Global Atmospheric and Oceanic Modelling

Combining rigorous theory with practical application, this book provides a unified and detailed account of the fundamental equations governing atmospheric and oceanic fluid flow on which global, quantitative models of weather and climate prediction are founded. It lays the foundation for more accurate models by making fewer approximations and imposing dynamical and thermodynamical consistency, moving beyond the assumption that the Earth is perfectly spherical. A general set of equations is developed in a standard notation with clearly stated assumptions, limitations, and important properties. Some exact, non-linear solutions are developed to promote further understanding and for testing purposes. This book contains a thorough consideration of the fundamental equations for atmospheric and oceanic models, and is therefore invaluable to both theoreticians and numerical modellers. It also stands as an accessible source for reference purposes.

Andrew N. Staniforth – now retired – led the development of dynamical cores for weather and climate prediction at two national centres (Canada and the UK). He has published over 100 peer-reviewed journal articles and is the recipient of various prizes and awards including the Editor's Award (American Meteorological Society, 1990); the Andrew Thompson Prize (Canadian Meteorological and Oceanographic Society, 1993); and the Buchan and Adrian Gill Prizes (Royal Meteorological Society, 2007 and 2009).

'Well, this is an impressive book. It covers both the equations of motion and how those equations and their approximations can be used in models of the ocean and atmosphere. It is clearly written, careful and thorough, with a range and a depth that is unmatched elsewhere. It will be of immense value both to those interested in the fundamentals and to those wishing to build models that have a sound foundation. It will be a standard for years to come.'

Geoffrey K. Vallis, University of Exeter

'This is the textbook I wish I'd had as a graduate student and course instructor! This is an incredibly comprehensive resource for students and researchers alike. I am confident the book will become the go-to reference on atmospheric and oceanic modelling for the 2020s and beyond.'

Andrew Weaver, University of Victoria

'Andrew Staniforth has produced a comprehensive and insightful book on the mathematical foundation of global atmosphere and oceanic modelling. For different geophysical fluid applications, he guides us masterfully from the first principles of fluid physics to their evolution equations. The book covers all the fundamental aspects of these equations including conservation laws and exact non-linear solutions. This brilliant book is ideal for introducing graduate students to the subject matter as much as it is relevant for experts as a reference book.'

Gilbert Brunet, Bureau of Meteorology, Melbourne

Global Atmospheric and Oceanic Modelling

Fundamental Equations

Andrew N. Staniforth

CAMBRIDGE
UNIVERSITY PRESS

University Printing House, Cambridge CB2 8BS, United Kingdom

One Liberty Plaza, 20th Floor, New York, NY 10006, USA

477 Williamstown Road, Port Melbourne, VIC 3207, Australia

314–321, 3rd Floor, Plot 3, Splendor Forum, Jasola District Centre, New Delhi – 110025, India

103 Penang Road, #05–06/07, Visioncrest Commercial, Singapore 238467

Cambridge University Press is part of the University of Cambridge.

It furthers the University's mission by disseminating knowledge in the pursuit of education, learning, and research at the highest international levels of excellence.

www.cambridge.org
Information on this title: www.cambridge.org/9781108838337

DOI: 10.1017/9781108974431

First published 2022

Printed in the United Kingdom by TJ Books Limited, Padstow Cornwall

A catalogue record for this publication is available from the British Library.

Library of Congress Cataloging-in-Publication Data
Names: Staniforth, Andrew N., author.
Title: Global atmospheric and oceanic modelling : fundamental equations / Andrew N. Staniforth.
Description: New York : Cambridge University Press, 2022. | Includes bibliographical references and index.
Identifiers: LCCN 2021038570 (print) | LCCN 2021038571 (ebook) | ISBN 9781108838337 (hardback) | ISBN 9781108974431 (ebook)
Subjects: LCSH: Geophysics–Fluid models. | Atmospheric models–Mathematics. | Fluid mechanics. | Oceanography. | Atmospheric thermodynamics.
Classification: LCC QC809.F5 S73 2022 (print) | LCC QC809.F5 (ebook) | DDC 551.5101/1–dc23/eng/20211021
LC record available at https://lccn.loc.gov/2021038570
LC ebook record available at https://lccn.loc.gov/2021038571

ISBN 978-1-108-83833-7 Hardback

To the memory of my parents,
who so strongly encouraged me to pursue the educational opportunities available to me, but not to them.

Contents

Preface

Purpose For novices and experts alike, this book aims to provide a unified, consolidated and detailed account of the underlying geophysical-fluid-dynamical equations on which global, quantitative, atmospheric, and oceanic models for climate and weather prediction are founded. Everything is built upon, or around, these governing equations. To accomplish this goal:

- A *general* set of governing equations is developed in a standard notation, with clearly stated assumptions, limitations, and important properties (in Part I).
- *Approximated* equation sets are related to this general set, again with clearly stated assumptions, limitations, and important properties (in Part II).
- Some *exact*, steady and unsteady, non-linear solutions are developed for testing purposes and to promote further understanding (in Part III).

A significant feature of this book that distinguishes it from other textbooks on the subject is that Earth is assumed to be *ellipsoidal* in shape instead of spherical. Gravity can then vary meridionally as it does physically. This then provides the firm foundation for formulating more accurate, quantitative, atmospheric, and oceanic prediction models in the future. This book also aims to provide an easily accessible source for reference purposes. Part I can be considered to be a long book covering the fundamentals, and Parts II and III to be two shorter ones following on from Part I.

Themes I have endeavoured to convey the importance of:

- Generality.
- Scientific rigour.
- Dynamical and thermodynamical consistency.
- Unification of atmospheric and oceanic modelling.

Readership The only prerequisite to reading this book is a basic mastery of vector calculus and partial differentiation. This makes the book accessible to graduate students. Prior knowledge of fluid dynamics and thermodynamics is helpful, but not essential. How much detail a reader needs to understand a subject depends very much on their prior knowledge. I have deliberately erred on the side of giving more rather than less detail. I believe that students will appreciate this – insufficient detail impedes understanding – and that experienced readers can easily filter out details already familiar to them.

A thorough understanding of the governing equations for atmospheric and oceanic models and their properties is essential to the design of accurate and efficient numerical methods for their solution. This book is therefore intended to appeal not only to theoreticians but also to numerical modellers. It is in fact the book that I wished for when I first embarked on numerical modelling in the early 1970s, albeit some of the knowledge for this had not yet been developed.

Scope Although there is some (necessary) overlap of content, this book aims to complement other textbooks on atmospheric and oceanic fluid dynamics. One important way in which it does so is by representing gravity throughout in a more general and physically realistic manner.

Gravity is a dominant force in the geophysical-fluid-dynamical equations governing atmospheric and oceanic motion. It is mathematically specified in terms of the gradient of a potential, termed the geopotential. Textbooks generally assume from the very beginning that geopotential isosurfaces are spherical in shape. This is termed the spherical-geopotential approximation. Governing equations are then expressed in spherical-polar coordinates from that point on. This unduly restricts generality and also introduces inconsistencies into quantitative prediction systems.

For mildly oblate planets (such as Earth), geopotential surfaces are more realistically approximated as spheroids (i.e. slightly squashed spheres). In principle, this then allows gravity to vary from pole to equator (as it does in reality) – this is excluded by the spherical-geopotential approximation – and also to vary realistically in the vertical. Doing so in practice, however, is highly challenging. Based on recent advances, three chapters are therefore dedicated to:

1 The rigorous derivation from first principles (systematically developed in Chapters 7 and 8) of a physically realistic representation of geopotential (and therefore of gravity).
2 The development (in Chapter 12) of an associated, mutually consistent, *spheroidal*, coordinate system in terms of axial-orthogonal-curvilinear coordinates.

This then allows expression (in other chapters) of the governing equations in these more general (spheroidal) coordinates. The classical, spherical-geopotential approximation is a special case of this more general approach to the representation of gravity. As a side benefit, it can then be mathematically justified as a zeroth-order asymptotic approximation of the more general representation.

Expression of the governing equations in terms of axial-orthogonal-curvilinear coordinates instead of spherical-polar coordinates is slightly more complicated. It is, however, a small price to pay for maintaining generality of the governing equations. At any stage it is straightforward to express any equation in terms of spherical-polar coordinates by setting the metric factors to their spherical form. To aid clarity, to link to the literature, and for reference purposes, sets of governing equations are also given explicitly for the special case of spherical-polar coordinates.

Many textbooks focus on *maximum* simplification of the governing equations to better understand, in the simplest possible framework, basic physical processes. This also facilitates the development of accurate and efficient numerical methods for modelling them in practice. Here, the complementary approach of *minimum* simplification is taken instead to derive a *general* set of governing equations; just enough simplification of physical reality to be practically viable, but no more than is necessary. The goal is to ultimately improve the accuracy of quantitative atmospheric and oceanic models by using a more accurate set of governing equations.

Various approximated and mostly familiar equation sets are derived from this general set in later chapters, with a focus on preserving analogues of the fundamental conservation laws for mass, axial angular momentum, total energy, and potential vorticity. This is termed dynamical consistency. Variational methods provide a useful and complementary tool to examine this. An introduction to these methods is therefore given in Chapter 14.

The atmosphere and oceans exchange mass, momentum, and energy. Inconsistencies in the thermodynamics of atmospheric and oceanic models can lead to spurious climate drift in quantitative predictions. This is best avoided by specifying a Gibbs potential (or other thermodynamic potential), and obtaining *all* thermodynamic quantities from it. The thermodynamic framework for this important subject is developed in Chapter 9. This framework is essential for a physically realistic representation of ocean thermodynamics. As a side benefit, an acceptably realistic, analytically tractable representation of phase changes of water substance also is developed. This provides a useful introduction to more advanced representations.

Chapter Organisation The underpinning rationale for the ordering of chapters is given at the end of the Introduction (Chapter 1). Chapters can (almost) be read sequentially if one wishes. I say 'almost' because there are many interrelated facets involved in the formulation of governing equations for realistic atmospheric and oceanic forecast models. To accommodate this, and to understand how the many pieces of the formulational jigsaw fit together to form a complete picture, links are made in chapters to both earlier and later ones. These links may be ignored on a first reading.

Chapters do not, however, necessarily have to be read sequentially. Each chapter aims to be a 'scientific adventure', motivated by a stated goal, with a sign-posted journey to the destination. With a willingness to accept certain affirmations, or to refer to part of another chapter, each chapter can be read almost independently of the others. This is to help the reader interested in a particular subject to more readily access the relevant material without needing to first carefully read other chapters. This has led to some repetition, but this is no bad thing – it is how we acquire many skills in life.

Some chapters (particularly later ones) contain some original material specifically developed to answer various questions that arose during writing. This material is aimed at readers interested in exploring the current boundaries of research, particularly regarding the generality of governing equation sets, their exact solution for testing purposes, and their properties.

Course Examples Some suggestions on using this book for courses follow.
A basic course on:

- Formulation of Governing Equations for Atmospheric and Oceanic Models (or Dynamical Cores) – based on the chapters of Part I, with a flexible choice of material to be covered according to interest.

Specialist courses on:

- Representation of Gravity in Atmospheric and Oceanic Models – Chapters 7, 8, and 12.
- Introduction to Basic Thermodynamics for Atmospheric and Oceanic Dynamical Cores – Part of Chapter 4 plus Chapters 9–11, possibly supplemented by other material.
- Variational Methods and Hamilton's Principle of Stationary Action – Chapter 14.
- Conservation Principles of Atmospheric and Oceanic Dynamical Cores and Dynamical Consistency – Chapters 15–17.
- Shallow-Water Equations and the Non-divergent Barotropic Potential-Vorticity Equation – Chapters 18 and 19.
- Exact, Steady and Unsteady, Non-linear Solutions of Atmospheric and Oceanic Dynamical Cores – Chapters 20–23.

Notation Notation is for the most part 'standard'. However, the book draws on different disciplines of physics and mathematics. Each has its own conventions, and each generally makes full use of familiar Roman and Greek symbols, of which there are a limited number. This has inevitably led to some notational compromises. Within a subject area, I have mostly used its standard notation to facilitate access to its specialised literature, whilst identifying and clarifying potential notational ambiguity.

References To help readers more readily access related scientific literature, particularly for specialised aspects, I have provided more references than often found in textbooks. These references, nevertheless limited in both number and scope, are simply those that I am familiar with, that I have found helpful, that came to mind whilst writing this book and that were accessible to me after my retirement. They are intended to provide a portal to further references, of which there are many of great value.

Technical This book was written using the LyX Document Processor (www.lyx.org). LyX provides a comprehensive, user-friendly, graphical interface, with an integrated equation editor, to the underlying power and flexibility of the TeX/LaTeX typesetting system. Without this open-source software, writing this book would have been a far more daunting challenge. Many of the figures were created using the xfig (www.xfig.org) open-source software. I am grateful to Markus Gross for his valuable help to use LyX with TeX/LaTeX document classes, and also to use xfig. This got me off to a very good start.

Acknowledgements I am greatly indebted to my very good friend and former colleague, Andy White; his help and good humour have proven invaluable. Over our two decades of scientific collaboration, Andy has not only stimulated my interest in the subject matter of this book but also brought great clarity to it. He has carefully reviewed drafts of many chapters, identified various errors therein, and suggested many improvements. He also collaborated with me on various journal papers that have been adapted for use herein. Without Andy's help and encouragement over the years, particularly when morale was flagging, this book would never have been written!

I am also indebted to many other people, too numerous to list, with whom I have interacted over the past four decades or so. This includes former colleagues at Environment Canada and the UK Meteorological Office, national and international colleagues, and journal editors and reviewers. Collectively, they have all greatly contributed to and influenced my understanding of, and way of viewing, the formulation of governing equations for atmospheric and oceanic models. I am particularly grateful to two of my collaborators, John Thuburn and Nigel Wood. John carefully reviewed several chapters and, in particular, identified some misconceptions and the means to address them; and Chapter 13 extends collaborative work with Nigel. Various chapters in Part I of this book were much influenced by Geoff Vallis' excellent book entitled *Atmospheric and Oceanic Fluid Dynamics*. I am also indebted to Geoff for providing valuable advice on the book publication process and also for providing some LaTeX macros.

I am very grateful to: Matt Lloyd, Sarah Lambert, and Elle Ferns at CUP for their guidance and encouragement to a novice book author; Bret Workman for copyediting the manuscript; and Vidya Ashwin and her team at Integra Software Services for typesetting. They all greatly contributed to bringing a challenging project to fruition.

I would have liked to thank my former employer, the UK Meteorological Office, for its encouragement and support in writing this book. Sadly, I cannot since my invitations to do so were declined. C'est la vie, et on tourne la page; goodbye bureaucracy, hello retirement! Time will tell whether my early retirement years have been well spent ...

I hope that readers of this book will find it both interesting and useful. At the very least, I found writing it to be challenging and fulfilling.

Finally, I thank my wife Lenore for her enduring love, patience, and support during the lengthy writing process.

Notation and Acronyms

A partial list of the more important and frequently used symbols employed herein follows. They are grouped as Roman symbols, Greek symbols, and operators, and are organised within groups in roughly alphabetical order. A list of acronyms also is provided.

Notation is fairly standard, with relatively few exceptions. Vectors are generally written in bold, upright Roman, characters (as in $\mathbf{V}$); and scalars and indices in italic, Roman, and Greek ones. An overbar (as in $\overline{S}$) usually indicates a basic-state or horizontally averaged quantity; an overhat (as in $\widehat{S}$), a vertically averaged quantity; and a zero subscript (as in S_0), a representative value of a quantity.

Due to inevitable overlap of meaning between standard usage in different disciplines, some characters have different meanings in different chapters herein. In such instances, different meanings are separated by a semicolon (';') in their entry in the lists here, with the relevant meaning usually clear from the context in which the symbol appears.

Roman Symbols

a, c	Equatorial and polar radii, respectively, of axially symmetric spheroids/ellipsoids.
$\overline{a}$	Earth's mean radius.
$\mathbf{a}$	Label coordinates $\mathbf{a} = (a_1, a_2, a_3)$.
$\mathbf{A}, \mathbf{B}, \mathbf{C}$	Generic vectors in vector identities.
$A(R), C(R)$	Equatorial and polar radii, respectively, of geopotential surface $\Phi(\chi, R) = \Phi(\chi = 0, r = R) \equiv \Phi_R = \text{constant}$.
$A(\xi_2), B(\xi_2)$	Two key functions in the derivation of the quasi-shallow equation set.
$\mathscr{A}$	Action of a physical system.
c_p, c_v	Specific heats at constant pressure and constant volume, respectively.
c_s	Local sound speed.
$\mathbb{C}$	A prototypical, materially conserved quantity for $\mathbb{C} = M, E, \Pi$.
ds, dS, dV	Distance metric, surface element, and volume element, respectively.
$đQ, đW, đC$	Infinitesimal changes in internal energy due to heat supplied, work done, and chemical composition, respectively.
e	First eccentricity $\left(a^2 - c^2\right)^{1/2}/a$, often shortened to eccentricity, of an ellipse or ellipsoid.
e'	Second eccentricity $\left(a^2 - c^2\right)^{1/2}/c$ of an ellipse or ellipsoid.
$(\mathbf{e}_1, \mathbf{e}_2, \mathbf{e}_3)$	Unit-vector triad in three, mutually orthogonal, curvilinear directions (ξ_1, ξ_2, ξ_3).
$E, \mathscr{E}$	Total and internal energies, respectively.
f	Coriolis parameter; a generic scalar function; Helmholtz energy.
$\mathbf{F}$	Force or sum of forces.
g	Gravity; Gibbs potential (or Gibbs free energy).
g_0	Value of 'standard gravity' for Earth ($g_0 \equiv 9.80665$ m s^{-2}).

g_S, g_S^E, g_S^P	Value of g at a planet's surface, equator, and a pole, respectively.
g^d, g^v, g^{av}	Gibbs potentials for dry air, water vapour, and humid air, respectively.
g^W, g^S	Gibbs potentials for pure liquid water and saline correction, respectively.
$g_S^P/g_S^E - 1$	Clairaut's fraction $\equiv \left(g_S^P - g_S^E\right)/g_S^E$.
$\mathbf{g}^N$	Newtonian gravitational force.
h	Elevation for geodetic coordinates.
h, h^0	Enthalpy and potential enthalpy, respectively.
(h_1, h_2, h_3)	Metric (or scale) factors for orthogonal-curvilinear coordinates (ξ_1, ξ_2, ξ_3).
H, B	Heights above a reference level of the top and bottom surfaces of a shallow layer of fluid, respectively.
$\widetilde{H}$	Depth of a shallow layer of fluid $\equiv H - B$.
$(\mathbf{i}_1, \mathbf{i}_2, \mathbf{i}_3)$	Unit-vector triad in three mutually orthogonal, Cartesian directions (x_1, x_2, x_3).
k	Integer index; wave number.
K	Kinetic energy.
$L, \mathscr{L}, \widehat{\mathscr{L}}$	Lagrangian, Lagrangian density, and vertically averaged Lagrangian density, respectively.
L_0^v, L_0^f	Latent heats of vaporisation and fusion, respectively.
$\mathbb{L}$	Latent energy.
m	Mixing ratio; $\Omega^2 a^3/(\gamma M)$, the ratio of centrifugal force to gravitational attraction; a point mass.
m, n	Order and degree of associated Legendre functions, respectively.
M, M_P	Total mass of a planet and its atmosphere (if present); axial absolute angular momentum.
M_E, M_A	Masses of Earth and of its atmosphere, respectively.
$\mathbb{M}$	Mass of an ideal gas or fluid parcel.
n_d	Number of independent degrees of freedom for a single molecule of an ideal gas.
$\mathbf{n}$	Unit vector normal to a surface.
$\mathbb{N}^D, \mathbb{N}^C, \mathbb{N}^P$	Number of independent degrees of freedom, components, and coexisting phases, respectively.
$\mathscr{N}$	Non-hydrostatic switch.
$p, \widehat{p}$	Pressure and vertically averaged pressure, respectively.
p^d, p^v	Partial pressures of dry air and water vapour, respectively.
p_{sat}^v	Saturation vapour pressure.
p_0	Constant reference pressure, usually set to 1 000 hPa or 1 1013.25 hPa.
p_0^d	Reference pressure for dry air.
p_0^{sat}	Saturation vapour pressure for pure water at $T = T_0$.
P_n	Legendre polynomial of degree n.
P_n^m, Q_n^m	Associated Legendre functions of the first and second kinds, respectively, of order m and degree n.
q	Mass fraction of water substance; $\ln p$.
$\dot{q}$	Rate of change of mass fraction.
$\dot{Q}, \dot{Q}_E$	Heating rate and rate of total-energy input, respectively.
$\mathcal{Q}$	Quasi-hydrostatic switch.
$\mathbf{q}(t), \dot{\mathbf{q}}(t)$	Vector of generalised coordinates and its rate of change, respectively.
r	Radial distance from an origin (e.g. in spherical-polar coordinates).
$\mathtt{r}$	Radial distance measured perpendicular to a reference axis in cylindrical-polar coordinates.
$r_\perp$	Perpendicular distance from a rotation axis.
$\mathbf{r}$	Position vector.
R, R^d, R^v	Gas constants per unit mass of air, dry air, and water vapour, respectively.

$s, \dot{s}$	Generalised vertical coordinate and generalised vertical velocity, respectively.
S	Mass fraction of substance (e.g. of dry air, of water substance, or of salinity); a surface.
S_A	Mass fraction of Absolute Salinity.
S_P	Practical Salinity.
S_0	Constant reference value for Absolute Salinity.
$\dot{S}$	Rate of change of substance.
$\mathscr{S}$	Surface area of a planet.
t	Time.
$\mathbf{t}$	Unit vector tangent to a curve or surface.
T, T_v	Temperature and virtual temperature, respectively.
T_0	Temperature at the triple point of water for pure-water substance (in the absence of dry air); constant reference value for T.
T_*	Temperature at the triple point of water in the presence of dry air.
$\mathbf{u}$	Velocity vector in a uniformly rotating frame of reference.
$\mathbf{u}_{hor}$	Horizontal velocity.
V	Newtonian gravitational potential (i.e. the potential due solely to Newtonian gravity); potential energy in Newtonian particle mechanics; a volume.
V^C	Potential of the centrifugal force.
V_E	Volume of Earth.
$\mathbb{V}$	Volume of an ideal gas or a mixture of ideal gases.
$\mathscr{V}$	Volume of a planet.
$\dot{W}$	Rate of work done per unit mass.
X	Type of constituent.
(x_1, x_2, x_3)	Cartesian coordinates.
z	Altitude (for $z > 0$), $\|z\|$ is depth (for $z < 0$); axial position in cylindrical-polar coordinates.
$\widetilde{z}$	Basic terrain-following coordinate.
$\mathbf{Z}$	Absolute vorticity (i.e. the sum of planetary and relative vorticities).
$\mathbf{Z}_r, \mathbf{Z}_a$	Absolute vorticity in deep and shallow spherical-polar coordinates, respectively.

Greek Symbols

α	Specific volume.
β_T, β_T^*	First and second thermal-expansion coefficients, respectively.
$\widehat{\beta}^T, \widehat{\beta}^S$	Thermal-expansion and saline-contraction coefficients with respect to in situ temperature T, respectively.
$\beta_S, \beta_S^*, \beta_p$	Saline-contraction, heat-capacity, and compressibility coefficients, respectively.
γ	Ratio of specific heats, c_p/c_v; universal gravitational constant.
γM	Standard gravitational parameter for Earth + its atmosphere.
$\gamma M_E, \gamma M_A$	Standard gravitational parameters for Earth and its atmosphere, respectively.
γ^*	Thermobaric parameter.
Γ	Temperature lapse rate.
$\delta f, \Delta f$	Increments of a generic scalar f.
ε	Ellipticity $(a - c)/a$, sometimes called first flattening parameter; R^d/R^v.
$\widetilde{\varepsilon}$	$\left(a^2 - c^2\right) / \left(2c^2\right)$.
ζ	Relative vorticity.
η	Entropy.
θ	Potential temperature; parametric latitude.
Θ	Conservative temperature.
κ	R/c_p.

λ	Zonal (or azimuthal) coordinate in axially symmetric coordinate systems (e.g. in spherical-polar coordinates).
Λ	A materially conserved scalar field.
μ	Chemical potential in thermodynamics; reduced mass in particle mechanics; viscosity.
ν	Kinematic viscosity.
μ, ν	Two coefficients appearing in geopotential representation for GREAT coordinates.
(ξ_1, ξ_2, ξ_3)	Orthogonal-curvilinear coordinates in the $(\mathbf{e}_1, \mathbf{e}_2, \mathbf{e}_3)$ directions.
π	Ratio of a circle's circumference to its diameter; hydrostatic pressure.
$\Pi, \overline{\Pi}, \overline{\Pi}_{\xi_2}$	Potential vorticity, basic-state potential vorticity, and meridional gradient of $\overline{\Pi}$, respectively.
ρ, ρ^θ	Density and potential density, respectively.
σ	Pressure-based, terrain-following coordinate; frequency.
τ	Time, associated with label coordinates $\mathbf{a} = (a_1, a_2, a_3)$.
φ	Generic latitude.
ϕ	Geographic latitude.
Φ	Potential of apparent gravity, often termed geopotential.
χ	Geocentric latitude.
ψ	Stream function; a coordinate.
Ψ	Stream function; thermodynamic potential.
ω	Frequency.
$\mathbf{\Omega}, \widehat{\mathbf{\Omega}}$	Angular velocity (assumed constant), and unit vector aligned with its direction, respectively.

Operators

D/Dt	Material derivative in a rotating frame of reference.
$(D/Dt)_I$	Material derivative in an inertial frame of reference.
D_S/Dt	Material derivative in shallow (non-Euclidean) geometry, in a rotating frame of reference.
D_r/Dt	Material derivative in spherical-polar coordinates.
D_a/Dt	Material derivative in shallow (non-Euclidean) spherical-polar coordinates.
$\widehat{F}$	Vertical average of F.
$\mathfrak{L}$	A second-order differential operator associated with the non-divergent barotropic PV equation.
O	Order of a quantity.
$\delta F\left[\mathrm{q}(t)\right]$	Variation of functional $F\left[\mathrm{q}(t)\right]$.
∂	Partial-derivative operator.
$\partial(\mathbf{r})/\partial(\mathbf{a})$	Jacobian of the transformation from coordinates $\mathbf{r} = (x_1, x_2, x_3)$ to coordinates $\mathbf{a} = (a_1, a_2, a_3)$.
∇, ∇_S	Gradient operator in Euclidean and shallow non-Euclidean geometries, respectively.
∇_r, ∇_a	Gradient operator in deep and shallow spherical-polar coordinates, respectively.
∇_{hor}	Horizontal gradient operator.
∇^2, ∇_S^2	Laplacian operator in Euclidean and shallow non-Euclidean geometries, respectively.

Acronyms

BPV	Barotropic Potential Vorticity
BTF	Basic Terrain Following

COS	Confocal-Oblate-Spheroidal
GPS	Global Positioning System
GREAT	Geophysically Realistic, Ellipsoidal, Analytically Tractable
GSW	Gibbs Sea Water
IAPSO	International Association for the Physical Sciences of the Oceans
IAPWS	International Association for the Properties of Water and Steam
IOC	Intergovernmental Oceanographic Commission
PV	Potential Vorticity
SCOR	Scientific Committee on Oceanic Research
SIA	Seawater Ice Air
SOS	Similar-Oblate-Spheroidal
STF	Smooth Terrain Following
TEOS	Thermodynamic Equation Of Seawater
WGS	World Geodetic System

Part I

FOUNDATIONS

1

Introduction

ABSTRACT

To offer a perspective, the genesis of Earth is very briefly reviewed. Since Earth rotates on its axis, its atmosphere and oceans behave as rotating fluids to leading order. This greatly helps unify their representation. The mathematical equations governing the flow of each of these two geophysical fluids then define an associated dynamical core. To obtain a realistic atmospheric or oceanic model for prediction purposes, the governing equations of the corresponding dynamical core need to be supplemented with forcing terms that represent other important physical processes (e.g. solar radiative heating, which drives atmospheric and oceanic circulations). Since gravity is such a dominant physical force, its representation is identified as a key factor for accurate, predictive, atmospheric, and oceanic modelling. Earth's shape is determined by its rotation rate and by gravity; it is not spherical, but spheroidal. Consequently, gravity varies as a function of latitude, but this variation is traditionally neglected in atmospheric and oceanic models. Although Earth's atmosphere and oceans behave to leading order as fluids, they nevertheless behave somewhat differently in detail. This is because their fluid composition is inherently different, with different properties, particularly with regard to thermodynamics. The fluid composition of Earth's atmosphere and oceans is briefly reviewed, both separately and comparatively. Both fluids are stratified, but in different ways according to their composition. This review helps relate later mathematical developments to physical reality. Finally, a summary guide to the organisation of the chapters is given.

1.1 A VERY BRIEF HISTORY OF TIME

Once upon a time there was – according to current scientific theory – a **BIG BANG**. A very big and very hot bang. Instantaneously, unimaginable amounts of hydrogen, helium, and other gases started whizzing around at breakneck speeds within rapidly expanding space. Over billions of years:

- These gases cooled and condensed under the influence of gravity, first into gas clouds, and then into stars.
- Thermonuclear reactions took place.
- Different forms of matter resulted.
- Yet more stars were created, with planets and their moons orbiting around them.

Fast forward more billions of years, and there was a far more modest bang. A celestial body crashed into a minor planet, in orbit around a minor star, of a minor constellation of stars, in a minor galaxy of the universe; and a moon was born.

Fast forward again and we are where we are now, 13.8 billion years (give or take) after the very big, very hot bang. The Moon orbits the rotating planet, Earth, which in turn orbits the Sun. This concludes our very brief history of time.[1] For a very readable, far more comprehensive history of time, see Hawking (2011).

This minor, rotating planet has both an atmosphere and oceans. The purpose of this book is to mathematically describe them in a fairly rigorous and physically realistic manner, based on their representation as fluids. This then provides the firm foundation for the construction of realistic, quantitative, atmospheric, and oceanic models for weather and climate prediction.

1.2 DYNAMICAL CORES

Earth's atmosphere and its oceans, *to leading order*, behave as fluids. From a modelling perspective, this is of crucial importance, since it allows us to translate the *physics* governing the motion of a fluid into *mathematics*. The resulting geophysical-fluid-dynamical, partial differential equations are the governing equations for the *dynamical core* of an atmospheric or oceanic model. In reality, these equations have to be supplemented by (often complicated) mathematical representations of other important physical phenomena (e.g. radiative heating by the Sun, which drives the global circulations of Earth's atmosphere and oceans). These representations then appear as *forcing terms* for the governing equations of a dynamical core.

To illustrate this, the governing equations of an atmospheric or oceanic model may be formally expressed in much simplified form as

The Dynamical Core Equations

$$\frac{\partial X}{\partial t} + Y(X) = F(X), \tag{1.1}$$

where t is time and:

- X is the vector of *dependent forecast variables* (such as velocity components, pressure, density, internal energy, water substance, salinity).
- Vector $Y(X)$ represents the *dynamical coupling* of these forecast variables in the absence of forcings other than gravitational attraction.
- $F(X)$ is the vector of all other *forcing functions* (such as solar radiative heating and friction).[2]

The left-hand side of (1.1) then represents the governing (geophysical-fluid-dynamical) equations of a dynamical core, and the right-hand side the forcing terms needed to turn a dynamical core into a realistic, quantitative, atmospheric, or oceanic forecast model.

Numerical approximation of the governing mathematical equations (including forcing terms, $F(X)$), coupled with use of many observations to determine an *initial state* of the atmosphere and/or of an ocean, then allows (approximate) *future states* to be determined using supercomputers. For numerical weather forecasting, Roulstone and Norbury (2013) and Lynch (2014) give very readable accounts of this process, intertwined with its historical development. Similar considerations also apply to oceanic forecasting.

The dynamical core is, as the name suggests, the cornerstone of an atmospheric or oceanic model; everything is built upon or around it. It is therefore very important that its governing equations not be oversimplified, otherwise an undesirable loss of fidelity with respect to reality results. This book focuses attention on this by:

[1] Well, you were advised that this history would be *very* brief... unlike the remainder of this book, you may say.

[2] Gravity is treated as part of the dynamical core because it has a known form, and because it guides the choice of coordinate systems. All other forcings are considered not to belong to the dynamical core. These forcings are important both practically and conceptually, and their accurate formulation represents a major scientific challenge; but the focus here is on the equally important and challenging problem of accurately formulating dynamical cores.

Part I Describing how to minimally approximate these governing equations, without prejudicing practical viability.

Part II Providing a means to understand the significance and potential impact of making further approximations, with emphasis on the importance of maintaining analogues in any approximated equation set of physical conservation laws embodied in the original set.

Part III Developing some exact, steady and unsteady, non-linear solutions of governing dynamical-core equation sets for testing purposes.

This book also places these important aspects into a broader context.

1.3 GRAVITY

Key to achievement of the three goals just presented is recognition that rotating planets are more accurately represented as oblate spheroids than spheres. A planet's shape is due to the combined influence of rotation about its axis and gravity; see Chapters 7 and 8. The faster a planet rotates, the more oblate it is. A deformable planet can only be spherical if it does not rotate. But Earth *is* deformable and it *does* rotate.

Atmospheric and oceanic dynamical cores traditionally assume the *spherical-geopotential approximation*; see Sections 7.1.5 and 8.6 for details. This practice dates back to the very beginnings of numerical modelling when:

- Computer power and modelling know-how were very limited indeed.
- Spatio-temporal resolution was very low.
- The representation of physical forcings was very crude.
- Observational data were sparse.
- Initial conditions had very large errors.

It was then clear that any error incurred by the spherical-geopotential approximation was negligible compared to other errors. Many decades later, with the advent of ever-more-powerful computers coupled with enormous improvements in data availability and its assimilation, numerical methods, and the representation of physical forcings, this is no longer evident. Looking to the future, the only convincing way to assess the validity of the spherical-geopotential approximation is to perform careful, controlled experiments using a model with a more accurate, *spheroidal* representation and an appropriate coordinate system. But first the governing equations of such a model have to be properly formulated!

It is relatively straightforward to describe the spherical-geopotential approximation and how it is applied in practice. However, it is surprisingly difficult to *rigorously* and convincingly derive it; see van der Toorn and Zimmerman (2008)'s critique of arguments advanced by previous authors for this. The underlying difficulty is that if a deformable planet is truly spherical and rotating, then the corresponding geopotential surfaces cannot be spherical, and vice versa. Also, to respect physical conservation principles, gravity cannot vary meridionally (i.e. as a function of latitude) with the spherical-geopotential approximation. However, it has been known for centuries that it does. Furthermore, artificial satellites have, over the past several decades, become a crucially important source of observational data for atmospheric and oceanic prediction systems. Locations of satellite observing platforms are reported (using the Global Positioning System, or GPS for short) with respect to a reference *ellipsoid*, and *not* to a spherical one; see Section 12.2. This then leads to *inconsistencies* in the use of satellite data.

Gravity is a dominant physical force for Earth's atmospheric and oceanic circulations. For the reasons just given, it seems perverse to represent Earth's geopotentials as *spheres*; mathematical simplicity and convenience should not outweigh physical fidelity. This practice, with the risk of creating cumulative, systematic, forecast errors, is shown to be unnecessary. A suitable coordinate system for formulating the governing equations of the dynamical cores of mildly oblate planets

Gas	Chemical symbol	Atomicity	Mass fraction with respect to dry air
Nitrogen	N_2	Diatomic	0.755 184 73
Oxygen	O_2	Diatomic	0.231 318 60
Argon	Ar	Monatomic	0.012 870 36
Carbon dioxide	CO_2	Triatomic	0.000 607 75
Neon	Ne	Monatomic	0.000 012 68
Krypton	Kr	Monatomic	0.000 003 18
Methane	CH_4	Polyatomic	0.000 000 83
Helium	He	Monatomic	0.000 000 72
Nitrous oxide	N_2O	Triatomic	0.000 000 46
Xenon	Xe	Monatomic	0.000 000 45
Carbon monoxide	CO	Diatomic	0.000 000 19
Hydrogen	H_2	Diatomic	0.000 000 03

Table 1.1 Composition of Earth's dry atmosphere by mass fraction of dry air. Source: Feistel et al. (2010a). By definition, molecules of monatomic, diatomic, and triatomic gases have one, two, and three atoms, respectively. Methane is a polyatomic gas with five atoms.

in *ellipsoidal* geometry is developed in Chapter 12. Adopting this ellipsoidal coordinate system should result in only a relatively modest increase in the complexity and computational cost of a forecast model compared to using *spherical* coordinates. Appropriately taking an asymptotic limit of this more general, ellipsoidal formulation, recovers the familiar set of governing equations in spherical geometry as a special case. As a side benefit, this sheds light on the validity and interpretation of the classical spherical-geopotential approximation.

1.4 FLUID COMPOSITION OF EARTH'S ATMOSPHERE AND OCEANS

To state the obvious, Earth's atmosphere is primarily *gaseous*, whilst its oceans are primarily *liquid*. This then influences the precise form of their respective governing equations. Their similarities unify many aspects, whereas their differences necessitate separate treatments of other aspects, particularly regarding equations of thermodynamic state. Before going into further details (in the following chapters), and to prepare the way for this, a brief overview is now given of the composition of Earth's atmosphere and of its oceans. This helps to relate later mathematical developments to physical reality.

1.4.1 The Atmosphere

Dry Gases

Earth's atmosphere is composed primarily of a mixture of dry gases; see Table 1.1. Just two dry (diatomic) gases (nitrogen and oxygen) comprise almost 99% of the total mass of dry gas in the atmosphere. The mass of carbon dioxide in Earth's atmosphere has gradually increased from pre-industrial levels (and currently continues to increase); this mass has an annual fluctuation, primarily due to the Northern Hemisphere's agricultural growing season.

Each dry gas (to an excellent approximation) individually behaves as an ideal gas. For most purposes, the mixture of dry gases can be considered to be of constant composition.[3] Because an individual dry gas behaves as an ideal gas, and because the mixture is of constant composition,[4]

[3] In other words, the individual masses of these gases, for any sample, are everywhere in the same proportion to one another.

[4] Depending on the application, there are exceptions to this rule. These include ozone (e.g. in the stratosphere over polar regions) and near-surface noxious gases (e.g. emissions from vehicles and factories). This is because some gases can

the mixture behaves as if it were a *single* ideal gas (see Section 2.3.3). This greatly simplifies matters. Instead of having to represent a dozen or more dry gases and their mutual interactions, only a single, equivalent, composite, (dry) ideal gas needs representation.

Water Vapour

A very important constituent of Earth's atmosphere is water vapour, the gaseous state of water substance (H_2O). Water vapour, a triatomic gas, occupies about 0.4% of the mass of the entire atmosphere and typically 1–4% of the mass of air near Earth's surface. Unlike the dry gas constituents of Earth's atmosphere, water vapour varies considerably both in space and time.

It can change from its vapour state to both its liquid state (e.g. cloud droplets, and rain droplets) and its frozen state (e.g. ice crystals, hail, and frost). When it does so, it can strongly interact with its environment (e.g. by heating the atmosphere via the release of latent heat, and by moisture loss from the atmosphere through precipitation). At Earth's surface, water vapour is also both continuously lost (e.g. via precipitation and condensation) and replenished (e.g. via evaporation over land and water, and via evapotranspiration by plants and trees). These various complex processes are part of the *hydrological cycle* – that is, the cycling of water substance (in its various forms):

- Within the atmosphere.
- At Earth's surface.
- Below its surface (e.g. the root systems of plants and trees, aquifers, and underground streams).

Furthermore, water vapour is a so-called *greenhouse gas* (in fact, the most important one by far, ahead of carbon dioxide, methane, and ozone), and it thereby considerably influences Earth's weather and climate through its radiative properties.[5] It also interacts chemically with various atmospheric constituents (e.g. sulphur dioxide and nitrogen oxides interact with water substance to produce acid rain).

Because of its many different properties, water vapour in physically realistic, quantitative weather and climate prediction models needs to be represented separately from dry gases. That said, for a lot of theoretical work, aimed at improving understanding by studying simplified models, it can often be ignored!

Various Liquids and Solids

Earth's atmosphere not only contains dry gases and water vapour, but also various liquids and solids. These are particularly important for representing clouds, and the physical processes taking place within them. In physically realistic, quantitative, weather and climate prediction models, aerosols,[6] cloud liquid water, and cloud frozen water are typically represented. In air-quality models, dozens (or more) of chemically active species may be needed.

chemically and photochemically (via solar radiation) interact with other atmospheric constituents. Furthermore, dry gases can interact with vegetation at Earth's surface, thereby changing the amounts of dry gas present in the atmosphere. For example, plants process oxygen and carbon dioxide via photosynthesis. If any such phenomenon is important for a particular application (e.g. modelling the destruction of stratospheric ozone over the Poles in wintertime, or for air-quality studies, or for surface emissions), then any such gas should not form part of the mixture of dry ideal gases, but instead be represented separately; see the '$+\cdots$' terms of (3.1) in Chapter 3.

5 In the absence of water vapour in the atmosphere, the temperature of Earth's atmosphere would be considerably colder (by an estimated 20° Celsius) and life, as we know it, would be considerably different (and possibly non-existent, albeit life forms have been found in the deep ocean at temperatures near 0° Celsius).

6 Aerosols are small particles, suspended in the air, with diameters typically smaller than a micrometre. They include dust, pollen, sea-salt particles, volcanic ash, and by-products of combustion. Although very small, they provide a nucleus for condensation to take place and thereby facilitate cloud and fog formation, and chemical reactions associated with air pollution and ozone depletion. They also influence radiative transfer within the atmosphere. Small though they are, their properties and vast number make them important.

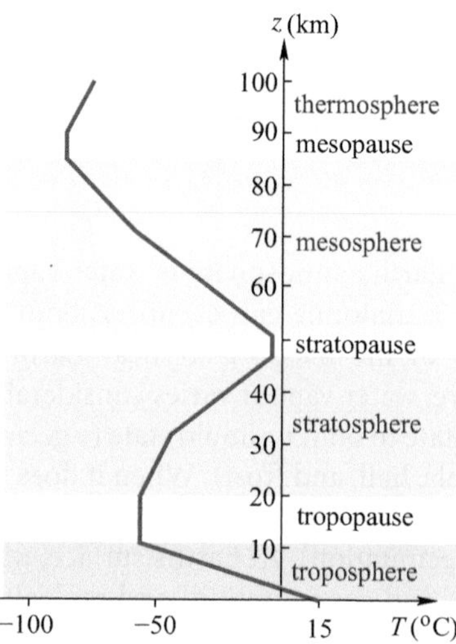

Figure 1.1 Temperature profile for the International Standard Atmosphere (1976), extended to include the mesopause and the lowest part of the thermosphere. Abscissa is temperature T (° C); ordinate is altitude z (km). Lowest four layers ('spheres') of the atmosphere (troposphere, stratosphere, mesosphere, and thermosphere) are shaded; locations of the tropopause, stratopause, and mesopause (isothermal) layers separating them are also depicted.

Stratification

To leading order, observations show that Earth's atmosphere can be considered to be a stratified fluid with several fairly distinct layers. Pressure (p) and density (ρ) in a column of air generally decrease monotonically as a function of increasing altitude (z).[7] Temperature (T), however, does not. It is the vertical variation of T that facilitates classification of the atmosphere into layers; see Fig. 1.1 for a schematic of the lowest layers. This figure is based on the International Standard Atmosphere (1976) for a representative mid-latitude column of (assumed dry) air extending upwards from Earth's surface. In reality, significant departures from this standard atmosphere are observed, even in mid-latitudes, but much more so in polar and tropical regions.

Based on time and space averages of observed temperature profiles in mid-latitudes, the International Standard Atmosphere (1976) defines a piecewise-linear, vertical temperature profile as a function of altitude (z). This is depicted in red in Fig. 1.1. Four (shaded) 'spheres' are shown (troposphere, stratosphere, mesosphere, and thermosphere), with a fifth 'sphere' (the exosphere) above them not shown.[8] For each of these 'spheres', T (by construction) either decreases or increases at a constant rate (measured in °C per km), termed the *temperature lapse rate*. Interleaved with these 'spheres' are 'pauses' (tropopause, stratopause, mesopause, thermopause), where T is isothermal (i.e. the temperature is constant, and consequently the temperature lapse rate is zero).

Assuming a stationary, hydrostatic atmosphere composed of ideal gases at constant composition, pressure and density profiles corresponding to a given temperature profile may be deduced. See Section 13.9 regarding the methodology for this.

[7] This decrease is approximately exponential – see Section 13.9.1.

[8] Since Earth is spheroidal, *troposphere* might whimsically be relabelled *tropospheroid*, and so on.

Solute	Ion	Mass fraction with respect to total sea salt
Chloride	Cl^-	0.550 3396
Sodium	Na^+	0.306 5958
Sulphate	SO_4^{--}	0.077 1319
Magnesium	Mg^{++}	0.036 5055
Calcium	Ca^{++}	0.011 7186
Potassium	K^+	0.011 3495
Bicarbonate	$HC0_3^-$	0.002 9805
Bromide	Br^-	0.001 9134

Table 1.2 Composition of Earth's oceans by mass fraction of total sea salt. Source: Feistel et al. (2010a).

1.4.2 The Oceans

Pure Liquid Water

Seawater is a dilute solution of salts. The oceanic analogue of dry air is pure liquid water – the water content of the oceans in the absence of any dissolved salts (also referred to as solutes) or other substances. Seawater and pure liquid water are much denser than dry and moist air. For example, the density of seawater at the atmosphere/ocean interface is about 800 times larger than that of the overlying air.

Salinity

Solutes collectively comprise approximately 3.5% of the total mass of the oceans. See Table 1.2 for the composition of Earth's oceans by mass fraction of total sea salt. Similarly to dry gases in Earth's atmosphere, the solutes in the oceans can also usually be considered to be of constant composition. This means that instead of having to represent many different solutes and their mutual interactions, only a single, equivalent, composite quantity, namely *salinity*, needs representation.

For many years salinity was defined to be *Practical Salinity* (denoted S_P). This definition is based on measured electrical conductivity; Practical Salinity is nevertheless dimensionless. In 2010 a new standard for the properties of seawater was developed and adopted by the international oceanographic community. It is called the Thermodynamic Equation of Seawater 2010 (or TEOS-10, for short); see IOC et al. (2010) for (lots of) details. In particular, a new measure of salinity, termed *Absolute Salinity* (S_A), has been defined. This new measure represents the mass fraction of solutes with respect to the combined mass of pure liquid water plus solutes. Since it is a true mass fraction, Absolute Salinity is also dimensionless. However, it is usually expressed in units of g kg^{-1} (i.e. as parts per thousand).

Stratification

Density (ρ) in an ocean is (with a few exceptions) generally constant to within 2 per cent. It has the value $\rho_0 = 1\,035 \text{ kg m}^{-3}$. This quasi-constancy reflects the fact that whereas air is highly compressible, liquid water is almost incompressible. That ρ is nearly constant can be exploited to determine, to a very good approximation, how pressure (p) in an ocean varies with respect to ocean depth ($-z$).

For a motionless (static) ocean, a vertical column of water satisfies the hydrostatic-balance equation,

$$\frac{1}{\rho}\frac{\partial p}{\partial z} = -g, \tag{1.2}$$

where g is gravity.[9] For given ρ and g, (1.2) is a first-order differential equation that can be integrated downwards from a given pressure, say $p(z=0) = p_0 = 1.01325 \times 10^5$ Pa $\equiv$ 1 atm $\equiv$ 1.01325 bar $\equiv$ 1013.25 hPa $\equiv$ 1013.25 mb (i.e. the value of 'standard sea-level pressure', a representative value at an ocean's upper surface).[10] Making the approximations $\rho \approx \rho_0 = 1035 \text{kgm}^{-3}$ and $g \approx g_0 = 9.80665 \text{ms}^{-2}$ (i.e. the value of 'standard gravity') in (1.2), and integrating downwards, then yields

The Approximate Pressure within a Water Column

$$p \approx p_0 - \rho_0 g_0 z. \tag{1.3}$$

Depth ($\equiv -z$) in (1.3) is measured *downwards* from an ocean's surface at $z = 0$. Thus pressure increases approximately linearly downwards (with respect to depth) from an ocean's surface at a rate of $\rho_0 g_0 = 1.015 \times 10^4$ kg m^{-2} s^{-2} $= 1.015 \times 10^4$ Pa m^{-1}, or approximately 1 atm $\equiv$ 1.01325×10^5 Pa (approximate pressure at an ocean's upper surface) for every 10 m of descent. This makes it easy to roughly estimate the pressure at a particular ocean depth; just divide the depth, measured in metres, by ten and then add unity to obtain how many times greater the pressure is than that at an ocean's upper surface. For example, at the bottom of the Mariana Trench (the lowest point of Earth's oceans) in the Western Pacific Ocean, the depth is about 11 km = 11 000 m. Thus the pressure there is approximately 1 100 times greater than that at the ocean's upper surface. And life still exists down there – wow!

Although density is approximately constant throughout the ocean, *density variation* nevertheless has an important influence on the *thermohaline* circulation of Earth's oceans.[11] Cold, dense water sinks in polar regions (this is termed *downwelling*) and feeds a deep-ocean return flow of the poleward transport of heat from the tropics by upper-ocean currents, such as the West Atlantic Gulf Stream. This circulation is part of what is termed the *ocean conveyor belt*. This circulates a huge amount of heat within Earth's oceans on much longer timescales than those for the atmospheric circulation. It is why oceanic circulation is so important for climate modelling and prediction.

For the atmosphere, the vertical variation of temperature (T) facilitates classification of the atmosphere into layers; see Section 1.4.1. This is also true for the oceans, but in a different way. For the oceans there are three fairly distinct layers:

- The mixed (or surface) layer, with small, vertical temperature gradient.
- The thermocline, with large gradient.
- The deep ocean, with small gradient.

This stratification is depicted in Fig. 1.2 for a representative mid-latitude temperature profile. For the atmosphere (see Fig. 1.1), it is the *sign* of the vertical temperature gradient that characterises

9 Equation (1.2) follows from staticity in the vertical component of the momentum equation. Physically, the vertical component of the pressure-gradient force then exactly balances gravity.

10 Many units for pressure have been used in meteorology and oceanography over the years, with standard SI (Système International) units now in frequent use. But old habits die hard, and there are many old publications that are still relevant today, so it is good to be aware of other units.

11 *Thermo* in 'thermohaline' refers to temperature, and *haline* to salt content, both of which jointly determine seawater density. The lower the temperature and the higher the salinity (e.g. due to an excess of surface evaporation versus precipitation over a region), the higher is the density. Conversely, the higher the temperature (e.g. due to solar heating) and the lower the salinity (e.g. due to dilution by fresh water from river outflow, or melting of icebergs), the lower is the density.

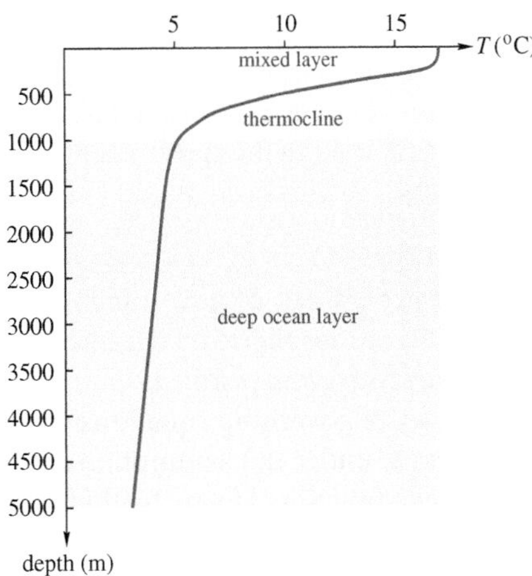

Figure 1.2 A representative, mid-latitude, oceanic temperature profile T (° C), plotted as a function of depth (m) measured downwards from an ocean's upper surface. Three layers are depicted. From top to bottom; mixed layer (light shading), thermocline (medium shading), and deep ocean (dark shading).

the layers. For the oceans (see Fig. 1.2), it is instead the *magnitude* of the vertical gradient, with two low-gradient layers (mixed and deep-ocean) separated by a large-gradient one (the thermocline).

Incoming solar radiation during daylight hours heats the near-surface of an ocean, primarily in its upper few centimetres; this is governed by the radiative properties of the ocean and the air above it. Wave activity (often wind-driven at an ocean's surface) and turbulent mixing then redistribute this heat fairly uniformly in the *mixed layer*. The depth of this layer depends upon a number of factors, particularly latitude and season. For the tropics and mid-latitudes, this depth is typically 100–200 m, but may be barely discernible, and even absent, at high latitudes. The temperature of this layer is generally higher in the tropics and subtropics (where solar heating is strong) than in mid- and high-latitudes (where it is much weaker). It is also somewhat colder during night-time due to the absence of incoming, short-wave, solar radiative heating, and the presence of outgoing, long-wave radiative cooling. In wintertime, the depth of the mixed layer is generally greater than in summertime, albeit it changes relatively little in the tropics, where the seasonal variation of T is relatively small. Mass, momentum, and energy are exchanged between the mixed layer and the atmosphere.

The *thermocline* is a transition layer between the mixed layer and the deep-ocean layer.[12] Its distinguishing feature is a strong temperature gradient between the top and bottom of the layer. This is generally strongest in the tropics and subtropics, reducing to non-existent in polar regions in wintertime. Its depth is highly variable, but typically of order 1 km. Associated with the thermocline (for temperature) are the *pycnocline* (for density) and *halocline* (for dissolved salt). The thermocline and pycnocline are generally quite well defined and correlated with one another. The halocline is generally less well-defined and only weakly correlated with the thermocline and pycnocline. A large density gradient can originate from a large temperature gradient, a very large salinity gradient, or a combination of the two.

The *deep-ocean layer* extends from the bottom of the thermocline down to an ocean's bathymetry. Conditions are generally fairly calm within this layer, with gradually reducing temperature

[12] *'Cline'* in oceanography denotes a relatively shallow layer of seawater with a strong vertical gradient of an associated quantity. Thus, *thermo*cline denotes a strong vertical gradient of temperature; *pycno*cline, a strong vertical gradient of density; and *halo*cline, a strong vertical gradient of dissolved salt.

and gradually increasing density as a function of depth. It comprises the vast majority of the total volume of the oceans; of order 90%. Because of its distance from an ocean's upper surface, the daily and seasonal variation of the deep-ocean layer is negligible.

1.5 ORGANISATION OF CHAPTERS

Part I

This part describes how to obtain a general set of governing equations, in a unified manner, for the dynamical core of a quantitative atmospheric or oceanic prediction model. To this end, *approximations are made only when necessary and justified.*

By the end of Chapter 4, a unified set of governing equations (in vector form) has been given for both the atmosphere and the oceans, under the assumptions that both the geopotential (Φ) and the internal energy ($\mathscr{E}$) are of known functional form. Nothing is said therein about a suitable coordinate system.

To prepare the way for this eventuality, definitions and relevant properties of various orthogonal-curvilinear coordinate systems are documented in Chapter 5. To accommodate the later description of some popular, approximate, governing equation sets (e.g. the hydrostatic primitive equations), shallow versions of these coordinates also are included.[13]

By the end of Chapter 6, the governing equations (in vector form) of Chapter 4 for the motion of geophysical fluids have been re-expressed in general axial-orthogonal-curvilinear coordinates. These equations are also given explicitly for the special cases of spherical-polar and cylindrical-polar coordinates. However, functional forms for Φ for the atmosphere and the oceans, and $\mathscr{E}$ for the oceans, remain to be specified. (Specification of $\mathscr{E}$ for a dry atmosphere, and also for a simple moist atmosphere, has already been taken care of in Chapters 2 and 3, respectively, since this is both analytically tractable and instructive.)

In Chapters 7 and 8, suitable functional forms for Φ are systematically developed from first principles. Doing so defines the mean shape of the planet (classically called its *figure*) – which is *ellipsoidal* rather than spherical – and the planet's gravitational attraction, both above *and* below its geoid. These functional forms are deliberately chosen to be consistent with the Reference *Ellipsoid* used for the Global Positioning System.

At this juncture, we have the basic ingredients for a realistic, *atmospheric*, dynamical core. This is not the case, however, for a realistic, *oceanic*, dynamical core since we still do not have a suitable functional form for $\mathscr{E}$, not even for a pure-water ocean with no salt.[14] To address this deficiency requires a more comprehensive and far more complicated treatment of thermodynamics than the relatively simple one – described in Chapters 2 and 3 – that suffices for the atmosphere. This is given in Chapters 9–11, with the bonus of also describing phase changes of water substance between its vapour, liquid, and frozen forms. This is an important aspect of both atmospheric and oceanic models.

By the end of Chapter 12, a self-consistent, geopotential coordinate system for a *smooth*, ellipsoidal (as opposed to spherical) planet has been developed. Planets are not, however, generally smooth; an atmosphere or an ocean usually has an underlying orography or bathymetry, respectively. To address this, generalised vertical coordinates – with suitable lower and upper boundary conditions – are developed in Chapter 13. This includes terrain-following coordinates as a special case. These provide a natural representation of orography and bathymetry in atmospheric and oceanic models, respectively.

Chapter 14, on variational methods and Hamilton's principle of stationary action, introduces an alternative methodology for deriving governing equation sets. This often-overlooked subject

[13] Shallow coordinates are a non-Euclidean geometric distortion of corresponding, deep, unapproximated Euclidean coordinates.

[14] Whereas an incompressible pure-water ocean can be very useful in theoretical studies designed to aid conceptual development and interpretation of model output, it is insufficiently accurate for realistic quantitative modelling.

helps to set the scene for Part II on dynamical consistency. Chapter 15 answers the question: what do various conservation principles look like:

- In vector form and in axial-orthogonal-curvilinear coordinates?
- In the presence of internal and external (conservative and non-conservative) forcings?

This prepares the way for constructing diagnostic budget relations for monitoring model integrations. It too helps to set the scene for Part II on dynamical consistency.

Part II

This part provides a means to understand the significance and potential impact of approximating the general set of governing dynamical-core equations, with emphasis on the importance of maintaining analogues in any approximated equation set of physical conservation laws embodied in the general, unapproximated set. Chapters in Part II are ordered from least approximated (3D) equation sets (when compared to the most general one developed in earlier chapters) to most approximated (2D) ones. Chapters 18 and 19 on 2D shallow-water equation sets and the 2D barotropic-potential-vorticity (BPV) equation then naturally follow on from Chapters 16 and 17 on dynamical consistency in 3D.

Part III

In this part, some exact, steady and unsteady, non-linear solutions of governing dynamical-core equation sets are developed for testing dynamical cores, once constructed. Chapters in Part III are ordered from the simplest exact solutions to the most complex ones. Some steady non-linear solutions are developed in Chapters 20 and 21. Some unsteady ones are similarly developed in Chapters 22 and 23.

An Appendix on Vectors

Vectors are used in many chapters, particularly in the early ones. For convenience, and for reference purposes, a comprehensive set of vector identities is assembled in the Appendix at the end of this book.

2

Governing Equations for Motion of a Dry Atmosphere: Vector Form

ABSTRACT

The governing equations for modelling Earth's global atmospheric and oceanic fluid flow are based on conservation principles for momentum, mass and energy, and specification of the state of the fluid in local, thermodynamic equilibrium. To develop a unified set of governing equations based on these principles, applicable to both the atmosphere and the oceans, is not straightforward. This is because the composition of the two fluids (predominantly gaseous for the atmosphere and liquid for the oceans) is very different. The differences in composition necessitate different representations of the associated physics, particularly for thermodynamic processes. A step-by-step approach to formulating the governing equations is taken. A governing equation set is classically developed in this chapter for the comparatively simple case of a dry atmosphere composed of a mixture of ideal gases of constant composition. This set is expressed in vector form in a rotating frame of reference. Various commonly used equivalent forms for the associated thermodynamic-energy equation are given. The dry-atmospheric formulation of this chapter is then extended to the following:

- A cloudy atmosphere (in Chapter 3).
- A general geophysical fluid (in Chapter 4), for application not only to the atmosphere, but also to the oceans.

2.1 PREAMBLE

The governing equations for modelling Earth's global atmospheric and oceanic fluid flow are based on three conservation principles, namely those for:

1. Momentum (as expressed by Newton's second law of motion).
2. Mass.
3. Energy.

It is noteworthy that these same principles also apply in the context of the atmospheres and oceans of other planets and, where applicable, also their moons. Indeed much of the content of this book has been written with this broader context in mind – albeit often more in the background than in the foreground! Vallis (2019) provides a concise and informative overview of planetary atmospheres and their dynamics, and Ingersoll (2013) a more comprehensive one. Although the same principles apply, other planetary atmospheres can, and usually do, behave in detail in a significantly different manner to Earth's. The chemical composition of their atmospheres and distance from their star – which influences atmospheric temperature – are important factors for this.

To develop a unified set of governing equations based on these three conservation principles (for momentum, mass, and energy), applicable to Earth's atmosphere and oceans, is not straightforward. As discussed in Chapter 1, the atmosphere is essentially gaseous, whereas the oceans are essentially liquid. That they are both fluids greatly simplifies matters. In principle, all we have to do is to write down (in vector form and/or in a suitable coordinate system) well-established equations for fluid flow. In practice, however, differences in the composition of the two fluids greatly complicate matters. These differences in composition necessitate different representations of the associated physics, particularly for the specification of equations of thermodynamic state. A step-by-step approach to formulating the governing equations is therefore taken herein.

We begin (in the present chapter) by defining a relatively simple, mathematical idealisation of the atmosphere. Complexity is then progressively increased in the following two chapters to ultimately obtain (in Chapter 4) a unified, general set of governing equations written in vector form for quantitative modelling of geophysical fluids such as Earth's atmosphere and oceans.

A significant virtue of proceeding in this manner is that the development of the governing equations for a dry or cloudy atmosphere – in this chapter and in Chapter 3, respectively – is analytically a lot more tractable than similarly doing so for the oceans. This eases conceptual understanding and, practically speaking, allows one to go much further in a realistic manner. The underlying reason for this is related to the thermodynamics of air versus that of seawater. For both dry and cloudy air it is relatively straightforward to formulate excellent, realistic approximations to an equation of state. For the oceans, the thermodynamic properties of seawater mean that no such approximation of comparable simplicity and breadth of applicability is available.

The remainder of the present chapter is organised as follows. Some well-known fundamental concepts and equations of fluid dynamics are assembled in Section 2.2. These include:

- Definition of fluid parcels, velocity, and material derivatives.
- Statement of the momentum equation in vector form in an inertial frame of reference, followed by transformation to a rotating frame.
- Definition of various functional forms of the mass-continuity equation.

Various thermodynamic properties of an ideal gas are stated in Section 2.3. These include:

- Definition of the ideal-gas law and its associated thermodynamic-energy equation.
- Application of the ideal-gas law to a mixture of ideal gases of constant composition.
- Definition of various equivalent forms of the thermodynamic-energy equation for an ideal gas.

A set of equations (summarised in Section 2.4) governing the motion of a *dry atmosphere* then results from the developments of Sections 2.2 and 2.3. The ideal gas assumption, together with the assumption that the atmosphere is of constant composition, significantly simplify matters.

2.2 FLUID DYNAMICS

2.2.1 Fluid Parcels, Velocity, and Material Derivatives

2.2.1.1 *Fluid Parcels*

Classical mechanics describes the motion of *point particles* moving under the influence of forces acting upon them, with mass *discretely* distributed at a set of *points* (Goldstein et al., 2001). Fluid mechanics, however, generally considers mass to instead be *continuously* distributed within a domain composed of the infinite union of *infinitesimal volumes*, called *fluid parcels*,[1] that collectively span the domain. A fluid parcel also moves under the influence of forces acting upon it.

[1] Fluid parcels are also often termed *fluid elements* or *material elements*.

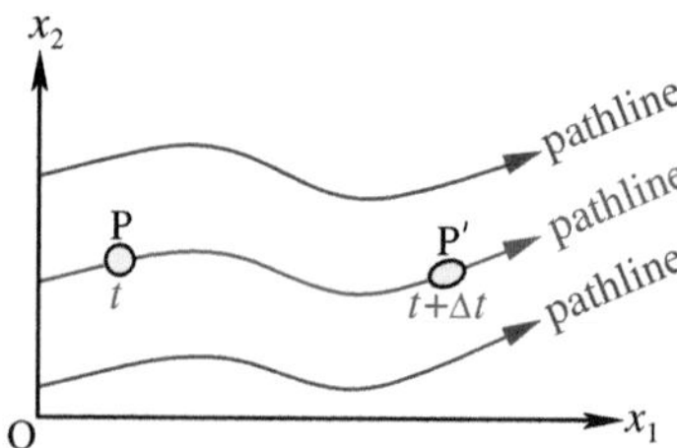

Figure 2.1 Path taken by a fluid parcel (depicted in yellow) embedded within a 2D flow field. The parcel moves along the red pathline; at time t it is located at point P, and by time $t + \Delta t$ it has arrived at point P′.

For present purposes, each (infinitesimal) fluid parcel is considered to:

- Be composed of indivisible fluid within its closed boundary.
- Have fixed mass as a consequence.
- Be continuous with its encompassing fluid environment.
- Deform, with possible change of volume, as it is transported by its encompassing fluid flow.

The path taken by such a fluid parcel – embedded within a two-dimensional (2D) flow field for simplicity – is depicted in Fig. 2.1. The parcel moves along the red pathline; at time t the parcel is located at point P, and by time $t + \Delta t$ it has arrived at point P′.[2] As the fluid parcel moves downstream along its pathline, its circular form at time t is gradually deformed to become oval by time $t + \Delta t$. This is due to the gradual convergence of the flow field in this illustrative example, leading to a convergence of the spacing between neighbouring pathlines downstream. Because, as the name suggests, a fluid parcel is composed of fluid material, it has associated properties:

- Volume (as illustrated in Fig. 2.1).
- Density.
- Pressure.
- Temperature.
- Velocity.

The associated properties of a fluid parcel in motion are functions of both time and space, where t is time and $\mathbf{r} = \mathbf{r}(t)$ is the position vector of the parcel at time t, measured from a fixed point in space; see Fig. 2.2. In three-dimensional (3D) Cartesian coordinates, for example, $\mathbf{r}(t) = [x_1(t), x_2(t), x_3(t)]$, with similar expressions in 2D and 1D.

To fully describe the motion of a fluid parcel, we need to be able to measure the time rate of change of the parcel's associated properties as the parcel is transported by the fluid flow in which it is embedded. This is where the *material derivative* enters the picture; it is very important in fluid dynamics.[3]

2.2.1.2 *The Velocity of a Fluid Parcel*

We begin by examining the time rate of change of a fluid parcel's position $\mathbf{r} = \mathbf{r}(t)$; this defines the parcel's velocity, which is a vector. Referring to Fig. 2.2, the fluid parcel at time t is located at point P, with position vector $\mathbf{r} = \mathbf{r}(t)$ (in green) with respect to a fixed origin at point O. The parcel is then transported by the flow in which it is embedded to arrive at point P′ at time $t + \Delta t$, with position vector $\mathbf{r} = \mathbf{r}(t + \Delta t)$ (in blue). Thus

$$\mathbf{r}(t + \Delta t) = \mathbf{r}(t) + \Delta \mathbf{r}, \tag{2.1}$$

[2] A pathline is the solution of $d\mathbf{r}/dt = \mathbf{u}$ over a given time period that passes through a given point at a given instant in time, where $\mathbf{r} = \mathbf{r}(t)$ is the position vector and $\mathbf{u} = \mathbf{u}(\mathbf{r}, t)$ defines the flow field.

[3] It is also termed total derivative, derivative following the fluid, substantive derivative, and Lagrangian derivative.

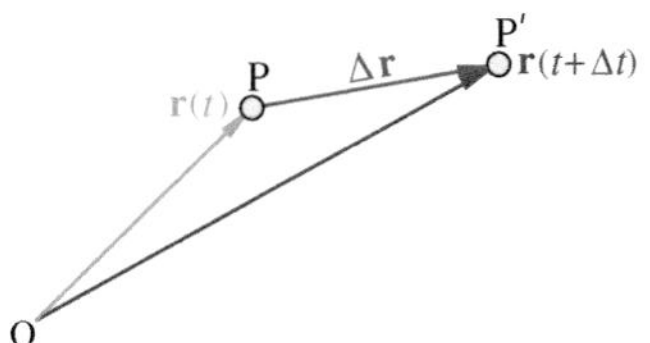

Figure 2.2 Displacement of a fluid parcel's position between times t and $t + \Delta t$. At time t, the parcel (depicted in yellow) is located at point P, with position vector $\mathbf{r} = \mathbf{r}(t)$ (in green) with respect to fixed origin at point O. By time $t + \Delta t$, it has moved to point P′, with position vector $\mathbf{r} = \mathbf{r}(t + \Delta t)$ (in blue). Its displacement vector between times t and $t + \Delta t$ is $\Delta\mathbf{r}$ (in red), so that $\mathbf{r}(t + \Delta t) = \mathbf{r}(t) + \Delta\mathbf{r}$. This diagram is drawn in the plane passing through (necessarily coplanar) points O, P, and P′.

where $\Delta\mathbf{r}$ is the displacement vector (in red) of the parcel's position during the time interval Δt. The time rate of change of the fluid parcel's position is then obtained by rearranging (2.1), dividing by Δt, and taking the limit $\Delta t \to 0$. Thus:

The Velocity of a Fluid Parcel

- The *velocity* of an (infinitesimal) fluid parcel is given by

$$\mathbf{u}(t) \equiv \lim_{\Delta t \to 0} \frac{\Delta\mathbf{r}}{\Delta t} \equiv \lim_{\Delta t \to 0} \frac{\mathbf{r}(t + \Delta t) - \mathbf{r}(t)}{\Delta t} \equiv \frac{d\mathbf{r}}{dt}. \tag{2.2}$$

- Velocity vector $\mathbf{u}$ is aligned with the direction of $\Delta\mathbf{r}$.

In Cartesian coordinates (x_1, x_2, x_3), (2.2) may be rewritten as

$$\mathbf{u} = (u_1, u_2, u_3) = u_1\mathbf{i}_1 + u_2\mathbf{i}_2 + u_3\mathbf{i}_3 = \frac{dx_1}{dt}\mathbf{i}_1 + \frac{dx_2}{dt}\mathbf{i}_2 + \frac{dx_3}{dt}\mathbf{i}_3 = \left(\frac{dx_1}{dt}, \frac{dx_2}{dt}, \frac{dx_3}{dt}\right) = \frac{d\mathbf{r}}{dt}, \tag{2.3}$$

where components are taken with respect to the Cartesian unit-vector triad $(\mathbf{i}_1, \mathbf{i}_2, \mathbf{i}_3)$. These three unit vectors are aligned with the three coordinate axes Ox_1, Ox_2, and Ox_3, respectively, in the usual way.

2.2.1.3 *The Material Derivative of a Scalar Fluid Property*

We now examine the time rate of change of a generic, scalar, fluid property $F = F[\mathbf{r}(t), t]$ – such as volume, density, pressure, or temperature – associated with a fluid parcel as it is transported by the fluid flow within which it is embedded. Referring to Fig. 2.2, we again assume that the fluid parcel moves from point P, located at $\mathbf{r} = \mathbf{r}(t)$ at time t, to point P′, located at $\mathbf{r} = \mathbf{r}(t + \Delta t)$ at time $t + \Delta t$. Let the *total change* in F by the end of this time interval be $\Delta F \equiv F[\mathbf{r}(t + \Delta t), t + \Delta t] - F[\mathbf{r}(t), t]$.

For simplicity, let us use Cartesian coordinates $\mathbf{r}(t) = [x_1(t), x_2(t), x_3(t)]$. Using the chain rule of differential calculus, the total change in $F = F[x_1(t), x_2(t), x_3(t), t]$ during time Δt is given by

$$\Delta F[x_1(t), x_2(t), x_3(t), t] = \frac{\partial F}{\partial t}\Delta t + \frac{\partial F}{\partial x_1}\Delta x_1 + \frac{\partial F}{\partial x_2}\Delta x_2 + \frac{\partial F}{\partial x_3}\Delta x_3, \tag{2.4}$$

where Δx_1, Δx_2, and Δx_3 are the corresponding individual changes in x_1, x_2, and x_3, respectively. It follows that the total rate of change in F is

$$\frac{dF}{dt} \equiv \lim_{\Delta t \to 0} \frac{\Delta F}{\Delta t} = \lim_{\Delta t \to 0} \left(\frac{\partial F}{\partial t} + \frac{\partial F}{\partial x_1}\frac{\Delta x_1}{\Delta t} + \frac{\partial F}{\partial x_2}\frac{\Delta x_2}{\Delta t} + \frac{\partial F}{\partial x_3}\frac{\Delta x_3}{\Delta t} \right)$$

$$= \frac{\partial F}{\partial t} + \frac{\partial F}{\partial x_1}\frac{dx_1}{dt} + \frac{\partial F}{\partial x_2}\frac{dx_2}{dt} + \frac{\partial F}{\partial x_3}\frac{dx_3}{dt}. \tag{2.5}$$

Using (2.3) to eliminate dx_1/dt, dx_2/dt, and dx_3/dt from (2.5) in favour of u_1, u_2, and u_3, followed by rearrangement, then leads to

$$\frac{dF}{dt} = \frac{\partial F}{\partial t} + u_1\frac{\partial F}{\partial x_1} + u_2\frac{\partial F}{\partial x_2} + u_3\frac{\partial F}{\partial x_3} = \frac{\partial F}{\partial t} + (\mathbf{u}\cdot\nabla)\,F, \tag{2.6}$$

where $\mathbf{u}\cdot\nabla$ is the *advection operator*, and

$$\mathbf{u}\cdot\nabla \equiv u_1\frac{\partial}{\partial x_1} + u_2\frac{\partial}{\partial x_2} + u_3\frac{\partial}{\partial x_3}, \tag{2.7}$$

in Cartesian coordinates.

Equation (2.6) expresses the total rate of change, dF/dt, of a generic scalar fluid property (F) associated with a fluid parcel. Because the d/dt operator of (2.6) appears so frequently in fluid dynamics, and also to distinguish it from the ordinary-derivative operator, operator d/dt in (2.6) is frequently denoted by D/Dt in the literature. We will follow this practice herein. Thus:

The Material Derivative

- The material derivative D/Dt is defined to be

$$\frac{D}{Dt} \equiv \frac{\partial}{\partial t} + (\mathbf{u}\cdot\nabla). \tag{2.8}$$

- It is also termed *derivative following the fluid, substantive derivative*, or *Lagrangian derivative* in the literature, according to author preference.

Defined this way, the material derivative can be applied to a scalar quantity (as in this section) or to a vector quantity (as in the next section). Definition (2.8) holds not only in Cartesian coordinates but also more generally; for example, in the orthogonal-curvilinear coordinate systems given in Chapter 5.

The preceding development of the material-derivative operator is primarily mathematical and somewhat abstract. To help make things clearer, it is instructive to visualise and interpret it in 1D and 2D. Referring to Fig. 2.3, a fluid parcel (denoted by a yellow circle) is initially located at point P at time t, and then transported by the encompassing fluid flow along vector $\overrightarrow{PP'}$ (in blue) to arrive at point P′ at time $t+\Delta t$. The time rate of change of scalar fluid property F along $\overrightarrow{PP'}$ is the material derivative of F (i.e. DF/Dt). It is *as if* the fluid parcel:

1. First moved along the green vector $\overrightarrow{PN}$ from point P to point N – carrying scalar fluid property F along with it – under the influence of partial time derivative $\partial F/\partial t$.[4]
2. Then moved along the red vector $\overrightarrow{NP'}$ from point N to point P′ – carrying scalar fluid property F along with it – under the influence of advection $\mathbf{u}\cdot\nabla F$ at constant $t+\Delta t$.

(An alternative route would be along the dashed vectors $\overrightarrow{PM}$ (in red) and $\overrightarrow{MP'}$ (in green) in Fig. 2.3.)

[4] This corresponds to a *local* time rate of change of F at a *fixed location* in space.

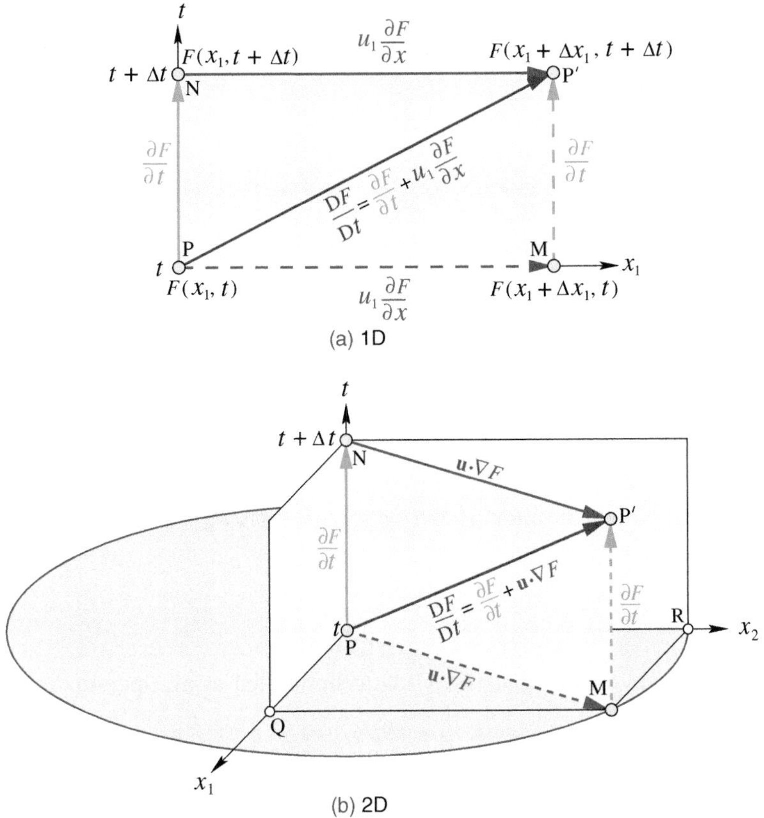

Figure 2.3 Visualisation of material derivative $DF/Dt \equiv \partial F/\partial t + \mathbf{u} \cdot \nabla F$ in (a) 1D and (b) 2D. Time t is depicted in the vertical direction, and space in the horizontal direction. A fluid parcel (denoted by a yellow circle) is initially located at point P at time t and then transported by the fluid flow along vector $\overrightarrow{PP'}$ (in blue) to arrive at point P′ at time $t + \Delta t$. The time rate of change of scalar fluid property F along $\overrightarrow{PP'}$ is the material derivative of F, (i.e. DF/Dt). It can be decomposed into the sum of its partial time derivative $\partial F/\partial t$ along $\overrightarrow{PN}$ (in green) and its advection $\mathbf{u} \cdot \nabla F$ along $\overrightarrow{NP'}$ (in red) at constant $t + \Delta t$. See main text for further details.

In reality, the fluid parcel moves along the blue vector $\overrightarrow{PP'}$ – carrying scalar fluid property F along with it – under the *simultaneous* influence of the partial time derivative $\partial F/\partial t$ and advection operator $\mathbf{u} \cdot \nabla F$, so that $DF/Dt = \partial F/\partial t + \mathbf{u} \cdot \nabla F$. The actual path along $\overrightarrow{PP'}$ is just the vector sum of the separate paths along $\overrightarrow{PN}$ and $\overrightarrow{NP'}$, that is,

$$\overrightarrow{PP'} = \overrightarrow{PN} + \overrightarrow{NP'}. \tag{2.9}$$

As an illustrative example of the usefulness of the material derivative, consider the special case of a fluid that is of constant density everywhere within the domain, including within any fluid parcel. Since we have assumed that the mass of a (closed) fluid parcel is constant for all time, its volume (which is equal to its mass divided by its density) must also be constant for all time for a fluid of constant density. This then means that

$$\frac{DF}{Dt} = 0, \tag{2.10}$$

where F, *for this special case*, can be the parcel's mass, its volume, or its density. This is an example of what is termed *material conservation of a scalar* or *passive advection of a scalar*. Physically, the associated quantity (F) within a fluid parcel that is advected by the fluid flow field $\mathbf{u}$ within which it is embedded preserves its initial value for all time. As we will see, other examples for F – in the absence of forces acting on the fluid parcel – that satisfy (2.10) include specific humidity in the atmosphere and salinity in the oceans, to name but two.

2.2.1.4 The Material Derivative of a Vector

The concept of the material derivative of a scalar property of a fluid parcel similarly extends to the material derivative of a vector property, such as velocity $\mathbf{u} = \mathbf{u}(\mathbf{r}, t)$. Applying material-derivative operator (2.8) to a generic vector property, $\mathbf{A} = \mathbf{A}(\mathbf{r}, t)$, gives:

The Material Derivative of a Vector

The material derivative of a vector $\mathbf{A}$ is

$$\frac{D\mathbf{A}}{Dt} \equiv \frac{\partial \mathbf{A}}{\partial t} + (\mathbf{u} \cdot \nabla)\,\mathbf{A}, \tag{2.11}$$

where the parentheses signify that $(\mathbf{u} \cdot \nabla)$ is to be treated as an operator that acts on vector $\mathbf{A}$.

The use of parentheses when $\mathbf{u} \cdot \nabla$ acts on a scalar property, F, is optional. Treating $\mathbf{u} \cdot \nabla$ as an operator acting on F unambiguously gives exactly the same result as first computing ∇F and then taking its scalar product with $\mathbf{u}$. This behaviour does not carry over to a vector property $\mathbf{A}$.

It is emphasised here that care needs to be taken when evaluating material derivatives of *vector* quantities in *general* coordinate systems, such as orthogonal, curvilinear ones. To illustrate this, we first apply (2.11) in *Cartesian coordinates* (x_1, x_2, x_3) to obtain

$$\begin{aligned}
\frac{D\mathbf{A}}{Dt} &\equiv \left[\frac{\partial}{\partial t} + (\mathbf{u} \cdot \nabla)\right]\mathbf{A} = \left(\frac{\partial}{\partial t} + u_1\frac{\partial}{\partial x_1} + u_2\frac{\partial}{\partial x_2} + u_3\frac{\partial}{\partial x_3}\right)(A_1\mathbf{i}_1 + A_2\mathbf{i}_2 + A_3\mathbf{i}_3) \\
&= \left[\left(\frac{\partial}{\partial t} + u_1\frac{\partial}{\partial x_1} + u_2\frac{\partial}{\partial x_2} + u_3\frac{\partial}{\partial x_3}\right)A_1\right]\mathbf{i}_1 + \left[\left(\frac{\partial}{\partial t} + u_1\frac{\partial}{\partial x_1} + u_2\frac{\partial}{\partial x_2} + u_3\frac{\partial}{\partial x_3}\right)A_2\right]\mathbf{i}_2 \\
&\quad + \left[\left(\frac{\partial}{\partial t} + u_1\frac{\partial}{\partial x_1} + u_2\frac{\partial}{\partial x_2} + u_3\frac{\partial}{\partial x_3}\right)A_3\right]\mathbf{i}_3 \\
&= \left(\frac{DA_1}{Dt}\right)\mathbf{i}_1 + \left(\frac{DA_2}{Dt}\right)\mathbf{i}_2 + \left(\frac{DA_3}{Dt}\right)\mathbf{i}_3,
\end{aligned} \tag{2.12}$$

where $(\mathbf{i}_1, \mathbf{i}_2, \mathbf{i}_3)$ is the unit-vector triad of Cartesian coordinates. The second-to-fourth lines of (2.12) hold since

$$\frac{\partial \mathbf{i}_j}{\partial t} = \frac{\partial \mathbf{i}_j}{\partial x_1} = \frac{\partial \mathbf{i}_j}{\partial x_2} = \frac{\partial \mathbf{i}_j}{\partial x_3} = 0, \quad j = 1, 2, 3, \tag{2.13}$$

in Cartesian coordinates (i.e. the unit-vector triad $(\mathbf{i}_1, \mathbf{i}_2, \mathbf{i}_3)$ for these coordinates is fixed in both time and space).

For *more general coordinate systems*, with coordinates (ξ_1, ξ_2, ξ_3) and unit-vector triad $(\mathbf{e}_1, \mathbf{e}_2, \mathbf{e}_3)$ (say), (2.13) (with $\mathbf{i}$'s replaced by $\mathbf{e}$'s) no longer holds and, consequently, neither does (2.12). For such coordinates

$$\frac{D\mathbf{A}}{Dt} = \frac{D}{Dt}(A_1\mathbf{e}_1 + A_2\mathbf{e}_2 + A_3\mathbf{e}_3) = \frac{DA_1}{Dt}\mathbf{e}_1 + \frac{DA_2}{Dt}\mathbf{e}_2 + \frac{DA_3}{Dt}\mathbf{e}_3 + \underbrace{A_1\frac{D\mathbf{e}_1}{Dt} + A_2\frac{D\mathbf{e}_2}{Dt} + A_3\frac{D\mathbf{e}_3}{Dt}}_{\neq 0}, \tag{2.14}$$

with *extra contributions* (compared to Cartesian coordinates) from $D\mathbf{e}_1/Dt$, $D\mathbf{e}_2/Dt$, and $D\mathbf{e}_3/Dt$. See Chapter 5 for various examples in orthogonal-curvilinear coordinate systems, such as spherical-polar coordinates and cylindrical-polar coordinates. Thus, *in general*,

$$\frac{D\mathbf{A}}{Dt} \neq \frac{DA_1}{Dt}\mathbf{e}_1 + \frac{DA_2}{Dt}\mathbf{e}_2 + \frac{DA_3}{Dt}\mathbf{e}_3. \tag{2.15}$$

Equality only holds for very special coordinates, such as Cartesian coordinates.

2.2.2. The Momentum Equation

2.2.2.1 *Preamble*

Newton's laws of motion are valid in an *inertial* (or *absolute*, or '*star-fixed*', or *non-rotating*) frame of reference (Goldstein et al., 2001, p. 2).[5] They are *not* directly applicable in a rotating frame, such as the one that rotates about Earth's rotation axis with its rotation rate. The starting point for the development of the momentum equation is therefore to apply Newton's second law of motion in an inertial frame of reference, where it is valid. This equation is then (exactly) transformed to express it in a *rotating* frame. This latter frame is more convenient for many purposes. In particular, important physical/dynamical balances are more naturally expressed in a rotating frame (see Section 7.3). Furthermore, most meteorological and oceanographical observations for Earth are taken with respect to its rotating frame (see Section 12.2).

2.2.2.2 *In an Inertial Frame*

Newton's second law of motion for a body of mass m is

$$\frac{d}{dt}(m\mathbf{u}) = \mathfrak{F}, \tag{2.16}$$

where $\mathbf{u}$ is its velocity, $m\mathbf{u}$ is its linear momentum, t is time, $\mathfrak{F}$ is the vector sum of all forces acting on the body, and $\mathfrak{F}$ and $\mathbf{u}$ are measured in an *inertial* frame of reference.

Application of this law, in this frame of reference, to a *fluid parcel* (or *material volume element* or *fluid particle*) of *unit mass* leads to the momentum equation

$$\left(\frac{D\mathbf{u}}{Dt}\right)_I = \left(\frac{1}{\rho}\mathfrak{F}\right)_I = \left(-\frac{1}{\rho}\nabla p - \nabla V + \mathbf{F}\right)_I. \tag{2.17}$$

Here:

- Subscript 'I' denotes measurement by an observer in an inertial frame of reference.
- $(\mathbf{r})_I$ is the position vector (relative to a fixed origin) of the fluid parcel.
- $(\mathbf{u} \equiv D\mathbf{r}/Dt)_I$ is its velocity.
- $\left(D\mathbf{u}/Dt \equiv D^2\mathbf{r}/Dt^2\right)_I$ is its acceleration.
- ρ is its density (i.e. mass per unit volume).
- p is pressure exerted, normal to its surface, by the encompassing fluid.
- ∇ is the 3D (three-dimensional) gradient operator.
- V is the potential of Newtonian gravity (a lot more on this in Chapters 7, 8, and 12).

5 After Isaac Newton (1643–1726).

Furthermore,

$$\left(\frac{D}{Dt}\right)_I \equiv \left(\frac{\partial}{\partial t} + \mathbf{u} \cdot \nabla\right)_I \tag{2.18}$$

is the *material rate of change* (of an operand) measured by an observer in the inertial frame; see Section 2.2.1.

The first two terms on the rightmost side of (2.17) are the pressure-gradient and gravitational forces (per unit mass), respectively, acting on a fluid parcel, and $\mathbf{F}$ is the sum of any other forces (per unit mass), such as friction:[6]

- The pressure applied to an infinitesimal fluid parcel by the fluid encompassing it (or, conversely, by the fluid parcel on the encompassing flow) generally varies as a function of position on the parcel's surface. This then gives rise to a vector force, $(-1/\rho)\,\nabla p$, proportional to the gradient of the pressure acting (inwards or outwards) on the surface of the fluid parcel.
- The potential, V, of Newtonian gravity is assumed to be of *known, specified form*. Determining a suitable, approximate, functional form for V that meets all requirements is highly challenging, and not at all obvious; see Chapters 7, 8, and 12.

Terms in (2.17) are arranged so that acceleration is placed on the left-hand side, with all forces placed on the right-hand side.

2.2.2.3 *In a Rotating Frame*

Let subscript 'R' denote measurement in a *rotating* frame of reference. For a *scalar* field, its material rate of change, $(D/Dt)_R$, observed in a frame rotating with uniform angular velocity $\boldsymbol{\Omega}$ relative to an inertial frame, is the same as its rate of change, $(D/Dt)_I$, observed in an inertial frame (see e.g. Section 2.1.4 of Vallis, 2017). (For an example of a rotating frame of reference embedded within an inertial frame, see Fig. 2.4.) However, this property does not (in general) hold for *vector* fields, due to different rates of change of direction being observed in the inertial and rotating frames.[7]

To transform (2.17), valid in an inertial frame, into an equivalent equation in the (uniformly) rotating frame, we need to relate $(D\mathbf{u}/Dt)_I$ to $(D\mathbf{u}/Dt)_R$. To do so, we first note that

$$\left(\frac{D}{Dt}\right)_I \mathbf{A} = \left(\frac{D}{Dt}\right)_R \mathbf{A} + \boldsymbol{\Omega} \times \mathbf{A}, \tag{2.19}$$

for arbitrary vector $\mathbf{A}$ (Goldstein et al., 2001, Chapter 4.9). Let $\mathbf{r}$ be the position vector, relative to a fixed point O on the axis of rotation,[8] of a fluid parcel. Setting $\mathbf{A} \equiv \mathbf{r} = (\mathbf{r})_I = (\mathbf{r})_R$ in (2.19) then gives:[9]

$$\begin{aligned} \left(\frac{D\mathbf{r}}{Dt}\right)_I &= \left(\frac{D}{Dt}\right)_I (\mathbf{r})_I = \left(\frac{D}{Dt}\right)_R (\mathbf{r})_I + \boldsymbol{\Omega} \times (\mathbf{r})_I = \left(\frac{D}{Dt}\right)_R (\mathbf{r})_R + \boldsymbol{\Omega} \times (\mathbf{r})_R \\ &= \left(\frac{D\mathbf{r}}{Dt} + \boldsymbol{\Omega} \times \mathbf{r}\right)_R . \end{aligned} \tag{2.20}$$

6 For Newtonian fluids, such as air and water, the frictional force can be well approximated by $\mathbf{F}^{\text{friction}} = \nu\left[\nabla^2\mathbf{u} + \nabla\left(\nabla \cdot \mathbf{u}\right)/3\right]$, where $\nu \equiv \mu/\rho$ is *kinematic viscosity* and μ is *viscosity*. The closer the flow is to being incompressible (for which $\nabla \cdot \mathbf{u} \equiv 0$), the less important is the second contribution to $\mathbf{F}^{\text{friction}}$. In quantitative global models of Earth's atmosphere and oceans, frictional effects are, however, usually parametrised in terms of other variables.

7 For example, to an observer in the rotating frame, the unit vector $\mathbf{e}_\lambda$ on Fig. 2.4 is perceived to be stationary; it does not move. However, for an observer in the inertial frame, it is perceived to be rotating about the axis of rotation (Oz on Fig. 2.4), with the rotation rate $\boldsymbol{\Omega}$ of the rotating frame.

8 This point – see Fig. 2.4 – is usually located at a planet's centre of mass.

9 Position vector $\mathbf{r}$ is the same in both frames since it does not involve D/Dt, the rate of change, with respect to time. It is the time rate of change that causes $\mathbf{u} \equiv D\mathbf{r}/Dt$ to behave differently to $\mathbf{r}$ under transformation from an inertial frame to a rotating one.

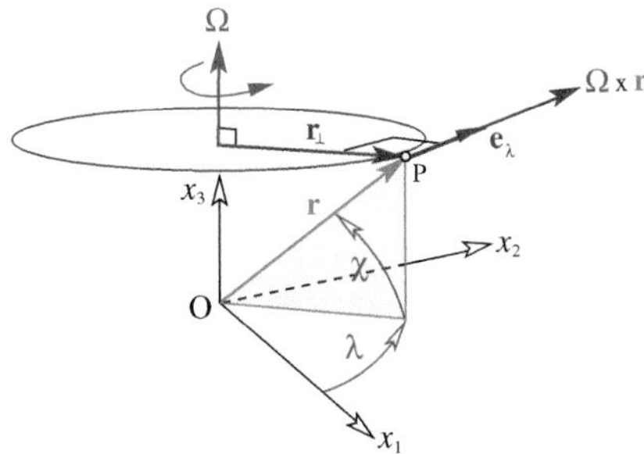

Figure 2.4 Frame $Ox_1x_2x_3$ rotates with uniform angular velocity $\boldsymbol{\Omega}$ about its x_3 axis. Point P is fixed in $Ox_1x_2x_3$ and has position vector $\mathbf{r}$ relative to O. $\mathbf{r}_\perp$ is the perpendicular from rotation axis Ox_3 to P; $\mathbf{e}_\lambda$ is the unit vector in the zonal direction at P (perpendicular to the plane containing $\boldsymbol{\Omega}$, $\mathbf{r}$, and $\mathbf{r}_\perp$). The velocity of point P relative to the inertial frame in which $Ox_1x_2x_3$ rotates is $\boldsymbol{\Omega} \times \mathbf{r}_\perp = |\boldsymbol{\Omega}|\,|\mathbf{r}_\perp|\,\mathbf{e}_\lambda = |\boldsymbol{\Omega}|\,|\mathbf{r}|\cos\chi\,\mathbf{e}_\lambda = \boldsymbol{\Omega} \times \mathbf{r}$, where χ is the angle subtended by $\mathbf{r}$ with respect to the Ox_1x_2 plane. In a spherical-polar coordinate system in which Ox_3 is the polar axis, χ is the latitude of P, and λ is the longitude of P relative to Ox_1 as zero. After Fig. 1.1 of Staniforth et al. (2006) and Fig. 2.1 of Vallis (2017).

Now $(\mathbf{u})_I$ and $(\mathbf{u})_R$ are defined by the kinematic equations (cf. (2.2))

$$(\mathbf{u})_I \equiv \left(\frac{D\mathbf{r}}{Dt}\right)_I, \quad (\mathbf{u})_R \equiv \left(\frac{D\mathbf{r}}{Dt}\right)_R, \tag{2.21}$$

where $(D/Dt)_I$ is defined by (2.18), and

$$\left(\frac{D}{Dt}\right)_R \equiv \left(\frac{\partial}{\partial t} + \mathbf{u}\cdot\nabla\right)_R \tag{2.22}$$

is the material rate of change observed in the *rotating* frame. Using kinematic identities (2.21), (2.20) may therefore be rewritten as

$$(\mathbf{u})_I = (\mathbf{u} + \boldsymbol{\Omega} \times \mathbf{r})_R\,. \tag{2.23}$$

Setting $\mathbf{A} \equiv (\mathbf{u})_I$ in (2.19) and using (2.21) and (2.23), the accelerations $(D\mathbf{u}/Dt)_R$ and $(D\mathbf{u}/Dt)_I$ in the rotating and inertial frames, respectively, are related to one another by

$$\begin{aligned}
\left(\frac{D\mathbf{u}}{Dt}\right)_I &= \left(\frac{D}{Dt}\right)_I (\mathbf{u})_I = \left(\frac{D}{Dt} + \boldsymbol{\Omega}\times\right)_R (\mathbf{u})_I = \left(\frac{D}{Dt} + \boldsymbol{\Omega}\times\right)_R (\mathbf{u} + \boldsymbol{\Omega}\times\mathbf{r})_R \\
&= \left[\left(\frac{D}{Dt} + \boldsymbol{\Omega}\times\right)(\mathbf{u} + \boldsymbol{\Omega}\times\mathbf{r})\right]_R \\
&= \left[\frac{D\mathbf{u}}{Dt} + \boldsymbol{\Omega}\times\mathbf{u} + \cancelto{0}{\frac{D\boldsymbol{\Omega}}{Dt}} \times \mathbf{r} + \boldsymbol{\Omega}\times\frac{D\mathbf{r}}{Dt} + \boldsymbol{\Omega}\times(\boldsymbol{\Omega}\times\mathbf{r})\right]_R \\
&= \left[\frac{D\mathbf{u}}{Dt} + 2\boldsymbol{\Omega}\times\mathbf{u} + \boldsymbol{\Omega}\times(\boldsymbol{\Omega}\times\mathbf{r})\right]_R .
\end{aligned} \tag{2.24}$$

The $(D\boldsymbol{\Omega}/Dt)\times\mathbf{r}$ term has been dropped to obtain the last line of (2.24), since the rotation vector, $\boldsymbol{\Omega}$, is assumed independent of time.[10]

[10] In reality, Earth wobbles slightly about its axis, so that $D\boldsymbol{\Omega}/Dt$ is not identically zero (Barnes et al., 1983). This wobble is, however, generally considered to be negligible in the context of modelling Earth's atmosphere and oceans. $\boldsymbol{\Omega}$ is therefore assumed here to be independent of time.

The forces on the right-hand side of (2.17), observed in an inertial frame, remain unchanged in the rotating frame. Substitution of (2.24) into (2.17) then yields

$$\frac{D\mathbf{u}}{Dt} + 2\boldsymbol{\Omega} \times \mathbf{u} + \boldsymbol{\Omega} \times (\boldsymbol{\Omega} \times \mathbf{r}) = -\frac{1}{\rho}\nabla p - \nabla V + \mathbf{F}, \tag{2.25}$$

where $2\boldsymbol{\Omega} \times \mathbf{u}$ and $\boldsymbol{\Omega} \times (\boldsymbol{\Omega} \times \mathbf{r})$ are called the *Coriolis* and *centripetal accelerations* per unit mass, respectively. The subscript 'R' has been dropped from (2.25) in the interests of brevity.

Nomenclature

Henceforth, in the absence of an 'I' or 'R' subscript, subscript 'R' is implied, with evaluation in the *rotating* frame of reference understood.

Terms in (2.25) are still arranged, as in (2.17), so that the *entire acceleration* is placed on the left-hand side, with all of the real (i.e. physical) *forces* on the right-hand side. That there are now *three* acceleration terms on the left-hand side of (2.25) – rather than the *single*, original one on the left-hand side of (2.17) – is entirely due to the change of an observer's frame of reference from an inertial frame to a rotating one. However, it is traditional in the atmospheric and oceanic sciences (and also beneficial, as we will see) to move the second and third terms of the left-hand side of the equation to the right-hand side to obtain

$$\frac{D\mathbf{u}}{Dt} = -2\boldsymbol{\Omega} \times \mathbf{u} - \boldsymbol{\Omega} \times (\boldsymbol{\Omega} \times \mathbf{r}) - \frac{1}{\rho}\nabla p - \nabla V + \mathbf{F}. \tag{2.26}$$

The first two terms on the right-hand side of (2.26) are then termed the *Coriolis* ($-2\boldsymbol{\Omega} \times \mathbf{u}$) and *centrifugal* ($-\boldsymbol{\Omega} \times (\boldsymbol{\Omega} \times \mathbf{r})$) *forces* per unit mass, respectively. These two forces are *not* real (i.e. physical) forces, but *apparent* (or *pseudo*) forces introduced by the change of an observer's frame of reference from an inertial frame to a rotating one. Both originate from acceleration, and moving them from the left-hand side of (2.25) to the right-hand side does not change this. What does change is the *interpretation* of these two terms in the rearranged equation; see White (2016) for a related, illuminating discussion.

Their appearance on the right-hand side of (2.26) is *as if* they are real *forces*, even though they are not (Phillips, 1973; Goldstein et al., 2001; Vallis, 2017). The *real* forces, observed in both the inertial and rotating frames of reference, are the last three terms on the right-hand side. The remaining term on the left-hand side of (2.26) is the *apparent* acceleration, as observed in the rotating frame of reference. It is different from the real acceleration, as evidenced by the left-hand side of (2.24). To an observer in the rotating frame, *it appears as if* a fluid parcel has an acceleration $(D\mathbf{u}/Dt)_R$ moving under the influence of the combined (apparent plus real) forces appearing on the right-hand side of (2.26).

The (apparent) Coriolis force, $-2\boldsymbol{\Omega} \times \mathbf{u}$, is (by definition) minus twice the vector product of the planet's rotation vector, $\boldsymbol{\Omega}$, and a fluid parcel's velocity vector, $\mathbf{u}$. Consequently the Coriolis force:

1. Acts in a direction perpendicular to both $\boldsymbol{\Omega}$ and $\mathbf{u}$, in a plane perpendicular to the planet's rotation axis (and therefore parallel with its equatorial plane, i.e. the plane containing its equator).
2. Depends on the velocity, $\mathbf{u}$, of the fluid parcel.
3. Does no work on a fluid parcel (because it acts perpendicular to the velocity vector, $\mathbf{u}$).
4. Is inactive when either $\boldsymbol{\Omega} \equiv 0$ (i.e. when observed in an inertial frame of reference), or $\mathbf{u} \equiv 0$ (i.e. the fluid parcel is motionless when observed in the rotating frame), or $\mathbf{u}$ is parallel to the planet's rotation axis.

By way of contrast, the (apparent) centrifugal force, $-\boldsymbol{\Omega} \times (\boldsymbol{\Omega} \times \mathbf{r})$:

1. Acts outwards in the direction $\mathbf{r}_\perp$, perpendicular to the planet's rotation axis (see Fig. 2.4).

2. Depends on the position vector, $\mathbf{r}$, of the fluid parcel, but is independent of its velocity, $\mathbf{u}$.
3. Appears to do work on a fluid parcel.
4. Is inactive when either $\boldsymbol{\Omega} \equiv 0$ (i.e. when observed in an inertial frame of reference) or the fluid parcel lies on the planet's rotation axis (i.e. at or vertically above its two poles).

The Placement of Terms in (2.26)

A crucially important advantage of the placement of terms in (2.26) is that the (apparent) centrifugal force, $-\boldsymbol{\Omega} \times (\boldsymbol{\Omega} \times \mathbf{r})$, and the (real) Newtonian gravitational force, $-\nabla V$, can be combined into a single (apparent) *conservative* force.[a]

[a] A force is conservative if no work is done by it when moving a fluid parcel around a closed path. The force is then expressible as the gradient of a *potential*.

To do so, the centrifugal force, $-\boldsymbol{\Omega} \times (\boldsymbol{\Omega} \times \mathbf{r})$, is first rewritten as the gradient of a potential. Thus (see Fig. 2.4):

$$-\boldsymbol{\Omega} \times (\boldsymbol{\Omega} \times \mathbf{r}) = -\boldsymbol{\Omega} \times (\boldsymbol{\Omega} \times \mathbf{r}_\perp) = \Omega^2 \mathbf{r}_\perp = \nabla \left(\frac{\Omega^2 r_\perp^2}{2} \right) = \nabla \left[\frac{(\boldsymbol{\Omega} \times \mathbf{r}) \cdot (\boldsymbol{\Omega} \times \mathbf{r})}{2} \right], \quad (2.27)$$

where $\mathbf{r}_\perp$ points outwards from the axis of $\boldsymbol{\Omega}$ and has magnitude $r_\perp$ equal to perpendicular distance from this axis. This then allows (2.26) to be expressed more concisely. Thus:

The Momentum Equation in a Rotating Frame of Reference

The momentum equation in a rotating frame of reference is

$$\frac{D\mathbf{u}}{Dt} = -2\boldsymbol{\Omega} \times \mathbf{u} - \frac{1}{\rho}\nabla p - \nabla \Phi + \mathbf{F}, \quad (2.28)$$

where

$$\Phi \equiv V + V^C \equiv V - \frac{(\boldsymbol{\Omega} \times \mathbf{r}) \cdot (\boldsymbol{\Omega} \times \mathbf{r})}{2} \quad (2.29)$$

is the *potential of apparent gravity*;

$$V^C \equiv -\frac{(\boldsymbol{\Omega} \times \mathbf{r}) \cdot (\boldsymbol{\Omega} \times \mathbf{r})}{2} \quad (2.30)$$

is the *potential of the centrifugal force*;

$$\mathbf{g} \equiv -\nabla \Phi \quad (2.31)$$

is the gravitational force per unit mass;[a] and Φ is usually referred to as the *geopotential.*

[a] Strictly speaking, $\mathbf{g}$ is the force due to *apparent* gravity, but most published works simply refer to $\mathbf{g}$ as the gravitational force, or gravity.

The surfaces for which Φ = constant are usually termed *geopotential surfaces* or, occasionally, *level surfaces*. They are *very important* in meteorology and oceanography; see Chapters 7, 8, and 12, particularly Sections 7.3 and 12.2, for the reasons why this is so.

2.2.2.4 *The Vector-Invariant Form*

Use of (2.22) (material rate of change in a rotating frame) and vector identity (A.23) leads to

$$\frac{D\mathbf{u}}{Dt} + 2\boldsymbol{\Omega} \times \mathbf{u} \equiv \frac{\partial \mathbf{u}}{\partial t} + (\mathbf{u} \cdot \nabla)\,\mathbf{u} + 2\boldsymbol{\Omega} \times \mathbf{u} \equiv \frac{\partial \mathbf{u}}{\partial t} + \nabla\left(\frac{\mathbf{u} \cdot \mathbf{u}}{2}\right) + (2\boldsymbol{\Omega} + \nabla \times \mathbf{u}) \times \mathbf{u}$$
$$= \frac{\partial \mathbf{u}}{\partial t} + \mathbf{Z} \times \mathbf{u} + \nabla K, \tag{2.32}$$

where

$$\mathbf{Z} \equiv 2\boldsymbol{\Omega} + \nabla \times \mathbf{u}, \quad K \equiv \frac{\mathbf{u} \cdot \mathbf{u}}{2}. \tag{2.33}$$

In (2.32) and (2.33):

$2\boldsymbol{\Omega}$ = planetary vorticity,
$\nabla \times \mathbf{u}$ = relative vorticity,
$\mathbf{Z}$ = absolute vorticity (the sum of planetary and relative vorticities), and
K = kinetic energy.

These are all specific quantities (i.e. quantities *per unit mass*).

Substitution of (2.32) into momentum equation (2.28) then gives:

The Vector-Invariant Form of the Momentum Equation

$$\frac{\partial \mathbf{u}}{\partial t} = -\mathbf{Z} \times \mathbf{u} - \nabla\,(K + \Phi) - \frac{1}{\rho}\nabla p + \mathbf{F}. \tag{2.34}$$

Since this form has the same appearance in both the rotating and inertial frames of reference, it is usually termed the *vector-invariant form* of the momentum equation.[11] This form facilitates the derivation of an equation for the conservation of potential vorticity; see Section 15.3. It has also been used in the formulation of many numerical models over the years.

2.2.3 The Mass-Continuity Equation (Conservation of Mass)

Momentum equation (2.28) has three components in three dimensions. This provides three equations for the five, unknown, dependent variables $\mathbf{u} = (u_1, u_2, u_3)$, ρ, and p, that appear in momentum equation (2.28). (It is assumed here that the geopotential, Φ, is a known field of prescribed form and that the forces, $\mathbf{F}$, do not bring in any further unknowns.[12]) The mass-continuity equation, given here in various forms, provides a *prognostic equation* (i.e. a predictive equation, with a time rate of change) for the density, ρ. There are then four equations for the five unknown variables $\mathbf{u} = (u_1, u_2, u_3)$, ρ, and p. The pressure, p, then still needs to be related to the four variables $\mathbf{u} = (u_1, u_2, u_3)$ and ρ. To do so is generally quite complicated and brings in thermodynamics, as discussed in Section 2.3.

The conservation equation for mass is usually referred to as the *mass-continuity equation* or, for short, the *continuity equation*. For a thorough exposition of this equation, see Section 1.2 of Vallis (2017), where three derivations are given; one for the Lagrangian[13] form, and two for the Eulerian[14] one.[15]

11 This form is not, however, *frame invariant*. This is due to the presence of $\partial\mathbf{u}/\partial t$, which is not frame invariant; and also to the hidden absence, or presence, of $\boldsymbol{\Omega}$ in definition (2.33), according to whether it is expressed in the inertial or rotating frames, respectively.

12 If they do – as is often the case for comprehensive models of Earth's atmosphere and oceans – then further equations are required to account for this.

13 After Joseph-Louis Lagrange (1736–1813).

14 After Leonhard Euler (1707–83).

15 See also Section 14.2.1 herein for a comparative discussion of Lagrangian and Eulerian frameworks.

2.2.3.1 *Lagrangian Form*

Consider an arbitrary, *closed* parcel of fluid in motion, of volume $\iiint_V dV$, density ρ, and mass $\iiint_V \rho dV$. Because the parcel is confined (by its enclosing material surface), its mass must remain constant for all time. In other words, the mass (i.e. material) contained within the parcel is *materially conserved*, so that $D\left(\iiint_V \rho dV\right)/Dt = 0$. It then follows from the Reynolds' transport theorem (A.53) (with substitution of $f = \rho$ therein) that

$$\frac{D}{Dt}\iiint_V \rho dV = \iiint_V \left(\frac{D\rho}{Dt} + \rho\nabla\cdot\mathbf{u}\right) dV = 0. \tag{2.35}$$

Since the volume is arbitrary, the integrand of the middle term must be zero. Thus, expressed in an inertial frame of reference, the mass-continuity equation in *Lagrangian form* is

$$\left(\frac{D\rho}{Dt} + \rho\nabla\cdot\mathbf{u}\right)_I = 0. \tag{2.36}$$

Now the material derivative of a scalar is independent of the reference frame of observation. Applying (2.35) in a rotating frame of reference, instead of an inertial one, then leads in a similar manner to:

The Lagrangian Form of the Mass-Continuity Equation in a Rotating Frame of Reference

$$\frac{D\rho}{Dt} + \rho\nabla\cdot\mathbf{u} = 0, \tag{2.37}$$

where the absence of a subscript 'I' or 'R' implies evaluation in the rotating frame of reference.

Dividing through by ρ, (2.37) can be rewritten in logarithmic form as

$$\frac{D(\ln\rho)}{Dt} + \nabla\cdot\mathbf{u} = 0. \tag{2.38}$$

This form has been used in semi-implicit, semi-Lagrangian discretisations of the governing equations (Staniforth and Côté, 1991).[16]

2.2.3.2 *Eulerian Form*

With use of (2.22) (for the material rate of change in a rotating frame) and vector identity (A.17), (2.37) can be rewritten in the rotating frame. This gives:

The Eulerian Form of the Mass-Continuity Equation in a Rotating Frame of Reference

$$\frac{\partial\rho}{\partial t} + \nabla\cdot(\rho\mathbf{u}) = 0. \tag{2.39}$$

[16] It facilitates elimination of $\nabla\cdot\mathbf{u}$ at the new timestep between the continuity equation and the divergence of the momentum equation, ultimately leading to an elliptic boundary-value problem for pressure.

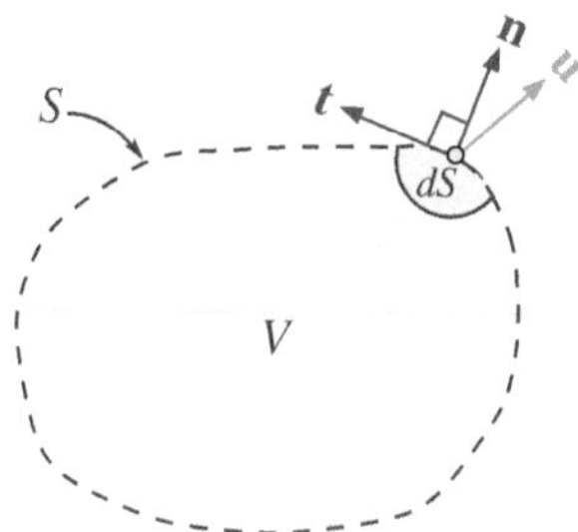

Figure 2.5 A fixed arbitrary volume V (shaded in light blue), with open surface S (dashed dark blue contour) and surface element dS (shaded in mid-blue). Normal and tangential unit vectors (in red) for dS are **n** and **t**, respectively. Velocity vector **u** for fluid exiting V across dS is denoted in green.

This form is written in terms of the fluxes of momentum ($\rho\mathbf{u}$) across the bounding surface of a volume *fixed in space*. As such, it is particularly useful for numerical models of the atmosphere and oceans discretised in an Eulerian manner. However, (2.37) is arguably the more natural form of the mass-continuity equation compared to (2.39), since its individual terms are all *frame invariant*. This is not so for (2.39) because $\partial\rho/\partial t$ is not frame invariant; see (2.21) of Vallis (2017).

An alternative derivation of (2.39) is as follows. For an arbitrary, notional volume V, *fixed in space* (see Fig. 2.5):

$$\text{The rate of change in the mass contained within it} = \frac{\partial}{\partial t}\iiint_V \rho dV = \iiint_V \frac{\partial\rho}{\partial t} dV. \qquad (2.40)$$

The rightmost term of this equation holds only because the volume is assumed to be fixed in space. The rate of change is positive when mass is increasing and negative when it is decreasing. Now the mass contained within V can only change due to mass entering or leaving across its *fixed, but open*, surrounding boundary surface, S.[17] Let dS be an infinitesimal surface element of boundary surface S, so that:

$$\text{The rate of change of mass per unit time across } dS = \mathbf{u}\cdot\mathbf{n}dS = \rho\mathbf{u}\cdot d\mathbf{S}, \qquad (2.41)$$

where $d\mathbf{S} \equiv \mathbf{n}dS$, and $\mathbf{n}$ is the outward-pointing normal from dS. For $\mathbf{u}\cdot\mathbf{n} > 0$, this corresponds to an outflow (decrease) of mass and, similarly, it corresponds to an inflow (increase) of mass for $\mathbf{u}\cdot\mathbf{n} < 0$. Integrating over the entire surface S, with use of Gauss's divergence theorem (A.33), then gives:

$$\text{The net outflow of mass per unit time across } S = \oiint_S \rho\mathbf{u}\cdot d\mathbf{S} = \iiint_V \nabla\cdot(\rho\mathbf{u})\, dV. \qquad (2.42)$$

If this integrated quantity is positive, then the mass contained within V decreases and, similarly, if it is negative, then the mass increases.

Equations (2.40) and (2.42) are two different ways of computing the rate of change of mass within V. After taking proper consideration of signs, they must agree with one another to ensure that mass is neither created nor destroyed (i.e. to ensure that mass is conserved). Thus

$$\iiint_V \left[\frac{\partial\rho}{\partial t} + \nabla\cdot(\rho\mathbf{u})\right] dV = 0. \qquad (2.43)$$

[17] This situation should not be confused with that examined earlier of a *moving* fluid element, *with a deformable, closed boundary*. The two situations are physically very different.

Since the volume (V) has been assumed to be arbitrary, for (2.43) to hold, its integrand must be identically zero. Thus

$$\frac{\partial \rho}{\partial t} + \nabla \cdot (\rho \mathbf{u}) = 0, \tag{2.44}$$

in agreement with (2.39).

2.2.3.3 *Specific Volume Form*

Specific volume, α, is defined to be the inverse of mass density, ρ. Thus:

The Definition of Specific Volume

$$\alpha \equiv \frac{1}{\rho}. \tag{2.45}$$

Physically, specific volume corresponds to the volume occupied by unit mass of a substance, and it is a natural and convenient variable for thermodynamics; see Sections 2.3, 3.2, 4.2, and 4.3, and Chapters 9–11.

Substitution of definition (2.45) into (2.37) yields:

The Mass-Continuity Equation in Specific Volume Form

$$\frac{D\alpha}{Dt} - \alpha \nabla \cdot \mathbf{u} = 0. \tag{2.46}$$

An alternative derivation of this equation may be found in Section 14.3.

Dividing through by α, (2.46) can also be rewritten in logarithmic form – cf. (2.38) – as

$$\frac{D(\ln \alpha)}{Dt} - \nabla \cdot \mathbf{u} = 0. \tag{2.47}$$

2.3 THERMODYNAMICS OF AN IDEAL GAS

2.3.1 Preamble

Momentum equation (2.28) provides three equations for its five unknown variables $\mathbf{u} = (u_1, u_2, u_3)$, ρ, and p. Mass-continuity equation (2.37) provides a fourth equation.[18] This means that at least one more equation is required. If pressure depends *only* on density (and vice versa), then an equation of the functional form

$$p = p(\rho) \quad \Leftrightarrow \quad \rho = \rho(p) \tag{2.48}$$

provides the missing equation, and we are done.[19]

Equation (2.48) is the definition of a *barotropic/homentropic* fluid.[20] (More generally, if (2.48) is not satisfied, then the fluid is termed *baroclinic/non-homentropic*.) Although the atmosphere

18 One of the alternative forms (2.38), (2.39), (2.46), or (2.47) can instead be used. If (2.46) or (2.47) is used, then $1/\rho$ in the momentum equation, (2.28), is replaced by α; see (2.45).

19 For example, consider an ideal gas that everywhere has the same constant (isothermal) temperature, $\overline{T}$. Then (from (2.50) of Section 2.3.2) $p = \rho R\overline{T} = \text{constant} * \rho$, and (2.48) is satisfied.

20 Terminology; barotropic is sometimes termed homentropic in oceanography.

and oceans can, in restricted circumstances, behave as barotropic fluids, in general they do not. Pressure (p) and density (ρ) are related not only to one another, but also to *temperature* (T). (In general, they are also related to other variables such as water vapour (in the atmosphere) and salinity (in the ocean); we return to this point in Chapters 3 and 4, and elsewhere.) Instead of (2.48), we therefore need an equation of the more general form

$$p = p\left(\rho, T, \ldots\right) \quad \Leftrightarrow \quad \rho = \rho\left(p, T, \ldots\right), \tag{2.49}$$

where '$, \ldots$' signifies the possibility of further variables being involved. The addition of T to the list of dependent variables means that we then need a further equation (more if there are additional variables in (2.49) or in this further equation). Keeping track of the accounting, this gives six equations (three momentum components + mass continuity + equation of state + thermodynamic energy), for the six fields $\mathbf{u} = (u_1, u_2, u_3)$, ρ, p, and T.

To fix ideas, we first discuss (in this chapter) the atmospheric case under the assumption that the atmosphere behaves as an *ideal gas*, which is both frictionless and dry (i.e. without any water content whatsoever). By assuming for now that the gas is *dry*, we avoid having to bring in further variables (such as water vapour, liquid water, frozen water, aerosols). This greatly simplifies matters. Furthermore, to avoid (again for now) getting into the complexities of thermodynamics, we simply write down various thermodynamic equations for ideal gases as and when needed. Forward references to other chapters for more detailed information are, however, given. These may be ignored on a first reading. The emphasis in this chapter is on assembling a set of governing equations suitable for describing the motion of a dry atmosphere. This then sets the scene for more detailed and more general developments in later chapters.

2.3.2 The Ideal-Gas Law

Although this is certainly not the case for the oceans, dry gases in Earth's atmosphere collectively behave, to an excellent approximation, as an ideal gas. The ideal-gas law may be obtained in various ways. Historically, it evolved empirically from Boyle's law ($p\alpha = \text{constant}$),[21] and Charles' law ($\alpha = \text{constant} * T$).[22] It was later obtained theoretically using the kinetic theory of gases and statistical mechanics. Thus:

The Ideal-Gas Law

The equation of state for an ideal gas is the *ideal-gas law*

$$p = \rho R T, \tag{2.50}$$

where R is the gas constant of the ideal gas.

Using definition (2.45) of specific volume, (2.50) can be equivalently rewritten as:

An Alternative Form of the Ideal-Gas Law

$$p\alpha = RT. \tag{2.51}$$

Taking the logarithm of both sides of (2.50) and (2.51), the ideal-gas law can also be written in logarithmic form as:

21 After Robert Boyle (1629–91).

22 After Jacques Alexandre César Charles (1746–1823).

A Further Alternative Form of the Ideal-Gas Law

$$\ln p = \ln \rho + \ln R + \ln T = -\ln \alpha + \ln R + \ln T. \tag{2.52}$$

These latter two forms of the ideal-gas law are useful for manipulating and interrelating various thermodynamic quantities; see Section 4.3, for example.

2.3.3 A Mixture of Inert Ideal Gases of Constant Composition

Equations (2.50)–(2.52) are valid not only for an *individual*, ideal gas but also for a *mixture, of constant composition*,[23] of *inert*,[24] ideal gases. To see this, consider a fixed volume, $\mathbb{V}^g$, occupied by a mixture of inert, ideal gases at temperature, T.[25] For this mixture of ideal gases, it is assumed that:

1. Each gas completely occupies the fixed *gaseous* volume, $\mathbb{V}^g$, and shares this volume with the other gases of the mixture.
2. Each gas obeys its own equation of state, (2.50), so that for the ideal gas of index i,
$$p^i = \rho^i R^i T. \tag{2.53}$$
3. The total pressure (p) exerted by the mixture of ideal gases is the sum of the partial pressures (p^i) exerted by the individual gases (of index i)[26], so that
$$p = \sum_i p^i. \tag{2.54}$$

Let $\mathbb{M}^{(i)}$ be the mass of an individual ideal gas, of index i, and let $\mathbb{M}^g$ be the total mass of the mixture of ideal gases, so that

$$\mathbb{M}^g \equiv \sum_i \mathbb{M}^{(i)}. \tag{2.55}$$

Now, by definition of density (i.e. mass per unit volume),

$$\rho^i \equiv \frac{\mathbb{M}^{(i)}}{\mathbb{V}^g}, \quad \rho^g \equiv \frac{\mathbb{M}^g}{\mathbb{V}^g}. \tag{2.56}$$

Substitution of (2.56) into (2.53) gives the partial pressures

$$p^i = \frac{\mathbb{M}^{(i)}}{\mathbb{V}^g} R^i T. \tag{2.57}$$

Using (2.54) to sum over the partial pressures (2.57) and using identities (2.55) and (2.56) then yields the total pressure

$$p = \sum_i p^i = \sum_i \left(\frac{\mathbb{M}^{(i)}}{\mathbb{V}^g} R^i T \right) = \left(\frac{\mathbb{M}^g}{\mathbb{V}^g} \right) \left(\sum_i \frac{\mathbb{M}^{(i)}}{\mathbb{M}^g} R^i \right) T = \rho^g R T, \tag{2.58}$$

23 In other words, the composition of the mixture is such that the individual masses of these gases, for any volume, are everywhere in the same proportion to one another.

24 An inert gas is considered here to be a gas that does not chemically react with other gases.

25 Although redundant for the situation considered here, superscript 'g' is appended to $\mathbb{V}$ to signify that volume $\mathbb{V}^g$ is occupied *solely by gases*. This is consistent with later use of $\mathbb{V}$ (in Section 3.3, and in the Appendix to Chapter 3) to include not only gaseous volume $\mathbb{V}^g$, *shared* by all gases, but also volumes $\mathbb{V}^l$ and $\mathbb{V}^f$, *separately occupied* by liquid water and frozen water, respectively. Distinguishing here between $\mathbb{V}$ and $\mathbb{V}^g$ avoids later ambiguity. Similar comments also apply for $\mathbb{M}^g$ – see (2.55).

26 This is Dalton's Law of Partial Pressures, after John Dalton (1766–1844).

where

$$R \equiv \sum_i \left(\frac{\mathbb{M}^{(i)}}{\mathbb{M}^g} R^i \right). \tag{2.59}$$

Now (see footnote 23) a mixture of gases is of *constant* composition if the individual masses of its constituent gases are everywhere in the same proportion for any sample volume. Thus the ratio $\mathbb{M}^{(i)}/\mathbb{M}^g$, termed a *mass fraction*, has the same value anywhere, and everywhere, within the domain. Consequently R, as defined by (2.59), is *constant throughout the domain*. Therefore, comparing (2.58) and (2.59) (for a *mixture* of ideal gases of constant composition) with (2.50) (for an *individual* ideal gas), it is seen that:

The Ideal-Gas Law for a Mixture of Inert Ideal Gases of Constant Composition

- The ideal-gas law (2.50) holds not only for an individual ideal gas but also for a mixture of (inert) ideal gases *of constant composition*.
- Constant R – as defined by (2.59) – is then the gas constant for the mixture, *of constant composition*, of (inert) ideal gases, and $\rho = \rho^g$ – defined by (2.56) – is the corresponding density of this mixture.

For the composition of Earth's dry air – see Table 1.1 – the experimentally measured value of R is approximately $R = 287\,\mathrm{J\,kg^{-1}\,K^{-1}}$.

2.3.4 Work Done by Expansion or Compression of a Gas

Consider, as depicted in Fig. 2.6, a piston that freely moves within a cylinder, closed at one end. Assume that a gas, confined within the cylinder by the piston, is undergoing expansion at a given instant in time, and let:

- $\mathbb{A}$ be the cylinder's cross-sectional area.
- $\mathbb{M}$ be the (constant) mass of the confined gas.
- x be the distance of the piston from the cylinder's closed end.
- $\mathbb{V}$ be the volume of the confined gas, at the instant in time.
- p be the pressure exerted *by the gas on the piston*, at the instant in time.

With these definitions, we now argue as follows:

- The force exerted by the gas on the piston is $p\mathbb{A}$ (= pressure $\times$ area).
- The work done *by* the gas to move the piston an infinitesimal distance, dx, along the cylinder is therefore $p\mathbb{A}dx$ (= force $\times$ distance). But $\mathbb{A}dx$ is the infinitesimal change, $d\mathbb{V}$, in volume of the gas in moving the piston an infinitesimal distance dx. Thus the work done *by* the gas is $p\mathbb{A}dx = pd\mathbb{V}$.
- But, by definition (2.45), $\alpha \equiv 1/\rho = \mathbb{V}/\mathbb{M}$, where α is the specific volume of the gas. Thus $d\mathbb{V} = d(\mathbb{M}\alpha) = \mathbb{M}d\alpha$ (since the confined gas has *constant* mass, $\mathbb{M}$), and so the work done *by* the confined gas *on* the piston is $pd\mathbb{V} = p\mathbb{M}d\alpha$.
- Dividing by the constant mass, $\mathbb{M}$, the work done *by* the gas, *per unit mass*, to move the piston an infinitesimal distance dx by expansion is then $pd\alpha$.
- The work, $đW$, done *on* the gas *by* the piston is therefore the negative of this, and so

$$đW = -pd\alpha. \tag{2.60}$$

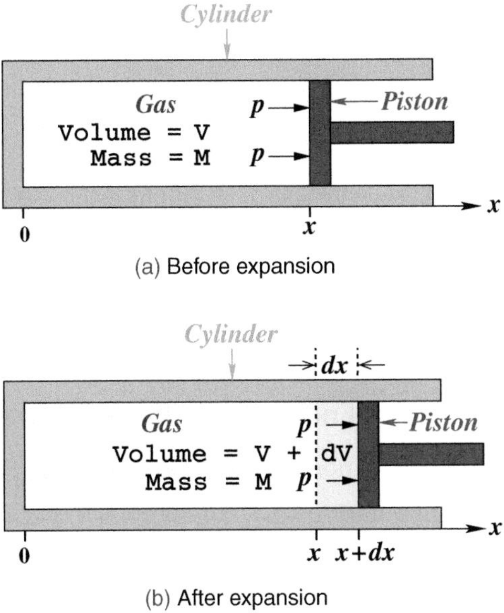

Figure 2.6 Work done by expansion of a gas on a piston that moves freely within a cylinder. Cross sections: (a) before gas expands; and (b) after gas has expanded and moved the piston a distance dx along the cylinder. The gas occupies volumes: (a) $V = \mathbb{V}$, before expansion; and (b) $V + dV = \mathbb{V} + d\mathbb{V}$, after expansion. See text for further details.

From (2.60):

The Rate of Work Done on a Gaseous Parcel of Air

The rate of work done[a] (per unit mass) *on* a gaseous parcel of air (transported by the flow, **u**) *by* its surrounding environment is

$$\dot{W} = -p\frac{D\alpha}{Dt}. \tag{2.61}$$

a A thermodynamic subtlety: $\text{đ}W$ in (2.60) is (see Section 4.2.4) an *imperfect* (as opposed to *exact*) differential that depends upon the thermodynamic path taken to go from one thermodynamic state of equilibrium to another. Therefore $\dot{W} \neq DW/Dt$ in (2.61). $\dot{W}$ here is simply shorthand for the work, $\text{đ}W$, done on the gas by the piston, *per unit time*, with the dot notation signifying this. A similar comment applies to $\dot{Q} \neq DQ/Dt$ in (2.63); again see Section 4.2.4.

Importantly, this equation holds for a general, compressible fluid.[27] If $D\alpha/Dt$ is positive, then the fluid parcel is undergoing expansion (and the fluid parcel therefore does work on its surrounding environment), but, if it is negative, then the fluid parcel is undergoing compression (and the surrounding environment therefore does work on the fluid parcel).

2.3.5 The Thermodynamic-Energy Equation for a Dry Ideal Gas

From the kinetic theory of gases and statistical mechanics:

[27] It also holds if different phases of a substance, such as water substance, are present within a mixture of substances. The key reason for this is that although a change of phase changes the *individual masses* of the phases, the *total mass*, $\mathbb{M}$, nevertheless remains unchanged. Thus $d\mathbb{V} = d\left(\mathbb{M}\alpha\right) = \mathbb{M}d\alpha$ still holds and still leads to (2.60) and (2.61).

The Internal Energy for a Dry Ideal Gas

- The *internal energy* of a dry ideal gas is

$$\mathscr{E} \equiv c_v T, \tag{2.62}$$

where T is temperature and c_v (a constant) is *specific heat* at constant *volume*; see (3.71)–(3.73) with $q^d = 1$ and $q^v = q^l = q^f = 0$.

Inserting this definition into (3.70):

The Thermodynamic-Energy Equation for a Dry Ideal Gas

- The thermodynamic-energy equation for a dry ideal gas undergoing reversible expansion or compression is

$$c_v \frac{DT}{Dt} + p \frac{D\alpha}{Dt} \equiv c_v \frac{DT}{Dt} + p \frac{D}{Dt}\left(\frac{1}{\rho}\right) = \dot{Q}, \tag{2.63}$$

where $\dot{Q}$ is *heating rate* per unit mass.
- This equation relates T to p and ρ, and thereby provides the missing equation – see Section 2.3.1 – to close the equation set for the six fields $\mathbf{u} = (u_1, u_2, u_3)$, ρ, p, and T.

The physical interpretation of the first two terms in (2.63) is as follows:

- $c_v DT/Dt = D(c_v T)/Dt = D\mathscr{E}/Dt$ is the rate of change of internal energy (per unit mass) of a dry ideal gas during reversible expansion or compression.
- $pD\alpha/Dt$ – see (2.61) – is the work done (per unit mass) *by* the gas *on* its surrounding environment or, equivalently, $-pD\alpha/Dt$ is the work done (per unit mass) *on* the gas *by* its surrounding environment.

An *adiabatic process* is, by definition, one in which no heat is transferred between a fluid parcel and the environment in which it is embedded. This corresponds to the cylinder's walls and the piston in Fig. 2.6 being perfect insulators, with no heat transfer between them and the confined gas. The heating rate, $\dot{Q}$, is then identically zero. Thus (2.61)–(2.63) show that:

- For an ideal gas undergoing an adiabatic process (so that $\dot{Q} \equiv 0$), the change in internal energy exactly balances the work done during the process.

2.3.6 Equivalent Forms for the Thermodynamic-Energy Equation of a Dry Ideal Gas

Using (2.45) (definition of ρ in terms of α), (2.50) or (2.51) (ideal-gas law), and Carnot's Law[28] – see (4.93) with $\widetilde{R} \equiv R$ for a dry ideal gas –

$$R = c_p - c_v, \tag{2.64}$$

where R is the gas constant, and c_p (another constant) is specific heat at constant *pressure*, (2.63) may be rewritten as:

[28] After Nicolas Léonard Sadi Carnot (1796–1832).

An Alternative Form of the Thermodynamic-Energy Equation for a Dry Ideal Gas

$$c_p \frac{DT}{Dt} - \alpha \frac{Dp}{Dt} = c_p \frac{DT}{Dt} - \frac{1}{\rho} \frac{Dp}{Dt} = \dot{Q}. \qquad (*) \qquad (2.65)$$

The *specific enthalpy* of a dry ideal gas is $h = c_p T$; see Table 9.4 of Chapter 9. In (2.65), the term $c_p DT/Dt = D\left(c_p T\right)/Dt = Dh/dt$ is therefore the rate of change of specific enthalpy (per unit mass) of a dry ideal gas.

The asterisk $(*)$ in (2.65) indicates that although this form of the thermodynamic-energy equation is *exact* for a dry ideal gas, it does not straightforwardly generalise in an exact manner to a cloudy atmosphere; see Section 3.6. A similar comment applies to the forms (2.68)–(2.71) that also bear an asterisk. These forms are included here since they frequently appear in the literature and it is good to be aware of them. Since they are exact for a dry ideal gas, they are entirely appropriate for studies of dry dynamics.

Using the ideal-gas law – in one or other of the forms (2.50)–(2.52) – and Carnot's law (2.64), thermodynamic-energy equation (2.65) can be rewritten in a number of alternative ways (any one of which can replace (2.63) or (2.65) as the thermodynamic-energy equation). Thus:

$$c_v \frac{DT}{Dt} - \frac{RT}{\rho} \frac{D\rho}{Dt} = \dot{Q}, \qquad (2.66)$$

$$\frac{D \ln T}{Dt} + \frac{R}{c_v} \frac{D \ln \alpha}{Dt} = \frac{\dot{Q}}{c_v T}, \qquad (2.67)$$

$$c_p \frac{DT}{Dt} - \alpha \frac{Dp}{Dt} = \dot{Q}, \qquad (*) \qquad (2.68)$$

$$\frac{Dp}{Dt} - \gamma RT \frac{D\rho}{Dt} = \frac{\rho R \dot{Q}}{c_v}, \qquad (*) \qquad (2.69)$$

$$\frac{D \ln T}{Dt} - \kappa \frac{D \ln p}{Dt} = \frac{\dot{Q}}{c_p T}, \qquad (*) \qquad (2.70)$$

$$\frac{D\theta}{Dt} = \frac{\theta \dot{Q}}{c_p T}, \qquad (*) \qquad (2.71)$$

$$c_p \frac{D \ln \theta}{Dt} = \frac{\dot{Q}}{T}. \qquad (2.72)$$

In these equations:

- $\kappa \equiv R/c_p =$ constant.
- R is the gas constant of the dry, ideal gas.
- c_p and c_v are (constant) specific heats at constant pressure and volume, respectively, of the ideal gas.
- $\dot{Q}$ is heating rate per unit mass.

Also, $\gamma \equiv c_p/c_v$ (another constant) is the ratio of specific heats at constant pressure and volume of the ideal gas. For a diatomic ideal gas (the dry component of Earth's atmosphere is composed almost entirely of diatomic gases – see Table 1.1 of Chapter 1), $\gamma = 7/5$ (as discussed in Section 4.3). The factor γRT in (2.69) turns out to be the square of the local sound speed, c_s, for an ideal gas; cf. (4.94).

Furthermore,

$$\theta \equiv T\left(\frac{p}{p_r}\right)^{-\kappa} \tag{2.73}$$

is *dry potential temperature*, where p_r is a constant reference pressure (usually set to 1 000 hPa or 1 013.25 hPa, according to context). If all thermodynamic processes take place *adiabatically* (i.e. they take place without heat transfer between a fluid parcel and its surrounding environment), then $\dot{Q} \equiv 0$, and the right-hand sides of (2.65)–(2.72) are all identically zero. Potential temperature, θ, of a fluid parcel at temperature, T, is the temperature obtained by adiabatic expansion or compression of a fluid parcel (so that no heat is transferred between the fluid parcel and its surrounding environment, and therefore $\dot{Q} \equiv 0$) to a constant reference pressure, p_r; see Section 10.7.

The forms (2.63) and (2.65)–(2.72) of the thermodynamic-energy equation (obtained assuming constant values of c_p, c_v, R, γ, and κ) are valid only for a dry ideal gas or, more generally, for a mixture of ideal gases of *constant* composition. Cloudy air contains water vapour plus other constituents, and both dry air and water vapour individually behave as ideal gases. However, the ratio of water vapour to dry air in the atmosphere *varies* as a function of both space and time. A gaseous mixture of dry air and water vapour is consequently (in general) *not* of constant composition, but of *variable* composition. Thus the forms (2.63) and (2.65)–(2.72) of the thermodynamic-energy equation are not generally valid for cloudy air; see Sections 3.5 and 3.6.1 for forms that are.

2.4 THE GOVERNING EQUATIONS FOR MOTION OF AN IDEAL GAS

The preceding developments lead to the governing equations in the summary box that follows.

Doing the accounting, the three components of the vector momentum equation (2.74), plus the three scalar equations (2.75)–(2.77), comprise six equations for the six, unknown, dependent variables $\mathbf{u} = (u_1, u_2, u_3)$, ρ, T, and p.

Equations (2.74)–(2.76) are prognostic equations for the five unknown variables $\mathbf{u} = (u_1, u_2, u_3)$, ρ, and T. This means that, given values of all variables at an instant in time, the five prognostic equations (i.e. those with a D/Dt term) can be used to obtain values of the five variables $\mathbf{u} = (u_1, u_2, u_3)$, ρ, and T at a slightly later instant in time. (Because of the non-linear nature of these equations, this usually has to be done numerically for all but the simplest situations.) Pressure, p, at the later instant in time can then be obtained using the diagnostic equation (2.77), and the prognosed values of ρ and T at the later instant in time. The entire process can then be repeated to obtain all variables at an even later instant in time, and so on.

As discussed many other equivalent combinations of equations can be formed to obtain a complete set. The set of equations (2.74)–(2.77) is particularly simple and concise, and easily related to the underlying physics.

The Governing Equations for Motion of an Ideal Gas

A set of governing equations (expressed in a rotating frame of reference) for the motion of an ideal gas, or for a mixture of ideal gases *of constant composition*, is comprised of the momentum equation (2.28), the mass-continuity equation (2.37), thermodynamic-energy equation (2.63), and the state equation (2.50), namely:

Momentum

$$\frac{D\mathbf{u}}{Dt} = -2\boldsymbol{\Omega} \times \mathbf{u} - \frac{1}{\rho}\nabla p - \nabla\Phi + \mathbf{F}. \tag{2.74}$$

Continuity of Mass Density

$$\frac{D\rho}{Dt} + \rho\nabla\cdot\mathbf{u} = 0. \tag{2.75}$$

Thermodynamic Energy

$$c_V\frac{DT}{Dt} + p\frac{D}{Dt}\left(\frac{1}{\rho}\right) = \dot{Q}. \tag{2.76}$$

State

$$p = \rho RT. \tag{2.77}$$

2.5 CONCLUDING REMARKS

From the preceding developments, we conclude that all is well for a planet with a *dry* atmosphere *of constant composition.* Such planets do exist, at least approximately (e.g. Mars has an almost dry, rarefied atmosphere composed of approximately 95.5% carbon dioxide, 2.7% nitrogen, and 1.6% argon, with only trace amounts of other gases, such as water vapour).[29] That said, the composition of a planetary atmosphere is dependent not only on the cycling of water substance between its vapour, liquid, and frozen states – as is the case for Earth – but also (in different temperature regimes, particularly cold ones) on other substances such as carbon dioxide and methane (Ingersoll, 2013). This then needs to be taken into consideration in the formulation of a suitable set of governing equations for other planetary atmospheres.

Water substance in its various forms (vapour, liquid, and frozen) in Earth's atmosphere (and other planetary atmospheres) has, to this point herein, been put to one side, but, with appropriate generalisations to the formulation, it can be accommodated – see Chapter 3 and, for a more detailed account, Chapter 10. Earth's oceans are another story, and this is taken up in Chapter 4 and, in further detail, in Chapter 11.

[29] Although the atmosphere of Mars at the present time is almost dry – with only trace amounts of water substance in it – observations of its surface strongly suggest that this was not always the case. It is currently believed that Mars in the distant past – more than 3.5 billion years ago – had a thick enough atmosphere to have standing and running water on its surface.

3

Governing Equations for Motion of a Cloudy Atmosphere: Vector Form

ABSTRACT

A set of governing equations – expressed in vector form – was assembled in Chapter 2 to describe the motion of a dry atmosphere composed of a mixture of ideal gases of constant composition. That formulation is extended here to the motion of a cloudy atmosphere. This latter atmosphere additionally includes water substance (in its gaseous, liquid, and frozen forms) plus other substances (such as aerosols). To prepare the way for this extended formulation, a cloudy-air parcel is defined together with various measures of substance content. Attention is then turned to Lagrangian and Eulerian forms of substance transport within a cloudy atmosphere. Whereas dry air can be considered to be of constant composition, cloudy air is of variable composition. This significantly complicates matters. The presence of water substance and other substances then not only introduces the need for additional equations to transport them, it also necessitates modification of the dry forms of the equation of state, and momentum and thermodynamic-energy equations. It is shown how to consistently accomplish this. Various algebraically equivalent but superficially different forms of the thermodynamic-energy equation for cloudy air are obtained and contrasted. This provides some insight into the representation of latent heating in the thermodynamic-energy equation. Finally, based on the preceding developments of the present chapter, a set of governing equations – expressed in vector form – is assembled to describe the motion of a cloudy atmosphere.

3.1 PREAMBLE

In Chapter 2, we assembled a set of governing equations – expressed in vector form – to describe the motion of a *dry* atmosphere composed of a mixture of ideal gases of constant composition. The goal of the present chapter is to extend this formulation to a *cloudy* atmosphere. By cloudy atmosphere is meant an atmosphere comprised of:

- Dry air.
- Water substance in its three forms, namely vapour, liquid, and frozen.
- Other possible substances (e.g. aerosols).

The ideal-gas law is a simple and highly realistic representation of a dry gas. As shown in Chapter 2, it is straightforward to define a dry atmosphere of *constant composition*, comprised of a set of *inert* dry gases. This composite dry atmosphere then behaves *as if* it were a single dry gas. This greatly simplifies matters.

If water vapour also behaved as an inert gas, then it could be straightforwardly included as any other gas in the mixture of ideal gases of constant composition described in

Chapter 2. Unfortunately the behaviour of water vapour is significantly anomalous,[1] particularly with regard to:

- Its significant spatial and temporal variation. When combined with dry air, the resulting mixture is no longer of *constant* composition, but of *variable* composition.
- Its behaviour with respect to temperature and pressure. A change in environmental temperature or pressure can trigger a change from the vapour form of water substance to liquid and frozen forms.

These anomalies very much complicate matters. Despite this, it turns out to be possible – to a very good approximation – to represent water vapour as an ideal gas provided that (a lot of) care is taken to accommodate how water vapour responds to changes in environmental temperature and pressure and how water vapour may interact with other substances (including liquid and frozen forms of water substance).[2] These aspects will become clearer later, both in this chapter and in Chapters 9 and 10.

The remainder of this chapter is organised as follows.

Representation of water substance, and other substances, in the atmosphere is described in Section 3.2. This includes:

- Definition of cloudy air and its substance content.
- Equations for the transport of substances.
- Various equivalent forms of the mass-continuity equation for cloudy air.

The presence of substances – other than dry air – in a cloudy atmosphere necessitates modification of the governing equations for a dry atmosphere. The following modified equations for cloudy air are developed:

- Equation of state (in Section 3.3).
- Momentum equation (in Section 3.4).
- Thermodynamic-energy equation (in Section 3.5).

Various alternative forms of the thermodynamic-energy equation for cloudy air are developed in Section 3.6. They fall into two classes – namely *exact* and *almost-exact* – with many of the forms frequently used in the literature. Various equation forms are compared for a simple model problem. It is shown that the same physical, latent heating appears very differently in equivalent forms. The reasons for this are explained.

Finally, a set of equations based on the preceding developments is assembled in Section 3.7 to describe the motion of a cloudy atmosphere. In a similar manner to that of Chapter 2 for a *dry* atmosphere, the emphasis in this chapter is on assembling a set of governing equations suitable for describing the motion of a *cloudy* atmosphere. To avoid getting into the complexities of thermodynamics here, we again make forward references to other chapters for more detailed accounts. These references may similarly be ignored on a first reading.

3.2 REPRESENTATION OF WATER AND OTHER SUBSTANCES IN THE ATMOSPHERE

As indicated in Section 3.1, the presence of water substance (in its gaseous, liquid, and frozen forms) in the atmosphere significantly complicates matters. Not only are further equations required to represent water substance, various modifications to the dry equations (2.74)–(2.77) are also needed.

[1] When viewed from a purely mathematical standpoint, but not from a physical one since water substance in the atmosphere is why Earth's atmosphere is as warm and as conducive to life as we know it.

[2] In-depth discussion of this topic is deferred until Chapters 9 and 10, since a prerequisite to doing so is the introduction of a significantly more advanced treatment of thermodynamics than the simplified one employed here.

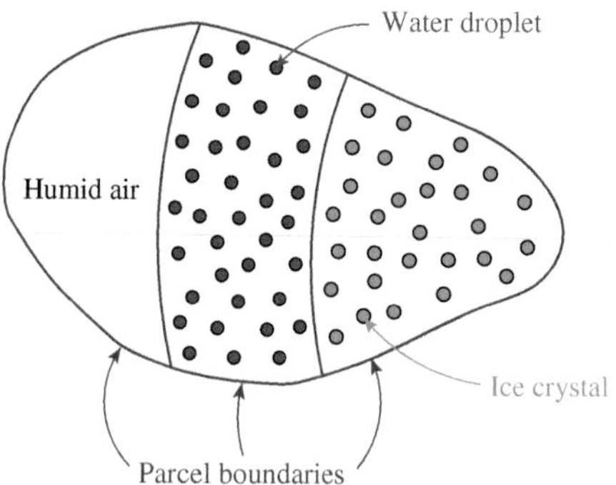

Figure 3.1 Three cloudy-air parcels of (from left to right): *humid air*, composed of a gaseous mixture of dry air and water vapour within a shared volume; *liquid cloud*, composed of liquid-water droplets embedded within humid air; and *frozen cloud*, composed of ice crystals embedded within humid air. Humid air is depicted in light blue shading, liquid-water droplets in magenta, and ice crystals in medium blue. The unshared volumes occupied by liquid-water droplets and ice crystals are greatly exaggerated for clarity.

In what follows we need to distinguish between *dry*-air quantities, and *cloudy*-air quantities that depend on water substance. It seems natural to use unqualified symbols (such as p, ρ, κ, c_p) for cloudy air, since *cloudy air* (i.e. dry air plus three phases of water substance, plus possibly other substances) is the multicomponent system that we wish to model in this chapter. To identify dry quantities, a superscript '*d*' is appended to them as necessary. We note in what follows where the new 'superscript-*d*' notation must be applied to earlier equations.

3.2.1 Cloudy Air and Measures of Substance Content

3.2.1.1 A Cloudy-Air Parcel

Consider a closed,[3] cloudy-air parcel of volume $\mathbb{V}$, containing a mixture of dry air and of water substance. Water substance has three possible phases (i.e. forms): vapour, liquid, and frozen. Three examples of cloudy-air parcels are depicted in Fig. 3.1. From left to right, they are parcels of:

- *Humid air*, composed of a gaseous mixture of dry air plus water vapour within a shared volume.
- *Liquid cloud*, composed of humid air plus embedded liquid-water droplets.
- *Frozen cloud*, composed of humid air plus embedded ice crystals.

It is also possible for a cloudy-air parcel to be composed of humid air plus liquid-water droplets *and* ice crystals, that is, with all four constituents simultaneously present (as opposed to two or three in the preceding examples). This, however, can only happen at a very special temperature that is very close to zero degrees Celsius; see Section 10.6 for details.

A cloudy-air parcel's total mass $\mathbb{M}$ is the sum of the masses of the individual constituents contained within it. Thus:

Definition of Masses within a Cloudy-Air Parcel

$$\mathbb{M} = \mathbb{M}^d + \mathbb{M}^v + \mathbb{M}^l + \mathbb{M}^f + \cdots = \sum_{X=(d,v,l,f,\ldots)} \mathbb{M}^X, \tag{3.1}$$

[3] By definition, no mass is macroscopically transferred across the boundary between a *closed* fluid parcel and its encompassing environment; see Section 4.2. Microscopic mass transfer, occurring randomly across a parcel's boundary and summing to zero, is nevertheless permissible.

where:[a]

$$\mathbb{M} = \text{total mass of the parcel,} \tag{3.2}$$

$$\mathbb{M}^d = \text{mass of } \mathbf{d}\text{ry air in the parcel,} \tag{3.3}$$

$$\mathbb{M}^v = \text{mass of water } \mathbf{v}\text{apour in the parcel,} \tag{3.4}$$

$$\mathbb{M}^l = \text{mass of } \mathbf{l}\text{iquid water in the parcel,} \tag{3.5}$$

$$\mathbb{M}^f = \text{mass of } \mathbf{f}\text{rozen water in the parcel.} \tag{3.6}$$

a See footnote 4 of Chapter 1 and Point 4 of Section 3.3 of this chapter, regarding possible decomposition of dry air into a set of dry gases of *constant* composition (treated as a single, equivalent ideal gas, with a composite gas constant) and other dry gases of *variable* composition (treated individually, with their individual gas constants).

In (3.1)–(3.6), the superscripts appended to $\mathbb{M}$ denote the type $X = (d, v, l, f, \ldots)$ of constituent transported by the flow, **u**. In (3.1), '. . .' signifies the possible inclusion of constituents other than dry air and phases of water substance (e.g. dust, condensation nuclei, aerosols, volatile organic compounds).

The focus herein is very much on the four dominant named types of constituent (**d**ry air, water **v**apour, **l**iquid water, and **f**rozen water), but with occasional comments made regarding further possibilities that it is good to be aware of. Although there are four named types of *constituent*, there are only two named *substances*; dry air (with a single phase – vapour), and water substance (with three phases – vapour, liquid, and frozen). As described in detail in Sections 10.3–10.5, the three phases of water substance are not independent of one another. Additionally, dry air and water vapour share a common volume, whereas other constituents do not. These aspects significantly complicate matters.

3.2.1.2 *Measures of Substance Content*

Two measures of substance content and associated mass content are in common use for modelling Earth's atmosphere: *mass fraction* (q) and *mixing ratio* (m). These measures first need to be defined, and equations then derived for their transport.

Both measures normalise substance content, and both involve ratios and are non-dimensional. Furthermore, both are very similar to one another. So similar in fact that they are frequently (but incorrectly) used interchangeably when the substance content is water (e.g. water vapour, liquid water, frozen water). The difference between them is that they are normalised with respect to slightly different quantities (of at most only a few per cent difference in value):

- *Mass fraction* is normalised with respect to *total mass*, $\mathbb{M}$, within a cloudy-air parcel.
- *Mixing ratio* is normalised with respect to the amount of *dry mass*, $\mathbb{M}^d$, within the parcel.

As we will see (in Section 3.2.1.5), it is straightforward to convert one set of mass fractions, or of mixing ratios to and from the other set of mixing ratios, or of mass fractions, respectively. The two sets are equivalent to one another.

3.2.1.3 *Mass Fraction*

Mass fraction, q^X, is defined to be the mass of a constituent, of type $X = (d, v, l, f, \ldots)$, per unit mass of total substance in a cloudy-air parcel, comprised of a mixture of dry air and water substance (and any other substances, if present). Thus:

Definition of Mass Fraction

$$q^X \equiv \frac{\mathbb{M}^X}{\mathbb{M}}, \quad X = (d, v, l, f, \ldots), \tag{3.7}$$

where (from (3.1)), $\mathbb{M} = \sum_{X=(d,v,l,f,\ldots)} \mathbb{M}^X$.

When the constituent is water vapour, q^v is usually termed *specific humidity*. To be explicit:

$$q^d \equiv \frac{\mathbb{M}^d}{\mathbb{M}} = \text{mass fraction of } \textbf{d}\text{ry air}, \tag{3.8}$$

$$q^v \equiv \frac{\mathbb{M}^v}{\mathbb{M}} = \text{mass fraction of water } \textbf{v}\text{apour}, \tag{3.9}$$

$$q^l \equiv \frac{\mathbb{M}^l}{\mathbb{M}} = \text{mass fraction of } \textbf{l}\text{iquid water}, \tag{3.10}$$

$$q^f \equiv \frac{\mathbb{M}^f}{\mathbb{M}} = \text{mass fraction of } \textbf{f}\text{rozen water}, \tag{3.11}$$

and so on. Summing (3.7) over $X = (d, v, l, f, \ldots)$ and using definition (3.1) of total mass gives the constraints

$$\sum_{X=(d,v,l,f,\ldots)} q^X = \frac{1}{\mathbb{M}} \sum_{X=(d,v,l,f,\ldots)} \mathbb{M}^X = \frac{\mathbb{M}}{\mathbb{M}} = 1 \quad \Rightarrow \quad q^d = 1 - \sum_{X=(v,l,f,\ldots)} q^X. \tag{3.12}$$

Thus, $q^d, q^v, q^l, q^f, \ldots$ depend (linearly) on one another and are *not* independent of one another.

3.2.1.4 *Mixing Ratio*

Similarly, *mixing ratio*, m^X, is defined to be the mass of a constituent, of type $X = (d, v, l, f, \ldots)$, per unit mass of *dry air* in a cloudy-air parcel comprised of a mixture of dry air and water substance (and any other substances, if present). Thus:

Definition of Mixing Ratio

$$m^X \equiv \frac{\mathbb{M}^X}{\mathbb{M}^d}, \quad X = (d, v, l, f, \ldots). \tag{3.13}$$

To be explicit:

$$m^d \equiv \frac{\mathbb{M}^d}{\mathbb{M}^d} = \text{mixing ratio of } \textbf{d}\text{ry air} \equiv 1, \tag{3.14}$$

$$m^v \equiv \frac{\mathbb{M}^v}{\mathbb{M}^d} = \text{mixing ratio of water } \textbf{v}\text{apour}, \tag{3.15}$$

$$m^l \equiv \frac{\mathbb{M}^l}{\mathbb{M}^d} = \text{mixing ratio of } \textbf{l}\text{iquid water}, \tag{3.16}$$

$$m^f \equiv \frac{\mathbb{M}^f}{\mathbb{M}^d} = \text{mixing ratio of } \textbf{f}\text{rozen water}, \tag{3.17}$$

and so on. The mixing ratio of dry air is trivially given by (3.14). This constraint is the analogue of constraint (3.12) on $q^X, X = (d, v, l, f, \ldots)$; if n is the total number of constituents in a parcel, at most only $n - 1$ of them may be independently specified.

3.2.1.5 *Conversion between Mass Fractions and Mixing Ratios*

Given a set of mass fractions, $\{q^X,\ X = (d, v, l, f, \ldots)\}$, it is possible to obtain an equivalent set of mixing ratios, $\{m^X,\ X = (d, v, l, f, \ldots)\}$, and vice versa. This may be needed in certain circumstances (e.g. for physical-parametrisation purposes).

With use of (3.7) (definition of mass fraction), equation (3.1) (for the total mass of all constituents in a cloudy-air parcel) can be rewritten as

$$\frac{\mathbb{M}}{\mathbb{M}^d} \equiv \frac{\mathbb{M}}{\mathbb{M} - \sum_{Y=(v,l,f,\ldots)} \mathbb{M}^Y} \equiv \frac{1}{1 - \sum_{Y=(v,l,f,\ldots)} \mathbb{M}^Y/\mathbb{M}} \equiv \frac{1}{1 - \sum_{Y=(v,l,f,\ldots)} q^Y}, \tag{3.18}$$

where Y is a dummy index of summation. Using (3.7) (definition of mass fraction) and (3.18), (3.13) (definition of mixing ratio) can then be rewritten as

$$m^X \equiv \frac{\mathbb{M}^X}{\mathbb{M}^d} \equiv \frac{\mathbb{M}^X}{\mathbb{M}} \frac{\mathbb{M}}{\mathbb{M}^d} \equiv q^X \frac{\mathbb{M}}{\mathbb{M}^d} \equiv \frac{q^X}{1 - \sum_{Y=(v,l,f,\ldots)} q^Y}. \tag{3.19}$$

Similarly, with use of (3.13) (definition of mixing ratio), equation (3.1) (for the total mass of all constituents in a cloudy-air parcel) can be rewritten as

$$\frac{\mathbb{M}}{\mathbb{M}^d} \equiv \sum_{Y=(d,v,l,f,\ldots)} \frac{\mathbb{M}^Y}{\mathbb{M}^d} \equiv \sum_{Y=(d,v,l,f,\ldots)} m^Y \equiv 1 + \sum_{Y=(v,l,f,\ldots)} m^Y. \tag{3.20}$$

Using (3.13) (definition of mixing ratio) and (3.20), (3.7) (definition of mass fraction) can then be rewritten as

$$q^X \equiv \frac{\mathbb{M}^X}{\mathbb{M}} \equiv \frac{\mathbb{M}^X}{\mathbb{M}^d} \frac{\mathbb{M}^d}{\mathbb{M}} \equiv \frac{m^X \mathbb{M}^d}{\mathbb{M}} \equiv \frac{m^X \mathbb{M}^d}{\mathbb{M}^d + \sum_{Y=(v,l,f,\ldots)} \mathbb{M}^Y} \equiv \frac{m^X}{1 + \sum_{Y=(v,l,f,\ldots)} m^Y}. \tag{3.21}$$

Summarising:

Conversion between Mixing Ratios and Mass Fractions

If required (e.g. for physical-parametrisation purposes), the two equation sets

$$m^d \equiv 1, \quad m^X = \frac{q^X}{1 - \sum_{Y=(v,l,f,\ldots)} q^Y}, \quad X = (v, l, f, \ldots), \tag{3.22}$$

$$q^X = \frac{m^X}{1 + \sum_{Y=(v,l,f,\ldots)} m^Y}, \quad X = (d, v, l, f, \ldots), \tag{3.23}$$

provide the means to *exactly* convert between a set of mass fractions, $\{q^X, X = (d, v, l, f, \ldots)\}$, and a set of mixing ratios, $\{m^X, X = (d, v, l, f, \ldots)\}$.

3.2.2 Substance Transport

Having defined cloudy air and introduced measures of substance content, we now turn our attention to substance transport. Consider (as in Section 3.2.1.1) an arbitrary, *closed* parcel of cloudy air, of volume $\mathbb{V}$, and of total mass $\mathbb{M} \equiv \sum_{X=(d,v,l,f,\ldots)} \mathbb{M}^X$, but now materially transported by a flow field, **u**. Because the parcel is closed, there is no (macroscopic) mass flux (i.e. transport) of dry air, water substance, or other substance across the parcel's bounding surface; see Section 4.2. Note that dry air is assumed here to have a single (namely vapour) phase.[4]

3.2.2.1 *Material Transport of Dry Air*

Because there is no flux of dry air across the parcel's bounding surface, the mass of dry air contained within it is constant during material transport by flow field **u**. It thus obeys the conservation law

$$\frac{D\mathbb{M}^d}{Dt} = 0, \tag{3.24}$$

where, from (2.8), $D/Dt \equiv \partial/\partial t + \mathbf{u} \cdot \nabla$ is material derivative.

3.2.2.2 *Material Transport of Water Substance and Other Substances*

Similarly to dry air, the *total* mass of water substance contained within a closed, cloudy-air parcel is also constant during material transport.[5] Two very important differences, however, are that:

1. The three phases (vapour, liquid, and frozen) of water substance are all physically realisable in Earth's atmosphere.
2. They can coexist within a cloudy-air parcel.

This then means (depending upon, for example, the temperature and pressure within the parcel) that a change of phase can occur during material transport. When this happens, the mass of one phase increases at the expense of a corresponding decrease in mass of another phase, leaving unchanged the total mass of the three phases of water substance within the parcel. This very much complicates matters. It leads to the appearance of sources and sinks – see $F^{\mathbb{M}^X}$ in (3.25) – of the masses ($\mathbb{M}^X$) of *individual* constituents within a *closed* cloudy-air parcel; these correspond to a *redistribution* of water substance between its three phases, with no net change in the total mass $\left(\sum_{X=(v,l,f)} \mathbb{M}^X\right)$ of water substance contained within the parcel.

The equations governing material transport of an *individual* constituent, of mass $\mathbb{M}^X$, contained within a cloudy-air parcel are

$$\frac{D\mathbb{M}^X}{Dt} = F^{\mathbb{M}^X}, \quad X = \left(d, v, l, f, \ldots\right), \quad F^{\mathbb{M}^d} \equiv 0. \tag{3.25}$$

The first equation holds for each type $X = \left(d, v, l, f, \ldots\right)$ of constituent transported by the flow, **u** – see (3.3)–(3.6).[6] Note the presence of a mass source/sink term, $F^{\mathbb{M}^X}$, on the right-hand side

[4] Dry air is considered to be a mixture (of constant composition) of gases; see Table 1.1 of Chapter 1. Physically, each of these gases can exist in gaseous, liquid, and frozen form, but only under conditions that are unrealisable in Earth's atmosphere. For the atmospheres of other celestial bodies, liquid and frozen phases of dry gases may need to be represented.

[5] This assumes the absence of chemical reactions between water substance and other substances. When such reactions occur (e.g. in air-quality models), the total mass of water substance within a closed parcel can change, albeit usually by very little.

[6] Falling precipitation, such as rain, snow, and hail, is not transported by the flow. This phenomenon is usually represented in atmospheric models via additional equations. Some bookkeeping is then involved for the partition of water substance into:

of this equation. This source/sink term represents changes of water phase (vapour, liquid, frozen), precipitation formation (and also evaporation and sublimation), chemical and photochemical reactions, and so on. As discussed in detail in Sections 10.3–10.5, equations (3.25) for $X = v, l, f$ are not independent of one another,[7] but constrained by thermodynamic equilibrium. Note also that for the special case $X = d$ (with $F^{\mathbb{M}^d} \equiv 0$, since it is assumed that there are no sources or sinks of dry air), (3.25) subsumes (3.24).

Summing (3.25) over all constituents of type $X = (d, v, l, f, \ldots)$, and using definition (3.1), shows that total mass ($\mathbb{M}$) contained within a cloudy-air parcel obeys

$$\frac{D\mathbb{M}}{Dt} = \sum_{X=(d,v,l,f,\ldots)} F^{\mathbb{M}^X} \equiv F^{\mathbb{M}} \ (\text{say}), \tag{3.26}$$

where $F^{\mathbb{M}}$ is the *net* source/sink of mass of the mixture. Note that (3.25) and (3.26) formally hold not only for a closed parcel, but also for an open one (with mass flux across its bounding surface, and with an appropriate definition of $F^{\mathbb{M}^X}$ that reflects this). For the *closed* parcels of cloudy air under present consideration, however, $F^{\mathbb{M}} \equiv \sum_{X=(d,v,l,f,\ldots)} F^{\mathbb{M}^X} \equiv 0$ since, by definition, no mass (macroscopically) crosses the bounding surface of a *closed* parcel.

Furthermore, (3.26) is a *consequence* of equations (3.25), and therefore it is *not* an independent equation. It can be used *instead of* (3.24), albeit (3.24) is simpler.

3.2.2.3 *Substance Transport Using Mixing Ratios and Mass Fractions*

Equation set (3.25) governs material transport of constituents contained within a closed parcel of cloudy air. It is written in terms of masses, $\mathbb{M}^X$, of the parcel's individual constituents of type $X = (d, v, l, f, \ldots)$. However, a practical disadvantage of expressing material transport in this way is that values of $\mathbb{M}^X$ depend upon the volume ($\mathbb{V}$) of the fluid parcel. This motivates the development of equivalent transport equations, written in terms of mass fractions (q^X) or mixing ratios (m^X); both of these are normalised, *non-dimensional* quantities that are *independent of* $\mathbb{V}$.

Mass Fractions

Taking the material derivative of definition (3.7) of mass fraction, q^X, of type $X = (d, v, l, f, \ldots)$ gives

$$\frac{Dq^X}{Dt} \equiv \frac{D}{Dt}\left(\frac{\mathbb{M}^X}{\mathbb{M}}\right) = \frac{1}{\mathbb{M}}\frac{D\mathbb{M}^X}{Dt} - \frac{\mathbb{M}^X}{\mathbb{M}^2}\frac{D\mathbb{M}}{Dt}, \quad X = (d, v, l, f, \ldots). \tag{3.27}$$

Note that $Dq^d/Dt, Dq^v/Dt, Dq^l/Dt, Dq^f/Dt, \ldots$ are *not* independent of one another – see (3.12). Substitution of mass-transport equations (3.25) and (3.26) for parcel constituents into (3.27), with use of definition (3.7) of mass fraction, then yields

1. Gaseous, liquid and solid phases contained within the parcel and transported by the flow, **u**.
2. Liquid and frozen forms of precipitation that enter and exit the parcel with a velocity different to the parcel velocity, **u**. This aspect is not pursued herein but is important in realistic weather-prediction and climate-prediction models.

[7] *Substances* (such as dry air and water substance) *are* (in the absence of chemical reactions) independent of one another. However, *phases* of substances (such as water vapour, liquid water, and frozen water, for water substance) are not.

$$\frac{Dq^X}{Dt} = \frac{F^{\mathbb{M}^X}}{\mathbb{M}} - \frac{\mathbb{M}^X}{\mathbb{M}^2}\left(\sum_{Y=(d,v,l,f,\dots)} F^{\mathbb{M}^Y}\right) = \frac{F^{\mathbb{M}^X}}{\mathbb{M}} - \frac{q^X}{\mathbb{M}} \sum_{Y=(d,v,l,f,\dots)} F^{\mathbb{M}^Y}$$
$$\equiv \dot{q}^X \text{ (say)}, \quad X = (d, v, l, f, \dots). \tag{3.28}$$

Thus

$$\frac{Dq^d}{Dt} = \dot{q}^d, \quad \frac{Dq^v}{Dt} = \dot{q}^v, \quad \frac{Dq^l}{Dt} = \dot{q}^l, \quad \frac{Dq^f}{Dt} = \dot{q}^f, \dots \tag{3.29}$$

For the *closed* parcels under consideration here,

$$\dot{q}^d \equiv 0, \quad \dot{q}^w \equiv \dot{q}^v + \dot{q}^l + \dot{q}^f \equiv 0, \tag{3.30}$$

where

$$q^w \equiv q^v + q^l + q^f, \tag{3.31}$$

is *mass fraction of total water substance* (i.e. the sum of the three constituents; water vapour, liquid water, and frozen water). Equations (3.30) hold since substances (dry air and water substance here) are independent of one another and are materially transported for a closed parcel. Although $\dot{q}^w \equiv 0$, the *individual* contributions (namely $\dot{q}^v$, $\dot{q}^l$, and $\dot{q}^f$) to it may nevertheless be non-zero due to a phase change of water substance (e.g. between water vapour and liquid water) occurring during transport.

Mixing Ratios

Similarly, taking the material derivative of definition (3.13) of mixing ratio, m^X, of constituent of type $X = (d, v, l, f, \dots)$ gives

$$\frac{Dm^X}{Dt} \equiv \frac{D}{Dt}\left(\frac{\mathbb{M}^X}{\mathbb{M}^d}\right) = \frac{1}{\mathbb{M}^d}\frac{D\mathbb{M}^X}{Dt} - \frac{\mathbb{M}^X}{\left(\mathbb{M}^d\right)^2}\frac{D\mathbb{M}^d}{Dt}, \quad X = (d, v, l, f, \dots). \tag{3.32}$$

Substitution of mass transport equations (3.25) for parcel constituents into (3.32) then yields

$$\frac{Dm^X}{Dt} = \frac{F^{\mathbb{M}^X}}{\mathbb{M}^d} \equiv \dot{m}^X \text{ (say)}, \quad X = (d, v, l, f, \dots). \tag{3.33}$$

Thus

$$\frac{Dm^d}{Dt} = \dot{m}^d, \quad \frac{Dm^v}{Dt} = \dot{m}^v, \quad \frac{Dm^l}{Dt} = \dot{m}^l, \quad \frac{Dm^f}{Dt} = \dot{m}^f, \dots \tag{3.34}$$

For the closed parcels under consideration here – and noting that dry air and water substance are independent of one another –

$$\dot{m}^d \equiv 0, \quad \dot{m}^w \equiv \dot{m}^v + \dot{m}^l + \dot{m}^f \equiv 0, \tag{3.35}$$

where

$$m^w \equiv \frac{\mathbb{M}^v + \mathbb{M}^l + \mathbb{M}^f}{\mathbb{M}} \equiv m^v + m^l + m^f \tag{3.36}$$

is the *mixing ratio of total water substance.*

Summary

Substance Transport Using Mixing Ratios and Mass Fractions

- From (3.25), material transport of constituents of type $X = (d, v, l, f, \ldots)$ and of mass $\mathbb{M}^X$, contained within a *closed* parcel of cloudy air, is governed by equations of the form

$$\frac{D\mathbb{M}^X}{Dt} = F^{\mathbb{M}^X}, \quad X = (d, v, l, f, \ldots), \quad F^{\mathbb{M}^d} \equiv 0. \tag{3.37}$$

- From (3.28) and (3.33), the two equation sets

$$\frac{Dq^X}{Dt} = \frac{F^{\mathbb{M}^X}}{\mathbb{M}} - \frac{q^X}{\mathbb{M}} \sum_{Y=(d,v,l,f,\ldots)} F^{\mathbb{M}^Y} \equiv \dot{q}^X, \quad X = (d, v, l, f, \ldots), \tag{3.38}$$

$$\frac{Dm^X}{Dt} = \frac{F^{\mathbb{M}^X}}{\mathbb{M}^d} \equiv \dot{m}^X, \quad X = (d, v, l, f, \ldots), \tag{3.39}$$

  are equivalent to equation set (3.37), but expressed instead in terms of (non-dimensional) mass fractions, q^X, and mixing ratios, m^X, respectively.
- Equation sets (3.38) and (3.39) formally hold not only for a closed parcel but also for an open one (with macroscopic mass flux across its bounding surface, and with an appropriate definition of $F^{\mathbb{M}^X}$ that reflects this).
- For a *closed* parcel of cloudy air, however, the right-hand sides of (3.38) and (3.39) are constrained by (3.30) and (3.35), and so

$$\dot{q}^d \equiv 0, \quad \dot{q}^w \equiv \dot{q}^v + \dot{q}^l + \dot{q}^f \equiv 0, \tag{3.40}$$

$$\dot{m}^d \equiv 0, \quad \dot{m}^w \equiv \dot{m}^v + \dot{m}^l + \dot{m}^f \equiv 0. \tag{3.41}$$

- From (3.38), q^X may be forecast so long as the current values of q^X, $\dot{q}^X$, and flow field $\mathbf{u}$ (for evaluation of the material derivative) are known.
- Similarly, from (3.39), m^X may be forecast so long as the current values of m^X, $\dot{m}^X$, and flow field $\mathbf{u}$ are known.

3.2.3 The Mass-Continuity Equation for Cloudy Air

From (2.37), the Lagrangian form of the dry mass-continuity equation is

$$\frac{D\rho^d}{Dt} + \rho^d \nabla \cdot \mathbf{u} = 0, \tag{3.42}$$

where superscript '*d*' signifies that ρ^d is ***dry*** mass density. Now the derivation of (2.37) (obtained by applying the Reynolds' transport theorem) holds not only for a closed parcel of *dry* air but also for a closed parcel of *cloudy* air. Thus:

The Mass-Continuity Equation for Cloudy Air

- The *mass-continuity equation* for a *closed* parcel of cloudy air is

$$\frac{D\rho}{Dt} + \rho \nabla \cdot \mathbf{u} = 0, \tag{3.43}$$

  where $\rho \equiv \mathbb{M}/\mathbb{V}$, $\mathbb{M}$, and $\mathbb{V}$ are the density, total mass, and total volume of the parcel, respectively.

- For an *open* parcel, a source/sink term, F^{ρ}, needs to be inserted on the right-hand side of (3.43) to account for any macroscopic mass flux across the parcel's boundary.

To alternatively derive (3.43), note that although $D\mathbb{M}/Dt = 0$ for a closed fluid parcel (i.e. its total mass remains unchanged when materially transported), it is not (in general) true that its volume, $\mathbb{V}$, analogously satisfies $D\mathbb{V}/Dt = 0$. This is because the volume of a parcel depends upon the flow field, $\mathbf{u}$, in which it is embedded and, specifically, on the flow divergence, $\nabla \cdot \mathbf{u}$. For $\nabla \cdot \mathbf{u} > 0$, the volume of the parcel increases as time goes by and, similarly, it decreases for $\nabla \cdot \mathbf{u} < 0$. It can be shown (cf. (2.46) herein and equation (1.15) of Vallis (2017)) that

$$\frac{D\mathbb{V}}{Dt} = \mathbb{V}\nabla \cdot \mathbf{u}. \tag{3.44}$$

With use of $\rho \equiv \mathbb{M}/\mathbb{V}$ (definition of density within the parcel), $D\mathbb{M}/Dt = 0$ (conservation of mass within the parcel), and (3.44) (expansion/compression of the parcel's volume by the flow field), we obtain

$$\frac{D\rho}{Dt} \equiv \frac{D}{Dt}\left(\frac{\mathbb{M}}{\mathbb{V}}\right) = \frac{1}{\mathbb{V}}\cancelto{0}{\frac{D\mathbb{M}}{Dt}} - \frac{\mathbb{M}}{\mathbb{V}^2}\frac{D\mathbb{V}}{Dt} = -\frac{\mathbb{M}}{\mathbb{V}}\nabla \cdot \mathbf{u} = -\rho\nabla \cdot \mathbf{u}. \tag{3.45}$$

This equation is just a rearrangement of (3.43).

3.2.4 Eulerian Forms of Substance Transport

In Section 3.2.2.3, transport of the constituents of a closed, cloudy-air parcel is expressed in Lagrangian form, using either mass fractions (q^X) or mixing ratios (m^X). For numerical modelling and budget purposes, particularly with regard to exact conservation, it is useful to alternatively express substance transport in Eulerian form. Mass-continuity equation (3.43) for cloudy air facilitates this, as now shown.

Mass Fractions

Summing $\rho\times$(3.38) and $q^X\times$(3.43) gives

$$\frac{D\left(\rho q^X\right)}{Dt} + \rho q^X\nabla \cdot \mathbf{u} = \rho\dot{q}^X, \quad X = \left(d, v, l, f, \ldots\right). \tag{3.46}$$

Using (2.8) (definition of material derivative) and rearranging, (3.46) may be rewritten as:

Substance Transport in Eulerian Flux Form Using Mass Fractions

$$\frac{\partial\left(\rho q^X\right)}{\partial t} + \nabla \cdot \left(\rho q^X\mathbf{u}\right) = \rho\dot{q}^X, \quad X = \left(d, v, l, f, \ldots\right). \tag{3.47}$$

Mixing Ratios

Similarly, summing $\rho\times$(3.39) and $m^X\times$(3.43) gives

$$\frac{D\left(\rho m^X\right)}{Dt} + \rho m^X\nabla \cdot \mathbf{u} = \rho\dot{m}^X, \quad X = \left(d, v, l, f, \ldots\right). \tag{3.48}$$

Using (2.8) (definition of material derivative) and rearranging, (3.48) may be rewritten as:

Substance Transport in Eulerian Flux Form Using Mixing Ratios

$$\frac{\partial\left(\rho m^X\right)}{\partial t} + \nabla\cdot\left(\rho m^X \mathbf{u}\right) = \rho \dot{m}^X, \quad X = \left(d, v, l, f, \ldots\right). \tag{3.49}$$

3.3 THE EQUATION OF STATE FOR CLOUDY AIR

3.3.1 Full Form

Recall that ideal-gas law (2.50) is

$$p = \rho R T, \tag{3.50}$$

where:

- p is the pressure exerted (on its surrounding environment) by a parcel of ideal gas, of density ρ, and of temperature T.
- R is the gas *constant* of the ideal gas.

This law is valid for any ideal gas.[8] As shown in Chapter 2.3.3, it is also valid for a mixture of ideal gases of *constant* composition.[9] Gas constant R is then a mass-weighted average of the individual gas constants – see (2.59).

The challenge now is to generalise the ideal-gas law for a *dry* atmosphere of *constant* composition to a *cloudy* atmosphere of *variable* composition comprised of:

- *Gaseous* constituents of dry air and water vapour (occupying a common, gaseous volume).
- *Non-gaseous* constituents of liquid water and frozen water (occupying two, separate, unshared volumes).

(For simplicity, attention is restricted here to the presence of only two substances, namely dry air and water substance.) This challenge is met in the Appendix to this chapter, where an equation of state is derived from first principles for a *cloudy*-air parcel comprised of a mixture of dry air, water vapour, liquid water, and frozen water. The results of this analysis are summarised in the shaded box.

The Equation of State for Cloudy Air

For a closed parcel of cloudy air, of total mass $\mathbb{M}$ and total volume $\mathbb{V}$, composed of the four named constituents (dry air, water vapour, liquid water, and frozen water) appearing in (3.1), the *equation of state for cloudy air* is

$$p = p^d + p^v = \rho R T \equiv \frac{RT}{\alpha}, \tag{3.51}$$

[8] The value of the gas constant, R, does, however, depend upon the gas.

[9] This actually includes water vapour, *but only provided there are no sources or sinks of water vapour, and no change of its state* (between water vapour and liquid water or frozen water) since the mixture would then no longer be of constant composition.

where:

- The partial pressures individually exerted by dry air and water vapour are

$$p^d \equiv \rho^d R^d T = \rho \left[\frac{R^d q^d}{1 - (\alpha^l/\alpha)\, q^l - (\alpha^f/\alpha)\, q^f} \right] T, \tag{3.52}$$

$$p^v \equiv \rho^v R^v T = \rho \left[\frac{R^v q^v}{1 - (\alpha^l/\alpha)\, q^l - (\alpha^f/\alpha)\, q^f} \right] T, \tag{3.53}$$

  respectively.
- The composite gas 'constant' for the mixture is

$$R = \frac{R^d q^d + R^v q^v}{1 - (\alpha^l/\alpha)\, q^l - (\alpha^f/\alpha)\, q^f} \equiv \frac{\widetilde{R}}{1 - (\alpha^l/\alpha)\, q^l - (\alpha^f/\alpha)\, q^f}, \quad \widetilde{R} \equiv R^d q^d + R^v q^v. \tag{3.54}$$

- $\rho \equiv \mathbb{M}/\mathbb{V}$ is density of total substance (per unit volume) within the parcel.
- R^d and R^v are the gas constants (per unit mass) of dry air and water vapour, respectively.
- $q^d \equiv \mathbb{M}^d/\mathbb{M}$ and $q^v \equiv \mathbb{M}^v/\mathbb{M}$ are the mass fractions of dry air and water vapour, respectively, contained within the parcel (this includes liquid and frozen water, in addition to dry air and water vapour).
- α^l and α^f are the (constant) volumes occupied by unit mass of liquid water and frozen water, respectively.
- $\alpha \equiv 1/\rho$ is the volume occupied by unit mass of total substance within the parcel.

Regarding (3.51)–(3.54), note that:

1. Equation (3.51) for a *cloudy* atmosphere of *variable* composition retains the same form as (3.50) for a *dry* atmosphere of *constant* composition. However, there is an important difference; whereas R is *constant* in (3.50), R in (3.51) *varies* as a function of both space and time due to the presence of $q^d = q^d(\mathbf{r}, t)$ and $q^v = q^v(\mathbf{r}, t)$. This has important consequences when evaluating Dp/Dt, since R no longer commutes with D/Dt as it does for an ideal gas of constant composition.
2. In (3.54), $\widetilde{R}$ has been introduced to make the forms of various terms in the analysis of Section 3.6 more compact; see, for example, Forms 2–4 in Table 3.1. It is an approximation of R that neglects the volumes separately occupied by liquid water and frozen water. As noted in Staniforth and White (2019) (but for the more restricted case with no frozen water) $\widetilde{R}$ is not the gas 'constant' for the gaseous part of the mixture. That quantity is

$$R^g \equiv \frac{R^d q^d + R^v q^v}{q^d + q^v} \equiv \frac{\widetilde{R}}{q^d + q^v}. \tag{3.55}$$

   For later use, (3.54) leads to

$$R - \widetilde{R} = R - \left[1 - \frac{\left(\alpha^l q^l + \alpha^f q^f\right)}{\alpha} \right] R = \frac{\left(\alpha^l q^l + \alpha^f q^f\right)}{\alpha} R. \tag{3.56}$$

3. From (3.51) (the equation of state for cloudy air) and (3.54) (the composite gas 'constant'),

$$\rho T = \frac{p}{R} = \frac{\left[1 - \left(\alpha^l/\alpha\right) q^l - \left(\alpha^f/\alpha\right) q^f\right] p}{R^d q^d + R^v q^v}. \tag{3.57}$$

Substitution of (3.57) into (3.52) and (3.53), with use of (3.54), then gives the alternative expressions

$$p^d = \left(\frac{R^d q^d}{R^d q^d + R^v q^v}\right) p = \frac{R^d q^d}{\widetilde{R}} p, \tag{3.58}$$

$$p^v = \left(\frac{R^v q^v}{R^d q^d + R^v q^v}\right) p = \frac{R^v q^v}{\widetilde{R}} p, \tag{3.59}$$

for the partial pressures p^d and p^v. Summing (3.58) and (3.59) recovers $p = p^d + p^v$; see (3.51).

4. In deriving (3.51)–(3.54), it has been implicitly assumed that:

- Dry air is a mixture of dry gases of *constant* composition.
- There are no other dry gases in the mixture.

If, however, some of the dry gases are of *variable* composition,[10] then ρ^d in (3.52) needs to be decomposed into $\rho^d = \rho^d_{cc} + \sum_i \rho^d_i$. Here:

- $\rho^d_{cc} \equiv \mathbb{M}^d_{cc}/\mathbb{V}^g$ is mass density of the composite mixture of dry gases of constant composition in volume $\mathbb{V}^g$.
- $\rho^d_i \equiv \mathbb{M}^d_i/\mathbb{V}^g$, $i = 1, 2, \ldots$, are mass densities of the dry gases of variable composition (of type i).
- $\mathbb{M}^d \equiv \mathbb{M}^d_{cc} + \sum_i \mathbb{M}^d_i$, where $\mathbb{M}^d_{cc}$ and $\mathbb{M}^d_i$ are the masses of dry gas of constant composition, and of variable composition (of type i), respectively, in volume $\mathbb{V}^g$.

Equations (3.52) and (3.54) are then replaced by

$$p^d = \rho \left[\frac{R^d_{cc} q^d_{cc} + \sum_i R^d_i q^d_i}{1 - \left(\alpha^l/\alpha\right) q^l - \left(\alpha^f/\alpha\right) q^f}\right] T, \tag{3.60}$$

$$R = \frac{R^d_{cc} q^d_{cc} + \sum_i R^d_i q^d_i + R^v q^v}{1 - \left(\alpha^l/\alpha\right) q^l - \left(\alpha^f/\alpha\right) q^f}, \tag{3.61}$$

where:

- R^d_{cc} and R^d_i are the gas constants for the dry gases of constant composition, and of variable composition (of type i), respectively.
- $q^d_{cc} \equiv \mathbb{M}^d_{cc}/\mathbb{M}$ and $q^d_i \equiv \mathbb{M}^d_i/\mathbb{M}$ are the mass fractions of the dry gases of constant composition, and of variable composition (of type i), respectively, contained within the parcel of total volume $\mathbb{V}$.

3.3.2 Approximated Form

Consider now a form of the equation of state for cloudy air that frequently appears in textbooks and in the meteorological literature. Since:

- α^l/α and α^f/α are both of order 10^{-3} (being essentially the ratios of the densities of air to water, and of air to ice, respectively); and
- q^l and q^f (mass fractions of liquid water and frozen water, respectively) are considerably less than unity;

[10] This, for example, is so for a mixture of chemically reactive gases in air-pollution and air-chemistry studies.

both $\left(\alpha^l/\alpha\right) q^l$ and $\left(\alpha^f/\alpha\right) q^f$ can be considered to be quantitatively negligible with respect to unity.

Neglecting the $\left(\alpha^l/\alpha\right) q^l$ and $\left(\alpha^f/\alpha\right) q^f$ terms with respect to unity, unapproximated equations (3.51)–(3.54) simplify to

$$p = p^d + p^v = \rho RT \approx \rho \widetilde{R} T, \tag{3.62}$$

$$p^d \equiv \rho^d R^d T \approx \rho R^d q^d T, \tag{3.63}$$

$$p^v \equiv \rho^v R^v T \approx \rho R^v q^v T, \tag{3.64}$$

$$R \approx R^d q^d + R^v q^v \equiv \widetilde{R}. \tag{3.65}$$

It is seen that the composite gas 'constant' $R = \widetilde{R} / \left[1 - \left(\alpha^l/\alpha\right) q^l - \left(\alpha^f/\alpha\right) q^f\right]$ is approximated by the much simpler $\widetilde{R}$.[11] *This amounts to assuming that liquid water and frozen water occupy negligible volume in a parcel of cloudy air when compared to the gaseous volume.*

One can, if one wishes, systematically neglect α^l/α and α^f/α terms in the analysis of Section 3.3, and also in other analyses. In the interests of completeness, and also for accuracy in a numerical model of Earth's atmosphere, this is not done here. Furthermore, including these extra terms allows us to assess different forms of the thermodynamic-energy equation and of latent heating, some of which are more natural and more accurate than others; see Section 3.6 and also Section 10.6.5.

Using definitions (3.8) and (3.9) for q^d and q^v, respectively, (3.62) is often rewritten as

$$p = \rho RT = \rho R^d \left(\frac{R}{R^d} T\right) = \rho R^d T_v, \tag{3.66}$$

where

$$T_v \equiv \left(q^d + \frac{q^v}{\varepsilon}\right) T \tag{3.67}$$

is termed *virtual temperature*, and $\varepsilon \equiv R^d / R^v$ (≈ 0.622) is the ratio of gas constants for dry air and water vapour. Physically, T_v is the temperature that *dry* air would have to have, at a given density, in order to exert the same pressure as a mixture of dry air and water vapour at temperature T. Mathematically, the spatial and temporal variation of R in (3.62) is absorbed into the temperature variable T_v, and R is replaced by R^d, the gas *constant* per unit mass for *dry* air.

The motivation for rewriting (3.62) as (3.66) is that it gets around the problem (identified in Point 1 for the unapproximated form of the equation of state for cloudy air) of R not commuting with D/Dt. However, as noted in Akmaev and Juang (2008), this is something of an illusion since it effectively transfers the problem elsewhere (specifically into the derivation of an unapproximated thermodynamic energy equation, consistent with other equations). In fact, Akmaev and Juang (2008) conclude that '*the use of virtual temperature apparently offers no benefits for atmospheric modelling with variable composition*'. Their arguments for this have merit.

Note that index 'v' has now accumulated three different meanings: 'virtual' (as in T_v); 'vapour' (as in R^v); and 'constant volume' (as in c_v)! No ambiguity should arise as long as the possibility of it is appreciated.

3.4 THE MOMENTUM EQUATION FOR CLOUDY AIR

The momentum equation for a cloudy atmosphere has exactly the same form as (2.74) for a dry atmosphere. Thus:

[11] For the special case of humid air (a purely gaseous mixture of dry air and water vapour, so that $q^l = q^f \equiv 0$), (3.62)–(3.65) are no longer approximate, but exact.

The Momentum Equation for Cloudy Air

The momentum equation for cloudy air is

$$\frac{D\mathbf{u}}{Dt} = -2\mathbf{\Omega} \times \mathbf{u} - \frac{1}{\rho}\nabla p - \nabla\Phi + \mathbf{F}. \tag{3.68}$$

What is different, however, are the definitions of density (ρ) and pressure (p) used for the evaluation of the pressure-gradient term, $(1/\rho)\,\nabla p$, in this equation.

For the special case of a *dry* atmosphere of *constant* composition, recall that ρ and p are simply the dry quantities, ρ^d and p^d (these are expressed here in the new notation, which distinguishes between types of constituent). These quantities (ρ^d and p^d) satisfy the ideal-gas law $p^d = \rho^d R^d T$ for a mixture of dry gases of constant composition; see (2.58) with ρ^d and p^d made explicit therein. Thus, for a dry atmosphere (for which $\rho \equiv \rho^d$), the term $(1/\rho)\,\nabla p$ in (3.68) is evaluated as $\left(1/\rho^d\right)\nabla p^d$.

For a cloudy atmosphere of *variable* composition:

- $\rho \equiv \mathbb{M}/\mathbb{V}$ is mass density of cloudy air (including all of its constituents) – see Table 3.3.
- p is the sum of the partial pressures, as given by (3.51).
- These expressions for ρ and p are used to evaluate the pressure-gradient term, $(1/\rho)\,\nabla p$, in (3.68).
- p, ρ, and T satisfy the equation of state ($p = \rho RT$) for cloudy air – see (3.51).

Setting all masses, except $\mathbb{M}^d$, to zero in (3.1) and also setting p^v, q^v, q^l, and q^f to zero in (3.51)–(3.54) recovers the special case of a dry atmosphere – for which $(1/\rho)\,\nabla p$ in (3.68) is evaluated as $\left(1/\rho^d\right)\nabla p^d$.

3.5 THE THERMODYNAMIC-ENERGY EQUATION FOR CLOUDY AIR

In a realistic atmospheric model for weather and climate prediction, the representation of water substance is of crucial importance. Without the presence of water substance in Earth's atmosphere, its climate would be very different due to the radiative properties of water substance in its different phases. In particular, it would be much colder; circa 20 K at Earth's surface. It is therefore important to represent water substance in a realistic manner.

Existing representations can be extremely complex. Furthermore, they are the subject of current research and are still evolving. Attention is therefore restricted herein to a representation of water substance that is quite realistic from a physical perspective, yet reasonably simple and tractable from a mathematical standpoint. This approach aims to promote understanding of basic principles.

Consider therefore *cloudy air*. This is defined herein – see Sections 3.2.1, 4.3, and 10.5 – to be composed of a mixture of dry air, water vapour, liquid water, and frozen water, where other substances are neglected here for simplicity and tractability. Thus (3.1) simplifies to

$$\mathbb{M} = \mathbb{M}^d + \mathbb{M}^v + \mathbb{M}^l + \mathbb{M}^f = \sum_{X=(d,v,l,f)} \mathbb{M}^X, \tag{3.69}$$

where $\mathbb{M}^d$, $\mathbb{M}^v$, $\mathbb{M}^l$, and $\mathbb{M}^f$ are the masses of dry air, water vapour, liquid water, and frozen water, respectively, contained within a cloudy-air parcel.

Consider now a parcel of cloudy air for which no mass is (macroscopically) transferred between it and its encompassing environment across their mutual boundary. Such an air parcel constitutes a *closed* thermodynamic system. Whereas *mass* is not transferred across its boundary, *energy* can nevertheless be transferred through heat supplied/removed (e.g. by radiation), or work done (by mechanical expansion/compression). Furthermore, water substance within the air parcel can change its phase (between gaseous, liquid, and frozen forms). When it does so, latent internal energy is released or absorbed, leading to latent heating or cooling of the parcel (see Chapter 10). *It is very important to include this process in the representation of water substance.* The following analysis generalises Staniforth and White (2019)'s for an air parcel (of dry air + water vapour + liquid water) by additionally including frozen water for added realism – as advocated in Thuburn (2017).

In the present context, the first law of thermodynamics for a cloudy-air parcel is (see Section 4.2.5)

$$\frac{D\mathscr{E}}{Dt} + p\frac{D\alpha}{Dt} = \frac{D\mathscr{E}}{Dt} - \frac{p}{\rho^2}\frac{D\rho}{Dt} = \dot{Q}, \tag{3.70}$$

where (3.70) corresponds to (4.13), with the chemical potential terms absent (since the parcel is assumed closed, there is no mass flux across its boundary). Here:

- $\mathscr{E}$ is *total specific internal energy* (i.e. internal energy per unit mass) contained within the air parcel, *including latent internal energy.*
- $\dot{W} \equiv -pD\alpha/D$ (see (2.61)) is *rate at which work is done* (per unit mass) on the air parcel by its surrounding environment, during reversible expansion (or compression) of its gaseous constituents.
- $\dot{Q}$ is *total energy input rate* (per unit mass) supplied to (or removed from) the air parcel, *excluding any latent heating (or cooling).*[12]

$\dot{Q}$ includes heat-related changes in internal energy of an air parcel due to:

- Radiative heating and cooling.
- Diffusion of heat to or from the surrounding environment.
- Dissipation of kinetic energy into heat by viscous forces.
- Heating and cooling at Earth's surface.

Equation (3.70) expresses the principle of *conservation of internal energy*. Internal energy of a closed air parcel is conserved in the absence of:

1. Energy supplied to/removed from the parcel (so that $\dot{Q} = 0$).
2. Work done between the fluid parcel and its surrounding environment (so that $pD\alpha/Dt = 0$).

For the assumed mass distribution (3.69), and assuming that the gases (including water vapour) in the air parcel behave as ideal gases (see Section 4.3 for details), the internal energy satisfies (4.71), that is,

$$\mathscr{E} = c_v T + \mathbb{L}_0 = \mathscr{E}^d q^d + \mathscr{E}^v q^v + \mathscr{E}^l q^l + \mathscr{E}^f q^f. \tag{3.71}$$

In this equation,

$$c_v\left(q^d, q^v, q^l, q^f\right) = c_v^d q^d + c_v^v q^v + c^l q^l + c^f q^f, \tag{3.72}$$

is specific heat capacity at constant ***volume*** of the air parcel, where c_v^d, c_v^v, c^l, and c^f are constants. Furthermore,

$$\mathbb{L}_0\left(q^v, q^l\right) \equiv \left(L_0^v + L_0^f\right) q^v + L_0^f q^l \tag{3.73}$$

[12] Latent heating/cooling is handled separately via inclusion of latent internal energy in the definition of total specific internal energy, $\mathscr{E}$.

is latent internal energy of water substance, where the constants L_0^v and L_0^f are the latent heats of vaporisation and fusion, respectively, extrapolated to $T = 0$ and $p = 0$. ($\mathbb{L}_0\left(q^v, q^l\right)$ acts as an energy reservoir, available to change water phase between vapour, liquid, and solid forms. How much energy is in the reservoir depends upon how much water vapour and liquid water there is in the parcel, that is, it depends upon the values of the mass fractions q^v and q^l.)

Both $\mathscr{E}$ and c_v are mass-weighted quantities, with mass fractions q^d, q^v, q^l, and q^f; see (3.7). The quantities

$$\mathscr{E}^d = c_v^d T, \quad \mathscr{E}^v = c_v^v T + L_0^v + L_0^f, \quad \mathscr{E}^l = c^l T + L_0^f, \quad \mathscr{E}^f = c^f T, \tag{3.74}$$

are the *individual* internal energies of the constituents of type (d, v, l, f). The expression in (3.74) for internal energy of a dry ideal gas corresponds to that of (2.62). (For this special case, $q^d = 1$, $q^v = q^l = 0$, and (3.71)–(3.73) reduce to $\mathscr{E} = c_v^d T$, $c_v = c_v^d$, and $\mathbb{L}_0 = 0$, respectively.)

For the composition of cloudy air considered here (and in more detail in Section 4.3 and Chapter 10) only water vapour (q^v) and liquid water (q^l) contain latent internal energy; it is assumed that dry air (q^d) and frozen water (q^f) do not. The constants c_v^d, c_v^v, c^l, and c^f are the corresponding specific heat capacities (at constant *volume*) of the associated constituents. (Note that specific heat capacities c^l and c^f for the *non-gaseous* constituents do not bear a subscript. This is because their values are assumed to be independent of whether heat is transferred at constant volume or at constant pressure.) These specific heat capacities are also assumed to be independent of temperature, T, everywhere in the atmosphere.[13]

In contradistinction to a *dry-air* parcel, for which $c_v =$ constant – see (2.63) – c_v in (3.72) (for a *cloudy-air* parcel) varies both spatially and temporally due to the presence of the mass fractions q^d, q^v, q^l, and q^f. A similar comment also applies to

$$c_p\left(q^d, q^v, q^l, q^f\right) = c_p^d q^d + c_p^v q^v + c^l q^l + c^f q^f, \tag{3.75}$$

the specific heat capacity of an air parcel at constant ***pressure***, where c_p^d, c_p^v, c^l, and c^f are constants. It further applies to

$$R\left(q^d, q^v, q^l, q^f\right) = \frac{R^d q^d + R^v q^v}{1 - \left(\alpha^l/\alpha\right) q^l - \left(\alpha^f/\alpha\right) q^f} = \frac{\widetilde{R}\left(q^d, q^v\right)}{1 - \left(\alpha^l/\alpha\right) q^l - \left(\alpha^f/\alpha\right) q^f}, \tag{3.76}$$

the mass weighting of the gas constants R^d and R^v for dry air and water vapour, respectively, associated with the partial pressures, p^d and p^v; see (3.51)–(3.54).

Carnot's Law for a *dry* air parcel is given by (2.64). To obtain its counterpart for a *cloudy-air* parcel, subtraction of (3.72) from (3.75), with use of $R^d \equiv c_p^d - c_v^d$ and $R^v \equiv c_p^v - c_v^v$ (see (2.64)), and also of (3.54), yields

$$\begin{aligned} c_p - c_v &= c_p^d q^d + c_p^v q^v + \cancel{c^l q^l + c^f q^f} - \left(c_v^d q^d + c_v^v q^v + \cancel{c^l q^l + c^f q^f}\right) \\ &= \left(c_p^d - c_v^d\right) q^d + \left(c_p^v - c_v^v\right) q^v = R^d q^d + R^v q^v \equiv \widetilde{R} \neq R. \end{aligned} \tag{3.77}$$

In the last line, that $\widetilde{R} \neq R$ (in general) follows from comparison with (3.76).

Two substances are present in (3.69)–(3.77): dry air and water substance. Under our assumption that there is no net mass transfer across the bounding surface of the air parcel, the mass

[13] For the temperatures and pressures of Earth's atmosphere, this is generally an excellent approximation to reality. Broadly speaking, it is increasingly true, the higher the temperature, the lower the pressure, albeit at high-enough altitudes the atmosphere becomes so rarefied that ideal gas assumptions break down.

fractions of dry air and water substance within it must remain unchanged for all time. Dry air cannot change into water substance, nor vice versa. Thus

$$\frac{Dq^d}{Dt} = 0, \tag{3.78}$$

$$\frac{Dq^w}{Dt} \equiv \frac{D\left(q^v + q^l + q^f\right)}{Dt} = 0 \quad \Rightarrow \quad \frac{Dq^l}{Dt} = \dot{q}^l = -\left(\frac{Dq^v}{Dt} + \frac{Dq^f}{Dt}\right) = -\left(\dot{q}^v + \dot{q}^f\right), \tag{3.79}$$

where

$$q^w \equiv q^v + q^l + q^f = 1 - q^d = \text{constant} \tag{3.80}$$

is mass fraction of total water substance contained within the parcel.

Inserting (3.71) (definition of internal energy) into (3.70) (thermodynamic-energy equation), and using (3.38) (transport of q^X), (3.73) (definition of latent internal energy), and (3.79) (transport of q^l for a closed parcel), leads to:

The Thermodynamic-Energy Equation for Cloudy Air

$$\frac{D(c_V T)}{Dt} + p\frac{D\alpha}{Dt} \equiv \frac{D(c_V T)}{Dt} + p\frac{D}{Dt}\left(\frac{1}{\rho}\right)$$
$$= \dot{Q} - \left(L_0^v + L_0^f\right)\dot{q}^v - L_0^f \dot{q}^l = \dot{Q} - L_0^v \dot{q}^v + L_0^f \dot{q}^f. \tag{3.81}$$

The first expression on the last line holds for any parcel (open or closed), whereas the second holds only for a closed parcel.

Formally comparing (3.81) for a *cloudy*-air parcel to (2.63) for a *dry* one, they are very similar. There are, however, two notable differences:

1. c_V in (3.81) *is no longer constant*. It now appears as $D(c_V T)/Dt$ rather than as $c_V DT/Dt$.
2. Latent heating (cooling) terms, with rates $-L_0^v \dot{q}^v$ and $+L_0^f \dot{q}^f$, now appear on the right-hand side. These reflect a change in latent internal energy resulting from a change of phase of water substance between vapour, liquid, and frozen forms. For negative (positive) $\dot{q}^v$, latent heating (cooling) of the air parcel results, and similarly for positive (negative) $\dot{q}^f$. See Chapter 10 for further details.

3.6 ALTERNATIVE FORMS FOR THE THERMODYNAMIC-ENERGY EQUATION

For the present model problem (of cloudy air), thermodynamic-energy equation (3.81) has been derived *exactly*. Some alternative forms are now developed. They are grouped into two classes: equivalent to (3.81), and therefore *exact*; and *almost-exact*, with their equations bearing an asterisk to identify this at a glance.

For reference and comparison purposes, a representative subset of these forms are collated in Table 3.1. They correspond to a subset of those given in Table 1 of Staniforth and White (2019) for the simpler case of cloudy air (but with no frozen water). Indeed, setting $\dot{q}^f \equiv 0$ in Table 3.1 herein recovers a subset of their Table 1. *Form 0 is the base form* (3.81) *from which all other forms are obtained.*

#	Accuracy	Thermodynamic-energy equation	Equation
0	Exact	$\frac{D(c_v T)}{Dt} + p\frac{D\alpha}{Dt} = \dot{Q} - L_0^v \dot{q}^v + L_0^f \dot{q}^f$	(3.81)
1	Exact	$c_v \frac{DT}{Dt} + p\frac{D\alpha}{Dt} = \dot{Q} - \left[L_0^v + \left(c_v^v - c^l\right) T\right] \dot{q}^v + \left[L_0^f + \left(c^l - c^f\right) T\right] \dot{q}^f$	(3.84)
2	Exact	$\frac{D}{Dt}\left(c_p T\right) - \alpha \frac{Dp}{Dt} + \frac{D}{Dt}\left[\left(R - \widetilde{R}\right) T\right] = \dot{Q} - L_0^v \dot{q}^v + L_0^f \dot{q}^f$	(3.87)
3	Exact	$c_p \frac{DT}{Dt} - \alpha \frac{Dp}{Dt} + \frac{D}{Dt}\left[\left(R - \widetilde{R}\right) T\right] = \dot{Q} - \left[L_0^v + \left(c_p^v - c^l\right) T\right] \dot{q}^v + \left[L_0^f + \left(c^l - c^f\right) T\right] \dot{q}^f$	(3.88)
4	Exact	$c_p \frac{DT}{Dt} - \frac{\widetilde{R}T}{p}\frac{Dp}{Dt} = \dot{Q} - \left[L_0^v + \left(c_p^v - c^l\right) T - \alpha^l p\right] \dot{q}^v + \left[L_0^f + \left(c^l - c^f\right) T - \left(\alpha^f - \alpha^l\right) p\right] \dot{q}^f$	(3.90)
5	Exact	$\frac{Dh}{Dt} - \alpha \frac{Dp}{Dt} = \dot{Q}$	(3.95)
2A	Almost exact	$\frac{D\left(c_p T\right)}{Dt} - \alpha \frac{Dp}{Dt} = \dot{Q} - L_0^v \dot{q}^v + L_0^f \dot{q}^f$	(3.97)
3A	Almost exact	$c_p \frac{DT}{Dt} - \alpha \frac{Dp}{Dt} = \dot{Q} - \left[L_0^v + \left(c_p^v - c^l\right) T\right] \dot{q}^v + \left[L_0^f + \left(c^l - c^f\right) T\right] \dot{q}^f$	(3.98)

Table 3.1 Some exact and almost-exact forms of the thermodynamic-energy equation. Form 0 is the base form from which other forms are obtained. To obtain the two almost-exact forms, $\left(R - \widetilde{R}\right)$ is set to zero in exact Forms 2 and 3; the resulting Forms 2A and 3A are equivalent to one another. To avoid possible confusion, note that c_v^v appears in Form 1, but c_p^v in Forms 3, 4, and 3A. Equation numbers in the last column refer to equations in this chapter.

3.6.1 Exact

Despite c_v being variable, it is nevertheless possible to *exactly* rewrite (3.81) in a form similar to that of (2.63). Using (3.72), we first rewrite $D(c_vT)/Dt$ as

$$\frac{D(c_vT)}{Dt} = c_v\frac{DT}{Dt} + \frac{Dc_v}{Dt}T = c_v\frac{DT}{Dt} + \left(c_v^d\frac{Dq^d}{Dt} + c_v^v\frac{Dq^v}{Dt} + c^l\frac{Dq^l}{Dt} + c^f\frac{Dq^f}{Dt}\right)T. \tag{3.82}$$

Insertion of substance-transport equations (3.78) and (3.79) into (3.82) gives

$$\begin{aligned}\frac{D(c_vT)}{Dt} &= c_v\frac{DT}{Dt} + \left[\left(c_v^v - c^l\right)\frac{Dq^v}{Dt} - \left(c^l - c^f\right)\frac{Dq^f}{Dt}\right]T \\ &= c_v\frac{DT}{Dt} + \left[\left(c_v^v - c^l\right)\dot{q}^v - \left(c^l - c^f\right)\dot{q}^f\right]T.\end{aligned} \tag{3.83}$$

Elimination of $D(c_vT)/Dt$ between (3.81) and (3.83), followed by rearrangement, then yields

$$\begin{aligned}c_v\frac{DT}{Dt} + p\frac{D\alpha}{Dt} &\equiv c_v\frac{DT}{Dt} + p\frac{D}{Dt}\left(\frac{1}{\rho}\right) \\ &= \dot{Q} - \left[L_0^v + \left(c_v^v - c^l\right)T\right]\dot{q}^v + \left[L_0^f + \left(c^l - c^f\right)T\right]\dot{q}^f.\end{aligned} \tag{3.84}$$

Comparing (3.84) for a *cloudy*-air parcel to (2.63) for a *dry* one, they both have the same form, but with different right-hand sides. Both right-hand sides have the $\dot{Q}$ heating rate term in common, but (3.84) for cloudy air additionally has latent heating/cooling terms.

Comparing (3.84) now with (3.81), it is seen that latent heating appears in a slightly different way. This is because the $TD(c_v)/Dt$ contribution to $D(c_vT)/Dt$, and its influence, has migrated from the left-hand side of thermodynamic-energy equation (3.81) to the right-hand side. *No approximation has been introduced*, the underlying physics remains unchanged, and although the two mathematical forms – (3.81) versus (3.84) – are different, *they are equivalent*.

Using cloudy-air equation of state (3.51), definition $\alpha \equiv 1/\rho$, and Carnot's law for a cloudy-air parcel, (3.77), other forms equivalent to (3.81) (i.e. without any approximation) are:

$$\frac{D(c_vT)}{Dt} + \frac{RT}{\alpha}\frac{D\alpha}{Dt} = \frac{D(c_vT)}{Dt} - \frac{RT}{\rho}\frac{D\rho}{Dt} = \dot{Q} - L_0^v\dot{q}^v + L_0^f\dot{q}^f, \tag{3.85}$$

$$\frac{D\ln(c_vT)}{Dt} + \frac{R}{c_v}\frac{D\ln\alpha}{Dt} = \frac{\dot{Q} - L_0^v\dot{q}^v + L_0^f\dot{q}^f}{c_vT}, \tag{3.86}$$

$$\frac{D}{Dt}(c_pT) - \alpha\frac{Dp}{Dt} + \frac{D}{Dt}\left[\left(R - \widetilde{R}\right)T\right] = \dot{Q} - L_0^v\dot{q}^v + L_0^f\dot{q}^f, \tag{3.87}$$

$$c_p\frac{DT}{Dt} - \alpha\frac{Dp}{Dt} + \frac{D}{Dt}\left[\left(R - \widetilde{R}\right)T\right] = \dot{Q} - \left[L_0^v + \left(c_p^v - c^l\right)T\right]\dot{q}^v + \left[L_0^f + \left(c^l - c^f\right)T\right]\dot{q}^f. \tag{3.88}$$

(Note that the right-hand side of (3.88) differs from the right-hand side of (3.84); c_p^v appears in (3.88), whereas it is c_v^v that appears in (3.84).)

Equations (3.85)–(3.88) are the *cloudy*-air analogues (where R, $\widetilde{R}$, c_p, and c_v are all *variable*) of *dry*-air equations (2.66)–(2.68) (where $R = \widetilde{R} = R^d$, $c_p = c_p^d$, and $c_v = c_v^d$ are all *constant*). (As a cross-check, setting R, c_p, and c_V to their constant dry values in cloudy-air equations (3.85)–(3.88), and also setting $\dot{q}^v \equiv \dot{q}^f \equiv 0$ to eliminate latent heating, recovers dry-air equations (2.66)–(2.68).) Although (3.87) and (3.88) are equivalent to (3.81), a small term $D\left[\left(R-\widetilde{R}\right)T\right]/Dt$, absent in the analogous dry equation (2.68), appears in them.

Using (3.38), (3.51), (3.76), and (3.79), the third term on the left-hand side of (3.88) may be rewritten as

$$\begin{aligned}\frac{D}{Dt}\left[\left(R-\widetilde{R}\right)T\right] &= \frac{D}{Dt}\left[\frac{\left(\alpha^l q^l + \alpha^f q^f\right)}{\alpha}RT\right] = \frac{D}{Dt}\left[\left(\alpha^l q^l + \alpha^f q^f\right)p\right] \\ &= \left(\alpha^l q^l + \alpha^f q^f\right)\frac{Dp}{Dt} + \left(\alpha^l \frac{Dq^l}{Dt} + \alpha^f \frac{Dq^f}{Dt}\right)p \\ &= \frac{\alpha}{R}\left(R-\widetilde{R}\right)\frac{Dp}{Dt} + \left(\alpha^l \dot{q}^l + \alpha^f \dot{q}^f\right)p \\ &= \alpha\frac{Dp}{Dt} - \frac{\widetilde{R}T}{p}\frac{Dp}{Dt} - \left[\alpha^l \dot{q}^v + \left(\alpha^l - \alpha^f\right)\dot{q}^f\right]p. \end{aligned} \tag{3.89}$$

Insertion of (3.89) into (3.88), followed by rearrangement, then delivers the following, further, equivalent form to (3.81), namely

$$\begin{aligned} c_p\frac{DT}{Dt} - \frac{\widetilde{R}T}{p}\frac{Dp}{Dt} &= \dot{Q} - \left[L_0^v + \left(c_p^v - c^l\right)T - \alpha^l p\right]\dot{q}^v + \left[L_0^f + \left(c^l - c^f\right)T - \left(\alpha^f - \alpha^l\right)p\right]\dot{q}^f \\ &\equiv \dot{Q} - \mathbb{L}^{l\to v}(T,p)\,\dot{q}^v + \mathbb{L}^{f\to l}(T,p)\,\dot{q}^f. \end{aligned} \tag{3.90}$$

In (3.90),

$$\mathbb{L}^{l\to v}(T,p) \equiv h^v - h^l = L_0^v + \left(c_p^v - c^l\right)T - \alpha^l p, \tag{3.91}$$

$$\mathbb{L}^{f\to l}(T,p) \equiv h^l - h^f = L_0^f + \left(c^l - c^f\right)T - \left(\alpha^f - \alpha^l\right)p \tag{3.92}$$

are the latent heats (also known as enthalpies) of vaporisation and fusion, respectively; see (10.25) and (10.26) of Section 10.4.5, (44) of Staniforth and White (2019), and (61) of Thuburn (2017). These latent heats have important physical significance. Latent heat $\mathbb{L}^{A\to B}(T,p)$ measures the amount of heat required, at temperature T and pressure p, to completely change unit mass of phase A of a substance to phase B; see Section 10.4.4.

Akmaev and Juang (2008) propose the use of *specific enthalpy*,[14]

$$h \equiv \mathscr{E} + \frac{p}{\rho} \equiv \mathscr{E} + p\alpha, \tag{3.93}$$

instead of internal energy, $\mathscr{E}$; see (3.71)–(3.74).[15] Introducing (3.71) into (3.93) gives

$$h = c_V T + \mathbb{L}_0 + p\alpha = c_V T + p\alpha + \left(L_0^v + L_0^f\right)q^v + L_0^f q^l. \tag{3.94}$$

Eliminating $c_V T$ between (3.81) and (3.94), and using (3.38) and (3.79), then yields (their equation (5))

[14] Enthalpy is a thermodynamic potential (see Section 9.2) with important physical significance regarding latent heating (see Section 10.4.4).

[15] They do not, however, include latent potential internal energy $\mathbb{L}_0$ in their definition of $\mathscr{E}$. Instead, they implicitly (in words rather than equations) include its impact as an extra contribution to the heating term $\dot{Q}$.

$$\frac{Dh}{Dt} - \alpha \frac{Dp}{Dt} \equiv \frac{Dh}{Dt} - \frac{1}{\rho}\frac{Dp}{Dt} = \dot{Q}. \tag{3.95}$$

This is another thermodynamic-energy equation for cloudy air that is equivalent to (3.81). It is of particularly simple form.

3.6.2 Almost Exact

Following Staniforth and White (2019), but additionally including frozen water, almost-exact forms of the thermodynamic-energy equation for a cloudy atmosphere can be obtained from (3.87) and (3.88) by exploiting the fact that R and $\widetilde{R}$ are almost (but not quite) identical. Using (3.56), the ratio of the third and first terms in (3.87) behaves as

$$\frac{\left(R - \widetilde{R}\right) T}{c_p T} = \frac{\left(\alpha^l q^l + \alpha^f q^f\right)}{\alpha} \frac{R}{c_p} \sim \left[2 \times \left(\frac{5}{4} \times 10^{-3} \times 3 \times 10^{-3}\right)\right] \times \frac{2}{7} \sim 2 \times 10^{-6} \tag{3.96}$$

The justification for these estimates is as follows. The ratios α^l/α and α^f/α are essentially the ratios of the densities of air to liquid water and to frozen water, respectively, and these are approximately equal. They attain their maximum values of approximately $5/4 \times 10^{-3}$ at sea level (since air density diminishes as a function of altitude). The ratio R/c_p is approximately that of dry air, which is usually taken to be 2/7, the ideal-gas value for an ideal, diatomic gas (about 99% of Earth's composite dry gas is composed of diatomic gases; see Section 1.4). Mass fractions q^l and q^f (which measure the relative amounts of liquid water and frozen water, respectively, in a cloudy-air parcel) are very small, with maximum values estimated to be less than 3×10^{-3}, but usually much less. Precise estimates are unimportant here. The important point is that $\left(R - \widetilde{R}\right) T$ is negligibly small with respect to $c_p T$, and therefore the $\left(R - \widetilde{R}\right) T$ terms appearing in (3.87) and (3.88) are quantitatively negligible.

Setting $R \equiv \widetilde{R}$ in (3.87) and (3.88) gives the approximate '*almost-exact*' forms

$$\frac{D\left(c_p T\right)}{Dt} - \alpha \frac{Dp}{Dt} = \dot{Q} - L_0^v \dot{q}^v + L_0^f \dot{q}^f, \qquad (*) \tag{3.97}$$

$$c_p \frac{DT}{Dt} - \alpha \frac{Dp}{Dt} = \dot{Q} - \left[L_0^v + \left(c_p^v - c^l\right) T\right] \dot{q}^v + \left[L_0^f + \left(c^l - c^f\right) T\right] \dot{q}^f. \qquad (*) \tag{3.98}$$

Comparing (3.97) for a cloudy-air parcel to (2.65) for a dry one, they are quite similar. As for the comparison of (3.81) with (2.63), there are the same two notable differences except that for the first difference it is c_p in (3.97) that *is no longer constant*. It now appears as $D\left(c_p T\right)/Dt$ rather than as $c_p DT/Dt$.

Comparing (3.98) with (3.90), it is seen that the dependence on pressure (p) of the latent-heating-related terms on the right-hand side of (3.90) is lost in (3.98) as a consequence of approximating (3.90) by (3.98). See Section 10.6.5 for further discussion of this, particularly with regard to thermodynamical consistency.

Other forms of the thermodynamic-energy equation (equivalent to (3.97), and therefore *almost equivalent* to (3.81)) may be obtained by setting $R \equiv \widetilde{R}$ and using the equation of state ($p\alpha = RT = \widetilde{R}T$) for cloudy air. Thus

$$\frac{D\left(c_p T\right)}{Dt} - \frac{1}{\rho}\frac{Dp}{Dt} = \dot{Q} - L_0^v \dot{q}^v + L_0^f \dot{q}^f, \qquad (*) \tag{3.99}$$

$$\frac{Dp}{Dt} - (\gamma RT)\left[\frac{1}{\kappa}\frac{D}{Dt}(\kappa\rho)\right] = \frac{\rho R}{c_v}\left(\dot{Q} - L_0^v\dot{q}^v + L_0^f\dot{q}^f\right), \qquad (*) \qquad (3.100)$$

$$\frac{D\ln(c_p T)}{Dt} - \kappa\frac{D\ln p}{Dt} = \frac{\dot{Q} - L_0^v\dot{q}^v + L_0^f\dot{q}^f}{c_p T}, \qquad (*) \qquad (3.101)$$

$$\frac{D}{Dt}\left(\frac{T_v}{\kappa}\right) - \frac{T_v}{p}\frac{Dp}{Dt} = \frac{\dot{Q} - L_0^v\dot{q}^v + L_0^f\dot{q}^f}{R_d}, \qquad (*) \qquad (3.102)$$

$$\frac{D}{Dt}\ln\left(\frac{T_v}{\kappa}\right) - \kappa\frac{D\ln p}{Dt} = \frac{\kappa\left(\dot{Q} - L_0^v\dot{q}^v + L_0^f\dot{q}^f\right)}{R^d T_v}, \qquad (*) \qquad (3.103)$$

where

$$T_v \equiv \frac{R}{R^d}T, \qquad (3.104)$$

is *virtual temperature*, and $\gamma \equiv c_p/c_v$.

Equations (3.99)–(3.101) are the *cloudy*-air analogues (where R, c_p, c_v, γ, and κ are all *variable*) of *dry*-air equations (2.65), (2.69), and (2.70) (where $R = R^d$, $c_p = c_p^d$, $c_v = c_v^d$, $\gamma = \gamma^d \equiv c_p^d/c_v^d$, and $\kappa = \kappa^d \equiv R^d/c_d^p$ are all *constant*). (As a cross-check, setting R, c_p, c_v, γ, and κ to their constant dry values in cloudy-air equations (3.99)–(3.101), and also setting $\dot{q}^v \equiv \dot{q}^f \equiv 0$ to eliminate latent heating, recovers dry-air equations (2.65), (2.69), and (2.70).)

3.6.3 Discussion

Various *exact* and *almost-exact* forms of the thermodynamic-energy equation for the model problem (of cloudy air) are collated in Table 3.1. Form 0 is the base form from which the other forms (1-5, 2A, and 3A) are obtained. Exact Forms 1–5 are equivalent to Form 0. Almost-exact Forms 2A and 3A are equivalent to one another, and almost equivalent to Form 0 (and, therefore, to Forms 1–5); these almost-exact forms only neglect the minute fractional volume occupied by liquid water and frozen water.

As discussed in Staniforth and White (2019) (but for the simpler case of cloudy air in the absence of frozen water), there is considerable variation in the way in which latent-heating contributions appear on the right-hand sides of *equivalent* forms. One might expect that latent heating should appear in the same way for equivalent forms. So why don't they? Depending on form, it is because some elements of the latent-heating terms have cancelled or have combined with other terms during the derivation. Variation of specific heat capacity (at constant volume or at constant pressure) with vapour content is an important contributor for this.

All of the right-hand sides of the forms appearing in Table 3.1 (with the exception of Form 5) can be described as:

'*including latent heating and other physical processes*'.

This is a common practice in the literature, with the virtue of being succinct ... albeit with the vice of being imprecise. Whilst this description is true, it can be seen from Table 3.1 that the corresponding mathematical expressions can be very different between different forms. One therefore has to be very careful when interpreting *verbal* descriptions of the formulation of a thermodynamic-energy equation. For a particular formulation, it is insufficient to know that a right-hand side 'includes latent heating'; one additionally needs to know precisely how, *mathematically*, it is included. Unfortunately, this information is not always provided by authors, and this can lead to misinterpretation and misleading conclusions; see Section 3.6.4 for an example of this.

Table 3.1 provides a practical guide and useful reference for assessing the fidelity of the representation of latent heating for the well-defined problem of cloudy air.

3.6.4 A Clarification

Akmaev and Juang (2008) derive two equations – their (10) and (11) – whose left-hand sides correspond to those of Forms 2A and 3A, respectively, displayed in Table 3.1.[16] They *assume* that the corresponding right-hand sides are the *same* quantity (their Q, with *verbal* definition '*Q represents non-adiabatic contributions to internal energy per unit mass (heating rates) by such processes as radiative and chemical heating, phase transitions, ...*', which includes latent heating due to phase transitions). They then argue that the common use in meteorology of their (11) (i.e. Form 3A herein) is highly questionable, since c_p has to be constant (which is not generally true) for it to be valid.

Now it has been shown earlier that Forms 2A and 3A – whose left-hand sides correspond to Akmaev and Juang (2008)'s (10) and (11), respectively – are *equivalent* forms. However, examination of their right-hand sides (Table 3.1) shows that they are *mathematically different* (and not identical, as assumed in Akmaev and Juang (2008)). As pointed out in Staniforth and White (2019), but for the simpler case of a cloudy-air parcel with no frozen water, this means that use of Form 3A is fully justified (with variable c_p) *provided that* the latent-heating contribution to its right-hand side is evaluated appropriately. This evaluation corresponds to that commonly used in meteorology, with latent heats of vaporisation and fusion that depend on temperature, T.

There is nothing wrong with Akmaev and Juang (2008)'s logic. It is their *assumption* that the right-hand sides of their (10) and (11) are equal that is problematic. This then led to their unfortunate and misleading conclusion, repeated in Akmaev (2011)'s comprehensive and authoritative review of whole-atmosphere modelling.

This clarification does not, however, affect the validity of Akmaev and Juang (2008)'s principal conclusions:

- Enthalpy as a prognostic variable in the thermodynamic-energy equation is both viable and convenient.
- Proper account needs to be taken of the variability of specific heat capacity in the formulation of a thermodynamic-energy equation.

3.7 THE GOVERNING EQUATIONS FOR MOTION OF A CLOUDY ATMOSPHERE

Summarising the developments of the preceding sections, a set of governing equations for the motion of a cloudy-air parcel (comprised of dry air, water vapour, liquid water, frozen water, plus other substances – if present) can be assembled in many different ways.

Similarly to the case of a dry atmosphere, any one of many different forms of the thermodynamic energy equation can be adopted. In what follows, the particular form (3.81) is chosen. This is a very natural form since it is a prognostic equation for internal energy, $\mathcal{E} = c_v T + \mathbb{L}_0$, and closely related to the underlying physics.

For modelling purposes, there are two natural representations of constituents:

1. Mass fractions, q^X.
2. Mixing ratios, m^X.

As shown in Section 3.2.1.5, it is straightforward to convert between these two representations. Basing a model primarily on one representation does not, therefore, preclude secondary use of the other representation for special purposes (e.g. for parametrising unresolved, or unrepresented, physical processes); one can transform between them when necessary.

[16] Taking into account notational differences, their term $(RT/p)\,\omega \equiv (RT/p)\,Dp/Dt = \alpha Dp/Dt$.

- So which representation is preferable?

There are pros and cons for either choice, depending upon application (e.g. for the atmosphere only, or for the ocean only, or for *both*). The present work aims to formulate the basic governing equations for atmospheric and oceanic modelling *within a unified framework*. A key aspect is the representation of the thermodynamics for a mixture of substances, particularly for the formulation of an associated equation of state. Now the thermodynamics for a mixture of substances (such as dry air, water substance, sea salts) is most naturally formulated in terms of the mass fractions of its constituents; see Chapter 9. This is one good reason to use mass fractions.

Another good reason is that *mixing ratios are susceptible to singular behaviour*. Constituent masses – see (3.13) – are normalised by the *dry* mass, $\mathbb{M}^d$, in a fluid parcel. This is fine for the atmosphere, where dry air is the major constituent. However, it is not fine when dry air is absent (e.g. for oceans or for pure water substance), since the normalisation is then singular due to division by zero when $\mathbb{M}^d \equiv 0$. This precludes a unified, *overarching*, representation – as in Feistel et al. (2010a,b) and Wright et al. (2010) – of the thermodynamics of a mixture of constituents that span the atmosphere and the ocean, whereby one simply specifies the values of the mass fractions (with some being identically zero) according to the (atmospheric, oceanic, or coupled) application. Contrastingly, mass fractions are normalised with respect to *total* mass ($\mathbb{M}$) – see (3.7) – and they can therefore never be singular since $\mathbb{M} \geq \mathbb{M}^X$ for *any* constituent mass, $\mathbb{M}^X$.

For these reasons, mass fractions are used in developments from here onwards. A suitable set of governing equations, using mass fractions, for motion of a cloudy-air parcel is summarised in the shaded box. For simplicity, the (four) constituents of the parcel are assumed to be dry air and the (three) phases of water substance, with other substances absent. With appropriate definitions of the right-hand side forcing functions, this set is valid for both open and closed parcels. Simplifications that arise for closed parcels are noted below the set of governing equations.

For $X = (d, v, l, f)$, there are four constituents (dry air, water, vapour, liquid water and frozen water) with associated mass fractions q^d, q^v, q^l, and q^f. Doing the accounting for equation set (3.105)–(3.112), the three components of vector equation (3.105), plus the six scalar equations (3.106)–(3.108) and (3.110), comprise nine equations for the nine unknown variables $\mathbf{u} = (u_1, u_2, u_3)$, ρ, q^v, q^l, q^f, T, and p.[17]

Equations (3.105)–(3.108) are prognostic equations for the eight unknown variables $\mathbf{u} = (u_1, u_2, u_3)$, ρ, q^v, q^l, q^f, and T. This means that given values of all variables at an instant in time, the eight prognostic equations (i.e. those with a D/Dt term) can be used to obtain values of the eight variables $\mathbf{u} = (u_1, u_2, u_3)$, ρ, q^v, q^l, q^f, and T at a later instant in time. (Because of the non-linear nature of these equations, this usually has to be done numerically for all but the simplest situations).

Pressure (p) at the later instant in time can then be obtained using diagnostic equation (3.110) and the prognosed values of ρ and T at the later instant in time.

3.8 CONCLUDING REMARKS

A set of governing equations for a realistic representation of a cloudy atmosphere has been assembled in this chapter. A key ingredient for the construction of this equation set was the use of ideal-gas law (3.110) as the equation of thermodynamic state. This greatly simplified developments, whilst avoiding the need to delve too deeply into the challenges of thermodynamic theory.

[17] q^d is omitted from the accounting, and also from the outlined solution procedure. If needed, it can be obtained diagnostically at the end of the solution procedure using (3.12); i.e. it can be obtained from $q^d = 1 - \left(q^v + q^l + q^f\right)$, using known values of q^v, q^l, and q^f.

As the name suggests, the ideal-gas law is not valid for the representation of seawater in the oceans. Furthermore, no suitable thermodynamic equation of state of comparable simplicity and accuracy is available for the oceans. A far more general, unified formulation of thermodynamics than the greatly simplified one considered in Chapter 2 and the present chapter is therefore needed. This is developed in a step-by-step manner in Chapters 4 and 9–11 and is valid for both quantitative atmospheric *and* oceanic modelling.

The Governing Equations for Motion of a Cloudy-Air Parcel

A set of governing equations for motion of a cloudy-air parcel, expressed in a rotating frame of reference, is comprised of momentum equation (3.68), mass-continuity equation (3.43), water-substance-transport equations (3.38), thermodynamic-energy equation (3.81), and state equation (3.51), namely:

Momentum

$$\frac{D\mathbf{u}}{Dt} = -2\boldsymbol{\Omega} \times \mathbf{u} - \frac{1}{\rho}\nabla p - \nabla\Phi + \mathbf{F}. \tag{3.105}$$

Continuity of Total Mass Density

$$\frac{D\rho}{Dt} + \rho\nabla\cdot\mathbf{u} = F^{\rho}. \tag{3.106}$$

Water Substance Transport

$$\frac{Dq^X}{Dt} = \dot{q}^X, \quad X = (v, l, f). \tag{3.107}$$

Thermodynamic-Energy

$$\frac{D(c_v T)}{Dt} + p\frac{D}{Dt}\left(\frac{1}{\rho}\right) = \dot{Q} - \left(L_0^v + L_0^f\right)\dot{q}^v - L_0^f \dot{q}^l, \tag{3.108}$$

where

$$c_v = c_v^d q^d + c_v^v q^v + c^l q^l + c^f q^f. \tag{3.109}$$

State

$$p = \rho R T, \tag{3.110}$$

where

$$R = \frac{R^d q^d + R^v q^v}{1 - \left(\alpha^l/\alpha\right) q^l - \left(\alpha^f/\alpha\right) q^f}, \tag{3.111}$$

$$\alpha \equiv \frac{1}{\rho}, \quad \alpha^l \equiv \frac{1}{\rho^l} = \text{constant}, \quad \alpha^f \equiv \frac{1}{\rho^f} = \text{constant}. \tag{3.112}$$

For a closed parcel

(with therefore no net mass transfer across its bounding surface):

- $F^{\rho} \equiv 0.$
- $\sum_{X=(v,l,f)} \dot{q}^X \equiv 0.$
- **Consequently, the right-hand side of (3.108) may then be rewritten as** $\dot{Q} - L_0^v \dot{q}^v + L_0^f \dot{q}^f$.

Symbol	Description
$\mathbb{V}^g$	Volume occupied by *gas* (**shared** between *dry air + water vapour*)
$\mathbb{V}^l$	Volume occupied (**exclusively**) by *liquid water*
$\mathbb{V}^f$	Volume occupied (**exclusively**) by *frozen water*
$\mathbb{V}$	Total volume occupied by mixture = $\mathbb{V}^g + \mathbb{V}^l + \mathbb{V}^f$ (*dry air + water vapour + liquid water + frozen water*)
$\mathbb{M}^d$	Mass of *dry air* (in volume $\mathbb{V}^g$)
$\mathbb{M}^v$	Mass of *water vapour* (in volume $\mathbb{V}^g$)
$\mathbb{M}^g$	Mass of *gas* (in volume $\mathbb{V}^g$) $= \mathbb{M}^d + \mathbb{M}^v$
$\mathbb{M}^l$	Mass of *liquid water* (in volume $\mathbb{V}^l$)
$\mathbb{M}^f$	Mass of *frozen water* (in volume $\mathbb{V}^f$)
$\mathbb{M}$	Total mass (in volume $\mathbb{V} = \mathbb{V}^g + \mathbb{V}^l + \mathbb{V}^f$) of mixture (*dry air + water vapour + liquid water + frozen water*) $= \mathbb{M}^g + \mathbb{M}^l + \mathbb{M}^f$ $= \mathbb{M}^d + \mathbb{M}^v + \mathbb{M}^l + \mathbb{M}^f$

Table 3.2 Nomenclature and definitions for various volumes and masses.

APPENDIX: DERIVATION OF THE EQUATION OF STATE FOR CLOUDY AIR FROM FIRST PRINCIPLES

Nomenclature and Definitions

Consider a parcel of a mixture (termed *cloudy air*) composed of dry air, water vapour, liquid water, and frozen water. The nomenclature used to define various volumes, masses, mass densities, and specific volumes associated with this parcel is tabulated in Tables 3.2 and 3.3. The fact that gases (dry air and water vapour) share a *common* volume ($\mathbb{V}^g$), whereas liquid water and frozen water occupy *separate* (unshared) volumes ($\mathbb{V}^l$ and $\mathbb{V}^f$), complicates the derivation of the equation of state for cloudy air.

Partial and Total Pressures

Only gases (dry air and water vapour here) exert a pressure; liquid water and frozen water do not. According to Dalton's Law of Partial Pressures,[18] (and consistent with the ideal-gas law (3.50)), the pressure exerted by a mixture of dry air and water vapour is equal to the sum of the partial pressures which would be exerted by the dry air and water vapour fractions separately, *both occupying the same volume* $\mathbb{V}^g$.[19] The volume $\mathbb{V}^l + \mathbb{V}^f$ occupied by liquid water and frozen water is both very small and distinct from $\mathbb{V}^g$. (Volumes $\mathbb{V}^l$ and $\mathbb{V}^f$ are commonly neglected in textbooks on atmospheric dynamics, which is an approximation, albeit a quantitatively excellent one for Earth's atmosphere. Doing so can, however, lead to inconsistencies in the thermodynamics; see e.g. Section 10.6.5.)

If R^d and R^v are the gas constants (per unit mass) for dry air and water vapour, respectively, then (using ideal-gas law (3.50), and definitions given in Table 3.2) the total pressure exerted by the gases in volume $\mathbb{V}^g$ is

[18] After John Dalton (1766–1844).

[19] Underlying Dalton's Law are the ideal gas assumptions that the molecules of the two gaseous components have negligible volume, and that *molecules do not interact with one another*. In reality, these two assumptions do not hold exactly, only to a very good approximation over the range of temperatures and pressures observed in Earth's atmosphere. Molecular interactions can be accounted for in the representation of gases, but at the cost of increased complexity of the thermodynamics.

Symbol	Description
ρ^d	Density of *dry air* (per unit volume in $\mathbb{V}^g$) $= \dfrac{\mathbb{M}^d}{\mathbb{V}^g} = \dfrac{1}{\alpha^d}$
ρ^v	Density of *water vapour* (per unit volume in $\mathbb{V}^g$) $= \dfrac{\mathbb{M}^v}{\mathbb{V}^g} = \dfrac{1}{\alpha^v}$
ρ^g	Density of *gas* (per unit volume in $\mathbb{V}^g$) $= \dfrac{\mathbb{M}^d + \mathbb{M}^v}{\mathbb{V}^g} = \dfrac{\mathbb{M}^g}{\mathbb{V}^g}$ $= \rho^d + \rho^v = \dfrac{1}{\alpha^d} + \dfrac{1}{\alpha^v} = \dfrac{1}{\alpha^g}$
ρ^l	Density of *liquid water* (per unit volume in $\mathbb{V}^l$) $= \dfrac{\mathbb{M}^l}{\mathbb{V}^l} = \dfrac{1}{\alpha^l}$
ρ^f	Density of *frozen water* (per unit volume in $\mathbb{V}^f$) $= \dfrac{\mathbb{M}^f}{\mathbb{V}^f} = \dfrac{1}{\alpha^f}$
ρ	Density of total substance (*gas* + *liquid* + *frozen*) (per unit volume in $\mathbb{V} = \mathbb{V}^g + \mathbb{V}^l + \mathbb{V}^f$) $= \dfrac{\mathbb{M}}{\mathbb{V}} = \dfrac{1}{\alpha}$
α^d	Volume (in $\mathbb{V}^g$) occupied by unit mass of dry air $= \dfrac{\mathbb{V}^g}{\mathbb{M}^d} = \dfrac{1}{\rho^d}$
α^v	Volume (in $\mathbb{V}^g$) occupied by unit mass of water vapour $= \dfrac{\mathbb{V}^g}{\mathbb{M}^v} = \dfrac{1}{\rho^v}$
α^g	Volume (in $\mathbb{V}^g$) occupied by unit mass of *gas* (*dry air* + *water vapour*) $= \dfrac{\mathbb{V}^g}{\mathbb{M}^g} = \dfrac{\mathbb{V}^g}{\mathbb{M}^d + \mathbb{M}^v} = \dfrac{1}{\rho^g}$
α^l	Volume (in $\mathbb{V}^l$) occupied by unit mass of *liquid water* $= \dfrac{\mathbb{V}^l}{\mathbb{M}^l} = \dfrac{1}{\rho^l}$
α^f	Volume (in $\mathbb{V}^f$) occupied by unit mass of *frozen water* $= \dfrac{\mathbb{V}^f}{\mathbb{M}^f} = \dfrac{1}{\rho^f}$
α	Volume (in $\mathbb{V} = \mathbb{V}^g + \mathbb{V}^l + \mathbb{V}^f$) occupied by unit mass of total substance (*dry air* + *water vapour* + *liquid water* + *frozen water*) $= \dfrac{\mathbb{V}}{\mathbb{M}} = \dfrac{1}{\rho}$

Table 3.3 Nomenclature and definitions for various mass densities (mass contained within a volume divided by this volume) and specific volumes (volume occupied by unit mass of a substance within the stated volume).

$$p = p^d + p^v = \left(\rho^d R^d + \rho^v R^v\right) T = \left(R^d \frac{\mathbb{M}^d}{\mathbb{V}^g} + R^v \frac{\mathbb{M}^v}{\mathbb{V}^g}\right) T = \frac{\mathbb{M}}{\mathbb{V}} \left(R^d \frac{\mathbb{M}^d}{\mathbb{M}} + R^v \frac{\mathbb{M}^v}{\mathbb{M}}\right) \frac{\mathbb{V}}{\mathbb{V}^g} T$$
$$= \rho \left(R^d q^d + R^v q^v\right) \frac{\mathbb{V}}{\mathbb{V}^g} T, \tag{3.113}$$

where:

- $p^d = \rho^d R^d T$ and $p^v = \rho^v R^v T$ are the partial pressures exerted by dry air and water vapour, respectively.
- $q^d \equiv \mathbb{M}^d/\mathbb{M}$ and $q^v \equiv \mathbb{M}^v/\mathbb{M}$ are the mass fractions of dry air and water vapour, respectively, contained within the parcel (which latter includes liquid water and frozen water in addition to dry air and water vapour).

Now (using Tables 3.2 and 3.3, and definition (3.7) of mass fraction)

$$\frac{\mathbb{V}^g}{\mathbb{V}} = \frac{\mathbb{V} - \mathbb{V}^l - \mathbb{V}^f}{\mathbb{V}} = 1 - \frac{\mathbb{V}^l}{\mathbb{V}} - \frac{\mathbb{V}^f}{\mathbb{V}} = 1 - \frac{\mathbb{V}^l}{\mathbb{M}^l}\frac{\mathbb{M}}{\mathbb{V}}\frac{\mathbb{M}^l}{\mathbb{M}} - \frac{\mathbb{V}^f}{\mathbb{M}^f}\frac{\mathbb{M}}{\mathbb{V}}\frac{\mathbb{M}^f}{\mathbb{M}} = 1 - \frac{\alpha^l}{\alpha}q^l - \frac{\alpha^f}{\alpha}q^f. \quad (3.114)$$

Substituting (3.114) into (3.113) finally yields the total pressure

$$p = \rho\left[\frac{R^d q^d + R^v q^v}{1 - \left(\alpha^l/\alpha\right)q^l - \left(\alpha^f/\alpha\right)q^f}\right]T = \rho R T. \quad (3.115)$$

Here

$$R \equiv \frac{R^d q^d + R^v q^v}{1 - \left(\alpha^l/\alpha\right)q^l - \left(\alpha^f/\alpha\right)q^f} \quad (3.116)$$

is the 'gas constant' of the mixture. This is generally not a constant but (due to its dependence on q^d and q^v) varies as a function of the composition of the mixture of substances (dry air + water vapour + liquid water + frozen water).

Equations (3.115) and (3.116) correspond to (3.51) and (3.54), respectively, of Section 3.3.

4

Governing Equations for Motion of Geophysical Fluids: Vector Form

ABSTRACT

A set of governing equations – expressed in vector form – to describe the motion of a cloudy atmosphere was assembled in Chapter 3. Due to use of the ideal-gas law, this equation set is unsuitable for modelling oceanic flows. To prepare the way to address this shortcoming, a more general approach to thermodynamics is developed in the present chapter. Thermodynamic vocabulary and fundamental principles are reviewed. Applying thermodynamic theory, prognostic equations for the internal energy and entropy of a geophysical fluid – be it primarily gaseous or liquid – are obtained. The thermodynamic representation of a cloudy-air parcel is revisited since it is analytically tractable to do so, leading to explicit functional forms for various thermodynamic quantities of interest. This provides valuable insight into the application of thermodynamic theory in the context of geophysical fluid dynamics. It also illustrates the theoretical principle that all thermodynamic properties of a fluid parcel can be obtained from a prescribed, functional form for its internal energy, or from one for its entropy. Finally, a unified set of governing equations for the motion of geophysical fluids (predominantly gaseous for the atmosphere, and liquid for the oceans) is assembled using the developments of the present chapter and those of the two preceding chapters. This unified set is expressed in vector form in a rotating frame of reference.

4.1 PREAMBLE

The thermodynamics described in Section 2.3 (for a dry atmosphere) and Chapter 3 (for a cloudy one) assume the ideal-gas law (2.50). This approach has the virtues of simplicity, analytical tractability and realism for atmospheric applications. It is, however, unduly restrictive. In particular, the equation of state for an ideal gas is invalid for the representation of seawater.

To prepare the way for the representation of seawater, a more general approach to thermodynamics is therefore described in Section 4.2. A general overview of thermodynamic vocabulary and principles is given first. This includes discussion of:

- Microscopic and macroscopic scales of matter.
- Thermodynamic systems, equilibrium, and state variables.
- Internal energy and its conservation.
- The fundamental equation of state for substances.
- The fundamental thermodynamic relation.
- Prognostic equations for the internal energy ($\mathscr{E}$) and entropy (η) of a fluid, be it primarily gaseous *or* liquid.[1]
- Specific heat capacity.

[1] An important further benefit of the development of these equations is that they facilitate later application (in Chapter 14) – in a *unified* manner (for both the atmosphere *and* the oceans) – of Hamilton's principle of stationary action, after William Rowan Hamilton (1805–1865).

Next, the thermodynamic representation of a cloudy-air parcel – described in Chapter 3 – is revisited in Section 4.3. This provides a concrete example of the application of the thermodynamic theory described in Section 4.2 of the present chapter. This theory is quite abstract. Furthermore, it leaves the reader wondering what practical and realistic functional forms for $\mathscr{E}$ and η might look like. A cloudy-air parcel facilitates thermodynamic understanding by providing an *analytically tractable* example of significant, practical applicability. *Explicit* functional forms are given for the thermodynamic quantities of interest in the present context. Doing so makes it easier to follow the developments – in Chapter 11 – for the representation of ocean thermodynamics.

Having developed explicit, functional forms of $\mathscr{E}$ and η for a cloudy-air parcel, various thermodynamic properties of cloudy air are then explicitly derived from the functional form of $\mathscr{E}$. This illustrates the theoretical principle – stated in Section 4.2 – that *all* thermodynamic properties of a fluid parcel can be obtained from a prescribed, functional form for $\mathscr{E}$ or, alternatively and equivalently, from that for η.

Putting together the thermodynamic developments of this chapter, together with the fluid-dynamical ones of Chapters 2 and 3, a *unified set of governing equations* for geophysical fluids (predominantly gaseous for the atmosphere, and liquid for the oceans) is assembled in Section 4.4. This set is expressed in vector form in a rotating frame of reference. Concluding remarks are made in Section 4.5. Finally, theoretical values for the specific heats of an ideal gas – derived from statistical mechanics – are given in the Appendix to this chapter.

4.2 MORE GENERAL THERMODYNAMICS

Thermodynamics is a broad and challenging subject.[2] What follows is a description of the basic thermodynamic principles needed to represent Earth's atmosphere and oceans as fluids. Readers interested in acquiring a broader and deeper knowledge of this important subject may wish to consult specialist textbooks such as Callen (1985), Curry and Webster (1999), and Ambaum (2010).

4.2.1 Some Thermodynamic Vocabulary

Thermodynamics has a lot of vocabulary, so we begin by defining some of it.

4.2.1.1 Matter

Matter is composed of independent *substances*. Each substance is termed a *component* of a thermodynamic system. Examples of independent substances/components in Earth's atmosphere and oceans are:

- Dry air.
- Water substance.
- Ocean solute.

Each substance may have one or more *phases*. Examples of individual substances and their associated phases (vapour, liquid, solid) are:

- Dry air (one phase – vapour).
- Water substance (three phases – vapour, liquid, solid).
- Ocean solute (one phase – water-soluble solid).

A *constituent* of a thermodynamic system is any substance in a particular phase. Examples of constituents are:

[2] As Gill (1982) wryly notes (on page 41): '*Textbooks differ considerably in their approach and there are difficulties obtaining a completely logical treatment of thermodynamics*'!

- Dry air.
- Water vapour.
- Liquid water.
- Frozen water.
- Ocean solute.

One can give more formal definitions for these quantities, but the informal ones presented in the preceding list suffice for present purposes. Illustrative examples of the composition of two-component thermodynamic systems of relevance to atmospheric and oceanic modelling may be found in Table 9.6 of Chapter 9.

4.2.1.2 *Microscopic and Macroscopic Scales*

Substances are comprised of huge numbers of molecules, invisible to the naked eye. Each of these molecules whizzes around and interacts with other molecules. The molecules themselves are comprised of atoms and subatomic particles to further complicate things. In principle, one could track each molecule in terms of its mass, position, momentum, and other characteristics as a function of time. In practice this is completely unfeasible due to the huge number of molecules involved. To address this, thermodynamic theory essentially defines two spatial scales:

1. *Microscopic* (visible only with the aid of a powerful magnifying instrument, such as an optical or electron *microscope*).
2. *Macroscopic* (a scale that is orders of magnitude larger than microscopic).

The microscopic scale relates to that of individual molecules and their mean separation, whilst the macroscopic scale relates to that of traditional measurements of bulk flow and properties such as temperature, pressure, and density.

To illustrate this concept of scale separation and how it can be exploited, consider the cylinder and piston depicted in Fig. 2.6 and described in Section 2.3.4. Microscopically, molecules move around within the cylinder in an incoherent manner, interacting with one another and bouncing off the walls of the cylinder and piston. The characteristic time of these interactions is extremely short, and they take place so quickly that they are unmeasurable using traditional thermometers (for temperature) and barometers (for pressure).

However, although the molecules *individually* behave randomly and incoherently from the microscopic viewpoint, they *collectively* behave coherently at the macroscopic scale. When their motions are statistically averaged in both space and time, the molecules collectively define a (macroscopic) temperature (i.e. T, an average kinetic energy per unit mass), and a (macroscopic) pressure (i.e. p, an average force per unit area, exerted on the cylinder walls and on the piston). The characteristic timescale of these two quantities (T and p) is sufficiently long for the gas to be considered to be in a state of (thermodynamic) equilibrium, and for traditional thermometers and barometers to measure them.

Thus the huge number of variables required to represent a huge number of molecules, when viewed at microscopic scale, reduces here to just two (T and p) when viewed macroscopically! Equilibrium thermodynamics essentially filters out the details of what is *randomly* happening at microscopic scales, and focuses attention on what is *coherently* happening macroscopically, in a statistically averaged sense. See Chapter 1 of Callen (1985) for a comprehensive overview of the underpinning principles and assumptions of equilibrium thermodynamics.

4.2.1.3 *Closed and Open Thermodynamic Systems*

A *thermodynamic system* is composed of matter (i.e. substances) enclosed within its bounding walls. These *walls* can be either physical (e.g. the cylinder and piston depicted in Fig. 2.6 of Section 2.3.4) or notional (e.g. the hypothetical boundaries of a fluid parcel embedded within a much

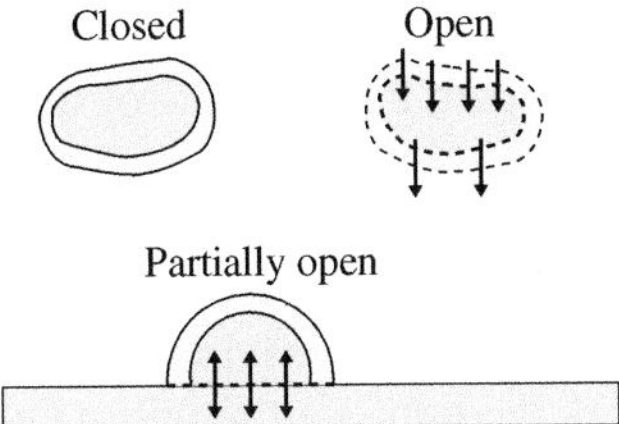

Figure 4.1 Examples of cross sections of closed, partially open, and open fluid parcels. Solid (dashed) contours; impermeable (permeable) to macroscopic mass transfer. Doubled-up contours; possible expansion or compression of the parcel. Arrows; possible macroscopic mass transfer across a parcel boundary. Shaded rectangle; a reservoir of fluid. See text for further details.

larger body of fluid, as depicted in Fig. 2.1 of Section 2.2.1). In our fluid-dynamical context, the walls might be the bounding material surface of a parcel of cloudy air in the atmosphere or a parcel of seawater in the ocean. Walls constrain the thermodynamic operations that can take place within a thermodynamic system. They are classed according to their permeability with respect to *matter*, that is, to the mass of substance(s) contained within the system (e.g. within a fluid parcel).

There is no macroscopic transfer of mass between a *closed thermodynamic system* and its environment (i.e. its walls are *closed* and macroscopically impermeable to mass transfer).[3] Microscopic exchange of mass, occurring randomly across its walls but summing to zero, is nevertheless allowed.[4] There is no change in the composition of the thermodynamic system (i.e. in the amount(s) of substance(s) contained within it). In our fluid-dynamical context, this means no change in the amounts of dry air and water substance in a closed cloudy-air parcel, nor in the amounts of water substance and salinity in a closed seawater parcel.[5] Although the walls of a closed thermodynamic system are impermeable to macroscopic *mass* transfer, *heat* can nevertheless be transferred between the system and its surrounding environment. Furthermore, *work* can also be done by one on the other, through expansion or compression of the system's volume. For example, an inflated balloon, heated by the Sun, will expand and do work on the surrounding air; work is also done against the surface tension of the balloon's elastic surface.

An *open thermodynamic system* additionally allows net *mass* transfer with its adjacent environment. This then results in a change in *composition* of the system (i.e. a change in the net amount of substance(s) contained within it). Some examples of cross sections of closed, partially open, and open fluid parcels are depicted in Fig. 4.1.

A *closed* fluid parcel, embedded entirely *within* a fluid, is depicted on the left-hand side of Fig. 4.1. Solid contours denote no macroscopic mass transfer across a parcel boundary (i.e. impermeability to macroscopic mass transfer). The parcel's two contours denote the shape and location of its boundary, before and after any work done expanding or compressing it; see Section 2.3.4.

A similar, but *open* parcel of fluid embedded entirely within a fluid is depicted on the right-hand side of Fig. 4.1. Its dashed contours denote the possibility of macroscopic mass transfer across the parcel boundary (i.e. a possible permeability to mass transfer). Arrows represent such a transfer

[3] This definition of closure is the one given in the footnote on page 17 of Callen (1985). It implies a wall that is *only* restrictive with respect to (net) transfer of mass. The wall is *not* additionally restrictive to energy and volume, as in the alternative definition given by Callen in his main text.

[4] If the microscopic exchange of mass does not sum to zero, then the parcel does not constitute a perfectly closed system. Such a situation arises for systems having two (or more) components if diffusion of one component within the other is allowed. For simplicity, this effect will be neglected here; it can be included if required for completeness, but the system is no longer closed.

[5] A change of phase of a substance *within* a *closed* thermodynamic system is nevertheless possible. For example, water substance can change between its vapour, liquid, and frozen phases, but the total amount of water substance within the system is constant before, during, and after the change.

of mass; inward (outward) pointing arrows depict mass inflow (outflow). So, for example, the four topmost arrows could represent precipitating raindrops, or ice pellets, entering a cloudy-air parcel from above under the influence of gravity. The remaining two arrows then represent liquid or frozen precipitation leaving the parcel. The number of arrows indicates the relative amounts of mass (water substance in this example) entering and leaving the fluid parcel. For the depicted parcel, more mass is entering from above than is leaving below.

Strictly speaking, a thermodynamic system is either closed or open; if it is not closed, then it must be open. However, it is sometimes convenient to consider a fluid parcel for which part of its boundary is impermeable to macroscopic mass transfer, with the remainder being permeable, and to term this a *partially open* fluid parcel. Such a parcel, in contact with a fluid reservoir, is depicted at the bottom of Fig. 4.1. No macroscopic mass transfer is allowed across its closed contours, but is permitted across its (dashed) open one. So, for example, the fluid parcel could be an air parcel overlying and in contact with an underlying ocean or lake. At the open boundary between them, a macroscopic mass transfer (in the form of water vapour) to the air parcel can take place via evaporation of liquid water (denoted by the three upward-pointing arrows) from the underlying reservoir.[6] Conversely, a macroscopic mass transfer (in the form of liquid water) from the air parcel to an underlying surface can take place via condensation of water vapour (denoted by the three downward-pointing arrows). *The focus herein, however, is very much on closed thermodynamic systems in the form of closed fluid parcels.*

4.2.1.4 Thermodynamic Equilibrium and State Variables

For a thermodynamic system to be in *thermodynamic equilibrium*, it must be simultaneously in thermal, mechanical, and chemical equilibrium, with dynamical balance between the system and its surrounding environment holding for each of these three kinds of equilibrium; see Sections 10.3 and 10.4 for detailed discussion. Classical thermodynamics (as applied herein) assumes that a thermodynamic system is either in local thermodynamic equilibrium with its surrounding environment or it is very close to being so. Implicit in this assumption is that changes take place suitably slowly during a thermodynamic process. Such changes are described as *quasi-static*.

A *thermodynamic state variable* is a quantity that serves to specify the *state* of a thermodynamic system. Familiar examples of state variables are:

- Temperature (T).
- Pressure (p).
- Density (ρ).
- Specific volume (α).

Less-familiar ones include:

- Internal energy ($\mathscr{E}$).
- Entropy (η).
- Various other thermodynamic potentials – see Section 9.2 for detailed discussion of these.

Thermodynamic state variables are not independent of one another, but interrelated in many different ways. A particularly simple example is the ideal-gas law (2.50). It constrains p, ρ, and T; given any two of these, the third is uniquely determined from (2.50), and specific volume (α) is obtained from (2.45) by taking the reciprocal of density (ρ).

Thermodynamic variables come in two flavours:

1. *Extensive*: they depend on the mass of the system.
2. *Intensive*: they are independent of the mass of the system.

[6] This is a very important mechanism for Earth's climate. Evaporation of seawater in the tropics drives the circulation of the Hadley cell – after George Hadley (1685–1768) – via moist convection.

Thus mass and volume are extensive variables, whereas temperature, pressure, density, and specific volume are intensive ones. The usual convention in thermodynamics is that *extensive variables* are denoted by *upper-case* letters, and *intensive variables* by *lower-case* ones, albeit with some exceptions; notably T for temperature, and $\dot{Q}$ for heating rate per unit mass. For the most part we use intensive variables, since this keeps things local and simplifies keeping track of thermodynamic quantities. Dividing an extensive variable (usually denoted in upper case) by the mass of the system converts it into an intensive variable (denoted in lower case). The resulting intensive variable is then termed a *specific* variable, and it is the amount of the variable *per unit mass.*

4.2.2 Internal Energy and Its Conservation

4.2.2.1 *Definition of Internal Energy*

For a thermodynamic system, *internal energy* is the microscopic energy contained within it. For present purposes, this system is taken to be a fluid parcel; of cloudy air (including dry air, water vapour, liquid water, and frozen water) for the atmosphere, and of seawater (including liquid and frozen water, and solutes) for the oceans. The internal energy within a fluid parcel includes:

- Microscopic *kinetic energy*, due to translations, rotations, and vibrations of its molecules.
- Microscopic *potential energy*, due to chemical bonds within molecules and to forces between them.

It excludes:

- Kinetic energy of the motion of a fluid parcel (considered as a macroscopic entity, with microscopic (internal) variations averaged out).
- Potential energy of a fluid parcel (considered as a macroscopic entity) due to external forces (such as gravity) acting on it.

These excluded energies are represented in the momentum equation (2.34) (expressed in vector-invariant form). Concretely, they appear as $\nabla(K + \Phi)$, where K is specific kinetic energy, and Φ is geopotential (i.e. potential of apparent gravity).

4.2.2.2 *Conservation of Internal Energy*

As discussed in further detail later, conservation of internal energy dictates that internal energy of an *open* fluid parcel can only change due to:

1. Heat transfer between the fluid parcel and its surrounding environment.
2. Work done between the fluid parcel and its surrounding environment.
3. Change of composition of the fluid parcel.

For a *closed* fluid parcel, only the first two conditions apply, since there is then no change in composition of the parcel. This greatly simplifies things. *It is emphasised that attention in this book is almost exclusively focused on closed fluid parcels* to make things simpler and easier to understand.

4.2.3 The Fundamental Equation of State

Specific internal energy is a function of its natural variables, namely:

1. Specific volume (α).
2. Specific entropy (η).
3. Mass fractions ($S^1, S^2, \ldots$) of the fluid's components (i.e. substances).[7]

[7] See Table 9.6 of Section 9.4.4 for some two-component examples of atmospheric and oceanic thermodynamic systems.

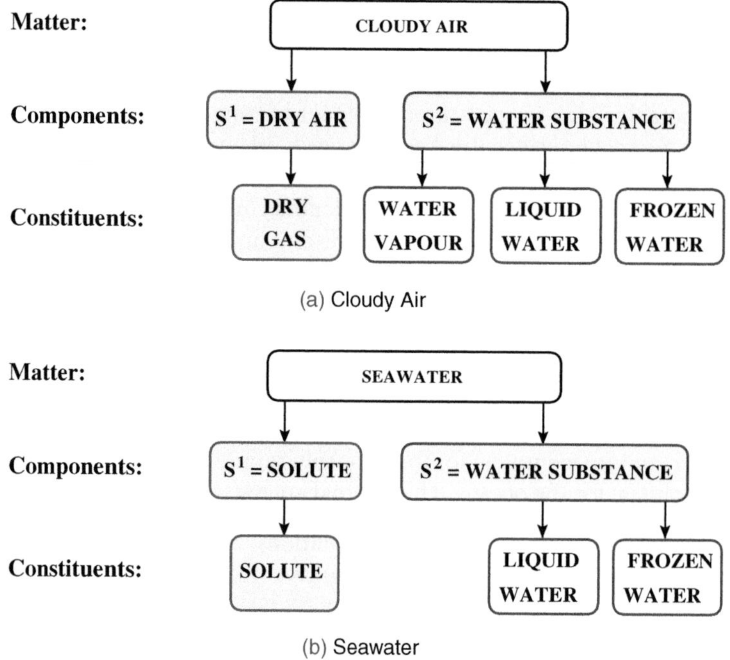

Figure 4.2 Matter, components, and constituents for: (a) cloudy air; and (b) seawater. See text for further details.

Thus:

The Fundamental (or Characteristic) Equation of State

The *fundamental (or characteristic) equation of state* is

$$\mathscr{E} = \mathscr{E}\left(\alpha, \eta, S^1, S^2, \ldots\right), \tag{4.1}$$

where:

- $\mathscr{E}$ is *specific internal energy* (per unit mass of fluid, including all of its components).
- η is *specific entropy*.
- $S^1, S^2, \ldots$ are the mass fractions of the fluid's components (i.e. substances). These mass fractions collectively define the fluid's *composition*.

The precise functional form of $\mathscr{E} = \mathscr{E}\left(\alpha, \eta, S^1, S^2, \ldots\right)$ *depends only on the physical properties of the fluid.*

As the name suggests, the fundamental equation of state is a very important equation since it embodies everything needed to *completely describe the thermodynamic equilibrium state of a fluid parcel* (or, more generally, the equilibrium state of a thermodynamic system).

Mass fractions $S^1, S^2, \ldots$ in (4.1) are now related to the constituents of Earth's atmosphere and oceans. In what follows, arrows ($\rightarrow$) denote mapping from a quantity or a set of quantities to another. So, in words, '$\rightarrow$' reads 'maps to'.

Atmosphere For the atmosphere, the two principal components are *dry air* (with a single phase) and *water substance* (with three phases – gaseous, liquid, and frozen). See Fig. 4.2a.

Attention herein is (mostly) restricted to these two substances, that is, the atmosphere is assumed to be a two-component thermodynamic system (termed *cloudy air*) comprised of dry air + water substance. Thus $\left(S^1, S^2, \ldots\right) \rightarrow \left(q^d, q^w\right)$, where q^d and $q^w \equiv q^v + q^l + q^f$ are the mass fractions of dry air and (total) water substance, respectively, with respect to unit mass of air (comprised of dry air + water substance); see Section 3.2. The dry-gas component is assumed to be of constant composition (i.e. all of the dry gases in any sample volume are in the same proportion to one another). The dry component (q^d) is then related to the mass fractions of water substance by $q^d = 1 - q^w \equiv 1 - \left(q^v + q^l + q^f\right)$; see (3.12) and (3.31).[8]

Oceans For the oceans, the two principal components are *solute* (i.e. dissolved salts, with a single phase) and *water substance* (with two phases – liquid and frozen). See Fig. 4.2b. Attention herein is restricted to these two substances, that is, the oceans are assumed to be a two-component thermodynamic system (termed *seawater*) comprised of solute + water substance. Thus $\left(S^1, S^2, \ldots\right) \rightarrow \left(S_A, q^w\right)$, where S_A (termed *Absolute Salinity*) and $q^w \equiv q^l + q^f$ are the mass fractions of solute and water substance, respectively, with respect to unit mass of seawater, comprised of solute + water substance. The solute component (often termed *saline component*) is the sum of all solutes. This component is assumed to be of constant composition (i.e. the mass fractions of all of the solutes in any sample are in the same proportion to one another).[9] The solute (saline) component (S_A) is then related to the mass fractions of water substance by $S_A = 1 - q^w \equiv 1 - \left(q^l + q^f\right)$.

In what follows, when taking partial derivatives of functions, the arguments of these functions will usually be given explicitly. For example, $\partial\mathscr{E}\left(\alpha, \eta, S^1, S^2\right)/\partial\eta$ is the first partial derivative of the function $\mathscr{E} = \mathscr{E}\left(\alpha, \eta, S^1, S^2\right)$ with respect to η, taken whilst holding fixed its other arguments, namely α, S^1, and S^2. Although this may seem somewhat pedestrian and, at times, unnecessary, it has the important virtue of avoiding ambiguity. *What varies, and what is held fixed, is very important indeed in thermodynamics.*

4.2.4 The Fundamental Thermodynamic Relation

Conservation of internal energy (see Section 4.2.2.2) means that if internal energy changes by an infinitesimal amount, $d\mathscr{E}$, due to infinitesimal changes to the state of a fluid parcel, then

$$d\mathscr{E} = đQ + đW + đC, \tag{4.2}$$

where:

1. $đQ$ (if positive) is the infinitesimal change in internal energy due to an infinitesimal amount of *heat* supplied to a fluid parcel by its surrounding environment, or (if negative) released from a fluid parcel to its surrounding environment.
2. $đW$ (if positive) is the infinitesimal change in internal energy due to an infinitesimal amount of *work* done by the surrounding environment on a fluid parcel, or (if negative) by a fluid parcel on its surrounding environment.

[8] To be clear, q^d is composed *solely* of dry gases. Water vapour (q^v), the gaseous form of water substance, is a constituent of total water substance (q^w), the other two constituents being q^l and q^f.

[9] For a realistic, quantitative model of the oceans, the single, saline component, S_A (Absolute Salinity of total solute), varying in space and in time, is sufficient for most global applications.

3. $đC$ is the infinitesimal change in internal energy due to an infinitesimal change in a fluid parcel's *chemical composition* (i.e. a change in the amount of one or both of the substances contained within a fluid parcel).

Regarding notation, thermodynamic theory distinguishes between exact differentials and imperfect (or inexact) differentials. For an *exact differential* (such as $d\mathcal{E}$), its value is *independent of the thermodynamic path taken* to go from an initial thermodynamic state to a final thermodynamic state. As such, it can be treated using the usual rules of differential calculus, which implicitly assume this. By way of contrast, the value of an *imperfect differential* (such as $đQ$, $đW$, or $đC$) *can depend on the thermodynamic path taken* to go from an initial state to a final state; the symbol 'đ' denotes this dependence on path taken. Therefore an imperfect differential – in general – does not obey the usual rules of differential calculus.

An imperfect differential should be thought of as an increment of something (such as an incremental change to internal energy due to heating ($đQ$), work done ($đW$), or compositional change ($đC$)), whose value depends upon the thermodynamic path taken. However, in special circumstances, imperfect differentials can sometimes be converted into exact ones by multiplication by an integrating factor; see (4.3) and (4.4) for two examples of this.

Further details for the imperfect differentials $đQ$, $đW$, and $đC$ are now given; see Callen (1985), Salmon (1998), Curry and Webster (1999), Ambaum (2010), and Vallis (2017).

Heat From thermodynamic theory

$$đQ = Td\eta, \tag{4.3}$$

for an infinitesimal *quasi-static* or *reversible*[10] thermodynamic process at constant composition, where T is temperature of the fluid parcel (assumed to be uniform within it, as implied by thermodynamic equilibrium of the parcel throughout a quasi-static or reversible thermodynamic process), and η is its specific entropy.[11]

Work For a quasi-static or reversible thermodynamic process, the work done (per unit mass) on a fluid parcel by its surrounding environment due to a change $d\alpha$ in its volume (i.e. by mechanical *expansion* or *compression*) is

$$đW = -pd\alpha, \tag{4.4}$$

see (2.60) in Section 2.3.4.[12] If $d\alpha$ is negative, then this corresponds to reducing the volume of the fluid parcel (i.e. to mechanical *compression*), and the surrounding environment does work *on* the fluid parcel. If $d\alpha$ is positive, then this instead corresponds to increasing the volume of the fluid parcel (i.e. to mechanical *expansion*), and the fluid parcel does work on its surrounding environment.

[10] Here, a *reversible* thermodynamic process is one such that a change in state of a fluid parcel, from initial to final, can be *exactly* reversed to recover the initial state, *without increasing total entropy* of the *combined system* comprised of the fluid parcel *and* its surrounding environment. Note though that a reversible process is a theoretical concept that can never quite be achieved physically. In physical reality, *any* change results in an increase in *total* entropy (of the fluid parcel and of its surrounding environment), no matter how small this may be. The initial state (including entropy) can therefore never be exactly recovered, at best only almost so.

[11] This is an example of an imperfect differential being converted to an exact one. The imperfect differential, $đQ$, is multiplied by the integrating factor, $1/T$, to give the exact differential, $d\eta = đQ/T$. The special circumstance is that the thermodynamic process is assumed here to be *quasi-static* or *reversible*, and the usual rules of differential calculus then apply to $d\eta$.

[12] This is a second example of an imperfect differential being converted to an exact one. The imperfect differential, $đW$, is multiplied by the integrating factor, $1/p$, to give the exact differential, $d\alpha = -đW/p$. The special circumstance is again that the thermodynamic process is assumed to be *quasi-static* or *reversible,* and the usual rules of differential calculus then apply to $d\alpha$.

Composition The change in internal energy, $đC$ (often termed *chemical energy* for want of a better term), per unit mass due to a change in composition of the fluid parcel is

$$đC = \sum_{i=1}^{2} \mu^i dS^i, \tag{4.5}$$

where μ^1 and μ^2 are the *chemical potentials* of the two components (i.e. substances), with mass fractions S^1 and S^2, respectively (with respect to unit mass of the fluid parcel), and with specific volume (α) and specific entropy (η) held fixed. For a fluid of *constant* composition, $dS^1 = dS^2 \equiv 0$, and both sides of (4.5) are identically zero. For a fluid of *variable* composition, either or both of dS^1 and dS^2 may be non-zero, depending upon whether a particular fluid parcel within the fluid is undergoing change in its composition, or not, and to what extent.

Substitution of (4.3)–(4.5) into (4.2) finally yields:

The Fundamental Thermodynamic Relation (or First Law of Thermodynamics)

The *fundamental thermodynamic relation* (or *the first law of thermodynamics*) is

$$d\mathscr{E} = T d\eta - p d\alpha + \sum_{i=1}^{2} \mu^i dS^i. \tag{4.6}$$

As developed, (4.6) holds for quasi-static or reversible thermodynamic changes of a two-component thermodynamic system. However, since $\mathscr{E}$, η, and α are thermodynamic functions of state, it also holds for arbitrary irreversible changes.

Using (4.3) and (4.5), fundamental thermodynamic relation (4.6) can be rewritten as

$$d\mathscr{E} + p d\alpha = đQ + đC = đQ + \sum_{i=1}^{2} \mu^i dS^i. \tag{4.7}$$

This energy-budget equation describes what happens when an infinitesimal fluid parcel is heated (or cooled) whilst simultaneously undergoing a possible change in its composition.

4.2.5 The Prognostic Internal Energy Equation

Forming material derivatives in fundamental thermodynamic relation (4.7) leads to

$$\frac{D\mathscr{E}}{Dt} + p\frac{D\alpha}{Dt} = \dot{Q} + \sum_{i=1}^{2} \mu^i \frac{DS^i}{Dt}, \tag{4.8}$$

where $\dot{Q}$ is heating rate (per unit mass).

- But what is DS^i/Dt in this equation?

For the atmosphere (see Section 4.2.3), $\left(S^1, S^2\right) \rightarrow \left(q^d, q^w\right)$, where q^d and q^w are the mass fractions of dry air and water substance, respectively. For the oceans, $\left(S^1, S^2\right) \rightarrow \left(S_A, q^w\right)$, where S_A (Absolute Salinity) and q^w are the mass fractions of solute and water substance, respectively. In either case, there is a prognostic equation for S^i of the form

$$\frac{DS^i}{Dt} = \dot{S}^i, \quad i = 1, 2, \tag{4.9}$$

where $\dot{S}^i$ is the rate of change of substance S^i contained within the bounding surface of the fluid parcel.

To be more precise, for the *atmosphere*, (4.9) has the forms (see (3.38))

$$\frac{Dq^d}{Dt} = \dot{q}^d, \quad \frac{Dq^w}{Dt} = \frac{Dq^v}{Dt} + \frac{Dq^l}{Dt} + \frac{Dq^f}{Dt} = \dot{q}^v + \dot{q}^l + \dot{q}^f = \dot{q}^w, \tag{4.10}$$

where $\dot{q}^v$, $\dot{q}^l$, and $\dot{q}^f$ are the rates of change of constituents q^v, q^l, and q^f, respectively, for the air parcel. Similarly, for the *oceans*, (4.9) has the forms

$$\frac{DS_A}{Dt} = \dot{S}_A, \quad \frac{Dq^w}{Dt} = \frac{Dq^l}{Dt} + \frac{Dq^f}{Dt} = \dot{q}^l + \dot{q}^f = \dot{q}^w, \tag{4.11}$$

where $\dot{S}_A$, $\dot{q}^l$, $\dot{q}^f$, and $\dot{q}^w$ are the rates of change of S_A (Absolute Salinity), q^l, q^f, and q^w, respectively, of the seawater parcel (water vapour, q^v, being absent for the ocean).

For the *closed* fluid parcels (atmospheric or oceanic) of primary interest herein,

$$\frac{DS^1}{Dt} = \dot{S}^1 = 0, \quad \frac{Dq^w}{Dt} = \dot{q}^w = 0, \tag{4.12}$$

where $S^1 = q^d$ for the atmosphere, and $S^1 = S_A$ for the ocean.

Substitution of (4.9) into (4.8) yields:

The Prognostic Internal-Energy Equation for a Fluid

The *prognostic internal-energy equation for a fluid* is

$$\frac{D\mathscr{E}}{Dt} + p\frac{D\alpha}{Dt} = \dot{Q} + \sum_{i=1}^{2} \mu^i \dot{S}^i \equiv \dot{Q}_E, \tag{4.13}$$

where:

- $\mathscr{E} = \mathscr{E}\left(\alpha, \eta, S^1, S^2\right)$ = specific internal energy of the fluid parcel.
- $\alpha \equiv 1/\rho$ = total specific volume of the fluid parcel.
- η = specific entropy of the fluid parcel.
- $\dot{Q}$ = heating rate (per unit mass) of the fluid parcel.
- μ^1 and μ^2 = chemical potentials of substances S^1 and S^2, respectively.
- $\dot{S}^1$ and $\dot{S}^2$ = rate of change of substances S^1 and S^2, respectively.
- $\dot{Q}_E$ = rate of total energy input to the fluid parcel.

As derived here:

- Prognostic internal-energy equation (4.13) holds for both the atmosphere and the oceans.

All that differs between them is the precise definition of specific internal energy $\mathscr{E} = \mathscr{E}\left(\alpha, \eta, S^1, S^2\right)$, according to whether the atmosphere (dry air + water substance) is being represented, or the oceans (solute + water substance). In either case, *everything needed to represent their thermodynamic states is embodied in the appropriate definition of* $\mathscr{E}$. To help clarify this aspect, and to make things more tangible, explicit functional forms for $\mathscr{E}$ for a cloudy-air parcel are given in Section 4.3. It is also shown there how to derive important thermodynamic quantities from the fundamental equation of state.

4.2.6 An Alternative Fundamental Equation of State

Specific internal energy $\mathscr{E} = \mathscr{E}\left(\alpha, \eta, S^1, S^2\right)$ – see Section 4.2.3 – completely defines the thermodynamic equilibrium state of a fluid. Specific entropy,

$$\eta = \eta\left(\alpha, \mathscr{E}, S^1, S^2\right), \tag{4.14}$$

also does so. This equation is just a rearrangement of (4.1) whereby $\mathscr{E} = \mathscr{E}\left(\alpha, \eta, S^1, S^2\right)$ is solved to obtain η as a function of the remaining variables, including $\mathscr{E}$.[13] As such, it is an alternative, equivalent form of the fundamental equation of state, (4.1), with exactly the same information content.

If internal energy ($\mathscr{E}$) is the dependent variable and entropy (η) is an independent variable (as in (4.1)), then this is termed as being in the *energy representation* (Callen, 1985, p. 41). Similarly, if entropy (η) is the dependent variable and internal energy ($\mathscr{E}$) is an independent variable (as in (4.14)), then it is in the *entropy representation*. Although either representation can be used, according to convenience, vacillating between them for a specific problem can lead to confusion (Callen, 1985, p. 41) and is therefore best avoided.

4.2.7 The Prognostic Entropy Equation

A prognostic entropy equation is now developed as an alternative thermodynamic-energy equation[14] to prognostic internal-energy equation (4.13). This is obtained by dividing (4.3) by T and taking the material derivative. Thus:

The Prognostic Entropy Equation for a Fluid

The *prognostic entropy equation for a fluid* is

$$\frac{D\eta}{Dt} = \frac{\dot{Q}}{T}, \tag{4.15}$$

where $\dot{Q}$ is heating rate (per unit mass) of the fluid parcel.

Equation (4.15) for η is of simpler form, with a single term on each side, than (4.13) for $\mathscr{E}$. In the absence of heating (i.e. when $\dot{Q} \equiv 0$), $D\eta/Dt \equiv 0$, and entropy (η) of the fluid parcel is then materially conserved following the fluid.

[13] For some fluids (e.g. seawater), solving (4.1) for η may very well be analytically intractable. It may therefore be necessary to instead do so numerically. For an ideal gas, it is, however, straightforward to obtain η from $\mathscr{E}$, analytically; see Section 4.3.2.

[14] It is an *alternative* equation since the two equations are not independent of one another, but are obtained from the same fundamental equation of state, albeit expressed slightly differently.

4.2.8 Equations of State

4.2.8.1 Internal Energy Forms

Since $d\mathscr{E}$ is an *exact* differential (as opposed to an imperfect one), and using the fundamental equation of state, (4.1), and the chain rule for differentiation, $d\mathscr{E}$ can be alternatively written as

$$d\mathscr{E} = \frac{\partial \mathscr{E}\left(\alpha, \eta, S^1, S^2\right)}{\partial \eta} d\eta + \frac{\partial \mathscr{E}\left(\alpha, \eta, S^1, S^2\right)}{\partial \alpha} d\alpha + \sum_{i=1}^{2} \frac{\partial \mathscr{E}\left(\alpha, \eta, S^1, S^2\right)}{\partial S^i} dS^i. \tag{4.16}$$

Comparing (4.16) with (4.6) term by term, the following relations must hold for consistency:

Equations of State from Internal Energy

$$T = \frac{\partial \mathscr{E}\left(\alpha, \eta, S^1, S^2\right)}{\partial \eta} = T\left(\alpha, \eta, S^1, S^2\right), \tag{4.17}$$

$$p = -\frac{\partial \mathscr{E}\left(\alpha, \eta, S^1, S^2\right)}{\partial \alpha} = p\left(\alpha, \eta, S^1, S^2\right), \tag{4.18}$$

$$\mu^i = \frac{\partial \mathscr{E}\left(\alpha, \eta, S^1, S^2\right)}{\partial S^i} = \mu^i\left(\alpha, \eta, S^1, S^2\right). \tag{4.19}$$

Thus if $\mathscr{E} = \mathscr{E}\left(\alpha, \eta, S^1, S^2\right)$ is of given functional form (and therefore completely defines the thermodynamic equilibrium state of the fluid), then temperature (T), pressure (p), and chemical potentials (μ^i) are defined by (4.17)–(4.19), respectively; they are all functions of $\left(\alpha, \eta, S^1, S^2\right)$. Equations (4.17)–(4.19), derived from the fundamental thermodynamic relation, (4.6), are termed *equations of state.*

Elimination of specific entropy, η, from (4.17) and (4.18) gives an equation of the form

$$p = p\left(\alpha, T, S^1, S^2\right). \tag{4.20}$$

This derived equation is also known as an equation of state. For an ideal gas, it takes the familiar forms $p\alpha = RT$ or (since $\alpha \equiv 1/\rho$) $p = \rho RT$; see (2.51) and (2.50), respectively. It is then conventionally called *the* equation of state.

In a similar manner to elimination of specific entropy (η), elimination of specific volume (α) from (4.17) and (4.18) gives an equation of the form

$$p = p\left(\eta, T, S^1, S^2\right). \tag{4.21}$$

So we now have *three* functional forms for p, given by (4.18), (4.20), and (4.21). These are expressed in terms of three pairs of thermodynamic variables, namely (α, η), (η, T), and (α, T) (these are cyclic permutations of α, η, and T), plus the set of composition variables S^1 and S^2. From these three functional forms, we can obtain many alternative forms. For example, rearrangement of (4.20) yields

$$\alpha = \alpha\left(p, T, S^1, S^2\right), \tag{4.22}$$

$$T = T\left(\alpha, p, S^1, S^2\right). \tag{4.23}$$

(To help fix ideas, for an ideal gas, (4.22) and (4.23) correspond to simply rewriting $p\alpha = RT$ as $\alpha = RT/p$, and $T = p\alpha/R$, respectively.)

Similarly, (4.17) for temperature (T), and (4.18) and (4.21) for pressure (p), can be rearranged to give the alternative forms

$$\eta = \eta\left(\alpha, T, S^1, S^2\right), \quad \alpha = \alpha\left(\eta, T, S^1, S^2\right), \tag{4.24}$$

$$\eta = \eta\left(\alpha, p, S^1, S^2\right), \quad \alpha = \alpha\left(\eta, p, S^1, S^2\right), \tag{4.25}$$

$$\eta = \eta\left(p, T, S^1, S^2\right), \quad T = T\left(\eta, p, S^1, S^2\right). \tag{4.26}$$

From these examples, we see that there are many possible ways of expressing equations of state.

4.2.8.2 *Entropy Forms*

To add to the confusion, equations of state, similar to those just developed using internal energy ($\mathscr{E}$), can also be developed using entropy (η)!

Rearrangement of fundamental thermodynamic relation (4.6) gives

$$d\eta = \frac{1}{T}d\mathscr{E} + \frac{p}{T}d\alpha - \frac{1}{T}\sum_{i=1}^{2}\mu^i dS^i. \tag{4.27}$$

Since $d\eta$ is an *exact* differential, and using the alternative form, (4.14), of the fundamental equation of state, $d\eta$ can also be written (using the chain rule for differentiation) as

$$d\eta = \frac{\partial \eta\left(\alpha, \mathscr{E}, S^1, S^2\right)}{\partial \mathscr{E}}d\mathscr{E} + \frac{\partial \eta\left(\alpha, \mathscr{E}, S^1, S^2\right)}{\partial \alpha}d\alpha + \sum_{i=1}^{2}\frac{\partial \eta\left(\alpha, \mathscr{E}, S^1, S^2\right)}{\partial S^i}dS^i. \tag{4.28}$$

Comparing (4.27) with (4.28), term by term, the following relations must hold for consistency:

Equations of State from Entropy

$$T = \left[\frac{\partial \eta\left(\alpha, \mathscr{E}, S^1, S^2\right)}{\partial \mathscr{E}}\right]^{-1} = T\left(\alpha, \mathscr{E}, S^1, S^2\right), \tag{4.29}$$

$$p = T\frac{\partial \eta\left(\alpha, \mathscr{E}, S^1, S^2\right)}{\partial \alpha} = p\left(\alpha, \mathscr{E}, S^1, S^2\right), \tag{4.30}$$

$$\mu^i = -T\frac{\partial \eta\left(\alpha, \mathscr{E}, S^1, S^2\right)}{\partial S^i} = \mu^i\left(\alpha, \mathscr{E}, S^1, S^2\right). \tag{4.31}$$

Thus if $\eta = \eta\left(\alpha, \mathscr{E}, S^1, S^2\right)$ is of given functional form (and therefore completely defines the thermodynamic equilibrium state of the fluid), then temperature (T), pressure (p), and chemical potentials (μ^i) are alternatively defined by (4.29)–(4.31), respectively. Equations (4.29)–(4.31), derived using entropy (η), are also termed *equations of state*. They are the analogues of the equations of state (4.17)–(4.19), derived using internal energy ($\mathscr{E}$).

Manipulation of (4.29)–(4.31), in a similar manner to that for (4.17)–(4.19), yields yet more analogous equations of state with different functional forms. These are not, however, given here, but the interested reader may derive them.

4.2.8.3 *Concluding Remarks*

So what can we conclude from the preceding analysis of the various forms of the equations of state? Firstly, there are many such forms, and which particular form or forms to use depends upon the application. Secondly, and to (almost) quote Vallis (2017) (the parenthetical comment in what follows has been added):

> Given the fundamental equation of state, the thermodynamic state of a body is fully specified by a knowledge of any two of p, ρ (or, equivalently, α, since $\alpha \equiv 1/\rho$), T, η, and $\mathscr{E}$, plus its composition.

4.2.9 Specific Heat Capacity

Assume no change in composition (i.e. $dS^i \equiv 0$ in (4.5) and (4.6)) during the reversible transfer of an infinitesimal amount of heat, $đQ$, between a fluid parcel and its surrounding environment. From (4.3), the change in entropy resulting from heating this fluid parcel (for $đQ > 0$), or cooling it (for $đQ < 0$), is

$$d\eta = \frac{đQ}{T}. \tag{4.32}$$

The induced change in temperature, dT, satisfies

$$đQ = c dT, \tag{4.33}$$

where c is *specific heat capacity* (often shortened to *specific heat*).[15] Elimination of $đQ$ from (4.32) and (4.33) then gives

$$c dT = T d\eta. \tag{4.34}$$

The precise value of c depends upon *how* this heat is transferred. If it is transferred at *constant volume*, then $d\alpha \equiv 0$ and $c \equiv c_V$. If, instead, it is transferred at *constant pressure*, then $dp \equiv 0$ and $c \equiv c_p$. The value of c_p is generally larger than that of c_V. This is because c_p includes additional energy related to work done by the fluid parcel against the constant, maintained pressure of the surrounding environment.

4.2.9.1 *Specific Heat Capacity at Constant Volume*

Taking η to be a function of specific volume (α) and temperature (T) gives

$$d\eta = \frac{\partial \eta(\alpha, T)}{\partial \alpha} d\alpha + \frac{\partial \eta(\alpha, T)}{\partial T} dT. \tag{4.35}$$

Transferring heat at *constant volume* (so that $d\alpha \equiv 0$) eliminates the first term on the right-hand side of this equation, so that $d\eta = [\partial \eta(\alpha, T)/\partial T] dT$. Insertion of this simplified equation into (4.34) then yields:

Specific Heat Capacity at Constant Volume (1)

Specific heat capacity at constant volume is given by

$$c_V \equiv T \frac{\partial \eta(\alpha, T)}{\partial T}. \tag{4.36}$$

Definition (4.36) of c_V is expressed in terms of the partial derivative of specific entropy (η) with respect to temperature (T). It can be equivalently expressed in terms of the partial

15 Physically, c is the amount of heat ($đQ$) required to raise the temperature of unit mass of fluid by an amount dT; c has units J kg^{-1} K, as does gas constant R.

derivative (again with respect to T) of specific internal energy, $\mathscr{E}$. To see this, we proceed as follows. From the fundamental thermodynamic relation, (4.6), and recalling that dS^i is assumed zero here,

$$d\mathscr{E} = Td\eta - pd\alpha. \tag{4.37}$$

Using (4.35) to eliminate $d\eta$ from (4.37) gives

$$d\mathscr{E} = \left[T\frac{\partial \eta (\alpha, T)}{\partial \alpha} - p \right] d\alpha + T\frac{\partial \eta (\alpha, T)}{\partial T} dT. \tag{4.38}$$

Taking $\mathscr{E}$ to be a function of specific volume (α) and temperature (T), $d\mathscr{E}$ may be alternatively written as (see (4.16) with $dS^i \equiv 0$)

$$d\mathscr{E} = \frac{\partial \mathscr{E} (\alpha, T)}{\partial \alpha} d\alpha + \frac{\partial \mathscr{E} (\alpha, T)}{\partial T} dT. \tag{4.39}$$

Comparing the coefficient of dT in these two equations for $d\mathscr{E}$ (i.e. (4.38) and (4.39)), it is seen that

$$T\frac{\partial \eta (\alpha, T)}{\partial T} = \frac{\partial \mathscr{E} (\alpha, T)}{\partial T}. \tag{4.40}$$

Insertion of this equation into (4.36) gives an alternative definition of c_v, expressed in terms of internal energy ($\mathscr{E}$) instead of entropy (η). Thus:

Specific Heat Capacity at Constant Volume (2)

Specific heat capacity at constant volume may be alternatively written as

$$c_v \equiv \frac{\partial \mathscr{E} (\alpha, T)}{\partial T}. \tag{4.41}$$

4.2.9.2 *Specific Heat Capacity at Constant Pressure*

Taking η now to instead be a function of pressure (p) and temperature (T) gives

$$d\eta = \frac{\partial \eta (p, T)}{\partial p} dp + \frac{\partial \eta (p, T)}{\partial T} dT. \tag{4.42}$$

Transferring heat at *constant pressure* (so that $dp \equiv 0$) eliminates the first term on the right-hand side of this equation, leaving $d\eta = \left[\partial \eta (p, T) / \partial T \right] dT$. Insertion of this simplified equation into (4.34) then yields:

Specific Heat Capacity at Constant Pressure

Specific heat capacity at constant pressure is given by

$$c_p \equiv T\frac{\partial \eta (p, T)}{\partial T}. \tag{4.43}$$

4.2.9.3 *Difference in Specific Heats, $c_p - c_v$*

Taking specific volume (α) to be a function of pressure (p) and temperature (T), the infinitesimal change in specific volume may be written as

$$d\alpha = \frac{\partial \alpha (p, T)}{\partial p} dp + \frac{\partial \alpha (p, T)}{\partial T} dT. \tag{4.44}$$

Substituting this into (4.35) gives

$$d\eta = \frac{\partial\eta\,(\alpha,T)}{\partial\alpha}\frac{\partial\alpha\,(p,T)}{\partial p}dp + \left[\frac{\partial\eta\,(\alpha,T)}{\partial T} + \frac{\partial\eta\,(\alpha,T)}{\partial\alpha}\frac{\partial\alpha\,(p,T)}{\partial T}\right]dT. \tag{4.45}$$

Equating coefficients of dp and dT in (4.42) and (4.45) then yields

$$\frac{\partial\eta\,(p,T)}{\partial p} = \frac{\partial\eta\,(\alpha,T)}{\partial\alpha}\frac{\partial\alpha\,(p,T)}{\partial p}, \tag{4.46}$$

$$\frac{\partial\eta\,(p,T)}{\partial T} = \frac{\partial\eta\,(\alpha,T)}{\partial T} + \frac{\partial\eta\,(\alpha,T)}{\partial\alpha}\frac{\partial\alpha\,(p,T)}{\partial T}. \tag{4.47}$$

From definitions (4.36) and (4.43) of c_v and c_p, respectively, the difference in specific heats at constant pressure and volume is given by

$$c_p - c_v = T\left[\frac{\partial\eta\,(p,T)}{\partial T} - \frac{\partial\eta\,(\alpha,T)}{\partial T}\right]. \tag{4.48}$$

Insertion of (4.47) into (4.48) finally yields:

The Difference in Specific Heat Capacities at Constant Pressure and Constant Volume

The *difference in specific heat capacities at constant pressure and constant volume* is given by

$$c_p - c_v = T\frac{\partial\eta\,(\alpha,T)}{\partial\alpha}\frac{\partial\alpha\,(p,T)}{\partial T}. \tag{4.49}$$

For an ideal gas, it is shown in Section 4.3.3 that (4.49) reduces to Carnot's relation[16] – see (4.93) with $\widetilde{R} \equiv R$ for a dry ideal gas –

$$c_p - c_v = R. \tag{4.50}$$

4.2.9.4 *Specific Heat Capacities for an Ideal Gas*

Values of specific heat capacities, c_v and c_p, depend upon the properties of the molecules that comprise the gas. For an ideal gas, statistical mechanics theory leads to *explicit formulae* for these in terms of the number of independent degrees of freedom associated with a single molecule of gas. See the Appendix to this Chapter for further details.

4.2.10 The Local Speed of Sound

Assume that sound waves propagate:

1. Adiabatically (so that $đQ = 0$ in (4.3), which implies $d\eta = 0$, which implies absence of the $Td\eta$ term in fundamental thermodynamic relation (4.6)).
2. Without change of composition (so that (4.5) reduces to $\sum_{i=1,2}\mu^i dS^i = đC = 0$, which implies that fundamental thermodynamic relation (4.6) further reduces to $d\mathcal{E} = -pd\alpha$).

[16] After Nicolas Léonard Sadi Carnot (1796–1832).

Thus it is assumed that sound waves physically propagate due to adiabatic expansion and compression of the fluid; see Section 2.3.4. This is generally a very good approximation for both Earth's atmosphere and its oceans. An expression for the *local sound speed* (relative to the local flow velocity, **u**) may be found by linearising the governing fluid-dynamical equations about a representative state of the fluid; see, for example, Salmon (1998), p. 14, and Vallis (2017), p. 40. Thus:

The Local Speed of Sound

- The *local speed of sound* (relative to the local flow velocity, **u**), c_s, may be obtained from

$$c_s^2 \equiv \frac{\partial p\left(\rho, \eta\right)}{\partial \rho}. \tag{4.51}$$

- Using the definitions $\alpha \equiv 1/\rho$ (see (2.45)) and $p \equiv -\partial \mathscr{E}\left(\alpha, \eta\right)/\partial \alpha$ (see (4.18)), (4.51) can be equivalently rewritten as

$$c_s^2 \equiv -\alpha^2 \frac{\partial p\left(\alpha, \eta\right)}{\partial \alpha} \equiv \alpha^2 \frac{\partial^2 \mathscr{E}\left(\alpha, \eta\right)}{\partial \alpha^2}. \tag{4.52}$$

- This alternative expression for c_s^2 agrees with that given at the bottom of the third column of Table 9.2 in Chapter 9.

The local sound speed depends upon the properties of the medium of propagation. To an excellent approximation, the atmosphere generally behaves as an ideal gas, and so $c_s^2 \approx \left(c_p/c_v\right) RT$ (with equality holding if it were indeed an ideal, diatomic gas); see (4.94). Thus, physically, c_s in the atmosphere depends almost linearly on temperature (T), but hardly at all on density (ρ) or pressure (p.) A typical mid-tropospheric value (for $c_p/c_v = 7/5$, $R = 287$ J kg^{-1} K^{-1} and $T = 273.15$ K) is therefore of order $c_s \approx 330$ m s^{-1}.

For the oceans, no such analytic approximation is available. Measurements show, however, that c_s depends on temperature (T), density (ρ), and pressure (p). Because the oceans are liquids (and much less compressible than gases), c_s is much larger for the oceans than for the atmosphere. A representative value for the oceans is instead $c_s \approx 1500$ m s^{-1} (see e.g. Table A3-1 of Gill, 1982, and p. 40 of Vallis, 2017).

The fastest speed of propagation of disturbances in Earth's atmosphere and oceans is the sound speed, c_s. This has important implications for a numerical model. For a model with an *explicit* time scheme, the maximum permissible timestep (for computational stability) is inversely proportional to the fastest-propagating signal speed (i.e. the faster the propagation speed, the smaller the timestep). All other things being equal (e.g. spatial resolution), the maximum permissible timestep in an oceanic model is therefore four to five times smaller than in an atmospheric model; cf. 330 m s^{-1} vs. 1500 m s^{-1} for c_s.

However, in both atmospheric and oceanic models, acoustic (i.e. sound) oscillations generally propagate negligible amounts of energy, and an accurate representation of them therefore has negligible impact on the overall accuracy of a numerical model. Consequently, it is numerically a lot more efficient to instead treat the terms responsible for the propagation of acoustic oscillations in a *time-implicit* manner. This then allows much longer timesteps by greatly slowing down their speed of numerical propagation, with no discernible impact on accuracy. Consequently, computational efficiency is greatly enhanced; it takes far fewer timesteps to achieve the same accuracy.

This is an example of how a thorough understanding of the fundamental properties of the fluid-dynamical representation of Earth's atmosphere and oceans can be exploited in the design of accurate and efficient numerical models. Such an understanding is highly beneficial, not only in theory but also in practice!

4.3 FUNCTIONAL FORMS FOR A CLOUDY-AIR PARCEL

Up to this point, the fundamental equation of state, in its two forms (4.1) (for internal energy, $\mathcal{E}$) and (4.14) (for entropy, η), has been abstractly presented as some *known*, *prescribed function* of its state parameters. From this assumed, prescribed function, everything thermodynamic can then be deduced through the use of various definitions.

Now Earth's atmosphere behaves very much like a mixture of ideal gases, water droplets, and ice crystals. It is therefore instructive to construct *explicit* functional forms for both $\mathcal{E}$ and η for a cloudy-air parcel. This then helps fix ideas by turning the abstract into the practical – at least for Earth's atmosphere! A further benefit is that it helps prepare the way for the representation of the equations of state for the oceans; see Chapter 11.

4.3.1 Construction of Functional Forms for Internal Energy ($\mathcal{E}$)

To construct explicit functional forms for a cloudy-air parcel, we make the following assumptions:

1. Dry air and water vapour behave as ideal gases. A *cloudy-air parcel* is composed of dry air, water vapour, liquid water, and frozen water, so that (from (3.69) and (3.80))

$$\mathbb{M} = \mathbb{M}^d + \mathbb{M}^v + \mathbb{M}^l + \mathbb{M}^f, \tag{4.53}$$

$$q^d + q^v + q^l + q^f = q^d + q^w = 1, \tag{4.54}$$

where

$$q^w \equiv q^v + q^l + q^f \tag{4.55}$$

is mass fraction of total water substance in a cloudy-air parcel. Since mass fractions q^d, q^v, q^l, and q^f are (linearly) related by (4.54), any function of these four variables equivalently reduces to a function of any three of them.

Notation For brevity in what follows, functional dependence on the four mass fractions, q^d (for **d**ry air) and q^v, q^l, and q^f (for the three phases – **v**apour, **l**iquid and **f**rozen, respectively – of water substance) is abbreviated to functional dependence on $\mathbf{q} \equiv \left(q^d, q^v, q^l, q^f\right)$. For example, $F = F\left(q^d, q^v, q^l, q^f\right)$ is abbreviated to $F = F\left(\mathbf{q}\right)$ for some quantity, F. In words, F is a function of both dry air and water substance, but not of specific volume (α), nor of specific entropy (η).

2. The equation of state for a cloudy-air parcel is given by (3.51), that is, by

$$p\alpha = RT, \tag{4.56}$$

where – from (3.54), (4.54), and (4.55) –

$$R\left(\mathbf{q}\right) = \frac{\widetilde{R}\left(q^d, q^v\right)}{1 - \left(\alpha^l/\alpha\right) q^l - \left(\alpha^f/\alpha\right) q^f}, \tag{4.57}$$

$$\widetilde{R}\left(q^d, q^v\right) \equiv R^d q^d + R^v q^v = R^d\left(1 - q^w\right) + R^v q^v = R^d\left(1 - q^v - q^l - q^f\right) + R^v q^v, \tag{4.58}$$

$\alpha \equiv 1/\rho$; and α^l, α^f, R^d, and R^v are all constants.
(For a dry atmosphere, (4.57) and (4.58) reduce to $R = \widetilde{R} = R^d =$ constant.)

3. For a cloudy-air parcel, definitions (4.17) and (4.18) of temperature (T) and pressure (p) are then given by

$$T = \frac{\partial \mathscr{E}(\alpha, \eta, \mathbf{q})}{\partial \eta}, \quad p = -\frac{\partial \mathscr{E}(\alpha, \eta, \mathbf{q})}{\partial \alpha}. \tag{4.59}$$

(In (4.59), $\mathscr{E}(\alpha, \eta, \mathbf{q})$ is shorthand for $\mathscr{E}\left(\alpha, \eta, q^d, q^v, q^l, q^f\right)$.)

4. The specific heat at constant volume (c_v) is given by (4.41) (but generalised here to a cloudy atmosphere). Thus

$$c_v(\mathbf{q}) = \frac{\partial \mathscr{E}(\alpha, T, \mathbf{q})}{\partial T}, \tag{4.60}$$

where it is assumed that c_v only depends on the mass fractions of dry air and water substance of the fluid parcel and not, for example, on T.[17] In (4.60), $\mathscr{E}$ is equivalently expressed as a function of α and T instead of as a function of α and η (cf. (4.59)), with dependence on $\mathbf{q}$ (i.e. on q^d, q^v, q^l, and q^f) appended to account for contributions of dry air and water substance to internal energy. (In (4.60), $c_v(\mathbf{q})$ is shorthand for $c_v = c_v\left(q^d, q^v, q^l, q^f\right)$. For a dry atmosphere, $c_v = c_v(\mathbf{q})$ reduces to $c_v = c_v^d =$ constant.)

5. Water vapour and liquid water contain *latent internal energy*

$$\mathbb{L}_0\left(q^v, q^l\right) = \left(L_0^v + L_0^f\right) q^v + L_0^f q^l, \tag{4.61}$$

where constants L_0^v and L_0^f are latent heats of vaporisation and fusion, respectively, extrapolated to $T = 0$ and $p = 0$. (These latent internal energies are an energy reservoir, available to change a parcel's water state between vapour, liquid, and frozen phases. See Chapter 10.)

From these five assumptions, explicit functional forms for $\mathscr{E} = \mathscr{E}(\alpha, \eta, \mathbf{q})$, $\mathscr{E} = \mathscr{E}(\alpha, T, \mathbf{q})$, $\mathscr{E} = \mathscr{E}(p, T, \mathbf{q})$, and $\mathscr{E} = \mathscr{E}(\alpha, p, \mathbf{q})$ are now derived.

Substituting (4.59) into (4.56), and using (4.57), yields

$$\left(\alpha - \alpha^l q^l - \alpha^f q^f\right) \frac{\partial \mathscr{E}(\alpha, \eta, \mathbf{q})}{\partial \alpha} = -\widetilde{R}\left(q^d, q^v\right) \frac{\partial \mathscr{E}(\alpha, \eta, \mathbf{q})}{\partial \eta}. \tag{4.62}$$

This first-order, partial-differential equation is solved for $\mathscr{E} = (\alpha, \eta, \mathbf{q})$ by:

- Expanding $\mathscr{E}(\alpha, \eta, \mathbf{q})$ as

$$\mathscr{E}(\alpha, \eta, \mathbf{q}) = \mathscr{E}^\alpha(\alpha, \mathbf{q})\, \mathscr{E}^\eta(\eta, \mathbf{q}) + \mathbb{L}_0\left(q^v, q^l\right), \tag{4.63}$$

  where $\mathbb{L}_0\left(q^v, q^l\right)$ is defined by (4.61).
- Substituting (4.63) into (4.62).
- Separating variables.
- Integrating two, first-order, ordinary-differential equations, one for $\mathscr{E}^\alpha = \mathscr{E}^\alpha(\alpha, \mathbf{q})$, the other for $\mathscr{E}^\eta = \mathscr{E}^\eta(\eta, \mathbf{q})$.

Applying this procedure then leads to the solution,

$$\mathscr{E}(\alpha, \eta, \mathbf{q}) = \left(\alpha - \alpha^l q^l - \alpha^f q^f\right)^{-B(\mathbf{q})} \exp\left\{\frac{[\eta - A(\mathbf{q})]\, B(\mathbf{q})}{\widetilde{R}\left(q^d, q^v\right)}\right\} + \mathbb{L}_0\left(q^v, q^l\right), \tag{4.64}$$

of (4.62), where:

[17] Thus $\mathscr{E}$ depends linearly on T for functional form (4.60) to hold; this is a consequence of Assumption 1 (that dry air and water vapour behave as ideal gases). In reality, $\mathscr{E}$ does have a weak, non-linear dependence on T.

- $B = B(\mathbf{q})$ is the separation 'constant' (i.e. it depends on q^d, q^v, q^l, and q^f, but not on α and η).
- $A = A(\mathbf{q})$ is a 'constant' of integration (i.e. it also depends on q^d, q^v, q^l, and q^f, but not on α and η).
- $\mathbb{L}_0 = \mathbb{L}_0\left(q^v, q^l\right)$ is latent internal potential energy (this appears here due to Assumption 5).

It is easily verified that (4.64) is indeed a solution of (4.62).

The function $A = A(\mathbf{q})$ is arbitrary. It simply serves to specify a reference state for measuring specific entropy, η – see (4.74).[18]

To determine $B = B(\mathbf{q})$, the fourth assumption (definition (4.60) of c_v) is exploited. To prepare for this, solution (4.64) of partial-differential equation (4.62) is inserted into definitions (4.59) of temperature (T) and pressure (p). Thus

$$T = \frac{\partial \mathscr{E}(\alpha, \eta, \mathbf{q})}{\partial \eta} = \frac{B(\mathbf{q})}{\widetilde{R}(q^d, q^v)}\left[\mathscr{E}(\alpha, \eta, \mathbf{q}) - \mathbb{L}_0\left(q^v, q^l\right)\right], \tag{4.65}$$

$$p = -\frac{\partial \mathscr{E}(\alpha, \eta, \mathbf{q})}{\partial \alpha} = \frac{B(\mathbf{q})}{\left(\alpha - \alpha^l q^l - \alpha^f q^f\right)}\left[\mathscr{E}(\alpha, \eta, \mathbf{q}) - \mathbb{L}_0\left(q^v, q^l\right)\right]. \tag{4.66}$$

Using these two equations, (4.64) can be equivalently rewritten as

$$\mathscr{E}(\alpha, T, \mathbf{q}) = \mathscr{E}(p, T, \mathbf{q}) = \frac{\widetilde{R}\left(q^d, q^v\right)}{B(\mathbf{q})} T + \mathbb{L}_0\left(q^v, q^l\right), \tag{4.67}$$

$$\mathscr{E}(\alpha, p, \mathbf{q}) = \frac{\left(\alpha - \alpha^l q^l - \alpha^f q^f\right)}{B(\mathbf{q})} p + \mathbb{L}_0\left(q^v, q^l\right). \tag{4.68}$$

With this preparation, (4.67) is now inserted into (4.60) to obtain

$$c_v(\mathbf{q}) = \frac{\partial \mathscr{E}(\alpha, T, \mathbf{q})}{\partial T} = \frac{\widetilde{R}\left(q^d, q^v\right)}{B(\mathbf{q})} \quad \Rightarrow \quad B(\mathbf{q}) = \frac{\widetilde{R}\left(q^d, q^v\right)}{c_v(\mathbf{q})}. \tag{4.69}$$

Thus the separation 'constant', B, must have the form $B(\mathbf{q}) = \widetilde{R}\left(q^d, q^v\right)/c_v(\mathbf{q})$. Insertion of this function back into (4.64), (4.67), and (4.68) yields the following four alternative forms for $\mathscr{E}$:

Explicit Functional Forms for the Internal Energy of a Cloudy Air Parcel

$$\mathscr{E}(\alpha, \eta, \mathbf{q}) = \left(\alpha - \alpha^l q^l - \alpha^f q^f\right)^{-\widetilde{R}\left(q^d, q^v\right)/c_v(\mathbf{q})} \exp\left[\frac{\eta - A(\mathbf{q})}{c_v(\mathbf{q})}\right] + \mathbb{L}_0\left(q^v, q^l\right), \tag{4.70}$$

$$\mathscr{E}(\alpha, T, \mathbf{q}) = c_v(\mathbf{q})\, T + \mathbb{L}_0\left(q^v, q^l\right), \tag{4.71}$$

$$\mathscr{E}(p, T, \mathbf{q}) = c_v(\mathbf{q})\, T + \mathbb{L}_0\left(q^v, q^l\right), \tag{4.72}$$

$$\mathscr{E}(\alpha, p, \mathbf{q}) = \frac{c_v(\mathbf{q})}{\widetilde{R}\left(q^d, q^v\right)}\left(\alpha - \alpha^l q^l - \alpha^f q^f\right) p + \mathbb{L}_0\left(q^v, q^l\right), \tag{4.73}$$

where $A = A(\mathbf{q})$ is an arbitrary 'constant' (i.e. a function that is independent of α and η).

[18] Only changes, $d\eta$, to η have physical significance.

The first form, (4.70), corresponds to the fundamental equation of state, (4.1), for the particular case of a cloudy-air parcel (with dry air and water vapour assumed to behave as ideal gases). Thus (4.70) completely defines the thermodynamic equilibrium state of such an air parcel. *From it, all other thermodynamic relations may be derived.* See Section 4.3.3 for illustrative examples of this property.

The last form, (4.73), could have been obtained more directly by simply using the equation of state for a cloudy-air parcel, (4.56), to eliminate T from (4.71) or (4.72) in favour of $p\alpha$, with use of (4.57). The alternative forms (4.70)–(4.73) illustrate, once again, that there is a plethora of interrelated thermodynamic variables, and many different ways of mathematically expressing the same physical principles.

4.3.2 Functional Forms for Specific Entropy (η)

Having constructed explicit functional forms (4.70)–(4.73) for specific internal energy ($\mathscr{E}$), we can now use them to obtain corresponding explicit functional forms for specific entropy (η). Thus solving (4.70) for η gives:

The Basic Functional Form for the Entropy of a Cloudy-Air Parcel

$$\begin{aligned}\eta\left(\alpha,\mathscr{E},\mathbf{q}\right) &= c_v\left(\mathbf{q}\right)\ln\left\{\left(\alpha-\alpha^l q^l-\alpha^f q^f\right)^{\widetilde{R}\left(q^d,q^v\right)/c_v(\mathbf{q})}\left[\mathscr{E}-\mathbb{L}_0\left(q^v,q^l\right)\right]\right\}+A\left(\mathbf{q}\right)\\ &= c_v\left(\mathbf{q}\right)\ln\left[\mathscr{E}-\mathbb{L}_0\left(q^v,q^l\right)\right]+\widetilde{R}\left(q^d,q^v\right)\ln\left(\alpha-\alpha^l q^l-\alpha^f q^f\right)+A\left(\mathbf{q}\right),\end{aligned} \tag{4.74}$$

where $A = A\left(\mathbf{q}\right)$ is an arbitrary function that serves to determine a reference state for η, and is independent of α and η.

Functional form (4.74) corresponds to the fundamental equation of state, (4.14), for the particular case of a cloudy-air parcel (with dry air and water vapour assumed to behave as ideal gases). Similarly to (4.70), (4.74) also completely defines the thermodynamic equilibrium state of a fluid. As such, all other thermodynamic relations may be derived from (4.74).

Three alternative explicit forms for η are now developed, namely:

1. $\eta = \eta\left(\alpha, T, \mathbf{q}\right)$; see (4.75).
2. $\eta = \eta\left(p, T, \mathbf{q}\right)$; see (4.79).
3. $\eta = \eta\left(\alpha, p, \mathbf{q}\right)$; see (4.80).

Eliminating $\mathscr{E}-\mathbb{L}_0\left(q^v,q^l\right)$ between (4.71) and (4.74) delivers the first alternative entropy form

$$\eta\left(\alpha,T,\mathbf{q}\right) = c_v\left(\mathbf{q}\right)\ln T+\widetilde{R}\left(q^d,q^v\right)\ln\left(\alpha-\alpha^l q^l-\alpha^f q^f\right)+\underbrace{c_v\left(\mathbf{q}\right)\ln\left[c_v\left(\mathbf{q}\right)\right]+A\left(\mathbf{q}\right)}_{\text{independent of }\alpha,T}. \tag{4.75}$$

From the equation of state for a cloudy-air parcel, (4.56), and using (4.57) and (4.58),

$$1 = \frac{R(\mathbf{q})T}{p\alpha} = \frac{\widetilde{R}\left(q^d, q^v\right)T}{\left[1 - \left(\alpha^l/\alpha\right)q^l - \left(\alpha^f/\alpha\right)q^f\right]p\alpha} = \frac{\widetilde{R}\left(q^d, q^v\right)T}{\left(\alpha - \alpha^l q^l - \alpha^f q^f\right)p}$$

$$\Downarrow$$

$$\alpha - \alpha^l q^l - \alpha^f q^f = \frac{\widetilde{R}\left(q^d, q^v\right)T}{p}. \tag{4.76}$$

Elimination of $\alpha - \alpha^l q^l - \alpha^f q^f$ between (4.75) and (4.76) then yields

$$\eta(p, T, \mathbf{q}) = \left[c_v(\mathbf{q}) + \widetilde{R}\left(q^d, q^v\right)\right]\ln T - \widetilde{R}\left(q^d, q^v\right)\ln p$$
$$+ \widetilde{R}\left(q^d, q^v\right)\ln\left[\widetilde{R}\left(q^d, q^v\right)\right] + c_v(\mathbf{q})\ln\left[c_v(\mathbf{q})\right] + A(\mathbf{q}). \tag{4.77}$$

Using definition (4.43) (but generalised here to a cloudy-air parcel) and (4.75) delivers

$$c_p(\mathbf{q}) \equiv T\frac{\partial\eta(p, T, \mathbf{q})}{\partial T} = c_v(\mathbf{q}) + \widetilde{R}\left(q^d, q^v\right), \tag{4.78}$$

that is, Carnot's Law (which holds for the *gaseous* component of total substance). From (4.78), c_p must be a function of $\mathbf{q}$ only (i.e. of q^d, q^v, q^l, and q^f only), since c_v and $\widetilde{R}$ are functions of $\mathbf{q}$ only (and not of p, nor of T). Substituting (4.78) into (4.77) gives the second alternative entropy form

$$\eta(p, T, \mathbf{q}) = c_p(\mathbf{q})\ln T - \widetilde{R}\left(q^d, q^v\right)\ln p$$
$$+ \underbrace{\widetilde{R}\left(q^d, q^v\right)\ln\left[\widetilde{R}\left(q^d, q^v\right)\right] + c_v(\mathbf{q})\ln\left[c_v(\mathbf{q})\right] + A(\mathbf{q})}_{\text{independent of } p, T}. \tag{4.79}$$

Eliminating T between (4.76) and (4.79), and using (4.78), yields the third alternative entropy form

$$\eta(\alpha, p, \mathbf{q}) = c_p(\mathbf{q})\ln\left(\alpha - \alpha^l q^l - \alpha^f q^f\right) + c_v(\mathbf{q})\ln p$$
$$+ \underbrace{c_v(\mathbf{q})\left\{\ln\left[c_v(\mathbf{q})\right] - \ln\left[\widetilde{R}\left(q^d, q^v\right)\right]\right\} + A(\mathbf{q})}_{\text{independent of } \alpha, p}. \tag{4.80}$$

4.3.3 Thermodynamic Properties of a Cloudy-Air Parcel from $\mathscr{E}$

The functional form for $\mathscr{E}(\alpha, \eta, \mathbf{q})$ is given by (4.70). This equation may be written in condensed form as

$$\mathscr{E}(\alpha, \eta, \mathbf{q}) = \left(\alpha - \alpha^l q^l - \alpha^f q^f\right)^{-\widetilde{R}/c_v}\exp\left(\frac{\eta - A}{c_v}\right) + \mathbb{L}_0, \tag{4.81}$$

where, in (4.81) and the equations that follow, it is understood that

$$\widetilde{R} = \widetilde{R}\left(q^d, q^v\right), \quad \mathbb{L}_0 = \mathbb{L}_0\left(q^v, q^l\right), \quad A = A(\mathbf{q}) = A\left(q^d, q^v, q^l, q^f\right), \tag{4.82}$$

$$c_v = c_v(\mathbf{q}) = c_v\left(q^d, q^v, q^l, q^f\right), \quad c_p = c_p(\mathbf{q}) = c_p\left(q^d, q^v, q^l, q^f\right). \tag{4.83}$$

Given functional form (4.81) for $\mathscr{E}(\alpha, \eta, \mathbf{q})$, the entire equilibrium state of a cloudy-air parcel (with dry air and water vapour assumed to behave as ideal gases) may be determined. To illustrate this property, the following thermodynamic quantities result from use of (4.81) and various definitions:

1. T and p from (4.81) and definitions (4.17) and (4.18), so that

$$\begin{aligned} T &= \frac{\partial \mathscr{E}(\alpha, \eta, \mathbf{q})}{\partial \eta} = \frac{1}{c_v}\left(\alpha - \alpha^l q^l - \alpha^f q^f\right)^{-\widetilde{R}/c_v} \exp\left(\frac{\eta - A}{c_v}\right) \\ &= \frac{1}{c_v}\left[\mathscr{E}(\alpha, \eta, \mathbf{q}) - \mathbb{L}_0\left(q^v, q^l\right)\right], \end{aligned} \tag{4.84}$$

$$\begin{aligned} p &= -\frac{\partial \mathscr{E}(\alpha, \eta, \mathbf{q})}{\partial \alpha} = \frac{\widetilde{R}}{c_v}\left(\alpha - \alpha^l q^l - \alpha^f q^f\right)^{-(\widetilde{R}+c_v)/c_v} \exp\left(\frac{\eta - A}{c_v}\right) \\ &= \frac{\widetilde{R}}{c_v\left(\alpha - \alpha^l q^l - \alpha^f q^f\right)}\left[\mathscr{E}(\alpha, \eta, \mathbf{q}) - \mathbb{L}_0\left(q^v, q^l\right)\right]. \end{aligned} \tag{4.85}$$

2. The equation of state for a cloudy-air parcel by elimination of $\left[\mathscr{E}(\alpha, \eta, \mathbf{q}) - \mathbb{L}_0\left(q^v, q^l\right)\right]$ from (4.84) and (4.85), so that

$$p\alpha = RT, \tag{4.86}$$

where

$$R = \frac{\widetilde{R}\left(q^d, q^v\right)}{\left[1 - \left(\alpha^l/\alpha\right) q^l - \left(\alpha^f/\alpha\right) q^f\right]}. \tag{4.87}$$

3. $\mathscr{E}(\alpha, T, \mathbf{q})$ by rearrangement of (4.84), so that

$$\mathscr{E}(\alpha, T, \mathbf{q}) = c_v T + \mathbb{L}_0\left(q^v, q^l\right). \tag{4.88}$$

4. $\eta(\alpha, \mathscr{E}, \mathbf{q})$ by solution of (4.81) for $\eta(\alpha, \mathscr{E}, \mathbf{q})$, so that

$$\eta(\alpha, \mathscr{E}, \mathbf{q}) = c_v \ln\left[\mathscr{E}(\alpha, \eta, \mathbf{q}) - \mathbb{L}_0\left(q^v, q^l\right)\right] + \widetilde{R}\ln\left(\alpha - \alpha^l q^l - \alpha^f q^f\right) + A. \tag{4.89}$$

5. $\eta(\alpha, T, \mathbf{q})$ by elimination of $\left[\mathscr{E}(\alpha, \eta, \mathbf{q}) - \mathbb{L}_0\left(q^v, q^l\right)\right]$ between (4.88) and (4.89), so that

$$\eta(\alpha, T, \mathbf{q}) = c_v \ln T + \widetilde{R}\ln\left(\alpha - \alpha^l q^l - \alpha^f q^f\right) + c_v \ln c_v + A. \tag{4.90}$$

6. $\eta(p, T, \mathbf{q})$ by elimination of α between (4.86) and (4.90), with use of (4.87), so that

$$\eta(p, T, \mathbf{q}) = \left(c_v + \widetilde{R}\right)\ln T - \widetilde{R}\ln p + \widetilde{R}\ln\widetilde{R} + c_v \ln c_v + A. \tag{4.91}$$

7. c_v (trivially) from definition (4.41) and (4.88), so that

$$c_v = \frac{\partial \mathscr{E}(\alpha, T, \mathbf{q})}{\partial T} = \frac{\partial\left(c_v T + \mathbb{L}_0\right)}{\partial T} = c_v. \tag{4.92}$$

8. c_p from definition (4.43) and (4.91), so that

$$\begin{aligned} c_p &= T\frac{\partial \eta(p, T, \mathbf{q})}{\partial T} = T\frac{\partial}{\partial T}\left[\left(c_v + \widetilde{R}\right)\ln T - \widetilde{R}\ln p + \widetilde{R}\ln\widetilde{R} + c_v \ln c_v + A\right] \\ &= c_v + \widetilde{R}. \end{aligned} \tag{4.93}$$

9. c_s from definition (4.52), so that

$$c_s^2 = \alpha^2 \frac{\partial^2 \mathscr{E}(\alpha,\eta,\mathbf{q})}{\partial \alpha^2} = \frac{\widetilde{R}}{c_V}\left(\frac{\widetilde{R}}{c_V}+1\right)\left[\mathscr{E}(\alpha,\eta,\mathbf{q}) - \mathbb{L}_0\left(q^v,q^l\right)\right]$$
$$= \frac{\widetilde{R}}{c_V}\left(\frac{c_p}{c_V}\right)\left[\mathscr{E}(\alpha,\eta,\mathbf{q}) - \mathbb{L}_0\left(q^v,q^l\right)\right] = \frac{c_p}{c_V}\widetilde{R}T = \gamma \widetilde{R}T, \tag{4.94}$$

where $\gamma \equiv c_p/c_V$.[19]

This is not an exhaustive list, but an illustrative one. Other thermodynamic relations may be similarly obtained. For example, equations (4.72), (4.73), and (4.80) for $\mathscr{E} = \mathscr{E}(p,T,\mathbf{q})$, $\mathscr{E} = \mathscr{E}(\alpha,p,\mathbf{q})$, and $\eta = \eta(\alpha,p,\mathbf{q})$, respectively, may also be derived; see Sections 4.3.1 and 4.3.2.

4.3.4 Thermodynamic Properties of a Cloudy Air Parcel from η

In Section 4.3.3, various thermodynamic properties of a cloudy-air parcel were obtained, starting from the functional form (4.81) for internal energy, $\mathscr{E} = \mathscr{E}(\alpha,\eta,\mathbf{q})$. A similar exercise, but instead starting from the functional form (4.74) for entropy, $\eta = \eta(\alpha,\mathscr{E},\mathbf{q})$, may be undertaken. The entire equilibrium state of a cloudy-air parcel may then be similarly determined from (4.74) and various definitions. This is left as an exercise for the interested reader.

4.4 THE GOVERNING EQUATIONS FOR MOTION OF A GEOPHYSICAL FLUID

A set of governing equations for general, rotating, geophysical fluids (primarily gaseous for the atmosphere, and liquid for the oceans), expressed in a rotating frame of reference, is assembled in the summary box that follows. As a cross-check, setting $\mathbb{L}_0 \equiv 0$ in (4.88), so that $\mathscr{E} \equiv c_V T$, and restricting attention to a mixture of ideal gases of constant composition, the ideal-gas equation set (2.74)–(2.77) can be recovered from (4.95)–(4.102). Many other equivalent equation sets are possible, due to the many different equivalent forms of various thermodynamic relations.

The Governing Equations for Motion of a Geophysical Fluid: Expressed in Vector Form

Momentum
From (3.68):

$$\frac{D\mathbf{u}}{Dt} = -2\boldsymbol{\Omega}\times\mathbf{u} - \frac{1}{\rho}\nabla p - \nabla\Phi + \mathbf{F}. \tag{4.95}$$

Continuity of Total Mass Density
From (2.37) with a possible source/sink term, F^ρ, for an open fluid parcel:

$$\frac{D\rho}{Dt} + \rho\nabla\cdot\mathbf{u} = F^\rho. \tag{4.96}$$

[19] For a dry atmosphere, for which $q^v = q^l = q^f = 0$, (4.94) reduces to $c_s^2 = \left(c_p^d/c_V^d\right) R^d T$. Equation (4.94) is then known as Laplace's formula, after Pierre-Simon Laplace (1749–1827). It corrects Newton's formula, $c_s^2 = p/\rho = R^d T$, after Isaac Newton (1643–1726). Newton had assumed that sound waves propagate *isothermally*, when in reality they propagate *adiabatically*, whence the factor c_p/c_V in (4.94). Earth's atmosphere is primarily composed of diatomic gases, and so $c_p/c_V \approx 7/5$; see Table 4.3 of the Appendix to this chapter.

Note the appearance of $\widetilde{R} = \widetilde{R}\left(q^d,q^v\right)$ in (4.94), rather than $R = R(\mathbf{q})$. Physically, this is because the speed of propagation depends on the compressibility of the gaseous component of the air mixture; liquid and frozen water do not contribute to compressibility since they have been assumed to be incompressible.

Substance Transport
From (4.9):

$$\frac{DS^i}{Dt} = \dot{S}^i, \quad i = 1, 2, \ldots, \tag{4.97}$$

where S^i and $\dot{S}^i$ are as in (4.10) (for the atmosphere) and (4.11) (for oceans).

Thermodynamic-Energy Equation
From (4.13) and (4.15):

$$\frac{D\mathcal{E}}{Dt} + p\frac{D\alpha}{Dt} = \dot{Q}_E \quad \text{or} \quad \frac{D\eta}{Dt} = \frac{\dot{Q}}{T}, \tag{4.98}$$

where $\dot{Q}$ is heating rate (per unit mass), and $\dot{Q}_E \equiv \dot{Q} + \sum_i \mu^i \dot{S}^i$ is total rate of energy input (per unit mass).

Fundamental Equation of State
From (4.1) and (4.14):

$$\mathcal{E} = \mathcal{E}\left(\alpha, \eta, S^1, S^2, \ldots\right) \quad \text{or} \quad \eta = \eta\left(\alpha, \mathcal{E}, S^1, S^2, \ldots\right), \tag{4.99}$$

where $\alpha \equiv 1/\rho$.

Definition of T, p, and μ^i
From (4.17)–(4.19) or (4.29)–(4.31):

$$T = \frac{\partial \mathcal{E}\left(\alpha, \eta, S^1, S^2, \ldots\right)}{\partial \eta} \quad \text{or} \quad T = \left[\frac{\partial \eta\left(\alpha, \mathcal{E}, S^1, S^2, \ldots\right)}{\partial \mathcal{E}}\right]^{-1}, \tag{4.100}$$

$$p = -\frac{\partial \mathcal{E}\left(\alpha, \eta, S^1, S^2, \ldots\right)}{\partial \alpha} \quad \text{or} \quad p = T\frac{\partial \eta\left(\alpha, \mathcal{E}, S^1, S^2, \ldots\right)}{\partial \alpha}, \tag{4.101}$$

$$\mu^i = \frac{\partial \mathcal{E}\left(\alpha, \eta, S^1, S^2, \ldots\right)}{\partial S^i} \quad \text{or} \quad \mu^i = -T\frac{\partial \eta\left(\alpha, \mathcal{E}, S^1, S^2, \ldots\right)}{\partial S^i}. \tag{4.102}$$

4.5 CONCLUDING REMARKS

At this juncture we have developed a unified set of governing equations for the motion of a geophysical fluid; primarily gaseous for the atmosphere, and primarily liquid for the oceans. This has been achieved under the twin assumptions that:

1. The geopotential of apparent gravity (Φ) is of known prescribed form. Once prescribed, this defines both the shape (classically termed *Figure*) of Earth's mean surface, and its associated gravitational attraction.
2. The internal energy ($\mathcal{E}$) or, equivalently, the entropy (η) – as defined by a fundamental equation of state, see (4.99) – is also of known prescribed form. Once either $\mathcal{E}$ or η is prescribed, the entire thermodynamic state of a fluid parcel can be obtained from it (at a given instant in time) via partial differentiation and algebra.

Prescribing these forms is, in practice, quite challenging. The first challenge (prescribing Φ) is met in Chapters 7 and 8.

For the second challenge, two such functional forms, one for $\mathcal{E}$, the other for η, have been given for the atmosphere, namely (4.70) and (4.74), respectively. These two forms are prescribed analytically and are equivalent to one another. They assume that the dry-gas and water-vapour

Motion	Monatomic ($N = 1$)	Diatomic ($N = 2$)	Triatomic ($N = 3$)
Translation	3	3	3
Rotation	0	3	3
Vibration	0	0	3
Total (n_d)	3	6	9

Table 4.1 Number of *theoretical* degrees of freedom (n_d) available for ideal monatomic, diatomic, and triatomic molecules; N is number of atoms per molecule.

constituents of Earth's atmosphere are well approximated by ideal gases, and that the vapour and liquid phases of water substance contain latent, internal, potential energy, expressed in a certain, physically realistic manner. These two forms provide a good representation for Earth's *atmosphere*, albeit accuracy can be improved at the price of increased complexity. For the *oceans*, however, we have remained silent about exactly *how* to suitably prescribe $\mathscr{E}$. This important, and highly challenging, issue is deferred until Chapter 11, since it is intimately linked to the further thermodynamic developments of Chapters 9 and 10.

The unified set of governing equations for the motion of a geophysical fluid, assembled in Section 4.4, has been expressed in vector form. For both practical applications and theoretical investigations – for example, the development of the geopotential approximations of Chapters 7 and 8 – it is desirable to also express this unified set in appropriate coordinate systems. This will be done in Chapter 6. To prepare the way for this, we define various orthogonal-curvilinear coordinate systems in Chapter 5.

APPENDIX: SPECIFIC HEAT CAPACITIES FOR AN IDEAL GAS

Values of specific heats, c_v and c_p, depend upon the properties of the molecules that comprise the gas. For an ideal gas, statistical-mechanics theory leads to

$$c_v = \frac{n_d}{2}R, \quad c_p = c_v + R = \left(\frac{n_d + 2}{2}\right)R, \tag{4.103}$$

where n_d is the number of independent degrees of freedom associated with a single molecule of gas (see e.g. Salmon, 1998, pp. 45–48; Curry and Webster, 1999, Section 2.9; and Ambaum, 2010, pp. 47–48). Each (independent) degree of freedom can contribute to storage of thermal energy within a molecule. The number of available degrees of freedom depends not only on the atomicity of the gas (i.e. the number of atoms per molecule) but also on how atoms are arranged within a molecule.

For an N-atomic molecule (i.e. a molecule comprised of N atoms), there are, theoretically (from the equipartition theorem), $3N$ degrees of freedom available (Table 4.1); three for each atom of the molecule, regardless of whether the molecule is monatomic or polyatomic.[20] These degrees of freedom are, depending upon the atomicity and structure of the molecule, divided between translational, rotational, and vibrational modes of motion (see Fig. 2.7 of Curry and Webster (1999) for a graphical depiction of diatomic and triatomic modes of motion). Thus:

1. A molecule of a *monatomic* gas, such as argon (Ar) and helium (He), has a single atom (i.e. $N = 1$). It then behaves as a point mass with three available degrees of freedom (i.e. $n_d = 3$); these correspond to translation in three independent directions in three-dimensional space. A monatomic molecule has no rotational or vibrational degrees of freedom available for heat storage.

[20] A polyatomic molecule has two or more atoms. For the special cases of molecules with just two or three molecules, however, the phraseology 'diatomic' or 'triatomic', respectively, is generally used instead.

2. A molecule of a *diatomic* gas, such as nitrogen (N_2) and oxygen (O_2), has two atoms (i.e. $N = 2$). A diatomic molecule also has three translational degrees of freedom, but it additionally has three rotational ones. These correspond to rotation about three independent axes. A diatomic molecule has no vibrational degrees of freedom available for heat storage.
3. A molecule of a *triatomic* gas, such as water vapour (H_2O) and carbon dioxide (CO_2), has three atoms (i.e. $N = 3$).[21] A triatomic molecule not only has three translational degrees of freedom available for heat storage, plus three rotational ones, it also has three vibrational ones.

If c_v and c_p are computed using (4.103) with the values of n_d displayed in Table 4.1, then their values can be compared with values measured, under realistic conditions, for the various gases in Earth's atmosphere. For monatomic gases (with $n_d = 3$), theoretical and measured values of c_v and c_p agree very well indeed. However, they do not agree at all well for diatomic and triatomic gases (with $n_d = 6$ and $n_d = 9$, respectively; see Table 4.1). At first sight this appears to be a very serious problem, but appearances can be deceptive.

The values of n_d given in Table 4.1 are the *maximum* theoretical numbers of degrees of freedom available for heat storage. However, they are not necessarily the most appropriate ones for Earth's atmosphere. For real gases in Earth's atmosphere, the following two considerations come into play:

- For diatomic gases, there are generally only *two* (instead of three) rotational modes of motion, since rotation about the axis aligned with its two atoms is negligible. Thus n_d reduces from $n_d = 6$ to $n_d = 5$ (three for translation plus two for rotation). This is a non-negligible reduction by one sixth, with a corresponding reduction in the predicted values of c_v and c_p; see (4.103). It is particularly important given that 99% of the dry component of Earth's atmosphere is composed of just two diatomic gases, namely nitrogen (N_2) and oxygen (O_2); see Table 1.1 of Chapter 1.
- Furthermore, vibrational degrees of freedom contribute negligibly to heat capacity. This is because atmospheric temperatures are generally insufficiently hot to excite vibrational states. Consequently, n_d reduces from $n_d = 9$ to $n_d = 6$ for triatomic gases, a significant reduction of one third, again with a corresponding reduction in the predicted values of c_v and c_p; see (4.103). This reduction is particularly important since water vapour (H_2O) is a triatomic gas of crucial importance for modelling Earth's atmosphere.

These two reductions of n_d lead to the simplified situation, summarised in Table 4.2, for the number of *accessible* (as opposed to the *maximum* number of theoretically possible) degrees of freedom available for heat storage in real gases at Earth's atmospheric temperatures. The corresponding values of c_v, c_p, κ, and γ are displayed in Table 4.3. They are computed using (4.103) and the definitions $\kappa \equiv R/c_p$ and $\gamma \equiv c_p/c_v$. The values displayed in Table 4.3 then agree very well indeed with measured values. Note in particular the value $\kappa = 2/7$ for a diatomic gas; this is the value given in many textbooks and journal publications for the dry component of Earth's atmosphere. This value is justified by the fact that 99% of the dry gases in Earth's atmosphere are diatomic gases.

From the preceding analysis, it is concluded that the significant individual gases in Earth's atmosphere can be represented, *to an excellent approximation*, as ideal gases according to:

- Their atomicity (i.e. number of atoms per molecule).
- The associated number of *accessible* degrees of freedom, where $n_d = 3, 5, 6$ for monatomic, diatomic, and triatomic gases, respectively.

[21] These are the two most important 'greenhouse' gases in Earth's atmosphere.

Motion	Monatomic ($N = 1$)	Diatomic ($N = 2$)	Triatomic ($N = 3$)
Translation	3	3	3
Rotation	0	2	3
Vibration	0	0	0
Total (n_d)	3	5	6

Table 4.2 Same as Table 4.1, but for number of *accessible* degrees of freedom for real gases in the range of Earth's atmospheric temperatures.

	Monatomic ($n_d = 3$)	Diatomic ($n_d = 5$)	Triatomic ($n_d = 6$)
$c_v = \frac{n_d}{2}R$	$\frac{3}{2}R$	$\frac{5}{2}R$	$3R$
$c_p \equiv c_v + R = \frac{n_d+2}{2}R$	$\frac{5}{2}R$	$\frac{7}{2}R$	$4R$
$\kappa \equiv \frac{R}{c_p} = \frac{2}{n_d+2}$	$\frac{2}{5}$	$\frac{2}{7}$	$\frac{1}{4}$
$\gamma \equiv \frac{c_p}{c_v} = \frac{n_d+2}{n_d}$	$\frac{5}{3}$	$\frac{7}{5}$	$\frac{4}{3}$

Table 4.3 Values of c_v, c_p, $\kappa \equiv R/c_p$, and $\gamma \equiv c_p/c_v$, for monatomic, diatomic, and triatomic molecules of Earth's atmosphere, as a function of *accessible* number of degrees of freedom, n_d, displayed in Table 4.2. Values are computed using (4.103) and definitions $\kappa \equiv R/c_p$ and $\gamma \equiv c_p/c_v$.

Nevertheless, as noted in Salmon (1998), for quantitative applications one can instead use a measured value of c_v or c_p for each individual gas, and then obtain the other from Carnot's Law, $c_p = c_v + R$.

5

Orthogonal-Curvilinear Coordinate Systems

ABSTRACT

Using vectors, one can develop the governing equations for atmospheric and oceanic modelling, and examine some of their properties. Coordinate systems, however, are needed for both practical applications and theoretical investigations. Orthogonal-curvilinear coordinates are a standard tool of mathematical physics. They are extensively employed herein. Definitions and relevant properties of various orthogonal-curvilinear coordinate systems are documented in the present chapter. This prepares the way for developments in later chapters and is also useful for reference purposes. These coordinate systems come in two flavours – deep and shallow – according to their application (in later chapters) to the governing equations. Shallow coordinates are a frequently used, non-Euclidean, geometric distortion of deep coordinates. They are based on the assumption that the depth of the atmosphere or an ocean can be considered to be shallow with respect to planetary radius. Using shallow coordinates then leads to simpler forms of the governing equations, albeit with some reduction of accuracy according to how well the shallowness assumption holds for a particular application. General forms for both deep and shallow orthogonal coordinate systems are presented, as well as some of their properties for representation of standard differential operators. Specific examples of such coordinate systems are also given (e.g. deep and shallow forms of both spherical-polar coordinates and cylindrical-polar coordinates).

5.1 PREAMBLE

One can make significant progress using vectors both to develop the governing equations for atmospheric and oceanic modelling (see Chapters 2–4) and to examine some of their fundamental properties (e.g. conservation of various physical quantities; see Chapter 15). Coordinate systems, however, are needed for both practical applications and theoretical investigations. For example, spherical-polar coordinates are employed (in Chapters 7 and 8) to determine the Figure (shape) of the Earth, and to develop various models of gravitational potential of differing accuracy and complexity.

Orthogonal-curvilinear coordinates are a standard tool of mathematical physics, derived in textbooks on vector analysis; for example, Spiegel and Lipschutz (2009). Indeed, these coordinates are extensively employed herein. This chapter therefore sets the scene for later chapters by providing pertinent documentation. For the purposes of this book, orthogonal-curvilinear coordinate systems come in two flavours.[1] These are labelled *deep* and *shallow*, and refer to the intended application of the coordinate system:

[1] Both delicious, albeit this is something of an acquired taste.

1. *'Deep'* refers to application of standard, unmodified orthogonal-curvilinear coordinates to the unapproximated governing equations of a deep atmosphere or ocean;[2] see Chapter 6. A description of the general form of *deep* orthogonal-curvilinear coordinates is given in Section 5.2, followed by specific examples (spherical-polar, cylindrical-polar, geodetic, and confocal-oblate-spheroidal) relevant to later chapters.
2. *'Shallow'* refers to application of *modified* orthogonal-curvilinear coordinates to approximate and simplify the unapproximated equations in a dynamically consistent manner. Such modifications are made under the assumption that the depth of the atmosphere or of an ocean can be considered to be shallow with respect to planetary radius; see Chapters 16–19. A description of the general form of *shallow* orthogonal-curvilinear coordinates is given in Section 5.3, followed by two specific examples (spherical-polar and cylindrical-polar) relevant to later chapters.

Following the developments of Sections 5.2 and 5.3 for deep and shallow coordinate systems, respectively, some concluding remarks are made in Section 5.4.

5.2 DEEP ORTHOGONAL-CURVILINEAR COORDINATES

5.2.1 General Deep Orthogonal-Curvilinear Coordinates (ξ_1, ξ_2, ξ_3)

Consider: an orthogonal-curvilinear coordinate system with coordinates (ξ_1, ξ_2, ξ_3) and associated unit-vector triad $(\mathbf{e}_1, \mathbf{e}_2, \mathbf{e}_3)$; and a Cartesian coordinate system with coordinates (x_1, x_2, x_3) and associated unit-vector triad $(\mathbf{i}_1, \mathbf{i}_2, \mathbf{i}_3)$. See Fig. 5.1.

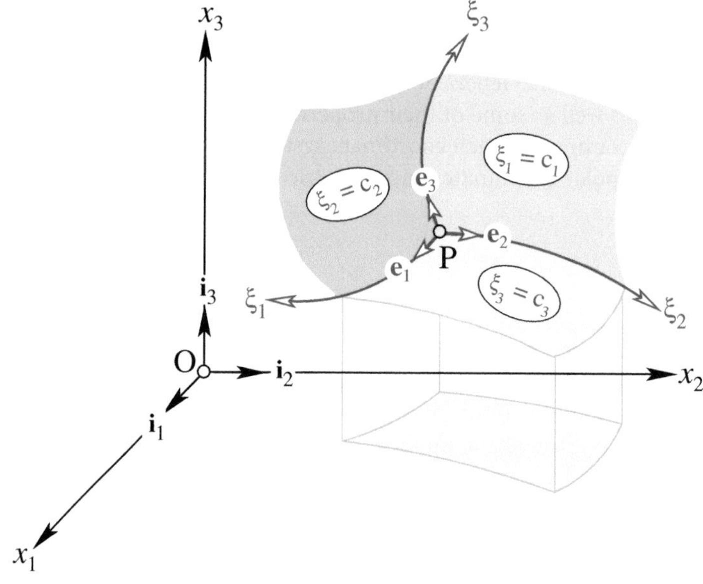

Figure 5.1 Orthogonal-curvilinear coordinates (ξ_1, ξ_2, ξ_3) with associated unit-vector triad $(\mathbf{e}_1, \mathbf{e}_2, \mathbf{e}_3)$ at an arbitrary point P; and Cartesian coordinates (x_1, x_2, x_3) with associated unit-vector triad $(\mathbf{i}_1, \mathbf{i}_2, \mathbf{i}_3)$ at origin O. Three (shaded) mutually orthogonal, curvilinear surfaces, on which $\xi_i = c_i = \text{constant}, i = 1, 2, 3$, are also shown. To help visualisation in 3D, the $\xi_3 = c_3$ plane is projected onto the Ox_1x_2 plane.

[2] Unapproximated here means no further approximation beyond that needed to obtain the equation set assembled in Section 4.4.

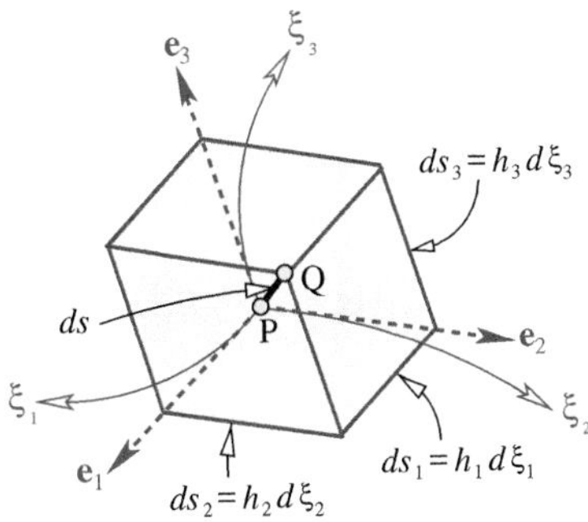

Figure 5.2 Distance and volume elements at an arbitrary point, P, of an orthogonal-curvilinear coordinate system (ξ_1, ξ_2, ξ_3). Distance elements $ds_i \equiv h_i d\xi_i, i = 1, 2, 3$, are the lengths of the sides of the rectangular cuboid (in magenta), and $dV \equiv ds_1 ds_2 ds_3 \equiv h_1 h_2 h_3 d\xi_1 d\xi_2 d\xi_3$ is its volume. Unit vectors of unit-vector triad $(\mathbf{e_1}, \mathbf{e_2}, \mathbf{e_3})$ are tangent to the three coordinate lines passing through point P. Distance element $ds \equiv \left[(ds_1)^2 + (ds_2)^2 + (ds_3)^2\right]^{1/2}$ is the length of diagonal PQ of the cuboid.

These two sets of coordinates are assumed to be related to one another by invertible relations of the form $x_i = x_i(\xi_j), i, j = 1, 2, 3$. Thus:

Position Vector r in Cartesian and Orthogonal-Curvilinear Coordinates

Position vector $\mathbf{r}$ may be equivalently written in Cartesian and orthogonal-curvilinear coordinates as

$$\begin{aligned} \mathbf{r} &= (\mathbf{r} \cdot \mathbf{i}_1)\,\mathbf{i}_1 + (\mathbf{r} \cdot \mathbf{i}_2)\,\mathbf{i}_2 + (\mathbf{r} \cdot \mathbf{i}_3)\,\mathbf{i}_3 \equiv x_1\mathbf{i}_1 + x_2\mathbf{i}_2 + x_3\mathbf{i}_3 \\ &= (\mathbf{r} \cdot \mathbf{e}_1)\,\mathbf{e}_1 + (\mathbf{r} \cdot \mathbf{e}_2)\,\mathbf{e}_2 + (\mathbf{r} \cdot \mathbf{e}_3)\,\mathbf{e}_3. \end{aligned} \tag{5.1}$$

Referring to Fig. 5.2:

The Distance Metric and Volume Element of an Orthogonal-Curvilinear Coordinate System

The distance metric (ds) and volume element (dV) of an orthogonal-curvilinear coordinate system are given by

$$(ds)^2 \equiv (ds_1)^2 + (ds_2)^2 + (ds_3)^2 \equiv (h_1 d\xi_1)^2 + (h_2 d\xi_2)^2 + (h_3 d\xi_3)^2, \tag{5.2}$$

$$dV \equiv ds_1 ds_2 ds_3 \equiv (h_1 d\xi_1)(h_2 d\xi_2)(h_3 d\xi_3) \equiv h_1 h_2 h_3 d\xi_1 d\xi_2 d\xi_3, \tag{5.3}$$

respectively, where the (positive) *metric* (or *scale*) *factors* (h_1, h_2, h_3) satisfy

$$(h_i)^2 \equiv \left|\frac{\partial \mathbf{r}}{\partial \xi_i}\right|^2 \equiv \frac{\partial \mathbf{r}}{\partial \xi_i} \cdot \frac{\partial \mathbf{r}}{\partial \xi_i} \equiv \left(\frac{\partial x_1}{\partial \xi_i}\right)^2 + \left(\frac{\partial x_2}{\partial \xi_i}\right)^2 + \left(\frac{\partial x_3}{\partial \xi_i}\right)^2, \quad i = 1, 2, 3. \tag{5.4}$$

The velocity vector in an orthogonal-curvilinear system can then be written in terms of its components as

$$\mathbf{u} = u_1\mathbf{e}_1 + u_2\mathbf{e}_2 + u_3\mathbf{e}_3 \equiv h_1\frac{D\xi_1}{Dt}\mathbf{e}_1 + h_2\frac{D\xi_2}{Dt}\mathbf{e}_2 + h_3\frac{D\xi_3}{Dt}\mathbf{e}_3 = \frac{Dx_1}{Dt}\mathbf{i}_1 + \frac{Dx_2}{Dt}\mathbf{i}_2 + \frac{Dx_3}{Dt}\mathbf{i}_3. \tag{5.5}$$

For a generic vector $\mathbf{A} = (A_1, A_2, A_3) \equiv A_1\mathbf{e}_1 + A_2\mathbf{e}_2 + A_3\mathbf{e}_3$, and a generic scalar f, the *gradient* (∇f), *divergence* ($\nabla \cdot \mathbf{A}$), *curl* ($\nabla \times \mathbf{A}$), *Laplacian* ($\nabla^2 f$), and *advection* ($\mathbf{A} \cdot \nabla f$) operations may be respectively written in orthogonal-curvilinear coordinates as:[3]

$$\nabla f = \frac{1}{h_1}\frac{\partial f}{\partial \xi_1}\mathbf{e}_1 + \frac{1}{h_2}\frac{\partial f}{\partial \xi_2}\mathbf{e}_2 + \frac{1}{h_3}\frac{\partial f}{\partial \xi_3}\mathbf{e}_3, \tag{5.6}$$

$$\nabla \cdot \mathbf{A} = \frac{1}{h_1 h_2 h_3}\left[\frac{\partial}{\partial \xi_1}(A_1 h_2 h_3) + \frac{\partial}{\partial \xi_2}(A_2 h_3 h_1) + \frac{\partial}{\partial \xi_3}(A_3 h_1 h_2)\right], \tag{5.7}$$

$$\begin{aligned}\nabla \times \mathbf{A} = \frac{1}{h_1 h_2 h_3}\begin{vmatrix} h_1\mathbf{e}_1 & h_2\mathbf{e}_2 & h_3\mathbf{e}_3 \\ \frac{\partial}{\partial \xi_1} & \frac{\partial}{\partial \xi_2} & \frac{\partial}{\partial \xi_3} \\ h_1 A_1 & h_2 A_2 & h_3 A_3 \end{vmatrix} &= \frac{1}{h_2 h_3}\left[\frac{\partial}{\partial \xi_2}(h_3 A_3) - \frac{\partial}{\partial \xi_3}(h_2 A_2)\right]\mathbf{e}_1 \\ &+ \frac{1}{h_3 h_1}\left[\frac{\partial}{\partial \xi_3}(h_1 A_1) - \frac{\partial}{\partial \xi_1}(h_3 A_3)\right]\mathbf{e}_2 \\ &+ \frac{1}{h_1 h_2}\left[\frac{\partial}{\partial \xi_1}(h_2 A_2) - \frac{\partial}{\partial \xi_2}(h_1 A_1)\right]\mathbf{e}_3, \end{aligned} \tag{5.8}$$

$$\nabla^2 f = \frac{1}{h_1 h_2 h_3}\left[\frac{\partial}{\partial \xi_1}\left(\frac{h_2 h_3}{h_1}\frac{\partial f}{\partial \xi_1}\right) + \frac{\partial}{\partial \xi_2}\left(\frac{h_3 h_1}{h_2}\frac{\partial f}{\partial \xi_2}\right) + \frac{\partial}{\partial \xi_3}\left(\frac{h_1 h_2}{h_3}\frac{\partial f}{\partial \xi_3}\right)\right], \tag{5.9}$$

$$\begin{aligned}\mathbf{A} \cdot \nabla f &= (A_1\mathbf{e}_1 + A_2\mathbf{e}_2 + A_3\mathbf{e}_3) \cdot \left(\frac{1}{h_1}\frac{\partial f}{\partial \xi_1}\mathbf{e}_1 + \frac{1}{h_2}\frac{\partial f}{\partial \xi_2}\mathbf{e}_2 + \frac{1}{h_3}\frac{\partial f}{\partial \xi_3}\mathbf{e}_3\right) \\ &= \frac{A_1}{h_1}\frac{\partial f}{\partial \xi_1} + \frac{A_2}{h_2}\frac{\partial f}{\partial \xi_2} + \frac{A_3}{h_3}\frac{\partial f}{\partial \xi_3}. \end{aligned} \tag{5.10}$$

Using definition (2.8) of material derivative and (5.10), the *material derivative of a scalar*, f, expressed in orthogonal-curvilinear coordinates, is

$$\frac{Df}{Dt} \equiv \frac{\partial f}{\partial t} + \mathbf{u} \cdot \nabla f = \left(\frac{\partial}{\partial t} + \frac{u_1}{h_1}\frac{\partial}{\partial \xi_1} + \frac{u_2}{h_2}\frac{\partial}{\partial \xi_2} + \frac{u_3}{h_3}\frac{\partial}{\partial \xi_3}\right) f. \tag{5.11}$$

Using (5.6) and (5.8) in vector identity (A.23), the *material derivative of the velocity vector*, $\mathbf{u}$ (i.e. the acceleration vector) in an orthogonal-curvilinear system is thus

$$\begin{aligned}\frac{D\mathbf{u}}{Dt} &\equiv \frac{\partial \mathbf{u}}{\partial t} + (\mathbf{u} \cdot \nabla)\mathbf{u} \equiv \frac{\partial \mathbf{u}}{\partial t} + \frac{1}{2}\nabla(\mathbf{u} \cdot \mathbf{u}) + (\nabla \times \mathbf{u}) \times \mathbf{u} \\ &= \left[\frac{Du_1}{Dt} + \frac{u_2}{h_1 h_2}\left(u_1\frac{\partial h_1}{\partial \xi_2} - u_2\frac{\partial h_2}{\partial \xi_1}\right) + \frac{u_3}{h_1 h_3}\left(u_1\frac{\partial h_1}{\partial \xi_3} - u_3\frac{\partial h_3}{\partial \xi_1}\right)\right]\mathbf{e}_1 \\ &+ \left[\frac{Du_2}{Dt} + \frac{u_3}{h_2 h_3}\left(u_2\frac{\partial h_2}{\partial \xi_3} - u_3\frac{\partial h_3}{\partial \xi_2}\right) + \frac{u_1}{h_2 h_1}\left(u_2\frac{\partial h_2}{\partial \xi_1} - u_1\frac{\partial h_1}{\partial \xi_2}\right)\right]\mathbf{e}_2 \\ &+ \left[\frac{Du_3}{Dt} + \frac{u_1}{h_3 h_1}\left(u_3\frac{\partial h_3}{\partial \xi_1} - u_1\frac{\partial h_1}{\partial \xi_3}\right) + \frac{u_2}{h_3 h_2}\left(u_3\frac{\partial h_3}{\partial \xi_2} - u_2\frac{\partial h_2}{\partial \xi_3}\right)\right]\mathbf{e}_3, \end{aligned} \tag{5.12}$$

[3] The three vector components of (5.6) may be found from one another by cyclic permutation of the three indices, and similarly for the three vector components of (5.8) and of (5.12). This is also the case for the three scalar contributions to the right-hand side of (5.7), and similarly for those of the right-hand sides of (5.9)–(5.11).

where D/Dt (of a scalar) is defined by (5.11).[4] The terms not involving D/Dt on the last three lines of (5.12) are usually referred to as *metric terms* and are not to be confused with the *metric (scale) factors* h_1, h_2, and h_3.

For later convenience and for reference purposes, explicit expressions are now given for various quantities needed to express the governing atmospheric and oceanographic equations in various orthogonal-curvilinear coordinate systems.

5.2.2 Spherical-Polar Coordinates (λ, ϕ, r)

Consider: a spherical-polar coordinate system with coordinates (λ, ϕ, r) and associated unit-vector triad $(\mathbf{e}_\lambda, \mathbf{e}_\phi, \mathbf{e}_r)$; and a Cartesian coordinate system with coordinates (x_1, x_2, x_3) and associated unit-vector triad $(\mathbf{i}_1, \mathbf{i}_2, \mathbf{i}_3)$. See Fig. 5.3. Thus the general orthogonal-curvilinear coordinates (ξ_1, ξ_2, ξ_3) of Section 5.2.1 are replaced by the spherical-polar coordinates (λ, ϕ, r) of the present section, that is,

$$(\xi_1, \xi_2, \xi_3) \rightarrow (\lambda, \phi, r) . \tag{5.13}$$

The spherical-polar coordinates of point P are:

- *Longitude* (λ), measuring angular distance eastwards, from the (dark-shaded) reference meridional plane, along the latitude circle passing through point P.
- *Latitude* (ϕ), measuring angular distance northward (for positive ϕ) from the equator, along the meridian containing point P, and southward for negative ϕ.
- *Radial distance* (r), measured from origin O at the sphere's centre.

Each coordinate (λ, ϕ or r) increases in the direction of its associated unit vector ($\mathbf{e}_\lambda$, $\mathbf{e}_\phi$ or $\mathbf{e}_r$, respectively).[5]

Examination of Fig. 5.3 shows that the Cartesian and spherical-polar coordinate systems are related to one another by the coordinate-transformation equations

$$(x_1, x_2, x_3) = (r \cos\lambda \cos\phi, r \sin\lambda \cos\phi, r \sin\phi) . \tag{5.14}$$

Using this equation in (5.4), the (positive) metric (or scale) factors are

[4] In Cartesian coordinates, (5.11) (valid for a scalar, f) also holds for the individual components $u_i, i = 1, 2, 3$, of the vector $\mathbf{u}$, i.e.

$$\frac{D\mathbf{u}}{Dt} = \frac{Du_1}{Dt}\mathbf{i}_1 + \frac{Du_2}{Dt}\mathbf{i}_2 + \frac{Du_3}{Dt}\mathbf{i}_3.$$

This is because unit-vector triad $(\mathbf{i}_1, \mathbf{i}_2, \mathbf{i}_3)$ has the same orientation (aligned with the Cartesian coordinate axes) at *all* points in space. It is very important, however, to realise that this property does *not* hold, in general, for orthogonal-curvilinear coordinates because the orientation of the associated unit-vector triad $(\mathbf{e}_1, \mathbf{e}_2, \mathbf{e}_3)$ generally varies as a function of its position $\mathbf{r}$, so that

$$\frac{D\mathbf{u}}{Dt} = \frac{Du_1}{Dt}\mathbf{e}_1 + \frac{Du_2}{Dt}\mathbf{e}_2 + \frac{Du_3}{Dt}\mathbf{e}_3 + \underbrace{u_1\frac{D\mathbf{e}_1}{Dt} + u_2\frac{D\mathbf{e}_2}{Dt} + u_3\frac{D\mathbf{e}_3}{Dt}}_{\neq 0}.$$

[5] The spherical-polar coordinates defined and used herein are those generally employed in the atmospheric and oceanic sciences, where latitude ϕ is the meridionally varying coordinate, i.e. the coordinate that varies along a meridian for fixed λ and r. In other fields, colatitude $\theta \equiv (\pi/2) - \phi$ is often used instead of latitude. Colatitude also measures angular distance along a meridian, but southwards from the north pole of the coordinate system instead of northwards from the equator. To maintain a conventional *right*-handed coordinate triad, (λ, ϕ, r) is then replaced by (θ, λ, r); note the different ordering of coordinates with λ then being the second coordinate instead of the first. Latitude (as opposed to colatitude) is used everywhere herein.

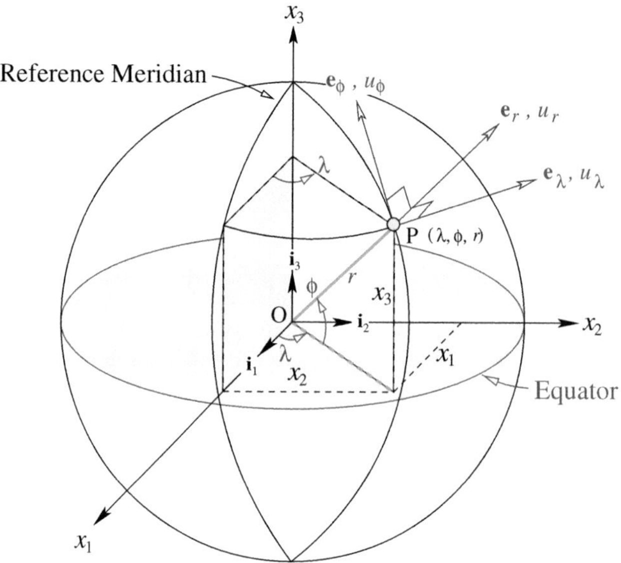

Figure 5.3 Spherical-polar coordinates (λ, ϕ, r) with associated unit-vector triad $(\mathbf{e}_\lambda, \mathbf{e}_\phi, \mathbf{e}_r)$ at an arbitrary point P; and Cartesian coordinates (x_1, x_2, x_3) with associated unit-vector triad $(\mathbf{i}_1, \mathbf{i}_2, \mathbf{i}_3)$ at origin O. The reference meridional plane $(\lambda = 0)$ is dark shaded (in blue), and the meridional plane containing point P is light shaded (in blue). The equatorial plane is shaded in magenta. See text for further details.

$$\left(h_\lambda, h_\phi, h_r\right) = (r\cos\phi, r, 1)\,. \tag{5.15}$$

From (5.2), (5.3), (5.13), and (5.15), the distance metric (ds) and volume element (dV) of a spherical-polar coordinate system are then given by

$$(ds)^2 = \left(h_\lambda d\lambda\right)^2 + \left(h_\phi d\phi\right)^2 + \left(h_r dr\right)^2 = r^2\cos^2\phi\,(d\lambda)^2 + r^2\,(d\phi)^2 + (dr)^2\,, \tag{5.16}$$

$$dV = \left(h_\lambda d\lambda\right)\left(h_\phi d\phi\right)\left(h_r dr\right) = r^2\cos\phi\, d\lambda\, d\phi\, dr. \tag{5.17}$$

Inserting (5.13) and metric factors (5.15) into (5.5), the velocity vector in a spherical-polar coordinate system can then be written in terms of its components as

$$\begin{aligned}\mathbf{u} = u_\lambda\mathbf{e}_\lambda + u_\phi\mathbf{e}_\phi + u_r\mathbf{e}_r &\equiv h_\lambda\frac{D\lambda}{Dt}\mathbf{e}_\lambda + h_\phi\frac{D\phi}{Dt}\mathbf{e}_\phi + h_r\frac{D\phi}{Dt}\mathbf{e}_r\\ &= r\cos\phi\frac{D\lambda}{Dt}\mathbf{e}_\lambda + r\frac{D\phi}{Dt}\mathbf{e}_\phi + \frac{Dr}{Dt}\mathbf{e}_r.\end{aligned} \tag{5.18}$$

For a generic vector $\mathbf{A} = \left(A_\lambda, A_\phi, A_r\right) \equiv A_\lambda\mathbf{e}_\lambda + A_\phi\mathbf{e}_\phi + A_r\mathbf{e}_r$, and a generic scalar f, and using (5.13) and metric factors (5.15) in (5.6)–(5.10), the *gradient* (∇f), *divergence* $(\nabla\cdot\mathbf{A})$, *curl* $(\nabla\times\mathbf{A})$, *Laplacian* $(\nabla^2 f)$, and *advection* $(\mathbf{A}\cdot\nabla f)$ operations may be respectively written in spherical-polar coordinates as:

$$\nabla f = \frac{1}{r\cos\phi}\frac{\partial f}{\partial\lambda}\mathbf{e}_\lambda + \frac{1}{r}\frac{\partial f}{\partial\phi}\mathbf{e}_\phi + \frac{\partial f}{\partial r}\mathbf{e}_r, \tag{5.19}$$

$$\nabla \cdot \mathbf{A} = \frac{1}{r^2 \cos\phi} \left[\frac{\partial}{\partial \lambda} (r A_\lambda) + \frac{\partial}{\partial \phi} \left(r \cos\phi A_\phi \right) + \frac{\partial}{\partial r} \left(r^2 \cos\phi A_r \right) \right], \tag{5.20}$$

$$\nabla \times \mathbf{A} = \frac{1}{r} \left[\frac{\partial A_r}{\partial \phi} - \frac{\partial}{\partial r} \left(r A_\phi \right) \right] \mathbf{e}_\lambda + \frac{1}{r\cos\phi} \left[\frac{\partial}{\partial r} \left(r \cos\phi A_\lambda \right) - \frac{\partial A_r}{\partial \lambda} \right] \mathbf{e}_\phi + \frac{1}{r \cos\phi} \left[\frac{\partial A_\phi}{\partial \lambda} - \frac{\partial}{\partial \phi} \left(\cos\phi A_\lambda \right) \right] \mathbf{e}_r, \tag{5.21}$$

$$\nabla^2 f = \frac{1}{r^2 \cos^2\phi} \frac{\partial^2 f}{\partial \lambda^2} + \frac{1}{r^2 \cos\phi} \frac{\partial}{\partial \phi} \left(\cos\phi \frac{\partial f}{\partial \phi} \right) + \frac{1}{r^2} \frac{\partial}{\partial r} \left(r^2 \frac{\partial f}{\partial r} \right), \tag{5.22}$$

$$\mathbf{A} \cdot \nabla f = \frac{A_\lambda}{r \cos\phi} \frac{\partial f}{\partial \lambda} + \frac{A_\phi}{r} \frac{\partial f}{\partial \phi} + A_r \frac{\partial f}{\partial r}. \tag{5.23}$$

Using (2.8) and (5.23), the *material derivative of a scalar*, f, expressed in spherical-polar coordinates, is

$$\frac{Df}{Dt} = \left(\frac{\partial}{\partial t} + \frac{u_\lambda}{r \cos\phi} \frac{\partial}{\partial \lambda} + \frac{u_\phi}{r} \frac{\partial}{\partial \phi} + u_r \frac{\partial}{\partial r} \right) f. \tag{5.24}$$

Using (5.19) and (5.21) in vector identity (A.23), the *material derivative of the velocity vector*, $\mathbf{u}$ (i.e. the acceleration vector), in spherical-polar coordinates is thus

$$\frac{D\mathbf{u}}{Dt} = \left(\frac{Du_\lambda}{Dt} + \frac{u_\lambda u_r}{r} - \frac{u_\lambda u_\phi \tan\phi}{r} \right) \mathbf{e}_\lambda + \left(\frac{Du_\phi}{Dt} + \frac{u_\phi u_r}{r} + \frac{u_\lambda^2 \tan\phi}{r} \right) \mathbf{e}_\phi + \left[\frac{Du_r}{Dt} - \frac{\left(u_\lambda^2 + u_\phi^2 \right)}{r} \right] \mathbf{e}_r. \tag{5.25}$$

Outside a rotating spherical planet of radius a, and making the classical spherical-geopotential approximation (in deep form) (see Sections 7.1.5 and 8.6),

$$\Phi\left(r\right) = -\, g_a \frac{a^2}{r} = -\frac{\gamma M}{r}, \tag{5.26}$$

$$\Downarrow$$

$$\mathbf{g}\left(r\right) \equiv -\, \nabla \Phi\left(r\right) = -\frac{\gamma M}{r^2} \mathbf{e}_r = -g_a \left(\frac{a}{r} \right)^2 \mathbf{e}_r. \tag{5.27}$$

Here:

- Φ is geopotential.
- $g_a \equiv |\nabla \Phi|_{r=a}$ is the absolute value of gravity on the surface ($r = a$) of the rotating spherical planet.
- γ is Newton's universal gravitational constant.
- M is the total mass of the planet (including its atmosphere and oceans).

5.2.3 Cylindrical-Polar Coordinates $(\mathtt{r}, \lambda, z)$ or $(\lambda, z, \mathtt{r})$

The usual convention in the scientific literature for ordering cylindrical-polar coordinates is $(\mathtt{r}, \lambda, z)$, where $\mathtt{r}$, λ, and z are defined later in this section. It turns out, however, that cyclically reordering them instead as $(\lambda, z, \mathtt{r})$ is more convenient for the developments presented herein.[6] Reordering does not change anything fundamentally. Both orderings share the same physical,

[6] Putting the azimuthal coordinate (λ) first and the radial coordinate ($\mathtt{r}$) last facilitates the development of a unified framework for expressing governing equation sets in ellipsoidal, spherical, *and* cylindrical geometries. This is because these developments assume axial symmetry of the coordinate system, and that surfaces for which the third coordinate is constant are geopotential surfaces. See Sections 6.4 and 6.5.

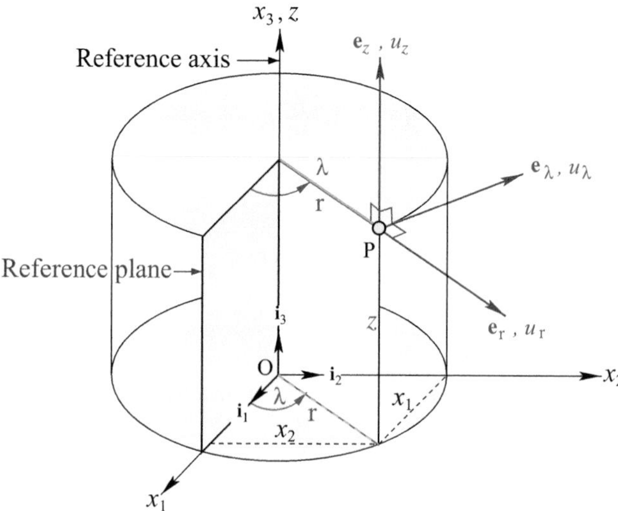

Figure 5.4 Cylindrical-polar coordinates $(\lambda, z, \mathtt{r})$ or $(\mathtt{r}, \lambda, z)$ with associated unit-vector triad $(\mathbf{e}_\lambda, \mathbf{e}_z, \mathbf{e}_\mathtt{r})$ or $(\mathbf{e}_\mathtt{r}, \mathbf{e}_\lambda, \mathbf{e}_z)$, respectively, at an arbitrary point P; and Cartesian coordinates (x_1, x_2, x_3) with associated unit-vector triad $(\mathbf{i}_1, \mathbf{i}_2, \mathbf{i}_3)$ at origin O. The reference axis is $Oz = Ox_3$, the reference plane $(\lambda = 0)$ is dark shaded (in blue), and the $\lambda =$ constant plane through point P is light shaded (in blue). See text for further details.

right-handed, unit-vector triad, as depicted in Fig. 5.4. Using either ordering leads to exactly the same expressions for operators such as gradient, divergence, curl, Laplacian, and material derivatives of scalars and vectors – see (5.34)–(5.40). The only real difference is that terms in the expressions for such operators appear in a different (cyclically permuted) order. To maintain the link between the present work, with $(\lambda, z, \mathtt{r})$ ordering, and other works, with standard ordering $(\mathtt{r}, \lambda, z)$, both are developed in parallel with one another. The overhead of doing so is small, and it is almost all up front. In what follows, the ordering for the present $(\lambda, z, \mathtt{r})$ convention is given first, followed by the alternative ordering $(\mathtt{r}, \lambda, z)$ that corresponds to the standard convention.

Consider therefore: a cylindrical-polar coordinate system with coordinates ordered as $(\lambda, z, \mathtt{r})$ or $(\mathtt{r}, \lambda, z)$, and associated unit-vector triad, $\left(\mathbf{e}_\lambda, \mathbf{e}_\varphi, \mathbf{e}_\mathtt{r}\right)$ or $(\mathbf{e}_\mathtt{r}, \mathbf{e}_\lambda, \mathbf{e}_z)$, respectively; and a Cartesian coordinate system with coordinates (x_1, x_2, x_3) and associated unit-vector triad $(\mathbf{i}_1, \mathbf{i}_2, \mathbf{i}_3)$. See Fig. 5.4. Thus the general orthogonal-curvilinear coordinates (ξ_1, ξ_2, ξ_3) of Section 5.2.1 are replaced by the cylindrical-polar coordinates $(\lambda, z, \mathtt{r})$ or $(\mathtt{r}, \lambda, z)$ of the present section, that is,

$$(\xi_1, \xi_2, \xi_3) \to (\lambda, z, \mathtt{r}) \quad \text{or} \quad (\xi_1, \xi_2, \xi_3) \to (\mathtt{r}, \lambda, z) . \tag{5.28}$$

The cylindrical-polar coordinates of point P (see Fig. 5.4) are:[7]

- *Azimuthal position* [λ, measuring angular distance from the (dark-shaded) reference plane containing the reference axis Oz].
- *Axial position* (z, measuring distance along the reference axis Oz from origin O on that axis).
- *Radial distance* ($\mathtt{r}$, measured in a direction normal to the reference axis Oz).[8]

Each coordinate (λ, z, or $\mathtt{r}$) increases in the direction of its associated unit vector ($\mathbf{e}_\lambda$, $\mathbf{e}_z$, or $\mathbf{e}_\mathtt{r}$, respectively). The direction associated with z is aligned with the reference axis of a cylinder.

[7] The definitions of the coordinates themselves are independent of their ordering as a triad.

[8] Note the distinction that $\mathtt{r}$ denotes the radial coordinate of *cylindrical-polar* coordinates $(\lambda, z, \mathtt{r})$ or $(\mathtt{r}, \lambda, z)$, whereas r denotes the radial coordinate of *spherical-polar* coordinates (λ, ϕ, r); $\mathtt{r}$ versus r.

Examination of Fig. 5.4 shows that Cartesian and cylindrical-polar coordinate systems are related to one another by the coordinate-transformation equations

$$(x_1, x_2, x_3) = (\mathrm{r}\cos\lambda, \mathrm{r}\sin\lambda, z). \tag{5.29}$$

Using this equation and either mapping of (5.28) in (5.4), the (positive) metric (or scale) factors are

$$(h_\lambda, h_z, h_\mathrm{r}) = (\mathrm{r}, 1, 1). \tag{5.30}$$

From (5.2), (5.3), and either mapping of (5.28), the distance metric (ds) and volume element (dV) of a cylindrical-polar coordinate system are then given by

$$(ds)^2 = (h_\lambda d\lambda)^2 + (h_z dz)^2 + (h_\mathrm{r} d\mathrm{r})^2 = \mathrm{r}^2 (d\lambda)^2 + (dz)^2 + (d\mathrm{r})^2, \tag{5.31}$$

$$dV = (h_\lambda d\lambda)(h_z dz)(h_\mathrm{r} d\mathrm{r}) = \mathrm{r} d\lambda dz d\mathrm{r}. \tag{5.32}$$

Inserting either mapping of (5.28) together with metric factors (5.30) into (5.5), the velocity vector in a cylindrical-polar coordinate system can then be written in terms of its components as

$$\mathbf{u} = u_\lambda \mathbf{e}_\lambda + u_z \mathbf{e}_z + u_\mathrm{r} \mathbf{e}_\mathrm{r} \equiv h_\lambda \frac{D\lambda}{Dt}\mathbf{e}_\lambda + h_z \frac{Dz}{Dt}\mathbf{e}_z + h_\mathrm{r}\frac{D\mathrm{r}}{Dt}\mathbf{e}_\mathrm{r} = \mathrm{r}\frac{D\lambda}{Dt}\mathbf{e}_\lambda + \frac{Dz}{Dt}\mathbf{e}_z + \frac{D\mathrm{r}}{Dt}\mathbf{e}_\mathrm{r}. \tag{5.33}$$

For a generic vector $\mathbf{A} = (A_\lambda, A_z, A_\mathrm{r}) \equiv A_\lambda \mathbf{e}_\lambda + A_z \mathbf{e}_z + A_\mathrm{r}\mathbf{e}_\mathrm{r}$ (or, equivalently, $\mathbf{A} = (A_\mathrm{r}, A_\lambda, A_z) \equiv A_\mathrm{r}\mathbf{e}_\mathrm{r} + A_\lambda \mathbf{e}_\lambda + A_z \mathbf{e}_z$), and a generic scalar f, and using (5.28) and metric factors (5.30) in (5.6)–(5.10), the *gradient* (∇f), *divergence* ($\nabla \cdot \mathbf{A}$), *curl* ($\nabla \times \mathbf{A}$), *Laplacian* ($\nabla^2 f$), and *advection* ($\mathbf{A}\cdot\nabla f$) operations, may be respectively written in cylindrical-polar coordinates as:

$$\nabla f = \frac{1}{\mathrm{r}}\frac{\partial f}{\partial \lambda}\mathbf{e}_\lambda + \frac{\partial f}{\partial z}\mathbf{e}_z + \frac{\partial f}{\partial \mathrm{r}}\mathbf{e}_\mathrm{r}, \tag{5.34}$$

$$\nabla \cdot \mathbf{A} = \frac{1}{\mathrm{r}}\frac{\partial A_\lambda}{\partial \lambda} + \frac{\partial A_z}{\partial z} + \frac{1}{\mathrm{r}}\frac{\partial}{\partial \mathrm{r}}(\mathrm{r}A_\mathrm{r}), \tag{5.35}$$

$$\nabla \times \mathbf{A} = \left(\frac{\partial A_\mathrm{r}}{\partial z} - \frac{\partial A_z}{\partial \mathrm{r}}\right)\mathbf{e}_\lambda + \frac{1}{\mathrm{r}}\left[\frac{\partial}{\partial \mathrm{r}}(\mathrm{r}A_\lambda) - \frac{\partial A_\mathrm{r}}{\partial \lambda}\right]\mathbf{e}_z + \left(\frac{1}{\mathrm{r}}\frac{\partial A_z}{\partial \lambda} - \frac{\partial A_\lambda}{\partial z}\right)\mathbf{e}_\mathrm{r}, \tag{5.36}$$

$$\nabla^2 f = \frac{1}{\mathrm{r}^2}\frac{\partial^2 f}{\partial \lambda^2} + \frac{\partial^2 f}{\partial z^2} + \frac{1}{\mathrm{r}}\frac{\partial}{\partial \mathrm{r}}\left(\mathrm{r}\frac{\partial f}{\partial \mathrm{r}}\right), \tag{5.37}$$

$$\mathbf{A}\cdot\nabla f = \frac{A_\lambda}{\mathrm{r}}\frac{\partial f}{\partial \lambda} + A_z\frac{\partial f}{\partial z} + A_\mathrm{r}\frac{\partial f}{\partial \mathrm{r}}. \tag{5.38}$$

Using (2.8) and (5.38), the *material derivative of a scalar*, f, expressed in cylindrical-polar coordinates, is

$$\frac{Df}{Dt} = \left(\frac{\partial}{\partial t} + \frac{u_\lambda}{\mathrm{r}}\frac{\partial}{\partial \lambda} + u_z\frac{\partial}{\partial z} + u_\mathrm{r}\frac{\partial}{\partial \mathrm{r}}\right)f. \tag{5.39}$$

Using (5.34) and (5.36) in vector identity (A.23), the *material derivative of velocity vector*, **u** (i.e. the acceleration vector), in cylindrical-polar coordinates is thus

$$\frac{D\mathbf{u}}{Dt}=\left(\frac{Du_\lambda}{Dt}+\frac{u_\lambda u_\mathrm{r}}{\mathrm{r}}\right)\mathbf{e}_\lambda+\frac{Du_z}{Dt}\mathbf{e}_z+\left(\frac{Du_\mathrm{r}}{Dt}-\frac{u_\lambda^2}{\mathrm{r}}\right)\mathbf{e}_\mathrm{r}. \tag{5.40}$$

Outside a uniform circular cylinder of radius $\mathrm{r}=\mathrm{r}_S$, of infinite length, and rotating with constant angular frequency Ω about its axis,

$$\Phi(\mathrm{r})=V(\mathrm{r})-\frac{\Omega^2\mathrm{r}^2}{2}=\mathrm{r}_S g_S^N\ln\left(\frac{\mathrm{r}}{\mathrm{r}_S}\right)-\frac{\Omega^2\left(\mathrm{r}^2-\mathrm{r}_S^2\right)}{2}, \tag{5.41}$$

$$\Downarrow$$

$$\mathbf{g}(\mathrm{r})\equiv-\frac{d\Phi}{d\mathrm{r}}\mathbf{e}_\mathrm{r}=-\left(g_S^N\frac{\mathrm{r}_S}{\mathrm{r}}-\Omega^2\mathrm{r}\right)\mathbf{e}_\mathrm{r}. \tag{5.42}$$

Here:

- $\Phi(\mathrm{r})$ is geopotential.
- $V(\mathrm{r})$ is Newtonian gravitational potential.
- $g_S^N\equiv|\nabla V|_{\mathrm{r}=\mathrm{r}_S}$ is the absolute value of Newtonian gravity on the cylinder's surface, $\mathrm{r}=\mathrm{r}_S$.[9]

5.2.4 Geodetic Coordinates $(\lambda,\phi,\mathrm{h})$

Consider a geodetic coordinate system (WGS 84, 2004; Torge, 2001; White and Inverarity, 2012) with coordinates $(\lambda,\phi,\mathrm{h})$ and associated unit-vector triad $\left(\mathbf{e}_\lambda,\mathbf{e}_\phi,\mathbf{e}_\mathrm{h}\right)$; and a Cartesian coordinate system with coordinates (x_1,x_2,x_3) and associated unit-vector triad $(\mathbf{i}_1,\mathbf{i}_2,\mathbf{i}_3)$. See Fig. 5.5. Thus the general orthogonal-curvilinear coordinates (ξ_1,ξ_2,ξ_3) of Section 5.2.1 are replaced by the geodetic coordinates $(\lambda,\phi,\mathrm{h})$ of the present section, that is,

$$(\xi_1,\xi_2,\xi_3)\rightarrow(\lambda,\phi,\mathrm{h}). \tag{5.43}$$

This coordinate system is employed in many geodetic applications; see Section 12.2. For example, it is used to specify the position of satellite platforms that provide remotely sensed observations of Earth's atmosphere and oceans. Earth's WGS 84 reference ellipsoid[10] (WGS 84, 2004) provides the zero-elevation surface for the definition of geodetic coordinates $(\lambda,\phi,\mathrm{h})$, where (Fig. 5.5):

- λ is *longitude*, measured with respect to the WGS 84 reference meridian.
- ϕ is *geographic latitude*,[11] that is, the angle between the normal at a point on the reference ellipsoid and the equatorial plane.
- h is *elevation*, that is, distance along the normal from a point on the reference ellipsoid with (horizontal, two-dimensional) coordinates (λ,ϕ).

Each coordinate (λ, ϕ, or h) increases in the direction of its associated unit vector ($\mathbf{e}_\lambda$, $\mathbf{e}_\phi$, or $\mathbf{e}_\mathrm{h}$, respectively).

It can be shown (Torge, 2001; White and Inverarity, 2012; and Section 12.2.5 herein) that the Cartesian and geodetic coordinate systems are related to one another by the coordinate-transformation equations

[9] Equations (5.41) and (5.42) are exact for this problem; no geopotential approximation is required. To obtain them, $\nabla^2V(\mathrm{r})=(1/\mathrm{r})\,d\left(\mathrm{r}dV/d\mathrm{r}\right)/d\mathrm{r}=0$ is solved for $V(\mathrm{r})$ outside the cylinder (for $\mathrm{r}\geq\mathrm{r}_S$). This introduces an arbitrary constant; this constant is determined by imposing the arbitrary condition $\Phi(\mathrm{r}=\mathrm{r}_S)=0$. The procedure summarised here is the adaptation to cylindrical geometry of that outlined in Section 8.4 for spheroidal and spherical geometry.

[10] Nomenclature: Various conventions regarding the use of the terms spheroid and ellipsoid may be found in the literature. The convention adopted herein is that a *spheroid* is an approximately spherical solid of revolution, having an *almost*, or *precisely*, elliptic cross section in any meridional plane. When this cross section is *precisely* elliptic, the solid is then termed an *ellipsoid*.

[11] Geographic latitude is also known as geodetic latitude.

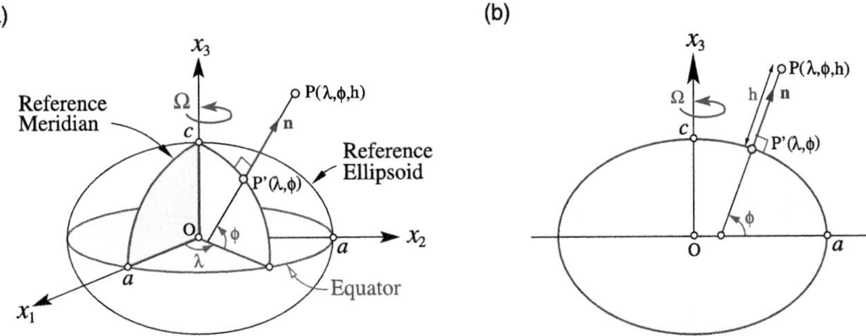

Figure 5.5 Geodetic coordinates $(\lambda, \phi, \mathrm{h})$, where λ is longitude, ϕ is geographic (geodetic) latitude, and h is elevation (distance from the WGS 84 reference ellipsoid along a straight-line normal, $\mathbf{n}(\lambda, \phi)$). Point P $(\lambda, \phi, \mathrm{h})$ is a general point. Line P′P is oriented along the normal to the reference ellipsoid that passes through point P $(\lambda, \phi, \mathrm{h})$. Point P′ (λ, ϕ) lies on the surface of the reference ellipsoid. The axes of the WGS 84 coordinate system (x_1, x_2, x_3) and its reference meridian are also depicted. The reference-meridional plane $(\lambda = 0)$ is dark shaded (in blue), and the meridional plane containing point P is light shaded (in blue). The equatorial plane is shaded in magenta. See text for further details.

$$x_1 = \left[\frac{a}{\left(1 - e^2 \sin^2 \phi\right)^{1/2}} + \mathrm{h} \right] \cos \lambda \cos \phi, \tag{5.44}$$

$$x_2 = \left[\frac{a}{\left(1 - e^2 \sin^2 \phi\right)^{1/2}} + \mathrm{h} \right] \sin \lambda \cos \phi, \tag{5.45}$$

$$x_3 = \left[\frac{a\left(1 - e^2\right)}{\left(1 - e^2 \sin^2 \phi\right)^{1/2}} + \mathrm{h} \right] \sin \phi, \tag{5.46}$$

where (Section 12.2.4) a and c are the major and minor semi-axes, respectively, of the WGS 84 reference ellipsoid[12] that represents the 'Figure of the Earth', and

$$e \equiv \frac{\left(a^2 - c^2\right)^{1/2}}{a} \tag{5.47}$$

is its eccentricity[13]. Coordinate-transformation equations (5.44)–(5.46) are equivalent to (12.7) of Section 12.2.5, but written here in White and Inverarity (2012)'s notation. (Their form is more convenient for computing metric factors.)

Using (5.43)–(5.46) in (5.4), the (positive) metric (or scale) factors are (White and Inverarity, 2012):

$$h_\lambda = \left[\frac{a}{\left(1 - e^2 \sin^2 \phi\right)^{1/2}} + \mathrm{h} \right] \cos \phi, \quad h_\phi = \frac{a\left(1 - e^2\right)}{\left(1 - e^2 \sin^2 \phi\right)^{3/2}} + \mathrm{h}, \quad h_{\mathrm{h}} = 1. \tag{5.48}$$

12 Only Earth's surface is ellipsoidal; all other horizontal coordinate surfaces of geodetic coordinates are spheroidal (i.e. approximately ellipsoidal), rather than precisely ellipsoidal (White and Inverarity, 2012).

13 e is sometimes called the *first* eccentricity, the *second* eccentricity being $e' \equiv \left(a^2 - c^2\right)^{1/2} / c$.

From (5.2), (5.3), and (5.48), the distance metric (ds) and volume element (dV) of the geodetic coordinate system are then given by

$$(ds)^2 = \left[\frac{a}{\left(1-e^2\sin^2\phi\right)^{1/2}} + \mathsf{h}\right]^2 \cos^2\phi\,(d\lambda)^2 + \left[\frac{a\left(1-e^2\right)}{\left(1-e^2\sin^2\phi\right)^{3/2}} + \mathsf{h}\right]^2 d\phi^2 + (d\mathsf{h})^2, \tag{5.49}$$

$$dV = \left[\frac{a}{\left(1-e^2\sin^2\phi\right)^{1/2}} + \mathsf{h}\right]\left[\frac{a\left(1-e^2\right)}{\left(1-e^2\sin^2\phi\right)^{3/2}} + \mathsf{h}\right]\cos\phi\, d\lambda\, d\phi\, d\mathsf{h}. \tag{5.50}$$

Inserting (5.43) and metric factors (5.48) into (5.5), the velocity vector in a geodetic coordinate system can then be written in terms of its components as

$$\begin{aligned}\mathbf{u} &= u_\lambda \mathbf{e}_\lambda + u_\phi \mathbf{e}_\phi + u_{\mathsf{h}} \mathbf{e}_{\mathsf{h}} \equiv h_\lambda \frac{D\lambda}{Dt}\mathbf{e}_\lambda + h_\phi \frac{D\phi}{Dt}\mathbf{e}_\phi + h_{\mathsf{h}} \frac{D\mathsf{h}}{Dt}\mathbf{e}_{\mathsf{h}} \\ &= \left[\frac{a}{\left(1-e^2\sin^2\phi\right)^{1/2}} + \mathsf{h}\right]\cos\phi \frac{D\lambda}{Dt}\mathbf{e}_\lambda + \left[\frac{a\left(1-e^2\right)}{\left(1-e^2\sin^2\phi\right)^{3/2}} + \mathsf{h}\right]\frac{D\phi}{Dt}\mathbf{e}_\phi + \frac{D\mathsf{h}}{Dt}\mathbf{e}_{\mathsf{h}}. \end{aligned} \tag{5.51}$$

For a generic vector $\mathbf{A} = \left(A_\lambda, A_\phi, A_{\mathsf{h}}\right) \equiv A_\lambda \mathbf{e}_\lambda + A_\phi \mathbf{e}_\phi + A_{\mathsf{h}} \mathbf{e}_{\mathsf{h}}$, and a generic scalar f, the *gradient* (∇f), *divergence* ($\nabla \cdot \mathbf{A}$), *curl* ($\nabla \times \mathbf{A}$), *Laplacian* ($\nabla^2 f$), and *advection* ($\mathbf{A} \cdot \nabla f$) operations, may be obtained in geodetic coordinates by inserting (5.43) and metric factors (5.48) into (5.6)–(5.10). The resulting expressions are somewhat messy and for this reason, and other reasons given later, they are not given here.

These expressions are not required in geodesy and navigation applications. However, by assuming that surfaces of constant h are geopotential surfaces (White and Inverarity, 2012), geodetic coordinates could be used as a geopotential coordinate system to model the atmosphere. This possibility is, however, of limited interest. White and Inverarity (2012) note that the magnitude of apparent gravity may be represented in a physically realistic manner as

$$g(\mathsf{h}) = \frac{g_a}{\left(1 + \mathsf{h}/a\right)^2}, \tag{5.52}$$

where g_a is a constant surface value of g at $\mathsf{h} = 0$. However, g cannot vary meridionally since the geopotential surfaces (on which $\mathsf{h} =$ constant) have equal normal separation at all latitudes. This is unphysical (albeit no worse than when the spherical-geopotential approximation is adopted, and better than using COS coordinates – see Section 5.2.5 and Table 8.2). As discussed in detail in Chapters 7, 8, and 12, everything boils down in practice to two 'extreme' options:

1. The simplicity, and limitations, of spherical-polar coordinates coupled with the spherical-geopotential approximation.
2. The less simple – but physically more realistic – use of the geopotential coordinate system and ellipsoidal-geopotential approximation developed in Chapter 12.[14]

[14] Other options of intermediate complexity are available as discussed in Chapters 7, 8, and 12. However, they offer no practical advantage, being more complex than Option 1, whilst being less realistic than Option 2. If Option 1 is insufficiently accurate, then Option 2 is the best way forward.

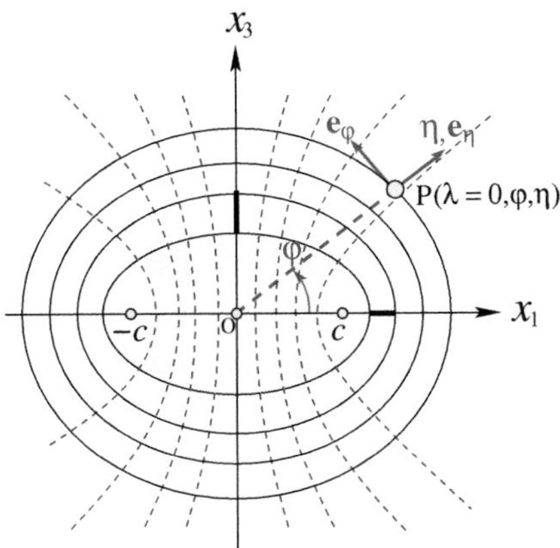

Figure 5.6 Confocal-oblate-spheroidal (COS) coordinates (λ, φ, η) in the meridional reference plane $(\lambda = 0)$ with associated unit-vector triad $(\mathbf{e}_\lambda, \mathbf{e}_\varphi, \mathbf{e}_\eta)$ at an arbitrary point P in this plane; and Cartesian coordinates (x_1, x_2, x_3) with associated unit-vector triad $(\mathbf{i}_1, \mathbf{i}_2, \mathbf{i}_3)$ at origin O. The common focal distance of the ellipses (solid contours in black) and hyperbolae (short-dashed contours in blue) is c. The long-dashed line (in red) denotes the asymptote of the hyperbola that passes through point P. The angle between this line and the equatorial plane is φ. The cross section of the planet is shaded (in light blue). The three-dimensional coordinate system is obtained by rotating the described two-dimensional family about the x_3 axis. See text for further details.

In a nutshell:

> Geodetic coordinates are essential for reporting and processing observational data (particularly remotely sensed observations from satellite platforms) of Earth's atmosphere and oceans, but of little or no practical use for modelling purposes since there are better alternatives.

5.2.5 Confocal-Oblate-Spheroidal (COS) Coordinates (λ, φ, η)

Consider: a confocal-oblate-spheroidal[15] (COS) coordinate system (Gates, 2004) with coordinates (λ, φ, η), and an associated unit-vector triad $(\mathbf{e}_\lambda, \mathbf{e}_\varphi, \mathbf{e}_\eta)$; and a Cartesian coordinate system with coordinates (x_1, x_2, x_3) and an associated unit-vector triad $(\mathbf{i}_1, \mathbf{i}_2, \mathbf{i}_3)$.[16] Thus the general orthogonal-curvilinear coordinates (ξ_1, ξ_2, ξ_3) of Section 5.2.1 are replaced by the COS coordinates (λ, φ, η) of the present section, that is,

$$(\xi_1, \xi_2, \xi_3) \rightarrow (\lambda, \varphi, \eta) . \tag{5.53}$$

This coordinate system is obtained by rotating a two-dimensional family of confocal ellipses and hyperbolae, with common focal distance c, about the non-focal axis of the ellipses. See Fig. 5.6.

The coordinates are:

- *Longitude* (λ), measuring angular distance eastwards from a reference meridian – as for spherical-polar and geodetic coordinates.
- *Asymptotic angle* (φ), measuring the angle between the equatorial plane and the asymptote of the hyperbola in the meridional plane that passes through a given point – northward from the equator of the coordinate system when positive, and southward when negative.

[15] Since the horizontal surfaces are ellipsoidal (i.e. of elliptic cross section), confocal-oblate-*spheroidal* coordinates could alternatively be called confocal-oblate-*ellipsoidal* coordinates.

[16] To avoid confusion, Gates (2004)'s coordinate ξ is replaced herein by η, and his (x, y, z) by (x_1, x_2, x_3).

- η, an indirect measure of outward distance along the hyperbola passing through a given point.

Each coordinate (λ, φ, or η) increases in the direction of its associated unit vector ($\mathbf{e}_\lambda$, $\mathbf{e}_\varphi$, or $\mathbf{e}_\eta$, respectively).

The Cartesian and COS coordinate systems are related to one another by the coordinate-transformation equations

$$(x_1, x_2, x_3) = \left(c \cosh\eta \cos\lambda \cos\varphi, c \cosh\eta \sin\lambda \cos\varphi, c \sinh\eta \sin\varphi\right), \tag{5.54}$$

where c is the common focal distance of the ellipses and hyperbolae.

Using (5.53) and (5.54) in (5.4), the (positive) metric (or scale) factors are:

$$\left(h_\lambda, h_\varphi, h_\eta\right) = \left[c \cosh\eta \cos\varphi, c\left(\sin^2\varphi + \sinh^2\eta\right)^{\frac{1}{2}}, c\left(\sin^2\varphi + \sinh^2\eta\right)^{\frac{1}{2}}\right]. \tag{5.55}$$

From (5.2), (5.3), (5.53), and (5.55), the distance metric (ds) and volume element (dV) of the COS coordinate system are then given by

$$(ds)^2 = c^2 \cosh^2\eta \cos^2\varphi \,(d\lambda)^2 + c^2\left(\sin^2\varphi + \sinh^2\eta\right)\left[(d\varphi)^2 + (d\eta)^2\right], \tag{5.56}$$

$$dV = c^3 \cosh\eta \cos\varphi \left(\sin^2\varphi + \sinh^2\eta\right) d\lambda d\varphi d\eta. \tag{5.57}$$

Inserting (5.53) and metric factors (5.55) into (5.5), the velocity vector in the COS coordinate system can then be written in terms of its components as

$$\begin{aligned}\mathbf{u} = u_\lambda \mathbf{e}_\lambda + u_\varphi \mathbf{e}_\varphi + u_\eta \mathbf{e}_\eta &\equiv h_\lambda \frac{D\lambda}{Dt}\mathbf{e}_\lambda + h_\varphi \frac{D\varphi}{Dt}\mathbf{e}_\varphi + h_\eta \frac{D\eta}{Dt}\mathbf{e}_\eta \\ &= c\cosh\eta\cos\varphi \frac{D\lambda}{Dt}\mathbf{e}_\lambda + c\left(\sin^2\varphi + \sinh^2\eta\right)^{\frac{1}{2}}\left(\frac{D\varphi}{Dt}\mathbf{e}_\varphi + \frac{D\eta}{Dt}\mathbf{e}_\eta\right).\end{aligned} \tag{5.58}$$

For a generic vector $\mathbf{A} = \left(A_\lambda, A_\varphi, A_\text{h}\right) \equiv A_\lambda \mathbf{e}_\lambda + A_\varphi \mathbf{e}_\varphi + A_\text{h}\mathbf{e}_\text{h}$, and a generic scalar f, the *gradient* (∇f), *divergence* ($\nabla \cdot \mathbf{A}$), *curl* ($\nabla \times \mathbf{A}$), *Laplacian* ($\nabla^2 f$), and *advection* ($\mathbf{A} \cdot \nabla f$) operations, may be obtained in COS coordinates by inserting (5.53) and metric factors (5.55) into (5.6)–(5.10). The resulting expressions are not given here.

At first sight, COS coordinates appear to be a very attractive proposition:

- They are analytically tractable.
- Earth's surface is approximated by an ellipsoid of revolution.

However, they are intrinsically based on a family of *confocal* ellipses. As a direct (and unfortunate) consequence, their horizontal coordinate surfaces (with the exception of Earth's surface) very poorly represent, *even qualitatively*, Earth's geopotential surfaces (White et al., 2008). In particular, because the polar separation of the coordinate surfaces is greater than the equatorial separation (see the thick bars on Fig. 5.6), the latitudinal variation of apparent gravity is *opposite in sign* to that observed. This fails the test of properly representing the underlying physics.

COS coordinates for atmospheric and oceanic modelling are consequently impractical and only of historical interest.[17]

5.3 SHALLOW ORTHOGONAL-CURVILINEAR COORDINATES

5.3.1 General Shallow Orthogonal-Curvilinear Coordinates (ξ_1, ξ_2, ξ_3)

Consider: a *shallow* orthogonal-curvilinear coordinate system with coordinates (ξ_1, ξ_2, ξ_3) and associated unit-vector triad $(\mathbf{e}_1, \mathbf{e}_2, \mathbf{e}_3)$; and a Cartesian coordinate system with coordinates (x_1, x_2, x_3) and associated unit-vector triad $(\mathbf{i}_1, \mathbf{i}_2, \mathbf{i}_3)$. See Fig. 5.1.

For *position (location) purposes,* these two sets of coordinates are assumed to be related to one another by invertible relations of the form $x_i = x_i(\xi_j), i, j = 1, 2, 3$ in Euclidean space. Just as for the general *deep* orthogonal case of Section 5.2.1, position vector $\mathbf{r}$ may be equivalently written in the two systems as (5.1), that is, as

$$\mathbf{r} = (\mathbf{r} \cdot \mathbf{i}_1)\,\mathbf{i}_1 + (\mathbf{r} \cdot \mathbf{i}_2)\,\mathbf{i}_2 + (\mathbf{r} \cdot \mathbf{i}_3)\,\mathbf{i}_3 \equiv x_1\mathbf{i}_1 + x_2\mathbf{i}_2 + x_3\mathbf{i}_3 = (\mathbf{r} \cdot \mathbf{e}_1)\,\mathbf{e}_1 + (\mathbf{r} \cdot \mathbf{e}_2)\,\mathbf{e}_2 + (\mathbf{r} \cdot \mathbf{e}_3)\,\mathbf{e}_3. \tag{5.59}$$

Thus

Navigation in shallow orthogonal-curvilinear coordinates is exactly the same as in deep orthogonal-curvilinear coordinates.

The distinguishing feature between deep and shallow geometries is the definition of the distance metric.

The distance metric determines *all* of the local, geometrical properties of space. These include:

- Curvature.
- The behaviour of geodesics.
- The definition of differential operations such as gradient, divergence, curl, Laplacian, and advection.

As discussed in Thuburn and White (2013), simplifying the definition of the distance metric results in the *Euclidean* space of the *deep* case becoming *non-Euclidean* for the *shallow* case. They pointed out that some of the consequences of the curved geometry implied by the shallow-atmosphere approximation are surprising and, importantly, that

'Euclidean space intuition cannot be relied upon in shallow-atmosphere geometry.'

And similarly, of course, for shallow-ocean geometry.

To put things on a sound footing, Thuburn and White (2013) showed that although shallow geometry is a curved, *non-Euclidean, three*-dimensional space, it can be embedded within a flat, *Euclidean, four*-dimensional space. The shallow, three-dimensional subspace then has the shallow metric (5.60).

[17] This does not mean, per se, that COS coordinates could not, *in principle*, be used as a geophysical coordinate system. They could, but then a realistic representation of apparent gravity would give rise to very large components in the horizontal (see Section 7.3.4). These would consequently have to be balanced by a correspondingly large, horizontal pressure gradient, and if this were not handled very accurately, the accuracy of the physical geostrophic balance would be correspondingly compromised. This is highly undesirable.

For the deep case (in Section 5.2.1), the metric factors (h_1, h_2, h_3) appearing in the *deep* distance metric (5.2) are obtained directly – by use of (5.4) – from the transformational relations, $x_i = x_i(\xi_j), i, j = 1, 2, 3$, between the orthogonal-curvilinear and Cartesian coordinate systems. By way of contrast, the distance metric in *shallow* orthogonal-curvilinear coordinates is obtained by simply setting the corresponding (*deep*) metric factors (h_1, h_2, h_3) in (5.2) to their values (h_1^S, h_2^S, h_3^S) *evaluated at the surface of the planet.*[18] This means that

$h_i^S = h_i^S(\xi_1, \xi_2), i = 1, 2, 3$, with no dependence of h_i^S on ξ_3, so that $\partial h_i^S / \partial \xi_3 \equiv 0$.

Thus:

The Distance Metric and Volume Element of a Shallow Orthogonal-Curvilinear Coordinate System

The distance metric (ds) and volume element (dV) of a *shallow* orthogonal-curvilinear coordinate system are given by

$$(ds)^2 \equiv (ds_1)^2 + (ds_2)^2 + (ds_3)^2 \equiv \left(h_1^S d\xi_1\right)^2 + \left(h_2^S d\xi_2\right)^2 + \left(h_3^S d\xi_3\right)^2, \tag{5.60}$$

$$dV \equiv ds_1 ds_2 ds_3 \equiv \left(h_1^S d\xi_1\right)\left(h_2^S d\xi_2\right)\left(h_3^S d\xi_3\right) \equiv h_1^S h_2^S h_3^S d\xi_1 d\xi_2 d\xi_3, \tag{5.61}$$

where $h_i^S = h_i^S(\xi_1, \xi_2), i = 1, 2, 3$.

See Sections 5.3.2 and 5.3.3 for two illustrative examples of shallow orthogonal-curvilinear coordinate systems.

It is seen from (5.60) and (5.61) that, contrary to their counterparts in deep geometry, the distance and volume metrics of shallow geometry are both *independent* of vertical coordinate ξ_3. This phenomenon is depicted pictorially *in Euclidean geometry* in Fig. 5.7. For *deep* geometry (Fig. 5.7a), vertical columns of fluid diverge as a function of increasing vertical coordinate ξ_3, with ever-increasing cross-sectional area $(h_1 h_2 d\xi_1 d\xi_2)$. However, in *shallow* geometry (Fig. 5.7b) they do not; vertical columns instead have *uniform* cross-sectional area $(h_1^S h_2^S d\xi_1 d\xi_2)$. Although there appear to be gaps between adjacent fluid columns in Fig. 5.7b for shallow geometry, this is an artefact of their depiction in deep (Euclidean) geometry. In geometrically deformed (non-Euclidean) geometry, there are no such gaps. This illustrates how intuitive reasoning for shallow (non-Euclidean) geometry can mislead. See Thuburn and White (2013) for further examples.

Note that Fig. 5.7 grossly exaggerates matters by depicting a domain whose vertical extent is several times that of the sphere's radius a. By contrast, the depth of Earth's atmosphere is instead but a small fraction – of $O\left(10^{-2}\right)$ – of a; and an even smaller one for an ocean. The fluid depth on Fig. 5.7 then reduces to the order of the thickness of the blue contour! Thus, quantitatively, shallow-fluid approximation for Earth's atmosphere and oceans is actually quite good for many (but not all) purposes.

Using (5.60), the velocity vector in a shallow orthogonal-curvilinear system can be written in terms of its components as

$$\mathbf{u} = u_1 \mathbf{e}_1 + u_2 \mathbf{e}_2 + u_3 \mathbf{e}_3 \equiv h_1^S \frac{D\xi_1}{Dt} \mathbf{e}_1 + h_2^S \frac{D\xi_2}{Dt} \mathbf{e}_2 + h_3^S \frac{D\xi_3}{Dt} \mathbf{e}_3 = \frac{Dx_1}{Dt} \mathbf{i}_1 + \frac{Dx_2}{Dt} \mathbf{i}_2 + \frac{Dx_3}{Dt} \mathbf{i}_3. \tag{5.62}$$

[18] (h_1, h_2, h_3) only need to be set to values $\left(h_1^{\text{ref}}, h_2^{\text{ref}}, h_3^{\text{ref}}\right)$ evaluated on some representative reference surface. For atmospheric and oceanic modelling purposes, it is both traditional and natural to choose this reference surface to be the (idealised) bounding surface of the planet.

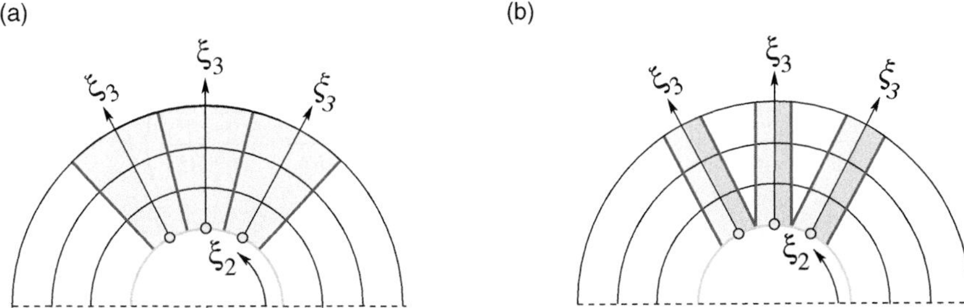

Figure 5.7 Grossly exaggerated depiction – using spherical-polar coordinates – of a hemispheric cross-section for: (a) deep (*Euclidean*) geometry, and (b) shallow (*non-Euclidean*) geometry. Three columns (in magenta) are shown for each geometry. Each column is centred around the vertical coordinate emanating from a yellow point lying on the surface of the (blue) hemisphere. See text for further details.

A crucial property of shallow orthogonal-curvilinear coordinates is that the usual vector identities involving gradient, divergence, and curl (as in the Appendix on Vector Identities at the end of this book) continue to hold (White et al., 2005), *but with use of shallow metric factors* $\left(h_1^S, h_2^S, h_3^S\right)$ *instead of deep ones* $\left(h_1, h_2, h_3\right)$. Thus, for a generic vector $\mathbf{A} = (A_1, A_2, A_3) \equiv A_1\mathbf{e}_1 + A_2\mathbf{e}_2 + A_3\mathbf{e}_3$ and a generic scalar f, and taking into consideration that $h_i^S = h_i^S(\xi_1, \xi_2)$, the *gradient* $(\nabla_S f)$, *divergence* $(\nabla_S \cdot \mathbf{A})$, *curl* $(\nabla_S \times \mathbf{A})$, *Laplacian* $(\nabla_S^2 f)$, and *advection* $(\mathbf{A} \cdot \nabla_S f)$ operations may be respectively written in shallow orthogonal-curvilinear coordinates as:

$$\nabla_S f = \frac{1}{h_1^S}\frac{\partial f}{\partial \xi_1}\mathbf{e}_1 + \frac{1}{h_2^S}\frac{\partial f}{\partial \xi_2}\mathbf{e}_2 + \frac{1}{h_3^S}\frac{\partial f}{\partial \xi_3}\mathbf{e}_3, \tag{5.63}$$

$$\nabla_S \cdot \mathbf{A} = \frac{1}{h_1^S h_2^S h_3^S}\left[\frac{\partial}{\partial \xi_1}\left(A_1 h_2^S h_3^S\right) + \frac{\partial}{\partial \xi_2}\left(A_2 h_3^S h_1^S\right)\right] + \frac{1}{h_3^S}\frac{\partial A_3}{\partial \xi_3}, \tag{5.64}$$

$$\begin{aligned}
\nabla_S \times \mathbf{A} = \frac{1}{h_1^S h_2^S h_3^S}\begin{vmatrix} h_1^S\mathbf{e}_1 & h_2^S\mathbf{e}_2 & h_3^S\mathbf{e}_3 \\ \frac{\partial}{\partial \xi_1} & \frac{\partial}{\partial \xi_2} & \frac{\partial}{\partial \xi_3} \\ h_1^S A_1 & h_2^S A_2 & h_3^S A_3 \end{vmatrix} &= \frac{1}{h_2^S h_3^S}\left[\frac{\partial}{\partial \xi_2}\left(h_3^S A_3\right) - h_2^S\frac{\partial A_2}{\partial \xi_3}\right]\mathbf{e}_1 \\
&+ \frac{1}{h_3^S h_1^S}\left[h_1^S\frac{\partial A_1}{\partial \xi_3} - \frac{\partial}{\partial \xi_1}\left(h_3^S A_3\right)\right]\mathbf{e}_2 \\
&+ \frac{1}{h_1^S h_2^S}\left[\frac{\partial}{\partial \xi_1}\left(h_2^S A_2\right) - \frac{\partial}{\partial \xi_2}\left(h_1^S A_1\right)\right]\mathbf{e}_3,
\end{aligned} \tag{5.65}$$

$$\nabla_S^2 f = \frac{1}{h_1^S h_2^S h_3^S}\left[\frac{\partial}{\partial \xi_1}\left(\frac{h_2^S h_3^S}{h_1^S}\frac{\partial f}{\partial \xi_1}\right) + \frac{\partial}{\partial \xi_2}\left(\frac{h_3^S h_1^S}{h_2^S}\frac{\partial f}{\partial \xi_2}\right)\right] + \frac{1}{\left(h_3^S\right)^2}\frac{\partial^2 f}{\partial \xi_3^2}, \tag{5.66}$$

$$\begin{aligned}
\mathbf{A} \cdot \nabla_S f &= (A_1\mathbf{e}_1 + A_2\mathbf{e}_2 + A_3\mathbf{e}_3) \cdot \left(\frac{1}{h_1^S}\frac{\partial f}{\partial \xi_1}\mathbf{e}_1 + \frac{1}{h_2^S}\frac{\partial f}{\partial \xi_2}\mathbf{e}_2 + \frac{1}{h_3^S}\frac{\partial f}{\partial \xi_3}\mathbf{e}_3\right) \\
&= \frac{A_1}{h_1^S}\frac{\partial f}{\partial \xi_1} + \frac{A_2}{h_2^S}\frac{\partial f}{\partial \xi_2} + \frac{A_3}{h_3^S}\frac{\partial f}{\partial \xi_3}.
\end{aligned} \tag{5.67}$$

Equations (5.63)–(5.67) are the shallow analogues of the corresponding deep relations (5.6)–(5.10).

Using (2.8) and (5.67), the *material derivative of a scalar, f*, expressed in shallow orthogonal-curvilinear coordinates, is

$$\frac{D_S f}{Dt} \equiv \frac{\partial f}{\partial t} + \mathbf{u} \cdot \nabla_S f = \left(\frac{\partial}{\partial t} + \frac{u_1}{h_1^S} \frac{\partial}{\partial \xi_1} + \frac{u_2}{h_2^S} \frac{\partial}{\partial \xi_2} + \frac{u_3}{h_3^S} \frac{\partial}{\partial \xi_3} \right) f. \tag{5.68}$$

Using (5.63) and (5.65) in the shallow version of vector identity (A.23), the *material derivative of the velocity vector*, **u** (i.e. the acceleration vector), in a shallow orthogonal-curvilinear system is thus

$$\begin{aligned}
\frac{D_S \mathbf{u}}{Dt} &= \frac{\partial \mathbf{u}}{\partial t} + (\mathbf{u} \cdot \nabla_S)\, \mathbf{u} \equiv \frac{\partial \mathbf{u}}{\partial t} + \frac{1}{2} \nabla_S (\mathbf{u} \cdot \mathbf{u}) + (\nabla_S \times \mathbf{u}) \times \mathbf{u} \\
&= \left[\frac{D_S u_1}{Dt} + \frac{u_2}{h_1^S h_2^S} \left(u_1 \frac{\partial h_1^S}{\partial \xi_2} - u_2 \frac{\partial h_2^S}{\partial \xi_1} \right) - \frac{u_3^2}{h_1^S h_3^S} \frac{\partial h_3^S}{\partial \xi_1} \right] \mathbf{e}_1 \\
&+ \left[\frac{D_S u_2}{Dt} - \frac{u_3^2}{h_2^S h_3^S} \frac{\partial h_3^S}{\partial \xi_2} + \frac{u_1}{h_2^S h_1^S} \left(u_2 \frac{\partial h_2^S}{\partial \xi_1} - u_1 \frac{\partial h_1^S}{\partial \xi_2} \right) \right] \mathbf{e}_2 \\
&+ \left[\frac{D_S u_3}{Dt} + \frac{u_3}{h_3^S} \left(\frac{u_1}{h_1^S} \frac{\partial h_3^S}{\partial \xi_1} + \frac{u_2}{h_2^S} \frac{\partial h_3^S}{\partial \xi_2} \right) \right] \mathbf{e}_3,
\end{aligned} \tag{5.69}$$

where D_S/Dt (of a scalar) is defined by (5.68). The terms not involving D_S/Dt on the last three lines of (5.12) are usually referred to as *metric terms* and are not to be confused with the *metric (scale) factors* h_1^S, h_2^S, and h_3^S.

Comparing (5.63)–(5.69) with their deep counterparts (5.6)–(5.12), it is seen that they are very similar, with the former being generally simpler and more compact than the latter.

For later convenience, and also for reference purposes, explicit expressions are now given for various quantities needed to express the governing atmospheric and oceanographic equations in shallow spherical-polar and cylindrical-polar coordinate systems.

5.3.2 Shallow Spherical-Polar Coordinates (λ, ϕ, r)

Consider: a *shallow* spherical-polar coordinate system with coordinates (λ, ϕ, r) and associated unit-vector triad $(\mathbf{e}_\lambda, \mathbf{e}_\phi, \mathbf{e}_r)$; and a Cartesian coordinate system with coordinates (x_1, x_2, x_3) and associated unit-vector triad $(\mathbf{i}_1, \mathbf{i}_2, \mathbf{i}_3)$. See Fig. 5.3. Thus the general, shallow, orthogonal-curvilinear coordinates (ξ_1, ξ_2, ξ_3) of Section 5.3.1 are replaced by the shallow, spherical-polar coordinates (λ, ϕ, r) of the present section, that is,

$$(\xi_1, \xi_2, \xi_3) \rightarrow (\lambda, \phi, r)\,. \tag{5.70}$$

The shallow, spherical-polar coordinates of point P are (see Fig. 5.3):

- *Longitude* (λ), measuring angular distance eastwards, from the (dark-shaded) reference meridional plane, along the latitude circle passing through point P.
- *Latitude* (ϕ), measuring angular distance northward (for positive ϕ) from the equator, along the meridian containing point P, and southward for negative ϕ.
- *Radial distance* (r), measured from origin O at the sphere's centre.

Each coordinate (λ, ϕ, or r) increases in the direction of its associated unit vector ($\mathbf{e}_\lambda$, $\mathbf{e}_\phi$, or $\mathbf{e}_r$, respectively). For navigation purposes, the Cartesian and spherical-polar coordinate systems are related to one another by the coordinate transformation equations (5.14) of the deep case, that is, by

$$(x_1, x_2, x_3) = (r \cos\lambda \cos\phi, r \sin\lambda \cos\phi, r \sin\phi). \tag{5.71}$$

The (positive) metric (or scale) factors for *deep* spherical-polar coordinates are given by (5.15), that is, by $(h_\lambda, h_\phi, h_r) = (r\cos\phi, r, 1)$. Setting r to its surface value $r = a$ then gives the corresponding shallow metric (or scale) factors:

$$\left(h_\lambda^S, h_\phi^S, h_r^S\right) = (a\cos\phi, a, 1), \tag{5.72}$$

for shallow spherical-polar coordinates. Using (5.70) and (5.72) in (5.60) and (5.61), the distance metric (ds) and volume element (dV) of a shallow, spherical-polar, coordinate system are thus

$$(ds)^2 = a^2\cos^2\phi\,(d\lambda)^2 + a^2\,(d\phi)^2 + (dr)^2, \tag{5.73}$$

$$dV = a^2\cos\phi\, d\lambda\, d\phi\, dr. \tag{5.74}$$

Inserting (5.70) and metric factors (5.72) into (5.62), the velocity vector in a shallow spherical-polar coordinate system can then be written in terms of its components as

$$\begin{aligned}\mathbf{u} = u_\lambda \mathbf{e}_\lambda + u_\phi \mathbf{e}_\phi + u_r \mathbf{e}_r &\equiv h_\lambda^S \frac{D\lambda}{Dt}\mathbf{e}_\lambda + h_\phi^S \frac{D\phi}{Dt}\mathbf{e}_\phi + h_r^S \frac{Dr}{Dt}\mathbf{e}_r \\ &= a\cos\phi\frac{D\lambda}{Dt}\mathbf{e}_\lambda + a\frac{D\phi}{Dt}\mathbf{e}_\phi + \frac{Dr}{Dt}\mathbf{e}_r.\end{aligned} \tag{5.75}$$

For a generic vector $\mathbf{A} = \left(A_\lambda, A_\phi, A_r\right) \equiv A_\lambda \mathbf{e}_\lambda + A_\phi \mathbf{e}_\phi + A_r \mathbf{e}_r$ and a generic scalar f, and using (5.70) and metric factors (5.72) in (5.63)–(5.67), the *gradient* ($\nabla_S f$), *divergence* ($\nabla_S \cdot \mathbf{A}$), *curl* ($\nabla_S \times \mathbf{A}$), *Laplacian* ($\nabla_S^2 f$), and *advection* ($\mathbf{A}\cdot\nabla_S f$) operations may be respectively written in shallow spherical-polar coordinates as:

$$\nabla_S f = \frac{1}{a\cos\phi}\frac{\partial f}{\partial\lambda}\mathbf{e}_\lambda + \frac{1}{a}\frac{\partial f}{\partial\phi}\mathbf{e}_\phi + \frac{\partial f}{\partial r}\mathbf{e}_r, \tag{5.76}$$

$$\nabla_S \cdot \mathbf{A} = \frac{1}{a\cos\phi}\left[\frac{\partial A_\lambda}{\partial\lambda} + \frac{\partial}{\partial\phi}\left(\cos\phi A_\phi\right)\right] + \frac{\partial A_r}{\partial r}, \tag{5.77}$$

$$\begin{aligned}\nabla_S \times \mathbf{A} = &\left(\frac{1}{a}\frac{\partial A_r}{\partial\phi} - \frac{\partial A_\phi}{\partial r}\right)\mathbf{e}_\lambda + \left(\frac{\partial A_\lambda}{\partial r} - \frac{1}{a\cos\phi}\frac{\partial A_r}{\partial\lambda}\right)\mathbf{e}_\phi \\ &+ \frac{1}{a\cos\phi}\left[\frac{\partial A_\phi}{\partial\lambda} - \frac{\partial}{\partial\phi}\left(\cos\phi A_\lambda\right)\right]\mathbf{e}_r,\end{aligned} \tag{5.78}$$

$$\nabla_S^2 f = \frac{1}{a^2\cos^2\phi}\frac{\partial^2 f}{\partial\lambda^2} + \frac{1}{a^2\cos\phi}\frac{\partial}{\partial\phi}\left(\cos\phi\frac{\partial f}{\partial\phi}\right) + \frac{\partial^2 f}{\partial r^2}, \tag{5.79}$$

$$\mathbf{A}\cdot\nabla_S f = \frac{A_\lambda}{a\cos\phi}\frac{\partial f}{\partial\lambda} + \frac{A_\phi}{a}\frac{\partial f}{\partial\phi} + A_r\frac{\partial f}{\partial r}. \tag{5.80}$$

Using (5.70) and (5.72) in (5.68), the *material derivative of a scalar, f*, expressed in shallow spherical-polar coordinates, is

$$\frac{D_S f}{Dt} \equiv \frac{\partial f}{\partial t} + \mathbf{u}\cdot\nabla_S f = \left(\frac{\partial}{\partial t} + \frac{u_\lambda}{a\cos\phi}\frac{\partial}{\partial\lambda} + \frac{u_\phi}{a}\frac{\partial}{\partial\phi} + u_r\frac{\partial}{\partial r}\right)f. \tag{5.81}$$

Using (5.70) and (5.72) in (5.69), the *material derivative of the velocity vector,* **u** (i.e. the acceleration vector), in shallow spherical-polar coordinates is thus

$$\frac{D_S \mathbf{u}}{Dt} = \left(\frac{D_S u_\lambda}{Dt} - \frac{u_\lambda u_\phi \tan\phi}{a} \right) \mathbf{e}_\lambda + \left(\frac{D_S u_\phi}{Dt} + \frac{u_\lambda^2 \tan\phi}{a} \right) \mathbf{e}_\phi + \frac{D_S u_r}{Dt} \mathbf{e}_r. \tag{5.82}$$

Outside a rotating spherical planet of radius a, and making the classical spherical-geopotential approximation (in shallow form) (see Sections 7.1.5 and 8.6),

$$\Phi(r) = g_a (r - a), \tag{5.83}$$

$$\Downarrow$$

$$\mathbf{g} \equiv -\nabla_S \Phi(r) = -\frac{d\Phi}{dr} \mathbf{e}_r = -g_a \mathbf{e}_r, \tag{5.84}$$

where:

- Φ is geopotential.
- $g_a \equiv |\nabla \Phi|_{r=a}$ is the absolute value of gravity on the surface ($r = a$) of the rotating planet.

5.3.3 Shallow Cylindrical-Polar Coordinates (r, λ, z) or (λ, z, r)

Consider (cf. the introductory paragraph of Section 5.2.3): a cylindrical-polar coordinate system with coordinates ordered as (λ, z, r) or (r, λ, z), and associated unit-vector triad $(\mathbf{e}_\lambda, \mathbf{e}_z, \mathbf{e}_\mathrm{r})$ or $(\mathbf{e}_\mathrm{r}, \mathbf{e}_\lambda, \mathbf{e}_z)$, respectively; and a Cartesian coordinate system with coordinates (x_1, x_2, x_3) and associated unit-vector triad $(\mathbf{i}_1, \mathbf{i}_2, \mathbf{i}_3)$. See Fig. 5.4. Thus the general shallow orthogonal-curvilinear coordinates (ξ_1, ξ_2, ξ_3) of Section 5.3.1 are replaced by the shallow cylindrical-polar coordinates (λ, z, r) or (r, λ, z) of the present section, that is,

$$(\xi_1, \xi_2, \xi_3) \rightarrow (\lambda, z, \mathrm{r}) \quad \text{or} \quad (\xi_1, \xi_2, \xi_3) \rightarrow (\mathrm{r}, \lambda, z). \tag{5.85}$$

The shallow cylindrical-polar coordinates of point P (see Fig. 5.4) are:[19]

- *Azimuthal position* (λ), measuring angular distance from the (dark-shaded) reference plane containing the reference axis Oz.
- *Axial position* (z), measuring distance along the reference axis Oz from origin O on that axis.
- *Radial distance* (r), measured in a direction normal to the reference axis Oz.

Each coordinate (λ, z, or r) increases in the direction of its associated unit vector ($\mathbf{e}_\lambda$, $\mathbf{e}_z$, or $\mathbf{e}_\mathrm{r}$, respectively). The direction associated with z is aligned with the reference axis of a cylinder.

For navigation purposes, Cartesian and cylindrical-polar coordinate systems are related to one another by the coordinate-transformation equations (5.29) of the deep case, that is, by

$$(x_1, x_2, x_3) = (\mathrm{r}\cos\lambda, \mathrm{r}\sin\lambda, z). \tag{5.86}$$

The (positive) metric (or scale) factors for *deep* cylindrical-polar coordinates are given by (5.30), that is, by $(h_\lambda, h_z, h_\mathrm{r}) = (\mathrm{r}, 1, 1)$. Setting r to its surface value $\mathrm{r} = \mathrm{r}_S$ then gives the corresponding *shallow* metric (or scale) factors:

$$\left(h_\lambda^S, h_z^S, h_\mathrm{r}^S\right) = (\mathrm{r}_S, 1, 1), \tag{5.87}$$

[19] The definitions of the coordinates themselves are independent of their ordering as a triad.

for *shallow* cylindrical coordinates. Using either mapping of (5.85) together with (5.87) in (5.60) and (5.61), the distance metric (ds) and volume element (dV) of a shallow cylindrical-polar coordinate system are thus

$$ds^2 \equiv \mathrm{r}_S^2 d\lambda^2 + dz^2 + d\mathrm{r}^2, \tag{5.88}$$

$$dV \equiv \mathrm{r}_S d\lambda dz d\mathrm{r}. \tag{5.89}$$

Inserting either mapping of (5.85) and metric factors (5.87) into (5.62), the velocity vector in a shallow cylindrical-polar coordinate system can then be written in terms of its components as

$$\mathbf{u} = u_\lambda \mathbf{e}_\lambda + u_z \mathbf{e}_z + u_\mathrm{r} \mathbf{e}_\mathrm{r} \equiv h_\lambda^S \frac{D\lambda}{Dt}\mathbf{e}_\lambda + h_z^S \frac{Dz}{Dt}\mathbf{e}_z + h_\mathrm{r}^S \frac{D\mathrm{r}}{Dt}\mathbf{e}_\mathrm{r} = \mathrm{r}_S \frac{D\lambda}{Dt}\mathbf{e}_\lambda + \frac{Dz}{Dt}\mathbf{e}_z + \frac{D\mathrm{r}}{Dt}\mathbf{e}_\mathrm{r}. \tag{5.90}$$

For a generic vector $\mathbf{A} = (A_\lambda, A_z, A_\mathrm{r}) \equiv A_\lambda \mathbf{e}_\lambda + A_z \mathbf{e}_z + A_\mathrm{r} \mathbf{e}_\mathrm{r}$ (or, equivalently, $\mathbf{A} = (A_\mathrm{r}, A_\lambda, A_z) \equiv A_\mathrm{r} \mathbf{e}_\mathrm{r} + A_\lambda \mathbf{e}_\lambda + A_z \mathbf{e}_z$) and a generic scalar f, and using (5.85) and metric factors (5.87) in (5.63)–(5.67), the *gradient* ($\nabla_S f$), *divergence* ($\nabla_S \cdot \mathbf{A}$), *curl* ($\nabla_S \times \mathbf{A}$), *Laplacian* ($\nabla_S^2 f$), and *advection* ($\mathbf{A} \cdot \nabla_S f$) operations may be respectively written in shallow cylindrical-polar coordinates as:

$$\nabla_S f = \frac{1}{\mathrm{r}_S}\frac{\partial f}{\partial \lambda}\mathbf{e}_\lambda + \frac{\partial f}{\partial z}\mathbf{e}_z + \frac{\partial f}{\partial \mathrm{r}}\mathbf{e}_\mathrm{r}, \tag{5.91}$$

$$\nabla_S \cdot \mathbf{A} = \frac{1}{\mathrm{r}_S}\frac{\partial A_\lambda}{\partial \lambda} + \frac{\partial A_z}{\partial z} + \frac{\partial A_\mathrm{r}}{\partial \mathrm{r}}, \tag{5.92}$$

$$\nabla_S \times \mathbf{A} = \left(\frac{\partial A_\mathrm{r}}{\partial z} - \frac{\partial A_z}{\partial \mathrm{r}}\right)\mathbf{e}_\lambda + \left(\frac{\partial A_\lambda}{\partial \mathrm{r}} - \frac{1}{\mathrm{r}_S}\frac{\partial A_\mathrm{r}}{\partial \lambda}\right)\mathbf{e}_z + \left(\frac{1}{\mathrm{r}_S}\frac{\partial A_z}{\partial \lambda} - \frac{\partial A_\lambda}{\partial z}\right)\mathbf{e}_\mathrm{r}, \tag{5.93}$$

$$\nabla_S^2 f = \frac{1}{\mathrm{r}_S^2}\frac{\partial^2 f}{\partial \lambda^2} + \frac{\partial^2 f}{\partial z^2} + \frac{\partial^2 f}{\partial \mathrm{r}^2}, \tag{5.94}$$

$$\mathbf{A} \cdot \nabla_S f = \frac{A_\lambda}{\mathrm{r}_S}\frac{\partial f}{\partial \lambda} + A_z \frac{\partial f}{\partial z} + A_\mathrm{r}\frac{\partial f}{\partial \mathrm{r}}. \tag{5.95}$$

Using (5.85) and (5.87) in (5.68), the *material derivative of a scalar*, f, expressed in shallow cylindrical-polar coordinates, is

$$\frac{D_S f}{Dt} = \left(\frac{\partial}{\partial t} + \frac{u_\lambda}{\mathrm{r}_S}\frac{\partial}{\partial \lambda} + u_z \frac{\partial}{\partial z} + u_\mathrm{r}\frac{\partial}{\partial \mathrm{r}}\right) f. \tag{5.96}$$

Using (5.85) and (5.87) in (5.69), the *material derivative of the velocity vector*, $\mathbf{u}$ (i.e. the acceleration vector), in shallow cylindrical-polar coordinates is thus

$$\frac{D_S \mathbf{u}}{Dt} = \left(\frac{D_S u_\lambda}{Dt} + \frac{u_\lambda u_\mathrm{r}}{\mathrm{r}_S}\right)\mathbf{e}_\lambda + \frac{D_S u_z}{Dt}\mathbf{e}_z + \frac{D_S u_3}{Dt}\mathbf{e}_\mathrm{r}. \tag{5.97}$$

Outside a uniform circular cylinder, of radius r_S, of infinite length, and rotating with constant angular frequency Ω about its axis,

$$\Phi(\mathrm{r}) = \left(g_S^N - \Omega^2 \mathrm{r}_S\right)(\mathrm{r} - \mathrm{r}_S) = g_S(\mathrm{r} - \mathrm{r}_S), \tag{5.98}$$

$$\Downarrow$$

$$\mathbf{g} \equiv -\nabla\Phi(\mathrm{r}) = -\frac{d\Phi}{d\mathrm{r}}\mathbf{e}_\mathrm{r} = -\left(g_S^N - \Omega^2 \mathrm{r}_S\right)\mathbf{e}_\mathrm{r} = -g_S \mathbf{e}_\mathrm{r}. \tag{5.99}$$

Here:

- $\Phi(\mathrm{r})$ is geopotential.

- $V(\mathtt{r})$ is Newtonian gravitational potential.
- g_S^N and $g_S \equiv g_S^N - \Omega^2 \mathtt{r}_S$ are the absolute values of Newtonian gravity and (total) gravity, respectively, on the cylinder's surface, $\mathtt{r} = \mathtt{r}_S$.

To obtain (5.98) and (5.99), the procedure given in footnote 9 is adapted to shallow geometry.[20] The impact of shallow geometry is to everywhere replace Newtonian gravity and the centrifugal force by their values at the cylinder's surface.

5.4 CONCLUDING REMARKS

In principle, GREAT (Geophysically Realistic, Ellipsoidal, Analytically Tractable) coordinates could have been included in Section 5.2. Instead they are developed in great(!) detail in Chapter 12.

Furthermore, similar-oblate-spheroidal (SOS) coordinates (White et al., 2008) could also have been included in Section 5.2. However, as formulated in White et al. (2008), these coordinates are singular at the equator of the coordinate system. This is not a fundamental problem; it is simply an artefact of the White et al. (2008) implementation, due to the introduction and use of an auxiliary coordinate. This limitation is addressed at source in the development of GREAT coordinates (in Chapter 12). Non-singular SOS coordinates then arise as a special case (namely with parameter ν set identically zero) of (non-singular) GREAT coordinates.

Having documented various orthogonal-curvilinear coordinate systems, we are now ready to recall the governing equations – expressed in vector form – of Chapter 4 for general geophysical fluids, and to re-express them in axial-orthogonal-curvilinear coordinates. This challenge is met in Chapter 6.

[20] $\nabla_S^2 V(\mathtt{r}) = d^2 V / d\mathtt{r}^2 = 0$ is solved (in *shallow* geometry) for $V(\mathtt{r})$ outside the cylinder (for $\mathtt{r} \geq \mathtt{r}_S$). This introduces an arbitrary constant; this constant is again determined by imposing the arbitrary condition $\Phi(\mathtt{r} = \mathtt{r}_S) = 0$. Equations (5.98) and (5.99) are approximate for this physical problem; see footnote 9 for this problem in deep geometry.

6

Governing Equations for Motion of Geophysical Fluids: Curvilinear Form

ABSTRACT

A set of governing equations for geophysical fluids (primarily gaseous for the atmosphere, and liquid for the oceans) is recalled from Chapter 4. It is expressed in vector form in a rotating frame of reference. To express this set in general orthogonal-curvilinear coordinates is, for the most part, straightforward. Quantities appearing in these equations, with one exception, can be rewritten in these general coordinates using expressions given in Chapter 5 for this purpose. The exception is the Coriolis force of the momentum equation. To maintain generality and flexibility, a general representation of the Coriolis force is needed, rather than a coordinate-specific one. Using metric (scale) factors, such a representation is developed for general axial-orthogonal-curvilinear coordinates (obtained by rotation of a 2D orthogonal system about an axis). Using this representation – together with expressions given in Chapter 5 – the governing equations are re-expressed in general axial-orthogonal-curvilinear coordinates. As corollaries, the governing equations are given for the special cases of spherical-polar coordinates and cylindrical-polar coordinates. In the absence of forcings other than gravity, the momentum-component equations are more compactly written as Euler–Lagrange equations.

6.1 PREAMBLE

A set of governing equations for general fluids (primarily gaseous for the atmosphere, and liquid for the oceans) is given in vector form in Section 4.4. To express this set in general orthogonal-curvilinear coordinates is, for the most part, straightforward. Quantities appearing in these equations, *with one exception*, can be rewritten in these general coordinates by using expressions given in Chapter 5.2.1 for this purpose. The exception is the Coriolis force ($-2\boldsymbol{\Omega}\times\mathbf{u}$) that appears on the right-hand side of momentum equation (4.95).

For some orthogonal-curvilinear coordinate systems, it is straightforward to express $-2\boldsymbol{\Omega}\times\mathbf{u}$ in terms of their coordinates. In spherical-polar coordinates (λ, ϕ, r), for example, $-2\boldsymbol{\Omega}\times\mathbf{u}$ can be written (in the notation of Section 5.2.2) as

$$-2\boldsymbol{\Omega}\times\mathbf{u} = 2\Omega\left[\left(u_\phi \sin\phi - u_r\cos\phi\right)\mathbf{e}_\lambda - u_\lambda \sin\phi\,\mathbf{e}_\phi + u_\lambda\cos\phi\,\mathbf{e}_r\right]. \qquad (6.1)$$

The corresponding expression in cylindrical-polar coordinates (λ, z, r) is (in the notation of Section 5.2.3)

$$-2\boldsymbol{\Omega}\times\mathbf{u} = 2\Omega\left(-u_{\mathrm{r}}\mathbf{e}_\lambda + u_\lambda\mathbf{e}_{\mathrm{r}}\right). \qquad (6.2)$$

However, for other (important) orthogonal-curvilinear coordinate systems of interest (such as ellipsoidal coordinates, obtained by rotation of a 2D, elliptic, coordinate system about a planet's rotation axis), things are much less straightforward. To maintain generality and flexibility, what is needed is a *general* representation of $-2\boldsymbol{\Omega}\times\mathbf{u}$, rather than a *coordinate-specific* one such as (6.1) or (6.2). Using metric (scale) factors, White and Wood (2012) have shown how to accomplish this for general *axial*-orthogonal-curvilinear coordinates (obtained by rotation of a 2D orthogonal system about an axis).

Their assumption that the coordinate system be *axially symmetric* is crucially important. This is related to the representation of the geopotential, and to the importance of geopotential-coordinate systems; see Sections 6.5 and 7.3 and Chapter 12. In practice, *the geopotential should be axially symmetric.* If it is not, then gravity (which is normal to geopotential surfaces) does not (in general) act within meridional planes. This significantly complicates the development of a geopotential coordinate system, particularly regarding the coordinates to be used within its geopotential surfaces.

The remainder of this chapter is organised as follows. To set the scene (in Section 6.2), the set of governing equations for general fluids is recalled from Section 4.4. To prepare the way forward, axial-orthogonal-curvilinear coordinates are described in Section 6.3. Next (in Section 6.4), the crucial step of representing the Coriolis force ($-2\boldsymbol{\Omega}\times\mathbf{u}$) in terms of the metric (scale) factors of general axial-orthogonal-curvilinear coordinates is developed. Armed with this representation, and using expressions given in Section 5.2.1 for various differential operations, the governing equations are then written (in Section 6.5) in terms of general axial-orthogonal-curvilinear coordinates. As a corollary (and also to relate the present, general representation to the specific one usually adopted in geophysical-fluid-dynamics textbooks), the governing equations are expressed (in Appendix A to the present chapter) in spherical-polar coordinates. They are also similarly expressed in cylindrical-polar coordinates in Appendix B. In the absence of forcings other than gravity, the momentum-component equations are written more compactly in Section 6.7 as Euler–Lagrange equations. Finally, some concluding remarks are made in Section 6.8.

6.2 THE GOVERNING EQUATIONS IN VECTOR FORM

A set of governing equations for general, rotating, geophysical fluids, expressed in the notation of Chapter 4 and in vector form in a rotating frame of reference – see (4.95)–(4.102), (2.21), and (2.22) – is as follows:

Momentum

$$\frac{D\mathbf{u}}{Dt} = -2\boldsymbol{\Omega}\times\mathbf{u} - \frac{1}{\rho}\nabla p - \nabla\Phi + \mathbf{F}. \tag{6.3}$$

Continuity of Total Mass Density

$$\frac{D\rho}{Dt} + \rho\nabla\cdot\mathbf{u} = F^{\rho}. \tag{6.4}$$

Substance Transport

$$\frac{DS^i}{Dt} = \dot{S}^i, \quad i = 1, 2, \ldots, \tag{6.5}$$

where S^i and $\dot{S}^i$ are as in (4.10) (for the atmosphere) and (4.11) (for oceans).

Thermodynamic-Energy Equation

$$\frac{D\mathscr{E}}{Dt} + p\frac{D\alpha}{Dt} = \dot{Q}_E \quad \text{or} \quad \frac{D\eta}{Dt} = \frac{\dot{Q}}{T}, \tag{6.6}$$

where $\dot{Q}$ is heating rate (per unit mass), and $\dot{Q}_E \equiv \dot{Q} + \sum_i \mu^i \dot{S}^i$ is total rate of energy input.

Fundamental Equation of State

$$\mathscr{E} = \mathscr{E}\left(\alpha, \eta, S^1, S^2, \ldots\right) \quad \text{or} \quad \eta = \eta\left(\alpha, \mathscr{E}, S^1, S^2, \ldots\right), \tag{6.7}$$

where $\alpha \equiv 1/\rho$.

Definition of T, p, and μ^i

$$T = \frac{\partial \mathscr{E}\left(\alpha, \eta, S^1, S^2, \ldots\right)}{\partial \eta} \quad \text{or} \quad T = \left[\frac{\partial \eta\left(\alpha, \mathscr{E}, S^1, S^2, \ldots\right)}{\partial \mathscr{E}}\right]^{-1}, \tag{6.8}$$

$$p = -\frac{\partial \mathscr{E}\left(\alpha, \eta, S^1, S^2, \ldots\right)}{\partial \alpha} \quad \text{or} \quad p = T\frac{\partial \eta\left(\alpha, \mathscr{E}, S^1, S^2, \ldots\right)}{\partial \alpha}, \tag{6.9}$$

$$\mu^i = \frac{\partial \mathscr{E}\left(\alpha, \eta, S^1, S^2, \ldots\right)}{\partial S^i} \quad \text{or} \quad \mu^i = -T\frac{\partial \eta\left(\alpha, \mathscr{E}, S^1, S^2, \ldots\right)}{\partial S^i}. \tag{6.10}$$

Kinematic Equation

$$\mathbf{u} = \frac{D\mathbf{r}}{Dt}. \tag{6.11}$$

Material Derivative

$$\frac{D}{Dt} = \frac{\partial}{\partial t} + \mathbf{u} \cdot \nabla. \tag{6.12}$$

6.3 AXIAL-ORTHOGONAL-CURVILINEAR COORDINATES

An axial-orthogonal-curvilinear coordinate system (ξ_1, ξ_2, ξ_3) is defined by rotating a 2D orthogonal-curvilinear coordinate system (ξ_2, ξ_3) about an axis. See Fig. 6.1. In the present atmospheric and oceanic context, this axis is either a planet's rotation axis or the axis of a cylinder. The ξ_1 coordinate (usually denoted by λ) then varies around this axis from 0 to 2π. The other two coordinates (ξ_2 and ξ_3) vary in any $\xi_1 =$ constant plane containing the rotation axis.

Rotation of a 2D, elliptic coordinate system, (ξ_2, ξ_3), defined in such a plane – as in Fig. 6.1b – leads to an ellipsoidal-coordinate system. Similarly, rotation of Cartesian (x_1, x_3) coordinates – as in Fig. 6.1c – leads to cylindrical-polar coordinates. Unit vectors $\mathbf{e}_2$ and $\mathbf{e}_3$ are tangential and normal, respectively, to $\xi_3 =$ constant surfaces (the latter surfaces usually coincide with the surface of a planet). In Fig. 6.1b, the $\xi_3 =$ constant surfaces are ellipsoidal, whereas they are cylindrical in Fig. 6.1c.

For ellipsoidal (including spherical) geometry, a $\xi_1 =$ constant plane is termed a *meridional plane* – see the light blue shaded plane in Figs. 6.1a and b – since it contains not only the rotation axis but also a meridian. For spherical-polar coordinates (λ, ϕ, r), a meridian is a line of longitude (i.e. a great, semicircular arc between the coordinate system's south and north poles) along which

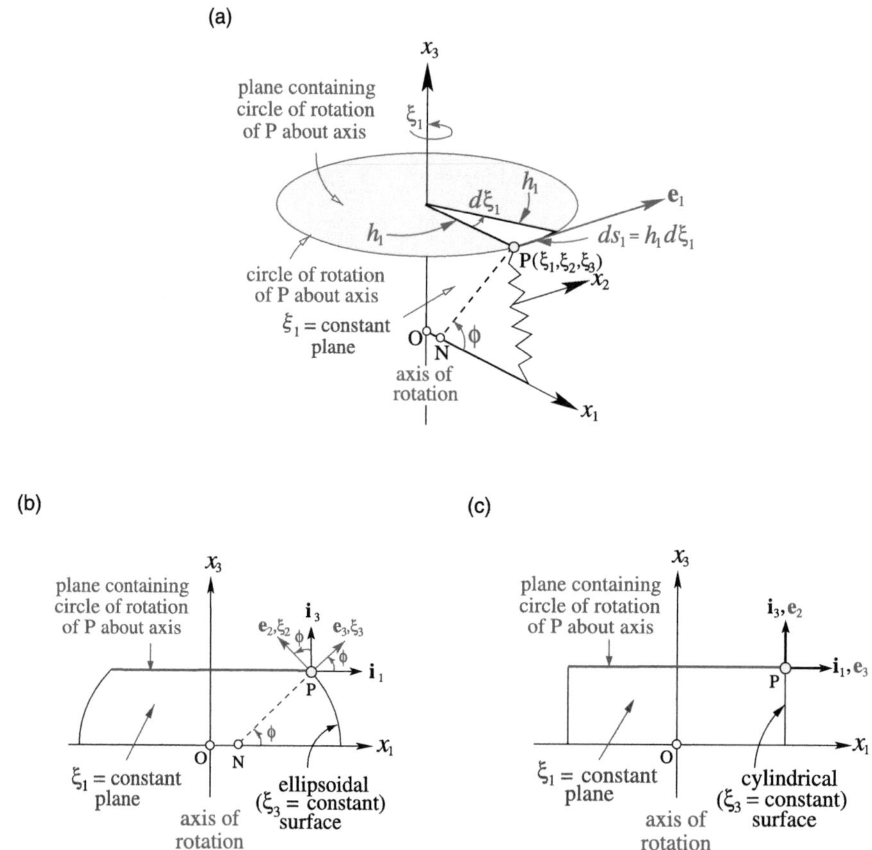

Figure 6.1 Axial-orthogonal-curvilinear coordinates (ξ_1, ξ_2, ξ_3), obtained by rotation of a 2D orthogonal coordinate system (ξ_2, ξ_3) about an axis (Ox_3).
(a) 3D view. The circle traced by rotation of an arbitrary point P about Ox_3, and the plane containing it, are in magenta. The $\xi_1 =$ constant plane containing P is light shaded (in blue). N is the point at which the normal to the $\xi_3 =$ constant surface through P intersects Ox_1. Cartesian coordinate x_3 is aligned with the rotation axis; x_1 and x_3 are embedded in the $\xi_1 =$ constant plane through P. The precise (smooth) shape that replaces the jagged line depends upon the type of coordinate system:
(b) 2D view. Cross section of the $\xi_1 =$ constant plane through P for ellipsoidal coordinates.
(c) Ditto, but for cylindrical coordinates (λ, z, r).
See text for further details.

λ and r are held constant, and latitude (ϕ) varies from $-\pi/2$ to $+\pi/2$. For ellipsoidal (including spherical) geometry, the magenta plane in Figs. 6.1a and b is termed a *latitudinal plane* since it contains the *latitude circle* (of rotation about the axis) passing through point P, at geographic latitude ϕ.

6.4 CORIOLIS TERMS

The developments in this section describe how to represent the Coriolis terms in an ellipsoidal-coordinate system, as depicted in Figs. 6.1a,b. This is by far the more difficult of the two cases; ellipsoidal coordinates versus cylindrical-polar coordinates. Similar arguments apply to cylindrical-polar coordinates (λ, z, r), as depicted in Figs. 6.1a,c. These arguments are not given here but may be developed in an analogous manner to those that follow for ellipsoidal coordinates.

6.4.1 Relating Geographic Latitude to Metric Factors

To express the components of the Coriolis force in terms of metric factors (and also velocity components), geographic latitude ϕ (the angle that the local vertical $\mathbf{e}_3$ makes with the equatorial plane[1] in ellipsoidal geometry) is first expressed in terms of metric factor h_1.[2] *This approach assumes, and relies on, the coordinate system being axial* (i.e. constructed by rotation of a 2D, orthogonal coordinate system about an axis); see Figs. 6.1a,b and 6.2. Thus:

The Axial Assumption of Axial-Orthogonal-Curvilinear Coordinates

For *axial-orthogonal-curvilinear coordinates,* it is assumed that h_1, h_2, and h_3 are all independent of ξ_1, that is,

$$h_i = h_i(\xi_2, \xi_3), \quad i = 1, 2, 3. \tag{6.13}$$

For general orthogonal-curvilinear coordinates in 3D, distance metric ds is given by (5.2). Thus

$$(ds)^2 = (ds_1)^2 + (ds_2)^2 + (ds_3)^2, \tag{6.14}$$

where

$$ds_1 = h_1 d\xi_1, \quad ds_2 = h_2 d\xi_2, \quad ds_3 = h_3 d\xi_3, \tag{6.15}$$

are the 1D distance metrics, and h_1, h_2, and h_3 are the associated 1D metric (scale) factors.

Consider the situation when ξ_2 and ξ_3 are held fixed; then $d\xi_2 = d\xi_3 = 0$, and only ξ_1 varies. Consequently, (6.14) and (6.15) reduce to $ds = ds_1 = h_1 d\xi_1$. Defining ξ_1 to be the *azimuthal angle* (i.e. angular distance) around the coordinate system's rotation axis, ξ_1 varies from 0 to 2π (just as λ does in spherical-polar coordinates). For fixed ξ_2 and ξ_3, and taking into consideration constraints (6.13) for axial-orthogonal-curvilinear coordinates, h_1 = constant. Solving $ds_1 = h_1 d\xi_1$ for constant h_1 then gives $s_1 = h_1 \xi_1$, where the resulting arbitrary constant of integration has been set to zero.[3] Thus s_1 *measures distance around a circle of revolution of radius* h_1, where h_1 is measured perpendicular from the axis of rotation to point P; see Figs. 6.1a and 6.2.[4]

Referring to Fig. 6.2a, point Q is separated from point P by distance $d\xi_3$, as measured by coordinate ξ_3 whilst holding coordinates ξ_1 and ξ_2 fixed. From (6.15), this corresponds to the physical separation $ds_3 = h_3 d\xi_3$ between points P and Q. The associated change in h_1 (holding ξ_1 and ξ_2 fixed), namely dh_1, when moving from P to Q along the coordinate curve between them is obtained by elementary trigonometry from triangle PQS. Note that point S is located at distance $h_1 + dh_1$ from point R. Thus

$$dh_1 = (\text{PS}) = (\text{PQ}) \cos\phi = ds_3 \cos\phi = h_3 d\xi_3 \cos\phi, \tag{6.16}$$

where dh_1, ds_3, and $d\xi_3$ are all increments measured whilst holding ξ_1 and ξ_2 fixed. Solving (6.16) for $\cos\phi$, and noting that increments dh_1, ds_1, and $d\xi_3$ correspond to holding ξ_1 and ξ_2 fixed, then yields

1 The equatorial plane is the plane that contains the equator of the coordinate system. It is parallel to a latitudinal plane (coloured in magenta in Figs. 6.1a and b), which contains a latitude circle.

2 Andy White kindly provided the analysis on which this section is based.

3 The arbitrary constant merely determines where, on the circle of revolution around the rotation axis, the measurement of distance s_1 begins. In other words, it determines the location of the reference point on the circle for this.

4 For the latitude-longitude representation of Earth's surface: s_1 measures distance along the latitude circle passing through point P, from the point at 0° longitude on this latitude circle; and h_1 is the radius of this latitude circle.

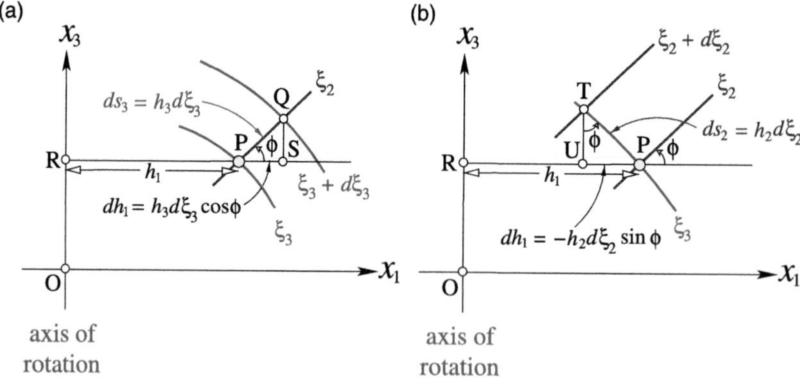

Figure 6.2 Relating $\cos\phi$ and $\sin\phi$, where ϕ is geographic latitude, to metric factors (h_1, h_2, h_3) for axial-orthogonal-curvilinear coordinates (ξ_1, ξ_2, ξ_3), obtained by rotating 2D orthogonal-coordinate system (ξ_2, ξ_3) about rotation axis Ox_3.
Both panels depict the $\xi_1 = $ constant meridional cross section passing through arbitrary point P (ξ_1, ξ_2, ξ_3); (x_1, x_3) are Cartesian coordinates embedded within this plane. Blue curves are surfaces of constant ξ_2, red curves of constant ξ_3. P lies at perpendicular distance h_1 away from Ox_3. Point R is at the intersection of Ox_3 and the latitudinal plane containing P.
(a) Determination of $\cos\phi$.
Point Q $(\xi_1, \xi_2, \xi_3 + d\xi_3)$ lies on isosurface $\xi_3 + d\xi_3$. It also lies on the $\xi_2 = $ constant curve normal to this surface that passes through P. Point Q is at physical distance $ds_3 = h_3 d\xi_3$ from P. Point S lies on the latitudinal plane containing P, and line QS is perpendicular to this plane. Point S is at physical distance $dh_1 = ds_3 \cos\phi = h_3 d\xi_3 \cos\phi$ from P.
(b) Determination of $\sin\phi$.
Point T $(\xi_1, \xi_2 + d\xi_2, \xi_3)$ lies on isosurface $\xi_2 + d\xi_2$. It also lies on the ξ_3 isosurface passing through P, at physical distance $ds_2 = h_2 d\xi_2$ from P. Point U lies on the latitudinal plane containing P, and line TU is perpendicular to this plane. Point U is at physical distance $dh_1 = -ds_2 \sin\phi = -h_2 d\xi_2 \sin\phi$ from P.

$$\cos\phi = \frac{1}{h_3}\left(\frac{dh_1}{d\xi_3}\right)_{\xi_1,\xi_2 \text{ held fixed}} = \frac{1}{h_3}\frac{\partial h_1}{\partial \xi_3}. \tag{6.17}$$

Referring now to Fig. 6.2b, point T is separated from point P by distance $d\xi_2$, as measured by coordinate ξ_2 whilst holding coordinates ξ_1 and ξ_3 fixed. From (6.15), this corresponds to the physical separation $ds_2 = h_2 d\xi_2$ between points P and T. The associated change in h_1 (holding ξ_1 and ξ_3 fixed), namely dh_1, when moving from P to T along the coordinate curve between them, is obtained by elementary trigonometry from triangle PTU. Note that U is located at distance $h_1 + dh_1$ from point R, and dh_1 is negative since U is closer to R than P is to R. Thus

$$dh_1 = -\,(\text{PU}) = -\,(\text{PT})\sin\phi = -ds_2 \sin\phi = -h_2 d\xi_2 \sin\phi, \tag{6.18}$$

where dh_1, ds_2, and $d\xi_2$ are all increments measured whilst holding ξ_1 and ξ_3 fixed. Solving (6.18) for $\sin\phi$, and noting that increments dh_1, ds_2, and $d\xi_2$ correspond to holding ξ_1 and ξ_3 fixed, then yields

$$\sin\phi = -\frac{1}{h_2}\left(\frac{dh_1}{d\xi_2}\right)_{\xi_1,\xi_3 \text{ held fixed}} = -\frac{1}{h_2}\frac{\partial h_1}{\partial \xi_2}. \tag{6.19}$$

Summarising the preceding analysis:

Expression of cos ϕ and sin ϕ in Terms of Metric Factors

For *axial*-orthogonal-curvilinear coordinates (ξ_1, ξ_2, ξ_3), $\cos\phi$ and $\sin\phi$ are related to metric factors (h_1, h_2, h_3) by the relations

$$\cos\phi = \frac{1}{h_3}\frac{\partial h_1}{\partial \xi_3} = \frac{1}{h_3(\xi_2,\xi_3)}\frac{\partial h_1(\xi_2,\xi_3)}{\partial \xi_3}, \tag{6.20}$$

$$\sin\phi = -\frac{1}{h_2}\frac{\partial h_1}{\partial \xi_2} = -\frac{1}{h_2(\xi_2,\xi_3)}\frac{\partial h_1(\xi_2,\xi_3)}{\partial \xi_2}. \tag{6.21}$$

Equations (6.20) and (6.21) hold for any *axial*-orthogonal-curvilinear coordinate system, constructed by rotation of a 2D orthogonal-coordinate system about an axis. For such a coordinate system (but *not* more generally), constraint equations between h_1, h_2, and h_3 may be obtained as follows. Squaring and summing (6.20) and (6.21), followed by rearrangement, gives

$$h_2 = \left[1 - \left(\frac{1}{h_3}\frac{\partial h_1}{\partial \xi_3}\right)^2\right]^{-1/2}\frac{\partial h_1}{\partial \xi_2}, \quad h_3 = \left[1 - \left(\frac{1}{h_2}\frac{\partial h_1}{\partial \xi_2}\right)^2\right]^{-1/2}\frac{\partial h_1}{\partial \xi_3}. \tag{6.22}$$

Dividing the first of these two equations by the second then yields

$$\frac{h_2}{h_3} = \frac{\partial h_1/\partial \xi_2}{\partial h_1/\partial \xi_3} \quad \Rightarrow \quad h_2 = \left(\frac{\partial h_1/\partial \xi_2}{\partial h_1/\partial \xi_3}\right)h_3, \quad h_3 = \left(\frac{\partial h_1/\partial \xi_3}{\partial h_1/\partial \xi_2}\right)h_2. \tag{6.23}$$

Thus, given metric factors h_1 and h_3, metric factor h_2 may be obtained from either the first equation of (6.22) or from the second equation of (6.23). Similarly, given metric factors h_1 and h_2, metric factor h_3 may be obtained from either the second equation of (6.22) or from the third equation of (6.23). Doing so provides a cross-check on independently computed metric factors. Note that the constraint equations given in (6.22) and (6.23) are *not* independent of one another. Care must also be exercised to avoid division by zero for the special cases $h_1 = h_1(\xi_2)$ or $h_1 = h_1(\xi_3)$.

6.4.2 Expressing Coriolis Components in Terms of Metric Factors

The rotation vector $\mathbf{\Omega}$ *is assumed to be aligned with the rotation axis of the axial-orthogonal-curvilinear coordinate system.* Now construct a (x_1, x_2, x_3) Cartesian coordinate system with x_1 and x_3 lying in the $\xi_1 =$ constant plane going through point P, and with x_3 aligned with the axis of rotation. See Fig. 6.1. Thus $\mathbf{\Omega}$ can be written in the (x_1, x_2, x_3) Cartesian coordinate system as

$$\mathbf{\Omega} = \Omega \mathbf{i}_3. \tag{6.24}$$

For a planet in ellipsoidal geometry, the next step is to express Cartesian unit vector $\mathbf{i}_3$ in terms of unit vectors $\mathbf{e}_2$ and $\mathbf{e}_3$ of the axial-orthogonal-curvilinear coordinate system, and geographic latitude, ϕ. Resolving $\mathbf{i}_3$ into its two components with respect to $\mathbf{e}_2$ and $\mathbf{e}_3$ – see Fig. 6.1b and also Fig. 17.2 of Chapter 17 – and multiplying by Ω gives

$$\mathbf{\Omega} = \Omega \mathbf{i}_3 = \Omega\left[(\cos\phi)\,\mathbf{e}_2 + (\sin\phi)\,\mathbf{e}_3\right]. \tag{6.25}$$

Eliminating $\cos\phi$ and $\sin\phi$ from this equation using (6.20) and (6.21) then yields

$$\mathbf{\Omega} = \Omega\left(\frac{1}{h_3}\frac{\partial h_1}{\partial \xi_3}\mathbf{e}_2 - \frac{1}{h_2}\frac{\partial h_1}{\partial \xi_2}\mathbf{e}_3\right) = \Omega\left(0, \frac{1}{h_3}\frac{\partial h_1}{\partial \xi_3}, -\frac{1}{h_2}\frac{\partial h_1}{\partial \xi_2}\right). \tag{6.26}$$

Using this representation of the rotation vector, $\boldsymbol{\Omega}$, then leads to:

Expression of the Coriolis Force in Terms of Metric Factors

The Coriolis force, $-2\boldsymbol{\Omega} \times \mathbf{u}$, may be expressed in component form in terms of the metric factors h_1, h_2, and h_3 of *axial*-orthogonal-curvilinear coordinates (ξ_1, ξ_2, ξ_3) as

$$-2\boldsymbol{\Omega} \times \mathbf{u} = -2\Omega\left(0, \frac{1}{h_3}\frac{\partial h_1}{\partial \xi_3}, -\frac{1}{h_2}\frac{\partial h_1}{\partial \xi_2}\right) \times (u_1, u_2, u_3) = -2\Omega \begin{vmatrix} \mathbf{e}_1 & \mathbf{e}_2 & \mathbf{e}_3 \\ 0 & \dfrac{1}{h_3}\dfrac{\partial h_1}{\partial \xi_3} & -\dfrac{1}{h_2}\dfrac{\partial h_1}{\partial \xi_2} \\ u_1 & u_2 & u_3 \end{vmatrix}$$

$$= -2\Omega\left[\left(\frac{u_2}{h_2}\frac{\partial h_1}{\partial \xi_2} + \frac{u_3}{h_3}\frac{\partial h_1}{\partial \xi_3}\right)\mathbf{e}_1 - \frac{u_1}{h_2}\frac{\partial h_1}{\partial \xi_2}\mathbf{e}_2 - \frac{u_1}{h_3}\frac{\partial h_1}{\partial \xi_3}\mathbf{e}_3\right]. \qquad (6.27)$$

As a cross-check, inserting (5.13) and (5.15) into (6.27) recovers expression (6.1) for the components of the Coriolis force in spherical-polar coordinates. Similarly, inserting (5.28) and (5.30) into (6.27) recovers expression (6.2) for cylindrical-polar coordinates.

6.5 THE GOVERNING EQUATIONS IN AXIAL-ORTHOGONAL-CURVILINEAR COORDINATES

Governing-equation set (6.3)–(6.12) is written in vector form. Having derived representation (6.27) of the Coriolis force, we now re-express this equation set in axial-orthogonal-curvilinear coordinates. To do so, we exploit axial assumption (6.13) and insert (6.27) and identities (5.5)–(5.12) into (6.3)–(6.12). After some algebra, this yields the equation set given in the summary box that follows. The grouping of terms in (6.29)–(6.31) is for later convenience (in Chapter 16). Absence of metric terms involving $\partial h_i/\partial \xi_1, i = 1, 2, 3$, is due to axial symmetry of the coordinate system, as expressed by (6.13).

As discussed in Section 7.3 and Chapter 12, because apparent gravity is such a dominant force, it is highly desirable (and arguably essential) that geophysical coordinate systems be chosen such that two of the three spatial coordinates are embedded in the geopotential surface passing through a point, with the third coordinate aligned with the direction of the local vertical, normal to the geopotential surface. Such a coordinate system is termed a *geopotential coordinate system*. This then ensures that there is no component of apparent, gravitational force in any direction tangential to geopotential surfaces. It thereby avoids difficulties with the representation of the horizontal and vertical balance of forces, such as geostrophic and (quasi-)hydrostatic balance, respectively.

For a geopotential coordinate system, Φ is (by definition) constant on geopotential surfaces $\xi_3 =$ constant, and therefore $\Phi = \Phi(\xi_3)$. The horizontal component of $\nabla\Phi$, namely

$$\nabla_{\text{hor}}\Phi \equiv \frac{1}{h_1}\frac{\partial \Phi}{\partial \xi_1}\mathbf{e}_1 + \frac{1}{h_2}\frac{\partial \Phi}{\partial \xi_2}\mathbf{e}_2, \qquad (6.28)$$

is then identically zero, and the two horizontal components $(1/h_1)\,\partial\Phi/\partial\xi_1$ and $(1/h_2)\,\partial\Phi/\partial\xi_2$ of $\nabla\Phi$ are then absent from (6.29) and (6.30), respectively. Since one is not obliged to use a geopotential coordinate system (despite it nevertheless being highly advisable), these two horizontal components have – for now – been retained in (6.29) and (6.30). In later chapters they will often be left out in the interests of brevity; they can anyway be straightforwardly reinserted at any time, if desired.

The precise functional form of the geopotential (Φ) is determined by the geometry of the problem (spheroidal, ellipsoidal, spherical, cylindrical, etc.), together with a suitable model of gravity. See Chapters 7, 8, and 12 for general theory and spheroidal geometry, and Sections 5.2.2 and 5.2.3

for the special cases of (deep) spherical and cylindrical geometries, respectively. For an axially symmetric planet,[5] $\Phi = \Phi(\xi_2, \xi_3)$ and $(1/h_1)\,\partial\Phi/\partial\xi_1$ is anyway absent from (6.29), irrespective of whether a geopotential-coordinate system is employed or not.

The Governing Equations for Motion of a Geophysical Fluid: Expressed in Axial-Orthogonal-Curvilinear Coordinates $(\xi_1\xi_2, \xi_3)$

Momentum

$$\frac{Du_1}{Dt} + \left(\frac{u_1}{h_1} + 2\Omega\right)\left(\frac{u_2}{h_2}\frac{\partial h_1}{\partial \xi_2} + \frac{u_3}{h_3}\frac{\partial h_1}{\partial \xi_3}\right) + \frac{1}{\rho h_1}\frac{\partial p}{\partial \xi_1} + \frac{1}{h_1}\frac{\partial \Phi}{\partial \xi_1} = F^{u_1}, \tag{6.29}$$

$$\frac{Du_2}{Dt} - \left(\frac{u_1}{h_1} + 2\Omega\right)\frac{u_1}{h_2}\frac{\partial h_1}{\partial \xi_2} - \frac{u_3^2}{h_2 h_3}\frac{\partial h_3}{\partial \xi_2} + \frac{u_2}{h_2}\frac{u_3}{h_3}\frac{\partial h_2}{\partial \xi_3} + \frac{1}{\rho h_2}\frac{\partial p}{\partial \xi_2} + \frac{1}{h_2}\frac{\partial \Phi}{\partial \xi_2} = F^{u_2}, \tag{6.30}$$

$$\frac{Du_3}{Dt} + \frac{u_2}{h_2}\frac{u_3}{h_3}\frac{\partial h_3}{\partial \xi_2} - \left(\frac{u_1}{h_1} + 2\Omega\right)\frac{u_1}{h_3}\frac{\partial h_1}{\partial \xi_3} - \frac{u_2^2}{h_2 h_3}\frac{\partial h_2}{\partial \xi_3} + \frac{1}{\rho h_3}\frac{\partial p}{\partial \xi_3} + \frac{1}{h_3}\frac{\partial \Phi}{\partial \xi_3} = F^{u_3}. \tag{6.31}$$

Continuity of Total Mass Density

$$\frac{D\rho}{Dt} + \frac{\rho}{h_1 h_2 h_3}\left[\frac{\partial}{\partial \xi_1}(u_1 h_2 h_3) + \frac{\partial}{\partial \xi_2}(u_2 h_3 h_1) + \frac{\partial}{\partial \xi_3}(u_3 h_1 h_2)\right] = F^{\rho}. \tag{6.32}$$

Substance Transport

$$\frac{DS^i}{Dt} = \dot{S}^i, \quad i = 1, 2, \ldots, \tag{6.33}$$

where S^i and $\dot{S}^i$ are as in (4.10) (for the atmosphere) and (4.11) (for oceans).

Thermodynamic-Energy Equation

$$\frac{D\mathscr{E}}{Dt} + p\frac{D\alpha}{Dt} = \dot{Q}_E \quad \text{or} \quad \frac{D\eta}{Dt} = \frac{\dot{Q}}{T}, \tag{6.34}$$

where $\dot{Q}$ is heating rate (per unit mass), and $\dot{Q}_E \equiv \dot{Q} + \sum_i \mu^i \dot{S}^i$ is total rate of energy input.

Fundamental Equation of State

$$\mathscr{E} = \mathscr{E}\left(\alpha, \eta, S^1, S^2, \ldots\right) \quad \text{or} \quad \eta = \eta\left(\alpha, \mathscr{E}, S^1, S^2, \ldots\right), \tag{6.35}$$

where $\alpha \equiv 1/\rho$.

Definition of T, p, and μ^i

$$T = \frac{\partial \mathscr{E}\left(\alpha, \eta, S^1, S^2, \ldots\right)}{\partial \eta} \quad \text{or} \quad T = \left[\frac{\partial \eta\left(\alpha, \mathscr{E}, S^1, S^2, \ldots\right)}{\partial \mathscr{E}}\right]^{-1}, \tag{6.36}$$

[5] See Section 12.2.6 for why this assumption is a good one for atmospheric and oceanic modelling.

$$p = -\frac{\partial \mathscr{E}\left(\alpha, \eta, S^1, S^2, \ldots\right)}{\partial \alpha} \quad \text{or} \quad p = T\frac{\partial \eta\left(\alpha, \mathscr{E}, S^1, S^2, \ldots\right)}{\partial \alpha}, \tag{6.37}$$

$$\mu^i = \frac{\partial \mathscr{E}\left(\alpha, \eta, S^1, S^2, \ldots\right)}{\partial S^i} \quad \text{or} \quad \mu^i = -T\frac{\partial \eta\left(\alpha, \mathscr{E}, S^1, S^2, \ldots\right)}{\partial S^i}. \tag{6.38}$$

Kinematic Equation

$$\mathbf{u} \equiv u_1\mathbf{e}_1 + u_2\mathbf{e}_2 + u_3\mathbf{e}_3 = h_1\frac{D\xi_1}{Dt}\mathbf{e}_1 + h_2\frac{D\xi_2}{Dt}\mathbf{e}_2 + h_3\frac{D\xi_3}{Dt}\mathbf{e}_3. \tag{6.39}$$

Material Derivative of a Scalar

$$\frac{Df}{Dt} = \frac{\partial f}{\partial t} + \frac{u_1}{h_1}\frac{\partial f}{\partial \xi_1} + \frac{u_2}{h_2}\frac{\partial f}{\partial \xi_2} + \frac{u_3}{h_3}\frac{\partial f}{\partial \xi_3}. \tag{6.40}$$

6.6 THE GOVERNING EQUATIONS IN SPHERICAL-POLAR AND CYLINDRICAL-POLAR COORDINATES

To make things more concrete and familiar, and also to relate the present general formulation to the specific one usually adopted in geophysical-fluid-dynamics textbooks, the special case of spherical-polar coordinates is considered in Appendix A to this chapter. As a second special case, cylindrical-polar coordinates are considered in Appendix B.

6.7 EULER–LAGRANGE FORMS OF THE MOMENTUM COMPONENTS

6.7.1 In Axial-Orthogonal-Curvilinear Coordinates

In the absence of right-hand-side forcings, momentum-component equations (6.29)–(6.31) may be more compactly written as:

The Euler–Lagrange Equations for Momentum Components

$$\frac{D}{Dt}\left(\frac{\partial \mathscr{L}}{\partial \dot{\xi}_i}\right) - \frac{\partial \mathscr{L}}{\partial \xi_i} = -\alpha\frac{\partial p}{\partial \xi_i}, \quad i = 1, 2, 3, \tag{6.41}$$

where $\dot{\xi}_i \equiv D\xi_i/Dt$, and $\mathscr{L}$ is

The Lagrangian Density in Axial-Orthogonal-Curvilinear Coordinates

$$\mathscr{L}\left(\xi_1, \xi_2, \xi_3, \dot{\xi}_1, \dot{\xi}_2, \dot{\xi}_3, \alpha, \eta, S^1, S^2, \ldots\right)$$
$$= \frac{h_1^2\dot{\xi}_1^2 + h_2^2\dot{\xi}_2^2 + h_3^2\dot{\xi}_3^2}{2} + \Omega h_1^2\dot{\xi}_1 - \mathscr{E}\left(\alpha, \eta, S^1, S^2, \ldots\right) - \Phi\left(\xi_2, \xi_3\right). \tag{6.42}$$

See Section 14.10 for a variational derivation of these equations,[6] and verification that they do indeed lead to (6.29)–(6.31) in the absence of right-hand-side forcings. See also Chapters 14–18 for further discussion of variational methods and the Euler–Lagrange equations, particularly regarding practical applications.

6.7.2 In Spherical-Polar Coordinates

In spherical-polar coordinates (λ, ϕ, r), ξ_i and h_i are given by (6.45). Inserting these into (6.42) then yields:

The Lagrangian Density in Spherical-Polar Coordinates

$$\begin{aligned}&\mathscr{L}\left(\lambda, \phi, r, \dot{\lambda}, \dot{\phi}, \dot{r}, \alpha, \eta, S^1, S^2, \ldots\right) \\ &= \frac{r^2 \cos^2 \phi \dot{\lambda}^2 + r^2 \dot{\phi}^2 + \dot{r}^2}{2} + \Omega r^2 \cos^2 \phi \dot{\lambda} - \mathscr{E}\left(\alpha, \eta, S^1, S^2, \ldots\right) - \Phi\left(\phi, r\right). \quad (6.43)\end{aligned}$$

6.7.3 In Cylindrical-Polar Coordinates

In cylindrical-polar coordinates (λ, z, r), ξ_i and h_i are given by (6.58). Inserting these into (6.42) then yields:

The Lagrangian Density in Cylindrical-Polar Coordinates

$$\mathscr{L}\left(\lambda, z, \mathrm{r}, \dot{\lambda}, \dot{z}, \dot{\mathrm{r}}, \alpha, \eta, S^1, S^2, \ldots\right) = \frac{\mathrm{r}^2 \dot{\lambda}^2 + \dot{z}^2 + \dot{\mathrm{r}}^2}{2} + \Omega \mathrm{r}^2 \dot{\lambda} - \mathscr{E}\left(\alpha, \eta, S^1, S^2, \ldots\right) - \Phi\left(z, \mathrm{r}\right). \quad (6.44)$$

6.8 CONCLUDING REMARKS

Recall from Section 4.5 that although a unified set of governing equations for the motion of a geophysical fluid has been assembled, it was assumed when doing so that:

1. The geopotential (Φ) is of known prescribed form.
2. The internal energy $(\mathscr{E})$ of the fluid is also of known prescribed form.

This is the case regardless of whether the governing equations are expressed in vector form (as in Section 4.4) or in axial-orthogonal-curvilinear coordinates (as in Section 6.5).

The challenge of prescribing a suitable form for the geopotential (according to a chosen geometry) is met in the next two chapters (Chapters 7 and 8) and in Chapter 12. The second challenge (i.e. prescribing a suitable form for the internal energy of a fluid according to whether it is primarily gaseous or liquid) is pursued in Chapters 9–11.

[6] This is not the only way to derive these equations. Historically, they were derived *before* the advent of the (elegant and powerful) variational methodology.

APPENDIX A: THE GOVERNING EQUATIONS IN SPHERICAL-POLAR COORDINATES

Consider the special case of spherical-polar coordinates for which – from (5.13) and (5.15) –

$$(\xi_1, \xi_2, \xi_3) = (\lambda, \phi, r), \quad (h_1, h_2, h_3) = (r\cos\phi, r, 1). \tag{6.45}$$

Introducing (6.45) into (6.29)–(6.40) leads to the following equations:

Momentum

$$\frac{Du_\lambda}{Dt} - \left(\frac{u_\lambda}{r\cos\phi} + 2\Omega\right)(u_\phi \sin\phi - u_r\cos\phi) + \frac{1}{r\cos\phi}\left(\frac{1}{\rho}\frac{\partial p}{\partial \lambda} + \frac{\partial \Phi}{\partial \lambda}\right) = F^{u_\lambda}, \tag{6.46}$$

$$\frac{Du_\phi}{Dt} + \left(\frac{u_\lambda}{r\cos\phi} + 2\Omega\right)u_\lambda \sin\phi + \frac{u_\phi u_r}{r} + \frac{1}{r}\left(\frac{1}{\rho}\frac{\partial p}{\partial \phi} + \frac{\partial \Phi}{\partial \phi}\right) = F^{u_\phi}, \tag{6.47}$$

$$\frac{Du_r}{Dt} - \left(\frac{u_\lambda^2}{r} + \frac{u_\phi^2}{r} + 2\Omega u_\lambda \cos\phi\right) + \frac{1}{\rho}\frac{\partial p}{\partial r} + \frac{\partial \Phi}{\partial r} = F^{u_r}. \tag{6.48}$$

Continuity of Total Mass Density

$$\frac{D\rho}{Dt} + \frac{\rho}{r^2\cos\phi}\left[\frac{\partial}{\partial \lambda}(u_\lambda r) + \frac{\partial}{\partial \phi}(u_\phi r\cos\phi) + \frac{\partial}{\partial r}(u_r r^2\cos\phi)\right] = F^{\rho}. \tag{6.49}$$

Substance Transport

$$\frac{DS^i}{Dt} = \dot{S}^i, \quad i = 1, 2, \ldots, \tag{6.50}$$

where S^i and $\dot{S}^i$ are as in (4.10) (for the atmosphere) and (4.11) (for oceans).

Thermodynamic-Energy Equation

$$\frac{D\mathscr{E}}{Dt} + p\frac{D\alpha}{Dt} = \dot{Q}_E \quad \text{or} \quad \frac{D\eta}{Dt} = \frac{\dot{Q}}{T}, \tag{6.51}$$

where $\dot{Q}$ is heating rate (per unit mass), and $\dot{Q}_E \equiv \dot{Q} + \sum_i \mu^i \dot{S}^i$ is total rate of energy input.

Fundamental Equation of State

$$\mathscr{E} = \mathscr{E}\left(\alpha, \eta, S^1, S^2, S^3, \ldots\right) \quad \text{or} \quad \eta = \eta\left(\alpha, \mathscr{E}, S^1, S^2, S^3, \ldots\right), \tag{6.52}$$

where $\alpha \equiv 1/\rho$.

Definition of T, p, and μ^i

$$T = \frac{\partial \mathscr{E}\left(\alpha, \eta, S^1, S^2, S^3, \ldots\right)}{\partial \eta} \quad \text{or} \quad T = \left[\frac{\partial \eta\left(\alpha, \mathscr{E}, S^1, S^2, S^3, \ldots\right)}{\partial \mathscr{E}}\right]^{-1}, \tag{6.53}$$

$$p = -\frac{\partial \mathscr{E}\left(\alpha, \eta, S^1, S^2, S^3, \ldots\right)}{\partial \alpha} \quad \text{or} \quad p = T\frac{\partial \eta\left(\alpha, \mathscr{E}, S^1, S^2, S^3, \ldots\right)}{\partial \alpha}, \tag{6.54}$$

$$\mu^i = \frac{\partial \mathscr{E}\left(\alpha, \eta, S^1, S^2, S^3, \ldots\right)}{\partial S^i} \quad \text{or} \quad \mu^i = -T\frac{\partial \eta\left(\alpha, \mathscr{E}, S^1, S^2, S^3, \ldots\right)}{\partial S^i}. \tag{6.55}$$

Kinematic Equation

$$\mathbf{u} = u_\lambda \mathbf{e}_\lambda + u_\phi \mathbf{e}_\phi + u_r \mathbf{e}_r \equiv r\cos\phi \frac{D\lambda}{Dt}\mathbf{e}_\lambda + r\frac{D\phi}{Dt}\mathbf{e}_\phi + \frac{Dr}{Dt}\mathbf{e}_r. \tag{6.56}$$

Material Derivative of a Scalar

$$\frac{Df}{Dt} = \frac{\partial f}{\partial t} + \frac{u_\lambda}{r\cos\phi}\frac{\partial f}{\partial \lambda} + \frac{u_\phi}{r}\frac{\partial f}{\partial \phi} + u_r\frac{\partial f}{\partial r}. \tag{6.57}$$

Outside a rotating spherical planet of radius a and making the spherical geopotential approximation, $\Phi = \Phi(r)$ in (6.46)–(6.48) is given by (5.26). The horizontal components of $\nabla\Phi$ in (6.46) and (6.47) are then absent.

APPENDIX B: THE GOVERNING EQUATIONS IN CYLINDRICAL-POLAR COORDINATES

Consider the special case of cylindrical-polar coordinates for which – from (5.28) and (5.30) –

$$(\xi_1, \xi_2, \xi_3) = (\lambda, z, \mathrm{r}), \quad (h_1, h_2, h_3) = (\mathrm{r}, 1, 1). \tag{6.58}$$

Introducing (6.58) into (6.29)–(6.40) leads to the following equations.

Momentum

$$\frac{Du_\lambda}{Dt} + \left(\frac{u_\lambda}{\mathrm{r}} + 2\Omega\right)u_\mathrm{r} + \frac{1}{\rho \mathrm{r}}\frac{\partial p}{\partial \lambda} + \frac{1}{\mathrm{r}}\frac{\partial \Phi}{\partial \lambda} = F^{u_\lambda}, \tag{6.59}$$

$$\frac{Du_z}{Dt} + \frac{1}{\rho}\frac{\partial p}{\partial z} + \frac{\partial \Phi}{\partial z} = F^{u_z}, \tag{6.60}$$

$$\frac{Du_\mathrm{r}}{Dt} - \left(\frac{u_\lambda}{\mathrm{r}} + 2\Omega\right)u_\lambda + \frac{1}{\rho}\frac{\partial p}{\partial \mathrm{r}} + \frac{\partial \Phi}{\partial \mathrm{r}} = F^{u_\mathrm{r}}. \tag{6.61}$$

Continuity of Total Mass Density

$$\frac{D\rho}{Dt} + \rho\left[\frac{1}{\mathrm{r}}\frac{\partial u_\lambda}{\partial \lambda} + \frac{\partial u_z}{\partial z} + \frac{1}{\mathrm{r}}\frac{\partial}{\partial \mathrm{r}}(\mathrm{r}u_\mathrm{r})\right] = F^\rho. \tag{6.62}$$

Substance Transport

$$\frac{DS^i}{Dt} = \dot{S}^i, \quad i = 1, 2, \ldots, \tag{6.63}$$

where S^i and $\dot{S}^i$ are as in (4.10) (for the atmosphere) and (4.11) (for oceans).

Thermodynamic-Energy Equation

$$\frac{D\mathscr{E}}{Dt} + p\frac{D\alpha}{Dt} = \dot{Q}_\mathrm{E} \quad \text{or} \quad \frac{D\eta}{Dt} = \frac{\dot{Q}}{T}, \tag{6.64}$$

where $\dot{Q}$ is heating rate (per unit mass), and $\dot{Q}_\mathrm{E} \equiv \dot{Q} + \sum_i \mu^i \dot{S}^i$ is total rate of energy input.

Fundamental Equation of State

$$\mathscr{E} = \mathscr{E}\left(\alpha, \eta, S^1, S^2, S^3, \ldots\right) \quad \text{or} \quad \eta = \eta\left(\alpha, \mathscr{E}, S^1, S^2, S^3, \ldots\right), \tag{6.65}$$

where $\alpha \equiv 1/\rho$.

Definition of T, p, and μ^i

$$T = \frac{\partial \mathscr{E}\left(\alpha, \eta, S^1, S^2, S^3, \ldots\right)}{\partial \eta} \quad \text{or} \quad T = \left[\frac{\partial \eta\left(\alpha, \mathscr{E}, S^1, S^2, S^3, \ldots\right)}{\partial \mathscr{E}}\right]^{-1}, \tag{6.66}$$

$$p = -\frac{\partial \mathscr{E}\left(\alpha, \eta, S^1, S^2, S^3, \ldots\right)}{\partial \alpha} \quad \text{or} \quad p = T\frac{\partial \eta\left(\alpha, \mathscr{E}, S^1, S^2, S^3, \ldots\right)}{\partial \alpha}, \tag{6.67}$$

$$\mu^i = \frac{\partial \mathscr{E}\left(\alpha, \eta, S^1, S^2, S^3, \ldots\right)}{\partial S^i} \quad \text{or} \quad \mu^i = -T\frac{\partial \eta\left(\alpha, \mathscr{E}, S^1, S^2, S^3, \ldots\right)}{\partial S^i}. \tag{6.68}$$

Kinematic Equation

$$\mathbf{u} = u_\lambda \mathbf{e}_\lambda + u_z \mathbf{e}_z + u_{\mathrm{r}} \mathbf{e}_{\mathrm{r}} \equiv \mathrm{r}\frac{D\lambda}{Dt}\mathbf{e}_\lambda + \frac{Dz}{Dt}\mathbf{e}_z + \frac{D\mathrm{r}}{Dt}\mathbf{e}_{\mathrm{r}}. \tag{6.69}$$

Material Derivative of a Scalar

$$\frac{Df}{Dt} = \left(\frac{\partial f}{\partial t} + \frac{u_\lambda}{\mathrm{r}}\frac{\partial f}{\partial \lambda} + u_z\frac{\partial f}{\partial z} + u_{\mathrm{r}}\frac{\partial f}{\partial \mathrm{r}}\right). \tag{6.70}$$

Outside a uniform circular cylinder, of radius $\mathrm{r} = \mathrm{r}_S$, of infinite length, and rotating with constant angular frequency about its axis, $\Phi = \Phi(\mathrm{r})$ in (6.59)–(6.61) is given by (5.41). The horizontal components of $\nabla\Phi$ in (6.59) and (6.60) are then absent.

7

Representation of Gravity: Basic Theory and Spherical Planets

ABSTRACT

The principal characteristics of Earth's shape (classically termed its 'Figure') are summarised. A brief history of the development of the mathematical description of Earth's gravitational field follows. Why Earth has the shape it does is explained in terms of cause (Earth rotates about its axis) and effect (Earth becomes, to leading order, a mildly oblate ellipsoid rather than a sphere). A framework (based on Gauss's reformulation of Newtonian gravity) is given for the development (in terms of gravitational potential, termed geopotential) of a mathematical representation for the gravitational field of mildly oblate, idealised planets. The described framework is applied to four idealised planets of increasing geometrical and mass-distribution complexity:

1. Spherical, with constant density (in this chapter).
2. Spherical, with variable density (in this chapter).
3. Mildly ellipsoidal, with constant density (in Chapter 8).
4. Mildly ellipsoidal, with variable density (in Chapter 8).

Doing so allows various mathematical challenges to be identified and successively addressed. This ultimately culminates (in Chapter 8) in a geophysically realistic representation of a mildly oblate planet's shape and its associated gravitational field. A somewhat surprising, intermediate result (of this chapter) is that the classical, ubiquitous, spherical-geopotential approximation does not correspond to a physically realisable representation of gravity for a spherical planet. A side benefit of the further, more general analysis of Chapter 8 – for the representation of gravity for mildly oblate, spheroidal planets – is that it clarifies, in a rigorous manner, the applicability and validity of the classical spherical-geopotential approximation.

7.1 PREAMBLE

7.1.1 Latitudinal and Vertical Variation of Gravity

Now we have the governing equations for the motion of a geophysical fluid both in vector form (in Section 4.4) and in axial-orthogonal-curvilinear coordinates (in Section 6.5), we need, for practical applications, to specify the location of Earth's mean surface and the geopotential of apparent gravity.[1] As we will see, this is a challenging problem.

Earth's mean surface is historically called the *Figure of the Earth*. It has been known since the time of Newton (Todhunter, 1873; Chandrasekhar, 1967) that:

[1] *Geopotential of apparent gravity* is usually abbreviated in the literature to *geopotential*. This practice is followed herein.

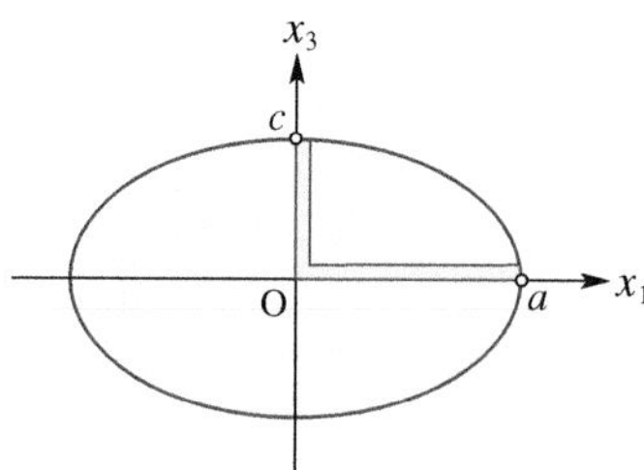

Figure 7.1 An exaggerated cross section in the Ox_1x_3 meridional plane of the Figure of the Earth, where a and c are its semi-major and semi-minor axes, respectively. Newton's two conceptual 'canals', bored to Earth's centre from the North Pole and a point on the Equator, are also shown (dark shaded).

- Earth's mean surface is more accurately approximated by an oblate spheroidal[2] surface of revolution (see Fig. 7.1), having an approximately elliptic cross section in a meridional plane,[3] than by a sphere.
- Gravitational attraction, exerted exterior to Earth by it, varies as a function of geographic latitude, as well as distance from Earth's surface.

Quantitatively, the semi-major (a) and semi-minor (c) axes of the quasi-elliptic meridional cross section are approximately 6378 km and 6357 km, respectively. See Table 7.1 for more precise values of these and other parameters for Earth's representation.

Thus Earth is a *mildly oblate spheroid* (i.e. slightly flattened sphere) with an equatorial bulge of approximately 21 km ($= 6378\,\text{km} - 6357\,\text{km}$). This is more than twice the peak height of Everest above mean sea level. The Pole-to-Equator variation of apparent gravity at Earth's surface is approximately 0.5 %. The vertical variation can be much larger than this, however. See Table 7.2.

Whilst these variations (at least in the lower atmosphere) are not large, they are systematic. They could therefore conceivably give rise to significant cumulative effects, both in climate-scale integrations and (via use of background fields in data assimilation) in numerical weather prediction (White et al., 2008; Bénard, 2014, 2015; Staniforth, 2014b; Staniforth and White, 2015a).

Furthermore, the total depth of the atmosphere represented in numerical models keeps increasing as technology advances. For *whole-atmosphere models*, used to model *space weather*, the atmospheric depth can be 600 km or more (Akmaev, 2011). The vertical variation of gravity over the whole atmosphere is then of order 20%, which is far from negligible. Today's operational numerical weather-prediction and climate-simulation models typically include a mesosphere, and the vertical variation is then of order 3%. Whilst this is not extraordinarily large, it is not necessarily negligible. In particular, this vertical variation systematically affects the mean stratification of Earth's atmosphere, thereby affecting the global atmospheric circulation.

7.1.2 Cause and Effect

- So why is Earth's mean surface spheroidal in shape rather than spherical?

[2] Nomenclature: There is some confusion in the literature regarding the use of the terms spheroid and ellipsoid. These are often used interchangeably, or with conflicting meanings between different authors. The convention adopted herein is that a *spheroid* is a solid of revolution that is approximately spherical, having an *almost*, or *precisely*, elliptic cross section in any meridional plane. When this cross section is *precisely* elliptic, the solid is then termed an *ellipsoid*.

[3] A meridional plane is one embedding a meridian, i.e. one that embeds an arc of constant longitude. This plane can equivalently be described as any plane containing Earth's rotation axis.

Parameter	Quantity	Value	Units
a	Equatorial radius	6378.1370000	km
c	Polar radius [$\equiv (1-\varepsilon)\,a$, computed]	6356.7523142	km
$1/\varepsilon$	Inverse of ellipticity (flattening parameter)	298.257223563	–
$1/m$	Inverse of $m \equiv \Omega^2 a^3/(\gamma M)$ (computed)	288.901121086	–
Ω	Earth's angular speed in inertial space	7.292115×10^{-5}	rad s^{-1}
γ	Universal gravitational constant	6.673×10^{-11}	$m^3\,kg^{-1}\,s^{-2}$
γM	Standard gravitational parameter (for Earth + its atmosphere)	$3.986004418 \times 10^{14}$	$m^3\,s^{-2}$
γM_E	Standard gravitational parameter (for Earth only)	3.9860009×10^{14}	$m^3\,s^{-2}$
γM_A	Standard gravitational parameter (for atmosphere only)	3.5×10^{8}	$m^3\,s^{-2}$
M_E	Mass of Earth (computed)	5.97333×10^{24}	kg
M_A	Mass of Earth's atmosphere (computed)	5.2×10^{18}	kg
V_E	Volume of Earth (computed)	$1.08320732 \times 10^{21}$	m^3

Table 7.1 Values of various parameters for Earth, obtained from the WGS 84 (World Geodetic System 84) documentation. See WGS 84 (2004), Chapter 3. The four defining parameters of the WGS 84 reference ellipsoid are a, $1/\varepsilon$, Ω, and γM. $\gamma M \equiv \gamma M_E + \gamma M_A$ is known to higher accuracy than either of its two constituents, and γM_A to much lower accuracy than γM_E. ε and m are dimensionless.

Approximate location	$r - \bar{a}$ (km)	Variation of $\lvert\mathbf{g}\rvert$ from its mean surface value (%)
Thermopause	600	19.7
Mesopause	85	2.7
Stratopause	55	1.7
Tropopause	10	0.3
Mean sea level	0	0.0

Table 7.2 Percentage variation of $\lvert\mathbf{g}\rvert$ as a function of atmospheric height $r - \bar{a}$, where r is distance from Earth's centre and $\bar{a}$ ($\approx$ 6371 km) is mean distance of Earth's mean sea level from Earth's centre.

It is because Earth approximately behaves as if it were a uniformly rotating fluid spheroid in motionless (relative to a frame of reference rotating synchronously with Earth) hydrostatic equilibrium. As will be seen, and as discussed in Chandrasekhar (1967, 1969), two small non-dimensional quantities govern this behaviour, one measuring *cause*, the other *effect*.

Cause

The non-dimensional parameter for the *cause* is

$$m \equiv \frac{\Omega^2 a^3}{\gamma M} \equiv \frac{\Omega^2 a}{\gamma M/a^2} \approx \frac{\text{centrifugal force}}{\text{gravitational attraction}}. \tag{7.1}$$

Here:

- Ω is Earth's (assumed constant) angular-rotation rate about its axis.
- a is Earth's equatorial radius (see Fig. 7.1).
- γ is Newton's universal gravitational constant.
- M is the combined mass of Earth and its atmosphere.[4]

Thus the faster the planet rotates, and the larger it is, the stronger the cause, whereas the more massive it is, the weaker the cause. Physically, the rightmost expression in (7.1) is the ratio of the (apparent) *centrifugal force* anywhere on Earth's Equator and approximately (with negligible difference) the *Newtonian gravitational attraction* there. Thus m is a measure of the relative importance of these two forces, centrifugal to Newtonian gravitational attraction.

Effect

The non-dimensional parameter for the *effect* is *ellipticity*

$$\varepsilon \equiv \frac{a-c}{a}. \tag{7.2}$$

Here, a is again equatorial radius, and c is polar radius.[5] See Fig. 7.1. Thus the larger ε is, the more oblate is the spheroid. Furthermore, if $\varepsilon \equiv 0$, then the spheroid degenerates into a perfect sphere.

Both m and ε are small for Earth, with approximately equal values of about $1/300$.[6] See Table 7.1. That they are both so small is key to mathematical tractability, as we shall see.

If Earth did not rotate, then Ω, and therefore m, would both be identically zero. This would then suppress the cause, thereby also suppressing the effect. Consequently ε would be identically zero, and Earth's mean shape would be a sphere. But Earth does rotate – once per terrestrial day – and so it is approximately spheroidal in shape rather than spherical.

7.1.3 Some History

As an idealisation of Earth, Isaac Newton (1643–1727) assumed it to be (in modern parlance) a fluid spheroid that is:

- Almost spherical in shape (i.e. $m, \varepsilon \ll 1$).
- Of homogeneous composition (i.e. its density is constant within its interior).
- Motionless within its interior relative to a frame of reference rotating synchronously with Earth (i.e. it rotates about its axis as if it were a solid body).
- Subject to constant pressure (taken to be zero) everywhere on its surface.
- In perfect hydrostatic balance everywhere within its interior.

Aside In Newton's day (*Principia Mathematica* was published in 1687) the equations for fluid dynamics had not yet been established. He therefore had to instead make some ingenious arguments that amount to an assumption of hydrostatic balance.

As reviewed in Chandrasekhar (1967, 1969), he imagined two 'canals' of unit cross section (see Fig. 7.1), bored to Earth's centre from the North Pole and a point on the Equator, and then filled with fluid. The 'weights' of the two fluids must balance at Earth's centre for a state of equilibrium. In the absence of Earth's rotation, this would give two columns

4 In the WGS 84 documentation, the mass of Earth's atmosphere is assumed to be 'condensed' to lie on Earth's reference surface (WGS 84, 2004).

5 ε is also known as the *flattening* parameter, and as the *first ellipticity*, with $(a-c)/c$ being the *second ellipticity*.

6 This is an astronomical coincidence.

of fluid of equal depth (i.e. Earth would be spherical). However, Earth's rotation gives rise to a weak (compared with gravity) centrifugal force acting outwards from Earth's rotation axis, which then slightly 'dilutes' the 'weight' in the column of fluid bored from the Equator. To compensate for this 'dilution' and to maintain equilibrium, the column of fluid bored from the Equator must therefore be slightly deeper than that bored from the North Pole. Therefore Earth's mean surface must be slightly oblate (i.e. slightly flattened) in shape, rather than spherical.

As briefly summarised in Chandrasekhar (1967, 1969) – based on Todhunter (1873)'s historical account – Newton's theory was nevertheless doubted by some scientists of the time, who believed Earth's shape to be a *prolate* spheroid rather than an *oblate* one. The matter was unequivocally settled in Newton's favour by geodetic measurements made during a scientific expedition to Lapland in 1736–1737, organised by l'Académie des Sciences (Paris) for this purpose.

The stated assumptions, after some truly extraordinary work and the neglect of various terms of $O\left(\varepsilon^2, \varepsilon m, m^2\right)$, led Newton to conclude that the Figure of the Earth (i.e. its shape) is approximately an ellipsoid of revolution (i.e. the solid formed by rotating an ellipse, defined in a meridional plane, about Earth's rotation axis). Furthermore, he showed that under these conditions

$$\varepsilon = \frac{5}{4}m. \tag{7.3}$$

However, from the consolidation of available measurements it was found that $\varepsilon \approx m$ (as opposed to $\varepsilon \approx 5m/4$) and that (7.3) is therefore quantitatively incorrect. But it was not known why.

In 1742, Colin Maclaurin (1698–1746) repeated Newton's work, using the same assumptions but, remarkably, *without* approximation. This however still leads to (7.3), thereby corroborating Newton's work, rather than to $\varepsilon \approx m$. Whereas Newton's derivation relies on m and ε being sufficiently small, Maclaurin's does not, and it is therefore of more general applicability to other planetary atmospheres (but still limited by the assumption of constant density).[7] Maclaurin also showed that apparent gravity[8] acts in a direction *normal* (i.e. perpendicular) to geopotential surfaces.[9]

Alexis Claude Clairaut (1713–65) finally solved the mystery of the quantitative disagreement between Newton's theory and observations (Clairaut, 1743). The key reason is that *the assumption of constant density for Earth's composition is insufficiently true.* It turns out that Earth's density is an order of magnitude larger in its core than it is at its crust (for quantitative discussions, see e.g. Jeffreys, 1976 and Dziewonski and Anderson, 1981). Furthermore, its density varies meridionally (i.e. as a function of latitude), being denser at the Poles than at the Equator. Clairaut relaxed Newton's constant-density assumption by instead assuming that Earth is composed of concentric spheroidal shells – see Fig. 7.2 – each of which is bounded by two concentric geopotential surfaces. The density within each shell is constant, but the density can vary from shell to shell, with jumps in density between adjacent shells.[10] Taking all this into consideration then reconciled theory with observations.

[7] A surprising feature of Maclaurin's analysis, discovered in 1743 by Thomas Simpson (1710–61), is that for a given small value of the rotation rate Ω, *two* oblate spheroids are possible. The first corresponds (approximately) to Newton's slightly flattened ellipsoid, whereas the second is a highly flattened spheroid. The existence of this supplementary equilibrium solution ultimately led to Karl Jacobi (1804–51) showing (in 1834) that a homogeneous spheroid in equilibrium does not *necessarily* have to be a solid of revolution. Such spheroids are termed *triaxial*, because they have three unequal axes (a, b, and c) rather than the two (a and c) of spheroids of revolution. See Chandrasekhar (1969, 1967) and Tassoul (1978) for both this and subsequent developments for planetary atmospheres.

[8] This is aligned with a plumb line and defines the direction of the local vertical.

[9] A geopotential surface is one on which the geopotential, Φ, is everywhere constant.

[10] The density can also vary continuously by considering shells of infinitesimal thickness and then integrating over shells.

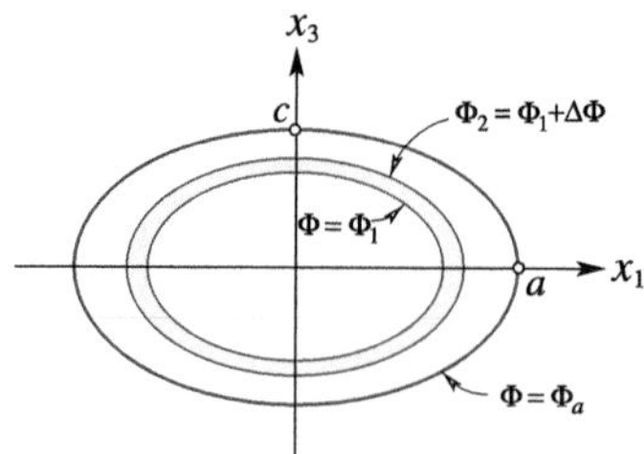

Figure 7.2 An exaggerated cross section in the Ox_1x_3 meridional plane of the Figure of the Earth, where a and c are its semi-major and semi-minor axes, respectively. A shell (dark shaded) is shown that is bounded by two concentric geopotential surfaces $\Phi = \Phi_1$ and $\Phi = \Phi_2 \equiv \Phi_1 + \Delta\Phi$, where Φ_1, Φ_2, and $\Delta\Phi$ are constants.

As discussed more fully in Chandrasekhar (1967, 1969) and Tassoul (1978), the theory developed by the pioneers was gradually refined during the latter part of the eighteenth century and the early part of the nineteenth century:

- Adrien-Marie Legendre (1752–1833) introduced the concept of gravitational potential and the condition that the sum of the gravitational and centrifugal potentials is constant on geopotential surfaces. He also developed the polynomials that bear his name.
- Meanwhile, Pierre-Simon de Laplace (1749–1827) developed spherical harmonics and the differential equation that bears his name. Furthermore, he showed that the surfaces of equal pressure, equal density, and geopotential all coincide for a uniformly rotating fluid spheroid in equilibrium.
- Siméon-Denis Poisson (1781–1840) developed the differential equation that bears his name. This equation relates the gravitational potential at a point inside Earth with the density at that point.

The use of concentric spheroidal shells is quite cumbersome mathematically. Fortunately Johann Carl Friedrich Gauss (1777–1855) developed (in the early nineteenth century) an alternative, equivalent, formulation of gravitational attraction to that proposed by Newton.[11] This then led to the development of modern potential theory.[12] This framework, together with the incorporation of the developments due to Legendre, Laplace, and Poisson, provides the basis for the present exposition of the theory of gravitational attraction for atmospheric and oceanic modelling.

7.1.4 Apparent Gravity and the Geopotential

Recall from Section 2.2.2 that the apparent gravitational force $\mathbf{g}$, per unit mass, is the combined force of Newtonian gravitational attraction and the apparent centrifugal force due to Earth's diurnal rotation about its axis.[13] Furthermore, $\mathbf{g}$ is expressible in terms of the geopotential, Φ, of apparent gravity as (see (2.29)–(2.31))

$$\mathbf{g} \equiv \mathbf{g}^N + \mathbf{g}^C \equiv -\nabla\left(V + V^C\right) \equiv -\nabla\Phi, \tag{7.4}$$

[11] Gauss's formulation states that the flux due to Newtonian gravity across any closed surface is proportional to the enclosed mass; see Section 7.5, and Chapter 8 for application to spheroidal planets.

[12] Potential theory is widely used in many disciplines.

[13] Also recall – from Section 2.2.2 – that Newtonian gravitational attraction is a *real* force, observed both in an inertial frame of reference and one rotating synchronously with Earth. The centrifugal force however is an *apparent* force. It must be included when the momentum equation is expressed in a rotating frame of reference in order for Newton's law to hold in an inertial frame. See Section 2.2.2.3.

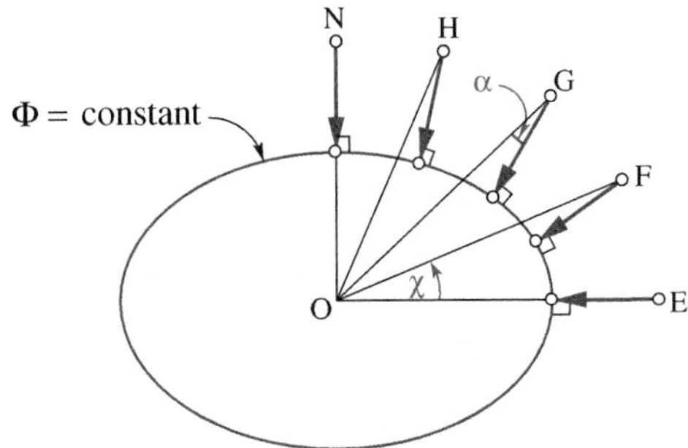

Figure 7.3 Meridional cross section of a spheroidal geopotential surface, Φ = constant, and directions (downward-pointing arrows) of apparent vertical normal to it. O is the centre of Earth, which rotates with angular velocity $\boldsymbol{\Omega}$ about its polar axis ON. OE lies in the equatorial plane. Apparent vertical is the direction of apparent gravity, locally aligned with a pendulum bob hanging at rest relative to rotating Earth. At the Poles and at the Equator, the direction of apparent vertical coincides with the radii (ON and OE), but at other (*geocentric*) latitudes (χ) centrifugal force leads to a deviation α of apparent vertical from the local radial direction. α attains its maximum value at about $\pm 45°$ of latitude. For pictorial clarity: geopotential eccentricity; typical atmospheric values of α; and distances of points E, F, G, H, and N from the geopotential surface Φ = constant are all greatly exaggerated. After Fig. 1 of White et al. (2005).

where

$$\mathbf{g}^N \equiv -\nabla V, \quad \mathbf{g}^C \equiv -\nabla V^C \equiv -\nabla \left[-\frac{(\boldsymbol{\Omega} \times \mathbf{r}) \cdot (\boldsymbol{\Omega} \times \mathbf{r})}{2} \right], \tag{7.5}$$

$$\Phi \equiv V + V^C = V - \frac{(\boldsymbol{\Omega} \times \mathbf{r}) \cdot (\boldsymbol{\Omega} \times \mathbf{r})}{2}, \quad V^C \equiv -\frac{(\boldsymbol{\Omega} \times \mathbf{r}) \cdot (\boldsymbol{\Omega} \times \mathbf{r})}{2}. \tag{7.6}$$

In these equations:

- V is potential of Newtonian gravity.[14]
- V^C is potential of centrifugal force.
- $\mathbf{r}$ is position vector.
- $\boldsymbol{\Omega}$ is Earth's rotation vector, aligned with its axis of rotation.

Potential theory therefore provides a suitable mathematical framework for representing and suitably approximating Earth's geopotential in a tractable manner. See Sections 7.5 and 8.4.

Since apparent gravity is expressible as the gradient of the geopotential, Φ, it acts in a direction normal to a geopotential surface (i.e. normal to a surface on which Φ is everywhere constant). See Fig. 7.3.

The direction of apparent gravitational force, $\mathbf{g}$, is aligned with a plumb line (this defines the direction of the local vertical) and *not* with the direction of the Newtonian gravitational force, $\mathbf{g}^N$. The difference in direction between the two (α in Fig. 7.3) is entirely due to the centrifugal force, which acts in a direction *normal to Earth's rotation axis*. Because the centrifugal force is two orders of magnitude smaller than Newtonian gravitational attraction,[15] the difference in direction of Newtonian and apparent gravitational forces is, however, less than 0.2°. The largest difference occurs at a latitude of approximately $\pm 45°$.

14 Physically, Newtonian gravitational potential $V(\mathbf{r})$ is the work done by the force of gravity to move a particle of unit mass from position $\mathbf{r}$ (with respect to Earth's centre) to a fixed reference location. Conventionally, the fixed location (for a deep atmosphere) is taken to be at infinite distance from Earth's centre, and $V(|\mathbf{r}| \to \infty)$ is arbitrarily set to zero; $V(\mathbf{r})$ is then negative everywhere, except at infinity where it is zero. No work is done by Newtonian gravity when a particle is moved on an equipotential surface of V, since V = constant on such a surface by definition.

15 Recall from Section 7.1.2 that m is a measure of the ratio of these two forces, and $m \approx 1/300$.

7.1.5 The Classical Spherical-Geopotential Approximation

It is standard practice in the atmospheric and oceanic sciences to model Earth's gravitational field using the *the classical spherical-geopotential approximation*. This is motivated by two facts:

1. Earth is approximately spherical.
2. Spherical geometry is simpler than spheroidal geometry.

Earth's geopotential surfaces (i.e. surfaces on which Φ is constant) are then taken to be *perfectly spherical*. This includes Earth's mean sea level surface, which is implicitly assumed to coincide with a geopotential surface.

In standard texts on the subject, Φ is usually approximated (to within an arbitrary additive constant) by:

The Classical Spherical-Geopotential Approximation (in Shallow Form)

$$\Phi(r) = g_a(r - a), \tag{7.7}$$

where r is distance from the centre of a sphere of radius a, and g_a is a constant representative value for $|\mathbf{g}|$ on the sphere's surface, $r = a$.[a] The geopotential surfaces are perfectly spherical.

a That $\Phi(r) \geq 0$ for $r \geq a$ is due to the arbitrary, fixed, reference location – mentioned in footnote 14 – having (for convenience) been relocated from $r \to \infty$ (for a deep atmosphere) to $r = a$ (for a shallow atmosphere). The term $r - a$ in (7.7) is often denoted in the literature by z, altitude above Earth's mean sea level, reference surface.

Using (7.4), apparent gravity is then given by[16]

$$\mathbf{g} = -\nabla\Phi(r) = -\frac{d\Phi}{dr}\mathbf{e}_r = -g_a\mathbf{e}_r, \tag{7.8}$$

where $\mathbf{e}_r$ is the outward-pointing normal to the sphere. Thus apparent gravity is not only constant on the surface of the sphere for spherical-geopotential approximation (7.7), but also *everywhere outside it*.

Some texts (e.g. Haltiner and Martin, 1957; White et al., 2005; Wood et al., 2014) instead approximate Φ (to within an arbitrary additive constant) by:

The Classical Spherical-Geopotential Approximation (in Deep Form)

$$\Phi(r) = -g_a\frac{a^2}{r} = -\frac{\gamma M}{r}, \tag{7.9}$$

where $g_a \equiv \gamma M/a^2$, γ is Newton's universal gravitational constant, and M is the total mass of the sphere. The geopotential surfaces are again perfectly spherical.

[16] The negative sign in the rightmost expression of (7.8) reflects the fact that apparent gravity (above the sphere's surface i.e. for the atmosphere) points *towards* the sphere's surface rather than away from it. This keeps everyone's feet on the ground.

For (7.9), apparent gravity is then given by

$$\mathbf{g}(r) = -\nabla\Phi(r) = -\frac{d\Phi}{dr}\mathbf{e}_r = -g_a\frac{a^2}{r^2}\mathbf{e}_r = -\frac{\gamma M}{r^2}\mathbf{e}_r. \tag{7.10}$$

For this alternative form of the classical spherical-geopotential approximation, apparent gravity is still constant on spherical surfaces. However, instead of being constant *everywhere* outside the sphere, it now falls off radially as $1/r^2$. This agrees much better with what one would expect on physical grounds for Newtonian gravity[17] outside a sphere with a radially symmetric mass distribution; note, though, that for a sufficiently shallow layer of fluid over a sphere, the radial variation is expected to be negligible.

When using either of the two forms (7.7) and (7.9) of the spherical-geopotential approximation, the equations of motion are expressed (using a shallow or a deep metric) in terms of (shallow or deep) spherical-polar coordinates. Everything from that point on is then done within the simplified framework of (shallow or deep) *spherical* geometry rather than that of (deep) *spheroidal* geometry.

Two forms for $\Phi = \Phi(r)$ have been given, namely (7.7) and (7.9).

- Is there any reason to prefer one over the other?

Yes, indeed there is, depending upon the chosen form of the governing equations for the motion of a geophysical fluid; see Section 8.6 for the reasons why. If the governing fluid-dynamical equations are written (after approximation using a *shallow* metric) in shallow form (see Chapter 16 for details), then the (shallow) spherical-geopotential approximation (7.7) should be used. Similarly, if the governing fluid-dynamical equations are written (using a *deep* metric) in deep form, then the (deep) spherical-geopotential approximation (7.9) should be used.

It is relatively straightforward to describe – as we have just done – the spherical-geopotential approximation and how it is applied in practice. Although this approximation is – almost universally – well accepted by the atmospheric and oceanic modelling communities, it is nevertheless surprisingly difficult to rigorously derive it in a convincing manner for a *rotating* planet. In this regard, van der Toorn and Zimmerman (2008)'s critique of the arguments advanced by previous authors to justify the classical spherical-geopotential approximation is an interesting read!

7.2 A GUIDE TO THIS CHAPTER AND TO THE NEXT ONE

The approach taken here on geopotential representation is to examine – in a systematic and complementary manner – a quartet of idealised problems of increasing complexity. Each idealisation isolates and examines a specific challenge for the derivation of geopotential representations. To help fix ideas and to help the reader keep track of things, Fig. 7.4 schematically displays cross sections of the geopotential surfaces associated with four idealised planets. This provides a visual overview of the influence a planet's shape and composition have on the induced gravitational field.

A guide to the content of the remainder of this chapter and to the next one now follows. This is to help the reader understand how the individual pieces of the geopotential jigsaw fit together to form a picture. Most of the theory presented here applies not only to planet Earth, but also to other mildly oblate planets and, furthermore, to mildly oblate moons. Except when the context is specific to Earth, the generality of generic planets (and moons) is preserved during the development of the theory.

Sections 7.3–7.5 of this chapter prepare the way for the analysis of the geopotential fields induced by idealised rotating planets:

1. Section 7.3 examines some important equilibrium states of geophysical fluids and discusses the importance of accurately representing intrinsic horizontal and vertical balances.

[17] For (near) Earth, and as previously noted, Newtonian gravity is the dominant contribution to apparent gravity. It is two orders of magnitude larger than the centrifugal contribution.

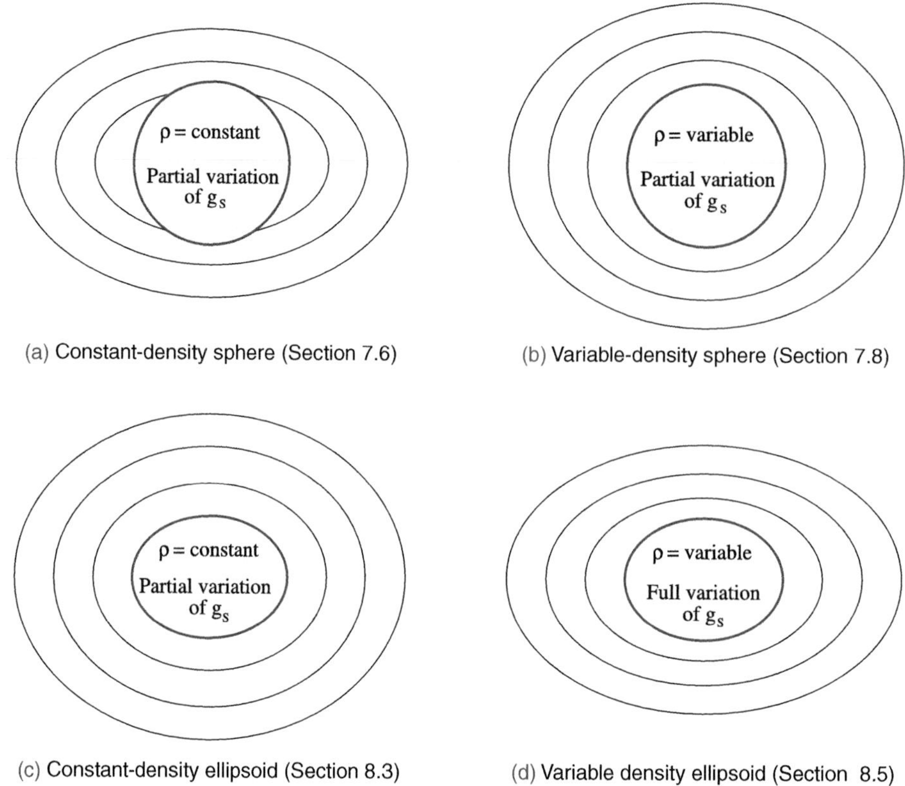

Figure 7.4 Cross sections of geopotential surfaces for constant- and variable-density spherical and ellipsoidal planets. Planets are shaded. The associated partial or full variation of surface gravity (g_S) is annotated on the planets' cross sections.

2. Section 7.4 describes how Earth can be approximately represented as a rotating fluid spheroid in hydrostatic equilibrium. It also introduces the concept of a reference ellipsoid and its importance for geopotential representation.
3. Section 7.5 describes how potential theory can be used to determine the Newtonian gravitational potential induced by a planet having a specified, known density distribution. This provides the analysis framework for:
   (a) Section 7.6 on gravity for a *spherical* planet of *constant* density.
   (b) Section 7.8 on gravity for a *spherical* planet of *variable* density.
   (c) Section 8.3 on gravity for a *spheroidal* planet of *constant* density.
   (d) Section 8.5 on gravity for a *spheroidal* planet of *variable* density.

The analysis (in Section 7.6) for a *constant*-density *spherical* planet is highly tractable. It provides insight into how the analysis framework (described in Section 7.5) works in practice. It is found that the twin assumptions of constant density and sphericity for a *rotating* planet are such that the planet's surface does *not* coincide with a geopotential surface. This demonstrates that *representing Earth as a constant-density sphere would be a bad idea.* To address this deficiency, three possible avenues for investigation are identified in Section 7.7.

The analysis that follows (in Section 7.8) for a *variable*-density *spherical* planet is also highly tractable. It provides valuable insight into the circumstances under which the planet's internal density distribution leads to its spherical surface coinciding, as desired, with a geopotential surface. *These circumstances are surprisingly limited.*

Some possible functional forms for *spheroidal* planets, expressed in various coordinate systems, are developed in Section 8.2. This then prepares the way for the analysis (in Section 8.3) of a *constant*-density *ellipsoidal* planet.[18] Because of the more complicated geometry, this analysis is significantly less straightforward than the two preceding spherical ones. It provides insight into how to handle the additional challenges posed by the change of geometry from spherical to ellipsoidal.

The two analyses for *constant*-density planets, one spherical (in Section 7.6), the other ellipsoidal (in Section 8.3), may be contrasted. The *spherical* analysis of Section 7.6 assumes a prescribed, *rigid*, precisely spherical planet of constant density. This then uniquely defines the Newtonian gravitational potential which, when combined with the known centrifugal potential, gives the geopotential. The planet's surface does *not* then correspond to a geopotential surface. This is because the planet is *assumed* to be a *rigid* solid body that cannot be deformed by forces acting on it.

By way of contrast, the corresponding analysis of Section 8.3 assumes that the planet is a *fluid* (in hydrostatic equilibrium) of *ellipsoidal* form, but of *unknown ellipticity*. Its surface, rather than being rigid, is deformed by and adjusts to the forces acting on it. The precise value of its ellipticity is then determined by a balance of forces at equilibrium such that the planet's ellipsoidal surface coincides with a geopotential surface. This is a much more difficult, and *more strongly coupled* mathematical problem to solve than that for a rigid sphere. The underlying reason for this is that the precise shape of the planet (as measured by its ellipticity) influences the geopotential (via Newtonian gravity), but then the geopotential and equilibrium force balance influences the precise shape of the planet, and so on ad infinitum. (For the rigid spherical planet of Section 7.6, these two aspects nicely decouple.)

As suggested by the analysis (in Section 7.8) for a *variable*-density spherical planet, if all that is of interest is the gravitational field *outside* a planet, and if one knows (by independent means) the total mass of the planet and its precise shape (including the precise value of its ellipticity), then it is possible to obtain this field without knowing anything whatsoever about the planet's *internal density distribution* or *internal gravitational field*. This greatly simplifies the problem. The simplified potential framework to achieve this is presented in Section 8.4.

This prepares the way for the analysis (in Section 8.5) for a *variable*-density ellipsoidal planet. The resulting analysis is of crucial importance to the development (in Chapter 12) of a self-consistent, physically realistic, ellipsoidal, orthogonal, geopotential coordinate system for atmospheric and oceanic modelling. A further (and arguably important) benefit is that the classical spherical-geopotential approximation can be justified (in Section 8.6) as an asymptotic limit of the ellipsoidal geopotential approximation developed in Section 8.5.

When reading the remainder of this chapter and the next one, the reader may wish to consider consulting the present section from time to time. This may help to separate, and keep track of, the various aspects and subtleties regarding the mathematical representation of gravity for idealised rotating planets.

7.3 EQUILIBRIUM STATES FOR UNACCELERATED FLOW

7.3.1 Unaccelerated Flow

Consider unaccelerated flow in Earth's rotating frame of reference (i.e. flow for which the acceleration $D\mathbf{u}/Dt$ is identically zero *when observed in this frame*). In the absence of any external forcing ($\mathbf{F}$), momentum equation (2.28) then reduces to

[18] This analysis is essentially equivalent to Newton's original analysis, but achieved using mathematical techniques developed after his time.

$$\begin{array}{ccccccc} 0 & = & - & 2\boldsymbol{\Omega}\times\mathbf{u} & - & \dfrac{1}{\rho}\nabla p & - & \nabla\Phi. \\ & & & 10^{-3} & & 10 & & 10 \end{array} \tag{7.11}$$

Some approximate magnitudes (m s^{-2}), to be discussed, for each of the terms have been inserted below them. Equation (7.11) is a *three*-way balance between:

1. The Coriolis force, $-2\boldsymbol{\Omega}\times\mathbf{u}$.
2. The pressure-gradient force, $-(1/\rho)\,\nabla p$.
3. The force due to apparent gravity, $-\nabla\Phi$.

Now $g \equiv |\nabla\Phi| \approx 10\,\mathrm{m\,s^{-2}}$, as is well known. Assuming (for sake of argument) that $|\mathbf{u}| \approx 7\,\mathrm{m\,s^{-1}}$, then $|2\boldsymbol{\Omega}\times\mathbf{u}| \approx 10^{-3}\,\mathrm{m\,s^{-2}}$, where (from Table 7.1) $|\boldsymbol{\Omega}| \approx 7.3\times10^{-5}\,\mathrm{s^{-1}}$. Thus, with these (physically realistic) values for Earth's atmosphere, gravitational attraction ($\mathbf{g} \equiv -\nabla\Phi$) is 10^4 times stronger than the Coriolis force ($-2\boldsymbol{\Omega}\times\mathbf{u}$), and *gravitational attraction must therefore be balanced almost entirely by the pressure-gradient force* $-(1/\rho)\,\nabla p$ *in* (7.11).

So far we have only considered the relative *magnitude* of these three forces, but their *direction* is also very important. Apparent gravity, because it is expressible as the gradient of a potential, only acts in the direction *normal* to a geopotential surface (i.e. normal to a surface on which the geopotential Φ is *constant*). See Fig. 7.3. This direction defines the *local vertical*, and it is aligned with a *plumb line*. Importantly, apparent gravity does *not* act in any direction that is *tangential* to a geopotential surface.

One can express (7.11) in *any* convenient coordinate system. However, for global atmospheric and oceanic modelling, the physics is much better represented in a *geopotential coordinate system* (i.e. a coordinate system whose horizontal surfaces are surfaces of *constant* geopotential, aligned with an embedded *spirit level*) than in other coordinate systems. This is because of:

- The dominant nature of gravity, both within Earth and within its atmosphere and oceans.
- The importance of well representing physically important equilibrium states.

To see this, first consider spherical-polar coordinates (λ, χ, r), whose locally 'horizontal' (r = constant) coordinate surfaces do *not* coincide with (spheroidal) geopotential surfaces. See Fig. 7.5 for a meridional cross section. One can then decompose (7.11) into 'horizontal' and 'vertical' components, where 'horizontal' (with quotation marks) means surfaces on which r is constant (i.e. spherical surfaces), and 'vertical' (with quotation marks) means in the direction normal to these 'horizontal' surfaces (i.e. in the direction of $\mathbf{r}$). This then gives

$$\begin{array}{ccccccc} 0 & = & - & (2\boldsymbol{\Omega}\times\mathbf{u})_{\text{`hor'}} & - & \dfrac{1}{\rho}\nabla_{\text{`hor'}}p & - & \nabla_{\text{`hor'}}\Phi, \\ & & & 10^{-3} & & 3\times10^{-2} & & 3\times10^{-2} \end{array} \tag{7.12}$$

$$\begin{array}{ccccccc} 0 & = & - & (2\boldsymbol{\Omega}\times\mathbf{u})\cdot\mathbf{e}_r & - & \dfrac{1}{\rho}\nabla p\cdot\mathbf{e}_r & - & \nabla\Phi\cdot\mathbf{e}_r, \\ & & & 10^{-3} & & 10 & & 10 \end{array} \tag{7.13}$$

where subscript 'hor' denotes the 'horizontal' component (on the surface of a sphere), and $\mathbf{e}_r$ is a unit vector in the 'vertical' (i.e. radially, in the direction of $\mathbf{r}$). The approximate magnitudes given in (7.12) are discussed in Section 7.3.2.

Thus far, nothing much has really changed, and the force balance in both (7.12) and (7.13) is still a *three*-way one, both in the 'horizontal' and in the 'vertical' (i.e. along a *spherical* surface and normal to it, respectively).

7.3.2 Geostrophic and Quasi-Hydrostatic Balance

Consider now how the situation changes if we instead choose a *geopotential coordinate system*, depicted in blue on Fig. 7.5. Because Φ is constant on a (truly horizontal) *geopotential* surface

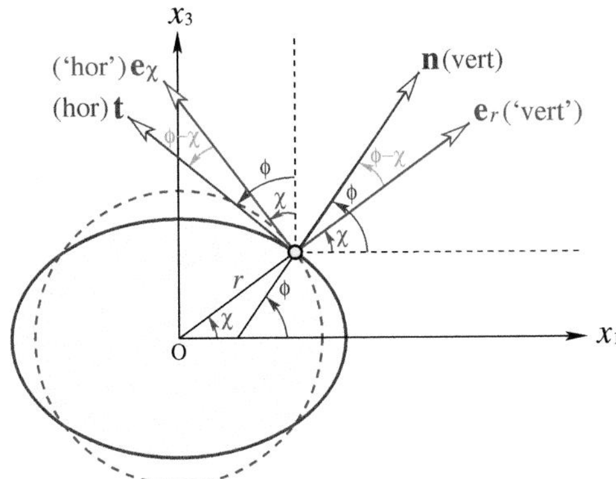

Figure 7.5 Unit vectors at an arbitrary point (depicted in yellow) of a meridional cross section; χ and ϕ correspond to geocentric and geographic latitudes, respectively. Spherical-polar coordinates (depicted in red) have unit vectors $\mathbf{e}_r$ and $\mathbf{e}_\chi$, normal and tangential, respectively, to the (red-dashed) spherical surface that passes through the point. Similarly; geopotential coordinates (depicted in blue) have unit vectors $\mathbf{n}$ and $\mathbf{t}$, normal and tangential, respectively, to the (continuous-blue) spheroidal, geopotential surface that passes through the point. The angle (depicted in green) between the two sets of unit vectors is $\phi - \chi \lesssim 0.2° \approx (1/300)$ rad, with maximum difference at $\phi, \chi \approx \pm 45°$. See text for further details.

(perpendicular to a plumb line, and aligned with a spirit level), so $\nabla_{\text{hor}} \Phi \equiv 0$ on it, where hor (without the quotation marks) denotes the (truly) horizontal component within a geopotential surface. Decomposing (7.11) in the (true and blue) horizontal and vertical directions, respectively, then gives:

The Geostrophic Balance Equation

$$\begin{array}{ccccccc} 0 & = & - & (2\boldsymbol{\Omega} \times \mathbf{u})_{\text{hor}} & - & \dfrac{1}{\rho}\nabla_{\text{hor}} p, \\ & & & 10^{-3} & & 10^{-3} \end{array} \tag{7.14}$$

and

The Quasi-Hydrostatic Balance Equation

$$\begin{array}{ccccccccc} 0 & = & - & (2\boldsymbol{\Omega} \times \boldsymbol{u}) \cdot \boldsymbol{n} & - & \dfrac{1}{\rho}\nabla p \cdot \boldsymbol{n} & - & \nabla \Phi \cdot \boldsymbol{n}. \\ & & & 10^{-3} & & 10 & & 10 \end{array} \tag{7.15}$$

In (7.15), $\mathbf{n}$ is the unit vector in the truly vertical direction (aligned with a plumb line), which is normal to the geopotential surface (aligned with an embedded spirit level).

Equations (7.14) and (7.15) embody the principal horizontal and vertical balances, respectively, in Earth's atmosphere and oceans. Note the absence of a term $\nabla_{\text{hor}} \Phi$ in (7.14) since $\Phi =$ constant on a geopotential surface. This then means that the horizontal pressure-gradient term $(- (1/\rho)\, \nabla_{\text{hor}} p,)$ is approximately $10^{-3}\ \text{m s}^{-2}$ in magnitude in order to balance the only other term, the Coriolis force $(- (2\boldsymbol{\Omega} \times \mathbf{u})_{\text{hor}})$.

The geostrophic balance equation (7.14) has the great virtue that it is a *two*-way force balance between the *Coriolis force* and the *(truly) horizontal, pressure-gradient force*, rather than the *three*-way force balance of (7.12). Reducing a *three*-way force balance in the horizontal to a *two*-way one has not changed the physics in any way. It has merely changed the way in which it is expressed; in a well-chosen (geopotential) coordinate system versus in a less well-chosen (spherical-polar) one. Using the assumptions of Section 7.3.1 for the magnitude of the forces, the two terms in (7.14) necessarily balance one another; both are then approximately equal in value to $10^{-3}\,\mathrm{m\,s^{-2}}$ in magnitude.

Contrast this now with the force balance in (7.12) if spherical-polar coordinates are used instead of geopotential coordinates. See Fig. 7.5 with its colour coding; red for spherical-polar coordinates, blue for geopotential coordinates, and green for the angular difference between geocentric and geographic latitudes. This helps to distinguish between the two closely related coordinate systems. For Earth, the angle between spherical and geopotential surfaces can be as large as (but no more than) $\alpha \equiv \phi - \chi \approx 0.2^\circ = \pi/900$ (this maximum value occurs at a latitude of approximately $\pm 45^\circ$). Projecting $\nabla\Phi$ (which is aligned with unit vector $\mathbf{n}$) into the 'horizontal' of spherical-polar coordinates (aligned with unit vector $\mathbf{e}_\chi$ in the meridional cross section) then yields

$$\begin{aligned} |\nabla_{\text{'hor'}}\Phi| &= |\nabla\Phi \cdot \mathbf{e}_\chi| = |\mathbf{g} \cdot \mathbf{e}_\chi| = g\,|\mathbf{n} \cdot \mathbf{e}_\chi| = g \sin(\phi - \chi) \\ &\approx g \sin\left(\frac{\pi}{900}\right) \approx \frac{g\pi}{900} \approx 3 \times 10^{-2}\,\mathrm{m\,s^{-2}}, \end{aligned} \tag{7.16}$$

where $g \approx 10\,\mathrm{m\,s^{-2}}$ and $\sin x \approx x$ (for small x) have been used.

Thus – from (7.12) and (7.16) – the component of gravity in the 'horizontal' (i.e. on a *spherical* surface) dominates the Coriolis term $-(2\boldsymbol{\Omega} \times \mathbf{u})_{\text{'hor'}}$ by more than a factor of thirty; cf. $3\times10^{-2}\,\mathrm{m\,s^{-2}}$ of (7.16) with 10^{-3} of (7.12). The component of gravity in the 'horizontal' $(\nabla_{\text{'hor'}}\Phi)$ in (7.12) must then be compensated by a correspondingly large contribution from the 'horizontal' pressure-gradient force $-(1/\rho)\,\nabla_{\text{'hor'}}p$.

The use of spherical-polar coordinates in the analysis to this point greatly masks the simplicity of the *physical* balance by effectively adding and subtracting two terms (which only occur because of a suboptimal choice of coordinate system). These terms are *more than an order of magnitude larger* than those that *physically* balance one another. This can introduce serious inaccuracies in a numerical model. For example, assume that the numerical evaluation of the 'horizontal' pressure-gradient and 'horizontal' gravitational forces in (7.12) introduce individual errors of 1% (i.e. errors of $O\left(10^{-4}\right)\,\mathrm{m\,s^{-2}}$). For a flow in perfect (physical) geostrophic balance, this would spuriously create a 'horizontal' acceleration almost as large in magnitude as that of the Coriolis term. The exact geostrophic balance would then be spuriously destroyed. Even decreasing the individual errors in the numerical evaluation of the 'horizontal' pressure-gradient and 'horizontal' gravitational forces by an order of magnitude in this example (from one per cent to one tenth of a per cent) would still unacceptably perturb geostrophic balance.

These illustrative estimates are for the atmosphere. Similar considerations also apply for the oceans. By using different estimates for the magnitude of $\mathbf{u}$, the poor representation of geostrophic balance in spherical-polar (i.e. non-geopotential) coordinates can be made more or less striking.[19]

7.3.3 Hydrostatic Balance

If $\mathbf{u} \equiv 0$ (as it is assumed to be within Earth – see Section 7.1.3), then (7.15) reduces to

[19] Balances for other mildly oblate, spheroidal celestial bodies can also be examined by appropriately varying the estimated values of the parameters for rotation rate and gravity.

The Hydrostatic Balance Equation

$$0 = -\underbrace{\frac{1}{\rho}\nabla p \cdot \mathbf{n}}_{10} - \underbrace{\nabla\Phi \cdot \mathbf{n}}_{10}. \tag{7.17}$$

Gravity is then *exactly* balanced by a vertical pressure gradient. Because, with the illustrative parameter values used for the atmosphere in the preceding analysis, apparent gravitational attraction is 10^4 times stronger than the Coriolis force, and the quasi-hydrostatic balance of (7.15) is very close to the hydrostatic balance of (7.17). For *large-scale* (but not small-scale) atmospheric and oceanic flows, the hydrostatic balance equation is highly accurate, both qualitatively and quantitatively.

7.3.4 Representation of Physical Balances

Consistent with previous work – for example, by Phillips (1973), Gill (1982), White et al. (2008), van der Toorn and Zimmerman (2008), Staniforth (2014b), Staniforth and White (2015a), and Vallis (2017) – we conclude from our discussion and analysis that:

Geopotential Coordinate Systems and Representation of Physical Balances

- By definition of a *geopotential coordinate system*, two of its three spatial coordinates are embedded in the geopotential surface passing through a point, with the third coordinate aligned with the direction of the local vertical, normal to the geopotential surface.
- Because apparent gravity is such a dominant force, it is highly desirable (and arguably essential) for quantitative modelling of Earth's atmosphere and oceans that a geopotential coordinate system be used in order to accurately represent the principal horizontal and vertical force balances of geostrophy and quasi-hydrostaticity, respectively.
- This then ensures that there is no component of apparent gravitational force in any direction tangential to geopotential surfaces. It thereby avoids difficulties with the representation of the horizontal and vertical balance of forces, such as geostrophic and (quasi-)hydrostatic balance, respectively.

7.4 THE GEOPOTENTIAL AT AND NEAR EARTH'S SURFACE

The assumption that Earth is, to a very good approximation, a uniformly rotating fluid spheroid in equilibrium is crucially important for mathematical tractability. It is indeed fortunate that it holds so well in an approximate sense, even though it does not do so in the sense of finest detail. The approach taken for atmospheric and oceanic modelling is very similar to that adopted in *geodesy*.[20] A *reference ellipsoid* (see Fig. 7.6) is first defined to represent Earth's surface of mean sea level, and then the finer detail (such as the superposition of orography) can be incorporated in terms of *small* perturbations about this reference ellipsoid. For terrestrial atmospheric and oceanic modelling purposes, the impact on the gravitational field of these small perturbations can be safely ignored.[21] (When Earth's orography is included, the gravitational

[20] Geodesy is the branch of science concerned with the determination of the size and shape of Earth, and the exact positions of points on its surface; and with the description of variations of its gravitational field.

[21] This is not, however, the case for satellite navigation and Global Positioning System (GPS) applications.

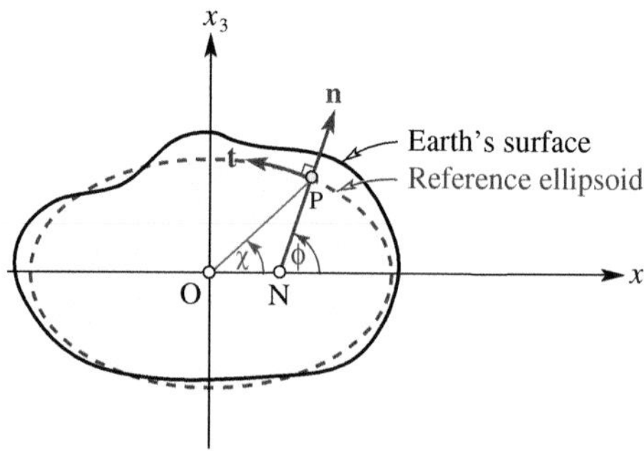

Figure 7.6 An exaggerated cross section in the Ox_1x_3 meridional plane. Earth's surface is depicted by the solid (black) contour, and that of the reference ellipsoid by the dashed (blue) contour. The interior of the reference ellipsoid is shaded; N is the point at which the normal to it at the general point P (yellow dot) intersects Ox_1. At P, χ is *geocentric* latitude, ϕ is *geographic* latitude, and **n** and **t** are normal and tangential unit vectors to the reference ellipsoid there, respectively. After Fig. 1 of Staniforth and White (2015b).

influence of the layer of mass contained between it and the reference ellipsoid is assumed to be 'condensed' to the ellipsoid; see WGS 84 (2004).) The reference ellipsoid is very important, since the position of observed data (particularly satellite data) is reported with respect to it; see Section 12.2 for more information on this aspect. Approximating Earth as a sphere rather than as a spheroid introduces inconsistencies into the handling of observational data. This is undesirable and may lead to systematic errors in model forecasts (Bénard, 2015).

Before proceeding further, we first establish an important property of a *rotating fluid spheroid*. Assume therefore that a planet behaves as a fluid spheroid uniformly rotating about its axis as if it were a solid body, so that $\mathbf{u} \equiv 0$ (where $\mathbf{u}$ is velocity measured in the planet's rotating frame of reference). Now seek equilibrium solutions such that D/Dt and $\partial/\partial t$ (as observed in the planet's rotating frame of reference) of all hydrodynamic and thermodynamic variables are identically zero within the planet's interior. The thermodynamic and continuity equations given in Section 4.4 are then all trivially satisfied,[22] and momentum equation (4.95) reduces to

$$\frac{1}{\rho}\nabla p + \nabla\Phi = 0. \tag{7.18}$$

This equation holds everywhere *within* the planet, including at its surface.[23] Taking the scalar (i.e. dot) product of (7.18), evaluated at the planet's surface, with **t** gives

$$(\nabla\Phi)_S \cdot \mathbf{t} = -\left(\frac{1}{\rho}\nabla p\right)_S \cdot \mathbf{t}, \tag{7.19}$$

where **t** is any unit vector tangent to the planet's surface, and subscript 'S' denotes evaluation there. Now assume that pressure p is everywhere constant at the planet's surface.[24] Applying this condition to (7.19) then gives

$$(\nabla\Phi)_S \cdot \mathbf{t} = 0, \tag{7.20}$$

[22] Although the geophysical governing equations of Section 4.4 were developed with a focus on planetary atmospheres and oceans, they also hold within the interior of rotating fluid spheroids.

[23] The precise shape of the planet's surface at this point in the argument still remains to be determined in some way.

[24] This is a very important assumption that greatly aids tractability.

(i.e. Φ = constant everywhere on the planet's surface). Therefore:

Coincidence of a Rotating Fluid Planet's Bounding Surface with a Geopotential Surface

For a rotating fluid spheroidal planet in equilibrium, having constant pressure on its bounding surface, this surface coincides with a geopotential surface. This is a crucially important property.

7.5 NEWTONIAN GRAVITY AND POTENTIAL THEORY

The determination of the potential, V, for Newtonian gravity can be formulated in terms of either Newton's law of universal gravitation or, equivalently, Gauss's reformulation of this law. For present purposes, the latter is the mathematically simpler approach and is therefore adopted herein. Thus:

Gauss's Procedure to Determine Newtonian Gravity (V)

This (see Fig. 7.7) is comprised of the following three steps:

1. Solve Laplace's equation

$$\nabla^2 V = 4\pi\gamma\rho = 0, \tag{7.21}$$

outside the planet, down to its surface. It is assumed that:

(a) A vacuum[a] exists everywhere outside the planet.[b]

(b) V goes to zero at infinite distance from the planet.

Solution of (7.21) introduces constants, whose values need to be determined from application of boundary conditions – see Step 1(b) and Step 3.

2. Solve Poisson's equation

$$\nabla^2 V = 4\pi\gamma\rho, \tag{7.22}$$

inside the planet and up to its surface. It is assumed that:

(a) ρ is known everywhere within the planet[c] as a function of position.

(b) V is bounded everywhere within the planet.

Solution of (7.22) also introduces constants, whose values need to be determined from application of boundary conditions – see Step 2(b) and Step 3.

3. Match the two solutions of Steps 1 and 2, respectively, for V across the planet's surface, such that V and ∇V are continuous across it.[d]

a In principle, however, ρ should be specified everywhere in the Universe as a function of both position *and* time!

b In making the assumption $\rho \equiv 0$ outside Earth, it is assumed that self-gravitation of Earth's atmosphere is negligible, as are the gravitational influences of the Moon, the Sun, and the remainder of the entire Universe. Neglect of self-gravitation of Earth's atmosphere can be justified on the basis that its density is four orders of magnitude smaller than that of Earth. Although Earth's oceanic tides are driven by the gravitational influence of the Moon, its gravitational influence on the mean state of the ocean (i.e. mean sea level) is negligible. Despite its great mass, the Sun's gravitational influence can be ignored for present purposes, due to its great distance from Earth.

c Thus self-gravitation within the planet is taken into consideration.

d $\nabla^2 V$ is not continuous across the planet's surface since ρ is assumed to vary discontinuously across it and $\nabla^2 V = 4\pi\gamma\rho$.

It remains to discuss the location of the planet's surface, where boundary conditions are applied. A planet is assumed herein to be of either rigid or fluid composition:

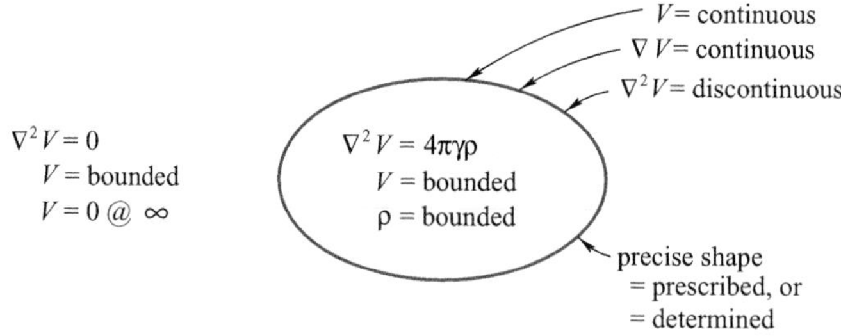

Figure 7.7 The differential equations and boundary conditions that determine Newtonian gravity, V, inside and outside a spheroid (shaded). Poisson's equation, $\nabla^2 V = 4\pi\gamma\rho$, holds inside the spheroid. Laplace's equation, $\nabla^2 V = 0$, holds outside. V and ρ are bounded everywhere. V is arbitrarily set to zero at infinite distance (in any direction) away from the spheroid. V and ∇V are continuous across the spheroid's surface, whereas $\nabla^2 V$ is discontinuous across it. For an assumed *rigid* planet its precise shape is prescribed ab initio, whereas for an assumed *fluid* planet its precise shape is determined simultaneously with determination of V.

- For a *rigid* planet, its shape is assumed prescribed from the very beginning; for example, the constant-density and variable-density spheres examined in Sections 7.6 and 7.8, respectively.
- For a *fluid* planet, its geopotential field (including the contribution of the Newtonian gravitational potential) is determined via an equilibrium of forces at its surface. Further details regarding this are given in Chapter 8.

7.6 A SPHERICAL PLANET OF CONSTANT DENSITY

7.6.1 Preliminaries

The purpose of this section is to provide insight into the challenge of developing a suitable representation of gravity for atmospheric and oceanic modelling. This then motivates the developments that follow.

Since Earth is only mildly oblate, it is natural to consider representing it in the simplest possible manner, namely as a *rigid* spherical planet of constant density. The simplicity of this representation makes it a highly tractable problem analytically. Strengths and weaknesses of this representation for atmospheric and oceanic modelling can then be straightforwardly assessed.

Assume therefore a perfectly spherical, rigid planet, of radius a and of *constant density* $\rho = \overline{\rho}$. The natural choice of coordinate system is then spherical-polar coordinates (λ, χ, r), where λ is longitude, χ is latitude,[25] and r is distance from the planet's centre. See Fig. 7.8.

Because of the radial symmetry for this geometry, $V = V(r)$. Thus the Newtonian potential of gravity, V, does not depend on direction, only on distance from the planet's centre. From (5.22), the Laplacian operator ∇^2 acting on V then simplifies to

$$\nabla^2 V = \frac{1}{r^2}\frac{d}{dr}\left(r^2\frac{dV}{dr}\right). \tag{7.23}$$

The assumption of a spherical planet, with constant density, allows its mass to be straightforwardly computed. Thus, since the volume of a sphere of radius a is $4\pi a^3/3$, the mass of the planet is

$$M = \frac{4\pi a^3\overline{\rho}}{3}. \tag{7.24}$$

[25] There is no need to specify what kind of latitude here since, for a sphere, all latitudes (e.g. geocentric, geographic, and parametric – see Section 8.2 for their definitions) reduce to exactly the same thing.

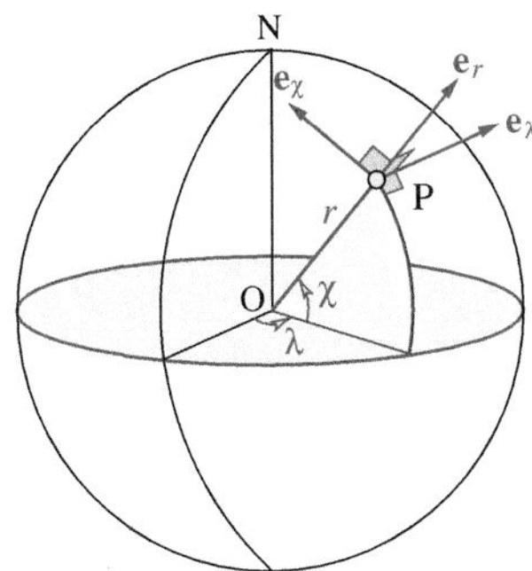

Figure 7.8 Spherical-polar coordinates. P is a general point with coordinates (λ, χ, r) and mutually orthogonal unit vectors $(\mathbf{e}_\lambda, \mathbf{e}_\chi, \mathbf{e}_r)$.

In spherical-polar coordinates, the potential of the centrifugal force, V^C – see (7.6) and Fig. 2.4 – is

$$V^C \equiv -\frac{(\boldsymbol{\Omega} \times \mathbf{r}) \cdot (\boldsymbol{\Omega} \times \mathbf{r})}{2} = -\frac{\Omega^2 r^2 \cos^2 \chi}{2}. \tag{7.25}$$

With this preparation, we can now apply Steps 1–3 of Section 7.5 to explicitly determine the Newtonian gravitational potential $V(r)$.

7.6.2 The Exterior Problem for V (Step 1)

For Step 1, ρ is identically zero (by assumption) everywhere *outside* the planet (i.e. for $r > a$). Inserting definition (7.23) into Laplace's equation (7.21) gives

$$\frac{1}{r^2}\frac{d}{dr}\left[r^2 \frac{dV(r)}{dr}\right] = 0, \quad r \geq a. \tag{7.26}$$

Integrating this equation twice with respect to r and applying the condition that V goes to zero at infinite distance from the planet (i.e. $V \to 0$ as $r \to \infty$) then yields

$$V(r) = \frac{C_1}{r}, \quad r \geq a. \tag{7.27}$$

This equation gives the Newtonian gravitational potential $V(r)$ *outside* the planet and down to its surface $r = a$. The constant C_1 remains to be determined by application (in Step 3) of the boundary conditions at the planet's surface.

7.6.3 The Interior Problem for V (Step 2)

For Step 2, inserting definition (7.23) into Poisson's equation (7.22) (with $\rho = \overline{\rho} =$ constant) gives

$$\frac{1}{r^2}\frac{d}{dr}\left[r^2 \frac{dV(r)}{dr}\right] = 4\pi\gamma\overline{\rho}, \quad r \leq a. \tag{7.28}$$

Integrating this equation twice with respect to r and applying the boundedness of V at $r = 0$ (i.e. at the planet's centre) then yields

$$V(r) = A_0 + \frac{2\pi\gamma\overline{\rho}}{3} r^2, \quad r \leq a. \tag{7.29}$$

This equation gives the Newtonian gravitational potential $V(r)$ *inside* the planet and up to its surface $r = a$. The constant A_0 remains to be determined by application (in Step 3) of the boundary conditions at the planet's surface.

7.6.4 Application of Boundary Conditions at the Planet's Surface (Step 3)

The final step to determine $V(r)$ is to apply the boundary conditions at the planet's surface $r = a$. From (7.27) and (7.29), continuity of V at $r = a$ gives

$$V(r = a) = A_0 + \frac{2\pi\gamma a^2\overline{\rho}}{3} = \frac{C_1}{a}. \tag{7.30}$$

Since $V = V(r)$, continuity of ∇V at the planet's surface simplifies to continuity of dV/dr at $r = a$. Thus differentiating (7.27) and (7.29), followed by evaluation at $r = a$, gives

$$\left(\frac{dV}{dr}\right)_{r=a} = \frac{4\pi\gamma a\overline{\rho}}{3} = -\frac{C_1}{a^2}. \tag{7.31}$$

Solving (7.30) and (7.31) for A_0 and C_1 and using (7.24) to eliminate $\overline{\rho}$ in favour of M then yields

$$A_0 = -2\pi\gamma a^2\overline{\rho} = -\frac{3\gamma M}{2a}, \quad C_1 = -\frac{4\pi\gamma a^3\overline{\rho}}{3} = -\gamma M. \tag{7.32}$$

7.6.5 V for a Spherical Planet of Constant Density

Putting together the preceding results:

The Newtonian Gravitational Potential V for a Spherical Planet of Constant Density

The Newtonian gravitational potential *inside* and *outside* a spherical planet of *constant* density, with radius a and mass M, is given by

$$V(r) = -\frac{\gamma M}{a}\frac{1}{2}\left(3 - \frac{r^2}{a^2}\right), \quad r \leq a, \tag{7.33}$$

$$= -\frac{\gamma M}{r}, \quad r \geq a. \tag{7.34}$$

7.6.6 Newtonian Gravity for a Spherical Planet of Constant Density

Differentiating (7.33) and (7.34) with respect to r, using definition (7.5) of $\mathbf{g}^N$ and definition (5.19) of gradient in spherical-polar coordinates, then gives

$$\mathbf{g}^N \equiv -\nabla V(r) = -\frac{dV}{dr}\mathbf{e}_r = -\frac{\gamma M r}{a^3}\mathbf{e}_r, \quad r \leq a, \tag{7.35}$$

$$= -\frac{\gamma M}{r^2}\mathbf{e}_r, \quad r \geq a, \tag{7.36}$$

where $\mathbf{e}_r$ is a unit vector pointing from the planet's centre to the point with position vector $\mathbf{r}$.[26] The negative signs in (7.35) and (7.36) signify that the Newtonian gravitational force exerted by the planet on a particle of unit mass is directed *towards* the planet's centre, as one expects on physical grounds.[27]

Expressions (7.34) and (7.36) are just the familiar ones for the Newtonian gravitational potential and Newtonian gravitational attraction, respectively, between two particles of unit mass and

[26] From (7.35) and (7.36), it is trivial to verify that $\mathbf{g}^N \equiv -\nabla V$ is continuous across the spherical surface $r = a$.

[27] Caveat: This is *only* true for the *precisely spherical* planet assumed here for now. It is *not* true for a *spheroidal* planet; if this property (of the Newtonian gravitational force being directed to the planet's centre) is incorrectly assumed for a spheroidal planet, then it leads to *incorrect* meridional (latitudinal) variation of gravity.

of mass M, separated by distance r. Thus the Newtonian attraction exerted by a spherical planet of constant density on a particle of unit mass located *outside* the planet is the same as if all of the planet's mass were located at the planet's centre.[28]

7.6.7 The Geopotential Φ for a Spherical Planet of Constant Density

The complete potential (i.e. the geopotential Φ), is obtained by inserting (7.25) – for the centrifugal potential – and (7.33) and (7.34) – for the Newtonian gravitational potential – into definition (7.6) of geopotential. Thus

$$\Phi(\chi, r) = -\frac{\gamma M}{a}\frac{1}{2}\left(3 - \frac{r^2}{a^2}\right) - \frac{\Omega^2 r^2 \cos^2\chi}{2}, \qquad r \leq a, \tag{7.37}$$

$$= -\frac{\gamma M}{r} - \frac{\Omega^2 r^2 \cos^2\chi}{2}, \qquad r \geq a, \tag{7.38}$$

where (7.37) and (7.38) are the geopotentials *inside* and *outside* the spherical planet, respectively. Inserting definition (7.1) of small parameter m into (7.37) and (7.38) then leads to:

The Geopotential Φ for a Spherical Planet of Constant Density

The geopotential *inside* and *outside* a spherical planet of *constant* density, with radius a and mass M, is given by

$$\Phi(\chi, r) = -\frac{\gamma M}{a}\left[\frac{1}{2}\left(3 - \frac{r^2}{a^2}\right) + \frac{m}{2}\left(\frac{r}{a}\right)^2 \cos^2\chi\right], \qquad r \leq a, \tag{7.39}$$

$$= -\frac{\gamma M}{a}\left[\frac{a}{r} + \frac{m}{2}\left(\frac{r}{a}\right)^2 \cos^2\chi\right], \qquad r \geq a, \tag{7.40}$$

where

$$m \equiv \frac{\Omega^2 a^3}{\gamma M}. \tag{7.41}$$

7.6.8 Gravity outside and on a Spherical Planet of Constant Density

The apparent gravitational force *outside* the spherical planet is obtained by inserting (7.40) into (7.4), and using definition (5.19) of gradient in spherical-polar coordinates. Thus

$$\mathbf{g} \equiv -\nabla\Phi = -\frac{1}{r}\frac{\partial\Phi}{\partial\chi}\mathbf{e}_\chi - \frac{\partial\Phi}{\partial r}\mathbf{e}_r \tag{7.42}$$

$$= -\frac{\gamma M}{a^2}\left[m\frac{r}{a}\sin\chi\cos\chi\,\mathbf{e}_\chi + \left(\frac{a^2}{r^2} - m\frac{r}{a}\cos^2\chi\right)\mathbf{e}_r\right], \tag{7.43}$$

$$|\mathbf{g}| = \frac{\gamma M}{a^2}\left[\frac{a^4}{r^4} - 2m\left(\frac{a}{r} - \frac{m}{2}\frac{r^2}{a^2}\right)\cos^2\chi\right]^{\frac{1}{2}}. \tag{7.44}$$

Evaluation of (7.44) at the planet's spherical surface $r = a$ then gives

$$|\mathbf{g}|_{r=a} = \frac{\gamma M}{a^2}\left[1 - 2m\left(1 - \frac{m}{2}\right)\cos^2\chi\right]^{\frac{1}{2}} = \frac{\gamma M}{a^2}\left[1 - m\cos^2\chi + O\left(m^2\right)\right] \tag{7.45}$$

$$= g_S^E\left[1 + m\sin^2\chi + O\left(m^2\right)\right], \tag{7.46}$$

[28] This well-known result is also true if the planet's density *for a spherical planet* varies as a function of radial distance only (i.e. if $\rho = \rho(r)$ instead of $\rho = \overline{\rho} =$ constant).

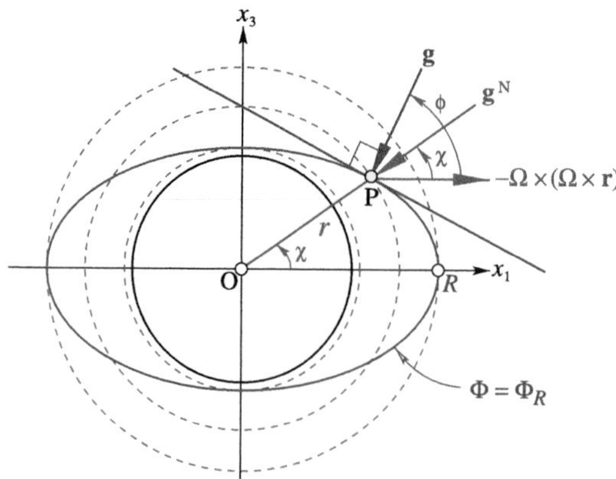

Figure 7.9 An exaggerated cross section in the Ox_1x_3 meridional plane showing the geopotential surface $\Phi = \Phi_R$ (blue curve) for a spherical planet (shaded disc), where R is the surface's equatorial radius. For a general point P on the geopotential surface, $r \equiv |\mathbf{r}|$ is the absolute value of the position vector $\mathbf{r} \equiv \overrightarrow{OP}$ relative to the origin O, χ is *geocentric* latitude, and ϕ is *geographic* latitude. At P, the Newtonian gravitational force $\mathbf{g}^N$ is aligned with $\mathbf{r}$, and the centrifugal force acts in the direction normal to the planet's rotation axis (Ox_3). The apparent gravitational force $\mathbf{g} \equiv \mathbf{g}^N - \boldsymbol{\Omega} \times (\boldsymbol{\Omega} \times \mathbf{r})$ acts in the (inward-pointing) direction normal to the geopotential surface at P. Surfaces of constant r are denoted by dashed circles (in red).

where

$$g_S^E \equiv \frac{\gamma M}{a^2}(1-m) \tag{7.47}$$

is the value of g at the planet's equator.

From (7.45) and (7.46), it is seen that gravity at the surface of a *spherical* planet of constant density is larger at its poles than at its equator.[29]

7.6.9 Geopotential Surfaces outside a Spherical Planet of Constant Density

The geopotential *outside* a spherical planet of constant density is given by (7.40). Consider now the geopotential surface that passes identically through a point with equatorial radius R (i.e. a point lying in the equatorial plane such that $\chi = 0$ and $r = R$). See Fig. 7.9. Label this geopotential surface Φ_R (a constant). Thus, from (7.40),

$$\Phi_R \equiv \Phi(\chi = 0, r = R) = -\frac{\gamma M}{a}\left[\frac{a}{R} + \frac{m}{2}\left(\frac{R}{a}\right)^2\right]. \tag{7.48}$$

The equation that defines this particular geopotential surface is $\Phi(\chi, r) = \Phi_R =$ constant, so that – equating (7.40) and (7.48) –

$$\frac{1}{r}\left[1 + \frac{m}{2}\left(\frac{r}{a}\right)^3 \cos^2\chi\right] = \frac{1}{R}\left[1 + \frac{m}{2}\left(\frac{R}{a}\right)^3\right]. \tag{7.49}$$

This is a cubic equation for r. To exactly solve this equation is complicated so, to provide further insight, we instead solve it approximately for $m \ll 1$. The square-bracketed term on the left-hand side of this equation is first rewritten as

[29] This *qualitatively* agrees with reality for Earth but *quantitatively* underestimates the pole-to-equator variation by about one third. See Table 8.2 for further details.

$$1+\frac{m}{2}\left(\frac{r}{a}\right)^3\cos^2\chi = 1+\frac{m}{2}\left(\frac{R}{a}\right)^3\left(1+\frac{r-R}{R}\right)^3\cos^2\chi = 1+\frac{m}{2}\left(\frac{R}{a}\right)^3\cos^2\chi + O\left(m^2\right), \tag{7.50}$$

where it has been assumed that $(r-R)/R \lesssim O(m)$, that is, the geopotential surface $\Phi = \Phi_R$ is almost spherical.[30] Substituting (7.50) into (7.49) and expanding binomially then gives

$$\begin{aligned} r &= R\left[1+\frac{m}{2}\left(\frac{R}{a}\right)^3\cos^2\chi + O\left(m^2\right)\right]\left[1+\frac{m}{2}\left(\frac{R}{a}\right)^3\right]^{-1} \\ &= R\left[1-\frac{m}{2}\left(\frac{R}{a}\right)^3\sin^2\chi + O\left(m^2\right)\right]. \end{aligned} \tag{7.51}$$

This, for a spherical planet, is an $O(m)$-accurate equation for the geopotential surface $\Phi = \Phi_R$ *outside* the planet. For Earth – see Table 7.1 – $m \approx 1/289 \Rightarrow O\left(m^2\right) \sim 10^{-5} \ll 1$. Thus (7.51) is an excellent approximation of the solution of cubic equation (7.49) with Earth's parameters.

7.6.10 Assessment of the Model of Gravity for a Spherical Planet of Constant Density

Assuming a spherical planet of constant density and examining the situation *outside* the planet:

- Newtonian gravitational potential (V) is given by (7.34).
- Geopotential (Φ) by (7.40).
- Gravity ($\mathbf{g}$) by expressions (7.42)–(7.47).
- The equation for geopotential surfaces by (7.49).

7.6.10.1 Non-Rotating Planet

Consider first the situation where the planet does not rotate about its axis (i.e. its rotation rate Ω is identically zero). From definition (7.41), this implies that $m \equiv 0$ as well. There is then no centrifugal force (since Ω is now zero). The geopotential Φ, given by (7.40), reduces to the Newtonian gravitational potential V, given by (7.34), so that $\Phi = V$. Setting $m \equiv 0$ in (7.49) and solving for r then gives $r = R$ for the equation of the geopotential surfaces. The geopotential surfaces therefore reduce to spheres. In particular, the geopotential surface $\Phi_{R=a}$ coincides with the reference surface of this spherical, non-rotating planet. Setting $m \equiv 0$ in (7.43) reduces it to $\mathbf{g} = -\gamma M/r^2$, so that $\mathbf{g}$ is constant on any geopotential surface $r = R$ and varies inversely as the square of the distance from the planet's centre. Thus, overall, this would be an excellent model of gravity for a *non-rotating* spherical planet of constant density.

7.6.10.2 Rotating Planet

Consider now the more realistic situation where the planet rotates about its axis with non-zero angular frequency Ω. There is now a (non-zero) centrifugal force, and the geopotential Φ, given by (7.40), no longer reduces to the Newtonian gravitational potential V, given by (7.34).

If the planet were truly spherical and homogeneous (i.e. of constant density), then the direction of Newtonian gravity would be normal to the planet's (perfectly spherical) surface, but apparent gravity (and therefore a plumb line) would not be aligned with this normal. Consequently – see Fig. 7.9 – there is:

[30] Do not be misled by Fig. 7.9 which, for illustrative purposes, grossly exaggerates the relative importance of the centrifugal force with respect to the Newtonian gravitational force for terrestrial applications.

A Serious Deficiency of the Model of Gravity for a Rotating Spherical Planet of Constant Density

- The planet's perfectly spherical surface would not coincide with a constant geopotential surface.
- This would then cause possible/probable difficulties with accurate representation of horizontal and vertical force balances when using spherical-polar coordinates, the natural coordinate system for a perfectly spherical planet.

This would be bad news indeed. Fortunately, Earth is not *spherical* with *constant density* but is instead (approximately) *spheroidal* with *variable density*.

7.7 AVENUES FOR INVESTIGATION

To address this important deficiency (namely non-coincidence of a constant-density spherical planet's surface with a geopotential surface), three possible avenues are explored. In summary, they are to:

1. Retain the simplicity of a spherical planet, but relax the assumption of constant density (in Section 7.8).
2. Retain the simplicity of constant density, but relax the assumption of sphericity (in Section 8.3).
3. Simultaneously relax the two assumptions of constant density and of sphericity (in Section 8.5).

Common to all three avenues is the requirement that the planet's surface coincides with a geopotential surface. What distinguishes the three avenues from one another are the conditions under which this is achieved.

The first avenue provides additional insight into the challenge of representing Earth's geopotential in a physically realistic manner. However, it results in only two-thirds of the meridional variation of gravity at Earth's surface being represented. This is attributable to the sphericity assumption. The second avenue provides further insight, but it also leads to underestimation of the observed meridional variation of gravity at Earth's surface. This is attributable to the assumption of constant density. The third avenue, which *simultaneously* relaxes the assumptions of both constant density and sphericity, leads to capture of the full variation.

The reader is reminded that Fig. 7.4 provides a visual summary of the influence a planet's shape and composition have on the induced gravitational field.

7.8 A SPHERICAL PLANET OF VARIABLE DENSITY

7.8.1 Preamble

In Section 7.6, we found that the surface of a constant-density, rotating, spherical planet does not coincide with a geopotential surface. *This may then create serious difficulties for accurate representation of horizontal and vertical force balances in atmospheric and oceanic models.*

The purpose of this section is therefore to explore the extent to which it is possible to simultaneously retain the simplicity of a spherical planet, whilst nevertheless obtaining coincidence of the planet's surface with a geopotential surface. This would address, at least partially, the deficiency identified in Section 7.6.

If we wish to maintain sphericity of the planet, there is then no alternative but to relax the assumption that the planet's density is constant. The planet is still assumed to be *rigid* – with its bounding (spherical) surface prescribed a priori – but its composition is now allowed to vary within its interior. Consideration of this possibility raises the following questions:

- Is it actually possible to vary the density $\rho(\chi, r)$ in such a way that the planet's spherical surface exactly coincides with a geopotential surface?
- If so:
  - How should density vary?
  - Does gravity at the spherical planet's surface vary meridionally?
    - If so, does it do so in a physically realistic manner?

To prepare the way for what follows, some useful, standard, spherical relations are collated in the Appendix to this chapter. These are frequently used in what follows (and are also used elsewhere).

We are now ready to apply the three-step procedure of Section 7.5 to determine the Newtonian gravitational potential $V(\chi, r)$ for a *rigid, spherical* planet of *variable* density.

7.8.2 The Exterior Problem for V (Step 1)

7.8.2.1 *Problem Definition and Functional Form for $V(\chi, r)$*

Outside a spherical planet of radius $r = a$, the exterior solution for $V(\chi, r)$ must satisfy Laplace's equation (7.21). Using zonally averaged Laplacian (7.118) then gives

$$\nabla^2 V(\chi, r) \equiv \frac{1}{r^2 \cos\chi}\frac{\partial}{\partial\chi}\left(\cos\chi\frac{\partial V}{\partial\chi}\right) + \frac{1}{r^2}\frac{\partial}{\partial r}\left(r^2\frac{\partial V}{\partial r}\right) = 0, \quad r \geq a. \tag{7.52}$$

Since $V = V(\chi, r)$ only enters the definition of gravity in differentiated form (as a gradient), it is only determined to within an arbitrary additive constant. Applying the condition at infinity that

$$V(\chi, r) \to 0 \text{ as } r \to \infty \tag{7.53}$$

then fixes the value of this arbitrary constant.[31] The solution must also be bounded everywhere outside the planet. These requirements mean that the exterior solution must have the form of a weighted sum of *spherical solid harmonics*.[32] Thus

$$V(\chi, r) = C_0\frac{a}{r} + C_1\frac{a^2}{r^2}P_1(\sin\chi) + C_2\frac{a^3}{r^3}P_2(\sin\chi) + C_3\frac{a^4}{r^4}P_3(\sin\chi)\ldots, \quad r \geq a, \tag{7.54}$$

where C_0, C_1, C_2, C_3, ... are constants to be determined, and $P_0 \equiv 1$ (see (7.115)) has been exploited to simplify the first term on the right-hand side.

Partially differentiating (7.54) with respect to r gives

$$\frac{\partial V(\chi, r)}{\partial r} = -C_0\frac{a}{r^2} - 2C_1\frac{a^2}{r^3}P_1(\sin\chi) - 3C_2\frac{a^3}{r^4}P_2(\sin\chi) - 4C_3\frac{a^4}{r^5}P_3(\sin\chi)\ldots, \quad r \geq a. \tag{7.55}$$

Evaluation of (7.54) and (7.55) at the planet's surface $r = a$ then yields

$$V(\chi, r = a) = C_0 + C_1P_1(\sin\chi) + C_2P_2(\sin\chi) + C_3P_3(\sin\chi) + \cdots, \quad r \geq a, \tag{7.56}$$

$$\frac{\partial V(\chi, r = a)}{\partial r} = -\frac{1}{a}\left[C_0 + 2C_1P_1(\sin\chi) + 3C_2P_2(\sin\chi) + 4C_3P_3(\sin\chi) + \cdots\right]. \quad r \geq a. \tag{7.57}$$

[31] Another natural way of doing so would be to set $V(\chi, r = a) = 0$. This would not change the ensuing representation of gravity.

[32] Spherical solid harmonics are eigen solutions of Laplace's equation in spherical geometry. For problems with no variation in λ (as here), they take the forms (see (7.119) and (7.120)) $r^n P_n(\sin\chi)$, $n = 0, 1, 2, \ldots$ and $r^{-(n+1)}P_n(\sin\chi)$, $n = 0, 1, 2, \ldots$. For the present problem, solutions of the first form are excluded by boundary condition (7.53), leaving only those of the second form.

The complete potential (i.e. the geopotential Φ) is obtained by inserting (7.25) for the centrifugal potential, V^C, into (7.6). Thus, both inside and outside the planet,

$$\Phi(\chi, r) = V(\chi, r) - \frac{\Omega^2}{2} r^2 \cos^2 \chi = V(\chi, r) - \frac{\Omega^2}{3} r^2 \left[1 - P_2(\sin \chi)\right], \quad r \geq 0, \tag{7.58}$$

where the second identity of (7.116) has been used to re-express $\cos^2 \chi$. Outside the planet, $V(\chi, r)$ has the form (7.54). Inside the planet, $V(\chi, r)$ remains to be determined – see Section 7.8.4.

Now from (7.58) and using (7.56), the geopotential at the planet's surface is given by

$$\begin{aligned}\Phi(\chi, r = a) &= V(\chi, r = a) - \frac{\Omega^2}{3} a^2 \left[1 - P_2(\sin \chi)\right] \\ &= C_0 + C_1 P_1(\sin \chi) + C_2 P_2(\sin \chi) + C_3 P_3(\sin \chi) + \cdots - \frac{\Omega^2}{3} a^2 \left[1 - P_2(\sin \chi)\right].\end{aligned} \tag{7.59}$$

We demand that the planet's surface coincides with a geopotential surface. For this to be so, $\Phi(\chi, r = a)$ must be constant. Because the Legendre polynomials $P_n(\sin \chi)$ are mutually independent, examination of (7.59) shows that $\Phi(\chi, r = a)$ can only be constant if

$$C_1 = 0, \quad C_2 = -\frac{\Omega^2 a^2}{3}, \quad C_3 = C_4 = C_5 = \cdots = 0. \tag{7.60}$$

Thus only C_0 and C_2 in (7.54) can be non-zero. Substituting the coefficient values given in (7.60) into (7.54), the exterior solution can therefore be more succinctly rewritten as

$$V(\chi, r) = C_0 \frac{a}{r} - \frac{\Omega^2 a^2}{3} \frac{a^3}{r^3} P_2(\sin \chi), \quad r \geq a, \tag{7.61}$$

where only C_0 remains to be determined. This will be achieved by matching V and $\partial V / \partial r$ to the interior solution at the planet's surface.

7.8.2.2 *Exterior Matching Conditions*

To prepare for this matching, partial differentiation of (7.61) yields

$$\frac{\partial V(\chi, r)}{\partial r} = -C_0 \frac{a}{r^2} + \Omega^2 a^2 \frac{a^3}{r^4} P_2(\sin \chi), \quad r \geq a. \tag{7.62}$$

Evaluating exterior solution (7.61) and its derivative $\partial V / \partial r$ (given by (7.62)) at $r = a$ then gives the exterior matching conditions:

$$V(\chi, r = a) = C_0 - \frac{\Omega^2 a^2}{3} P_2(\sin \chi), \tag{7.63}$$

$$\frac{\partial V(\chi, r = a)}{\partial r} = -\frac{1}{a} \left[C_0 - \Omega^2 a^2 P_2(\sin \chi)\right]. \tag{7.64}$$

The corresponding interior solutions, evaluated at $r = a$, must match these.

7.8.3 Gravity at the Planet's Surface

Before examining the interior problem, it is instructive to obtain an expression for apparent gravity, $g_S = g_S(\chi) \equiv \left|\mathbf{g}(\chi, r = a)\right|$, at the planet's surface. Now – from definitions (7.4) and (5.19) –

$$\mathbf{g}(\chi, r) \equiv -\nabla\Phi(\chi, r) = -\frac{\partial\Phi}{\partial r}\mathbf{e}_r - \frac{1}{r}\frac{\partial\Phi}{\partial\chi}\mathbf{e}_\chi, \tag{7.65}$$

where $\mathbf{e}_r$ and $\mathbf{e}_\chi$ are unit vectors in the r and χ directions respectively. Evaluating (7.65) at the planet's spherical surface at r = a and noting that *this is assumed to be a geopotential surface* (for which $\Phi = \text{constant} \Rightarrow \partial\Phi(\chi, r = a)/\partial\chi = 0$) then leads to

$$\mathbf{g}_S(\chi) \equiv \mathbf{g}(\chi, r = a) = -\frac{\partial\Phi}{\partial r}(\chi, r = a)\mathbf{e}_r - \frac{1}{a}\cancelto{0}{\frac{\partial\Phi}{\partial\chi}(\chi, r = a)}\mathbf{e}_\chi = -\frac{\partial\Phi}{\partial r}(\chi, r = a)\mathbf{e}_r. \tag{7.66}$$

This simplification (of no contribution to $\mathbf{g}_S(\chi)$ from $\nabla\Phi$ in the direction of $\mathbf{e}_\chi$) is because the planet's (geopotential) surface coincides with a (spherical-polar) coordinate surface. As we will see, this simplification does not occur for the ellipsoidal cases examined in Sections 8.3 and 8.5 where the (ellipsoidal) planet's (geopotential) surface no longer coincides with a (spherical-polar) coordinate surface.[33]

From the last expression of (7.66), $\partial\Phi(\chi, r = a)/\partial r$ must be strictly positive otherwise gravity at the planet's surface would not point inwards as it must. Partially differentiating (7.58) with respect to r, followed by evaluation at $r = a$ and use of (7.115), (7.64), and (7.66), then yields

$$\begin{aligned} g_S(\chi) &\equiv |\mathbf{g}(\chi, r = a)| = \frac{\partial\Phi}{\partial r}(\chi, r = a) = \frac{\partial V}{\partial r}(\chi, r = a) - \frac{2\Omega^2 a}{3}[1 - P_2(\sin\chi)] \\ &= -\frac{1}{a}\left[C_0 + \frac{2\Omega^2 a^2}{3} - \frac{5\Omega^2 a^2}{3}P_2(\sin\chi)\right] = -\frac{1}{a}\left(C_0 + \frac{3\Omega^2 a^2}{2} - \frac{5\Omega^2 a^2}{2}\sin^2\chi\right). \end{aligned} \tag{7.67}$$

Evaluating (7.67) at $\chi = 0$ gives

$$g_S^E \equiv g_S(\chi = 0) = -\frac{1}{a}\left(C_0 + \frac{3\Omega^2 a^2}{2}\right) \quad \Rightarrow \quad C_0 = -\left(ag_S^E + \frac{3\Omega^2 a^2}{2}\right), \tag{7.68}$$

where g_S^E is apparent gravity at the planet's equator ($\chi = 0, r = a$). Using (7.68), (7.67) is more compactly written as

$$g_S(\chi) = g_S^E + \frac{5\Omega^2 a}{2}\sin^2\chi. \tag{7.69}$$

Without any need to examine the interior problem, it is seen from (7.67) and (7.69) that:

Gravity outside and on a Rotating Spherical Planet of Variable Density

- Gravity at the surface of a rotating ($\Omega \neq 0$) spherical planet must vary meridionally.

[33] Even for the present case (of a spherical planet of variable density) this simplification only occurs at the planet's surface (at $r = a$). Elsewhere, as we will find, the geopotential surfaces are no longer spherical in shape, and the simplification is lost.

- Gravity will be strongest at the planet's poles (where $\sin^2\chi = 1$) and weakest around its equator (where $\sin^2\chi = 0$).[a]
- The only impact that the planet's density distribution has on gravity *external to the planet* (other than constraining the planet's surface to coincide with a geopotential surface) is through the value of g_S^E, i.e. through the value of C_0 since g_S^E and C_0 are proxies of one another – see (7.68).

These are important results.

The underlying reason for them is the requirement that the planet's surface coincides with a geopotential surface. This is a very strong (and crucially important) constraint.

a This qualitatively agrees with observations for Earth, albeit not quantitatively.

7.8.4 The Interior Problem for V (Step 2)

Returning to the main argument, we now examine the interior problem for V.

7.8.4.1 *Problem Definition*

Inside a spherical planet of radius $r = a$, with density distribution $\rho = \rho(\chi, r)$, the interior solution for $V(\chi, r)$ must satisfy Poisson's equation (7.22). Using zonally averaged Laplacian (7.118) this gives

$$\nabla^2 V(\chi, r) \equiv \frac{1}{r^2 \cos\chi}\frac{\partial}{\partial\chi}\left(\cos\chi\frac{\partial V}{\partial\chi}\right) + \frac{1}{r^2}\frac{\partial}{\partial r}\left(r^2\frac{\partial V}{\partial r}\right) = 4\pi\gamma\rho(\chi, r), \quad r \leq a, \qquad (7.70)$$

where γ is Newton's universal gravitational constant. We wish to specify $\rho = \rho(\chi, r)$ in such a way that:

- Poisson equation (7.70) is satisfied.
- Both $\rho(\chi, r \leq a)$ and $V(\chi, r \leq a)$ are bounded.
- Exterior matching conditions (7.63) and (7.64) are satisfied.
- The planet's surface corresponds to both a geopotential surface and an isopycnal surface (i.e. a surface on which density is constant).

As we will see, *these conditions do not uniquely determine the density distribution $\rho(\chi, r)$, although they do very much constrain it.* Construction of a model of $\rho(\chi, r)$ that satisfies these conditions is now investigated.

7.8.4.2 *Functional Forms for $\rho(\chi, r)$ and $V(\chi, r)$*

The question now is:

- What functional form should the density distribution $\rho = \rho(\chi, r)$ have?

We are principally interested in *simple representations* that characterise the most important features of Earth's density distribution.[34] The simplest representation is to assume that ρ is everywhere constant. This is certainly too simple. As seen in Section 7.6.9, *the planet's surface cannot then coincide with a geopotential surface*, due to the implied radial symmetry of $V = V(r)$.

To address this deficiency we can infer from exterior matching conditions (7.63) and (7.64) that there must be a contribution to ρ that varies as $P_2(\sin\chi)$ so that, when solving Poisson's equation (7.70) for $V = V(\chi, r)$, these conditions are satisfied. Now we know that Earth – to

[34] Similar considerations also apply to other spherical, or near-spherical, celestial bodies.

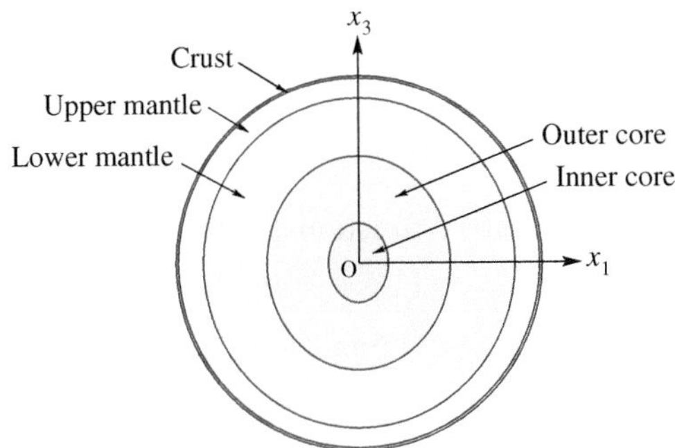

Figure 7.10 Cross section in the Ox_1x_3 meridional plane, of some constant-density (isopycnal) surfaces inside a rigid, spherical planet of variable composition; darker shading indicates higher density. The outermost contour corresponds to the planet's rigid, spherical boundary, which is assumed to coincide with a geopotential surface. Isopycnal contours have been chosen to broadly correspond with those associated with Earth's internal layers. See text for further details.

leading order – has an onion-like structure, with its mass distributed in a series of five concentric shells from Earth's crust to its inner core (Jeffreys, 1976; Dziewonski and Anderson, 1981). See Fig. 7.10 (for the spherical planet under consideration here) and 8.5 (for an ellipsoidal planet) for meridional cross sections of density distribution. We also know that Earth's inner core is much denser than its crust.[35] Therefore, to be physically realistic:

- ρ should vary as a function of distance r from the planet's centre.
- The planet's boundary (i.e. thin crust) should correspond to an isopycnal surface.
- Furthermore, for V in the interior to satisfy boundary conditions (7.63) and (7.64) at $r = a$, and to therefore match the exterior solution (7.61) there, we expect ρ to vary meridionally solely as a function of $P_2(\sin\chi)$.

Although these considerations very much limit the possibilities for the functional form of ρ, some further detective work is needed. This is provided by examining the *inverse* problem:

- Given particular functional forms for $V(\chi, r)$, what do these imply for the functional form of $\rho(\chi, r)$?

A contribution to $V(\chi, r)$ of the form $r^2 P_2(\sin\chi)$ cannot contribute to $\rho(\chi, r)$ since $r^2 P_2(\sin\chi)$ is a spherical solid harmonic, and the Laplacian, ∇^2, of it is therefore identically zero; see (7.119). Contributions involving r alone, or $P_2(\sin\chi)$ alone, or $rP_2(\sin\chi)$ are excluded because, after taking ∇^2 of them, they introduce a singularity in ρ at $r = 0$; see (7.121)–(7.123).

Taking these considerations into account, the simplest possible functional form for $V = V(\chi, r)$ that meets our requirements turns out to be

$$V(\chi, r) = A_0 + A_2\frac{r^2}{a^2} + A_3\frac{r^3}{a^3} - \frac{\Omega^2 r^2}{3}P_2(\sin\chi) + \left(B_2\frac{r^2}{a^2} + B_3\frac{r^3}{a^3} + B_4\frac{r^4}{a^4}\right)P_2(\sin\chi), \quad r \leq a. \tag{7.71}$$

(The separation of the $r^2 P_2(\sin\chi)$ term into two contributions is for later convenience.) We can add as many higher-order polynomial terms in r as we wish to the A and B series of terms without adversely affecting mathematical tractability, but this only adds complexity. We should not,

[35] This is consistent with Earth's interior behaving as a fluid in hydrostatic balance (see Section 7.4); the nearer the centre, the greater the hydrostatic pressure, and the greater the compression of matter. It is also consistent with denser matter having sunk centreward.

however, add any terms involving $P_n(\sin\chi)$, $n \neq 0$ or 2. This would violate exterior matching condition (7.63), with the consequence that the planet's spherical surface would no longer coincide with a geopotential surface.

The question now is:

- What does this proposed, minimalist, functional form for $V(\chi, r)$ imply for the sought functional form for $\rho(\chi, r)$?

To determine this, we insert expansion (7.71) into Poisson's equation (7.70) to obtain

$$\rho(\chi, r) = \frac{1}{4\pi\gamma}\nabla^2 V(\chi, r) = \frac{3}{2\pi\gamma a^2}\left[A_2 + 2A_3\frac{r}{a} + \left(B_3\frac{r}{a} + \frac{7}{3}B_4\frac{r^2}{a^2}\right)P_2(\sin\chi)\right], \quad r \leq a, \tag{7.72}$$

as a physically reasonable, leading-order model of $\rho(\chi, r)$. Conceptually, we can start with density distribution (7.72) and expect the corresponding interior solution for $V(\chi, r)$ to then have the form (7.71).

We are now in a position to apply the constraint that the planet's spherical surface at $r = a$ corresponds to an isopycnal surface (i.e. a surface on which density is constant). To do so, we impose the condition that

$$\rho(\chi, r = a) \equiv \rho_a = \text{constant}. \tag{7.73}$$

Evaluating density distribution (7.72) at $r = a$ and applying constraint (7.73) then gives

$$\rho(\chi, r = a) = \frac{3}{2\pi\gamma a^2}\left[A_2 + 2A_3 + \left(B_3 + \frac{7}{3}B_4\right)P_2(\sin\chi)\right] = \rho_a = \text{constant}. \tag{7.74}$$

For this equation to hold, the $P_2(\sin\chi)$ term must be zero, which implies that

$$B_4 = -\frac{3}{7}B_3. \tag{7.75}$$

Insertion of (7.75) into (7.71) and (7.72) then leads to

$$V(\chi, r) = A_0 + A_2\frac{r^2}{a^2} + A_3\frac{r^3}{a^3} - \frac{\Omega^2 r^2}{3}P_2(\sin\chi) + \left[B_2\frac{r^2}{a^2} + B_3\frac{r^3}{a^3}\left(1 - \frac{3}{7}\frac{r}{a}\right)\right]P_2(\sin\chi), \quad r \leq a, \tag{7.76}$$

$$\rho(\chi, r) = \frac{3}{2\pi\gamma a^2}\left[A_2 + 2A_3\frac{r}{a} + B_3\frac{r}{a}\left(1 - \frac{r}{a}\right)P_2(\sin\chi)\right], \quad r \leq a. \tag{7.77}$$

Equation (7.77) captures the most important aspects of the physics, namely radial and meridional variation of density:

- The A_2 term provides a basic, constant contribution.
- The A_3 term allows the planet's core to be denser than its crust, including in the absence of planetary rotation (when B_3 turns out to be zero).
- The B_3 term allows meridional variation; *without this term, the planet's surface could not be an isopycnal surface nor a geopotential surface.*

A_2 and A_3 can be considered to be free parameters that (almost) define the density distribution (B_3 is also needed). In (7.76) and (7.77), A_0, B_2, and B_3 are *not* free parameters – their values are obtained by applying appropriate boundary conditions at $r = a$.

7.8.4.3 An Equivalent Functional Form for $\rho(\chi, r)$

To help further fix ideas, the simple functional form (7.77) for $\rho(\chi, r)$ is now equivalently rewritten by re-expressing A_2 and A_3 in terms of two new parameters, namely ρ_0 and ρ_a, that are easier to interpret physically.

Specifically, let ρ_0 and ρ_a be the constant values of ρ at the planet's centre and at its spherical boundary, respectively. Evaluating (7.77) at $r = 0$ and $r = a$, respectively, gives

$$\rho_0 \equiv \rho(r=0) = \frac{3}{2\pi\gamma a^2} A_2, \quad \rho_a \equiv \rho(r=a) = \frac{3}{2\pi\gamma a^2}(A_2 + 2A_3). \tag{7.78}$$

Solving these two equations for A_2 and A_3 then leads to

$$A_2 = \frac{2\pi\gamma a^2}{3}\rho_0, \quad A_3 = -\frac{\pi\gamma a^2}{3}(\rho_0 - \rho_a). \tag{7.79}$$

Inserting these expressions for A_2 and A_3 into (7.77) results in the following equivalent functional form for density distribution (7.77):

$$\rho(\chi, r) = \underbrace{\rho_0 - (\rho_0 - \rho_a)\frac{r}{a}}_{\text{linear - radial variation}} + \underbrace{\frac{3}{2\pi\gamma a^2} B_3 \frac{r}{a}\left(1 - \frac{r}{a}\right) P_2(\sin\chi)}_{\text{quadratic - radial \& meridional variation}}, \quad r \leq a. \tag{7.80}$$

In (7.80), ρ_0 and ρ_a are free parameters, and B_3 is determined by application of boundary conditions at the planet's assumed spherical surface at $r = a$; see Section 7.8.5. It will turn out that $B_3 \propto \Omega^2 \propto m$ – see (7.93) – and, in the present context, B_3 is small. This also means that the B_3 term in (7.80) is entirely absent in the absence of planetary rotation, since B_3 is then identically zero. Examination of (7.80) shows that $\rho_0 > \rho_a$ then guarantees that ρ monotonically increases linearly (with respect to r) from the planet's crust to its core along any radius. Setting $\rho_a = \rho_0/10$ would, for example, make the density at the planet's core an order of magnitude larger than at its crust. For Earth, this would be qualitatively correct and, at the current level of understanding of Earth's composition, quantitatively in the right ball park.

7.8.4.4 Solving the Poisson Problem

Returning to the main argument, using (7.77) as the model for density distribution $\rho(\chi, r)$ – rather than the equivalent (7.80) – the interior solution for $V(\chi, r)$ can now be obtained. To do so, we must solve Poisson's equation (7.70) for $V(\chi, r)$ – with $\rho(\chi, r)$ defined by (7.77) – subject to the conditions:

- $V(\chi, r \leq a)$ is bounded.
- V and $\partial V/\partial r$ match the exterior solution (7.61) at $r = a$ (i.e. (7.63) and (7.64) are simultaneously satisfied).

By construction, and before application of the boundary conditions, the solution of the Poisson problem therefore has the form (7.76).[36]

[36] It might be thought that (7.76) is missing a term or terms. By solving the Poisson problem directly rather than inferring the solution from the inverse problem, it can, however, be shown that this is not the case. (The solution is a sum of spherical solid harmonics plus a particular solution. A particular solution can be straightforwardly obtained, inspired by (7.76). Application of the boundary conditions to the sum of the general and particular solutions then gives the complete solution.) Alternatively, once the final solution (after application of the matching conditions) has been found, it can be verified a posteriori that each and every equation and condition of the problem is satisfied.

7.8.4.5 *Interior Matching Conditions*

To prepare for the matching, (7.76) for $V = V(\chi, r)$ is evaluated at $r = a$ to obtain

$$V(\chi, r = a) = A_0 + A_2 + A_3 - \frac{\Omega^2 a^2}{3} P_2(\sin\chi) + \left(B_2 + \frac{4}{7}B_3\right) P_2(\sin\chi). \quad (7.81)$$

Partially differentiating (7.76) with respect to r gives

$$\frac{\partial V(\chi, r)}{\partial r} = 2A_2\frac{r}{a^2} + 3A_3\frac{r^2}{a^3} - \frac{2\Omega^2 r}{3}P_2(\sin\chi) + \left[2B_2\frac{r}{a^2} + 3B_3\frac{r^2}{a^3}\left(1 - \frac{4}{7}\frac{r}{a}\right)\right] P_2(\sin\chi), \quad (7.82)$$

and evaluating this at $r = a$ then yields

$$\frac{\partial V(\chi, r = a)}{\partial r} = \frac{1}{a}(2A_2 + 3A_3) - \frac{2\Omega^2 a}{3}P_2(\sin\chi) + \frac{1}{a}\left(2B_2 + \frac{9}{7}B_3\right) P_2(\sin\chi). \quad (7.83)$$

Equations (7.81) and (7.83) are the two interior matching conditions. Everything is now in place to match the interior and exterior solutions at $r = a$.

7.8.5 Matching the Interior and Exterior Solutions (Step 3)

Interior conditions (7.81) and (7.83) must agree with exterior conditions (7.63) and (7.64), respectively. Comparing (7.63) with (7.81) for $V(\chi, r = a)$, and (7.64) with (7.83) for $\partial V(\chi, r = a)/\partial r$, then gives:

$$\begin{aligned} V(\chi, r = a) &= A_0 + A_2 + A_3 - \frac{\Omega^2 a^2}{3}P_2(\sin\chi) + \left(B_2 + \frac{4}{7}B_3\right)P_2(\sin\chi) \\ &= C_0 - \frac{\Omega^2 a^2}{3}P_2(\sin\chi), \end{aligned} \quad (7.84)$$

$$\begin{aligned} \frac{\partial V(\chi, r = a)}{\partial r} &= \frac{1}{a}(2A_2 + 3A_3) - \frac{2\Omega^2 a}{3}P_2(\sin\chi) + \frac{1}{a}\left(2B_2 + \frac{9}{7}B_3\right)P_2(\sin\chi) \\ &= -\frac{1}{a}\left[C_0 - \Omega^2 a^2 P_2(\sin\chi)\right]. \end{aligned} \quad (7.85)$$

These equations imply that:

$$A_0 + A_2 + A_3 - C_0 = 0, \quad (7.86)$$

$$B_2 + \frac{4}{7}B_3 = 0, \quad (7.87)$$

$$2A_2 + 3A_3 + C_0 = 0, \quad (7.88)$$

$$2B_2 + \frac{9}{7}B_3 = \frac{5\Omega^2 a^2}{3}. \quad (7.89)$$

Solving (7.86) and (7.88) for C_0 and A_0, and (7.87) and (7.89) for B_2 and B_3, then yields:

$$C_0 = -(2A_2 + 3A_3), \tag{7.90}$$

$$A_0 = -(3A_2 + 4A_3), \tag{7.91}$$

$$B_2 = -\frac{20}{3}\Omega^2 a^2 = -\frac{20m}{3}\frac{\gamma M}{a}, \tag{7.92}$$

$$B_3 = \frac{35}{3}\Omega^2 a^2 = \frac{35m}{3}\frac{\gamma M}{a}, \tag{7.93}$$

where

$$m \equiv \frac{\Omega^2 a^3}{\gamma M} \tag{7.94}$$

is the previously defined ratio – see (7.1) and (7.41) – of centrifugal force to gravitational attraction.

7.8.6 The Complete Solution for V for a Spherical Planet of Variable Density

Substituting into (7.77) and (7.76), respectively, the values for the constants given in (7.90)–(7.93) leads to the density distribution

$$\rho(\chi, r) = \frac{3}{2\pi\gamma a^2}\left[A_2 + 2A_3\frac{r}{a} + \frac{35m}{3}\frac{\gamma M}{a}\frac{r}{a}\left(1 - \frac{r}{a}\right)P_2(\sin\chi)\right], \quad r \le a, \tag{7.95}$$

and to the interior solution

$$V(\chi, r) = -A_2\left(3 - \frac{r^2}{a^2}\right) - A_3\left(4 - \frac{r^3}{a^3}\right) - 7\frac{\gamma M}{a}m\left[1 - \frac{5}{3}\frac{r}{a}\left(1 - \frac{3}{7}\frac{r}{a}\right)\right]\frac{r^2}{a^2}P_2(\sin\chi), \quad r \le a. \tag{7.96}$$

The presence of the two free parameters A_2 and A_3 in (7.96) reflects the impact that density distribution (7.95) has on the Newtonian gravitational potential $V(\chi, r)$ in the planet's interior. The presence of the $P_2(\sin\chi)$ term in (7.95) is what allows the planet's surface to coincide with a geopotential surface. Specifying the values of parameters A_2 and A_3 in (7.95) determines the precise form of the planet's internal density distribution and, thereby, its internal Newtonian potential (7.96). For example, by specifying the values of ρ_0 and ρ_a (and also setting B_3 to its value in (7.93)), one could use the equivalent representation of Section 7.8.4.3. These values would then, using (7.79), define the corresponding values of A_2 and A_3 to be used there and onwards.

The Newtonian gravitational potential $V(\chi, r)$ external to the planet is given by (7.61), where C_0 is now known – see (7.90). Thus

$$V(\chi, r) = -(2A_2 + 3A_3)\frac{a}{r} - \frac{m}{3}\frac{\gamma M}{a}\frac{a^3}{r^3}P_2(\sin\chi), \quad r \ge a. \tag{7.97}$$

As written, the Newtonian gravitational potential $V(\chi, r)$ external to the planet responds to the specified values of A_2 and A_3 that appear in definition (7.95) of the density distribution. So far, so unsurprising; one intuitively expects the Newtonian gravitational potential to respond to the planet's mass, including how it is distributed. There is, however, an at first sight, somewhat surprising result left to reveal. It turns out that *it is crucial to the development (in Section 8.5) of a spheroidal model of gravity for atmospheric and oceanic modelling.*

7.8.7 A Somewhat Surprising Result and a Speculation

It would be disastrous for atmospheric and oceanic modelling if we needed an accurate representation of Earth's internal density distribution, since we do not have this (Jeffreys, 1976; Dziewonski and Anderson, 1981).

Given a planet's density distribution, ρ, its total mass can be obtained by integration of ρ over its volume. It is instructive to carry out this computation for density distribution (7.95) and see what we get. Thus:

$$
\begin{aligned}
M &= \int_0^{2\pi} \int_{-\frac{\pi}{2}}^{+\frac{\pi}{2}} \int_0^{a} \rho\left(\chi, r\right) r^2 \cos\chi \, d\lambda d\chi dr \\
&= \frac{3}{2\pi\gamma a^2} \int_0^{2\pi} \int_{-\frac{\pi}{2}}^{+\frac{\pi}{2}} \int_0^{a} \left[A_2 + 2A_3 \frac{r}{a} + \frac{35m}{3} \frac{\gamma M}{a} \frac{r}{a} \left(1 - \frac{r}{a}\right) P_2\left(\sin\chi\right) \right] \\
&\qquad \times r^2 \cos\chi \, d\lambda d\chi dr \\
&= \frac{a}{\gamma} \left(2A_2 + 3A_3\right), \\
&\Downarrow
\end{aligned}
$$

$$
2A_2 + 3A_3 = \frac{\gamma M}{a}, \tag{7.98}
$$

where M is the planet's total mass, and definition (7.115) of $P_2\left(\sin\chi\right)$ has been used in the evaluation of the volume integral.

Lo and behold, the quantity $2A_2 + 3A_3$ that appears in (7.97) is, from (7.98), directly proportional to the planet's mass! This means that (7.97) – which defines the Newtonian gravitational potential $V\left(\chi, r\right)$ external to the planet – can be rewritten as

$$
V\left(\chi, r\right) = -\frac{\gamma M}{r} \left[1 + \frac{m}{3} \frac{a^2}{r^2} P_2\left(\sin\chi\right) \right], \quad r \geq a. \tag{7.99}
$$

Provided that one knows the values for the spherical planet's:

- Total mass (M).
- Radius (a).
- Rotation rate (Ω).
- Newton's universal gravitational constant (γ).

then the Newtonian gravitational potential $V\left(\chi, r\right)$ external to the variable-density spherical planet is given by (7.99).

This suggests that, although knowledge of a *particular* density distribution – namely (7.95) – was used to obtain this result, perhaps this knowledge may not actually be necessary if the planet's total mass is known. For Earth, its total mass is independently known to great accuracy from astronomical, geological, and satellite observations (Pavlis et al., 2012, 2013). This speculation is now examined.

7.8.8 Rederivation of $V(\chi, r)$ Exterior to the Planet

Recapitulating, Laplace's equation (7.21) was solved in Section 7.8.2 subject to:

- Boundary condition (7.53) at infinity.
- Boundedness of the solution.
- The constraint that the planet's surface coincides with a geopotential surface.

This then led to (7.61), that is, to

$$V(\chi, r) = C_0 \frac{a}{r} - \frac{\Omega^2 a^2}{3} \frac{a^3}{r^3} P_2(\sin\chi), \quad r \geq a, \tag{7.100}$$

for the Newtonian gravitational potential $V(\chi, r)$ *exterior to the planet.*

In Section 7.8.3, apparent gravity *at the surface of the planet*, $g_S(\chi)$, was shown to satisfy

$$g_S(\chi) = \frac{\partial \Phi}{\partial r}(\chi, r = a) = \frac{\partial V}{\partial r}(\chi, r = a) - \frac{2\Omega^2 a}{3}\left[1 - P_2(\sin\chi)\right] = g_S^E + \frac{5\Omega^2 a}{2}\sin^2\chi, \tag{7.101}$$

where

$$g_S^E \equiv g_S(\chi = 0) = -\frac{1}{a}\left(C_0 + \frac{3\Omega^2 a^2}{2}\right) \quad \Rightarrow \quad C_0 = -\left(a g_S^E + \frac{3\Omega^2 a^2}{2}\right), \tag{7.102}$$

and g_S^E is the value of g at the planet's equator.

To obtain (7.100)–(7.102), *no information whatsoever regarding the planet's composition was required.* This only came into play to determine the value of C_0 that appears in (7.100). To achieve this, Poisson's equation (7.22) was solved subject to a number of conditions, one of which was to assume a density distribution of the specific form (7.72). *We now demonstrate that this assumption is unnecessary if one knows the total mass of the planet.*

The total mass of the planet, M, may be obtained by integrating Poisson's equation (7.22) over the volume of the planet. Thus:

The Mass of a Planet

$$M \equiv \iiint_{\mathscr{V}} \rho \, d\mathscr{V} = \frac{1}{4\pi\gamma}\iiint_{\mathscr{V}} \nabla^2 V d\mathscr{V} = \frac{1}{4\pi\gamma}\oiint_{\mathscr{S}} \nabla V \cdot \mathbf{n} d\mathscr{S}, \tag{7.103}$$

where:

- $\mathscr{V}$ and $\mathscr{S}$ are the volume and surface area of the planet, respectively.
- $d\mathscr{V}$ and $d\mathscr{S}$ are the corresponding infinitesimal elements.
- $\mathbf{n}$ is the outward-pointing normal at the planet's surface.
- Gauss's divergence theorem (A.33) (with $\mathbf{A} \equiv \nabla V$) has been used to transform a volume integral into a surface one.

Note that *no assumption about the form of the density distribution ρ is needed, nor has been used, to obtain* (7.103).

Now from (7.6), the Newtonian potential (V) is related to the geopotential (Φ) and to the centrifugal potential (V^C) by

$$V = \Phi - V^C, \tag{7.104}$$

where

$$V^C = -\frac{(\boldsymbol{\Omega} \times \mathbf{r}) \cdot (\boldsymbol{\Omega} \times \mathbf{r})}{2}. \tag{7.105}$$

We may therefore rewrite (7.103) as

$$M=\frac{1}{4\pi\gamma}\oiint_{\mathscr{S}}\left(\nabla\Phi-\nabla V^{C}\right)\cdot\mathbf{n}d\mathscr{S}=\frac{1}{4\pi\gamma}\oiint_{\mathscr{S}}\left(-\mathbf{g}-\nabla V^{C}\right)\cdot\mathbf{n}d\mathscr{S},\tag{7.106}$$

where $\mathbf{g}\equiv-\nabla\Phi$ is the (inward-pointing) vector of apparent gravity. Very importantly, (7.103) and (7.106) are *general* results; *they are valid not only for a spherical planet but also for a spheroidal one.*

For the *particular* case of spherical geometry, $-\left(\mathbf{g}\cdot\mathbf{n}\right)_{\mathscr{S}}=g_S\left(\chi\right)$ (where $g_S\left(\chi\right)$ is given by (7.101)), and $V^C=-\Omega^2r^2\cos^2\chi/2$. Using these relations in (7.106) then yields

$$\begin{aligned}M&=\frac{1}{4\pi\gamma}\oiint_{\mathscr{S}}\left(-\mathbf{g}-\nabla V^{C}\right)\cdot\mathbf{n}d\mathscr{S}\\&=\frac{1}{4\pi\gamma}\int_0^{2\pi}\int_{-\frac{\pi}{2}}^{+\frac{\pi}{2}}\left[g_S^E+\frac{5\Omega^2a}{2}\sin^2\chi-\frac{\partial}{\partial r}\left(-\frac{\Omega^2r^2\cos^2\chi}{2}\right)\right]_{r=a}a^2\cos\chi\,d\lambda\,d\chi\\&=-\frac{a}{\gamma}C_0,\end{aligned}\tag{7.107}$$

where (7.102) has been used to eliminate g_S^E in favour of C_0. Inserting (7.107) into (7.100), with use of (7.94), finally gives

$$V\left(\chi,r\right)=-\frac{\gamma M}{a}\left[\frac{a}{r}+\frac{m}{3}\frac{a^3}{r^3}P_2\left(\sin\chi\right)\right],\quad r\geq a,\tag{7.108}$$

which is precisely (7.99). This *particular* case explicitly demonstrates that *to obtain the Newtonian gravitational potential $V\left(\chi,r\right)$ exterior to the (assumed spherical) planet, it is not necessary to know anything at all about the planet's mass distribution, only its total mass.*

It has also been implicitly assumed that the planet's surface coincides with a geopotential surface. This assumption was made to obtain (7.102), which has been used to obtain (7.108). Not making this important assumption leads to the serious practical difficulties discussed in Section 7.3 for global atmospheric and oceanic modelling. *It is therefore considered herein to be an essential assumption.*

7.8.9 Assessment of the Model of Gravity for a Spherical Planet of Variable Density

In the context of modelling Earth's atmosphere and oceans, it is natural to wonder:

- How physically realistic is the (non-classical)[37] spherical-geopotential approximation, developed in this section (i.e. Section 7.8), for a spherical planet of variable density?

Now the standard assessment test for meridional variation of gravity (Todhunter, 1873; Staniforth and White, 2015a) is to compute *Clairaut's famous fraction*, namely $\left(g_S^P-g_S^E\right)/g_S^E$, and to then compare it against the actual (asymptotic) value $(5m/2)-\varepsilon$ obtained from spheroidal geopotential theory (see Section 8.5 herein). So we now proceed to do so. Note that Clairaut's fraction is a measure of the meridional variation of gravity between a planet's equator and a pole.

[37] The use of '*non-classical*' here is to distinguish this spherical-geopotential approximation from the '*classical*' one described in Section 7.1.5. *They are different*, and not to be confused with one another. The present 'non-classical' spherical geopotential is, by construction, physically realisable. As argued in Section 7.9, the classical spherical geopotential is not.

Assessment Criteria for Clairaut's Fraction for a Mildly Oblate Planet

- Clairaut's fraction $\left(g_S^P - g_S^E\right)/g_S^E$ should be *positive* for any realistic geopotential approximation for a mildly oblate planet such as Earth, since gravity is measurably larger at a Pole for Earth than at its Equator.
- The closer Clairaut's fraction is to the asymptotic value of $(5m/2) - \varepsilon$, the better the geopotential approximation.

Using (7.58), (7.94), and (7.99) to recapitulate and set the scene:

The Geopotential and Gravity outside a Spherical Planet of Variable Density

The geopotential *outside* a spherical planet of *variable* density, with radius a and mass M, is

$$\begin{aligned}\Phi(\chi, r) &= -\frac{\gamma M}{r}\left\{1 + \frac{m}{3}\frac{a^2}{r^2}P_2(\sin\chi) + \frac{m}{3}\frac{r^3}{a^3}\left[1 - P_2(\sin\chi)\right]\right\} \\ &= -\frac{\gamma M}{r}\left[1 - \frac{m}{6}\frac{a^2}{r^2} + \frac{m}{2}\frac{r^3}{a^3} + \frac{m}{2}\left(\frac{a^2}{r^2} - \frac{r^3}{a^3}\right)\sin^2\chi\right], \qquad (7.109)\end{aligned}$$

where $m \equiv \Omega^2 a^3/(\gamma M)$; see (7.94). Gravity at the planet's spherical surface is then given by

$$\begin{aligned}g_S(\chi) &= \left(\frac{\partial\Phi}{\partial r}\right)_{r=a} = -\gamma M\left\{\frac{\partial}{\partial r}\left[\frac{1}{r} - \frac{m}{6}\frac{a^2}{r^3} + \frac{m}{2}\frac{r^2}{a^3} + \frac{m}{2}\left(\frac{a^2}{r^3} - \frac{r^2}{a^3}\right)\sin^2\chi\right]\right\}_{r=a} \\ &= \frac{\gamma M}{a^2}\left(1 - \frac{3m}{2} + \frac{5m}{2}\sin^2\chi\right) \approx \frac{\gamma M}{a^2}\left(1 - \frac{3m}{2}\right)\left(1 + \frac{5m}{2}\sin^2\chi\right) \\ &= g_S^E\left(1 + \frac{5m}{2}\sin^2\chi\right). \qquad (7.110)\end{aligned}$$

In (7.110):

- Binomial expansion has been used.
- Terms of $O\left(m^2\right)$ have been neglected ($m \approx 1/300$ for Earth, and so $m^2 \sim 10^{-5}$).
- The value of g at the planet's equator is

$$g_S^E \equiv g_S(\chi = 0) = \frac{\gamma M}{a^2}\left(1 - \frac{3m}{2}\right). \qquad (7.111)$$

Evaluating (7.110) at $\chi = \pm\pi/2$ gives

$$g_S^P \equiv g_S\left(\chi = \pm\frac{\pi}{2}\right) = g_S^E\left(1 + \frac{5m}{2}\right), \qquad (7.112)$$

where g_S^P is the value of g at the planet's poles. Using (7.112),

$$\text{Clairaut's fraction} \equiv \frac{g_S^P - g_S^E}{g_S^E} = \frac{5m}{2} \text{ for geopotential representation (7.109).} \qquad (7.113)$$

Comparing (7.113) against the asymptotic value $(5m/2) - \varepsilon$, it is seen that the term in ε is missing. This can be attributed to the fact that ε *is zero for a sphere* but non-zero for a spheroid. Quantitatively, $\varepsilon \approx m$ for Earth (from observed data), and so the actual value $(5m/2) - \varepsilon$ for Clairaut's fraction is approximately equal to $3m/2$. This means that the value given in (7.113) for (non-classical) spherical-geopotential representation (7.109) *over*estimates the actual value by approximately m; $5m/2$ versus $3m/2$.

By way of further comparison, the classical (shallow and deep) spherical-geopotential approximations (7.7) and (7.9) *under*estimate the actual value by approximately $3m/2$; 0 versus $3m/2$. Thus:

Non-Classical versus Classical Spherical-Geopotential Representation for Mildly Oblate Planets such as Earth

Non-classical spherical-geopotential representation (7.109) for a rotating spherical planet of variable density is quantitatively more accurate than classical spherical-geopotential approximations (7.7) and (7.9), with the added virtue of permitting meridional variation of gravity (albeit with some exaggeration).

This is the *theoretical* upside of nonclassical spherical-geopotential representation (7.109) when compared to classical spherical-geopotential approximations (7.7) and (7.9). There is, however, an important *practical* downside. This downside is that the *only* perfectly spherical geopotential surface associated with (non-classical) spherical-geopotential representation (7.109) is the one coincident with the planet's assumed bounding spherical surface.[38] See Fig. 7.11 for a schematic that graphically depicts this behaviour (in a much exaggerated manner).

The reason for this is that substituting $r = R =$ constant (the equation that defines a spherical surface of radius R, whose centre coincides with the planet's) into (7.109) shows that Φ is only constant (and therefore a geopotential surface) when $R = a$ (i.e. only for the planet's assumed spherical surface). This behaviour is in contradistinction to that of the classical (shallow and deep) spherical-geopotential approximations (7.7) and (7.9), for which the geopotential surfaces are *all* assumed to be precisely spherical. Thus, whereas the classical spherical-geopotential approximations can take advantage of the simplifications afforded by spherical geometry, the (non-classical) spherical-geopotential representation (7.109) cannot. This is a serious practical drawback.

Despite this difficulty, it is natural to wonder:

1. Could the (non-classical) spherical-geopotential representation (7.109) nevertheless be used to develop an orthogonal, curvilinear, *geopotential* coordinate system?
2. Would it be advantageous to do so, practically speaking?

The answers are:

1. Yes, it certainly could.
2. But no, it would not really be advantageous to do so.

It turns out that the associated coordinate system is a special case of GREAT (Geophysically Realistic, Ellipsoidal, Analytically Tractable) coordinates (to be developed in Chapter 12), so it is certainly possible. However, there would be little point in proceeding in this manner since one

[38] It is nevertheless a significant improvement over what was found in Section 7.6 for a *homogeneous* spherical planet, where *none* of the geopotential surfaces were spherical. In particular, that the planet's surface does *not* coincide with a geopotential surface precludes its use as the basis of a geopotential coordinate system. This is in contradistinction to the (non-classical) spherical-geopotential representation (7.109) which, because the planet's surface *does* coincide with a geopotential surface, *can* be used as the basis of a geopotential coordinate system.

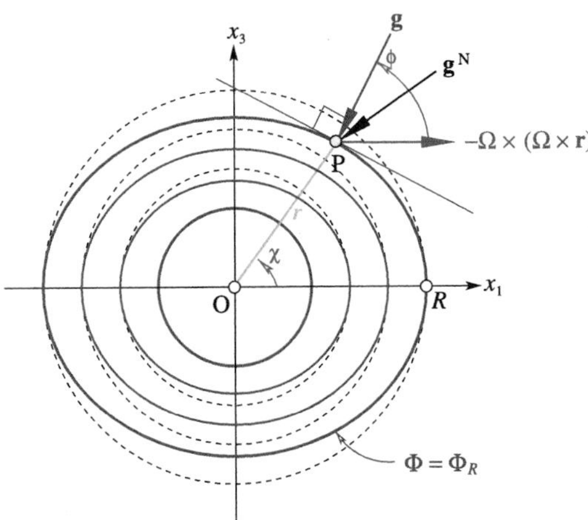

Figure 7.11 An exaggerated cross section in the Ox_1x_3 meridional plane showing the geopotential surfaces (blue curves) for a spherical planet (shaded disc) of variable density. Surfaces of constant r are denoted by dashed circles. For a general point P on the geopotential surface $\Phi = \Phi_R$ (where R is the surface's equatorial radius); $r \equiv |\mathbf{r}|$ is the absolute value of position vector $\mathbf{r} \equiv \overrightarrow{OP}$ relative to origin O, χ is *geocentric* latitude, and ϕ is *geographic* latitude. At P, the Newtonian gravitational force $\mathbf{g}^N$ is no longer aligned with $\mathbf{r}$ since the density is no longer constant. The centrifugal force acts in the direction normal to the planet's rotation axis (Ox_3 axis). The apparent gravitational force $\mathbf{g} \equiv \mathbf{g}^N - \boldsymbol{\Omega} \times (\boldsymbol{\Omega} \times \mathbf{r})$ acts in the direction normal to the geopotential surface at P.

could, for no extra complexity and negligible extra computational cost, anyway use the more accurate and physically realistic GREAT representation.

The importance of the (non-classical) spherical-geopotential representation (7.109) is not so much that it is, practically speaking, that useful, but rather that it provides important insight, and a conceptually useful stepping stone, towards the development of a more accurate *ellipsoidal* representation. That said, *if* the planet were precisely spherical, and *if* its internal mass distribution were such that its surface coincided with a geopotential surface, then it would unequivocally be the *best* geopotential representation, since it would be *exact*. It is only because Earth is closer to being an ellipsoid than a sphere that (non-classical) spherical-geopotential representation (7.109) is not a particularly useful practical proposition. It essentially falls between two stools; the simpler, classical, spherical-geopotential one, and the more realistic ellipsoidal one.

The following is offered as a clarification. One might think that the analysis for a rotating spherical planet of *constant density* should be recoverable as a special case of that for a rotating spherical planet of *variable density*. This is not true, however. The variable-density case *assumes* – with no ifs, ands, or buts – that the planet's bounding surface *must* coincide with a constant geopotential surface. This is then incompatible with the geopotential field for a rotating *constant-density* spherical planet, for which the planet's surface *cannot* coincide with a constant geopotential surface. Taking the limit that the variable density becomes constant is thus a *singular limit*, with the consequence that *the uniform-density analysis cannot be recovered as a special case of the variable-density one.*

7.9 CONCLUDING REMARKS

For a rotating spherical planet of *constant density*, it was found – in Section 7.6 – that its perfectly spherical bounding surface *cannot* coincide with a constant geopotential surface. This would then cause possible/probable difficulties with accurate representation of horizontal and vertical force balances when using spherical-polar coordinates – the natural coordinate system for a perfectly spherical planet. The underlying cause of this serious deficiency is the assumption of *constant density*.

This motivated the development – in Section 7.8 – of a representation of gravity for a rotating spherical planet of *variable density*, subject to the crucial constraint that the planet's perfectly spherical bounding surface *must* coincide with a constant geopotential surface. This representation of gravity is consequently much better than that for a rotating spherical planet of constant density. In particular:

- Gravity varies meridionally, albeit in a slightly exaggerated manner with respect to reality.
- A geopotential coordinate system can be constructed, a highly desirable property.

A limitation then, however, is that the *only* geopotential that is purely spherical is the one that coincides with the planet's surface. This is in contradistinction to the situation for the classical spherical-geopotential approximation – described in Section 7.1.5 – for which *all* constant geopotential surfaces are assumed to be purely spherical, with ensuing, practical simplifications. A price paid for these simplifications, however, is that gravity cannot vary meridionally.

For a rotating *spherical* planet with variable density, there is one, *and only one*, spherical-geopotential approximation possible *outside* the planet that respects the essential constraint that the planet's bounding surface coincides with a constant geopotential surface. And it is *not* the classical spherical-geopotential approximation. This means that *the classical spherical-geopotential approximation is not physically realisable.* This is disappointing from a conceptual standpoint, and has led to much confusion over the years regarding rigorous justification of this widely used approximation (see e.g. van der Toorn and Zimmerman (2008)'s overview). This issue is revisited in Section 8.6, where it is shown that this approximation can be justified as a mathematical *asymptotic limit* of a physically realisable *ellipsoidal* geopotential approximation.

The facts that:

- None of the spherical-geopotential approximations properly capture the observed meridional variation of gravity.
- Earth's shape (its Figure) is closer to being ellipsoidal than spherical.

motivate the analyses of Chapter 8 for ellipsoidal planets of both constant and variable density. The end result is then two physically realistic, ellipsoidal, geopotential approximations – one above Earth's geoid, the other below – that do properly represent the observed meridional variation of gravity. They are subsequently used – in Chapter 12 – as the basis for the development of a unified, ellipsoidal, geopotential coordinate system for atmospheric and oceanic modelling.

APPENDIX: SOME SPHERICAL RELATIONS

Some spherical relations used in this chapter, and elsewhere, are given in this appendix – see, for example, Lebedev (1972).

Legendre Polynomials

Legendre polynomials, $P_n(\sin\chi)$, are the zonally averaged eigenfunctions of the Laplacian operator, ∇^2, in spherical-polar coordinates (λ, χ, r). As such, they satisfy Legendre's equation

$$\frac{1}{\cos\chi}\frac{d}{d\chi}\left[\cos\chi\frac{dP_n(\sin\chi)}{d\chi}\right]+n(n+1)P_n(\sin\chi)=0,\quad n\geq 0 \text{ and integer.} \tag{7.114}$$

Of particular importance in this chapter are those of low degree, namely

$$P_0\equiv 1,\quad P_1\equiv\sin\chi,\quad P_2(\sin\chi)\equiv\frac{3\sin^2\chi-1}{2},\quad P_3(\sin\chi)\equiv\frac{5\sin^3\chi-3\sin\chi}{2}. \tag{7.115}$$

Using (7.115), it is straightforward to verify the following identities:

$$\sin^2\chi \equiv \frac{1+2P_2(\sin\chi)}{3}, \quad \cos^2\chi \equiv \frac{2\left[1-P_2(\sin\chi)\right]}{3}, \tag{7.116}$$

$$\frac{dP_1(\sin\chi)}{d\chi} \equiv \cos\chi, \quad \frac{dP_2(\sin\chi)}{d\chi} \equiv 3\sin\chi\cos\chi, \quad \frac{dP_3(\sin\chi)}{d\chi} \equiv \frac{3\left(5\sin^2\chi-1\right)\cos\chi}{2}. \tag{7.117}$$

Laplacians of Zonally Averaged Functions

In geocentric spherical-polar coordinates (λ, χ, r), the Laplacian of a zonally averaged function $f = f(\chi, r)$ takes the form (from (5.22))

$$\nabla^2 f(\chi, r) \equiv \frac{1}{r^2\cos\chi}\frac{\partial}{\partial\chi}\left(\cos\chi\frac{\partial f}{\partial\chi}\right) + \frac{1}{r^2}\frac{\partial}{\partial r}\left(r^2\frac{\partial f}{\partial r}\right). \tag{7.118}$$

Using (7.114)–(7.118), it can be shown that

$$\nabla^2\left[r^n P_n(\sin\chi)\right] = 0, \quad n \geq 0 \text{ and integer}, \tag{7.119}$$

$$\nabla^2\left[r^{-(n+1)} P_n(\sin\chi)\right] = 0, \quad n \geq 0 \text{ and integer}, \tag{7.120}$$

$$\nabla^2 r^n = n(n+1)\,r^{n-2}, \quad n \text{ integer}, \tag{7.121}$$

$$\nabla^2 P_2(\sin\chi) = -\frac{6}{r^2}P_2(\sin\chi), \tag{7.122}$$

$$\nabla^2\left[r^n P_2(\sin\chi)\right] = (n+3)(n-2)\,r^{n-2}P_2(\sin\chi), \quad n \geq 0 \text{ and integer}. \tag{7.123}$$

Zonally Averaged Spherical Solid Harmonics

Functions in spherical geometry that satisfy Laplace's equation,

$$\nabla^2 f = 0, \tag{7.124}$$

are known as *spherical solid harmonics*. Thus the functions $r^n P_n(\sin\chi)$ and $r^{-(n+1)} P_n(\sin\chi)$ appearing in (7.119) and (7.120), respectively, are zonally averaged spherical solid harmonics.

8

Representation of Gravity: Further Theory and Spheroidal Planets

ABSTRACT

A framework for the development of a mathematical representation for the gravitational field of rotating, mildly oblate, idealised planets was given in Chapter 7. This framework was applied there to two rigid, rotating, spherical, idealised planets; one with constant density, the other with variable density. The spherical context simplified things conceptually and eased analytic tractability. The resulting meridional variation of gravity at the planet's bounding spherical surface does not, however, agree with that observed for Earth, regardless of whether density is held constant or allowed to vary. This is because the planet's surface has been constrained to be precisely spherical. This artificial constraint – which breaks a cause-and-effect relationship – is therefore removed in the present chapter. Two fluid, rotating, ellipsoidal, idealised planets are instead considered; one again with constant density, the other with variable density. Doing so allows further mathematical challenges to be identified and successively addressed. This culminates in a geophysically realistic representation of a rotating planet's (mildly oblate) shape and its gravitational field, not only above its geoid but also below it. The associated geopotential representation provides the basis for the development (in Chapter 12) of a self-consistent, geophysically realistic, ellipsoidal, analytically tractable, geopotential coordinate system, above and below the geoid of a mildly oblate, rotating planet. It is also shown in the present chapter how the less accurate, classical, spherical-geopotential approximation may be obtained as an asymptotic limit of that developed here for a mildly oblate, rotating, ellipsoidal planet of variable density.

8.1 PREAMBLE

The representation of gravity outside and inside a rotating *spherical* planet was examined in Chapter 7. To well represent important force balances in quantitative atmospheric and oceanic models – such as quasi-hydrostatic and geostrophic balance – it was argued that it is essential that a planet's bounding surface coincide with a geopotential surface (i.e. a surface on which the geopotential is constant).

A somewhat surprising consequence of this constraint is that the geopotential *outside* a rotating spheroidal planet (which includes a spherical planet as a special case) is insensitive to the precise distribution of mass within the planet. This does not mean that a planet's mass distribution is of no importance. Arbitrary mass (or, equivalently, density) distributions generally lead to constraint violation. Respect of the constraint then restricts how density may vary within the planet. It does not, however, restrict mass distribution sufficiently to make it unique, only to strongly constrain the possibilities.[1]

[1] For example, two *different* mass distributions within a rotating *ellipsoidal* planet, whose bounding surface coincides with both a geopotential surface and an isopycnal (constant-density) surface, are explicitly constructed in Section 8.5.3. Both of them lead to *exactly the same* geopotential *outside* the planet.

As discussed in Sections 7.8.7 and 7.8.8, the geopotential *outside* a rotating ellipsoidal planet, *subjected to the constraint that the planet's bounding surface coincides with a geopotential surface*, depends only upon:

- The planet's total mass (M).
- Its precise shape, as measured by its ellipticity (ε).
- Its angular rotation rate (Ω).
- Newton's universal gravitational constant (γ).

It is *independent* of internal mass distribution. For the determination of the gravitational potential *outside* a planet *of known shape*, it is the planet's total mass that is important rather than its internal distribution.

For the special case of a rotating *spherical* planet of variable density – and obviously of known shape – it was found in Section 7.8 that meridional variation of gravity at the planet's bounding surface does not agree with that observed for Earth. The underlying reason for this is the assumption that the planet be *precisely spherical*. This assumption breaks the *cause* and *effect* relationship described in Section 7.1.2. Specifically, *artificially* constraining the planet to be precisely spherical prevents it from properly responding to the cause – namely planetary rotation – in a physically realistic manner.

To address this important deficiency, instead of insisting that the planet is *rigid* and *perfectly spherical* – as assumed in Chapter 7 for conceptual simplicity and analytic tractability – we now allow it to be *fluid* and *spheroidal*. In particular, we examine the situation where it is precisely *ellipsoidal*.[2] Changing the geometry from spherical to spheroidal/ellipsoidal does, of course, complicate things. To set the scene, various functional forms for spheroidal/ellipsoidal planets are therefore defined in Section 8.2.

With this preparation, the representation of gravity for an ellipsoidal planet of *constant density* is examined in Section 8.3. This problem is equivalent to that examined in Newton's seminal work. The planet's precise ellipticity (i.e. its precise shape) is determined (as opposed to prescribed) via respect of equilibrium force balances at the planet's bounding, ellipsoidal, geopotential surface. This restores the cause and effect relationship that was broken in Chapter 7 in the spherical context. The resulting meridional variation of gravity is better than that obtained for a spherical planet of variable density, but it nevertheless still does not agree with that observed for Earth. This is due to the assumption of *constant density*, which does not correspond well enough to Earth's actual internal density distribution.

To address this deficiency we need to allow the planet's density distribution to vary – but how? – whilst still respecting the constraint that the planet's bounding surface coincides with a constant geopotential surface. Mathematically, this (coupled) problem is a lot more challenging to solve than the analogous problem examined in Section 7.8 for a rigid *spherical* planet of *variable* density, where the *precise* (i.e. spherical) shape of the planet is imposed, which breaks the coupling. Furthermore, an accurate representation of Earth's internal density distribution is unavailable (Jeffreys, 1976; Dziewonski and Anderson, 1981).

- What to do?

Based on the analyses of Sections 7.8.7 and 7.8.8 and *exploiting the fact that Earth's ellipticity and total mass are known very accurately* – see Table 7.1 and WGS 84 (2004) – we turn the problem around. Independently knowing the precise value of an ellipsoidal planet's ellipticity together with its total mass, angular rotation rate, and Newton's universal gravitational constant

[2] This choice is motivated by the benefits of compatibility with the use of the WGS 84 (2004) Reference *Ellipsoid* for geodetic applications of importance to quantitative atmospheric and oceanic modelling. See Section 12.2.2 for detailed discussion.

allows us to uniquely determine the geopotential (and hence gravity) *outside* a rotating ellipsoidal planet. This separates cause and effect but, very importantly, *without detrimentally impacting accuracy with respect to reality*. The methodology for accomplishing this is summarised in Section 8.4. It implicitly assumes that the observed shape of the planet – as measured by its observed ellipticity – is consistent with respect of an equilibrium force balance at a fluid planet's bounding surface. This is a very reasonable and physically realistic assumption.

In Section 8.5, the methodology of Section 8.4 is applied to obtain the geopotential *outside* a rotating ellipsoidal planet. This then defines gravity above a planet's/Earth's geoid (i.e. for an atmosphere), and it is shown that the resulting meridional variation of gravity agrees with that observed for Earth. Phew, finally!

- But what about below the geoid (i.e. for oceans)?[3]

To answer this question, a model of a planet's density distribution is needed. Fortunately, for shallow oceans such as Earth's – with a maximum depth of about 11 km compared to Earth's mean radius of about 6400 km – it does not need to be very sophisticated for two reasons:

1. The solution of the Poisson equation for Newtonian gravitational potential just below the planet's bounding surface/geoid – where we need it – is very strongly controlled by continuity with the (known) solution just above it. This latter solution is independently known to good accuracy from application of the methodology of Section 8.4.
2. Solutions of Poisson equations generally respond weakly to the small scales of their right-hand-side forcing functions. Any small-scale detail in the definition of the density distribution then anyway ends up being heavily smoothed in the corresponding solution.

A fairly simple minimalist model of density distribution is therefore constructed, based on the broad hypothesised features of Earth's density distribution, as inferred from limited indirect observational data. Applying the methodology of Section 8.4 then results in the definition of the geopotential and gravity *inside an ellipsoidal planet of variable density*. Via application of appropriate boundary conditions at the planet's bounding surface, these properly match those already obtained *outside* the planet. Due to the strong influence of the accurate outside solution – down to the geoid – the resulting representation of gravity in a shallow layer beneath the geoid can be expected to be acceptably accurate, albeit progressively less so downwards from an ocean's bathymetry.

It is shown in Section 8.6 that the classical, spherical, geopotential approximation may be rigorously obtained and justified as an asymptotic limit of that developed in Section 8.5 for a mildly oblate, rotating, ellipsoidal planet of variable density. Unsurprisingly, the zeroth-order, classical, *spherical*, geopotential approximation is less accurate than the first-order, *ellipsoidal*, geopotential approximation. Concluding remarks are made in Section 8.7.

8.2 FUNCTIONAL FORMS FOR SPHEROIDAL PLANETS

8.2.1 Spheroidal Form

Consider now a *spheroidal* planet that is symmetric with respect to both its rotation axis and its equatorial plane – see Fig. 8.1 for a three-dimensional schematic of the planet, and Fig. 8.2 for its cross section in a meridional plane. Now Legendre polynomials, $P_n(\sin\chi)$, are the zonally averaged eigenfunctions of the Laplacian, ∇^2, in spherical-polar coordinates. By potential theory (see Section 7.5), the potential for Newtonian gravity outside the planet satisfies Laplace's equation.

[3] Although Earth is the only planet in the Solar System to have oceans, remote observations suggest that several moons in the Solar System – e.g. Jupiter's moons Callisto, Europa, and Ganymede – have liquid-water oceans capped by a layer of ice. The theoretical representation of gravity described in this chapter is applicable not only to mildly oblate planets *and* moons within the Solar System, but also beyond it.

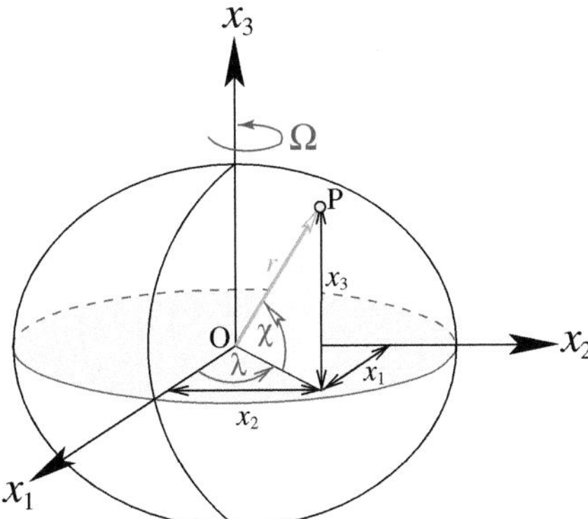

Figure 8.1 A spheroid rotating about its axis with angular frequency Ω, and two synchronously rotating coordinate systems having origin O at the spheroid's centre. The Ox_1, Ox_2 axes of the Cartesian system both lie in the equatorial plane of the spheroid, and Ox_3 is aligned with the spheroid's axis of symmetry/rotation. Also indicated are the corresponding rotating geocentric-polar coordinates of an arbitrary point P on the spheroid's surface, at distance r from O, with longitude λ and geocentric latitude χ.

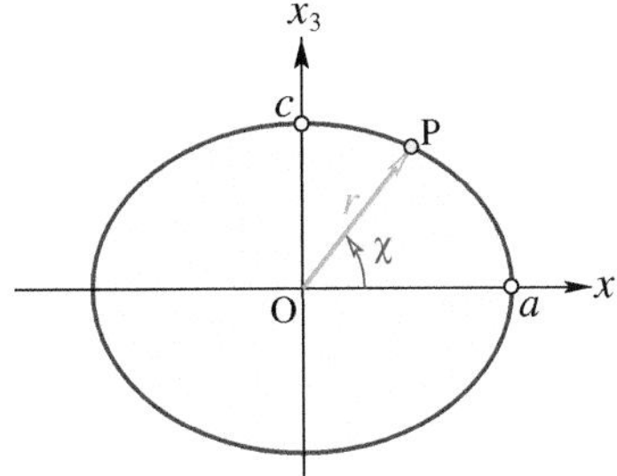

Figure 8.2 A cross section in the Ox_1x_3 meridional plane of a spheroid, where $r \equiv |\mathbf{r}|$ is the absolute value of position vector $\mathbf{r}$ relative to the origin O, χ is *geocentric* latitude, and a and c are the semi-major and semi-minor axes, respectively.

It is therefore natural, in the first instance, to represent an almost-spherical planet in terms of Legendre polynomials. Thus, in a geocentric, spherical-polar, coordinate system, synchronously rotating about the planet's axis of rotation with angular frequency Ω, the equation for the surface of a spheroidal planet can be expressed as the infinite sum[4]

$$r_S(\chi) = \sum_{n=0}^{\infty} r_S^n P_n(\sin\chi), \tag{8.1}$$

where χ is *geocentric* latitude, $r_S(\chi)$ is the magnitude of position vector $\mathbf{r}_S$ at the planet's surface, and r_S^n are (constant) coefficients.

Because of the assumed symmetry about the equatorial plane, and the symmetry/asymmetry properties of Legendre polynomials, $r_S^n \equiv 0$ for n odd. Furthermore, because the planet will be assumed to be almost spherical, and also for reasons of analytical tractability, it suffices here to

[4] More generally, if no symmetry assumptions are made, it can be expressed in terms of an infinite series of surface spherical harmonics $Y_n^m(\lambda,\chi) \equiv P_n^m(\sin\chi)\exp(im\lambda)$, where $P_n^m(\sin\chi)$ is an associated Legendre function of order m and degree n, and λ is longitude (Jeffreys, 1976).

consider only the two most significant (hemispherically symmetric) terms of the infinite sum in (8.1). Thus, using the definitions – see (7.115) –

$$P_0 \equiv 1, \quad P_2(\sin\chi) \equiv \frac{3\sin^2\chi - 1}{2}, \tag{8.2}$$

(8.1) simplifies to

$$r_S(\chi) = r_S^0 + r_S^2\left(\frac{3\sin^2\chi - 1}{2}\right). \tag{8.3}$$

Note that the superscript in r_S^2 denotes an index (rather than a square); and r_S^0 is just the mean value of $r_S(\chi)$ since

$$\frac{\int_{-\pi/2}^{+\pi/2} r_S(\chi)\cos\chi\, d\chi}{\int_{-\pi/2}^{+\pi/2}\cos\chi\, d\chi} = \frac{1}{\int_{-\pi/2}^{+\pi/2}\cos\chi\, d\chi}\int_{-\pi/2}^{+\pi/2}\left[r_S^0 + r_S^2\left(\frac{3\sin^2\chi - 1}{2}\right)\right]\cos\chi\, d\chi = r_0^S. \tag{8.4}$$

(That r_S^2 does not contribute to the right-hand side of (8.4) is implicitly due to the orthogonality of P_0 and $P_2(\sin\chi)$. Explicitly computing the last integral in (8.4) simply confirms this.)

Further assume that a and c are the equatorial and polar semi-axes, respectively, of the planet's assumed spheroidal surface. Thus – from (8.3) –

$$a \equiv r_S(\chi = 0) = r_S^0 - \frac{1}{2}r_S^2, \quad c \equiv r_S\left(\chi = \frac{\pi}{2}\right) = r_S^0 + r_S^2. \tag{8.5}$$

Solving (8.5) for r_S^0 and r_S^2 then allows them to be expressed in terms of the semi-major and semi-minor axes a and c. Thus

$$r_S^0 = \frac{2a + c}{3} = a\left(1 - \frac{\varepsilon}{3}\right), \quad r_S^2 = -\frac{2(a - c)}{3} = -\frac{2a\varepsilon}{3}, \tag{8.6}$$

where

$$\varepsilon \equiv \frac{a - c}{a} \quad \Rightarrow \quad c \equiv a(1 - \varepsilon), \tag{8.7}$$

and ε is a measure of the ellipticity of the planet's assumed spheroidal surface. Only two of the five parameters r_S^0, r_S^2, a, c, and ε are independent. Which pair to use is a matter of convenience for what is being done.

Substitution of (8.5)–(8.7) into (8.3), with use of (8.2), then gives the following equivalent forms for the equation in geocentric-polar coordinates that defines the surface of the spheroid:

$$r = r_S(\chi) = a\left(1 - \varepsilon\sin^2\chi\right) = a\left[\left(1 - \frac{\varepsilon}{3}\right) - \frac{2\varepsilon}{3}P_2(\sin\chi)\right]. \tag{8.8}$$

The first form shows the simple $\sin^2\chi$, equator-to-pole variation of r_S with respect to geocentric latitude χ. The second form, in terms of $P_2(\sin\chi)$, facilitates various $O(\varepsilon)$ approximations to both the representation of the planet's surface and the planet's associated geopotential field.

Equation (8.8) may be rewritten in rotating Cartesian coordinates (x_1, x_2, x_3) as (see Fig. 8.1)

$$x_1 = a\left(1 - \varepsilon\sin^2\chi\right)\cos\lambda\cos\chi, \tag{8.9}$$

$$x_2 = a\left(1 - \varepsilon\sin^2\chi\right)\sin\lambda\cos\chi, \tag{8.10}$$

$$x_3 = a\left(1 - \varepsilon\sin^2\chi\right)\sin\chi. \tag{8.11}$$

The origin for these coordinates is located at the ellipsoidal planet's centre. This coordinate system rotates synchronously about the planet's rotation axis Ox_3 with the planet's angular rotation rate Ω.

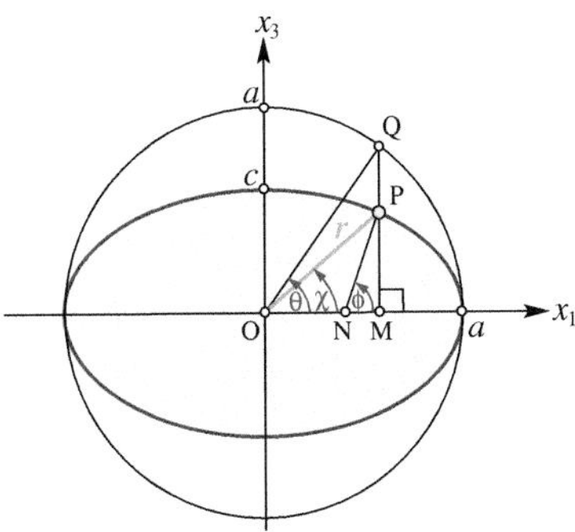

Figure 8.3 A cross section in the Ox_1x_3 meridional plane of an ellipsoid (shaded) having semi-major and semi-minor axes a and c, respectively, that is inscribed in a circle of radius a. P is a general point on the ellipse with coordinates (x_1, x_3). N is the point at which the normal to the ellipse at the general point P (x_1, x_3) intersects the x_1 axis. θ, χ, and ϕ are *parametric, geocentric,* and *geographic* latitudes, respectively. For a spherical planet these latitudes are identical to one another, but for an ellipsoidal planet they are all distinct. The straight line QPM is parallel to Ox_3 and intersects the circle and Ox_1 at Q and M respectively. The equation of the ellipse is $\left(x_1^2/a^2\right) + \left(x_3^2/c^2\right) = 1$, and $r = \left(x_1^2 + x_3^2\right)^{1/2}$ is the distance of P from origin O along the green line. After Fig. 3 of Staniforth and White (2015a).

8.2.2 Ellipsoidal Form

8.2.2.1 *Cartesian Representation of an Ellipsoid*

For mildly oblate (almost spherical) planets, the spheroid defined by (8.9)–(8.11) in rotating Cartesian coordinates (and equivalently by (8.8) in rotating geocentric-polar coordinates) may be approximated as an ellipsoid. To see this, square equations (8.9)–(8.11), expand them in ε, and then neglect terms of $O\left(\varepsilon^2\right)$ to obtain

$$x_1^2 = a^2\left(1 - 2\varepsilon \sin^2\chi\right)\cos^2\lambda\cos^2\chi, \tag{8.12}$$

$$x_2^2 = a^2\left(1 - 2\varepsilon \sin^2\chi\right)\sin^2\lambda\cos^2\chi, \tag{8.13}$$

$$x_3^2 = a^2\left(1 - \varepsilon\right)^2\left(1 + 2\varepsilon \cos^2\chi\right)\sin^2\chi. \tag{8.14}$$

Dividing (8.12) and (8.13) by a^2, dividing (8.14) by $c^2 \equiv a^2\left(1-\varepsilon\right)^2$, summing all three results, and neglecting terms of $O\left(\varepsilon^2\right)$ then gives (to $O\left(\varepsilon\right)$ accuracy)

$$\frac{x_1^2 + x_2^2}{a^2} + \frac{x_3^2}{c^2} = 1. \tag{8.15}$$

This is the equation for an *ellipsoid* of revolution about the Ox_3 axis, having semi-major and semi-minor axes a and c, respectively. See Fig. 8.3 for a cross section in the Ox_1x_3 meridional plane.

8.2.2.2 *Geocentric Polar Representation of an Ellipsoid*

For later convenience, the ellipsoid described by (8.15) in rotating Cartesian coordinates (x_1, x_2, x_3) is now exactly rewritten in terms of rotating geocentric-polar coordinates (λ, χ, r), where λ is longitude, χ is *geocentric* latitude, and $r \equiv \left(x_1^2 + x_2^2 + x_3^2\right)^{1/2}$ is distance from the origin at the ellipsoid's centre. Now rotating Cartesian coordinates are related to rotating geocentric coordinates by (see Fig. 8.1)

$$x_1 = r\cos\lambda\cos\chi, \quad x_2 = r\sin\lambda\cos\chi, \quad x_3 = r\sin\chi. \tag{8.16}$$

Substitution of (8.16) into (8.15) gives

$$r = r_S(\chi) = \frac{ac}{\left(a^2\sin^2\chi + c^2\cos^2\chi\right)^{\frac{1}{2}}} = \frac{a}{\left(1 + 2\widetilde{\varepsilon}\sin^2\chi\right)^{\frac{1}{2}}}, \tag{8.17}$$

where

$$\widetilde{\varepsilon} \equiv \frac{a^2 - c^2}{2c^2} \equiv \frac{1 - (1-\varepsilon)^2}{2(1-\varepsilon)^2} = \varepsilon + O\left(\varepsilon^2\right). \tag{8.18}$$

Equation (8.17) is the geocentric-polar representation of the ellipsoid's surface. It is equivalent to the Cartesian representation (8.15). The parameter $\widetilde{\varepsilon}$ is introduced here in anticipation of the development – in Chapter 12 – of a geophysically realistic, ellipsoidal, analytically tractable, geopotential coordinate system.[5]

Using (8.17) in (8.16), the equation for the surface of the ellipsoid may be equivalently written as

$$x_1 = \frac{ac\cos\lambda\cos\chi}{\left(a^2\sin^2\chi + c^2\cos^2\chi\right)^{\frac{1}{2}}} = \frac{a\cos\lambda\cos\chi}{\left(1 + 2\widetilde{\varepsilon}\sin^2\chi\right)^{\frac{1}{2}}}, \tag{8.19}$$

$$x_2 = \frac{ac\sin\lambda\cos\chi}{\left(a^2\sin^2\chi + c^2\cos^2\chi\right)^{\frac{1}{2}}} = \frac{a\sin\lambda\cos\chi}{\left(1 + 2\widetilde{\varepsilon}\sin^2\chi\right)^{\frac{1}{2}}}, \tag{8.20}$$

$$x_3 = \frac{ac\sin\chi}{\left(a^2\sin^2\chi + c^2\cos^2\chi\right)^{\frac{1}{2}}} = \frac{a\sin\chi}{\left(1 + 2\widetilde{\varepsilon}\sin^2\chi\right)^{\frac{1}{2}}}. \tag{8.21}$$

8.2.2.3 *Parametric Polar Representation of an Ellipsoid*

Equation (8.15) for an ellipsoid can also be rewritten parametrically as

$$x_1 = a\cos\lambda\cos\theta, \quad x_2 = a\sin\lambda\cos\theta, \quad x_3 = c\sin\theta = a(1-\varepsilon)\sin\theta, \tag{8.22}$$

where θ is *parametric* latitude (Bénard, 2015; Staniforth and White, 2015b). See Fig. 8.3. Squaring the equations of (8.22) and summing the results leads to the equivalent parametric-polar forms,

$$r = r_S(\theta) = \left(a^2\cos^2\theta + c^2\sin^2\theta\right)^{\frac{1}{2}} = a\left(1 - 2\widetilde{\varepsilon}\sin^2\theta\right)^{\frac{1}{2}}, \tag{8.23}$$

for the representation of the ellipsoid's surface, where $\widetilde{\varepsilon}$ is defined by (8.18).

8.2.2.4 *Geographic Polar Representation of an Ellipsoid*

Equation (8.15) for an ellipsoid may also be further rewritten parametrically (Bénard, 2015; Staniforth and White, 2015b) as

$$x_1 = \frac{a^2\cos\lambda\cos\phi}{r_{ac}^{\phi}(\phi)}, \quad x_2 = \frac{a^2\sin\lambda\cos\phi}{r_{ac}^{\phi}(\phi)}, \quad x_3 = \frac{c^2\sin\phi}{r_{ac}^{\phi}(\phi)}, \tag{8.24}$$

where

$$r_{ac}^{\phi}(\phi) \equiv \left(a^2\cos^2\phi + c^2\sin^2\phi\right)^{\frac{1}{2}}, \tag{8.25}$$

[5] In some texts $e' \equiv \left(a^2 - c^2\right)^{1/2}/c$ is called the *second* eccentricity, the first eccentricity being $e \equiv \left(a^2 - c^2\right)^{1/2}/a$. This means that $\widetilde{\varepsilon} \equiv e'/2$. One could proceed from here onwards using e' instead of $\widetilde{\varepsilon}$, taking into account the difference of a factor of two in their definitions. To maintain consistency with the notation employed in relevant journal publications, in particular Staniforth and White (2015a), this however has not been done.

φ	$r(\varphi)$	$x_1(\lambda,\varphi)$	$x_2(\lambda,\varphi)$	$x_3(\varphi)$	$\dfrac{x_3(\varphi)}{x_1(\lambda=0,\varphi)}$
χ	$\dfrac{ac}{\left(a^2\sin^2\chi+c^2\cos^2\chi\right)^{\frac{1}{2}}}$	$\dfrac{ac\cos\lambda\cos\chi}{r_{ac}^{\chi}(\chi)}$	$\dfrac{ac\sin\lambda\cos\chi}{r_{ac}^{\chi}(\chi)}$	$\dfrac{ac\sin\chi}{r_{ac}^{\chi}(\chi)}$	$\tan\chi$
θ	$\left(a^2\cos^2\theta+c^2\sin^2\theta\right)^{\frac{1}{2}}$	$a\cos\lambda\cos\theta$	$a\sin\lambda\cos\theta$	$c\sin\theta$	$\dfrac{c}{a}\tan\theta$
ϕ	$\dfrac{\left(a^4\cos^4\phi+c^4\sin^4\phi\right)^{\frac{1}{2}}}{\left(a^2\cos^2\phi+c^2\sin^2\phi\right)^{\frac{1}{2}}}$	$\dfrac{a^2\cos\lambda\cos\phi}{r_{ac}^{\phi}(\phi)}$	$\dfrac{a^2\sin\lambda\cos\phi}{r_{ac}^{\phi}(\phi)}$	$\dfrac{c^2\sin\phi}{r_{ac}^{\phi}(\phi)}$	$\dfrac{c^2}{a^2}\tan\phi$

Table 8.1 Expressions for $r(\varphi)$, $x_1(\lambda,\varphi)$, $x_2(\lambda,\varphi)$, $x_3(\varphi)$, and the quotient $x_3(\varphi)/x_1(\lambda=0,\varphi)$ for the representation of an ellipsoid using meridional coordinates; χ (*geocentric* latitude), θ (*parametric* latitude), and ϕ (*geographic* latitude). In this table, $r_{ac}^{\chi}(\chi)\equiv\left(a^2\sin^2\chi+c^2\cos^2\chi\right)^{\frac{1}{2}}$ and $r_{ac}^{\phi}(\phi)\equiv\left(a^2\cos^2\phi+c^2\sin^2\phi\right)^{\frac{1}{2}}$.

and ϕ is *geographic* latitude. See Fig. 8.3. Squaring the equations of (8.24) and summing the results leads to the equivalent geographic-polar form,

$$r=r_S(\phi)=\frac{\left(a^4\cos^4\phi+c^4\sin^4\phi\right)^{\frac{1}{2}}}{\left(a^2\cos^2\phi+c^2\sin^2\phi\right)^{\frac{1}{2}}},\tag{8.26}$$

for the representation of the ellipsoid's surface.

8.2.2.5 *Summary Table*

For reference purposes, the expressions (given in Sections 8.2.2.1–8.2.2.4 in various coordinate systems) for the representation of an ellipsoid's surface are consolidated in Table 8.1.

8.2.2.6 *Volume and Surface Area*

Anticipating the need for this, expressions are now given for the volume and surface area of the ellipsoid defined by (8.15) (and, equivalently, by (8.22)). Thus the volume $\mathscr{V}$ and surface area $\mathscr{S}$, respectively, of an ellipsoidal planet are

$$\mathscr{V}=\pi\int_{-\pi/2}^{\pi/2}\left(x_1^2+x_2^2\right)\frac{dx_3}{d\theta}d\theta=\frac{4\pi}{3}a^3(1-\varepsilon),\tag{8.27}$$

$$\begin{aligned}\mathscr{S}&=2\pi\int_{-\pi/2}^{\pi/2}\left(x_1^2+x_2^2\right)^{\frac{1}{2}}\left[\left(\frac{dx_1}{d\chi}\right)^2+\left(\frac{dx_2}{d\chi}\right)^2+\left(\frac{dx_3}{d\chi}\right)^2\right]^{\frac{1}{2}}d\chi\\&=2\pi a^2\left[1+\left(1-e^2\right)\frac{\tanh^{-1}e}{e}\right]=4\pi a^2\left[1-\frac{e^2}{3}+O\left(e^4\right)\right]=4\pi a^2\left[1-\frac{2\varepsilon}{3}+O\left(\varepsilon^2\right)\right],\end{aligned}\tag{8.28}$$

where (using (8.7))

$$e^2\equiv1-\frac{c^2}{a^2}=2\varepsilon+O\left(\varepsilon^2\right),\tag{8.29}$$

and e is the ellipsoid's *eccentricity*.

Result (8.27) for $\mathcal{V}$ is exact. The first two expressions of (8.28) for $\mathcal{S}$ are also exact, whereas the last two are approximative. Setting $\varepsilon \equiv e \equiv 0$ in (8.27) and (8.28) recovers the exact, familiar expressions for a spherical planet that $\mathcal{V} = 4\pi a^3/3$ and $\mathcal{S} = 4\pi a^2$.

8.3 AN ELLIPSOIDAL PLANET OF CONSTANT DENSITY

8.3.1 Preliminaries

For compatibility with later developments, it is convenient to specify the planet's ellipsoidal surface in the form – see (8.17) and (7.116) –

$$r = r_S(\chi) = a\left(1 + 2\widetilde{\varepsilon}\sin^2\chi\right)^{-\frac{1}{2}} = a\left[1 + \frac{2\widetilde{\varepsilon}}{3} + \frac{4\widetilde{\varepsilon}}{3}P_2(\sin\chi)\right]^{-\frac{1}{2}}, \tag{8.30}$$

where – see (8.7) and (8.18) –

$$\widetilde{\varepsilon} \equiv \frac{a^2 - c^2}{2c^2} \equiv \frac{1 - (1-\varepsilon)^2}{2(1-\varepsilon)^2} = \varepsilon + O\left(\varepsilon^2\right), \quad \varepsilon \equiv \frac{a-c}{a}. \tag{8.31}$$

For simplicity, the theory is only developed here to $O(\widetilde{\varepsilon}, m)$ accuracy, that is, to first order in $\widetilde{\varepsilon} \sim \varepsilon$ and m, where $m \equiv \Omega^2 a^3/(\gamma M)$; see (7.41).[6] Terms of higher order are therefore consistently dropped throughout.

The mass M of an ellipsoid of *constant* density $\rho = \overline{\rho}$, whose surface is defined by (8.30), is

$$M \equiv \iiint \overline{\rho}\, d\mathcal{V} = \overline{\rho} \iiint d\mathcal{V} = \frac{4\pi a^3 \overline{\rho}}{3}(1-\varepsilon) \approx \frac{4\pi a^3\overline{\rho}}{3}(1-\widetilde{\varepsilon}), \tag{8.32}$$

where (8.27) has been used to evaluate the volume $\mathcal{V} \equiv \iiint d\mathcal{V}$, and (8.31) to replace ε by $\widetilde{\varepsilon}$. Thus, to $O(\widetilde{\varepsilon})$ and expanding $(1-\widetilde{\varepsilon})$ binomially,

$$4\pi\gamma\overline{\rho} = 3\left(\frac{1+\widetilde{\varepsilon}}{a^2}\right)\frac{\gamma M}{a}. \tag{8.33}$$

In geocentric-polar coordinates, the potential of the centrifugal force – see (7.6), (7.116), and Fig. 2.4 – becomes

$$V^C \equiv -\frac{(\boldsymbol{\Omega}\times\mathbf{r})\cdot(\boldsymbol{\Omega}\times\mathbf{r})}{2} = -\frac{\Omega^2 r^2\cos^2\chi}{2} = -\frac{\Omega^2 r^2}{3}\left[1 - P_2(\sin\chi)\right]. \tag{8.34}$$

With this preparation, we are now ready to apply the three-step procedure of Section 7.5 to determine the Newtonian gravitational potential $V(\chi, r)$ for a rotating, ellipsoidal planet of constant density.

8.3.2 Step 1: *V* outside the Planet

Because it is assumed that the ellipsoid has very small ellipticity ($\varepsilon \approx 1/300$ for Earth), spherical-polar coordinates (λ, χ, r) are used, where λ is longitude, χ is latitude, and r is distance from the

[6] As mentioned in Section 7.1.3, Maclaurin (in 1742) determined the geopotential *exactly* for an ellipsoidal planet of *constant* density. This is not done here since, whilst interesting, it does not generalise to an ellipsoidal planet of *variable* density. A first-order theory is instead developed that can be generalised. Although it is tractable to also do so to second order – as in e.g. Jeffreys (1976) – this greatly complicates matters for little, if any, practical benefit. Because, for Earth, $\widetilde{\varepsilon} \approx m \approx 1/300$ are so small, second-order terms are negligibly small. Furthermore – for tractability reasons – it could very well turn out to be impossible to develop a second-order analogue of the first-order, self-consistent, geopotential coordinate system developed in Chapter 12.

planet's centre. See Figs 8.1 and 8.2. For Step 1 of Section 7.5, and with use of zonally averaged Laplacian (7.118), Laplace's equation (7.21) then becomes

$$\nabla^2 V(\chi, r) \equiv \frac{1}{r^2 \cos\chi}\frac{\partial}{\partial\chi}\left(\cos\chi\frac{\partial V}{\partial\chi}\right) + \frac{1}{r^2}\frac{\partial}{\partial r}\left(r^2\frac{\partial V}{\partial r}\right) = 0. \tag{8.35}$$

Applying the condition that $V \to 0$ as $r \to \infty$, the Newtonian gravitational potential *outside* the planet is the (highly truncated) series of spherical solid harmonics

$$V(\chi, r) = C_0\frac{a}{r}P_0 + C_2\frac{a^3}{r^3}P_2(\sin\chi) = C_0\frac{a}{r} + C_2\frac{a^3}{r^3}P_2(\sin\chi), \quad r \geq r_S(\chi), \tag{8.36}$$

where C_0 and C_2 are constants to be determined by application (in Step 3) of boundary conditions at the planet's ellipsoidal surface.[7]

8.3.3 Step 2: V inside the Planet

For Step 2 of Section 7.5, Poisson's equation (7.22), with use of (7.118) and (8.33), then becomes

$$\nabla^2 V \equiv \frac{1}{r^2 \cos\chi}\frac{\partial}{\partial\chi}\left(\cos\chi\frac{\partial V}{\partial\chi}\right) + \frac{1}{r^2}\frac{\partial}{\partial r}\left(r^2\frac{\partial V}{\partial r}\right) = 4\pi\gamma\overline{\rho} = 3\left(\frac{1+\widetilde{\varepsilon}}{a^2}\right)\frac{\gamma M}{a}. \tag{8.37}$$

A particular solution $V = V_{\text{part}}(r)$ of (8.37) can be found by seeking a bounded solution, in the planet's interior, of

$$\frac{1}{r^2}\frac{d}{dr}\left[r^2\frac{dV_{\text{part}}(r)}{dr}\right] = 3\left(\frac{1+\widetilde{\varepsilon}}{a^2}\right)\frac{\gamma M}{a}. \tag{8.38}$$

Thus, integrating twice with respect to r,

$$V_{\text{part}}(r) = \left(\frac{1+\widetilde{\varepsilon}}{2}\right)\frac{\gamma M}{a}\frac{r^2}{a^2}. \tag{8.39}$$

The complete solution is the sum of this solution and a (highly truncated) series of spherical solid harmonics that satisfy Laplace's equation, $\nabla^2 V = 0$. Thus the Newtonian gravitational potential *inside* the planet is

$$\begin{aligned} V(\chi, r) &= V_{\text{part}}(r) + A_0 P_0 + B_2\frac{r^2}{a^2}P_2(\sin\chi) \\ &= A_0 + \left[\left(\frac{1+\widetilde{\varepsilon}}{2}\right)\frac{\gamma M}{a} + B_2 P_2(\sin\chi)\right]\frac{r^2}{a^2}, \quad r \leq r_S(\chi), \end{aligned} \tag{8.40}$$

where P_0 and $P_2(\sin\chi)$ are defined by (8.2), and A_0 and B_2 are constants to be determined by application (in Step 3) of boundary conditions at the planet's ellipsoidal surface.

[7] We are free to express Laplace's equation in any convenient coordinate system. For small $\widetilde{\varepsilon}$, an ellipsoidal planet is almost spherical. It is therefore convenient to use spherical-polar coordinates. We can then expand $V(\chi, r)$ as an infinite series of solid harmonics that exactly satisfies $\nabla^2 V = 0$ everywhere outside the planet. Because $\widetilde{\varepsilon}$ is assumed (very) small, it turns out that the series is rapidly convergent and so, to $O(\widetilde{\varepsilon}, m)$, we only need its first two terms, as in (8.36). Although a *spherical*-harmonic expansion is used, continuity conditions are nevertheless applied (in Section 8.3.4) at the planet's *ellipsoidal* surface, and *not* on a spherical surface. A similar rationale to that described in this footnote also applies to Step 2 for solving Poisson's equation (8.37) to obtain $V(\chi, r)$ *inside* the planet.

8.3.4 Step 3: Application of Boundary Conditions at the Planet's Surface

For Step 3 of Section 7.5, the interior and exterior solutions for V and ∇V should match on the planet's ellipsoidal surface $r = r_S(\chi)$, defined by (8.30).

Comparing the functional forms of (8.36) and (8.40) for an *ellipsoidal* planet, with the corresponding functional forms of (7.34) and (7.33) for a *spherical* planet, and noting that they should agree when $\widetilde{\varepsilon} \equiv 0$ (with no variation in χ), we see that the constants A_0, B_2, C_0, and C_2 in (8.40) and (8.36) have the following orders:

$$A_0 = O(1), \quad B_2 = O(\widetilde{\varepsilon}), \quad C_0 = O(1), \quad C_2 = O(\widetilde{\varepsilon}). \tag{8.41}$$

In what follows, much use is made of asymptotic expressions (8.41) to consistently drop $O(\widetilde{\varepsilon}^2)$ terms throughout. When done, this is indicated by use of the approximation sign ($\approx$), with neglect of terms of $O(\widetilde{\varepsilon}^2)$ implied.

Continuity of $V(\chi, r)$ on $r = r_S(\chi)$

Evaluating (8.36) and (8.40) on the planet's ellipsoidal surface using (8.30) and (8.41) gives

$$\begin{aligned} V^{\text{ext}}_{r=r_S(\chi)} &= C_0\left[1 + \frac{2\widetilde{\varepsilon}}{3} + \frac{4\widetilde{\varepsilon}}{3}P_2(\sin\chi)\right]^{\frac{1}{2}} + C_2\left[1 + \frac{2\widetilde{\varepsilon}}{3} + \frac{4\widetilde{\varepsilon}}{3}P_2(\sin\chi)\right]^{\frac{3}{2}} P_2(\sin\chi) \\ &\approx C_0\left(1 + \frac{\widetilde{\varepsilon}}{3}\right) + \left(C_2 + \frac{2\widetilde{\varepsilon}}{3}C_0\right)P_2(\sin\chi), \end{aligned} \tag{8.42}$$

$$\begin{aligned} V^{\text{int}}_{r=r_S(\chi)} &= A_0 + \left[\left(\frac{1+\widetilde{\varepsilon}}{2}\right)\frac{\gamma M}{a} + B_2 P_2(\sin\chi)\right]\left[1 + \frac{2\widetilde{\varepsilon}}{3} + \frac{4\widetilde{\varepsilon}}{3}P_2(\sin\chi)\right]^{-1} \\ &\approx A_0 + \frac{1}{2}\left(1 + \frac{\widetilde{\varepsilon}}{3}\right)\frac{\gamma M}{a} + \left(B_2 - \frac{2\widetilde{\varepsilon}}{3}\frac{\gamma M}{a}\right)P_2(\sin\chi), \end{aligned} \tag{8.43}$$

where superscripts 'ext' and 'int' denote solutions exterior and interior to the planet, respectively, and subscript 'S' in $r = r_S(\chi)$ denotes evaluation on the planet's surface. The approximation sign ($\approx$) in (8.43) and (8.42) denotes neglect of terms of $O(\widetilde{\varepsilon}^2)$ using binomial expansions together with use of (8.41). Successively matching the constant and $P_2(\sin\chi)$ terms of (8.42) and (8.43) then yields

$$A_0 = \left(C_0 - \frac{1}{2}\frac{\gamma M}{a}\right)\left(1 + \frac{\widetilde{\varepsilon}}{3}\right), \quad B_2 = C_2 + \frac{2\widetilde{\varepsilon}}{3}\left(C_0 + \frac{\gamma M}{a}\right). \tag{8.44}$$

Using (8.44) together with (8.36) and (8.40), the solutions outside and inside the planet are

$$V^{\text{ext}}(\chi, r) = C_0\frac{a}{r} + C_2\frac{a^3}{r^3}P_2(\sin\chi), \qquad r \geq r_S(\chi), \tag{8.45}$$

$$\begin{aligned} V^{\text{int}}(\chi, r) &= \left(C_0 - \frac{1}{2}\frac{\gamma M}{a}\right)\left(1 + \frac{\widetilde{\varepsilon}}{3}\right) \\ &+ \left\{\left(\frac{1+\widetilde{\varepsilon}}{2}\right)\frac{\gamma M}{a} + \left[C_2 + \frac{2\widetilde{\varepsilon}}{3}\left(C_0 + \frac{\gamma M}{a}\right)\right]P_2(\sin\chi)\right\}\frac{r^2}{a^2}, \qquad r \leq r_S(\chi). \end{aligned} \tag{8.46}$$

Continuity of $\nabla V(\chi, r)$ *on* $r = r_S(\chi)$

From (5.19), $\nabla V(\chi, r)$ in axially symmetric, spherical-polar coordinates (λ, χ, r) is

$$\nabla V(\chi, r) = \frac{1}{r}\frac{\partial V}{\partial \chi}\mathbf{e}_\chi + \frac{\partial V}{\partial r}\mathbf{e}_r, \tag{8.47}$$

where $\mathbf{e}_r$ is a unit vector pointing from the planet's centre to the point with position vector $\mathbf{r}$, and $\mathbf{e}_\chi$ is a unit vector normal to it in the meridional plane containing $\mathbf{r}$.[8] Differentiation of definition (8.2) for $P_2(\sin\chi)$ gives

$$\frac{d}{d\chi}P_2(\sin\chi) = 3\sin\chi\cos\chi. \tag{8.48}$$

With this preparation we are now ready to apply continuity of ∇V on $r = r_S(\chi)$.

Using (8.45), (8.46), and (8.48) in (8.47) gives

$$\nabla V^{\text{ext}}(\chi, r) = \frac{3}{a}C_2\frac{a^4}{r^4}\sin\chi\cos\chi\,\mathbf{e}_\chi - \frac{1}{a}\left[C_0\frac{a^2}{r^2} + 3C_2\frac{a^4}{r^4}P_2(\sin\chi)\right]\mathbf{e}_r, \tag{8.49}$$

$$\begin{aligned}\nabla V^{\text{int}}(\chi, r) = {}& \frac{3}{a}\left[C_2 + \frac{2\tilde{\varepsilon}}{3}\left(C_0 + \frac{\gamma M}{a}\right)\right]\frac{r}{a}\sin\chi\cos\chi\,\mathbf{e}_\chi \\ &+ \frac{2}{a}\left\{\left(\frac{1+\tilde{\varepsilon}}{2}\right)\frac{\gamma M}{a} + \left[C_2 + \frac{2\tilde{\varepsilon}}{3}\left(C_0 + \frac{\gamma M}{a}\right)\right]P_2(\sin\chi)\right\}\frac{r}{a}\mathbf{e}_r.\end{aligned} \tag{8.50}$$

With use of (8.30) and (8.41), evaluation of these two equations at $r = r_S(\chi)$ then yields

$$\begin{aligned}\nabla V^{\text{ext}}_{r=r_S(\chi)} = {}& \frac{3}{a}C_2\left[1 + \cancel{\frac{2\tilde{\varepsilon}}{3}} + \cancel{\frac{4\tilde{\varepsilon}}{3}P_2(\sin\chi)}\right]^2 \sin\chi\cos\chi\,\mathbf{e}_\chi \\ &- \frac{1}{a}C_0\left[1 + \frac{2\tilde{\varepsilon}}{3} + \frac{4\tilde{\varepsilon}}{3}P_2(\sin\chi)\right]\mathbf{e}_r - \frac{1}{a}3C_2\left[1 + \cancel{\frac{2\tilde{\varepsilon}}{3}} + \cancel{\frac{4\tilde{\varepsilon}}{3}P_2(\sin\chi)}\right]^2 P_2(\sin\chi)\,\mathbf{e}_r \\ \approx {}& \frac{3}{a}C_2\sin\chi\cos\chi\,\mathbf{e}_\chi - \frac{1}{a}\left[\left(1 + \frac{2\tilde{\varepsilon}}{3}\right)C_0 + \left(\frac{4\tilde{\varepsilon}}{3}C_0 + 3C_2\right)P_2(\sin\chi)\right]\mathbf{e}_r,\end{aligned} \tag{8.51}$$

$$\begin{aligned}\nabla V^{\text{int}}_{r=r_S(\chi)} = {}& \frac{3}{a}\left[C_2 + \frac{2\tilde{\varepsilon}}{3}\left(C_0 + \frac{\gamma M}{a}\right)\right]\left[1 + \cancel{\frac{2\tilde{\varepsilon}}{3}} + \cancel{\frac{4\tilde{\varepsilon}}{3}P_2(\sin\chi)}\right]^{-\frac{1}{2}}\sin\chi\cos\chi\,\mathbf{e}_\chi \\ &+ \frac{2}{a}\left(\frac{1+\tilde{\varepsilon}}{2}\right)\frac{\gamma M}{a}\left[1 + \frac{2\tilde{\varepsilon}}{3} + \frac{4\tilde{\varepsilon}}{3}P_2(\sin\chi)\right]^{-\frac{1}{2}}\mathbf{e}_r \\ &+ \frac{2}{a}\left[C_2 + \frac{2\tilde{\varepsilon}}{3}\left(C_0 + \frac{\gamma M}{a}\right)\right]P_2(\sin\chi)\left[1 + \cancel{\frac{2\tilde{\varepsilon}}{3}} + \cancel{\frac{4\tilde{\varepsilon}}{3}P_2(\sin\chi)}\right]^{-\frac{1}{2}}\mathbf{e}_r \\ \approx {}& \frac{1}{a}\left[3C_2 + 2\tilde{\varepsilon}\left(C_0 + \frac{\gamma M}{a}\right)\right]\sin\chi\cos\chi\,\mathbf{e}_\chi \\ &+ \frac{1}{a}\left\{\left(1 + \frac{2\tilde{\varepsilon}}{3}\right)\frac{\gamma M}{a} + 2\left[C_2 + \frac{\tilde{\varepsilon}}{3}\left(2C_0 + \frac{\gamma M}{a}\right)\right]P_2(\sin\chi)\right\}\mathbf{e}_r.\end{aligned} \tag{8.52}$$

Successively matching the components of $\mathbf{e}_\chi$ and $\mathbf{e}_r$ for (8.51) and (8.52) finally leads to

[8] The contribution in the direction of $\mathbf{e}_\lambda$ is absent since $V(\chi, r)$ is independent of λ.

$$C_0 = -\frac{\gamma M}{a}, \quad C_2 = \frac{2\widetilde{\varepsilon}}{5}\frac{\gamma M}{a}. \tag{8.53}$$

8.3.5 V for an Ellipsoidal Planet of Constant Density

Substitution of (8.53) into (8.45) and (8.46), with use of (8.32), leads to:

The Newtonian Gravitational Potential V for an Ellipsoidal Planet of Constant Density

$$V(\chi, r) = -\frac{\gamma M}{r}\left[1 - \frac{2\widetilde{\varepsilon}}{5}\frac{a^2}{r^2}P_2(\sin\chi)\right], \qquad r \geq r_S(\chi), \tag{8.54}$$

$$= -\frac{\gamma M}{a}\left[1 + \left(\frac{1+\widetilde{\varepsilon}}{2}\right)\left(1 - \frac{r^2}{a^2}\right) - \frac{2\widetilde{\varepsilon}}{5}\frac{r^2}{a^2}P_2(\sin\chi)\right], \quad r \leq r_S(\chi). \tag{8.55}$$

8.3.6 Φ for an Ellipsoidal Planet of Constant Density (1)

The complete potential (i.e. the geopotential Φ) is obtained by inserting (8.34) for the centrifugal potential, V^C, and (8.54)–(8.55), for the Newtonian gravitational potential, V, into (7.6). Thus, to $O(\widetilde{\varepsilon}, m)$, this leads to:

$$\Phi(\chi, r) = -\frac{\gamma M}{r}\left[1 - \frac{2\widetilde{\varepsilon}}{5}\frac{a^2}{r^2}P_2(\sin\chi)\right] - \frac{m}{2}\frac{\gamma M}{a}\frac{r^2}{a^2}\cos^2\chi, \qquad r \geq r_S(\chi), \tag{8.56}$$

$$= -\frac{\gamma M}{a}\left[1 + \left(\frac{1+\widetilde{\varepsilon}}{2}\right)\left(1 - \frac{r^2}{a^2}\right) - \frac{2\widetilde{\varepsilon}}{5}\frac{r^2}{a^2}P_2(\sin\chi) + \frac{m}{2}\frac{r^2}{a^2}\cos^2\chi\right], \quad r \leq r_S(\chi), \tag{8.57}$$

where (from (7.41) and (8.31))

$$m \equiv \frac{\Omega^2 a^3}{\gamma M}, \quad \widetilde{\varepsilon} \equiv \frac{a^2 - c^2}{2c^2}. \tag{8.58}$$

8.3.7 Newton's Constraint between ε and m for an Ellipsoid of Constant Density

Thus far, our analysis is valid for both a *rigid*, rotating, ellipsoidal planet and a *fluid* one.

For a *rigid* ellipsoidal planet of known mass (M), of known rotational frequency (Ω), and of known shape (i.e. a and c are of known value), the values of m and $\widetilde{\varepsilon}$ are known from (8.58). It then follows from (8.54)–(8.57) that $V = V(\chi, r)$ and $\Phi = \Phi(\chi, r)$ are known functions, with all of their coefficients being of known value. However (in general) *a rigid planet's ellipsoidal bounding surface will not coincide with a geopotential surface* – as indeed we found in Section 7.6 for the special case of a rigid, rotating, *spherical* planet of constant density.

For a deformable, *fluid* planet of constant density – as assumed by Newton, and as assumed here for now – its shape is determined by the constraint that the planet's bounding, fluid surface is in equilibrium. This implies that the planet's bounding surface coincides with a geopotential surface, as described in Sections 7.1 and 7.4. Doing so then determines the functional relation $\widetilde{\varepsilon} = \widetilde{\varepsilon}(m)$ between $\widetilde{\varepsilon}$ and m for equilibrium to hold at and across the planet's bounding fluid surface. This constraint for a rotating fluid ellipsoid of constant density is known as *Newton's constraint*. The object of the exercise now is to apply this constraint – just as Newton did – to determine the functional form $\widetilde{\varepsilon} = \widetilde{\varepsilon}(m)$ and thereby determine the fluid planet's precise, heretofore unknown, shape.

Inserting definition (8.2) of $P_2(\sin\chi)$ into (8.56) gives

$$\Phi(\chi, r) = -\frac{\gamma M}{r}\left[1 - \frac{2\widetilde{\varepsilon}}{5}\frac{a^2}{r^2}\left(\frac{3\sin^2\chi - 1}{2}\right)\right] - \frac{m}{2}\frac{\gamma M}{a}\frac{r^2}{a^2}\cos^2\chi, \quad r \geq r_S(\chi), \tag{8.59}$$

for the geopotential outside the planet. In particular, (8.59) holds at the planet's surface $r = r_S(\chi)$ which – for equilibrium to hold there – must coincide with a geopotential surface (i.e. a surface on which the geopotential (Φ) is constant). This condition is what allows us to determine the functional relation between $\widetilde{\varepsilon} \sim \varepsilon$ and m.

Evaluation of (8.59) at an ellipsoidal planet's equator (where $r = r_S(\chi = 0) = a$ from (8.30)) and at its poles (where $r = r_S(\chi = \pm\pi/2) = a(1 + 2\widetilde{\varepsilon})^{-1/2}$) gives

$$\Phi[\chi = 0, r = r_S(\chi = 0)] = -\frac{\gamma M}{a}\left(1 + \frac{\widetilde{\varepsilon}}{5} + \frac{m}{2}\right), \tag{8.60}$$

$$\Phi[\chi = \pi/2, r = r_S(\chi = \pm\pi/2)] = -\frac{\gamma M(1 + 2\widetilde{\varepsilon})^{\frac{1}{2}}}{a}\left[1 - \frac{2\widetilde{\varepsilon}}{5}(1 + 2\widetilde{\varepsilon})\right] \approx -\frac{\gamma M}{a}\left(1 + \frac{3\widetilde{\varepsilon}}{5}\right). \tag{8.61}$$

But, since Φ = constant on $r = r_S(\chi)$, which includes the two points $[\chi = 0, r = r_S(\chi = 0)]$ and $[\chi = \pi/2, r = r_S(\chi = \pm\pi/2)]$, the last-most expressions of (8.60) and (8.61) must – to $O(\widetilde{\varepsilon}, m)$ – be equal. Setting them equal and using (8.31) then gives:

Newton's Constraint for a Constant-Density Ellipsoidal Planet

$$\varepsilon \approx \widetilde{\varepsilon} = \frac{5m}{4}. \tag{8.62}$$

Thus:

Cause and Effect

The planet's rotation (the **cause**, characterised by the non-dimensional parameter m) gives rise to an ellipsoidal bulge (the **effect**, characterised by the non-dimensional parameter $\widetilde{\varepsilon} \approx \varepsilon = 5m/4$).

8.3.8 Φ for an Ellipsoidal Planet of Constant Density (2)

Inserting Newton's constraint (8.62) into (8.56) and (8.57), with use of (8.2) for $P_2(\sin\chi)$ and of the identity $\cos^2\chi \equiv 1 - \sin^2\chi$, leads to:

The Geopotential Φ for an Ellipsoidal Planet of Constant Density

$$\Phi(\chi, r) = -\frac{\gamma M}{r}\left\{1 + \frac{\widetilde{\varepsilon}}{5}\left[\frac{a^2}{r^2} + 2\frac{r^3}{a^3} - \left(3\frac{a^2}{r^2} + 2\frac{r^3}{a^3}\right)\sin^2\chi\right]\right\}, \quad r \geq r_S(\chi), \quad (8.63)$$

$$= -\frac{\gamma M}{a}\left[\left(\frac{3+\widetilde{\varepsilon}}{2}\right) - \frac{1}{2}\left(1 - \frac{\widetilde{\varepsilon}}{5} + 2\widetilde{\varepsilon}\sin^2\chi\right)\frac{r^2}{a^2}\right], \quad r \leq r_S(\chi). \quad (8.64)$$

8.3.9 Surface Gravity for an Ellipsoidal Planet of Constant Density

The apparent gravitational force (i.e. gravity) outside an *ellipsoidal* planet of constant density is obtained by inserting (8.63) into (7.4) and using (5.19) to evaluate the gradient. Thus

$$\begin{aligned}\mathbf{g} \equiv -\nabla\Phi &= -\frac{1}{r}\frac{\partial\Phi}{\partial\chi}\mathbf{e}_\chi - \frac{\partial\Phi}{\partial r}\mathbf{e}_r \\ &= -\frac{2\widetilde{\varepsilon}}{5}\frac{\gamma M}{a^2}\left(3\frac{a^4}{r^4} + 2\frac{r}{a}\right)\sin\chi\cos\chi\,\mathbf{e}_\chi \\ &\quad - \frac{\gamma M}{a^2}\left\{\frac{a^2}{r^2} - \frac{\widetilde{\varepsilon}}{5}\left[3\frac{a^4}{r^4}\left(3\sin^2\chi - 1\right) + 4\frac{r}{a}\left(1 - \sin^2\chi\right)\right]\right\}\mathbf{e}_r. \end{aligned} \quad (8.65)$$

From (8.30), the planet's ellipsoidal surface is given by

$$r = r_S(\chi) = a\left(1 + 2\widetilde{\varepsilon}\sin^2\chi\right)^{-\frac{1}{2}} = a\left[1 - \widetilde{\varepsilon}\sin^2\chi + O\left(\widetilde{\varepsilon}^2\right)\right]. \quad (8.66)$$

Substitution of (8.66) into (8.65) evaluated at $r = r_S$, and dropping terms of $O\left(\widetilde{\varepsilon}^{\,2}\right)$, the apparent gravitational force on an ellipsoidal fluid planet of constant density is

$$(\mathbf{g})_{r=r_S(\chi)} \approx -2\widetilde{\varepsilon}\frac{\gamma M}{a^2}\sin\chi\cos\chi\,\mathbf{e}_\chi - \frac{\gamma M}{a^2}\left(1 - \frac{\widetilde{\varepsilon}}{5} + \widetilde{\varepsilon}\sin^2\chi\right)\mathbf{e}_r. \quad (8.67)$$

Thus, to first order in $\widetilde{\varepsilon}$ and using Newton's constraint (8.62), the variation of gravity g at the planet's surface is given by

$$(g)_{r=r_S(\chi)} \equiv \left(|\mathbf{g}|\right)_{r=r_S(\chi)} \approx \frac{\gamma M}{a^2}\left(1 - \frac{\widetilde{\varepsilon}}{5} + \widetilde{\varepsilon}\sin^2\chi\right) \quad (8.68)$$

$$\approx g_S^E\left(1 + \widetilde{\varepsilon}\sin^2\chi\right) \approx g_S^E\left(1 + \frac{5m}{4}\sin^2\chi\right), \quad (8.69)$$

where

$$g_S^E \equiv (g)_{r=r_S(\chi=0)} = \frac{\gamma M}{a^2}\left(1 - \frac{\widetilde{\varepsilon}}{5}\right) \approx \frac{\gamma M}{a^2}\left(1 - \frac{m}{4}\right), \quad (8.70)$$

is the value of g at the planet's equator.

Comparing (8.69) for a rotating, fluid, ellipsoidal planet of constant density with (7.46) for a rigid, spherical one, it is seen that:

- Surface gravity for a rotating, *fluid, ellipsoidal* planet of *constant density* varies more (by 25%) than for a *rigid, spherical* one of *constant density.*

8.3.10 Clairaut's Fraction for an Ellipsoidal Planet of Constant Density

Evaluating (8.69) at $\chi = \pm\pi/2$, the value of g at an ellipsoidal planet's poles is

$$g_S^P \equiv (g)_{r=r_S(\chi=\pm\pi/2)} \approx g_S^E \left(1 + \frac{5m}{4}\right). \tag{8.71}$$

Thus, to first order in m, (8.71) leads to:

Clairaut's Fraction for an Ellipsoidal Planet of Constant Density

$$\frac{g_S^P - g_S^E}{g_S^E} \equiv \text{Clairaut's fraction} = \frac{5m}{4}. \tag{8.72}$$

As we will see (in Section 8.5), the asymptotic value (to first order in ε and m) for Clairaut's fraction for a rotating, fluid, ellipsoidal planet of *variable density* is $(5m/2) - \varepsilon$. This is the gold standard for Clairaut's fraction, since it well matches observations for Earth. As discussed in Section 7.1, $\widetilde{\varepsilon} \approx \varepsilon$ is observed to be approximately equal to m for Earth rather than to the value of $5m/4$ deduced by Newton under the assumption of constant density, and rederived here. Using the realistic value $\widetilde{\varepsilon} \approx \varepsilon \approx m$, the correct asymptotic value (that corresponds to observations) for Clairaut's fraction is then $(5m/2) - \widetilde{\varepsilon} \approx (5m/2) - m = 3m/2$. This can be compared with the value $5m/4$ given by (8.72), which is an underestimate of that observed. The fundamental reason for this discrepancy is that Earth's density is not well enough approximated by a constant.

- How good, despite its deficiencies, is the representation of gravity for an ellipsoidal Earth of constant density when compared with the classical, ubiquitous, spherical-geopotential approximation?

The answer (White et al., 2008) is very good. The spherical-geopotential approximation captures none of the Equator-to-Pole variation of gravity, whereas the representation of gravity for an ellipsoidal Earth of constant density captures it quite well, albeit it is underestimated.[9]

8.4 REFORMULATION OF THE PROCEDURE TO DETERMINE NEWTONIAN GRAVITY OUTSIDE A PLANET

Under the assumptions that:

- The planet's surface coincides with a geopotential surface.
- A vacuum exists everywhere outside the planet.

it was shown in Section 7.8.8 that the potential V of Newtonian gravity *outside* a *spherical* planet of *variable* density can be determined from the planet's:

- Radius (a).
- Rotation rate (Ω).
- Total mass (M).

Furthermore, this was achieved without:

- Any knowledge whatsoever of the planet's density distribution ρ.

[9] The classical spherical-geopotential approximation is usually applied in shallow geometry, albeit it can also be applied in deep geometry; see Sections 7.1.5 and 8.6. When applied in shallow geometry, gravity is constant everywhere, with no vertical variation whatsoever. Gravity then no longer decreases as a function of increasing altitude, as it does physically, and as it does with the ellipsoidal model of gravity developed in this section in deep geometry.

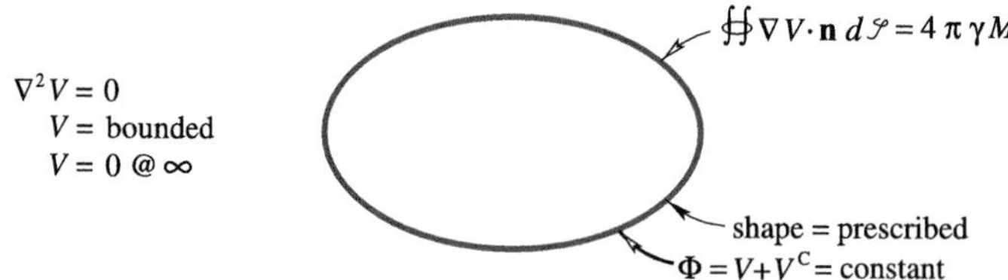

$\oiint \nabla V \cdot \mathbf{n}\, d\mathcal{S} = 4\pi\gamma M$
$\nabla^2 V = 0$
V = bounded
$V = 0$ @ ∞
shape = prescribed
$\Phi = V + V^C$ = constant

Figure 8.4 The differential equation and boundary conditions that determine Newtonian gravity, V, *outside* a spheroid (shaded), when the spheroid's surface is of prescribed shape and its total mass, M, is known. Laplace's equation, $\nabla^2 V = 0$, holds outside the spheroid. V is bounded everywhere and is arbitrarily set to zero at infinite distance (in any direction) away from the spheroid.

- Any need to determine the form of V inside the planet.
- Any need to match interior and exterior solutions for V across the planet's surface.

This greatly simplified the overarching, three-step solution procedure described in Section 7.5. Steps 2 (the solution of a Poisson problem to determine the form of V inside the planet) and 3 (the matching of interior and exterior solutions across the planet's surface) of this solution procedure were essentially replaced by use of the integrated Newtonian gravitational flux condition (7.103) at the planet's surface. This then closed the problem (see Step 1 of the solution procedure) for solving Laplace's equation for V everywhere *outside* the spherical planet down to and including its surface.

The questions now are:

1. What is the key ingredient that allowed simplification of the three-step procedure of Section 7.5?
2. Is it possible to also analogously simplify this three-step procedure to determine the potential V of Newtonian gravity *outside* a *spheroidal* planet of *variable* density?

The answer to the first question is that the key ingredient is the assumption that the location of the planet's surface is known *precisely*. Specifically, the planet's surface is assumed to be a sphere of known radius.[10] This situation may be contrasted with that examined in Section 8.3 for an *ellipsoidal* planet of *constant* density. For this latter case, the planet is assumed to be of ellipsoidal *form* of known (constant) density, but of *unknown ellipticity*. The location of its surface is therefore not known precisely and has to be determined via the consistent matching of the interior and exterior solutions for V. To achieve this, the planet is assumed to be a fluid in hydrostatic equilibrium; the precise shape of the planet is determined by the balance of forces at its assumed *deformable* surface. By way of contrast, the spherical planet of variable density is assumed to be *rigid*, with a *rigid*, spherical, bounding surface.

The answer to the second question is that, yes and with two provisos, it is indeed possible to analogously simplify the three-step procedure of Section 7.5 to determine the potential V of Newtonian gravity *outside* a *spheroidal* planet of *variable* density. The two provisos are related to the answer to the first question. They are that:

- The location of the planet's assumed spheroidal surface is known *precisely*, with known values for its parameters, rather than just in functional form, with unknown values for its parameters.
- The value of the planet's total mass M is also known *precisely* (to within measurement error).

The generalisation of this methodology from spherical to spheroidal geometry can be concisely summarised (see Fig. 8.4) as follows.

[10] It is also assumed that its total mass M is known, and that its surface coincides with a geopotential surface. No assumption is made regarding its internal density distribution ρ; this does not need to be known.

Reformulated Procedure to Determine Newtonian Gravity (V) *outside* a Planet

Solve Laplace's equation

$$\nabla^2 V = 4\pi\gamma\rho = 0, \tag{8.73}$$

outside the planet, and down to its surface, subject to the boundary conditions:

1. V goes to zero at infinite distance from the planet, that is,
$$V \to 0 \quad \text{as } r \to \infty. \tag{8.74}$$
2. The planet's surface $\mathscr{S}$, whose location is known precisely, coincides with a geopotential surface, that is,
$$\Phi = \text{constant, on } \mathscr{S}. \tag{8.75}$$
Thus, using (7.6),
$$V = -V^C + \text{constant, on } \mathscr{S}, \tag{8.76}$$
where V^C is centrifugal potential.
3. The integrated Newtonian gravitational flux at the planet's surface satisfies the constraint
$$\oiint_{\mathscr{S}} \nabla V \cdot \mathbf{n} d\mathscr{S} = 4\pi\gamma M, \tag{8.77}$$
where $\mathbf{n}$ is the outward-pointing normal at the planet's surface $\mathscr{S}$, and γM (standard gravitational parameter – see Table 7.1) is of known value.

This simplified procedure for determining Newtonian gravity outside a planet is applied in Section 8.5 to obtain the Newtonian gravitational potential *exterior* to an ellipsoidal planet of variable density and, thereby, the associated exterior geopotential.

Defining a bounded density distribution (with parameters to be determined) inside the planet, solving a Poisson equation therein, and appropriately matching interior and exterior solutions across the planet's boundary to determine values of the parameters then leads to the Newtonian gravitational potential *interior* to an ellipsoidal planet of variable density and, thereby, the associated interior geopotential. This prepares the way for the development – in Chapter 12 – of a unified, ellipsoidal, geopotential coordinate system for modelling Earth's atmosphere and oceans, as well as those of other, mildly oblate, uniformly rotating planets and moons.

8.5 AN ELLIPSOIDAL PLANET OF VARIABLE DENSITY

8.5.1 Preliminaries

In the analysis of Section 8.3 for a *constant*-density *ellipsoidal* planet, the precise value of the planet's ellipticity was initially *unknown*. Its value was then determined by matching the solution exterior to the planet with the derived interior solution under an assumption of force balance at the planet's surface. Here a *variable*-density *ellipsoidal* planet of *known ellipticity* is instead assumed. As remarked earlier, the assumption that the value of the planet's ellipticity is known means that the geopotential outside the planet can be determined once and for all – as described in Section 8.4 – *without the need to simultaneously determine it inside the planet.* This greatly simplifies matters. That the methodology simplifies in this way is very fortunate indeed, since Earth's mass distribution is relatively poorly known, but its total mass is known to considerable accuracy.

The planet's ellipsoidal surface is again specified in the form (8.30). Thus

$$r = r_S(\chi) = a\left(1 + 2\widetilde{\varepsilon}\sin^2\chi\right)^{-\frac{1}{2}} = a\left[1 + \frac{2\widetilde{\varepsilon}}{3} + \frac{4\widetilde{\varepsilon}}{3}P_2(\sin\chi)\right]^{-\frac{1}{2}}, \tag{8.78}$$

where – see (8.31) –

$$\widetilde{\varepsilon} \equiv \frac{a^2 - c^2}{2c^2} \equiv \frac{1 - (1-\varepsilon)^2}{2(1-\varepsilon)^2} = \varepsilon + O\left(\varepsilon^2\right), \tag{8.79}$$

with the value of $\widetilde{\varepsilon}$ *assumed known.* For simplicity, the theory is again only developed here to $O(\widetilde{\varepsilon}, m)$ accuracy (i.e. to first order in $\widetilde{\varepsilon}$ and m) where

$$m \equiv \frac{\Omega^2 a^3}{\gamma M} \equiv \frac{\Omega^2 a}{\gamma M/a^2} \approx \frac{\text{centrifugal force}}{\text{gravitational attraction}}, \tag{8.80}$$

is a measure of the ratio of the centrifugal force on the planet's surface to Newtonian gravitational attraction there; see (7.1). Terms of higher order than first in $\widetilde{\varepsilon}$ and m are therefore consistently dropped throughout, as in the analysis of Section 8.3 for a *constant*-density, ellipsoidal planet.

Noting that (7.103) holds not only for a spherical planet but also for an ellipsoidal one, we see that the mass M of an ellipsoidal planet of variable density ρ is

$$M \equiv \iiint_{\mathscr{V}} \rho\, d\mathscr{V} = \frac{1}{4\pi\gamma}\iiint_{\mathscr{V}} \nabla^2 V d\mathscr{V} = \frac{1}{4\pi\gamma}\oiint_{\mathscr{S}} \nabla V \cdot \mathbf{n}\, d\mathscr{S}, \tag{8.81}$$

where:

- $\mathscr{V}$ and $\mathscr{S}$ are the volume and surface area of the ellipsoidal planet, respectively.
- $d\mathscr{V}$ and $d\mathscr{S}$ are the corresponding infinitesimal elements.
- The planet's ellipsoidal surface $\mathscr{S}$ is defined by (8.78).
- $\mathbf{n}$ is the outward-pointing normal to this surface.
- Gauss's divergence theorem (A.33) (with $\mathbf{A} \equiv \nabla V$) has been used to transform a volume integral into a surface one.

From (8.34), the potential of the centrifugal force in geocentric-polar coordinates is

$$V^C \equiv -\frac{(\boldsymbol{\Omega} \times \mathbf{r}) \cdot (\boldsymbol{\Omega} \times \mathbf{r})}{2} = -\frac{\Omega^2 r^2 \cos^2\chi}{2} = -\frac{\Omega^2 r^2}{3}\left[1 - P_2(\sin\chi)\right]. \tag{8.82}$$

With this preparatory introduction, we now apply the procedure of Section 8.4 to obtain the Newtonian gravitational potential, $V(\chi, r)$, and the geopotential, $\Phi(\chi, r)$, *outside* the ellipsoidal planet defined by (8.78).

8.5.2 The Exterior Problem

8.5.2.1 *Laplace's Equation*

Laplace's equation in zonally averaged, geocentric-polar coordinates is – see (8.35) –

$$\nabla^2 V(\chi, r) = \frac{1}{r^2\cos\chi}\frac{\partial}{\partial\chi}\left(\cos\chi\frac{\partial V}{\partial\chi}\right) + \frac{1}{r^2}\frac{\partial}{\partial r}\left(r^2\frac{\partial V}{\partial r}\right) = 0. \tag{8.83}$$

8.5.2.2 *Application of the First Boundary Condition (at Infinity)*

Applying condition (8.74) that $V \rightarrow 0$ as $r \rightarrow \infty$, we find that the Newtonian gravitational potential *outside* the planet is the (highly truncated) series of spherical solid harmonics

$$V(\chi, r) = C_0 \frac{a}{r} + C_2 \frac{a^3}{r^3} P_2(\sin\chi), \quad r \geq r_S(\chi), \tag{8.84}$$

where C_0 and C_2 are constants to be determined by application of boundary conditions at the planet's ellipsoidal surface. Comparing the functional forms of (8.84), for an ellipsoidal planet with *variable* density, with (8.54) for one with *constant* density, and noting that they should agree when density is constant, we see that the constants C_0 and C_2 satisfy

$$C_0 = O(1), \quad C_2 = O(m, \widetilde{\varepsilon}); \tag{8.85}$$

compare, for example, (8.41). In what follows, much use of asymptotic expressions similar in form to (8.85) is made to consistently drop $O\left(\widetilde{\varepsilon}^2, \widetilde{\varepsilon}m, m^2\right)$ terms throughout. When done, this is indicated by use of the approximation sign ($\approx$), with neglect of terms of $O\left(\widetilde{\varepsilon}^2, \widetilde{\varepsilon}m, m^2\right)$ implied.

8.5.2.3 *Application of the Second Boundary Condition (at the Planet's Surface)*

The geopotential, $\Phi(\chi, r)$ is obtained by inserting (8.82) for centrifugal potential $V^C(\chi, r)$, and (8.84), for Newtonian gravitational potential $V(\chi, r)$, into definition (7.6) of the geopotential. Thus, to $O(\widetilde{\varepsilon}, m)$,

$$\begin{aligned}\Phi(\chi, r) \equiv V(\chi, r) + V^C(\chi, r) &= C_0 \frac{a}{r} + C_2 \frac{a^3}{r^3} P_2(\sin\chi) - \frac{\Omega^2 r^2}{3}\left[1 - P_2(\sin\chi)\right] \\ &= C_0 \frac{a}{r} + C_2 \frac{a^3}{r^3} P_2(\sin\chi) - \frac{\gamma M}{a}\frac{m}{3}\frac{r^2}{a^2}\left[1 - P_2(\sin\chi)\right], \quad r \geq r_S(\chi),\end{aligned} \tag{8.86}$$

where definition (8.80) for m has been used. Inserting (8.78) for the planet's ellipsoidal surface into (8.86) and then expanding binomially to $O(m, \widetilde{\varepsilon})$ with use of (8.85), the geopotential at the planet's surface is

$$\begin{aligned}\Phi\left[\chi, r_S(\chi)\right] = {} & C_0\left[1 + \frac{2\widetilde{\varepsilon}}{3} + \frac{4\widetilde{\varepsilon}}{3}P_2(\sin\chi)\right]^{\frac{1}{2}} + C_2\left[1 + \cancel{\frac{2\widetilde{\varepsilon}}{3}} + \cancel{\frac{4\widetilde{\varepsilon}}{3}P_2(\sin\chi)}\right]^{\frac{3}{2}} P_2(\sin\chi) \\ & - \frac{\gamma M}{a}\frac{m}{3}\left[1 + \cancel{\frac{2\widetilde{\varepsilon}}{3}} + \cancel{\frac{4\widetilde{\varepsilon}}{3}P_2(\sin\chi)}\right]^{-1}\left[1 - P_2(\sin\chi)\right] \\ \approx {} & C_0\left(1 + \frac{\widetilde{\varepsilon}}{3}\right) - \frac{\gamma M}{a}\frac{m}{3} + \left(C_0\frac{2\widetilde{\varepsilon}}{3} + C_2 + \frac{m}{3}\frac{\gamma M}{a}\right)P_2(\sin\chi).\end{aligned} \tag{8.87}$$

For $\Phi\left[\chi, r_S(\chi)\right]$ to be a geopotential surface, as required – see (8.75) – its right-hand side must equal a constant. For this to be so, the coefficient of the term in $P_2(\sin\chi)$ must be identically zero. Thus:

$$C_2 = -\frac{1}{3}\left(2C_0\widetilde{\varepsilon} + \frac{\gamma M}{a}m\right). \tag{8.88}$$

8.5.2.4 *Application of the Third Boundary Condition (at the Planet's Surface)*

From Newtonian gravitational potential (8.84), and using (8.47) and (8.48), the gradient of $V(\chi, r)$ outside the planet and down to its surface is

$$\begin{aligned}\nabla V(\chi, r) &= \frac{1}{r}\frac{\partial V}{\partial \chi}\mathbf{e}_\chi + \frac{\partial V}{\partial r}\mathbf{e}_r = \frac{1}{r}\frac{\partial}{\partial \chi}\left[C_0\frac{a}{r} + C_2\frac{a^3}{r^3}P_2(\sin\chi)\right]\mathbf{e}_\chi \\ &\quad + \frac{\partial}{\partial r}\left[C_0\frac{a}{r} + C_2\frac{a^3}{r^3}P_2(\sin\chi)\right]\mathbf{e}_r \\ &= \frac{C_2}{a}\frac{a^4}{r^4}3\sin\chi\cos\chi\,\mathbf{e}_\chi - \frac{1}{a}\left[C_0\frac{a^2}{r^2} + 3C_2\frac{a^4}{r^4}P_2(\sin\chi)\right]\mathbf{e}_r. \end{aligned} \tag{8.89}$$

Evaluating this equation at the planet's ellipsoidal surface $r = r_S(\chi)$ and using (8.78) and (8.85) then gives

$$\begin{aligned}(\nabla V \cdot \mathbf{n})_{r=r_S(\chi)} &= \frac{C_2}{a}\left[1 + \frac{2\widetilde{\varepsilon}}{3} + \frac{4\widetilde{\varepsilon}}{3}P_2(\sin\chi)\right]^2 3\sin\chi\cos\chi\left(\mathbf{e}_\chi \cdot \mathbf{n}\right) \\ &\quad - \left\{\frac{C_0}{a}\left[1 + \frac{2\widetilde{\varepsilon}}{3} + \frac{4\widetilde{\varepsilon}}{3}P_2(\sin\chi)\right]\right. \\ &\quad \left. + 3\frac{C_2}{a}\left[1 + \frac{2\widetilde{\varepsilon}}{3} + \frac{4\widetilde{\varepsilon}}{3}P_2(\sin\chi)\right]^2 P_2(\sin\chi)\right\}(\mathbf{e}_r \cdot \mathbf{n}) \\ &\approx -\frac{1}{a}\left[C_0\left(1 + \frac{2\widetilde{\varepsilon}}{3}\right) + \left(3C_2 + \frac{4\widetilde{\varepsilon}}{3}C_0\right)P_2(\sin\chi)\right]. \end{aligned} \tag{8.90}$$

To obtain the last line, the approximations

$$\mathbf{e}_\chi \cdot \mathbf{n} = \sin(\phi - \chi) = O(\widetilde{\varepsilon}), \quad \mathbf{e}_r \cdot \mathbf{n} = \cos(\phi - \chi) = 1 + O\left(\widetilde{\varepsilon}^2\right) \tag{8.91}$$

have been exploited, where – see Figs. 7.5 and 8.3 – χ and ϕ are geocentric and geographic latitudes, respectively.

Substitution of (8.90) into (8.81) – see (8.77) – yields

$$\begin{aligned}4\pi\gamma M &= \oiint_{\mathscr{S}} \nabla V \cdot \mathbf{n}\, d\mathscr{S} = \int_0^{2\pi}\int_{-\frac{\pi}{2}}^{+\frac{\pi}{2}} (\nabla V \cdot \mathbf{n})_{r=r^S}\left(r^S\right)^2 \cos\chi\, d\lambda\, d\chi \\ &= -2\pi a\int_{-1}^{+1}\left[C_0\left(1 + \frac{2\widetilde{\varepsilon}}{3}\right) + \left(3C_2 + \frac{4\widetilde{\varepsilon}}{3}C_0\right)P_2(\mu)\right]\left[1 + \frac{2\widetilde{\varepsilon}}{3} + \frac{4\widetilde{\varepsilon}}{3}P_2(\mu)\right]^{-1} d\mu \\ &\approx -2\pi a\int_{-1}^{+1}\left[C_0\left(1 + \frac{2\widetilde{\varepsilon}}{3}\right) + \left(3C_2 + \frac{4\widetilde{\varepsilon}}{3}C_0\right)P_2(\mu)\right]\left[1 - \frac{2\widetilde{\varepsilon}}{3} - \frac{4\widetilde{\varepsilon}}{3}P_2(\mu)\right] d\mu \\ &\approx -2\pi a\int_{-1}^{+1}\left[C_0 + 3C_2\left(\frac{3\mu^2 - 1}{2}\right)\right] d\mu = -4\pi a C_0, \end{aligned} \tag{8.92}$$

where $\mu \equiv \sin\chi$ is a dummy variable of integration. For given values of M and a, (8.92) determines C_0, and so

$$C_0 = -\frac{\gamma M}{a}. \tag{8.93}$$

This value of C_0 is then used to determine C_2 from (8.88). Thus

$$C_2 = \frac{\gamma M}{a}\left(\frac{2\widetilde{\varepsilon} - m}{3}\right). \tag{8.94}$$

8.5.2.5 *The Newtonian Gravitational Potential Exterior to an Ellipsoid of Variable Density*

Using (8.93) and (8.94) to eliminate C_0 and C_2 from (8.84), and (8.2) to rewrite $P_2(\sin\chi)$ finally yields the first-order-accurate representation of the Newtonian gravitational potential, $V(\chi, r)$, exterior to a rotating, variable-density, ellipsoidal planet of known ellipticity and mass. Thus

$$\begin{aligned} V(\chi, r) &= -\frac{\gamma M}{r}\left[1 - \left(\frac{2\widetilde{\varepsilon} - m}{3}\right)\frac{a^2}{r^2}P_2(\sin\chi)\right] \\ &= -\frac{\gamma M}{r}\left[1 - \left(\frac{2\widetilde{\varepsilon} - m}{2}\right)\frac{a^2}{r^2}\left(\sin^2\chi - \frac{1}{3}\right)\right], \quad r \geq r_S(\chi). \end{aligned} \tag{8.95}$$

8.5.2.6 *The Geopotential Φ Exterior to and on an Ellipsoidal Planet of Variable Density*

Similarly, using (8.93) and (8.94) to eliminate C_0 and C_2 from (8.86) yields the associated first-order-accurate representation of the geopotential. Thus to $O(\widetilde{\varepsilon}, m)$:

The Geopotential Φ *outside* an Ellipsoidal Planet of Variable Density

For an *ellipsoidal* planet of *variable* density, with semi-major and semi-minor axes a and c, respectively, and mass M, the geopotential *outside* it is

$$\begin{aligned} \Phi(\chi, r) &= -\frac{\gamma M}{r}\left\{1 - \left(\frac{2\widetilde{\varepsilon} - m}{3}\right)\frac{a^2}{r^2}P_2(\sin\chi) + \frac{m}{3}\frac{r^3}{a^3}\left[1 - P_2(\sin\chi)\right]\right\} \\ &= -\frac{\gamma M}{r}\left\{1 - \left(\frac{2\widetilde{\varepsilon} - m}{3}\right)\frac{a^2}{r^2}\left(\frac{3\sin^2\chi - 1}{2}\right) + \frac{m}{3}\frac{r^3}{a^3}\left[1 - \left(\frac{3\sin^2\chi - 1}{2}\right)\right]\right\} \\ &= -\frac{\gamma M}{r}\left\{1 + \left(\frac{2\widetilde{\varepsilon} - m}{6}\right)\frac{a^2}{r^2} + \frac{m}{2}\frac{r^3}{a^3} - \left[\left(\frac{2\widetilde{\varepsilon} - m}{2}\right)\frac{a^2}{r^2} + \frac{m}{2}\frac{r^3}{a^3}\right]\sin^2\chi\right\}, \\ & \qquad r \geq r_S(\chi), \end{aligned} \tag{8.96}$$

where (from (8.79) and (8.80))

$$\widetilde{\varepsilon} \equiv \frac{a^2 - c^2}{2c^2}, \quad m \equiv \frac{\Omega^2 a^3}{\gamma M}. \tag{8.97}$$

8.5.2.7 *Gravity Exterior to and on an Ellipsoidal Planet of Variable Density*

The apparent gravitational force, $\mathbf{g}(\chi, r)$, outside an ellipsoidal planet of *variable* density is obtained by inserting (8.96) into (7.4), and using (8.47) to evaluate the gradient. Thus:

Gravity *outside* an Ellipsoidal Planet of Variable Density

Gravity *outside* a spherical planet of *variable* density is given by

$$\mathbf{g}(\chi, r) \equiv -\nabla\Phi = -\frac{1}{r}\frac{\partial\Phi}{\partial\chi}\mathbf{e}_\chi - \frac{\partial\Phi}{\partial r}\mathbf{e}_r$$

$$= -\frac{\gamma M}{r^2}\left[(2\widetilde{\varepsilon} - m)\frac{a^2}{r^2} + m\frac{r^3}{a^3}\right]\sin\chi\cos\chi\,\mathbf{e}_\chi$$
$$-\frac{\gamma M}{r^2}\left\{1 + \left(\frac{2\widetilde{\varepsilon} - m}{2}\right)\frac{a^2}{r^2} - m\frac{r^3}{a^3} - \left[3\left(\frac{2\widetilde{\varepsilon} - m}{2}\right)\frac{a^2}{r^2} - m\frac{r^3}{a^3}\right]\sin^2\chi\right\}\mathbf{e}_r,$$
$$r \geq r_S(\chi), \quad (8.98)$$

where – see (8.78) –

$$r_S(\chi) = a\left(1 + 2\widetilde{\varepsilon}\sin^2\chi\right)^{-\frac{1}{2}}. \quad (8.99)$$

Substituting (8.99) into (8.98), evaluated at the planet's ellipsoidal surface $r = r_S(\chi)$, and dropping second-order terms, we see that the apparent gravitational force on the surface of an ellipsoidal planet of variable density is

$$(\mathbf{g})_{r=r_S(\chi)} \approx -2\widetilde{\varepsilon}\frac{\gamma M}{a^2}\sin\chi\cos\chi\,\mathbf{e}_\chi - \frac{\gamma M}{a^2}\left[1 + \widetilde{\varepsilon} - \frac{3m}{2} + \left(\frac{5m}{2} - \widetilde{\varepsilon}\right)\sin^2\chi\right]\mathbf{e}_r. \quad (8.100)$$

Thus, to first order in $\widetilde{\varepsilon}$ and m, the variation of gravity g at the planet's surface is given by

$$(g)_{r=r_S(\chi)} \equiv \left(|\mathbf{g}|\right)_{r=r_S(\chi)} = \frac{\gamma M}{a^2}\left\{4\widetilde{\varepsilon}^2\sin^2\chi\cos^2\chi + \left[1 + \widetilde{\varepsilon} - \frac{3m}{2} + \left(\frac{5m}{2} - \widetilde{\varepsilon}\right)\sin^2\chi\right]^2\right\}^{\frac{1}{2}}$$
$$\approx \frac{\gamma M}{a^2}\left\{\left[1 + \widetilde{\varepsilon} - \frac{3m}{2} + \left(\frac{5m}{2} - \widetilde{\varepsilon}\right)\sin^2\chi\right]^2\right\}^{\frac{1}{2}}$$
$$= \frac{\gamma M}{a^2}\left[1 + \widetilde{\varepsilon} - \frac{3m}{2} + \left(\frac{5m}{2} - \widetilde{\varepsilon}\right)\sin^2\chi\right]$$
$$\approx g_S^E\left[1 + \left(\frac{5m}{2} - \widetilde{\varepsilon}\right)\sin^2\chi\right], \quad (8.101)$$

where

$$g_S^E \equiv (g)_{r=r_S(\chi=0)} = \frac{\gamma M}{a^2}\left(1 + \widetilde{\varepsilon} - \frac{3m}{2}\right), \quad (8.102)$$

is the value of g at the planet's equator.

As a cross-check, applying Newton's constraint (8.62) for an ellipsoidal planet of constant density (that $\varepsilon \approx \widetilde{\varepsilon} = 5m/4$) recovers the equations given in Section 8.3.9 as a special case.

8.5.2.8 *Clairaut's Fraction for an Ellipsoidal Planet of Variable Density*

Evaluating (8.101) at $\chi = \pm\pi/2$, the value of g at the planet's poles is

$$g_S^P = (g)_{r=r_S(\chi=\pm\pi/2)} \approx g_S^E\left(1 + \frac{5m}{2} - \widetilde{\varepsilon}\right)$$
$$= \frac{\gamma M}{a^2}\left(1 + \widetilde{\varepsilon} - \frac{3m}{2}\right)\left(1 + \frac{5m}{2} - \widetilde{\varepsilon}\right) \approx \frac{\gamma M}{a^2}(1 + m), \quad (8.103)$$

where – dropping second-order terms in m and $\widetilde{\varepsilon}$ – (8.102) has been used to obtain the last-most expression. Thus (8.103) leads to:

Clairaut's Fraction for a Rotating Ellipsoidal Planet of Variable Density

$$\frac{g_S^P - g_S^E}{g_S^E} \equiv \text{Clairaut's fraction} \approx \frac{5m}{2} - \widetilde{\varepsilon} \approx \frac{5m}{2} - \varepsilon, \tag{8.104}$$

to first order in m and $\widetilde{\varepsilon}$ or, equivalently, to first order in m and ε.

The parameters m and ε in (8.104) are determined by a combination of:

- The planet's rotation rate, Ω.
- Its equatorial radius, a.
- Its ellipticity, ε (or $\widetilde{\varepsilon}$).
- Its mass, M.

Included as special cases are:

- The variable-density sphere examined in Section 7.8 (for which $\varepsilon \equiv \widetilde{\varepsilon} \equiv 0$).
- The constant-density ellipsoid examined in Section 8.3 (for which $\varepsilon = 5m/4$).

8.5.3 The Interior Problem

8.5.3.1 *Problem Definition*

If we only wish to model gravity exterior to a mildly oblate, ellipsoidal planet of variable density – for example, to model Earth's atmosphere for weather-prediction purposes – then we can simply stop here since we already have everything we need:

- Equation (8.99) defines the planet's ellipsoidal surface, $r = r_S(\chi)$.
- Newtonian gravitational potential, $V(\chi, r)$; geopotential, $\Phi(\chi, r)$; and apparent gravitational force vector, $\mathbf{g}(\chi, r)$; are given by (8.95), (8.96), and (8.98), respectively.
- Values of parameters γ, M, $\widetilde{\varepsilon}$, and m therein are all known to high accuracy.

However, if we also wish to model gravity *below* the planet's reference surface (i.e. below its geoid) – for example, to model Earth's deep oceans – then we need to go further by defining a suitable density distribution for the planet's interior and then developing the associated geopotential. *This geopotential should be consistent at the planet's surface with the outer model of gravity developed in Section 8.5.2* (valid outside the planet and down to its surface). This inner geopotential (valid inside the planet and up to its surface) should also be accurate to first order in $\widetilde{\varepsilon}$ and m everywhere within an ocean.

Application of the following two-step procedure allows accomplishment of these goals in a conceptually straightforward manner (but with some complications regarding the details):

Determination of Newtonian Gravitational Potential *below* a Planet's Geoid

To determine Newtonian gravitational potential below a planet's geoid:

1. Define a bounded, physically reasonable density distribution, $\rho(\chi, r)$, inside the planet (i.e. for $r \leq r_S(\chi)$).
2. Solve Poisson's equation

$$\nabla^2 V(\chi, r) = 4\pi\gamma\rho(\chi, r), \tag{8.105}$$

inside the planet subject to:

(a) Boundedness of $V(\chi, r)$ everywhere inside the planet.
(b) Continuity of $V(\chi, r)$ and $\nabla V(\chi, r)$ at the planet's surface, $r = r_S(\chi)$, with the corresponding outer solutions there.

Noting that centrifugal potential $V^C(\chi, r)$ and its gradient are continuous at the planet's surface, and – from (7.6) – that

$$\Phi(\chi, r) \equiv V(\chi, r) + V^C(\chi, r), \tag{8.106}$$

boundary conditions of Step 2(b) ensure that $\Phi(\chi, r)$, and the apparent gravitational force vector,

$$\mathbf{g}(\chi, r) \equiv -\nabla\Phi(\chi, r) \equiv -\nabla V(\chi, r) - \nabla V^C(\chi, r), \tag{8.107}$$

are both continuous at and across the planet's surface.

8.5.3.2 *Exterior Matching Conditions*

From (8.95), the Newtonian gravitational potential exterior to an ellipsoid of variable density is (to first order in m and $\widetilde{\varepsilon}$)

$$\begin{aligned} V(\chi, r) &= -\frac{\gamma M}{r}\left[1 - \left(\frac{2\widetilde{\varepsilon} - m}{3}\right)\frac{a^2}{r^2}P_2(\sin\chi)\right] \\ &= -\frac{\gamma M}{r}\left[1 - \left(\frac{2\widetilde{\varepsilon} - m}{2}\right)\frac{a^2}{r^2}\left(\sin^2\chi - \frac{1}{3}\right)\right], \quad r \geq r_S(\chi). \end{aligned} \tag{8.108}$$

To prepare the way for application of the boundary conditions of Step 2(b) of the reformulated procedure of Section 8.5.3.1, (8.108) is used to compute V and ∇V at the planet's surface.

Evaluating (8.108) at the planet's ellipsoidal surface $r = r_S(\chi) = a\left(1 + 2\widetilde{\varepsilon}\sin^2\chi\right)^{-\frac{1}{2}}$ gives, to first order in m and $\widetilde{\varepsilon}$,

$$\begin{aligned} V(\chi, r = r_S) &= -\frac{\gamma M}{a}\left(1 + 2\widetilde{\varepsilon}\sin^2\chi\right)^{\frac{1}{2}}\left[1 - \left(\frac{2\widetilde{\varepsilon} - m}{2}\right)\left(1 + 2\widetilde{\varepsilon}\sin^2\chi\right)\left(\sin^2\chi - \frac{1}{3}\right)\right] \\ &\approx -\frac{\gamma M}{a}\left(1 + \widetilde{\varepsilon}\sin^2\chi\right)\left[1 - \left(\frac{2\widetilde{\varepsilon} - m}{2}\right)\left(\sin^2\chi - \frac{1}{3}\right)\right] \\ &\approx -\frac{\gamma M}{a}\left[1 + \left(\frac{2\widetilde{\varepsilon} - m}{6}\right) + \frac{m}{2}\sin^2\chi\right] \\ &= -\frac{\gamma M}{a}\left\{1 + \left(\frac{2\widetilde{\varepsilon} - m}{6}\right) + \frac{m}{2}\left[\frac{1 + 2P_2(\sin\chi)}{3}\right]\right\} \\ &= -\frac{\gamma M}{a}\left[1 + \frac{\widetilde{\varepsilon}}{3} + \frac{m}{3}P_2(\sin\chi)\right], \end{aligned} \tag{8.109}$$

where binomial expansion has been used. Equation (8.109) is the first, exterior, matching condition.

From (8.108) and (8.47), $\nabla V(\chi, r)$ outside the planet is given by

$$\nabla V(\chi, r) = \frac{1}{r}\frac{\partial V}{\partial \chi}\mathbf{e}_\chi + \frac{\partial V}{\partial r}\mathbf{e}_r$$

$$
\begin{aligned}
&= -\frac{1}{r}\frac{\partial}{\partial \chi}\left\{\frac{\gamma M}{r}\left[\not\chi - \left(\frac{2\widetilde{\varepsilon}-m}{2}\right)\frac{a^2}{r^2}\left(\sin^2\chi - \frac{1}{3}\right)\right]\right\}\mathbf{e}_\chi \\
&\quad - \frac{\partial}{\partial r}\left\{\frac{\gamma M}{r}\left[1 - \left(\frac{2\widetilde{\varepsilon}-m}{2}\right)\frac{a^2}{r^2}\left(\sin^2\chi - \frac{1}{3}\right)\right]\right\}\mathbf{e}_r \\
&= \frac{\gamma M}{a^2}\left(2\widetilde{\varepsilon}-m\right)\frac{a^4}{r^4}\sin\chi\cos\chi\,\mathbf{e}_\chi + \frac{\gamma M}{a^2}\left[\frac{a^2}{r^2} - 3\left(\frac{2\widetilde{\varepsilon}-m}{2}\right)\frac{a^4}{r^4}\left(\sin^2\chi - \frac{1}{3}\right)\right]\mathbf{e}_r.
\end{aligned}
\tag{8.110}
$$

Evaluation of this at the planet's surface $r = r_S(\chi) = a\left(1 + 2\widetilde{\varepsilon}\sin^2\chi\right)^{-\frac{1}{2}}$, together with binomial expansion, then gives

$$
\begin{aligned}
\nabla V(\chi, r = r_S) &\approx \frac{\gamma M}{a^2}\left(2\widetilde{\varepsilon}-m\right)\sin\chi\cos\chi\,\mathbf{e}_\chi \\
&\quad + \frac{\gamma M}{a^2}\left[1 + 2\widetilde{\varepsilon}\sin^2\chi - 3\left(\frac{2\widetilde{\varepsilon}-m}{2}\right)\left(\sin^2\chi - \frac{1}{3}\right)\right]\mathbf{e}_r \\
&= \frac{\gamma M}{a^2}\left(2\widetilde{\varepsilon}-m\right)\sin\chi\cos\chi\,\mathbf{e}_\chi + \frac{\gamma M}{a^2}\left[1 + \left(\frac{2\widetilde{\varepsilon}-m}{2}\right) + \left(\frac{3}{2}m - \widetilde{\varepsilon}\right)\sin^2\chi\right]\mathbf{e}_r.
\end{aligned}
\tag{8.111}
$$

Thus – from (8.109) and (8.111) – the two exterior, matching conditions for application of Step 2(b) of the procedure outlined in Section 8.5.3.1 are

$$
V(\chi, r = r_S) = -\frac{\gamma M}{a}\left[1 + \frac{\widetilde{\varepsilon}}{3} + \frac{m}{3}P_2(\sin\chi)\right], \tag{8.112}
$$

$$
\begin{aligned}
\nabla V(\chi, r = r_S) &= \frac{\gamma M}{a^2}\left(2\widetilde{\varepsilon}-m\right)\sin\chi\cos\chi\,\mathbf{e}_\chi \\
&\quad + \frac{\gamma M}{a^2}\left[1 + \left(\frac{2\widetilde{\varepsilon}-m}{2}\right) + \left(\frac{3}{2}m - \widetilde{\varepsilon}\right)\sin^2\chi\right]\mathbf{e}_r.
\end{aligned}
\tag{8.113}
$$

8.5.3.3 *Functional Forms for $\rho(\chi, r)$ and $V(\chi, r)$ inside the Planet*

The question now is:

- What functional form should density distribution $\rho(\chi, r)$ have?

The answer, broadly speaking, is that:

- It should be a *simple* representation that captures meridional and radial variation of Earth's density distribution.
- It should lead to a Newtonian gravitational potential within the planet that agrees with matching conditions (8.112) and (8.113) at the planet's surface.

A very similar argument to that given in Section 7.8.4.2 for a *spherical* planet of variable density indicates that the simplest possible functional form for $V(\chi, r)$ inside the planet that meets requirements is – cf. (7.71) –

$$
V(\chi, r) = A_0 + A_2\frac{r^2}{a^2} + A_3\frac{r^3}{a^3} - \frac{\Omega^2 r^2}{3}P_2(\sin\chi) + \left(B_2\frac{r^2}{a^2} + B_3\frac{r^3}{a^3} + B_4\frac{r^4}{a^4}\right)P_2(\sin\chi), \quad r \leq r_S(\chi). \tag{8.114}
$$

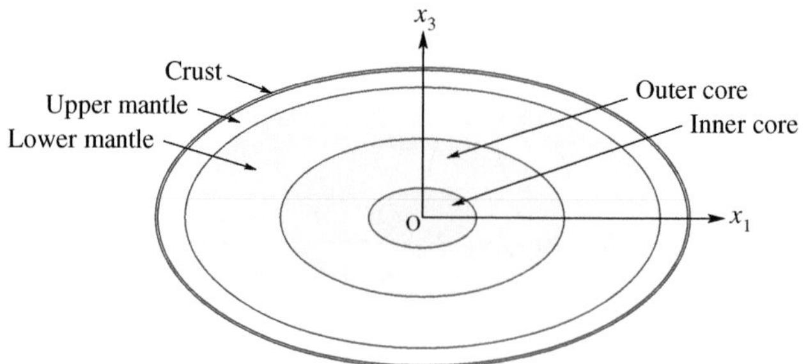

Figure 8.5 Cross section in the Ox_1x_3 meridional plane of some constant-density (isopycnal) surfaces inside an ellipsoidal, fluid planet of variable composition; darker shading indicates higher density. The outermost contour corresponds to the planet's deformable, ellipsoidal, fluid boundary, which is assumed to coincide with a geopotential surface. Isopycnal contours have been chosen to broadly correspond with those associated with Earth's internal layers. See text for further details.

The associated density distribution is – cf. (7.72) –

$$\rho(\chi, r) = \frac{1}{4\pi\gamma}\nabla^2 V(\chi, r) = \frac{3}{2\pi\gamma a^2}\left[A_2 + 2A_3\frac{r}{a} + \left(B_3\frac{r}{a} + \frac{7}{3}B_4\frac{r^2}{a^2}\right)P_2(\sin\chi)\right],$$
$$r \leq r_S(\chi). \quad (8.115)$$

See Fig. 8.5 for a graphical depiction (greatly exaggerated for visual clarity) of a cross section of a simple, physically reasonable density distribution for mildly oblate, rotating planets, based on that for Earth.

We are now in a position to apply the constraint that the planet's ellipsoidal surface at $r = r_S(\chi)$ corresponds to an isopycnal surface (i.e. a surface on which density is constant). To do so, we impose the condition that

$$\rho[\chi, r = r_S(\chi)] = \rho_a \equiv \rho(\chi = 0, r = a) = \text{constant}, \quad (8.116)$$

where – from (8.78) – the planet's ellipsoidal surface is located at $r = r_S(\chi) = a\left(1 + 2\widetilde{\varepsilon}\sin^2\chi\right)^{-\frac{1}{2}}$. Evaluating density distribution (8.115) at $r = r_S(\chi)$ and applying constraint (8.116) then gives

$$\begin{aligned}\rho[\chi, r_S(\chi)] &= \frac{3}{2\pi\gamma a^2}\left[A_2 + 2A_3\frac{r_S}{a} + \left(B_3\frac{r_S}{a} + \frac{7}{3}B_4\frac{r_S^2}{a^2}\right)P_2(\sin\chi)\right] \\ &= \frac{3}{2\pi\gamma a^2}\left\{A_2 + \frac{2A_3}{\left(1 + 2\widetilde{\varepsilon}\sin^2\chi\right)^{\frac{1}{2}}}\right. \\ &\quad \left. + \left[\frac{B_3}{\left(1 + 2\widetilde{\varepsilon}\sin^2\chi\right)^{\frac{1}{2}}} + \frac{7}{3}\frac{B_4}{\left(1 + 2\widetilde{\varepsilon}\sin^2\chi\right)}\right]P_2(\sin\chi)\right\} \\ &= \rho_a = \text{constant}.\end{aligned} \quad (8.117)$$

Assume (cf. (8.41) for the much simpler case of an ellipsoidal planet of constant density) that

$$A_0, A_2 = O(1), \quad A_3 = O(1) \text{ or } O(m, \widetilde{\varepsilon}), \quad B_2, B_3, B_4 = O(m, \widetilde{\varepsilon}). \tag{8.118}$$

The orders of the four coefficients A_0, A_2, B_2, and B_3 in (8.118) follow naturally from the earlier analyses for spherical planets – of both constant and variable density – and for an ellipsoidal planet of *constant* density. For the case under present consideration (of an ellipsoidal planet of variable density), the order of coefficient A_3 is less obvious. As we will see, there are generality versus simplicity and tractability trade-offs between them. We therefore leave the order of A_3 open for now by simultaneously maintaining the possibility of it being of either $O(1)$ or of $O(m, \widetilde{\varepsilon})$, so that:

- For $A_3 = O(1)$, all terms are retained in the equations, *including all terms coloured in red.*
- For $A_3 = O(m, \widetilde{\varepsilon})$, *all terms coloured in red are omitted.*

This way, we obtain two analyses for the bargain price of one!

In what follows, much use of asymptotic expressions (8.118) is made to consistently drop second-order terms in m and $\widetilde{\varepsilon}$. Again, when done, this is indicated by the use of the approximation sign ($\approx$), with neglect of second-order terms implied. Thus – cf. (8.117) –

$$\begin{aligned} & A_2 + \frac{2A_3}{\left(1 + 2\widetilde{\varepsilon}\sin^2\chi\right)^{\frac{1}{2}}} + \left[\frac{B_3}{\left(1 + \cancel{2\widetilde{\varepsilon}\sin^2\chi}\right)^{\frac{1}{2}}} + \frac{7}{3}\frac{B_4}{\left(1 + \cancel{2\widetilde{\varepsilon}\sin^2\chi}\right)}\right] P_2(\sin\chi) \\ & \approx A_2 + 2A_3\left(1 - \widetilde{\varepsilon}\sin^2\chi\right) + \left(B_3 + \frac{7}{3}B_4\right) P_2(\sin\chi) \\ & = A_2 + 2A_3\left(1 - \frac{\widetilde{\varepsilon}}{3}\right) + \left(B_3 + \frac{7}{3}B_4 - \frac{4\widetilde{\varepsilon}}{3}A_3\right) P_2(\sin\chi), \end{aligned} \tag{8.119}$$

to $O(m, \widetilde{\varepsilon})$, where (7.116) has been used to obtain the last line. Insertion of (8.119) into (8.117) then gives

$$\rho\left[\chi, r_S(\chi)\right] \approx \frac{3}{2\pi\gamma a^2}\left[A_2 + 2A_3\left(1 - \frac{\widetilde{\varepsilon}}{3}\right) + \left(B_3 + \frac{7}{3}B_4 - \frac{4\widetilde{\varepsilon}}{3}A_3\right) P_2(\sin\chi)\right]. \tag{8.120}$$

Now the right-hand side of this equation is constant when the coefficient of $P_2(\sin\chi)$ is zero, that is, when

$$B_4 = -\frac{3}{7}B_3 + \frac{4\widetilde{\varepsilon}}{7}A_3. \tag{8.121}$$

To first-order accuracy, $\rho(\chi, r = r_S)$ is then an isopycnal surface as desired. Using (8.121), we can therefore rewrite (8.114) and (8.115) as[11]

[11] Elimination of B_4 from the subsequent analysis has the great virtue that it shortens lengthy equations, thereby simplifying things. A minor inconvenience, however, is that this excludes a homogeneous (i.e. constant-density) ellipsoidal planet as being a special case of the ensuing analysis. This subtlety is because the surface of a homogeneous ellipsoidal planet is, by definition, an isopycnal surface, and constraint (8.121) is then no longer a necessary condition for this to be so; it already is. Parameter B_4 is then *arbitrary* and can be chosen to satisfy (8.121) if one wishes. This does not, however, correspond to the planet being of constant density, but to something else.

$$V(\chi,r) = A_0 + A_2\frac{r^2}{a^2} + A_3\frac{r^3}{a^3} - \frac{\Omega^2 r^2}{3}P_2(\sin\chi)$$
$$+\left[B_2\frac{r^2}{a^2} + B_3\frac{r^3}{a^3}\left(1 - \frac{3}{7}\frac{r}{a}\right) + \frac{4\widetilde{\varepsilon}}{7}A_3\frac{r^4}{a^4}\right]P_2(\sin\chi), \quad r \le r_S(\chi), \quad (8.122)$$

$$\rho(\chi,r) = \frac{3}{2\pi\gamma a^2}\left\{A_2 + 2A_3\frac{r}{a} + \left[B_3\frac{r}{a}\left(1 - \frac{r}{a}\right) + \frac{4\widetilde{\varepsilon}}{3}A_3\frac{r^2}{a^2}\right]P_2(\sin\chi)\right\}, \quad r \le r_S(\chi). \quad (8.123)$$

The A_2 term in density distribution (8.123) provides a basic, constant contribution; the A_3 term allows the planet's core to be denser than its crust; and the $P_2(\sin\chi)$ term introduces meridional variation (without this term, the planet's surface could not coincide with a geopotential surface nor with an isopycnal surface). The parameters A_0, A_2, A_3, B_2, and B_3 in (8.122) and (8.123) are *not* free parameters. Their values are determined – see Sections 8.5.3.5 and 8.5.3.6 – as functions of the following parameters, *with assumed known values*:

- The planet's total mass, M.
- Newton's universal gravitational constant, γ.
- The planet's ellipticity, as measured by $\widetilde{\varepsilon}$.
- The ratio of centrifugal force to Newtonian gravitational attraction, as measured by m.

Density distribution (8.123) satisfies the boundedness condition of Step 1 of the procedure outlined in Section 8.5.3.1. Poisson's equation for Step 2 can be written as

$$\nabla^2 V(\chi,r) \equiv \frac{1}{r^2\cos\chi}\frac{\partial}{\partial\chi}\left(\cos\chi\frac{\partial V}{\partial\chi}\right) + \frac{1}{r^2}\frac{\partial}{\partial r}\left(r^2\frac{\partial V}{\partial r}\right) = 4\pi\gamma\rho(\chi,r), \quad (8.124)$$

where $\rho(\chi,r)$ is defined by (8.123). Equation (8.122) is, by construction, a bounded solution of (8.124), as required by Step 2. Coefficients A_0, A_2, A_3, B_2, and B_3 in (8.122) are determined – see Step 2(b) and Section 8.5.3.6 – by matching exterior conditions (8.112) and (8.113) to the corresponding interior ones developed in Section 8.5.3.5.

8.5.3.4 *An Equivalent Functional Form for $\rho(\chi,r)$*

To help further our understanding, the functional form (8.123) for $\rho(\chi,r)$ is now equivalently rewritten – as in Section 7.8.4.3 – by re-expressing A_2 and A_3 in terms of ρ_0 and ρ_a. These two parameters are easier to interpret physically.

Specifically, let ρ_0 and ρ_a be the values of ρ at the planet's centre and at its isopycnal, ellipsoidal boundary, respectively. Evaluating (8.123) at $r = 0$ immediately gives

$$\rho_0 \equiv \rho(r=0) = \frac{3}{2\pi\gamma a^2}A_2. \quad (8.125)$$

Applying isopycnal condition (8.116) to (8.123), with use of (8.2), then yields – to first order in $\widetilde{\varepsilon}$ –

$$\rho_a \equiv \rho(r = r_S) \approx \frac{3}{2\pi\gamma a^2}\left[A_2 + 2A_3\left(1 - \frac{\widetilde{\varepsilon}}{3}\right)\right]. \quad (8.126)$$

Solving (8.125) and (8.126) for A_2 and A_3 results in

$$A_2 = \frac{2\pi\gamma a^2}{3}\rho_0, \quad A_3 = -\frac{\pi\gamma a^2}{(3-\widetilde{\varepsilon})}(\rho_0 - \rho_a) \approx -\frac{\pi\gamma a^2}{3}\left(1 + \frac{\widetilde{\varepsilon}}{3}\right)(\rho_0 - \rho_a). \quad (8.127)$$

Inserting these expressions for A_2 and A_3 into (8.123) finally leads to the following, equivalent (to first order) functional form for density distribution (8.123):

$$\rho(\chi, r) = \underbrace{\rho_0 - \left(1 + \frac{\widetilde{\varepsilon}}{3}\right)(\rho_0 - \rho_a)\frac{r}{a}}_{\text{linear-radial variation}} + \underbrace{\frac{3}{2\pi\gamma a^2}\left[B_3\frac{r}{a}\left(1 - \frac{r}{a}\right) + \frac{4\widetilde{\varepsilon}}{3}A_3\frac{r^2}{a^2}\right]P_2(\sin\chi)}_{\text{quadratic-radial and meridional variation}}, \quad r \le r_S(\chi). \tag{8.128}$$

In (8.128), ρ_0 and ρ_a are free parameters, and B_3 and A_3 are determined by application of boundary conditions at the planet's assumed, ellipsoidal surface at $r = r_S(\chi)$; see Section 8.5.3.5. Examination of the leading-order (first two) terms on the right-hand side of (8.128) indicates that ρ monotonically increases linearly (with respect to r) from the planet's crust to its core when $\rho_0 > \rho_a$. Setting $\rho_a = \rho_0/10$ would, for example, make the density at the planet's core an order of magnitude larger than at its crust. For Earth, this would be qualitatively correct and, at the current (crude) level of understanding of Earth's composition, quantitatively in the right ballpark. For present purposes, however, we only need an acceptably accurate representation of composition in the vicinity of Earth's crust for oceanic applications.

8.5.3.5 *Interior Matching Conditions*

Using (8.97), (8.118), and (7.116) to evaluate the interior potential (8.122) on the planet's surface $r = r_S(\chi) = a\left(1 + 2\widetilde{\varepsilon}\sin^2\chi\right)^{-\frac{1}{2}}$ gives (to first order in m and $\widetilde{\varepsilon}$)

$$\begin{aligned}
V(\chi, r = r_S) &= A_0 + A_2\frac{r_S^2}{a^2} + A_3\frac{r_S^3}{a^3} - \frac{\Omega^2 r_S^2}{3}P_2(\sin\chi) \\
&\quad + \left[B_2\frac{r_S^2}{a^2} + B_3\frac{r_S^3}{a^3}\left(1 - \frac{3}{7}\frac{r_S}{a}\right) + \frac{4\widetilde{\varepsilon}}{7}A_3\frac{r_S^4}{a^4}\right]P_2(\sin\chi) \\
&\approx A_0 + A_2\left(1 + 2\widetilde{\varepsilon}\sin^2\chi\right)^{-1} + A_3\left(1 + 2\widetilde{\varepsilon}\sin^2\chi\right)^{-\frac{3}{2}} - \frac{\Omega^2 a^2}{3}P_2(\sin\chi) \\
&\quad + \left\{B_2 + B_3\left[1 - \frac{3}{7}\left(1 + \cancel{2\widetilde{\varepsilon}\sin^2\chi}\right)^{-\frac{1}{2}}\right]\right. \\
&\quad \left. + \frac{4\widetilde{\varepsilon}}{7}A_3\left(1 + \cancel{2\widetilde{\varepsilon}\sin^2\chi}\right)^{-2}\right\}P_2(\sin\chi) \\
&\approx A_0 + A_2\left(1 - 2\widetilde{\varepsilon}\sin^2\chi\right) + A_3\left(1 - 3\widetilde{\varepsilon}\sin^2\chi\right) - \frac{\gamma M}{a}\frac{m}{3}P_2(\sin\chi) \\
&\quad + \left(B_2 + \frac{4}{7}B_3 + \frac{4\widetilde{\varepsilon}}{7}A_3\right)P_2(\sin\chi) \\
&= A_0 + A_2\left[1 - \frac{2\widetilde{\varepsilon}}{3} - \frac{4\widetilde{\varepsilon}P_2(\sin\chi)}{3}\right] + A_3\left[1 - \widetilde{\varepsilon} - 2\widetilde{\varepsilon}P_2(\sin\chi)\right] \\
&\quad + \left(B_2 + \frac{4}{7}B_3 + \frac{4\widetilde{\varepsilon}}{7}A_3 - \frac{\gamma M}{a}\frac{m}{3}\right)P_2(\sin\chi) \\
&= A_0 + \left(1 - \frac{2\widetilde{\varepsilon}}{3}\right)A_2 + (1 - \widetilde{\varepsilon})A_3
\end{aligned}$$

$$
\begin{aligned}
&+\left[B_2+\frac{4}{7}B_3+\frac{4\widetilde{\varepsilon}}{7}A_3-\left(\frac{4\widetilde{\varepsilon}}{3}A_2+2\widetilde{\varepsilon}A_3+\frac{\gamma M}{a}\frac{m}{3}\right)\right]P_2(\sin\chi)\\
&=A_0+\left(1-\frac{2\widetilde{\varepsilon}}{3}\right)A_2+(1-\widetilde{\varepsilon})A_3\\
&+\left(B_2+\frac{4}{7}B_3-\frac{4\widetilde{\varepsilon}}{3}A_2-\frac{10\widetilde{\varepsilon}}{7}A_3-\frac{\gamma M}{a}\frac{m}{3}\right)P_2(\sin\chi). \qquad (8.129)
\end{aligned}
$$

Using (8.122), the gradient of V inside the planet is then

$$
\begin{aligned}
\nabla V(\chi,r)=&\ \frac{1}{r}\frac{\partial V}{\partial\chi}\mathbf{e}_\chi+\frac{\partial V}{\partial r}\mathbf{e}_r\\
=&\ \frac{1}{r}\left[B_2\frac{r^2}{a^2}+B_3\frac{r^3}{a^3}\left(1-\frac{3}{7}\frac{r}{a}\right)+\frac{4\widetilde{\varepsilon}}{7}A_3\frac{r^4}{a^4}-\frac{\Omega^2r^2}{3}\right]3\sin\chi\cos\chi\,\mathbf{e}_\chi\\
&+\frac{1}{a}\left\{2A_2\frac{r}{a}+3A_3\frac{r^2}{a^2}+\left[2B_2\frac{r}{a}+3B_3\frac{r^2}{a^2}\left(1-\frac{4}{7}\frac{r}{a}\right)\right.\right.\\
&\left.\left.+\frac{16\widetilde{\varepsilon}}{7}A_3\frac{r^3}{a^3}-\frac{2\Omega^2ar}{3}\right]P_2(\sin\chi)\right\}\mathbf{e}_r. \qquad (8.130)
\end{aligned}
$$

Noting from (8.118) that $B_2, B_3 = O(m,\widetilde{\varepsilon})$ and evaluating (8.130) at $r = r_S(\chi) = a\left(1+2\widetilde{\varepsilon}\sin^2\chi\right)^{-\frac{1}{2}}$, with use of (7.115), gives

$$
\begin{aligned}
\nabla V(\chi,r_S)\approx&\ \frac{1}{a}\left(B_2+\frac{4}{7}B_3+\frac{4\widetilde{\varepsilon}}{7}A_3-\frac{\Omega^2a^2}{3}\right)3\sin\chi\cos\chi\,\mathbf{e}_\chi\\
&+\frac{1}{a}\left[2A_2\left(1+2\widetilde{\varepsilon}\sin^2\chi\right)^{-\frac{1}{2}}+3A_3\left(1+2\widetilde{\varepsilon}\sin^2\chi\right)^{-1}\right]\mathbf{e}_r\\
&+\frac{1}{a}\left(2B_2+\frac{9}{7}B_3+\frac{16\widetilde{\varepsilon}}{7}A_3-\frac{2\Omega^2a^2}{3}\right)P_2(\sin\chi)\,\mathbf{e}_r\\
\approx&\ \frac{1}{a}\left(B_2+\frac{4}{7}B_3+\frac{4\widetilde{\varepsilon}}{7}A_3-\frac{\gamma M}{a}\frac{m}{3}\right)3\sin\chi\cos\chi\,\mathbf{e}_\chi\\
&+\frac{1}{a}\left[2A_2\left(1-\widetilde{\varepsilon}\sin^2\chi\right)+3A_3\left(1-2\widetilde{\varepsilon}\sin^2\chi\right)\right]\mathbf{e}_r\\
&+\frac{1}{a}\left(2B_2+\frac{9}{7}B_3+\frac{16\widetilde{\varepsilon}}{7}A_3-\frac{\gamma M}{a}\frac{2m}{3}\right)\left(\frac{3\sin^2\chi-1}{2}\right)\mathbf{e}_r\\
=&\ \frac{1}{a}\left(B_2+\frac{4}{7}B_3+\frac{4\widetilde{\varepsilon}}{7}A_3-\frac{\gamma M}{a}\frac{m}{3}\right)3\sin\chi\cos\chi\,\mathbf{e}_\chi\\
&+\frac{1}{a}\left(2A_2+3A_3-B_2-\frac{9}{14}B_3-\frac{8\widetilde{\varepsilon}}{7}A_3+\frac{\gamma M}{a}\frac{m}{3}\right)\mathbf{e}_r\\
&-\frac{1}{a}\left(2\widetilde{\varepsilon}A_2+\frac{18\widetilde{\varepsilon}}{7}A_3-3B_2-\frac{27}{14}B_3+\frac{\gamma M}{a}m\right)\sin^2\chi\,\mathbf{e}_r. \qquad (8.131)
\end{aligned}
$$

Thus – from (8.129) and (8.131) – the two interior, matching conditions for application of Step 2(b) of the procedure outlined in Section 8.5.3.1 are

$$\begin{aligned}
V(\chi, r_S) = {} & A_0 + \left(1 - \frac{2\tilde{\varepsilon}}{3}\right) A_2 + (1 - \tilde{\varepsilon})\, A_3 \\
& + \left(B_2 + \frac{4}{7} B_3 - \frac{4\tilde{\varepsilon}}{3} A_2 - \frac{10\tilde{\varepsilon}}{7} A_3 - \frac{\gamma M}{a}\frac{m}{3}\right) P_2(\sin\chi), \qquad (8.132)
\end{aligned}$$

$$\begin{aligned}
\nabla V(\chi, r_S) = {} & \frac{1}{a}\left(B_2 + \frac{4}{7}B_3 + \frac{4\tilde{\varepsilon}}{7}A_3 - \frac{\gamma M}{a}\frac{m}{3}\right) 3 \sin\chi \cos\chi\, \mathbf{e}_\chi \\
& + \frac{1}{a}\left(2A_2 + 3A_3 - B_2 - \frac{9}{14}B_3 - \frac{8\tilde{\varepsilon}}{7}A_3 + \frac{\gamma M}{a}\frac{m}{3}\right)\mathbf{e}_r \\
& - \frac{1}{a}\left(2\tilde{\varepsilon}A_2 + \frac{18\tilde{\varepsilon}}{7}A_3 - 3B_2 - \frac{27}{14}B_3 + \frac{\gamma M}{a} m\right)\sin^2\chi\, \mathbf{e}_r. \qquad (8.133)
\end{aligned}$$

8.5.3.6 *Matching the Interior and Exterior Solutions*

Setting the outer equation (8.112) for $V(\chi, r = r_S)$ equal to the corresponding inner equation (8.132) gives

$$\begin{aligned}
-\frac{\gamma M}{a}\left[1 + \frac{\tilde{\varepsilon}}{3} + \cancel{\frac{m}{3}P_2(\sin\chi)}\right] = {} & A_0 + \left(1 - \frac{2\tilde{\varepsilon}}{3}\right)A_2 + (1 - \tilde{\varepsilon})\, A_3 \\
& + \left(B_2 + \frac{4}{7}B_3 - \frac{4\tilde{\varepsilon}}{3}A_2 - \frac{10\tilde{\varepsilon}}{7}A_3 - \cancel{\frac{\gamma M}{a}\frac{m}{3}}\right) P_2(\sin\chi), \qquad (8.134)
\end{aligned}$$

and rearranging this equation leads to

$$\begin{aligned}
& A_0 + \left(1 - \frac{2\tilde{\varepsilon}}{3}\right)A_2 + (1 - \tilde{\varepsilon})\, A_3 + \frac{\gamma M}{a}\left(1 + \frac{\tilde{\varepsilon}}{3}\right) \\
& = -\left(B_2 + \frac{4}{7}B_3 - \frac{4\tilde{\varepsilon}}{3}A_2 - \frac{10\tilde{\varepsilon}}{7}A_3\right) P_2(\sin\chi). \qquad (8.135)
\end{aligned}$$

Independently setting to zero the constant (left-hand-side) and $P_2(\sin\chi)$ (right-hand-side) terms of (8.135) then yields the two constraint equations:

$$A_0 + \left(1 - \frac{2\tilde{\varepsilon}}{3}\right)A_2 + (1 - \tilde{\varepsilon})\, A_3 = -\frac{\gamma M}{a}\left(1 + \frac{\tilde{\varepsilon}}{3}\right), \qquad (8.136)$$

$$B_2 + \frac{4}{7}B_3 - \frac{4\tilde{\varepsilon}}{3}A_2 - \frac{10\tilde{\varepsilon}}{7}A_3 = 0. \qquad (8.137)$$

Similarly matching outer equation (8.113) for $\nabla V(\chi, r_S)$ to the corresponding inner one (8.133) gives

$$\begin{aligned}
\nabla V(\chi, r_S) = {} & \frac{1}{a}\left(B_2 + \frac{4}{7}B_3 + \frac{4\tilde{\varepsilon}}{7}A_3 - \cancel{\frac{\gamma M}{a}\frac{m}{3}}\right) 3 \sin\chi \cos\chi\, \mathbf{e}_\chi \\
& + \frac{1}{a}\left(2A_2 + 3A_3 - B_2 - \frac{9}{14}B_3 - \frac{8\tilde{\varepsilon}}{7}A_3 + \frac{\gamma M}{a}\frac{m}{3}\right)\mathbf{e}_r \\
& - \frac{1}{a}\left(2\tilde{\varepsilon}A_2 + \frac{18\tilde{\varepsilon}}{7}A_3 - 3B_2 - \frac{27}{14}B_3 + \frac{\gamma M}{a} m\right)\sin^2\chi\, \mathbf{e}_r
\end{aligned}$$

$$= \frac{\gamma M}{a^2}(2\widetilde{\varepsilon} - m)\sin\chi\cos\chi\,\mathbf{e}_\chi + \frac{\gamma M}{a^2}\left[1 + \left(\frac{2\widetilde{\varepsilon} - m}{2}\right) + \left(\frac{3}{2}m - \widetilde{\varepsilon}\right)\sin^2\chi\right]\mathbf{e}_r. \tag{8.138}$$

The (independent) components for $\mathbf{e}_\chi$ and $\mathbf{e}_r$ in (8.138) then yield the three further constraint equations

$$B_2 + \frac{4}{7}B_3 + \frac{4\widetilde{\varepsilon}}{7}A_3 = \frac{\gamma M}{a}\frac{2\widetilde{\varepsilon}}{3}, \tag{8.139}$$

$$2A_2 + \left(3 - \frac{8\widetilde{\varepsilon}}{7}\right)A_3 - B_2 - \frac{9}{14}B_3 = \frac{\gamma M}{a}\left(1 + \widetilde{\varepsilon} - \frac{5m}{6}\right), \tag{8.140}$$

$$2\widetilde{\varepsilon}A_2 + \frac{18\widetilde{\varepsilon}}{7}A_3 - 3B_2 - \frac{27}{14}B_3 = -\frac{\gamma M}{a}\left(\frac{5}{2}m - \widetilde{\varepsilon}\right), \tag{8.141}$$

where the $\mathbf{e}_r$ component has been split into its two independent components.

For the two choices:

1. $A_3 = O(1)$. This is physically realistic because it allows density at Earth's core to be ten times larger than at its crust. The downside is that the extra terms adversely affect tractability for the development (in Chapter 12) of GREAT coordinates.
2. $A_3 = O(m, \widetilde{\varepsilon})$. This simpler form is physically a little less realistic, since density at Earth's core is only allowed to differ from that at Earth's crust by an amount of $O(m, \widetilde{\varepsilon})$. The upside, however, is that it is acceptably accurate in the vicinity of Earth's crust – where needed – and its simplicity aids tractability for the development of GREAT coordinates (in Chapter 12) below the geoid.

Equations (8.136) and (8.137) together with (8.139)–(8.141) comprise a set of five linear equations for the five coefficients A_0, A_2, A_3, B_2, and B_3 of (8.122) and (8.123). There are *two* sets of solutions according to the presence or absence of the red terms within them. After individually solving the two equation sets, we obtain:

The Coefficient Values

1. For $A_3 = O(1)$, that is, in the *presence* of the red terms, the coefficient values are

$$A_0 = -2\frac{\gamma M}{a}, \quad A_2 = 2\frac{\gamma M}{a}, \quad A_3 = -\frac{\gamma M}{a}, \tag{8.142}$$

$$B_2 = 10\left(\frac{3\widetilde{\varepsilon} - 2m}{3}\right)\frac{\gamma M}{a}, \quad B_3 = -\left(\frac{46\widetilde{\varepsilon} - 35m}{3}\right)\frac{\gamma M}{a}. \tag{8.143}$$

2. For $A_3 = O(\widetilde{\varepsilon}, m)$, that is, in the *absence* of the red terms, the coefficient values are

$$A_0 = -\frac{3}{2}\left(1 + \frac{2\widetilde{\varepsilon}}{9}\right)\frac{\gamma M}{a}, \quad A_2 = \frac{1}{2}\frac{\gamma M}{a}, \quad A_3 = \frac{\widetilde{\varepsilon}}{3}\frac{\gamma M}{a}, \tag{8.144}$$

$$B_2 = -\frac{2}{3}(10m - 9\widetilde{\varepsilon})\frac{\gamma M}{a}, \quad B_3 = \frac{7}{3}(5m - 4\widetilde{\varepsilon})\frac{\gamma M}{a}. \tag{8.145}$$

Two *different* density distributions have been constructed within a rotating, *ellipsoidal* planet, whose bounding surface coincides with a geopotential surface, namely:

1. Distribution (8.123) with its coefficients defined by (8.142) and (8.143).

2. Distribution (8.123) with its coefficients defined by (8.144) and (8.145).

Both of these are consistent with *exactly the same* geopotential *outside* the planet, namely (8.96). This explicitly demonstrates that:

> Different density distributions within a planet can lead to exactly the same geopotential outside it.

8.5.3.7 *Cross-Check on the Five Coefficients*

In (8.142)–(8.145), the mass (M) of the ellipsoid is a parameter of *assumed known value.* It can, however, be independently computed in terms of the coefficients A_2, A_3, and B_3 that appear in the assumed density distribution (8.123). This is achieved by integrating $\rho\,(\chi, r)$ over the volume of the ellipsoid. The end result should then be consistent – to $O\,(m, \tilde{\varepsilon})$ – with the values of A_2, A_3, and B_3 given in (8.142)–(8.145). Thus, assuming asymptotic behaviours (8.118) and integrating (8.123) over the volume of the ellipsoid, the total mass of the planet is given by

$$
\begin{aligned}
M &= \int_0^{2\pi} d\lambda \int_{-\pi/2}^{\pi/2} d\chi \int_0^{r_S(\chi)} dr\rho\,(\chi, r)\,r^2\cos\chi = 2\pi \int_{-\pi/2}^{\pi/2} d\chi \int_0^{r_S(\chi)} dr\rho\,(\chi, r)\,r^2\cos\chi \\
&= 2\pi \int_{-1}^{1} d\,(\sin\chi) \int_0^{r_S(\chi)} dr \frac{3}{2\pi\gamma a^2}\left\{A_2 + 2A_3\frac{r}{a} + \left[B_3\frac{r}{a}\left(1-\frac{r}{a}\right) + \frac{4\tilde{\varepsilon}}{3}A_3\frac{r^2}{a^2}\right]P_2\,(\sin\chi)\right\}r^2 \\
&= \frac{3}{\gamma a^2}\int_{-1}^{1} d\mu \int_0^{r_S(\mu)} dr\left\{A_2 r^2 + 2A_3\frac{r^3}{a} + \left[B_3\frac{r^3}{a}\left(1-\frac{r}{a}\right) + \frac{4\tilde{\varepsilon}}{3}A_3\frac{r^4}{a^2}\right]P_2\,(\mu)\right\} \\
&= \frac{3}{\gamma a^2}\int_{-1}^{1} d\mu\left\{A_2\frac{r_S^3\,(\mu)}{3} + \frac{2A_3}{a}\frac{r_S^4\,(\mu)}{4} + \left[\frac{B_3}{a}\left(\frac{r_S^4}{4} - \frac{r_S^5}{5a}\right) + \frac{4\tilde{\varepsilon}}{3}\frac{A_3}{a}\frac{r_S^5}{5a}\right](3\mu^2-1)\right\} \\
&= \frac{3}{\gamma a^2}\int_{-1}^{1} d\mu\left[\frac{A_2 a^3}{3\left(1+2\tilde{\varepsilon}\mu^2\right)^{\frac{3}{2}}} + \frac{A_3}{a}\frac{a^4}{2\left(1+2\tilde{\varepsilon}\mu^2\right)^2}\right] \\
&\quad + \frac{3}{\gamma a^2}\int_{-1}^{1} d\mu\left\{\frac{B_3}{a}\left[\frac{a^4}{4\left(1+\cancel{2\tilde{\varepsilon}\mu^2}\right)^2} - \frac{a^4}{5\left(1+\cancel{2\tilde{\varepsilon}\mu^2}\right)^{\frac{5}{2}}}\right]\right. \\
&\qquad\left. + \frac{4\tilde{\varepsilon}}{3}\frac{A_3}{a}\frac{a^4}{5\left(1+\cancel{2\tilde{\varepsilon}\mu^2}\right)^{\frac{5}{2}}}\right\}(3\mu^2-1) \\
&\approx \frac{3a}{\gamma}\int_0^1 d\mu\left[\frac{\left(1-3\tilde{\varepsilon}\mu^2\right)}{3}A_2 + \frac{A_3}{2}\left(1-4\tilde{\varepsilon}\mu^2\right)\right] \\
&\quad + \frac{3a}{\gamma}\int_0^1 d\mu\left[\left(\frac{1}{4}-\frac{1}{5}\right)B_3 + \frac{4\tilde{\varepsilon}}{15}A_3\right](3\mu^2-1) \\
&= \frac{3a}{\gamma}\left[\frac{2\,(1-\tilde{\varepsilon})}{3}A_2 + A_3\left(1-\frac{4\tilde{\varepsilon}}{3}\right)\right].
\end{aligned}
\tag{8.146}
$$

After rearrangement, this then gives the constraint

$$2\left(1-\widetilde{\varepsilon}\right)A_2+\left(3-4\widetilde{\varepsilon}\right)A_3=\frac{\gamma M}{a}. \tag{8.147}$$

Substitution of the values for A_2 and A_3 – obtained from either (8.142) or from (8.144) – into (8.147), with use of (8.118) for the two individual cases (i.e. for $A_3=O(1)$ or for $A_3=O(m,\widetilde{\varepsilon})$), verifies that (8.147) is indeed satisfied for both cases.

8.5.3.8 *The Newtonian Gravitational Potential $V(\chi,r)$ inside an Ellipsoidal Planet of Variable Density $\rho(\chi,r)$*

Recall that the Newtonian gravitational potential $V(\chi,r)$ inside an ellipsoidal planet of variable density $\rho(\chi,r)$, defined by (8.123), is given by (8.122). Specifically,

$$\rho(\chi,r)=\frac{3}{2\pi\gamma a^2}\left\{A_2+2A_3\frac{r}{a}+\left[B_3\frac{r}{a}\left(1-\frac{r}{a}\right)+\frac{4\widetilde{\varepsilon}}{3}A_3\frac{r^2}{a^2}\right]P_2(\sin\chi)\right\},\quad r\leq r_S(\chi), \tag{8.148}$$

$$\begin{aligned}V(\chi,r)&=A_0+A_2\frac{r^2}{a^2}+A_3\frac{r^3}{a^3}-\frac{\Omega^2r^2}{3}P_2(\sin\chi)\\&+\left[B_2\frac{r^2}{a^2}+B_3\frac{r^3}{a^3}\left(1-\frac{3}{7}\frac{r}{a}\right)+\frac{4\widetilde{\varepsilon}}{7}A_3\frac{r^4}{a^4}\right]P_2(\sin\chi),\quad r\leq r_S(\chi),\end{aligned} \tag{8.149}$$

where coefficients A_0, A_2, A_3, B_2, and B_3 are given by either (8.142) and (8.143) (when red terms are retained in (8.148) and (8.149)), or by (8.144) and (8.145) (when they are dropped), according to whether $A_3=O(1)$ or $A_3=O(m,\widetilde{\varepsilon})$, respectively.

8.5.3.9 *The Geopotential $\Phi(\chi,r)$ inside an Ellipsoidal Planet of Variable Density $\rho(\chi,r)$*

The geopotential $\Phi(\chi,r)$ inside an ellipsoidal planet of variable density $\rho(\chi,r)$, defined by (8.148), is obtained by adding (see (8.106)) the centrifugal potential, $V^C(\chi,r)$, of (8.82) to the Newtonian gravitational potential, $V(\chi,r)$, of (8.149) to obtain

$$\begin{aligned}\Phi(\chi,r)&\equiv V(\chi,r)+V^C(\chi,r)\\&=A_0+A_2\frac{r^2}{a^2}+A_3\frac{r^3}{a^3}-\cancel{\frac{\Omega^2r^2}{3}P_2(\sin\chi)}-\frac{\Omega^2r^2}{3}\left[1-\cancel{P_2(\sin\chi)}\right]\\&+\left[B_2\frac{r^2}{a^2}+B_3\frac{r^3}{a^3}\left(1-\frac{3}{7}\frac{r}{a}\right)+\frac{4\widetilde{\varepsilon}}{7}A_3\frac{r^4}{a^4}\right]P_2(\sin\chi)\\&=A_0+A_2\frac{r^2}{a^2}+A_3\frac{r^3}{a^3}-\frac{\Omega^2r^2}{3}\\&+\left[B_2\frac{r^2}{a^2}+B_3\frac{r^3}{a^3}\left(1-\frac{3}{7}\frac{r}{a}\right)+\frac{4\widetilde{\varepsilon}}{7}A_3\frac{r^4}{a^4}\right]\left(\frac{3\sin^2\chi-1}{2}\right)\\&=A_0+\left(A_2-\frac{B_2}{2}-\frac{\gamma M}{a}\frac{m}{3}\right)\frac{r^2}{a^2}+\left[A_3\left(1-\frac{2\widetilde{\varepsilon}}{7}\frac{r}{a}\right)-\frac{B_3}{2}\left(1-\frac{3}{7}\frac{r}{a}\right)\right]\frac{r^3}{a^3}\\&+\frac{3}{2}\left[B_2\frac{r^2}{a^2}+B_3\frac{r^3}{a^3}\left(1-\frac{3}{7}\frac{r}{a}\right)+\frac{4\widetilde{\varepsilon}}{7}A_3\frac{r^4}{a^4}\right]\sin^2\chi,\quad r\leq r_S(\chi).\end{aligned} \tag{8.150}$$

Summarising:

The Geopotential inside an Ellipsoidal Planet of Variable Density

For an *ellipsoidal* planet of *variable* density, with semi-major and semi-minor axes a and c, respectively, and mass M, the geopotential within it is

$$\Phi\left(\chi,r\right)=A_0+\left(A_2-\frac{B_2}{2}-\frac{\gamma M}{a}\frac{m}{3}\right)\frac{r^2}{a^2}+\left[A_3\left(1-\frac{2\widetilde{\varepsilon}}{7}\frac{r}{a}\right)-\frac{B_3}{2}\left(1-\frac{3}{7}\frac{r}{a}\right)\right]\frac{r^3}{a^3}$$
$$+\frac{3}{2}\left[B_2\frac{r^2}{a^2}+B_3\frac{r^3}{a^3}\left(1-\frac{3}{7}\frac{r}{a}\right)+\frac{4\widetilde{\varepsilon}}{7}A_3\frac{r^4}{a^4}\right]\sin^2\chi,\quad r\leq r_S\left(\chi\right), \tag{8.151}$$

where

$$m\equiv\frac{\Omega^2a^3}{\gamma M},\quad \widetilde{\varepsilon}\equiv\frac{a^2-c^2}{2c^2}, \tag{8.152}$$

and coefficients A_0, A_2, A_3, B_2, and B_3 are given either by (8.142) and (8.143) (when red terms are retained in (8.148) and (8.149)), or by (8.144) and (8.145) (when they are dropped), according to whether $A_3=O\left(1\right)$ or $A_3=O\left(m,\widetilde{\varepsilon}\right)$, respectively.

8.5.3.10 *Gravity inside and on an Ellipsoidal Planet of Variable Density $\rho\left(\chi,r\right)$*

The apparent gravitational force, $\mathbf{g}\left(\chi,r\right)$, inside an ellipsoidal planet of *variable* density is obtained by inserting geopotential (8.151) into definition (8.107) of gravity and using (5.19) to evaluate the gradient. Thus

$$\mathbf{g}\left(\chi,r\right)\equiv-\nabla\Phi=-\frac{1}{r}\frac{\partial\Phi}{\partial\chi}\mathbf{e}_\chi-\frac{\partial\Phi}{\partial r}\mathbf{e}_r$$
$$=-\frac{3}{a}\left[B_2\frac{r}{a}+B_3\frac{r^2}{a^2}\left(1-\frac{3}{7}\frac{r}{a}\right)+\frac{4\widetilde{\varepsilon}}{7}A_3\frac{r^3}{a^3}\right]\sin\chi\cos\chi\,\mathbf{e}_\chi$$
$$-\frac{1}{a}\left\{\left(2A_2-B_2-\frac{\gamma M}{a}\frac{2m}{3}\right)\frac{r}{a}+3\left[A_3\left(1-\frac{8\widetilde{\varepsilon}}{21}\frac{r}{a}\right)-\frac{B_3}{2}\left(1-\frac{4}{7}\frac{r}{a}\right)\right]\frac{r^2}{a^2}\right\}\mathbf{e}_r$$
$$-\frac{3}{a}\left[B_2\frac{r}{a}+\frac{3B_3}{2}\left(1-\frac{4}{7}\frac{r}{a}\right)\frac{r^2}{a^2}+\frac{8\widetilde{\varepsilon}}{7}A_3\frac{r^3}{a^3}\right]\sin^2\chi\,\mathbf{e}_r,\quad r\leq r_S\left(\chi\right), \tag{8.153}$$

where (from (8.78))

$$r=r_S\left(\chi\right)=a\left(1+2\widetilde{\varepsilon}\sin^2\chi\right)^{-\frac{1}{2}}\approx a\left(1-\widetilde{\varepsilon}\sin^2\chi\right). \tag{8.154}$$

Substituting (8.154) into (8.153), evaluated at the planet's ellipsoidal surface $r=r_S\left(\chi\right)$, and only retaining terms to $O\left(m,\widetilde{\varepsilon}\right)$, the apparent gravitational force on an ellipsoidal planet of variable density is

$$\left(\mathbf{g}\right)_{r=r_S\left(\chi\right)}=-\frac{3}{a}\left[B_2\cancel{\frac{r_S}{a}}+B_3\cancel{\frac{r_S^2}{a^2}}\left(1-\frac{3}{7}\cancel{\frac{r_S}{a}}\right)+\frac{4\widetilde{\varepsilon}}{7}A_3\cancel{\frac{r_S^3}{a^3}}\right]\sin\chi\cos\chi\,\mathbf{e}_\chi$$
$$-\frac{1}{a}\left[2A_2\frac{r_S}{a}-\left(B_2+\frac{\gamma M}{a}\frac{2m}{3}\right)\cancel{\frac{r_S}{a}}\right.$$
$$\left.+3A_3\left(1-\frac{8\widetilde{\varepsilon}}{21}\cancel{\frac{r_S}{a}}\right)\frac{r_S^2}{a^2}-\frac{3B_3}{2}\left(1-\frac{4}{7}\cancel{\frac{r_S}{a}}\right)\cancel{\frac{r_S^2}{a^2}}\right]\mathbf{e}_r$$

$$
\begin{aligned}
&-\frac{3}{a}\left[B_2 \xcancel{\frac{r_S}{a}} + \frac{3B_3}{2}\left(1-\frac{4}{7}\xcancel{\frac{r_S}{a}}\right)\xcancel{\frac{r_S^2}{a^2}} + \frac{8\widetilde{\varepsilon}}{7}A_3\xcancel{\frac{r_S^3}{a^3}}\right]\sin^2\chi\,\mathbf{e}_r \\
\approx &-\frac{3}{a}\left(B_2+\frac{4}{7}B_3+\frac{4\widetilde{\varepsilon}}{7}A_3\right)\sin\chi\cos\chi\,\mathbf{e}_\chi \\
&-\frac{1}{a}\left[2A_2+3\left(1-\frac{8\widetilde{\varepsilon}}{21}\right)A_3-\left(B_2+\frac{9}{14}B_3+\frac{\gamma M}{a}\frac{2m}{3}\right)\right]\mathbf{e}_r \\
&-\frac{1}{a}\left(3B_2+\frac{27}{14}B_3-2\widetilde{\varepsilon}A_2-\frac{18\widetilde{\varepsilon}}{7}A_3\right)\sin^2\chi. \qquad (8.155)
\end{aligned}
$$

With use of (8.142) and (8.143) (when red terms are retained), and of (8.144) and (8.145) (when they are not), it can be shown that

$$B_2+\frac{4}{7}B_3+\frac{4\widetilde{\varepsilon}}{7}A_3=\frac{2}{3}\widetilde{\varepsilon}\frac{\gamma M}{a}, \qquad (8.156)$$

$$2A_2+3\left(1-\frac{8\widetilde{\varepsilon}}{21}\right)A_3-\left(B_2+\frac{9}{14}B_3+\frac{\gamma M}{a}\frac{2m}{3}\right)=\left(1+\widetilde{\varepsilon}-\frac{3m}{2}\right)\frac{\gamma M}{a}, \qquad (8.157)$$

$$3B_2+\frac{27}{14}B_3-2\widetilde{\varepsilon}A_2-\frac{18\widetilde{\varepsilon}}{7}A_3=\left(\frac{5m}{2}-\widetilde{\varepsilon}\right)\frac{\gamma M}{a}. \qquad (8.158)$$

This means that the right-hand sides of (8.156)–(8.158) are *independent* of whether $A_3 = O(1)$ (when red terms are retained) or $A_3 = O(m,\widetilde{\varepsilon})$ (when they are not). Substitution of (8.156)–(8.158) into (8.155) – to replace coefficients by their values – then leads to

$$(\mathbf{g})_{r=r_S(\chi)} \approx -2\widetilde{\varepsilon}\frac{\gamma M}{a^2}\sin\chi\cos\chi\,\mathbf{e}_\chi-\frac{\gamma M}{a^2}\left[1+\widetilde{\varepsilon}-\frac{3m}{2}+\left(\frac{5m}{2}-\widetilde{\varepsilon}\right)\sin^2\chi\right]\mathbf{e}_r, \qquad (8.159)$$

regardless of whether $A_3 = O(1)$ or $A_3 = O(m,\widetilde{\varepsilon})$. Therefore, to first order in $\widetilde{\varepsilon}$ and m, the variation of gravity g at the planet's surface is given by:

$$
\begin{aligned}
(g)_{r=r_S(\chi)} \equiv (|\mathbf{g}|)_{r=r_S(\chi)} &\approx \frac{\gamma M}{a^2}\left[1+\widetilde{\varepsilon}-\frac{3m}{2}+\left(\frac{5m}{2}-\widetilde{\varepsilon}\right)\sin^2\chi\right] \\
&\approx g_S^E\left[1+\left(\frac{5m}{2}-\widetilde{\varepsilon}\right)\sin^2\chi\right], \qquad (8.160)
\end{aligned}
$$

where

$$g_S^E \equiv (g)_{r=r_S(\chi=0)}=\frac{\gamma M}{a^2}\left(1+\widetilde{\varepsilon}-\frac{3m}{2}\right) \qquad (8.161)$$

is the value of g at the planet's equator.

8.5.3.11 *Cross-Check on Clairaut's Fraction for an Ellipsoidal Planet of Variable Density*

Evaluating (8.160) at $\chi = \pm\pi/2$, the value of g at the planet's poles is

$$g_S^P=(g)_{r=r_S(\chi=\pm\pi/2)}\approx g_S^E\left(1+\frac{5m}{2}-\widetilde{\varepsilon}\right)\approx\frac{\gamma M}{a^2}(1+m), \qquad (8.162)$$

where (8.161) has been used to obtain the last-most expression. Thus (8.162) leads to:

Clairaut's Fraction for a Rotating Ellipsoidal Planet of Variable Density

$$\frac{g_S^P - g_S^E}{g_S^E} \equiv \text{Clairaut's fraction} \approx \frac{5m}{2} - \widetilde{\varepsilon} \approx \frac{5m}{2} - \varepsilon, \tag{8.163}$$

to first order in m and $\widetilde{\varepsilon}$ or, equivalently, to first order in m and ε.

This asymptotic result, obtained by approaching the planet's surface from below, agrees (as it should) with (8.104), obtained by approaching the planet's surface from above. This serves as a useful cross-check that the analysis for gravity within the planet is correct.

8.6 SPHERICAL GEOPOTENTIAL APPROXIMATION AS AN ASYMPTOTIC LIMIT

Simultaneously taking the limits $\widetilde{\varepsilon} \to 0$ and $m \to 0$ in (8.96), (8.98), and (8.99) asymptotically results in – cf. (7.9) and (7.10) –

$$\Phi(r) = -\frac{\gamma M}{r}, \quad r \geq a, \tag{8.164}$$

$$\mathbf{g}(r) = -\nabla\Phi(r) = -\frac{\gamma M}{r^2}\mathbf{e}_r, \quad r \geq a \tag{8.165}$$

$$r_S = a. \tag{8.166}$$

This is the form – see, for example, Haltiner and Martin, 1957, and White et al., 2005 – of the (classical) spherical geopotential approximation where gravity falls off as $1/r^2$ outside the sphere rather than remaining constant.[12] Not only is the surface of the planet spherical – from (8.166) – *all* geopotential surfaces are spherical.[13]

Thus the consequences of the (classical) spherical-geopotential approximation are:

- The ellipsoidal geopotential surfaces deform to become perfectly spherical.
- The centrifugal force $-\Omega \times (\Omega \times \mathbf{r})$ is neglected.
- The geopotential field Φ no longer varies as a function of latitude, but only of radial distance r from a planet's centre.
- The apparent acceleration due to gravity varies as the inverse square of r, with no latitudinal variation.

Taking the limit $\widetilde{\varepsilon} \sim \varepsilon \to 0$ is a geometric approximation, whereby spheroids are distorted/mapped into spheres. See Fig. 8.6 for a graphical representation of this distortion/mapping. Taking the limit $m \to 0$ results in neglect of the centrifugal force $-\Omega \times (\Omega \times \mathbf{r})$. Thus:

12 This is consistent with having worked throughout in deep geometry; the shallow assumption (used, for example, in the derivation of the hydrostatic primitive equations) has not been made anywhere in the derivation of the geopotential for a mildly oblate, ellipsoidal planet of variable density.

13 This can be seen by considering an arbitrary geopotential surface, $\Phi =$ constant. Inserting this constant value into the left-hand side of (8.164) and solving for r then gives $r =$ constant as the equation for an arbitrary geopotential surface; i.e. this surface is perfectly spherical. This may be contrasted with the situation (in Section 7.8) for a spherical planet with variable density, for which the *only* geopotential surface that is perfectly spherical is the one coincident with the planet's surface. Thus the classical spherical-geopotential approximation is *not* equivalent to the physical situation of a spherical planet with variable density. In fact, it is not equivalent to *any* physically realisable situation, which possibly helps explain why attempts to justify the classical spherical-geopotential approximation have been so controversial (van der Toorn and Zimmerman, 2008).

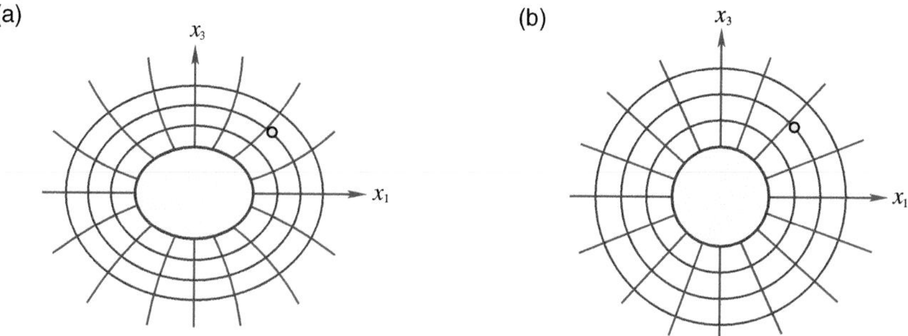

Figure 8.6 Exaggerated meridional cross sections of two geopotential coordinate systems for modelling a planet's atmosphere: (a) spheroidal; and (b) spherical. The planet is shaded, geopotential surfaces are depicted by concentric (blue) curves, and vertical coordinate lines by (red) radiating curves. The solid circle (in yellow) on each panel denotes the location of an arbitrary point on a *spheroidal*-geopotential coordinate surface (panel (a)), and where it maps to, and from, on the corresponding *spherical*-geopotential coordinate surface (panel (b)).

- The classical spherical-geopotential approximation amounts to *simultaneously* distorting the geometry from spheroidal to spherical, whilst neglecting the centrifugal force.

This interpretation is consistent with the comments made in Section 7.1.2, where m was identified with *cause*, and ε with *effect*, with ε being a fixed multiple (approximately equal to unity) of m. Recall the interdependence between m and ε. A planet's shape (as measured by ε) depends on the equilibrium (as measured by m) between Newtonian gravity and a planet's (apparent) centrifugal force, but Newtonian gravity itself depends on a planet's shape, which depends on gravity and a planet's rotation rate, and so on *ad infinitum*. Because of this *intrinsic* physical interdependence, it is natural to handle m and ε *together* when approximating the geopotential via the asymptotic limiting process just described.

The classical spherical-geopotential approximation (8.164)–(8.166) *is completely independent of a planet's rotation rate,* Ω. This is consistent with neglect of the centrifugal force $-\Omega \times (\Omega \times \mathbf{r})$, which implies (see (7.6)) that Φ *(potential of apparent gravity) then reduces to* V *(potential of Newtonian gravity)*. If a planet rotates faster or slower, this has no impact whatsoever on the location of geopotential surfaces, nor on the representation of gravity on them; they remain unchanged.[14] This behaviour can be contrasted with geopotential approximation (8.96)–(8.99) for an ellipsoidal planet of variable density. The location of geopotential surfaces and the representation of gravity on them do then depend on a planet's rotation rate (directly through m, the cause, and indirectly through $\widetilde{\varepsilon}$, the effect).

It remains to explain the assertion made in Section 7.1.5 that spherical-geopotential form (7.9) (for which $\Phi \sim V \sim 1/r$) is appropriate for models expressed in *deep* geometry, whereas form (7.7) (for which $\Phi \sim V \sim r$) is appropriate for models expressed in *shallow* geometry. Central to this is that – see the following highlighted box – whichever of the two forms of spherical geopotential is used, *it should result in the corresponding Newtonian gravitational flux divergence outside a planet being identically zero.*

[14] This does not, of course, mean that a planet's rotation rate has no impact on the governing equations and solutions of them; a faster or slower rotation rate changes the magnitude of the Coriolis terms and thereby solutions of the governing equations.

Recall from Section 7.1.5 that the classical spherical-geopotential approximation comes in two flavours, namely *deep* and *shallow*. Recapitulating:

The Classical Spherical-Geopotential Approximation

Expressed in Deep Spherical Coordinates

- From (7.9) and (7.10), the classical spherical-geopotential approximation, expressed in *deep* spherical-polar coordinates (λ, ϕ, r), is

$$\Phi(r) = -g_a \frac{a^2}{r} = -\frac{\gamma M}{r} \quad \Rightarrow \quad \mathbf{g}(r) = -\nabla\Phi(r) = -\frac{d\Phi}{dr}\mathbf{e}_r = -\frac{\gamma M}{r^2}\mathbf{e}_r = -\frac{g_a}{r}\mathbf{e}_r, \tag{8.167}$$

where g_a is the constant value of gravity at the planet's surface, $r = a$.

- In (8.167), the gravity vector has been *consistently computed* in *deep* coordinates as

$$\mathbf{g} = -\nabla\Phi. \tag{8.168}$$

Expressed in Shallow Spherical Coordinates

- From (7.7) and (7.8), the classical spherical-geopotential approximation, expressed in *shallow* spherical-polar coordinates (λ, ϕ, r), is

$$\Phi(r) = g_a(r - a) \quad \Rightarrow \quad \mathbf{g} = -\nabla_a\Phi(r) = -\frac{d\Phi}{dr}\mathbf{e}_r = -g_a\mathbf{e}_r, \tag{8.169}$$

where g_a is the constant value of gravity at the planet's surface, $r = a$.

- In (8.169), the gravity vector has been *consistently computed* in *shallow* coordinates as

$$\mathbf{g} = -\nabla_a\Phi. \tag{8.170}$$

In what follows, use is made of the facts that:

1. Φ reduces to V for the classical spherical-geopotential approximation.
2. V outside the planet should satisfy Laplace's equation (8.83), expressed in the appropriate (deep or shallow) geometry.

The Newtonian gravitational flux divergence may then be evaluated as follows:

The Newtonian Gravitational Flux Divergence

In Deep Spherical Coordinates

- Using (7.6) and (5.22), the Newtonian gravitational flux divergence outside a planet, expressed in *deep* spherical-polar coordinates, is

$$\nabla\cdot\mathbf{g}^N \equiv -\nabla\cdot\nabla V(r) = -\nabla\cdot\nabla\Phi(r) = -\nabla^2\Phi(r) = -\frac{1}{r^2}\frac{d}{dr}\left(r^2\frac{d\Phi}{dr}\right) = 0. \tag{8.171}$$

- This verifies that the Newtonian gravitational flux divergence, *consistently computed* in *deep* spherical-polar coordinates, is identically zero as it should be.

In Shallow Spherical Coordinates

- Using (5.76), (5.77), and (5.79), the corresponding Newtonian gravitational flux divergence, expressed in *shallow* spherical-polar coordinates, is

$$\nabla_a \cdot \mathbf{g}^N \equiv -\nabla_a \cdot \nabla_a V(r) = -\nabla_a \cdot \nabla_a \Phi(r) = -\nabla_a^2 \Phi(r) = -\frac{d^2\Phi}{dr^2} = 0. \quad (8.172)$$

- This verifies that the Newtonian gravitational flux divergence, *consistently computed* in *shallow* spherical-polar coordinates, is identically zero as it should be.

Important Corollary

- As noted in White et al. (2005), it is inconsistent and potentially dangerous to use *shallow* spherical-geopotential approximation (8.169) (which implies constant gravity) in a *deep*-atmosphere model.
- This amounts to adding a *spurious* and *systematic* source term, namely $-g_a\left(1 - a^2/r^2\right)$, to the right-hand side of the vertical component of the momentum equation.
- *This term is always of the same sign and increases in absolute value as a function of geometric height.* It will adversely impact the vertical stratification of the atmosphere and, because it acts *systematically*, it is likely to adversely affect the global atmospheric circulation. In practice this would then mean that the physical parametrisations in a model would end up being tuned to partially compensate for a systematic error in the dynamics, something that is unjustifiable and highly undesirable.

8.7 CONCLUDING REMARKS

The geopotentials outside and inside four idealised planets – two spherical ones in Chapter 7, and two ellipsoidal ones in the present chapter – of increasing geometrical and mass-distribution complexity have been successively developed. In doing so, various mathematical challenges were identified and addressed as they arose.

For geopotential representations to be geophysically realistic, they must lead to Clairaut's fraction of $(5m/2 - \varepsilon)$. For comparative purposes, this fraction is displayed in Table 8.2 for the geopotential representations associated with various, mildly oblate, idealised planets, of different shapes and density distributions. These representations include those developed in this chapter (for ellipsoidal planets) and Chapter 7 (for spherical planets), plus some others that have appeared in the literature.

The exterior and interior geopotential representations (8.96) and (8.151), respectively, were developed in Section 8.5 for a rotating, mildly oblate, ellipsoidal planet, such as Earth, of variable density. What remains to be done is to develop an associated, analytically tractable, *coordinate system* for atmospheric and oceanic modelling. Based on representations (8.96) and (8.151), this challenge is met in Chapter 12 in a unified, continuous manner.

But first we develop (in Chapters 9–11) a more comprehensive treatment of thermodynamics than the simple one, described in Chapter 3, that suffices for the atmosphere. Doing so then provides (in Chapter 11) the heretofore missing representation of ocean thermodynamics. A realistic representation of phase changes of water substance is also developed (in Chapter 10). This is an important aspect of both atmospheric and oceanic models.

#	Geopotential representations (underpinning basis)	Clairaut's fraction $\equiv \dfrac{g_S^P - g_S^E}{g_S^E}$
1	Classical variable-density ellipsoid (Section 8.5)	$\dfrac{5m}{2} - \varepsilon$ (correct)
2	Classical spherical-geopotential approximation (Section 7.1.5 and Section 8.6)	0
3	Constant-density sphere (surface not a geopotential surface) (Section 7.6)	m
4	Variable-density sphere (Section 7.8)	$\dfrac{5m}{2}$
5	Constant-density ellipsoid (Section 8.3 and special case of Section 8.5)	$\dfrac{5m}{4}$ $(= \varepsilon)$
6	Ellipsoid with confocal-oblate-spheroidal surfaces (Gates, 2004)	$-\varepsilon$
7	Ellipsoid with geodetic coordinate surfaces (White and Inverarity, 2012)	0
8	Variable-density spheroid (Staniforth, 2014b)	$\dfrac{5m}{2} - \varepsilon$ (correct)
9	(a) Variable-density ellipsoid, and (b) Variable-density spheroid (Staniforth and White, 2015a)	$\dfrac{5m}{2} - \varepsilon$ (correct)
10	Variable-density spheroid (numerical) (Bénard, 2014)	$\dfrac{5m}{2} - \varepsilon$ (correct)

Table 8.2 Clairaut's fraction for various geopotential representations, with references for the detailed description of the representations given in parentheses. The planet's surface is a geopotential surface for all representations except #3. For comparison purposes, the correct result (to $O(m, \varepsilon)$) is that of representation #1. At the time of writing, representation #2 – the classical spherical-geopotential approximation – was, to the best of the author's knowledge, used in all global operational atmospheric and oceanic models.

9

Thermodynamic Potentials and Thermodynamical Consistency

ABSTRACT

Inconsistencies in the representation of thermodynamics in atmospheric and oceanic models can lead to spurious climate drift when they are integrated. This is best avoided by specifying a Gibbs potential (or other thermodynamic potential) and obtaining *all* thermodynamic quantities from it. To do so, our hitherto limited treatment of thermodynamics is therefore broadened and deepened here. The definition and fundamental properties of four commonly used thermodynamic potentials are reviewed. To illustrate practical application of thermodynamic potentials, some basic Gibbs potentials are first derived individually and then combined to form a composite one for the atmosphere. The individual Gibbs potentials for the three constituents of water substance – namely water vapour, liquid water, and frozen water – are normalised in such a way as to straightforwardly respect the constraints for these constituents to coexist at the triple point of water. The constructed composite Gibbs potential physically corresponds to the representation of cloudy air examined in some detail in Chapters 3 and 4. Thermodynamical consistency is very important not only within an atmospheric or oceanic model, but also at their mutual interface where significant mass, momentum, and energy exchanges physically take place in a zero-sum (i.e. conserved) manner. Failure to respect this physical constraint in coupled quantitative atmosphere-ocean models can be expected to lead to climate drift during forecast simulations.

9.1 PREAMBLE

A unified set of governing equations for modelling Earth's global atmospheric and oceanic fluid flow was developed in a step-by-step manner in Chapters 2–4. To avoid being overwhelmed by a complicated and highly challenging subject, the thermodynamics for this was introduced gradually in a limited way. The starting point was to examine the thermodynamics of an ideal gas (derivable from the kinetic theory of gases) and thereby obtain a set of governing equations for a dry atmosphere. By making the excellent approximation that water vapour in the atmosphere also behaves as an ideal gas, this equation set was then extended to include the presence of water substance in the atmosphere.

For the ocean, however, there is no theory of realism comparable to that of an ideal gas for the atmosphere. Two questions then are:

- How does one represent the thermodynamics of an ocean?
- Can this be done in a unified manner that also includes the thermodynamics of the atmosphere?

To answer these questions, the specialised thermodynamic framework based on ideal gas theory (described in Chapters 2 and 3) was replaced by a somewhat more general one (in Chapter 4).

This was sufficient to obtain a unified set of governing equations (in Section 4.4) to model not only the atmosphere but also the oceans. It was, however, *assumed* that the fundamental equation of state (which describes the local properties of a fluid in thermodynamic equilibrium) is of *known*, prescribed form, expressed in terms of the fluid's thermodynamic state parameters.[1] Making this assumption (temporarily) avoided the practical problem of actually prescribing the form!

Assuming ideal gas theory, explicit functional forms for the fundamental equation of state were constructed (in Section 4.3.1) for a cloudy-air parcel composed of dry air, water vapour, liquid-water droplets, and ice crystals. This works very well for the atmosphere, but not at all for an ocean. Chapter 4 left open two important questions:

1. How can phase transitions of water substance be represented in atmospheric and oceanic models?
2. How can suitable representations of the fundamental equation of state for an ocean be constructed?

To answer these two important questions, we have to first broaden and deepen our hitherto limited treatment of thermodynamics. Central to this is the important concept of *thermodynamical consistency*.

The complexity of atmospheric and oceanic thermodynamics has led to the necessity of a plethora of thermodynamic approximations being made in the formulations of practical quantitative models. As reviewed in IOC et al. (2010) (for the ocean) and Thuburn (2017) (for the atmosphere) – with concrete examples – a highly undesirable consequence of making such approximations is that it often leads to inconsistencies. These inconsistencies can further lead to unforeseen consequences such as the introduction of spurious systematic errors in numerical integrations ... and to the possibility of subsequently drawing false conclusions from such integrations. Furthermore, due to the ever-increasing complexity of thermodynamic representations in quantitative atmospheric and oceanic models – particularly when such models are coupled at their mutual interface – it is very difficult to even identify all such inconsistencies, let alone quantify their importance in model integrations.

- What to do?

The answer to this vexing question is remarkably simple. It is to derive *everything thermodynamic* from a thermodynamic potential using the fundamental laws and rules of thermodynamics. Everything then boils down to formulating a thermodynamic potential that is suitable for the envisaged application. This potential can be as simple or as complex as one wishes; see Section 9.3 for some very simple ones for the atmosphere, and Section 11.5 for a highly complex one for the oceans. The crucially important point here is that no matter how simple or complex the thermodynamic potential may be, everything thermodynamic is consistently obtained from it. To formalise this important principle:

Definition of Thermodynamical Consistency

- A model of a global atmosphere or of an ocean is considered to be *thermodynamically consistent* if everything thermodynamic within the model is derived in a self-consistent manner from a prescribed thermodynamic potential – such as internal energy or a Gibbs potential – following the fundamental laws and rules of equilibrium thermodynamics.
- Doing so then avoids introducing inconsistencies that can lead to spurious systematic errors and inaccurate model integrations.

[1] These parameters depend upon the precise composition of the fluid and are different for the ocean than for the atmosphere.

The remainder of this chapter is organised as follows. To set the scene, the basic thermodynamic theory given in Section 4.2 is further developed here in Sections 9.2–9.4. Thermodynamic potentials, their importance, and their interrelationships are introduced and reviewed in Section 9.2. These potentials are crucially important tools to ensure thermodynamical consistency. Internal energy $\mathcal{E} = \mathcal{E}\left(\alpha, \eta, S^1, S^2, \ldots\right)$ – where α is specific volume, η is specific entropy, and $\left\{S^1, S^2, \ldots\right\}$ are mass fractions of the fluid's components – is the cardinal (i.e. fundamental) equation of state from which other potentials are derived. This thermodynamic potential directly embodies the physical principle of conservation of energy. It is well suited to examining the theoretical properties of approximated equation sets and, in particular, to examination of dynamical consistency (in later chapters). However, neither internal energy nor entropy (one of its state variables) is directly measurable. For practical purposes the Gibbs potential $g = g\left(p, T, S^1, S^2, \ldots\right)$ is therefore to be preferred since its state variables $\left(p, T, S^1, S^2, \ldots\right)$ *are* directly measurable. One can anyway, when needed, exactly transform between various thermodynamic potentials using fundamental thermodynamic definitions.

Some basic Gibbs potentials for dry air, water vapour, pure liquid water, and pure frozen water are derived (in Section 9.3) from first principles. It is then shown in Section 9.4 (and later in Sections 10.2 and 10.5) how to use these to form composite Gibbs potentials for the atmosphere. This provides insight into the construction and use of Gibbs thermodynamic potentials for a now familiar, analytically tractable and highly relevant problem. Concluding remarks are made in Section 9.5.

9.2 THERMODYNAMIC POTENTIALS

9.2.1 Overview

Thermodynamics is based on the physical principle of conservation of energy of a thermodynamic system; see Section 4.2. In our fluid-dynamical context, *specific internal energy* ($\mathcal{E}$) is a thermodynamic potential of the form (4.1):

$$\mathcal{E} = \mathcal{E}\left(\alpha, \eta, S^1, S^2, \ldots\right). \tag{9.1}$$

Recall from Section 4.2.3 that:

- $\mathcal{E}$ is specific internal energy (per unit mass of fluid, including all of its components, i.e. substances).
- η is *specific entropy*.
- $S^1, S^2, \ldots$ are *mass fractions* of the fluid's components of type i (these mass fractions collectively define the fluid's composition).
- Thermodynamic potential (9.1) depends only on the physical properties of the fluid, and it completely defines the fluid's thermodynamic equilibrium state.
- This potential is expressed in terms of its *natural (canonical) variables*, α, η, and $\left\{S^i\right\} \equiv \left\{S^1, S^2, \ldots\right\}$, the set of mass fractions of the fluid's components.

Specific internal energy is the *cardinal* (i.e. *fundamental*) thermodynamic potential, from which other thermodynamic potentials may be obtained. See Table 9.1 for:

- The definition and interrelationship of four thermodynamic potentials in common usage:
  - Specific internal energy ($\mathcal{E}$).
  - Specific enthalpy (h).

Potential	Definition	Natural variables	Fundamental thermodynamic relation	First partial derivatives	Maxwell relations
$\mathscr{E}$	$\mathscr{E}$	$\alpha, \eta, \{S^i\}$	$d\mathscr{E} = Td\eta - pd\alpha + \sum_i \mu^i dS^i$	$T = \frac{\partial \mathscr{E}(\alpha, \eta, \{S^i\})}{\partial \eta}$ $p = -\frac{\partial \mathscr{E}(\alpha, \eta, \{S^i\})}{\partial \alpha}$	$\frac{\partial T(\alpha, \eta, \{S^i\})}{\partial \alpha} = -\frac{\partial p(\alpha, \eta, \{S^i\})}{\partial \eta}$
h	$\mathscr{E} + p\alpha$	$\eta, p, \{S^i\}$	$dh = Td\eta + \alpha dp + \sum_i \mu^i dS^i$	$T = \frac{\partial h(\eta, p, \{S^i\})}{\partial \eta}$ $\alpha = \frac{\partial h(\eta, p, \{S^i\})}{\partial p}$	$\frac{\partial T(\eta, p, \{S^i\})}{\partial p} = \frac{\partial \alpha(\eta, p, \{S^i\})}{\partial \eta}$
f	$\mathscr{E} - T\eta$	$\alpha, T, \{S^i\}$	$df = -pd\alpha - \eta dT + \sum_i \mu^i dS^i$	$p = -\frac{\partial f(\alpha, T, \{S^i\})}{\partial \alpha}$ $\eta = -\frac{\partial f(\alpha, T, \{S^i\})}{\partial T}$	$\frac{\partial p(\alpha, T, \{S^i\})}{\partial T} = \frac{\partial \eta(\alpha, T, \{S^i\})}{\partial \alpha}$
g	$\mathscr{E} + p\alpha - T\eta$	$p, T, \{S^i\}$	$dg = \alpha dp - \eta dT + \sum_i \mu^i dS^i$	$\eta = -\frac{\partial g(p, T, \{S^i\})}{\partial T}$ $\alpha = \frac{\partial g(p, T, \{S^i\})}{\partial p}$	$\frac{\partial \eta(p, T, \{S^i\})}{\partial p} = -\frac{\partial \alpha(p, T, \{S^i\})}{\partial T}$

Table 9.1 Four thermodynamic potentials in common usage; their definitions and interrelationships, natural variables, fundamental thermodynamic relation, first partial derivatives, and Maxwell relations. Potentials are $\mathscr{E}$ (specific internal energy); h (specific enthalpy); f (specific Helmholtz free energy); and g (specific Gibbs energy). Natural variables are α (specific volume); η (specific entropy); p (pressure); T (temperature); and $\{S^i\} \equiv \{S^1, S^2, \ldots\}$ (set of mass fractions of the fluid's components). μ^i is the chemical potential of component S^i.

  - Specific Helmholtz free energy (f).[2]
  - Specific Gibbs[3] energy (g).[4]
- Their associated natural variables.
- Their fundamental thermodynamic relations.
- Their first partial derivatives with respect to their first two natural variables.
- Their associated Maxwell relations,[5] obtained by cross differentiation of terms given in the penultimate column.

Each of these thermodynamic potentials is based on internal energy ($\mathscr{E}$). Each potential also completely and equivalently defines *all* of the thermodynamic properties of a fluid in terms of the potential's natural (canonical) variables. To do so, partial derivatives of the potential are taken *with respect to its natural variables*, and then combined in various ways. For example, expressions to obtain various thermodynamic properties from four commonly used potentials are displayed in Table 9.2; cf. Tables 18 and 20 of Feistel (2008), Tables G2 and G3 of Feistel et al. (2010a), Table P.1 of IOC et al. (2010), and Appendix A of Thuburn (2017). Note that if a thermodynamic potential is not given as a function of its natural variables, but instead of other variables, it will in general yield many, *but not necessarily all* of the thermodynamic properties of the fluid (Callen, 1985, Section 3–3).

- So why not keep things simple, just use specific internal energy ($\mathscr{E}$) as the thermodynamic potential, and forget about the other thermodynamic potentials (such as h, f, and g)? Specific internal energy is, after all, the *cardinal* thermodynamic potential and the one that is most closely related to, and most clearly expresses, the underlying physical principle of conservation of energy.

For practical quantitative applications, such as modelling Earth's atmosphere and oceans, thermodynamic equations need to be expressible in terms of available physically measurable quantities such as pressure (p) and temperature (T). For ideal gases (including mixtures of ideal gases), it is possible to *theoretically* relate internal energy ($\mathscr{E}$) to p and T; see Section 4.3. This then means that it is possible, to an excellent approximation, to represent the gaseous constituents of Earth's atmosphere, since they behave very similarly to a variable composition of (dry + moist) ideal gases.

For Earth's oceans no such *theoretical* representation of acceptable accuracy exists. An *empirical* representation, of assumed functional form, has instead to be made, with coefficients determined from best fits to observational data (Feistel et al., 2010a; IOC et al., 2010). Of the alternative thermodynamic potentials, the *Gibbs potential* (or *Gibbs function*, or *Gibbs free energy*) is arguably the most practical. This is because its natural thermodynamic variables are pressure (p), temperature (T), and mass fractions (S^i) of substances, for all of which experimental data are available. The Gibbs potential is also convenient for the representation of phase transitions; see Sections 10.3 and 10.4. In particular, the Gibbs potentials of two phases are equal during a phase transition, whereas other potentials are discontinuous across the transition (Callen, 1985, p. 221).

Furthermore, the IOC et al. (2010) representation of Earth's atmosphere and oceans is expressed in terms of a single mathematical function, the Gibbs potential. From this potential, *all* thermodynamic quantities can be obtained – including those such as entropy that cannot be measured – in a self-consistent manner via partial differentiation and algebraic manipulation. This representation, as discussed in Section 11.2, is a consolidation of those obtained from various sources of highly accurate experimental data.

[2] After Hermann Ludwig Ferdinand von Helmholtz (1821–1894).

[3] After Josiah Willard Gibbs (1839–1903).

[4] The use of g for Gibbs potential – a standard notation in thermodynamics – should not be confused with the use of g to denote acceleration due to gravity.

[5] After James Clerk Maxwell (1831–1879).

Quantity	Symbol	From Internal Energy $\mathscr{E}(\alpha,\eta,S^1,S^2,\ldots)$	From Enthalpy $h(\eta,p,S^1,S^2,\ldots)$	From Helmholtz energy $f(\alpha,T,S^1,S^2,\ldots)$	From Gibbs energy $g(p,T,S^1,S^2,\ldots)$	Unit
Specific volume	α	–	h_p	–	g_p	$\mathrm{m^3\,kg^{-1}}$
Specific entropy	η	–	–	$-f_T$	$-g_T$	$\mathrm{J\,kg^{-1}\,K^{-1}}$
Pressure	p	$-\mathscr{E}_\alpha$	–	$-f_\alpha$	–	Pa
Temperature	T	$\mathscr{E}_\eta$	h_η	–	–	K
Chemical potential	μ^i	$\mathscr{E}_{S^i}$	h_{S^i}	f_{S^i}	g_{S^i}	$\mathrm{J\,kg^{-1}}$
Specific internal energy	$\mathscr{E}$	$\mathscr{E}$	$h - p h_p$	$f - T f_T$	$g - T g_T - p g_p$	$\mathrm{J\,kg^{-1}}$
Specific enthalpy	h	$\mathscr{E} - \alpha\mathscr{E}_\alpha$	h	$f - \alpha f_\alpha - T f_T$	$g - T g_T$	$\mathrm{J\,kg^{-1}}$
Specific Helmholtz energy	f	$\mathscr{E} - \eta\mathscr{E}_\eta$	$h - \eta h_\eta - p h_p$	f	$g - p g_p$	$\mathrm{J\,kg^{-1}}$
Specific Gibbs energy	g	$\mathscr{E} - \alpha\mathscr{E}_\alpha - \eta\mathscr{E}_\eta$	$h - \eta h_\eta$	$f - \alpha f_\alpha$	g	$\mathrm{J\,kg^{-1}}$
Specific heat capacity at constant pressure	c_p	$\dfrac{\mathscr{E}_\eta\mathscr{E}_{\alpha\alpha}}{\left(\mathscr{E}_{\alpha\alpha}\mathscr{E}_{\eta\eta} - \mathscr{E}_{\alpha\eta}^2\right)}$	$\dfrac{h_\eta}{h_{\eta\eta}}$	$T\dfrac{\left(f_{\alpha T}^2 - f_{\alpha\alpha}f_{TT}\right)}{f_{\alpha\alpha}}$	$-T g_{TT}$	$\mathrm{J\,kg^{-1}\,K^{-1}}$
Specific heat capacity at constant volume	c_v	$\dfrac{\mathscr{E}_\eta}{\mathscr{E}_{\eta\eta}}$	$\dfrac{h_\eta h_{pp}}{\left(h_{\eta\eta}h_{pp} - h_{\eta p}^2\right)}$	$-T f_{TT}$	$T\dfrac{\left(g_{pT}^2 - g_{pp}g_{TT}\right)}{g_{pp}}$	$\mathrm{J\,kg^{-1}\,K^{-1}}$
Local sound speed squared	c_s^2	$\alpha^2\mathscr{E}_{\alpha\alpha}$	$-\dfrac{h_p^2}{h_{pp}}$	$\dfrac{\alpha^2}{f_{TT}}\left(f_{\alpha\alpha}f_{TT} - f_{\alpha T}^2\right)$	$\dfrac{g_p^2 g_{TT}}{g_{pT}^2 - g_{TT}g_{pp}}$	$\mathrm{m^2\,s^{-2}}$

Table 9.2 Some general thermodynamic properties obtainable via partial differentiation of the four potentials defined in Table 9.1: $\mathscr{E} = \mathscr{E}(\alpha,\eta,S^1,S^2,\ldots)$ = specific internal energy; $h = h(\eta,p,S^1,S^2,\ldots)$ = specific enthalpy; $f = f(\alpha,T,S^1,S^2,\ldots)$ = specific Helmholtz free energy; and $g = g(p,T,S^1,S^2,\ldots)$ = specific Gibbs energy. Arguments of these potentials are their natural variables. Subscripts (other than for c_p, c_v, and c_s) denote partial differentiation with respect to that independent variable, with remaining independent variables held fixed. Thus $\mathscr{E}_\alpha \equiv \partial\mathscr{E}(\alpha,\eta,S^1,S^2,\ldots)/\partial\alpha$, etc.

Because Gibbs potentials provide *all* thermodynamic properties as functions of directly measurable quantities, they are very convenient for many purposes. For some applications, however, other potentials may be more convenient. For example, enthalpy (h) is well suited for examination of isentropic processes and for latent heating/cooling (see Sections 10.3, 10.4, and 10.7), whilst internal energy ($\mathscr{E}$) is convenient for examination of dynamical consistency (see e.g. Chapters 16 and 17). As noted in Feistel et al. (2010a):

'All these potential functions are mathematically and physically equivalent; the choice of which to use depends on application requirements or numerical simplicity.'

Because of this equivalence between thermodynamic potential functions $\mathscr{E}$, h, f, and g, we can use whichever one is most convenient for a particular application. In fact we can use more than one, *with the very important proviso that we do so in a self-consistent manner.*

9.2.2 Fundamental Thermodynamic Relations

9.2.2.1 *Specific Internal Energy*

In our fluid-dynamical context, the fundamental thermodynamic relation for *specific internal energy* is

$$d\mathscr{E} = Td\eta - pd\alpha + \sum_i \mu^i dS^i. \tag{9.2}$$

This equation is derived in Section 4.2.4; see (4.6).

9.2.2.2 *Specific Enthalpy*

The definition of *specific enthalpy*, h, is (see Table 9.1)

$$h \equiv \mathscr{E} + p\alpha. \tag{9.3}$$

Taking the infinitesimal differential of definition (9.3) and using the fundamental thermodynamic relation (9.2) for internal energy to eliminate $d\mathscr{E}$ yields

$$\begin{aligned} dh = d\left(\mathscr{E} + p\alpha\right) = d\mathscr{E} + pd\alpha + \alpha dp &= Td\eta - \cancel{pd\alpha} + \sum_i \mu^i dS^i + \cancel{pd\alpha} + \alpha dp \\ &= Td\eta + \alpha dp + \sum_i \mu^i dS^i. \end{aligned} \tag{9.4}$$

Thus (9.4) is *the fundamental thermodynamic relation for specific enthalpy*; see Table 9.1. It corresponds to (9.2), the fundamental thermodynamic relation for internal energy.

9.2.2.3 *Specific Helmholtz Free Energy*

The definition of *specific Helmholtz free energy*, f, is (see Table 9.1)

$$f \equiv \mathscr{E} - T\eta. \tag{9.5}$$

Taking the infinitesimal differential of definition (9.5) and using the fundamental thermodynamic relation (9.2) for internal energy to eliminate $d\mathscr{E}$ yields

$$\begin{aligned} df = d\left(\mathscr{E} - T\eta\right) = d\mathscr{E} - Td\eta - \eta dT &= \cancel{Td\eta} - pd\alpha + \sum_i \mu^i dS^i - \cancel{Td\eta} - \eta dT \\ &= -pd\alpha - \eta dT + \sum_i \mu^i dS^i. \end{aligned} \tag{9.6}$$

Thus (9.6) is *the fundamental thermodynamic relation for specific Helmholtz free energy*; see Table 9.1. It corresponds to (9.2), the fundamental thermodynamic relation for internal energy.

9.2.2.4 *Specific Gibbs Energy*

The definition of *specific Gibbs thermodynamic potential* (or *specific Gibbs free energy*, or *specific Gibbs function*), g, is (see Table 9.1)

$$g \equiv \mathscr{E} + p\alpha - T\eta. \tag{9.7}$$

Taking the infinitesimal differential of definition (9.7) and using the fundamental thermodynamic relation (9.2) for internal energy to eliminate $d\mathscr{E}$ yields

$$\begin{aligned} dg &= d\left(\mathscr{E} + p\alpha - T\eta\right) = d\mathscr{E} + pd\alpha + \alpha dp - Td\eta - \eta dT \\ &= \cancel{Td\eta} - \cancel{pd\alpha} + \sum_i \mu^i dS^i + \cancel{pd\alpha} + \alpha dp - \cancel{Td\eta} - \eta dT \\ &= \alpha dp - \eta dT + \sum_i \mu^i dS^i. \end{aligned} \tag{9.8}$$

Thus (9.8) is *the fundamental thermodynamic relation for the Gibbs potential*; see Table 9.1. It corresponds to (9.2), the fundamental thermodynamic relation for internal energy.

9.2.2.5 *Summary Table*

For reference purposes, definitions of the four thermodynamic potentials given in Sections 9.2.2.1–9.2.2.4, together with their associated fundamental thermodynamic relations, are assembled in the second and fourth columns of Table 9.1, respectively.

9.3 BASIC GIBBS THERMODYNAMIC POTENTIALS

9.3.1 Thermodynamic Properties from a Gibbs Potential

Recall that from a given Gibbs potential of functional form $g = g\left(p, T, S^1, S^2, \ldots\right)$ (where the arguments are the natural variables of g), *all* of the thermodynamic properties of the fluid may be obtained. Expressions to obtain various commonly used thermodynamic properties from a Gibbs potential are displayed in the penultimate column of Table 9.2; and similarly – in three further columns – for three other potentials, expressed in terms of their natural variables.

To make things less abstract and more concrete, Gibbs thermodynamic potentials for dry air, water vapour, pure liquid water, and pure frozen water are given explicitly in Sections 9.3.2–9.3.5. A further important benefit of doing so is that this prepares the way for the construction (in Section 9.4) of composite Gibbs potentials from these four basic thermodynamic potentials.

9.3.2 Dry Air

Various functional forms for cloudy air (composed of dry air and water vapour, treated as ideal gases, plus liquid water and frozen water) are given in Section 4.3. From these, we can construct the Gibbs potential for dry air (i.e. the Gibbs potential for a mixture of ideal gases *of constant composition*). From (4.72), (4.56), (4.77), and (4.78), with c_v, c_p, and R set to their constant, dry values, c_v^d, c_p^d, and R^d, respectively, and $\mathbb{L}_0$ set identically zero, we have

$$\mathscr{E}^d\left(p^d, T\right) = c_v^d T, \tag{9.9}$$

$$p^d\alpha^d = R^d T, \tag{9.10}$$

$$\eta^d\left(p^d, T\right) = c_p^d \ln T - R^d \ln p^d + R^d \ln R^d + c_v^d \ln c_v^d + A, \tag{9.11}$$

$$c_p^d = c_v^d + R^d, \tag{9.12}$$

where A is an arbitrary constant, and an additive arbitrary constant in (9.9) has been set to zero so that $\mathscr{E}^d = 0$ when $T = 0$. Inserting these four equations into definition (9.7) of the Gibbs potential yields

$$\begin{aligned} g^d\left(p^d, T\right) &\equiv \mathscr{E}^d + p^d\alpha^d - T\eta^d = c_v^d T + R^d T - T\left(c_p^d \ln T - R^d \ln p^d + R^d \ln R^d + c_v^d \ln c_v^d + A\right) \\ &= -c_p^d T \ln T + R^d T \ln p^d + \left(c_p^d - R^d \ln R^d - c_v^d \ln c_v^d - A\right) T. \end{aligned} \tag{9.13}$$

The value of constant A in (9.13) is arbitrary. Choosing it, as in Thuburn (2017), to satisfy

$$g^d\left(p_0^d, T_0\right) = 0 \quad \Rightarrow \quad A = -c_p^d\left(\ln T_0 - 1\right) - c_v^d \ln c_v^d - R^d\left(\ln R^d - \ln p_0^d\right), \tag{9.14}$$

where T_0 and p_0^d are two further constants, then leads to his equations (44) and (46), and to:

The Basic Gibbs Potential for Dry Air

$$g^d\left(p^d, T\right) = -c_p^d T \ln\left(\frac{T}{T_0}\right) + R^d T \ln\left(\frac{p^d}{p_0^d}\right). \tag{9.15}$$

Setting $p^d = p_0^d$ and $T = T_0$ in (9.15) verifies that $g^d\left(p_0^d, T_0\right) = 0$ is indeed satisfied by (9.15).

Equation (9.13) is the *Gibbs potential for a mixture ('dry air') of dry gases of constant composition*, where A is an arbitrary constant that serves to determine a reference value for entropy (η). Equation (9.15) is the same Gibbs potential but with A chosen such that $g^d\left(p_0^d, T_0\right) = 0$ for constant reference values $p^d = p_0^d$ and $T = T_0$.

9.3.3 Water Vapour

Water vapour can also, to an excellent approximation, be represented as an ideal gas. The Gibbs potential for water vapour therefore has the same functional form as (9.15) for dry air. Thus

$$g^v\left(p^v, T\right) = -c_p^v T \ln\left(\frac{T}{T_0}\right) + R^v T \ln\left(\frac{p^v}{p_0^v}\right), \tag{9.16}$$

where dry superscripts ('d') have been replaced by vapour ones ('v'), and c_p^v, R^v, and p_0^v are the constant vapour analogues of c_p^d, R^d, and p_0^d.[6]

As noted in IOC et al. (2010) and Thuburn (2017), we are free to add any linear function $C_1 + C_2 T$ to the definition of any Gibbs potential, where C_1 and C_2 are arbitrary constants. In the present context this means the freedom to add a linear function of T to the right-hand side of (9.16). Doing so has no effect on any quantity measurable by physical experiment.[7] Thus $g^v\left(p^v, T\right)$ in (9.16) is only defined to within an arbitrary, additive, linear function of T.

Inspired by Thuburn (2017), a term $\left(L_0^v + L_0^f\right)\left(1 - T/T_0\right)$ is added to the right-hand side of (9.16), and arbitrary constant p_0^v is set equal to p_0^{sat}. This leads to:

[6] For the range of pressures and temperatures observed in Earth's atmosphere, setting these quantities constant is an excellent approximation.

[7] Constant C_1 offsets the origin of Gibbs potential (g^v) or (equivalently) of internal energy ($\mathscr{E}^v$). Similarly, constant C_2 offsets the origin of entropy (η^v).

The Basic Gibbs Potential for Water Vapour

$$g^v\left(p^v, T\right) = -c_p^v T \ln\left(\frac{T}{T_0}\right) + R^v T \ln\left(\frac{p^v}{p_0^{sat}}\right) + \left(L_0^v + L_0^f\right)\left(1 - \frac{T}{T_0}\right), \tag{9.17}$$

where L_0^v and L_0^f are two further constants.[8]

The motivation (as described in Sections 10.4 and 10.5) for our choices is to represent latent heating and cooling when there is a phase transition between the vapour, liquid, and frozen forms of water substance. (It turns out that L_0^v and L_0^f are the latent heats of *vaporisation* and *fusion*, respectively, when both are extrapolated to $T = 0$ and $p = 0$.) By construction, and analogously to the dry-air case of Section 9.3.2, Gibbs potential (9.17) for water vapour has been normalised to satisfy $g^v\left(p_0^{sat}, T_0\right) = 0$ for constant reference values $p^v = p_0^{sat}$ and $T = T_0$. (It turns out that p_0^{sat} is equal to the saturation vapour pressure of pure water vapour at $T = T_0$, the temperature at the triple point of water; see Sections 10.3–10.5.)

9.3.4 Pure Liquid Water

Assume that pure liquid water is an *incompressible* fluid of constant mass density, ρ^l. Thus

$$\alpha^l \equiv \frac{1}{\rho^l} = \text{constant}, \tag{9.18}$$

where superscript '*l*' denotes a liquid-water quantity.[9] Further assume no change in composition (i.e. $dS^i \equiv 0$ in (4.5) and (4.6)) during the reversible transfer of an infinitesimal amount of heat, $đQ$, between a closed parcel of pure liquid water and its surrounding environment. From (4.32), the change in entropy resulting from heating this liquid-water parcel (for $đQ > 0$) or cooling it (for $đQ < 0$) is

$$d\eta^l = \frac{đQ}{T}. \tag{9.19}$$

From (4.33), the induced change in temperature, dT, satisfies

$$đQ = c^l dT, \tag{9.20}$$

where c^l (assumed constant here) is *specific heat capacity of pure liquid water*. (Because of the incompressibility assumption (9.18), heat is implicitly transferred at constant volume.) Elimination of $đQ$ from (9.19) and (9.20) then gives

$$c^l dT = T d\eta^l. \tag{9.21}$$

From (9.2), the fundamental thermodynamic relation for pure liquid water is

$$d\mathcal{E}^l = T d\eta^l - p d\alpha^l + \sum_i \mu^i dS^i. \tag{9.22}$$

Under the assumptions of incompressibility of pure liquid water and no change in composition during the reversible thermodynamic process, $d\alpha^l = 0$ (from (9.18)) and $dS^i = 0$. Using these values, (9.22) simplifies to

[8] Compared to Thuburn (2017)'s analysis, frozen water is additionally included in the present one, leading to the introduction of L_0^f into (9.17).

[9] In reality, $\alpha^l = \alpha^l\left(p, T\right)$. Assuming α^l = constant here greatly benefits analytical tractability. Doing so is an excellent approximation for atmospheric applications, but not for oceanic ones and, in particular, for the representation of deep oceans. This assumption is relaxed in Chapter 11.

$$d\mathscr{E}^l = Td\eta^l. \tag{9.23}$$

Eliminating $Td\eta^l$ between (9.21) and (9.23) then gives

$$d\mathscr{E}^l = c^l dT. \tag{9.24}$$

Assume now that both internal energy ($\mathscr{E}^l$) and entropy (η^l) depend only upon temperature (T), and that both are independent of pressure (p). Equations (9.21) and (9.24) can then be rewritten as the pair of first-order differential equations

$$\frac{d\eta^l}{dT} = \frac{c^l}{T}, \quad \frac{d\mathscr{E}^l}{dT} = c^l. \tag{9.25}$$

Integrating these two equations then gives

$$\eta^l = c^l \ln T + C_2, \quad \mathscr{E}^l = c^l T + C_1, \tag{9.26}$$

where C_1 and C_2 are two constants to be specified.

Having constructed functional forms for internal energy ($\mathscr{E}^l$) and specific entropy (η^l), that is, (9.26), we now use these to construct the Gibbs potential for pure liquid water. Recall that the Gibbs potential is defined by (9.7) as follows:

$$g^l \equiv \mathscr{E}^l + p\alpha^l - T\eta^l. \tag{9.27}$$

In (9.27), p is the pressure exerted on a closed parcel of pure liquid water by its encompassing environment.[10] Substitution of functional forms (9.26) for $\mathscr{E}^l$ and η^l into definition (9.27) of the Gibbs potential and use of incompressibility assumption (9.18) yields

$$g^l(p,T) = c^l T + C_1 + p\alpha^l - T\left(c^l \ln T + C_2\right). \tag{9.28}$$

It remains to specify the two constants C_1 and C_2. To do so, we first demand that

$$g^l\left(p = p_0^{sat}, T = T_0\right) = 0, \tag{9.29}$$

where T_0 and p_0^{sat} are reference values. (It again turns out that p_0^{sat} is equal to the saturation vapour pressure of pure water vapour at $T = T_0$, the temperature at the triple point of water – see Sections 10.3–10.5.) Thus, introducing (9.29) into (9.28) and solving for C_2 leads to

$$C_2 = -c^l \ln T_0 + c^l + \alpha^l \frac{p_0^{sat}}{T_0} + \frac{C_1}{T_0}. \tag{9.30}$$

Elimination of C_2 between (9.28) and (9.30) then yields

$$g^l(p,T) = -c^l T \ln\left(\frac{T}{T_0}\right) + \alpha^l\left(p - p_0^{sat}\frac{T}{T_0}\right) + C_1\left(1 - \frac{T}{T_0}\right). \tag{9.31}$$

The remaining constant, C_1, is set to L_0^f to obtain

The Basic Gibbs Potential for Pure Liquid Water

$$g^l(p,T) = -c^l T \ln\left(\frac{T}{T_0}\right) + \alpha^l\left(p - p_0^{sat}\frac{T}{T_0}\right) + L_0^f\left(1 - \frac{T}{T_0}\right). \tag{9.32}$$

[10] This environment could, for example, be composed of other similar parcels of pure liquid water. However, it could instead be a gaseous mixture of dry air and water vapour, encompassing a water droplet. There are many possibilities.

The motivation for this is to represent latent heating and cooling when there is a phase transition to/from frozen water; see the additional term introduced into (9.17).[11]

For the temperatures (T) and pressures (p) observed in Earth's atmosphere, the middle term on the right-hand side of (9.32) is typically four orders of magnitude smaller than the first term ($< 10^2\,\mathrm{J kg^{-1}}$ versus $\sim 1.2 \times 10^6\,\mathrm{J kg^{-1}}$). It is therefore often neglected in the meteorological literature. We will nevertheless retain it since this allows assessment (in Section 10.5) of its influence on the representation of cloudy air, and also on thermodynamical consistency.

9.3.5 Pure Frozen Water

Assume that pure frozen water (i.e. ice formed from pure liquid water) is an *incompressible* solid of constant mass density, ρ^f. Thus

$$\alpha^f \equiv \frac{1}{\rho^f} = \text{constant}, \tag{9.33}$$

where superscript 'f' denotes a frozen-water quantity. Note that ice is less dense (by about 8%) than liquid water,[12] and so, from (9.18) and (9.33),

$$\rho^f < \rho^l \quad \Rightarrow \quad \alpha^f > \alpha^l. \tag{9.34}$$

A similar analysis to that given in Section 9.3.4 for the Gibbs potential (g^l) of pure liquid water, but with c^l replaced by c^f and C_1 set identically zero, leads to:

The Basic Gibbs Potential for Pure Frozen Water

$$g^f(p,T) = -c^f T \ln\left(\frac{T}{T_0}\right) + \alpha^f\left(p - p_0^{sat}\frac{T}{T_0}\right), \tag{9.35}$$

where c^f (assumed constant here) is *specific heat capacity of pure frozen water.*

Since α^f is of the same order as α^l ($(1/917)\ \mathrm{m^3 kg^{-1}}$ versus $(1/1000)\ \mathrm{m^3 kg^{-1}}$, from Table 9.3), and c^f is of the same order as c^l ($2106\,\mathrm{J kg^{-1} K^{-1}}$ versus $4186\,\mathrm{J kg^{-1} K^{-1}}$, again from Table 9.3), the argument given for the possible neglect of the α^l term in (9.32) also justifies the possible neglect of $\alpha^f\left(p - p_0^{sat} T/T_0\right)$ in (9.35).

9.3.6 Normalisation of the Gibbs Potentials g^v, g^l, and g^f for Water Substance

As noted in Sections 9.3.2–9.3.5, there is some arbitrariness involved in choosing values for certain constants when constructing the Gibbs potentials g^v, g^l, and g^f for water substance. However, when these *individual* Gibbs potentials are combined to obtain *composite* Gibbs potentials (see Sections 9.4 and 10.2–10.5), maintenance of thermodynamic equilibrium between the three phases of water substance then restricts these choices.

To avoid unnecessary proliferation of constants and to facilitate later developments, this need has been anticipated in the construction of Gibbs potentials (9.17), (9.32), and (9.35) for g^v, g^l, and g^f, respectively. Specifically, the three phases of water substance must be in chemical

[11] Equation (9.32) agrees with Thuburn (2017)'s equation (50) for the Gibbs potential of pure liquid water. To obtain this agreement, L_0^f has to be set identically zero. This is because Thuburn (2017)'s analysis excludes frozen water and, thereby, any change of phase to or from frozen water, with an associated change of latent internal energy.

[12] This is why ice cubes in a glass of water and icebergs in the ocean float, frozen water pipes burst, and lakes freeze from the top down, rather than from the bottom up.

equilibrium at the triple point of water (where $p = p^v = p_0^{sat}$ and $T = T_0$) and must therefore satisfy constraint (10.15). Thus:

Constraints on the Basic Gibbs Potentials at the Triple Point of Water

For coexistence of water vapour, liquid water, and frozen water at the triple point of water to be possible, the basic Gibbs potentials for water substance must simultaneously satisfy the constraints

$$g^v = g^l = g^f. \tag{9.36}$$

Satisfaction of these two constraints is, by construction, built into Gibbs potentials (9.17), (9.32), and (9.35) since g^v, g^l, and g^f are all, by definition, identically zero at the triple point of water.

9.3.7 Representative Values for the Physical Constants in the Basic Gibbs Potentials

Representative values are displayed in Table 9.3 for the physical constants appearing in the basic Gibbs potentials (9.15), (9.17), (9.32), and (9.35) for dry air, water vapour, pure liquid water, and pure frozen water, respectively. They are based on the values given in the Appendix of Bryan and Fritsch (2002) – which are repeated in Table 1 of Thuburn (2017) – supplemented by some representative values for frozen-water quantities.[13] Values from Table 9.3 have been used for the computations reported in Section 10.6 for the triple point of water in the presence of dry air.

9.3.8 Thermodynamic Properties from the Gibbs Potentials g^d, g^v, g^l, and g^f

Having derived Gibbs potentials (9.15), (9.17), (9.32), and (9.35) for dry air, water vapour, pure liquid water, and pure frozen water, respectively, various thermodynamic quantities are obtained from them for later use. These quantities are displayed in Tables 9.4 and 9.5. They were obtained using the general expressions given in the penultimate column of Table 9.2 and are duplicated for convenience in the last columns of Tables 9.4 and 9.5.

9.4 COMPOSITE GIBBS POTENTIALS

9.4.1 Preamble

Composite Gibbs potentials can be constructed from the individual ones, given in Section 9.3, for dry air, water vapour, pure liquid water, and pure frozen water. In what follows, care needs to be exercised regarding the definition of any thermodynamic system under examination. The focus here will be almost exclusively on *closed* thermodynamic systems in the form of a closed air parcel. For such an air parcel, no *matter* (i.e. substance) macroscopically crosses its boundary. Nevertheless, *energy* can still be transferred between a parcel and its encompassing environment through heat supplied or removed (e.g. by radiation) or through work done (by mechanical expansion or compression); see Section 4.2.

Within a particular environment, for example, cloudy air (see Fig. 9.1 and Section 10.5), parcels of different composition may exist, each one of which can nevertheless be considered to be a closed thermodynamic system. Thus some parcels might be composed of humid air (dry air + water vapour, in a shared volume, depicted in light blue in Fig. 9.1), others of humid air + water droplets (in their own volume, depicted in magenta), and yet others of humid air + ice crystals (in

[13] The negative value for L_0^f is an artifice of the mathematical extrapolation of formula (10.26) outside its domain of validity; that this value is negative has no physical significance.

Symbol	Description	Value	Units
c_p^d	Specific heat of dry air at constant pressure	1004	$\mathrm{J\,kg^{-1}\,K^{-1}}$
R^d	Gas constant for dry air	287	$\mathrm{J\,kg^{-1}\,K^{-1}}$
$c_v^d = c_p^d - R^d$	Specific heat of dry air at constant volume	717	$\mathrm{J\,kg^{-1}\,K^{-1}}$
c_p^v	Specific heat of water vapour at constant pressure	1885	$\mathrm{J\,kg^{-1}\,K^{-1}}$
R^v	Gas constant for water vapour	461	$\mathrm{J\,kg^{-1}\,K^{-1}}$
$c_v^v = c_p^v - R^v$	Specific heat of water vapour at constant volume	1424	$\mathrm{J\,kg^{-1}\,K^{-1}}$
c^l	Specific heat of liquid water	4186	$\mathrm{J\,kg^{-1}\,K^{-1}}$
c^f	Specific heat of frozen water	2106	$\mathrm{J\,kg^{-1}\,K^{-1}}$
$L_0^v = \mathbb{L}^{l \to v}(T=0, p=0)$	Latent heat of vaporisation at $T = 0, p = 0$	3.1285×10^6	$\mathrm{J\,kg^{-1}}$
$L_0^f = \mathbb{L}^{f \to l}(T=0, p=0)$	Latent heat of fusion at $T = 0, p = 0$	-0.2341×10^6	$\mathrm{J\,kg^{-1}}$
$L_0^s = L_0^f + L_0^v$	Latent heat of sublimation at $T = 0, p = 0$	2.8944×10^6	$\mathrm{J\,kg^{-1}}$
$\mathbb{L}^{l \to v}\left(T_0, p_0^{sat}\right) = L_0^v + \left(c_p^v - c^l\right) T_0 - \alpha^l p_0^{sat}$	Latent heat of vaporisation at $T = T_0, p = p_0^{sat}$	2.501×10^6	$\mathrm{J\,kg^{-1}}$
$\mathbb{L}^{f \to l}\left(T_0, p_0^{sat}\right) = L_0^f + \left(c^l - c^f\right) T_0 + \left(\alpha^l - \alpha^f\right) p_0^{sat}$	Latent heat of fusion at $T = T_0, p = p_0^{sat}$	0.334×10^6	$\mathrm{J\,kg^{-1}}$
$\mathbb{L}^{f \to v}\left(T_0, p_0^{sat}\right) = \mathbb{L}^{f \to l}\left(T_0, p_0^{sat}\right) + \mathbb{L}^{l \to v}\left(T_0, p_0^{sat}\right)$	Latent heat of sublimation at $T = T_0, p = p_0^{sat}$	2.835×10^6	$\mathrm{J\,kg^{-1}}$
α^l	Specific volume of liquid water	1/1000	$\mathrm{m^3\,kg^{-1}}$
α^f	Specific volume of frozen water	1/917	$\mathrm{m^3\,kg^{-1}}$
p_0^d	Reference pressure for dry air	10^5	Pa
p_0^{sat}	Saturation vapour pressure for pure water at $T = T_0$	611.657	Pa
T_0	Temperature at triple point of pure water	273.16	K

Table 9.3 Representative values (after Appendix of Bryan and Fritsch (2002) and Table 1 of Thuburn (2017)) for the physical constants appearing in basic Gibbs potentials (9.15), (9.17), (9.32), and (9.35) for dry air, water vapour, pure liquid water, and pure frozen water, respectively. Values for specific heats and gas constants are at $T = T_0 = 273.16$ K. Latent heats of vaporisation ($\mathbb{L}^{l \to v}\left(T_0, p_0^{sat}\right)$) and fusion ($\mathbb{L}^{f \to l}\left(T_0, p_0^{sat}\right)$) at $T = T_0, p = p_0^{sat}$ have been computed using equations (10.25) and (10.26), respectively. See footnote 13 for why $L_0^f = \mathbb{L}^{f \to l}(T = 0, p = 0)$ is negative. For units, $1\ \mathrm{Pa} = 1\ \mathrm{m^{-1}\,s^{-2}\,kg} = 1\ \mathrm{J\,m^{-3}}$.

Quantity	Dry air	Water vapour	Definition
$g\,(p,T)$	$g^d = -c_p^d T \ln\left(\frac{T}{T_0}\right) + R^d T \ln\left(\frac{p^d}{p_0^d}\right)$	$g^v = -c_p^v T \ln\left(\frac{T}{T_0}\right) + R^v T \ln\left(\frac{p^v}{p_0^{sat}}\right) + \left(L_0^v + L_0^f\right)\left(1 - \frac{T}{T_0}\right)$	g
$\alpha\,(p,T)$	$\alpha^d = \frac{R^d T}{p^d} = \frac{1}{\rho^d}$	$\alpha^v = \frac{R^v T}{p^v} = \frac{1}{\rho^v}$	g_p
$\eta\,(p,T)$	$\eta^d = c_p^d \left[1 + \ln\left(\frac{T}{T_0}\right)\right] - R^d \ln\left(\frac{p^d}{p_0^d}\right)$	$\eta^v = c_p^v \left[1 + \ln\left(\frac{T}{T_0}\right)\right] - R^v \ln\left(\frac{p^v}{p_0^{sat}}\right) + \frac{L_0^v + L_0^f}{T_0}$	$-g_T$
$\mathscr{E}\,(T)$	$\mathscr{E}^d = c_v^d T$	$\mathscr{E}^v = c_v^v T + \left(L_0^v + L_0^f\right)$	$g - Tg_T - pg_p$
$h\,(T)$	$h^d = c_p^d T$	$h^v = c_p^v T + \left(L_0^v + L_0^f\right)$	$g - Tg_T$
$f\,(p,T)$	$f^d = -c_p^d T \ln\left(\frac{T}{T_0}\right) + R^d T \ln\left(\frac{p^d}{p_0^d}\right) - R^d T$	$f^v = -c_p^v T \ln\left(\frac{T}{T_0}\right) + R^v T \ln\left(\frac{p^v}{p_0^{sat}}\right) - R^v T + \left(L_0^v + L_0^f\right)\left(1 - \frac{T}{T_0}\right)$	$g - pg_p$

Table 9.4 Some thermodynamic quantities obtained via partial differentiation of the gaseous Gibbs potentials $g^d\left(p^d, T\right)$ and $g^v\left(p^v, T\right)$ for dry air and water vapour, respectively. Subscripts (other than for constants c_p, c_v, p_0^{sat}, and T_0) denote partial differentiation with respect to that independent variable (p or T), with the remaining independent variable (T or p) held fixed. In the entries for $\alpha\,(p,T)$, use has been made of the definitions $\alpha^d \equiv 1/\rho^d$ and $\alpha^v \equiv 1/\rho^v$.

Quantity	Pure liquid water	Pure frozen water	Definition
$g\left(p,T\right)$	$g^{l}=-c^{l}T\ln\left(\frac{T}{T_0}\right)+\alpha^{l}\left(p-p_0^{sat}\frac{T}{T_0}\right)+L_0^{f}\left(1-\frac{T}{T_0}\right)$	$g^{f}=-c^{f}T\ln\left(\frac{T}{T_0}\right)+\alpha^{f}\left(p-p_0^{sat}\frac{T}{T_0}\right)$	g
α	α^{l}	α^{f}	g_p
$\eta\left(T\right)$	$\eta^{l}=c^{l}\left[1+\ln\left(\frac{T}{T_0}\right)\right]+\alpha^{l}\frac{p_0^{sat}}{T_0}+\frac{L_0^{f}}{T_0}$	$\eta^{f}=c^{f}\left[1+\ln\left(\frac{T}{T_0}\right)\right]+\alpha^{f}\frac{p_0^{sat}}{T_0}$	$-g_T$
$\mathscr{E}\left(T\right)$	$\mathscr{E}^{l}=c^{l}T+L_0^{f}$	$\mathscr{E}^{f}=c^{f}T$	$g-Tg_T-pg_p$
$h\left(p,T\right)$	$h^{l}=c^{l}T+\alpha^{l}p+L_0^{f}$	$h^{f}=c^{f}T+\alpha^{f}p$	$g-Tg_T$
$f\left(T\right)$	$f^{l}=-c^{l}T\ln\left(\frac{T}{T_0}\right)-\alpha^{l}p_0^{sat}\frac{T}{T_0}+L_0^{f}\left(1-\frac{T}{T_0}\right)$	$f^{f}=-c^{f}T\ln\left(\frac{T}{T_0}\right)-\alpha^{f}p_0^{sat}\frac{T}{T_0}$	$g-pg_p$

Table 9.5 As in Table 9.4, but for non-gaseous Gibbs potentials $g^{l}\left(p,T\right)$ and $g^{f}\left(p,T\right)$ for pure liquid water and pure frozen water, respectively.

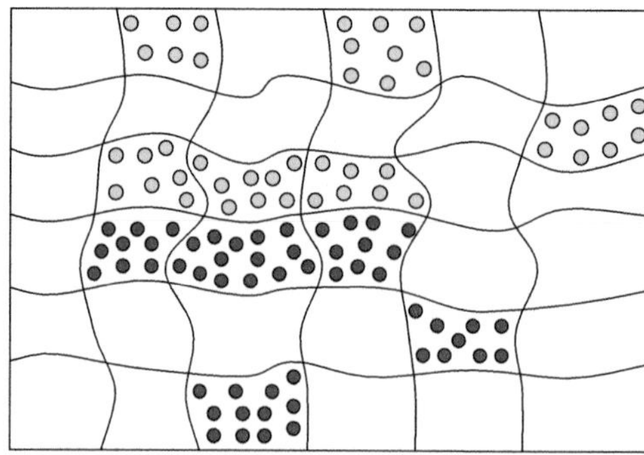

Figure 9.1 Schematic for cloudy air: humid air (dry air + water vapour, in a shared volume) is depicted in light blue, liquid water droplets in magenta, and ice crystals in dark blue. A mixed-phase cloud, composed of six contiguous cloud parcels (three liquid, three frozen) is shown. Also shown are five further cloud parcels (two liquid, three frozen) separated by humid air from both one another and the mixed-phase cloud. Water droplets and ice crystals are absent in the parcels of the humid-air environment in which the liquid-cloud and frozen-cloud parcels are embedded.

their own volume, depicted in dark blue). Energy can be transferred (via heat supplied/removed, and by work done) between each parcel and its surrounding environment (i.e. with immediately adjacent parcels), but *no matter can be macroscopically transferred* (since each parcel is assumed to be a closed thermodynamic system). The union of these parcels then allows liquid and frozen clouds, embedded within a domain of otherwise humid air, to be present and represented.

However, if precipitation occurs (e.g. after coalescence of water droplets into raindrops within a parcel which – as depicted in the rightmost parcel of Fig. 4.1 in Chapter 4 – then fall out of the parcel into an underlying parcel, due to gravity), then neither parcel can be considered to be a closed thermodynamic system, but rather they are open ones due to mass transfer. Other considerations then come into play, such as irreversible changes of entropy. For further information on open thermodynamic systems, the interested reader should consult a specialist book on thermodynamics, for example, Callen (1985), Curry and Webster (1999), and Ambaum (2010).

9.4.2 Thermodynamic Equilibrium

Thermodynamic equilibrium is assumed in classical thermodynamic theory and herein. It is the *stable* stationary state that is reached when a thermodynamic system interacts with its environment over a long period of time. If a system in equilibrium is subjected to a change of imposed conditions, then it will respond in such a way as to reduce those changes. This is known as Le Chatelier's Principle (Callen, 1985).[14] If a system does not respond in this way, then it is not, by definition, in thermodynamic equilibrium.

A system in thermodynamic equilibrium is thus stable to *small* fluctuations of its state variables. Fluctuations occur continually and spontaneously from the random behaviour of individual molecules in substances. Statistically, however, these random *microscopic* perturbations average out *macroscopically*. Thus the equilibrium is dynamic when viewed microscopically, but static when viewed macroscopically or statistically.

9.4.3 Thermodynamic Degrees of Freedom

For a single substance, such as an ideal gas, its entire equilibrium thermodynamic state can be determined by specifying the values of just two thermodynamic variables; for example,

[14] After Henry Louis Le Chatelier (1850–1936).

pressure (p) and temperature (T). From values of these two thermodynamic variables, values of all other thermodynamic variables can then be obtained using diagnostic relations; for example, $\rho = RT/p$ from the equation of state for an ideal gas, where ρ is mass density. Thus there are two *independent* degrees of freedom associated with specification of the equilibrium thermodynamic state of an ideal gas. The Gibbs potential $g = g(p, T)$ provides a natural representation of fluid parcels, including those of an ideal gas.

For *composite mixtures* of substances, such as

- Cloudy air (composed of dry air + water vapour + cloud droplets + ice crystals).
- Seawater with sea ice (composed of liquid water + solute + frozen water).

we need to know the number of *independent* degrees of freedom that uniquely defines the equilibrium thermodynamic state of the mixture. This brings in the *Gibbs phase rule*. Before stating this rule, however, we first define some further thermodynamic vocabulary.

9.4.4 Components and Phases

Many substances can exist in different forms; gaseous, liquid, or solid in the context of Earth's atmosphere and oceans. A *phase* of a substance is any one of these forms. Thus water substance has three phases; water vapour, liquid water, and frozen water.

The number of *components* in a thermodynamic system in equilibrium is the number of chemically different substances present, irrespective of their phase (gaseous, liquid, or solid).

A substance with a single phase present, for example, dry air (considered as a single ideal gas), contributes a single component to the total number of components. A substance with two or more coexisting phases (gaseous, liquid, solid), for example, water substance (with the three phases of water vapour, liquid water, and frozen water) also only contributes a single component to the total number of components. Thus:

- A thermodynamic system consisting of dry air + water vapour + liquid droplets + ice crystals is a two-component system; dry air (with a single phase) + water substance (with three phases). See Fig. 4.2a.
- A thermodynamic system consisting of liquid water + solute + ice is also a two-component system; solute (with a single phase) + water substance (with two phases). See Fig. 4.2b.

Some one- and two-component thermodynamic systems of relevance to meteorology and oceanography are listed in the second column of Table 9.6, grouped according to their composition.

9.4.5 The Gibbs Phase Rule

The Gibbs phase rule allows us to determine the number of thermodynamic state variables that can be independently varied without changing the number of components and/or phases. Underlying this rule is the fact that thermodynamic equilibrium between phases imposes a constraint on the variation of thermodynamic variables: see Section 10.3.2. This then reduces the number of *independent* degrees of freedom available to uniquely determine the equilibrium thermodynamic state of a fluid parcel. Thus:

The Gibbs Phase Rule

The *Gibbs phase rule* is

$$\mathbb{N}^D = \mathbb{N}^C - \mathbb{N}^P + 2, \tag{9.37}$$

where $\mathbb{N}^D$, $\mathbb{N}^C$, and $\mathbb{N}^P$ are numbers of independent degrees of freedom, components, and coexisting phases, respectively.

	Thermodynamic system	Components ($\mathbb{N}^C$)	Phases ($\mathbb{N}^P$)	Degrees of freedom ($\mathbb{N}^D = \mathbb{N}^C - \mathbb{N}^P + 2$)
I.1	Dry air	1	1	2
I.2	Water vapour	1	1	2
I.3	Liquid water	1	1	2
I.4	Frozen water	1	1	2
W.1	Liquid water + frozen water	1	2	1
W.2	Water vapour + water droplets	1	2	1
W.3.	Water vapour + ice crystals	1	2	1
W.4	Water vapour + water droplets + ice crystals (all at the triple point of water)	1	3	0
H	Humid air (dry air + water vapour)	2	1	3
C.1	Liquid cloud (dry air + water vapour + water droplets)	2	2	2
C.2	Ice cloud (dry air + water vapour + ice crystals)	2	2	2
C.3	Mixed-phase cloud (dry air + water vapour + water droplets + ice crystals)	2	3	1
S.1	Seawater without sea ice (liquid water + solute)	2	1	3
S.2	Seawater with sea ice (liquid water + solute + sea ice)	2	2	2

Table 9.6 Examples (after Curry and Webster, 1999, Section 4.2) of Gibbs phase rule $\mathbb{N}^D = \mathbb{N}^C - \mathbb{N}^P + 2$ for thermodynamic systems, where $\mathbb{N}^D$, $\mathbb{N}^C$, and $\mathbb{N}^P$ are numbers of independent degrees of freedom, components, and coexisting phases, respectively. Thermodynamic systems are grouped according to composition of a fluid parcel: I (individual); W (water substance); H (humid air); C (cloudy air); and S (seawater).

In (9.37), the number '2' can be considered to refer to the two degrees of freedom associated with pressure (p) and temperature (T), both of which are common to all substances and phases present in a fluid parcel in thermodynamic equilibrium.[15] Values of $\mathbb{N}^D$, $\mathbb{N}^C$, and $\mathbb{N}^P$ for various thermodynamic systems of relevance to meteorology and oceanography are displayed in the last three columns of Table 9.6.

We have just described the situation when there are no phase transitions (i.e. no change in the number of substances in a fluid parcel, nor in the number of phases present, nor in the amount of mass present for any phase). The representation of phase transitions, which are very important in meteorology and (in polar regions) in oceanography, is described in Sections 10.3 and 10.4.

9.4.6 Composite Gibbs Potentials for Fluid Parcels

The Gibbs potentials (9.15), (9.17), (9.32), and (9.35), developed separately for dry air, water vapour, pure liquid water, and pure frozen water, respectively, can be combined to provide *composite Gibbs potentials*. Doing so enables us to represent the thermodynamics of various mixtures of dry gas and water substance (in its vapour, liquid, and frozen forms). For the four individual Gibbs potentials, everything thermodynamic can be obtained from them separately, as illustrated in Tables 9.4 and 9.5. Similarly, everything thermodynamic for a mixture can also be obtained in a self-consistent manner from its composite Gibbs potential via partial differentiation and algebraic manipulation; see the penultimate column of Table 9.2 for examples of how this is done.

Consider now a fluid parcel composed of a mixture of dry air, water vapour, liquid water, and frozen water. From (4.53), the total mass of such a parcel is

$$\mathbb{M} = \mathbb{M}^d + \mathbb{M}^v + \mathbb{M}^l + \mathbb{M}^f, \tag{9.38}$$

where $\mathbb{M}^d$, $\mathbb{M}^v$, $\mathbb{M}^l$, and $\mathbb{M}^f$ are the partial masses of dry air, water vapour, liquid water, and frozen water, respectively. From definitions (3.7), the corresponding mass fractions are

$$q^d = \frac{\mathbb{M}^d}{\mathbb{M}}, \quad q^v = \frac{\mathbb{M}^v}{\mathbb{M}}, \quad q^l = \frac{\mathbb{M}^l}{\mathbb{M}}, \quad q^f = \frac{\mathbb{M}^f}{\mathbb{M}}. \tag{9.39}$$

Dividing (9.38) by $\mathbb{M}$ with use of definitions (9.39), these mass fractions therefore sum to unity:

$$q^d + q^v + q^l + q^f = 1. \tag{9.40}$$

For a fluid parcel with composition (9.38), we can formally define a composite Gibbs potential as the mass-weighted sum of the four basic Gibbs potentials (9.15), (9.17), (9.32), and (9.35). Thus:

[15] Equation (9.37) assumes that equilibrium between phases only depends upon pressure, temperature, and mass fractions of substances. It does not, for example, hold for situations where surface tension of raindrops is important. Such effects are ignored herein but can be important in cloud physics.

Composite Gibbs Potentials for Fluid Parcels

Composite Gibbs potentials $g = g\left(p, T, q^d, q^v, q^l, q^f\right)$ may be constructed as a mass-weighted sum of the four, individual, basic, Gibbs potentials (9.15), (9.17), (9.32), and (9.35) to obtain

$$g\left(p, T, q^d, q^v, q^l, q^f\right) = q^d g^d\left(p^d, T\right) + q^v g^v\left(p^v, T\right) + q^l g^l\left(p, T\right) + q^f g^f\left(p, T\right), \quad (9.41)$$

where (as will be shown later) $p^d = p^d\left(p\right)$ and $p^v = p^v\left(p\right)$; see (10.6) and (10.7). In (9.41), the values of mass fractions q^d, q^v, q^l, and q^f have to be specified in some way, subject to appropriate physical constraints determined by the particular thermodynamic system under consideration.

9.4.7 Three Illustrative Examples of Composite Gibbs Potentials

In Chapter 10, we examine – in some detail – three illustrative examples of the formal composite Gibbs potential (9.41):

1. **Humid air**
   Setting $q^l = q^f \equiv 0$ in (9.41), there are two substances (dry air + water substance) sharing a common volume, with a single (gaseous) phase. This facilitates examination of a mixture of two gases (dry air + water vapour) *only*, in the absence of liquid and frozen water; see Section 10.2.
2. **Water substance**
   Setting $q^d \equiv 0$ in (9.41), there is a single substance (water substance) with three possible phases (water vapour, liquid water, and frozen water). This facilitates a qualitative and quantitative examination of phase transitions in the simplest possible context; see Sections 10.3 and 10.4.
3. **Cloudy air, possibly containing liquid and frozen water**
   This is the general case that combines the building blocks developed in Sections 10.2 and 10.4 for humid air and water substance, respectively. This then describes cloudy air; see Section 10.5.

9.4.8 A Very Useful Property of Compositions of the Four Basic Gibbs Potentials

For cloudy air, we have already seen (in Section 3.5) that internal energy ($\mathcal{E}$) and specific heats at constant volume (c_v) and constant pressure (c_p) can be expressed as mass-weighted sums of their corresponding basic quantities; see (3.71), (3.72), and (3.75), respectively. For example, $\mathcal{E} = \mathcal{E}^d q^d + \mathcal{E}^v q^v + \mathcal{E}^l q^l + \mathcal{E}^f q^f$ from (3.71), where $\mathcal{E}^d$, $\mathcal{E}^v$, $\mathcal{E}^l$, and $\mathcal{E}^f$ are as in (3.74) and Tables 9.4 and 9.5. Similarly, (9.41) defines a composite Gibbs potential that is a mass-weighted sum of the basic Gibbs potentials g^d, g^v, g^l, and g^f.

Would it not be nice if other thermodynamic potentials, such as specific enthalpy (h), specific Helmholtz potential (f), and specific entropy (η), also had this property? This would mean we could straightforwardly obtain expressions for them using the basic quantities given in Tables 9.4 and 9.5 instead of:

- Summing the individual Gibbs potentials.
- Laboriously computing various partial derivatives from the sum.
- Algebraically combining the resulting quantities.

Very good news; it turns out that (consistent with the assumptions made to obtain the basic Gibbs potentials)[16] other associated thermodynamic potentials can indeed be obtained as the mass-weighted sum of the corresponding (already available) basic quantities! Ain't life grand? This property will be exploited later; see Sections 10.5.6 and 10.5.7.

9.5 CONCLUDING REMARKS

The importance of using a Gibbs potential (or another thermodynamic potential) to obtain everything thermodynamic cannot be overemphasised. Doing so *guarantees thermodynamical consistency*. This avoids introducing spurious sources and sinks of energy that can then lead to spurious climate drift in atmospheric and oceanic models. It also facilitates the development of improved representation of thermodynamic processes. A striking example of this is the development of the comprehensive TEOS-10 Gibbs potential, which embodies many high-quality observational databases (IOC et al., 2010).

Having now broadened and deepened our treatment of thermodynamics – by examining thermodynamical consistency and thermodynamic potentials – we are now in a position to address the two important questions identified in Section 9.1:

1. How can phase transitions of water substance be represented in atmospheric and oceanic models?
2. How can suitable forms of the fundamental equation of state for an ocean be constructed?

The first of these is answered in Chapter 10, and the second in Chapter 11.

[16] In particular, it is assumed that certain virial interaction terms are absent. Real gases behave, to an excellent approximation, as ideal gases for large values of specific volume (α). For diminishing values of α, however, they gradually depart from this behaviour. This is because intermolecular forces are negligible at low density (i.e. for large α), but not at high density, where molecules are much closer to one another. To account for this small departure from ideal gas behaviour (Callen, 1985, Section 13-3), the ideal-gas law ($p/T = R/\alpha$) is replaced by $p/T = (R/\alpha)\left[1 + \sum_{i=1,2,\ldots} c_i(T)\,\alpha^i\right]$. The term in square brackets is termed a virial expansion, $c_i(T)$ is the ith virial coefficient, and its functional form depends on the form of the intermolecular forces of the gas.

10

Moist Thermodynamics

ABSTRACT

Phase transitions of water substance – between its vapour, liquid, and frozen forms – involve transfers of energy that significantly influence atmospheric and oceanic circulations, and thereby Earth's weather and climate. A realistic representation of phase transitions for pure-water substance in a vacuum is developed. It combines the basic Gibbs potentials for water vapour, liquid water, and frozen water, developed in Chapter 9. Exploiting this composite potential provides valuable insight into phase transitions, latent heating and cooling, and the 'triple point of water' – the only point on a 2D (T-p^v) phase diagram where all three water states can coexist. An important missing ingredient for atmospheric applications is dry air – the atmosphere's primary constituent. Analysis for the triple point of water is therefore extended to include:

- The presence of dry-air substance.
- The determination of the distribution of water vapour, liquid water, and frozen water at the only point of a 3D (T-p^v-p^d) phase diagram where water vapour, liquid water, frozen water, *and dry air* can coexist.

The basic Gibbs potentials for liquid water and frozen water may be simplified, whilst nevertheless maintaining thermodynamical consistency. The triple-point temperature in the presence of dry air then reduces to that in its absence; this corresponds to a widely used approximation in meteorology. The ability to account for the approximately 0.01 °C difference between the triple-point temperature in the absence and presence of dry air is, however, lost. Finally, potential temperature, potential enthalpy, conservative temperature, and potential density are all defined – in terms of thermodynamic potentials – for subsequent use.

10.1 PREAMBLE

For Earth's atmosphere, water substance exists in three forms; vapour, liquid, and frozen. For the oceans, it is limited to liquid and frozen forms. Water substance can change from one form to another. When it does, a *phase transition* is said to take place. This involves significant transfers of energy, primarily in the form of heat. These phase transitions then greatly influence – both directly and indirectly – the circulations of Earth's atmosphere and oceans, and thereby its weather and climate.[1]

Important, related, hydrological processes include cloud formation and precipitation – such as rain, hail, snow, and sleet – in the atmosphere; and the formation, melting, and sublimation of sea ice in lakes and in the Arctic and Southern Oceans. These hydrological processes influence

[1] For other celestial bodies – with different régimes of temperature and pressure – phase transitions of other substances, such as carbon dioxide and methane, may instead be important.

Earth's albedo (i.e. the amount of light that gets reflected back from a surface). For example, high-level clouds in the atmosphere reflect incoming solar radiation back to Space, as do snow and ice at Earth's surface over both land and water. If the total amount and distribution of cloud cover, or that of sea ice, changes, then this will change the amount of heat that is absorbed by Earth's atmosphere and oceans – both locally and globally – thereby changing their circulations.

Suffice it to say that:

- Phase transitions are very important for atmospheric and oceanic circulations.
- They significantly influence Earth's weather and climate.

The focus of this chapter is therefore to tackle the first question[2] posed in Section 9.5:

- How can phase transitions of water substance be represented in atmospheric and oceanic models?

A first step towards answering this question has already been taken in Chapter 9, with the introduction and definition of various thermodynamic potentials, together with their associated properties. In particular, some basic Gibbs potentials for dry air, water vapour, pure liquid water, and pure frozen water were derived (in Section 9.3) from first principles. These basic potentials are exploited in this chapter to form composite Gibbs potentials. A simple, composite Gibbs potential – constructed in Section 10.2 – is that for humid air, a gaseous mixture of dry air and water vapour that is transparent to the human eye.

The next step (in Section 10.3) is to *qualitatively* describe the physics of phase transitions. This prepares the way to *quantitatively* translate (in Section 10.4) this physical behaviour into mathematical equations for the phase transitions of water substance *in a vacuum*. This (temporarily) keeps things a little simpler than would otherwise be the case if dry air were to be immediately included.

Building on the valuable insight provided by this analysis, the mathematical formulation of phase transitions for pure-water substance is extended (in Section 10.5) to additionally allow for the presence of dry air. The end result is a realistic representation of dry air and water substance in the atmosphere that includes phase transitions and latent heating and cooling. The triple point of water *in the presence of dry air* is examined in Section 10.6. This includes determination of the distribution of water vapour, liquid water, and frozen water at this very special point of a 3D temperature-pressure (T-p^v-p^d) phase diagram,

General definitions – in terms of thermodynamic potentials – are given in Section 10.7 for potential temperature, potential enthalpy, conservative temperature, and potential density; these hold for both the atmosphere and the oceans. This prepares the way for use of these variables in Chapter 11. It is also useful in other contexts. Concluding remarks are made in Section 10.8.

10.2 HUMID AIR

Humid air is a (purely gaseous) mixture of dry air and water vapour, with no liquid or frozen water present (i.e. $q^l = q^f \equiv 0$ in composite Gibbs potential (9.41)).[3] As discussed in the Appendix to Chapter 3, dry air and water vapour *share the same gaseous volume*. Following Thuburn (2017), who himself followed Feistel et al. (2010a), let a be the mass fraction of *dry air* in this shared volume.

Aside One could instead let a be the mass fraction of *water vapour* in this shared, gaseous volume. In the absence of other forms of water substance (i.e. liquid and frozen water), a

[2] The second question will be addressed in Chapter 11.

[3] For simplicity, it is assumed in this section that saturation of a humid-air parcel does not occur. Both dry air and water vapour coexist within the parcel, but there is insufficient water vapour to allow a change of its phase. Saturation of air parcels is examined in Sections 10.3–10.5.

would then be specific humidity (q^v) with small values of order $0 - 0.04$. All other things being equal, it is natural to define a this way for the atmosphere. However, Thuburn (2017) argues that it is nevertheless desirable to instead let a be the mass fraction of *dry air* since '*this leads to a formal symmetry between humid air and saline water*'. This (somewhat cryptic) comment alludes to the fact that water substance is common to both humid air and saline water. For humid air it is the *minor* constituent (with dry air being the major one), whereas for saline water it is the *major* constituent (with solute being the minor one).

Now Feistel et al. (2010a) obtained a *universally valid*, highly accurate, composite Gibbs potential, applicable not only to Earth's oceans, but also to its atmosphere and cryosphere.[4] The only difference between the application of their composite Gibbs potential to these various domains is in the specification of appropriate mass fractions for the various gaseous, liquid, and frozen constituents. For example, the mass fraction of salinity would be set to zero for global atmospheric models, and that for dry air would similarly be set to zero for global ocean models. Using a *universal* Gibbs potential offers the opportunity to intrinsically ensure thermodynamical consistency at and across air/sea/ice boundaries. In particular, this then allows computation of *consistent* thermodynamic fluxes across these boundaries.

One does not have to adopt the Feistel et al. (2010a) composite Gibbs potential; one can make it simpler or more complicated, according to application. However, by defining a to be the mass fraction of dry air (instead of water vapour) within the shared gaseous volume, the resulting Gibbs potential should then have approximately the same functional form as the Feistel et al. (2010a) one, thereby facilitating validation against it.

- For the reasons given, it is beneficial for meteorologists to adopt the oceanographic convention. a is then the mass fraction of dry air, rather than that of water vapour.

Returning to the main argument, and recalling that a is the mass fraction of dry air *in the shared, gaseous volume*,[5] the Gibbs potential for a humid-air parcel may be written as a mass-weighted combination of the basic Gibbs potentials (9.15) and (9.17) for dry air and water vapour, respectively. Setting $q^l = q^f \equiv 0$ in composite Gibbs potential (9.41) then leads to:

The Gibbs Potential for Humid Air

$$g^{av}\left(p^d, p^v, T, a\right) = a g^d\left(p^d, T\right) + (1-a)\, g^v\left(p^v, T\right). \tag{10.1}$$

In (10.1), and as in previous chapters, superscripts 'd' and 'v' denote values associated with **d**ry air and water **v**apour, respectively, and p^d and p^v are the partial pressures of dry air and water vapour, respectively.[6] Furthermore, q^d has been written as a, and q^v as $1-a$ (cf. (10.1) for humid air with prototypical equation (9.41) for fluid parcels). This has been done to prepare

[4] The cryosphere is that region of Earth, including, for example, Antarctica, where the surface is frozen.

[5] For the more general case of cloudy air, this volume excludes that occupied by liquid and frozen water.

[6] In writing (10.1), certain 'virial interaction terms' (these represent intermolecule interactions) are absent. This is a *consequence* of our assumption that dry air and water vapour behave as ideal gases, with the underlying assumption that their molecules do not interact with one another. (In physical reality they do, and these interactions lead to weak virial terms; these are usually neglected in meteorology.)

the way for examination (in Sections 10.4 and 10.5) of saturation and phase transitions of water substance.

Because a is the mass fraction of dry air in the *shared* gaseous volume, and $\mathbb{M}^d + \mathbb{M}^v$ is the total mass within this shared volume, it follows from definitions (9.38) (with $\mathbb{M}^l = \mathbb{M}^f = 0$) and (9.39) that

$$a = \frac{\mathbb{M}^d}{\mathbb{M}^d + \mathbb{M}^v} = \frac{\mathbb{M}^d/\mathbb{V}^g}{\left(\mathbb{M}^d + \mathbb{M}^v\right)/\mathbb{V}^g} = \frac{\rho^d}{\rho^d + \rho^v}, \quad 1 - a = 1 - \frac{\rho^d}{\rho^d + \rho^v} = \frac{\rho^v}{\rho^d + \rho^v}, \tag{10.2}$$

for humid air. (This is also true when liquid and/or frozen water are additionally present; see Sections 10.4 and 10.5.)[7]

As written, g^{av} in (10.1) is expressed as a function of p^d and p^v (as well as of T and a). However, partial pressures p^d and p^v are related to total pressure (p) by $p = p^d + p^v$ (i.e. by Dalton's Law of Partial Pressures).[8] Since p (as opposed to p^d and p^v) is a natural variable of a *composite* Gibbs potential, we now express p^d and p^v in terms of p. Note that p^v is often denoted by e in the meteorological literature. This practice is not followed herein; this avoids any confusion regarding the use of e for other purposes (e.g. for the base of natural logarithms or for the eccentricity of an ellipse). From Table 9.4 of Chapter 9, and using (10.2), the ideal-gas laws for dry air and water vapour, respectively, are

$$p^d = \rho^d R^d T = \left(\frac{\rho^d}{\rho^d + \rho^v}\right) R^d \left(\rho^d + \rho^v\right) T = aR^d \left[\left(\rho^d + \rho^v\right) T\right], \tag{10.3}$$

$$p^v = \rho^v R^v T = \left(\frac{\rho^v}{\rho^d + \rho^v}\right) R^v \left(\rho^d + \rho^v\right) T = (1 - a) R^v \left[\left(\rho^d + \rho^v\right) T\right]. \tag{10.4}$$

Elimination of the square-bracketed expression between these two equations then relates p^d to p^v by

$$p^d = \frac{aR^d}{(1 - a) R^v} p^v \equiv \frac{a\varepsilon}{(1 - a)} p^v, \tag{10.5}$$

where $\varepsilon \equiv R^d/R^v$.[9] Using $p = p^d + p^v$ (i.e. the total pressure is the sum of the partial pressures) to eliminate p^v in (10.5) yields

$$p^d = \frac{a\varepsilon}{(1 - a)} \left(p - p^d\right) \quad \Rightarrow \quad p^d = \frac{a\varepsilon}{1 + a(\varepsilon - 1)} p. \tag{10.6}$$

[7] Whereas $\mathbb{M} = \mathbb{M}^d + \mathbb{M}^v$ for humid air, $\mathbb{M} = \mathbb{M}^d + \mathbb{M}^v + \mathbb{M}^l + \mathbb{M}^f$ for cloudy air. Care then has to be taken when defining and using mass fractions (which normalise partial masses by $\mathbb{M}$), to properly take into account the presence of liquid and frozen water. See Sections 3.3, 10.4, and 10.5. In general, $a \neq q^d$; $a = q^d$ *only* when $q^l = q^f \equiv 0$ (i.e. *only* in the absence of both liquid water and frozen water).

[8] After John Dalton (1766–1844).

[9] Caution: do not be misled by the notation (which is classical); $\varepsilon \approx 287/461 \approx 0.6226$, and it is consequently *not* a small, negligible quantity with respect to unity.

Therefore

$$p^v = p - p^d = p - \frac{a\varepsilon}{1 + a(\varepsilon - 1)}p = \frac{1 - a}{1 + a(\varepsilon - 1)}p, \tag{10.7}$$

where $p = p^d + p^v$ has been reused to obtain (10.7).[10]

Thus (10.6) and (10.7) relate p^d and p^v to p, as desired. This permits p^d and p^v to be eliminated in favour of p whenever convenient to do so. In particular, substitution of (10.6) and (10.7) into Gibbs potentials (9.15) and (9.17) gives

$$g^d\left(p^d, T\right) = -c_p^d T \ln\left(\frac{T}{T_0}\right) + R^d T \ln\left(\frac{p^d}{p_0^d}\right) \tag{10.8}$$

$$= g^d(p, T) = -c_p^d T \ln\left(\frac{T}{T_0}\right) + R^d T \ln\left\{\frac{a\varepsilon}{[1 + a(\varepsilon - 1)]}\frac{p}{p_0^d}\right\}, \tag{10.9}$$

$$g^v\left(p^v, T\right) = -c_p^v T \ln\left(\frac{T}{T_0}\right) + R^v T \ln\left(\frac{p^v}{p_0^{sat}}\right) + \left(L_0^v + L_0^f\right)\left(1 - \frac{T}{T_0}\right) \tag{10.10}$$

$$= g^v(p, T) = -c_p^v T \ln\left(\frac{T}{T_0}\right) + R^v T \ln\left\{\frac{(1 - a)}{[1 + a(\varepsilon - 1)]}\frac{p}{p_0^{sat}}\right\} + \left(L_0^v + L_0^f\right)\left(1 - \frac{T}{T_0}\right). \tag{10.11}$$

In the absence of saturation (as assumed in this section), the latent internal energy term $\left(L_0^v + L_0^f\right)(1 - T/T_0)$ in (10.10) and (10.11) is inactive. Latent internal energy is then passively transported by a humid-air parcel, with no change of phase of the water vapour contained within it.

Using (10.9) and (10.11) in (10.1) – instead of the equivalent (10.8) and (10.10) – allows the Gibbs potential (10.1) for humid air to be simplified to:

An Alternative, Equivalent Form of the Gibbs Potential for Humid Air

$$g^{av}(p, T, a) = a g^d(p, T) + (1 - a) g^v(p, T). \tag{10.12}$$

Equation (10.12) corresponds to Thuburn (2017)'s (45), and equations (10.8)–(10.11) to his (46)–(49).

The derivation of composite Gibbs potential (10.12) for *humid air* prepares the way for the development (in Section 10.5) of a composite Gibbs potential for the more complicated case of *cloudy air*.

[10] Setting $a \equiv 0$ in (10.6) and (10.7) gives $p^d = 0$ and $p^v = p$, respectively, i.e. the parcel consists entirely of water vapour. Similarly, setting $a \equiv 1$ gives $p^d = p$ and $p^v = 0$, respectively, i.e. the parcel consists entirely of dry air. Summing (10.6) and (10.7) recovers Dalton's Law of Partial Pressures $p = p^d + p^v$.

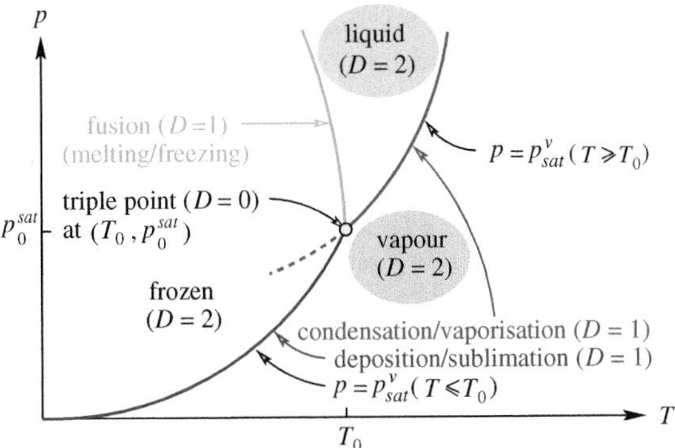

Figure 10.1 Phase diagram (not to scale) for water substance for different conditions of temperature (T) and pressure (p); $D \equiv \mathbb{N}^D$ is number of independent degrees of freedom. There are three regions (for vapour, liquid, and frozen phases, for which $D = 2$), separated by three (solid) saturation/coexistence curves (for change of phase via condensation/vaporisation, fusion (melting/freezing) and deposition/sublimation, for which $D = 1$). These three curves intersect at the triple point of water (for which $D = 0$), the only point on the diagram where all three phases of water can coexist in equilibrium. The dashed curve denotes the possibility of supercooled liquid. See text for further interpretation of this diagram.

10.3 LATENT INTERNAL ENERGY AND PHASE TRANSITIONS

10.3.1 Qualitative Overview of Phase Transitions

The development and application of composite Gibbs potentials based on (9.41) is complicated by the fact that although we only have (by the assumptions made in Section 9.3) two substances (dry air and water substance), water substance has three different phases (vapour, liquid, and frozen) with very different properties.[11] Depending on the physical situation, there can be a transition of water phase from one form to another (i.e. a *phase transition*). This then results in what is termed *latent heating* or *cooling* of an air parcel. Phase transitions have a very important impact on the formation of clouds and precipitation, and on the freezing and melting of sea ice. Through related transfers of heat and energy, these transitions significantly influence the global circulation of Earth's atmosphere and oceans, with a consequent impact on Earth's climate.

To set the scene for subsequent developments, a qualitative overview of phase transitions is now given. A *phase diagram* for the physical forms of a substance, be it water substance or some other substance, provides a useful graphical representation. Such a phase diagram (not to scale) for water substance is depicted in Fig. 10.1 for different combinations of temperature (T) and pressure (p).

Consider a parcel of pure-water vapour at fixed temperature, T. By assumption, there is no dry air ($q^d \equiv 0$), and so $p^d \equiv 0 \Rightarrow p = p^d + p^v = p^v$. There is a maximum amount of water vapour that the parcel can hold without condensation or deposition taking place.[12] When this maximum amount is attained, the parcel is said to be *saturated*, and a change of phase is then initiated for any excess water vapour.

[11] Dry air is a mixture of gases. At sufficiently (i.e. very-) low temperatures these gases can also change phase to liquid and frozen forms. However they do so at temperatures (and pressures) outside the range of those observed in Earth's atmosphere. We therefore consider dry air herein to have a single (vapour) phase. For planets with very cold atmospheres, this simplification is however invalid.

[12] This is true not only for water vapour, but also for the vapour phase of other substances.

Let $p = p^v_{sat}(T)$ be the *saturation vapour pressure* at temperature T (i.e. the pressure exerted by the parcel's water vapour on its encompassing environment *at saturation of the parcel*). Referring to Fig. 10.1, for any point (T, p^v) that lies anywhere beneath the blue and red *saturation curves* (also known as *coexistence curves*) $p = p^v_{sat}(T)$ on the diagram, the parcel is in a purely gaseous, subsaturated phase. Within this region both T and p^v can be varied independently, and the number of independent degrees of freedom is $\mathbb{N}^D \equiv D = 2$. ($\mathbb{N}^D$ is denoted by D on Fig. 10.1.)

If, at fixed temperature, T, more water vapour is added to the parcel (e.g. from its underlying surface), then the vapour pressure, p^v, increases. If enough water vapour is added, p^v eventually reaches its saturation value, $p^v_{sat}(T)$.[13] The point (T, p^v_{sat}) – which represents the temperature and vapour pressure of the parcel at saturation – then lies on either the blue *saturation curve* or the red one. Along the two saturation curves, the number of independent degrees of freedom is $\mathbb{N}^D \equiv D = 1$. This is one less than anywhere below these two curves and signifies that, at saturation, *only one of T and $p = p^v_{sat}$ can be varied independently*. The question now is:

- What happens at saturation of the water vapour parcel?

The answer to this question depends upon the value of temperature (T) at saturation vapour pressure ($p = p^v_{sat}$). This brings in the *triple point of a substance*. For a given substance (water content in the atmospheric context), there is only a *single* point in the (T, p) plane – as depicted by the yellow point in Fig. 10.1 – where all *three* phases (vapour, liquid, and frozen) of a substance can *coexist* in thermodynamic equilibrium. This very special point, with coordinates $(T, p) = (T_0, p^{sat}_0)$, is known as the *triple point of a substance*.[14] At the triple point (see composition W.4 in Table 9.6), the number of independent degrees of freedom $\mathbb{N}^D \equiv D = 0$, that is, there is no freedom whatsoever, *without losing a phase*, to vary (even slightly) either T or $p = p^v = p^{sat}_0$ whilst maintaining thermodynamic equilibrium.

For a point lying on the saturation curve $p = p^v_{sat}(T \geq T_0)$ (i.e. on the red curve of Fig. 10.1), any excess water vapour in the parcel, beyond the saturation amount, changes phase and condenses to liquid water.[15] As a consequence, some latent heat (i.e. some latent internal energy) is released.

Similarly, for a point lying on the saturation curve $p = p^v_{sat}(T \leq T_0)$ (i.e. on the blue curve of Fig. 10.1), any excess water vapour in the parcel instead freezes.[16] [17] As a consequence, some latent heat is again released.

The preceding paragraphs describe what happens at saturation of the gaseous phase of water substance. Any excess water vapour either condenses (for $T \geq T_0$, the temperature at the triple point of water substance) or it freezes (for $T \leq T_0$). Condensation or freezing of excess water vapour continues until a new state of thermodynamic equilibrium is established. In what follows, it is assumed for simplicity that this happens almost instantaneously after saturation. Supplying additional water vapour at saturation does not increase vapour pressure. Any additional water

[13] Adding water vapour is not the only way to saturate the parcel; decreasing T at fixed vapour pressure, p^v, also does so. The point (T, p^v_{sat}) then moves leftward on the diagram (instead of upwards), and if T is decreased sufficiently, the point then lies on one of the two (blue or red) saturation curves.

[14] For pure-water substance, $(T_0, p^{sat}_0) = (273.16\ \text{K}, 611.657\ \text{Pa}) = (273.16\ \text{K}, 6.11657\ \text{hPa}) = (0.01\ ^\circ\text{C}, 6.11657\ \text{mb})$, precisely. Note that, by international convention and for historical reasons, T_0 is *not* equal to precisely 0 °C, but to precisely 0.01 °C ≡ 0.01 K. The intervals of the Kelvin and Celsius temperature scales are, by definition, *identical*, and 1 K ≡ 1 °C.

[15] The parcel remains saturated, and the vapour and condensed liquid are in thermodynamic equilibrium with one another after completion of the transition.

[16] The parcel again remains saturated, but this time it is the vapour and frozen phases that are in mutual thermodynamic equilibrium.

[17] Physically it is possible for vapour at temperature $T < T_0$ to condense to supercooled liquid form instead of to frozen form; see the dashed curve on Fig. 10.1. This phenomenon can be important for the detailed representation and modelling of clouds and fog in Earth's atmosphere. However, for simplicity, it is ignored herein.

vapour (i.e. excess water vapour) condenses or freezes instead, leaving vapour pressure unchanged at its saturation value ($p^v = p^v_{sat}(T)$).

Consider now what happens if the temperature (T) of the parcel increases at constant vapour pressure (p^v), or the vapour pressure (p^v) decreases at constant temperature (T). In the first case, a point $\left(p^v_{sat}, T\right)$ on either of the two (blue or red) saturation curves of Fig. 10.1 moves rightwards. In the second case it moves downwards. In either case, the result is that the water vapour in the parcel is no longer at saturation, but at subsaturation. Furthermore, the water vapour in the parcel is no longer in thermodynamic equilibrium with the parcel's liquid- or frozen-water content. A phase transition from liquid to gaseous (for $T \geq T_0$, see red curve) or from frozen to gaseous (for $T \leq T_0$, see blue curve) is then initiated for the excess liquid or frozen water. This continues until:

1. *Either* a new state of thermodynamic equilibrium is established between the parcel's water vapour, and either the parcel's remaining liquid water or its remaining frozen water (depending upon whether $T \geq T_0$ or $T \leq T_0$, respectively).
2. *Or* there is no longer any liquid or frozen water left, only water vapour.

In any of these circumstances, latent cooling of the parcel occurs. It is again assumed for simplicity that establishment of a new state of thermodynamic equilibrium happens almost instantaneously.

For a parcel of pure-water substance (as considered here), the only way the area above the red and blue vapour-saturation curves of Fig. 10.1 can be accessed is for *all* of a parcel's water vapour to condense to liquid and/or frozen form. For this to happen, the specific volume occupied by water vapour (α^v) must reduce to zero (via compression by the parcel's environment), leaving only the specific volumes α^l and α^v occupied by liquid and/or frozen water. When *both* forms are present, the point lies on the green fusion curve.

In a similar manner to that described earlier in this section for phase transitions between water vapour and liquid or frozen water, any point that is pushed away from the (green) fusion curve by a change in T or p is pushed back to it in order to re-establish thermodynamic equilibrium. This again results in either latent heating (for transition from liquid water to frozen water, i.e. freezing of liquid water) or cooling (for transition from frozen water to liquid water, i.e. melting of ice) of the parcel until equilibrium is re-established.

The terminology used for transitions between vapour, liquid, and frozen phases of a substance is graphically displayed in Fig. 10.2, with the spacing of h^f, h^l, and h^v being approximately to scale (for water substance); see Sections 10.4.4 and 10.4.5 for definition and discussion of these quantities. Fig. 10.2 graphically illustrates (for water substance) that (per unit mass):

- Much less energy is required to melt frozen water than to evaporate liquid water ($\mathbb{L}^{f\to l} \equiv h^l - h^f \approx 0.334 \times 10^6$ J kg^{-1} versus $\mathbb{L}^{l\to v} \equiv h^v - h^l \approx 2.501 \times 10^6$ J kg^{-1}).
- The same amount of energy is required to first melt frozen water, and to then evaporate the resulting liquid water to obtain water vapour ($\mathbb{L}^{f\to l} + \mathbb{L}^{l\to v} \equiv \left(h^l - h^f\right) + \left(h^v - h^l\right) = h^v - h^f$), as it does to directly sublimate frozen water to obtain water vapour ($\mathbb{L}^{f\to v} \equiv h^v - h^f$).

To place the present discussion in a broader context, almost all of the water substance in the atmosphere (in its vapour, liquid, and frozen phases) originates from evaporation and sublimation at Earth's surface (primarily over its oceans). The energy required for evaporation and sublimation is mostly supplied by (short-wave) radiation from the Sun. This supplied energy overcomes the intermolecular attraction between water molecules that holds them together in the liquid and frozen-water reservoirs at Earth's surface. Some molecules then escape from these reservoirs, thereby increasing the water vapour content and internal energy of air parcels adjacent to them. The more water vapour is added to an air parcel, the higher its vapour pressure and internal energy become.

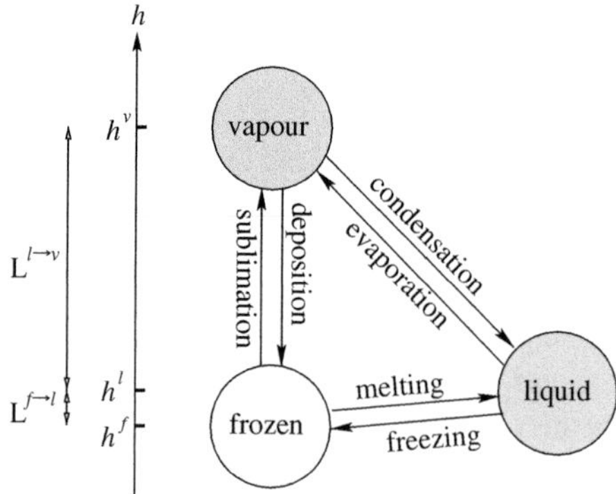

Figure 10.2 Terminology for phase transitions. Each arrow points towards the phase resulting from the transition. Colour coding is as in Fig. 10.1. h is specific enthalpy; h^f, h^l, and h^v are values for frozen water, liquid water, and water vapour. $\mathrm{L}^{l\to v} \equiv \mathbb{L}^{l\to v} \equiv h^v - h^l$ and $\mathrm{L}^{f\to l} \equiv \mathbb{L}^{f\to l} \equiv h^l - h^f$ are enthalpies of vaporisation and fusion. No physical significance is attached to the abscissa, only to the ordinate.

When such a parcel saturates for whatever reason (typically by cooling of the parcel by adiabatic expansion as its rises in the atmosphere), a change of phase (to either liquid or frozen form) of excess water vapour takes place. Latent internal energy (heat) is then released to the air parcel via condensation or deposition of the excess. (It is termed 'latent' since it is 'hidden' within a subsaturated air parcel. This energy was supplied during evaporation and sublimation at Earth's surface and is passively carried elsewhere by the parcel as water vapour until saturation occurs (Dutton, 1986, p. 269).) The released energy, in some combination, simultaneously increases the temperature (T) of the parcel (via thermal heating) and causes the parcel to expand (thereby doing mechanical work). Under the right conditions, accumulated liquid and frozen water can fall out of an air parcel as precipitation. Again, if conditions are right, this water will fall all the way to Earth's surface.

This continual processing and recycling of water substance in the atmosphere is the dominant part of Earth's *hydrological cycle*. About 97% of Earth's total water substance is locked into its oceans, and about 2% into ice and glaciers, with much less than 1% actually residing in the atmosphere at any one time.

This concludes the general qualitative overview of phase transitions. Although couched in fairly general terms, the reader is advised that phase transitions are, in reality, more complicated than those described here. For in-depth analysis of phase transitions, a specialised book on thermodynamics and cloud physics should be consulted; e.g. Callen (1985) and Curry and Webster (1999).

10.3.2 Coexistence of Phases

Let $A = v, l, f$ denote a particular phase of a substance, where v, l, and f signify vapour, liquid, and frozen phases, respectively, and similarly for B.

Coexistence of Phases

For two phases of a substance to be in thermodynamic equilibrium at a common boundary, three conditions must be simultaneously satisfied:

1. **Thermal equilibrium,**

$$T^B = T^A, \tag{10.13}$$

so that no heat flows between the two phases.

2. **Mechanical equilibrium,**

$$p^B = p^A, \tag{10.14}$$

so that neither phase does mechanical work on the other via expansion or compression.

3. **Chemical equilibrium,**

$$g^B = g^A. \tag{10.15}$$

so that there is no change in chemical composition (i.e. no *net* change in the number of molecules of each phase). [a]

a More generally, the condition for chemical equilibrium is that $\mu^B = \mu^A$, where μ is chemical potential. However, consistent with our assumptions herein (in particular, the absence of virial interaction terms due to the ideal gas assumption), this condition reduces to (10.15).

10.4 WATER SUBSTANCE IN A VACUUM

10.4.1 The Gibbs Potential for Water Substance

Consider a fluid parcel of pure-water substance, so that $q^d \equiv 0$, that is, *there is no dry air*, only water substance with its three possible phases (water vapour, liquid water, and frozen water). This facilitates qualitative (as in Section 10.3) and quantitative (as in this section) examination of phase transitions in the simplest possible context. Since $q^d \equiv 0 \Rightarrow p^d \equiv 0 \Rightarrow p \equiv p^d + p^v = p^v$, setting $q^d \equiv 0$ and $p = p^v$ in (9.41) gives:

The Mass-Weighted Gibbs Potential for Pure-Water Substance

$$g\left(p^v, T, q^v, q^l, q^f\right) = q^v g^v\left(p^v, T\right) + q^l g^l\left(p^v, T\right) + q^f g^f\left(p^v, T\right). \tag{10.16}$$

In (10.16), $g^v\left(p^v, T\right)$, $g^l\left(p^v, T\right)$, and $g^f\left(p^v, T\right)$ are obtained from basic Gibbs potentials (9.17), (9.32), and (9.35) for water vapour, liquid water, and frozen water, respectively.

10.4.2 Saturation, Instability, and Phase Transition

Referring to Fig. 10.1, recall that for an arbitrary point in the vapour region of this (T, p) phase diagram (i.e. below the blue and red saturation curves), thermodynamic equilibrium holds. A small fluctuation may induce a tiny droplet (for $T \geq T_0$) or ice pellet (for $T \leq T_0$) to momentarily form, but this then almost instantaneously vaporises to re-establish thermodynamic equilibrium. Only water vapour can exist (in equilibrium) within the vapour region. Liquid and/or frozen water cannot; they can only coexist with water vapour at a point on the blue (for $T \leq T_0$) and red (for $T \geq T_0$) saturation (also known as *coexistence*) curves of Fig. 10.1.

However, as a point (T, p) approaches (for whatever reason) either of these coexistence curves, the parcel becomes less stable to any fluctuation of its state variables. For a small enough fluctuation it still remains stable (in its purely vapour form), but a larger fluctuation can lead to saturation. At saturation, when a point (T, p) lies on either of the two coexistence curves, a dramatic *instability* is triggered within the air parcel. This instability causes a phase transition of excess water vapour, to either liquid or frozen form, to take place; see Chapter 9 of Callen (1985) for details. A transition from water vapour to liquid water physically takes place almost instantaneously. A transition from water vapour to frozen water can take place more slowly, but for simplicity it is assumed herein that it takes place almost instantaneously and also in the absence of any supercooled liquid water.

After phase transition, the parcel is again in a state of thermodynamic equilibrium, but a different one to that prior to transition. Prior to transition, water vapour was present in the parcel, but liquid and frozen water were entirely absent. After transition, both water vapour and liquid or frozen water coexist, in thermodynamic equilibrium, within the parcel.

10.4.3 Discontinuity of Thermodynamic Potentials during Phase Transition

Phase transitions can only take place at a point (T_c, p_c) that lies on one of the three coexistence curves depicted on Fig. 10.1, where subscript 'c' denotes coexistence.

During a phase transition at such a point, conditions (10.13)–(10.15) for coexistence of two phases continue to hold. Thus – from (10.13) and (10.14) – T_c and p_c remain constant during phase transition; any change of latent internal energy (heat) is entirely expended to effectuate phase transition, with none used to change T_c or p_c. Furthermore – from (10.15) – the specific Gibbs potential (g) for the two phases (A and B, say) that can coexist there are equal. Thus $g^B(T_c, p_c) = g^A(T_c, p_c)$ at such a point *before*, *during*, and *after* phase transition. Therefore:

The Continuity of the Gibbs Potential during Phase Transition

The Gibbs potential (g) maintains its continuity between phases A and B during a phase transition.

However, for other thermodynamic potentials, such as specific internal energy ($\mathcal{E}$), specific Helmholtz free energy (f), and specific enthalpy (h), this is not the case:

The Discontinuity of Other Thermodynamic Potentials

Discontinuity between phases A and B of thermodynamic potentials (the Gibbs potential excepted) at a point (T_c, p_c) on a coexistence curve is the signature of a phase transition taking place there (Callen, 1985, p. 221).

Before transition at a point (T_c, p_c) on a coexistence curve, thermodynamic equilibrium holds. *All* thermodynamic potentials Ψ (say) then satisfy $\Psi^B(T_c, p_c) = \Psi^A(T_c, p_c)$. *During* transition, *only* the Gibbs potential, g, maintains continuity of its phases. Thus, during phase transition, $g^B(T_c, p_c) = g^A(T_c, p_c)$, but $\Psi^B(T_c, p_c) \neq \Psi^A(T_c, p_c)$ for other thermodynamic potentials. For example, $h^B(T_c, p_c) \neq h^A(T_c, p_c)$ – see (10.19) – and so the enthalpies of the two phases are discontinuous *during* (but only during) phase transition.

After phase transition, thermodynamic equilibrium is re-established. Consequently all thermodynamic potentials regain continuity between their phases (i.e. $\Psi^B(T_c, p_c) = \Psi^A(T_c, p_c)$ again for all Ψ).

10.4.4 Discontinuity of Enthalpy/Entropy and Latent Internal Energy/Heat

Of particular interest and importance is the discontinuity in *specific enthalpy* (h) that arises *during* phase transition; this discontinuity measures the amount (per unit mass) of latent internal energy (i.e. latent heat) released to, or absorbed by a parcel due to a phase transition. This, at first sight, is somewhat counter-intuitive. One might expect it would be the discontinuity in *specific internal energy* ($\mathcal{E}$), rather than that of *specific enthalpy*, that would be the key, physical quantity. The underlying reason why this is not the case is that any release or absorption of latent internal energy not only results (as intuitively expected) in a change of specific internal energy, but also (via mechanical work done) in a change in *specific volume* (α); see Section 4.2.4. Specific enthalpy includes the *additional* impact of a change in volume caused by a change in specific internal energy. This property is also related to, and consistent with the fact that in the total-energy budget equation (15.71) of Section 15.3.7, it is the flux of *specific enthalpy* (h) that appears rather than that of *specific internal energy* ($\mathcal{E}$).

By our assumption of thermodynamic equilibrium, conditions (10.13)–(10.15) simultaneously hold for all phases of a substance. They also hold during a phase transition between phases A and B of a substance but, additionally, the temperature (T) and pressure (p) in (10.13) and (10.14), respectively, remain constant during the transition. Taking the difference of definitions (9.3) and (9.7) for specific enthalpy (h) and specific Gibbs potential (g), respectively, gives

The Gibbs Potential in Terms of h, T and η

$$g \equiv h - T\eta. \tag{10.17}$$

Applying condition (10.15) at constant T (the constant temperature *during* transition), with use of identity (10.17), then yields

$$g^B \equiv h^B - T\eta^B = h^A - T\eta^A \equiv g^A, \tag{10.18}$$

$$\Downarrow$$

$$\mathbb{L}^{A\to B} \equiv h^B - h^A = T\left(\eta^B - \eta^A\right) \neq 0. \tag{10.19}$$

In (10.19), $\mathbb{L}^{A\to B}$ is *specific latent heat* for the transition from phase A of a substance to phase B. It measures the amount of heat required to completely change unit mass of phase A of a substance to phase B. The terminology ('enthalpy of' or 'latent heat of') used for $\mathbb{L}^{A\to B}$ depends upon what A and B denote; see Table 10.1, and also Fig. 10.2, for this. From (10.19) and the fact that $T = T_c =$ constant during phase transition, it is seen that the discontinuity $(h^B - h^A)$ in specific enthalpy corresponds to a discontinuity $(\eta^B - \eta^A)$ in specific entropy.

Note that

$$\mathbb{L}^{f\to v} \equiv h^v - h^f \equiv \left(h^l - h^f\right) + \left(h^v - h^l\right) \equiv \mathbb{L}^{f\to l} + \mathbb{L}^{l\to v} \equiv -\mathbb{L}^{v\to f}, \tag{10.20}$$

$$\mathbb{L}^{v\to f} \equiv h^f - h^v \equiv \left(h^l - h^v\right) + \left(h^f - h^l\right) \equiv \mathbb{L}^{v\to l} + \mathbb{L}^{l\to f} \equiv -\mathbb{L}^{f\to v}. \tag{10.21}$$

Thus, from (10.20), the enthalpy of sublimation ($\mathbb{L}^{f\to v}$) is the sum of the enthalpies of fusion ($\mathbb{L}^{f\to l}$) and vaporisation ($\mathbb{L}^{l\to v}$). Physically, this means that it takes the same amount of energy (heat) to first turn unit mass of frozen water into liquid water (via melting), and then turn this liquid water into water vapour (via evaporation), as it does to directly turn unit mass of frozen water into water vapour (via sublimation).

Similarly, from (10.21), the enthalpy of deposition ($\mathbb{L}^{v\to f}$) is the sum of the enthalpies of condensation ($\mathbb{L}^{v\to l}$) and freezing ($\mathbb{L}^{l\to f}$). Physically, this means that the same amount of energy (latent heat) is released to the parcel by first turning unit mass of water vapour into liquid water (via condensation) and then turning this liquid water into frozen water (via freezing), as is released to the parcel by directly turning unit mass of water vapour into frozen water (via deposition).

$\mathbb{L}$		Definition(s)			Terminology		
$\mathbb{L}^{l\rightarrow v}$	$\equiv$	$h^{v}-h^{l}$		$=$	enthalpy of vaporisation	$=$	latent heat of vaporisation
						$=$	latent heat of evaporation
$\mathbb{L}^{f\rightarrow l}$	$\equiv$	$h^{l}-h^{f}$		$=$	enthalpy of fusion	$=$	latent heat of fusion
$\mathbb{L}^{f\rightarrow v}$	$\equiv$	$h^{v}-h^{f}$		$=$	enthalpy of sublimation	$=$	latent heat of sublimation
$\mathbb{L}^{v\rightarrow l}$	$\equiv$	$h^{l}-h^{v}$	$\equiv -\mathbb{L}^{l\rightarrow v}$	$=$	enthalpy of condensation	$=$	latent heat of condensation
$\mathbb{L}^{l\rightarrow f}$	$\equiv$	$h^{f}-h^{l}$	$\equiv -\mathbb{L}^{f\rightarrow l}$	$=$	enthalpy of freezing	$=$	latent heat of freezing
$\mathbb{L}^{v\rightarrow f}$	$\equiv$	$h^{f}-h^{v}$	$\equiv -\mathbb{L}^{f\rightarrow v}$	$=$	enthalpy of deposition	$=$	latent heat of deposition

Table 10.1 Terminology for enthalpies and latent heats associated with phase transitions between vapour (v), liquid (l), and frozen (f) phases of substances. They are grouped into two sets of three, the second acting in the opposite sense to that of the first.

10.4.5 Enthalpies of Vaporisation, Fusion, and Condensation

For the basic Gibbs potentials, developed in Section 9.3 and appearing in the composite Gibbs potential (10.16) for water substance, the specific enthalpies for water vapour, liquid water, and frozen water are given in Tables 9.4 and 9.5. Thus:

Specific Enthalpies of Water Vapour, Liquid Water, and Frozen Water for the Basic Gibbs Potentials of Pure-Water Substance

The specific enthalpies of water vapour, liquid water, and frozen water for pure-water substance are

$$h^{v}\left(T\right)=L_{0}^{v}+L_{0}^{f}+c_{p}^{v}T, \tag{10.22}$$

$$h^{l}\left(T,p\right)=L_{0}^{f}+c^{l}T+\alpha^{l}p, \tag{10.23}$$

$$h^{f}\left(T,p\right)=c^{f}T+\alpha^{f}p, \tag{10.24}$$

respectively.[a]

a In (10.23) and (10.24), p is the total pressure exerted by the environment on a closed fluid parcel. For a fluid parcel composed only of liquid or frozen water, enclosed by an environment similarly composed, $p=p^{v}$. For a fluid parcel composed only of liquid or frozen water, enclosed by an environment that additionally includes dry air, $p=p^{d}+p^{v}$.

Inserting these expressions for h^{v}, h^{l}, and h^{f} into the definitions of the enthalpies of vaporisation ($\mathbb{L}^{l\rightarrow v}$), fusion ($\mathbb{L}^{f\rightarrow l}$), and condensation ($\mathbb{L}^{f\rightarrow v}$) – see Table 10.1 – yields:

The Enthalpies of Vaporisation, Fusion, and Condensation for the Basic Gibbs Potentials of Pure-Water Substance

The enthalpies of vaporisation, fusion, and condensation for pure-water substance are

$$\mathbb{L}^{l\rightarrow v}\left(T,p\right)\equiv h^{v}-h^{l}=L_{0}^{v}+\left(c_{p}^{v}-c^{l}\right)T-\alpha^{l}p, \tag{10.25}$$

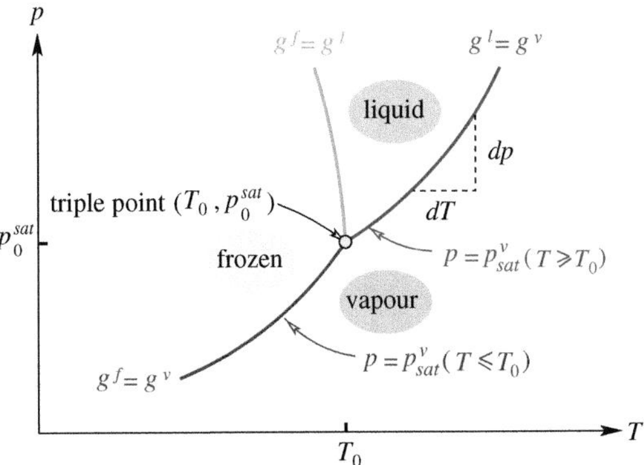

Figure 10.3 Phase diagram (not to scale) for water substance for different conditions of temperature (T) and pressure (p). There are three regions (for vapour, liquid, and frozen phases) where only one phase can exist, separated by three coexistence curves along which two phases can coexist. These three curves intersect at the triple point of water, where $(p^v, T) = (p_0^{sat}, T_0)$, the only point where all three phases of water can coexist in equilibrium. Increment dp denotes the change in p when T is changed by an amount dT for a point constrained to lie on the (red) liquid–vapour coexistence curve. See text for further interpretation of this diagram.

$$\mathbb{L}^{f \to l}(T,p) \equiv h^l - h^f = L_0^f + \left(c^l - c^f\right) T + \left(\alpha^l - \alpha^f\right) p, \tag{10.26}$$

$$\mathbb{L}^{f \to v}(T,p) \equiv h^v - h^f = L_0^v + L_0^f + \left(c_p^v - c^f\right) T - \alpha^f p, \tag{10.27}$$

respectively.

As a cross-check, it is seen that (10.25)–(10.27) satisfy (10.20) and (10.21).

10.4.6 Slopes of Coexistence Curves and Clapeyron's Equation

The form of the three coexistence curves (depicted on the phase diagrams of Figs. 10.1 and 10.3) is determined by the properties of the two phases that coexist along each of them. All three curves emanate from the triple point of water, *the only point where all three phases can coexist.* At the triple point (i.e. at the intersection of the three coexistence curves), $g^v = g^l = g^f$ from (9.36). As noted in Section 9.3.6, these two conditions are automatically satisfied by constructions (9.17), (9.32), and (9.35) for the Gibbs potentials g^v, g^l, and g^f, respectively.[18] Slopes along all three of these coexistence curves are constrained by Clapeyron's equation,[19] as now shown.

Consider a coexistence curve that separates the region of pure phase B from that of pure phase A, and along which phases A and B coexist. For example, the (red) liquid–vapour curve on Fig. 10.3 corresponds to $(A, B) = (l, v)$; liquid water (denoted by l) and water vapour (by v) coexist along this curve, with (only) liquid water existing to the left, and (only) water vapour to the right of the curve. For thermodynamic equilibrium to hold along such a curve, the condition $g^B = g^A$ must be satisfied (as depicted on Fig. 10.3); see (10.15). Consequently the infinitesimal changes dg^A and dg^B of g^A and g^B, respectively, for two neighbouring points an infinitesimal distance apart along the curve, must be equal. Thus

[18] By these constructions, $g^v = g^l = g^f = 0$.

[19] After Benoît Paul Émile Clapeyron (1799–1864).

$$dg^B = dg^A, \tag{10.28}$$

along the coexistence curve for which phases A and B coexist. These changes correspond to infinitesimal changes dT and dp in T and p, respectively, as depicted in Fig. 10.3 for the (red) liquid–vapour coexistence curve.

Applying $dg = \alpha dp - \eta dT$ for each phase – see fundamental thermodynamic relation (9.8) with $dS^i \equiv 0$ since the parcel is assumed closed – gives

$$dg^B = \alpha^B dp - \eta^B dT, \quad dg^A = \alpha^A dp - \eta^A dT. \tag{10.29}$$

Insertion of these two equations into (10.28), with use of definition (10.19) of specific latent heat ($\mathbb{L}^{A\to B}$), then yields Clapeyron's equation for dp/dT, as given in the highlighted box.

Clapeyron's Equation for *dp/dT*

- Clapeyron's equation for dp/dT is

$$\left(\frac{dp}{dT}\right)^{A\to B} = \frac{\eta^B - \eta^A}{\alpha^B - \alpha^A} = \frac{h^B - h^A}{T\left(\alpha^B - \alpha^A\right)} = \frac{\mathbb{L}^{A\to B}}{T\left(\alpha^B - \alpha^A\right)}. \tag{10.30}$$

- It holds along the coexistence curve for phases A and B of a substance, and $\mathbb{L}^{A\to B}$ is termed *specific latent heat for transition from phase A to phase B*.
- Rewriting (10.30) explicitly for each of the three coexistence curves for water substance, followed by use of (10.25)–(10.27) for the basic Gibbs potentials defined in Section 9.3, delivers

$$\underbrace{\left(\frac{dp}{dT}\right)^{l\to v} = \frac{\mathbb{L}^{l\to v}}{T\left(\alpha^v - \alpha^l\right)} = \frac{L_0^v + \left(c_p^v - c^l\right)T - \alpha^l p}{T\left(\alpha^v - \alpha^l\right)},}_{\text{liquid to vapour}} \tag{10.31}$$

$$\underbrace{\left(\frac{dp}{dT}\right)^{f\to l} = \frac{\mathbb{L}^{f\to l}}{T\left(\alpha^l - \alpha^f\right)} = \frac{L_0^f + \left(c^l - c^f\right)T + \left(\alpha^l - \alpha^f\right)p}{T\left(\alpha^l - \alpha^f\right)},}_{\text{frozen to liquid}} \tag{10.32}$$

$$\underbrace{\left(\frac{dp}{dT}\right)^{f\to v} = \frac{\mathbb{L}^{f\to v}}{T\left(\alpha^v - \alpha^f\right)} = \frac{L_0^v + L_0^f + \left(c_p^v - c^f\right)T - \alpha^f p}{T\left(\alpha^v - \alpha^f\right)}.}_{\text{frozen to vapour}} \tag{10.33}$$

- In (10.31)–(10.33), the middle expressions are *general*, but the rightmost ones are *particular* to the basic potentials defined in Section 9.3, for which c_p^v, c^l, c^f, α^l, and α^f (but *not* α^v) are all constants.

10.4.7 Determination of Coexistence Curves and the Clausius–Clapeyron Approximation

Each of (10.31)–(10.33) is a first-order, ordinary-differential equation that can, in principle, be solved using the boundary condition that the triple point lies on the coexistence curve; see Fig. 10.3. In other words, (10.31)–(10.33) can be solved *outwards* from the triple point to determine the three coexistence curves. In *general* (using the middle expressions of (10.31)–(10.33)), this may or may not be straightforward (or even analytically tractable), depending upon the properties of any particular representation of the phases of water substance. The coexistence curves for a *particular* (analytically tractable) case (using the rightmost expressions of (10.31)–(10.33), with the basic potentials defined as in Section 9.3) are now examined. We start with the easiest case before moving on to the two more complicated ones.

10.4.7.1 Liquid–Frozen Coexistence Curve

For the *particular* liquid–frozen case, (10.32) (obtained using the basic potentials defined in Section 9.3) can be solved *exactly* subject to the triple-point condition $p(T_0) = p_0^{sat}$. As is easily verified, this gives:

The Exact Equation for the Particular Liquid–Frozen Coexistence Curve

$$p(T) = p_0^{sat}\frac{T}{T_0} + \frac{1}{(\alpha^f - \alpha^l)}\left[L_0^f\left(1 - \frac{T}{T_0}\right) - \left(c^l - c^f\right)T\ln\frac{T}{T_0}\right]. \qquad (10.34)$$

This solution can be obtained by using standard techniques for solving a first-order differential equation of the form of (10.32). Alternatively, it can be obtained more directly by simply applying $g^f = g^l$ along the green coexistence curve depicted in Fig. 10.3, where g^l and g^f are given by (9.32) and (9.35), respectively; the resulting equation is then solved for $p(T)$ to give (10.34).

The slope of the liquid–frozen coexistence curve, expressed in terms of T (only), can be obtained by either substituting (10.34) into (10.32) and rearranging, or by differentiating (10.34) with respect to T. Either way, this leads to:

The Exact Slope of the Particular Liquid–Frozen Coexistence Curve

$$\left(\frac{dp}{dT}\right)^{f\to l} = \underset{\substack{2.24\\ \mathrm{m^{-3}\,J\,K^{-1}}}}{\frac{p_0^{sat}}{T_0}} - \underset{\substack{1.1\times10^4\\ \mathrm{m^{-3}\,kg}}}{\frac{1}{(\alpha^f - \alpha^l)}}\left\{\underset{\substack{-0.86\times10^3\\ \mathrm{J\,kg^{-1}\,K^{-1}}}}{\frac{L_0^f}{T_0}} + \underset{\substack{2\times10^3\\ \mathrm{J\,kg^{-1}\,K^{-1}}}}{\left(c^l - c^f\right)}\left[1 + \ln\left(\frac{T}{T_0}\right)\right]\right\}, \qquad (10.35)$$

where some rough estimates for the values of various terms on the right-hand side have been inserted below them using values given in Table 9.3. See footnote 13 of Chapter 9 for why L_0^f/T_0 is negative.

Since (from Table 9.3) $\alpha^f > \alpha^l$ and $c^l > c^f$, (10.35) shows that the liquid–frozen coexistence curve (defined by (10.34)) is almost vertical with a slight backwards tilt to it; see the green curve on Fig. 10.3. Note that *water substance behaves anomalously in this regard.* For most other substances, the specific volume of their frozen phase is generally *less* than that of their liquid phase (i.e. $\alpha^f < \alpha^l$), whereas for water substance it is the converse. This difference in sign of $\left(\alpha^f - \alpha^l\right)$ means that other substances (e.g. carbon dioxide or methane for some cold planets and moons) usually have a *forwards* (instead of *backwards*) tilt, with some substances having stronger tilts than others.

An interesting consequence of this (anomalous) backwards tilt is what happens to a liquid-water parcel in the liquid region of Fig. 10.3 when it is held at constant temperature (T), whilst its pressure (p) is continually reduced. (This corresponds to a point moving vertically downwards on the figure.) For $T \geq T_0$, the point $\left(T \geq T_0, p\right)$ intersects the (red) vapour–liquid coexistence curve; consequently the liquid water then *evaporates* to water vapour. For $T < T_0$, however, the point $\left(T < T_0, p\right)$ first intersects the (green) liquid–frozen coexistence curve, whereupon the liquid water instead *freezes*. Further reducing p then leads to intersection with the (blue) vapour–frozen coexistence curve; consequently the frozen water then *sublimates* to water vapour (without passing through the liquid phase). In both circumstances (for constant $T \geq T_0$ or $T < T_0$, with a continual reduction in p), the parcel starts off as liquid water and ends up as water vapour. However, in the first case (for $T \geq T_0$) it does so *directly*, whereas for the second one ($T < T_0$), it does so by first freezing.

10.4.7.2 *Vapour–Liquid Coexistence Curve*

For the *particular* vapour–liquid case, things are not as simple as for the particular liquid–frozen case developed in Section 10.4.7.1. The principal reason for this is the appearance of α^v in (10.31); α^v is *not* constant, in contradistinction to α^f and α^l in (10.32). However – see Table 9.4 – α^v satisfies the ideal-gas law

$$\alpha^v = \frac{R^v T}{p^v}. \tag{10.36}$$

For the pure-water-substance case under consideration here, $p = p_s^v$ in (10.31) since $p^d = 0 \Rightarrow p = p^d + p_s^v = p_s^v$. (For notational brevity, p_{sat}^v (the value of p^v on the vapour–liquid coexistence curve) has been abbreviated to p_s^v, both here and onwards.) Substitution of (10.36) into (10.31) (with $p = p_s^v$) then gives:

The Differential Equation for the Particular Vapour–Liquid Coexistence Curve

$$\left(\frac{1}{p_s^v}\frac{dp_s^v}{dT}\right)^{l\to v} = \frac{L_0^v + \left(c_p^v - c^l\right)T - \alpha^l p_s^v}{T\left(R^v T - \alpha^l p_s^v\right)}, \quad T \geq T_0. \tag{10.37}$$

This first-order differential equation is more challenging to solve than (10.32). Rather than solving it directly as a differential equation (which is complicated, but nevertheless feasible), we instead adopt the alternative approach given earlier for solving (10.32). Since vapour and liquid are in thermal equilibrium along the vapour–liquid coexistence curve (see the red curve on Fig. 10.3), $g^l = g^v$ along it, where g^v and g^l are given by (9.17) and (9.32), respectively. Substituting these two equations into $g^l = g^v$, followed by rearrangement, then yields:

The Exact Equation for the Particular Vapour–Liquid Coexistence Curve

$$p_s^v = p_0^{sat}\left(\frac{T}{T_0}\right)^{\frac{\left(c_p^v - c^l\right)}{R^v}} \exp\left[-\frac{L_0^v}{R^v T}\left(1 - \frac{T}{T_0}\right) + \frac{\alpha^l}{R^v T}\left(p_s^v - p_0^{sat}\frac{T}{T_0}\right)\right], \quad T \geq T_0. \tag{10.38}$$

Equation (10.38) for the vapour–liquid coexistence curve can be compared with its counterpart, (10.34), for the liquid–frozen one. It is seen that whereas (10.34) is an *explicit* equation for $p(T)$, (10.38) is an *implicit* transcendental equation for $p_s^v(T \geq T_0)$. By differentiating (10.38), followed by rearrangement, it can be verified that (10.38) does indeed satisfy first-order differential equation (10.37) and boundary condition $p_s^v(T = T_0) = p_0^{sat}$ at the triple point.

To obtain $p_s^v(T \geq T_0)$, transcendental equation (10.38) has to be solved *numerically*. This can be done to machine precision using a suitable numerical algorithm, such as Newton–Raphson iteration[20] (with generally rapid, quadratic convergence); see, for example, Chapter 9 of Press et al. (1992). Although this is less convenient than having an explicit equation available – cf. (10.34) for the liquid–frozen case – there is nevertheless no loss of accuracy.

10.4.7.3 *Vapour–Frozen Coexistence Curve*

Because of the functional similarity between (10.31) and (10.33), the procedures used in Section 10.4.7.2 for the vapour–liquid coexistence curve can also be employed for the vapour–frozen one. Doing so then leads to:

The Differential Equation for the Particular Vapour–Frozen Coexistence Curve

$$\left(\frac{1}{p_s^v}\frac{dp_s^v}{dT}\right)^{f\to v} = \frac{\left(L_0^v + L_0^f\right) + \left(c_p^v - c^f\right)T - \alpha^f p_s^v}{T\left(R^v T - \alpha^f p_s^v\right)}, \quad T \leq T_0, \tag{10.39}$$

and to:

The Exact Equation for the Particular Vapour–Frozen Coexistence Curve

$$p_s^v = p_0^{sat}\left(\frac{T}{T_0}\right)^{\frac{\left(c_p^v - c^f\right)}{R^v}} \exp\left[-\frac{\left(L_0^v + L_0^f\right)}{R^v T}\left(1 - \frac{T}{T_0}\right) + \frac{\alpha^f}{R^v T}\left(p_s^v - p_0^{sat}\frac{T}{T_0}\right)\right], \quad T \leq T_0. \tag{10.40}$$

These two equations are the counterparts to (10.37) and (10.38), respectively. Solution (10.40) satisfies the boundary condition $p_s^v(T = T_0) = p_0^{sat}$ at the triple point, as it must.

10.4.7.4 *Clausius–Clapeyron Approximation for the Vapour–Liquid Coexistence Curve*

Examination of (10.38) shows that its implicitness is due to the presence of the term $\alpha^l\left(p_s^v - p_0^{sat}T/T_0\right)/\left(R^v T\right)$ on its right-hand side. *If this term could be neglected*, then (10.38) would simplify to the *explicit* equation

20 After Isaac Newton (1643–1726) and Joseph Raphson (circa 1648–circa 1715).

$$p_s^v = p_0^{sat}\left(\frac{T}{T_0}\right)^{\frac{(c_p^v - c^l)}{R^v}} \exp\left[-\frac{L_0^v}{R^v T}\left(1-\frac{T}{T_0}\right)\right], \quad T \geq T_0. \tag{10.41}$$

(A similar remark also applies to (10.40) for the term $\alpha^f \left(p_s^v - p_0^{sat} T/T_0\right) / (R^v T)$.) The question now is:

- Under what circumstances (if any) is neglect of this term justifiable?

To examine this question we return to (10.31). Observe that:

- $\alpha^l/\alpha^v \equiv \rho^v/\rho^l$ is the ratio of the density of water vapour to that of liquid water, with an approximate value at the triple point of water of $\alpha^l/\alpha^v \approx 10^{-3}/206 \approx 0.5 \times 10^{-5} \ll 1$.
- Using this estimate for the value of α^l/α^v, and also using (10.36), gives

$$\frac{\alpha^l p_s^v}{\left|c_p^v - c^l\right| T} = \left(\frac{\alpha^l}{\alpha^v}\right)\frac{R^v}{\left|c_p^v - c^l\right|} \approx \left(\frac{1}{2}\times 10^{-5}\right)\times\frac{1}{5} = 10^{-6} \ll 1. \tag{10.42}$$

This then justifies neglect of $\alpha^l p$ in (10.31).[21] Doing so then leads to:

The Clausius–Clapeyron Approximation of the Slope of the Vapour–Liquid Coexistence Curve

- The Clausius–Clapeyron approximation[a] of the slope of the vapour–liquid coexistence curve (10.31) is

$$\left(\frac{dp_s^v}{dT}\right)^{l\to v} \approx \frac{\mathbb{L}^{l\to v}}{R^v T^2} p_s^v \approx \frac{\left[L_0^v + \left(c_p^v - c^l\right)T\right]}{R^v T^2} p_s^v, \quad T \geq T_0, \tag{10.43}$$

  where use has been made of (10.36) to eliminate α^v in favour of p_s^v and T.
- The middle expression in (10.43) is *general*, but the rightmost one is *particular* to the basic potentials defined in Section 9.3.

a After Rudolf Clausius (1822–1888) and Benoît Paul Émile Clapeyron (1799–1864).

Solution of (10.43) (using its rightmost expression), subject to the triple-point condition $p_s^v(T_0) = p_0^{sat}$, yields (10.41). Success! Taking into account differences in notation,[22] (10.43) and (10.41) correspond to equations (5.16) and (5.17), respectively, of Ambaum (2010).

10.4.7.5 *Clausius–Clapeyron Approximation for the Vapour–Frozen Coexistence Curve*

Similar approximations to those just given may also be made to (10.33) for the vapour–frozen coexistence curve. Doing so then leads to:

[21] Neglect of $\alpha^l p$ in $\mathbb{L}^{l\to v}$ (enthalpy of vaporisation) in (10.31) (originating from (10.25)) is a frequently used approximation in the meteorological literature (Staniforth and White, 2019). Its neglect is a crucially important step for the development (using the Clausius–Clapeyron approximation) of a highly accurate *explicit* representation of the vapour–liquid curve for pure-water substance.

[22] $p_s^v \to e_s$, $p_0^{sat} \to e_s^0$, $L_0^v \to L_0 + \left(c^l - c_p^v\right) T_0$.

The Clausius–Clapeyron Approximation of the Slope of the Vapour–Frozen Coexistence Curve

- The Clausius–Clapeyron approximation of the slope of the vapour–frozen coexistence curve (10.33) is

$$\left(\frac{dp_s^v}{dT}\right)^{f\to v} \approx \frac{\mathbb{L}^{f\to v}}{R^v T^2}p_s^v \approx \frac{\left[\left(L_0^v + L_0^f\right) + \left(c_p^v - c^f\right)T\right]}{R^v T^2}p_s^v, \quad T \le T_0, \tag{10.44}$$

where use has been made of (10.36) to eliminate α^v in favour of p_s^v and T.
- The middle expression in (10.44) is *general*, but the rightmost one is *particular* to the basic potentials defined in section 9.3.

Solution of (10.44) (using its rightmost expression), subject to the triple-point condition $p_s^v(T_0) = p_0^{sat}$, yields

$$p_s^v = p_0^{sat}\left(\frac{T}{T_0}\right)^{\frac{(c_p^v - c^f)}{R^v}} \exp\left[-\frac{\left(L_0^v + L_0^f\right)}{R^v T}\left(1 - \frac{T}{T_0}\right)\right], \quad T \le T_0. \tag{10.45}$$

Equations (10.44) and (10.45) for the vapour–frozen coexistence curve are the counterparts of (10.43) and (10.41), respectively, for the vapour–liquid one.

10.4.7.6 *Slopes of the Vapour–Frozen and Vapour–Liquid Coexistence Curves at the Triple Point*

On Figs. 10.1 and 10.3, the slope of the vapour–frozen coexistence curve at the triple point has been depicted as being steeper than that of the vapour–liquid one there.

- Is this in fact true?

To answer this question we evaluate Clausius–Clapeyron approximations (10.43) and (10.44) at the triple point and compare them.

Evaluating (10.43) and (10.44) at the triple point $(p_s^v, T) = \left(p_0^{sat}, T_0\right)$ leads to

$$\left(\frac{dp_s^v}{dT}\right)^{l\to v} \approx \frac{p_0^{sat}}{R^v T_0^2}\left[L_0^v + \left(c_p^v - c^l\right)T_0\right], \tag{10.46}$$

$$\left(\frac{dp_s^v}{dT}\right)^{f\to v} \approx \frac{p_0^{sat}}{R^v T_0^2}\left[\left(L_0^v + L_0^f\right) + \left(c_p^v - c^f\right)T_0\right]. \tag{10.47}$$

Taking the difference of (10.46) and (10.47) then gives

$$\left(\frac{dp_s^v}{dT}\right)^{f\to v} - \left(\frac{dp_s^v}{dT}\right)^{l\to v} \approx \frac{p_0^{sat}}{R^v T_0^2}\left[L_0^f + \left(c^l - c^f\right)T_0\right] > 0, \tag{10.48}$$

since L_0^f, $\left(c^l - c^f\right)$, and T_0 are all strictly positive. Thus:

Slopes of the Vapour–Frozen and Vapour–Liquid Coexistence Curves at the Triple Point of Water Substance

The slope of the vapour–frozen coexistence curve at the triple point is steeper than that of the vapour–liquid one there.

This is an example of the usefulness of the Clausius–Clapeyron approximation of Clapeyron's equation. To obtain this result from (10.37) and (10.39) would be less straightforward.

10.4.7.7 Thermodynamical Consistency

Recall (from Section 9.1) that to guarantee thermodynamical consistency, approximations should be made, *once and for all*, to the Gibbs potential, with everything thermodynamic then being obtained from this approximated potential, *without further approximation*. Making ad hoc approximations here and there – *even quantitatively excellent ones* – risks introducing inconsistency and *systematic cumulative errors*.

The development (using the Clausius–Clapeyron approximation) of the explicit representations (10.41) and (10.45) of the vapour–liquid and vapour–frozen coexistences curves, respectively, has, however, been achieved in an ad hoc manner. Thermodynamical consistency with other thermodynamic relations is therefore not guaranteed (and is, in fact, violated – as shown later in this section). The question thus arises:

- Can (10.41) and (10.45) be obtained in a thermodynamically consistent manner via suitable approximation of the basic Gibbs potentials (g^v, g^l, and g^f)?

Indeed they can. This is achieved by neglecting up front the $\left[p - p_0^{sat}\left(T/T_0\right)\right]$ terms in both (9.32) and (9.35) – for the basic Gibbs potentials of pure liquid water and pure frozen water, respectively – by formally setting $\alpha^l \equiv \alpha^f \equiv 0$.[23] Thus (with (9.17) for $g^v\left(p^v, T\right)$ left unchanged)

$$g^l_{\text{Approximated}}\left(T\right) = -c^l T \ln\left(\frac{T}{T_0}\right) + L_0^f\left(1 - \frac{T}{T_0}\right), \tag{10.49}$$

$$g^f_{\text{Approximated}}\left(T\right) = -c^f T \ln\left(\frac{T}{T_0}\right), \tag{10.50}$$

$$g^v\left(p^v, T\right) = -c_p^v T \ln\left(\frac{T}{T_0}\right) + R^v T \ln\left(\frac{p^v}{p_0^{sat}}\right) + \left(L_0^v + L_0^f\right)\left(1 - \frac{T}{T_0}\right). \tag{10.51}$$

Setting $g^l_{\text{Approximated}}\left(T\right) = g^v\left(p^v, T\right)$ and $g^f_{\text{Approximated}}\left(T\right) = g^v\left(p^v, T\right)$ (the conditions for coexistence when using (10.49)–(10.51)) leads directly, as desired, to (10.41) and (10.45) for the vapour–liquid and vapour–frozen coexistence curves, respectively.[24]

Although the two coexistence curves (10.41) and (10.45) have been obtained directly from the approximated Gibbs potentials (10.49) and (10.50), the same cannot be said for the derivation (from the *unapproximated* Gibbs potentials (9.32) and (9.35)) of the third coexistence equation,

[23] Although it appears that setting $\alpha^l \equiv \alpha^f \equiv 0$ in this context means that the volumes occupied by liquid and/or frozen water are identically zero, this is not the case. The correct interpretation is that these occupied volumes are so minute that the terms they appear in can be neglected with respect to much larger terms. Formally setting $\alpha^l \equiv \alpha^f \equiv 0$ is simply a convenient means of neglecting minutely small terms instead of achieving the same result via asymptotic analysis.

[24] An alternative procedure is to set $\alpha^l \equiv \alpha^f \equiv 0$ in (10.22)–(10.27), and similarly in the equations used to obtain (10.38) and (10.40). This is not only a lot more complicated, it introduces a spurious singularity into the derivation of (10.32), which then has to be addressed. This issue is best avoided by using the non-singular method described in the main body of the text!

(10.34). This is *thermodynamically inconsistent*. To address this violation of consistency, the condition $g^l_{\text{Approximated}}(T) = g^f_{\text{Approximated}}(T)$ is applied to obtain the liquid–frozen coexistence curve that *is* consistent with coexistence curves (10.41) and (10.45). Thus

$$L_0^f \left(1 - \frac{T}{T_0}\right) = \left(c^l - c^f\right) T \ln\left(\frac{T}{T_0}\right) \quad \Rightarrow \quad T = T_0, \tag{10.52}$$

instead of the more complicated (10.34). The liquid–frozen coexistence curve $T = T_0$, which passes through the triple-point $\left(T_0, p_0^{sat}\right)$, is now perfectly vertical and has lost its tilt (cf. Fig. 10.3). This loss of tilt is due to setting $\alpha^l \equiv \alpha^f \equiv 0$ whilst respecting thermodynamical consistency.

Consistently using approximate Gibbs potentials (10.49) and (10.50) throughout (instead of (9.32) and (9.35)) also affects other thermodynamic relations. For example, all terms involving α^l or α^f in Table 9.5 are absent, thereby redefining the thermodynamic potentials η^l, η^f, h^l, h^f, f^l, and f^f. Since latent heats are defined in terms of enthalpies (h) – see Table 10.1 – this means that latent heats are also redefined for thermodynamical consistency; see Section 10.6.5.1.

The present simple example illustrates that the procedure of *only approximating the Gibbs potential* and then obtaining everything thermodynamic from its approximated form is a powerful tool for ensuring thermodynamical consistency. A further welcome benefit is that it brings order to, and discrimination between what would otherwise be a bewildering range of possible combinations of approximations with unpredictable consequences. *This rationalises the complicated whilst enhancing understanding.*

10.5 CLOUDY AIR, POSSIBLY CONTAINING LIQUID AND/OR FROZEN WATER

The present analysis extends that of Thuburn (2017) for moist air (composed of humid air + liquid water) by additionally including frozen water. This follows on from his remark that doing so is a minimum requirement for weather forecasting and climate modelling.

10.5.1 Definition of Cloudy Air

For present purposes, and as in Section 4.3, *cloudy air* (also known as *mixed-phase cloud*) is defined to be composed of a mixture of humid air (dry air + water vapour, as in Section 10.2) + liquid water (droplets) + frozen water (ice crystals).

From (4.53) or (9.38), the total mass of a parcel of cloudy air is

$$\mathbb{M} = \mathbb{M}^d + \mathbb{M}^v + \mathbb{M}^l + \mathbb{M}^f, \tag{10.53}$$

where $\mathbb{M}^d$, $\mathbb{M}^v$, $\mathbb{M}^l$, and $\mathbb{M}^f$ are the individual masses of its dry-air, water-vapour, liquid-water, and frozen-water constituents, respectively. From (3.7) or (9.39), the corresponding mass fractions are

$$q^d = \frac{\mathbb{M}^d}{\mathbb{M}}, \quad q^v = \frac{\mathbb{M}^v}{\mathbb{M}}, \quad q^l = \frac{\mathbb{M}^l}{\mathbb{M}}, \quad q^f = \frac{\mathbb{M}^f}{\mathbb{M}}. \tag{10.54}$$

Using (10.53), these mass fractions satisfy (see (4.54) and (4.55))

$$q^d + q^v + q^l + q^f = q^d + q^w = 1, \tag{10.55}$$

where

$$q^w \equiv q^v + q^l + q^f \tag{10.56}$$

is the mass fraction of total water substance in a cloudy-air parcel. Furthermore, from (10.2), (10.54), and (10.55), the mass fraction, a, of dry air in the *gaseous* part of a parcel's total volume[25] satisfies

[25] A parcel's *total* volume additionally includes the volumes (separately) occupied by liquid water and frozen water.

$$a = \frac{\mathbb{M}^d}{\mathbb{M}^d + \mathbb{M}^v} = \frac{q^d}{q^d + q^v} = \frac{1 - q^w}{1 - q^l - q^f}, \tag{10.57}$$

and so

$$1 - a = \frac{\mathbb{M}^v}{\mathbb{M}^d + \mathbb{M}^v} = \frac{q^v}{q^d + q^v} = \frac{q^w - q^l - q^f}{1 - q^l - q^f}. \tag{10.58}$$

10.5.2 The Gibbs Potential for Cloudy Air

For a cloudy-air parcel, possibly containing liquid and/or frozen water, (9.41) may be rewritten as

The Composite Gibbs Potential for Cloudy Air

$$g\left(p, T, q^w\right) = \left(1 - q^l - q^f\right) g^{av}\left(p, T, a\right) + q^l g^l\left(p, T\right) + q^f g^f\left(p, T\right), \tag{10.59}$$

where – from (10.55) and (10.56) –

$$1 - q^l - q^f = 1 - q^w + q^v = q^d + q^v. \tag{10.60}$$

Furthermore:

- $g^{av}\left(p, T, a\right)$ is defined by (10.8)–(10.12).
- $g^l\left(p, T\right)$ and $g^f\left(p, T\right)$ are defined by (9.32) and (9.35), respectively (see also Table 9.5).

Equation (10.59) is a mass-weighted sum of the Gibbs potentials, $g^{av}\left(p, T, a\right)$, $g^l\left(p, T\right)$, and $g^f\left(p, T\right)$, for humid air (dry air + water vapour), liquid water, and frozen water, respectively; the associated mass fractions are $\left(1 - q^l - q^f\right)$, q^l, and q^f, respectively. For the special case of no frozen water (i.e. $q^f \equiv 0$), (10.59) here corresponds to Thuburn (2017)'s equation (51).

As discussed in Sections 10.3.1 and 10.4.7.7, there is only one point (the triple point) on the 2D temperature-pressure (T-p) phase diagram for water substance – see Figs. 10.1 and 10.3 – where water vapour, liquid water, and frozen water can coexist in thermodynamic equilibrium. Furthermore, water vapour must be present for liquid or frozen water to also be present.[26] This simplifies things by reducing the number of possible combinations of phases of water substance that need to be considered. It leads to the realisable combinations (of q^v, q^l, and q^f) summarised in Table 10.2 for the composition of a cloudy-air parcel; humid air and liquid or frozen cloud are special cases. The expressions appearing in this table for a (mass fraction of dry air *in the gaseous part of the parcel*), $1 - a$ (mass fraction of water vapour *in the gaseous part of the parcel*), and g (composite Gibbs potential) are obtained from (10.57), (10.58), and (10.59), respectively. Humid air (by its definition given in Section 10.2) is unsaturated air (i.e. $q^l = q^f = 0$), whereas cloudy air is (in general) saturated (i.e. one or both of q^l and q^f are non-zero). Subscripts 's' on q^v and 'sat' on a emphasise that the values of q^v and a are limited to those occurring at saturation of water vapour.

Importantly, a, $(1 - a)$ and g for *frozen cloud* have the same form as their counterparts for *liquid cloud*: q^l for liquid cloud is replaced by q^f for frozen cloud – see the last three columns of Table 10.2. This means that, with relatively few modifications, frozen cloud can be handled

[26] To maintain thermodynamic equilibrium within the parcel, some liquid or frozen water is necessarily converted to water vapour via a phase transition.

(a)

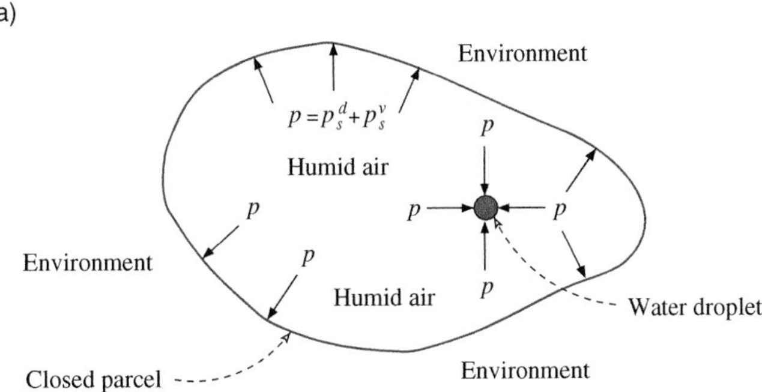

(b)

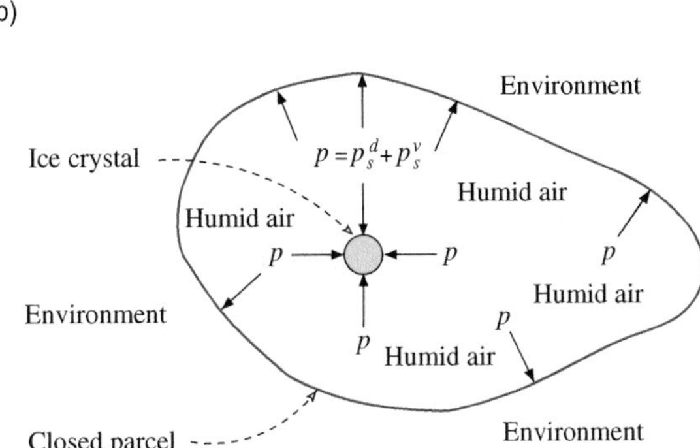

Figure 10.4 Two closed, cloudy-air parcels, composed of a mixture of humid air (dry air + water vapour in a shared volume, in light blue) with (occupying separate volumes): (a) a liquid-water droplet (in magenta); and (b) an ice crystal (in dark blue).
Total pressure, acting within a parcel and also on the ice-crystal or water-droplet boundaries, is $p = p_s^d + p_s^v$, where p_s^d and p_s^v are dry-air and water-vapour partial pressures at saturation, respectively. Arrow heads for p denote total pressure acting on a surface at that point. For a parcel of pure-water substance, $p_s^d \equiv 0$ and $p = p_s^v$.

in a similar manner to that described in Thuburn (2017) for liquid cloud. (This simplification would not be possible if all three phases of water substance could coexist at temperatures and pressures different to those of the triple point of water. That they cannot is due to the assumption of thermodynamic equilibrium.) The triple point of water *in the presence of dry air* does, however, complicate matters and requires special consideration; see Section 10.6.

10.5.3 Saturation for $T \geq T_*$

Consider a closed, *liquid-cloud* parcel; see Fig. 10.4a, with a single water droplet. Note that Fig. 10.4a is not to scale; the volume occupied by a water droplet is many orders of magnitude smaller than that occupied by humid air.

As discussed in Sections 10.3.1 and 10.4.7.2, if a parcel of pure-water substance saturates for $T \geq T_0$, then the saturation vapour pressure, $p_s^v\left(p, T \geq T_0\right)$, lies on the red curve of Figs. 10.1 and 10.3. A similar situation arises for a cloudy-air parcel, except that T_0 (with known value) is replaced by T_* (with value that depends on dry-air pressure, p^d). Liquid cloud exists for $T \geq T_*$, and frozen cloud for $T \leq T_*$. The value of T_* (in the presence of dry air) is very close indeed

Composition	q^v	q^l	q^f	q^w	a	$1-a$	g
Humid air	✓	0	0	q^v	$1-q^w$	q^w	$g^{av}(p,T,a)$
Liquid cloud	✓	✓	0	$q_s^v+q^l$	$\dfrac{1-q^w}{1-q^l}$	$\dfrac{q^w-q^l}{1-q^l}$	$\left(1-q^l\right)g^{av}(p,T,a_{sat})+q^lg^l(p,T)$
Frozen cloud	✓	0	✓	$q_s^v+q^f$	$\dfrac{1-q^w}{1-q^f}$	$\dfrac{q^w-q^f}{1-q^f}$	$\left(1-q^f\right)g^{av}(p,T,a_{sat})+q^fg^f(p,T)$
Liquid & frozen cloud	✓	✓	✓	$q_s^v+q^l+q^f$	$\dfrac{1-q^w}{1-q^l-q^f}$	$\dfrac{q^w-q^l-q^f}{1-q^l-q^f}$	$\left(1-q^l-q^f\right)g^{av}(p_*,T_*,a_{sat})+q^lg^l(p_*,T_*)+q^fg^f(p_*,T_*)$

Table 10.2 Realisable combinations of phases of water substance for a cloudy-air parcel with mass fractions q^v, q^l and q^f; $q^w \equiv q^v + q^l + q^f$ is mass fraction of total water substance in the parcel. Presence of a quantity is denoted by a tick mark (✓), and its absence by zero for its associated mass fraction. Associated expressions for a (mass fraction of dry air *in the gaseous part of the parcel*), $1-a$ (mass fraction of water vapour *in the gaseous part of the parcel*), and g (composite Gibbs potential) are also tabulated; they are obtained from (10.57), (10.58), and (10.59), respectively. Subscripts '*s*' and '*sat*' denote values at saturation of water vapour. Subscript '$*$' denotes evaluation at the triple point of water in the presence of dry air. See text for further details.

to that of T_0 (in the absence of dry air). Equality of T_* and T_0 is in fact assumed in much of the meteorological literature; see Section 10.6 for why, although not precisely true, this is an excellent approximation. The determination of the precise (unapproximated) value of T_* at a specified value, $p^d = p^d_*$ (say), of dry-air pressure, is deferred to Section 10.6. For now we simply assume that T_* can be determined and therefore has known value.

At saturation, condensation of excess water vapour to liquid-water droplets takes place through a phase transition, with an associated release of latent internal energy (heat). Two questions now arise. For specified values of p (total pressure, i.e. the sums of the partial pressures p^d and p^v for dry air and water vapour, respectively) and $T \geq T_*$:

1. At what value of saturation vapour pressure, $p^v_s\left(p, T \geq T_*\right)$, will a humid-air parcel saturate?
2. What is the value of the associated mass fraction of dry air, $a_{sat}\left(p, T \geq T_*\right)$, at saturation?

10.5.3.1 Saturation Vapour Pressure $p^v_s\left(p, T \geq T_*\right)$

For the first question, before saturation the air parcel only contains humid air (i.e. dry air plus a single phase (water vapour) of water substance). At saturation, liquid-water droplets start to form; the first droplet to form is depicted in Fig. 10.4a. The humid-air parcel then becomes a liquid-cloud parcel (since $T \geq T_*$ by hypothesis here) with two phases (water vapour and liquid water) of water substance present.

As discussed in Section 10.3.2, conditions (10.13)–(10.15) for thermodynamic equilibrium at a common boundary between two phases (here, water vapour and liquid water) must be satisfied. Thus (from (10.15))

$$g^v\left(p^v, T\right) = g^l\left(p, T\right), \tag{10.61}$$

at saturation of the liquid-cloud parcel, where $g^v\left(p^v, T\right)$ is defined by (10.10) or (equivalently) by (10.11), and $g^l\left(p, T\right)$ by (9.32). (Since the input arguments of both g^v and g^l are $\left(p, T\right)$, conditions (10.13) and (10.14) are automatically satisfied.) Substitution of (10.10) (for g^v) and (9.32) (for g^l) into (10.61) gives the saturation condition

$$R^v T \ln\left[\frac{p^v_s\left(p, T\right)}{p^{sat}_0}\right] = \left(c^v_p - c^l\right) T \ln\left(\frac{T}{T_0}\right) - L^v_0\left(1 - \frac{T}{T_0}\right) + \alpha^l\left(p - p^{sat}_0 \frac{T}{T_0}\right). \tag{10.62}$$

In (10.62), $p^v_s\left(p, T\right)$ is the value of saturation vapour pressure at the boundary between a liquid-water droplet and its surrounding gaseous volume – see Fig. 10.4a – and $p = p^d + p^v$ is the total gaseous pressure there (i.e. the sum of dry-air and water-vapour pressures). For specified values of p and $T \geq T_*$, solving (10.62) for $p^v_s\left(p, T\right)$ yields:

The Saturation Vapour Pressure for Liquid Cloud ($T \geq T_*$)

$$p^v_s\left(p, T\right) = p^{sat}_0 \left(\frac{T}{T_0}\right)^{\frac{\left(c^v_p - c^l\right)}{R^v}} \exp\left[-\frac{L^v_0}{R^v T}\left(1 - \frac{T}{T_0}\right) + \frac{\alpha^l}{R^v T}\left(p - p^{sat}_0 \frac{T}{T_0}\right)\right]. \tag{10.63}$$

With the exception of p and T, all quantities on the right-hand side of (10.63) are constants of known value. Thus, for specified values of p and $T \geq T_*$, the corresponding saturation vapour pressure, $p^v_s\left(p, T \geq T_*\right)$, can be explicitly determined from (10.63).

As a cross-check, consider now the special case (examined in Section 10.4.7.2) where the parcel is composed *solely* of water substance. Thus: $q^d \equiv 0 \Rightarrow p^d \equiv 0$ (i.e. dry air is absent, so it cannot exert pressure); p in (10.63) reduces to p_s^v; and (reassuringly) transcendental equation (10.38) for p_s^v is recovered.

If, instead, $q^d \neq 0 \Rightarrow p^d \neq 0$ (i.e. dry air is present with an associated non-zero pressure), but $\alpha^l \left(p - p_0^{sat} T/T_0\right) / \left(R^v T\right)$ is neglected in (10.63), then (10.63) reduces to (10.41). This means that, *with this approximation*, saturation vapour pressure (p_s^v) is then *independent* of dry-air pressure (p^d). It is conventional wisdom in the meteorological literature that this is indeed so. However (as just shown), it is only *precisely true* if $\alpha^l \left(p - p_0^{sat} T/T_0\right) / \left(R^v T\right)$ is *absent* from (10.63). As argued in Sections 10.4.7.4 and 10.4.7.7, neglecting this term is an excellent approximation, but it is an approximation nevertheless, with implications for thermodynamical consistency.

10.5.3.2 *Mass Fraction $a_{sat}\left(p, T \geq T_*\right)$ of Dry Air*

For the second question, $p_s^v\left(p, T\right)$ may be alternatively written by evaluating (10.7) at saturation to obtain

$$p_s^v\left(p, T\right) = \frac{\left[1 - a_{sat}\left(p, T\right)\right] p}{1 + a_{sat}\left(p, T\right)\left(\varepsilon - 1\right)}. \tag{10.64}$$

For specified values of p and $T \geq T_*$, solving this equation for a_{sat} then yields:

The Mass Fraction of Dry Air for Liquid Cloud ($T \geq T_*$)

$$a_{sat}\left(p, T\right) = \frac{p - p_s^v\left(p, T\right)}{p + \left(\varepsilon - 1\right) p_s^v\left(p, T\right)}, \tag{10.65}$$

where $p_s^v\left(p, T \geq T_*\right)$ is known from (10.63).

From (10.64), and noting that $p_s^d\left(p, T\right) = p - p_s^v\left(p, T\right)$, the corresponding partial pressure of dry air at saturation is given by

$$p_s^d\left(p, T\right) = \frac{\varepsilon a_{sat}\left(p, T\right) p}{1 + a_{sat}\left(p, T\right)\left(\varepsilon - 1\right)}. \tag{10.66}$$

(Alternatively, (10.6) can be evaluated at saturation to obtain (10.66).)

From (10.65), the corresponding mass fraction $1 - a_{sat}\left(p, T\right)$ of water vapour in the liquid-cloud parcel is given by

$$1 - a_{sat}\left(p, T\right) = \frac{\varepsilon p_s^v\left(p, T\right)}{p + \left(\varepsilon - 1\right) p_s^v\left(p, T\right)}, \tag{10.67}$$

where $p_s^v\left(p, T \geq T_*\right)$ is known from (10.63).

10.5.4 Saturation for $T \leq T_*$

In the previous subsection, a closed, *liquid-cloud* parcel was examined for $T \geq T_*$. Similar considerations and conditions also apply for a closed, *frozen-cloud* parcel (for $T \leq T_*$) – see Fig. 10.4b, with the water droplet of Fig. 10.4a replaced by a single ice crystal – but with a few differences.

If a humid-air parcel saturates for $T \leq T_*$, then the saturation vapour pressure, $p_s^v(p, T \leq T_*)$, lies on the *blue* (instead of red) curve of Figs. 10.1 and 10.3. Condensation of excess water vapour to *ice crystals* (instead of to water droplets) then takes place through a phase transition, again with an associated release of latent internal energy. Two similar questions again arise. For specified values of p (total pressure) and $T \leq T_*$:

1. At what value of saturation vapour pressure, $p_s^v(p, T \leq T_*)$, will a humid-air parcel saturate?
2. What is the value of the associated mass fraction of dry air, $a_{sat}(p, T \leq T_*)$, at saturation?

10.5.4.1 Saturation Vapour Pressure $p_s^v(p, T \leq T_*)$

For the first question, before saturation the air parcel only contains humid air (i.e. dry air plus a single phase (water vapour) of water substance). At saturation, ice crystals start to form; the first crystal to form is depicted in Fig. 10.4b. The humid-air parcel then becomes a frozen-cloud parcel (since $T \leq T_*$ by hypothesis here) with two phases (water vapour and frozen water) of water substance present.

For thermodynamic equilibrium at a common boundary between two phases (here, water vapour and frozen water) conditions (10.13)–(10.15) must be satisfied. Thus (from (10.15))

$$g^v\left(p^v, T\right) = g^f\left(p, T\right), \tag{10.68}$$

at saturation of the frozen-cloud parcel, where $g^v\left(p^v, T\right)$ is defined by (10.10) or (equivalently) by (10.11), and $g^f\left(p, T\right)$ by (9.35). (Since the input arguments of both g^v and g^f are $\left(p, T\right)$, conditions (10.13) and (10.14) are automatically satisfied.) Substitution of (10.10) (for g^v) and (9.35) (for g^f) into (10.68) gives the saturation condition

$$R^v T \ln\left[\frac{p_s^v(p, T)}{p_0^{sat}}\right] = \left(c_p^v - c^f\right) T \ln\left(\frac{T}{T_0}\right) - \left(L_0^v + L_0^f\right)\left(1 - \frac{T}{T_0}\right) + \alpha^f\left(p - p_0^{sat}\frac{T}{T_0}\right). \tag{10.69}$$

For specified values of p and $T \leq T_*$, solving this equation for $p_s^v\left(p, T\right)$ then yields:

The Saturation Vapour Pressure for Frozen Cloud ($T \leq T_*$)

$$p_s^v\left(p, T\right) = p_0^{sat}\left(\frac{T}{T_0}\right)^{\frac{\left(c_p^v - c^f\right)}{R^v}} \exp\left[-\frac{\left(L_0^v + L_0^f\right)}{R^v T}\left(1 - \frac{T}{T_0}\right) + \frac{\alpha^f}{R^v T}\left(p - p_0^{sat}\frac{T}{T_0}\right)\right]. \tag{10.70}$$

Thus for specified values of p and $T \leq T_*$, the corresponding saturated vapour pressure $p_s^v\left(p, T \leq T_*\right)$ can be explicitly determined from (10.70).

As a further cross-check, consider again the special case (examined in Section 10.4.7.3) where the parcel is composed *solely* of water substance. Thus: $q^d \equiv 0 \Rightarrow p^d \equiv 0$ (i.e. dry air is absent, so it cannot exert pressure); p in (10.70) reduces to p_s^v; and transcendental equation (10.40) for p_s^v is recovered.

If, instead, $q^d \neq 0 \Rightarrow p^d \neq 0$ (i.e. dry air is present with an associated non-zero pressure), but $\alpha^f\left(p - p_0^{sat}T/T_0\right) / \left(R^v T\right)$ is neglected in (10.70), then (10.70) reduces to (10.45). This means that, *with this approximation*, saturation vapour pressure (p_s^v) is then *independent* of dry-air pressure (p^d). It is conventional wisdom in the meteorological literature that this is indeed so. However (as just shown), it is only *precisely true* if $\alpha^f\left(p - p_0^{sat}T/T_0\right) / \left(R^v T\right)$ is absent from (10.70). With a similar argument to that used in Section 10.4.7.4, neglecting this term is an excellent approximation, but it is an approximation nevertheless with, again, implications for thermodynamical consistency.

10.5.4.2 *Mass Fraction $a_{sat}(p, T \leq T_*)$ of Dry Air*

For the second question, (10.64) and (10.65) hold not only for $T \geq T_*$ but also for $T \leq T_*$. Thus for specified values of p and $T \leq T_*$, the corresponding value of $a_{sat}(p, T \leq T_*)$ can be explicitly determined from (10.65), where $p_s^v(p, T \leq T_*)$ is known from (10.70).

Equations (10.66) and (10.67) also continue to hold for $T \leq T_*$. They can be similarly used to obtain $p_s^d(p, T)$ and $1 - a_{sat}(p, T)$ for $T \leq T_*$, but this time using (10.70) instead of (10.63).

10.5.5 The Gibbs Potential for a Saturated Cloudy Air Parcel

Recall that for a closed parcel of cloudy air, possibly containing liquid and/or frozen water, its composite Gibbs potential is given by (10.59). As summarised in Table 10.2, there are four realisable combinations of the phases of water substance that respect thermodynamic equilibrium. They are termed:

1. Humid air.
2. Liquid cloud.
3. Frozen cloud.
4. Liquid and frozen cloud.

Each combination is examined individually in Sections 10.5.5.2–10.5.5.6, but before doing so, we give a criterion for saturation.

10.5.5.1 *A Criterion for Saturation*

Because the cloudy-air parcel is assumed *closed*, the total amount of water substance (with mass fraction q^w) within it remains constant for all time (as also does the mass fraction q^d of dry air within it). What can change, however, is the *distribution* of total water substance as a function of its three possible phases, where – from (10.55) and (10.56) – their associated mass fractions are constrained to satisfy

$$q^w \equiv q^v + q^l + q^f \equiv 1 - q^d = \text{constant}, \tag{10.71}$$

and q^v, q^l, and q^f are the mass fractions of water vapour, liquid water, and frozen water, respectively.

To respect thermodynamic equilibrium, saturation can occur only if there is sufficient (total) water substance in the parcel to achieve this. The question now is:

- How much (as measured by mass fraction q^w) is sufficient to achieve saturation?

Given that q^w = constant is fixed (for all time) for a closed parcel, the answer depends upon temperature (T) and pressure (p) of the parcel, as described in Sections 10.5.3 and 10.5.4. The key quantities (one follows from the other) are:

- $a_{sat} = a_{sat}(p, T)$ (mass fraction of dry air *in the gaseous part of the parcel at saturation*).
- $1 - a_{sat} = 1 - a_{sat}(p, T)$ (mass fraction of water vapour *in the gaseous part of the parcel at saturation*).

For specified p and T, a_{sat} is given by (10.65), where $p_s^v(p, T \geq T_*)$ and $p_s^v(p, T \leq T_*)$ satisfy (10.63) and (10.70), respectively. Thus:

A Criterion for Saturation of a Cloudy Air Parcel

A criterion for saturation of a cloudy-air parcel is

$$q^w \equiv q^v + q^l + q^f \geq 1 - a_{sat}(p, T), \tag{10.72}$$

that is, saturation can occur only if the mass fraction of total water substance (q^w) in the parcel is greater than the mass fraction of water vapour, $(1 - a_{sat})$, *in the gaseous part of the parcel at saturation.*

10.5.5.2 *Humid (Subsaturated) Air*

At subsaturation, there is (by definition) insufficient total water substance in the parcel to achieve saturation (i.e. $q^w < 1 - a_{sat}$); see (10.72). Condensation of water vapour to liquid water, and/or deposition of water vapour to frozen water, therefore cannot take place. Thus $q^l = q^f = 0$ (i.e. there is no liquid or frozen water in the parcel, only water vapour). Therefore, using (10.57) and (10.71),

$$q^l = q^f = 0 \quad \Rightarrow \quad q^v = q^w, \quad a = 1 - q^w = q^d, \quad 1 - a = q^w = q^v. \tag{10.73}$$

Using (10.73), Gibbs potential (10.59) for cloudy air then reduces to that for humid air (i.e. to (10.12)). This leads to:

The Gibbs Potential for Humid (Subsaturated) Air

$$g(p, T, q^w) = g^{av}(p, T, 1 - q^w) = (1 - q^w)\, g^d(p, T) + q^w g^v(p, T), \tag{10.74}$$

where $q^w = q^v = 1 - q^d =$ constant, and $g^d(p, T)$ and $g^v(p, T)$ are given by (10.9) and (10.11), respectively.

10.5.5.3 *Cloudy (Saturated) Air*

For a saturated cloudy-air parcel (containing liquid and/or frozen water), the composite Gibbs potential (10.59) holds, but with a replaced by its saturated value, a_{sat}. This leads to:

The Gibbs Potential for Cloudy Saturated Air

$$g(p, T, q^w) = \left(1 - q^l - q^f\right) g^{av}(p, T, a_{sat}) + q^l g^l(p, T) + q^f g^f(p, T), \tag{10.75}$$

where q^l and/or q^f are non-zero.

There are three possibilities (examined in detail in Sections 10.5.5.4–10.5.5.6) for cloudy (saturated) air (see Table 10.2):

1. Liquid cloud, for which $T \geq T_*$, and $q^f \equiv 0$ in (10.75).

2. Frozen cloud, for which $T \le T_*$, and $q^l \equiv 0$ in (10.75).
3. Liquid and frozen cloud, for which $T = T_*$, and both q^l and q^f are non-zero in (10.75).

Determination of the value of T_* (the triple-point temperature of water substance at dry-air pressure p_*^d) is described in Section 10.6. This value is very close to T_0, the triple-point temperature of water substance in a vacuum. In fact – see Section 10.6 – if the terms involving α^l and α^f are omitted from basic Gibbs potentials (9.32) and (9.35), then $T_* \equiv T_0$. As previously mentioned, this approximation, which greatly simplifies things, is an excellent one that is frequently made in meteorology.

10.5.5.4 Liquid Cloud ($T \ge T_*$)

For a liquid-cloud parcel to exist:

- $T \ge T_*$ (for condensation to be possible).
- $q^f \equiv 0$ (i.e. there is no frozen water).
- q^v and q^l are non-zero (for water vapour and liquid water to coexist).

Furthermore, the parcel must be saturated (for condensation to actually take place). Therefore (from (10.72) with $q^f \equiv 0$)

$$q^w = q_s^v + q^l \ge 1 - a_{sat}\left(p, T\right), \tag{10.76}$$

where, for specified p and T, $1 - a_{sat}\left(p, T\right)$ is given by (10.67), and $p_s^v\left(p, T \ge T_*\right)$ satisfies (10.63).

Setting $q^f \equiv 0$ in (10.57) (evaluated at saturation) gives

$$a = a_{sat} = \frac{1 - q^w}{1 - q^l} \quad \Rightarrow \quad q^l = \frac{q^w + a_{sat} - 1}{a_{sat}}, \quad 1 - q^l = \frac{1 - q^w}{a_{sat}}. \tag{10.77}$$

Setting $q^f \equiv 0$ in (10.71) with use of (10.77) then yields

$$q^v = q_s^v = q^w - q^l = \left(1 - q^w\right) \frac{\left(1 - a_{sat}\right)}{a_{sat}}. \tag{10.78}$$

Since q^w = constant (because the parcel is closed), q^v (from (10.78)) and q^l (from (10.77)) are functions of $a_{sat} = a_{sat}\left(p, T\right)$ only.

The composite Gibbs potential for liquid cloud is given by substitution of (10.77) and (10.78) into (10.75) (with q^f set identically zero). This leads to:

The Gibbs Potential for Liquid Cloud ($T \ge T_*$)

$$\begin{aligned} g\left(p, T, q^w\right) &= \left(1 - q^l\right) g^{av}\left(p, T, a_{sat}\right) + q^l g^l\left(p, T\right) \\ &= \left(\frac{1 - q^w}{a_{sat}}\right) g^{av}\left(p, T, a_{sat}\right) + \left(\frac{q^w + a_{sat} - 1}{a_{sat}}\right) g^l\left(p, T\right), \end{aligned} \tag{10.79}$$

where

$$g^{av}\left(p,T,a_{sat}\right) = a_{sat}g^{d}\left(p,T\right) + \left(1-a_{sat}\right)g^{v}\left(p,T\right), \tag{10.80}$$

and $g^d\left(p,T\right)$, $g^v\left(p,T\right)$, and $g^l\left(p,T\right)$ are given by (10.9) (with $a = a_{sat}$), (10.11) (with $a = a_{sat}$), and (9.32), respectively.

The present analysis corresponds to that given in Section 3.2 of Thuburn (2017), but in more detail here.

10.5.5.5 Frozen Cloud ($T \leq T_*$)

For a frozen-cloud parcel to exist:

- $T \leq T_*$ (for deposition to be possible).
- $q^l \equiv 0$ (i.e. there is no liquid water).
- q^v and q^f are non-zero (for water vapour and frozen water to coexist).

Furthermore, the parcel must be saturated (for deposition to actually take place). Therefore (from (10.72) with $q^l \equiv 0$)

$$q^w \equiv q_s^v + q^f \geq 1 - a_{sat}\left(p,T\right), \tag{10.81}$$

where, for specified p and T, $1-a_{sat}\left(p,T\right)$ is given by (10.67), and $p_s^v\left(p,T\leq T_*\right)$ satisfies (10.70).

Setting $q^l \equiv 0$ in (10.57) (evaluated at saturation) gives

$$a = a_{sat} = \frac{1-q^w}{1-q^f} \quad \Rightarrow \quad q^f = \frac{q^w + a_{sat} - 1}{a_{sat}}, \quad 1-q^f = \frac{1-q^w}{a_{sat}}. \tag{10.82}$$

Setting $q^l \equiv 0$ in (10.71), with use of (10.82), then yields

$$q^v = q_s^v = q^w - q^f = \left(1-q^w\right)\frac{\left(1-a_{sat}\right)}{a_{sat}}. \tag{10.83}$$

Since q^w = constant (because the parcel is closed), q^v (from (10.83)) and q^f (from (10.82)) are functions of $a_{sat} = a_{sat}\left(p,T\right)$, only.

The composite Gibbs potential for frozen cloud is given by substitution of (10.82) into (10.75) (with q^l set identically zero therein). This leads to:

The Gibbs Potential for Frozen Cloud ($T \leq T_*$)

$$\begin{aligned} g\left(p,T,q^w\right) &= \left(1-q^f\right)g^{av}\left(p,T,a_{sat}\right) + q^f g^f\left(p,T\right) \\ &= \left(\frac{1-q^w}{a_{sat}}\right)g^{av}\left(p,T,a_{sat}\right) + \left(\frac{q^w + a_{sat} - 1}{a_{sat}}\right)g^f\left(p,T\right), \end{aligned} \tag{10.84}$$

where $g^{av}\left(p,T,a_{sat}\right)$, $g^d\left(p,T\right)$, $g^v\left(p,T\right)$, and $g^f\left(p,T\right)$ are given by (10.80), (10.9), (10.11), and (9.35), respectively.

The present analysis extends Thuburn (2017)'s for liquid cloud to frozen cloud. This addresses his remark that doing so is a minimum requirement for weather forecasting and climate modelling.

Note the (expected) similarity between (10.76)–(10.79) for liquid cloud and (10.81)–(10.84) for frozen cloud; q^l in the former equations are replaced by q^f in the latter ones. A difference, however, is that $g^l(p,T)$ and $g^f(p,T)$ have different functional forms; cf. (9.32) with (9.35). Apart from differences in coefficients (c^l versus c^f, and α^l versus α^f), a latent-energy contribution appears in (9.32), but is absent from (9.35).

10.5.5.6 *Liquid and Frozen Cloud ($T = T_*$)*

For a liquid-and-frozen cloud parcel to exist:

- $T = T_*$ and $p = p_*$ (for liquid water and frozen water to coexist).
- q^v, q^l, and q^f are all non-zero (for water vapour, liquid water, and frozen water to coexist).

Furthermore, the parcel must be saturated (for water vapour, liquid water, and frozen water to coexist). Therefore (from (10.72) with $(p,T) = (p_*, T_*)$)

$$q^w \equiv q_s^v + q^l + q^f \geq 1 - a_{sat}(p_*, T_*). \tag{10.85}$$

In (10.85), $1 - a_{sat}(p,T)$ is given by (10.67) with $(p,T) = (p_*, T_*)$, that is, by

$$1 - a_{sat}(p_*, T_*) = \frac{\varepsilon p_s^v(p_*, T_*)}{p_*^d + \varepsilon p_s^v(p_*, T_*)}, \tag{10.86}$$

where $p_s^v(p_*, T_*)$ satisfies both (10.63) (for liquid cloud) *and* (10.70) (for frozen cloud). Furthermore, $p_* = p_*^d + p_*^v$ is the sum of dry-air pressure and water-vapour pressure at the triple point of water in the presence of dry air with specified dry-air pressure $p^d = p_*^d$. (That $p_s^v(p_*, T_*)$ must satisfy both (10.63) and (10.70) is used in Section 10.6.2 to determine the precise values of T_* and p_* for specified dry-air pressure $p^d = p_*^d$; doing so also leads to (in Section 10.6.3) the equation for the triple-point coexistence curve in the presence of dry air.)

From (10.57), evaluated at saturation,

$$a = a_{sat} = \frac{1 - q^w}{1 - q^l - q^f} \quad \Rightarrow \quad q^l + q^f = \frac{q^w + a_{sat} - 1}{a_{sat}}, \quad 1 - q^l - q^f = \frac{1 - q^w}{a_{sat}}. \tag{10.87}$$

Solving (10.71) at saturation for q_s^v, with use of (10.87), yields

$$q^v = q_s^v = q^w - q^l - q^f = (1 - q^w)\frac{(1 - a_{sat})}{a_{sat}}. \tag{10.88}$$

Evaluating Gibbs potential (10.59) at $(p,T) = (p_*, T_*)$ and using (10.87) leads to:

The Gibbs Potential for Liquid-and-Frozen Cloud ($T = T_*$)

$$g(p_*, T_*, q^w) = \left(\frac{1 - q^w}{a_{sat}}\right) g^{av}(p_*, T_*, a_{sat}) + q^l g^l(p_*, T_*) + q^f g^f(p_*, T_*). \tag{10.89}$$

It remains to determine the values of (p_*, T_*), q^l, and q^f. To do so is non-trivial; see Section 10.6.2 for (p_*, T_*), and Section 10.6.4 for q^l and q^f.

10.5.6 Other Thermodynamic Potentials for Liquid Cloud ($T \geq T_*$)

10.5.6.1 Methodologies

Other thermodynamic potentials may be obtained from a Gibbs potential. We now examine how to do so for liquid cloud (in this section) and for frozen cloud (in Section 10.5.7). Using the relations given in Section 10.5.5.4, Gibbs potential (10.79) for liquid cloud may be rewritten in the form

$$g\left(p,T,q^w\right)=\left(1-q^w\right)g^d\left(p,T\right)+q_s^v g^v\left(p,T\right)+\left(q^w-q_s^v\right)g^l\left(p,T\right), \tag{10.90}$$

where

$$1-q^w=q^d=\text{constant},\quad q^w-q_s^v=q^l,\quad q_s^v=\left(1-q^w\right)\frac{\left(1-a_{sat}\right)}{a_{sat}},\quad a_{sat}=\frac{1-q^w}{1-q^l}, \tag{10.91}$$

and $g^d\left(p,T\right)$, $g^v\left(p,T\right)$, and $g^l\left(p,T\right)$ are defined in Tables 9.4 and 9.5 and by (10.9) and (10.11). Inserting (10.90) into the various relations given in column 6 of Table 9.2 then leads to explicit expressions for other thermodynamic potentials. This is hard work.

There is an alternative, less labour-intensive method available to that just described; see Section 9.4.8. It exploits the fact that Gibbs potential (10.90) is *a mass-weighted sum* of the basic potentials $g^d\left(p,T\right)$, $g^v\left(p,T\right)$, and $g^l\left(p,T\right)$. This allows us to write other thermodynamic potentials as similar mass-weighted sums of their own basic potentials. Although this method can be used in the present circumstances, it is not, however, of general applicability; see footnote 16 of Chapter 9.

Using either of the two methods just described leads to the following results.

10.5.6.2 Specific Internal Energy ($\mathcal{E}$) for Liquid Cloud ($T \geq T_*$)

Specific internal energy for liquid cloud may be written as the mass-weighted sum

$$\mathcal{E}=\left(1-q^w\right)\mathcal{E}^d\left(T\right)+q_s^v\mathcal{E}^v\left(T\right)+\left(q^w-q_s^v\right)\mathcal{E}^l\left(T\right), \tag{10.92}$$

where (from Tables 9.4 and 9.5)

$$\mathcal{E}^d\left(T\right)=c_v^dT,\quad \mathcal{E}^v\left(T\right)=c_v^vT+L_0^v+L_0^f,\quad \mathcal{E}^l\left(T\right)=c^lT+L_0^f. \tag{10.93}$$

Using (10.91) and (10.93), (10.92) may be alternatively written as

$$\mathcal{E}=\left(1-q^w\right)c_v^dT+q_s^v\left[L_0^v+\left(c_v^v-c^l\right)T\right]+q^w\left(c^lT+L_0^f\right), \tag{10.94}$$

$$\mathcal{E}=c_v^{sat,l}T+\left(L_0^v+L_0^f\right)q_s^v+L_0^f\left(q^w-q_s^v\right), \tag{10.95}$$

where

$$c_v^{sat,l}\equiv\left(1-q^w\right)c_v^d+q_s^vc_v^v+\left(q^w-q_s^v\right)c^l. \tag{10.96}$$

The preceding equations correspond to (3.71)–(3.74) of Section 3.5, with q^f set identically zero therein to be applicable for liquid cloud.

10.5.6.3 Specific Enthalpy (h) for Liquid Cloud ($T \geq T_*$)

Specific enthalpy for liquid cloud may be written as the mass-weighted sum

$$h=\left(1-q^w\right)h^d\left(T\right)+q_s^vh^v\left(T\right)+\left(q^w-q_s^v\right)h^l\left(p,T\right), \tag{10.97}$$

where (from Tables 9.4 and 9.5)

$$h^d\left(T\right)=c_p^dT,\quad h^v\left(T\right)=c_p^vT+L_0^v+L_0^f,\quad h^l\left(p,T\right)=c^lT+\alpha^lp+L_0^f. \tag{10.98}$$

Using (10.91) and (10.98), (10.97) may be alternatively written as

$$h = \left(1 - q^w\right) h^d(T) + q_s^v \left[h^v(T) - h^l(p,T)\right] + q^w h^l(p,T), \tag{10.99}$$

$$h = \left(1 - q^w\right) h^d(T) + q_s^v \mathbb{L}^{l\to v}(T,p) + q^w h^l(p,T), \tag{10.100}$$

$$h = c_p^{sat,l} T + \left(q^w - q_s^v\right) \alpha^l p + L_0^v q_s^v + L_0^f q^w, \tag{10.101}$$

where – cf. (10.25) –

$$\mathbb{L}^{l\to v}(T,p) \equiv h^v(T) - h^l(p,T) = L_0^v + \left(c_p^v - c^l\right) T - \alpha^l p, \tag{10.102}$$

is the enthalpy (latent heat) of vaporisation, and – cf. (3.75) –

$$c_p^{sat,l} \equiv \left(1 - q^w\right) c_p^d + q_s^v c_p^v + \left(q^w - q_s^v\right) c^l. \tag{10.103}$$

10.5.6.4 Specific Helmholtz Energy (f) for Liquid Cloud ($T \geq T_*$)

Specific Helmholtz energy for liquid cloud may be written as the mass-weighted sum

$$f = \left(1 - q^w\right) f^d(p,T) + q_s^v f^v(p,T) + \left(q^w - q_s^v\right) f^l(p,T), \tag{10.104}$$

where (from Tables 9.4 and 9.5)

$$f^d(p,T) = -c_p^d T \ln\left(\frac{T}{T_0}\right) + R^d T \ln\left(\frac{p^d}{p_0^d}\right) - R^d T, \tag{10.105}$$

$$f^v(p,T) = -c_p^v T \ln\left(\frac{T}{T_0}\right) + R^v T \ln\left(\frac{p^v}{p_0^{sat}}\right) - R^v T + \left(L_0^v + L_0^f\right)\left(1 - \frac{T}{T_0}\right), \tag{10.106}$$

$$f^l(p,T) = -c^l T \ln\left(\frac{T}{T_0}\right) - \alpha^l p_0^{sat} \frac{T}{T_0} + L_0^f \left(1 - \frac{T}{T_0}\right), \tag{10.107}$$

and – from (10.6) and (10.7) –

$$p^d = \frac{a_{sat}\,\varepsilon p}{1 + a_{sat}(\varepsilon - 1)}, \quad p^v = \frac{(1 - a_{sat})\, p}{1 + a_{sat}(\varepsilon - 1)}. \tag{10.108}$$

10.5.6.5 Specific Entropy (η) for Liquid Cloud ($T \geq T_*$)

Specific entropy for liquid cloud may be written as the mass-weighted sum

$$\eta = \left(1 - q^w\right) \eta^d(p,T) + q_s^v \eta^v(p,T) + \left(q^w - q_s^v\right) \eta^l(p,T), \tag{10.109}$$

where (from Tables 9.4 and 9.5)

$$\eta^d(p,T) = c_p^d \left[1 + \ln\left(\frac{T}{T_0}\right)\right] - R^d \ln\left(\frac{p^d}{p_0^d}\right), \tag{10.110}$$

$$\eta^v(p,T) = c_p^v \left[1 + \ln\left(\frac{T}{T_0}\right)\right] - R^v \ln\left(\frac{p^v}{p_0^{sat}}\right) + \frac{L_0^v + L_0^f}{T_0}, \tag{10.111}$$

$$\eta^l(p,T) = c^l \left[1 + \ln\left(\frac{T}{T_0}\right)\right] + \frac{\alpha^l p_0^{sat}}{T_0} + \frac{L_0^f}{T_0}, \tag{10.112}$$

and (10.108) relates p^d and p^v to p.

Elimination of $\mathcal{E} + p\alpha$ from (9.3) and (9.7), followed by evaluation for water vapour and liquid water, gives

$$g^v = h^v - T\eta^v, \quad g^l = h^l - T\eta^l \quad \Rightarrow \quad g^v - g^l = h^v - h^l - T\left(\eta^v - \eta^l\right). \tag{10.113}$$

Now for a liquid-cloud parcel, water vapour must be in thermodynamic equilibrium with liquid water. Thus, from (10.15), $g^v = g^l$. Using this in the last equation of (10.113), with use of (10.102), then yields

$$\eta^v - \eta^l = \frac{h^v - h^l}{T} = \frac{\mathbb{L}^{l\to v}(T,p)}{T} = \frac{L_0^v + \left(c_p^v - c^l\right)T - \alpha^l p}{T}. \quad (10.114)$$

Substitution of (10.102) into (10.109) finally yields the alternative forms

$$\eta = \left(1 - q^w\right)\eta^d(p,T) + q^w\eta^l(p,T) + q_s^v\frac{\left[L_0^v + \left(c_p^v - c^l\right)T - \alpha^l p\right]}{T}, \quad (10.115)$$

$$\eta = \left(1 - q^w\right)\eta^d(p,T) + q^w\eta^l(p,T) + q_s^v\frac{\mathbb{L}^{l\to v}(T,p)}{T}. \quad (10.116)$$

Form (10.116) corresponds to (6.12) of Ambaum (2010) and to (18.157) of Vallis (2017).

10.5.7 Other Thermodynamic Potentials for Frozen Cloud ($T \leq T_*$)

In a similar manner to that given in Section 10.5.6.5 for liquid cloud, other thermodynamic potentials may be obtained from Gibbs potential (10.84) for frozen cloud. This potential may be rewritten as

$$g\left(p,T,q^w\right) = \left(1 - q^w\right)g^d(p,T) + q_s^v g^v(p,T) + \left(q^w - q_s^v\right)g^f(p,T), \quad (10.117)$$

where

$$1 - q^w = q^d = \text{constant}, \quad q^w - q_s^v = q^f, \quad q_s^v = \left(1 - q^w\right)\frac{\left(1 - a_{sat}\right)}{a_{sat}}, \quad a_{sat} = \frac{1 - q^w}{1 - q^f}, \quad (10.118)$$

and $g^d(p,T)$, $g^v(p,T)$, and $g^f(p,T)$ are defined in Tables 9.4 and 9.5 and by (10.9) and (10.11). Using either of the two methods outlined in Section 10.5.6.1 leads to the following results for frozen cloud.

10.5.7.1 *Specific Internal Energy ($\mathcal{E}$) for Frozen Cloud ($T \leq T_*$)*

Specific internal energy for frozen cloud may be written as the mass-weighted sum

$$\mathcal{E} = \left(1 - q^w\right)\mathcal{E}^d(T) + q_s^v\mathcal{E}^v(T) + \left(q^w - q_s^v\right)\mathcal{E}^f(T), \quad (10.119)$$

where (from Tables 9.4 and 9.5)

$$\mathcal{E}^d(T) = c_v^d T, \quad \mathcal{E}^v(T) = c_v^v T + L_0^v + L_0^f, \quad \mathcal{E}^f(T) = c^f T. \quad (10.120)$$

Using (10.118) and (10.120), (10.119) may be alternatively written as

$$\mathcal{E} = \left(1 - q^w\right)c_v^d T + q_s^v\left[L_0^v + L_0^f + \left(c_v^v - c^f\right)T\right] + q^w c^f T, \quad (10.121)$$

$$\mathcal{E} = c_v^{sat,f}T + \left(L_0^v + L_0^f\right)q_s^v, \quad (10.122)$$

where

$$c_v^{sat,f} \equiv \left(1 - q^w\right)c_v^d + q_s^v c_v^v + \left(q^w - q_s^v\right)c^f. \quad (10.123)$$

Equations (10.119)–(10.123) for frozen cloud correspond to (3.71)–(3.74) of Section 3.5, with q^l set identically zero therein to be applicable for frozen cloud. Equations (10.119)–(10.123) are fairly similar in form to their counterparts (10.92)–(10.96) for liquid cloud.

10.5.7.2 *Specific Enthalpy (h) for Frozen Cloud ($T \le T_*$)*

Specific enthalpy for frozen cloud may be written as the mass-weighted sum

$$h = \left(1 - q^w\right) h^d \left(T\right) + q_s^v h^v \left(T\right) + \left(q^w - q_s^v\right) h^f \left(p, T\right), \tag{10.124}$$

where (from Tables 9.4 and 9.5)

$$h^d \left(T\right) = c_p^d T, \quad h^v \left(T\right) = c_p^v T + L_0^v + L_0^f, \quad h^f \left(p, T\right) = c^f T + \alpha^f p. \tag{10.125}$$

Using (10.118) and (10.125), (10.124) may be alternatively written as

$$h = \left(1 - q^w\right) h^d \left(T\right) + q_s^v \left[h^v \left(T\right) - h^f \left(p, T\right)\right] + q^w h^f \left(p, T\right) \tag{10.126}$$

$$h = \left(1 - q^w\right) h^d \left(T\right) + q_s^v \mathbb{L}^{f \to v} \left(T, p\right) + q^w h^f \left(p, T\right), \tag{10.127}$$

$$h = c_p^{sat,f} T + \left(q^w - q_s^v\right) \alpha^f p + \left(L_0^v + L_0^f\right) q_s^v, \tag{10.128}$$

where – cf. (10.27) –

$$\mathbb{L}^{f \to v} \left(T, p\right) \equiv h^v - h^f = L_0^v + L_0^f + \left(c_p^v - c^f\right) T - \alpha^f p, \tag{10.129}$$

is the enthalpy (latent heat) of sublimation, and – cf. (3.75) –

$$c_p^{sat,f} \equiv \left(1 - q^w\right) c_p^d + q_s^v c_p^v + \left(q^w - q_s^v\right) c^f. \tag{10.130}$$

Equations (10.124)–(10.130) for frozen cloud are fairly similar in form to their counterparts (10.97)–(10.103) for liquid cloud.

10.5.7.3 *Specific Helmholtz Energy (f) for Frozen Cloud ($T \le T_*$)*

Specific Helmholtz energy for frozen cloud may be written as the mass-weighted sum

$$f = \left(1 - q^w\right) f^d \left(p, T\right) + q_s^v f^v \left(p, T\right) + \left(q^w - q_s^v\right) f^f \left(p, T\right), \tag{10.131}$$

where (from Tables 9.4 and 9.5)

$$f^d \left(p, T\right) = - c_p^d T \ln \left(\frac{T}{T_0}\right) + R^d T \ln \left(\frac{p^d}{p_0^d}\right) - R^d T, \tag{10.132}$$

$$f^v \left(p, T\right) = - c_p^v T \ln \left(\frac{T}{T_0}\right) + R^v T \ln \left(\frac{p^v}{p_0^{sat}}\right) - R^v T + \left(L_0^v + L_0^f\right) \left(1 - \frac{T}{T_0}\right), \tag{10.133}$$

$$f^f \left(p, T\right) = - c^f T \ln \left(\frac{T}{T_0}\right) - \alpha^f p_0^{sat} \frac{T}{T_0}, \tag{10.134}$$

and (10.108) relates p^d and p^v to p.

10.5.7.4 *Specific Entropy (η) for Frozen Cloud ($T \le T_*$)*

Specific entropy for frozen cloud may be written as the mass-weighted sum

$$\eta = \left(1 - q^w\right) \eta^d \left(p, T\right) + q_s^v \eta^v \left(p, T\right) + \left(q^w - q_s^v\right) \eta^f \left(p, T\right), \tag{10.135}$$

where (from Tables 9.4 and 9.5)

$$\eta^d \left(p, T\right) = c_p^d \left[1 + \ln \left(\frac{T}{T_0}\right)\right] - R^d \ln \left(\frac{p^d}{p_0^d}\right), \tag{10.136}$$

$$\eta^v \left(p, T\right) = c_p^v \left[1 + \ln \left(\frac{T}{T_0}\right)\right] - R^v \ln \left(\frac{p^v}{p_0^{sat}}\right) + \frac{L_0^v + L_0^f}{T_0}, \tag{10.137}$$

$$\eta^f(p,T) = c^f\left[1+\ln\left(\frac{T}{T_0}\right)\right] + \frac{\alpha^f p_0^{sat}}{T_0}, \tag{10.138}$$

and (10.108) relates p^d and p^v to p.

Elimination of $\mathcal{E}+p\alpha$ from (9.3) and (9.7), followed by evaluation for water vapour and frozen water, gives

$$g^v = h^v - T\eta^v, \quad g^f = h^f - T\eta^f \quad \Rightarrow \quad g^v - g^f = h^v - h^f - T\left(\eta^v - \eta^f\right). \tag{10.139}$$

Now for a frozen-cloud parcel, water vapour must be in thermodynamic equilibrium with frozen water. Thus, from (10.15), $g^v = g^f$. Using this in the last equation of (10.139), with use of (10.129), then yields

$$\eta^v - \eta^f = \frac{h^v - h^f}{T} = \frac{\mathbb{L}^{f\to v}(T,p)}{T} = \frac{L_0^v + L_0^f + \left(c_p^v - c^f\right)T - \alpha^f p}{T}. \tag{10.140}$$

Substitution of (10.129) into (10.135) finally yields the alternative forms

$$\eta = \left(1-q^w\right)\eta^d(p,T) + q^w\eta^f(p,T) + q_s^v\frac{\left[L_0^v + L_0^f + \left(c_p^v - c^f\right)T - \alpha^f p\right]}{T}, \tag{10.141}$$

$$\eta = \left(1-q^w\right)\eta^d(p,T) + q^w\eta^f(p,T) + q_s^v\frac{\mathbb{L}^{f\to v}(T,p)}{T}. \tag{10.142}$$

10.6 THE TRIPLE POINT OF WATER IN THE PRESENCE OF DRY AIR

10.6.1 Graphical Overview

Recall (from Sections 10.3 and 10.4) that the three phases (vapour, liquid, frozen) of pure-water substance (*in the absence of dry air*) can only coexist at the triple point $(T,p) = (T,p^v) = (T_0, p_0^{sat})$ on the 2D (T,p) phase diagram for water substance; see Figs. 10.1 and 10.3. The question now is:

- Can the three phases (vapour, liquid, frozen) of pure-water substance also coexist *in the presence of dry air*?

Indeed they can, as graphically depicted on the 3D $\left(T,p^v,p^d\right)$ phase diagram of Fig. 10.5, and as quantitatively described in Sections 10.6.2 and 10.6.3.

The horizontal coordinates in Fig. 10.5 are (T,p^v), the vertical coordinate is p^d, and the total pressure is given by $p = p^d + p^v$, the sum of the two partial pressures. In the absence of dry air, $p^d \equiv 0 \Rightarrow p \equiv p^v + p^d = p^v$. The $p^d = 0$ plane in Fig. 10.5 therefore corresponds to the 2D (T,p) phase diagram for pure-water substance, depicted in Figs. 10.1 and 10.3, and discussed in detail in Sections 10.3 and 10.4. Coexistence curves are shown on the $p^d = 0$ plane for water vapour and frozen water (in dark blue); water vapour and liquid water (in red); and frozen water and liquid water (in green). The point $\left(T,p^v,p^d\right) = \left(T_0, p_0^{sat}, 0\right)$ (depicted in yellow) corresponds to the triple point of pure-water substance (i.e. in the absence of dry air), where water vapour, liquid water, and frozen water can coexist.

In the presence of dry air, $p^d \neq 0 \Rightarrow p \equiv p^v + p^d \neq p^v$. For a specified value $p^d = p_*^d$ of dry-air pressure, the coexistence curves on the $p^d = p_*^d$ plane and the triple point of water substance (in the presence of dry air) are all slightly displaced horizontally with respect to their counterparts (in the absence of dry air) on the $p^d = 0$ plane. For physically realisable values of p^d in Earth's atmosphere, this horizontal displacement is very small indeed and, for clarity, it is

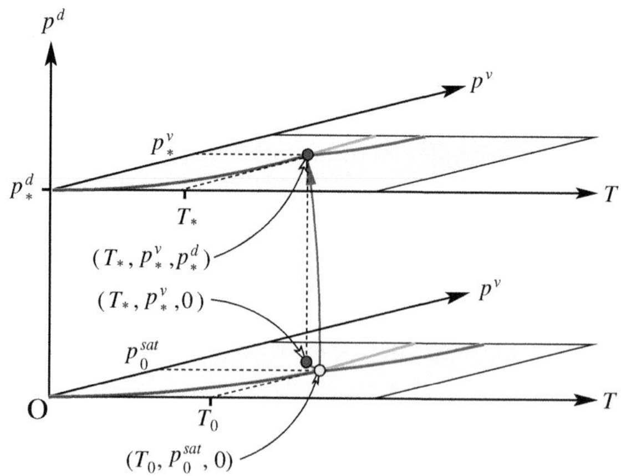

Figure 10.5 A 3D (T, p^v, p^d) phase diagram (not to scale) for water substance for different conditions of temperature (T), water-vapour pressure (p^v), and dry-air pressure (p^d). Two planes (in light blue) are shown; for $p^d = 0$ and $p^d = p^d_*$. The plane $p^d = 0$ corresponds to the 2D (T, p^v) phase diagrams depicted in Figs. 10.1 and 10.3 for pure-water substance (in the absence of dry air). Coexistence curves on the $p^d = 0$ and $p^d = p^d_*$ planes are shown for water vapour and frozen water (in dark blue); water vapour and liquid water (in red); and frozen water and liquid water (in green). The triple point of water substance (yellow circle) *in the absence of dry air* is located at $(T, p^v, p^d) = (T_0, p^{sat}_0, 0)$. The triple point of dry air and water substance (upper magenta circle) for $p^d = p^d_*$ is located at $(T, p^v, p^d) = (T_*, p^v_*, p^d_*)$. The lower magenta circle is the projection of the upper one onto the $p^d = 0$ plane. As p^d increases from $p^d = 0$ to $p^d = p^d_*$, the triple point of water *in the presence of dry air* follows the magenta curve. Dashed lines are parallel to coordinate axes.

grossly exaggerated on Fig. 10.5. The triple point of water substance (in the presence of dry air) is located at the intersection point (T_*, p^v_*, p^d_*) of the three coexistence curves on the $p^d = p^d_*$ plane. At this special point, the total pressure is $p_* = p^d_* + p^v_*$ (i.e. the sum of the partial pressures there). As p^d increases from $p^d = 0$ to some specified value $p^d = p^d_*$, the triple point of water *in the presence of dry air* follows the arrowed curve shown in magenta on Fig. 10.5.

The $p^d = 0$ plane depicted (but not to scale) in Fig. 10.5, is also shown (but now to scale) in Fig. 10.6a. Although (as discussed in Sections 10.3 and 10.4) the slopes of the vapour–frozen (in dark blue) and vapour–liquid (in red) coexistence curves differ from one another at the triple point (yellow circle) of water substance, the difference between these slopes is barely visible, if indeed at all.

There are in fact *two* sets of coexistence curves depicted on Fig. 10.6a: the set shown on the $p^d = 0$ plane of Fig. 10.5, and that of the $p^d = p^d_*$ plane, projected downwards onto the $p^d = 0$ plane. At the scale of Fig. 10.6a, however, the two sets almost perfectly overlay one another and are visually indistinguishable from one another. This means that the arrowed magenta curve on Fig. 10.5 is almost vertical.

To address this lack of visible separation between the two sets of curves, Fig. 10.6b is a zoom of Fig. 10.6a about the triple point (yellow circle) of pure-water substance. The separation of the two sets of curves is now visible. The solid curves correspond to those on the $p^d = 0$ plane for pure-water substance (in the absence of dry air); the dashed curves correspond to those on the $p^d = p^d_* = 1\,000$ hPa $= 10^5$ Pa plane (in the presence of dry air at a typical pressure at Earth's surface). Careful examination of Fig. 10.6b, with the aid of a ruler, confirms that the slopes of

(a)

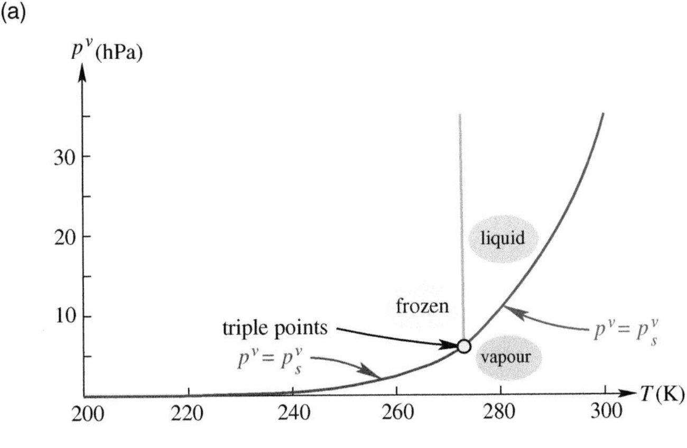

(b)

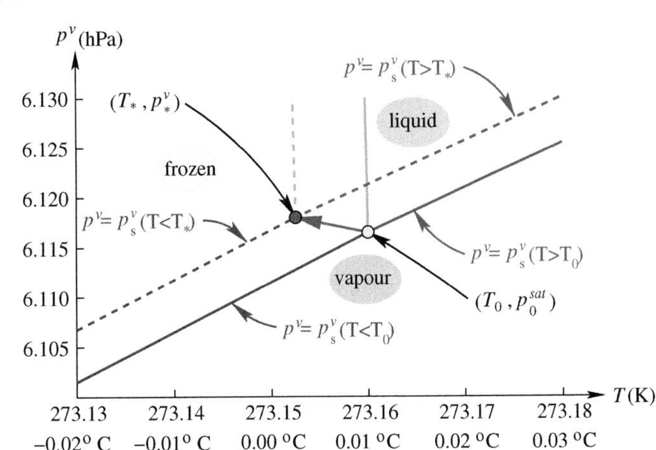

Figure 10.6 2D phase diagrams (to scale) for water substance for different conditions of temperature (T) and water vapour pressure (p^v). For both panels, (colour-coded) coexistence (dashed) curves on the $p^d = p^d_*$ plane of Fig. 10.5 are projected onto the $p^d = 0$ plane and superposed on the corresponding (solid) curves of the $p^d = 0$ plane. The scale of panel (a) is such that the two sets of curves are indistinguishable from one another. Panel (b) is a zoom of panel (a) about the triple (circled) points of water substance in the absence of dry air (for $p^d = 0$) and in its presence (for $p^d_* = 1000$ hPa $= 10^5$ Pa); the separation of the two sets of curves is now visible. The triple point of water substance in the presence of dry air follows the arrowed magenta curve on panel (b) as p^d increases from $p^d = 0$ to $p^d = p^d_*$; this curve is the projection onto the $p^d = 0$ plane of the arrowed magenta curve on Fig. 10.5.
(This figure is based on data computed and kindly provided by John Thuburn.)

the vapour–frozen (in dark blue) and vapour–liquid (in red) coexistence curves do indeed differ slightly from one another at the triple point of water substance, both in the absence (solid curves) and presence (dashed curves) of dry air. See Section 10.4.7.6 for why this is so in the absence of dry air.

The triple point of water substance in the presence of dry air follows the arrowed magenta curve on Fig. 10.6b as p^d increases from $p^d = 0$ to $p^d = p^d_*$; this curve is the projection onto the $p^d = 0$ plane of the arrowed magenta curve of Fig. 10.5. The location of the triple point (yellow circle) on Fig. 10.6b, in the absence of dry air, is fixed; it is entirely defined by the properties of pure-water substance (in the absence of dry air, i.e. in a vacuum). It is located (see footnote 14) at *precisely* $T = 273.16\text{ K} \equiv 0.01\ ^\circ\text{C}$ and $p^v = p_0^{sat} = 6.11657\text{ hPa} \equiv 611.657\text{ Pa}$. Thus, in the *absence* of dry air, the temperature at the triple point of water (where water freezes and ice melts) is not 0 °C but (by international convention and contrary to what one might expect) at *precisely* 0.01 °C.

The location of the triple point (magenta circle) of water substance in the *presence* of dry air depends upon the pressure exerted by the dry air; this varies from next to nothing in Earth's upper atmosphere, to approximately $p^d = 1\,000\text{ hPa} = 10^5\text{ Pa}$ at Earth's surface. The circled magenta point on Fig. 10.6b is therefore close to the maximum (2D) displacement of the triple point in the presence of dry air from its (2D) location in the absence of dry air. This displacement is very small indeed, and it is frequently neglected in the meteorological literature.

It is seen from Fig. 10.6b that the triple point of water in the *presence* of dry air at Earth's surface occurs at a temperature *almost* exactly equal to 0 °C, as opposed to exactly 0.01 °C in the *absence* of dry air. How closely to 0 °C, in the presence of dry air, depends upon the value of the dry-air pressure (p^d). Historically, experiments (to empirically deduce classical gas laws) during the early development of thermodynamics were performed in the laboratory at atmospheric pressure *at Earth's surface* rather than in a vacuum. This accounts for the very small difference of approximately 0.01 °C between the triple-point temperature in the absence, and in the presence (near Earth's surface) of dry air.

10.6.2 Determination of T_* and p_*^v for Specified $p^d = p_*^d$

As shown graphically on Figs. 10.5 and 10.6, the triple point of water in the presence of dry air, with specified pressure $p^d = p_*^d$, has 2D coordinates $(T, p^v) = (T_*, p_*^v)$. This point corresponds to the intersection of the coexistence curves defined by (10.63) and (10.70) for water vapour and liquid water (in red), and water vapour and frozen water (in blue), respectively, on the $p^d = p_*^d$ plane of Fig. 10.5. Evaluation of the saturation vapour-pressure equations (10.63) and (10.70) at $(T, p) = (T_*, p_*) \equiv \left(T_*, p_*^d + p_*^v\right)$, that is, at the magenta point $\left(T, p^v, p^d\right) = \left(T_*, p_*^v, p_*^d\right)$ on Fig. 10.5, then gives

$$p_*^v \equiv p_s^v(p_*, T_*) = p_0^{sat}\left(\frac{T_*}{T_0}\right)^{\frac{\left(c_p^v - c^l\right)}{R^v}} \exp\left[-\frac{L_0^v}{R^v T_*}\left(1 - \frac{T_*}{T_0}\right) + \frac{\alpha^l}{R^v T_*}\left(p_*^d + p_*^v - p_0^{sat}\frac{T_*}{T_0}\right)\right], \tag{10.143}$$

$$p_*^v \equiv p_s^v(p_*, T_*) = p_0^{sat}\left(\frac{T_*}{T_0}\right)^{\left(\frac{c_p^v - c^f}{R^v}\right)} \exp\left[-\frac{\left(L_0^v + L_0^f\right)}{R^v T_*}\left(1 - \frac{T_*}{T_0}\right) + \frac{\alpha^f}{R^v T_*}\left(p_*^d + p_*^v - p_0^{sat}\frac{T_*}{T_0}\right)\right]. \tag{10.144}$$

- For a specified value of dry-air pressure, $p^d = p_*^d$, and (10.143) and (10.144) are two non-linearly coupled equations for the two sought quantities, T_* and p_*^v.

Because of the non-linear coupling, they have to be solved numerically; this can be done to machine precision (e.g. using Newton–Raphson iteration). This then yields the triple-point temperature (T_*) and water-vapour pressure (p_*^v) (or, equivalently, total pressure $p_* \equiv p_*^d + p_*^v$) *at dry-air pressure* $p^d = p_*^d$.

10.6.3 The Triple-Point Coexistence Curve in the Presence of Dry Air

Setting the right-hand sides of (10.143) and (10.144) equal and rearranging yields

The Triple Point Coexistence Curve in the Presence of Dry Air with Specified Dry Partial Pressure $p^d = p_*^d$

$$\left(c^l - c^f\right) T_* \ln\left(\frac{T_*}{T_0}\right) - L_0^f\left(1 - \frac{T_*}{T_0}\right) + \left(\alpha^f - \alpha^l\right)\left(p_*^d + p_*^v - p_0^{sat}\frac{T_*}{T_0}\right) = 0. \quad (10.145)$$

For specified dry partial pressure $p^d = p_*^d$, this equation constrains the 3D coordinates T_*, p_*^v, and p_*^d of the 3D $\left(T, p^v, p^d\right)$ phase diagram of Fig. 10.5.

The triple-point coexistence curve is depicted on Fig. 10.5 by the arrowed curve in magenta that passes through the end points $\left(T, p^v, p^d\right) = \left(T_0, p_0^{sat}, 0\right)$ and $\left(T, p^v, p^d\right) = \left(T_*, p_*^v, p_*^d\right)$. Increasing p_*^d continuously as a parameter, away from $p_*^d = 0$, traces out this curve. We can obtain further insight regarding the triple-point coexistence curve (10.145) via approximation.

In the *absence* of dry air (i.e. $p^d \equiv 0$), $\left(T_*, p_*^v\right) = \left(T_0, p_0^{sat}\right)$ satisfies (10.145); this corresponds to the triple point of pure-water substance, depicted in Fig. 10.5 by the yellow point on the $p^d = 0$ plane. In the *presence* of dry air (i.e. $p^d \neq 0$), $\left(T_*, p_*^v\right) = \left(T_0, p_0^{sat}\right)$ no longer satisfies (10.145), due to the (now) non-zero contribution of p_*^d. For Earth's atmosphere, however, the $\left(\alpha^f - \alpha^l\right)$ term in (10.145) is orders of magnitude smaller than the other terms. Neglecting this term, (10.145) reduces to

$$\left(c^l - c^f\right) T_* \ln\left(\frac{T_*}{T_0}\right) - L_0^f\left(1 - \frac{T_*}{T_0}\right) = 0, \quad (10.146)$$

with solution $T_* = T_0$. This means that, *with this approximation*, it is then possible for liquid cloud and frozen cloud to coexist *at triple-point temperature T_0 of pure-water substance* for *any* value of dry-air pressure, p^d. This behaviour can be contrasted with the unapproximated situation, where $T_* = T_0$ *only holds* at $p^d = 0$ (i.e. in the total absence of dry air). *With* this approximation (namely neglect of the $\left(\alpha^f - \alpha^l\right)$ term in (10.145)), the triple-point coexistence curve on Fig. 10.5 is perfectly vertical, whereas *without* this approximation, it has a slight, backwards vertical tilt.

10.6.4 The Distribution of Water Substance at the Triple Point of Water

Recall that for some specified dry-air pressure, $p^d = p_*^d$, water vapour, liquid water, and frozen water can only coexist in a parcel of liquid-and-frozen cloud at the triple point of water $\left(T, p^v, p^d\right) = \left(T_*, p_*^v, p_*^d\right)$. The corresponding values of T_* and p_*^v are obtained from simultaneous solution of (10.143) and (10.144). The question now is:

- What is the distribution of water substance within a parcel at the triple point; that is, what are the values of mass fractions q^v, q^l, and q^f there?

Velasco and Fernández-Pineda (2007) have examined this question in detail for *pure-water substance* (i.e. in the absence of dry air, for which $q^d \equiv 0$ and $p^d = p_*^d \equiv 0$). For a specified mass fraction q^d of dry air in a closed parcel of liquid and frozen cloud, it is fairly straightforward to extend Velasco and Fernández-Pineda (2007)'s analysis[27] to additionally include the presence and influence of dry air. Without going into too much detail, how to do so is now outlined.

For a closed parcel of cloudy air, mass fractions q^d, q^v, q^l, and q^f satisfy

$$q^v + q^l + q^f \equiv q^w \equiv 1 - q^d = \text{constant}, \tag{10.147}$$

$$\alpha^v q^v + \alpha^l q^l + \alpha^f q^f = \alpha - \alpha^d q^d, \tag{10.148}$$

$$\eta^v q^v + \eta^l q^l + \eta^f q^f = \eta - \eta^d q^d. \tag{10.149}$$

The first of these three equations follows from (10.71); mass fractions of all constituents of a parcel must sum to unity. The second and third equations exploit the property, given in Section 9.4.8, that various thermodynamic quantities may be written as the mass-weighted sum of the corresponding basic quantities given in Tables 9.4 and 9.5. In particular, this applies to specific volume (α) and specific entropy (η); these are two state variables that specify the thermodynamic state of the system at the triple point.

Equations (10.147)–(10.149) comprise a set of three linear equations for the three mass fractions q^v, q^l, and q^f. The coefficients on the left-hand sides are determined from Tables 9.4 and 9.5 by evaluating quantities at the triple point $\left(T, p^v, p^d\right) = \left(T_*, p^v_*, p^d_*\right)$. For known values of α and η, the right-hand sides of these equations are also known, since q^d, α^d, and η^d are all of known value.

Solving linear equation set (10.147)–(10.149) for q^v, q^l, and q^f leads to:

The Distribution of Water Substance at the Triple Point of Water

- The distribution of water substance within a closed cloudy-air parcel at the triple point of water is uniquely determined by the values of the two state parameters, α and η, within the parcel.
- The corresponding mass fractions q^v, q^l, and q^f can then be determined from solution of (10.147)–(10.149) so that

$$q^v = \frac{\alpha^{lf}\left(\eta - \eta^l - \eta^{ld} q^d\right) - \eta^{lf}\left(\alpha - \alpha^l - \alpha^{ld} q^d\right)}{\alpha^{lf}\eta^{lv} - \alpha^{lv}\eta^{lf}}, \tag{10.150}$$

$$q^l = \frac{\alpha^{fv}\left(\eta - \eta^f - \eta^{fd} q^d\right) - \eta^{fv}\left(\alpha - \alpha^f - \alpha^{fd} q^d\right)}{\alpha^{lf}\eta^{lv} - \alpha^{lv}\eta^{lf}}, \tag{10.151}$$

$$q^f = \frac{\alpha^{vl}\left(\eta - \eta^v - \eta^{vd} q^d\right) - \eta^{vl}\left(\alpha - \alpha^v - \alpha^{vd} q^d\right)}{\alpha^{lf}\eta^{lv} - \alpha^{lv}\eta^{lf}}, \tag{10.152}$$

where, for arbitrary F,

$$F^{ij} \equiv F^j - F^i. \tag{10.153}$$

Note that setting $q^d \equiv 0$ in (10.150)–(10.152) recovers Velasco and Fernández-Pineda (2007)'s solution (their (5)–(7)) for the phases of pure-water substance, as it should.

Further useful information may be obtained from (10.150)–(10.152) by constructing the specific volume-specific entropy ($\alpha - \eta$) phase diagram at the triple point; see Fig. 10.7. To construct

[27] John Thuburn kindly brought their analysis to the present author's attention.

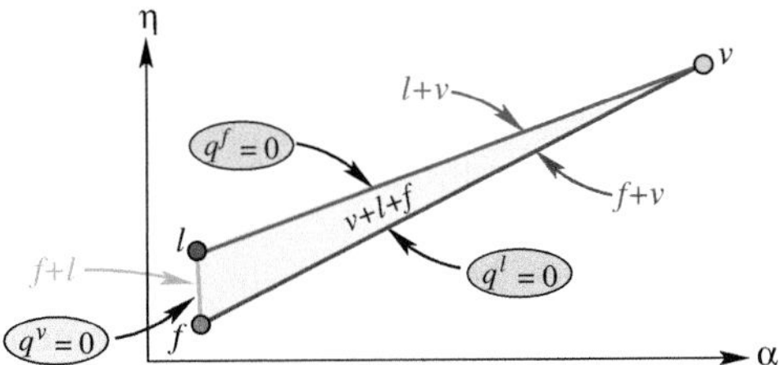

Figure 10.7 Specific volume-specific entropy ($\alpha - \eta$) phase diagram at the triple point of water substance. Water–vapour (v), liquid–water (l), and frozen–water (f) phases coexist everywhere within the yellow triple-point triangle; this is indicated by $v+l+f$. Similarly, only two phases coexist along triangle sides, and a single phase at triangle vertices. Triangle sides are defined by the equations shown within ellipses. See text for further details.

Fig. 10.7, consider the conditions under which one of the three phases of water substance at the triple point is lost. This occurs when the corresponding mass fraction goes from being non-zero to zero. Thus:

- Water vapour is lost when $q^v \to 0$.
- Liquid water is lost when $q^l \to 0$.
- Frozen water is lost when $q^f \to 0$.

Setting $q^v \equiv 0$, $q^l \equiv 0$, and $q^f \equiv 0$ in (10.150)–(10.152), respectively, and rearranging then leads to

$$q^v \equiv 0 \quad \Rightarrow \quad \eta = \frac{\eta^{lf}}{\alpha^{lf}} \left(\alpha - \alpha^l - \alpha^{ld} q^d \right) + \eta^l + \eta^{ld} q^d, \qquad \text{(green line)} \tag{10.154}$$

$$q^l \equiv 0 \quad \Rightarrow \quad \eta = \frac{\eta^{fv}}{\alpha^{fv}} \left(\alpha - \alpha^f - \alpha^{fd} q^d \right) + \eta^f + \eta^{fd} q^d, \qquad \text{(blue line)} \tag{10.155}$$

$$q^f \equiv 0 \quad \Rightarrow \quad \eta = \frac{\eta^{vl}}{\alpha^{vl}} \left(\alpha - \alpha^v - \alpha^{vd} q^d \right) + \eta^v + \eta^{vd} q^d. \qquad \text{(red line)} \tag{10.156}$$

Each of these three equations defines a straight line in the $\alpha - \eta$ plane. Plotted together on the $\alpha - \eta$ phase diagram, they form the *triple-point triangle* of Fig. 10.7.

Three situations occur, according to the position of a point (α, η). When it:

1. Falls within the triangle, all three phases of water substance are present.
2. Lies on a triangle side, only two phases are present.
3. Is located at a vertex, only a single phase is present.

Which phases are present, according to situation, is summarised on Fig. 10.7. Corresponding mass fractions are obtained by substitution of the values of (α, η) into (10.150)–(10.152).

The reader may wonder why the triple-point triangle has the shape and orientation depicted on Fig. 10.7. It is because of Clapeyron's equation (10.30) for slopes dp/dT along coexistence curves on the $T - p$ phase diagram; see Section 10.4.6. Rewriting (10.30), using (10.153), gives

$$\frac{dp}{dT} = \frac{\eta^B - \eta^A}{\alpha^B - \alpha^A} \equiv \frac{\eta^{AB}}{\alpha^{AB}}, \tag{10.157}$$

where $A = v, l, f$ is a particular phase and $B = v, l, f$ is a different phase. Note that η^{lf}/α^{lf}, η^{fv}/α^{fv}, and η^{vl}/α^{vl} are also the slopes of the lines defined by (10.154)–(10.156), respectively. The slopes $d\eta/d\alpha$ of the sides of the triple-point triangle on the $\alpha - \eta$ phase diagram are therefore exactly the same as the slopes dp/dT for the coexistence curves at the triple point on the $T-p$ phase diagram (i.e. the two sets of slopes are parallel to one another). Examining Figs. 10.6b and 10.7, it is seen that the slopes of blue and red lines are quite close to one another, and those of green lines have a slight backwards vertical tilt. These observations explain why the triple-point triangle has the shape and orientation that it does.

Velasco and Fernández-Pineda (2007) show that further examination of the triple-point triangle provides additional useful information. They provide a geometrical interpretation of this triangle; this leads to generalisation (to *three* coexisting phases) of so-called *'lever rules'* for determining the mass fractions of *two* coexisting phases (Callen, 1985).

They also examine the transfer of energy to a parcel of water substance at the triple point by reversible heating/cooling and/or reversible compression/expansion, whilst holding temperature and pressure constant. Doing so results in a redistribution of water substance between its three phases (i.e. to a change in the mass fractions q^v, q^l, and q^f). They show (in the absence of dry air) that the amount present of each phase changes according to whether energy is transferred *isochorically* (i.e. at constant volume, α, corresponding to vertical displacement on Fig. 10.7), or *isentropically* (i.e. at constant entropy, η, corresponding to horizontal displacement on Fig. 10.7). Similar behaviour also holds at the triple point of water substance in the presence of dry air.

10.6.5 Thermodynamical Consistency

It was found in the preceding analysis that the determination of T_* in the presence of dry air is greatly simplified when the $\left(\alpha^f - \alpha^l\right)$ term in (10.145) is neglected; T_* then reduces to the known value T_0. However, *only* introducing this approximation into (10.145) but not elsewhere is *thermodynamically inconsistent* and to be avoided.

If one wishes to take advantage of the simplification afforded by neglect of the $\left(\alpha^f - \alpha^l\right)$ term in (10.145), one should first appropriately approximate the basic Gibbs potentials (9.32) and (9.35) for pure liquid water (g^l) and pure frozen water (g^f), respectively. Having done so, everything thermodynamic, *without further approximation*, should then be obtained using the two approximated Gibbs potentials. This (consistently) leads to approximate forms for any thermodynamic quantity or relation whose unapproximated form depends upon α^l and/or α^f and, in particular, to (10.146). *This procedure guarantees thermodynamical consistency* between *all* thermodynamic variables and relations and is highly recommended.

Neglect of the $\left(\alpha^f - \alpha^l\right)$ term in (10.145) can be achieved by a priori approximation of the basic Gibbs potentials (9.32) and (9.35). This can be accomplished in two different ways:

1. Neglecting the α^l and α^f terms in (9.32) and (9.35), respectively ('Approximation 1').
2. Retaining these terms, but making the weaker assumption that $\alpha^f = \alpha^l \, (\neq 0)$ ('Approximation 2').

Whilst respecting the principle of thermodynamical consistency, we now compare and contrast the ensuing impact each of these two approximations has on the thermodynamics. In particular, we examine the ensuing impact on the basic specific enthalpies (h^v, h^l, h^f) and on the associated latent heats (enthalpies) of water substance ($\mathbb{L}^{l \to v}$, $\mathbb{L}^{f \to l}$, $\mathbb{L}^{f \to v}$). Comparative results are summarised in Table 10.3; to facilitate comparison, all terms that depend on α^l and/or on α^f are coloured red. Although either approximation of the basic Gibbs potentials g^l and g^f reduces (10.145) to the simpler form (10.146), the ensuing impact on the basic specific enthalpies and latent heats (enthalpies) of water substance is nevertheless different.

Quantity	Definition	Unapproximated	Approximation 1 ($\alpha^f = \alpha^l \equiv 0$)	Approximation 2 ($\alpha^f = \alpha^l \neq 0$)
$g^v(T,p)$	–	$\left(L_0^v + L_0^f\right)\left(1 - \frac{T}{T_0}\right) - c_p^v T \ln\left(\frac{T}{T_0}\right) + R^v T \ln\left(\frac{p}{p_0^{sat}}\right)$	$\left(L_0^v + L_0^f\right)\left(1 - \frac{T}{T_0}\right) - c_p^v T \ln\left(\frac{T}{T_0}\right) + R^v T \ln\left(\frac{p}{p_0^{sat}}\right)$	$\left(L_0^v + L_0^f\right)\left(1 - \frac{T}{T_0}\right) - c_p^v T \ln\left(\frac{T}{T_0}\right) + R^v T \ln\left(\frac{p}{p_0^{sat}}\right)$
$g^l(T,p)$	–	$L_0^f\left(1 - \frac{T}{T_0}\right) - c^l T \ln\left(\frac{T}{T_0}\right) + \alpha^l\left(p - p_0^{sat}\frac{T}{T_0}\right)$	$L_0^f\left(1 - \frac{T}{T_0}\right) - c^l T \ln\left(\frac{T}{T_0}\right)$	$L_0^f\left(1 - \frac{T}{T_0}\right) - c^l T \ln\left(\frac{T}{T_0}\right) + \alpha^l\left(p - p_0^{sat}\frac{T}{T_0}\right)$
$g^f(T,p)$	–	$-c^f T \ln\left(\frac{T}{T_0}\right) + \alpha^f\left(p - p_0^{sat}\frac{T}{T_0}\right)$	$-c^f T \ln\left(\frac{T}{T_0}\right)$	$-c^f T \ln\left(\frac{T}{T_0}\right) + \alpha^l\left(p - p_0^{sat}\frac{T}{T_0}\right)$
$h^v(T)$	$g^v - Tg_T^v$	$L_0^v + L_0^f + c_p^v T$	$L_0^v + L_0^f + c_p^v T$	$L_0^v + L_0^f + c_p^v T$
$h^l(T,p)$	$g^l - Tg_T^l$	$L_0^f + c^l T + \alpha^l p$	$L_0^f + c^l T$	$L_0^f + c^l T + \alpha^l p$
$h^f(T,p)$	$g^f - Tg_T^f$	$c^f T + \alpha^f p$	$c^f T$	$c^f T + \alpha^l p$
$\mathbb{L}^{l\to v}(T,p)$	$h^v - h^l$	$L_0^v + \left(c_p^v - c^l\right) T - \alpha^l p$	$L_0^v + \left(c_p^v - c^l\right) T$	$L_0^v + \left(c_p^v - c^l\right) T - \alpha^l p$
$\mathbb{L}^{f\to l}(T,p)$	$h^l - h^f$	$L_0^f + \left(c^l - c^f\right) T + \left(\alpha^l - \alpha^f\right) p$	$L_0^f + \left(c^l - c^f\right) T$	$L_0^f + \left(c^l - c^f\right) T$
$\mathbb{L}^{f\to v}(T,p)$	$h^v - h^f$	$L_0^v + L_0^f + \left(c_p^v - c^f\right) T - \alpha^f p$	$L_0^v + L_0^f + \left(c_p^v - c^f\right) T$	$L_0^v + L_0^f + \left(c_p^v - c^f\right) T - \alpha^l p$

Table 10.3 Thermodynamical consistency of two approximations to the basic Gibbs potentials for water substance. Approximation 1 sets $\alpha^f \equiv \alpha^l \equiv 0$ in $g^l(T,p)$ and $g^f(T,p)$; Approximation 2 sets $\alpha^f = \alpha^l \neq 0$ in $g^f(T,p)$; and both approximations leave $g^v(T,p)$ unchanged. Specific enthalpies (h^v, h^l, h^f) and specific latent heats (enthalpies) ($\mathbb{L}^{l\to v}, \mathbb{L}^{f\to l}, \mathbb{L}^{f\to v}$) are then computed from the basic Gibbs potentials (g^v, g^l, g^f) in a thermodynamically consistent manner, using definitions (10.22)–(10.27) (repeated in Column 2). Results for the unapproximated and two approximated cases are displayed in Columns 3–5, respectively. All terms that depend on α^l and/or on α^f are in red. See text for further details.

10.6.5.1 Approximation 1 – Setting $\alpha^f = \alpha^l = 0$ in (9.32) and (9.35)

Setting $\alpha^l = \alpha^f = 0$ in (9.32) and (9.35) simplifies them to

$$g^l_{\text{Approximation 1}}(T) = -c^l T \ln\left(\frac{T}{T_0}\right) + L^f_0\left(1 - \frac{T}{T_0}\right), \tag{10.158}$$

$$g^f_{\text{Approximation 1}}(T) = -c^f T \ln\left(\frac{T}{T_0}\right). \tag{10.159}$$

Approximation 1 corresponds to that used to obtain the 'almost exact' approximations of the thermodynamic-energy equation given in Section 3.6.2; see also (10.49) and (10.50).

The dependence of g^l and g^f on p is lost by Approximation 1; g^l and g^f then depend on T alone. As shown in Section 10.6.3, setting $\alpha^l = \alpha^f = 0$ simplifies the triple-point coexistence curve (10.145) to (10.146), with solution $T_* = T_0$. Using (10.158) and (10.159) (with $T_* = T_0$) in the analysis of Sections 10.5.3 and 10.5.4 (instead of using unapproximated (9.32) and (9.35)) simplifies the *implicit* equations (for $p^v_s(p, T)$), (10.63) and (10.70), to the *explicit* equations

$$p^v_s(T) = p^{sat}_0\left(\frac{T}{T_0}\right)^{\frac{(c^v_p - c^l)}{R^v}} \exp\left[-\frac{L^v_0}{R^v T}\left(1 - \frac{T}{T_0}\right)\right], \qquad T \geq T_0, \tag{10.160}$$

$$p^v_s(T) = p^{sat}_0\left(\frac{T}{T_0}\right)^{\frac{(c^v_p - c^f)}{R^v}} \exp\left[-\frac{\left(L^v_0 + L^f_0\right)}{R^v T}\left(1 - \frac{T}{T_0}\right)\right], \qquad T \leq T_0, \tag{10.161}$$

respectively. Evaluation of (10.160) or (10.161) at $T = T_* = T_0$ delivers $p^v_* \equiv p^v_s(p_*, T_*) = p^{sat}_0$. Approximations (10.160) and (10.161), derived for water substance in the presence of dry air, are identical to the Clausius–Clapeyron approximations (10.41) and (10.45), respectively, derived in the absence of dry air.

The impact (see Table 10.3) of Approximation 1 on the basic specific enthalpies and latent heats (enthalpies) of water substance is to lose the dependency on p of h^l and h^f and, consequently, also of the latent heats (enthalpies) of vaporisation ($\mathbb{L}^{l\to v}$), fusion ($\mathbb{L}^{f\to l}$), and sublimation ($\mathbb{L}^{f\to v}$) (cf. columns 3 and 4).

10.6.5.2 Approximation 2 – Setting $\alpha^f = \alpha^l$ ($\neq 0$) in (9.35)

Making the weaker assumption $\alpha^f = \alpha^l$ ($\neq 0$) in (9.35) gives

$$g^f_{\text{Approximation 2}}(p, T) = -c^f T \ln\left(\frac{T}{T_0}\right) + \alpha^l\left(p - p^{sat}_0 \frac{T}{T_0}\right), \tag{10.162}$$

with (9.32) remaining unchanged, that is,

$$g^l(p, T) = -c^l T \ln\left(\frac{T}{T_0}\right) + \alpha^l\left(p - p^{sat}_0 \frac{T}{T_0}\right) + L^f_0\left(1 - \frac{T}{T_0}\right). \tag{10.163}$$

Clearly, (10.162) and (10.163) are of less simple form than (10.159) and (10.158), respectively, with dependence on p now being retained. With this approximation (i.e. $\alpha^f = \alpha^l \neq 0$), triple-point coexistence curve (10.145) (in the presence of dry air) nevertheless still simplifies to (10.146) and thereby still leads to $\left(T_*, p^v_*\right) = \left(T_0, p^{sat}_0\right)$. Using (10.162) and (10.163) (with $T_* = T_0$) in the analysis of Sections 10.5.3 and 10.5.4 (instead of using (9.32) and (9.35)) results in very little simplification. Both (10.63) and (10.70) remain *implicit* equations for saturation vapour pressure, $p_s^v(p,T)$, and the *only simplification* (apart from replacing T_* by T_0) is to replace constant α^f by α^l in (10.70).

In contradistinction to Approximation 1, Approximation 2 retains dependency on p of h^l, h^f, $\mathbb{L}^{l\to v}$, and $\mathbb{L}^{f\to v}$, albeit this is lost for $\mathbb{L}^{f\to l}$ and is only approximate for $\mathbb{L}^{f\to v}$ (cf. columns 3 and 5 of Table 10.3).

10.6.5.3 Comparative Assessment of the Two Approximations

The upside of using either of the two approximations – defined in Sections 10.6.5.1 and 10.6.5.2 – of the basic Gibbs potentials, g^l and g^f, is that it significantly simplifies matters; T_* reduces to known value T_0, which means that T_* no longer needs to be computed. The downside is that this is achieved at the cost of no longer being able to account for the approximately 0.01 °C difference (discussed in Section 10.6.1) between the triple-point temperature (T_*) in the absence, and in the presence (near Earth's surface) of dry air.

Note that although both approximations lead to the significant simplification $\left(T_*, p_*^v\right) = \left(T_0, p_0^{sat}\right)$, the thermodynamically consistent approximation of the saturation vapour-pressure curves (10.63) and (10.70) nevertheless depends upon which of the two approximations of the basic Gibbs potentials g^l and g^v is used. When setting $\alpha^l = \alpha^f = 0$ in (9.32) and (9.35), approximations (10.160) and (10.161) of the saturation vapour-pressure curves must be used; to do otherwise would be thermodynamically inconsistent. However, when α^f is instead set equal to α^l in (9.35), the appropriate forms of the saturation vapour-pressure curves are (10.63) and (10.70), with T_* set equal to T_0, and with α^f set equal to α^l in (10.70).

The reader may naturally wonder:

- Which of the two approximations of the basic Gibbs potentials g^l and g^v is better?

The trade-off is between simplicity and accuracy and is a judgement call:

- The first approximation (neglecting the α^l and α^f terms in (9.32) and (9.35)) is a *little simpler*, whereas the second (setting $\alpha^f = \alpha^l$) is *marginally more accurate*, with both being excellent approximations.
- With either approximation, the ability to account for the approximately 0.01 °C difference between the triple-point temperature in the absence and presence of dry air is lost.[28]
- Overall, one may as well opt for the added simplicity of Approximation 1 (as is frequently done in the meteorological literature) and forgo the marginal relative improvement in accuracy of Approximation 2.

10.6.5.4 Remarks

The analysis of this section (Section 10.6.5) illustrates application of the principle of thermodynamical consistency in the relatively simple context of a highly tractable, yet fairly realistic and understandable representation of cloudy air. More complete representations, such as those described in the general meteorological literature and in IOC et al. (2010) and Feistel et al. (2010a,b), are (unsurprisingly) far more complicated. It is therefore correspondingly much more difficult to assess the impact of making ad hoc approximations, here and there, for convenience. Such approximations are frequently made in meteorological modelling to both improve computational efficiency and to simplify, but often with unintended and poorly understood consequences (Thuburn, 2017).

[28] This can be very important in certain situations, e.g. to determine whether precipitation will fall as rain, freezing rain, or snow at near-freezing temperatures.

Application of the principle of thermodynamical consistency is thus a very valuable tool for ensuring that there are no spurious sources or sinks of energy or entropy that often occur with ad hoc approximations. This is particularly important for climate modelling, since thermodynamical inconsistency can lead to a spurious drift of the model's climate away from reality. For accuracy reasons, it is also important for weather forecasting. However, climate drift in this latter context is somewhat less problematic (albeit still undesirable) since:

1. The time periods of integration are much shorter (with correspondingly reduced time for errors to grow unacceptably large).
2. Observed data, used in data-assimilation cycles to provide initial conditions for weather models, continuously nudges any drift of climate in the cycle back towards reality, thereby limiting the growth of any such drift from cycle to cycle.

10.7 DEFINITION OF SOME THERMODYNAMIC QUANTITIES

10.7.1 Definition of Potential Temperature

Potential temperature is a frequently used thermodynamic variable in meteorology and oceanography. We begin with:

The Verbal Definition of Potential Temperature θ

Potential temperature θ is the temperature a fluid parcel would have if brought at constant entropy (equivalently termed '*isentropically*') and at constant composition to a reference pressure p_r.

IOC et al. (2010) note that, strictly speaking, one should add the proviso '*without dissipation of mechanical energy*' to this definition. Satisfaction of all three provisos then ensures reversibility of the notional thermodynamic process. In this definition, '*at constant entropy*' is often also alternatively phrased in the literature as '*adiabatically and reversibly*'. Because entropy is held constant, and assuming a reversible thermodynamic process, the parcel does not exchange any heat with its encompassing environment; see (4.3) with $d\eta \equiv 0$, which then implies $đQ \equiv 0$. The value of reference pressure p_r is arbitrary. It is usually chosen to be either 10^5 Pa $= 1,000$ hPa $= 1,000$ mb or 1.01325×10^5 Pa $= 1013.25$ hPa $= 1013.25$ mb, both being representative values of atmospheric pressure at or near Earth's oceanic surface. In oceanography, seawater is a solution of water substance and solutes (i.e. dissolved salts). *Absolute Salinity*, S_A, is the amount of solute per unit mass of seawater. A thermodynamic operation performed at '*constant composition*' is often phrased in the oceanographic literature as being performed in an '*isohaline*' manner (i.e. at a fixed value of Absolute Salinity S_A so that $dS_A \equiv 0$).

Physically one can imagine a closed fluid parcel being passively transported (downward for an air parcel, and upward for an ocean one) from a fluid's interior to its surface, with no exchange of heat or mass between it and its encompassing environment. It departs with pressure p, temperature T, and composition $\{S^1, S^2, \ldots\}$. It then arrives at the surface where the pressure is assumed to be p_r, a reference pressure. Since it is assumed that the composition is constant throughout the parcel's journey, the final composition is exactly the same as its initial composition:

- But what is the value of the parcel's remaining state variable, the temperature?

It is simply the potential temperature θ, which is uniquely determined via the two imposed constraints during the parcel's journey of:

1. Constant entropy.
2. Constant composition.

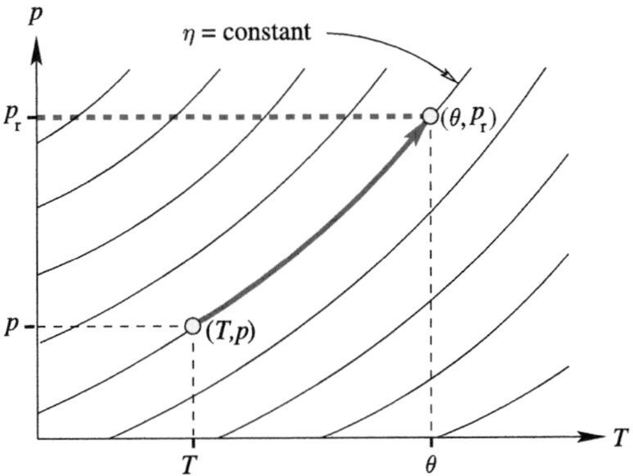

Figure 10.8 Isentropes (curves along which η = constant, in black) on a T-p diagram, at constant composition (i.e. with $S^1, S^2, \ldots$ held constant); after Fig. 3.5 of Ambaum (2010). A fluid parcel with temperature T, pressure p, and specific entropy η notionally undergoes an isentropic (i.e. with η held constant) thermodynamic process. Its temperature then changes to θ (potential temperature), with (constant) reference pressure p_r, and with (unchanged) specific entropy η. Graphically, arbitrary point (T, p) moves along the (red) isentrope to the point (θ, p_r).

We first translate the verbal definition – see the first highlighted box of this section – into an equation for a general fluid parcel. We then apply this equation to obtain an explicit expression for the potential temperature of the special case of a humid (unsaturated) air parcel.

As a preparatory step, specific entropy η can be obtained from a specific Gibbs potential $g = g\left(p, T, S^1, S^2, \ldots\right)$ by use of the identity – see Table 9.2 –

$$\eta\left(p, T, S^1, S^2, \ldots\right) \equiv -g_T\left(p, T, S^1, S^2, \ldots\right). \tag{10.164}$$

In (10.164), the arguments of η and g are the natural variables of g, and subscript 'T' denotes partial differentiation with respect to T whilst holding the other natural variables $(p, S^1, S^2, \ldots)$ constant.

Consider a fluid parcel that has pressure p, temperature T, and composition $\left\{S^1, S^2, \ldots\right\}$ at the beginning of the notional thermodynamic process that defines potential temperature θ; see Fig. 10.8. At the end of the process it has (by the verbal definition of potential temperature) pressure p_r, temperature θ, and unchanged composition $\left\{S^1, S^2, \ldots\right\}$. Since the fluid parcel's entropy is held constant during the process, its final value is equal to its initial value. This leads to:

The Mathematical Definition of Potential Temperature θ

Potential temperature θ of a fluid parcel with temperature T, pressure p, and composition $\left\{S^1, S^2, \ldots\right\}$ is the solution of

$$\eta\left(p_r, \theta, S^1, S^2, \ldots\right) = \eta\left(p, T, S^1, S^2, \ldots\right), \tag{10.165}$$

where specific entropy η is obtained from the Gibbs potential $g\left(p, T, S^1, S^2, \ldots\right)$ of the parcel using definition (10.164), and p_r is a constant, prescribed reference pressure.

(The left-hand side of (10.165) is the same as its right-hand side, but evaluated using the particular values $(p, T) = (p_r, \theta)$.)

Equation (10.165) represents the translation of the verbal definition of the potential temperature for a general fluid parcel into an equation using thermodynamic symbols that we have grown to know and love. Graphically, point (T, p) on Fig. 10.8 moves along the (red) isentrope that passes through it to the point (θ, p_r). Point (θ, p_r) lies at the intersection of the (red) isentrope passing through point (T, p) with the (red) reference line $p = p_r =$ constant.

For a given Gibbs thermodynamic potential, $g = g\left(p, T, S^1, S^2, \ldots\right)$:

- Equation (10.164) provides the means to evaluate specific entropy (η).
- Equation (10.165) then provides an implicit equation that can be solved to obtain potential temperature (θ) in terms of given values of $\left(p, T, S^1, S^2, \ldots\right)$ and of reference pressure p_r. In practice this usually needs to be done numerically[29] due to the implicit nature of (10.165). For a humid (unsaturated) air parcel it is, however, possible and instructive to *explicitly* solve (10.165) for θ, as now shown.

10.7.2 Potential Temperature for Humid (Subsaturated) Air

The specific Gibbs potential for a humid (subsaturated) air parcel is given by (10.74), that is, by

$$g\left(p, T, q^w\right) = \left(1 - q^w\right) g^d\left(p, T\right) + q^w g^v\left(p, T\right), \tag{10.166}$$

where (from (10.73)) $q^w = q^v = 1 - q^d =$ constant, and $g^d\left(p, T\right)$ and $g^v\left(p, T\right)$ are given by (10.9) and (10.11), respectively, with $a = 1 - q^w$ therein. The corresponding specific entropy $\eta = \eta\left(p, T, q^w\right)$ can then be obtained by insertion of (10.166) into (10.164). Thus the specific entropy for a humid (subsaturated) air parcel is

$$\eta\left(p, T, q^w\right) = \left(1 - q^w\right) \eta^d\left(p, T\right) + q^w \eta^v\left(p, T\right), \tag{10.167}$$

where (see (10.110) and (10.111), and (10.108) with $a_{sat} = 1 - q^w$)

$$\eta^d\left(p, T\right) = c_p^d\left[1 + \ln\left(\frac{T}{T_0}\right)\right] - R^d \ln\left\{\frac{\left(1 - q^w\right)\varepsilon}{\left[\left(1 - q^w\right)\varepsilon + q^w\right]}\frac{p}{p_0^d}\right\}, \tag{10.168}$$

$$\eta^v\left(p, T\right) = c_p^v\left[1 + \ln\left(\frac{T}{T_0}\right)\right] - R^v \ln\left\{\frac{q^w}{\left[\left(1 - q^w\right)\varepsilon + q^w\right]}\frac{p}{p_0^{sat}}\right\} + \frac{L_0^v + L_0^f}{T_0}. \tag{10.169}$$

Inserting (10.167)–(10.169) (expressions for specific entropy) into (10.165) (the mathematical definition of potential temperature) yields

$$\begin{aligned}
0 &= \eta\left(p_r, \theta, q^w\right) - \eta\left(p, T, q^w\right) \\
&= \left(1 - q^w\right)\left[\eta^d\left(p_r, \theta\right) - \eta^d\left(p, T\right)\right] + q^w\left[\eta^v\left(p_r, \theta\right) - \eta^v\left(p, T\right)\right] \\
&= \left(1 - q^w\right)\left[c_p^d \ln\left(\frac{\theta}{T}\right) - R^d \ln\left(\frac{p_0}{p}\right)\right] + q^w\left[c_p^v \ln\left(\frac{\theta}{T}\right) - R^v \ln\left(\frac{p_r}{p}\right)\right] \\
&= \widetilde{c}_p \ln\left(\frac{\theta}{T}\right) + \widetilde{R} \ln\left(\frac{p}{p_r}\right),
\end{aligned} \tag{10.170}$$

where

$$\widetilde{c}_p \equiv \left(1 - q^w\right) c_p^d + q^w c_p^v, \quad \widetilde{R} \equiv \left(1 - q^w\right) R^d + q^w R^v. \tag{10.171}$$

Solving (10.170) for θ then gives:

[29] This can be done to machine precision using a suitable numerical algorithm, such as Newton–Raphson iteration (with generally rapid, quadratic convergence); see e.g. Chapter 9 of Press et al. (1992).

The Potential Temperature of a Humid (Subsaturated) Air Parcel

$$\theta = T\left(\frac{p}{p_r}\right)^{-\widetilde{R}/\widetilde{c}_p}, \tag{10.172}$$

where $\widetilde{c}_p$ and $\widetilde{R}$ are defined by (10.171).

10.7.3 Potential Temperature for Dry Air

For the special case of a dry-air parcel, (10.172) reduces to:

The Potential Temperature of a Dry-Air Parcel

$$\theta = T\left(\frac{p}{p_r}\right)^{-R^d/c_p^d}, \tag{10.173}$$

since $\widetilde{R}$ in (10.172) then reduces to R^d, and $\widetilde{c}_p$ to c_p^d.

Note that R^d and c_p^d in (10.173) (for dry air) are *constants*, whereas $\widetilde{R}$ and $\widetilde{c}_p$ in (10.172) (for humid, unsaturated air) both depend on the mass fractions of water substance, which depend on location $\mathbf{r}$ and time t.

10.7.4 Potential Temperature versus Entropy

Potential temperature is intrinsically defined in terms of entropy; see (10.165). Ambaum (2010, pp. 62–63) gives an interesting and thought-provoking discussion of the relative merits of use of these two quantities in atmospheric science.

He notes that specific entropy is based on extensive total entropy, and therefore:

- Entropies can be added.
- Thermodynamic subsystems can be joined together.
- The total entropy of the composite system can be calculated.

He further notes that *these properties do not hold for potential temperature*. For example, potential temperature (10.172) for humid air does not analogously decompose into the sum of potential temperatures for dry air and water vapour.

He concludes that:

- Entropy is a fundamental thermodynamic quantity, whereas potential temperature is a derived one.
- Use of potential temperature may distract from consideration of more fundamental aspects of thermodynamics.
- It appears illogical to replace a variable (η) based on an extensive quantity, with an intrinsically intensive quantity (θ).

His conclusions do not mean that one should entirely avoid potential temperature for atmospheric applications. There is, after all, a lot of literature that uses it (and its many variants), and its dry form is particularly suitable for studies of dry dynamics. One should, however, be mindful that there is a more fundamental alternative available and, depending on application, one should consider using entropy instead of potential temperature. A further possibility is to instead use potential enthalpy.

10.7.5 Definition of Potential Enthalpy

- Why might one be interested in potential enthalpy?

Potential enthalpy is a function of potential temperature and, as discussed in Section 10.7.4, the use of potential temperature has some drawbacks. So one might think that potential enthalpy would suffer from similar drawbacks and therefore not be of any great interest. However, as discussed at length in Appendix A of IOC et al. (2010), potential enthalpy has some very good conservation properties, so good in fact that IOC et al. (2010) highly recommend its use! Furthermore, potential enthalpy is based on enthalpy, and enthalpy plays a crucial role during phase transitions of water substance – see Section 10.4.4. We start with:

The Verbal Definition of Potential Enthalpy h^0

Potential enthalpy h^0 is the enthalpy a fluid parcel would have if brought at constant entropy and constant composition to a reference pressure p_r.

Graphically, because both entropy and composition are held constant during the notional thermodynamic process, point (p, T) on Fig. 10.8 again moves along the (red) isentrope that passes through it to the point (θ, p_r).

As a preparatory step for the translation of the verbal definition to an equation, specific enthalpy is obtained from a specific Gibbs potential $g = g\left(p, T, S^1, S^2, \ldots\right)$ by use of the identity – see Table 9.2 –

$$h\left(p, T, S^1, S^2, \ldots\right) = g\left(p, T, S^1, S^2, \ldots\right) - Tg_T\left(p, T, S^1, S^2, \ldots\right). \tag{10.174}$$

In (10.174), the arguments of h and g are the natural variables of g, and subscript 'T' again denotes partial differentiation with respect to T whilst holding the other natural variables $(p, S^1, S^2, \ldots)$ constant.

Consider a fluid parcel that has pressure p, temperature T, and composition $\{S^1, S^2, \ldots\}$ at the beginning of the notional thermodynamic process that defines potential enthalpy $h^0 = h^0\left(p, T, S^1, S^2, \ldots\right)$. During the notional process its entropy and composition are both held constant, and at the end of the process its pressure is p_r. By the definition of potential temperature (which undergoes the same notional thermodynamic process at constant entropy and composition), the temperature at the end of the process is θ, and the composition remains unchanged. Thus the enthalpy at the end of the process is $h = h\left(p_r, \theta, S^1, S^2, \ldots\right)$. But, by the definition of potential enthalpy, the value of specific enthalpy at the end of the process is also $h^0 = h^0\left(p, T, S^1, S^2, \ldots\right)$. Equating these two expressions then leads to:

The Mathematical Definition of Potential Enthalpy h^0

- Potential enthalpy h^0 of a fluid parcel with temperature T, pressure p, and composition $\{S^1, S^2, \ldots\}$ is given by

$$h^0\left(p, T, S^1, S^2, \ldots\right) = h\left(p_r, \theta, S^1, S^2, \ldots\right), \tag{10.175}$$

where potential temperature θ is the solution of (10.165),

$$\eta\left(p_r, \theta, S^1, S^2, \ldots\right) = \eta\left(p, T, S^1, S^2, \ldots\right). \tag{10.176}$$

- In these two equations, specific enthalpy (h) and specific entropy (η) are obtained from the specific Gibbs potential $g = g\left(p, T, S^1, S^2, \ldots\right)$ of the parcel using definitions (10.174) and (10.164), respectively, and p_r is a constant reference pressure.

10.7.6 Definition of Conservative Temperature

Conservative temperature Θ is defined to be potential enthalpy divided by the precise value $c_{p0} \equiv 3991.86795711963\ \mathrm{J\,kg^{-1}\,K^{-1}}$ (IOC et al., 2010). Thus:

The Definition of Conservative Temperature Θ

Conservative temperature of a fluid parcel with temperature T, pressure p, composition $\{S, S_2, \ldots\}$, and potential enthalpy h_0 is given by

$$\Theta\left(p, T, S^1, S^2, \ldots\right) \equiv \frac{h^0\left(p, T, S^1, S^2, \ldots\right)}{c_{p0}}, \tag{10.177}$$

where $c_{p0} \equiv 3991.86795711963\ \mathrm{J\,kg^{-1}\,K^{-1}}$.

The unit of conservative temperature is then, as the name suggests, that (K) of temperature instead of those ($\mathrm{J\,kg^{-1}}$) of potential enthalpy.

One might naturally wonder:

- Why bother scaling potential enthalpy by the constant c_{p0}, when one could simply use potential enthalpy and be done with?

The answer is that with this scaling, conservative temperature (with its good conservation properties) approximately (as in Section 11.3.10 for an ocean) and sometimes exactly (as in the example that follows for the atmosphere) reduces to potential temperature (with its not-as-good properties). This is convenient for those who find potential temperature helpful for understanding what is going on physically.

To see why, consider a dry gas for which specific enthalpy (from Table 9.4) is $h^d = c_p^d T$ and (from (10.173)) $\theta^d = T\left(p/p_r\right)^{-R^d/c_p^d}$. Substitution of these into (10.175) then gives $h^0\left(p, T, S^1, S^2, \ldots\right) = h^d\left(p_r, \theta, S^1, S^2, \ldots\right) = c_p^d \theta^d$. Thus for dry air, potential enthalpy h^0 is simply potential temperature multiplied by the constant $c_p^d = c_{p0}$ and (from (10.177)), the corresponding conservative temperature Θ is exactly equal to potential temperature θ. This simple example illustrates the scaling of potential enthalpy h^0 by c_{p0} to obtain conservative temperature Θ. Scaling does not change the physics in any way, and it is done only for convenience of physical interpretation.

10.7.7 Definition of Potential Density

Potential density is another variable in widespread use in oceanography that intrinsically depends on potential temperature. We start with:

The Verbal Definition of Potential Density ρ^θ

Potential density ρ^θ is the density a fluid parcel would have if brought at constant entropy and constant composition to a reference pressure p_r.

Because both entropy and composition are held constant during the notional thermodynamic process, point (p, T) on Fig. 10.8 yet again moves along the (red) isentrope that passes through it to the point (θ, p_r).

As a preparatory step for the translation of the verbal definition of potential density to an equation, specific volume is obtained from a specific Gibbs potential $g = g(p, T, S)$ by use of the identity – see Table 9.2 –

$$\alpha(p, T, S_1, S_2, \ldots) \equiv \frac{\partial g(p, T, S_1, S_2, \ldots)}{\partial p} \tag{10.178}$$

$$\Downarrow$$

$$\rho(p, T, S_1, S_2, \ldots) \equiv \frac{1}{\alpha(p, T, S_1, S_2, \ldots)} \equiv \left[\frac{\partial g(p, T, S_1, S_2, \ldots)}{\partial p}\right]^{-1}, \tag{10.179}$$

where $\rho = \rho(p, T, S_1, S_2, \ldots)$ is density. Similar reasoning to that given in Section 10.7.5 for potential enthalpy leads to:

The Mathematical Definition of Potential Density ρ^θ

- Potential density ρ^θ of a fluid parcel with temperature T, pressure p, and composition $\{S_1, S_2, \ldots\}$ is given by

$$\rho^\theta(p, T, S_1, S_2, \ldots) = \rho(p_r, \theta, S_1, S_2, \ldots) = \left[\alpha(p_r, \theta, S_1, S_2, \ldots)\right]^{-1}, \tag{10.180}$$

where p_r is a constant reference pressure, and potential temperature θ is the solution of (10.165), that is, of

$$\eta(p_r, \theta, S^1, S^2, \ldots) = \eta(p, T, S^1, S^2, \ldots). \tag{10.181}$$

- Specific entropy $\eta = \eta(p, T, S^1, S^2, \ldots)$ is obtained from the specific Gibbs potential $g = g(p, T, S^1, S^2, \ldots)$ of the parcel by use of the identity $\eta(p, T, S^1, S^2, \ldots) \equiv -\partial g(p, T, S^1, S^2, \ldots)/\partial T$.

10.8 CONCLUDING REMARKS

Two issues were raised in Section 9.1:

1. How can phase transitions of water substance be represented in atmospheric and oceanic models?
2. How can suitable representations of the fundamental equation of state for an ocean be constructed?

To prepare to answer these two questions, we had to first broaden and deepen our hitherto limited treatment of thermodynamics. This was accomplished in the remainder of Chapter 9. Keys to this were:

- The definition of thermodynamical consistency.
- The introduction of thermodynamic potentials, together with an overview of their important properties.

This then facilitated the developments of this chapter to answer the first question raised in Section 9.1 and repeated in Section 10.1 and here. The purpose of the next chapter is to answer the second question.

11

Ocean Thermodynamics

ABSTRACT

For the oceans, there is no simple, physically realistic analogue of the ideal-gas law for the atmosphere. This means that things have to be done *empirically*. The thermodynamics is then most naturally represented by a Gibbs potential, which is a function of its physically measurable natural variables of pressure, temperature, and composition. To provide preliminary insight, two relatively simple, prototypical, oceanic Gibbs potentials are developed from first principles. The first corresponds to one given in a popular textbook. The second tweaks this potential by selectively modifying its functional form. The resulting 'polynomial-like' form then corresponds to that of the comprehensive Thermodynamic Equation of Seawater (TEOS-10) Gibbs potential, but truncated to far fewer terms. Its coefficients can be determined by either fitting them to experimental data – in a manner similar to that used to obtain TEOS-10 coefficients – or by setting them to representative values (e.g. for theoretical studies). This alternative, prototypical Gibbs potential then provides a simple model of the comprehensive TEOS-10 one adopted by the international oceanographic community. The actual definition of the TEOS-10 potential is summarised. An important virtue of this potential is that its pure-water and saline-correction parts are independent of one another. This prepares the way for future improvements when new experimental data become available for either part, with no need to modify the other part. Another important virtue is that two toolboxes of computer software are freely available for the computation of a vast array of thermodynamic quantities from the TEOS-10 Gibbs potential.

11.1 PREAMBLE

The issue to be addressed in this chapter is the second one raised in Section 9.1:

- How can suitable representations of the fundamental equation of state for an ocean be constructed?

There is, unfortunately, no theory comparable in realism to that of the ideal-gas law for the atmosphere to guide us. A realistic thermodynamic potential for the oceans therefore has to be constructed *empirically*. Of the four thermodynamic potentials defined and discussed in Section 9.2, the most natural choice for this purpose is the Gibbs potential since:

- A potential's coefficients have to be determined empirically by fitting them to sets of laboratory measurements plus other available data.
- Available datasets generally use pressure, temperature, and salinity as independent variables since these are all measurable quantities.
- These three independent variables are the natural variables of the Gibbs potential. See Table 9.1.

- Amongst thermodynamic potentials, the Gibbs potential is the only one whose natural variables (p, T, S^i) are all *intensive* quantities (i.e. independent of the total mass of the thermodynamic system); see Section 4.2.1.4. This helps simplify things.
- Only the Gibbs potential maintains its continuity between phases during a phase transition. See Section 10.4.3.

Feistel (2003), Feistel et al. (2008), and Feistel et al. (2010a) describe a highly accurate, unified Gibbs potential for the representation of the thermodynamics (including phase transitions) of:

1. The oceans.
2. The atmosphere (including dry air and cloudy air).
3. The cryosphere.

This comprehensive representation has been adopted by the international oceanic community (IOC et al., 2010). It is known as the *International Thermodynamic Equation of Seawater – 2010*, which is usually abbreviated to TEOS-10.

Depending on intended use, a Gibbs potential does not have to be as sophisticated as the TEOS-10 one. However, if a simplified Gibbs potential is adopted, then – as discussed in Chapter 9 – it is important that everything thermodynamic be obtained from the simplified potential *without further approximation* in order to ensure thermodynamical consistency. For atmospheric applications, Thuburn (2017) provides an illustrative, much simplified version of Feistel et al. (2010a)'s Gibbs potential. A minor generalisation of Thuburn (2017)'s has been derived in Chapter 10.5, for cloudy air. It additionally includes frozen water, as indeed advocated by Thuburn (2017).

For a saline ocean, two relatively simple Gibbs potentials are derived in this chapter from first principles. To prepare the way for this, a methodology to construct a prototypical Gibbs potential for an ocean is given (in Section 11.2). This methodology is first applied (in Section 11.3) to construct the Gibbs potential given in Chapter 1.7.2 of Vallis (2017). Doing so provides some insight into the specification of an appropriate functional form for the development of an empirical Gibbs potential.

The functional form of Vallis (2017)'s potential is quite similar in many ways to a low-order truncation of the highly accurate TEOS-10 one, with its many terms. There are, however, two notable differences, even at low order:

1. For both the pure-water and the saline-correction parts of Vallis (2017)'s potential, there are contributions of the form $T\left[\ln\left(T/T_0\right) - 1\right]$, where $T = T_0$ is a constant reference temperature. No such terms of this form appear in the TEOS-10 potential.
2. For the saline-correction part of the TEOS-10 potential and for compatibility with the theory of ideal solutions, there are two terms containing contributions of the form $x^2 \ln x$, where $x \equiv (S_A/S_u)^{1/2}$; S_A is Absolute Salinity; and S_u is a unit-related constant. No such terms appear in Vallis (2017)'s potential.

To provide a conceptual pathway to the highly sophisticated TEOS-10 Gibbs potential – with its many terms – an alternative, prototypical Gibbs potential for an ocean is therefore developed in Section 11.4, again using the methodology described in Section 11.2. It is of comparable order to the Vallis (2017) potential. However, its (low-order) functional form mimics that of the (high-order) TEOS-10 one more closely than the Vallis (2017) one does. In particular:

1. Contributions of the form $T\left[\ln\left(T/T_0\right) - 1\right]$ are absent.
2. A logarithmic contribution, of desired form, to the saline-correction part of the potential now appears.

To be clear, the goal here is not to improve the accuracy of the Vallis (2017) potential per se. It is instead to improve understanding of the underlying structure of the TEOS-10 potential by examining a low-order, analytically tractable prototype of more similar form than the Vallis (2017) one.

Following these preparatory steps, the TEOS-10 potential, with its many dozens of terms, is described in Section 11.5, including tabulation of its coefficients. An important virtue of both the alternative prototypical Gibbs potential and the TEOS-10 one is that they decompose into *independent* potentials for their pure-water and saline-correction parts. This means that they can be independently approximated or improved according to practical application and new experimental data, respectively. This is an important property of their definition. The availability of two comprehensive toolboxes of freely available computer software for the computation of a vast array of thermodynamic quantities from the TEOS-10 Gibbs potential is noted. Concluding remarks are made in Section 11.6.

11.2 OCEANIC GIBBS POTENTIALS

11.2.1 Decomposition of an Oceanic Gibbs Potential

The complete Gibbs potential for the TEOS-10 representation of seawater is, in the present notation, written as the sum of a pure-water part, $g^W = g^W(p, T)$, and a saline-correction part, $g^S = g^S(p, T, S_A)$. This decomposition is unique; to achieve this, it is assumed that $g^W(p, T)$ and $g^S(p, T, S_A)$ have particular functional forms and that the values of four arbitrary constants are fixed in a particular way. Thus:

Decomposition of an Oceanic Gibbs Potential into Pure-Water and Saline-Correction Parts

An oceanic Gibbs potential $g = g(p, T, S_A)$ may be decomposed into the sum of a pure-water part, $g^W = g^W(p, T)$, and a saline-correction part, $g^S = g^S(p, T, S_A)$, so that

$$g(p, T, S_A) = g^W(p, T) + g^S(p, T, S_A), \tag{11.1}$$

where p is pressure; T is temperature; $S \equiv S_A$ is Absolute Salinity; and

$$g^S(p, T, S_A = 0) = 0, \tag{11.2}$$

to ensure that the total Gibbs potential $g(p, T, S_A)$ reduces to that of pure water $(g^W(p, T))$ in the absence of solute (i.e. when $S_A \equiv 0$).

An important benefit of this decomposition is that *the two parts can be examined and modified independently of one another.*

11.2.2 A Methodology to Construct a Prototypical Oceanic Gibbs Potential

From (9.8) and assuming two substances (pure water and salt), the fundamental thermodynamic relation for an oceanic Gibbs potential is given by

$$dg(p, T, S) = \alpha(p, T, S)\,dp - \eta(p, T, S)\,dT + \mu(p, T, S)\,dS. \tag{11.3}$$

Using the chain rule for differentiation, $dg(p, T, S)$ may be alternatively written as

$$dg(p, T, S) = \frac{\partial g(p, T, S)}{\partial p}dp + \frac{\partial g(p, T, S)}{\partial T}dT + \frac{\partial g(p, T, S)}{\partial S}dS. \tag{11.4}$$

Equating coefficients of dp, dT, and dS between (11.3) and (11.4) then leads to

$$\frac{\partial g\left(p,T,S\right)}{\partial p}=\alpha\left(p,T,S\right),\quad \frac{\partial g\left(p,T,S\right)}{\partial T}=-\eta\left(p,T,S\right),\quad \frac{\partial g\left(p,T,S\right)}{\partial S}=\mu\left(p,T,S\right). \tag{11.5}$$

Cross differentiation of these three equations yields the Maxwell relations:

$$\frac{\partial \eta\left(p,T,S\right)}{\partial p}=-\frac{\partial \alpha\left(p,T,S\right)}{\partial T},\quad \frac{\partial \mu\left(p,T,S\right)}{\partial T}=-\frac{\partial \eta\left(p,T,S\right)}{\partial S},\quad \frac{\partial \mu\left(p,T,S\right)}{\partial p}=\frac{\partial \alpha\left(p,T,S\right)}{\partial S}. \tag{11.6}$$

Now if, from theoretical considerations, $\alpha=\alpha\left(p,T,S\right)$, $\eta=\eta\left(p,T,S\right)$, and $\mu=\mu\left(p,T,S\right)$ were *known functions*, then (subject to analytic tractability) the three first-order partial differential equations in (11.5) could be simultaneously solved (subject to appropriate conditions) to obtain the Gibbs potential $g=g\left(p,T,S\right)$. For the ocean, however, no such functions are known theoretically. Instead, one assumes that such functions can be approximated by series expansions, with associated coefficients. Values for these coefficients can then be determined *empirically* by fitting to a dataset of laboratory measurements.

This approach works very well for $\alpha=\alpha\left(p,T,S\right)$, the *conventional equation of state* for an ocean. A finite Taylor-series expansion of specific volume, $\alpha=\alpha\left(p,T,S\right)$, can be made in terms of polynomials about reference values $\left(p,T,S\right)=\left(p_0,T_0,S_0\right)$, with coefficients determined by data fitting. Generally, the more terms there are, the better the fit and the better the accuracy.[1] This approach, however, does not similarly work for entropy $\eta=\eta\left(p,T,S\right)$ since:

1. Entropy is not a measured quantity, so no data for η.
2. Entropy (η) is not independent of specific volume (α), but tied to it.

So how can these two problems be addressed? Examining each in turn:

1. Specific heat at constant pressure, $c_p=c_p\left(p,T,S\right)$, *is* a measured quantity. Using its definition (4.43), it can therefore be used as a substitute for entropy. Thus

$$\frac{\partial \eta\left(p,T,S\right)}{\partial T}=\frac{1}{T}c_p\left(p,T,S\right). \tag{11.7}$$

   For specified $c_p=c_p\left(p,T,S\right)$, this is a first-order partial differential equation for $\eta=\eta\left(p,T,S\right)$. In a similar manner to that described earlier in this section for $\alpha=\alpha\left(p,T,S\right)$, $c_p=c_p\left(p,T,S\right)$ can be represented as a finite series of polynomials.
2. Consistency between the representations of $\alpha=\alpha\left(p,T,S\right)$ and $\eta=\eta\left(p,T,S\right)$ can be obtained using Maxwell relations (11.6). This imposes constraints on some of the values of the coefficients of the polynomial expansions of $\alpha=\alpha\left(p,T,S\right)$ and $c_p=c_p\left(p,T,S\right)$, and on the form of some integration functions and constants that appear when integrating first-order partial differential equations to obtain $\eta=\eta\left(p,T,S\right)$ and $g=g\left(p,T,S\right)$.

Taking into account these two points, a simple Gibbs potential for an ocean may be constructed as shown in the summary box.

The resulting Gibbs potential then provides a (relatively) simple, complete, consistent thermodynamic description of the oceans. Everything thermodynamic can then be uniquely obtained from it via differentiation and algebra using the laws and rules of equilibrium thermodynamics.

The described methodology is not restricted to empirical constructions. When the functional forms for $\alpha=\alpha\left(p,T,S\right)$ and $c_p=c_p\left(p,T,S\right)$ are actually known theoretically – for example, $\alpha=R^dT/p$ and $c_p^d=$ constant for a dry atmosphere are known theoretically from the kinetic

[1] The maximum number of terms to be retained is, however, limited by the number of measurements available and their accuracy.

theory of gases – then their forms can be used in Steps 1 and 2 instead of polynomial expansions, thereby obviating the need for data fitting.

To illustrate how a Gibbs potential may be constructed using measured data, the methodology is now applied to derive Vallis (2017)'s relatively simple, yet quite accurate prototypical Gibbs potential for an ocean and to examine the assumptions that underlie it. The methodology could be applied in one go to obtain the complete potential directly. However, it is more instructive to instead construct its pure-water and saline-correction parts *separately*, so this is what we do.

Construction Procedure for a Simple Gibbs Potential of an Ocean

1. Expand $\alpha = \alpha\left(p, T, S\right)$ in a Taylor series, with a finite number of terms, about the constant reference values $\left(p, T, S\right) = \left(p_0, T_0, S_0\right)$ where, for the ocean, $S \equiv S_A$ is Absolute Salinity. For brevity, subscript 'A' is omitted from S_A, both here and onwards. Thus

$$\alpha\left(p, T, S\right) = \sum_{i,j,k=0,1,2,\ldots} A_{ijk}\left(p - p_0\right)^i \left(T - T_0\right)^j \left(S - S_0\right)^k, \tag{11.8}$$

where A_{ijk} are the coefficients of the Taylor series about $\left(p, T, S\right) = \left(p_0, T_0, S_0\right)$. We are free to retain as few or as many terms as we wish in this expansion, depending on accuracy requirements. At one extreme we could take a single term, so that $A_{000} = \alpha\left(p_0, T_0, S_0\right) = \alpha_0 =$ constant; this corresponds to a relatively inaccurate, constant-density representation of an ocean. At the other extreme is the Feistel et al. (2010a) expansion that has many dozens of coefficients; this leads to a highly accurate representation (to within experimental error) everywhere within an ocean.

2. Differentiating the first equation of (11.6) with respect to T, and (11.7) with respect to p, leads to

$$\frac{\partial^2 \eta\left(p, T, S\right)}{\partial p \partial T} = -\frac{\partial^2 \alpha\left(p, T, S\right)}{\partial T^2} = \frac{\partial}{\partial p}\left[\frac{1}{T} c_p\left(p, T, S\right)\right], \tag{11.9}$$

$$\Downarrow$$

$$\frac{\partial c_p\left(p, T, S\right)}{\partial p} = -\frac{\partial^2 \alpha\left(p, T, S\right)}{\partial T^2} T. \tag{11.10}$$

Equation (11.10) can be considered to be a consistency equation that links $c_p\left(p, T, S\right)$ and $\alpha\left(p, T, S\right)$. Inserting polynomial expansion (11.8) for $\alpha = \alpha\left(p, T, S\right)$ into (11.10) and integrating the functional form of $c_p = c_p\left(p, T, S\right)$ can be deduced. This introduces an integration function whose form has to be specified.

3. Simultaneously integrate the first equation of (11.6) and (11.7), that is, simultaneously integrate

$$\frac{\partial \eta\left(p, T, S\right)}{\partial p} = -\frac{\partial \alpha\left(p, T, S\right)}{\partial T}, \quad \frac{\partial \eta\left(p, T, S\right)}{\partial T} = \frac{1}{T} c_p\left(p, T, S\right), \tag{11.11}$$

to obtain $\eta = \eta\left(p, T, S\right)$. This also introduces an integration function whose form has to be specified.

4. Using the expressions for $\alpha = \alpha\left(p, T, S\right)$ and $\eta = \eta\left(p, T, S\right)$ of the first three steps, simultaneously integrate the first two equations of (11.5), that is, simultaneously integrate

$$\frac{\partial g\left(p, T, S\right)}{\partial p} = \alpha\left(p, T, S\right), \quad \frac{\partial g\left(p, T, S\right)}{\partial T} = -\eta\left(p, T, S\right), \tag{11.12}$$

to obtain $g = g(p, T, S)$. This introduces a further integration function whose form has to be specified.

5. Specify the form of the integration functions that appear in the expressions for $\eta = \eta(p, T, S)$ and $g = g(p, T, S)$ of Steps 2–4.
6. Determine the values of various constant coefficients (with the exception of any arbitrary ones) via a thermodynamically consistent fitting of functions to a dataset of laboratory measurements. Alternatively, they may be set to representative values for idealised studies.

11.3 DERIVATION OF VALLIS (2017)'S PROTOTYPICAL GIBBS POTENTIAL

11.3.1 Definition and Decomposition

Before deriving Vallis (2017)'s prototypical Gibbs potential for an ocean from first principles, we first:

1. State it.
2. Decompose it into the sum of its pure-water and saline-correction parts.

Doing so then defines the desired result for the derivation that follows. Thus:

Vallis (2017)'s Prototypical Gibbs Potential for an Ocean

Vallis (2017)'s prototypical Gibbs potential for an ocean – his equation (1.146) – may be compactly written as

$$g(p, T, S) = g_0 - \eta_0 T' + \mu_0 S' + c_{p0} T \left[1 - \ln\left(\frac{T}{T_0}\right)\right]\left(1 + \beta_S^* S'\right)$$
$$+ \alpha_0 \left[1 - \frac{\beta_P}{2} p' + \beta_T T' - \beta_S S' + \frac{\beta_T \gamma^*}{2} p' T' + \frac{\beta_T^*}{2}\left(T'\right)^2\right] p', \tag{11.13}$$

where

$$p' \equiv p - p_0, \quad T' \equiv T - T_0, \quad S' \equiv S - S_0, \tag{11.14}$$

are perturbations about reference values $(p, T, S) = (p_0, T_0, S_0)$, and S_A has been abbreviated to S.

A description of the physical significance of the parameters appearing in (11.13) is given in his Table 1.2, together with some representative values for them.[2] From this potential, expressions for various quantities of interest are derived, namely his equations (1.147)–(1.151), (1.171), and (1.172), respectively, for the corresponding conventional equation of state, entropy, heat capacity, thermal expansion coefficient, adiabatic lapse rate, enthalpy, and potential enthalpy.

Gibbs potential (11.13) may be decomposed in the form of (11.1). Thus:

Decomposition of Vallis (2017)'s Prototypical Gibbs Potential for an Ocean

$$g(p, T, S) = g^W(p, T) + g^S(p, T, S), \tag{11.15}$$

[2] Values for g_0, η_0, and μ_0 are not given, since they are of arbitrary value; more on this later.

where

$$g^W(p,T) = (g_0 - \mu_0 S_0) - \eta_0 T' + c_{p0}\left(1 - \beta_S^* S_0\right) T\left[1 - \ln\left(\frac{T}{T_0}\right)\right]$$
$$+ \alpha_0\left[(1+\beta_S S_0) - \frac{\beta_p}{2}p' + \beta_T T' + \frac{\beta_T\gamma^*}{2}p'T' + \frac{\beta_T^*}{2}\left(T'\right)^2\right]p', \tag{11.16}$$

$$g^S(p,T,S) = \left\{\mu_0 - \alpha_0\beta_S p' + c_{p0}\beta_S^* T\left[1 - \ln\left(\frac{T}{T_0}\right)\right]\right\}S. \tag{11.17}$$

Note that $g(p,T,S=0) = 0$, that is, constraint (11.2) is indeed satisfied (by construction). It is satisfaction of this constraint that accounts for the presence of S_0 in the pure-water part (11.16). However, one expects the *pure-water* part $g^W = g^W(p,T)$ of a Gibbs potential to be *independent of salinity*. This implies that S_0 should be set identically zero in (11.16).[3]

The pure-water part, $g^W = g^W(p,T)$, and the saline-correction part, $g^S = g^S(p,T,S)$, of Vallis (2017)'s prototypical Gibbs potential are individually constructed in Sections 11.3.2 and 11.3.3, respectively. This mimics the decomposition of the comprehensive TEOS-10 Gibbs potential into a pure-water part plus a saline-correction part. It is done here in a far simpler context to provide insight.

11.3.2 The Pure-Water Part $g^W = g^W(p,T)$

The pure-water part, $g^W = g^W(p,T)$ is now constructed in a step-by-step manner using the methodology described in Section 11.2.2. (For the pure-water part, there is no dependence of any quantity on S, since solute is entirely absent and therefore $S \equiv 0$.) Thus:

1. Assume the quadratic representation

$$\alpha(p,T) = A_{00} + A_{10}p' + A_{01}T' + A_{20}\left(p'\right)^2 + A_{11}p'T' + A_{02}\left(T'\right)^2, \tag{11.18}$$

where the (constant) coefficients in (11.18) are those of a Taylor series truncated to second order. To derive his Gibbs potential (1.146), Vallis (2017) has implicitly assumed that

$$A_{20} = 0, \tag{11.19}$$

that is, that the term $A_{20}\left(p'\right)^2$ in (11.18) is quantitatively negligible with respect to retained terms. This does not have to be done, but we do so here in the interests of keeping things simple, without losing anything of significance, and of maintaining compatibility with Vallis (2017)'s potential. Thus (11.18), with use of (11.19), simplifies to

$$\alpha(p,T) = A_{00} + A_{10}p' + A_{01}T' + A_{11}p'T' + A_{02}\left(T'\right)^2. \tag{11.20}$$

2. Insertion of (11.20) into consistency condition (11.10) gives

$$\frac{\partial c_p(p,T)}{\partial p} = -\frac{\partial^2\alpha(p,T,S)}{\partial T^2}T = -2A_{02}T \quad \Rightarrow \quad c_p(p,T) = F^{c_p}(T) - 2A_{02}p'T. \tag{11.21}$$

[3] This is not mandatory. However, if one does not, then Vallis (2017)'s prototypical Gibbs potential does not decompose as naturally into a pure-water part and a saline-correction part. The value $S_0 = 35\ \text{g kg}^{-1}$ is suggested in Table 1.2 of Vallis (2017). This, as shown in Vallis (2017), is a very good choice for quantitatively representing perturbations away from a well-mixed saline ocean in a simple, yet quite accurate way. The focus here is different; it is instead on further improving understanding.

We are free to choose the functional form for $F^{c_p} = F^{c_p}(T)$, with a Taylor-series expansion being a natural choice. For simplicity, and to maintain compatibility with Vallis (2017)'s prototypical Gibbs potential, we set $F^{c_p}(T) = \bar{c}_{p0} = \text{constant}$.[4] With this choice, (11.21) simplifies to

$$c_p\left(p, T\right) = \bar{c}_{p0} - 2A_{02}p^{'}T. \tag{11.22}$$

3. Inserting (11.20) into (11.11) and using (11.22) gives

$$\frac{\partial \eta\left(p, T\right)}{\partial p} = -\frac{\partial \alpha\left(p, T\right)}{\partial T} = -\left(A_{01} + A_{11}p^{'} + 2A_{02}T^{'}\right), \tag{11.23}$$

$$\frac{\partial \eta\left(p, T\right)}{\partial T} = \frac{c_p\left(p, T\right)}{T} = \frac{\bar{c}_{p0}}{T} - 2A_{02}p^{'}. \tag{11.24}$$

Simultaneously integrating these two equations then yields

$$\eta\left(p, T\right) = \eta_0 + \bar{c}_{p0}\ln\left(\frac{T}{T_0}\right) - \left(A_{01} + \frac{A_{11}}{2}p^{'} + 2A_{02}T^{'}\right)p^{'}, \tag{11.25}$$

where η_0 is an integration constant. Equation (11.25) is written in terms of perturbed quantities, with a constant term $(\bar{c}_{p0}\ln T_0)$ that would otherwise appear absorbed into η_0.[5]

4. Inserting (11.20) and (11.25) into (11.12) gives

$$\frac{\partial g^W\left(p, T\right)}{\partial p} = \alpha\left(p, T\right) = A_{00} + A_{10}p^{'} + A_{01}T^{'} + A_{11}p^{'}T^{'} + A_{02}\left(T^{'}\right)^2, \tag{11.26}$$

$$\frac{\partial g^W\left(p, T\right)}{\partial T} = -\eta\left(p, T\right) = -\eta_0 - \bar{c}_{p0}\ln\left(\frac{T}{T_0}\right) + \left(A_{01} + \frac{A_{11}}{2}p^{'} + 2A_{02}T^{'}\right)p^{'}. \tag{11.27}$$

Simultaneously integrating these two equations then yields

The Pure-Water Part of a Simple Gibbs Potential for an Ocean

$$\begin{aligned} g^W\left(p, T\right) =& \bar{g}_0 - \eta_0 T^{'} + \bar{c}_{p0}T\left[1 - \ln\left(\frac{T}{T_0}\right)\right] \\ &+ \left[A_{00} + \frac{A_{10}}{2}p^{'} + A_{01}T^{'} + \frac{A_{11}}{2}p^{'}T^{'} + A_{02}\left(T^{'}\right)^2\right]p^{'}, \end{aligned} \tag{11.28}$$

where $\bar{g}_0$ is another integration constant.

5. As noted in Section 9.3.3, a Gibbs potential is only defined to within an arbitrary linear function of the form $C_0 + C_1T$, where C_0 and C_1 are arbitrary constants. Neither of them can be measured by experiment; they simply serve to determine the arbitrary offsets of g and η. The constants $\bar{g}_0$ and η_0 in (11.28) therefore have arbitrary values that can be set for convenience.
6. Values for the remaining constants in (11.28), namely $\bar{c}_{p0}$, A_{00}, A_{10}, A_{01}, A_{11}, and A_{02}, are obtained via a thermodynamically consistent fitting to a dataset of laboratory measurements. Alternatively, they may be set to representative values for idealised studies.

[4] Setting $F^{c_p}(T) = \bar{c}_{p0}(T/T_0) = \bar{c}_{p0}\left[1 + \left(T^{'}/T_0\right)\right]$ instead leads to the pure-liquid part of the alternative prototypical Gibbs potential developed in Section 11.4.1.

[5] $\ln(T/T_0)$ is a perturbed quantity since $\ln(T/T_0) = \ln\left[\left(T_0 + T^{'}\right)/T_0\right] = \ln\left[1 + \left(T^{'}/T_0\right)\right] = \left(T^{'}/T_0\right) + O\left(T^{'}/T_0\right)^2$.

The absence of salinity in (11.28) is a welcome feature; one expects the pure-water part of a Gibbs potential for a saline ocean to be independent of salinity.

11.3.3 The Saline-Correction Part $g^S = g^S(p, T, S)$

The saline-correction part, $g^S = g^S(p, T, S)$, is also constructed in a step-by-step manner using the methodology of Section 11.2.2. Thus:

1. Assume the linear representation
$$\alpha(S) = A_0 + A_1 S. \tag{11.29}$$
2. Insertion of (11.29) into consistency condition (11.10) gives
$$\frac{\partial c_p}{\partial p} = -\frac{\partial^2 \alpha}{\partial T^2} T = 0 \quad \Rightarrow \quad c_p = F^{c_p}(S). \tag{11.30}$$
Here, for simplicity, it is assumed that the saline correction for c_p is independent not only of p but also of T, as is implicitly the case for Vallis (2017)'s potential. The function $F^{c_p} = F^{c_p}(S)$ is then an integration function whose form has to be specified. It represents the influence of salinity on the heat capacity of seawater.
3. Inserting (11.29) and (11.30) into (11.11) gives
$$\frac{\partial \eta(p, T, S)}{\partial p} = -\frac{\partial \alpha(S)}{\partial T} = 0, \quad \frac{\partial \eta(p, T, S)}{\partial T} = \frac{1}{T} c_p(p, T, S) = \frac{F^{c_p}(S)}{T}. \tag{11.31}$$
Simultaneously integrating these two equations then yields
$$\eta(T, S) = F^{\eta}(S) + F^{c_p}(S) \ln\left(\frac{T}{T_0}\right), \tag{11.32}$$
where $F^{\eta} = F^{\eta}(S)$ is another integration function whose form has to be specified. A term $(F^{c_p}(S) \ln T_0)$ that would otherwise appear in (11.32) has been absorbed into $F^{\eta}(S)$.
4. Inserting (11.29) and (11.32) into (11.12) gives
$$\frac{\partial g^S(p, T, S)}{\partial p} = \alpha(S) = A_0 + A_1 S, \tag{11.33}$$
$$\frac{\partial g^S(p, T, S)}{\partial T} = -\eta(T, S) = -F^{\eta}(S) - F^{c_p}(S) \ln\left(\frac{T}{T_0}\right). \tag{11.34}$$
Simultaneously integrating these two equations then yields
$$g^S(p, T, S) = F^g(S) - T' F^{\eta}(S) + (A_0 + A_1 S) p' + F^{c_p}(S) T \left[1 - \ln\left(\frac{T}{T_0}\right)\right], \tag{11.35}$$
where $F^g = F^g(S)$ is a further integration function whose form has to be specified.
5. For simplicity, expand the functions $F^{c_p} = F^{c_p}(S)$, $F^{\eta} = F^{\eta}(S)$, and $F^g = F^g(S)$ as linear functions of S so that
$$F^{c_p}(S) = F_0^{c_p} + F_1^{c_p} S, \quad F^{\eta}(S) = F_0^{\eta} + F_1^{\eta} S, \quad F^g(S) = F_0^g + F_1^g S, \tag{11.36}$$
where the coefficients are all constants. These functional forms, which depend only on salinity, facilitate understanding of how the thermodynamics of seawater is influenced by the presence of salinity. Inserting (11.36) into (11.35) and applying condition (11.2) then gives
$$g^S(p, T, S = 0) = F_0^g - F_0^{\eta} T' + A_0 p' + F_0^{c_p} T \left[1 - \ln\left(\frac{T}{T_0}\right)\right] = 0. \tag{11.37}$$

For this equation to hold for all values of $T \equiv T_0 + T'$ and $p \equiv p_0 + p'$, all four coefficients in the middle expression must be identically zero, that is,

$$F_0^g = F_0^\eta = A_0 = F_0^{c_p} = 0. \tag{11.38}$$

Inserting (11.36) and (11.38) into (11.35) then yields

The Saline-Correction Part of a Simple Gibbs Potential for an Ocean

$$g^S(p,T,S) = \left\{F_1^g - F_1^\eta T' + A_1 p' + F_1^{c_p} T\left[1 - \ln\left(\frac{T}{T_0}\right)\right]\right\} S. \tag{11.39}$$

As noted in IOC et al. (2010) – see their equation (2.6.2) – a Gibbs potential is only defined to within an arbitrary linear function of the form $(C_0 + C_1 T)\,S$. The constants F_1^g and F_1^η in (11.39) therefore have arbitrary values that can be set for convenience.

6. The values of the remaining constants, A_1 and $F_1^{c_p}$, are obtained via a thermodynamically consistent fitting to a dataset of laboratory measurements. Alternatively, they may be set to representative values for idealised studies.

11.3.4 The Complete Gibbs Potential

From (11.1), the complete, simple Gibbs potential for the ocean is the sum of the pure-water part (11.28) and the saline-correction part (11.39). This then leads to:

The Complete Simple Gibbs Potential for an Ocean

$$\begin{aligned} g(p,T,S) &= g^W(p,T) + g^S(p,T,S) \\ &= \left(\bar{g}_0 + F_1^g S_0\right) - \eta_0 T' + F_1^g S' \\ &\quad + \left[\left(\bar{c}_{p0} + F_1^{c_p} S_0\right) + F_1^{c_p} S'\right] T\left[1 - \ln\left(\frac{T}{T_0}\right)\right] - F_1^\eta S T' \\ &\quad + \left[(A_{00} + A_1 S_0) + \frac{A_{10}}{2} p' + A_{01} T' + A_1 S' + \frac{A_{11}}{2} p' T' + A_{02}\left(T'\right)^2\right] p'. \end{aligned} \tag{11.40}$$

In this equation, g_0, η^0, F_1^η, and F_1^g are constants of arbitrary value. The remaining constants are obtained via a thermodynamically consistent fitting to a dataset of laboratory measurements or, for idealised studies, by setting their values to representative ones.

11.3.5 Equivalence with Vallis (2017)'s Prototypical Gibbs Potential

Comparing (11.40) term-by-term with (11.13) for Vallis (2017)'s Gibbs potential for the ocean (or, equivalently, comparing (11.28) and (11.39) with (11.16) and (11.17), respectively) shows:

Equivalence of the Simple Gibbs Potential and Vallis (2017)'s

The simple Gibbs potential developed in Sections 11.3.2–11.3.4 is equivalent to Vallis (2017)'s Gibbs potential (1.146) when the two sets of coefficients are related to one another by

$$\bar{g}_0 = g_0 - \mu_0 S_0, \quad \bar{c}_{p0} = \left(1 - \beta_S^* S_0\right) c_{p0}, \quad A_{00} = \alpha_0 \left(1 + \beta_S S_0\right), \tag{11.41}$$

$$A_{10} = -\alpha_0 \beta_p, \quad A_{01} = \alpha_0 \beta_T, \quad A_{11} = \alpha_0 \beta_T \gamma^*, \quad A_{02} = \frac{\alpha_0 \beta_T^*}{2}, \tag{11.42}$$

$$F_1^g = \mu_0, \quad F_1^\eta = 0, \quad A_1 = -\alpha_0 \beta_S, \quad F_1^{c_p} = c_{p0} \beta_S^*. \tag{11.43}$$

Coefficients on the left-hand sides of (11.41)–(11.43) correspond to those in (11.40), and those on the right-hand sides to (11.13).

Some representative values for the parameters present in (11.13) and on the right-hand sides of (11.41)–(11.43) may be found in Table 1.2 of Vallis (2017), and in Table 1 of the earlier study of de Szoeke (2004).[6] For comparative and reference purposes, these two tables have been merged to form Table 11.1 herein. A suitable value for p_0 is either 10^5 Pa $= 1\,000$ hPa $= 1\,000$ mb or 1.01325×10^5 Pa $= 1\,013.25$ hPa $= 1\,013.25$ mb, both being representative values of atmospheric pressure at or near Earth's ocean surface. In the absence of any constraint (e.g. compatibility with some other study that used $p_0 = 10^5$ hPa), it is probably best to use $p_0 = 1.01325 \times 10^5$ Pa since this is the value used in the TEOS-10 standard.[7]

11.3.6 Some Derived Quantities

Everything thermodynamic can now be uniquely obtained in a self-consistent manner from the complete Gibbs potential (11.40) or from the equivalent (11.13). Via differentiation and algebraic manipulation, this is accomplished by using the definitions of thermodynamic quantities expressed as a function of Gibbs potential, g. For example, expressions for the following derived quantities are summarised in Tables 11.2 and 11.3, for the pure-water and saline-correction parts, respectively:

- $\alpha \equiv \partial g/\partial p$ [specific volume].
- $\eta \equiv -\partial g/\partial T$ [specific entropy].
- $\mu \equiv \partial g/\partial S$ [relative chemical potential, often called the chemical potential of seawater].[8]
- $c_p \equiv -T\partial^2 g/\partial T^2 \equiv T\partial\eta/\partial T$ [specific heat capacity at constant pressure].
- $\widehat{\beta}^T \equiv \left(\partial g/\partial p\right)^{-1} \partial^2 g/\partial p \partial T \equiv \alpha^{-1}\partial\alpha/\partial T$
  [thermal-expansion coefficient with respect to in situ temperature T].
- $\widehat{\beta}^S \equiv -\left(\partial g/\partial p\right)^{-1} \partial^2 g/\partial p \partial S \equiv \alpha^{-1}\partial\alpha/\partial S$
  [saline- or haline-contraction coefficient at constant in situ temperature T].
- $\Gamma \equiv -\left(1/\partial^2 g/\partial T^2\right) \partial^2 g/\left(\partial p \partial T\right) \equiv \left(T/c_p\right) \partial\alpha/\partial T$ [adiabatic lapse rate].

6 Vallis (2017)'s Gibbs potential evolved from de Szoeke (2004)'s.

7 To obtain quantities derived from his prototypical equation of state (1.146), Vallis (2017) uses $p_0 = 0$. One can infer from this that, although not stated explicitly, p in his (1.146) must be sea pressure, which is total pressure minus a reference pressure $p_r \equiv 1.01325 \times 10^5$ Pa (as defined in the TEOS-10 standard). Provided that everything is done consistently, and that it is clear which convention is being used, it does not matter whether one uses sea pressure or total pressure – they only differ by the constant value p_r.

8 Inserting $dp = dT = 0$ into (9.8) leads to $dg = \mu dS$. Physically, this describes the change in Gibbs potential for a sample of seawater if a small mass of water is replaced, at constant p and T, by an equal mass of solute.

Parameter	Description	Value (Vallis, 2017)	Value (de Szoeke, 2004)
$\rho_0 \equiv 1/\alpha_0$	Reference density	1.027×10^3 kg m^{-3}	1.0277×10^3 kg m^{-3}
α_0	Reference specific volume	9.738×10^{-4} m^3 kg^{-1}	9.731×10^{-4} m^3 kg^{-1}
T_0	Reference temperature	283 K $= 9.85$°C	278.15 K $= 5$°C
S_0	Reference salinity	35 ppt $= 35$ g kg^{-1}	35 psu ≈ 35 g kg^{-1}
c_{s0}	Reference sound speed	1490 m s^{-1}	1466 m s^{-1}
β_T	First thermal expansion coefficient	1.670×10^{-4} K^{-1}	1.067×10^{-4} K^{-1}
β_T^*	Second thermal expansion coefficient	1.000×10^{-5} K^{-2}	1.041×10^{-5} K^{-2}
β_S	Haline contraction coefficient	0.780×10^{-3} ppt^{-1}	0.754×10^{-3} psu^{-1}
β_p	Compressibility coefficient ($= \alpha_0/c_{s0}^2$)	4.390×10^{-10} m s^2 kg^{-1}	4.530×10^{-10} Pa^{-1}
γ^*	Thermobaric parameter	1.100×10^{-8} Pa^{-1}	1.860×10^{-8} Pa^{-1}
c_{p0}	Specific heat capacity at constant pressure	3986 J kg^{-1} K^{-1}	
β_S^*	Haline heat capacity coefficient	1.5×10^{-3} ppt^{-1}	

Table 11.1 Representative values for various thermodynamic parameters, drawn from Table 1.2 of Vallis (2017) and Table 1 of de Szoeke (2004). For units: 1 ppt = 1 part per thousand by weight = 1 g kg^{-1}; 1 psu = 1 Practical Salinity unit $\approx$ 1 g kg^{-1}; and 1 Pa $= 1$ m^{-1} s^{-2} kg $= 1$ J m^{-3}.

See the second column of Tables 11.2 and 11.3 herein and Section 2 of TEOS-10 regarding the definition of these quantities. Expressions for these quantities are displayed in the notation used to derive (11.40) (in column 3), and also in Vallis (2017)'s notation (in column 4). Most of the expressions in column 4 correspond to those given in Vallis (2017)'s equations (1.147)–(1.150), with one small difference. His expressions assume $p_0 \equiv 0$, whereas those in Tables 11.2 and 11.3 do not; those in these tables are consequently slightly more general. Further quantities can be similarly obtained via further partial differentiation and algebraic manipulation of the Gibbs potential. For example, $\mathscr{E} \equiv g - p\partial g/\partial p - T\partial g/\partial T \equiv g - p\alpha + T\eta$ (internal energy) and $h \equiv g - T\partial g/\partial T \equiv g + T\eta$ (specific enthalpy).

Thermodynamic quantities in Tables 11.2 and 11.3 are grouped according to whether they depend *linearly* on the Gibbs potential, or otherwise, with the two groupings separated by a horizontal line. The significance of these two groupings is that for the upper grouping, the *linearly dependent* pure-water part of a quantity can be *added* to its corresponding saline-correction part to form the complete quantity. For example, the total specific volume $\alpha = \alpha\left(p, T, S\right)$ is the sum of the pure-water part $\alpha^W = \alpha^W\left(p, T\right)$ (displayed in Table 11.2) and the saline-correction part $\alpha^S = \alpha^S\left(p, T, S\right)$ (displayed in Table 11.3). It is the *linear dependence* of $\alpha\left(p, T, S\right)$ on $g\left(p, T, S\right)$ that gives rise to this additive property. Specifically,

$$\begin{aligned}\alpha\left(p, T, S\right) &\equiv \frac{\partial g\left(p, T, S\right)}{\partial p} \equiv \frac{\partial}{\partial p}\left[g^W\left(p, T\right) + g^S\left(p, T, S\right)\right] \\ &\equiv \frac{\partial g^W\left(p, T\right)}{\partial p} + \frac{\partial g^S\left(p, T, S\right)}{\partial p} \equiv \alpha^W\left(p, T\right) + \alpha^S\left(p, T, S\right).\end{aligned} \tag{11.44}$$

In contradistinction, this additive property does *not* hold for the lower grouping of quantities due to the *non-linearity* of their definitions. For example, for the thermal expansion coefficient

$$\begin{aligned}\widehat{\beta}^T\left(p, T, S\right) &\equiv \frac{1}{\partial g/\partial p}\frac{\partial^2 g}{\partial p\partial T} \equiv \frac{1}{\partial g^W/\partial p + \partial g^S/\partial p}\left(\frac{\partial^2 g^W}{\partial p\partial T} + \frac{\partial^2 g^S}{\partial p\partial T}\right) \\ &\neq \frac{1}{\partial g^W/\partial p}\frac{\partial^2 g^W}{\partial p\partial T} + \frac{1}{\partial g^S/\partial p}\frac{\partial^2 g^S}{\partial p\partial T} \equiv \left(\widehat{\beta}^T\right)^W + \left(\widehat{\beta}^T\right)^S.\end{aligned} \tag{11.45}$$

The total thermal expansion coefficient $\widehat{\beta}^T = \widehat{\beta}^T\left(p, T, S\right)$ can nevertheless be obtained from Tables 11.2 and 11.3, albeit not as the simple sum of the pure-water and saline-correction parts. Inserting the definitions of α^W and α^S (extracted from the second columns of Tables 11.2 and 11.3, respectively) into the first line of (11.45), with use of the definitions of $\left(\widehat{\beta}^T\right)^W$ and $\left(\widehat{\beta}^T\right)^S$ (extracted from these two tables), leads to

$$\begin{aligned}\widehat{\beta}^T\left(p, T, S\right) &\equiv \frac{1}{\partial g/\partial p}\frac{\partial^2 g}{\partial p\partial T} \equiv \frac{1}{\alpha}\frac{\partial\alpha}{\partial T} \equiv \frac{1}{\alpha^W + \alpha^S}\frac{\partial\left(\alpha^W + \alpha^S\right)}{\partial T} \\ &\equiv \frac{1}{\alpha^W + \alpha^S}\left[\alpha^W\left(\frac{1}{\alpha^W}\frac{\partial\alpha^W}{\partial T}\right) + \alpha^S\left(\frac{1}{\alpha^S}\frac{\partial\alpha^S}{\partial T}\right)\right] \\ &\equiv \frac{\alpha^W\left(\widehat{\beta}^T\right)^W + \alpha^S\left(\widehat{\beta}^T\right)^S}{\alpha^W + \alpha^S} \neq \left(\widehat{\beta}^T\right)^W + \left(\widehat{\beta}^T\right)^S.\end{aligned} \tag{11.46}$$

Since only definitions have been used (as opposed to a particular specification of Gibbs potential) this is a general result. It shows that $\widehat{\beta}^T = \widehat{\beta}^T\left(p, T, s\right)$ is *not* the *simple sum* of $\left(\widehat{\beta}^T\right)^W$ and $\left(\widehat{\beta}^T\right)^S$, but instead a *weighted sum* with weights $\alpha^W/\left(\alpha^W + \alpha^S\right)$ and $\alpha^S/\left(\alpha^W + \alpha^S\right)$, respectively. Inserting the particular results for Vallis (2017)'s Gibbs potential (drawn from columns 3 and 4 of

Symbol	Definition	In Present Notation	With Vallis (2017)'s Notation for Parameters
g^W	g^W	$\overline{g}_0 \quad -\eta_0 T' + \overline{c}_{p0} T \left[1 - \ln\left(\frac{T}{T_0}\right)\right] + \left[A_{00} + \frac{A_{10}}{2} p' + A_{01} T' + \frac{A_{11}}{2} p' T' + A_{02} \left(T'\right)^2\right] p'$	$g_0 \quad -\mu_0 S_0 - \eta_0 T' + \left(1 - \beta_S^* S_0\right) c_{p0} T \left[1 - \ln\left(\frac{T}{T_0}\right)\right] + \alpha_0 \left[1 + \beta_S S_0 - \frac{\beta_p}{2} p' + \beta_T T' + \frac{\beta_T \gamma^*}{2} p' T' + \frac{\beta_T^*}{2} \left(T'\right)^2\right] p'$
α^W	$\frac{\partial g^W}{\partial p}$	$A_{00} + A_{10} p' + A_{01} T' + A_{11} p' T' + A_{02} \left(T'\right)^2$	$\alpha_0 \left[1 + \beta_S S_0 - \beta_p p' + \beta_T T' + \beta_T \gamma^* p' T' + \frac{\beta_T^*}{2} \left(T'\right)^2\right]$
η^W	$-\frac{\partial g^W}{\partial T}$	$\eta_0 + \overline{c}_{p0} \ln\left(\frac{T}{T_0}\right) - \left(A_{01} + \frac{A_{11}}{2} p' + 2 A_{02} T'\right) p'$	$\eta_0 + \left(1 - \beta_S^* S_0\right) c_{p0} \ln\left(\frac{T}{T_0}\right) - \alpha_0 \left(\beta_T + \frac{\beta_T \gamma^*}{2} p' + \beta_T^* T'\right) p'$
μ^W	$\frac{\partial g^W}{\partial S}$	0	0
c_p^W	$-T \frac{\partial^2 g^W}{\partial T^2}$	$\overline{c}_{p0} - 2 A_{02} p' T$	$\left(1 - \beta_S^* S_0\right) c_{p0} - \alpha_0 \beta_T^* p' T$
$\left(\widehat{\beta}^T\right)^W$	$\frac{1}{\partial g^W / \partial p} \frac{\partial^2 g^W}{\partial p \partial T}$	$\frac{1}{\alpha^W} \left(A_{01} + A_{11} p' + 2 A_{02} T'\right)$, α^W as above.	$\frac{\alpha_0}{\alpha^W} \left(\beta_T + \beta_T \gamma^* p' + \beta_T^* T'\right)$, α^W as above.
$\left(\widehat{\beta}^S\right)^W$	$-\frac{1}{\partial g^W / \partial p} \frac{\partial^2 g^W}{\partial p \partial S}$	0	0
Γ^W	$-\frac{1}{\partial^2 g^W / \partial T^2} \frac{\partial^2 g^W}{\partial p \partial T}$	$\frac{T}{c_p^W} \left(A_{01} + A_{11} p' + 2 A_{02} T'\right)$, c_p^W as above.	$\frac{T}{c_p^W} \alpha_0 \left(\beta_T + \gamma^* \beta_T p' + \beta_T^* T'\right)$, c_p^W as above.

Table 11.2 Various thermodynamic quantities (column 1) derived from the pure-water ($g^W = g^W(p, T)$) Gibbs potentials (11.28) (in the present notation, column 3) and (11.16) (with Vallis (2017)'s notation for parameters, column 4). These potentials are equivalent to one another, and also (after decomposition) to that given in equation (1.146) of Vallis (2017). Definitions of derived quantities in terms of Gibbs potential (g^W) are given in column 2. Quantities in the upper part of the table (above the dividing line) depend linearly on g^W, whereas those below do not. See text for further details.

Symbol	Definition	In Present Notation	With Vallis (2017)'s Notation for Parameters
g^S	g^S	$\left\{F_1^g + A_1 p' + F_1^{c_p} T\left[1 - \ln\left(\frac{T}{T_0}\right)\right]\right\} S$	$\left\{\mu_0 - \alpha_0 \beta_S p' + c_{p0}\beta_S^* T\left[1 - \ln\left(\frac{T}{T_0}\right)\right]\right\} S$
α^S	$\frac{\partial g^S}{\partial p}$	$A_1 S$	$-\alpha_0 \beta_S S$
η^S	$-\frac{\partial g^S}{\partial T}$	$F_1^{c_p} \ln\left(\frac{T}{T_0}\right) S$	$c_{p0}\beta_S^* \ln\left(\frac{T}{T_0}\right) S$
μ^S	$\frac{\partial g^S}{\partial S}$	$F_1^g + A_1 p' + F_1^{c_p} T\left[1 - \ln\left(\frac{T}{T_0}\right)\right]$	$\mu_0 - \alpha_0 \beta_S p' + c_{p0}\beta_S^* T\left[1 - \ln\left(\frac{T}{T_0}\right)\right]$
c_p^S	$-T\frac{\partial^2 g^S}{\partial T^2}$	$F_1^{c_p} S$	$c_{p0}\beta_S^* S$
$\left(\widehat{\beta^T}\right)^S$	$\frac{1}{\partial g^S/\partial p}\frac{\partial^2 g^S}{\partial p \partial T}$	0	0
$\left(\widehat{\beta^S}\right)^S$	$-\frac{1}{\partial g^S/\partial p}\frac{\partial^2 g^S}{\partial p \partial S}$	$-\frac{1}{\alpha^S}A_1$, α^S as above.	$\frac{\alpha_0}{\alpha^S}\beta_S$, α^S as above.
Γ^S	$-\frac{1}{\partial^2 g^S/\partial T^2}\frac{\partial^2 g^S}{\partial p \partial T}$	0	0

Table 11.3 As in Table 11.2 but derived from the saline-correction ($g^S = g^S(p, T, S)$) Gibbs potentials (11.39) and (11.17); $F_1^\eta \equiv 0$ for consistency with Vallis (2017). See text for further details.

Tables 11.2 and 11.3) into (11.46) leads to

$$\widehat{\beta}^{T}(p,T,S)=\frac{A_{01}+A_{11}p'+2A_{02}T'}{A_{00}+A_{10}p'+A_{01}T'+A_{11}p'T'+A_{02}\left(T'\right)^{2}+A_{1}S} \tag{11.47}$$

$$=\frac{\beta_{T}+\beta_{T}\gamma^{*}p'+\beta_{T}^{*}T'}{1-\beta_{p}p'+\beta_{T}T'+\beta_{T}\gamma^{*}p'T'+\frac{\beta_{T}^{*}}{2}\left(T'\right)^{2}-\beta_{S}S'}. \tag{11.48}$$

(From (11.41) and (11.43), $A_{00}+A_{1}S=\alpha_{0}\left(1+\beta_{S}S_{0}\right)-\alpha_{0}\beta_{S}\left(S_{0}+S'\right)=\alpha_{0}\left(1-\beta_{S}S'\right)$. This explains why an S appears in (11.47), but an S' in (11.48), something that at first sight looks suspicious.)

Some of the derived quantities in Tables 11.2 and 11.3 help explain the origin and physical significance of some of the (at first sight, mysterious) parameters present in Vallis (2017)'s Gibbs potential – and also in the equivalent (11.13) herein. Parameter μ_0 originates from $\mu=\mu\left(p,T,S\right)$, the relative chemical potential. Similarly, parameters β_T, $\beta_T\gamma^*$, and β_T^* originate from $\widehat{\beta}^T=\widehat{\beta}^T\left(p,T,S\right)$, the thermal expansion coefficient with respect to in situ temperature T; β_S from $\widehat{\beta}^S=\widehat{\beta}^S\left(p,T,S\right)$, the haline (or saline) contraction coefficient; and β_S^* and β_T^* from $c_p=c_p\left(p,T,S\right)$, the specific heat capacity at constant pressure.

As noted in Vallis (2017), parameters with asterisks multiply non-linear (higher-order) product terms in p', T', and S'. Setting the three asterisked parameters (β_T^*, β_S^*, and γ^*) to zero in Gibbs potential (11.13), and in all quantities derived from it, simplifies things a lot. For example, specific volume (which corresponds to the conventional equation of state) then only has linear terms. However, the convenience of this simplification comes at the price of significantly limiting applicability.

11.3.7 Potential Temperature $\theta=\theta\left(p,T,S\right)$

Potential temperature, θ, is defined and discussed in Section 10.7. In the present oceanic context, θ is the solution of – cf. (10.164) and (10.165) –

$$\frac{\partial g\left(p_{\mathrm{r}},\theta,S\right)}{\partial T}=\frac{\partial g\left(p,T,S\right)}{\partial T}\quad\Rightarrow\quad\eta\left(p_{\mathrm{r}},\theta,S\right)=\eta\left(p,T,S\right). \tag{11.49}$$

In these equations, p_{r} is a constant reference pressure, usually taken to be either 10^5 Pa $=$ 1 000 hPa $=$ 1 000 mb or 1.01325×10^5 Pa $=$ 1 013.25 hPa $=$ 1 013.25 mb, both being representative values of atmospheric pressure at Earth's surface that acts downwards on an ocean's upper surface. Substituting the expressions for $\eta^W=\eta^W\left(p,T\right)$ and $\eta^S=\eta^S\left(p,T,S\right)$ – given in column 3 of Tables 11.2 and 11.3, respectively – into (11.49) (these expressions are additive since $\eta\equiv\partial g/\partial T$ is a linear function of g) yields

$$\left(\bar{c}_{p0}+F_{1}^{c_p}S\right)\ln\left(\frac{\theta}{T_0}\right)-\left[A_{01}+\frac{A_{11}}{2}\left(p_{\mathrm{r}}-p_0\right)+2A_{02}\left(\theta-T_0\right)\right]\left(p_{\mathrm{r}}-p_0\right)$$
$$=\left(\bar{c}_{p0}+F_{1}^{c_p}S\right)\ln\left(\frac{T}{T_0}\right)-\left[A_{01}+\frac{A_{11}}{2}\left(p-p_0\right)+2A_{02}\left(T-T_0\right)\right]\left(p-p_0\right). \tag{11.50}$$

In general, this is a transcendental equation for θ of the form $\ln\theta+C\theta=$ known, where C is a constant of known value. As such, this equation usually has to be solved numerically. However, for the special case where $p_{\mathrm{r}}\equiv p_0$, C is then identically zero, and (11.50) then has the form $\ln\theta=$ known, which can be solved analytically. Thus setting $p_{\mathrm{r}}\equiv p_0$ in (11.50) leads to

$$\theta\left(p,T,S\right)=T\exp\left\{-\frac{\left[A_{01}+\left(A_{11}/2\right)p^{'}+2A_{02}T^{'}\right]p^{'}}{\left(\bar{c}_{p0}+F_1^{c_p}S\right)}\right\}, \tag{11.51}$$

where $p^{'}\equiv p-p_0\equiv p-p_{\rm r}$. Using (11.41)–(11.43), (11.51) may be equivalently written as

$$\theta\left(p,T,S\right)=T\exp\left\{-\frac{\alpha_0\left[\beta_T+\left(\beta_T\gamma^*/2\right)p^{'}+\beta_T^*T^{'}\right]p^{'}}{c_{p0}\left(1+\beta_S^*S^{'}\right)}\right\}. \tag{11.52}$$

This equation corresponds to Vallis (2017)'s equation (1.152), albeit (11.52) is slightly more general since it has only been assumed that $p_{\rm r}\equiv p_0$, rather than Vallis (2017)'s more restrictive assumption $p_{\rm r}\equiv p_0\equiv 0$.

For specified values of the state variables, namely pressure (p), temperature (T), and salinity (S), and specified values of various parameters, (11.51) and (11.52) give the value of the corresponding potential temperature θ. This is the temperature that a fluid parcel would have if it were to undergo a thermodynamic process at constant entropy (and therefore with no heating or cooling) and at constant composition (and therefore with no change in salinity).

11.3.8 Enthalpy $h=h\left(p,T,S\right)$

From the penultimate column of Table 9.2, $h\left(p,T,S\right)=g\left(p,T,S\right)+T\eta\left(p,T,S\right)$, where h is specific enthalpy. Substitution of the expressions for $g=g\left(p,T,S\right)$ and $\eta=\eta\left(p,T,S\right)$ given in the third and fourth columns of Tables 11.2 and 11.3 into this equation, followed by some algebraic manipulation, then yields the equivalent expressions

$$\begin{aligned}h\left(p,T,S\right)=&\bar{g}_0+\eta_0T_0+F_1^gS+\left(\bar{c}_{p0}+F_1^{c_p}S\right)T\\&+\left[A_{00}-T_0A_{01}+\frac{1}{2}\left(A_{10}-T_0A_{11}\right)p^{'}+A_1S-A_{02}\left(T^2-T_0^2\right)\right]p^{'},\end{aligned} \tag{11.53}$$

$$\begin{aligned}h\left(p,T,S\right)=&g_0+\eta_0T_0+\mu_0S^{'}+c_{p0}\left(1+\beta_S^*S^{'}\right)T\\&+\alpha_0\left[1-T_0\beta_T-\frac{1}{2}\left(\beta_p+T_0\beta_T\gamma^*\right)p^{'}-\beta_SS^{'}-\frac{\beta_T^*}{2}\left(T^2-T_0^2\right)\right]p^{'}.\end{aligned} \tag{11.54}$$

Setting $p_0\equiv 0$ in (11.54) (so that $p^{'}\equiv p-p_0$ reduces to p) recovers Vallis (2017)'s (1.171) to within the unimportant additive constant $g_0+\eta_0T_0-\mu_0S_0$.[9]

To obtain potential enthalpy, it is convenient to first evaluate (11.49) and (11.50) at a reference pressure such that $p_{\rm r}\equiv p_0\Rightarrow p^{'}=p_{\rm r}-p_0=0$. Thus (11.53) and (11.54) reduce to

[9] It is unimportant since g_0, η_0, and μ_0 are arbitrary constants that have no physical meaning.

$$h\left(p=p_{\mathrm{r}}=p_0,T,S\right)=\bar{g}_0+\eta_0T_0+F_1^gS+\left(\bar{c}_{p0}+F_1^{c_p}S\right)T, \tag{11.55}$$

$$h\left(p=p_{\mathrm{r}}=p_0,T,S\right)=g_0+\eta_0T_0+\mu_0S^{'}+c_{p0}\left(1+\beta_S^*S^{'}\right)T. \tag{11.56}$$

11.3.9 Potential Enthalpy $h^0\left(p,T,S\right)$

Potential enthalpy $h^0 = h^0\left(p,T,S\right)$ is defined in Section 10.7.5. It is obtained in the present context by:

1. Solving implicit equation (10.176) – alias (10.165) – for potential temperature $\theta=\theta\left(p,T,S\right)$, where entropy $\eta=\eta\left(p,T,S\right)=-\partial g\left(p,T,S\right)/\partial T$ (see Table 9.2) is obtained from the Gibbs potential $g=g\left(p,T,S\right)$. As noted in Section 10.7.1, this implicit equation in general has to be solved numerically. However, for Vallis (2017)'s Gibbs potential (11.13), it can be solved exactly, and (11.51) and (11.52) result under the assumption that $p_{\mathrm{r}}\equiv p_0$.
2. Obtaining enthalpy $h = h\left(p,T,S\right)$ from the Gibbs potential, either using $h\left(p,T,S\right) = g\left(p,T,S\right)-T\left[\partial g\left(p,T,S\right)/\partial T\right]$ (see Table 9.2) or the equivalent $h\left(p,T,S\right) = g\left(p,T,S\right)+T\eta\left(p,T,S\right)$. Either way, (11.53) and (11.54) result for Vallis (2017)'s Gibbs potential (11.13) under the assumption that $p_{\mathrm{r}}\equiv p_0$.
3. Evaluating (11.53) and (11.54) at reference pressure $p_{\mathrm{r}}\equiv p_0$ to obtain (11.55) and (11.56).
4. Using definition (10.175) to obtain potential enthalpy $h^0=h^0\left(p,T,S\right)\equiv h\left(p_{\mathrm{r}},\theta,S\right)$, where $\theta=\theta\left(p,T,S\right)$ and $h=h\left(p,T,S\right)$ are obtained as described in Steps 1 and 2, respectively.

Thus for Vallis (2017)'s Gibbs potential (11.13), use of (10.175) (definition of potential enthalpy) and (11.53) and (11.54) (enthalpy evaluated at reference pressure $p = p_{\mathrm{r}} \equiv p_0$) yields the equivalent expressions

$$h^0\left(p,T,S\right)=\bar{g}_0+\eta_0T_0+F_1^gS+\left(\bar{c}_{p0}+F_1^{c_p}S\right)\theta\left(p,T,S\right), \tag{11.57}$$

$$h^0\left(p,T,S\right)=g_0+\eta_0T_0+\mu_0S^{'}+c_{p0}\left(1+\beta_S^*S^{'}\right)\theta\left(p,T,S\right), \tag{11.58}$$

where $h^0\left(p,T,S\right) \equiv h\left(p=p_{\mathrm{r}},T=\theta,S\right)$, $\theta = \theta\left(p,T,S\right)$ is given by (11.51) (in (11.57)) or by (11.52) (in (11.58)), and $p_{\mathrm{r}}\equiv p_0$ is assumed.

Since $S^{'}\equiv S-S_0$, (11.58) recovers Vallis (2017)'s (1.172) to within the unimportant additive constant $g_0+\eta_0T_0-\mu_0S_0$.

11.3.10 Conservative Temperature $\Theta=\Theta\left(p,T,S\right)$

As discussed in Section 10.7.6, conservative temperature $\Theta = \Theta\left(p,T,S\right)$ is simply potential enthalpy scaled by $c_{p0} = 3991.86795711963\ \mathrm{J\,kg^{-1}\,K}$. Thus, from (10.177) (definition of conservative temperature), (11.57), and (11.58), conservative temperature for Vallis (2017)'s Gibbs potential (11.13) is given by the equivalent expressions

$$\Theta\left(p,T,S\right)=\frac{h^0\left(p,T,S\right)}{c_{p0}}=\frac{\bar{g}_0+\eta_0T_0+F_1^gS}{c_{p0}}+\frac{\left(\bar{c}_{p0}+F_1^{c_p}S\right)}{c_{p0}}\theta\left(p,T,S\right), \tag{11.59}$$

$$\Theta\left(p,T,S\right)=\frac{h^0\left(p,T,S\right)}{c_{p0}}=\frac{g_0+\eta_0T_0+\mu_0S^{'}}{c_{p0}}+\left(1+\beta_S^*S^{'}\right)\theta\left(p,T,S\right), \tag{11.60}$$

where $\theta = \theta(p, T, S)$ is given by (11.51) (in (11.59)) or by (11.52) (in (11.60)), and $p_r \equiv p_0$ is assumed.

Since g_0, η_0, and μ_0 are arbitrary constants (which is why values for them do not appear in Table 9.3) and F_1^g is also an arbitrary constant, we are free to set them identically zero if we wish. If we do so, then $\bar{g}_0 + \eta_0 T_0 + F_1^g S = g_0 + \eta_0 T_0 + \mu_0 S' = 0$ and (11.59) and (11.60) reduce to

$$\Theta(p, T, S) = \frac{\left(\bar{c}_{p0} + F_1^{c_p} S\right)}{c_{p0}} \theta(p, T, S) = \left(1 + \beta_S^* S'\right) \theta(p, T, S). \tag{11.61}$$

Conservative temperature $\Theta(p, T, S)$ is then equal to potential temperature $\theta(p, T, S)$ multiplied by the salinity-dependent 'correction' factor $\left[\left(\bar{c}_{p0} + F_1^{c_p} S\right)/c_{p0}\right] = \left(1 + \beta_S^* S'\right)$. Since $\left|\beta_S^* S'\right| \ll 1$, this means that $\Theta(p, T, S) \approx \theta(p, T, S)$ in these particular circumstances; that is, conservative temperature is then, to a very good approximation, equal to potential temperature.

11.3.11 Potential Density $\rho^\theta(p, T, S)$

Potential density $\rho^\theta = \rho^\theta(p, T, S)$ is defined in Section 10.7.7. In the present context, the potential temperature $\theta = \theta(p, T, S)$ is given by (11.51) and (11.52), which were obtained under the assumption that $p_r \equiv p_0$. Making this assumption and inserting the expressions for α^W and α^S given in Tables 11.2 and 11.3, respectively, into definition (10.180) for potential density then leads to the equivalent expressions

$$\rho^\theta(p, T, S) = \left[A_{00} + A_{01}(\theta - T_0) + A_{02}(\theta - T_0)^2 + A_1 S\right]^{-1}, \tag{11.62}$$

$$\rho^\theta(p, T, S) = \alpha_0^{-1}\left[1 + \beta_T(\theta - T_0) + \frac{\beta_T^*}{2}(\theta - T_0)^2 - \beta_S S'\right]^{-1}, \tag{11.63}$$

where $\rho^\theta(p, T, S) \equiv \rho(p_r, \theta, S) \equiv \left[\alpha(p_r, \theta, S)\right]^{-1}$, $\theta = \theta(p, T, S)$ is given by (11.51) (in (11.62)) or by (11.52) (in (11.63)), and $p_r \equiv p_0$ is assumed. Thus (11.62) and (11.63) are two expressions for the potential density that follow from (11.13) for Vallis (2017)'s Gibbs potential.

11.4 AN ALTERNATIVE PROTOTYPICAL GIBBS POTENTIAL FOR AN OCEAN

Vallis (2017)'s prototypical Gibbs potential (11.13) for an ocean provides a very useful and convenient approximation to the highly accurate, but highly complicated TEOS-10 one. In particular, his potential represents a very good compromise between simplicity and accuracy. Any simpler, and it would be insufficiently accurate for many applications. Significantly more accurate, and it would be significantly more complicated.

Two differences between the functional forms of Vallis (2017)'s potential and that of TEOS-10 were, however, identified in Section 11.1. It is therefore desirable to develop a Gibbs potential of comparable accuracy and simplicity to the Vallis (2017) one, but with a functional form that mimics that assumed for the TEOS-10 one. Since the specifications of g^W and g^S are *independent of one another*, we can modify them independently of one another. This illustrates the benefit of decomposing a complete Gibbs potential into its distinct pure-water and saline-correction parts.

11.4.1 The Pure-Water Part, $g^W = g^W(p, T)$

For the pure-water part, $g^W = g^W(p, T)$, it is straightforward to tweak Gibbs potential (11.16) and the equivalent (11.28). There is a $\ln(T/T_0)$ contribution in both equations. As will be seen in Section 11.5 – peek ahead if you wish – there is no corresponding contribution of this form in the TEOS-10 one. All we have to do to address this discrepancy is to expand $T\left[\ln\left(T/T_0\right)-1\right]$ in these two equations as a polynomial using Taylor series. With a sufficient number of terms, the resulting Gibbs potential will, quantitatively, give virtually identical results to Vallis (2017)'s. Noting that $T \equiv T_0 + T'$, expansion of $\ln\left(T/T_0\right) \equiv \ln\left[1+\left(T'/T_0\right)\right]$ for small T'/T_0 (with respect to unity) leads to

$$T\left[1-\ln\left(\frac{T}{T_0}\right)\right]=\left(T_0+T'\right)\left\{1-\left[\frac{T'}{T_0}-\frac{1}{2}\left(\frac{T'}{T_0}\right)^2+O\left(\left(\frac{T'}{T_0}\right)^3\right)\right]\right\}$$
$$=T_0\left[1-\frac{1}{2}\left(\frac{T'}{T_0}\right)^2+O\left(\left(\frac{T'}{T_0}\right)^3\right)\right]. \tag{11.64}$$

(Depending on the application, and on accuracy requirements, further terms could be included in (11.64). This would lengthen equations but would not unduly affect analytic tractability.) Substituting (11.64) into (11.28) and (11.16) then gives:

The Pure-Water Part of the Alternative Simple Gibbs Potential for an Ocean

Equivalent expressions for the pure-water part of the alternative simple Gibbs potential for an ocean are

$$g^W(p,T)=\overline{g}_0-\eta_0T'+\overline{c}_{p0}T_0\left[1-\frac{1}{2}\left(\frac{T'}{T_0}\right)^2\right]$$
$$+\left[A_{00}+\frac{A_{10}}{2}p'+A_{01}T'+\frac{A_{11}}{2}p'T'+A_{02}\left(T'\right)^2\right]p', \tag{11.65}$$

$$g^W(p,T)=g_0-\mu_0S_0-\eta_0T'+\left(1-\beta_S^*S_0\right)c_{p0}T_0\left[1-\frac{1}{2}\left(\frac{T'}{T_0}\right)^2\right]$$
$$+\alpha_0\left[1+\beta_SS_0-\frac{\beta_p}{2}p'+\beta_TT'+\frac{\beta_T\gamma^*}{2}p'T'+\frac{\beta_T^*}{2}\left(T'\right)^2\right]p'. \tag{11.66}$$

Various thermodynamic quantities, derived from (11.65) and (11.66), are displayed in Table 11.4. These may be compared to the corresponding ones in Table 11.2 – derived from (11.35) and (11.16) – for Vallis (2017)'s potential. It is seen that both the alternative prototypical Gibbs potential $g^W = g^W\left(p,T,S\right)$ for liquid water and the corresponding Vallis (2017) one lead to exactly the same: α (conventional equation of state); $\left(\widehat{\beta}^T\right)^W$ (thermal expansion coefficient); $\left(\widehat{\beta}^S\right)^W$ (haline contraction coefficient); and Γ^W (adiabatic lapse rate). However, η (entropy), μ (relative chemical potential), and c_p^W (specific heat capacity at constant pressure) all differ due to the replacement of $T\left[1-\ln\left(T/T_0\right)\right]$ by polynomial expansion (11.64).

The difference in the two construction procedures is that the assumed form for c_p^W is slightly different:

Symbol	Definition	In Present Notation	With Vallis (2017)'s Notation for Parameters
g^W	g^W	$\bar{g}_0 \quad -\eta_0 T' + \bar{c}_{p0} T_0 \left[1 - \frac{1}{2}\left(\frac{T'}{T_0}\right)^2\right] + \left[A_{00} + \frac{A_{10}}{2} p' + A_{01} T' + \frac{A_{11}}{2} p' T' + A_{02} \left(T'\right)^2\right] p'$	$g_0 \quad -\mu_0 S_0 - \eta_0 T' + (1 - \beta_S^* S_0)\, c_{p0} T_0 \left[1 - \frac{1}{2}\left(\frac{T'}{T_0}\right)^2\right] + \alpha_0 \left[1 - \frac{\beta_p}{2} p' + \beta_T T' + \frac{\beta_T \gamma^*}{2} p' T' + \frac{\beta_T^*}{2} \left(T'\right)^2\right] p'$
α^W	$\frac{\partial g^W}{\partial p}$	$A_{00} + A_{10} p' + A_{01} T' + A_{11} p' T' + A_{02} \left(T'\right)^2$	$\alpha_0 \left[1 + \beta_S S_0 - \beta_p p' + \beta_T T' + \beta_T \gamma^* p' T' + \frac{\beta_T^*}{2} \left(T'\right)^2\right]$
η^W	$-\frac{\partial g^W}{\partial T}$	$\eta_0 + \bar{c}_{p0} \frac{T'}{T_0} - \left(A_{01} + \frac{A_{11}}{2} p' + 2 A_{02} T'\right) p'$	$\eta_0 + (1 - \beta_S^* S_0)\, c_{p0} \frac{T'}{T_0} - \alpha_0 \left(\beta_T + \frac{\beta_T \gamma^*}{2} p' + \beta_T^* T'\right) p'$
μ^W	$\frac{\partial g^W}{\partial S}$	0	0
c_p^W	$-T \frac{\partial^2 g^W}{\partial T^2}$	$\bar{c}_{p0} \frac{T}{T_0} - 2 A_{02} p' T$	$(1 - \beta_S^* S_0)\, c_{p0} \frac{T}{T_0} - \alpha_0 \beta_T^* p' T$
$\left(\widehat{\beta^T}\right)^W$	$\frac{1}{\partial g^W/\partial p} \frac{\partial^2 g^W}{\partial p \partial T}$	$\frac{1}{\alpha^W} \left(A_{01} + A_{11} p' + 2 A_{02} T'\right)$, α^W as above.	$\frac{\alpha_0}{\alpha^W} \left(\beta_T + \beta_T \gamma^* p' + \beta_T^* T'\right)$, α^W as above.
$\left(\widehat{\beta^S}\right)^W$	$-\frac{1}{\partial g^W/\partial p} \frac{\partial^2 g^W}{\partial p \partial S}$	0	0
Γ^W	$-\frac{1}{\partial^2 g^W/\partial T^2} \frac{\partial^2 g^W}{\partial p \partial T}$	$\frac{T}{c_p^W} \left(A_{01} + A_{11} p' + 2 A_{02} T'\right)$, c_p^W as above.	$\frac{T}{c_p^W} \alpha_0 \left(\beta_T + \gamma^* \beta_T p' + \beta_T^* T'\right)$, c_p^W as above.

Table 11.4 As in Table 11.2, but derived from the alternative pure-water ($g^W = g^W(p, T)$) Gibbs potentials (11.65) and (11.66). See text for further details.

- Vallis (2017) implicitly assumes that c_p^W has the form

$$c_p^W = \bar{c}_{p0} - 2A_{02}p'T = \left(1 - \beta_S^* S_0\right) c_{p0} - \alpha_0 \beta_{TP}^* p' T. \tag{11.67}$$

See Table 11.2. This then leads to a $\ln(T/T_0)$ term in $g^W = g^W\left(p, T\right)$.

- For the present construction, c_p^W is instead implicitly assumed to have the slightly different form

$$c_p^W = \bar{c}_{p0}\left(T/T_0\right) - 2A_{02}p'T = \left(1 - \beta_S^* S_0\right) c_{p0}\left(T/T_0\right) - \alpha_0 \beta_{TP}^* p' T. \tag{11.68}$$

This form then no longer leads to a $\ln(T/T_0)$ term in $g^W = g^W\left(p, T\right)$ when integrating the modified (11.24), whereby $\bar{c}_{p0}/T$ is replaced in this equation by $\bar{c}_{p0}/T_0$, a constant.

A virtue of the alternative potential (11.65) – and its equivalent, (11.66) – for the pure-water part is that, as intended, it mimics (albeit in a much simplified manner) the functional form of the corresponding TEOS-10 one.

Although it is straightforward to modify the Vallis (2017) Gibbs potential for the pure-water part to obtain the alternative one, this is a somewhat convoluted way of doing things. One first constructs a potential with an assumed form that turns out not to match the TEOS one. And then one modifies it so it does! However, one could have started from scratch and immediately assumed a form for the Gibbs potential that is a simplified subset of the TEOS-10 one. Applying various thermodynamic rules would then still permit identification of the physical significance of various coefficients in an analogous manner to the construction of the Vallis (2017) one.

11.4.2 The Saline-Correction Part, $g^S = g^S\left(p, T, S\right)$

It is less straightforward to modify the saline-correction part, $g^S = g^S\left(p, T, S\right)$, to mimic the functional form of the TEOS-10 one. To do so, we restart from scratch. Guided by the form of the TEOS-10 potential – see Appendix H of IOC et al. (2010) and Section 11.5 herein:

The Saline-Correction Part
of the Alternative Simple Gibbs Potential for an Ocean

Let the saline-correction part of the alternative simple Gibbs potential for the ocean have the form

$$g^S\left(p, T, S\right) = \left[B_0 + B_1 p' + B_2\left(T'\right)^2\right] S + B_{\text{ln}} S \ln\left[\left(\frac{S}{S_{\text{u}}}\right)^{\frac{1}{2}}\right], \tag{11.69}$$

where S_{u} is a unit-related constant, and the constant coefficients B_0, B_1, B_2, and B_{ln} are to be determined by data fitting.[a]

a Mathematically it is arguably neater to write the logarithmic term as $\left(B_{\text{ln}}/2\right) S \ln\left(S/S_u\right)$. This is not done here for compatibility with the functional form of the TEOS-10 potential.

We are free to take more or fewer terms in (11.69) according to accuracy requirements. Those present in (11.69) suffice for present purposes. The last term (in red) on the right-hand side of (11.69) allows us to examine the influence of the logarithmic contribution; this term is absent from Vallis (2017)'s prototypical Gibbs potential (11.17) for the saline correction. The remaining three terms allow us to link his potential with (11.69) and thereby attach physical meaning to the constants in (11.69).

Symbol	Definition	In Present Notation	With Vallis (2017)'s Notation for Parameters
g^S	g^S	$\left[B_0 + B_1 p' + B_2 \left(T'\right)^2\right] S + B_{\ln} S \ln \left(\frac{S}{S_u}\right)^{\frac{1}{2}}$	$\left[\mu_0 + c_{p0}\beta_S^* T_0 - \alpha_0 \beta_S p' - \frac{c_{p0}\beta_S^*}{2T_0}\left(T'\right)^2\right] S + B_{\ln} S \ln \left(\frac{S}{S_u}\right)^{\frac{1}{2}}$
α^S	$\frac{\partial g^S}{\partial p}$	$B_1 S$	$-\alpha_0 \beta_S S$
η^S	$-\frac{\partial g^S}{\partial T}$	$-2B_2 T' S$	$c_{p0}\beta_S^* \frac{T'}{T_0} S$
μ^S	$\frac{\partial g^S}{\partial S}$	$B_0 + B_1 p' + B_2 \left(T'\right)^2 + B_{\ln}\left[\frac{1}{2} + \ln \left(\frac{S}{S_u}\right)^{\frac{1}{2}}\right]$	$\mu_0 + c_{p0}\beta_S^* T_0 - \alpha_0 \beta_S p' - \frac{c_{p0}\beta_S^*}{2T_0}\left(T'\right)^2 + B_{\ln}\left[\frac{1}{2} + \ln \left(\frac{S}{S_u}\right)^{\frac{1}{2}}\right]$
c_p^S	$-T \frac{\partial^2 g^S}{\partial T^2}$	$-2B_2 T S$	$c_{p0}\beta_S^* \frac{T}{T_0} S$
$\left(\widehat{\beta^T}\right)^S$	$\frac{1}{\partial g^S/\partial p} \frac{\partial^2 g^S}{\partial p \partial T}$	0	0
$\left(\widehat{\beta^S}\right)^S$	$-\frac{1}{\partial g^S/\partial p} \frac{\partial^2 g^S}{\partial p \partial S}$	$-\frac{1}{\alpha^S} B_1$, α^S as above.	$\frac{\alpha_0}{\alpha^S}\beta_S$, α^S as above.
Γ^S	$-\frac{1}{\partial^2 g^S/\partial T^2} \frac{\partial^2 g^S}{\partial p \partial T}$	0	0

Table 11.5 As in Table 11.3 but derived from the alternative saline-correction $g^S = g^S(p, T, S)$ Gibbs potential (11.69). Red terms are absent if the $S \ln (S/S_u)^{1/2}$ term in (11.69) is omitted. See text for further details.

Various thermodynamic quantities for the saline correction, $g^S = g^S(p, T, S)$, derived from (11.69), are displayed in column 3 of Table 11.5. These may be compared to the corresponding ones in Table 11.3 for Vallis (2017)'s potential (11.17). To do so, insert expansion (11.64) into (11.17) and (11.39) for Vallis (2017)'s potential to obtain

$$g^S(p, T, S) = \left\{ F_1^g - F_1^\eta T' + A_1 p' + F_1^{c_p} T_0 \left[1 - \frac{1}{2}\left(\frac{T'}{T_0}\right)^2 \right] + O\left(\frac{T'}{T_0}\right)^3 \right\} S, \tag{11.70}$$

$$g^S(p, T, S) = \left[\left(\mu_0 + c_{p0}\beta_S^* T_0\right) - \alpha_0 \beta s p' - \frac{c_{p0}\beta_S^*}{2T_0}\left(T'\right)^2 + O\left(\frac{T'}{T_0}\right)^3 \right] S. \tag{11.71}$$

Setting $B_{\text{ln}} \equiv 0$ in (11.69) (i.e. dropping the logarithmic term) and neglecting the $O\left(T'/T_0\right)^3$ terms in (11.70) and (11.71) yields three different expressions for $g^S(p, T, S)$. Comparing these three expressions term by term shows that they agree when

$$B_0 = F_1^g + F_1^{c_p} T_0 = \mu_0 + c_{p0}\beta_S^* T_0, \quad B_1 = A_1 = -\alpha_0 \beta_S, \quad B_2 = -\frac{F_1^{c_p}}{2T_0} = -\frac{c_{p0}\beta_S^*}{2T_0}. \tag{11.72}$$

Thus Vallis (2017)'s prototypical oceanic saline correction is asymptotically equivalent to the alternative one of this section when the (red) logarithmic term in (11.69) is absent. This means that the representative values given in Table 11.1 for Vallis (2017)'s prototypical oceanic Gibbs potential can also be considered to be representative values for the alternative saline correction of this section; they can then, if desired, be used in (11.72) as representative values for B_0, B_1, and B_2. Insertion of (11.72) into the expressions of the third column of Table 11.5 leads to the expressions given in the fourth one. Values for the constant coefficients B_0, B_1, B_2, and B_{ln} can be obtained via a thermodynamically consistent fitting to a dataset of laboratory values or by using the representative values given in Table 11.1, supplemented by a suitable representative value for B_{ln}.

The tweak (11.64) used to replace the $T\left[\ln\left(T/T_0\right) - 1\right]$ term in the pure-water Gibbs potentials (11.16) and (11.39) by a polynomial has been implicitly used to replace a similar term in the saline correction (11.69). This could have been anticipated by analogy.

11.4.3 Potential Temperature $\theta(p, T, S)$

Potential temperature, θ, is again defined by (11.49), but this time $g(p, T, S) \equiv g^W(p, T) + g^S(p, T, S)$ and $\eta(p, T, S) \equiv \eta^W(p, T) + \eta^S(p, T, S)$ are instead defined by the expressions given in Tables 11.4 and 11.5. Substituting these expressions into (11.49) then yields

$$\begin{aligned} \left[\frac{\bar{c}_{p0}}{T_0} - 2B_2 S - 2A_{02}\left(p_r - p_0\right)\right](\theta - T) = &-\left[A_{01} + \frac{A_{11}}{2}\left(p - p_0\right) + 2A_{02}\left(T - T_0\right)\right]\left(p - p_0\right) \\ &+ \left[A_{01} + \frac{A_{11}}{2}\left(p_r - p_0\right) + 2A_{02}\left(T - T_0\right)\right]\left(p_r - p_0\right). \end{aligned} \tag{11.73}$$

Although this equation can be solved explicitly for θ, it is easier to do so under the previous assumption that $p_r \equiv p_0$, and then

$$\theta\left(p,T,S\right) = T - T_0\frac{\left[A_{01} + (A_{11}/2)\,p' + 2A_{02}T'\right]p'}{\left(\bar{c}_{p0} - 2B_2T_0S\right)}. \tag{11.74}$$

Using (11.41), (11.42), and (11.72), (11.74) may be equivalently rewritten as

$$\theta\left(p,T,S\right) = T - T_0\frac{\alpha_0\left[\beta_T + (\beta_T\gamma^*/2)\,p' + \beta_T^*T'\right]p'}{c_{p0}\left(1 + \beta_S^*S'\right)}. \tag{11.75}$$

Examining (11.74) and (11.75) for the alternative Gibbs potential and comparing them with (11.51) and (11.52) for Vallis (2017)'s prototypical Gibbs potential:

1. Coefficient $B_{\ln}$ is absent from (11.74). This means that the $B_{\ln}S\ln\left(S/S_u\right)^{1/2}$ term in (11.69) does not influence potential temperature $\theta = \theta\left(p,T,S\right)$.
2. The polynomial forms of (11.74) and (11.75) for θ are simpler than the exponential ones of (11.51) and (11.52).
3. From (11.72), $F_1^{c_p} = -2B_2T_0$, and (11.51) may then be rewritten as

$$\theta\left(p,T,S\right) = T\exp\left\{-\frac{\left[A_{01} + (A_{11}/2)\,p' + 2A_{02}T'\right]p'}{\left(\bar{c}_{p0} - 2B_2T_0S\right)}\right\}. \tag{11.76}$$

Noting that $T \equiv T_0 + T'$ and expanding the exponential in (11.51) under the assumptions that its argument is small and that $\left|T'\right| \ll T_0$, leads to (11.74), but as an approximation. Thus to this order of approximation, (11.51) asymptotes to (11.74).

11.4.4 Enthalpy, Potential Enthalpy, Conservative Temperature, and Potential Density

The alternative prototypical Gibbs potential is the sum of the individual Gibbs potentials for pure water and for the saline correction, that is, $g\left(p,T,S\right) = g^W\left(p,T\right) + g^S\left(p,T,S\right)$. Proceeding as in Sections 11.3.8–11.3.11, expressions for enthalpy ($h = h\left(p,T,S\right)$), potential enthalpy ($h^0 = h^0\left(p,T,S\right)$), conservative temperature ($\Theta = \Theta\left(p,T,S\right)$), and potential density ($\rho^\theta = \rho^\theta\left(p,T,S\right)$) can be similarly derived from the alternative Gibbs potential. The resulting expressions can then be compared with the corresponding ones obtained from Vallis (2017)'s prototypical Gibbs potential. To facilitate this comparison, expressions for θ, h, h^0, Θ, and ρ^θ, derived from these two Gibbs potentials, are displayed side-by-side in Tables 11.4 and 11.5. [10]

It is seen that whereas the functional form for θ of the alternative potential is a bit simpler than that of Vallis (2017)'s, the corresponding forms for enthalpy (h), potential enthalpy (h^0), and conservative temperature (Θ) are a bit more complicated. Enthalpy, potential enthalpy, and conservative temperature are all influenced by the $B_{\ln}S\ln\left(S/S_u\right)^{1/2}$ term in (11.69), but potential density (ρ^θ) is not since this term is absent in α^S.

The conventional equation of state for Vallis (2017)'s potential may be obtained from $\alpha\left(p,T,S\right) = \alpha^W\left(p,T\right) + \alpha^S\left(p,T,S\right)$, where $\alpha^W\left(p,T\right)$ and $\alpha^S\left(p,T,S\right)$ are displayed in Tables 11.2 and 11.3, respectively. Similarly, that for the alternative potential may be found using Tables 11.4

[10] In doing so, (11.72) has been used to replace $F_1^{c_p}$, F_1^g, and A_1 by $-2B_2T_0$, $B_0 + 2B_2T_0^2$, and B_1, respectively.

and 11.5. Although the Vallis (2017) and alternative potentials both lead to the same conventional equation of state

$$\alpha\left(p,T,S\right)=\alpha_0\left[1-\beta_p p^{'}+\beta_T T^{'}-\beta_S S^{'}+\beta_T\gamma^* p^{'} T^{'}+\frac{\beta_T^*}{2}\left(T^{'}\right)^2\right], \tag{11.77}$$

they do nevertheless lead to differences in various other thermodynamic quantities derived from these potentials; see, for example, the green terms in Tables 11.6 and 11.7.

11.5 THE TEOS-10 GIBBS POTENTIAL

11.5.1 Background

The TEOS-10 specific Gibbs potential is designed to address various weaknesses of the International Equation of State of Seawater 1980 (known as EOS80) that were hindering progress in oceanography. See Pawlowicz et al. (2012) for a historical perspective and IOC et al. (2010) for comprehensive documentation and bibliography. Significant changes and improvements of TEOS-10 compared with EOS80 include:

- Definition of a specific Gibbs potential from which *all* thermodynamic properties (including, importantly, those that are not measured) of the ocean, atmosphere, and ice can, solely by differentiation and algebraic manipulation, be determined in a thermodynamically consistent manner.
- Replacement of Practical Salinity (based on electrical conductivity) as a state variable by Absolute Salinity (the mass fraction of solute in seawater).
- Use of the best available experimental data for heat capacities, freezing points, vapour pressures, and mixing heats at atmospheric pressure to improve accuracy.
- Facilitation of accurate representation of heat transport in the ocean and of thermodynamically consistent heat and mass exchange between the ocean, the atmosphere, and ice to improve modelling capability and accuracy.
- Availability of a comprehensive set of computer software to compute a broad suite of thermodynamic quantities from the TEOS-10 specific Gibbs potential.

11.5.2 Notation and Units

Notation and units can be something of a minefield, particularly when attempting to unify things between different disciplines. For consistency with the notation of Appendix A of Chapter 18 of Vallis (2017) and that of Thuburn (2017), we have thus far ordered the state variables as $\left(p,T,S\right)$, as in $g=g\left(p,T,S\right)$ in Sections 11.2–11.4 of the present chapter. This is the opposite ordering convention to that employed for TEOS-10. For compatibility with TEOS-10, we therefore now change our convention and adopt its reverse ordering, as in $g=g\left(S_{\mathrm{A}},T,p\right)$. However, we are not yet out of the notational woods.

In meteorology, absolute pressure, denoted by p, is simply atmospheric pressure (i.e. the total pressure exerted on a closed parcel of air by its surrounding environment). In oceanography, sea pressure is often used instead of absolute pressure. The difference between the two is that sea pressure is defined to be absolute pressure minus one standard atmosphere (with pressure $p_0 = 101\,325$ Pa $=$ constant); see Section 2.2 of IOC et al. (2010). Thus sea pressure is simply absolute pressure shifted by a constant value of pressure, typical of that exerted by the atmosphere on an ocean's surface. So far, so good. However, sea pressure (a relative pressure) is then usually also denoted by p (which in meteorology denotes absolute pressure, not relative pressure)! To avoid confusion, TEOS-10 therefore denotes absolute pressure by P, so that $P=p_0+p$ is absolute pressure, with p therein denoting sea pressure. In what follows, we nevertheless maintain the convention that p is absolute pressure (rather than sea pressure), and simply subtract p_0 from it

Quantity	From Vallis (2017)'s Gibbs Potential	From Alternative Gibbs potential
θ	$T\exp\left\{-\frac{\left[A_{01}+(A_{11}/2)\,p'+2A_{02}T'\right]p'}{\left(\bar{c}_{p0}-2B_2T_0S\right)}\right\}$	$T-T_0\frac{\left[A_{01}+(A_{11}/2)\,p'+2A_{02}T'\right]p'}{\left(\bar{c}_{p0}-2B_2T_0S\right)}$
h	$\bar{g}_0+\eta_0T_0+\left(B_0-2T_0B_2T'\right)S+\bar{c}_{p0}T$ $+\left[A_{00}-T_0A_{01}+\frac{1}{2}\left(A_{10}-T_0A_{11}\right)p'\right.$ $\left.+B_1S-A_{02}\left(T^2-T_0^2\right)\right]p'$	$\bar{g}_0+\eta_0T_0+\left[B_0-(T+T_0)B_2T'\right]S+\bar{c}_{p0}\frac{\left(T^2+T_0^2\right)}{2T_0}+B_{\ln}S\ln\left(\frac{S}{S_u}\right)^{\frac{1}{2}}$ $+\left[A_{00}-T_0A_{01}+\frac{1}{2}\left(A_{10}-T_0A_{11}\right)p'\right.$ $\left.+B_1S-A_{02}\left(T^2-T_0^2\right)\right]p'$
h^0	$\bar{g}_0+\eta_0T_0+\left[B_0-2T_0B_2\left(\theta-T_0\right)\right]S+\bar{c}_{p0}\theta$, θ as above.	$\bar{g}_0+\eta_0T_0+\left[B_0-(\theta+T_0)B_2\left(\theta-T_0\right)\right]S+\bar{c}_{p0}\frac{\left(\theta^2+T_0^2\right)}{2T_0}+B_{\ln}S\ln\left(\frac{S}{S_u}\right)^{\frac{1}{2}}$, θ as above.
Θ	$\frac{\bar{g}_0+\eta_0T_0}{c_{p0}}+\frac{\left[B_0-2T_0B_2\left(\theta-T_0\right)\right]S+\bar{c}_{p0}\theta}{c_{p0}}$, θ as above.	$\frac{\bar{g}_0+\eta_0T_0}{c_{p0}}+\frac{\left[B_0-(\theta+T_0)B_2\left(\theta-T_0\right)\right]S}{c_{p0}}+\frac{\bar{c}_{p0}}{c_{p0}}\frac{\left(\theta^2+T_0^2\right)}{2T_0}+\frac{B_{\ln}}{c_{p0}}S\ln\left(\frac{S}{S_u}\right)^{\frac{1}{2}}$, θ as above.
ρ^θ	$\left[A_{00}+A_{01}\left(\theta-T_0\right)+A_{02}\left(\theta-T_0\right)^2+B_1S\right]^{-1}$, θ as above.	$\left[A_{00}+A_{01}\left(\theta-T_0\right)+A_{02}\left(\theta-T_0\right)^2+B_1S\right]^{-1}$, θ as above.

Table 11.6 Various thermodynamic quantities (in column 1) derived from Vallis (2017)'s Gibbs potential (in column 2) and from the alternative Gibbs potential (in column 3); all quantities are expressed in the present notation. To facilitate comparison, differences between the two sets of expressions are highlighted in green, and logarithmic contributions for the alternative Gibbs potential in red.

Quantity	From Vallis (2017)'s Gibbs Potential	From Alternative Gibbs potential
θ	$T\exp\left\{-\frac{\alpha_0\left[\beta_T+(\beta_T\gamma^*/2)p'+\beta_T^*T'\right]p'}{c_{p0}(1+\beta_S^*S')}\right\}$	$T-T_0\frac{\alpha_0\left[\beta_T+(\beta_T\gamma^*/2)p'+\beta_T^*T'\right]p'}{c_{p0}(1+\beta_S^*S')}$
h	$g_0+\eta_0T_0+\mu_0S'+c_{p0}\left(1+\beta_S^*S'\right)T$ $+\alpha_0\left[1-T_0\beta_T-\frac{1}{2}(\beta_p+T_0\beta_T\gamma^*)p' -\beta_SS'-\frac{\beta_T^*}{2}(T^2-T_0^2)\right]p'$	$g_0+\eta_0T_0+\mu_0S'+c_{p0}\left(1+\beta_S^*S'\right)\frac{(T^2+T_0^2)}{2T_0}+B_{\ln}S\ln\left(\frac{S}{S_u}\right)^{\frac{1}{2}}$ $+\alpha_0\left[1-T_0\beta_T-\frac{1}{2}(\beta_p+T_0\beta_T\gamma^*)p' -\beta_SS'-\frac{\beta_T^*}{2}(T^2-T_0^2)\right]p'$
h^0	$g_0+\eta_0T_0+\mu_0S'+c_{p0}\left(1+\beta_S^*S'\right)\theta,$ θ as above.	$g_0+\eta_0T_0+\mu_0S'+c_{p0}\left(1+\beta_S^*S'\right)\frac{(\theta^2+T_0^2)}{2T_0}+B_{\ln}S\ln\left(\frac{S}{S_u}\right)^{\frac{1}{2}},$ θ as above.
Θ	$\frac{g_0+\eta_0T_0+\mu_0S'}{c_{p0}}+\left(1+\beta_S^*S'\right)\theta,$ θ as above.	$\frac{g_0+\eta_0T_0+\mu_0S'}{c_{p0}}+\left(1+\beta_S^*S'\right)\frac{(\theta^2+T_0^2)}{2T_0}+\frac{B_{\ln}}{c_{p0}}S\ln\left(\frac{S}{S_u}\right)^{\frac{1}{2}},$ θ as above.
ρ^θ	$\alpha_0^{-1}\left[1+\beta_T(\theta-T_0)+\frac{\beta_T^*}{2}(\theta-T_0)^2-\beta_SS'\right]^{-1},$ θ as above.	$\alpha_0^{-1}\left[1+\beta_T(\theta-T_0)+\frac{\beta_T^*}{2}(\theta-T_0)^2-\beta_SS'\right]^{-1},$ θ as above.

Table 11.7 As in Table 11.6, but expressed in Vallis (2017)'s notation.

when required. This is simply a notational issue; it does not fundamentally affect anything, only the way it is expressed.

There is also a notational issue with temperature. The TEOS-10 specific Gibbs potential is defined in terms of the Celsius temperature scale rather than that of absolute temperature. To make the distinction between these two temperature scales, t in TEOS-10 is used to denote temperature in degrees Celsius. This then causes a notational problem when, for example, one wishes to take a material derivative, D/Dt, where t then denotes time and not temperature. In what follows, and to avoid this problem, we continue to use absolute temperature T and to explicitly subtract the constant $T_0 = 273.15$ K from it whenever degrees Celsius appear in TEOS-10 formulae.

A related, general issue is that of units in TEOS-10, with a mix of SI units and traditional oceanographic ones! Ideally it would be best to work exclusively in SI units. However, this would go against traditional oceanographic practice, developed over many decades, with a vast literature and many datasets that make extensive use of non-SI units. As discussed in Section 1.5 of IOC et al. (2010), one consequently has to be vigilant regarding units when using TEOS-10 and to make appropriate adjustments when using formulae and software.

11.5.3 Unique Decomposition of the TEOS-10 Gibbs Potential into Pure-Water and Saline-Correction Parts

The functional form of the alternative, prototypical, specific Gibbs potential of Section 11.4 is, by design, similar to the TEOS-10 one, but with far fewer terms. As such it provides, in a much simpler context, a conceptual pathway towards the TEOS-10 Gibbs potential. Both the simple alternative Gibbs potential and the TEOS-10 one are composed of the sum of a pure-water part and a saline-correction part. Thus:

The TEOS-10 Gibbs Potential

The TEOS-10 Gibbs potential $g = g\left(S_A, T, p\right)$ is the sum of a pure-water part, $g^W = g^W\left(T, p\right)$, and a saline-correction part, $g^S = g^S\left(S_A, T, p\right)$, so that

$$g\left(S_A, T, p\right) = g^W\left(T, p\right) + g^S\left(S_A, T, p\right), \tag{11.78}$$

where S_A is Absolute Salinity, T is absolute temperature, and p is absolute pressure. Furthermore

$$g^S\left(S_A = 0, T, p\right) = 0, \tag{11.79}$$

to ensure that the total Gibbs potential, $g\left(S_A, T, p\right)$, reduces to that of pure water, $g^W\left(T, p\right)$, in the absence of solute (i.e. when $S_A \equiv 0$).

As written, this decomposition is not unique. This is because a specific Gibbs potential is only defined to within an additive function of temperature and salinity of the form $c_1 + c_2 T + \left(c_3 + c_4 T\right) S_A$, where c_1, c_2, c_3, and c_4 are constants of unknown, and unknowable, values. The first two coefficients (c_1 and c_2) are related to the pure-water part, $g^W = g^W\left(T, p\right)$, and the last two ($c_3$ and c_4) to the saline-correction part, $g^S = g^S\left(S_A, T, p\right)$.

To achieve uniqueness of the decomposition, it is standard practice to choose the first two coefficients to make the specific entropy (η) and the specific energy ($\mathcal{E}$) zero at the triple point of water.[11] Thus:

[11] See Section 10.3 for definition and discussion of the triple point of water and its importance.

Uniqueness of the TEOS-10 Gibbs Potential: the Pure-Water Part

To achieve uniqueness of the decomposition of the TEOS-10 Gibbs potential, the two constraints

$$\eta^W\left(T_t, p_t\right) = \mathscr{E}^W\left(T_t, p_t\right) = 0 \tag{11.80}$$

are applied to the pure-water part, where $T_t \equiv 273.16\ \text{K} \equiv 0.01\ ^\circ\text{C}$ and $p_t \equiv 611.657$ Pa are the values of T and p, respectively, at the triple point of water.[a]

[a] Note that the temperature at the triple point of water is *not* 0.00 °C but 0.01 °C. This is not a typographical error, but an international convention, as briefly discussed in Section 10.6!

The remaining two coefficients, c_3 and c_4, are chosen to make the specific enthalpy (h) and specific entropy (η) of a sample of standard seawater, with standard ocean properties

$$\left(S_{S0}, T_{S0}, p_{S0}\right) = \left(35.16504\,\text{g}\,\text{kg}^{-1} = 35\,\text{psu}, 273.15\,\text{K} = 0\,^\circ\text{C}, 101325\,\text{Pa}\right), \tag{11.81}$$

both zero, where (recall from Table 11.1) 1 psu=1 Practical Salinity unit $\approx 1\,\text{g}\,\text{kg}^{-1}$. Thus:

Uniqueness of the TEOS-10 Gibbs Potential: the Saline-Correction Part

To achieve uniqueness of the decomposition of the TEOS-10 Gibbs potential, the two constraints

$$h^S\left(S_{S0}, T_{S0}, p_{S0}\right) = \eta^S\left(S_{S0}, T_{S0}, p_{S0}\right) = 0 \tag{11.82}$$

are applied to the saline-correction part, where S_{S0}, T_{S0}, and p_{S0} satisfy (11.81).

Setting the four coefficients in this manner then makes the decomposition unique.

- Why is this important?

It is because *independence* of the pure-water part of the Gibbs potential from the saline-correction part means that we are free to combine *any* variant of the pure-water part with *any* variant of the saline-correction part. This is a very useful property. When more accurate measurements become available in the future for either the pure-water part or for the saline-correction part, then the corresponding TEOS-10 Gibbs potential can be modified and improved *independently of the other part.* This is not by chance, but by design of the TEOS-10 Gibbs potential. Furthermore one may, according to application, use simplified forms of either part, independently of the other part.

11.5.4 Definition of the Pure-Water Part $g^W(T, p)$

The pure-water part of the alternative prototypical Gibbs potential is given by (11.65). This is a low-order Taylor-series expansion about a reference state $(T, p) = (T_0, p_0)$.

The pure-water part of the TEOS-10 Gibbs potential is similarly defined as a Taylor-series expansion. There are many more terms (41 versus 8), the reference state is $(T, p) = (T_{S0}, p_{S0})$ instead of $(T, p) = (T_0, p_0)$, and perturbations $\left(T', p'\right) \equiv \left(T - T_{S0}, p - p_{S0}\right)$ are scaled to make them non-dimensional, but otherwise it is just more of the same; a higher-order Taylor-series expansion instead of a lower-order one.

Thus (from Appendix G of IOC et al. 2010, but in the present notation):

j	k		g_{jk}^W	j	k		g_{jk}^W
0	0		$0.101\,342\,743\,139\,674 \times 10^3$	3	2		$0.499\,360\,390\,819\,152 \times 10^3$
0	1		$0.100\,015\,695\,367\,145 \times 10^6$	3	3	−	$0.239\,545\,330\,654\,412 \times 10^3$
0	2	−	$0.254\,457\,654\,203\,630 \times 10^4$	3	4		$0.488\,012\,518\,593\,872 \times 10^2$
0	3		$0.284\,517\,778\,446\,287 \times 10^3$	3	5	−	$0.166\,307\,106\,208\,905 \times 10$
0	4	−	$0.333\,146\,754\,253\,611 \times 10^2$	4	0	−	$0.148\,185\,936\,433\,658 \times 10^3$
0	5		$0.420\,263\,108\,803\,084 \times 10$	4	1		$0.397\,968\,445\,406\,972 \times 10^3$
0	6	−	$0.546\,428\,511\,471\,039$	4	2	−	$0.301\,815\,380\,621\,876 \times 10^3$
1	0		$0.590\,578\,347\,909\,402 \times 10$	4	3		$0.152\,196\,371\,733\,841 \times 10^3$
1	1	−	$0.270\,983\,805\,184\,062 \times 10^3$	4	4	−	$0.263\,748\,377\,232\,802 \times 10^2$
1	2		$0.776\,153\,611\,613\,101 \times 10^3$	5	0		$0.580\,259\,125\,842\,571 \times 10^2$
1	3	−	$0.196\,512\,550\,881\,220 \times 10^3$	5	1	−	$0.194\,618\,310\,617\,595 \times 10^3$
1	4		$0.289\,796\,526\,294\,175 \times 10^2$	5	2		$0.120\,520\,654\,902\,025 \times 10^3$
1	5	−	$0.213\,290\,083\,518\,327 \times 10$	5	3	−	$0.552\,723\,052\,340\,152 \times 10^2$
2	0	−	$0.123\,577\,859\,330\,390 \times 10^5$	5	4		$0.648\,190\,668\,077\,221 \times 10$
2	1		$0.145\,503\,645\,404\,680 \times 10^4$	6	0	−	$0.189\,843\,846\,514\,172 \times 10^2$
2	2	−	$0.756\,558\,385\,769\,359 \times 10^3$	6	1		$0.635\,113\,936\,641\,785 \times 10^2$
2	3		$0.273\,479\,662\,323\,528 \times 10^3$	6	2	−	$0.222\,897\,317\,140\,459 \times 10^2$
2	4	−	$0.555\,604\,063\,817\,218 \times 10^2$	6	3		$0.817\,060\,541\,818\,112 \times 10$
2	5		$0.434\,420\,671\,917\,197 \times 10$	7	0		$0.305\,081\,646\,487\,967 \times 10$
3	0		$0.736\,741\,204\,151\,612 \times 10^3$	7	1	−	$0.963\,108\,119\,393\,062 \times 10$
3	1	−	$0.672\,507\,783\,145\,070 \times 10^3$				

Table 11.8 Non-zero coefficients g_{jk}^W of the pure-water Gibbs potential $g^W = g^W(T,p)$ of IAPWS-09, reproduced from Appendix G of IOC et al. (2010). See text for further details.

Definition of the Pure-Water Part of the TEOS-10 Gibbs Potential

The pure-water part of the TEOS-10 Gibbs potential is defined by the Taylor-series expansion

$$g^W(T,p) = g_u \sum_{j=0}^{7} \sum_{k=0}^{6} g_{jk}^W y^j z^k. \tag{11.83}$$

For this expansion:

$$y = \frac{T - T_{S0}}{T_u}, \quad z = \frac{p - p_{S0}}{p_u} \tag{11.84}$$

are dimensionless temperature and pressure perturbations, respectively, about the reference state

$$(T_{S0}, p_{S0}) \equiv (273.15\,\text{K} \equiv 0\,°\text{C}, 101325\,\text{Pa}), \tag{11.85}$$

and

$$(g_u, T_u, p_u) \equiv \left(1\,\text{J}\,\text{kg}^{-1}, 40\,°\text{C}, 10^8\,\text{Pa}\right) \tag{11.86}$$

are scaling units. Values for the 41 non-zero coefficients g_{jk}^W appearing in (11.83) were obtained by fitting them to high-quality sets of experimental data. These values are tabulated in Appendix G of IOC et al. (2010) which, for convenience, is reproduced here as Table 11.8.

The reference state defined by (11.85) is representative of that at the surface of a (pure-water) ocean in polar regions. This means that a Taylor expansion around this reference state is most accurate in near-surface polar regions, where perturbations are small, but less accurate equatorward (with much higher temperatures) and at depth (with much larger pressures), where perturbations are much larger. This influences the number of terms required to achieve high accuracy, with more terms needed where perturbations are larger. The number of terms retained in the TEOS-10 Gibbs potential for pure water has been chosen to ensure that accuracy *everywhere* within an ocean is within the accuracy of experimental measurements. This would not be the case with fewer terms. Similar considerations also apply to the Vallis (2017) and alternative prototypical Gibbs potentials, except that the value (283 K $= 9.85\,^\circ$C) given in Table 9.3 for T_0 corresponds instead to a mid-latitude ocean surface temperature instead of a polar one. Furthermore, the limited number of terms in these potentials reduces accuracy in some regions of an ocean, with errors there significantly greater than the measurement error of experimental data. Generally, the fewer the number of terms, the larger the error.

11.5.5 Definition of the Saline-Correction Part $g^S(S_A, T, p)$

The saline-correction part of the alternative prototypical Gibbs potential is given by (11.69). This is a very-low-order Taylor-series expansion about a reference state $(T, p) = (T_0, p_0)$, plus a logarithmic term in $S^{1/2}$, introduced to mimic the functional form used in TEOS-10, but at lower order.

The saline-correction part of the TEOS-10 Gibbs potential is similarly defined, and its functional form is termed *'polynomial-like'* by Feistel (2008). There are many more terms (64 versus 4, including two logarithmic terms instead of one), the reference state is $(S_A, T, p) = (0, T_{S0}, p_{S0})$ instead of $(S, T, p) = (0, T_0, p_0)$, and perturbations $\left(S_A', T', p'\right) \equiv (S_A, T - T_{S0}, p - p_{S0})$ are scaled to make them non-dimensional, but otherwise it is again just (much) more of the same; a higher-order Taylor-series expansion plus two logarithmic terms instead of a much-lower-order one with a single such term. The (two) logarithmic terms are consistent with Planck's theory of ideal solutions, and the quadratic scaling $x^2 = S_A/S_u$ results from the statistical theory of electrolytes (Feistel, 2008). The logarithmic terms are particularly important for weak solutions; S is then small, $\ln S^{1/2}$ is very large, and $S \ln S^{1/2}$ tends slowly to zero as S tends to zero.

Thus (from Appendix H of IOC et al. 2010, but in the present notation):

Definition of the Saline-Correction Part of the TEOS-10 Gibbs Potential

The saline-correction part of the TEOS-10 Gibbs potential is defined by the Taylor-series expansion

$$g^S(S_A, T, p) = g_u \sum_{j=0}^{6} \sum_{k=0}^{5} \left(g^S_{1jk} x^2 \ln x + \sum_{i=2}^{7} g^S_{ijk} x^i \right) y^j z^k. \tag{11.87}$$

For this expansion:

$$x^2 = \frac{S_A}{S_u}, \quad y = \frac{T - T_{S0}}{T_u}, \quad z = \frac{p - p_{S0}}{p_u}, \tag{11.88}$$

are dimensionless salinity, temperature, and pressure perturbations, respectively, about the reference state

$$(S_{A0}, T_{S0}, p_{S0}) \equiv (0, 273.15\,\mathrm{K} = 0\,^\circ\mathrm{C}, 101325\,\mathrm{Pa}), \tag{11.89}$$

and

$$(g_u, S_u, T_u, p_u) = \left(1\,\mathrm{J\,kg^{-1}}, \frac{40}{35} \times 35.16504\,\mathrm{g\,kg^{-1}}, 40\,^\circ\mathrm{C}, 10^8\,\mathrm{Pa}\right), \tag{11.90}$$

i	j	k	g^S_{ijk}	i	j	k	g^S_{ijk}
1	0	0	5812.81456626732	2	2	1	-860.764303783977
1	1	0	851.226734946706	3	2	1	383.058066002476
2	0	0	1416.27648484197	2	3	1	694.244814133268
3	0	0	-2432.14662381794	3	3	1	-460.319931801257
4	0	0	2025.80115603697	2	4	1	-297.728741987187
5	0	0	-1091.66841042967	3	4	1	234.565187611355
6	0	0	374.601237877840	2	0	2	384.794152978599
7	0	0	-48.5891069025409	3	0	2	-52.2940909281335
2	1	0	168.072408311545	4	0	2	-4.08193978912261
3	1	0	-493.407510141682	2	1	2	-343.956902961561
4	1	0	543.835333000098	3	1	2	83.1923927801819
5	1	0	-196.028306689776	2	2	2	337.409530269367
6	1	0	36.7571622995805	3	2	2	-54.1917262517112
2	2	0	880.031352997204	2	3	2	-204.889641964903
3	2	0	-43.0664675978042	2	4	2	74.7261411387560
4	2	0	-68.5572509204491	2	0	3	-96.5324320107458
2	3	0	-225.267649263401	3	0	3	68.0444942726459
3	3	0	-10.0227370861875	4	0	3	-30.1755111971161
4	3	0	49.3667694856254	2	1	3	124.687671116248
2	4	0	91.4260447751259	3	1	3	-29.4830643494290
3	4	0	.875600661808945	2	2	3	-178.314556207638
4	4	0	-17.1397577419788	3	2	3	25.6398487389914
2	5	0	-21.6603240875311	2	3	3	113.561697840594
4	5	0	2.49697009569508	2	4	3	-36.4872919001588
2	6	0	2.13016970847183	2	0	4	15.8408172766824
2	0	1	-3310.49154044839	3	0	4	-3.41251932441282
3	0	1	199.459603073901	2	1	4	-31.6569643860730
4	0	1	-54.7919133532887	2	2	4	44.2040358308000
5	0	1	36.0284195611086	2	3	4	-11.1282734326413
2	1	1	729.116529735046	2	0	5	-2.62480156590992
3	1	1	-175.292041186547	2	1	5	7.04658803315449
4	1	1	-22.6683558512829	2	2	5	-7.92001547211682

Table 11.9 Non-zero coefficients g^S_{ijk} of the saline specific Gibbs potential for seawater $g^S = g^S(S_A, T, p)$ of IAPWS-08, reproduced from Appendix H of IOC et al. (2010). See text for further details.

are scaling units. Values for the 64 non-zero coefficients g^S_{ijk} appearing in (11.87) were obtained by fitting to high-quality sets of experimental data. These values are tabulated in Appendix H of IOC et al. (2010) which, for convenience, is reproduced here as Table 11.9.

The reference state defined by (11.89) is again representative of that at the surface of a pure-water ocean in polar regions. Similar considerations to those for the pure-water part $g^W = g^W(T, p)$ of the Gibbs potential also apply to the saline-correction part $g^S = g^S(S_A, T, p)$. Since the reference value for salinity has been taken to be zero, many more terms are required in regions of an ocean with high salinity than those with low salinity. The number of terms retained in the saline-correction part of the TEOS-10 Gibbs potential has been chosen to ensure that accuracy everywhere in the ocean, including regions of high salinity, is within the accuracy of experimental measurements. This would not be the case with fewer terms.

11.5.6 Toolboxes

The TEOS-10 specific Gibbs potential addresses various weaknesses of the EOS80 Equation of State for Seawater. This is very good news. However, to benefit from this advance, computer software is needed to transfer theory into practical application. The development of such software is then a daunting task due to the complexity of the Gibbs potential and the significant number of quantities that need to be derived from it. Equally good news therefore is that two toolboxes of computer software are freely available for the computation of thermodynamic quantities from the TEOS-10 specific Gibbs potential. The authors of this software have done the hard work so that we fortunately do not have to do it! Because of the care that has gone into the development, optimisation, and – very importantly – testing of this software, it is advisable to use it rather than develop one's own.

The SIA (Seawater Ice Air) toolbox mostly uses SI units and is the more comprehensive and more accurate of the two toolboxes. It not only operates within the oceanographic range of state variables $\left(S_{\mathrm{A}}, T, p\right)$, but beyond it. The GSW (Gibbs-Sea-Water) toolbox mostly uses traditional oceanographic units and is more limited in scope. However, its subroutines typically execute an order of magnitude faster, or more, than those of the SIA toolbox, with little or no loss of accuracy within the oceanographic range of state variables. See Appendices M (for the SIA toolbox) and N (for the GSW one) of IOC et al. (2010), and references therein, for details on how to access and use these two toolboxes.

11.6 CONCLUDING REMARKS

Vallis (2017)'s prototypical Gibbs potential for an ocean has been derived from first principles and various thermodynamic quantities derived from it. It provides a very useful and convenient, low-order representation of an ocean's thermodynamics, plus some useful insight. Its functional form does not, however, correspond to a low-order approximation of the highly accurate but highly complicated high-order TEOS-10 one. An alternative prototypical Gibbs potential that does so has therefore been developed. This then provides a natural, conceptual pathway to understanding the structure of the empirical state-of-the-science TEOS-10 representation.

The TEOS-10 representation is not only of great benefit for oceanography, but also for meteorology. In particular, it is important that fluxes of mass, momentum, and energy at the interface between the atmosphere and oceans be consistent. This avoids introducing spurious sources and sinks of energy there that can lead to spurious climate drift in quantitative atmospheric and oceanic models. This then requires consistency between the thermodynamic representations of the atmosphere and oceans at their mutual interface. Use of the TEOS-10 representation, which is applicable not only to the ocean side of the interface but also to the atmospheric side, is thus a very good way of ensuring thermodynamical consistency at the interface.

Having addressed the challenges of:

- Geopotential representation (in Chapters 7 and 8) for mildly oblate, ellipsoidal planets.
- Thermodynamic representation (in Chapters 9–11) of the atmosphere and oceans.

we now turn our attention (in Chapter 12) to the development of a suitable geopotential coordinate system for quantitative global atmospheric and oceanic modelling. This coordinate system is termed GREAT, an abbreviated form of Geophysically Realistic, Ellipsoidal, Analytically Tractable.

12

Geopotential Coordinates for Modelling Planetary Atmospheres and Oceans

ABSTRACT

Physically, gravity varies both meridionally and vertically for Earth. It is therefore desirable that this behaviour be properly represented in atmospheric and oceanic models for Earth, and also for other rotating, mildly oblate, celestial bodies. To do so necessitates a spheroidal rather than spherical representation of the geopotential. To prepare the way for the development of GREAT (Geophysically Realistic, Ellipsoidal, Analytically Tractable) coordinates for both atmospheric and oceanic modelling, some important concepts of geodesy are reviewed. These include:

- Earth's geoid.
- Mean sea level.
- The World Geodetic System.
- Earth's reference ellipsoid.
- Two kinds of global, navigational coordinates.

Two spheroidal geopotential approximations (one above the geoid, the other below) for a rotating, variable-density, ellipsoidal planet are recalled. These are further developed, without loss of accuracy. The orthogonal trajectories to the geopotential surfaces are then determined analytically. By construction, GREAT coordinates, in concert with the two modified geopotential approximations, satisfy the ideal criteria for a geopotential model of a planet, and an associated, ellipsoidal, geopotential coordinate system. The GREAT coordinate system then opens the way to improve the fidelity of atmospheric and oceanic models not only for Earth but also for other rotating, mildly oblate celestial bodies, such as other planets and moons.

12.1 PREAMBLE

Physically, gravity varies both meridionally and vertically for Earth. It is therefore desirable that this behaviour be properly represented in atmospheric and oceanic models for Earth and other rotating, mildly oblate celestial bodies. However, it is common practice in such models to represent the geopotential surfaces of apparent gravity as spheres. This is known as the *spherical-geopotential approximation*. See Sections 7.1.5 and 8.6 for discussion of this approximation, the challenges in justifying it, and some of its limitations.

Use of this approximation dates back to the tentative beginnings of global atmospheric modelling in the 1960s and 1970s, when (Staniforth and White, 2015a):

- Computer power and modelling know-how were very limited.
- Spatio-temporal resolution was very low.

- Physical parametrisations were very crude.
- Initial conditions had very large errors.

At the time, it was evident that the error incurred by the spherical-geopotential approximation was negligible when compared to these other sources of error. However, the subsequent development of increasingly powerful computers, coupled with great advances in numerical methods, physical parametrisations, data availability, and data assimilation, mean that what was once negligible may very well no longer be so, particularly at the much higher spatio-temporal resolutions of today's numerical models (White et al., 2008; White and Wood, 2012; Bénard, 2014, 2015; Staniforth, 2014b; Tort and Dubos, 2014b; Staniforth and White, 2015a).

Proper representation of gravity's meridional variation necessitates a *spheroidal* rather than *spherical* representation of the geopotential (Todhunter, 1873; Ramsey, 1940; Chandrasekhar, 1967, 1969; Phillips, 1973; Jeffreys, 1976). To do otherwise would be inconsistent (White et al., 2005). Furthermore, use of the spherical-geopotential approximation is also incompatible with the World Geodetic System (WGS 84) reference *ellipsoid* used for reporting data position using the Global Positioning System (GPS).

Because apparent gravity is such a dominant force, it is highly desirable that geophysical coordinate systems be chosen such that two of the three spatial coordinates are embedded in the geopotential surface passing through a point (Phillips, 1973; Gill, 1982; White et al., 2008; van der Toorn and Zimmerman, 2008; Staniforth and White, 2015a; Vallis, 2017). This is termed a *geopotential coordinate system*. It ensures that there is no component of apparent gravity in any direction tangential to locally horizontal coordinate surfaces. Accurate representation of horizontal and vertical force balances is thereby facilitated, as discussed in Section 7.3.

For the preceding reasons, the availability of a suitable *spheroidal*-geopotential approximation and an associated geopotential coordinate system is highly desirable. Thus the purpose of this chapter is to develop these in a systematic and self-consistent manner.

Ideally the geopotential model and associated spheroidal, geopotential coordinate system should be (Staniforth and White, 2015a):

- Acceptably accurate.
- Self-consistent.
- Acceptably simple.
- Analytically tractable.
- Fully compatible with the WGS 84 reference ellipsoid.

Simplicity and analytical tractability are substantially enhanced if the spheroidal coordinate system is also *orthogonal*.

Various mathematical challenges for representing the gravitational fields of idealised, mildly oblate planets have been identified and addressed in Chapter 8. This culminated in the derivation (in Section 8.5) of two geopotential approximations for an ellipsoidal planet of variable density; one valid above the planet's surface, the other below it. These two geopotential approximations provide the essential basis for the development in the present chapter of a generalisation (to the oceans) of a revised version[1] of Staniforth and White (2015a)'s GREAT (Geophysically Realistic, Ellipsoidal, Analytically Tractable) coordinates for the atmosphere.

An overview of the remainder of this chapter is as follows. To prepare the way for the development of GREAT coordinates, some important concepts of geodesy are introduced in Section 12.2.

[1] Revisions include:

- Changing the sign convention when defining the geopotential – for consistency with that adopted in various published works, and in Chapters 2, 7, and 8 herein.
- Reworking the description of further geopotential approximation.
- No longer scaling distances by the planet's equatorial radius – for improved clarity.

Specifically, Earth's geoid, mean sea level, the World Geodetic System, Earth's reference ellipsoid, and two kinds of global navigational coordinates are all described and discussed. Examination (in Section 12.3) of geodetic coordinate surfaces then leads to a reinterpretation of the classical spherical-geopotential approximation.

The two spheroidal geopotential approximations for a variable-density ellipsoidal planet (developed in Section 8.5) are summarised in Section 12.4. They are then further developed (without loss of accuracy) in Sections 12.5 (above a planet's geoid) and 12.6 (below it). This step is essential to satisfaction of all of the desiderata listed for a geopotential model and an associated spheroidal, geopotential coordinate system.

After a brief interlude to consolidate our thoughts (in Section 12.7), the orthogonal trajectories to the geopotential surfaces are determined (in Section 12.8). This then opens the way for the development of GREAT coordinates (in Section 12.9) for both atmospheric and oceanic modelling. By construction, GREAT coordinates, in concert with the two modified geopotential approximations, satisfy all five of the ideal criteria listed as well as coordinate orthogonality.

Concluding remarks are given in Section 12.10.

12.2 GEODESY AND THE WORLD GEODETIC SYSTEM

12.2.1 Earth's Geoid and Mean Sea Level

Geodesy is the branch of science concerned with:

- The determination of the size and shape of Earth, and the exact positions of points on its surface.
- The description of variations of its three-dimensional gravitational field.

This is a very challenging and complicated subject for many reasons (Jeffreys, 1976; Torge, 2001). In particular:

- Earth's composition is inhomogeneous.
- Its surface is highly irregular over land.
- Its surface varies temporally over the oceans.

Consequently many assumptions and approximations have to be made to mathematically represent the shape and composition of Earth and its associated gravitational field. What approximations are made, and the resulting complexity of any mathematical model of the physical world, depends very much on the intended application. For the navigation and tracking of artificial satellites and for many applications of satellite data, a highly detailed and accurate mathematical model of Earth's gravitational field is required.[2] Fortunately, however, much simpler representations of gravity suffice to represent it in models of Earth's atmosphere and oceans. To facilitate the following discussion, a schematic of the relative positions of Earth's geoid, mean sea level, continental orography, and oceanic bathymetry is displayed in Fig. 12.1.

There does not appear to be a precise, agreed definition of Earth's geoid in the literature. For present purposes, the working definition is that Earth's *geoid* is a hypothetical global geopotential surface that:

- Is in equilibrium in the absence of forces other than Newtonian gravitational attraction and apparent centrifugal force.
- Well fits global mean sea level over Earth's oceans.
- Smoothly extends under its continents, using a mathematical description of gravitational potential.

[2] This can involve evaluation of spherical-harmonic expansions with thousands of terms or more.

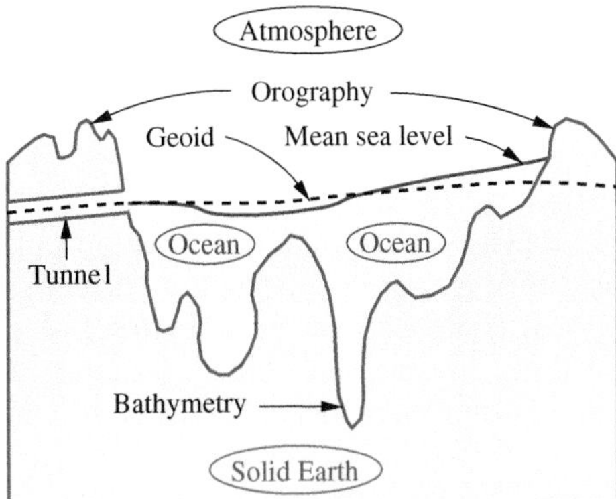

Figure 12.1 Relative positions of Earth's geoid, mean sea level, continental orography, and oceanic bathymetry. Solid Earth is denoted by brown shading, ocean by blue shading, and atmosphere by no shading. A hypothetical tunnel of small uniform cross section, centred about a curve embedded within the geoid and half filled with fluid at rest, is also shown. See text for further details.

The position of the geoid is determined empirically from a large observational database; measurements provided by instruments mounted on satellite platforms are of particular importance. Because the geoid is a geopotential surface (by its definition), a particle initially at rest on the geoid (as observed in an Earth-fixed, rotating frame of reference) remains at rest in the absence of forces other than Newtonian gravitational attraction and apparent centrifugal force. Apparent gravity (the sum of Newtonian gravitational attraction and apparent centrifugal force) acting on the geoid then acts only in the direction normal to it. This direction is aligned with a plumb line; and a tangent to any point on the geoid is aligned with a spirit level.

- But what is meant by *mean sea level*?

The surface of Earth's oceans is subject to approximately diurnal (i.e. daily) and semi-diurnal (i.e. twice-daily) tidal oscillations,[3] plus smaller-amplitude oscillations with other frequencies. These oscillations are due to the combined gravitational attraction of primarily the Moon, but secondarily of the Sun, both of which are in motion relative to Earth according to the laws of astronomy. Mean sea level is determined by averaging out these tidal oscillations over a period of many years. This eliminates most of the temporal variation of the position of the surface of Earth's oceans, but not all.

Even after averaging out tidal oscillations over a sufficiently long period of time to accomplish this, the surface of Earth's oceans is still not in an exact state of gravitational equilibrium. This is because the oceans are fluid, and they therefore evolve dynamically in the presence of forces. Mass, momentum, heat, and salinity are all physically transported on multiple time scales. This then results in slowly evolving changes to the spatial variation of sea level over multi-year time scales.[4] Furthermore, persistent atmospheric and oceanographic circulations, in response to physical forcings such as sea-surface heating and surface wind stress, also prevent the mean surface of Earth's oceans being in a state of gravitational equilibrium. As a consequence of these various factors, mean sea level of Earth's oceans does not correspond exactly with Earth's geoid, although agreement between the two is generally quite good.[5]

[3] The amplitude of tidal oscillations can be significant. For example, the largest difference between low and high tides in Earth's oceans occurs in the Bay of Fundy and – depending on the time of year – it can be as large as 16 m.

[4] For example, oceanic circulation and transport is a significant factor in the multi-year evolution of the El Nino/Southern Oscillation, with a significant impact on weather at the global scale.

[5] For many years it was believed that agreement between the two should be almost exact. It therefore came as a great surprise when high-quality satellite data showed agreement to be less precise than anticipated.

Discussion of the geoid has thus far focused on its position relative to the mean surface of Earth's *oceans*. The geoid is conceptually extended to Earth's *continents* by assuming that gravitational equilibrium holds within any hypothetical tunnel of small uniform cross section, bored within Earth and centred about any curve embedded within the geoid. See Fig. 12.1. Thus for any and all such hypothetical tunnels half filled with fluid at rest,[6] this fluid will remain undisturbed in the absence of forces other than Newtonian gravitational attraction and apparent centrifugal force. The precise positioning of the geoid underground depends on Earth's composition (i.e. on its density distribution) since this directly influences the geopotential and gravity below ground via a Poisson equation for the contribution of the Newtonian gravitational potential.[7] To determine the positioning of the geoid underground, a model of gravity is employed.

- How smooth are the geoid and mean sea level?

Both surfaces are much smoother than the underlying bathymetry or overlying orography, but they nevertheless do exhibit undulations. These undulations are caused by inhomogeneities in Earth's density distribution, primarily in its crust. Such inhomogeneities can vary quite rapidly locally, both horizontally and vertically. Rapid variations in density are, however, significantly smoothed out in the resulting gravitational potential, but not entirely so.[8] For a pictorial illustration of this phenomenon, see Fig. 12.1.

12.2.2 The World Geodetic System

The overview given in Section 12.2.1 of Earth's geoid and mean sea level is descriptive. Nothing is said therein about how the positions of these surfaces are specified, nor in what (global) coordinate system. Defining a suitable global coordinate system is a challenging problem of geodesy.

Until the 1950s there was no global, standard way of defining a smooth, zero-elevation surface (such as mean sea level, or the geoid), from which elevations (such as distance above mean sea level, or distance above the geoid) can be referenced. Each country or region did things their own way, resulting in a patchwork quilt of local coordinate systems, developed using different assumptions and incompatible with one another. The emerging need for global maps for navigation, aviation, and geography, and the advent of the satellite era,[9] provided the impetus to develop a coherent World Geodetic System (WGS 84, 2004).

The World Geodetic System system includes:

- A standard coordinate system.
- A standard reference ellipsoid.
- A geopotential surface (the geoid) that defines nominal sea level.

The Global Positioning System (GPS), used in many applications worldwide, is intimately linked to the World Geodetic System. Remotely sensed data from instruments mounted on satellite platforms are of crucial importance to the accuracy of the World Geodetic System. Without this previously unavailable data, accuracy and knowledge would be far less advanced.

[6] Imagine here a nineteenth-century transportation canal for barges, bored through a hill.

[7] Mathematically, density, ρ, is the forcing function of Poisson's equation $\nabla^2 V = 4\pi\gamma\rho$, where V is Newtonian gravitational potential, and γ is Newton's universal gravitational constant. See Section 7.5.

[8] Because of the properties of the ∇^2 operator, solving $\nabla^2 V = 4\pi\gamma\rho$ for V results in a smoothing of the scales present in ρ; the smaller the scale, the stronger the smoothing.

[9] This was initiated with the launch of the Sputnik satellite on 4 October 1957.

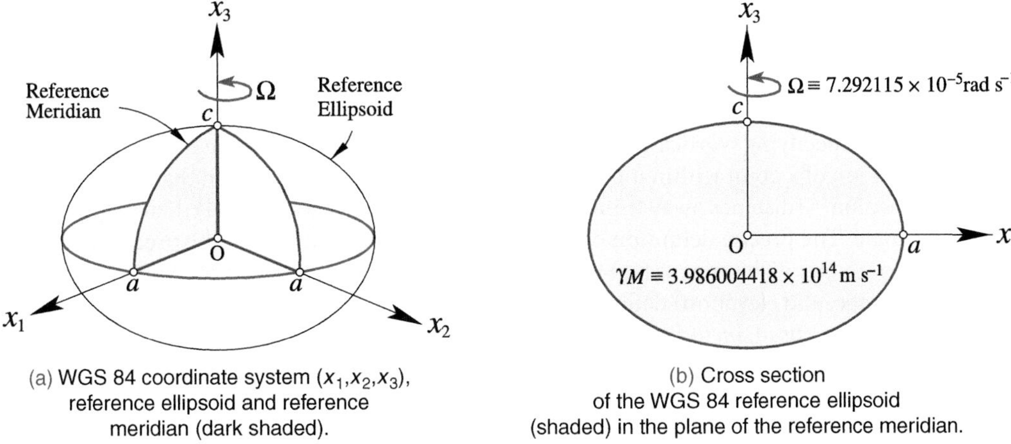

(a) WGS 84 coordinate system (x_1,x_2,x_3), reference ellipsoid and reference meridian (dark shaded).

(b) Cross section of the WGS 84 reference ellipsoid (shaded) in the plane of the reference meridian.

Figure 12.2 The WGS 84 (2004) coordinate system and reference ellipsoid. Earth's centre of mass is located at the system's origin, O. The system rotates about Earth's rotation axis, Ox_3, with angular speed $\Omega \equiv 7.292115 \times 10^{-5}$ rad s^{-1} (precisely). The reference ellipsoid has semi-major and semi-minor axes $a \equiv 6\,378.137$ km (precisely) and $c \equiv (1 - \varepsilon)\,a \approx 6\,356.752314$ km (approximately), respectively, where ellipticity has the value $\varepsilon \equiv 1/298.257\,223\,563$ (precisely). The WGS 84 reference meridian lies in the Ox_1x_3 plane. $\gamma M \equiv 3.986004418 \times 10^{14}$ ms^{-1} (precisely) is the product of Newton's universal gravitational constant, γ, and the total mass, M, of Earth, including its atmosphere and oceans. See text for further details.

12.2.3 The WGS 84 Coordinate System

An essential component of the World Geodetic System is its coordinate system, the purpose of which is to provide a practical framework to uniquely specify position in three-dimensional space.

The WGS 84 (2004) coordinate system is a right-handed, Earth-fixed, Cartesian (x_1, x_2, x_3), geocentric coordinate system, with origin located at Earth's centre of mass.[10] See Fig. 12.2. The x_3 axis coincides with Earth's rotation axis. The system rotates about this axis at Earth's angular speed $\Omega \equiv 7.292115 \times 10^{-5}$ rad s^{-1} in inertial space. The x_1 axis corresponds to the intersection of Earth's equatorial plane with the meridional plane of the IERS (International Earth Rotation Service) reference meridian (at zero degrees longitude). The reference meridian almost coincides with the traditional Greenwich prime meridian that passes through the courtyard of the Greenwich Royal Observatory in London, England.[11]

Using satellite data, together with data at a dozen reference surface stations of a global network, a consistent set of station coordinates (accurate to 5 cm) is inferred. The reference set of station coordinates anchors the WGS 84 coordinate system to Earth. This then allows the position of any location in three-dimensional space to be computed as coordinates in the WGS 84 coordinate system.

12.2.4 Earth's Reference Ellipsoid

As discussed in Section 12.2.1, Earth's surface is highly irregular, particularly over the continents, and it is also in perpetual motion over the oceans. This makes it very difficult to straightforwardly

[10] This mass includes that of Earth's atmosphere and oceans.

[11] Because of the slow movement of tectonic plates, the reference meridian, somewhat surprisingly, is not actually fixed to any point on Earth's surface but moves very slowly on geological time scales. The current eastward offset of the reference meridian with respect to the Greenwich prime meridian is approximately 5.31 seconds of arc; this corresponds to an eastward shift of about 102.5 m at the latitude of the Greenwich Royal Observatory.

and accurately specify three-dimensional position; for example, the location where an observation is made or the location of a grid point in a numerical model of Earth's atmosphere or oceans.

In practical applications, position is usually specified relative to some arbitrarily chosen reference surface, with two coordinates for (horizontal) navigation within the surface, and with a third coordinate specifying (vertical) navigation away from it. For example, longitude and latitude define the position of a point within the surface of a reference sphere, or within that of a reference ellipsoid; elevation[12] (distance away from the reference surface along a normal) then provides the third coordinate. The precise definition of a reference surface depends upon the precise values of its defining parameters. This means that when specifying position in terms of coordinates (such as longitude, latitude, and elevation) defined relative to a reference surface, one also needs to specify the reference surface used, *including the precise values of its defining parameters.*

A very important reference surface is the WGS 84 (2004), mildly oblate reference ellipsoid for Earth. It provides a smooth,[13] zero-elevation surface for geodetic applications. For example, many satellite observations, including raw GPS data, employ the surface of the WGS 84 reference ellipsoid as the zero-elevation surface. Although, in principle, the WGS 84 (Cartesian) coordinate system (described in Section 12.2.3) could be used to determine position for all geodetic applications, in practice this would be cumbersome. Elevation from a zero-elevation surface (needed for most geodetic applications) is most naturally expressed in terms of an ellipsoidal coordinate system that better matches Earth's geometry than a Cartesian system does. This motivated the development of the geodetic coordinate system (to be described in Section 12.2.5), which uses the WGS 84 (2004) reference ellipsoid to provide a smooth, zero-elevation surface. The two coordinate systems are mathematically equivalent since a position expressed in one system can be uniquely transformed to one in the other. Which coordinate system to use is then just a question of which is most convenient for what one wishes to do.

The WGS 84 (2004) Reference Ellipsoid

- The WGS 84 (2004) reference ellipsoid has four defining parameters; a, $1/\varepsilon$, Ω, and γM.
- The first two specify its size and shape, the third the angular speed of its uniform rotation, and the fourth its total mass.
- The centre of gravity of the reference ellipsoid is at the origin of the WGS 84 coordinate system. See Fig. 12.2.
- The reference ellipsoid has semi-major (equatorial) radius, a. Its semi-minor (polar) axis, c, is aligned with Earth's rotation axis. Thus Earth's reference ellipsoid is defined by

$$\frac{x_1^2 + x_2^2}{a^2} + \frac{x_3^2}{c^2} = 1, \tag{12.1}$$

where (x_1, x_2, x_3) are the coordinates of the WGS 84 coordinate system.

The precise value of a is $a \equiv 6\,378.137$ km. One might expect that c would also be specified as a precise value. However, for historical reasons (backward compatibility with a previously defined reference ellipsoid), this is not the case. Instead, the inverse of the ellipticity,[14] $1/\varepsilon \equiv a/(a-c)$, is specified to have the precise value $1/\varepsilon = 298.257\,223\,563$. The semi-minor axis, c, then has the (approximate) value $c \equiv (1-\varepsilon)\,a \approx 6\,356.752\,314$ km. This means that the semi-major axis is

[12] A synonym of elevation is altitude. When elevation/altitude is negative, as in the oceanic context, depth is usually used instead.

[13] Smoothness of the zero-elevation surface enhances accuracy for geodetic applications.

[14] The ellipticity, $\varepsilon \equiv (a-c)/a$, is also known as the flattening parameter, f.

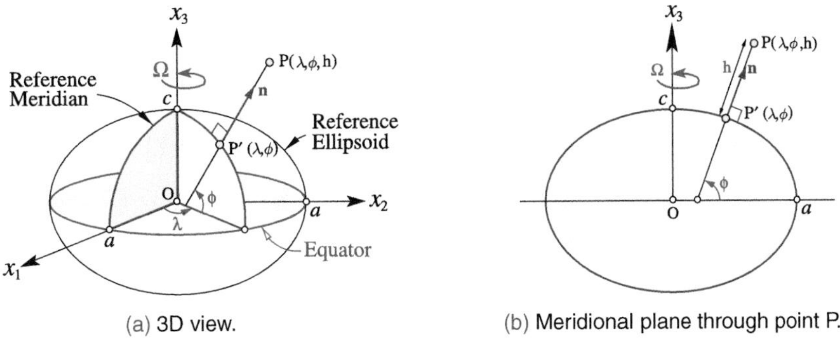

(a) 3D view. (b) Meridional plane through point P.

Figure 12.3 Geodetic coordinates $(\lambda, \phi, \text{h})$, where λ is longitude, ϕ is geographic (geodetic) latitude, and h is elevation (distance from the WGS 84 reference ellipsoid along a straight-line normal, $\mathbf{n}\,(\lambda, \phi)$). Point P $(\lambda, \phi, \text{h})$ is a general point. Line P′P is oriented along the normal to the reference ellipsoid that passes through point P $(\lambda, \phi, \text{h})$. Point P′ (λ, ϕ) lies on the surface of the reference ellipsoid. The axes of the WGS 84 coordinate system (x_1, x_2, x_3) and its reference meridian are also depicted. The reference meridional plane $(\lambda = 0)$ is dark shaded (in blue), and the meridional plane containing point P is light shaded (in blue). The equatorial plane is shaded in magenta. See text for further details.

a little over 21 km larger than the semi-minor axis. Thus the reference ellipsoid is mildly oblate with an equatorial bulge that quantitatively mimics Earth's.

The reference ellipsoid uniformly rotates (in inertial space) about its minor axis with the WGS coordinate system's angular speed of precisely $\Omega \equiv 7.292115 \times 10^{-5}$ rad s^{-1}. Thus the WGS 84 reference ellipsoid is stationary with respect to the WGS 84 coordinate system. The reference ellipsoid's total mass[15], M, is needed for gravity purposes and, in particular, to specify the position of Earth's geoid. Because the product γM is known to higher accuracy than both γ (Newton's universal gravitational constant) and M, and because the product γM appears naturally in gravity formulae, the mass of the WGS 84 reference ellipsoid is *indirectly specified* by setting γM to the precise value of $\gamma M \equiv 3.986004418 \times 10^{14}$ ms^{-1}.[16]

12.2.5 Geodetic Coordinates

Earth's WGS 84 reference ellipsoid provides the zero-elevation surface for the definition of *geodetic coordinates* $(\lambda, \phi, \text{h})$, where (see Fig. 12.3):

- λ is longitude, measured with respect to the WGS 84 reference meridian.
- ϕ is geographic latitude,[17] that is, the angle between the straight-line normal at a point on the reference ellipsoid and the equatorial plane.
- h is elevation, that is, distance along the straight-line normal from a point on the reference ellipsoid with (horizontal, two-dimensional) coordinates (λ, ϕ).

This coordinate system is employed in many geodetic applications; for example, to specify the position of satellite platforms used to provide remotely sensed observations of Earth's atmosphere and oceans.

Recall that the surface of the WGS 84 reference ellipsoid is defined by (12.1), where (x_1, x_2, x_3) are the Cartesian coordinates of the WGS 84 coordinate system. Equation (12.1) may be rewritten parametrically as (see Section 8.2.2.4)

[15] This includes the combined mass of Earth's atmosphere and oceans.

[16] Assuming a measured value for γ, of some known accuracy, the total mass M then follows to an accuracy determined by the accuracies of the values of γ and γM.

[17] Geographic latitude is also known as geodetic latitude.

$$x_1 = \frac{a^2 \cos\lambda \cos\phi}{r_{ac}^{\phi}(\phi)}, \quad x_2 = \frac{a^2 \sin\lambda \cos\phi}{r_{ac}^{\phi}(\phi)}, \quad x_3 = \frac{c^2 \sin\phi}{r_{ac}^{\phi}(\phi)}, \tag{12.2}$$

where

$$r_{ac}^{\phi}(\phi) \equiv \left(a^2 \cos^2\phi + c^2 \sin^2\phi\right)^{\frac{1}{2}} \equiv a\left(1 - e^2 \sin^2\phi\right)^{\frac{1}{2}}, \tag{12.3}$$

and

$$e^2 \equiv \frac{a^2 - c^2}{a^2} \equiv 1 - \frac{c^2}{a^2} \tag{12.4}$$

is the square of *eccentricity*, e. Unit normal vector $\mathbf{n}$ at point $P'(\lambda, \phi)$ on the reference ellipsoid is

$$\mathbf{n}(\lambda, \phi) = \cos\lambda \cos\phi \mathbf{e}_1 + \sin\lambda \cos\phi \mathbf{e}_2 + \sin\phi \mathbf{e}_3, \tag{12.5}$$

where $(\mathbf{e}_1, \mathbf{e}_2, \mathbf{e}_3)$ is the unit vector triad of the WGS 84 Cartesian coordinate system defined in Section 12.2.3.

Position vector $\mathbf{r}(\lambda, \phi, \mathrm{h})$ for the point $P(\lambda, \phi, \mathrm{h})$, at distance h along the *straight-line* normal[18] from point $P'(\lambda, \phi)$, is thus

$$\mathbf{r}(\lambda, \phi, \mathrm{h}) = \mathbf{r}_S(\lambda, \phi) + \mathrm{h}\,\mathbf{n}(\lambda, \phi), \tag{12.6}$$

where $\mathbf{r}_S(\lambda, \phi)$ is the position vector of point $P'(\lambda, \phi)$ on the surface of the reference ellipsoid, with Cartesian components given by (12.2). Taking the components of (12.6), and using (12.2) and (12.5), the transformation equations from geodetic coordinates $(\lambda, \phi, \mathrm{h})$ to Cartesian coordinates (x_1, x_2, x_3) are then

$$x_1 = \left[\frac{a^2}{r_{ac}^{\phi}(\phi)} + \mathrm{h}\right] \cos\lambda \cos\phi, \quad x_2 = \left[\frac{a^2}{r_{ac}^{\phi}(\phi)} + \mathrm{h}\right] \sin\lambda \cos\phi, \quad x_3 = \left[\frac{c^2}{r_{ac}^{\phi}(\phi)} + \mathrm{h}\right] \sin\phi. \tag{12.7}$$

When $c \equiv a$, (12.7) reduce to the familiar spherical-polar forms

$$x_1 = r \cos\lambda \cos\phi, \quad x_2 = r \sin\lambda \cos\phi, \quad x_3 = r \sin\phi, \tag{12.8}$$

where $r \equiv a + \mathrm{h}$.

As noted in White and Inverarity (2012), the inverse transformation from Cartesian coordinates (x_1, x_2, x_3) to geodetic coordinates $(\lambda, \phi, \mathrm{h})$ is not as straightforward. It is, however, a standard problem in satellite geodesy that can be efficiently solved numerically to arbitrary accuracy (Torge, 2001).

12.2.6 Assumptions for Representing Gravity in Atmospheric and Oceanic Models

In the preceding brief overview of geodesy and the World Geodetic System, various challenges have been discussed regarding the representation of gravity in atmospheric and oceanic models. The true physical situation is very complex, and simplifications necessarily have to be made. The question now is:

- What is the best compromise between simplicity and complexity?

Too simple prejudices physical realism, whereas too complex prejudices mathematical tractability.

The traditional answer has always been to assume that:

- Earth can be approximated as a sphere (with orography superimposed upon it).

[18] This *straight-line* normal diverges (as a function of increasing altitude) from the (curved) vertical coordinate line passing through point $P'(\lambda, \phi)$. This is because surfaces of constant h generally do *not* coincide with geopotential surfaces.

- Gravity can be represented using the classical spherical-geopotential approximation, with no meridional variation and, at most, radial variation (but usually no variation at all).

As discussed in the preamble of this chapter, this approach may now be too simple at the current state of atmospheric and oceanic modelling. In particular:

- It is inconsistent with the use of the WGS 84 reference ellipsoid for reporting the position of observations.
- This may lead to systematic accumulation of error in quantitative models.

Assuming that the spherical approach is indeed an oversimplification, then the next question is:

- How can this be addressed?

It is argued here that the key to this is provided by the reference ellipsoid.

By design, the reference ellipsoid:

- Approximates the Figure of the Earth much better than a sphere does.
- Has the same total mass as Earth (including the mass of its atmosphere and oceans).
- Is fully compatible with the Global Positioning System used to report data position for many kinds of observations.
- Agrees much better than a sphere does with the positions of both mean sea level and Earth's geoid.

It is therefore reasonable, as an excellent approximation for modelling Earth's atmosphere and oceans, to assume that Earth's geoid coincides with the WGS 84 reference ellipsoid. See Fig. 12.4. This is a far less drastic assumption than assuming that Earth's geoid is spherical.

Making this assumption then allows use of the two geopotential representations developed in Section 8.5 for an idealised, rotating, ellipsoidal planet of variable density. Thus:

Geopotential Representation above and below the Geoid

It is assumed that;

1. Geopotential representation (8.96) can be used *above* the geoid (for Earth's atmosphere).
2. Geopotential representation (8.151) can be used *below* it (for Earth's oceans).
3. The geoid coincides with the surface of the WGS 84 reference ellipsoid.

These assumptions are adopted later in this chapter but with an important added wrinkle. For consistency and tractability reasons, the two geopotential approximations are further approximated, *but with no loss of accuracy*. These further approximations retain first-order accuracy in the small parameters, m and ε; see Sections 12.5 and 12.6. This then leads to the development of a *unified*, Geophysically Realistic, Ellipsoidal, Analytically Tractable (GREAT) geopotential coordinate system for both Earth's atmosphere and its oceans.

In an analogous manner to the spherical case, orography is handled by superimposing it on the reference ellipsoid. Sea-surface height and mean sea level are represented as small-amplitude perturbations about the reference ellipsoid/geoid. The impact on Earth's gravitational field of any excursions between Earth's actual surface and the WGS 84 reference ellipsoid is assumed to be negligible. In the context of quantitatively modelling the evolution of Earth's atmosphere and oceans, this is an excellent approximation.

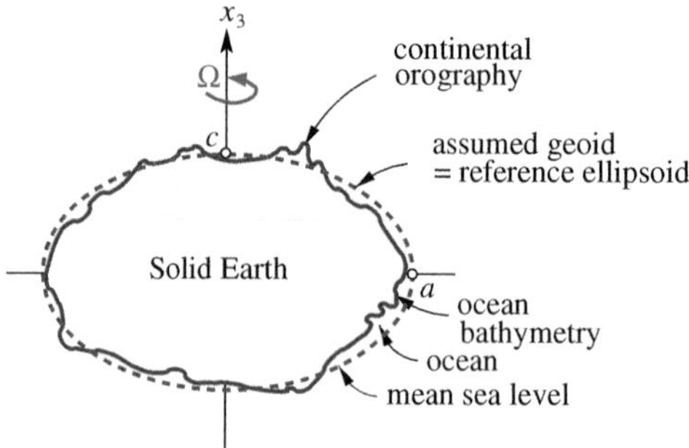

Figure 12.4 Depiction in a meridional plane of Earth's reference ellipsoid, geoid, continental orography, and ocean bathymetry, for atmospheric and oceanic modelling purposes. For tractability, Earth's geoid is assumed to everywhere coincide with the reference ellipsoid (blue dashed) that has semi-major and semi-minor axes a and c, respectively. Solid Earth is shaded brown, ocean is shaded blue.

12.3 THE CLASSICAL SPHERICAL GEOPOTENTIAL APPROXIMATION REVISITED

Geodetic coordinates were conceived to specify the position of points for geodetic applications such as satellite navigation and remote sensing. However, White and Inverarity (2012) observed that it is straightforward to obtain the metric factors h_1, h_2, and h_3 for a geodetic coordinate system, and proceeded to do so.[19] Knowing these factors allows various differential operators (such as gradient, divergence, and curl) to be determined using standard expressions for orthogonal-curvilinear coordinates. Thus the governing equations of global atmospheric and oceanic models can, if one wishes, be written in terms of geodetic coordinates. The question then arises:

- Is this possibility useful or not?

The answer is both no and yes.

Geodetic coordinates have the clear advantage over spherical-polar coordinates that they better represent Earth's shape, which is quasi-ellipsoidal rather than spherical. So far, so good. For the reasons discussed in Section 7.3, it is highly advantageous (and arguably essential) to use a geopotential coordinate system for atmospheric and oceanic modelling purposes, whereby horizontal coordinate surfaces coincide with geopotential surfaces. As discussed in White and Inverarity (2012), for this to be true for geodetic coordinates (*with surfaces of constant* h *assumed to coincide with geopotential surfaces*), gravity cannot vary meridionally,[20] even though it is known to do so physically. Thus geodetic coordinates, *when used as a geopotential coordinate system*,[21] represent something of a halfway house between two 'extreme' options:

1. A geopotential coordinate system with a spheroidal representation of gravity, that well captures meridional variation.
2. Spherical-polar coordinates with the classical spherical-geopotential approximation, that does not.

Geodetic coordinates, when used as a geopotential coordinate system, are therefore inferior to the first option, but nevertheless better than the second. Since, as shown later in this chapter, it

[19] These metric factors are reproduced in Section 5.2.4, but in the present notation ($\zeta \to \mathsf{h}$).

[20] This is because any two geopotential surfaces then have the same *constant* physical separation everywhere, independently of the horizontal coordinates λ and ϕ. The gradient of the geopotential, $\nabla\Phi \equiv -\mathbf{g}$, therefore has the same magnitude everywhere on a horizontal coordinate surface, and so therefore does gravity.

[21] Geodetic coordinates can, of course, be used to express the governing equations of global atmospheric and oceanic models *without* making the assumption that their horizontal coordinate surfaces coincide with geopotential surfaces. This, however, means that the apparent gravitational force then has *non-zero* components tangential to horizontal, geodetic coordinate surfaces, thereby causing difficulties with the accurate representation of force balances. Such behaviour is highly undesirable – see Section 7.3.

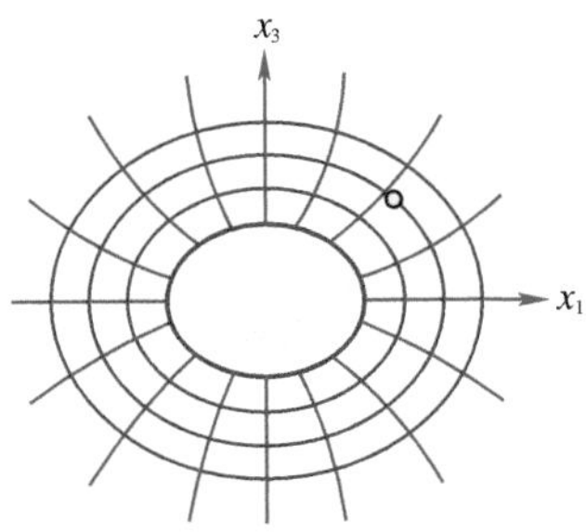

(a) Physically realistic, geopotential coordinate system, with a *first*-order (in *m* and ε), *spheroidal*, geopotential approximation.

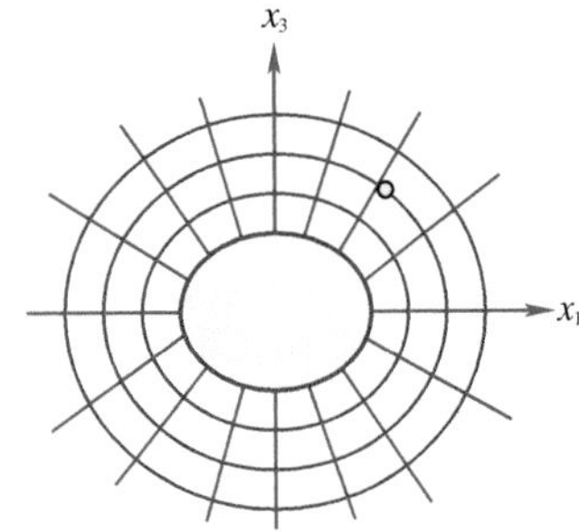

(b) Halfway-house configuration; a geodetic, geopotential coordinate system, with a *zeroth*-order (in *m* and ε), *spheroidal*, geopotential approximation.

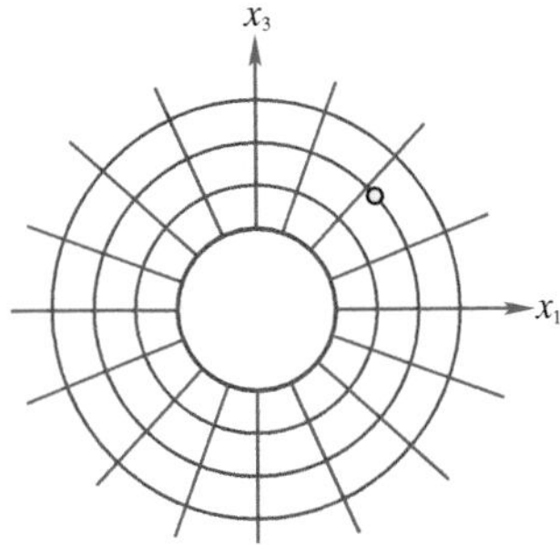

(b) Spherical-polar coordinates, with a *zeroth*-order (in *m* and ε), *spherical*, geopotential approximation.

Figure 12.5 Cross sections in the meridional plane of the reference meridian for the three configurations of geopotential approximation and geometry discussed in the text. Blue lines – horizontal (geopotential) surfaces. Red lines – vertical, orthogonal trajectories to the horizontal (geopotential) surfaces. Earth's cross section is shaded. Spacing of the geopotential surfaces along the x_1 axis is identical for all three panels. The yellow circle on each panel denotes the corresponding location of an arbitrarily chosen point on a horizontal (geopotential) surface.

is possible to affordably adopt the first option (using GREAT coordinates), it is of little practical interest to instead use geodetic coordinates as a geopotential coordinate system.[22] Thus the first answer to the question posed at the beginning of this section is *no*.

This does not, however, mean that use of geodetic coordinates as a geopotential coordinate system for modelling Earth's atmosphere and oceans is of no interest at all. For conceptual reasons, as now discussed, such an approach provides a helpful way of interpreting, and better understanding, the classical spherical-geopotential approximation.

As the starting point for the discussion, consider the first of the two extreme options, namely to use a geopotential coordinate system with a spheroidal representation of gravity that well captures its meridional variation. This can be considered to be the (almost) ideal, physically realistic situation, to which the second, degraded option (to use spherical-polar coordinates, with the classical spherical-geopotential approximation) is to be compared. To go from the first option to the second can be considered as making two successive *distinct* approximations to:

1. The *geopotential.*
2. The *geometry.*

With the aid of Fig. 12.5, we now examine these two approximations in turn.

[22] The cost of using geodetic coordinates is similar to that of GREAT coordinates, but the benefits are substantially less.

The first approximation goes from the ideal situation (depicted in Fig. 12.5a) to the (degraded) halfway house (depicted in Fig. 12.5b) of using geodetic coordinates as a geopotential coordinate system. It amounts to approximating the (assumed) ideal mathematical representation of the geopotential (accurate to first order in m and ε) by one that:

- Is less accurate (with similar formal accuracy to the classical spherical-geopotential approximation, both being of zeroth order in m and ε).
- Neglects meridional variation of gravity.
  but nevertheless
- Fully respects the (assumed) ellipsoidal form of Earth.

The ideal geopotential surfaces (depicted in Fig. 12.5a) – with the exception of the one coincident with Earth's ellipsoidal surface – are thus artificially distorted by this geopotential approximation to those depicted in Fig. 12.5b, and Clairaut's fraction is no longer respected. They nevertheless remain spheroidal in shape,[23] thereby (partially) preserving this property of the physical geopotential. However, whilst the orthogonal trajectories to the ideal geopotential surfaces (i.e. the curves normal to geopotential surfaces) are slightly curved (Fig. 12.5a), those after making this (geodetic) geopotential approximation are straight lines (Fig. 12.5b).

The second of the two approximations goes from the halfway house (of using geodetic coordinates as a geopotential coordinate system, as depicted in Fig. 12.5b) to the use of spherical-polar coordinates with the classical spherical-geopotential approximation (as depicted in Fig. 12.5c). This further approximation can be viewed as being a purely *geometric* mapping/degradation whereby all spheroidal surfaces are artificially distorted into spherical ones. The governing equations of atmospheric and oceanic models are then expressed in terms of spherical-polar coordinates. Whilst this is very convenient, it is accompanied by a loss of realism and accuracy.

As a result of the two successive approximations (firstly to the *geopotential*, and secondly to the *geometry*), the (almost) ideal, physically realistic, *spheroidal* world depicted in Fig. 12.5a is transformed into the less physically realistic, *spherical* world depicted in Fig. 12.5c, via the intermediate transformation to the *halfway-house* world depicted in Fig. 12.5b.

Thus, performing the thought experiment of formulating an atmospheric or oceanic model in terms of a geodetic, geopotential coordinate system provides additional insight into the two distinct approximations implicitly made (one *gravitational*, the other *geometric*) when adopting spherical geometry with the spherical-geopotential approximation. So the answer to the question posed at the beginning of this section is also *yes*!

12.4 GEOPOTENTIAL APPROXIMATION FOR ELLIPSOIDAL PLANETS

12.4.1 The Potential of Apparent Gravity

As discussed in Section 12.2.6, for the purposes of representing Earth's geopotential for atmospheric and oceanic modelling, it is assumed that:

- Earth's geoid coincides with the WGS 84 reference ellipsoid.
- The analysis of Section 8.5 for an idealised, ellipsoidal planet of variable density is applicable.
- Geopotential approximation (8.96), accurate to first order in m and ε, can therefore be used to represent gravity within Earth's *atmosphere* (*above* the geoid).
- Geopotential approximation (8.151), again accurate to first order in m and ε, can similarly be used to represent gravity within Earth's *oceans* (*below* the geoid).

Let a and c be the semi-major and semi-minor axes of the WGS 84 reference ellipsoid in a meridional plane, and let

$$\varepsilon \equiv \frac{a-c}{a} \quad \Rightarrow \quad c \equiv a(1-\varepsilon) \tag{12.9}$$

[23] They are not, however, ellipsoidal (White and Inverarity, 2012).

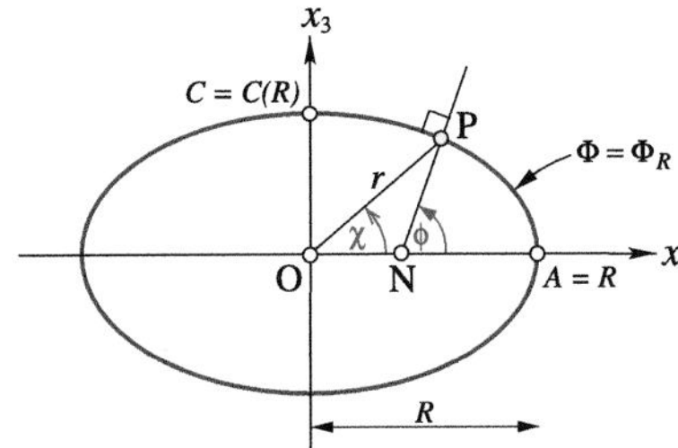

Figure 12.6 A cross section in the Ox_1x_3 meridional plane of the geopotential surface $\Phi = \Phi_R$, where R is the surface's label; $A = R$ is both equatorial radius and semi-major axis of the spheroid; and $C = C(R)$ is semi-minor axis. For a general point P on the spheroid, with coordinates (x_1, x_3): N is the point at which the normal to the spheroid at P intersects Ox_1; $r \equiv |\mathbf{r}|$ is absolute value of position vector $\mathbf{r}$ relative to the origin O; χ is *geocentric* latitude; and ϕ is *geographic* latitude.

be the associated ellipticity. See Section 12.2.4 and Fig. 12.2. Define – see (8.79) –

$$\widetilde{\varepsilon} \equiv \frac{a^2 - c^2}{2c^2} \equiv \frac{1 - (1-\varepsilon)^2}{2(1-\varepsilon)^2} = \varepsilon + O\left(\varepsilon^2\right). \tag{12.10}$$

As stated above, Earth's geoid is assumed to coincide with the WGS 84 reference ellipsoid. Written in geocentric polar coordinates, the geoid is thus given by – see (8.17) and (8.78) –

$$r = r_S(\chi) = a\left(1 + 2\widetilde{\varepsilon}\sin^2\chi\right)^{-\frac{1}{2}} = a\left(\cos^2\chi + \frac{a^2}{c^2}\sin^2\chi\right)^{-\frac{1}{2}}, \tag{12.11}$$

where χ is geocentric latitude.

The spheroidal-geopotential approximation (8.96) *outside* Earth (i.e. *above* the geoid $r = r_S(\chi)$)[24] – see Section 8.5.2 for its derivation – is

$$\Phi(\chi, r) = -\frac{\gamma M}{r}\left\{1 + \left(\frac{2\widetilde{\varepsilon} - m}{6}\right)\frac{a^2}{r^2} + \frac{m}{2}\frac{r^3}{a^3} - \left[\left(\frac{2\widetilde{\varepsilon} - m}{2}\right)\frac{a^2}{r^2} + \frac{m}{2}\frac{r^3}{a^3}\right]\sin^2\chi\right\}, \quad r \geq r_S(\chi), \tag{12.12}$$

where:

- A ratio of centrifugal and mass gravitational forces is defined by parameter

$$m \equiv \frac{\Omega^2 a^3}{\gamma M}. \tag{12.13}$$

- γ is the universal gravitational constant.
- M is the (assumed mildly oblate) planet's mass (including that of its atmosphere and oceans).
- r is distance from the (assumed ellipsoidal) planet's centre.

See Fig. 12.6 for a depiction of one such geopotential surface, $\Phi = \Phi_R =$ constant. Equation (12.12) defines Φ everywhere exterior to the planet in terms of parameters γ, M, a, m, and $\widetilde{\varepsilon}$, and geocentric coordinates (λ, χ, r). It is first-order accurate in the non-dimensional parameters

24 Taking into account a difference in sign convention, spheroidal geopotential approximation (8.96) is very similar to the classical geopotential approximation (2) given in Staniforth and White (2015a); ε in their (2) is replaced by $\widetilde{\varepsilon}$ in (8.96) herein. Since $\widetilde{\varepsilon} = \varepsilon + O\left(\varepsilon^2\right)$ – see (12.10) – these two approximations nevertheless agree with one another to their first-order accuracy. The use of $\widetilde{\varepsilon}$ herein – in preference to ε – eases the later development of further first-order accurate geopotential approximations for analytical tractability.

$\widetilde{\varepsilon}$ and m. Both parameters have terrestrial values of about 1/300, giving $O\left(\widetilde{\varepsilon}^2, \widetilde{\varepsilon}m, m^2\right) = O\left(10^{-5}\right)$ for neglected terms.

Similarly, spheroidal geopotential approximation (8.151) *inside* Earth (i.e. *below* the geoid $r = r_S(\chi)$) is

$$\Phi(\chi, r) = A_0 + \left(A_2 - \frac{B_2}{2} - \frac{\gamma M}{a}\frac{m}{3}\right)\frac{r^2}{a^2} + \left[A_3\left(1 - \frac{2\widetilde{\varepsilon}}{7}\frac{r}{a}\right) - \frac{B_3}{2}\left(1 - \frac{3}{7}\frac{r}{a}\right)\right]\frac{r^3}{a^3}$$
$$+ \frac{3}{2}\left[B_2\frac{r^2}{a^2} + B_3\frac{r^3}{a^3}\left(1 - \frac{3}{7}\frac{r}{a}\right) + \frac{4\widetilde{\varepsilon}}{7}A_3\frac{r^4}{a^4}\right]\sin^2\chi, \quad r \leq r_S(\chi). \tag{12.14}$$

In (8.151) – and therefore also in (12.14) – red terms are retained if $A_3 = O(1)$ is assumed, but they are omitted if $A_3 = O(m, \widetilde{\varepsilon})$ is assumed instead. As noted in Section 8.5.3.6, assuming $A_3 = O(m, \widetilde{\varepsilon})$ generally simplifies things somewhat and, in particular, it simplifies the present development of GREAT coordinates.[25] Thus, from here onwards, we assume that $A_3 = O(m, \widetilde{\varepsilon})$ and omit the red terms in (12.14) to obtain the geopotential representation

$$\Phi(\chi, r) = A_0 + \left(A_2 - \frac{B_2}{2} - \frac{\gamma M}{a}\frac{m}{3}\right)\frac{r^2}{a^2} + \left[A_3 - \frac{B_3}{2}\left(1 - \frac{3}{7}\frac{r}{a}\right)\right]\frac{r^3}{a^3}$$
$$+ \frac{3}{2}\left[B_2\frac{r^2}{a^2} + B_3\frac{r^3}{a^3}\left(1 - \frac{3}{7}\frac{r}{a}\right)\right]\sin^2\chi, \quad r \leq r_S(\chi), \tag{12.15}$$

below the geoid. Doing so then means – from (8.144) and (8.145) – that coefficients A_0, A_2, A_3, B_2, and B_3 in (12.15) are given by

$$A_0 = -\frac{3}{2}\left(1 + \frac{2\widetilde{\varepsilon}}{9}\right)\frac{\gamma M}{a}, \quad A_2 = \frac{1}{2}\frac{\gamma M}{a}, \quad A_3 = \frac{\widetilde{\varepsilon}}{3}\frac{\gamma M}{a}, \tag{12.16}$$

$$B_2 = -\frac{2}{3}\left(10m - 9\widetilde{\varepsilon}\right)\frac{\gamma M}{a}, \quad B_3 = \frac{7}{3}\left(5m - 4\widetilde{\varepsilon}\right)\frac{\gamma M}{a}. \tag{12.17}$$

In principle, everything we need to know about the geopotential is embodied in spheroidal-geopotential representations (12.12) and (12.15):

- The equation for geopotential surfaces.
- The equation for trajectories orthogonal to these surfaces.
- An orthogonal coordinate system based on these.

In practice, it is not quite this simple, as shown in Section 12.4.3. Analytic tractability and consistency are two stumbling blocks that need to be overcome.

12.4.2 Labelling of Geopotential Surfaces

Following White et al. (2008), Staniforth (2014b), and Staniforth and White (2015a), a geopotential surface is identified by its equatorial radius, R (i.e. the radius of its circular intersection with the equatorial plane $\chi = 0$). Thus:

- $R = a$ corresponds to Earth's geoid (which is assumed here to coincide with the surface of the reference ellipsoid).

[25] Assuming $A_3 = O(m, \widetilde{\varepsilon})$ is, however, somewhat more restrictive than instead assuming $A_3 = O(1)$, since it implicitly restricts the rate at which density can increase downward from the geoid towards Earth's core. This is probably (albeit not certainly!) unimportant in the present context. Physically, gravity just below Earth's geoid is strongly constrained by the matching conditions at the geoid to the outer representation of gravity (which is independent of either assumption for A_3). Furthermore, the representation of gravity below the geoid only needs to be acceptably accurate to a maximum depth of 11 km (i.e. that of the Mariana Trench) below the geoid, which is very small when compared to Earth's mean radius of about 6 400 km.

- $R > a$ corresponds to a geopotential surface *above* the geoid.
- $R < a$ corresponds to one *below* it.

Setting $\chi = 0$ and $r = R$ in (12.12) and (12.15) then gives

$$\Phi_R \equiv \Phi(\chi = 0, r = R) = -\frac{\gamma M}{R}\left[1 + \left(\frac{2\widetilde{\varepsilon} - m}{6}\right)\frac{a^2}{R^2} + \frac{m}{2}\frac{R^3}{a^3}\right], \quad R \geq a, \tag{12.18}$$

$$= A_0 + \left(A_2 - \frac{B_2}{2} - \frac{\gamma M}{a}\frac{m}{3}\right)\frac{R^2}{a^2} + \left[A_3 - \frac{B_3}{2}\left(1 - \frac{3}{7}\frac{R}{a}\right)\right]\frac{R^3}{a^3}, \quad R \leq a. \tag{12.19}$$

Further setting $R \equiv a$ in (12.18) – or in (12.19) with use of (12.16) and (12.17) to eliminate A_0, A_2, A_3, B_2, and B_3 – leads to (in both cases)

$$\Phi_a \equiv \Phi(\chi = 0, r = R = a) = -\frac{\gamma M}{a}\left(1 + \frac{m + \widetilde{\varepsilon}}{3}\right) \tag{12.20}$$

as the (constant) value of the geopotential on Earth's assumed geoid (for which $R \equiv a$).

12.4.3 Equations for Geopotential Surfaces above and below the Geoid

Above the Geoid

Above the geoid, the equation for the geopotential surface $\Phi = \Phi_R =$ constant (i.e. the functional relation that defines this surface for a specified value of R), is obtained by setting the right-hand sides of (12.12) and (12.18) equal, and dividing through by γM. Thus

$$\frac{1}{r}\left\{1 - \left[\left(\frac{2\widetilde{\varepsilon} - m}{2}\right)\frac{a^2}{r^2} + \frac{m}{2}\frac{r^3}{a^3}\right]\sin^2\chi\right\} + \frac{1}{a}\left[\left(\frac{2\widetilde{\varepsilon} - m}{6}\right)\frac{a^3}{r^3} + \frac{m}{2}\frac{r^2}{a^2}\right]$$
$$= \frac{1}{R}\left[1 + \left(\frac{2\widetilde{\varepsilon} - m}{6}\right)\frac{a^2}{R^2} + \frac{m}{2}\frac{R^3}{a^3}\right], \quad R \geq a. \tag{12.21}$$

Below the Geoid

Similarly, *below* the geoid, the equation for the geopotential surface $\Phi = \Phi_R =$ constant is obtained by setting the right-hand sides of (12.15) and (12.19) equal. Thus

$$\frac{r^2}{a^2}\left\{\left(A_2 - \frac{B_2}{2} - \frac{\gamma M}{a}\frac{m}{3}\right) + \left[A_3 - \frac{B_3}{2}\left(1 - \frac{3}{7}\frac{r}{a}\right)\right]\frac{r}{a} + \frac{3}{2}\left[B_2 + B_3\frac{r}{a}\left(1 - \frac{3}{7}\frac{r}{a}\right)\right]\sin^2\chi\right\}$$
$$= \frac{R^2}{a^2}\left\{\left(A_2 - \frac{B_2}{2} - \frac{\gamma M}{a}\frac{m}{3}\right) + \left[A_3 - \frac{B_3}{2}\left(1 - \frac{3}{7}\frac{R}{a}\right)\right]\frac{R}{a}\right\}, \quad R \leq a, \tag{12.22}$$

where coefficients A_2, A_3, B_2, and B_3 are given in (12.16) and (12.17).

12.4.4 Analytical Tractability and Consistency

We now see that there are difficulties with both analytical tractability and consistency:

1. Neither (12.21) nor (12.22) leads to nice, analytical solutions for r (which determines the geopotential surface associated with a given value of label R). This means that finding nice, analytic orthogonal trajectories to the geopotential surfaces also cannot be done tractably.
2. Furthermore, setting $R \equiv a$ in (12.21) or (12.22) does not exactly give the WGS 84 reference ellipsoid. This means that there is a mild inconsistency.[26]

[26] The inconsistency is mild because although the geopotential surfaces do not exactly overlie the reference ellipsoid, they almost do so (to first order in m and ε, but not to second order).

Analytical intractability and inconsistency therefore stand in the way of the development, *in a self-consistent manner*, of an analytic, geopotential coordinate system.

At first sight, these appear to be insuperable problems. Fortunately this is not the case. As shown in Staniforth (2014b) and Staniforth and White (2015a), the key to overcoming analytical intractability and inconsistency for the geopotential *above* the geoid is to further approximate it *without loss of accuracy*. It turns out that it is also possible to similarly do so *below* the geoid!

Geopotential approximations (12.12) and (12.15) are therefore further approximated, *without loss of accuracy*, in Sections 12.5 (for the geopotential above the geoid) and 12.6 (for that below it). The resulting approximations remain first-order accurate in the small parameters m and $\widetilde{\varepsilon}$. However, they have the very important advantage that they lead, in a *self-consistent*, *exact* manner, to *analytic* geopotential surfaces with associated *analytic* orthogonal trajectories.

- These analytic surfaces and trajectories are then used as the building blocks to develop (in Section 12.9) Geophysically Realistic, Ellipsoidal, Analytically Tractable (GREAT) coordinates for atmospheric and oceanic modelling.

12.4.5 Requirements for Improved Geopotential Approximation

To address the associated analytical intractability and inconsistency issues identified in Section 12.4.4, (12.12) and (12.15) are therefore approximated, *once and for all*, by expanding certain terms binomially and then neglecting $O\left(\widetilde{\varepsilon}^2, \widetilde{\varepsilon}m, m^2\right)$ terms. These approximations are based on the assumption that

$$\frac{r}{R} \equiv 1 + \frac{r-R}{R} = 1 + O\left(\widetilde{\varepsilon}, m\right), \tag{12.23}$$

so that r varies little over the geopotential surface $\Phi = \Phi_R$ (White et al., 2008; van der Toorn and Zimmerman, 2008; Staniforth and White, 2015a).[27]

As discussed in Sections 12.1 and 12.2, it is highly desirable to develop an orthogonal, *ellipsoidal* coordinate system to facilitate data consistency with respect to the WGS 84 reference *ellipsoid*. Thus the first challenge is to develop suitable geopotential approximations that lead to *precisely ellipsoidal* geopotential surfaces of the form

$$\frac{x_1^2 + x_2^2}{A^2\left(R\right)} + \frac{x_3^2}{C^2\left(R\right)} = 1, \tag{12.24}$$

where $A\left(R\right) \equiv R$ is equatorial semi-axis, and $C\left(R\right)$ is polar semi-axis. See Fig. 12.6.

Constraints on the Functional Form for $C\left(R\right)$

Constraints on the functional form for $C\left(R\right)$, which remains to be determined, are (as in Staniforth and White, 2015a, but extended here to further include geopotential surfaces below the geoid):

1. Clairaut's factor for the Pole-to-Equator variation of apparent gravitational acceleration at Earth's WGS 84 reference ellipsoid, $r = r_S\left(\chi\right)$, should be respected.
2. It should be possible to analytically determine the orthogonal trajectories (i.e. the curves normal to the geopotential surfaces) both above *and* below the geoid.
3. The surface value, $C\left(R = a\right)$, should *exactly* satisfy

$$C\left(R = a\right) = r_S\left(\chi = \pi/2\right) = c \equiv \left(1 - \varepsilon\right) a, \tag{12.25}$$

where c is the polar radius of Earth's WGS 84 reference ellipsoid – see (12.9).

[27] This is certainly true everywhere within 50 km or so of Earth's surface, albeit progressively less so much farther away.

Approximations of the geopotential are therefore sought that are as accurate as (12.12) and (12.15), that satisfy the three constraints given in the highlighted box, and that lead to ellipsoidal geopotential surfaces of the form (12.24) (or, equivalently, (12.27)). Such approximations are developed in a step-by-step manner (in Sections 12.5 and 12.6, above and below the geoid, respectively) by further approximating (12.12) and (12.15) to first order (in $\widetilde{\varepsilon}$ and m).

From here onwards, use of the approximation symbol ($\approx$) indicates that an approximation has been made such that the ensuing expression is accurate to first order in $\widetilde{\varepsilon}$ and m (i.e. a negligible error/difference of $O\left(\widetilde{\varepsilon}^2, \widetilde{\varepsilon}m, m^2\right)$ has been introduced).

12.4.6 Geocentric Polar Representation of Ellipsoids

Cartesian coordinates (x_1, x_2, x_3) are related to geocentric polar coordinates (λ, χ, r) by – see (8.16) –

$$x_1 = r \cos\lambda \cos\chi, \quad x_2 = r \sin\lambda \cos\chi, \quad x_3 = r \sin\chi, \tag{12.26}$$

where λ is longitude, χ is geocentric latitude, and $r \equiv \left(x_1^2 + x_2^2 + x_3^2\right)^{1/2}$ is distance from the mutual origin of the two coordinate systems. See Fig. 12.6 for a cross section in the Ox_1x_3 meridional plane.

Substituting (12.26) into (12.24) gives the *geocentric polar form*

$$r(\chi, R) = \frac{A(R)\,C(R)}{\left[A^2(R)\sin^2\chi + C^2(R)\cos^2\chi\right]^{\frac{1}{2}}} \equiv A(R)\left[\cos^2\chi + \frac{A^2(R)}{C^2(R)}\sin^2\chi\right]^{-\frac{1}{2}}, \tag{12.27}$$

for an ellipsoid whose Cartesian form is given by (12.24).

12.5 FURTHER GEOPOTENTIAL APPROXIMATION ABOVE EARTH'S GEOID

12.5.1 An Intermediate Geopotential Approximation

Expanding terms involving r/R binomially using (12.23) and neglecting all $O\left(\widetilde{\varepsilon}^2, \widetilde{\varepsilon}m, m^2\right)$ terms, we can rewrite (12.12) to first order in m and $\widetilde{\varepsilon}$ as

$$\begin{aligned}
\Phi(\chi, r) = &-\frac{\gamma M}{r} + \frac{\gamma M}{R}\left(1 + \cancel{\frac{r-R}{R}}\right)^{-1}\left[\left(\frac{2\widetilde{\varepsilon} - m}{2}\right)\frac{a^2}{R^2}\left(1 + \cancel{\frac{r-R}{R}}\right)^{-2}\right. \\
&\left. + \frac{m}{2}\frac{R^3}{a^3}\left(1 + \cancel{\frac{r-R}{R}}\right)^{3}\right]\sin^2\chi \\
&- \frac{\gamma M}{a}\left[\left(\frac{2\widetilde{\varepsilon} - m}{6}\right)\frac{a^3}{R^3}\left(1 + \cancel{\frac{r-R}{R}}\right)^{-3} + \frac{m}{2}\frac{R^2}{a^2}\left(1 + \cancel{\frac{r-R}{R}}\right)^{2}\right] \\
\approx &-\frac{\gamma M}{r} + \frac{\gamma M}{R}\left\{1 + \left[\left(\frac{2\widetilde{\varepsilon} - m}{2}\right)\frac{a^2}{R^2} + \frac{m}{2}\frac{R^3}{a^3}\right]\sin^2\chi\right\} \\
&- \frac{\gamma M}{R}\left[1 + \left(\frac{2\widetilde{\varepsilon} - m}{6}\right)\frac{a^2}{R^2} + \frac{m}{2}\frac{R^3}{a^3}\right] \\
\approx &-\frac{\gamma M}{r} + \frac{\gamma M}{R}\left\{1 + \left[(2\widetilde{\varepsilon} - m)\frac{a^2}{R^2} + m\frac{R^3}{a^3}\right]\sin^2\chi\right\}^{\frac{1}{2}}
\end{aligned} \tag{12.28}$$

$$-\frac{\gamma M}{R}\left[1+\left(\frac{2\widetilde{\varepsilon}-m}{6}\right)\frac{a^2}{R^2}+\frac{m}{2}\frac{R^3}{a^3}\right]. \tag{12.29}$$

For later convenience, a term $\gamma M/R$ has been added and subtracted from (12.28). *The reverse binomial expansion used in approximating* (12.28) *by* (12.29) *is crucial to obtaining ellipsoidal geopotential surfaces of the form* (12.27).

Solving

$$\Phi(\chi, r) = \Phi_R \equiv \Phi(\chi = 0, r = R), \tag{12.30}$$

with $\Phi(\chi, r)$ defined by geopotential approximation (12.29), then leads to

$$\begin{aligned} r(\chi, R) &= R\left\{1+\left[(2\widetilde{\varepsilon}-m)\frac{a^2}{R^2}+m\frac{R^3}{a^3}\right]\sin^2\chi\right\}^{-\frac{1}{2}} \\ &= R\left\{\cos^2\chi+\left[1+(2\widetilde{\varepsilon}-m)\frac{a^2}{R^2}+m\frac{R^3}{a^3}\right]\sin^2\chi\right\}^{-\frac{1}{2}}, \end{aligned} \tag{12.31}$$

for the corresponding equation of the geopotential surface labelled R. Setting

$$A(R) = R, \quad C(R) = R\left[1+(2\widetilde{\varepsilon}-m)\frac{a^2}{R^2}+m\frac{R^3}{a^3}\right]^{-\frac{1}{2}}, \tag{12.32}$$

in (12.27) yields (12.31). *Thus* (12.31) *is in the sought ellipsoidal form* (12.27). Nevertheless, the orthogonal trajectories to the ellipsoidal geopotential surfaces defined by (12.31) are not analytically accessible. Further first-order approximation of the geopotential approximation (12.29) is therefore sought.

12.5.2 The Sought Geopotential Approximation

To address this intractability problem, Staniforth and White (2015a) expanded a^2/R^2 and R^3/a^3 in terms of truncated Taylor series about $R^2 = a^2$ – this is accurate to $O(\widetilde{\varepsilon}, m)$ because of assumption (12.23) – to obtain

$$\frac{a^2}{R^2} \approx 1 - \frac{1}{a^2}\left(R^2 - a^2\right) = 2 - \frac{R^2}{a^2}, \tag{12.33}$$

$$\frac{R^3}{a^3} = \frac{\left(R^2\right)^{\frac{3}{2}}}{a^3} \approx 1 + \frac{3}{2}\frac{\left(a^2\right)^{\frac{1}{2}}}{a^3}\left(R^2 - a^2\right) = -\frac{1}{2} + \frac{3}{2}\frac{R^2}{a^2}. \tag{12.34}$$

Substitution of (12.33) and (12.34) into geopotential approximation (12.29) yields

$$\begin{aligned} \Phi(\chi, r) \approx& -\frac{\gamma M}{r} + \frac{\gamma M}{R}\left\{1+\left[\left(\frac{8\widetilde{\varepsilon}-5m}{2}\right)+\left(\frac{5m-4\widetilde{\varepsilon}}{2}\right)\frac{R^2}{a^2}\right]\sin^2\chi\right\}^{\frac{1}{2}} \\ &- \frac{\gamma M}{R}\left[1+\left(\frac{8\widetilde{\varepsilon}-7m}{12}\right)+\left(\frac{11m-4\widetilde{\varepsilon}}{12}\right)\frac{R^2}{a^2}\right]. \end{aligned} \tag{12.35}$$

Solving (12.30) with the form (12.35) then leads to

$$r(\chi, R) = R\left\{1+\left[\left(\frac{8\widetilde{\varepsilon}-5m}{2}\right)+\left(\frac{5m-4\widetilde{\varepsilon}}{2}\right)\frac{R^2}{a^2}\right]\sin^2\chi\right\}^{-\frac{1}{2}}, \tag{12.36}$$

for the corresponding equation of the (ellipsoidal) geopotential surfaces. Setting

$$A(R) = R, \quad C(R) = R\left[1 + \left(\frac{8\widetilde{\varepsilon} - 5m}{2}\right) + \left(\frac{5m - 4\widetilde{\varepsilon}}{2}\right)\frac{R^2}{a^2}\right]^{-\frac{1}{2}}, \tag{12.37}$$

in (12.27) yields (12.36). *Thus* (12.36) *is also in the sought ellipsoidal form* (12.27). Furthermore – as will be shown in Section 12.8 – *its orthogonal trajectories can be determined analytically* as required by the second constraint of Section 12.4.5.

Evaluating (12.36) at Earth's reference surface $R = a$ and using definition (12.10) then yields

$$r_S(\chi) \equiv r(\chi, R = a) = a\left(1 + 2\widetilde{\varepsilon}\sin^2\chi\right)^{-\frac{1}{2}} = a(1-\varepsilon)\left\{(1-\varepsilon)^2 + \left[1 - (1-\varepsilon)^2\right]\sin^2\chi\right\}^{-\frac{1}{2}}. \tag{12.38}$$

Thus

$$r_S(\chi = 0) = a, \quad r_S\left(\chi = \frac{\pi}{2}\right) = a(1-\varepsilon), \tag{12.39}$$

as required by the third constraint of Section 12.4.5.

12.5.3 Clairaut's Fraction

It is of paramount importance that any proposed spheroidal-geopotential approximation leads to Clairaut's fraction (White et al., 2008; Staniforth, 2014b; Staniforth and White, 2015a). The one developed in Section 12.5.2, namely (12.35), differs only to second order in ε and m from geopotential approximation (12.12). By construction, it should therefore also lead to Clairaut's fraction. This is now explicitly verified as in Staniforth and White (2015a).

Elimination of r between (12.35) and (12.36) allows (12.35) to be rewritten as

$$\Phi(R) = -\frac{\gamma M}{R}\left[1 + \left(\frac{8\widetilde{\varepsilon} - 7m}{12}\right) + \left(\frac{11m - 4\widetilde{\varepsilon}}{12}\right)\frac{R^2}{a^2}\right]. \tag{12.40}$$

The dependence of Φ on r and χ is now hidden in (12.36), which implicitly relates R to r and χ. Partially differentiating (12.40) with respect to r gives

$$\frac{\partial\Phi}{\partial r} = \frac{\partial R}{\partial r}\frac{d\Phi}{dR} = \frac{1}{\partial r/\partial R}\frac{\gamma M}{a^2}\left[\left(1 + \frac{8\widetilde{\varepsilon} - 7m}{12}\right)\frac{a^2}{R^2} - \left(\frac{11m - 4\widetilde{\varepsilon}}{12}\right)\right]. \tag{12.41}$$

Now, using (12.36),

$$\frac{\partial r}{\partial R} = \left[1 + \left(\frac{8\widetilde{\varepsilon} - 5m}{2}\right)\sin^2\chi\right]\left\{1 + \left[\left(\frac{8\widetilde{\varepsilon} - 5m}{2}\right) + \left(\frac{5m - 4\widetilde{\varepsilon}}{2}\right)\frac{R^2}{a^2}\right]\sin^2\chi\right\}^{-\frac{3}{2}}. \tag{12.42}$$

Evaluation at $(\chi = 0, R = a)$ and at $(\chi = \pi/2, R = a)$, respectively, then yields

$$\left(\frac{\partial r}{\partial R}\right)_{\chi=0,R=a} = 1, \quad \left(\frac{\partial r}{\partial R}\right)_{\chi=\pi/2,R=a} = \left(1 + 4\widetilde{\varepsilon} - \frac{5m}{2}\right)(1 + 2\widetilde{\varepsilon})^{-\frac{3}{2}}. \tag{12.43}$$

Thus, from (12.41) and (12.42),

$$g_S^P \equiv \left|\left(\frac{\partial\Phi}{\partial r}\right)_{\chi=\frac{\pi}{2},R=a}\right| = \frac{\gamma M}{a^2}\left(1 + \widetilde{\varepsilon} - \frac{3m}{2}\right)\frac{(1 + 2\widetilde{\varepsilon})^{\frac{3}{2}}}{(1 + 4\widetilde{\varepsilon} - 5m/2)}, \tag{12.44}$$

$$g_S^E \equiv \left|\left(\frac{\partial\Phi}{\partial r}\right)_{\chi=0,R=a}\right| = \frac{\gamma M}{a^2}\left(1 + \widetilde{\varepsilon} - \frac{3m}{2}\right), \tag{12.45}$$

where g_S^P and g_S^E are the accelerations of apparent gravity at the Poles and Equator, respectively, of Earth's assumed WGS 84 ellipsoidal reference surface. Clairaut's famous fraction (Todhunter, 1873) is then

$$\frac{g_S^P - g_S^E}{g_S^E} = \frac{(1+2\widetilde{\varepsilon})^{\frac{3}{2}}}{(1+4\widetilde{\varepsilon}-5m/2)} - 1 \approx \frac{5m}{2} - \widetilde{\varepsilon} \approx \frac{5m}{2} - \varepsilon, \tag{12.46}$$

where (12.10) has been used to obtain the rightmost expression. As noted in Staniforth and White (2015a), this result corresponds to 6.8(12) of Ramsey (1940), and to (18) of White et al. (2008), for the classical *spheroidal*-geopotential approximation. It also corresponds to (17) of Staniforth (2014b) for his spheroidal-geopotential approximation.

Thus ellipsoidal-geopotential approximation (12.35) *does indeed lead to Clairaut's fraction; see the first requirement of Section 12.4.5.*

12.5.4 Equilibrium Depth of an Ocean Covering a Solid Ellipsoidal Earth

van der Toorn and Zimmerman (2008) hypothesised (their Section 4.5) a solid ellipsoidal Earth covered by a shallow, non-self-gravitating global ocean. They implicitly placed this global ocean immediately *above* the geoid rather than immediately below it as done herein. Nevertheless the preceding analysis is still applicable provided that their hypothesised global ocean is indeed shallow and non-self-gravitating.

As a test of their spheroidal-geopotential approximation, van der Toorn and Zimmerman (2008) posed the question:

- How does the equilibrium depth of a very shallow fluid layer vary as a function of latitude and (in the present notation) the small parameters $\widetilde{\varepsilon}$ and m?

They then answered this question, after noting that:

- The construction of the final result is technically complicated.
- Full details of the derivation are therefore not given, but only an outline of the procedure used.

The following question then naturally arises:

- Can ellipsoidal-geopotential approximation (12.35) – which implies (12.36) and (12.38) – also be used to rederive van der Toorn and Zimmerman (2008)'s asymptotic result?

It is shown in the Appendix to this chapter that van der Toorn and Zimmerman (2008)'s quite lengthy derivation, using their spheroidal-geopotential approximation, very nicely reduces to just a few lines when using ellipsoidal-geopotential approximation (12.35).[28]

- This illustrates that suitably approximating the geopotential, *once and for all*, above Earth's geoid – as in Staniforth (2014a) and Staniforth and White (2015a) – not only facilitates analytical tractability and simplicity, but also can do so without sacrificing accuracy.

Following van der Toorn and Zimmerman (2008), (12.140) for the Pole-to-Equator relative equilibrium depth is plotted in Fig. 12.7 using physically realistic terrestrial values for $m \approx 0.00346$ and $\widetilde{\varepsilon} \approx 0.00337$, so that $(5m/2) - \widetilde{\varepsilon} \approx 0.00528$.

Assuming terrestrial values for $\widetilde{\varepsilon}$, m, and $a \approx 6\,400$ km, and also that

$$\frac{R-a}{a} \lesssim O\left(\widetilde{\varepsilon}^2, \widetilde{\varepsilon}m, m^2\right), \tag{12.47}$$

[28] Development of ellipsoidal-geopotential approximation (12.35) has, however, involved a fair amount of effort!

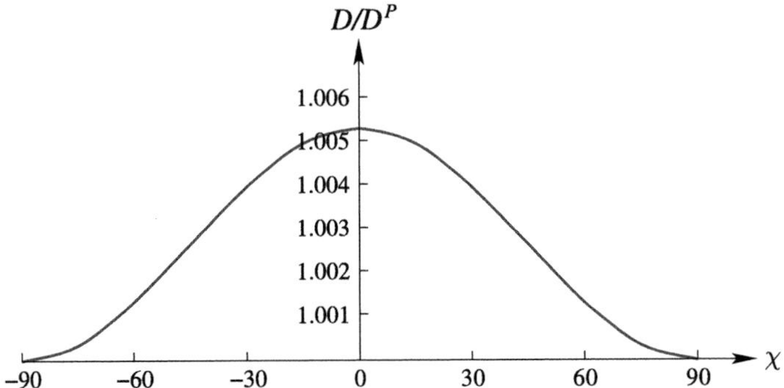

Figure 12.7 Relative equilibrium depth of a shallow global ocean covering a solid ellipsoidal Earth, as a function of geocentric latitude χ (in degrees). After Fig. 2 of van der Toorn and Zimmerman (2008).

it is seen from (12.140) , (12.47), and Fig. 12.7 that:

- Ocean depth monotonically decreases poleward from Earth's Equator.
- The parameter $(5m/2 - \widetilde{\varepsilon})$ appearing in (12.140) is Clairaut's fraction (12.46).
- $R - a \lesssim 20$ km, which restricts ocean depth to be somewhat larger than that ($\approx$ 11 km) of Earth's oceans.
- A global, equilibrium ocean of 1 km depth at Earth's Poles would be approximately 5 m deeper at its Equator.

12.6 FURTHER GEOPOTENTIAL APPROXIMATION BELOW EARTH'S GEOID

12.6.1 An Intermediate Geopotential Approximation

Noting from (12.16) and (12.17) that

$$A_0 = -\frac{3}{2}\left(1 + \frac{2\widetilde{\varepsilon}}{9}\right)\frac{\gamma M}{a} = O(1), \quad A_2 = \frac{1}{2}\frac{\gamma M}{a} = O(1), \quad A_3 = \frac{\widetilde{\varepsilon}}{3}\frac{\gamma M}{a} = O(m, \widetilde{\varepsilon}), \tag{12.48}$$

$$B_2 = -\frac{2}{3}(10m - 9\widetilde{\varepsilon})\frac{\gamma M}{a} = O(m, \widetilde{\varepsilon}), \quad B_3 = \frac{7}{3}(5m - 4\widetilde{\varepsilon})\frac{\gamma M}{a} = O(m, \widetilde{\varepsilon}), \tag{12.49}$$

geopotential approximation (12.15) below Earth's geoid can be rewritten as

$$\begin{aligned}\Phi(\chi, r) = \frac{1}{2}\frac{\gamma M}{a}\frac{r^2}{a^2} + A_0 + \frac{r^2}{a^2}\left\{-\frac{\gamma M}{a}\frac{m}{3} - \frac{B_2}{2} + \left[A_3 - \frac{B_3}{2}\left(1 - \frac{3}{7}\frac{r}{a}\right)\right]\frac{r}{a}\right\} \\ + \frac{3}{2}\frac{r^2}{a^2}\left[B_2 + B_3\frac{r}{a}\left(1 - \frac{3}{7}\frac{r}{a}\right)\right]\sin^2\chi,\end{aligned} \tag{12.50}$$

where the first two terms on the right-hand side are $O(1)$, and the remaining terms are $O(m, \widetilde{\varepsilon})$. Introducing assumption (12.23) into (12.50) and neglecting second-order terms in m and $\widetilde{\varepsilon}$ then gives

$$\begin{aligned}\Phi(\chi, r) &\approx \frac{1}{2}\frac{\gamma M}{a}\frac{r^2}{a^2} + A_0 + \frac{R^2}{a^2}\left\{-\frac{\gamma M}{a}\frac{m}{3} - \frac{B_2}{2} + \left[A_3 - \frac{B_3}{2}\left(1 - \frac{3}{7}\frac{R}{a}\right)\right]\frac{R}{a}\right\} \\ &\quad + \frac{3}{2}\frac{R^2}{a^2}\left[B_2 + B_3\frac{R}{a}\left(1 - \frac{3}{7}\frac{R}{a}\right)\right]\sin^2\chi \\ &= \frac{1}{2}\frac{\gamma M}{a}\left\langle\frac{r^2}{a^2} - \frac{R^2}{a^2}\left\{1 - \frac{3a}{\gamma M}\left[B_2 + B_3\frac{R}{a}\left(1 - \frac{3}{7}\frac{R}{a}\right)\right]\sin^2\chi\right\}\right\rangle\end{aligned}$$

$$+ A_0 + \frac{R^2}{a^2}\left\{\left(\frac{1}{2}\frac{\gamma M}{a} - \frac{\gamma M}{a}\frac{m}{3} - \frac{B_2}{2}\right) + \left[A_3 - \frac{B_3}{2}\left(1 - \frac{3}{7}\frac{R}{a}\right)\right]\frac{R}{a}\right\}, \tag{12.51}$$

$$\approx \frac{1}{2}\frac{\gamma M}{a}\left\langle \frac{r^2}{a^2} - \frac{R^2}{a^2}\left\{1 + \frac{3a}{\gamma M}\left[B_2 + B_3\frac{R}{a}\left(1 - \frac{3}{7}\frac{R}{a}\right)\right]\sin^2\chi\right\}^{-1}\right\rangle$$
$$+ A_0 + \frac{R^2}{a^2}\left\{\left(\frac{1}{2}\frac{\gamma M}{a} - \frac{\gamma M}{a}\frac{m}{3} - \frac{B_2}{2}\right) + \left[A_3 - \frac{B_3}{2}\left(1 - \frac{3}{7}\frac{R}{a}\right)\right]\frac{R}{a}\right\}, \tag{12.52}$$

where a term $(\gamma M/a)\left(R^2/a^2\right)/2$ has been added and subtracted for later convenience. *The reverse binomial expansion used in approximating* (12.51) *by* (12.52) *is crucial to obtaining ellipsoidal geopotential surfaces of the form* (12.27).

Now, from (12.52),

$$\Phi_R \equiv \Phi\left(\chi = 0, r = R\right) = A_0 + \frac{R^2}{a^2}\left\{\left(\frac{1}{2}\frac{\gamma M}{a} - \frac{\gamma M}{a}\frac{m}{3} - \frac{B_2}{2}\right) + \left[A_3 - \frac{B_3}{2}\left(1 - \frac{3}{7}\frac{R}{a}\right)\right]\frac{R}{a}\right\}. \tag{12.53}$$

Using (12.30), with $\Phi\left(\chi, r\right)$ defined by (12.52) and Φ_R defined by (12.53), leads to the equation for the geopotential surfaces, namely

$$0 = \frac{1}{2}\frac{\gamma M}{a}\left\langle \frac{r^2}{a^2} - \frac{R^2}{a^2}\left\{1 + \frac{3a}{\gamma M}\left[B_2 + B_3\frac{R}{a}\left(1 - \frac{3}{7}\frac{R}{a}\right)\right]\sin^2\chi\right\}^{-1}\right\rangle$$
$$+ \cancel{A_0} + \cancel{\frac{R^2}{a^2}\left\{\left(\frac{1}{2}\frac{\gamma M}{a} - \frac{\gamma M}{a}\frac{m}{3} - \frac{B_2}{2}\right) + \left[A_3 - \frac{B_3}{2}\left(1 - \frac{3}{7}\frac{R}{a}\right)\right]\frac{R}{a}\right\}}$$
$$- \cancel{A_0} - \cancel{\frac{R^2}{a^2}\left\{\left(\frac{1}{2}\frac{\gamma M}{a} - \frac{\gamma M}{a}\frac{m}{3} - \frac{B_2}{2}\right) + \left[A_3 - \frac{B_3}{2}\left(1 - \frac{3}{7}\frac{R}{a}\right)\right]\frac{R}{a}\right\}}$$
$$= \frac{1}{2}\frac{\gamma M}{a}\left\langle \frac{r^2}{a^2} - \frac{R^2}{a^2}\left\{1 + \frac{3a}{\gamma M}\left[B_2 + B_3\frac{R}{a}\left(1 - \frac{3}{7}\frac{R}{a}\right)\right]\sin^2\chi\right\}^{-1}\right\rangle. \tag{12.54}$$

Using (12.49) to eliminate B_2 and B_3 from (12.54) and solving for r then gives

$$r\left(\chi, R\right) = R\left\{1 + \frac{3a}{\gamma M}\left[B_2 + B_3\frac{R}{a}\left(1 - \frac{3}{7}\frac{R}{a}\right)\right]\sin^2\chi\right\}^{-\frac{1}{2}}$$
$$= R\left\{1 + \left[-2\left(10m - 9\widetilde{\varepsilon}\right) + 7\left(5m - 4\widetilde{\varepsilon}\right)\frac{R}{a}\left(1 - \frac{3}{7}\frac{R}{a}\right)\right]\sin^2\chi\right\}^{-\frac{1}{2}}. \tag{12.55}$$

Thus

$$r_S\left(\chi\right) \equiv r\left(\chi, R = a\right) = a\left(1 + 2\widetilde{\varepsilon}\sin^2\chi\right)^{-\frac{1}{2}} \tag{12.56}$$

at Earth's assumed geoid. Setting

$$A\left(R\right) = R, \quad C\left(R\right) = R\left[1 - 2\left(10m - 9\widetilde{\varepsilon}\right) + 7\left(5m - 4\widetilde{\varepsilon}\right)\frac{R}{a}\left(1 - \frac{3}{7}\frac{R}{a}\right)\right]^{-\frac{1}{2}} \tag{12.57}$$

in (12.27) yields (12.55). Thus (12.55) is in the sought ellipsoidal form (12.27).

However, we may or may not be able to analytically determine the orthogonal trajectories to the geopotential surfaces defined by (12.55). Even if we can, we also want the geopotential surfaces to have the same functional form just above and just below Earth's geoid. (As we will see, this unifies the two families of orthogonal trajectories – one above the geoid, the other below – into a single family that spans the geoid.) Comparing the form of (12.55) just below the geoid with that of (12.36) just above the geoid, we see that their functional forms are different; in (12.55) there are terms in both R/a and R^2/a^2 for $\sin^2\chi$, whereas in (12.36) there is a single term in R^2/a^2. Further first-order approximation of (12.52) is therefore sought.

12.6.2 The Sought Geopotential Approximation

To address this unification issue, R/a is expanded in a Taylor series about $R^2 = a^2$ as

$$\frac{R}{a} = \frac{\left(R^2\right)^{\frac{1}{2}}}{a} \approx 1 + \frac{1}{a}\frac{1}{2}\left[\left(R^2\right)^{-\frac{1}{2}}\right]_{R^2=a^2}\left(R^2 - a^2\right) = \frac{1}{2}\left(1 + \frac{R^2}{a^2}\right), \tag{12.58}$$

$$\Downarrow$$

$$\frac{R}{a}\left(1 - \frac{3}{7}\frac{R}{a}\right) = \frac{R}{a} - \frac{3}{7}\frac{R^2}{a^2} \approx \frac{1}{2}\left(1 + \frac{R^2}{a^2}\right) - \frac{3}{7}\frac{R^2}{a^2} = \frac{1}{2}\left(1 + \frac{1}{7}\frac{R^2}{a^2}\right). \tag{12.59}$$

Doing so here, *below* the geoid, is analogous to using expansions (12.33) and (12.34) *above* the geoid.

Introducing expansions (12.58) and (12.59) into (12.52) gives

$$\begin{aligned}\Phi(\chi, r) \approx\ & \frac{1}{2}\frac{\gamma M}{a}\left\langle \frac{r^2}{a^2} - \frac{R^2}{a^2}\left\{1 + \frac{3a}{\gamma M}\left[B_2 + \frac{B_3}{2}\left(1 + \frac{1}{7}\frac{R^2}{a^2}\right)\right]\sin^2\chi\right\}^{-1}\right\rangle \\ & + A_0 + \frac{R^2}{a^2}\left\{\left(\frac{1}{2}\frac{\gamma M}{a} - \frac{\gamma M}{a}\frac{m}{3} + \frac{A_3}{2} - \frac{B_2}{2} - \frac{B_3}{4}\right) + \frac{1}{2}\left(A_3 - \frac{B_3}{14}\right)\frac{R^2}{a^2}\right\},\end{aligned} \tag{12.60}$$

and using (12.16) and (12.17) to eliminate A_3, B_2 and B_3 then yields

$$\begin{aligned}\Phi(\chi, r) \approx\ & \frac{1}{2}\frac{\gamma M}{a}\left\langle \frac{r^2}{a^2} - \frac{R^2}{a^2}\left\{1 + \left[\left(\frac{8\widetilde{\varepsilon} - 5m}{2}\right) + \left(\frac{5m - 4\widetilde{\varepsilon}}{2}\right)\frac{R^2}{a^2}\right]\sin^2\chi\right\}^{-1}\right\rangle \\ & + A_0 + \frac{1}{2}\frac{\gamma M}{a}\frac{R^2}{a^2}\left[1 - \left(\frac{6\widetilde{\varepsilon} - m}{6}\right) + \left(\frac{6\widetilde{\varepsilon} - 5m}{6}\right)\left(\frac{R^2}{a^2}\right)\right].\end{aligned} \tag{12.61}$$

Now, from (12.61),

$$\Phi_R \equiv \Phi(\chi = 0, r = R) = A_0 + \frac{1}{2}\frac{\gamma M}{a}\frac{R^2}{a^2}\left[1 - \left(\frac{6\widetilde{\varepsilon} - m}{6}\right) + \left(\frac{6\widetilde{\varepsilon} - 5m}{6}\right)\left(\frac{R^2}{a^2}\right)\right]. \tag{12.62}$$

Using (12.30), with $\Phi(\chi, r)$ defined by (12.61) and Φ_R defined by (12.62), leads to the equation for the geopotential surfaces

$$\begin{aligned}0 =\ & \frac{1}{2}\frac{\gamma M}{a}\left\langle \frac{r^2}{a^2} - \frac{R^2}{a^2}\left\{1 + \left[\left(\frac{8\widetilde{\varepsilon} - 5m}{2}\right) + \left(\frac{5m - 4\widetilde{\varepsilon}}{2}\right)\frac{R^2}{a^2}\right]\sin^2\chi\right\}^{-1}\right\rangle \\ & + A_0 + \frac{1}{2}\frac{\gamma M}{a}\frac{R^2}{a^2}\left[1 - \left(\frac{6\widetilde{\varepsilon} - m}{6}\right) + \left(\frac{6\widetilde{\varepsilon} - 5m}{6}\right)\left(\frac{R^2}{a^2}\right)\right]\end{aligned}$$

$$
-\left\{A_0+\frac{1}{2}\frac{\gamma M}{a}\frac{R^2}{a^2}\left[1-\left(\frac{6\widetilde{\varepsilon}-m}{6}\right)+\left(\frac{6\widetilde{\varepsilon}-5m}{6}\right)\left(\frac{R^2}{a^2}\right)\right]\right\}
$$

$$
=\frac{\gamma M}{2a}\left\langle\frac{r^2}{a^2}-\frac{R^2}{a^2}\left\{1+\left[\left(\frac{8\widetilde{\varepsilon}-5m}{2}\right)+\left(\frac{5m-4\widetilde{\varepsilon}}{2}\right)\frac{R^2}{a^2}\right]\sin^2\chi\right\}^{-1}\right\rangle. \tag{12.63}
$$

Solving (12.63) for r then gives

$$
r(\chi,R)=R\left\{1+\left[\left(\frac{8\widetilde{\varepsilon}-5m}{2}\right)+\left(\frac{5m-4\widetilde{\varepsilon}}{2}\right)\frac{R^2}{a^2}\right]\sin^2\chi\right\}^{-\frac{1}{2}}. \tag{12.64}
$$

Thus

$$
r_S(\chi)\equiv r(\chi,R=a)=a\left(1+2\widetilde{\varepsilon}\sin^2\chi\right)^{-\frac{1}{2}} \tag{12.65}
$$

at Earth's assumed geoid. Setting

$$
A(R)=R,\quad C(R)=R\left[1+\left(\frac{8\widetilde{\varepsilon}-5m}{2}\right)+\left(\frac{5m-4\widetilde{\varepsilon}}{2}\right)\frac{R^2}{a^2}\right]^{-\frac{1}{2}} \tag{12.66}
$$

in (12.27) yields (12.64). Thus (12.64) is in the sought ellipsoidal form (12.27).

Since (12.65) is identical to (12.38), it follows that (12.39) holds as required by the third constraint of Section 12.4.5.

Both (12.64) and (12.65) – valid *below the geoid and up to it* – agree with the corresponding equations (12.36) and (12.38) for the geopotential surfaces *above and down to* the assumed ellipsoidal geoid. Thus:

- The orthogonal trajectories have the same functional form below the geoid as above it.

12.6.3 Clairaut's Fraction

It is important that geopotential approximation (12.61), derived above and valid below the geoid, satisfies Clairaut's fraction. This is now explicitly verified.

Differentiating (12.62) and evaluating the result at $R=a$ using (12.48) yields

$$
\left(\frac{d\Phi_R}{dR}\right)_{R=a}=\frac{\gamma M}{a^2}\left(1+\widetilde{\varepsilon}-\frac{3m}{2}\right). \tag{12.67}
$$

Partially differentiating (12.64) with respect to R gives

$$
\frac{\partial r}{\partial R}=\left[1+\left(\frac{8\widetilde{\varepsilon}-5m}{2}\right)\sin^2\chi\right]\left\{1+\left[\left(\frac{8\widetilde{\varepsilon}-5m}{2}\right)+\left(\frac{5m-4\widetilde{\varepsilon}}{2}\right)\frac{R^2}{a^2}\right]\sin^2\chi\right\}^{-\frac{3}{2}}, \tag{12.68}
$$

and so

$$
\left(\frac{\partial r}{\partial R}\right)_{\chi=0,R=a}=1,\quad \left(\frac{\partial r}{\partial R}\right)_{\chi=\pi/2,R=a}=\left(1+4\widetilde{\varepsilon}-\frac{5m}{2}\right)(1+2\widetilde{\varepsilon})^{-\frac{3}{2}}. \tag{12.69}
$$

Use of (12.67) and (12.69) then leads to

$$
g_S^P\equiv\left|\left(\frac{\partial\Phi}{\partial r}\right)_{\chi=\pi/2,R=a}\right|=\left|\left(\frac{\partial R}{\partial r}\right)_{\chi=\pi/2,R=a}\right|\left|\left(\frac{d\Phi}{dR}\right)_{R=a}\right|
$$

$$= \left(1 + 4\tilde{\varepsilon} - \frac{5m}{2}\right)^{-1} (1 + 2\tilde{\varepsilon})^{\frac{3}{2}} \frac{\gamma M}{a^2} \left(1 + \tilde{\varepsilon} - \frac{3m}{2}\right), \tag{12.70}$$

$$g_S^E \equiv \left|\left(\frac{\partial \Phi}{\partial r}\right)_{\chi=0,R=a}\right| = \left|\left(\frac{\partial R}{\partial r}\right)_{\chi=0,R=a}\right| \left|\left(\frac{d\Phi}{dR}\right)_{R=a}\right|$$

$$= \frac{\gamma M}{a^2} \left(1 + \tilde{\varepsilon} - \frac{3m}{2}\right). \tag{12.71}$$

Thus, from (12.70) and (12.71), Clairaut's fraction is

$$\frac{g_S^P - g_S^E}{g_S^E} = \frac{(1 + 2\tilde{\varepsilon})^{\frac{3}{2}}}{(1 + 4\tilde{\varepsilon} - 5m/2)} - 1 \approx \frac{5m}{2} - \tilde{\varepsilon} \approx \frac{5m}{2} - \varepsilon, \tag{12.72}$$

where (12.10) has been used to obtain the rightmost expression. This result corresponds to 6.8(12) of Ramsey (1940), and to (18) of White et al. (2008), for the classical spheroidal-geopotential approximation. Ellipsoidal-geopotential approximation (12.61) therefore does indeed lead to Clairaut's fraction; see the first requirement of Section 12.4.5.

12.7 INTERLUDE

We now pause to gather our thoughts. Ellipsoidal geopotential representations (12.35) (above the geoid) and (12.61) (below it) have been developed. They are as accurate as spheroidal-geopotential representations (12.12) and (12.15), respectively, whilst preserving both the ratio of Earth's Polar and Equatorial radii, and Clairaut's fraction for the Pole-to-Equator variation of apparent gravity at Earth's surface. Recapitulating:

The Geopotential Representations above and below the Geoid

Everything that needs to be known about the geopotential and gravity is embodied in the two representations (12.35) (above the geoid) and (12.61) (below it).

From them, the following can all be obtained analytically, that is, *without further approximation*:

- The equation for the associated ellipsoidal-geopotential surfaces, namely (12.36) and, equivalently, (12.64).
- The orthogonal trajectories to these surfaces (see Section 12.8).
- An orthogonal geopotential coordinate system (see Section 12.9).

The functional forms of geopotential representations (12.35) (above the geoid) and (12.61) (below it) are clearly different. The functional forms of gravity (obtained by taking minus the gradient of the geopotential) will therefore also be different. However – and this is a crucially important property – the equation for geopotential surfaces *below* the geoid, namely (12.64), is *identical* to that *above* the geoid, namely (12.36). Hence – see Fig. 12.8 for a graphical depiction –

There is:

- A single, *unified* family of *ellipsoidal-geopotential surfaces* that spans the geoid, rather than two separate families (one valid above the geoid, the other below it).
- A single, *unified* family of *orthogonal trajectories* to the ellipsoidal-geopotential surfaces (rather than two separate families, one valid above the geoid, the other below it).

This means that we can develop a *unified* geopotential coordinate system that spans the geoid and that can be used both above and below it. The only difference between the atmospheric and

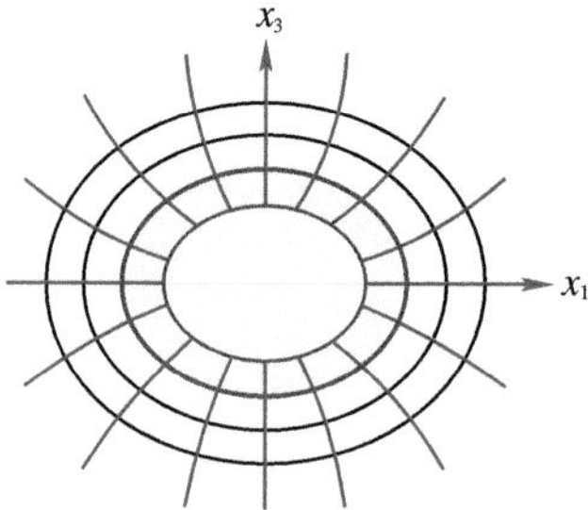

Figure 12.8 Cross section in the Ox_1x_3 meridional plane of the unified family of ellipsoidal-geopotential surfaces outside (in black) and inside (in blue) a mildly oblate, ellipsoidal planet, together with their orthogonal trajectories (in red). The planet's bounding surface is the dark blue ellipsoid, and the planet's interior is shaded (in two shades of light blue).

oceanic contexts is that gravity is evaluated using a different functional form for the geopotential. Above the geoid (for the atmospheric context) (12.35) is employed, whereas below it (for the oceanic context), (12.61) is employed instead. At the geoid, gravity can be evaluated using either of the two geopotential representations since they *and their gradients* are continuous across the geoid.

Now that the preparatory work has been done to develop ellipsoidal-geopotential representations (12.35) and (12.61), we are ready to determine the orthogonal trajectories to their geopotential surfaces (in Section 12.8). This then opens the way for the development of GREAT (Geophysically Realistic, Ellipsoidal, Analytically Tractable) coordinates (in Section 12.9) for both atmospheric *and* oceanic modelling.

In what follows, we very closely follow Staniforth and White (2015a)'s development of GREAT coordinates, but with various adaptations to the present context, such as:

- The sign convention for definition of the geopotential is changed – for consistency with that adopted in various published works, and in preceding chapters herein.
- Distances are not scaled herein by the planet's equatorial radius – for improved clarity.
- Staniforth and White (2015a)'s formulation, valid for an atmosphere *above* a planet's geoid, is generalised to also include oceans *below* a geoid.

12.8 ORTHOGONAL TRAJECTORIES TO THE GEOPOTENTIAL SURFACES

Following Staniforth and White (2015a), but *not* scaling lengths herein by the planet's equatorial radius a, the ellipsoidal-geopotential surfaces $\Phi = \Phi_R$ defined by (12.36) and (12.64) can be rewritten as

$$r(\chi, R) = R\left[1 + \left(\mu + \nu\frac{R^2}{a^2}\right)\sin^2\chi\right]^{-\frac{1}{2}} = \frac{A(R)\,C(R)}{\left[A^2(R)\sin^2\chi + C^2(R)\cos^2\chi\right]^{\frac{1}{2}}}, \tag{12.73}$$

where

$$\mu \equiv \frac{8\widetilde{\varepsilon} - 5m}{2}, \quad \nu \equiv \frac{5m - 4\widetilde{\varepsilon}}{2}, \tag{12.74}$$

$$A(R) = R, \quad C(R) = R\left(1 + \mu + \nu\frac{R^2}{a^2}\right)^{-\frac{1}{2}}. \tag{12.75}$$

Using (12.73) in the transformation equations (12.26) from geocentric-polar coordinates to Cartesian coordinates leads to (12.24), that is, to the equation for a family of ellipsoids with equatorial

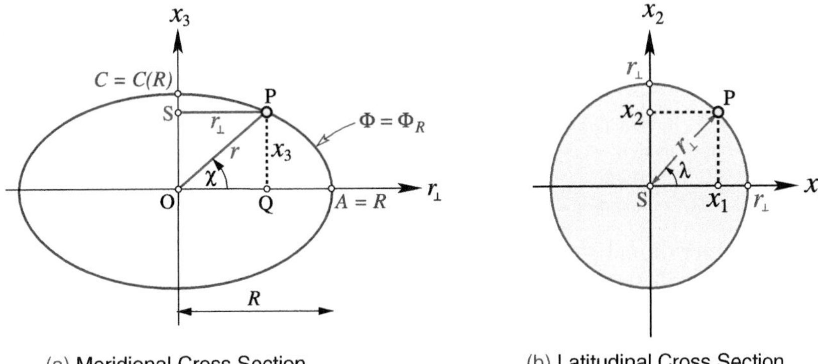

Figure 12.9 Meridional (panel a) and latitudinal (panel b) cross sections of prototypical ellipsoidal geopotential surface $\Phi = \Phi_R$, with equatorial and polar semi-axes $A(R)$ and $C(R)$ defined by (12.75). Point P is an arbitrary point on this surface; $r_\perp$ is perpendicular distance of point P from point S on the planet's rotation axis; the meridional cross-section is depicted in blue (on panel a); and the latitudinal one in magenta (on panel b). See text for further details.

and polar semi-axes $A(R)$ and $C(R)$ defined by (12.75). See Fig. 12.9 for a depiction of meridional and latitudinal cross sections of a prototypical, ellipsoidal-geopotential surface $\Phi = \Phi_R$.

Each ellipsoid of the family is a surface of revolution. Therefore an orthogonal trajectory that passes through an arbitrary point $(x_1, x_2, x_3) = (r\cos\lambda\cos\chi, r\sin\lambda\cos\chi, r\sin\chi)$ of an ellipsoid – cf. (12.26) – lies within the meridional plane that passes through that point. Equation (12.24) for the ellipsoidal-geopotential surfaces then simplifies to

$$\frac{r_\perp^2}{A^2(R)} + \frac{x_3^2}{C^2(R)} = 1, \tag{12.76}$$

where

$$r_\perp \equiv \left(x_1^2 + x_2^2\right)^{1/2} = r\cos\chi \tag{12.77}$$

is distance of a point (x_1, x_2, x_3) on the ellipsoid from its rotation axis of symmetry – see Fig. 12.9 – and

$$x_1 = r_\perp \cos\lambda, \quad x_2 = r_\perp \sin\lambda. \tag{12.78}$$

Expressing x_1 and x_2 in terms of $r_\perp$ and λ has the virtue that it shortens equations in what follows; for example, $r_\perp^2$ appears instead of $x_1^2 + x_2^2$.

Inserting (12.75) into (12.76) leads to

$$r_\perp^2 + \left(1 + \mu + \nu\frac{R^2}{a^2}\right)x_3^2 = R^2 \quad \Rightarrow \quad R^2 = \frac{r_\perp^2 + (1+\mu)\,x_3^2}{1 - \nu x_3^2/a^2}, \tag{12.79}$$

and then to the useful identity

$$1 + \mu + \nu\frac{r_\perp^2}{a^2} \equiv \left(1 - \nu\frac{x_3^2}{a^2}\right)\left(1 + \mu + \nu\frac{R^2}{a^2}\right). \tag{12.80}$$

Setting $a \equiv 1$ and $\nu \equiv 0$ in the first equation of (12.79) corresponds to White et al. (2008)'s equation (32) for their Similar-Oblate-Spheroidal (SOS) coordinates.[29] Thus:

[29] a has to be set to unity to take into account White et al. (2008)'s scaling of lengths by a. From (12.74), and as remarked in White et al. (2008), setting $\nu \equiv 0$ corresponds to $\widetilde{\varepsilon} = 5m/4$ and hence to Newton's *uniform*-density model of Earth. The

- White et al. (2008)'s SOS coordinates can be considered to be a special case of GREAT coordinates.[a]

[a] As formulated in White et al. (2008), SOS coordinates are singular at the Equator. This is, however, an artefact of their implementation, due to use of an auxiliary coordinate. This weakness is addressed at source here. Non-singular SOS coordinates then arise as a special case (namely with parameter ν set identically zero) of (non-singular) GREAT coordinates.

Tangents within a meridional plane to ellipsoidal-geopotential surfaces (12.76) are obtained by differentiating (12.76) with respect to $r_\perp$, at constant R (this labels and defines the geopotential surface $\Phi = \Phi_R$), followed by rearrangement, so that

$$\frac{dx_3}{dr_\perp} = -\frac{C^2}{A^2}\frac{r_\perp}{x_3}. \tag{12.81}$$

To obtain the orthogonal trajectories – that is, the curves *normal* to the family of ellipsoidal-geopotential surfaces – we exploit the fact that for mutual orthogonality, the product of the tangent and orthogonal-trajectory slopes must be equal to minus unity. Thus – from (12.75) and (12.81) – the orthogonal trajectories satisfy

$$\frac{dx_3}{dr_\perp} = -\left(-\frac{C^2}{A^2}\frac{r_\perp}{x_3}\right)^{-1} = \frac{A^2 x_3}{C^2 r_\perp} = \left(1 + \mu + \nu\frac{R^2}{a^2}\right)\frac{x_3}{r_\perp} = \left(1 + \mu + \nu\frac{r_\perp^2}{a^2}\right)\left(1 - \nu\frac{x_3^2}{a^2}\right)^{-1}\frac{x_3}{r_\perp}, \tag{12.82}$$

where use has been made of identity (12.80) to obtain the rightmost expression of (12.82). This crucially allowed R^2 to be eliminated and (12.82) to be written as a separable and *integrable* first-order differential equation in x_3 and $r_\perp$. Integrating (12.82) then gives

$$\frac{x_3}{a} = D\left(\frac{r_\perp}{a}\right)^{1+\mu}\exp\left[\frac{\nu}{2}\left(\frac{r_\perp^2 + x_3^2}{a^2}\right)\right], \tag{12.83}$$

where parameter D is the 'constant' of integration (i.e. it is independent of $r_\perp$ and x_3).

Setting $\nu \equiv 0$ in (12.83) corresponds to White et al. (2008)'s (33) for the special case of their SOS coordinates. Orthogonal-trajectory equation (12.82) was a key result of Staniforth and White (2015a) that allowed White et al. (2008)'s restriction to similar ellipsoids (for which $\nu \equiv 0$) to be lifted.

Parameter D in (12.83) takes a fixed value along each orthogonal trajectory and – as D is varied in the horizontal – it generates the family of orthogonal trajectories to the geopotential surfaces. Because of these properties, D must be a function of meridional coordinate φ. However, the precise form of the functional relation, say $D = D(\varphi)$, is a matter of choice. For the special case $\nu \equiv 0$ – examined in White et al. (2008) – (12.83) is *explicit* for x_3 in terms of $r_\perp$ (i.e. (12.83) reduces to the power-curve family $x_3/a = D(r_\perp/a)^{1+\mu}$). White et al. (2008) chose the functional relation $D(\varphi) = (1+\mu)^{-1/2}\tan^{1+\mu}\varphi$ for algebraic convenience. This, however, leads to singular behaviour at the coordinate system's equator. Consequently a different, non-singular approach was taken in Staniforth and White (2015a). Their approach is therefore adopted and adapted herein.

For this approach, each orthogonal trajectory is labelled in terms of its intersection with Earth's WGS 84 reference ellipsoid[30] rather than with a polar-tangent plane, thereby avoiding the singular

present formulation is more general. *Variable* density is now permitted, thereby leading to a *more physically realistic variation of gravity*. For Earth, $\widetilde{\varepsilon} \approx m \approx 1/300$ and so, from (12.74), $\nu \approx \mu/3 > 0$. This means that R^2 in (12.79) is potentially singular, which would then make R unsuitable as the vertical coordinate of a three-dimensional coordinate system. However, for realistic applications, $|x_3| \ll 2a$ and so $\nu x_3^2/a^2 \ll 1$ in (12.79). Thus this singularity would not be encountered in practice.

[30] Thus Earth's WGS 84 reference ellipsoid – which coincides with a geopotential surface – provides the reference surface for the construction of trajectories that are orthogonal at every point along them to the geopotential surface passing through that point.

behaviour at the equator of White et al. (2008)'s implementation of SOS coordinates. Importantly, Staniforth and White (2015a)'s procedure is applicable to the more complicated (and *implicit*) functional form of orthogonal-trajectory equation (12.83) when $\nu \neq 0$.

Evaluation of (12.83) at a point $(r_\perp, x_3)_S$ on a meridional cross section of Earth's WGS 84 reference ellipsoidal surface gives

$$\left(\frac{x_3}{a}\right)_S = D\left(\frac{r_\perp}{a}\right)_S^{1+\mu} \exp\left[\frac{\nu}{2}\left(\frac{r_\perp^2 + x_3^2}{a^2}\right)_S\right] \quad \Rightarrow \quad D = \frac{(x_3/a)_S}{(r_\perp/a)_S^{1+\mu}} \exp\left[-\frac{\nu}{2}\left(\frac{r_\perp^2 + x_3^2}{a^2}\right)_S\right]. \tag{12.84}$$

Thus, eliminating D from (12.83) and (12.84):

The Orthogonal Trajectories to Ellipsoidal Geopotential Surfaces

The (*generally implicit*) equation, in the prototypical $r_\perp$-x_3 meridional plane, for the orthogonal trajectories to ellipsoidal-geopotential surfaces is

$$\frac{x_3}{(x_3)_S} = \frac{r_\perp^{1+\mu}}{\left(r_\perp^{1+\mu}\right)_S} \exp\left\{\frac{\nu}{2}\left[\frac{(r_\perp^2 + x_3^2)}{a^2} - \frac{(r_\perp^2 + x_3^2)_S}{a^2}\right]\right\}, \tag{12.85}$$

where $r_\perp^2 = x_1^2 + x_2^2$ from (12.77) and (12.78).

This equation provides a critical component for the construction (in Section 12.9) of analytically defined, orthogonal, geopotential coordinates.

As a cross-check, taking the limit $\mu, \nu \to 0$ in (12.84) and (12.85) yields the spherical result $x_3 = Dr_\perp = (x_3/r_\perp)_S r_\perp$; that is, the orthogonal trajectories to concentric spheres are straight lines that emanate from the sphere's centre.

12.9 GREAT COORDINATES

12.9.1 Overview

Setting up an orthogonal coordinate system based on ellipsoidal-geopotential surfaces requires the ellipsoids and their orthogonal surfaces to be identified and labelled. In familiar spherical-polar coordinates (λ, ϕ, r) – and referring to Fig. 5.3 in Section 5.2.2 – the three orthogonal coordinates are:

- *Longitude* (λ), measuring angular distance eastwards, from the (dark-shaded) reference meridional plane, along the latitude circle passing through point P.
- *Latitude* (ϕ), measuring angular distance northward (for positive ϕ) from the equator, along the meridian containing point P, and southward for negative ϕ.
- *Radial distance* (r), measured from origin O at the sphere's centre.

Each spherical-polar coordinate (λ, ϕ, or r) increases in the direction of its associated unit vector ($\mathbf{e}_\lambda$, $\mathbf{e}_\phi$, or $\mathbf{e}_r$, respectively).

For a mildly oblate planet, such as Earth, m and $\tilde{\varepsilon}$ are small. Consequently GREAT coordinates (λ, φ, R) can be considered to be a relatively small geometrical distortion of spherical-polar coordinates (λ, ϕ, r). It helps understanding to keep this in mind in what follows. For GREAT coordinates:

- *Longitude* (λ) again measures angular distance eastwards – along a latitude circle – from a reference meridional plane, since ellipsoids are surfaces of revolution. The associated (λ = constant) coordinate surface is then the semi-infinite plane within which a meridian is embedded.

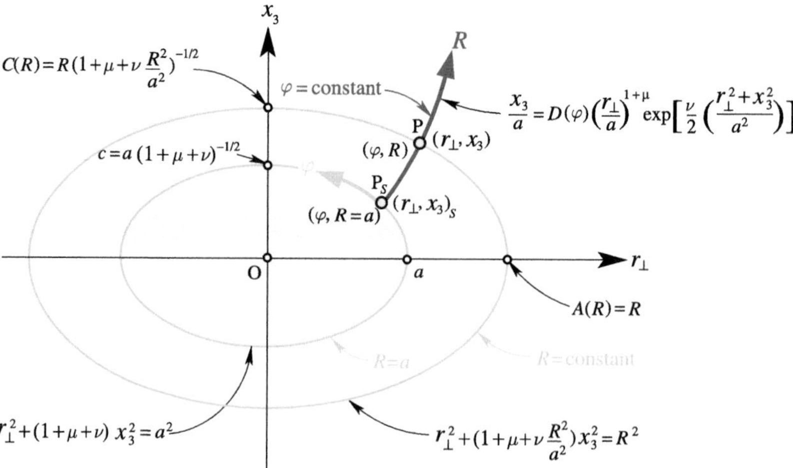

Figure 12.10 Key features of the GREAT coordinate system in a prototypical $r_\perp$-x_3 meridional cross-section; see text for details. After Fig. 2 of Staniforth and White (2015a), *but without length scaling.*

- *Latitude* (φ) measures angular distance (positive when northward, negative when southward) from the equator along a meridian.[31] It is, however, defined slightly differently (with three specific choices given in Section 12.9.5 for generic latitudinal coordinate φ) to account for meridians no longer being precisely semicircular in form, but instead elliptical. See Figs. 12.9–12.11. For a given $D(\varphi)$, defined by (12.84) in terms of surface values $[r_\perp(\varphi)]_S$ and $[x_3(\varphi)]_S$, the associated (φ = constant) coordinate surfaces are obtained by rotating the orthogonal trajectories – defined by (12.83) in a meridional plane – about the planet's polar axis.
- *Quasi-radial distance* (R), labelled by radial distance R in the equatorial plane, from origin O at the ellipsoid's centre. As identified in Section 12.4.2, this specifies the ellipsoid on which a given point lies, in the same way that radius r in spherical-polar coordinates specifies the sphere on which a given point lies. For fixed λ and φ – see Figs. 12.9 and 12.10 – the orthogonal trajectories to ellipsoidal geopotential surfaces are no longer straight, radial lines from a planet's spherical centre, but are instead slightly curved.

Staniforth and White (2015a) proposed a three-dimensional, orthogonal, ellipsoidal, coordinate system, brazenly termed GREAT (Geophysically Realistic, Ellipsoidal, Analytically Tractable), with coordinates λ, φ, and R. They scaled all distances by Earth's equatorial radius a. For clarity, this practice is not, however, followed herein; distances are *not* scaled. The distance ds between two points whose coordinates differ by $d\lambda$, $d\varphi$, and dR is given by – cf. (5.2) –

$$(ds)^2 = h_\lambda^2 (d\lambda)^2 + h_\varphi^2 (d\varphi)^2 + h_R^2 (dR)^2 . \tag{12.86}$$

Metric factors h_λ, h_φ, and h_R in (12.86) are related to Cartesian coordinates x_1, x_2, and x_3 by – cf. (5.4) –

$$h_\lambda^2 = \left(\frac{\partial x_1}{\partial \lambda}\right)^2 + \left(\frac{\partial x_2}{\partial \lambda}\right)^2 + \left(\frac{\partial x_3}{\partial \lambda}\right)^2 , \tag{12.87}$$

[31] Note that φ should not be confused with ϕ. Here, φ is a *generic* latitudinal coordinate, applicable to ellipsoids. It includes parametric (θ), geocentric (χ), and geographic (ϕ) latitudes as special cases. All three latitudes degenerate for the special case of a sphere to the usual latitude ϕ of spherical-polar coordinates. See Section 12.9.5 for further clarification.

$$h_\varphi^2 = \left(\frac{\partial x_1}{\partial \varphi}\right)^2 + \left(\frac{\partial x_2}{\partial \varphi}\right)^2 + \left(\frac{\partial x_3}{\partial \varphi}\right)^2, \tag{12.88}$$

$$h_R^2 = \left(\frac{\partial x_1}{\partial R}\right)^2 + \left(\frac{\partial x_2}{\partial R}\right)^2 + \left(\frac{\partial x_3}{\partial R}\right)^2. \tag{12.89}$$

As noted in White et al. (2008), *metric factors are the key to virtually all differential operations in any orthogonal coordinate system*. Since the ellipsoidal geopotentials correspond to surfaces of revolution, it suffices to restrict attention to a prototypical meridional plane to determine vertical (h_R) and meridional (h_φ) metric factors. Figure 12.10 illustrates key features of the GREAT coordinate system in a prototypical $r_\perp$-x_3 meridional cross-section and is applicable to all cases considered in Section 12.9.5.

12.9.2 Some Preparatory Steps for the Construction of Metric Factors

Intermediate Expressions for the Vertical Metric Factor

Partially differentiating (12.78) with respect to R, with φ held fixed, leads to

$$\left(\frac{\partial x_1}{\partial R}\right)^2 + \left(\frac{\partial x_2}{\partial R}\right)^2 = \left(\frac{\partial r_\perp}{\partial R}\cos\lambda\right)^2 + \left(\frac{\partial r_\perp}{\partial R}\sin\lambda\right)^2 = \left(\frac{\partial r_\perp}{\partial R}\right)^2. \tag{12.90}$$

Insertion of (12.90) into (12.89) then allows the square of the vertical metric factor to be re-expressed as

$$h_R^2 = \left(\frac{\partial r_\perp}{\partial R}\right)^2 + \left(\frac{\partial x_3}{\partial R}\right)^2. \tag{12.91}$$

Intermediate Expressions for the Meridional Metric Factor

Similarly, partially differentiating (12.78) with respect to φ, with R held fixed, leads to

$$\left(\frac{\partial x_1}{\partial \varphi}\right)^2 + \left(\frac{\partial x_2}{\partial \varphi}\right)^2 = \left(\frac{\partial r_\perp}{\partial \varphi}\cos\lambda\right)^2 + \left(\frac{\partial r_\perp}{\partial \varphi}\sin\lambda\right)^2 = \left(\frac{\partial r_\perp}{\partial \varphi}\right)^2. \tag{12.92}$$

Insertion of (12.92) into (12.88) then allows the square of the meridional metric factor to be re-expressed as

$$h_\varphi^2 = \left(\frac{\partial r_\perp}{\partial \varphi}\right)^2 + \left(\frac{\partial x_3}{\partial \varphi}\right)^2. \tag{12.93}$$

Two Logarithmic Forms

It is seen from (12.91) and (12.93) that to determine vertical metric factor h_R and meridional metric factor h_φ, we need appropriate expressions for $\partial r_\perp/\partial R$, $\partial x_3/\partial R$, $\partial r_\perp/\partial \varphi$, and $\partial x_3/\partial \varphi$. To prepare the way for determination of these, the first equation of (12.79) and (12.83) is rewritten in logarithmic form as

$$\ln x_3^2 = \ln\left(R^2 - r_\perp^2\right) - \ln\left(1 + \mu + \nu\frac{R^2}{a^2}\right), \tag{12.94}$$

$$\ln\left(\frac{x_3}{a}\right) = \ln\left[D(\varphi)\right] + (1+\mu)\ln\left(\frac{r_\perp}{a}\right) + \frac{\nu}{2}\left(\frac{r_\perp^2 + x_3^2}{a^2}\right), \tag{12.95}$$

where $D(\varphi)$ in (12.95) is defined by the second equation of (12.84), and $r_\perp$ and x_3 are functions of both meridional coordinate φ (see Section 12.9.5) and vertical (quasi-radial) coordinate R.

However, *explicit* relations for $r_\perp = r_\perp(\varphi, R)$ and $x_3 = x_3(\varphi, R)$ are not generally available, so h_R and h_φ cannot be found *directly* from (12.91) and (12.93). This complicates things somewhat, but the development of GREAT coordinates nevertheless remains analytically tractable, and we just have to work a little harder than would otherwise be the case!

Partial Derivatives with Respect to R

Partially differentiating (12.94) and (12.95) with respect to R, with φ held fixed, and using the first equation of (12.79) to eliminate $r_\perp^2$ yields

$$r_\perp \frac{\partial r_\perp}{\partial R} + \left(1 + \mu + \nu \frac{R^2}{a^2}\right) x_3 \frac{\partial x_3}{\partial R} = R\left(1 - \nu \frac{x_3^2}{a^2}\right), \tag{12.96}$$

$$\left(1 + \mu + \nu \frac{R^2}{a^2}\right) \frac{1}{r_\perp}\frac{\partial r_\perp}{\partial R} - \frac{1}{x_3}\frac{\partial x_3}{\partial R} = 0. \tag{12.97}$$

Solving coupled equations (12.96) and (12.97) for $\partial r_\perp/\partial R$ and $\partial x_3/\partial R$ then gives

$$\frac{\partial r_\perp}{\partial R} = \frac{R}{r_\perp}\left(1 - \nu \frac{x_3^2}{a^2}\right)\left[R^2 - \left(1 + \mu + \nu \frac{R^2}{a^2}\right) x_3^2\right]\left[R^2 + \left(\mu + \nu \frac{R^2}{a^2}\right)\left(1 + \mu + \nu \frac{R^2}{a^2}\right) x_3^2\right]^{-1}, \tag{12.98}$$

$$\frac{\partial x_3}{\partial R} = x_3 R\left(1 - \nu \frac{x_3^2}{a^2}\right)\left(1 + \mu + \nu \frac{R^2}{a^2}\right)\left[R^2 + \left(\mu + \nu \frac{R^2}{a^2}\right)\left(1 + \mu + \nu \frac{R^2}{a^2}\right) x_3^2\right]^{-1}. \tag{12.99}$$

Partial Derivatives with Respect to φ

Similarly, but instead partially differentiating (12.94) and (12.95) with respect to φ, with R held fixed, we get

$$r_\perp \frac{\partial r_\perp}{\partial \varphi} + \left(1 + \mu + \nu \frac{R^2}{a^2}\right) x_3 \frac{\partial x_3}{\partial \varphi} = 0, \tag{12.100}$$

$$\left(1 + \mu + \nu \frac{R^2}{a^2}\right) \frac{1}{r_\perp}\frac{\partial r_\perp}{\partial \varphi} - \frac{1}{x_3}\frac{\partial x_3}{\partial \varphi} = -\left(1 - \nu \frac{x_3^2}{a^2}\right)^{-1} \frac{d\left[\ln D(\varphi)\right]}{d\varphi}. \tag{12.101}$$

Solving coupled equations (12.100) and (12.101) for $\partial r_\perp/\partial \varphi$ and $\partial x_3/\partial \varphi$ then gives

$$\frac{\partial r_\perp}{\partial \varphi} = -r_\perp x_3^2\left(1 + \mu + \nu \frac{R^2}{a^2}\right)\left[R^2 + \left(\mu + \nu \frac{R^2}{a^2}\right)\left(1 + \mu + \nu \frac{R^2}{a^2}\right) x_3^2\right]^{-1}\left(1 - \nu \frac{x_3^2}{a^2}\right)^{-1} \frac{d(\ln D)}{d\varphi}, \tag{12.102}$$

$$\frac{\partial x_3}{\partial \varphi} = r_\perp^2 x_3\left[R^2 + \left(\mu + \nu \frac{R^2}{a^2}\right)\left(1 + \mu + \nu \frac{R^2}{a^2}\right) x_3^2\right]^{-1}\left(1 - \nu \frac{x_3^2}{a^2}\right)^{-1} \frac{d(\ln D)}{d\varphi}. \tag{12.103}$$

12.9.3 The Vertical Metric Factor

Using (12.98) and (12.99) in (12.91), the vertical metric factor is given by

$$h_R \equiv \left[\left(\frac{\partial r_\perp}{\partial R}\right)^2 + \left(\frac{\partial x_3}{\partial R}\right)^2\right]^{\frac{1}{2}} = R\left(1 - \nu \frac{x_3^2}{a^2}\right)\left[R^2 + \left(\mu + \nu \frac{R^2}{a^2}\right)\left(1 + \mu + \nu \frac{R^2}{a^2}\right) x_3^2\right]^{-\frac{1}{2}}. \tag{12.104}$$

12.9.4 The Meridional Metric Factor

Using (12.102) and (12.103) in (12.93), the meridional metric factor *in generic form* is then

$$h_\varphi \equiv \left[\left(\frac{\partial r_\perp}{\partial \varphi}\right)^2 + \left(\frac{\partial x_3}{\partial \varphi}\right)^2\right]^{\frac{1}{2}}$$
$$= r_\perp x_3 \left[R^2 + \left(\mu + \nu\frac{R^2}{a^2}\right)\left(1 + \mu + \nu\frac{R^2}{a^2}\right)x_3^2\right]^{-\frac{1}{2}}\left(1 - \nu\frac{x_3^2}{a^2}\right)^{-1}\frac{d\left(\ln D\right)}{d\varphi}. \tag{12.105}$$

Here, $D(\varphi)$ is defined by the second equation of (12.84), that is, by

$$D(\varphi) = \frac{\left[x_3(\varphi)/a\right]_S}{\left[r_\perp(\varphi)/a\right]_S^{1+\mu}} \exp\left\{-\frac{\nu}{2}\left[\frac{r_\perp^2(\varphi) + x_3^2(\varphi)}{a^2}\right]_S\right\}, \tag{12.106}$$

and φ is the generic meridional coordinate, yet to be specified (see Section 12.9.5).

Taking logarithms of both sides of (12.106) and differentiating the resulting equation with respect to φ yields

$$\frac{d\left(\ln D\right)}{d\varphi} = \left(1 - \nu\frac{x_3^2}{a^2}\right)_S \frac{d}{d\varphi}\left[\ln\left(\frac{x_3}{a}\right)\right]_S - \left(1 + \mu + \nu\frac{r_\perp^2}{a^2}\right)_S \frac{d}{d\varphi}\left[\ln\left(\frac{r_\perp}{a}\right)\right]_S. \tag{12.107}$$

This expression can be used to eliminate $d\left[\ln D(\varphi)\right]/d\varphi$ from (12.105) in favour of $\left[r_\perp(\varphi)\right]_S$ and $\left[x_3(\varphi)\right]_S$ and their derivatives.

12.9.5 Three Specific Choices of Meridional Coordinate

The role of φ is to act as a meridional, horizontal label, in the same way that R acts as a vertical label; and to characterise a particular orthogonal trajectory embedded in a meridional plane. Staniforth and White (2015a)'s characterisation involves the location of the intersection of the orthogonal trajectory with Earth's WGS 84 reference ellipsoid. Following Staniforth and White (2015a), evaluations of quantities on this ellipsoid are denoted by subscript or superscript 'S' – according to notational compactness and convenience – and the intersection is considered as a function of its geocentric (χ_S), parametric (θ_S), and geographic (ϕ_S) latitudes. Each of these three latitudes amounts to a different choice of generic meridional coordinate φ. See Fig. 12.11 for a graphical depiction of these in the x_1-x_3 meridional plane (this corresponds to the reference meridional plane of the WGS 84 reference ellipsoid). Expressions for the corresponding Cartesian expressions $\left[r_\perp(\varphi)\right]_S$ and $\left[x_3(\varphi)\right]_S$ in a $r_\perp$-x_3 meridional plane may be obtained from Table 8.1 of Chapter 8 for:

- Geocentric latitude (χ_S); by setting $\lambda \equiv 0$ and $\chi \to \chi_S$.
- Parametric latitude (θ_S); by setting $\lambda \equiv 0$ and $\theta \to \theta_S$.
- Geographic latitude (ϕ_S); by setting $\lambda \equiv 0$ and $\phi \to \phi_S$.

The resulting expressions are displayed in Table 12.1. In conjunction with the second equation of (12.84), these expressions can then be used to determine corresponding expressions for $D(\varphi)$ and $d\left[\ln D(\varphi)\right]/d\varphi$; results are summarised in Tables 12.2 and 12.3. The equatorial and polar singularities seen in $d\left[\ln D(\varphi)\right]/d\varphi$ in Table 12.3 are neutralised by the $r_\perp x_3$ factor that occurs in (12.105) when this equation is used to evaluate h_φ.

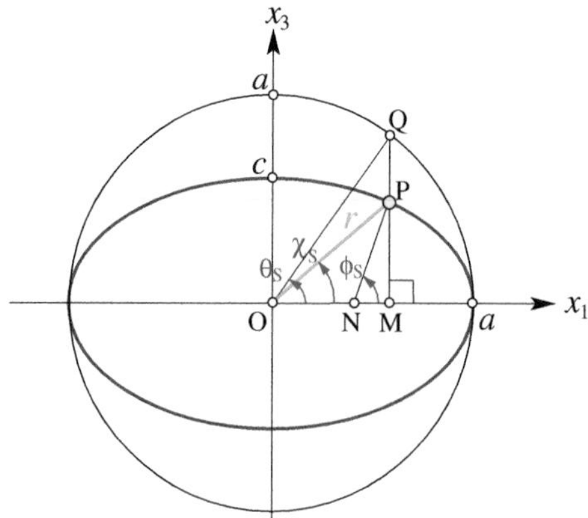

Figure 12.11 An ellipse in the Ox_1x_3 meridional plane, having semi-major and semi-minor axes a and c, respectively, and inscribed in a circle of radius a. P is a general point on the ellipse with coordinates (x_1, x_3). N is the point at which the normal to the ellipse at the general point P (x_1, x_3) intersects the x_1 axis. θ, χ, and ϕ are *parametric, geocentric*, and *geographic* latitudes, respectively. The line QPM is parallel to Ox_3 and intersects the circle and Ox_1 at Q and M respectively. The equation of the ellipse is $\left(x_1^2/a^2\right) + \left(x_3^2/c^2\right) = 1$, and $r = \left(x_1^2 + x_3^2\right)^{1/2}$ is the distance of P from the origin O along the green line. After Fig. 3 of Staniforth and White (2015a).

φ	$[r_\perp(\varphi)]_S$	$[x_3(\varphi)]_S$	$\dfrac{[x_3(\varphi)]_S}{[r_\perp(\varphi)]_S}$
χ_S	$\dfrac{ac\cos\chi_S}{\left(a^2\sin^2\chi_S + c^2\cos^2\chi_S\right)^{\frac{1}{2}}}$	$\dfrac{ac\sin\chi_S}{\left(a^2\sin^2\chi_S + c^2\cos^2\chi_S\right)^{\frac{1}{2}}}$	$\tan\chi_S$
θ_S	$a\cos\theta_S$	$c\sin\theta_S$	$\dfrac{c}{a}\tan\theta_S$
ϕ_S	$\dfrac{a^2\cos\phi_S}{\left(a^2\cos^2\phi_S + c^2\sin^2\phi_S\right)^{\frac{1}{2}}}$	$\dfrac{c^2\sin\phi_S}{\left(a^2\cos^2\phi_S + c^2\sin^2\phi_S\right)^{\frac{1}{2}}}$	$\dfrac{c^2}{a^2}\tan\phi_S$

Table 12.1 Expressions for $[r_\perp(\varphi)]_S$, $[x_3(\varphi)]_S$, and their quotient for the meridional coordinates: χ_S (geocentric latitude), θ_S (parametric latitude), and ϕ_S (geographic latitude). Subscript 'S' denotes evaluation on Earth's WGS 84 ellipsoidal reference surface, with semi-major and semi-minor axes a and c, respectively. After Table 1 of Staniforth and White (2015a).

It is emphasised that φ = constant everywhere along an orthogonal trajectory in GREAT coordinates (i.e. everywhere along the red curve on Figs. 12.10 and 12.12). Everywhere along this curve it has the value $\varphi = \chi_S$, $\varphi = \theta_S$, or $\varphi = \phi_S$, according to the specific choice made of meridional coordinate φ. Consequently it has exactly the same value at point P on this schematic as it does at point P_S. It does *not* have the value $\varphi = \chi$, $\varphi = \theta$, or $\varphi = \phi$ (without an 'S' subscript) at point P, but the value $\varphi = \chi_S$, $\varphi = \theta_S$, or $\varphi = \phi_S$. For example – and referring to Fig. 12.12 – for the choice $\varphi = \chi_S$, the physical interpretation of χ_S at point P is that it is the value of geocentric latitude at the intersection of the orthogonal trajectory (in red), passing through point P, with the WGS 84 reference ellipsoid $R = a$ (in blue), that is, it is its value (χ_S) at point P_S. It is *not* the value (χ) of geocentric latitude at point P, which is different; cf. the two green angles on Fig. 12.12. This means that one has to be very careful to use the correct geometrical relations when transforming between GREAT coordinates and Cartesian coordinates, and to make the distinction between surface values – for example, χ_S – and non-surface values – for example, χ.

φ	$D(\varphi)$
χ_S	$\dfrac{\left(a^2 \sin^2 \chi_S + c^2 \cos^2 \chi_S\right)^{\mu/2} \sin \chi_S}{c^{\mu} \cos^{1+\mu} \chi_S} \exp\left[-\dfrac{\nu c^2}{2\left(a^2 \sin^2 \chi_S + c^2 \cos^2 \chi_S\right)}\right]$
θ_S	$\dfrac{c \sin \theta_S}{a \cos^{1+\mu} \theta_S} \exp\left[-\dfrac{\nu}{2}\left(\dfrac{a^2 \cos^2 \theta_S + c^2 \sin^2 \theta_S}{a^2}\right)\right]$
ϕ_S	$\dfrac{c^2\left(a^2 \cos^2 \phi_S + c^2 \sin^2 \phi_S\right)^{\mu/2} \sin \phi_S}{a^{2+\mu} \cos^{1+\mu} \phi_S} \exp\left[-\dfrac{\nu}{2a^2}\left(\dfrac{a^4 \cos^2 \phi_S + c^4 \sin^2 \phi_S}{a^2 \cos^2 \phi_S + c^2 \sin^2 \phi_S}\right)\right]$

Table 12.2 $D(\varphi)$ for three meridional coordinates: χ_S (geocentric latitude), θ_S (parametric latitude), and ϕ_S (geographic latitude). Subscript 'S' denotes evaluation on Earth's WGS 84 ellipsoidal reference surface, with semi-major and semi-minor axes a and c, respectively. After Table 2 of Staniforth and White (2015a), but corrected here.

φ	$\dfrac{d\left[\ln D(\varphi)\right]}{d\varphi}$
χ_S	$\dfrac{1}{\sin \chi_S \cos \chi_S} + \mu \dfrac{a^2 \tan \chi_S}{\left(a^2 \sin^2 \chi_S + c^2 \cos^2 \chi_S\right)} + \nu \dfrac{\left(a^2 - c^2\right) c^2 \sin \chi_S \cos \chi_S}{\left(a^2 \sin^2 \chi_S + c^2 \cos^2 \chi_S\right)^2}$
θ_S	$\dfrac{1}{\sin \theta_S \cos \theta_S} + \mu \tan \theta_S + \nu \dfrac{\left(a^2 - c^2\right)}{a^2} \sin \theta_S \cos \theta_S$
ϕ_S	$\dfrac{1}{\sin \phi_S \cos \phi_S} + \mu \dfrac{c^2 \tan \phi_S}{\left(a^2 \cos^2 \phi_S + c^2 \sin^2 \phi_S\right)} + \nu \dfrac{\left(a^2 - c^2\right) c^2 \sin \phi_S \cos \phi_S}{\left(a^2 \cos^2 \phi_S + c^2 \sin^2 \phi_S\right)^2}$

Table 12.3 As in Table 12.2, but for $d\left[\ln D(\varphi)\right]/d\varphi$. After Table 3 of Staniforth and White (2015a).

12.9.6 The Zonal Metric Factor

It remains to give an expression for h_λ. Substituting (12.78) into (12.87) leads to

$$h_\lambda = \left(x_1^2 + x_2^2\right)^{\frac{1}{2}} = r_\perp. \tag{12.108}$$

As expected – see Fig. 12.9 – the zonal metric factor h_λ is simply the perpendicular distance, $r_\perp \equiv \left(x_1^2 + x_2^2\right)^{1/2}$, from the planet's rotation axis.

12.9.7 GREAT Metric Factors Summarised

From (12.108), (12.105), and (12.104):

The Metric Factors for GREAT Coordinates

The zonal, meridional, and vertical metric factors $\left(h_\lambda, h_\varphi, h_R\right)$ at any point $\left(\lambda_i, \varphi_j, R_k\right)$ for GREAT coordinates can be written as

$$h_\lambda = r_\perp = \left(x_1^2 + x_2^2\right)^{\frac{1}{2}}, \tag{12.109}$$

$$h_\varphi = \frac{h_\lambda h_R x_3}{R}\left(1 - \nu \frac{x_3^2}{a^2}\right)^{-2} \frac{d\left(\ln D\right)}{d\varphi}, \tag{12.110}$$

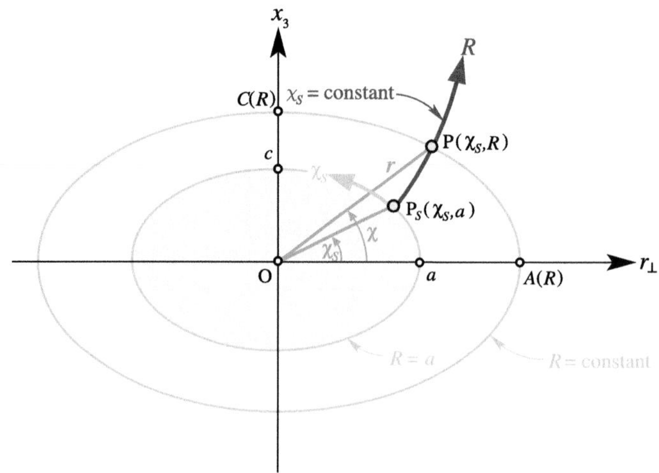

Figure 12.12 Distinguishing between geocentric latitudes χ and χ_S. Expressed in geocentric coordinates, Point P is located at (χ, r), whereas in GREAT coordinates it is located at (χ_S, R). See main text for further details.

$$h_R = R\left[R^2 + \left(\mu + \nu\frac{R^2}{a^2}\right)\left(1 + \mu + \nu\frac{R^2}{a^2}\right)x_3^2\right]^{-\frac{1}{2}}\left(1 - \nu\frac{x_3^2}{a^2}\right), \tag{12.111}$$

where h_φ in (12.110) has been rewritten more compactly using (12.108) for h_λ. In these expressions, all quantities are evaluated at a point $(\lambda_i, \varphi_j, R_k)$.

12.9.8 Transforming between the (λ, φ, R) and (x_1, x_2, x_3) Coordinate Systems

Since (12.109)–(12.111) involve (x_1, x_2, x_3) as well as (λ, φ, R), evaluating GREAT metric factors $(h_\lambda, h_\varphi, h_R)$ requires values of *both* sets of coordinates. If (λ, φ, R) are known at a set of gridpoints, then (x_1, x_2, x_3) must be calculated at the same points, and vice versa. A procedure to transform between the two coordinate systems is therefore required; this may also be of use in other contexts.

Transformation from GREAT (λ, φ, R) Coordinates to Cartesian (x_1, x_2, x_3) Coordinates

In the prototypical meridional plane, and referring to summary Fig. 12.10:

- The ellipsoidal-geopotential surfaces satisfy the first equation of (12.79).
- The orthogonal trajectories satisfy (12.83), where $D(\varphi)$ is defined by (12.106).

Squaring (12.83) and using the first equation of (12.79) to eliminate $r_\perp^2$ leads to the transcendental equation

$$\begin{aligned}\left(\frac{x_3}{a}\right)^2 &= D^2(\varphi)\left[\left(\frac{r_\perp}{a}\right)^2\right]^{1+\mu}\exp\left\{\nu\left[\frac{R^2}{a^2} - \left(\mu + \nu\frac{R^2}{a^2}\right)\left(\frac{x_3}{a}\right)^2\right]\right\},\\ &= D^2(\varphi)\left[\frac{R^2}{a^2} - \left(1 + \mu + \nu\frac{R^2}{a^2}\right)\left(\frac{x_3}{a}\right)^2\right]^{1+\mu}\exp\left\{\nu\left[\frac{R^2}{a^2} - \left(\mu + \nu\frac{R^2}{a^2}\right)\left(\frac{x_3}{a}\right)^2\right]\right\},\end{aligned} \tag{12.112}$$

for $(x_3/a)^2$ in terms of GREAT coordinates φ and R. For the intersection point of the $\varphi = \varphi_j$ orthogonal trajectory with the $R = R_k$ geopotential surface, that is, for the point (φ_j, R_k) in the prototypical meridional plane of Fig. 12.10, (12.112) may be solved iteratively (using an appropriate numerical algorithm) to obtain $(x_3)_{jk} \equiv x_3 \left(\varphi = \varphi_j, R = R_k\right)$ to machine precision. $\left(r_\perp^2\right)_{jk} \equiv r_\perp^2 \left(\varphi = \varphi_j, R = R_k\right)$ can then be obtained from the first equation of (12.79).

More generally, at a given point $\left(\lambda_i, \varphi_j, R_k\right)$ in three dimensions, (12.112) is numerically solved to machine precision for $(x_3)_{ijk} \equiv x_3 \left(\lambda = \lambda_i, \varphi = \varphi_j, R = R_k\right)$. Cartesian coordinates $(x_1)_{ijk}$ and $(x_2)_{ijk}$ can then be obtained from (12.78); that is, from

$$(x_1)_{ijk} = (r_\perp)_{ijk} \cos \lambda_i, \quad (x_2)_{ijk} = (r_\perp)_{ijk} \sin \lambda_i. \tag{12.113}$$

Thus, for every point $\left(\lambda_i, \varphi_j, R_k\right)$, the corresponding Cartesian coordinates $\left[(x_1)_{ijk}, (x_2)_{ijk}, (x_3)_{ijk}\right]$ can be found by successively using (12.112), the first equation of (12.79), and (12.113).

The complete procedure to transform from GREAT coordinates (λ, φ, R) to Cartesian coordinates (x_1, x_2, x_3) is summarised in Algorithm 12.1.

Transformation from Cartesian (x_1, x_2, x_3) Coordinates to (λ, φ, R) Coordinates

At a given point $\left[(x_1)_I, (x_2)_J, (x_3)_K\right]$ in the Cartesian coordinate system, the corresponding vertical coordinate, R_{IJK}, in the GREAT coordinate system can be obtained from the second equation of (12.79), that is, from

$$R_{IJK}^2 = \frac{\left(x_1^2\right)_I + \left(x_2^2\right)_J + (1+\mu)\left(x_3^2\right)_K}{1 - \nu\left(x_3^2/a^2\right)_K}, \tag{12.114}$$

where – from (12.77) – $r_\perp^2$ has been written explicitly as $r_\perp^2 \equiv x_1^2 + x_2^2$.

Eliminating $r_\perp$ from the two equations of (12.78), the zonal coordinate λ_{IJK} in the GREAT coordinate system can be obtained from

$$\lambda_{IJK} = \tan^{-1}\left[\frac{(x_2)_J}{(x_1)_I}\right] = \cot^{-1}\left[\frac{(x_1)_I}{(x_2)_J}\right]. \tag{12.115}$$

For small values of $(x_2)_J$, the middle expression of (12.115) should be used to avoid singular behaviour. Similarly, for small values of $(x_1)_I$, the last expression of (12.115) should be used instead to again avoid singular behaviour.

To determine φ_{IJK}, the orthogonal trajectory that passes through the point $\left[(x_1)_I, (x_2)_J, (x_3)_K\right]$ is tracked back to its intersection with the WGS 84 reference ellipsoid; see Fig. 12.10. To do so, (12.83) – with $r_\perp^2 \equiv \left(x_1^2 + x_2^2\right)$ – is first solved for $D\left(\varphi_{IJK}\right)$ to give

$$D\left(\varphi_{IJK}\right) = \frac{(x_3/a)_K}{\left\{\left[\left(x_1^2\right)_I + \left(x_2^2\right)_J\right]/a^2\right\}^{(1+\mu)/2}} \exp\left\{-\frac{\nu}{2}\left[\frac{\left(x_1^2\right)_I + \left(x_2^2\right)_J + \left(x_3^2\right)_K}{a^2}\right]\right\}. \tag{12.116}$$

It now suffices to examine what happens in the prototypical meridional plane of Fig. 12.10. The intersection of the orthogonal trajectory with the WGS 84 reference ellipsoid – this lies within the prototypical meridional plane – is obtained by solving (12.112), with R_{IJK} set to its surface value of $R = a$ – using an appropriate numerical algorithm – to obtain $(x_3)_S$ to machine precision. $\left[(r_\perp)_S\right]_{IJK}$ can then be recovered from (12.79) (with R_{IJK}^2 set to its surface value of $R^2 = a^2$, and with $r_\perp^2$ replaced by $\left(r_\perp^2\right)_S$). Using these values of $(r_\perp)_S$ and $(x_3)_S$ and noting that $\varphi =$ constant along an orthogonal trajectory, the value of φ can be obtained from the last column of Table 12.1 according to the specific meridional coordinate employed (χ_S, θ_S, or ϕ_S).

The complete procedure to transform from Cartesian coordinates (x_1, x_2, x_3) to GREAT coordinates (λ, φ, R) is summarised in Algorithm 12.2.

The Cartesian-to-ellipsoidal transformation is explicit except for the final step, in which φ is determined from $(x_3)_S$ and $(r_\perp)_S$. (Note that φ could be evaluated explicitly if $D(\varphi)$ were to be chosen as an easily invertible function of φ rather than via the procedure recommended in Section 12.9.5; see Table 12.2.)

Insofar as they are required for evaluation of the metric factors, these transformations need be done only once for a fixed grid of points in a numerical model.

12.9.9 The Spherical Case as an Asymptotic Limit

In a manner similar to that employed in Section 6 of Staniforth (2014b), Staniforth and White (2015a) found that simultaneously taking the limits $\widetilde{\varepsilon}, m \to 0$ throughout leads to

$$\Phi(R) = -\frac{\gamma M}{R}, \quad \mathbf{g} \equiv -\nabla\Phi = -\frac{\partial\Phi}{\partial r}\mathbf{e}_R = -\frac{\gamma M}{R^2}\mathbf{e}_R, \tag{12.117}$$

$$\varphi = \phi_S = \chi_S = \theta_S, \tag{12.118}$$

$$x_1 = R\cos\phi_S\cos\lambda, \quad x_2 = R\cos\phi_S\sin\lambda, \quad x_3 = R\sin\phi_S, \tag{12.119}$$

$$D(\varphi) = \tan\phi_S, \quad \frac{d(\ln D)}{d\varphi} = \frac{1}{\sin\phi_S\cos\phi_S}, \tag{12.120}$$

$$h_\lambda = R\cos\phi_S, \quad h_\varphi = R, \quad h_R = 1, \tag{12.121}$$

where $\mathbf{e}_R$ is a unit vector in the R direction.

Thus, in the asymptotic limit $\widetilde{\varepsilon}, m \to 0$:

- The ellipsoidal-geopotential approximation reduces to the classical (deep) spherical-geopotential approximation (12.117).
- The GREAT coordinate system reduces to the spherical-polar one (12.119), with coordinates (λ, ϕ_S, R).

12.10 CONCLUDING REMARKS

The exterior and interior geopotential representations developed in Section 8.5 for a mildly oblate, ellipsoidal planet of variable density have been further developed in Sections 12.5 and 12.6. Exploiting these modified geopotential representations, an orthogonal, ellipsoidal, geopotential coordinate system for atmospheric and oceanic modelling – termed Geophysically Realistic, Ellipsoidal, Analytically Tractable (GREAT) – was developed analytically in Section 12.9 in a self-consistent manner. Thus, by construction, GREAT coordinates, in concert with the modified geopotential representations, satisfy all five of the ideal criteria identified in Section 12.1 as well as orthogonality. They offer greater accuracy than that achievable using the ubiquitous, classical spherical-geopotential approximation and spherical-polar coordinate system.

Over the past fifty years or so, atmospheric and oceanic modelling has undergone a revolution. During the 1970s, the hydrostatic, shallow, primitive equations gradually became the mainstay of operational weather and climate prediction models, replacing the simpler equation sets used hitherto. These models were run at coarse resolutions, with horizontal meshlengths of hundreds of kilometres. As the term primitive suggests, it was believed at the time that it would forever be impractical to integrate more complete equation sets. However, as computing power greatly increased in concert with significant improvements in the efficiency of numerical methods, it has proven advantageous to use more general equation sets with greatly increased resolution.

The rationale for this evolution is that, as models become increasingly accurate, the assumptions of simplified equation sets at given resolution progressively lose their validity as resolution is increased. In particular, the hydrostatic and shallow assumptions of the primitive equations lose their validity, and many modelling centres are now using or developing more general, deep, non-hydrostatic models.

At the time of writing, almost all (and possibly all) such models continue to use the classical spherical-geopotential approximation. The limitations of this approximation are summarised in Section 12.1. They include:

- Neglect of meridional variation of gravity.
- Incompatibility with the WGS 84 reference ellipsoid widely used for specifying the position of atmospheric and oceanic data in three-dimensional space.

Algorithm 12.1 Transformation from GREAT (λ, φ, R) coordinates to Cartesian (x_1, x_2, x_3) coordinates.

For given GREAT coordinates $(\lambda_i, \varphi_j, R_k)$:

1. From Table 12.2, compute $D(\varphi_j)$ according to the specific meridional coordinate employed ($\varphi = \chi_S$, $\varphi = \theta_S$, or $\varphi = \phi_S$), that is,

$$D\left(\chi_j^S\right) = \frac{\left(a^2 \sin^2 \chi_j^S + c^2 \cos^2 \chi_j^S\right)^{\mu/2} \sin \chi_j^S}{a^\mu c^\mu \cos^{1+\mu} \chi_j^S} \exp\left[-\frac{\nu c^2}{2\left(a^2 \sin^2 \chi_j^S + c^2 \cos^2 \chi_j^S\right)}\right], \tag{12.122}$$

$$\text{or} \quad D\left(\theta_j^S\right) = \frac{c \sin \theta_j^S}{a^{1+\mu} \cos^{1+\mu} \theta_j^S} \exp\left[-\frac{\nu}{2}\left(\frac{a^2 \cos^2 \theta_j^S + c^2 \sin^2 \theta_j^S}{a^2}\right)\right], \tag{12.123}$$

$$\text{or} \quad D\left(\phi_j^S\right) = \frac{c^2\left(a^2 \cos^2 \phi_j^S + c^2 \sin^2 \phi_j^S\right)^{\mu/2} \sin \phi_j^S}{a^{2(1+\mu)} \cos^{1+\mu} \phi_j^S} \exp\left[-\frac{\nu}{2a^2}\left(\frac{a^4 \cos^2 \phi_j^S + c^4 \sin^2 \phi_j^S}{a^2 \cos^2 \phi_j^S + c^2 \sin^2 \phi_j^S}\right)\right]. \tag{12.124}$$

2. From (12.112), and using an appropriate iterative numerical algorithm, solve s

$$\left(\frac{x_3}{a}\right)_{ijk}^2 = D^2(\varphi_j)\left[\frac{R_k^2}{a^2} - \left(1+\mu+\nu\frac{R_k^2}{a^2}\right)\left(\frac{x_3}{a}\right)_{ijk}^2\right]^{1+\mu} \exp\left\{\nu\left[\frac{R_k^2}{a^2} - \left(\mu+\nu\frac{R_k^2}{a^2}\right)\left(\frac{x_3}{a}\right)_{ijk}^2\right]\right\}, \tag{12.125}$$

for $(x_3/a)_{ijk}^2$. To obtain $(x_3)_{ijk}$, take the positive root for $\varphi_j \geq 0$, and the negative one for $\varphi_j \leq 0$.

3. From the first equation of (12.79), compute

$$(r_\perp)_{ijk} = \left[R_k^2 - \left(1 + \mu + \nu\frac{R_k^2}{a^2}\right)(x_3)_{ijk}^2\right]^{\frac{1}{2}}, \tag{12.126}$$

taking the positive root since $r_\perp \geq 0$.

4. From (12.78), compute

$$(x_1)_{ijk} = (r_\perp)_{ijk} \cos \lambda_i, \quad x_2 = (r_\perp)_{ijk} \sin \lambda_i. \tag{12.127}$$

As written, this algorithm is applicable everywhere *except near the poles* due to the polar singularities in (12.122)–(12.124) for $D(\varphi_j = \pm\pi/2)$. For application in polar regions (poleward of some arbitrary near-polar latitude), the singularity in (12.125) may be removed by multiplying through by $\cos^{2(1+\mu)} \varphi_j$, followed by rearrangement. Doing so then gives the alternative form

$$\left[\frac{R_k^2}{a^2} - \left(1 + \mu + \nu\frac{R_k^2}{a^2}\right)\left(\frac{x_3}{a}\right)_{ijk}^2\right]^{1+\mu} = \frac{\cos^{2(1+\mu)} \varphi_j}{\cos^{2(1+\mu)} \varphi_j D^2(\varphi_j)}\left(\frac{x_3}{a}\right)_{ijk}^2 \exp\left\{-\nu\left[\frac{R_k^2}{a^2} - \left(\mu + \nu\frac{R_k^2}{a^2}\right)\left(\frac{x_3}{a}\right)_{ijk}^2\right]\right\}. \tag{12.128}$$

The division by $\cos^{2(1+\mu)} \varphi_j D^2(\varphi_j)$ on the right-hand side of this equation is non-singular at and near the poles. At the poles, the multiplication by $\cos^{2(1+\mu)} \varphi_j$ makes the right-hand zero, thereby reassuringly leading to the (exact, known) solution

$$(x_3)_{ijk}^2 = \left(1 + \mu + \nu\frac{R_k^2}{a^2}\right)^{-1} R_k^2. \tag{12.129}$$

Algorithm 12.2 Transformation from Cartesian (x_1, x_2, x_3) coordinates to GREAT (λ, φ, R) coordinates.

For given Cartesian coordinates $\left[(x_1)_I, (x_2)_J, (x_3)_K\right]$:

1. From (12.114), compute

$$R_{IJK} = \left[\frac{\left(x_1^2\right)_I + \left(x_2^2\right)_J + (1+\mu)\left(x_3^2\right)_K}{1 - \nu\left(x_3^2/a^2\right)_K}\right]^{\frac{1}{2}}. \tag{12.130}$$

2. From (12.115), compute

$$\lambda_{IJK} = \tan^{-1}\left[\frac{(x_2)_J}{(x_1)_I}\right] = \cot^{-1}\left[\frac{(x_1)_I}{(x_2)_J}\right]. \tag{12.131}$$

3. From (12.116), compute

$$D\left(\varphi_{IJK}\right) = \frac{(x_3/a)_K}{\left\{\left[\left(x_1^2\right)_I + \left(x_2^2\right)_J\right]/a^2\right\}^{(1+\mu)/2}} \exp\left\{-\frac{\nu}{2}\left[\frac{\left(x_1^2\right)_I + \left(x_2^2\right)_J + \left(x_3^2\right)_K}{a^2}\right]\right\}. \tag{12.132}$$

4. From (12.112), evaluated at $R = a$ – this corresponds to the surface of the WGS 84 reference ellipsoid – solve

$$(x_3)_S = aD\left(\varphi_{IJK}\right)\left[1 - (1+\mu+\nu)\left(\frac{x_3}{a}\right)_S^2\right]^{\frac{1+\mu}{2}} \exp\left\{\frac{\nu}{2}\left[1 - (\mu+\nu)\left(\frac{x_3}{a}\right)_S^2\right]\right\}, \tag{12.133}$$

for $(x_3)_S = \left[(x_3)_S\right]_{IJK}$.

5. From the first equation of (12.79), evaluated at $R = a$ – this corresponds to the surface of the WGS 84 reference ellipsoid – compute

$$\left[(r_\perp)_S\right]_{IJK} = \left\{a^2 - (1+\mu+\nu)\left[(x_3)_S^2\right]_{IJK}\right\}^{\frac{1}{2}}, \tag{12.134}$$

where $(r_\perp)_S$ is the value of $r_\perp$ at the surface of the WGS 84 reference ellipsoid.

6. From the last column of Table 12.1, compute $\varphi_{IJK} = (\varphi_S)_{IJK}$ according to the specific meridional coordinate employed ($\varphi = \chi_S$, $\varphi = \theta_S$ or $\varphi = \phi_S$); that is,

$$\chi_{IJK} = (\chi_S)_{IJK} = \tan^{-1}\left\{\frac{\left[(x_3)_S\right]_{IJK}}{\left[(r_\perp)_S\right]_{IJK}}\right\} = \cot^{-1}\left\{\frac{\left[(r_\perp)_S\right]_{IJK}}{\left[(x_3)_S\right]_{IJK}}\right\}, \tag{12.135}$$

$$\text{or} \quad \theta_{IJK} = (\theta_S)_{IJK} = \tan^{-1}\left\{\frac{a\left[(x_3)_S\right]_{IJK}}{c\left[(r_\perp)_S\right]_{IJK}}\right\} = \cot^{-1}\left\{\frac{c\left[(r_\perp)_S\right]_{IJK}}{a\left[(x_3)_S\right]_{IJK}}\right\}, \tag{12.136}$$

$$\text{or} \quad \phi_{IJK} = (\phi_S)_{IJK} = \tan^{-1}\left\{\frac{a^2\left[(x_3)_S\right]_{IJK}}{c^2\left[(r_\perp)_S\right]_{IJK}^2}\right\} = \cot^{-1}\left\{\frac{c^2\left[(r_\perp)_S\right]_{IJK}^2}{a^2\left[(x_3)_S\right]_{IJK}}\right\}. \tag{12.137}$$

The singularity in (12.132) when $\left(x_1^2\right)_I + \left(x_2^2\right)_J = 0$ may be similarly removed in an analogous manner to that employed to remove the $\cos^{2(1+\mu)}\varphi_j$ singularity in Algorithm 12.1.

- The introduction of systematic errors.

The next natural evolutionary step is therefore to replace the classical spherical-geopotential approximation by a more general, ellipsoidal one, together with an associated consistent, physically realistic, geopotential coordinate system, such as GREAT coordinates. This would then offer the opportunity to improve not only weather-prediction and climate-prediction models, but also planetary-atmosphere and whole-atmosphere models (Akmaev, 2011)[32].

At the current state of knowledge, it is not known whether, today or at some future time, the error induced by the classical spherical-geopotential approximation remains negligible with respect to other sources of error – different people have very different opinions on this. Development of atmospheric and oceanic models, using GREAT coordinates and associated geopotential representations, would permit quantification and assessment of this error. If this error already is or becomes non-negligible, then continued use of the classical spherical-geopotential approximation would have the perverse consequence that parametrisations of physical processes (e.g. 1 moist convection, 2 radiation) would end up being tuned to compensate for an addressable systematic deficiency in the representation of gravity! It would then be imperative to address the error at source by relaxing the classical spherical-geopotential approximation and adopting an ellipsoidal one instead.

Having now expressed the governing equations for quantitative atmospheric and oceanic modelling in axial-orthogonal-curvilinear/GREAT coordinates, our next task is to consider how to represent orography and bathymetry in such models. This can be accomplished by using generalised vertical coordinates together with appropriate lower and upper boundary conditions that respect fundamental conservation principles. Doing so is thus the subject of Chapter 13.

APPENDIX: THE EQUILIBRIUM DEPTH OF AN OCEAN COVERING A PLANET

van der Toorn and Zimmerman (2008) define the layer depth of an ocean covering a planet to be D.[33] In the present notation their definition corresponds to

$$D(\chi, R) \equiv r(\chi, R) - r_S(\chi) \equiv r(\chi, R) - r(\chi, R = a), \tag{12.138}$$

where $r(\chi, R)$ and $r_S(\chi)$ are defined by (12.36) and (12.38), respectively. Because the ocean is assumed to be in equilibrium, its surface must correspond to some geopotential surface. Let that surface be the geopotential surface $r = r(\chi, R)$. Expanding $r(\chi, R)$ in (12.138) as a 1D Taylor series in $(R - a)/a$ about $r(\chi, R = a)$ and using (12.36) gives

$$\begin{aligned} D(\chi, R) &= a\left[\left(\frac{\partial r}{\partial R}\right)_{(\chi, R=a)}\left(\frac{R-a}{a}\right) + O\left(\frac{R-a}{a}\right)^2\right] \\ &= a\left(\frac{R-a}{a}\right)\left[1 + \left(\widetilde{\varepsilon} - \frac{5m}{2}\right)\sin^2\chi + O\left(\widetilde{\varepsilon}^2, \widetilde{\varepsilon}m, m^2, \frac{R-a}{a}\right)\right]. \end{aligned} \tag{12.139}$$

Thus

$$\begin{aligned} \frac{D(\chi, R)}{D^P(R)} &= \frac{\left[1 + \left(\widetilde{\varepsilon} - \frac{5m}{2}\right)\sin^2\chi + O\left(\widetilde{\varepsilon}^2, \widetilde{\varepsilon}m, m^2, \frac{R-a}{a}\right)\right]}{\left[1 + \left(\widetilde{\varepsilon} - \frac{5m}{2}\right) + O\left(\widetilde{\varepsilon}^2, \widetilde{\varepsilon}m, m^2, \frac{R-a}{a}\right)\right]} \\ &= 1 + \left(\frac{5m}{2} - \widetilde{\varepsilon}\right)\cos^2\chi + O\left(\widetilde{\varepsilon}^2, \widetilde{\varepsilon}m, m^2, \frac{R-a}{a}\right), \end{aligned} \tag{12.140}$$

where

$$D^P(R) \equiv D\left(\chi = \pm\frac{\pi}{2}, R\right) \tag{12.141}$$

is the depth of the fluid at a pole.

[32] The atmospheres of such models are significantly deeper than those of current operational weather prediction models, and the vertical variation (from model bottom to top) of gravity is therefore correspondingly larger.

[33] D here is purely local to this appendix and should not be confused with the function $D(\phi)$ used elsewhere in this chapter.

Taking into account differences in notation $((e^2, \mu) \leftrightarrow (2\widetilde{\varepsilon}, m))$, (12.140) is equivalent to van der Toorn and Zimmerman (2008)'s (51). Thus their derivation, from their spheroidal-geopotential approximation, nicely reduces to the preceding few lines when using ellipsoidal-geopotential approximation (12.35).

For the classical *spherical*-geopotential approximation, simultaneously taking the limits as $\widetilde{\varepsilon}$ and m go to zero leads to loss of the latitudinal variation in (12.140), resulting in uniform depth everywhere. This is expected since geopotential surfaces then deform to become concentric spheres.

13

Vertical Coordinates and Boundary Conditions

ABSTRACT

Generalised vertical coordinates are developed for the deep, geophysical-fluid-dynamical equations. Some special vertical coordinates of interest are noted. These include height and isobaric coordinates, and height-based and pressure-based, terrain-following coordinates. Energy and axial-angular-momentum budgets are derived. It is shown that the implied principles of energy and axial-angular-momentum conservation depend on the lower and upper boundary conditions. For a deep atmosphere of finite extent, unforced within its domain, globally integrated energy conservation is only obtained for a rigid lid, fixed in space and time, or for an elastic lid with zero pressure there. Correspondingly, conservation of globally integrated axial angular momentum is obtained only when height at a rigid or elastic lid is independent of longitude, or (for a rigid lid) pressure is identically zero there. At the elastic interface between the atmosphere and an ocean, energy and axial angular momentum are exchanged in a zero-sum manner. At a rigid lower boundary, no work is done by or against the pressure field, but there is a mountain torque (for orography) or bottom torque (for bathymetry) if the boundary is not flat. Similar results to those summarised here also hold for shallow fluids (atmospheric or oceanic), but with shallow metric factors instead of deep ones. An energy-like invariant for elastic lids at non-zero pressure is derived. Finally, a stationary state is constructed for a deep atmosphere with zero pressure at finite height.

13.1 PREAMBLE

Now that we have the governing equations in axial-orthogonal-curvilinear/GREAT coordinates, we need to consider how to represent orography and bathymetry. This can be accomplished by *exactly* transforming from axial-orthogonal-curvilinear/GREAT coordinates to boundary-following coordinates; 'boundary' here corresponds to bathymetry for an ocean and, for the atmosphere, to its lower boundary with orography (over land) or water (over an ocean). To do so, a vertical coordinate is defined whose lowest 'horizontal' coordinate surface coincides with the boundary. This is depicted in Fig. 13.1 for a local vertical cross section over an underlying smooth orography/bathymetry beneath a flat, rigid upper boundary. More generally, an elastic upper boundary is also possible, albeit this has ramifications for conservation; see Section 13.6. Note that for conciseness, and for want of a better term, 'elastic boundary' is used informally here to simply mean a boundary that moves and deforms as a function of time, in contradistinction to a rigid boundary that does not.[1] Although the underlying primary motivation for a boundary-following coordinate is to naturally incorporate the *lower* boundaries of the atmosphere and oceans, somewhat similar considerations also apply at their *upper* boundaries.

[1] It is *not* used in the formal sense of physical elasticity, with associated material stresses and strains.

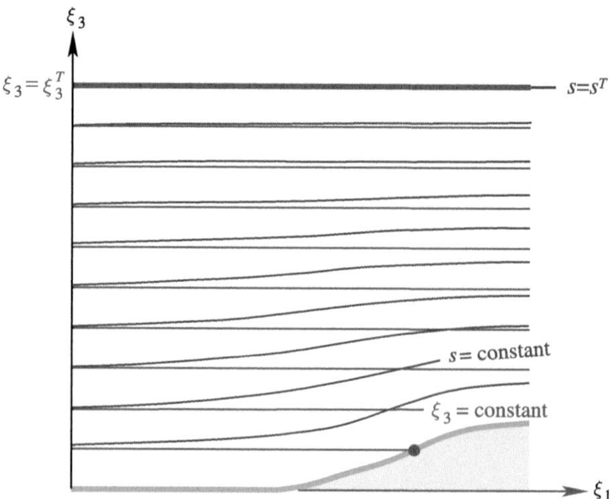

Figure 13.1 A local $\xi_1 - \xi_3$ vertical cross section over an underlying orography/bathymetry (in green) for geopotential (ξ_1, ξ_2, ξ_3, t) coordinates, and for vertically transformed, boundary-following (ξ_1, ξ_2, s, t) coordinates; $\xi_3 =$ constant (geopotential) surfaces are in red, and $s =$ constant surfaces in blue. The red point is the intersection of a geopotential surface with the bottom boundary. The top boundary (uppermost contour) is a rigid lid located at $\xi_3 = \xi_3^T =$ constant, coincident with $s = s^T =$ constant. See text for further details.

At this point, one could proceed to develop a particular boundary-following coordinate; this is often called a *terrain-following* coordinate, from when it was first devised by Phillips (1957) in the atmospheric context of an underlying orography/terrain. We do not, however, do so here since this unnecessarily restricts generality. We instead follow Kasahara (1974) (in his restricted context of the shallow, hydrostatic primitive equations) and Staniforth and Wood (2003) (in their restricted context of a deep atmosphere in spherical geometry), but here in the generalised context of a deep (atmospheric or oceanic) fluid in spheroidal geometry.[2] This approach uses a *generalised vertical coordinate*, defined to be any variable which is a single-valued monotonic function of geometric height. This includes not only boundary-following coordinates but also other (e.g. height, isobaric, and isentropic) coordinates.

It is assumed throughout that spatial variation of the gravitational influence of the mass associated with both the atmosphere and orography (above the geoid) is neglected, as discussed and justified in Section 12.2.6. This mass is assumed to have been 'condensed' to the surface of the WGS 84 reference ellipsoid, and Earth's geoid is assumed to coincide with this ellipsoid.

The plan of this chapter is as follows. In Section 13.2, the governing equations for a rotating, deep geophysical fluid are first given in axial-orthogonal-curvilinear coordinates; this includes GREAT coordinates. They are then transformed using a generalised vertical coordinate so that a variety of vertical coordinate systems ensue as special cases. Mass conservation in transformed coordinates is examined in Section 13.3. The equations governing the evolution of total energy and axial angular momentum are derived (in Sections 13.4 and 13.5 respectively), and the requirements on the lower and upper boundaries for conservation principles for each quantity to exist are determined. The impact of vertical boundary conditions on global conservation are analysed further in Section 13.6. Shallow-fluid equivalents of deep-fluid results are summarised in Section 13.7). An energy-like invariant for elastic lids at *non-zero pressure* is then derived in Section 13.8,

[2] This chapter contains public sector information licensed under the Open Government Licence v1.0. This information has been both adapted and generalised herein.

and a stationary atmospheric state with zero pressure *at finite height* is developed (in Section 13.9) for a deep atmosphere. Concluding remarks are made in Section 13.10. Finally, to avoid overburdening the mathematical developments in the remainder of the chapter, some useful identities are assembled in the Appendix to this chapter.

13.2 THE DEEP-FLUID EQUATIONS AND BOUNDARY CONDITIONS

13.2.1 The Governing Equations in Axial-Orthogonal-Curvilinear Coordinates

It is highly desirable to use a geopotential coordinate system; see Sections 7.3 and 12.1 for why. The horizontal components of $\nabla\Phi$ in the momentum equation are then absent. For brevity, these terms are omitted in the following equations (i.e. *we assume a geopotential coordinate system*).[3]

A set of governing equations for general (atmospheric or oceanic) fluids, expressed in axial-orthogonal-curvilinear coordinates (ξ_1, ξ_2, ξ_3) in a rotating frame of reference, may be written in the notation of Chapter 6. Thus, from (6.29)–(6.40), the governing equations in a geopotential coordinate system – so that $\Phi = \Phi(\xi_3)$ – are:

Momentum

$$\frac{Du_1}{Dt} + \left(\frac{u_1}{h_1} + 2\Omega\right)\frac{u_2}{h_2}\frac{\partial h_1}{\partial \xi_2} + \left(\frac{u_1}{h_1} + 2\Omega\right)\frac{u_3}{h_3}\frac{\partial h_1}{\partial \xi_3} + \frac{1}{\rho h_1}\frac{\partial p}{\partial \xi_1} = F^{u_1}, \tag{13.1}$$

$$\frac{Du_2}{Dt} - \left(\frac{u_1}{h_1} + 2\Omega\right)\frac{u_1}{h_2}\frac{\partial h_1}{\partial \xi_2} - \frac{u_3^2}{h_2 h_3}\frac{\partial h_3}{\partial \xi_2} + \frac{u_2}{h_2}\frac{u_3}{h_3}\frac{\partial h_2}{\partial \xi_3} + \frac{1}{\rho h_2}\frac{\partial p}{\partial \xi_2} = F^{u_2}, \tag{13.2}$$

$$\frac{Du_3}{Dt} + \frac{u_2}{h_2}\frac{u_3}{h_3}\frac{\partial h_3}{\partial \xi_2} - \left(\frac{u_1}{h_1} + 2\Omega\right)\frac{u_1}{h_3}\frac{\partial h_1}{\partial \xi_3} - \frac{u_2^2}{h_2 h_3}\frac{\partial h_2}{\partial \xi_3} + \frac{1}{\rho h_3}\frac{\partial p}{\partial \xi_3} + \frac{1}{h_3}\frac{d\Phi}{d\xi_3} = F^{u_3}. \tag{13.3}$$

Continuity of Total Mass Density

$$\frac{D\rho}{Dt} + \frac{\rho}{h_1 h_2 h_3}\left[\frac{\partial}{\partial \xi_1}(u_1 h_2 h_3) + \frac{\partial}{\partial \xi_2}(u_2 h_3 h_1) + \frac{\partial}{\partial \xi_3}(u_3 h_1 h_2)\right] = F^{\rho}. \tag{13.4}$$

Substance Transport

$$\frac{DS^i}{Dt} = \dot{S}^i, \quad i = 1, 2, \ldots \tag{13.5}$$

Thermodynamic-Energy Equation

$$\frac{D\mathcal{E}}{Dt} + p\frac{D\alpha}{Dt} = \dot{Q}_E \quad \text{or} \quad \frac{D\eta}{Dt} = \frac{\dot{Q}}{T}. \tag{13.6}$$

Definition of Specific Volume

$$\alpha \equiv \frac{1}{\rho}. \tag{13.7}$$

Definition of T, p, and μ^i

$$T = \frac{\partial \mathcal{E}(\alpha, \eta, S^1, S^2, \ldots)}{\partial \eta} \quad \text{or} \quad T = \left[\frac{\partial \eta(\alpha, \mathcal{E}, S^1, S^2, \ldots)}{\partial \mathcal{E}}\right]^{-1}, \tag{13.8}$$

$$p = -\frac{\partial \mathcal{E}(\alpha, \eta, S^1, S^2, \ldots)}{\partial \alpha} \quad \text{or} \quad p = T\frac{\partial \eta(\alpha, \mathcal{E}, S^1, S^2, \ldots)}{\partial \alpha}, \tag{13.9}$$

[3] If needed, the missing terms can be restored by inspection; after transformation, they have the same form that the pressure-gradient terms $(\partial p/\partial \xi_1)/(\rho h_1)$ and $(\partial p/\partial \xi_2)/(\rho h_2)$ of (13.1) and (13.2) have after transformation, but with p replaced by Φ and ρ by unity.

$$\mu^i = \frac{\partial \mathcal{E}\left(\alpha, \eta, S^1, S^2, \ldots\right)}{\partial S^i} \quad \text{or} \quad \mu^i = -T\frac{\partial \eta\left(\alpha, \mathcal{E}, S^1, S^2, \ldots\right)}{\partial S^i}. \tag{13.10}$$

Fundamental Equation of State

$$\mathcal{E} = \mathcal{E}\left(\alpha, \eta, S^1, S^2, \ldots\right) \quad \text{or} \quad \eta = \eta\left(\alpha, \mathcal{E}, S^1, S^2, \ldots\right). \tag{13.11}$$

Kinematic Equation

$$\mathbf{u} \equiv u_1\mathbf{e}_1 + u_2\mathbf{e}_2 + u_3\mathbf{e}_3 = h_1\frac{D\xi_1}{Dt}\mathbf{e}_1 + h_2\frac{D\xi_2}{Dt}\mathbf{e}_2 + h_3\frac{D\xi_3}{Dt}\mathbf{e}_3. \tag{13.12}$$

Material Derivative of a Scalar

$$\frac{Df}{Dt} = \frac{\partial f}{\partial t} + \frac{u_1}{h_1}\frac{\partial f}{\partial \xi_1} + \frac{u_2}{h_2}\frac{\partial f}{\partial \xi_2} + \frac{u_3}{h_3}\frac{\partial f}{\partial \xi_3}. \tag{13.13}$$

In these equations:

- (F^u, F^v, F^w) and F^ρ are (possibly parametrised) source/sink terms.
- $\Phi = \Phi(\xi_3)$ (since a geopotential coordinate system is assumed).
- S^i and $\dot{S}^i$ are as in (4.10) (for the atmosphere) and (4.11) (for oceans).
- $\dot{Q}$ is heating rate (per unit mass).
- $\dot{Q}_E \equiv \dot{Q} + \sum_i \mu^i \dot{S}^i$ is total rate of energy input (per unit mass).

Equation set (13.1)–(13.13) is expressed in general axial-orthogonal-curvilinear coordinates. The corresponding equation sets for the special cases of spherical-polar and cylindrical-polar coordinates are given explicitly in Appendices A and B, respectively, to Chapter 6.

13.2.2 Transformation to Generalised Vertical Coordinate s

Guided by Staniforth and Wood (2003) for the special case of the deep-atmosphere equations *in spherical-polar coordinates* (λ, ϕ, r), deep-fluid equations (13.1)–(13.13) *in axial-orthogonal-curvilinear coordinates* (ξ_1, ξ_2, ξ_3) are transformed using a *generalised vertical coordinate* $s = s(\xi_1, \xi_2, \xi_3, t)$ such that $(\xi_1, \xi_2, \xi_3, t) \rightarrow (\xi_1, \xi_2, s, t)$. Then:

Transformation Relations

$$\left(\frac{\partial}{\partial \xi_3}\right)_{\xi_1,\xi_2,t} = \frac{\partial s}{\partial \xi_3}\left(\frac{\partial}{\partial s}\right)_{\xi_1,\xi_2,t}, \tag{13.14}$$

$$\left(\frac{\partial}{\partial \psi}\right)_{\xi_3} = \left(\frac{\partial}{\partial \psi}\right)_s - \left(\frac{\partial \xi_3}{\partial \psi}\right)_s\left(\frac{\partial}{\partial \xi_3}\right)_{\xi_1,\xi_2,t} \tag{13.15}$$

$$= \left(\frac{\partial}{\partial \psi}\right)_s - \left(\frac{\partial \xi_3}{\partial \psi}\right)_s\left(\frac{\partial s}{\partial \xi_3}\right)\left(\frac{\partial}{\partial s}\right)_{\xi_1,\xi_2,t}, \tag{13.16}$$

hold for $\psi = \xi_1, \xi_2$ or t.

It is assumed that the transformation is strictly monotonic (i.e. $\partial s/\partial \xi_3$ is non-zero and single-signed (positive or negative) everywhere within the domain). Subscripts in (13.14)–(13.16) signify what is held constant whilst carrying out the operation contained within the associated parentheses. A local vertical cross section is depicted in Fig. 13.2.

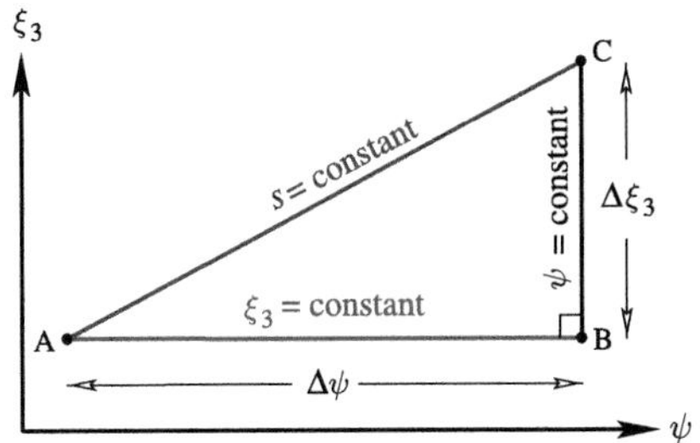

Figure 13.2 A local vertical cross section (containing the direction ψ): line BC is vertical (aligned with a plumb line); line AB is horizontal (aligned with a geopotential surface when $\psi = \xi_1$ or ξ_2); and generalised vertical coordinate s is constant on line AC.

Equation (13.14) represents a simple change of variable in the vertical (ξ_3) direction only, with no change of variable in the horizontal (ξ_1 and ξ_2) directions. For (13.15), consider the change in the ξ_3 direction of a differentiable scalar, f, along a surface of constant s. Thus (see Fig. 13.2):

$$\Delta f_{\mathrm{AC}} = \Delta f_{\mathrm{AB}} + \Delta f_{\mathrm{BC}} = \Delta\psi \left(\frac{\partial f}{\partial \psi}\right)_{\xi_3} + \Delta\xi_3 \left(\frac{\partial f}{\partial \xi_3}\right)_{\xi_1,\xi_2,t}, \tag{13.17}$$

$$\Downarrow$$

$$\left(\frac{\partial f}{\partial \psi}\right)_s = \left(\frac{\partial f}{\partial \psi}\right)_{\xi_3} + \left(\frac{\partial \xi_3}{\partial \psi}\right)_s \left(\frac{\partial f}{\partial \xi_3}\right)_{\xi_1,\xi_2,t}, \tag{13.18}$$

where Δf_{AC} is the incremental change in f when moving from point A to point C along line AC, and similarly for Δf_{AB} and Δf_{BC}. Rearrangement of (13.18) leads to (13.15) and thence – using (13.14) – to (13.16). When using (13.14)–(13.16) in what follows, explicit specifications that ξ_3 and s derivatives are taken at constant ξ_1, ξ_2, and t are omitted for brevity.

Transformation of the governing-equations set to generalised vertical coordinates can be broken down into substeps as follows.

13.2.2.1 *Pressure-Gradient and Gravity Terms*

With use of (13.16), the pressure-gradient terms in the *horizontal* components (13.1) and (13.2) of the momentum equation transform as

$$\frac{1}{\rho h_i}\frac{\partial p}{\partial \xi_i} \rightarrow \frac{1}{\rho h_i}\left(\frac{\partial p}{\partial \xi_i} - \frac{\partial \xi_3}{\partial \xi_i}\frac{\partial s}{\partial \xi_3}\frac{\partial p}{\partial s}\right), \quad i = 1, 2. \tag{13.19}$$

Using (13.14), the pressure-gradient and gravitational terms in the *vertical* component (13.3) of the momentum equation transform as

$$\frac{1}{\rho h_3}\frac{\partial p}{\partial \xi_3} \rightarrow \frac{1}{\rho h_3}\frac{\partial s}{\partial \xi_3}\frac{\partial p}{\partial s}, \quad \frac{1}{h_3}\frac{d\Phi}{d\xi_3} \rightarrow \frac{1}{h_3}\frac{\partial s}{\partial \xi_3}\frac{\partial \Phi}{\partial s}. \tag{13.20}$$

13.2.2.2 *Material Derivative of a Scalar and Vertical Motion*

Before the transformation $\xi_3 \rightarrow s$ of vertical coordinate, the material derivative of a scalar f is given by (13.13); Df/Dt now needs to be re-expressed in terms of the transformed coordinate, s. Repeatedly using (13.16) for $\psi = \xi_2, \xi_3, t$ in (13.13) to transform the first three terms of the right-hand side of (13.13) and using (13.14) for the fourth term leads to

$$\begin{aligned}\frac{Df}{Dt} &= \left(\frac{\partial f}{\partial t}\right)_{\xi_3} + \frac{u_1}{h_1}\left(\frac{\partial f}{\partial \xi_1}\right)_{\xi_3} + \frac{u_2}{h_2}\left(\frac{\partial f}{\partial \xi_2}\right)_{\xi_3} + \frac{u_3}{h_3}\frac{\partial f}{\partial \xi_3} \\ &= \left(\frac{\partial f}{\partial t}\right)_s + \frac{u_1}{h_1}\left(\frac{\partial f}{\partial \xi_1}\right)_s + \frac{u_2}{h_2}\left(\frac{\partial f}{\partial \xi_2}\right)_s \\ &\quad + \left[\frac{u_3}{h_3} - \left(\frac{\partial \xi_3}{\partial t}\right)_s - \frac{u_1}{h_1}\left(\frac{\partial \xi_3}{\partial \xi_1}\right)_s - \frac{u_2}{h_2}\left(\frac{\partial \xi_3}{\partial \xi_2}\right)_s\right]\frac{\partial s}{\partial \xi_3}\frac{\partial f}{\partial s}. \qquad (13.21)\end{aligned}$$

Setting $f \equiv s$ in this equation and noting that (ξ_1, ξ_2, s, t) are independent variables in the transformed coordinate system gives:

The Vertical Motion Equation in Transformed Coordinates (ξ_1, ξ_2, s, t)

$$\dot{s} \equiv \frac{Ds}{Dt} = \left[\frac{u_3}{h_3} - \left(\frac{\partial \xi_3}{\partial t}\right)_s - \frac{u_1}{h_1}\left(\frac{\partial \xi_3}{\partial \xi_1}\right)_s - \frac{u_2}{h_2}\left(\frac{\partial \xi_3}{\partial \xi_2}\right)_s\right]\frac{\partial s}{\partial \xi_3}, \qquad (13.22)$$

where $\dot{s}$ is *generalised vertical velocity*.

Noting that $\partial s/\partial \xi_3 \neq 0$ (otherwise the coordinate system would be singular), elimination of the square-bracketed terms between (13.21) and (13.22) then yields:

The Material Derivative Equation in Transformed Coordinates (ξ_1, ξ_2, s, t)

$$\frac{Df}{Dt} = \left(\frac{\partial f}{\partial t}\right)_s + \frac{u_1}{h_1}\left(\frac{\partial f}{\partial \xi_1}\right)_s + \frac{u_2}{h_2}\left(\frac{\partial f}{\partial \xi_2}\right)_s + \dot{s}\frac{\partial f}{\partial s}. \qquad (13.23)$$

This equation expresses the material derivative of a scalar function f after transformation to the generalised vertical coordinate system (ξ_1, ξ_2, s, t). The material derivative D/Dt appears in many of the transformed governing equations.

13.2.2.3 *The Mass-Continuity Equation*

Mass-continuity equation (13.4) for total mass density is less straightforward to transform than the other equations. Noting that $\partial(h_1h_2h_3)/\partial t \equiv 0$ for (untransformed) axial-orthogonal-curvilinear coordinates (ξ_1, ξ_2, ξ_3, t), it is convenient before transformation to multiply (13.4) through by $h_1h_2h_3$ (this is the Jacobian of the transformation from Cartesian coordinates to axial-orthogonal-curvilinear coordinates). The resulting equation can be rewritten as

$$\begin{aligned}&h_1h_2h_3\frac{D\rho}{Dt} + \rho\left[\frac{\partial}{\partial \xi_1}\left(\frac{u_1}{h_1}h_1h_2h_3\right) + \frac{\partial}{\partial \xi_2}\left(\frac{u_2}{h_2}h_1h_2h_3\right) + \frac{\partial}{\partial \xi_3}\left(\frac{u_3}{h_3}h_1h_2h_3\right)\right] \\ &= h_1h_2h_3\frac{D\rho}{Dt} + \rho\left[\frac{\partial}{\partial t} + \frac{u_1}{h_1}\frac{\partial}{\partial \xi_1} + \frac{u_2}{h_2}\frac{\partial}{\partial \xi_2} + \frac{u_3}{h_3}\frac{\partial}{\partial \xi_3}\right](h_1h_2h_3) \\ &\quad + \rho h_1h_2h_3\left[\frac{\partial}{\partial \xi_1}\left(\frac{u_1}{h_1}\right) + \frac{\partial}{\partial \xi_2}\left(\frac{u_2}{h_2}\right) + \frac{\partial}{\partial \xi_3}\left(\frac{u_3}{h_3}\right)\right] \\ &= \frac{D}{Dt}(\rho h_1h_2h_3) + \rho h_1h_2h_3\left[\frac{\partial}{\partial \xi_1}\left(\frac{u_1}{h_1}\right) + \frac{\partial}{\partial \xi_2}\left(\frac{u_2}{h_2}\right) + \frac{\partial}{\partial \xi_3}\left(\frac{u_3}{h_3}\right)\right] = h_1h_2h_3F^{\rho}.\end{aligned}$$

(13.24)

Using kinematic equation (13.12) in this equation then leads to

$$\frac{D}{Dt}\left(\rho h_1 h_2 h_3\right) + \rho h_1 h_2 h_3 \left(\frac{\partial \dot{\xi}_1}{\partial \xi_1} + \frac{\partial \dot{\xi}_2}{\partial \xi_2} + \frac{\partial \dot{\xi}_3}{\partial \xi_3}\right) = h_1 h_2 h_3 F^{\rho}, \tag{13.25}$$

where

$$\dot{\xi}_i \equiv \frac{D\xi_i}{Dt}, \quad i = 1, 2, 3. \tag{13.26}$$

Equation (13.25) is an alternative expression of the mass-continuity equation (13.4) in the (untransformed) (ξ_1, ξ_2, ξ_3, t) coordinate system.

With this preparation, (13.25) is now transformed from axial-orthogonal-curvilinear coordinates (ξ_1, ξ_2, ξ_3, t) to generalised vertical coordinates (ξ_1, ξ_2, s, t). From (13.15),

$$\left(\frac{\partial \dot{\xi}_i}{\partial \xi_i}\right)_{\xi_3} = \left(\frac{\partial \dot{\xi}_i}{\partial \xi_i}\right)_s - \left(\frac{\partial \xi_3}{\partial \xi_i}\right)_s \frac{\partial \dot{\xi}_i}{\partial \xi_3}, \quad i = 1, 2. \tag{13.27}$$

Rearranging vertical-motion equation (13.22), with use of (13.12) and (13.26), gives

$$\dot{\xi}_3 \equiv \frac{u_3}{h_3} = \left(\frac{\partial \xi_3}{\partial t}\right)_s + \dot{\xi}_1 \left(\frac{\partial \xi_3}{\partial \xi_1}\right)_s + \dot{\xi}_2 \left(\frac{\partial \xi_3}{\partial \xi_2}\right)_s + \dot{s}\frac{\partial \xi_3}{\partial s}. \tag{13.28}$$

From (13.14), (13.23) and (13.28),

$$\begin{aligned}
\frac{\partial \dot{\xi}_i}{\partial \xi_i} = \frac{\partial s}{\partial \xi_3}\frac{\partial \dot{\xi}_3}{\partial s} &= \frac{\partial s}{\partial \xi_3}\frac{\partial}{\partial s}\left[\left(\frac{\partial \xi_3}{\partial t}\right)_s + \dot{\xi}_1 \left(\frac{\partial \xi_3}{\partial \xi_1}\right)_s + \dot{\xi}_2 \left(\frac{\partial \xi_3}{\partial \xi_2}\right)_s + \dot{s}\frac{\partial \xi_3}{\partial s}\right] \\
&= \frac{\partial s}{\partial \xi_3}\frac{D}{Dt}\left(\frac{\partial \xi_3}{\partial s}\right) + \frac{\partial s}{\partial \xi_3}\left[\frac{\partial \dot{\xi}_1}{\partial s}\left(\frac{\partial \xi_3}{\partial \xi_1}\right)_s + \frac{\partial \dot{\xi}_2}{\partial s}\left(\frac{\partial \xi_3}{\partial \xi_2}\right)_s\right] + \frac{\partial \dot{s}}{\partial s} \\
&= \frac{\partial s}{\partial \xi_3}\frac{D}{Dt}\left(\frac{\partial \xi_3}{\partial s}\right) + \frac{\partial \dot{\xi}_1}{\partial \xi_3}\left(\frac{\partial \xi_3}{\partial \xi_1}\right)_s + \frac{\partial \dot{\xi}_2}{\partial \xi_3}\left(\frac{\partial \xi_3}{\partial \xi_2}\right)_s + \frac{\partial \dot{s}}{\partial s}.
\end{aligned} \tag{13.29}$$

Summing (13.27) for $i = 1, 2$, with (13.29) yields

$$\frac{\partial \dot{\xi}_1}{\partial \xi_1} + \frac{\partial \dot{\xi}_2}{\partial \xi_2} + \frac{\partial \dot{\xi}_3}{\partial \xi_3} = \frac{\partial s}{\partial \xi_3}\frac{D}{Dt}\left(\frac{\partial \xi_3}{\partial s}\right) + \left(\frac{\partial \dot{\xi}_1}{\partial \xi_1} + \frac{\partial \dot{\xi}_2}{\partial \xi_2}\right)_s + \frac{\partial \dot{s}}{\partial s}. \tag{13.30}$$

Substituting (13.30) into the untransformed mass-continuity equation (13.25), then leads to

$$\frac{D}{Dt}\left(\rho h_1 h_2 h_3\right) + \rho h_1 h_2 h_3 \left[\frac{\partial s}{\partial \xi_3}\frac{D}{Dt}\left(\frac{\partial \xi_3}{\partial s}\right) + \left(\frac{\partial \dot{\xi}_1}{\partial \xi_1} + \frac{\partial \dot{\xi}_2}{\partial \xi_2}\right)_s + \frac{\partial \dot{s}}{\partial s}\right] = h_1 h_2 h_3 F^{\rho}. \tag{13.31}$$

Multiplying (13.31) by $\partial \xi_3 / \partial s$, rearranging, and using (13.12) and (13.26) to restore u_1 and u_2 finally results in:

The Mass-Continuity Equation in Transformed Coordinates (ξ_1, ξ_2, s, t): Lagrangian Form

$$\frac{D\mathscr{F}}{Dt} + \mathscr{F}\left[\frac{\partial}{\partial \xi_1}\left(\frac{u_1}{h_1}\right) + \frac{\partial}{\partial \xi_2}\left(\frac{u_2}{h_2}\right) + \frac{\partial \dot{s}}{\partial s}\right] = \frac{\mathscr{F}}{\rho}F^{\rho}, \tag{13.32}$$

where

$$\mathscr{F} \equiv \rho h_1 h_2 h_3 \frac{\partial \xi_3}{\partial s}, \tag{13.33}$$

is total mass density (ρ) multiplied by the Jacobian ($h_1 h_2 h_3 \partial \xi_3 / \partial s$) of the transformation from Cartesian coordinates to generalised vertical coordinates (ξ_1, ξ_2, s, t).

Setting $f \equiv \mathscr{F}$ in material derivative equation (13.23), and inserting the resulting equation into (13.32) leads to:

The Mass Continuity Equation in Transformed Coordinates (ξ_1, ξ_2, s, t): Eulerian Form

$$\frac{\partial \mathscr{F}}{\partial t} + \frac{\partial}{\partial \xi_1}\left(\frac{u_1}{h_1}\mathscr{F}\right) + \frac{\partial}{\partial \xi_2}\left(\frac{u_2}{h_2}\mathscr{F}\right) + \frac{\partial}{\partial s}\left(\dot{s}\mathscr{F}\right) = \mathscr{F} F^{\rho}. \tag{13.34}$$

This form is convenient for examination (in Sections 13.3–13.6) of global conservation properties of the vertically transformed equation set.

Let $\mathcal{G}$ be an arbitrary scalar function. Setting $f \equiv \mathcal{G}$ in (13.23) and multiplying the resulting equation by $\mathscr{F}$ then leads to

$$\mathscr{F}\frac{D\mathcal{G}}{Dt} \equiv \mathscr{F}\frac{\partial \mathcal{G}}{\partial t} + \frac{u_1}{h_1}\mathscr{F}\frac{\partial \mathcal{G}}{\partial \xi_1} + \frac{u_2}{h_2}\mathscr{F}\frac{\partial \mathcal{G}}{\partial \xi_2} + \dot{s}\mathscr{F}\frac{\partial \mathcal{G}}{\partial s}. \tag{13.35}$$

Using (13.34), the right-hand side of (13.35) may be rewritten in flux form to give

$$\mathscr{F}\frac{D\mathcal{G}}{Dt} = \frac{\partial}{\partial t}\left(\mathcal{G}\mathcal{F}\right) + \frac{\partial}{\partial \xi_1}\left(\frac{u_1}{h_1}\mathcal{G}\mathcal{F}\right) + \frac{\partial}{\partial \xi_2}\left(\frac{u_2}{h_2}\mathcal{G}\mathcal{F}\right) + \frac{\partial}{\partial s}\left(\dot{s}\mathcal{G}\mathcal{F}\right) - \mathcal{G}\mathscr{F}F^{\rho}. \tag{13.36}$$

This equation will be used in Sections 13.4 and 13.5 on conservation of energy and axial angular momentum, respectively.

13.2.3 The Vertically Transformed Equations

Using (13.19) and (13.20) (for pressure-gradient and gravity terms in the horizontal-component equations of momentum), (13.22) (definition of generalised vertical velocity), (13.23) (material derivative in generalised vertical coordinates), and (13.32) (mass continuity in generalised vertical coordinates), the set of governing equations (13.1)–(13.13) in axial-orthogonal-curvilinear coordinates (ξ_1, ξ_2, ξ_3, t) transform to the following equation set in generalised vertical coordinates (ξ_1, ξ_2, s, t):

The Governing Equations in Vertically Transformed, Axial-Orthogonal-Curvilinear Coordinates (ξ_1, ξ_2, s, t)

Prognostic Equations

Momentum

$$\frac{Du_1}{Dt} + \left(\frac{u_1}{h_1} + 2\Omega\right)\frac{u_2}{h_2}\frac{\partial h_1}{\partial \xi_2} + \left(\frac{u_1}{h_1} + 2\Omega\right)\frac{u_3}{h_3}\frac{\partial h_1}{\partial \xi_3} + \frac{1}{\rho h_1}\left(\frac{\partial p}{\partial \xi_1} - \frac{\partial \xi_3}{\partial \xi_1}\frac{\partial s}{\partial \xi_3}\frac{\partial p}{\partial s}\right) = F^{u_1}, \tag{13.37}$$

$$\frac{Du_2}{Dt} - \left(\frac{u_1}{h_1} + 2\Omega\right)\frac{u_1}{h_2}\frac{\partial h_1}{\partial \xi_2} - \frac{u_3^2}{h_2 h_3}\frac{\partial h_3}{\partial \xi_2} + \frac{u_2}{h_2}\frac{u_3}{h_3}\frac{\partial h_2}{\partial \xi_3} + \frac{1}{\rho h_2}\left(\frac{\partial p}{\partial \xi_2} - \frac{\partial \xi_3}{\partial \xi_2}\frac{\partial s}{\partial \xi_3}\frac{\partial p}{\partial s}\right) = F^{u_2}, \tag{13.38}$$

$$\frac{Du_3}{Dt} + \frac{u_2}{h_2}\frac{u_3}{h_3}\frac{\partial h_3}{\partial \xi_2} - \left(\frac{u_1}{h_1} + 2\Omega\right)\frac{u_1}{h_3}\frac{\partial h_1}{\partial \xi_3} - \frac{u_2^2}{h_2 h_3}\frac{\partial h_2}{\partial \xi_3} + \frac{1}{\rho h_3}\frac{\partial s}{\partial \xi_3}\frac{\partial p}{\partial s} + \frac{1}{h_3}\frac{\partial s}{\partial \xi_3}\frac{\partial \Phi}{\partial s} = F^{u_3}. \tag{13.39}$$

Continuity of Total Mass Density

$$\frac{D}{Dt}\left(\rho h_1 h_2 h_3 \frac{\partial \xi_3}{\partial s}\right) + \left(\rho h_1 h_2 h_3 \frac{\partial \xi_3}{\partial s}\right)\left[\frac{\partial}{\partial \xi_1}\left(\frac{u_1}{h_1}\right) + \frac{\partial}{\partial \xi_2}\left(\frac{u_2}{h_2}\right) + \frac{\partial \dot{s}}{\partial s}\right] = h_1 h_2 h_3 \frac{\partial \xi_3}{\partial s} F^{\rho}. \tag{13.40}$$

Substance Transport

$$\frac{DS^i}{Dt} = \dot{S}^i, \quad i = 1, 2, \ldots \tag{13.41}$$

Thermodynamic-Energy Equation

$$\frac{D\mathcal{E}}{Dt} + p\frac{D\alpha}{Dt} = \dot{Q}_E \quad \text{or} \quad \frac{D\eta}{Dt} = \frac{\dot{Q}}{T}. \tag{13.42}$$

Definitions and Diagnostic Equations

Definition of Specific Volume

$$\alpha \equiv \frac{1}{\rho}. \tag{13.43}$$

Definition of T, p, and μ^i

$$T = \frac{\partial \mathcal{E}\left(\alpha, \eta, S^1, S^2, \ldots\right)}{\partial \eta} \quad \text{or} \quad T = \left[\frac{\partial \eta\left(\alpha, \mathcal{E}, S^1, S^2, \ldots\right)}{\partial \mathcal{E}}\right]^{-1}, \tag{13.44}$$

$$p = -\frac{\partial \mathcal{E}\left(\alpha, \eta, S^1, S^2, \ldots\right)}{\partial \alpha} \quad \text{or} \quad p = T\frac{\partial \eta\left(\alpha, \mathcal{E}, S^1, S^2, \ldots\right)}{\partial \alpha}, \tag{13.45}$$

$$\mu^i = \frac{\partial \mathcal{E}\left(\alpha, \eta, S^1, S^2, \ldots\right)}{\partial S^i} \quad \text{or} \quad \mu^i = -T\frac{\partial \eta\left(\alpha, \mathcal{E}, S^1, S^2, \ldots\right)}{\partial S^i}. \tag{13.46}$$

Fundamental Equation of State

$$\mathcal{E} = \mathcal{E}\left(\alpha, \eta, S^1, S^2, \ldots\right) \quad \text{or} \quad \eta = \eta\left(\alpha, \mathcal{E}, S^1, S^2, \ldots\right). \tag{13.47}$$

Kinematic Equation

$$\mathbf{u} \equiv u_1\mathbf{e}_1 + u_2\mathbf{e}_2 + u_3\mathbf{e}_3 = h_1\frac{D\xi_1}{Dt}\mathbf{e}_1 + h_2\frac{D\xi_2}{Dt}\mathbf{e}_2 + h_3\frac{D\xi_3}{Dt}\mathbf{e}_3. \tag{13.48}$$

Material Derivative of a Scalar

$$\frac{Df}{Dt} = \frac{\partial f}{\partial t} + \frac{u_1}{h_1}\frac{\partial f}{\partial \xi_1} + \frac{u_2}{h_2}\frac{\partial f}{\partial \xi_2} + \dot{s}\frac{\partial f}{\partial s}. \tag{13.49}$$

Vertical Motion

$$\dot{s} \equiv \frac{Ds}{Dt} = \left(\frac{u_3}{h_3} - \frac{\partial \xi_3}{\partial t} - \frac{u_1}{h_1}\frac{\partial \xi_3}{\partial \xi_1} - \frac{u_2}{h_2}\frac{\partial \xi_3}{\partial \xi_2}\right)\frac{\partial s}{\partial \xi_3}. \tag{13.50}$$

Setting $s \equiv \xi_3$ in this equation set recovers the untransformed equation set (13.1)–(13.13).

13.2.4 Time Integration

We now outline how, in principle, equation set (13.37)–(13.50) may be integrated forward in time. For simplicity, we assume two substances, S^1 and S^2, associated with substance transport equation (13.41).[4]

Doing the accounting, (13.37)–(13.42) comprise seven equations for the seven *prognostic* variables u_1, u_2, u_3, ρ, S^1, S^2, and $\mathcal{E}$ or η. This means that given values of all variables at an instant in time, the seven prognostic equations (13.37)–(13.42) (i.e. those with a D/Dt term) can be used to obtain values of the seven prognostic variables at a later instant in time. (Because of the non-linear nature of these equations, this has to be done numerically for all but the simplest situations.)

The remaining equations (13.43)–(13.50) are *definitions* and *diagnostic equations*. Once values for the seven prognostic variables at a later instant in time are available, values for the remaining variables at the later instant may then be obtained from these definitions and diagnostic equations. Having done so, values of all variables are then known at the later instant, and the procedure can be repeated to obtain all quantities at an even later instant.

Precisely how all this is done in practice, in an accurate, computationally stable, and efficient manner, is, however, a challenging subject in its own right and well beyond the scope of the present book!

13.2.5 Coordinate Surfaces, Unit Vectors, and Momentum Components

Velocity components u_1 and u_2 in transformed material derivative (13.49) are the usual horizontal components in axial-orthogonal-curvilinear coordinates (ξ_1, ξ_2, ξ_3, t) before vertical transformation. Thus:

Velocity Components u_1 and u_2 Are *Not* the Components of Velocity within/along Surfaces of Constant s

- The (orthogonal) unit vector triad $(\mathbf{e}_1, \mathbf{e}_2, \mathbf{e}_3)$ at a point remains *unchanged* by the transformation of vertical coordinates $(\xi_1, \xi_2, \xi_3, t) \rightarrow (\xi_1, \xi_2, s, t)$. This is depicted in Fig. 13.3 for a local vertical cross section over orography/bathymetry.

Derivatives with respect to ξ_1, ξ_2, and t in (13.49) are taken within constant s surfaces, so that increments of f are those seen as one moves in the relevant direction whilst constrained to remain on a constant s surface. The relevant distances, however, are those in the true horizontal (i.e. measured within a *geopotential surface*), and *not* those measured within a (quasi-horizontal) s surface. Also, $\partial / \partial s$ represents differentiation in the vertical, and *not* perpendicular to surfaces of constant s.[5]

Exactitude of Transformation

- The transformation $(\xi_1, \xi_2, \xi_3, t) \rightarrow (\xi_1, \xi_2, s, t)$ of coordinates is *exact*.

[4] For the atmosphere this corresponds to dry air plus water substance, and for the ocean to water substance plus solute.

[5] Representations in terms of velocity components and gradients within and perpendicular to s surfaces can, of course, be developed, but they are generally more complicated and consequently more difficult to use in numerical models.

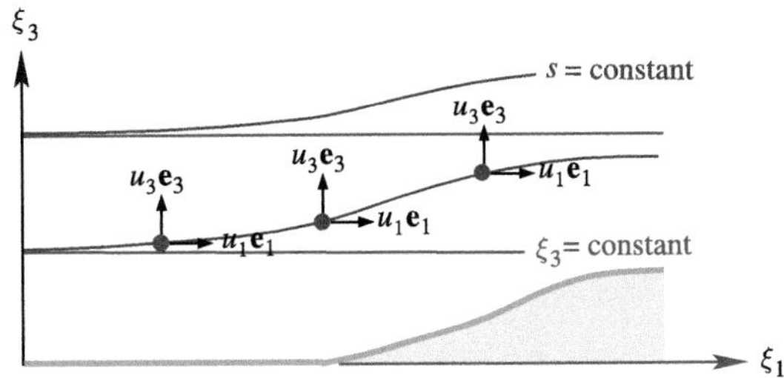

Figure 13.3 A local $\xi_1 - \xi_3$ vertical cross section over an underlying orography/bathymetry (in green) for geopotential (ξ_1, ξ_2, ξ_3, t) coordinates and for vertically transformed, boundary-following (ξ_1, ξ_2, s, t) coordinates; $\xi_3 =$ constant (geopotential) surfaces are in red, and $s =$ constant surfaces are in blue. Horizontal unit vector $\mathbf{e}_1$ is embedded within a geopotential ($\xi_3 =$ constant) surface; vertical unit vector $\mathbf{e}_3$ is aligned with a plumb line; momentum components u_1 and u_3 in the $\mathbf{e}_1$ and $\mathbf{e}_3$ directions, respectively, are shown at three points on a $s =$ constant surface. See text for further details.

The Generalised Vertical Coordinate System Is a Hybrid System

- It remains orthogonal in the sense that an *orthogonal* unit vector triad $(\mathbf{e}_1, \mathbf{e}_2, \mathbf{e}_3)$ is associated with every point within the domain. This triad corresponds to that of the *untransformed* equations – see Fig. 13.3.
- Furthermore, the momentum equation is still decomposed into three components (u_1, u_2, u_3) in the mutually orthogonal directions $(\mathbf{e}_1, \mathbf{e}_2, \mathbf{e}_3)$.
- In particular, the vertical-component equation of momentum remains a prognostic equation for u_3 (as opposed, for example, to one for $\dot{s}$).
- *However*, the generalised vertical coordinate system (ξ_1, ξ_2, s, t) is *not* orthogonal in the sense of its coordinate surfaces being mutually orthogonal to one another. Surfaces of constant s are *not* (in general) orthogonal to surfaces of constant ξ_1 or ξ_2.
- They can, however, be orthogonal in special circumstances; for example, when, for a particular column of fluid, the bottom and top boundaries both correspond to $\xi_3 =$ constant surfaces in the untransformed (geopotential) coordinate system (ξ_1, ξ_2, ξ_3, t).

Expression of the governing equations in the generalised vertical-coordinate system (ξ_1, ξ_2, s, t) facilitates examination (in Section 13.6) of the impact of boundary conditions in the vertical on the conservation (or otherwise) of energy and axial angular momentum.

13.2.6 Boundary Conditions in the Vertical

To conserve mass, both the lower and upper boundaries, $s = s^B(\xi_1, \xi_2, t)$ and $s = s^T(\xi_1, \xi_2, t)$, respectively, are assumed to be material surfaces *with no mass transport across them*. Thus, taking the material derivatives of $s = s^B(\xi_1, \xi_2, t)$ and $s = s^T(\xi_1, \xi_2, t)$ at the lower ('bottom') and upper ('top') material surfaces, respectively, and noting that s^B and s^T are independent of s leads to:

The Bottom and Top Boundary Conditions in the Vertical

$$\dot{s}^B = \frac{\partial s^B}{\partial t} + \frac{u_1^B}{h_1^B}\frac{\partial s^B}{\partial \xi_1} + \frac{u_2^B}{h_2^B}\frac{\partial s^B}{\partial \xi_2}, \tag{13.51}$$

$$\dot{s}^T = \frac{\partial s^T}{\partial t} + \frac{u_1^T}{h_1^T}\frac{\partial s^T}{\partial \xi_1} + \frac{u_2^T}{h_2^T}\frac{\partial s^T}{\partial \xi_2}, \tag{13.52}$$

where, analogously to Kasahara (1974) and Staniforth and Wood (2003), superscripts 'B' and 'T' denote evaluation at $s^B = s^B(\xi_1, \xi_2, t)$ and $s^T = s^T(\xi_1, \xi_2, t)$, respectively.

In particular, $\xi_3^B = \xi_3^B(\xi_1, \xi_2, t)$ and $\xi_3^T = \xi_3^T(\xi_1, \xi_2, t)$ are the corresponding locations of the lower and upper boundaries respectively in the (untransformed) axial-orthogonal-curvilinear, geopotential coordinate system (ξ_1, ξ_2, ξ_3, t).

The lower boundary is often assumed to be stationary, that is, $\xi_3^B = \xi_3^B(\xi_1, \xi_2)$. This is the case at an ocean's lower (bathymetric) surface and at the atmosphere's lower (orographic) surface. However, it is not so at the mutual boundary between the atmosphere and an ocean due to oceanic tides, wind stresses, temporal variation of atmospheric surface pressure, heat exchange, and so on there, albeit the temporal variation of ξ_3^B is usually quite small. See Sections 13.6.2.4 and 13.6.3.4 for the idealised case of an atmosphere–ocean boundary in the absence of physical forcings there.

Upper boundary condition (13.52) is a generalisation of that assumed by Kasahara (1974) in the more-restrictive context of the hydrostatic primitive equations. It reduces to his condition when s^T coincides with a coordinate surface s^T = constant, since then $\dot{s}^T = 0$. This is the case for most atmospheric models since they usually employ a boundary-following vertical coordinate designed to respect

$$\dot{s}^B = \dot{s}^T = 0, \tag{13.53}$$

where s^B and s^T are constants, often normalised to zero and unity. The lower and upper material surfaces then also correspond to (constant-s) coordinate surfaces. However, following Staniforth and Wood (2003), the generality of (13.51) and (13.52) is retained herein. Condition (13.53) will not be implicitly invoked in what follows.

13.2.7 Some Possible Choices of Vertical Coordinate

Boundary-following coordinates are widely used in atmospheric and oceanic models. They are usually either *height based* or *pressure based*, but isentropic-based, boundary-following coordinates are also possible (Kasahara, 1974).

13.2.7.1 *Height Based*

A simple example (Kasahara, 1974) of height-based coordinates in spherical geometry (λ, ϕ, r) is

$$s = \widetilde{z} \equiv \frac{r - r^B(\lambda, \phi)}{r^T - r^B(\lambda, \phi)}, \tag{13.54}$$

where r^T is constant.[6] This coordinate, termed Basic Terrain Following (BTF) by Klemp (2011), is indeed basic and (in the context of an atmosphere overlying orography) terrain following. However, it features 'wiggles' (when viewed in the untransformed (λ, ϕ, r) coordinate system) – of non-negligible amplitude – for constant-s surfaces at heights where they should ideally (for physical-balance reasons) be fairly flat. These 'wiggles' reflect the horizontal variation (with reducing amplitude for increasing height) of the underlying orography, $r^B = r^B(\lambda, \phi)$. This

[6] It is implicitly assumed in (13.54) that r^B is independent of time t. This is generally true for the atmosphere at a land point, but not exactly true at a sea point due to the presence of propagating surface ocean waves of various wavelengths and amplitudes.

then unduly reduces the accuracy of a numerical model and can create spurious atmospheric circulations.

Following Klemp (2011), an example of the 'wiggles' that occur when using (13.54) in Cartesian $x-z$ geometry (with $(\lambda,\phi)\rightarrow(x,z)$ and $s=\left[z-z^B(x)\right]/\left(\left[z^T-z^B(x)\right]\right)$ is displayed in Fig. 13.4a over a 30 km × 5 km ($x\times z$) subdomain of a larger domain. The lowest contour corresponds to an underlying orography of the form

$$z^B(x)=h_m\exp\left[-\left(\frac{x}{a}\right)^2\right]\cos^2\left(\pi\frac{x}{x_\lambda}\right),\quad a=5\,\text{km},\quad x_\lambda=4\,\text{km},\quad h_m=1\,\text{km},\tag{13.55}$$

where $z^B=z^B(x)$ is the height of the terrain above $z=0$, and parameters are set to the given values. The top of the domain is located at $z^T=20$ km, with only the lowest 5 km (of 20 km total height) of the domain displayed in Fig. 13.4. As noted in Klemp (2011), only about a quarter of the terrain influence has been removed at $z=5$km, which is inadequate for numerical-modelling purposes.

To address this deficiency of non-negligible 'wiggles' when using (13.54), various authors have proposed different ways of flattening the coordinate surfaces more rapidly as $s=\widetilde{z}$ increases (Arakawa and Lamb, 1977; Simmons and Burridge, 1981; Schär et al., 2002; Staniforth et al., 2006; Leuenberger et al., 2010; Klemp, 2011). For example, Klemp (2011)'s Smooth Terrain Following (STF) coordinate progressively flattens the coordinate surfaces more rapidly in a satisfactory manner; cf. Fig. 13.4b for STF coordinates, with Fig. 13.4a for BTF coordinates. As noted in Klemp (2011), the terrain influence with STF coordinates is not only essentially removed in the lowest 5 km of the domain but, additionally, the influence of the smaller-scale terrain structure is nearly absent above $z=2$ km.

13.2.7.2 *Pressure Based*

Setting $s=p$ gives the *isobaric*[7] coordinate system (ξ_1,ξ_2,p,t). In this system the vertical velocity is usually denoted by $\omega\equiv\dot{p}\equiv Dp/Dt$. Although useful for theoretical purposes, and a popular choice in the early days of numerical weather prediction, a serious drawback of isobaric coordinates is the practical difficulty of numerically approximating the governing equations at and near lower boundaries.

Thus a popular choice of vertical coordinate for atmospheric primitive-equation models[8] in spherical geometry is the pressure-based, *terrain-following* coordinate

$$s=\sigma\equiv\frac{p-p^T}{p^B(\lambda,\phi,t)-p^T},\tag{13.56}$$

where p^T (the pressure at $r=r^T(\lambda,\phi,t)$) is constant (Phillips, 1957; Kasahara, 1974; Staniforth and Wood, 2003).[9] For this coordinate, the surface $\sigma=0$ corresponds to the isobaric surface $p=p^T=$ constant. Similarly, the surface $\sigma=1$ corresponds to Earth's surface (assumed to be rigid), where $p=p^B(\lambda,\phi,t)$ and $r=r^B(\lambda,\phi)$.[10]

[7] Isobaric refers to surfaces of constant (whence *iso*) pressure (whence *bar*, a unit of pressure).

[8] As shown in Chapter 16, the primitive equations may be obtained from the unapproximated deep-fluid equations by making use of the *hydrostatic* and *shallow-fluid* approximations.

[9] Setting $p^T\equiv 0$ in (13.56) corresponds to Phillips (1957)'s original definition of σ coordinates.

[10] There is a typographic error in Staniforth and Wood (2003)'s equation (2.23). Bottom surface pressure $p_H(\lambda,\phi)$ therein should read $p_H(\lambda,\phi,t)$ since it also depends on time t, as in (13.56) for $p^B(\lambda,\phi,t)$. By way of contrast, the corresponding bottom surface $r_H=r_H(\lambda,\phi)$ therein is independent of time since it is assumed rigid.

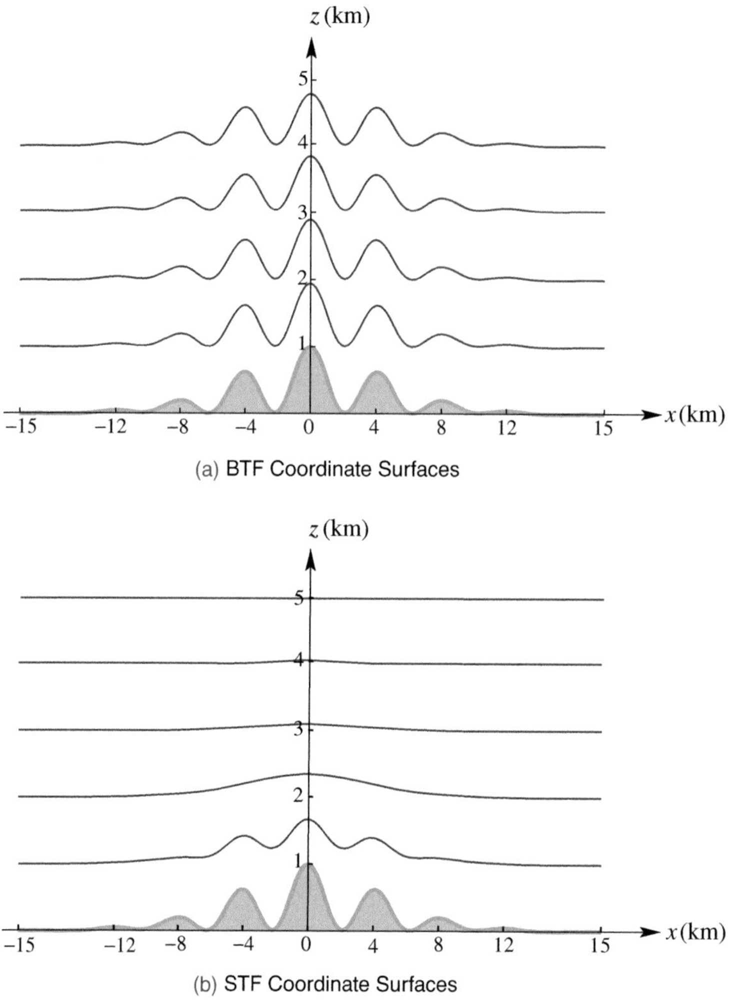

Figure 13.4 Constant-s surfaces (in blue) in Cartesian $x - z$ geometry, displayed over a $30\ \mathrm{km} \times 5\ \mathrm{km}$ subdomain, for: (a) BTF (Basic Terrain Following) coordinates; and (b) STF (Smoothed Terrain Following) coordinates. Lowest contour (in green) corresponds to the underlying orography/bathymetry, defined by (13.55); top of the domain is at $z = 20$ km. After Figs. 1a and 1e of Klemp (2011).

Laprise (1992) later extended this vertical coordinate to include *non-hydrostatic* but *nevertheless shallow* atmospheres by introducing a family of terrain-following, *hydrostatic-pressure* coordinates. A simple example of this coordinate is

$$s = \frac{\pi - \pi^T}{\pi^B(\lambda, \phi, t) - \pi^T}, \tag{13.57}$$

where (here, but not elsewhere)

$$\pi(\lambda, \phi, r, t) \equiv \pi^T + \int_r^{r^T} g\rho\left(\lambda, \phi, r', t\right) dr' \quad \Rightarrow \quad \frac{\partial \pi}{\partial \xi_3} = -\rho g, \tag{13.58}$$

π is 'hydrostatic pressure', g = constant (since the atmosphere is still assumed shallow) is the magnitude of gravitational acceleration, ρ is density, and π^T is constant.

Laprise (1992)'s 'hydrostatic-pressure' coordinate for shallow atmospheres has been further generalised by Wood and Staniforth (2003) to the *deep-atmosphere* case via the introduction of a *mass-based*, terrain-following, vertical coordinate. With this last choice for s, and in *spherical geometry*, equations (13.37)–(13.50) reduce to those derived in Wood and Staniforth (2003) for this special case.

13.2.8 The Transformed Equations in Spherical-Polar Coordinates

Consider now the special case of spherical-polar coordinates; see Section 5.2.2 for their definition. From (5.13) and (5.15) of Chapter 5,

$$(\xi_1, \xi_2, \xi_3) \rightarrow (\lambda, \phi, r), \quad (h_1, h_2, h_3) \rightarrow (r\cos\phi, r, 1). \tag{13.59}$$

Introducing (13.59) into vertically transformed equations (13.37)–(13.50) leads to the following equation set. Apart from a different thermodynamic treatment – which is more general here – this equation set agrees with that derived in Staniforth and Wood (2003) for spherical-polar coordinates.

The Governing Equations in Vertically Transformed Spherical-Polar Coordinates (λ, ϕ, s, t)

Prognostic Equations

Momentum

$$\frac{Du_\lambda}{Dt} - \left(\frac{u_\lambda}{r\cos\phi} + 2\Omega\right) u_\phi \sin\phi + \left(\frac{u_\lambda}{r\cos\phi} + 2\Omega\right) u_r \cos\phi + \frac{1}{\rho r\cos\phi}\left(\frac{\partial p}{\partial \lambda} - \frac{\partial r}{\partial \lambda}\frac{\partial s}{\partial r}\frac{\partial p}{\partial s}\right) = F^{u_\lambda}, \tag{13.60}$$

$$\frac{Du_\phi}{Dt} + \left(\frac{u_\lambda}{r\cos\phi} + 2\Omega\right) u_\lambda \sin\phi + \frac{u_\phi u_r}{r} + \frac{1}{\rho r}\left(\frac{\partial p}{\partial \phi} - \frac{\partial r}{\partial \phi}\frac{\partial s}{\partial r}\frac{\partial p}{\partial s}\right) = F^{u_\phi}, \tag{13.61}$$

$$\frac{Du_r}{Dt} - \left(\frac{u_\lambda}{r\cos\phi} + 2\Omega\right) u_\lambda \cos\phi - \frac{u_\phi^2}{r} + \frac{1}{\rho}\frac{\partial s}{\partial r}\frac{\partial p}{\partial s} + \frac{\partial s}{\partial r}\frac{\partial \Phi}{\partial s} = F^{u_r}. \tag{13.62}$$

Continuity of Total Mass Density

$$\frac{D}{Dt}\left(\rho r^2 \cos\phi \frac{\partial r}{\partial s}\right) + \left(\rho r^2 \cos\phi \frac{\partial r}{\partial s}\right)\left[\frac{\partial}{\partial \lambda}\left(\frac{u_\lambda}{r\cos\phi}\right) + \frac{\partial}{\partial \phi}\left(\frac{u_\phi}{r}\right) + \frac{\partial \dot{s}}{\partial s}\right] = r^2 \cos\phi \frac{\partial r}{\partial s} F^\rho. \tag{13.63}$$

Substance Transport

$$\frac{DS^i}{Dt} = \dot{S}^i, \quad i = 1, 2, \ldots \tag{13.64}$$

Thermodynamic-Energy Equation

$$\frac{D\mathcal{E}}{Dt} + p\frac{D\alpha}{Dt} = \dot{Q}_E \quad \text{or} \quad \frac{D\eta}{Dt} = \frac{\dot{Q}}{T}. \tag{13.65}$$

The Governing Equations in Vertically Transformed Spherical-Polar Coordinates (λ, ϕ, s, t)

Definitions and Diagnostic Equations

Definition of Specific Volume

$$\alpha \equiv \frac{1}{\rho}. \tag{13.66}$$

Definition of T, p, and μ^i

$$T = \frac{\partial \mathcal{E}\left(\alpha, \eta, S^1, S^2, \ldots\right)}{\partial \eta} \quad \text{or} \quad T = \left[\frac{\partial \eta\left(\alpha, \mathcal{E}, S^1, S^2, \ldots\right)}{\partial \mathcal{E}}\right]^{-1}, \tag{13.67}$$

$$p = -\frac{\partial \mathcal{E}\left(\alpha, \eta, S^1, S^2, \ldots\right)}{\partial \alpha} \quad \text{or} \quad p = T\frac{\partial \eta\left(\alpha, \mathcal{E}, S^1, S^2, \ldots\right)}{\partial \alpha}, \tag{13.68}$$

$$\mu^i = \frac{\partial \mathcal{E}\left(\alpha, \eta, S^1, S^2, \ldots\right)}{\partial S^i} \quad \text{or} \quad \mu^i = -T\frac{\partial \eta\left(\alpha, \mathcal{E}, S^1, S^2, \ldots\right)}{\partial S^i}. \tag{13.69}$$

Fundamental Equation of State

$$\mathcal{E} = \mathcal{E}\left(\alpha, \eta, S^1, S^2, \ldots\right) \quad \text{or} \quad \eta = \eta\left(\alpha, \mathcal{E}, S^1, S^2, \ldots\right). \tag{13.70}$$

Kinematic Equation

$$\mathbf{u} \equiv u_\lambda \mathbf{e}_\lambda + u_\phi \mathbf{e}_\phi + u_r \mathbf{e}_r = r\cos\phi\frac{D\lambda}{Dt}\mathbf{e}_\lambda + r\frac{D\phi}{Dt}\mathbf{e}_\phi + \frac{Dr}{Dt}\mathbf{e}_r. \tag{13.71}$$

Material Derivative of a Scalar

$$\frac{Df}{Dt} = \frac{\partial f}{\partial t} + \frac{u_\lambda}{r\cos\phi}\frac{\partial f}{\partial \lambda} + \frac{u_\phi}{r}\frac{\partial f}{\partial \phi} + \dot{s}\frac{\partial f}{\partial s}. \tag{13.72}$$

Vertical Motion

$$\dot{s} \equiv \frac{Ds}{Dt} = \left(u_r - \frac{\partial r}{\partial t} - \frac{u_\lambda}{r\cos\phi}\frac{\partial r}{\partial \lambda} - \frac{u_\phi}{r}\frac{\partial r}{\partial \phi}\right)\frac{\partial s}{\partial r}. \tag{13.73}$$

Vertical Boundary Conditions

$$\dot{s}^B = \frac{\partial s^B}{\partial t} + \frac{u_\lambda^B}{r^B\cos\phi}\frac{\partial s^B}{\partial \lambda} + \frac{u_\phi^B}{r^B}\frac{\partial s^B}{\partial \phi}, \tag{13.74}$$

$$\dot{s}^T = \frac{\partial s^T}{\partial t} + \frac{u_\lambda^T}{r^T\cos\phi}\frac{\partial s^T}{\partial \lambda} + \frac{u_\phi^T}{r^T}\frac{\partial s^T}{\partial \phi}. \tag{13.75}$$

13.2.9 The Transformed Equations in Cylindrical-Polar Coordinates

As a second special case, consider cylindrical-polar coordinates $(\lambda, z, \mathtt{r})$. With this ordering, the cylindrical surfaces $\mathtt{r}$ = constant correspond to geopotential surfaces; see Sections 5.2.3, 6.4, and 6.5. From (5.28), and (5.30),

$$(\xi_1, \xi_2, \xi_3) = (\lambda, z, \mathtt{r}), \quad \left(h_\lambda, h_z, h_\mathtt{r}\right) = (\mathtt{r}, 1, 1), \tag{13.76}$$

for these coordinates. Introducing (13.76) into the vertically transformed equations (13.37)–(13.50) leads to the following equation set.

The Governing Equations in Vertically Transformed Cylindrical-Polar Coordinates (λ, z, s, t)

Prognostic Equations

Momentum

$$\frac{Du_\lambda}{Dt}+\left(\frac{u_\lambda}{\mathrm{r}}+2\Omega\right)u_\mathrm{r}+\frac{1}{\rho\mathrm{r}}\left(\frac{\partial p}{\partial\lambda}-\frac{\partial\mathrm{r}}{\partial\lambda}\frac{\partial s}{\partial\mathrm{r}}\frac{\partial p}{\partial s}\right)=F^{u_\lambda}, \tag{13.77}$$

$$\frac{Du_z}{Dt}+\frac{1}{\rho}\left(\frac{\partial p}{\partial z}-\frac{\partial\mathrm{r}}{\partial z}\frac{\partial s}{\partial\mathrm{r}}\frac{\partial p}{\partial s}\right)=F^{u_z}, \tag{13.78}$$

$$\frac{Du_\mathrm{r}}{Dt}-\left(\frac{u_\lambda}{\mathrm{r}}+2\Omega\right)u_\lambda+\frac{1}{\rho}\frac{\partial s}{\partial\mathrm{r}}\frac{\partial p}{\partial s}+\frac{\partial s}{\partial\mathrm{r}}\frac{\partial\Phi}{\partial s}=F^{u_\mathrm{r}}. \tag{13.79}$$

Continuity of Total Mass Density

$$\frac{\mathrm{D}}{Dt}\left(\rho\mathrm{r}\frac{\partial\mathrm{r}}{\partial s}\right)+\left(\rho\mathrm{r}\frac{\partial\mathrm{r}}{\partial s}\right)\left[\frac{\partial}{\partial\lambda}\left(\frac{u_\lambda}{\mathrm{r}}\right)+\frac{\partial u_z}{\partial z}+\frac{\partial\dot{s}}{\partial s}\right]=\mathrm{r}\frac{\partial\mathrm{r}}{\partial s}F^\rho. \tag{13.80}$$

Substance Transport

$$\frac{DS^i}{Dt}=\dot{S}^i,\quad i=1,2,\ldots \tag{13.81}$$

Thermodynamic-Energy Equation

$$\frac{D\mathcal{E}}{Dt}+p\frac{D\alpha}{Dt}=\dot{Q}_E\qquad\text{or}\qquad\frac{D\eta}{Dt}=\frac{\dot{Q}}{T}. \tag{13.82}$$

Definitions and Diagnostic Equations

Definition of Specific Volume

$$\alpha\equiv\frac{1}{\rho}. \tag{13.83}$$

Definition of T, p, and μ^i

$$T=\frac{\partial\mathcal{E}\left(\alpha,\eta,S^1,S^2,\ldots\right)}{\partial\eta}\qquad\text{or}\quad T=\left[\frac{\partial\eta\left(\alpha,\mathcal{E},S^1,S^2,\ldots\right)}{\partial\mathcal{E}}\right]^{-1}, \tag{13.84}$$

$$p=-\frac{\partial\mathcal{E}\left(\alpha,\eta,S^1,S^2,\ldots\right)}{\partial\alpha}\qquad\text{or}\quad p=T\frac{\partial\eta\left(\alpha,\mathcal{E},S^1,S^2,\ldots\right)}{\partial\alpha}, \tag{13.85}$$

$$\mu^i=\frac{\partial\mathcal{E}\left(\alpha,\eta,S^1,S^2,\ldots\right)}{\partial S^i}\qquad\text{or}\quad\mu^i=-T\frac{\partial\eta\left(\alpha,\mathcal{E},S^1,S^2,\ldots\right)}{\partial S^i}. \tag{13.86}$$

Fundamental Equation of State

$$\mathcal{E} = \mathcal{E}\left(\alpha, \eta, S^1, S^2, \ldots\right) \qquad \text{or} \qquad \eta = \eta\left(\alpha, \mathcal{E}, S^1, S^2, \ldots\right). \tag{13.87}$$

Kinematic Equation

$$\mathbf{u} \equiv u_\lambda \mathbf{e}_\lambda + u_z \mathbf{e}_z + u_{\mathrm{r}} \mathbf{e}_{\mathrm{r}} = \mathrm{r}\frac{D\lambda}{Dt}\mathbf{e}_\lambda + \frac{Dz}{Dt}\mathbf{e}_z + \frac{D\mathrm{r}}{Dt}\mathbf{e}_{\mathrm{r}}. \tag{13.88}$$

Material Derivative of a Scalar

$$\frac{Df}{Dt} = \frac{\partial f}{\partial t} + \frac{u_\lambda}{\mathrm{r}}\frac{\partial f}{\partial \lambda} + u_z \frac{\partial f}{\partial z} + \dot{s}\frac{\partial f}{\partial s}. \tag{13.89}$$

Vertical Motion

$$\dot{s} \equiv \frac{Ds}{Dt} = \left(u_{\mathrm{r}} - \frac{\partial \mathrm{r}}{\partial t} - \frac{u_\lambda}{\mathrm{r}}\frac{\partial \mathrm{r}}{\partial \lambda} - u_z\frac{\partial \mathrm{r}}{\partial z}\right)\frac{\partial s}{\partial \mathrm{r}}. \tag{13.90}$$

Vertical Boundary Conditions

$$\dot{s}^B = \frac{\partial s^B}{\partial t} + \frac{u_\lambda^B}{\mathrm{r}^B}\frac{\partial s^B}{\partial \lambda} + u_z^B \frac{\partial s^B}{\partial z}, \tag{13.91}$$

$$\dot{s}^T = \frac{\partial s^T}{\partial t} + \frac{u_\lambda^T}{\mathrm{r}^T}\frac{\partial s^T}{\partial \lambda} + u_z^T \frac{\partial s^T}{\partial z}. \tag{13.92}$$

13.3 MASS CONSERVATION

Vertically integrating the mass-conservation equation – in its flux form (13.34) – with respect to s from s^B to s^T; applying (13.173) with $\mathcal{G} \equiv \mathscr{F}$; and using definition (13.33) leads to

$$\begin{aligned}\frac{\partial}{\partial t}\int_{s^B}^{s^T} \mathscr{F} ds = &- \frac{\partial}{\partial \xi_1}\int_{s^B}^{s^T} \frac{u_1 \mathscr{F}}{h_1} ds - \frac{\partial}{\partial \xi_2}\int_{s^B}^{s^T} \frac{u_2 \mathscr{F}}{h_2} ds + \int_{s^B}^{s^T} \mathscr{F} F^\rho ds \\ &+ \mathscr{F}^T \left(\frac{\partial s^T}{\partial t} + \frac{u_1^T}{h_1^T}\frac{\partial s^T}{\partial \xi_1} + \frac{u_2^T}{h_2^T}\frac{\partial s^T}{\partial \xi_2} - \dot{s}^T\right) - \mathscr{F}^B \left(\frac{\partial s^B}{\partial t} + \frac{u_1^B}{h_1^B}\frac{\partial s^B}{\partial \xi_1} + \frac{u_2^B}{h_2^B}\frac{\partial s^B}{\partial \xi_2} - \dot{s}^B\right),\end{aligned} \tag{13.93}$$

where – from (13.33) –

$$\mathscr{F} \equiv \rho h_1 h_2 h_3 \frac{\partial \xi_3}{\partial s}. \tag{13.94}$$

After applying lower and upper vertical-boundary conditions (13.51) and (13.52), respectively, (13.93) simplifies to

$$\frac{\partial}{\partial t}\int_{s^B}^{s^T} \mathscr{F} ds = - \frac{\partial}{\partial \xi_1}\int_{s^B}^{s^T} \frac{u_1 \mathscr{F}}{h_1} ds - \frac{\partial}{\partial \xi_2}\int_{s^B}^{s^T} \frac{u_2 \mathscr{F}}{h_2} ds + \int_{s^B}^{s^T} \mathscr{F} F^\rho ds. \tag{13.95}$$

Horizontally integrating (13.95) over the entire domain with respect to ξ_1 and ξ_2; applying periodicity in the ξ_1 direction; noting that when integrating in the ξ_2 direction, h_1 is distance from the rotation axis in axial-orthogonal-curvilinear coordinates (see Section 6.4.1), so that $h_1 \equiv 0$ at the poles; and using (13.94) then gives

The Evolution Equation for Globally Integrated Total Mass

$$\frac{\partial}{\partial t}\int_{\mathcal{V}} \rho d\mathcal{V} = \int_{\mathcal{V}} \rho F^{\rho} d\mathcal{V}, \tag{13.96}$$

where, for arbitrary $\mathcal{G}$, the global volume integral of $\mathcal{G}$ is defined by

$$\int_{\mathcal{V}} \mathcal{G} d\mathcal{V} \equiv \iint \left(\int_{s^B}^{s^T} \mathcal{G} h_1 h_2 h_3 \frac{\partial \xi_3}{\partial s} ds \right) d\xi_1 d\xi_2. \tag{13.97}$$

From Section 5.2:

- For spherical-polar coordinates;
  - $(h_1, h_2, h_3) = (r\cos\phi, r, 1)$,
  - $d\mathcal{V} = r^2 \cos\phi \, (\partial r/\partial s) \, d\lambda d\phi ds$.
- For cylindrical-polar coordinates;
  - $(h_1, h_2, h_3) = (\mathtt{r}, 1, 1)$,
  - $d\mathcal{V} = \mathtt{r} \, (\partial \mathtt{r}/\partial s) \, d\lambda dz d\mathtt{r}$.

- In these equations, only vertical-boundary conditions (13.51) and (13.52) were used to enforce no mass transport across the lower and upper boundaries of the domain, with no requirement to invoke (13.53) to accomplish this.

It is then evident from (13.96) that:

- In the absence of internal mass sources and sinks (i.e. $F^{\rho} \equiv 0$), the total mass of fluid is globally conserved provided that (13.51) and (13.52) are satisfied at the lower and upper boundaries, respectively.

13.4 ENERGETICS

13.4.1 Kinetic Energy Evolution

Individually multiplying component momentum equations (13.37)–(13.39) by their respective velocity component; summing the results; multiplying this by $\mathscr{F} \equiv \rho h_1 h_2 h_3 \partial\xi_3/\partial s$ (as defined by (13.33)); and then using (13.36), with $\mathscr{G} \equiv K$, to eliminate $\mathscr{F} DK/Dt$ from the resulting equation leads to

$$\begin{aligned}\frac{\partial}{\partial t}(K\mathscr{F}) = &-\frac{\mathscr{F}}{\rho}\left(\frac{u_1}{h_1}\frac{\partial p}{\partial \xi_1} + \frac{u_2}{h_2}\frac{\partial p}{\partial \xi_2}\right) + \frac{\mathscr{F}}{\rho}\left(\frac{u_1}{h_1}\frac{\partial \xi_3}{\partial \xi_1} + \frac{u_2}{h_2}\frac{\partial \xi_3}{\partial \xi_2} - \frac{u_3}{h_3}\right)\frac{\partial s}{\partial \xi_3}\frac{\partial p}{\partial s} \\ &- \frac{\partial}{\partial \xi_1}\left(\frac{u_1}{h_1}K\mathscr{F}\right) - \frac{\partial}{\partial \xi_2}\left(\frac{u_2}{h_2}K\mathscr{F}\right) - \frac{\partial}{\partial s}\left(\dot{s}K\mathscr{F}\right) - \frac{u_3}{h_3}\frac{\partial s}{\partial \xi_3}\frac{\partial \Phi}{\partial s}\mathscr{F} \\ &+ \left(u_1 F^{u_1} + u_2 F^{u_2} + u_3 F^{u_3} + K F^{\rho}\right)\mathscr{F},\end{aligned} \tag{13.98}$$

where

$$K \equiv \frac{u_1^2 + u_2^2 + u_3^2}{2}, \tag{13.99}$$

is specific kinetic energy. With use of vertical-motion equation (13.50), (13.98) may be rewritten as

$$\begin{aligned}\frac{\partial}{\partial t}(K\mathscr{F}) = &-\frac{\mathscr{F}}{\rho}\left(\frac{u_1}{h_1}\frac{\partial p}{\partial \xi_1} + \frac{u_2}{h_2}\frac{\partial p}{\partial \xi_2} + \dot{s}\frac{\partial p}{\partial s} + \frac{\partial \xi_3}{\partial t}\frac{\partial s}{\partial \xi_3}\frac{\partial p}{\partial s}\right)\\ &-\frac{\partial}{\partial \xi_1}\left(\frac{u_1}{h_1}K\mathscr{F}\right) - \frac{\partial}{\partial \xi_2}\left(\frac{u_2}{h_2}K\mathscr{F}\right) - \frac{\partial}{\partial s}(\dot{s}K\mathscr{F}) - \frac{u_3}{h_3}\frac{\partial s}{\partial \xi_3}\frac{\partial \Phi}{\partial s}\mathscr{F}\\ &+\left(u_1F^{u_1} + u_2F^{u_2} + u_3F^{u_3} + KF^{\rho}\right)\mathscr{F}. \qquad (13.100)\end{aligned}$$

Now

$$\begin{aligned}&\frac{\mathscr{F}}{\rho}\left(\frac{u_1}{h_1}\frac{\partial p}{\partial \xi_1} + \frac{u_2}{h_2}\frac{\partial p}{\partial \xi_2} + \dot{s}\frac{\partial p}{\partial s} + \frac{\partial \xi_3}{\partial t}\frac{\partial s}{\partial \xi_3}\frac{\partial p}{\partial s}\right)\\ &\equiv -p\left[\frac{\partial}{\partial \xi_1}\left(\frac{u_1}{h_1}\frac{\mathscr{F}}{\rho}\right) + \frac{\partial}{\partial \xi_2}\left(\frac{u_2}{h_2}\frac{\mathscr{F}}{\rho}\right) + \frac{\partial}{\partial s}\left(\frac{\mathscr{F}}{\rho}\dot{s}\right) + \frac{\partial}{\partial s}\left(\frac{\mathscr{F}}{\rho}\frac{\partial \xi_3}{\partial t}\frac{\partial s}{\partial \xi_3}\right)\right]\\ &+\frac{\partial}{\partial \xi_1}\left(\frac{u_1}{h_1}\frac{\mathscr{F}}{\rho}p\right) + \frac{\partial}{\partial \xi_2}\left(\frac{u_2}{h_2}\frac{\mathscr{F}}{\rho}p\right) + \frac{\partial}{\partial s}\left(\frac{\mathscr{F}}{\rho}\dot{s}p\right) + \frac{\partial}{\partial s}\left(\frac{\mathscr{F}}{\rho}\frac{\partial \xi_3}{\partial t}\frac{\partial s}{\partial \xi_3}p\right), \qquad (13.101)\end{aligned}$$

and substitution of this identity into (13.100) then yields

$$\begin{aligned}\frac{\partial}{\partial t}(K\mathscr{F}) = &\,p\left[\frac{\partial}{\partial \xi_1}\left(\frac{u_1}{h_1}\frac{\mathscr{F}}{\rho}\right) + \frac{\partial}{\partial \xi_2}\left(\frac{u_2}{h_2}\frac{\mathscr{F}}{\rho}\right) + \frac{\partial}{\partial s}\left(\frac{\mathscr{F}}{\rho}\dot{s}\right) + \frac{\partial}{\partial s}\left(\frac{\mathscr{F}}{\rho}\frac{\partial s}{\partial \xi_3}\frac{\partial \xi_3}{\partial t}\right)\right]\\ &-\frac{\partial}{\partial \xi_1}\left[\frac{u_1}{h_1}\left(K + \frac{p}{\rho}\right)\mathscr{F}\right] - \frac{\partial}{\partial \xi_2}\left[\frac{u_2}{h_2}\left(K + \frac{p}{\rho}\right)\mathscr{F}\right]\\ &-\frac{\partial}{\partial s}\left[\dot{s}\left(K + \frac{p}{\rho}\right)\mathscr{F}\right] - \frac{\partial}{\partial s}\left(p\frac{\mathscr{F}}{\rho}\frac{\partial s}{\partial \xi_3}\frac{\partial \xi_3}{\partial t}\right) - \frac{u_3}{h_3}\frac{\partial s}{\partial \xi_3}\frac{\partial \Phi}{\partial s}\mathscr{F}\\ &+\left(u_1F^{u_1} + u_2F^{u_2} + u_3F^{u_3} + KF^{\rho}\right)\mathscr{F}. \qquad (13.102)\end{aligned}$$

Setting $\psi = t$ and $f(\xi_1, \xi_2) = (\mathscr{F}/\rho)\,\partial s/\partial \xi_3 \equiv h_1(\xi_2, \xi_3)\,h_2(\xi_2, \xi_3)\,h_3(\xi_2, \xi_3)$ (from (13.94)) in identity (13.176) gives

$$\frac{\partial}{\partial s}\left(\frac{\mathscr{F}}{\rho}\frac{\partial s}{\partial \xi_3}\frac{\partial \xi_3}{\partial t}\right) \equiv \frac{\partial}{\partial s}\left(h_1h_2h_3\frac{\partial \xi_3}{\partial t}\right) \equiv \frac{\partial}{\partial t}\left(h_1h_2h_3\frac{\partial \xi_3}{\partial s}\right) \equiv \frac{\partial}{\partial t}\left(\frac{\mathscr{F}}{\rho}\right). \qquad (13.103)$$

Using (13.33) and (13.103) in (13.102) finally leads to:

The Kinetic Energy Evolution Equation

$$\begin{aligned}\frac{\partial (K\mathscr{F})}{\partial t} = &\,p\left[\frac{\partial}{\partial t}\left(\frac{\mathscr{F}}{\rho}\right) + \frac{\partial}{\partial \xi_1}\left(\frac{u_1}{h_1}\frac{\mathscr{F}}{\rho}\right) + \frac{\partial}{\partial \xi_2}\left(\frac{u_2}{h_2}\frac{\mathscr{F}}{\rho}\right) + \frac{\partial}{\partial s}\left(\dot{s}\frac{\mathscr{F}}{\rho}\right)\right]\\ &-\frac{\partial}{\partial \xi_1}\left[\frac{u_1}{h_1}\left(K + \frac{p}{\rho}\right)\mathscr{F}\right] - \frac{\partial}{\partial \xi_2}\left[\frac{u_2}{h_2}\left(K + \frac{p}{\rho}\right)\mathscr{F}\right]\\ &-\frac{\partial}{\partial s}\left[\dot{s}\left(K + \frac{p}{\rho}\right)\mathscr{F}\right] - \frac{\partial}{\partial s}\left(ph_1h_2h_3\frac{\partial \xi_3}{\partial t}\right) - \frac{u_3}{h_3}\frac{\partial s}{\partial \xi_3}\frac{\partial \Phi}{\partial s}\mathscr{F}\\ &+\left(u_1F^{u_1} + u_2F^{u_2} + u_3F^{u_3} + KF^{\rho}\right)\mathscr{F}. \qquad (13.104)\end{aligned}$$

13.4.2 Potential Gravitational Energy Evolution

Multiplying mass-continuity equation (13.34) through by Φ and exploiting transformation equation (13.16) and vertical-motion equation (13.50) leads to

$$\frac{\partial}{\partial t}(\Phi\mathscr{F}) = \left[\frac{\partial \Phi}{\partial t} + \frac{u_1}{h_1}\frac{\partial \Phi}{\partial \xi_1} + \frac{u_2}{h_2}\frac{\partial \Phi}{\partial \xi_2} - \left(\frac{\partial \xi_3}{\partial t} + \frac{u_1}{h_1}\frac{\partial \xi_3}{\partial \xi_1} + \frac{u_2}{h_2}\frac{\partial \xi_3}{\partial \xi_2}\right)\frac{\partial s}{\partial \xi_3}\frac{\partial \Phi}{\partial s}\right]\mathscr{F} + \frac{u_3}{h_3}\frac{\partial s}{\partial \xi_3}\frac{\partial \Phi}{\partial s}\mathscr{F}$$
$$- \frac{\partial}{\partial \xi_1}\left(\frac{u_1}{h_1}\Phi\mathscr{F}\right) - \frac{\partial}{\partial \xi_2}\left(\frac{u_2}{h_2}\Phi\mathscr{F}\right) - \frac{\partial}{\partial s}(\dot{s}\Phi\mathscr{F}) + \Phi\mathscr{F}F^{\rho}. \tag{13.105}$$

Using transformation equation (13.14) and noting that Φ is a function only of ξ_3 gives

$$\left(\frac{\partial \Phi}{\partial \psi}\right)_s - \left(\frac{\partial \xi_3}{\partial \psi}\right)_s \frac{\partial s}{\partial \xi_3}\frac{\partial \Phi}{\partial s} = \left(\frac{\partial \Phi}{\partial \psi}\right)_{\xi_3} = 0, \quad \psi = \xi_1, \xi_2, t. \tag{13.106}$$

Insertion of (13.106), for $\psi = \xi_1, \xi_2, t$, into (13.105) then yields:

The Potential Gravitational Energy Evolution Equation

$$\frac{\partial}{\partial t}(\Phi\mathscr{F}) = \frac{u_3}{h_3}\frac{\partial s}{\partial \xi_3}\frac{\partial \Phi}{\partial s}\mathscr{F} - \frac{\partial}{\partial \xi_1}\left(\frac{u_1}{h_1}\Phi\mathscr{F}\right) - \frac{\partial}{\partial \xi_2}\left(\frac{u_2}{h_2}\Phi\mathscr{F}\right) - \frac{\partial}{\partial s}(\dot{s}\Phi\mathscr{F}) + \Phi F^{\rho}\mathscr{F}. \tag{13.107}$$

13.4.3 Internal Energy Evolution

Setting $\mathcal{G} \equiv \mathcal{E}$ in (13.36), using thermodynamic-energy equation (13.42) to then eliminate $\mathscr{F}D\mathcal{E}/Dt$, and rearranging leads to

$$\frac{\partial}{\partial t}(\mathcal{E}\mathscr{F}) = -p\mathscr{F}\frac{D}{Dt}\left(\frac{1}{\rho}\right) - \frac{\partial}{\partial \xi_1}\left(\frac{u_1}{h_1}\mathcal{E}\mathscr{F}\right) - \frac{\partial}{\partial \xi_2}\left(\frac{u_2}{h_2}\mathcal{E}\mathscr{F}\right) - \frac{\partial}{\partial s}(\dot{s}\mathcal{E}\mathscr{F}) + \left(\dot{Q}_E + \mathcal{E}F^{\rho}\right)\mathscr{F}. \tag{13.108}$$

Applying (13.36) again, but with $\mathcal{G} \equiv \alpha \equiv 1/\rho$, and inserting the resulting equation into (13.108) then gives:

The Internal Energy Evolution Equation

$$\frac{\partial}{\partial t}(\mathcal{E}\mathscr{F}) = -p\left[\frac{\partial}{\partial t}\left(\frac{\mathscr{F}}{\rho}\right) + \frac{\partial}{\partial \xi_1}\left(\frac{u_1}{h_1}\frac{\mathscr{F}}{\rho}\right) + \frac{\partial}{\partial \xi_2}\left(\frac{u_2}{h_2}\frac{\mathscr{F}}{\rho}\right) + \frac{\partial}{\partial s}\left(\dot{s}\frac{\mathscr{F}}{\rho}\right)\right]$$
$$- \frac{\partial}{\partial \xi_1}\left(\frac{u_1}{h_1}\mathcal{E}\mathscr{F}\right) - \frac{\partial}{\partial \xi_2}\left(\frac{u_2}{h_2}\mathcal{E}\mathscr{F}\right) - \frac{\partial}{\partial s}(\dot{s}\mathcal{E}\mathscr{F}) + \left[\dot{Q}_E + \left(\mathcal{E} + \frac{p}{\rho}\right)F^{\rho}\right]\mathscr{F}. \tag{13.109}$$

13.4.4 Total Energy Evolution

Summing (13.104), (13.107), and (13.109), for the individual energy-evolution equations finally yields:

The Evolution Equation For Total Energy

$$\frac{\partial}{\partial t}(E\mathcal{F}) = -\frac{\partial}{\partial \xi_1}\left[\frac{u_1}{h_1}\left(E+\frac{p}{\rho}\right)\mathcal{F}\right] - \frac{\partial}{\partial \xi_2}\left[\frac{u_2}{h_2}\left(E+\frac{p}{\rho}\right)\mathcal{F}\right]$$
$$-\frac{\partial}{\partial s}\left[\dot{s}\left(E+\frac{p}{\rho}\right)\mathcal{F}\right] - \frac{\partial}{\partial s}\left(ph_1h_2h_3\frac{\partial \xi_3}{\partial t}\right)$$
$$+\left[u_1F^{u_1} + u_2F^{u_2} + u_3F^{u_3} + \dot{Q}_E + \left(E+\frac{p}{\rho}\right)F^{\rho}\right]\mathcal{F}, \tag{13.110}$$

where

$$E \equiv K + \Phi + \mathcal{E}, \tag{13.111}$$

is total energy per unit mass (i.e. specific total energy).

13.4.5 Global Conservation of Total Energy

Integrating (13.110) with respect to s from s^B to s^T; applying (13.174) with $\mathcal{G} \equiv E\mathcal{F}$, and (13.175) with $\mathcal{G} \equiv (p/\rho)\,\mathcal{F} \equiv h_1h_2h_3\partial\xi_3/\partial s$, where vertical boundary conditions (13.51) and (13.52) have been applied to obtain both (13.174) and (13.175); and using definition (13.33) then leads to

$$\frac{\partial}{\partial t}\int_{s^B}^{s^T} E\mathcal{F}\,ds = -\frac{\partial}{\partial \xi_1}\int_{s^B}^{s^T}\frac{u_1}{h_1}\left(E+\frac{p}{\rho}\right)\mathcal{F}\,ds - \frac{\partial}{\partial \xi_2}\int_{s^B}^{s^T}\frac{u_2}{h_2}\left(E+\frac{p}{\rho}\right)\mathcal{F}\,ds$$
$$-\left(ph_1h_2h_3\frac{\partial \xi_3}{\partial s}\right)^T\frac{\partial s^T}{\partial t} + \left(ph_1h_2h_3\frac{\partial \xi_3}{\partial s}\right)^B\frac{\partial s^B}{\partial t}$$
$$-\left(ph_1h_2h_3\frac{\partial \xi_3}{\partial t}\right)^T + \left(ph_1h_2h_3\frac{\partial \xi_3}{\partial t}\right)^B$$
$$+\int_{s^B}^{s^T}\left[u_1F^{u_1} + u_2F^{u_2} + u_3F^{u_3} + \dot{Q}_E + \left(E+\frac{p}{\rho}\right)F^{\rho}\right]\mathcal{F}\,ds. \tag{13.112}$$

Since s^T and s^B are generally functions of time, then, as noted by Kasahara (1974) but in the present context and notation,

$$\left[\left(\frac{\partial \xi_3}{\partial t}\right)_s\right]^T = \frac{\partial \xi_3^T}{\partial t} - \left(\frac{\partial \xi_3}{\partial s}\right)^T\frac{\partial s^T}{\partial t}, \tag{13.113}$$

$$\left[\left(\frac{\partial \xi_3}{\partial t}\right)_s\right]^B = \frac{\partial \xi_3^B}{\partial t} - \left(\frac{\partial \xi_3}{\partial s}\right)^B\frac{\partial s^B}{\partial t}, \tag{13.114}$$

where (13.114) is derived as (13.181) in the Appendix to this chapter. Hence, using (13.113) and (13.114), (13.112) reduces to

$$\frac{\partial}{\partial t}\int_{s^B}^{s^T} E\mathcal{F}\,ds = -\frac{\partial}{\partial \xi_1}\int_{s^B}^{s^T}\frac{u_1}{h_1}\left(E+\frac{p}{\rho}\right)\mathcal{F}\,ds - \frac{\partial}{\partial \xi_2}\int_{s^B}^{s^T}\frac{u_2}{h_2}\left(E+\frac{p}{\rho}\right)\mathcal{F}\,ds$$
$$+\int_{s^B}^{s^T}\left[u_1F^{u_1} + u_2F^{u_2} + u_3F^{u_3} + \dot{Q}_E + \left(E+\frac{p}{\rho}\right)F^{\rho}\right]\mathcal{F}\,ds$$
$$-\left(ph_1h_2h_3\right)^T\frac{\partial \xi_3^T}{\partial t} + \left(ph_1h_2h_3\right)^B\frac{\partial \xi_3^B}{\partial t}. \tag{13.115}$$

Finally, integrating (13.115) with respect to ξ_1 and ξ_2 over the entire global domain; applying periodicity in the ξ_1 direction; noting that when integrating in the ξ_2 direction, h_1 is distance from the rotation axis, so that $h_1 \equiv 0$ at the poles; and using definition (13.33) yields:

The Evolution Equation for Globally Integrated Total Energy

$$\frac{\partial}{\partial t}\int_{\mathcal{V}} \rho E d\mathcal{V} = \int_{\mathcal{V}} \rho \left[u_1 F^{u_1} + u_2 F^{u_2} + u_3 F^{u_3} + \dot{Q}_E + \left(E + \frac{p}{\rho} \right) F^{\rho} \right] d\mathcal{V}$$
$$+ \int_{\mathscr{A}^B} p^B h_3^B \frac{\partial \xi_3^B}{\partial t} d\mathscr{A}^B - \int_{\mathscr{A}^T} p^T h_3^T \frac{\partial \xi_3^T}{\partial t} d\mathscr{A}^T. \tag{13.116}$$

For arbitrary $\mathcal{G}$, the global volume integral of $\mathcal{G}$ is defined by (13.97), and boundary area integrals by

$$\int_{\mathscr{A}} \mathcal{G} d\mathscr{A} \equiv \iint \mathcal{G} h_1 h_2 d\xi_1 d\xi_2, \tag{13.117}$$

with evaluation at $s = s^B(\xi_1, \xi_2, t)$ for $\mathscr{A} = \mathscr{A}^B$, and at $s = s^T(\xi_1, \xi_2, t)$ for $\mathscr{A} = \mathscr{A}^T$.

From Section 5.3:

- For spherical-polar coordinates:
  - $(h_1, h_2) = (r\cos\phi, r)$,
  - $d\mathscr{A} = r^2 \cos\phi d\lambda d\phi$; and
- For cylindrical-polar coordinates:
  - $(h_1, h_2) = (\mathtt{r}, 1)$,
  - $d\mathscr{A} = \mathtt{r} d\lambda dz$.

Physical interpretation of (13.116) is given in Section 13.6.2. In particular, this equation is used to examine the impact of lower and upper boundary conditions on conservation of globally integrated total energy.

13.5 AXIAL-ANGULAR-MOMENTUM CONSERVATION

13.5.1 Evolution

By definition of an axial-orthogonal-curvilinear coordinate system, $h_1 = h_1(\xi_2, \xi_3)$; see (6.13). After multiplication by h_1, and use of (13.13) with $f \equiv h_1$, the u_1 component (13.37) of the momentum equation may be rewritten as

$$h_1 \frac{Du_1}{Dt} + \left(u_1 + 2\Omega h_1\right) \frac{Dh_1}{Dt} = -\frac{1}{\rho}\left(\frac{\partial p}{\partial \xi_1} - \frac{\partial \xi_3}{\partial \xi_1}\frac{\partial s}{\partial \xi_3}\frac{\partial p}{\partial s} \right) + h_1 F^{u_1}. \tag{13.118}$$

Using the identity

$$\frac{DM}{Dt} \equiv h_1 \frac{Du_1}{Dt} + \left(u_1 + 2\Omega h_1\right) \frac{Dh_1}{Dt}, \tag{13.119}$$

where

$$M \equiv \left(u_1 + \Omega h_1\right) h_1, \tag{13.120}$$

is *axial angular momentum per unit mass*, (13.118) may be rewritten as

$$\frac{DM}{Dt} = -\frac{1}{\rho}\left(\frac{\partial p}{\partial \xi_1} - \frac{\partial \xi_3}{\partial \xi_1}\frac{\partial s}{\partial \xi_3}\frac{\partial p}{\partial s} \right) + h_1 F^{u_1}. \tag{13.121}$$

Now evaluating (13.36) with $\mathcal{G} \equiv M$ gives

$$\mathscr{F}\frac{DM}{Dt} \equiv \frac{\partial}{\partial t}(M\mathscr{F}) + \frac{\partial}{\partial \xi_1}\left(\frac{u_1}{h_1}M\mathscr{F}\right) + \frac{\partial}{\partial \xi_2}\left(\frac{u_2}{h_2}M\mathscr{F}\right) + \frac{\partial}{\partial s}(\dot{s}M\mathscr{F}) - M\mathscr{F}F^{\rho}. \quad (13.122)$$

Eliminating DM/Dt between (13.121) and (13.122) then yields

$$\frac{\partial}{\partial t}(M\mathscr{F}) = -\frac{\partial}{\partial \xi_1}\left(\frac{u_1}{h_1}M\mathscr{F}\right) - \frac{\partial}{\partial \xi_2}\left(\frac{u_2}{h_2}M\mathscr{F}\right) - \frac{\partial}{\partial s}(\dot{s}M\mathscr{F}) - \frac{1}{\rho}\left(\frac{\partial p}{\partial \xi_1} - \frac{\partial \xi_3}{\partial \xi_1}\frac{\partial s}{\partial \xi_3}\frac{\partial p}{\partial s}\right)\mathscr{F} \\ + \left(h_1 F^{u_1} + MF^{\rho}\right)\mathscr{F}. \quad (13.123)$$

For the pressure-gradient terms in (13.123), use of definition (13.33) for $\mathscr{F}$ gives

$$\begin{aligned} \frac{1}{\rho}\left(\frac{\partial p}{\partial \xi_1} - \frac{\partial \xi_3}{\partial \xi_1}\frac{\partial s}{\partial \xi_3}\frac{\partial p}{\partial s}\right)\mathscr{F} &= \left(\frac{\partial p}{\partial \xi_1} - \frac{\partial \xi_3}{\partial \xi_1}\frac{\partial s}{\partial \xi_3}\frac{\partial p}{\partial s}\right)h_1 h_2 h_3 \frac{\partial \xi_3}{\partial s} \\ &= \frac{\partial}{\partial \xi_1}\left(p h_1 h_2 h_3 \frac{\partial \xi_3}{\partial s}\right) - p\frac{\partial}{\partial \xi_1}\left(h_1 h_2 h_3 \frac{\partial \xi_3}{\partial s}\right) \\ &\quad - \frac{\partial}{\partial s}\left(p h_1 h_2 h_3 \frac{\partial \xi_3}{\partial \xi_1}\right) + p\frac{\partial}{\partial s}\left(h_1 h_2 h_3 \frac{\partial \xi_3}{\partial \xi_1}\right). \end{aligned} \quad (13.124)$$

But – from (13.176) with $\psi = \xi_1$ and $f(\xi_2, \xi_3) = h_1(\xi_2, \xi_3)\, h_2(\xi_2, \xi_3)\, h_3(\xi_2, \xi_3)$ –

$$\frac{\partial}{\partial s}\left(h_1 h_2 h_3 \frac{\partial \xi_3}{\partial \xi_1}\right) = \frac{\partial}{\partial \xi_1}\left(h_1 h_2 h_3 \frac{\partial \xi_3}{\partial s}\right), \quad (13.125)$$

and so (13.124) reduces to

$$\frac{1}{\rho}\left(\frac{\partial p}{\partial \xi_1} - \frac{\partial \xi_3}{\partial \xi_1}\frac{\partial s}{\partial \xi_3}\frac{\partial p}{\partial s}\right)\mathscr{F} = \frac{\partial}{\partial \xi_1}\left(p h_1 h_2 h_3 \frac{\partial \xi_3}{\partial s}\right) - \frac{\partial}{\partial s}\left(p h_1 h_2 h_3 \frac{\partial \xi_3}{\partial \xi_1}\right). \quad (13.126)$$

Substituting (13.126) into (13.123) then yields:

The Evolution Equation for Axial Angular Momentum

$$\frac{\partial}{\partial t}(M\mathscr{F}) = -\frac{\partial}{\partial \xi_1}\left(\frac{u_1}{h_1}M\mathscr{F}\right) - \frac{\partial}{\partial \xi_2}\left(\frac{u_2}{h_2}M\mathscr{F}\right) - \frac{\partial}{\partial s}(\dot{s}M\mathscr{F}) \\ - \frac{\partial}{\partial \xi_1}\left(p h_1 h_2 h_3 \frac{\partial \xi_3}{\partial s}\right) + \frac{\partial}{\partial s}\left(p h_1 h_2 h_3 \frac{\partial \xi_3}{\partial \xi_1}\right) + \left(h_1 F^{u_1} + MF^{\rho}\right)\mathscr{F}. \quad (13.127)$$

13.5.2 Global Conservation of Axial Angular Momentum

Integrating (13.127) with respect to s from s^B to s^T; applying (13.172) with $G \equiv p\mathscr{F}/\rho$ and $\psi \equiv \xi_1$; applying (13.174) with $G \equiv M\mathscr{F}$; and using definition (13.33) for $\mathscr{F}$ leads to

$$\begin{aligned} \frac{\partial}{\partial t}\int_{s^B}^{s^T} M\mathscr{F}\, ds &= -\frac{\partial}{\partial \xi_1}\int_{s^B}^{s^T} \frac{u_1}{h_1}M\mathscr{F}\, ds - \frac{\partial}{\partial \xi_2}\int_{s^B}^{s^T} \frac{u_2}{h_2}M\mathscr{F}\, ds - \frac{\partial}{\partial \xi_1}\int_{s^B}^{s^T} p h_1 h_2 h_3 \frac{\partial \xi_3}{\partial s} ds \\ &\quad + \int_{s^B}^{s^T}\left(h_1 F^{u_1} + MF^{\rho}\right)\mathscr{F}\, ds + \left(p h_1 h_2 h_3 \frac{\partial \xi_3}{\partial \xi_1}\right)^T - \left(p h_1 h_2 h_3 \frac{\partial \xi_3}{\partial \xi_1}\right)^B \\ &\quad + \left(p h_1 h_2 h_3 \frac{\partial \xi_3}{\partial s}\right)^T \frac{\partial s^T}{\partial \xi_1} - \left(p h_1 h_2 h_3 \frac{\partial \xi_3}{\partial s}\right)^B \frac{\partial s^B}{\partial \xi_1}. \end{aligned} \quad (13.128)$$

Note that boundary conditions (13.51) and (13.52) have implicitly been applied due to use of (13.174), which assumes them.

Similarly as for (13.113) and (13.114),

$$\left(ph_1h_2h_3\frac{\partial \xi_3}{\partial \xi_1}\right)^T - \left(ph_1h_2h_3\frac{\partial \xi_3}{\partial \xi_1}\right)^B = (ph_1h_2h_3)^T\left[\frac{\partial \xi_3^T}{\partial \xi_1} - \left(\frac{\partial \xi_3}{\partial s}\right)^T \frac{\partial s^T}{\partial \xi_1}\right] - (ph_1h_2h_3)^B\left[\frac{\partial \xi_3^B}{\partial \xi_1} - \left(\frac{\partial \xi_3}{\partial s}\right)^B \frac{\partial s^B}{\partial \xi_1}\right], \tag{13.129}$$

and so (13.128) reduces to

$$\frac{\partial}{\partial t}\int_{s^B}^{s^T} M\mathscr{F}ds = -\frac{\partial}{\partial \xi_1}\int_{s^B}^{s^T}\frac{u_1}{h_1}M\mathscr{F}ds - \frac{\partial}{\partial \xi_2}\int_{s^B}^{s^T}\frac{u_2}{h_2}M\mathscr{F}ds - \frac{\partial}{\partial \xi_1}\int_{s^B}^{s^T} ph_1h_2h_3\frac{\partial \xi_3}{\partial s}ds + (ph_1h_2h_3)^T\frac{\partial \xi_3^T}{\partial \xi_1} - (ph_1h_2h_3)^B\frac{\partial \xi_3^B}{\partial \xi_1} + \int_{s^B}^{s^T}\left(h_1F^{u_1} + MF^{\rho}\right)\mathscr{F}ds. \tag{13.130}$$

Finally, integrating (13.130), with respect to ξ_1 and ξ_2, over the entire horizontal domain; applying periodicity in the ξ_1 direction; and noting that $h_1 \equiv 0$ at poles gives

The Evolution Equation for Globally Integrated Axial Angular Momentum

$$\frac{\partial}{\partial t}\int_{\mathcal{V}}\rho M d\mathcal{V} = \int_{\mathscr{A}^T} p^T h_3^T\frac{\partial \xi_3^T}{\partial \xi_1}d\mathscr{A}^T - \int_{\mathscr{A}^B} p^B h_3^B\frac{\partial \xi_3^B}{\partial \xi_1}d\mathscr{A}^B + \int_{\mathcal{V}}\rho\left(h_1F^{u_1} + MF^{\rho}\right)d\mathcal{V}, \tag{13.131}$$

where the volume and area integrals with respect to $d\mathcal{V}$ and $d\mathscr{A}$ are defined by (13.97) and (13.117), respectively.

Physical interpretation of (13.131) is given in Section 13.6.3. In particular, this equation is used to examine the impact of lower and upper boundary conditions on conservation of globally integrated axial angular momentum.

13.6 BOUNDARY CONDITIONS IN THE VERTICAL AND GLOBAL CONSERVATION

13.6.1 Preamble

Evolution equations for globally integrated total energy, and for globally integrated axial angular momentum, have been developed in Sections 13.4.5 and 13.5.2, respectively. These equations facilitate examination of how vertical boundary conditions influence conservation of these two quantities, considered both individually and together. See Fig. 13.5 for four prototypical configurations of interest, with various combinations of lower and upper, atmospheric and oceanic boundary conditions.

Two types of boundary are considered here; rigid and elastic:

1. A *rigid* boundary is fixed in space for all time, that is, $\xi_3 = \xi_3^{\text{rigid}}(\xi_1, \xi_2)$ – with no dependence on time t. Examples of rigid boundaries are (Figs. 13.5a–d):
   (a) The underlying orography of the atmosphere.
   (b) The underlying bathymetry of an ocean.
   (c) A rigid lid overlying the atmosphere or an ocean.

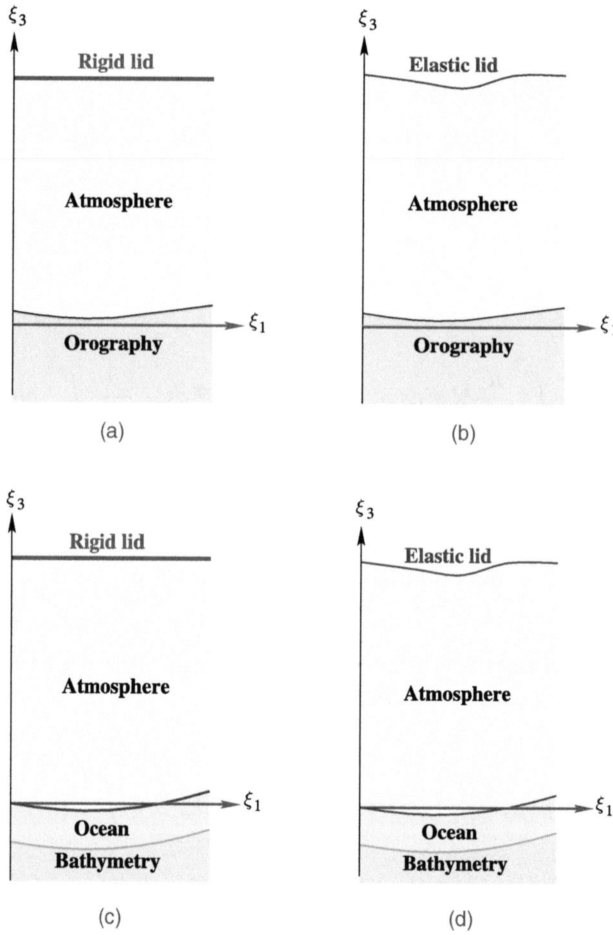

Figure 13.5 Local vertical slice depictions of: (a) atmosphere (with rigid lid), over orography; (b) atmosphere (elastic lid), over orography; (c) atmosphere (rigid lid) + ocean, over bathymetry; and (d) atmosphere (elastic lid) + ocean, over bathymetry. Orography and bathymetry are rigid; ocean-atmosphere interface is elastic.
Atmosphere is in light blue; ocean in darker blue; orography and bathymetry in green; geopotential surfaces in red; and the ξ_1 axis is embedded in the geoid.

Although a rigid boundary is fixed in space (with no dependence on time), other variables there (e.g. pressure p) can (and usually do) not only vary spatially, but also temporally.

2. An *elastic* boundary can move in space as a function of time t, that is, $\xi_3 = \xi_3^{\text{elastic}}(\xi_1, \xi_2, t)$. Its motion is nevertheless constrained; of particular interest here is when an imposed pressure is applied at the boundary. Examples of elastic boundaries are (Figs. 13.5b–d):
   (a) An elastic lid overlying the atmosphere, constrained by $p = p^{\text{elastic}}(\xi_1, \xi_2)$ (which, as a special case, could be equal to a constant) – with no dependence on time.
   (b) The boundary between the atmosphere and an ocean, constrained by continuity there of pressure $p = p^{\text{elastic}}(\xi_1, \xi_2, t)$ – with dependence on both space and time.

In what follows, atmospheric and oceanic forcings *within the fluid* (but *not* at its lower and upper boundaries) are assumed absent (i.e. $F^{u_1} = F^{u_2} = F^{u_3} = \dot{Q}_E = F^{\rho} = 0$).[11] This allows us to

[11] These forcings cannot, of course, be neglected in realistic atmospheric and oceanic models.

focus here on the impact of lower and upper boundary conditions on conservation of globally integrated total energy and axial angular momentum.

13.6.2 Total Energy

13.6.2.1 *Evolution Equation*

The evolution of globally integrated total energy of a fluid (be it the atmosphere or an ocean) is governed by (13.116). In the absence of atmospheric and oceanic forcings within the fluid (i.e. when $F^{u_1} = F^{u_2} = F^{u_3} = \dot{Q}_E = F^{\rho} = 0$), (13.116) reduces to

$$\frac{\partial}{\partial t}\int_{\mathcal{V}} \rho E d\mathcal{V} = \int_{\mathscr{A}^B} p^B h_3^B \frac{\partial \xi_3^B}{\partial t} d\mathscr{A}^B - \int_{\mathscr{A}^T} p^T h_3^T \frac{\partial \xi_3^T}{\partial t} d\mathscr{A}^T, \tag{13.132}$$

where superscripts 'B' and 'T' denote evaluation at the fluid's lower and upper boundaries, respectively. The two terms on the right-hand side of (13.132) represent a possible exchange of energy across the fluid's lower and upper boundaries, respectively. The mechanism of this possible exchange of energy is the rate of working of the pressure field at the boundary. If the two terms are identically zero (or fortuitously cancel one another), then globally integrated total energy of the fluid (be it for the atmosphere or for an ocean) is conserved.

13.6.2.2 *Rigid Boundary*

At a rigid boundary (as in Figs. 13.5a,b for orography, Figs. 13.5c,d for bathymetry, and Figs. 13.5a,c for a rigid lid), $\xi_3 = \xi_3^{\text{boundary}}(\xi_1, \xi_2)$, where superscript 'boundary' denotes evaluation at a lower or upper boundary. Thus

$$\frac{\partial \xi_3^{\text{boundary}}(\xi_1, \xi_2)}{\partial t} \equiv 0, \tag{13.133}$$

and the corresponding boundary term on the right-hand side of (13.132) vanishes.[12] There is then no work done by the pressure field at that (bottom, or top) boundary, and conservation of energy is entirely determined by whether work is done by the pressure field at the other (top, or bottom) boundary of the fluid (i.e. atmosphere or ocean).

13.6.2.3 *Elastic Boundary*

At an elastic boundary (as in Figs. 13.5b,d for an elastic atmospheric lid, and Figs. 13.5c,d for an elastic atmosphere/ocean boundary), $\partial \xi_3^{\text{boundary}}(\xi_1, \xi_2, t)/\partial t \neq 0$. This opens up the possibility of work being done by the pressure field at an elastic boundary (in contradistinction to no such

[12] Strictly speaking, (13.133) is not exactly satisfied at an orographic or bathymetric boundary. Physically, Earth exerts a pressure torque there, thereby changing the atmosphere's globally integrated axial angular momentum. This physical phenomenon is generally represented in atmospheric and oceanic models. However any increase or decrease of this quantity in the atmosphere or ocean results in a corresponding decrease or increase, respectively, in Earth's globally integrated axial angular momentum in order for the combined Earth-atmosphere-ocean system to respect conservation. This is associated with minute fluctuations in the length of day which are astronomically observable and well correlated with the fluctuations in the atmosphere's globally integrated axial angular momentum (Barnes et al., 1983). This physical phenomenon is generally neglected in atmospheric and oceanic models, and (13.133) is assumed to hold. We follow this practice herein, but it is good to be aware of this hidden assumption.

change at a rigid boundary, as noted earlier). Nevertheless, (13.132) implies that the pressure field does no work *in total* at an elastic boundary provided that

$$\int_{\mathscr{A}^{\text{boundary}}} \left(p h_3 \frac{\partial \xi_3}{\partial t} \right)^{\text{boundary}} d\mathscr{A}^{\text{boundary}} = 0. \tag{13.134}$$

If this is not the case, then energy is lost or gained by the system through the net work done by, or against, the pressure at the boundary in displacing it.

For (13.134) to hold generally, and hence for there to be no work done at the boundary, the pressure there must be identically zero (since $h_3 \partial \xi_3 / \partial t \neq 0$ for an elastic boundary). Physically, $p \equiv 0$ can only be realised (on planets with atmospheres) at the top boundary of the atmosphere, since the pressure exerted by an atmosphere on an ocean is always strictly positive.

13.6.2.4 Atmosphere–Ocean Boundary

For an atmosphere overlying an ocean (as in Figs. 13.5c,d), the globally integrated total energy is the sum of the individual atmospheric and oceanic energies. This raises two questions:

1. What happens at their interfacial surface (bottom of the atmosphere/top of the ocean)?
2. What impact does this have on the combined (atmospheric + oceanic) globally integrated total energy?

To examine these questions, we take the sum of (13.132), evaluated for the atmosphere, with its counterpart for an ocean. Thus

$$\begin{aligned}
\frac{\partial}{\partial t} & \left[\left(\int_{\mathcal{V}} \rho E d\mathcal{V} \right)^{\text{atmos}} + \left(\int_{\mathcal{V}} \rho E d\mathcal{V} \right)^{\text{ocean}} \right] \\
&= \cancel{\left(\int_{\mathscr{A}^B} p^B h_3^B \frac{\partial \xi_3^B}{\partial t} d\mathscr{A}^B \right)^{\text{atmos}}} - \left(\int_{\mathscr{A}^T} p^T h_3^T \frac{\partial \xi_3^T}{\partial t} d\mathscr{A}^T \right)^{\text{atmos}} \\
&+ \left(\int_{\mathscr{A}^B} p^B h_3^B \frac{\partial \xi_3^B}{\partial t} d\mathscr{A}^B \right)^{\text{ocean}} - \cancel{\left(\int_{\mathscr{A}^T} p^T h_3^T \frac{\partial \xi_3^T}{\partial t} d\mathscr{A}^T \right)^{\text{ocean}}} \\
&= \left(\int_{\mathscr{A}^B} p^B h_3^B \frac{\partial \xi_3^B}{\partial t} d\mathscr{A}^B \right)^{\text{ocean}} - \left(\int_{\mathscr{A}^T} p^T h_3^T \frac{\partial \xi_3^T}{\partial t} d\mathscr{A}^T \right)^{\text{atmos}},
\end{aligned} \tag{13.135}$$

where the superscripts 'atmos' and 'ocean' denote evaluation at an atmospheric or oceanic boundary, respectively. The mutual cancellation of two terms in (13.135) is due to the assumed continuity of pressure at the ocean-atmosphere interface; the pressures exerted by the atmosphere and ocean on either side of their mutual boundary are assumed equal.

Physically (under the given assumptions), energy can be exchanged across the ocean-atmosphere boundary, but only in a zero-sum way, otherwise continuity of pressure there would be compromised. It does not matter whether this boundary is rigid or elastic, since the net sum there is zero regardless. This means that the conservation of the combined (atmosphere + ocean) globally integrated total energy is determined by any net work done by or against the pressure field at the top of the atmosphere and at the bottom of the ocean. As shown in Section 13.6.2.2, there is no work done by the pressure field at the bottom of the ocean since the bathymetry is assumed rigid. Thus the conservation of the combined globally integrated total energy is entirely determined by what happens at the top of the atmosphere:

- For a rigid atmospheric lid, no work is done by or against the pressure field there – see Section 13.6.2.2 – and hence the globally integrated total energy of the combined (atmosphere + ocean) system is conserved. The atmospheric and oceanic contributions are not individually conserved, however, only their sum.

- For an elastic atmospheric lid, there is clearly no work done by the pressure field when $p = 0$ – see Section 13.6.2.3 – and the globally integrated total energy of the combined (atmosphere + ocean) system is again conserved. However, when $p \neq 0$, there may be net work done at an elastic lid, and hence the globally integrated total energy of the combined (atmosphere + ocean) system is then no longer conserved.

13.6.3 Axial Angular Momentum

13.6.3.1 *Evolution Equation*

The evolution of globally integrated axial angular momentum of a fluid (be it the atmosphere or an ocean) is governed by (13.131). In the absence of atmospheric and oceanic forcings within the fluid (i.e. when $F^{u_1} = F^{u_2} = F^{u_3} = \dot{Q}_E = F^{\rho} = 0$), (13.131) reduces to

$$\frac{\partial}{\partial t}\int_{\mathcal{V}} \rho M dV = \int_{\mathscr{A}^T} p^T h_3^T \frac{\partial \xi_3^T}{\partial \xi_1} d\mathscr{A}^T - \int_{\mathscr{A}^B} p^B h_3^B \frac{\partial \xi_3^B}{\partial \xi_1} d\mathscr{A}^B. \tag{13.136}$$

Equation (13.136) for the evolution of globally integrated axial angular momentum is of similar functional form to (13.132) for the evolution of global total energy; $\partial/\partial t$ in the surface integrals appearing in (13.132) for total energy are replaced by $-\partial/\partial \xi_1$ in the corresponding ones in (13.136) for axial angular momentum. Conditions under which the area integrals in (13.136) are zero are nevertheless different to conditions for the corresponding integrals in (13.132).

The tasks now are to determine under what conditions

$$\int_{\mathscr{A}^{\text{boundary}}} \left(p h_3 \frac{\partial \xi_3}{\partial \xi_1} \right)^{\text{boundary}} d\mathscr{A}^{\text{boundary}} = 0, \tag{13.137}$$

and to interpret what this means physically when (13.137) does not hold.

13.6.3.2 *Rigid Boundary*

At a rigid boundary, $\xi_3 = \xi_3^{\text{boundary}}(\xi_1, \xi_2)$ and consequently $\partial \xi_3^{\text{boundary}}/\partial \xi_1 \neq 0$ in general. Exceptionally, $\partial \xi_3^{\text{boundary}}/\partial \xi_1 \equiv 0$ if ξ_3^{boundary} is independent of ξ_1, that is, if $\xi_3^{\text{boundary}} = \xi_3^{\text{boundary}}(\xi_2)$. [This includes the special case $\xi_3^{\text{boundary}} =$ constant. Physically this corresponds to a rigid boundary that coincides with a geopotential surface; for example, the rigid atmospheric lid depicted in Figs. 13.5a,c, and a constant orography, bathymetry, or (rigid) ocean surface.]

Returning to the general case $\partial \xi_3^{\text{boundary}}/\partial \xi_1 \neq 0$; noting that p at a boundary generally depends on both space and time; and is generally non-zero, that is, $p^{\text{boundary}} = p^{\text{boundary}}(\xi_1, \xi_2, t) \neq 0$; and also noting that $h_3 = h_3(\xi_2, \xi_3) \neq 0$ for an axial-orthogonal-curvilinear coordinate system, it is seen that (13.137) is generally not satisfied at a rigid boundary. This then means that there will generally be a non-zero contribution at such a boundary unless $p \equiv 0$ there.

Such a change corresponds physically to the torque of the pressure field acting on any variation of the height of the boundary in the ξ_1-direction. When the boundary corresponds to orography, this torque is usually termed *mountain torque*; and for the ocean, *bottom torque*. For both torques, there is an exchange of axial angular momentum at the corresponding boundary. Any such exchange is, however, simply a zero-sum exchange of axial angular momentum between the atmosphere-ocean system encompassing a planet, and that of the planet itself (albeit the change of the planet's axial angular momentum is generally ignored – see footnote 12). Importantly, the globally integrated axial angular momentum of the *combined* atmosphere-ocean-planet system *is* physically conserved, even though those of the subsystems individually are not.

In the atmospheric context (but similarly for an ocean), the second term on the right-hand side of (13.136) is one form of mountain torque. However, since (by periodicity)

$$\int_{\mathscr{A}^B} \frac{\partial \left(p^B h_3^B \xi_3^B\right)}{\partial \xi_1} d\mathscr{A}^B \equiv 0, \tag{13.138}$$

and noting that h_3^B is independent of ξ_1 (by definition of an axial-orthogonal-curvilinear coordinate system – see (6.13)), integration by parts shows that the mountain-torque term can be equivalently written as

$$\begin{aligned} -\int_{\mathscr{A}^B} p^B h_3^B \frac{\partial \xi_3^B}{\partial \xi_1} d\mathscr{A}^B &= -\int_{\mathscr{A}^B} \frac{\partial \left(p^B h_3^B \xi_3^B\right)}{\partial \xi_1} d\mathscr{A}^B + \int_{\mathscr{A}^B} \xi_3^B \frac{\partial \left(h_3^B p^B\right)}{\partial \xi_1} d\mathscr{A}^B \\ &= +\int_{\mathscr{A}^B} \xi_3^B h_3^B \frac{\partial p^B}{\partial \xi_1} d\mathscr{A}^B. \end{aligned} \tag{13.139}$$

Expression (13.139) for the mountain torque makes the effect of a pressure difference across a mountain clearer, and is the preferred form (albeit in the restricted context of the hydrostatic primitive equations) of several authors, for example, Simmons and Burridge (1981) and Laprise and Girard (1990).

13.6.3.3 *Elastic boundary*

At an elastic boundary, $\partial \xi_3^{\text{boundary}} (\xi_1, \xi_2, t) / \partial \xi_1$ is generally non-zero. (It can exceptionally be zero if $\xi_3^{\text{boundary}} = \xi_3^{\text{boundary}} (\xi_2, t)$, something that is sometimes assumed in the restricted context of theoretical studies.) This means (in general) that whether or not condition (13.137) is satisfied depends upon the behaviour of the remaining two quantities in its integrand. To examine this, (13.137) is rewritten as

$$\int_{\mathscr{A}^{\text{boundary}}} \left[p \frac{\partial}{\partial \xi_1} (h_3 \xi_3) \right]^{\text{boundary}} d\mathscr{A}^{\text{boundary}} = 0. \tag{13.140}$$

This exploits the fact that for an axial-orthogonal-curvilinear coordinate system, $h_3 = h_3 (\xi_2, \xi_3)$. It is seen from (13.140) that when $p^{\text{boundary}} = p^{\text{boundary}} (\xi_2, t)$, it can be brought inside the derivative. Periodicity in the ξ_1 direction then leads to satisfaction of (13.140), and therefore also of (13.137). This then means that there is no change of globally integrated axial angular momentum due to a boundary torque when $p^{\text{boundary}} = p^{\text{boundary}} (\xi_2, t)$.

13.6.3.4 *Atmosphere–Ocean Boundary*

For an atmosphere overlying an ocean (as in Figs. 13.5c,d), the globally integrated axial angular momentum is the sum of the individual atmospheric and oceanic axial angular momenta. This raises two questions:

1. What happens at their interfacial surface (bottom of the atmosphere/top of the ocean)?
2. What impact does this have on the combined (atmospheric + oceanic) globally integrated axial angular momentum?

To examine these questions, we take the sum of (13.136), evaluated for the atmosphere, with its counterpart for an ocean. Thus

$$\begin{aligned} &\frac{\partial}{\partial t} \left[\left(\int_{\mathcal{V}} \rho M d\mathcal{V} \right)^{\text{atmos}} + \left(\int_{\mathcal{V}} \rho M d\mathcal{V} \right)^{\text{ocean}} \right] \\ &= \left(\int_{\mathscr{A}^T} p^T h_3^T \frac{\partial \xi_3^T}{\partial \xi_1} d\mathscr{A}^T \right)^{\text{atmos}} - \left(\int_{\mathscr{A}^B} p^B h_3^B \frac{\partial \xi_3^B}{\partial \xi_1} d\mathscr{A}^B \right)^{\text{atmos}} \end{aligned}$$

$$
\begin{aligned}
&+ \cancel{\left(\int_{\mathscr{A}^T} p^T h_3^T \frac{\partial \xi_3^T}{\partial \xi_1} d\mathscr{A}^T \right)^{\text{ocean}}} - \left(\int_{\mathscr{A}^B} p^B h_3^B \frac{\partial \xi_3^B}{\partial \xi_1} d\mathscr{A}^B \right)^{\text{ocean}} \\
&= \left(\int_{\mathscr{A}^T} p^T h_3^T \frac{\partial \xi_3^T}{\partial \xi_1} d\mathscr{A}^T \right)^{\text{atmos}} - \left(\int_{\mathscr{A}^B} p^B h_3^B \frac{\partial \xi_3^B}{\partial \xi_1} d\mathscr{A}^B \right)^{\text{ocean}} . \qquad (13.141)
\end{aligned}
$$

The mutual cancellation of two terms in (13.141) is due to the assumed continuity of pressure at the ocean-atmosphere interface; the pressures exerted by the atmosphere and ocean on either side of their mutual boundary are again assumed equal.

Physically (under the given assumptions), axial angular momentum can be exchanged across the ocean-atmosphere boundary, but only in a zero-sum way, otherwise continuity of pressure there would be compromised. It does not matter whether this boundary is rigid or elastic, since the net sum there is zero regardless. This means that the conservation of the combined (atmosphere + ocean) globally integrated axial angular momentum is determined by any pressure torques at the top of the atmosphere and at the bottom of the ocean:

- The bottom boundary of an ocean is rigid. Generally $\partial \xi_3 / \partial \xi_1 \neq 0$ there, and there will be a bottom torque; exceptionally there will not if $\xi_3 = \xi_3 (\xi_2)$ there. See Section 13.6.3.2.
- For a rigid boundary at the top of the atmosphere, there will generally be a torque; exceptionally there will not if either $\xi_3 = \xi_3^T (\xi_2)$ or $p \equiv 0$ there. See Section 13.6.3.2.
- For an elastic boundary at the top of the atmosphere, there will generally be a torque; exceptionally there will not if either $\xi_3 = \xi_3^T (\xi_2, t)$ or $p^T = p^T (\xi_2, t)$ there. See Section 13.6.3.3.

13.6.4 Prototypical Configurations

In Sections 13.6.2.4 and 13.6.3.4 it was found that there is no net contribution at an atmosphere–ocean boundary to either globally integrated total energy or globally integrated axial angular momentum, only a zero-sum, atmosphere-ocean exchange. Conservation (or otherwise) of these two quantities is therefore determined by pressure-work or pressure-torque contributions at:

- The top of the atmosphere.
- The bottom of the atmosphere (over orography).
- The bottom of the ocean (for an atmosphere over an ocean over bathymetry).

Using the results of Sections 13.6.2 and 13.6.3, we now examine changes at these boundaries for the four prototypical configurations depicted in Figs. 13.5a–d. The associated conservation properties for globally integrated energy and globally integrated axial angular momentum can then be deduced for each configuration.

13.6.4.1 Atmosphere over Orography

The *bottom* boundary of the atmosphere over orography is always *rigid* (see Figs. 13.5a,b). Thus:

- There is (from Section 13.6.2.2) no work done by the pressure field there.

  but
- There is generally (from Section 13.6.3.2) a (physical) mountain torque there, unless $\xi_3^B = \xi_3^B (\xi_2)$.

When the *top* boundary of the atmosphere is assumed *rigid* (see Fig. 13.5a):

- There is (from Section 13.6.2.2) no work done by the pressure field there.

  but

- There is generally (from Section 13.6.3.2) a (unphysical)[13] torque there, unless $\xi_3^T = \xi_3^T(\xi_2)$ or $p^T \equiv 0$.

When the *top* boundary of the atmosphere is assumed *elastic* (see Fig. 13.5b):

- There is generally (from Section 13.6.2.3) work done by the pressure field there, unless $p^T \equiv 0$.
- There is generally (from Section 13.6.3.3) a torque there, unless $\xi_3^T = \xi_3^T(\xi_2, t)$ or $p^T = p^T(\xi_2, t)$.

13.6.4.2 Atmosphere over Ocean over Bathymetry

The *bottom* boundary of an ocean over bathymetry is always *rigid* (see Figs. 13.5c,d). Thus:

- There is (from Section 13.6.2.2) no work done by the pressure field there.

  but
- There is generally (from Section 13.6.3.2) a (physical) bottom torque there, unless $\xi_3^B = \xi_3^B(\xi_2)$.

When the *top* boundary of the atmosphere is assumed *rigid* (see Fig. 13.5a):

- There is (from Section 13.6.2.2) no work done by the pressure field there.

  but
- There is generally (from Section 13.6.3.2) a (unphysical – see footnote 13) torque there, unless $\xi_3^T = \xi_3^T(\xi_2)$ or $p^T \equiv 0$.

When the *top* boundary of the atmosphere is assumed *elastic* (see Fig. 13.5b):

- There is generally (from Section 13.6.2.3) work done by the pressure field there, unless $p^T \equiv 0$.
- There is generally (from Section 13.6.3.3) a torque there, unless $\xi_3^T = \xi_3^T(\xi_2, t)$ or $p^T = p^T(\xi_2, t)$.

As seen here, for $p^T \neq 0$, conservation of globally integrated total energy requires $\xi_3^T = \xi_3^T(\xi_1, \xi_2)$, and so for such a model to conserve both globally integrated total energy *and* globally integrated axial angular momentum, the upper boundary must be fixed in space and time and can only be a function of latitude, $\xi_3^T = \xi_3^T(\xi_2)$. For models employing a height-based vertical coordinate, ξ_3^T is usually a constant in both space and time, and therefore such models conserve both globally integrated total energy and globally integrated axial angular momentum (in the absence of zonal mechanical forcing and mountain torque). As a corollary, models that impose a material surface with constant non-zero pressure at the upper boundary do not conserve total energy and axial angular momentum. They may, however, possess an energy-like invariant; see Section 13.8.

A way of interpreting these results is to consider the choice of an upper atmospheric boundary as dividing the entire atmosphere into two sub-atmospheres. The lower one corresponds to the model domain, and the upper one to the rest of the atmosphere. Mass conservation for each sub-atmosphere can be ensured by making the interface between the two a material surface. The entire atmosphere conserves total energy and axial angular momentum. Therefore if the lower sub-atmosphere is also to conserve these quantities, then so too must the upper sub-atmosphere. However, in general there will be an exchange of energy and axial angular momentum across the interface of the two sub-atmospheres (as at the interface of the atmosphere and an ocean – see Sections 13.6.2.4 and 13.6.3.4). The presented analysis shows that, at finite height, such an exchange is prevented only if the interface is fixed in space and time. Then both sub-atmospheres

[13] It is unphysical since the atmosphere is not in reality capped by a rigid lid. However, if it were, then it would be physical.

will independently conserve their own energy and axial angular momentum. Of course, in general in the real atmosphere, there will be no such fixed material surface other than that at the lower boundary. Therefore imposing this as an upper boundary will have an artificial impact on the simulated atmosphere (which is probably true for any type of upper boundary unless the behaviour of the upper sub-atmosphere is somehow known) and, in particular, can lead to spurious reflections.

A potential way out of the dilemma of the appearance of unphysical changes of globally integrated total energy and globally integrated axial angular momentum at an elastic atmospheric lid is to impose zero pressure there. This possibility is discussed in Section 13.9.

13.7 CONSERVATION WITH THE SHALLOW APPROXIMATION

Following a similar procedure to that just given, the shallow analogues of (13.116) and (13.131) for the conservation of globally integrated total energy and globally integrated axial angular momentum, respectively, can be obtained.

For globally integrated total energy, the deep result (13.116) is recovered except that the global volume and surface integrals (13.97) and (13.117) are redefined as

$$\int_{\mathcal{V}} \mathcal{G} d\mathcal{V} \equiv \iint \left(\int_{s^B}^{s^T} \mathcal{G} h_1^S h_2^S h_3^S \frac{\partial \xi_3}{\partial s} ds \right) d\xi_1 d\xi_2, \tag{13.142}$$

$$\int_{\mathscr{A}} \mathcal{G} d\mathscr{A} \equiv \iint \mathcal{G} h_1^S h_2^S d\xi_1 d\xi_2, \tag{13.143}$$

where superscript 'S' denotes evaluation at a planet's geoid, as in Section 5.3 for shallow approximation.

For globally integrated axial angular momentum, the shallow analogue of (13.131) is

$$\frac{\partial}{\partial t} \int_{\mathcal{V}} \rho M d\mathcal{V} = \int_{\mathscr{A}^T} p^T \left(h_3^S\right)^T \frac{\partial \xi_3^T}{\partial \xi_1} d\mathscr{A}^T - \int_{\mathscr{A}^B} p^B \left(h_3^S\right)^B \frac{\partial \xi_3^B}{\partial \xi_1} d\mathscr{A}^B + \int_{\mathcal{V}} \rho \left(h_1^S F^{u_1} + M F^{\rho}\right) d\mathcal{V}, \tag{13.144}$$

where the volume and surface integrals are again redefined by (13.142) and (13.143), and M and u_1 are also respectively redefined as

$$M \equiv \left(u_1 + \Omega h_1^S\right) h_1^S, \tag{13.145}$$

$$u_1 \equiv h_1^S \frac{D\xi_1}{Dt}. \tag{13.146}$$

Thus shallow approximation affects the constraints on globally integrated total energy and globally integrated axial angular momentum by fundamentally changing the definitions of the wind components, kinetic energy and axial angular momentum.

Because, for both globally integrated total energy and globally integrated axial angular momentum, the upper boundary contributions have the same functional form regardless of whether the fluid is shallow or deep, the conclusions given in Sections 13.3, 13.4.5, and 13.5.2 for a deep fluid also hold for a shallow one.

13.8 AN ENERGY-LIKE INVARIANT FOR ELASTIC LIDS AT FINITE PRESSURE

The results given in Section 13.6 show that globally integrated total energy is not – in general – conserved for an atmosphere with an elastic lid unless the pressure there is zero. However, under the assumption of such a finite, elastic, upper-boundary condition *at non-zero pressure*, Staniforth et al. (2003) have shown the existence of an invariant of the system. In the more restricted context of the hydrostatic primitive equations, this invariant had previously been associated with the energy of the system (Kasahara, 1974; Laprise and Girard, 1990; Arakawa and Konor, 1996).

In spherical-polar coordinates (λ, ϕ, r), with a rigid lower boundary $r = r^B(\lambda, \phi)$, and in the absence of forcing within the atmosphere, Staniforth et al. (2003) derived the conservation equation – their equation (3.3) –

$$\frac{\partial}{\partial t}\int_{\mathcal{V}}\left[\rho E + p^T(\lambda, \phi)\right] d\mathcal{V} = 0, \tag{13.147}$$

where ρ is density, E is specific total energy, $p^T = p^T(\lambda, \phi) \neq 0$ defines an assumed elastic lid, and $d\mathcal{V} = r^2 \cos\phi\, (\partial r/\partial s)\, d\lambda d\phi ds$ is the (transformed) volume element in generalised (λ, ϕ, s) coordinates. Their physical interpretation of (13.147) is that the global integral of $p^T(\lambda, \phi)$ represents the work done by the prescribed stationary pressure applied at the upper surface as the volume of the atmosphere below it changes.

It seems reasonable to expect that this same physical behaviour will also hold for the present, more general case, but expressed in axial-orthogonal-curvilinear coordinates instead of spherical-polar coordinates. The expected generalisation of (13.147), with an assumed, rigid, lower boundary $\xi_3 = \xi_3^B(\xi_1, \xi_2)$, is therefore

An Energy-Like Invariant for Elastic Lids at Finite Pressure

$$p^T = p^T(\xi_1, \xi_2)$$

$$\frac{\partial}{\partial t}\int_{\mathcal{V}}\left[\rho E + p^T(\xi_1, \xi_2)\right] d\mathcal{V} = 0, \tag{13.148}$$

where $d\mathcal{V} = h_1 h_2 h_3 (\partial\xi_3/\partial s)\, d\xi_1 d\xi_2 ds$, as in (13.97).

We now verify this conjecture. Using (13.97) for the volume integral in generalised axial-orthogonal-curvilinear coordinates (ξ_1, ξ_2, s), together with Leibniz's theorem (13.172), and noting that $s^B = s^B(\xi_1, \xi_2, t)$ and $s^T = s^T(\xi_1, \xi_2, t)$, the second term in (13.148) may be rewritten as

$$\begin{aligned}
\frac{\partial}{\partial t}\int_{\mathcal{V}} p^T(\xi_1, \xi_2)\, d\mathcal{V} &= \frac{\partial}{\partial t}\iint\left[\int_{s^B}^{s^T} p^T(\xi_1, \xi_2)\, h_1 h_2 h_3 \frac{\partial \xi_3}{\partial s} ds\right] d\xi_1 d\xi_2 \\
&= \iint\left\{\int_{s^B}^{s^T} \frac{\partial}{\partial t}\left[p^T(\xi_1, \xi_2)\, h_1 h_2 h_3 \frac{\partial \xi_3}{\partial s}\right] ds\right\} d\xi_1 d\xi_2 \\
&+ \iint\left[\frac{\partial s^T}{\partial t} p^T(\xi_1, \xi_2)\, h_1^T h_2^T h_3^T \left(\frac{\partial \xi_3}{\partial s}\right)^T\right] d\xi_1 d\xi_2 \\
&- \iint\left[\frac{\partial s^B}{\partial t} p^B(\xi_1, \xi_2)\, h_1^B h_2^B h_3^B \left(\frac{\partial \xi_3}{\partial s}\right)^B\right] d\xi_1 d\xi_2.
\end{aligned} \tag{13.149}$$

Exploiting identity (13.176) and vertically integrating from $s = s^B$ to $s = s^T$, the first term on the right-hand side of (13.149) may be rewritten as

$$\begin{aligned}
&\iint\left\{\int_{s^B}^{s^T} \frac{\partial}{\partial t}\left[p^T(\xi_1, \xi_2)\, h_1 h_2 h_3 \frac{\partial \xi_3}{\partial s}\right] ds\right\} d\xi_1 d\xi_2 \\
&= \iint\left\{\int_{s^B}^{s^T} \frac{\partial}{\partial s}\left[p^T(\xi_1, \xi_2)\, h_1 h_2 h_3 \frac{\partial \xi_3}{\partial t}\right] ds\right\} d\xi_1 d\xi_2
\end{aligned}$$

$$= \iint p^T(\xi_1,\xi_2)\left[h_1^T h_2^T h_3^T\left(\frac{\partial \xi_3}{\partial t}\right)^T - h_1^B h_2^B h_3^B\left(\frac{\partial \xi_3}{\partial t}\right)^B\right] d\xi_1 d\xi_2. \tag{13.150}$$

Inserting (13.150) into (13.149); collecting terms, using (13.113) and (13.114); noting that $\xi_3^B = \xi_3^B(\xi_1,\xi_2) \Rightarrow \partial \xi_3^B/\partial t \equiv 0$ for a rigid lower boundary; and using definition (13.117) for the area integral then yields

$$\begin{aligned}\frac{\partial}{\partial t}\int_{\mathcal{V}} p^T(\xi_1,\xi_2)\,d\mathcal{V} = & \iint p^T(\xi_1,\xi_2)\,h_1^T h_2^T h_3^T\left[\left(\frac{\partial \xi_3}{\partial t}\right)^T + \frac{\partial s^T}{\partial t}\left(\frac{\partial \xi_3}{\partial s}\right)^T\right] d\xi_1 d\xi_2 \\ & - \iint p^T(\xi_1,\xi_2)\,h_1^B h_2^B h_3^B\left[\left(\frac{\partial \xi_3}{\partial t}\right)^B + \frac{\partial s^B}{\partial t}\left(\frac{\partial \xi_3}{\partial s}\right)^B\right] d\xi_1 d\xi_2 \\ & = \iint p^T(\xi_1,\xi_2)\left(h_1^T h_2^T h_3^T\frac{\partial \xi_3^T}{\partial t} - h_1^B h_2^B h_3^B \underbrace{\frac{\partial \xi_3^B}{\partial t}}_{\equiv 0}\right) d\xi_1 d\xi_2 \\ & = \int_{\mathcal{A}^T} p^T(\xi_1,\xi_2)\,h_3^T\frac{\partial \xi_3^T}{\partial t}\,d\mathcal{A}^T. \end{aligned} \tag{13.151}$$

Evolution equation (13.132) for globally integrated total energy in the absence of forcing within the fluid (i.e. when $F^u = F^v = F^w = \dot{Q}_E = F^\rho = 0$), with a rigid lower boundary $\xi_3^B = \xi_3^B(\xi_1,\xi_2) \Rightarrow \partial \xi_3^B/\partial t \equiv 0$, and with an imposed elastic lid $p^T = p^T(\xi_1,\xi_2)$ simplifies to

$$\frac{\partial}{\partial t}\int_{\mathcal{V}} \rho E d\mathcal{V} + \int_{\mathcal{A}^T} p^T(\xi_1,\xi_2)\,h_3^T\frac{\partial \xi_3^T}{\partial t}\,d\mathcal{A}^T = 0. \tag{13.152}$$

Elimination of $\int_{\mathcal{A}^T} p^T(\xi_1,\xi_2)\,h_3^T\left(\partial \xi_3^T/\partial t\right) d\mathcal{A}^T$ between (13.151) and (13.152) then yields (13.148), thereby verifying the conjectured generalisation (13.148) of (13.147).

Conservation law (13.148) is valid for *deep* atmospheres provided $\xi_3^B = \xi_3^B(\xi_1,\xi_2)$, that is, the lower boundary is rigid, when the elastic-lid boundary condition $p^T = p^T(\xi_1,\xi_2)$ is applied. A similar result can also be shown to hold for *shallow* atmospheres, the only difference being that the volume integral in (13.148) is redefined using the shallow analogue (13.142). Thus (13.148) is general in the sense that it is independent of whether the atmosphere is assumed shallow or deep.

It is emphasised that, with an elastic upper boundary condition at non-zero pressure, whilst $\rho E + p^T$ *is* globally conserved (in the sense that its volume integral is an invariant of the system) the true total energy of the system, ρE, is *not* conserved. The difference between the two, namely the contribution of $p^T(\xi_1,\xi_2)$, represents the work done by the stationary pressure applied at the upper surface as the volume of the atmosphere below it changes.

13.9 AN ATMOSPHERIC STATE WITH ZERO PRESSURE AT FINITE HEIGHT

Closing the formulation of a realistic atmospheric model at, and near, an upper boundary is fraught with peril. The ideal-gas law is an excellent representation of the state of the combined troposphere, stratosphere and mesosphere (i.e. for altitudes up to $O(90\text{ km})$ above Earth's mean surface). At increasingly higher altitudes, however, the validity of the ideal-gas law gradually breaks down. The atmosphere becomes increasingly rarefied (i.e. air density becomes very small indeed) and gradually peters out with no clearly defined upper boundary. Consequently the atmosphere no longer behaves as a continuous fluid. In addition, other physical effects (electrodynamic, magnetic, radiative at high energy, ionisation, plasma, compositional change, etc.) become increasingly important (Akmaev, 2011).

Weather and climate forecast models usually only include the combined troposphere, stratosphere, and mesosphere within their domain; see Fig. 1.1. This practice assumes that extending the model domain upwards, with its many complications and computational cost, is impractical at the current state of the science and availability of data there.[14] This then poses the practical problem of how to acceptably close the model formulation at and near the top of a model's computational domain. This is usually done by introducing some kind of artificial, strongly dissipating/diffusing 'sponge' layer there to avoid spurious numerical reflection of upward-propagating disturbances, particularly at a rigid lid.

It was shown in Sections 13.6.2 and 13.6.3 that there is no pressure work done, and no axial torque due to the pressure field, at the upper boundary of an atmosphere, unforced within its domain, when $p \equiv 0$ there. However, it is conventional wisdom that p cannot be zero at *finite height* for Earth's atmosphere; see, for example, Rasch (1986), Staniforth et al. (2003), and Akmaev (2011). But is this actually true? If it is not, then employing an elastic upper boundary, with $p \equiv 0$ there, might significantly reduce the severity of spurious reflection without compromising conservation of energy and axial total angular momentum. Physically, dissipative and diffusive processes do act in the upper atmosphere. However, to the extent possible, they should be represented on a physical basis and not introduced as an artifice to compensate for fundamental deficiencies in formulation and numerics.

13.9.1 A Motionless Isothermal Shallow Atmosphere

The underlying argument (occasionally given, but usually implicitly assumed) for conventional wisdom is based on the derivation of an exact, motionless, isothermal solution of the unforced shallow-atmosphere equations in spherical-polar coordinates (λ, ϕ, r); small-amplitude perturbations can then be superimposed upon it. To prepare the way forward, we now summarise this derivation.

Applying the assumptions of no motion and isothermal temperature for the layer $r_B \leq r \leq r_T$, the only non-trivial equations that survive are the ideal-gas law (i.e. the equation of state)

$$p = \rho R T = \rho R T_B, \tag{13.153}$$

and (from the vertical-momentum equation) the (shallow-atmosphere) hydrostatic equation

$$\frac{1}{\rho}\frac{dp}{dr} = -g_a = \text{constant}. \tag{13.154}$$

Here:

- $a \leq r_B < r_T$ are all constants.
- $T_B =$ constant is isothermal temperature for $r_B \leq r \leq r_T$.
- Subscripts 'B' and 'T' denote evaluation at $r = r_B$ and $r = r_T$, respectively.
- $p_B = \rho_B R T_B$ from application of the ideal-gas law (13.153) at $r = r_B$.
- g_a is the absolute value of gravity at the planet's (spherical) surface $r = a$.
  That $g(r) = g_a =$ constant for $r \geq a$ is a (consistency) consequence of assuming a shallow atmosphere, as discussed in Section 8.6.

To solve these two coupled equations, only r_B, r_T, T_B, and p_B are prescribed; everything else is then determined as a mathematical consequence. Using (13.153) to eliminate ρ from hydrostatic equation (13.154) gives

[14] This may very well change in the future due to the influence and importance of the solar wind on, for example, telecommunications, satellite instrumentation, and electrical power networks. This emerging field is termed *space weather* (Akmaev, 2011).

$$\frac{d\ln p}{dr} = -\frac{1}{H_B}, \tag{13.155}$$

where $H_B \equiv RT_B/g_a$ = constant is a *scale height*. This first-order differential equation is solved subject to the boundary condition $p(r = r_B) \equiv p_B$ = prescribed constant. The solution is

$$p(r) = p_B \exp\left[-\frac{(r - r_B)}{H_B}\right], \tag{13.156}$$

and then (from (13.153))

$$\rho(r) = \frac{p}{RT_B} = \frac{p_B}{RT_B}\exp\left[-\frac{(r - r_B)}{H_B}\right] = \rho_B \exp\left[-\frac{(r - r_B)}{H_B}\right]. \tag{13.157}$$

- From (13.156), it is seen that $p(r)$ can only be zero as $r \to \infty$; this corresponds with conventional wisdom.
- From (13.157), $\rho(r)$ also asymptotically goes to zero (at the same rate) as $r \to \infty$.

13.9.2 A Motionless Deep Atmosphere Solution

The argument just presented, whilst mathematically correct, is however flawed. The atmosphere is *not* shallow, but *deep*, with the consequence that $g \neq$ constant. Furthermore, the assumption that a deep atmosphere composed of an assumed mixture of ideal gases is (even approximately) isothermal everywhere, *including at high altitude*, is highly questionable.[15]

To address these deficiencies, we follow the preceding methodology but with the more appropriate assumptions of:

1. A deep atmosphere, with the consequence that (see Section 8.6)

$$g(r) = \frac{g_a a^2}{r^2}, \tag{13.158}$$

instead of $g(r) = g_a$ = constant.

2. A linear temperature profile[a] that goes from $T = T_B$ at $r = r_B$ to $T = 0$ at $r = r_T$, that is (see the solid red line in Fig. 13.6 for $r \geq r_B$),

$$T(r) = \left(\frac{r_T - r}{r_T - r_B}\right) T_B, \quad r_B \leq r \leq r_T, \tag{13.159}$$

instead of T = constant (see the red dashed line in Fig. 13.6).

a The temperature does not have to go to zero linearly. What is important here is that *(absolute) temperature goes to zero*, rather than the way in which it does so. A linear profile has the virtue that it is the simplest generalisation of an isothermal profile, and this eases tractability and clarity of the analysis.

To make things yet more tangible, consider the temperature profile defined by the International Standard Atmosphere (1976) profile below the mesopause (i.e. for $r - a \leq r_B - a = 86$ km),

15 The internal energy of an ideal gas is proportional to T and, from the kinetic theory of gases, T is a measure of the average kinetic energy of an ensemble of microscopic particles moving randomly. If $p \equiv 0$, then no pressure is exerted by the random motion of these particles, i.e. they must all be at rest. This is consistent with $T = 0$ at $p = 0$, i.e. zero mean kinetic energy exerts zero pressure. These particles nevertheless have mass (albeit there may be no particles in a given volume) and therefore (conceivably at this stage of the argument) non-zero density.

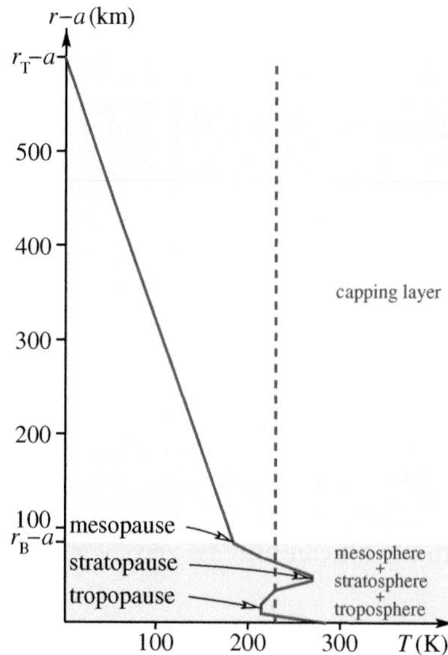

Figure 13.6 Temperature profiles: dotted line – isothermal; solid line – International Standard Atmosphere (1976) profile below $r = r_B$, and linear 'capping' profile above, with $T = 0$ at $r = r_T$. Abscissa is temperature T (K); ordinate is altitude $r - a$ (km), where r is distance from Earth's centre and a is Earth's mean radius. Lowest three layers ('spheres') of the atmosphere (troposphere, stratosphere, and mesosphere) are dark shaded; locations of the interleaved tropopause, stratopause, and mesopause (isothermal) layers separating them indicated by arrows; and idealised 'capping' (uppermost) layer is light shaded.

capped by the linear profile (13.159) above it that goes to zero at $r_T - a = 600$ km (say); see Fig. 13.6. (Note that r_B and r_T here are parameters that can have very different values to those shown on the figure. They define the lower and upper boundaries, respectively, of a 'capping' layer – lightly shaded – of fairly arbitrary depth. This capping layer is not *physical* – temperature actually increases rather than decreases in the thermosphere (i.e. in the layer directly above the mesopause). It is instead a *mathematical* artifice for avoiding the difficulties described of the atmosphere no longer behaving as a continuous fluid at high altitude, but gradually petering out.)

Applying the revised assumptions (13.158) and (13.159), plus the original assumptions of no motion and the ideal-gas law, the non-trivial equations in the 'capping' layer $r_B \leq r \leq r_T$ are then

$$p = \rho R T, \tag{13.160}$$

where $T = T(r)$ is now given by (13.159), and (from deep, vertical-momentum equation (13.3))

$$\frac{1}{\rho}\frac{dp}{dr} = -\frac{g_a a^2}{r^2}. \tag{13.161}$$

Coupled equations (13.160) and (13.161) are to be solved subject to the bottom boundary condition

$$p(r = r_B) = p_B = \text{prescribed constant,} \tag{13.162}$$

at the base of the 'capping' layer $r_B \le r \le r_T$.

The solution procedure is essentially the same as that used earlier for a shallow atmosphere, only with somewhat more complicated expressions. Eliminating ρ from (13.161) using (13.160), and eliminating T from the resulting equation using (13.159), then leads to

$$\frac{d\ln p(r)}{dr} = -\frac{a^2(r_T - r_B)}{H_B r_T^2}\left[\frac{r_T}{r^2} + \frac{1}{r} + \frac{1}{(r_T - r)}\right], \tag{13.163}$$

where $H_B \equiv RT_B/g_a$ is again a scale height. Solving differential equation (13.163) subject to boundary condition (13.162) then yields

$$p(r) = p_B\left[\frac{r_B(r_T - r)}{r(r_T - r_B)}\right]^{\frac{a^2(r_T - r_B)}{H_B r_T^2}} \exp\left[-\frac{a^2(r_T - r_B)}{H_B r_T}\left(\frac{1}{r_B} - \frac{1}{r}\right)\right], \quad r_B \le r \le r_T. \tag{13.164}$$

Setting $r = r_B$ in (13.164) recovers (as it must) boundary condition (13.162).

More interestingly, setting $r = r_T$ instead in (13.164) gives

$$p(r = r_T) = 0, \tag{13.165}$$

that is, $p = 0$ *at finite* $r = r_T$, instead of at infinite r for the shallow analysis of Section 13.9.1. This is a consequence of the assumptions made and of the governing equations.

It remains to determine the corresponding density profile. Inserting temperature profile (13.159) and pressure profile (13.164) into ideal-gas law (13.160) yields

$$\rho(r) = \rho_B\left[\frac{(r_T - r)}{(r_T - r_B)}\right]^{\frac{a^2(r_T - r_B)}{H_B r_T^2} - 1} \left(\frac{r_B}{r}\right)^{\frac{a^2(r_T - r_B)}{H_B r_T^2}} \exp\left[-\frac{a^2(r_T - r_B)}{H_B r_T}\left(\frac{1}{r_B} - \frac{1}{r}\right)\right], \tag{13.166}$$

where

$$\rho_B \equiv \rho(r = r_B) = \frac{p_B}{RT_B}. \tag{13.167}$$

Examination of (13.166) shows that the condition

$$r_T - r_B \ge H_B\frac{r_T^2}{a^2}, \tag{13.168}$$

must be respected for $\rho = \rho(r)$ to be bounded in the idealised 'capping' layer $r_B \le r \le r_T$. Furthermore, satisfaction of (13.168) implies that $\rho_T \equiv \rho(r = r_T) = 0$ except when $r_T - r_B = H_B(r_T^2/a^2)$, and then $\rho_T > 0$.

Condition (13.168) is not very restrictive in practice. It is satisfied for a broad range of parameter values, such as the following ones associated with Fig. 13.6:

$$\left. \begin{array}{c} a = 6\,400 \text{ km}, \quad r_B = a + 86 \text{ km} = 6\,486 \text{ km}, \quad r_T = a + 600 \text{ km} = 7\,000 \text{ km}, \\ r_T - r_B = 514 \text{ km}, \end{array} \right\} \tag{13.169}$$

$$T_B = 187 \text{ K}, \quad R = 287 \text{m}^2\text{s}^{-2}\text{K}^{-1}, \quad g_a = 9.81 \text{ m}^2\text{s}^{-2}, \quad H_B \frac{r_T^2}{a^2} = \frac{RT_B}{g_a} \frac{r_T^2}{a^2} = 6.54 \text{ km}. \tag{13.170}$$

Using these values gives

$$r_T - r_B = 514 \text{ km} \gg 6.54 \text{ km} = H_B \frac{r_T^2}{a^2}, \tag{13.171}$$

which easily satisfies boundedness condition (13.168).

The preceding analysis demonstrates, *by concrete example*, that it is indeed theoretically possible to have $p = 0$ at *finite* height in a deep-atmosphere model, unforced within its domain. This example suggests (but certainly does not prove) that it *might* be possible to employ an elastic upper boundary *at finite height* in a realistic, deep-atmosphere model in height-based coordinates and thereby reduce the deleterious effects of a computational 'sponge' layer.

One might wonder:

- Which of the two assumptions introduced at the beginning of this section is crucial to obtaining $p = 0$ at *finite* height?

It can be shown that:

1. Assuming T is isothermal in a deep atmosphere and solving the resulting equations *does not* lead to $p = 0$ at finite height.
2. Assuming T goes to zero at $r = r_T$ in a shallow atmosphere and solving the resulting equations *does* lead to $p = 0$ at finite height provided that $r_T - r_B \geq H_B$ (to ensure that ρ is bounded); this is a weaker condition than (13.168).

13.10 CONCLUDING REMARKS

A self-consistent, geopotential coordinate system – termed GREAT (Geophysically Realistic, Ellipsoidal, Analytically Tractable) – was developed in Chapter 12 for a (deep) geophysical fluid overlying a perfectly ellipsoidal planet. Planets, however, are generally not perfectly ellipsoidal, and their atmospheres and oceans usually overlie an orography or bathymetry, respectively, that varies in the horizontal. To accommodate this horizontal variation of a planet's orography and/or bathymetry, generalised vertical coordinates have therefore been developed in the present chapter.

At this juncture, we have now met the primary goal of Part I of this book, namely to develop a general, unified set of governing equations for the dynamical core of a quantitative, atmospheric or oceanic, prediction model in ellipsoidal geometry. In the course of doing so, approximations have only been made when necessary and justified.

The methodology used to develop this general, unified set of governing equations is the classical, familiar one employed in almost all textbooks in meteorology and oceanography. There is, however, a complementary variational methodology available. This oft-overlooked approach is mathematically quite challenging – which perhaps helps explain why it has not been used more frequently – but it has some very useful features, both in theory and, importantly, in practice. To prepare the way for its beneficial use from time to time in Part II of this book (on dynamically consistent approximation of the governing dynamical-core equations), the next chapter (Chapter 14) provides:

- An introduction to variational methods and Hamilton's principle of stationary action.[16]
- A complementary derivation – using this methodology – of the unified governing equations for an atmospheric or oceanic dynamical core.

APPENDIX: SOME USEFUL IDENTITIES

For convenience, some useful identities are gathered together here.

Leibniz's theorem for differentiation of an integral gives

Identity A:

$$\int_{s^B}^{s^T} \frac{\partial \mathcal{G}}{\partial \psi} ds \equiv \frac{\partial}{\partial \psi} \int_{s^B}^{s^T} \mathcal{G} ds - \left(\mathcal{G}^T \frac{\partial s^T}{\partial \psi} - \mathcal{G}^B \frac{\partial s^B}{\partial \psi} \right), \tag{13.172}$$

where $\mathcal{G}$ is a generic variable, ψ represents any of ξ_1, ξ_2, and t, $s^B = s^B(\xi_1, \xi_2)$, and $s^T = s^T(\xi_1, \xi_2)$.

It follows from (13.172) that

$$\begin{aligned} &\int_{s^B}^{s^T} \left[\frac{\partial \mathcal{G}}{\partial t} + \frac{\partial}{\partial \xi_1}\left(\frac{u_1}{h_1}\mathcal{G}\right) + \frac{\partial}{\partial \xi_2}\left(\frac{u_2}{h_2}\mathcal{G}\right) + \frac{\partial}{\partial s}(\dot{s}\mathcal{G}) \right] ds \\ &\equiv \frac{\partial}{\partial t} \int_{s^B}^{s^T} \mathcal{G} ds + \frac{\partial}{\partial \xi_1} \int_{s^B}^{s^T} \frac{u_1}{h_1} \mathcal{G} ds + \frac{\partial}{\partial \xi_2} \int_{s^B}^{s^T} \frac{u_2}{h_2} \mathcal{G} ds \\ &- \mathcal{G}^T \left(\frac{\partial s^T}{\partial t} + \frac{u_1^T}{h_1^T} \frac{\partial s^T}{\partial \xi_1} + \frac{u_2^T}{h_2^T} \frac{\partial s^T}{\partial \xi_2} - \dot{s}^T \right) + \mathcal{G}^B \left(\frac{\partial s^B}{\partial t} + \frac{u_1^B}{h_1^B} \frac{\partial s^B}{\partial \xi_1} + \frac{u_2^B}{h_2^B} \frac{\partial s^B}{\partial \xi_2} - \dot{s}^B \right). \end{aligned} \tag{13.173}$$

Invoking the lower and upper boundary conditions (13.51) and (13.52), this reduces to:

Identity B:

$$\begin{aligned} &\int_{s^B}^{s^T} \left[\frac{\partial \mathcal{G}}{\partial t} + \frac{\partial}{\partial \xi_1}\left(\frac{u_1}{h_1}\mathcal{G}\right) + \frac{\partial}{\partial \xi_2}\left(\frac{u_2}{h_2}\mathcal{G}\right) + \frac{\partial}{\partial s}(\dot{s}\mathcal{G}) \right] ds \\ &\equiv \frac{\partial}{\partial t} \int_{s^B}^{s^T} \mathcal{G} ds + \frac{\partial}{\partial \xi_1} \int_{s^B}^{s^T} \frac{u_1}{h_1} \mathcal{G} ds + \frac{\partial}{\partial \xi_2} \int_{s^B}^{s^T} \frac{u_2}{h_2} \mathcal{G} ds, \end{aligned} \tag{13.174}$$

with lower and upper boundary conditions (13.51) *and* (13.52) *for mass conservation assumed.*

Using $\psi \equiv t$ in (13.172), (13.174) is rewritten as:

[16] After William Rowan Hamilton (1805–1865).

Identity C:

$$\int_{s^B}^{s^T}\left[\frac{\partial}{\partial \xi_1}\left(\frac{u_1}{h_1}\mathcal{G}\right)+\frac{\partial}{\partial \xi_2}\left(\frac{u_2}{h_2}\mathcal{G}\right)\right]ds$$
$$\equiv\left(\mathcal{G}^T\frac{\partial s^T}{\partial t}-\mathcal{G}^B\frac{\partial s^B}{\partial t}\right)-\left(\dot{s}\mathcal{G}\right)_{s^B}^{s^T}+\frac{\partial}{\partial \xi_1}\int_{s^B}^{s^T}\frac{u_1}{h_1}\mathcal{G}ds+\frac{\partial}{\partial \xi_2}\int_{s^B}^{s^T}\frac{u_2}{h_2}\mathcal{G}ds, \tag{13.175}$$

with, again, lower and upper boundary conditions (13.51) *and* (13.52) *assumed.*

For $\psi=\xi_1$ or t, and arbitrary $f=f(\xi_2,\xi_3)$, the following holds:

Identity D:

$$\frac{\partial}{\partial s}\left[f(\xi_2,\xi_3)\frac{\partial \xi_3}{\partial \psi}\right]\equiv\frac{\partial}{\partial \psi}\left[f(\xi_2,\xi_3)\frac{\partial \xi_3}{\partial s}\right]. \tag{13.176}$$

Applying (13.15), with $\psi=t$, to $s=s(\xi_1,\xi_2,\xi_3,t)$ and noting that 'local' time derivatives in the s system are evaluated at constant s gives

$$\frac{\partial s(\xi_1,\xi_2,\xi_3,t)}{\partial t}=\cancelto{0}{\left(\frac{\partial s}{\partial t}\right)_s}-\frac{\partial \xi_3(\xi_1,\xi_2,s,t)}{\partial t}\frac{\partial s(\xi_1,\xi_2,\xi_3,t)}{\partial \xi_3}. \tag{13.177}$$

Evaluation of (13.177) at $\xi_3=\xi_3^B(\xi_1,\xi_2,t)$ then yields

$$\left[\frac{\partial s(\xi_1,\xi_2,\xi_3,t)}{\partial t}\right]^{\xi_3=\xi_3^B}=-\left[\frac{\partial \xi_3(\xi_1,\xi_2,s,t)}{\partial t}\right]^{s=s^B}\left[\frac{\partial s(\xi_1,\xi_2,\xi_3,t)}{\partial \xi_3}\right]^{\xi_3=\xi_3^B}, \tag{13.178}$$

where $s^B\equiv s\left(\xi_1,\xi_2,\xi_3=\xi_3^B,t\right)$ for the evaluation of the first factor on the right-hand side of (13.178). Using (13.178), $\partial s^B/\partial t$ may be rewritten as

$$\frac{\partial s^B}{\partial t}\equiv\frac{\partial s\left(\xi_1,\xi_2,\xi_3=\xi_3^B,t\right)}{\partial t}=\left[\frac{\partial s(\xi_1,\xi_2,\xi_3,t)}{\partial \xi_3}\right]^{\xi_3=\xi_3^B}\frac{\partial \xi_3^B}{\partial t}+\left[\frac{\partial s(\xi_1,\xi_2,\xi_3,t)}{\partial t}\right]^{\xi_3=\xi_3^B}$$
$$=\left[\frac{\partial s(\xi_1,\xi_2,\xi_3,t)}{\partial \xi_3}\right]^{\xi_3=\xi_3^B}\left\{\frac{\partial \xi_3^B}{\partial t}-\left[\frac{\partial \xi_3(\xi_1,\xi_2,s,t)}{\partial t}\right]^{s=s^B}\right\}. \tag{13.179}$$

Multiplying this equation by $\left[\partial \xi_3(\xi_1,\xi_2,s,t)/\partial s\right]^{s^B}$ then leads to

$$\left[\frac{\partial \xi_3(\xi_1,\xi_2,s,t)}{\partial s}\right]^{s^B}\frac{\partial s^B}{\partial t}$$
$$=\left[\frac{\partial \xi_3(\xi_1,\xi_2,s,t)}{\partial s}\right]^{s^B}\left[\frac{\partial s(\xi_1,\xi_2,\xi_3,t)}{\partial \xi_3}\right]^{\xi_3=\xi_3^B}\left\{\frac{\partial \xi_3^B}{\partial t}-\left[\frac{\partial \xi_3(\xi_1,\xi_2,s,t)}{\partial t}\right]^{s=s^B}\right\}$$
$$=\frac{\partial \xi_3^B}{\partial t}-\left[\frac{\partial \xi_3(\xi_1,\xi_2,s,t)}{\partial t}\right]^{s=s^B}. \tag{13.180}$$

Rearranging (13.180), and using the abbreviated notation of Section 13.2.6 that superscript 'B' denotes evaluation at $s^B=s^B(\xi_1,\xi_2,t)$, (13.180) may be rewritten as

$$\left[\left(\frac{\partial \xi_3}{\partial t}\right)_s\right]^B=\frac{\partial \xi_3^B}{\partial t}-\left(\frac{\partial \xi_3}{\partial s}\right)^B\frac{\partial s^B}{\partial t}, \tag{13.181}$$

with a similar equation holding at $s^T=s^T(\xi_1,\xi_2,t)$.

14

Variational Methods and Hamilton's Principle of Stationary Action

ABSTRACT

Using Hamilton's variational principle of stationary action, a unified set of governing equations for general fluids (primarily gaseous for the atmosphere, and liquid for the oceans) is developed in both vector form and in axial-orthogonal-curvilinear coordinates. The two forms correspond to the sets developed in Chapters 4 and 6. To prepare the way for this, the Eulerian and Lagrangian viewpoints of fluid dynamics are contrasted. This leads to the definition of *location* (Eulerian) and *label* (Lagrangian) spaces. Labels are assigned in a way that is particularly convenient for taking variations in label space. A brief introduction to functionals and variational methods is given using three simple, illustrative problems. Hamilton's variational principle of stationary action is then defined and its benefits discussed. The previously examined simple problem of gravitational attraction between two point particles is revisited as a first illustration of practical application of Hamilton's principle. As an intermediate illustration, a conservative system of point particles is then examined in a step-by-step manner. With these preparations, application of Hamilton's principle to a global atmosphere or ocean is developed, by making the transition from a set of discrete point particles to a fluid continuum. This is first done in vector form and then redone in axial-orthogonal-curvilinear coordinates. The governing equations developed in earlier chapters then result for an unforced atmosphere or ocean. The variational methods described in this chapter are applied in later chapters to obtain dynamically consistent, approximate equation sets.

14.1 PREAMBLE

A unified set of governing equations for general fluids (primarily gaseous for the atmosphere, and liquid for the oceans) has been developed in Chapters 4 and 6. The goals of the present chapter are to:

1. Provide a basic introduction to variational methods.
2. Apply a variational technique – namely *Hamilton's principle of stationary action* – to obtain the governing equations for a global atmosphere or ocean.

More general accounts of variational methods and the application of Hamilton's principle may be found elsewhere; for example, Lanczos (1970), Salmon (1998), Goldstein et al. (2001), Neuenschwander (2017), and Badin and Crisciani (2018).

As in earlier chapters, the approach herein is to start things off in a relatively simple way, and to then progressively increase complexity. To set the scene (in Section 14.2), the Eulerian and Lagrangian viewpoints of fluid dynamics are contrasted, both physically and mathematically. This leads to the definition of *location* (Eulerian) and *label* (Lagrangian) spaces. These are equivalent to one another, and are just two different useful ways of describing and examining the same things.

Labels are assigned (in Section 14.3) in a way that is convenient for taking variations in label space (as in Sections 14.8 and 14.9, and the Appendix to this chapter). It is shown how this leads to the familiar mass-continuity equation of fluid dynamics, given in various forms in Chapters 2–4 and 6.

To prepare the way for the definition of Hamilton's variational principle, a brief introduction is given (in Section 14.4) to functionals and variational methods. This is done through the use of three illustrative problems. The first two are static problems – to first find the shortest curve between two points on a plane, and to then similarly do so on the surface of a sphere. The third is a dynamical problem, namely gravitational attraction between two point particles. This problem is examined from both a mathematical and physical perspective to prepare the way for the definition and discussion (in Section 14.5) of Hamilton's variational principle of stationary action, including its beneficial properties.

As a first and very simple illustration of how Hamilton's principle can be applied in practice, the problem of gravitational attraction between two point particles is revisited (in Section 14.6). As an intermediate illustration, a conservative system of point particles is examined (in Section 14.7) in a step-by-step manner. With these preparations, the application of Hamilton's principle to a global atmosphere or ocean is developed – again in a step-by-step manner – by making the transition from a set of discrete point particles to a fluid continuum. This is done first in vector form (in Section 14.8), and then redone in axial-orthogonal-curvilinear coordinates (in Section 14.9). The governing equations developed in Chapters 4 (in vector form) and 6 (in axial-orthogonal-curvilinear coordinates) then result for an unforced atmosphere or ocean. The Euler–Lagrange equations for the momentum components of global geophysical fluids are derived in Section 14.10, and concluding remarks are made in Section 14.11.

14.2 EULERIAN VERSUS LAGRANGIAN VIEWPOINTS FOR FLUID DYNAMICS

14.2.1 Physical Representation

We first contrast the Eulerian and Lagrangian viewpoints of fluid dynamics from a physical perspective, under the assumption that the fluid is a continuum. Classical mechanics describes point particles moving under the influence of forces acting upon them (Goldstein et al., 2001), with mass *discretely* distributed at a set of *points*. Fluid dynamics instead considers mass to be *continuously* distributed within *infinitesimal volumes*. This leads to a mass-continuity equation and to a pressure-gradient force. Both of these are obtained by examining an infinitesimal volume, and then applying a limiting process. The distinguishing feature between the Eulerian and Lagrangian viewpoints is the precise definition of these infinitesimal volumes.

The *Eulerian* viewpoint (Fig. 14.1a) of fluid dynamics is that of a *stationary* observer. The stationary observer examines how the flow of the fluid varies (as a function of time) across the boundary of an infinitesimally small volume element, of *fixed size, shape, and location* in physical space.[1] In particular, the stationary observer examines the fluxes of various quantities (mass, momentum, energy, moisture, salinity, etc.) across the *fixed* (for all time) boundaries of the volume element. A protypical Eulerian volume element (shaded) is depicted in Fig. 14.1a.

The *Lagrangian* viewpoint (Fig. 14.1b) is instead that of an observer *moving with a fluid parcel.* The moving observer examines how an infinitesimally small volume element of fluid (i.e. a fluid parcel, shaded), of *fixed mass,* varies as it moves within the flow. Although the mass of the fluid parcel is fixed, its volume and shape are *not.* This leads to distortion in physical space of the (closed) surface of the fluid parcel as it moves. Fig. 14.1b depicts the Lagrangian volume element in physical space at both initial time and (instantaneously) at a later time.

[1] The volume element does not have to be a box, as depicted in Fig. 14.1a, but can be any shape of fixed infinitesimal size and of fixed location. It is just simpler to depict and describe it this way for presentational purposes.

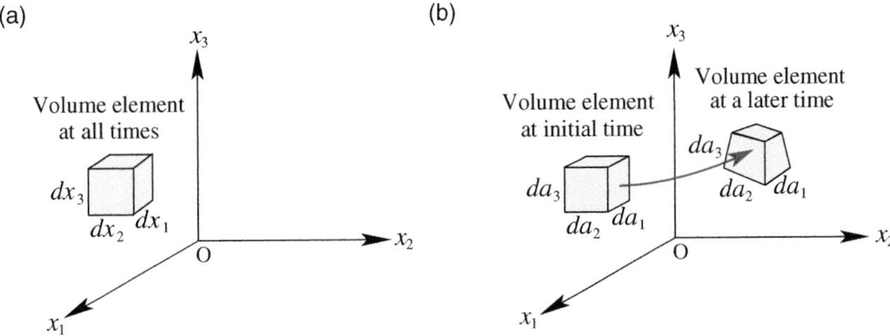

Figure 14.1 Infinitesimally small, prototypical, Eulerian and Lagrangian volume elements (blue shaded), depicted in Cartesian (x_1, x_2, x_3) coordinates in physical space.
(a) An Eulerian volume element, $dx_1 dx_2 dx_3$, fixed in space for all time.
(b) A moving Lagrangian volume element (fluid parcel), $da_1 da_2 da_3$, displayed at initial time and at a later time, where (a_1, a_2, a_3) are label coordinates. It moves with, and is deformed by, the fluid flow along the path depicted by the red arrowed curve. See main text for further details.

14.2.2 Mathematical Representation and Independent Variables

Mathematically, the Eulerian and Lagrangian physical viewpoints translate into two different sets of independent variables; see Chapter 1.1 of Salmon (1998). From the *Eulerian* viewpoint of a stationary observer, the independent variables are the *location coordinates* $\mathbf{r} = (x_1, x_2, x_3)$, plus time t. For simplicity, it is convenient to express location coordinates (position vector) $\mathbf{r}$ in terms of Cartesian coordinates (x_1, x_2, x_3) (fixed in physical space, and independent of time) as depicted in Fig. 14.1a. However $\mathbf{r}$ can equally well be expressed in terms of a set of curvilinear coordinates (again fixed in physical space, and independent of time), as in Section 14.9. Using Cartesian coordinates avoids – for now – the need to introduce scale factors (h_1, h_2, h_3), thereby simplifying the analysis. It is relatively straightforward to later adapt it to use curvilinear coordinates, via coordinate transformation, when needed.

From the *Lagrangian* viewpoint of an observer moving with a fluid parcel, the independent variables are instead the *label coordinates* $\mathbf{a} = (a_1, a_2, a_3)$ – these move in concert with fluid parcels – plus time τ. As the name suggests, an infinitesimal fluid parcel is assigned, *once and for all*, specific values, a_1, a_2, and a_3, of a three-dimensional *label* $\mathbf{a} = (a_1, a_2, a_3)$ that continuously varies throughout the fluid. The size and shape of the fluid parcel may (and usually does) change as a function of time, but the values of its label $\mathbf{a} = (a_1, a_2, a_3)$ do not. They are materially transported, *unchanged*, with the fluid parcel. The labelling of a fluid parcel, once done, is *permanent* and holds for all time. The designation of time as τ in the Lagrangian description, rather than as t as in the Eulerian one, is (see (14.5)) to facilitate taking partial derivatives correctly and unambiguously. The two times are taken to be identical; $\tau \equiv t$, and the initial instant in time is $\tau \equiv t = 0$.

The precise assignment of label coordinates to fluid parcels is quite arbitrary, and a matter of convenience and context. Label coordinates $\mathbf{a} = (a_1, a_2, a_3)$ vary continuously in space. Collectively they span the fluid's domain in the same way that location coordinates $\mathbf{r} = (x_1, x_2, x_3)$ do. Label coordinates $\mathbf{a} = (a_1, a_2, a_3)$ can be alternatively viewed as being curvilinear coordinates that are transported by the fluid flow in location space $\mathbf{r} = (x_1, x_2, x_3)$ (Salmon, 1998).

This is all rather abstract. To better make the connection with the literature on the subject and to be somewhat more concrete, note that label coordinates $\mathbf{a} = (a_1, a_2, a_3)$ are very often (arbitrarily) assigned to coincide at some initial time (but *only* at this time) with location coordinates $\mathbf{r} = (x_1, x_2, x_3)$. That said, this is not the way that label coordinates are assigned and used herein! Instead (see Section 14.3) we follow Salmon (1998) and assign them in a way that facilitates taking variations in label space. See Sections 14.8 and 14.9,and the Appendix to this chapter.[2] After the clock is started, location coordinates $\mathbf{r} = (x_1, x_2, x_3)$ remain unchanged in location space. However, label coordinates $\mathbf{a} = (a_1, a_2, a_3)$ move around in location space, in concert with the movement of associated fluid parcels within their enveloping fluid environment (Fig. 14.1b).

14.2.3 Transformation between Location Space and Label Space

Whatever the specifics of the chosen assignment of label coordinates, *location* coordinates $\mathbf{r} = (x_1, x_2, x_3)$, plus time t, are related to *label* coordinates $\mathbf{a} = (a_1, a_2, a_3)$, plus time τ, via relations of the form

$$x_1 = x_1(a_1, a_2, a_3, \tau), \quad x_2 = x_2(a_1, a_2, a_3, \tau), \quad x_3 = x_3(a_1, a_2, a_3, \tau), \quad t = \tau, \tag{14.1}$$

$$a_1 = a_1(x_1, x_2, x_3, t), \quad a_2 = a_2(x_1, x_2, x_3, t), \quad a_3 = a_3(x_1, x_2, x_3, t), \quad \tau = t. \tag{14.2}$$

Equations (14.1) and (14.2) can also be written in abbreviated form as

$$\mathbf{r} = \mathbf{r}(\mathbf{a}, \tau), \quad t = \tau, \tag{14.3}$$

$$\mathbf{a} = \mathbf{a}(\mathbf{r}, t), \quad \tau = t. \tag{14.4}$$

Whether written as (14.1) and (14.2), or as (14.3) and (14.4), each pair of equations represents a change of variables from a set of four *independent* variables to a different set of four *independent* variables. One set of variables ((14.1) or (14.3)) is intrinsically associated with *location* space and an *Eulerian* description of the fluid; the other ((14.2) or (14.4)) is intrinsically associated with *label* space and a *Lagrangian* description of the fluid.

By making a distinction between time variables t and τ, we can unambiguously distinguish, in a concise manner, between the two rates of change with respect to time, namely $\partial/\partial t$ and $\partial/\partial\tau$. It is simply a question of which independent variables are held constant when computing the partial derivative with respect to time. Because t is intrinsically associated with *location* coordinates $\mathbf{r} = (x_1, x_2, x_3)$, and with an *Eulerian* description of the flow, the partial time derivative $\partial/\partial t$ is to be taken whilst holding x_1, x_2, and x_3 all constant; this is consistent with the viewpoint of a *stationary observer* permanently located at fixed point $\mathbf{r} = (x_1, x_2, x_3)$. There is then no need to state this explicitly; it is implicit in the use of $\partial/\partial t$ rather than $\partial/\partial\tau$.

Similarly, because τ is intrinsically associated with *label* coordinates $\mathbf{a} = (a_1, a_2, a_3)$, and with a *Lagrangian* description of the flow, the partial time derivative $\partial/\partial\tau$ is to be taken whilst holding a_1, a_2, and a_3 all constant (i.e. it is taken along a fluid parcel's trajectory). This is consistent with the viewpoint of an *observer following a fluid parcel*, with the latter having *fixed*, permanent label, $\mathbf{a} = (a_1, a_2, a_3)$, so that

$$\frac{\partial \mathbf{a}}{\partial \tau} = 0. \tag{14.5}$$

Again there is no need to state this explicitly; it is implicit in the use of $\partial/\partial\tau$ rather than $\partial/\partial t$.

Because of the change of independent variables, the derivatives in location and label spaces are related in the usual way to one another, via the *chain rule* for partial differentiation. For any

[2] See also Section 18.9.3 for an example of label assignment in the context of the shallow-water equations.

function F that can be written in location space as $F = F(x_1, x_2, x_3, t)$ and in label space as $F = F(a_1, a_2, a_3, \tau)$, applying the chain rule for partial differentiation gives

$$\begin{aligned}\frac{\partial F(a_1,a_2,a_3,\tau)}{\partial \tau} &= \frac{\partial F(x_1,x_2,x_3,t)}{\partial t}\frac{\partial t}{\partial \tau} + \frac{\partial F(x_1,x_2,x_3,t)}{\partial x_1}\frac{\partial x_1}{\partial \tau} \\ &+ \frac{\partial F(x_1,x_2,x_3,t)}{\partial x_2}\frac{\partial x_2}{\partial \tau} + \frac{\partial F(x_1,x_2,x_3,t)}{\partial x_3}\frac{\partial x_3}{\partial \tau} \\ &= \frac{\partial F}{\partial t} + \frac{\partial x_1}{\partial \tau}\frac{\partial F}{\partial x_1} + \frac{\partial x_2}{\partial \tau}\frac{\partial F}{\partial x_2} + \frac{\partial x_3}{\partial \tau}\frac{\partial F}{\partial x_3}. \end{aligned} \tag{14.6}$$

Since $\partial \mathbf{r}/\partial \tau$ is the rate of change of $\mathbf{r}$ with respect to time (with $\mathbf{a}$ held fixed), as measured by an observer following a fluid parcel, it follows that

$$\mathbf{u} \equiv \frac{\partial \mathbf{r}}{\partial \tau} \equiv \left(\frac{\partial x_1}{\partial \tau}, \frac{\partial x_2}{\partial \tau}, \frac{\partial x_3}{\partial \tau}\right) \equiv (u_1, u_2, u_3) \tag{14.7}$$

is velocity as measured by an observer following a fluid parcel. Using (14.7) in (14.6) then gives

$$\frac{\partial F}{\partial \tau} = \frac{\partial F}{\partial t} + u_1\frac{\partial F}{\partial x_1} + u_2\frac{\partial F}{\partial x_2} + u_3\frac{\partial F}{\partial x_3} = \frac{\partial F}{\partial t} + \mathbf{u}\cdot\nabla F. \tag{14.8}$$

The right-hand side of (14.8) is immediately recognisable as being the usual *material* (or *substantial*, or *substantive*) *derivative* of F, as measured by an observer following a fluid parcel; see (2.6) and (2.8). Thus

$$\frac{\partial F}{\partial \tau} = \frac{DF}{Dt}, \tag{14.9}$$

where, in standard notation,

$$\frac{D}{Dt} \equiv \frac{\partial}{\partial t} + \mathbf{u}\cdot\nabla = \frac{\partial}{\partial t} + u_1\frac{\partial}{\partial x_1} + u_2\frac{\partial}{\partial x_2} + u_3\frac{\partial}{\partial x_3}. \tag{14.10}$$

14.3 MASS CONSERVATION

As mentioned earlier, assigning labels to fluid parcels is quite arbitrary. A very convenient way of doing so is as in Chapter 1.2 of Salmon (1998), whereby a volume element $d\mathbf{a} = da_1 da_2 da_3$ in *label space* $\mathbf{a} = (a_1, a_2, a_3)$ is equal to the mass contained within it.[3] Specifically, at some initial time, label coordinates $\mathbf{a} = (a_1, a_2, a_3)$ are assigned such that

$$d\mathbb{V}_{a_1 a_2 a_3} \equiv da_1 da_2 da_3 = d\,(mass), \tag{14.11}$$

where $d\mathbb{V}_{a_1 a_2 a_3}$ is the infinitesimal volume of a fluid parcel in *label* space $\mathbf{a} = (a_1, a_2, a_3)$, and $d\,(mass)$ is the mass enclosed within this fixed (in label space) volume (see Fig. 14.1b). Because, once assigned, a fluid parcel's label coordinates (a_1, a_2, a_3) are transported with it, *without change* (see (14.5)), (14.11) holds for all time.

[3] Although the volume of a fluid parcel is fixed in *label* space for all time, this volume generally varies when viewed in *location* (physical) space during material transport by the flow.

In *location* space $\mathbf{r} = (x_1, x_2, x_3)$, the same mass, $d\,(mass)$, of the same fluid parcel, is (at a later instant in time) enclosed within a corresponding, infinitesimal, physical volume $d\mathbb{V}_{x_1x_2x_3}$. Thus, by definition,

$$d\,(mass) = \rho d\mathbb{V}_{x_1x_2x_3}, \tag{14.12}$$

where ρ is the *mass density* of the fluid parcel (at the later instant in time). Eliminating $d\,(mass)$ between (14.11) and (14.12) then gives

$$\alpha \equiv \frac{1}{\rho} = \frac{d\mathbb{V}_{x_1x_2x_3}}{d\mathbb{V}_{a_1a_2a_3}}, \tag{14.13}$$

where α is *specific volume* of the fluid parcel – see (2.45). Specific volume, α, has been introduced here as it is a convenient variable for the thermodynamics of the fluid and, in particular, for the representation of the fluid's internal energy. It is also convenient for the derivation of the mass-continuity equation (14.23) in Lagrangian form.

Now infinitesimal volume $d\mathbb{V}_{x_1x_2x_3}$, in location coordinates $\mathbf{r} = (x_1, x_2, x_3)$, is related to the corresponding volume $d\mathbb{V}_{a_1a_2a_3} \equiv da_1da_2da_3$, in label coordinates $\mathbf{a} = (a_1, a_2, a_3)$, by the Jacobian of the transformation between these two coordinate systems. Thus

$$\frac{d\mathbb{V}_{x_1x_2x_3}}{d\mathbb{V}_{a_1a_2a_3}} \equiv \frac{\partial\,(x_1, x_2, x_3)}{\partial\,(a_1, a_2, a_3)}, \tag{14.14}$$

where

$$\frac{\partial\,(x_1, x_2, x_3)}{\partial\,(a_1, a_2, a_3)} \equiv \det\begin{bmatrix} \frac{\partial x_1}{\partial a_1} & \frac{\partial x_1}{\partial a_2} & \frac{\partial x_1}{\partial a_3} \\ \frac{\partial x_2}{\partial a_1} & \frac{\partial x_2}{\partial a_2} & \frac{\partial x_2}{\partial a_3} \\ \frac{\partial x_3}{\partial a_1} & \frac{\partial x_3}{\partial a_2} & \frac{\partial x_3}{\partial a_3} \end{bmatrix} \equiv \begin{vmatrix} \frac{\partial x_1}{\partial a_1} & \frac{\partial x_1}{\partial a_2} & \frac{\partial x_1}{\partial a_3} \\ \frac{\partial x_2}{\partial a_1} & \frac{\partial x_2}{\partial a_2} & \frac{\partial x_2}{\partial a_3} \\ \frac{\partial x_3}{\partial a_1} & \frac{\partial x_3}{\partial a_2} & \frac{\partial x_3}{\partial a_3} \end{vmatrix} \equiv \frac{\partial\,(\mathbf{r})}{\partial\,(\mathbf{a})} \tag{14.15}$$

is the Jacobian of the transformation from location coordinates, $\mathbf{r} = (x_1, x_2, x_3)$, to label coordinates, $\mathbf{a} = (a_1, a_2, a_3)$. Substitution of (14.14) into (14.13) then yields

$$\alpha \equiv \frac{1}{\rho} = \frac{\partial\,(x_1, x_2, x_3)}{\partial\,(a_1, a_2, a_3)} = \frac{\partial\,(\mathbf{r})}{\partial\,(\mathbf{a})}, \tag{14.16}$$

where $\partial\,(x_1, x_2, x_3)\,/\partial\,(a_1, a_2, a_3) \equiv \partial\,(\mathbf{r})\,/\partial\,(\mathbf{a})$ is given explicitly by evaluation of the determinant in (14.15).

Taking the material derivative $(\partial/\partial\tau)$ of (14.16) gives

$$\begin{aligned} \frac{\partial\alpha}{\partial\tau} = \frac{\partial}{\partial\tau}\left[\frac{\partial\,(x_1, x_2, x_3)}{\partial\,(a_1, a_2, a_3)}\right] &= \frac{\partial\,(\partial x_1/\partial\tau, x_2, x_3)}{\partial\,(a_1, a_2, a_3)} + \frac{\partial\,(x_1, \partial x_2/\partial\tau, x_3)}{\partial\,(a_1, a_2, a_3)} + \frac{\partial\,(x_1, x_2, \partial x_3/\partial\tau)}{\partial\,(a_1, a_2, a_3)} \\ &= \frac{\partial\,(u_1, x_2, x_3)}{\partial\,(a_1, a_2, a_3)} + \frac{\partial\,(x_1, u_2, x_3)}{\partial\,(a_1, a_2, a_3)} + \frac{\partial\,(x_1, x_2, u_3)}{\partial\,(a_1, a_2, a_3)}. \end{aligned} \tag{14.17}$$

Equation (14.1) means that $\partial/\partial\tau$ commutes with $\partial/\partial a_1$, $\partial/\partial a_2$, and $\partial/\partial a_3$, and this, together with the definition (14.15) of the Jacobian, has been exploited to obtain the right-hand side of the first line of (14.17). To obtain the second line, the definition (14.7) of $\mathbf{u}$ has been used.

Using the chain rule for Jacobians

$$\frac{\partial\,(\mathbf{b})}{\partial\,(\mathbf{a})} \equiv \frac{\partial\,(b_1, b_2, b_3)}{\partial\,(a_1, a_2, a_3)} = \frac{\partial\,(x_1, x_2, x_3)}{\partial\,(a_1, a_2, a_3)}\frac{\partial\,(b_1, b_2, b_3)}{\partial\,(x_1, x_2, x_3)} \equiv \frac{\partial\,(\mathbf{r})}{\partial\,(\mathbf{a})}\frac{\partial\,(\mathbf{b})}{\partial\,(\mathbf{r})}, \tag{14.18}$$

followed by use of (14.16), (14.17) can be rewritten as

$$\frac{\partial\alpha}{\partial\tau} = \frac{\partial(x_1,x_2,x_3)}{\partial(a_1,a_2,a_3)}\left[\frac{\partial(u_1,x_2,x_3)}{\partial(x_1,x_2,x_3)} + \frac{\partial(x_1,u_2,x_3)}{\partial(x_1,x_2,x_3)} + \frac{\partial(x_1,x_2,u_3)}{\partial(x_1,x_2,x_3)}\right]$$
$$= \alpha\left[\frac{\partial(u_1,x_2,x_3)}{\partial(x_1,x_2,x_3)} + \frac{\partial(x_1,u_2,x_3)}{\partial(x_1,x_2,x_3)} + \frac{\partial(x_1,x_2,u_3)}{\partial(x_1,x_2,x_3)}\right]. \quad (14.19)$$

Now (since x_1, x_2, and x_3 are all independent of one another)

$$\frac{\partial(u_1,x_2,x_3)}{\partial(x_1,x_2,x_3)} \equiv \begin{vmatrix} \frac{\partial u_1}{\partial x_1} & \frac{\partial u_1}{\partial x_2} & \frac{\partial u_1}{\partial x_3} \\ \frac{\partial x_2}{\partial x_1} & \frac{\partial x_2}{\partial x_2} & \frac{\partial x_2}{\partial x_3} \\ \frac{\partial x_3}{\partial x_1} & \frac{\partial x_3}{\partial x_2} & \frac{\partial x_3}{\partial x_3} \end{vmatrix} = \begin{vmatrix} \frac{\partial u_1}{\partial x_1} & \frac{\partial u_1}{\partial x_2} & \frac{\partial u_1}{\partial x_3} \\ 0 & 1 & 0 \\ 0 & 0 & 1 \end{vmatrix} = \frac{\partial u_1}{\partial x_1}, \quad (14.20)$$

and similarly

$$\frac{\partial(x_1,u_2,x_3)}{\partial(x_1,x_2,x_3)} = \frac{\partial u_2}{\partial x_2}, \quad \frac{\partial(x_1,x_2,u_3)}{\partial(x_1,x_2,x_3)} = \frac{\partial u_3}{\partial x_3}. \quad (14.21)$$

Substitution of (14.20) and (14.21) into (14.19), with use of (14.9), then yields – cf. (2.46) –

The Mass-Continuity Equation in Specific Volume Form

$$\frac{\partial\alpha}{\partial\tau} = \frac{D\alpha}{Dt} = \alpha\left(\frac{\partial u_1}{\partial x_1} + \frac{\partial u_2}{\partial x_2} + \frac{\partial u_3}{\partial x_3}\right) = \alpha\nabla\cdot\mathbf{u}. \quad (14.22)$$

Using (14.16) (i.e. $\alpha \equiv 1/\rho$) in (14.22) finally gives the familiar equation (2.37) of Chapter 2 for mass conservation, namely

$$\frac{D\rho}{Dt} + \rho\nabla\cdot\mathbf{u} = 0. \quad (14.23)$$

This equation is the *Lagrangian* form of the mass-continuity equation. Physically, it means that the mass density of a fluid parcel, moving within a flow, can only change when its volume either increases or decreases. Using (14.10), (14.23) can be rewritten as – cf. (2.39) of Chapter 2 –

$$\frac{\partial\rho}{\partial t} + \nabla\cdot(\rho\mathbf{u}) = \frac{\partial\rho}{\partial t} + \frac{\partial}{\partial x_1}(\rho u_1) + \frac{\partial}{\partial x_2}(\rho u_2) + \frac{\partial}{\partial x_3}(\rho u_3) = 0. \quad (14.24)$$

This is the *Eulerian* form of the mass-continuity equation for mass conservation. Physically, it means that the mass density of the fluid contained within a fixed (in both space and time) volume of fluid can only be changed by a net inflow or outflow of mass across its bounding surface.

14.4 FUNCTIONALS AND VARIATIONAL PRINCIPLES

14.4.1 The Shortest Curve between Two Points in the Plane

As a first introductory example of functionals and variational principles, consider the problem of determining the shortest curve between two points A and B in the $x_1 - x_2$ plane. See Fig. 14.2. Let these two points have Cartesian coordinates (x_1^A, x_2^A) and (x_1^B, x_2^B), respectively, and let $x_1^B > x_1^A$. Geometric intuition tells us that the shortest curve is a straight line (the solid black line in Fig. 14.2). This will now be confirmed analytically using *variational calculus*.

The differential of arc length, ds, in the $x_1 - x_2$ plane satisfies

$$(ds)^2 = (dx_1)^2 + (dx_2)^2 \quad \Rightarrow \quad ds = \left[1 + \left(\frac{dx_2}{dx_1}\right)^2\right]^{\frac{1}{2}} dx_1. \quad (14.25)$$

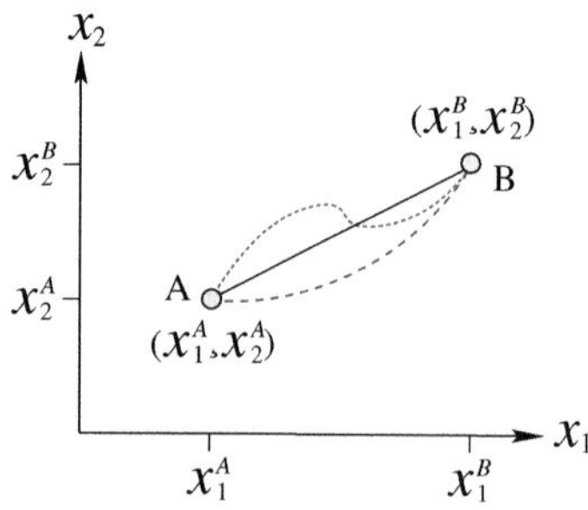

Figure 14.2 Three curves connecting points A and B with coordinates (x_1^A, x_2^A) and (x_1^B, x_2^B), respectively. Solid black line – shortest possible curve; blue dashed and red dotted lines – two arbitrary longer curves.

The total distance, S, between two points with coordinates (x_1^A, x_2^A) and (x_1^B, x_2^B) along the curve defined by $x_2 = x_2(x_1)$ is therefore given by the definite integral

$$S[x_2(x_1)] = \int_{x_1^A}^{x_1^B} \left[1 + \left(\frac{dx_2}{dx_1}\right)^2\right]^{\frac{1}{2}} dx_1, \tag{14.26}$$

where x_1^A and x_1^B are given constants.

In (14.26), $S[x_2(x_1)]$ is termed a *functional*, which may be considered herein to be a '*function of a function*'. For a given function $x_2 = x_2(x_1)$, that is, for a given curve in the Cartesian Ox_1x_2 plane, $S[x_2(x_1)]$ is a (real and positive) number that depends on *all* of the values of $x_2(x_1)$ for $x_1^A \leq x_1 \leq x_1^B$. By construction, this number is just the physical distance along the curve $x_2 = x_2(x_1)$ between two points with coordinates (x_1^A, x_2^A) and (x_1^B, x_2^B). The precise value of $S[x_2(x_1)]$ is determined by the precise functional form $x_2 = x_2(x_1)$, that is, by the precise definition of the curve $x_2 = x_2(x_1)$, with endpoints A and B. Three such curves are depicted on Fig. 14.2; what will turn out to be the shortest one (a straight line, in solid), plus two arbitrary longer ones (blue dashed and red dotted).

Of all possible functional forms $x_2 = x_2(x_1)$ – of which there is an infinity – we seek the one for which the value of the (real) number $S[x_2(x_1)]$ is smallest. The curve $x_2 = x_2(x_1)$ will then correspond to the shortest path between two given points A and B. From geometric considerations (Fig. 14.2), we anticipate (*but we do not assume*) that it will turn out to be the straight line that passes through the two endpoints with coordinates (x_1^A, x_2^A) and (x_1^B, x_2^B).

Let $x_2 = x_2(x_1)$ be the function (i.e. curve) that corresponds to the shortest path, for which the functional $S = S[x_1(x_2)]$ has its minimum value. At this point in the argument the precise functional form of $x_2 = x_2(x_1)$ is unknown; indeed it is the object of the exercise to determine this. Consider what happens when the sought function $x_2 = x_2(x_1)$ is *varied* by adding to it an *arbitrary* function $\delta x_2 = \delta x_2(x_1)$. This function – see the short-dashed blue curve on Fig. 14.3 – is assumed to be infinitesimally small everywhere. It is also constrained to be identically zero at its endpoints $x_1 = x_1^A$ and $x_1 = x_1^B$, so that

$$\delta x_2\left(x_1^A\right) = \delta x_2\left(x_1^B\right) = 0. \tag{14.27}$$

This condition corresponds to the assumption that the locations of points A and B are *fixed* in space and cannot be changed. The function $\delta x_2(x_1)$ is termed the *variation* of the *function* $x_2(x_1)$. Summing the function, $x_2(x_1)$, with its variation, $\delta x_2(x_1)$, then gives a new function, $x_2(x_1) + \delta x_2(x_1)$, namely the long-dashed red curve on Fig. 14.3). This new function (curve) also passes through the two points A and B, because of constraints (14.27) on the endpoint values of the variation $\delta x_2(x_1)$.

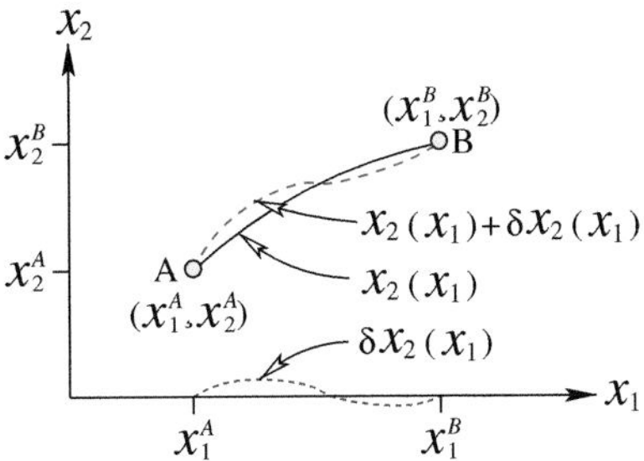

Figure 14.3 Schematic of three functions: $x_2 = x_2(x_1)$ – shortest curve (black-solid) between points A and B, with given coordinates (x_1^A, x_2^A) and (x_1^B, x_2^B), respectively; $\delta x_2 = \delta x_2(x_1)$, an arbitrary small variation (short-blue-dashed) of $x_2 = x_2(x_1)$; and the sum of these two functions (long-red-dashed), $x_2(x_1) + \delta x_2(x_1)$. See text for further details.

Consider now how the functional $S[x_2(x_1)]$ varies as the function $x_2(x_1)$ is varied by $\delta x_2(x_1)$. The resulting change in $S[x_2(x_1)]$ is then

$$\delta S[x_2(x_1)] \equiv S[x_2(x_1) + \delta x_2(x_1)] - S[x_2(x_1)], \tag{14.28}$$

where $\delta S[x_2(x_1)]$ is termed the *variation* of the *functional* $S[x_2(x_1)]$. To compute $\delta S[x_2(x_1)]$, we first evaluate the functional $S[x_2(x_1) + \delta x_2(x_1)]$. From definition (14.26) of S, this gives

$$\begin{aligned} S[x_2(x_1) + \delta x_2(x_1)] &= \int_{x_1^A}^{x_1^B} \left\{ 1 + \left[\frac{dx_2}{dx_1} + \frac{d(\delta x_2)}{dx_1} \right]^2 \right\}^{\frac{1}{2}} dx_1 \\ &= \int_{x_1^A}^{x_1^B} \left\{ 1 + \left(\frac{dx_2}{dx_1} \right)^2 + 2\frac{dx_2}{dx_1}\frac{d(\delta x_2)}{dx_1} + \left[\frac{d(\delta x_2)}{dx_1} \right]^2 \right\}^{\frac{1}{2}} dx_1. \end{aligned} \tag{14.29}$$

Assuming that $\delta x_2(x_1)$ is everywhere infinitesimally small and using Taylor's theorem to expand the integrand in (14.29) about $x_2(x_1)$ leads to

$$\begin{aligned} \left\{ 1 + \left(\frac{dx_2}{dx_1} \right)^2 + 2\frac{dx_2}{dx_1}\frac{d(\delta x_2)}{dx_1} + \left[\frac{d(\delta x_2)}{dx_1} \right]^2 \right\}^{\frac{1}{2}} &= \left[1 + \left(\frac{dx_2}{dx_1} \right)^2 \right]^{\frac{1}{2}} \\ &\quad + \left[1 + \left(\frac{dx_2}{dx_1} \right)^2 \right]^{-\frac{1}{2}} \frac{dx_2}{dx_1}\frac{d(\delta x_2)}{dx_1} \\ &\quad + \frac{1}{2}\left[1 + \left(\frac{dx_2}{dx_1} \right)^2 \right]^{-\frac{3}{2}} \left[\frac{d(\delta x_2)}{dx_1} \right]^2 \\ &\quad + O\left[(\delta x_2)^3 \right]. \end{aligned} \tag{14.30}$$

(Note that all $O\left[(\delta x_2)^2\right]$ terms have been explicitly retained in (14.30).) Substitution of (14.30) into (14.29), followed by integration by parts and use of (14.26)–(14.28), then yields

$$\begin{aligned} \delta S[x_2(x_1)] = &- \int_{x_1^A}^{x_1^B} \frac{d}{dx_1} \left\{ \left[1 + \left(\frac{dx_2}{dx_1} \right)^2 \right]^{-\frac{1}{2}} \frac{dx_2}{dx_1} \right\} (\delta x_2)\, dx_1 \\ &+ \frac{1}{2} \int_{x_1^A}^{x_1^B} \left[1 + \left(\frac{dx_2}{dx_1} \right)^2 \right]^{-\frac{3}{2}} \left[\frac{d(\delta x_2)}{dx_1} \right]^2 dx_1 + O\left[(\delta x_2)^3 \right], \end{aligned} \tag{14.31}$$

for *arbitrary*, infinitesimal variations $\delta x_2 = \delta x_2(x_1)$.

The last term on the right-hand side of (14.31) is negligible with respect to the other two and can be ignored. The middle term is guaranteed positive. For S to be a minimum when $x_2 = x_2(x_1)$ for *arbitrary* variations $\delta x_2 = \delta x_2(x_1)$ therefore requires the first term to be identically zero for *arbitrary* variations. Thus

$$\frac{d}{dx_1}\left\{\left[1+\left(\frac{dx_2}{dx_1}\right)^2\right]^{-\frac{1}{2}}\frac{dx_2}{dx_1}\right\}=0 \quad\Rightarrow\quad \frac{dx_2}{dx_1}=\text{constant}=m\ (\text{say}). \tag{14.32}$$

Integrating (14.32) and applying the boundary conditions that $x_2 = x_2(x_1)$ passes through points A and B, with coordinates (x_1^A, x_2^A) and (x_1^B, x_2^B), respectively, then gives

$$x_2 = mx_1 + c, \tag{14.33}$$

where

$$m = \frac{x_2^B - x_2^A}{x_1^B - x_1^A}, \qquad c = \frac{x_1^B x_2^A - x_1^A x_2^B}{x_1^B - x_1^A}. \tag{14.34}$$

Equation (14.33) is just the equation for the straight line passing through points A and B. As anticipated, this confirms that:

- The shortest curve between two points in the plane is the straight line passing through them.

14.4.2 The Shortest Arc between Two Points on a Unit Sphere

As a second, introductory example to functionals and variational principles, consider the problem of determining the shortest arc between two points A and B *on a unit sphere* (i.e. on a sphere of unit radius).[4] This problem is somewhat more challenging and interesting than the first one, with the result being less obvious (albeit well known).[5] To solve this problem we employ the same methodology as for the first one, but in *non-Euclidean* rather than Euclidean geometry. (In Euclidean geometry the shortest distance between two points on a sphere is along the straight line between them. This line (i.e. chord) lies *inside the sphere*. As such, it does not respect the present constraint that the sought arc must lie entirely *on the surface of the sphere*.)

In spherical-polar coordinates (λ, ϕ, r), where λ is longitude, ϕ is latitude, and r is radial distance from the sphere's centre, the differential of arc length on the surface of the unit sphere is (from (5.16), with dr set identically zero since the arc lies on $r \equiv 1$)

$$(ds)^2 = \cos^2\phi\,(d\lambda)^2 + (d\phi)^2 \quad\Rightarrow\quad ds = \left[1+\cos^2\phi\left(\frac{d\lambda}{d\phi}\right)^2\right]^{\frac{1}{2}} d\phi. \tag{14.35}$$

[4] There is no loss of generality in setting the sphere's radius to unity, since the actual radius can be reinserted at any stage simply by multiplying any quantity having units of length by this radius. Choosing it to be unity is simply a notational convenience that shortens the length of equations.

[5] Determining the shortest arc between two points on a sphere is in fact a special case of the more general problem of finding geodesics on the surface of a body. (A *geodesic* is a curve which gives the shortest distance between two points on a given surface.) Such problems, even for an ellipsoid of revolution, are a lot more complicated to solve than for the sphere. The behaviour of the geodesics is also a lot more complicated. For the special case of a sphere, a geodesic is an arc of a *closed* curve (namely an arc of a great circle). More generally, a geodesic is an arc of an *open* curve that winds multiple times (with no closed, periodic repetition) around the surface of a body.

The total distance, S, between two points on the unit sphere with two-dimensional[6] coordinates (λ^A, ϕ^A) and (λ^B, ϕ^B) is thus given by the definite integral

$$S[\lambda(\phi)] = \int_{\phi^A}^{\phi^B} \left[1 + \cos^2\phi \left(\frac{d\lambda}{d\phi}\right)^2\right]^{\frac{1}{2}} d\phi, \tag{14.36}$$

where ϕ^A and ϕ^B are given constants. Of all possible functional forms $\lambda = \lambda(\phi)$, we seek the one for which the (real) number $S[\lambda(\phi)]$ is smallest.

Following the methodology developed in the previous example, consider what happens when the sought function $\lambda = \lambda(\phi)$ is varied by adding to it an arbitrary function $\delta\lambda = \delta\lambda(\phi)$. This function is assumed to be infinitesimally small everywhere and also constrained to be identically zero at its (fixed) endpoints $\phi = \phi^A$ and $\phi = \phi^B$, so that

$$\delta\lambda(\phi^A) = \delta\lambda(\phi^B) = 0. \tag{14.37}$$

Taking the variation of the functional $S[\lambda(\phi)]$ in a manner analogous to that developed in the previous example leads to

$$\begin{aligned}
\delta S[\lambda(\phi)] &\equiv S[\lambda(\phi) + \delta\lambda(\phi)] - S[\lambda(\phi)] \\
&= \int_{\phi^A}^{\phi^B} \left\{1 + \cos^2\phi \left[\frac{d\lambda}{d\phi} + \frac{d(\delta\lambda)}{d\phi}\right]^2\right\}^{\frac{1}{2}} d\phi - \int_{\phi^A}^{\phi^B} \left[1 + \cos^2\phi \left(\frac{d\lambda}{d\phi}\right)^2\right]^{\frac{1}{2}} d\phi \\
&= \int_{\phi^A}^{\phi^B} \left\{1 + \cos^2\phi \left(\frac{d\lambda}{d\phi}\right)^2 + 2\cos^2\phi \frac{d\lambda}{d\phi}\frac{d(\delta\lambda)}{d\phi} + \cos^2\phi \left[\frac{d(\delta\lambda)}{d\phi}\right]^2\right\}^{\frac{1}{2}} d\phi \\
&\quad - \int_{\phi^A}^{\phi^B} \left[1 + \cos^2\phi \left(\frac{d\lambda}{d\phi}\right)^2\right]^{\frac{1}{2}} d\phi.
\end{aligned} \tag{14.38}$$

Assuming that $\delta\lambda(\phi)$ is everywhere infinitesimally small, the integrand of the first term on the right-hand side of (14.38) can be expanded in a Taylor series. Thus

$$\begin{aligned}
&\left\{1 + \cos^2\phi \left(\frac{d\lambda}{d\phi}\right)^2 + 2\cos^2\phi \frac{d\lambda}{d\phi}\frac{d(\delta\lambda)}{d\phi} + \cos^2\phi \left[\frac{d(\delta\lambda)}{d\phi}\right]^2\right\}^{\frac{1}{2}} \\
&= \left[1 + \cos^2\phi \left(\frac{d\lambda}{d\phi}\right)^2\right]^{\frac{1}{2}} + \left[1 + \cos^2\phi \left(\frac{d\lambda}{d\phi}\right)^2\right]^{-\frac{1}{2}} \cos^2\phi \frac{d\lambda}{d\phi}\frac{d(\delta\lambda)}{d\phi} \\
&\quad + \frac{1}{2}\left[1 + \cos^2\phi \left(\frac{d\lambda}{d\phi}\right)^2\right]^{-\frac{3}{2}} \cos^2\phi \left[\frac{d(\delta\lambda)}{d\phi}\right]^2 + O\left[(\delta\lambda)^3\right].
\end{aligned} \tag{14.39}$$

Substitution of (14.39) into (14.38), followed by integration by parts and use of (14.37), then yields

$$\delta S[\lambda(\phi)] = \int_{\phi^A}^{\phi^B} \left\{\left[1 + \cos^2\phi \left(\frac{d\lambda}{d\phi}\right)^2\right]^{-\frac{1}{2}} \cos^2\phi \frac{d\lambda}{d\phi}\frac{d(\delta\lambda)}{d\phi}\right\} d\phi$$

[6] The third coordinate, r, has been suppressed here, since every point on the unit sphere has $r \equiv 1$. Setting $r = 1 =$ constant ensures that the arc lies entirely *on the surface of the sphere*.

$$+\frac{1}{2}\int_{\phi^A}^{\phi^B}\left[1+\cos^2\phi\left(\frac{d\lambda}{d\phi}\right)^2\right]^{-\frac{3}{2}}\cos^2\phi\left[\frac{d(\delta\lambda)}{d\phi}\right]^2 d\phi + O\left[(\delta\lambda)^3\right]$$

$$=-\int_{\phi^A}^{\phi^B}\frac{d}{d\phi}\left\{\left[1+\cos^2\phi\left(\frac{d\lambda}{d\phi}\right)^2\right]^{-\frac{1}{2}}\cos^2\phi\frac{d\lambda}{d\phi}\right\}(\delta\lambda)\,d\phi$$

$$+\frac{1}{2}\int_{\phi^A}^{\phi^B}\left[1+\cos^2\phi\left(\frac{d\lambda}{d\phi}\right)^2\right]^{-\frac{3}{2}}\cos^2\phi\left[\frac{d(\delta\lambda)}{d\phi}\right]^2 d\phi + O\left[(\delta\lambda)^3\right]. \tag{14.40}$$

The last term on the right-hand side of (14.40) is negligible with respect to the other two and can be ignored. The middle term is guaranteed positive. For S to be a minimum when $\lambda = \lambda(\phi)$ for *arbitrary* variations $\delta\lambda = \delta\lambda(\phi)$ therefore requires the first term to be identically zero. Thus

$$\frac{d}{d\phi}\left\{\left[1+\cos^2\phi\left(\frac{d\lambda}{d\phi}\right)^2\right]^{-\frac{1}{2}}\cos^2\phi\frac{d\lambda}{d\phi}\right\}=0. \tag{14.41}$$

Equation (14.41) is a second-order, ordinary differential equation. Solving this equation subject to the boundary conditions that $\lambda(\phi)$ passes through points A and B, with coordinates $\left(\lambda^A,\phi^A\right)$ and $\left(\lambda^B,\phi^B\right)$, respectively, determines the arc of shortest length joining them. Integrating (14.41) with respect to ϕ gives

$$\left[1+\cos^2\phi\left(\frac{d\lambda}{d\phi}\right)^2\right]^{-\frac{1}{2}}\cos^2\phi\frac{d\lambda}{d\phi} = \text{constant} = \frac{1}{\left(1+c^2\right)^{\frac{1}{2}}}\ (\text{say}). \tag{14.42}$$

Solving this equation for $d\lambda/d\phi$ then yields

$$\frac{d\lambda}{d\phi}=\frac{1}{\left[\left(1+c^2\right)\cos^4\phi-\cos^2\phi\right]^{\frac{1}{2}}}=\frac{\sec^2\phi}{\left[\left(1+c^2\right)-\sec^2\phi\right]^{\frac{1}{2}}}, \tag{14.43}$$

which can be rewritten as

$$d\lambda=\frac{d(\tan\phi)}{\left[\left(1+c^2\right)-\left(1+\tan^2\phi\right)\right]^{\frac{1}{2}}}=\frac{d(\tan\phi)}{\left(c^2-\tan^2\phi\right)^{\frac{1}{2}}}. \tag{14.44}$$

Integration of (14.44) with respect to $\tan\phi$ delivers

$$\tan\phi = c\sin\left(\lambda-\lambda^c\right), \tag{14.45}$$

where λ^c is a constant. The constants c and λ^c in (14.45) are determined by application of the two boundary conditions. This is left as an exercise to the reader.

Although we now know that the equation for the arc that joins points A and B is (14.45), we nevertheless lack a geometrical interpretation. To provide this, we transform (14.45), expressed in spherical-polar coordinates $(\lambda,\phi,r\equiv 1)$, to Cartesian coordinates (x_1,x_2,x_3) using the transformation relations between these two coordinate systems. These are obtained by setting $r\equiv 1$ (the equation for the surface of the unit sphere) in (5.14) of Section 5.2.2. Thus

$$x_1=\cos\lambda\cos\phi,\quad x_2=\sin\lambda\cos\phi,\quad x_3=\sin\phi. \tag{14.46}$$

Squaring and summing these three equations gives the Cartesian representation

$$x_1^2 + x_2^2 + x_3^2 = 1, \tag{14.47}$$

of the unit sphere. Rewriting (14.45) as

$$c \sin \lambda_c \cos \lambda \cos \phi - c \cos \lambda_c \sin \lambda \cos \phi + \sin \phi = 0 \tag{14.48}$$

and using (14.46) then yields

$$\left(c \sin \lambda^c\right) x_1 - \left(c \cos \lambda^c\right) x_2 + x_3 = 0, \tag{14.49}$$

where the coefficients of x_1 and x_2 are known *constants*.

Equation (14.49) is just the equation for a plane that contains the sphere's centre with coordinates $(x_1, x_2, x_3) = (0, 0, 0)$. We can therefore geometrically interpret (14.45) as being the intersection of the unit sphere (see (14.47)) with a plane (see (14.49)) that passes through the sphere's centre. Therefore:

- The arc of shortest length between two points on the surface of a sphere lies on the great circle that passes through them.

14.4.3 Gravitational Attraction between Two Particles

The two examples of Sections 14.4.1 and 14.4.2 are *static* problems. They were introduced to help familiarise the reader with the concept of functionals and variational integrals in a fairly simple, concrete manner. By way of contrast, the example of this subsection is a *dynamical* problem. It is introduced here as a precursor to, and a special case of *Hamilton's principle of stationary action* – see Section 14.5 – for conservative, dynamical systems.

The preceding two problems determined the *minimum* value of a functional, defined as a definite integral. When taking variations, this led to inclusion of *second-order* terms to confirm that the extremum of the variational integral (obtained from *first-order* variation) is indeed a minimum. This complicated and lengthened the analysis.

To apply Hamilton's principle of *stationary* action *for dynamical problems* (see, for example, Sections 14.8 and 14.9 for the global atmosphere and oceans), it suffices to only examine *extrema*; there is no need to determine whether an extremum is a maximum, a minimum, or a saddle point. This means that only *first-order* variations are needed; higher-order ones can simply be ignored. This greatly simplifies matters. From here on (starting in Section 14.4.3.1) attention is therefore restricted to *first-order* variations, with the understanding that, by Hamilton's principle of stationary action (as stated in Section 14.5.3) for dynamical problems, second-order variations are redundant.

14.4.3.1 *Mathematical Perspective*

Consider the functional

$$I\left[r(t)\right] = \int_{t_1}^{t_2} \left\{ \frac{\mu}{2} \left[\frac{dr(t)}{dt} \right]^2 + \frac{\gamma m M}{r(t)} \right\} dt, \tag{14.50}$$

where:

- $r = r(t)$ is distance.
- t is time, with $t = t_1 < t_2$ and $t = t_2$ being two instants in time.
- μ, γ, m, and M are constants.

Of all possible functional forms $r = r(t)$, we seek the one for which the integrand in (14.50) corresponds to an extremum of the functional $I\left[r(t)\right]$, regardless of whether this extremum is a minimum, a maximum, or a point of inflection.

For now, we consider this problem from a purely *mathematical* perspective. Later (in Sections 14.4.3.2 and 14.6), we re-examine it from a *physical* perspective. Following the methodology developed in the two previous examples, consider what happens when the sought function $r = r(t)$ is varied by adding to it an arbitrary function $\delta r = \delta r(t)$. This function is assumed to be infinitesimally small everywhere. It is also constrained to be identically zero at times $t = t_1$ and $t = t_2$, so that

$$\delta r(t_1) = \delta r(t_2) = 0. \tag{14.51}$$

Taking the variation of functional $I[r(t)]$ in a manner analogous to that developed in the two previous examples leads to

$$\begin{aligned}\delta I[r(t)] &\equiv I[r(t) + \delta r(t)] - I[r(t)] \\ &= \int_{t_1}^{t_2} \left\langle \frac{\mu}{2}\left\{\frac{dr(t)}{dt} + \frac{d[\delta r(t)]}{dt}\right\}^2 + \frac{\gamma m M}{r(t) + \delta r(t)} - \frac{\mu}{2}\left[\frac{dr(t)}{dt}\right]^2 - \frac{\gamma m M}{r(t)}\right\rangle dt. \end{aligned} \tag{14.52}$$

Assuming that $\delta r(t)$ is everywhere infinitesimally small and expanding in a Taylor series, the terms involving $\delta r(t)$ in the integrand of the right-hand side of (14.52) can be rewritten as

$$\begin{aligned}\frac{\mu}{2}\left\{\frac{dr(t)}{dt} + \frac{d[\delta r(t)]}{dt}\right\}^2 + \frac{\gamma m M}{r(t) + \delta r(t)} &= \frac{\mu}{2}\left\{\left[\frac{dr(t)}{dt}\right]^2 + 2\frac{dr(t)}{dt}\frac{d[\delta r(t)]}{dt}\right\} \\ &\quad + \frac{\gamma m M}{r(t)} - \frac{\gamma m M}{r^2(t)}\delta r(t) + O[\delta r(t)]^2 . \end{aligned} \tag{14.53}$$

Substitution of (14.53) into (14.52), followed by integration by parts and use of constraint (14.51), then yields

$$\begin{aligned}\delta I[r(t)] &= \int_{t_1}^{t_2} \left\{\mu \frac{dr(t)}{dt}\frac{d[\delta r(t)]}{dt} - \frac{\gamma m M}{r^2(t)}\delta r(t)\right\} dt + O[\delta r(t)]^2 \\ &= \cancelto{0}{\left[\mu \frac{dr(t)}{dt}\delta r(t)\right]_{t_1}^{t_2}} - \int_{t_1}^{t_2}\left[\mu \frac{d^2 r(t)}{dt^2} + \frac{\gamma m M}{r^2(t)}\right]\delta r(t)\, dt + O[\delta r(t)]^2 \\ &= -\int_{t_1}^{t_2}\left[\mu \frac{d^2 r(t)}{dt^2} + \frac{\gamma m M}{r^2(t)}\right]\delta r(t)\, dt + O[\delta r(t)]^2 . \end{aligned} \tag{14.54}$$

Anticipating Hamilton's principle of stationary action (in Section 14.5), for $I[r(t)]$ to be an extremum for *arbitrary* variations $\delta r(t)$ then gives

$$\mu \frac{d^2 r(t)}{dt^2} + \frac{\gamma m M}{r^2(t)} = 0. \tag{14.55}$$

14.4.3.2 *Physical Perspective*

Equation (14.55) corresponds to application of Newton's second law to the standard *two-body problem* of the mutual gravitational attraction of two point particles of masses m and M. Mass M is located at $r = 0$, and mass m at a distance $r = r(t)$ from mass M; γ is Newton's universal gravitational constant. The two-body problem reduces to an equivalent *one-body, central-force problem*; see Chapter 3 of Goldstein et al. (2001). This results in the appearance of the *reduced mass*[7] $\mu \equiv mM/(M+m)$ in (14.55). For the reduced problem, a central force, $\gamma mM/r^2(t)$, *immovably* centred at $r=0$, acts on a point particle of mass μ at a distance $r=r(t)$ from the central point at $r=0$.

Equation (14.55) also corresponds to application of Newton's second law to the gravitational attraction of a *fluid parcel* (of *unit mass*) to a spherical planet (of mass M, centred on $r=0$, and of constant density). The fluid parcel is assumed to be located outside the planet. Applying Newton's second law, (2.17) (in an inertial frame of reference, and for a fluid parcel of *unit mass*), in the presence of Newtonian gravity (∇V) and in the absence of a pressure-gradient force ($(1/\rho)\nabla p$) and of other forcings ($\mathbf{F}$), gives $D^2\mathbf{r}/Dt^2 + \nabla V = 0$, where $V = V(r) = -\gamma M/r$ is given by (7.34) of Section 7.6.5. See also the comments of the last paragraph of Section 7.6.6.

The correspondence of (14.55) to a physical problem is not a coincidence. It is a special case of the application of Hamilton's principle of stationary action to a physical problem, as will be shown in Section 14.6.

14.5 HAMILTON'S PRINCIPLE OF STATIONARY ACTION

14.5.1 Background

Newtonian mechanics[8] describes point particles moving under the influence of forces acting upon them (Goldstein et al., 2001). Newton's laws of motion provide the underlying physical principles for this. They lead to important conservation laws for physical quantities such as linear momentum, angular momentum, and energy.

Lagrangian mechanics[9] provides an equivalent reformulation of Newtonian mechanics. It defines and exploits a mathematical function termed the *Lagrangian* (classically denoted by L). For *conservative systems*,[10] the Lagrangian is the difference between *kinetic energy* (classically denoted by T, but by K here since T is used for temperature) and *potential energy* (classically denoted by V, where V is the potential function, or potential for short). Thus

$$L = K - V. \tag{14.56}$$

For simplicity, and because of the importance of conservation of physical quantities for modelling atmospheric and oceanic flows, *attention is restricted herein to conservative systems*. (Non-conservative forces can be inserted a posteriori into the equations.)

Once constructed, the Lagrangian for a system of point particles then leads, via differentiation, to a set of differential equations that govern the motion. *Everything that one needs to know about the system (including the forces acting on its particles) is embodied in the Lagrangian.* It is usually relatively straightforward to construct the Lagrangian. For Newtonian mechanics (which is based on forces), one has to deal with many *vector* forces and accelerations to obtain the governing equations of motion. By way of contrast, for Lagrangian mechanics one only has to deal with two *scalar* functions, namely kinetic energy (K) and potential energy (V). This greatly simplifies matters.

Hamilton's principle[11] *of stationary action*[12] for conservative systems provides a complementary framework to Lagrangian mechanics for a conservative system of point particles. Lagrangian mechanics is based on a *differential* principle. By way of contrast, Hamilton's principle of

[7] μ is often alternatively written in the form $1/\mu \equiv 1/m + 1/M$ (Goldstein et al., 2001, p. 71). When mass M is much greater than mass m, i.e. $M \gg m$ (e.g. for an observational satellite orbiting Earth), then $1/\mu \equiv 1/m + 1/M \to 1/m$; μ in (14.55) can then be approximated by m.

[8] After Isaac Newton, 1643–1727.

[9] After Joseph-Louis Lagrange, 1736–1813.

[10] A conservative system is one where no work is done around a closed path by the forces. The forces are then expressible as the gradient of a potential.

[11] After William Rowan Hamilton, 1805–65.

[12] It is often called Hamilton's principle of *least* action because (a) this is the way it was developed historically, and (b) the stationary point often does turn out to be a minimum in practical applications.

stationary action is based on an *integral* principle. It involves determination of the extremum of a variational integral of a Lagrangian. Again:

- Everything that one needs to know about the system (including the forces acting on its particles) is embodied in the Lagrangian.

14.5.2 Benefits of Hamilton's Principle of Stationary Action

To motivate what follows, Hamilton's principle of stationary action for conservative systems has several benefits when compared to Newtonian and Lagrangian formulations of mechanics. These include (see e.g. Chapters 1 and 7 of Salmon (1998), and Chapter 2 of Goldstein et al. (2001)):

1. The mechanics of a conservative system of point particles can be constructed from Hamilton's principle as the basic postulate, instead of from Newton's laws of motion. The principle only involves physical quantities (kinetic and potential energies) that can be defined *without reference to a particular set of generalised coordinates*. This then has the benefit that the formulation is automatically invariant to the coordinate system used to express the Lagrangian (Goldstein et al., 2001, p. 51).
2. Via an appropriate definition of a Lagrangian, Hamilton's principle can be extended to describe various *nonmechanical* systems. *This includes fluid-dynamical systems, such as atmospheric and oceanic flows.* It also includes electromagnetic and quantum-dynamical systems. Techniques developed for one system can then be exploited for application to another system.
3. Hamilton's principle facilitates the incorporation of moving boundaries in fluid-dynamical problems.
4. Via Noether's theorem,[13] Hamilton's principle facilitates examination of the conservation properties of a system. This is achieved through examination of various symmetry properties of the Lagrangian which, if respected, guarantee conservation of associated physical quantities.
5. Of particular interest herein is that if one approximates the Lagrangian – see Chapters 16–18 – whilst maintaining respect of the symmetries of the Lagrangian, then the physical quantities conserved in the original system are also conserved in the approximated one. This provides a powerful tool for systematically developing approximated equation sets that respect the underlying conservation principles used to obtain the unapproximated equations.

14.5.3 Action and Hamilton's Principle of Stationary Action

We start by defining action and by stating Hamilton's principle of stationary action in general terms. This is rather abstract. It is, however, the nature of the subject and cannot be avoided. We then examine (in the following sections) three concrete problems of increasing complexity to illustrate how Hamilton's abstract principle can be tangibly translated into practical applications. This exercise culminates in the governing equations for a global atmosphere or ocean.

Action is an attribute of the dynamics of a physical system *from which the equations of motion of the system can be derived.*

The Action $\mathcal{A}\left[\mathbf{q}\left(t\right)\right]$ of a Physical System

The action, $\mathcal{A}\left[\mathbf{q}\left(t\right)\right]$, of a physical system is a functional defined in terms of a Lagrangian, $L\left[\mathbf{q}\left(t\right),\dot{\mathbf{q}}\left(t\right),t\right]$, as

$$\mathcal{A}\left[\mathbf{q}\left(t\right)\right] \equiv \int_{t_1}^{t_2} L\left[\mathbf{q}\left(t\right),\dot{\mathbf{q}}\left(t\right),t\right] dt, \tag{14.57}$$

[13] After Amalie Emmy Noether, 1882–1935.

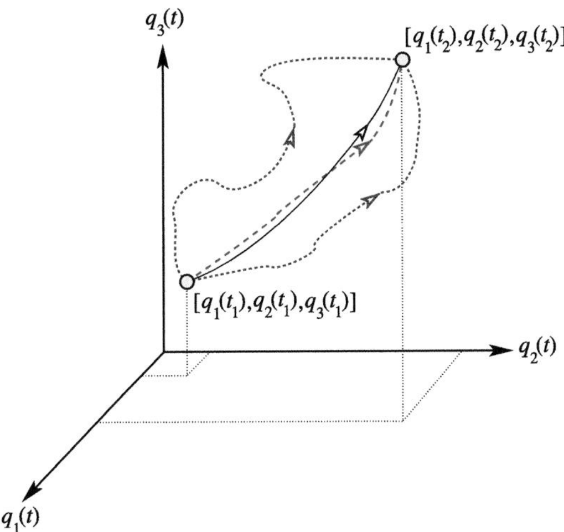

Figure 14.4 Configuration space in three dimensions ($n = 3$). $\mathbf{q}(t_1) = [q_1(t_1), q_2(t_1), q_3(t_1)]$ and $\mathbf{q}(t_2) = [q_1(t_2), q_2(t_2), q_3(t_2)]$ are beginning and end points, respectively, of four paths in configuration space; solid curve (black) – actual path; long-dashed curve (red) – a small variation from actual path; short-dashed curves (blue) – two random paths. Open arrows denote time increasing along a path, and dotted lines are parallel to coordinate axes.

where:

- $\mathbf{q}(t)$ is a vector of *generalised coordinates*, whose components are all *independent of one another.*
- $\dot{\mathbf{q}}(t) \equiv d\mathbf{q}/dt$ is its first derivative with respect to time, t.
- Lagrangian $L[\mathbf{q}(t), \dot{\mathbf{q}}(t), t]$ is independent of higher-order derivatives, such as $\ddot{\mathbf{q}}(t)$.
- $t = t_1$ and $t = t_2$ are two instants in time, with $t_1 < t_2$.

The vector space defined by $\mathbf{q}(t)$ is termed the *configuration space* of the physical system. A special case of configuration space is *location space*, that is, $\mathbf{q}(t) = \mathbf{r}(t)$; see Section 14.2.3. (For example, configuration space for the problems examined in Sections 14.6 and 14.7 is location space.) The value of the action $\mathcal{A}[\mathbf{q}(t)]$ (a real number) depends on the particular path taken in configuration space between times t_1 and t_2.

Assume that $\mathbf{q}(t)$ has n generalised (*independent*) components $q_1(t)$, $q_2(t)$, $q_3(t)$, …, $q_n(t)$, as depicted in Fig. 14.4 for $n = 3$. Time, t, here can be considered to be a parameter that varies between times t_1 and $t_2 > t_1$. The n components $q_i(t)$ form a vector hyperspace of dimension n. At any given instant in time, t, each component $q_i(t)$ has some value. At this instant in time, the vector of all n values $q_i(t)$, that is, $\mathbf{q}(t) \equiv [q_1(t), q_2(t), \ldots, q_n(t)]$, represents a point in this hyperspace (i.e. configuration space). Varying the parameter t from $t = t_1$ to $t = t_2$ then defines a continuous path between two points $\mathbf{q}(t_1)$ and $\mathbf{q}(t_2)$ in configuration space.

For the simple cases of $n = 2$ and $n = 3$, configuration space reduces to two and three dimensions respectively. Four paths, beginning at point $\mathbf{q}(t_1) = [q_1(t_1), q_2(t_1), q_3(t_1)]$ at time t_1, and ending at point $\mathbf{q}(t_2) = [q_1(t_2), q_2(t_2), q_3(t_2)]$ at time t_2, are depicted in Fig. 14.4 for $n = 3$. One can imagine a path between these two points as being a piece of (bent) wire joining them. For higher dimensions one has to exercise one's imagination!

Hamilton's principle of stationary action expressed verbally and mathematically, respectively, is as follows:

Hamilton's Principle of Stationary Action (Verbal)

- Of all possible paths that could be followed in configuration space, beginning at $\mathbf{q} = \mathbf{q}(t_1)$ at time t_1, and ending at $\mathbf{q} = \mathbf{q}(t_2)$ at time t_2, the path *actually* followed by a physical system is the one for which the action, $\mathcal{A}[\mathbf{q}(t)]$, is *stationary*.
- In other words, the path actually followed is the one that corresponds to an *extremum* of the integral functional, $\mathcal{A}[\mathbf{q}(t)]$, for the action.

Hamilton's Principle of Stationary Action (Mathematical)

- Mathematically,

$$\delta\mathcal{A}[\mathbf{q}(t)] \equiv \delta\int_{t_1}^{t_2} L[\mathbf{q}(t), \dot{\mathbf{q}}(t), t]\, dt = \int_{t_1}^{t_2} \delta L[\mathbf{q}(t), \dot{\mathbf{q}}(t), t]\, dt = 0, \tag{14.58}$$

  for *independent* variations $\delta\mathbf{q}(t)$ of the vector of generalised (independent) coordinates, $\mathbf{q}(t)$, such that

$$\delta\mathbf{q}(t_1) = \delta\mathbf{q}(t_2) = 0. \tag{14.59}$$

- Applying this procedure results in the governing equations of the physical system.

Note that variation and integration are commutative processes (e.g. Lanczos, 1970, p. 57), that is,

$$\delta\int_{t_1}^{t_2} F(t)\, dt = \int_{t_1}^{t_2} \delta F(t)\, dt. \tag{14.60}$$

This property has been exploited in rewriting the second expression in (14.58) as the third. A further useful property (again Lanczos, 1970, p. 57) is that variation and differentiation processes also commute – see also (14.72) herein – that is,

$$\delta\left[\frac{dF(t)}{dt}\right] = \frac{d}{dt}[\delta F(t)]. \tag{14.61}$$

Hamilton's principle has very wide application. Herein, as mentioned earlier, attention is restricted to its application for conservative systems. Central to Hamilton's principle is the definition of a Lagrangian for the physical system of interest. Lagrangians can, for example, be defined for (Salmon, 1998; Goldstein et al., 2001):

- A single point particle.
- A system of point particles of finite number.
- A *continuum*, such as a fluid composed of fluid parcels – of particular interest herein – with an infinite number of degrees of freedom.

One might think that the generalisation of a Lagrangian for a single point particle to a system of point particles can be straightforwardly accomplished by:

- Summing the kinetic energies of the individual particles to obtain the total kinetic energy, K.
- Summing the potential energies of the individual particles to obtain the total potential energy, V.
- Computing the Lagrangian, using (14.56), as $L = K - V$.

Sadly, things are not this simple. Although this procedure works just fine for rigid bodies, it does not do so more generally. The underlying physical reason for this is that, in general, point particles *mutually interact* (Goldstein et al., 2001, Chapter 1, p. 5). This gives rise to *internal forces.*[14] Total potential energy is then the sum of *internal* energy (due to internal conservative forces between point particles within the system), and *external* potential energy (imposed by external conservative forces acting on the system).[15]

The transition from a discrete system, with a finite number of point particles, to a continuum is accomplished via (see Section 14.8.2.2):

- Redistribution of point mass within the confines of an infinitesimal fluid parcel.
- Definition of an infinity of such parcels, continuously distributed throughout space and spanning it.
- Integration over the entire spatial domain.

Similar considerations regarding internal energy and external potential energy for systems of discrete particles also hold for a continuum of fluid parcels. The existence of internal energy very much complicates matters.

As is evident from the preceding discussion, Hamilton's principle of stationary action is a rather abstract concept. To make things more concrete, we now examine three problems of increasing complexity:

1. Motion of a point particle subject to the gravitational attraction of a second point particle (the two-body problem, Section 14.6).
   This simple problem, already examined in Section 14.4.3, is recast by defining a Lagrangian for it and then applying Hamilton's principle of stationary action using this Lagrangian.
2. Motion of a system of point particles (Section 14.7).
   This problem, of intermediate complexity, provides a bridge between the previous problem and the following one.
3. Motion of the global atmosphere or ocean, subject to conservative forces (Sections 14.8–14.10).
   This problem provides the basis for the derivation (in Chapters 16–18) of dynamically consistent, approximated equation sets for the global atmosphere or ocean.

14.6 GRAVITATIONAL ATTRACTION BETWEEN TWO PARTICLES REVISITED

The gravitational attraction between two point particles was introduced – in Section 14.4.3.1 – as a *mathematical* problem to help familiarise the reader with the concepts of functionals and variational integrals in a simple, concrete manner. It was only afterwards – in Section 14.4.3.2 – that it was linked to an actual *physical* problem. In this section we revisit this problem by instead starting from a physical perspective and using Hamilton's principle of stationary action. As discussed in Section 14.4.3.2, this two-body problem can be reduced to an equivalent one-body problem. This is achieved by simply using the reduced mass

$$\mu \equiv \frac{mM}{m+M} \tag{14.62}$$

to define the kinetic energy, instead of m (Goldstein et al., 2001, p. 71).

[14] *Internal* and *external* in the present context mean *internal to the system of point particles*, and *external to the system of point particles*, respectively.

[15] For rigid bodies, internal forces between particles cancel out. Internal potential energy is therefore absent, and this greatly simplifies matters.

With this introduction, a step-by-step procedure for solving this problem using Hamilton's principle of stationary action now follows. This procedure is subsequently adapted (in Sections 14.7–14.10) for more complicated problems.

14.6.1 Coordinates

The vector of generalised coordinates, $\mathbf{q}(t)$, consists of a single component

$$q(t) \equiv r(t), \tag{14.63}$$

where $r(t)$ is distance of a point particle, of mass m, from another point particle, of mass M. Thus configuration space is just location space, with a single 'generalised' coordinate.

14.6.2 Kinetic Energy (K)

From (14.63), the derivative of $q(t)$ with respect to time, t, is

$$\dot{q}(t) \equiv \frac{dq(t)}{dt} \equiv \frac{dr(t)}{dt} \equiv u. \tag{14.64}$$

Kinetic energy is thus

$$K\left[\dot{q}(t)\right] \equiv \frac{\mu}{2}u^2 \equiv \frac{\mu}{2}\left[\frac{dq(t)}{dt}\right]^2 \equiv \frac{\mu}{2}\left[\frac{dr(t)}{dt}\right]^2 \equiv \frac{\mu}{2}\left[\dot{r}(t)\right]^2 \equiv \frac{\mu}{2}\left[\dot{q}(t)\right]^2, \tag{14.65}$$

where μ is the reduced mass, defined by (14.62) and used to reduce the two-body problem to a one-body problem (see Section 14.4.3.2).

14.6.3 Potential Energy (V)

Potential energy is given by

$$V\left[q(t)\right] \equiv -\frac{\gamma mM}{r(t)} \equiv -\frac{\gamma mM}{q(t)}, \tag{14.66}$$

where $V\left[q(t)\right]$ is the potential for Newtonian gravitational attraction between two point particles, of masses m and M at distance $q(t) = r(t)$ apart, and γ is Newton's universal gravitational constant.

14.6.4 Lagrangian (L)

From (14.56), and using (14.65) and (14.66), the Lagrangian is

$$L\left[q(t),\dot{q}(t)\right] = K\left[\dot{q}(t)\right] - V\left[q(t)\right] = \frac{\mu}{2}\left[\dot{q}(t)\right]^2 + \frac{\gamma mM}{q(t)}. \tag{14.67}$$

14.6.5 Action ($\mathcal{A}$) and Hamilton's Principle

Inserting Lagrangian (14.67) into (14.57), the corresponding action is

$$\mathcal{A}\left[q(t)\right] = \int_{t_1}^{t_2} L\left[q(t),\dot{q}(t)\right]dt = \int_{t_1}^{t_2}\left\{\frac{\mu}{2}\left[\dot{q}(t)\right]^2 + \frac{\gamma mM}{q(t)}\right\}dt. \tag{14.68}$$

This is just (14.50) but rewritten in slightly different notation; $I \to \mathcal{A}$, $r \to q$, $dr(t)/dt \to \dot{q}(t)$.

Hamilton's principle of stationary action is to use (14.58) to set independent variations of the action to zero, subject to endpoint constraints (14.59) on the variation of the generalised coordinates. For the present problem, with a single generalised coordinate, $q(t)$, and using (14.68), this reduces to applying

$$\delta \mathcal{A}\left[q(t)\right] = \delta \int_{t_1}^{t_2} L\left[q(t), \dot{q}(t)\right] dt = \int_{t_1}^{t_2} \delta L\left[q(t), \dot{q}(t)\right] dt = 0, \tag{14.69}$$

subject to endpoint constraints

$$\delta q(t_1) = \delta q(t_2) = 0, \tag{14.70}$$

on the variation of $q(t)$.

Nothing has changed mathematically by rewriting (14.50) as (14.68) beyond introducing some definitions and nomenclature. What has changed is that (14.68) has now been derived from stated physical principles instead of just being written down as (14.50).

14.6.6 Variation of the Lagrangian

Rewriting (14.50) as (14.68) facilitates taking first-order variations without needing to *explicitly* expand the Lagrangian in a Taylor series. Explicit Taylor-series expansion is instead (and equivalently) replaced by partial differentiation; see (14.73). Doing things this way is of benefit when taking variations of complicated Lagrangians in several dimensions.

For this alternative method of taking variations, consider a Lagrangian of the form $L = L\left[q(t), \dot{q}(t)\right]$. Taking the variation $\delta L\left[q(t), \dot{q}(t)\right]$ and expanding to first order in a Taylor series in the two small quantities $\delta q(t)$ and $\delta \dot{q}(t)$ gives

$$\begin{aligned}
\delta L\left[q(t), \dot{q}(t)\right] &\equiv L\left[q(t) + \delta q(t), \dot{q}(t) + \delta \dot{q}(t)\right] - L\left[q(t), \dot{q}(t)\right] \\
&= \cancel{L\left[q(t), \dot{q}(t)\right]} + \frac{\partial L\left[q(t), \dot{q}(t)\right]}{\partial q} \delta q(t) + \frac{\partial L\left[q(t), \dot{q}(t)\right]}{\partial \dot{q}} \delta \dot{q}(t) - \cancel{L\left[q(t), \dot{q}(t)\right]} \\
&= \frac{\partial L\left[q(t), \dot{q}(t)\right]}{\partial q} \delta q(t) + \frac{\partial L\left[q(t), \dot{q}(t)\right]}{\partial \dot{q}} \delta \dot{q}(t).
\end{aligned} \tag{14.71}$$

Now $\delta \dot{q}(t)$ in (14.71) is not independent of $\delta q(t)$. Using $\dot{q}(t) \equiv d\left[q(t)\right]/dt$, the variation in $\dot{q}(t)$, that is, $\delta \dot{q}(t)$, is related to the variation in $q(t)$, that is, $\delta q(t)$, by

$$\delta \dot{q}(t) \equiv \delta \left\{ \frac{d\left[q(t)\right]}{dt} \right\} \equiv \frac{d\left[q(t) + \delta q(t)\right]}{dt} - \frac{d\left[q(t)\right]}{dt} = \frac{d}{dt}\left[\delta q(t)\right]. \tag{14.72}$$

(Equation (14.72) shows that δ commutes with d/dt, i.e. the processes of variation and differentiation commute; see (14.61).) Substituting (14.72) into (14.71) then yields

$$\delta L\left[q(t), \dot{q}(t)\right] = \frac{\partial L\left[q(t), \dot{q}(t)\right]}{\partial q} \delta q(t) + \frac{\partial L\left[q(t), \dot{q}(t)\right]}{\partial \dot{q}} \frac{d\left[\delta q(t)\right]}{dt}. \tag{14.73}$$

This equation expresses the variation $\delta L\left[q(t), \dot{q}(t)\right]$ of the Lagrangian in terms of the variation of the generalised coordinate $\delta q(t)$. Equation (14.73) is valid not only for the present problem, where the Lagrangian is specifically given by (14.67), but for *any* Lagrangian of the form $L = L\left[q(t), \dot{q}(t)\right]$.

When taking first-order variations, this means that one is no longer obliged to *explicitly* expand the Lagrangian, $L\left[q(t), \dot{q}(t)\right]$, in terms of a Taylor series. Instead, one can simply partially differentiate $L\left[q(t), \dot{q}(t)\right]$ to obtain the partial derivatives $\partial L\left[q(t), \dot{q}(t)\right]/\partial q$ and $\partial L\left[q(t), \dot{q}(t)\right]/\partial \dot{q}$, and then substitute these into (14.73) to obtain the variation $\delta L\left[q(t),\dot{q}(t)\right]$ in terms of the variation $\delta q(t)$. Due to its use in the derivation, Taylor-series expansion is still of course *implicit* in (14.73).

More generally, similar arguments for a Lagrangian of the form

$$L = L\left[\mathbf{q}(t),\dot{\mathbf{q}}(t)\right] \equiv L\left[q_1(t),q_2(t),\ldots,q_N(t),\dot{q}_1(t),\dot{q}_2(t),\ldots,\dot{q}_N(t)\right] \tag{14.74}$$

lead to:

Variations of a Lagrangian $L = L[\mathbf{q}(t),\dot{\mathbf{q}}(t)]$ via Partial Differentiation

$$\delta L\left[\mathbf{q}(t),\dot{\mathbf{q}}(t)\right] = \sum_{i=1}^{N} \frac{\partial L\left[q_i(t),\dot{q}_i(t)\right]}{\partial q_i}\delta q_i(t) + \sum_{i=1}^{N} \frac{\partial L\left[q_i(t),\dot{q}_i(t)\right]}{\partial \dot{q}_i}\frac{d\left[\delta q_i(t)\right]}{dt} \tag{14.75}$$

for the N independent variations, $\delta q_i(t), i = 1,2,\ldots,N$.

Returning to the present problem, the specific form of the Lagrangian, $L\left[q(t),\dot{q}(t)\right]$, is given by (14.67). Its two partial derivatives are thus

$$\frac{\partial L\left[q(t),\dot{q}(t)\right]}{\partial q} = -\frac{\gamma m M}{q^2(t)}, \quad \frac{\partial L\left[q(t),\dot{q}(t)\right]}{\partial \dot{q}} = \mu\dot{q}(t). \tag{14.76}$$

Substitution of (14.76) into (14.73) then yields the variation

$$\delta L\left[q(t),\dot{q}(t)\right] = \mu\dot{q}(t)\frac{d\left[\delta q(t)\right]}{dt} - \frac{\gamma m M}{q^2(t)}\delta q(t) \tag{14.77}$$

of the Lagrangian.

14.6.7 Variation of the Action and Governing Equation

Variation $\delta\mathcal{A}\left[q(t)\right]$ of action $\mathcal{A}\left[q(t)\right]$ is given by (14.69). Substitution of (14.77) – for variation $\delta L\left[q(t),\dot{q}(t)\right]$ of the Lagrangian – into (14.69) – for variation $\delta\mathcal{A}\left[q(t)\right]$ of the action – followed by integration by parts and use of (14.63), (14.64), and (14.70) then yields

$$\begin{aligned}
\delta\mathcal{A}\left[q(t)\right] &= \int_{t_1}^{t_2} \delta L\left[q(t),\dot{q}(t)\right]dt = \int_{t_1}^{t_2}\left[\mu\dot{q}\frac{d\left[\delta q(t)\right]}{dt} - \frac{\gamma m M}{q^2(t)}\delta q(t)\right]dt \\
&= \cancelto{0}{\left(\mu\dot{q}\delta q\right)_{t_1}^{t_2}} - \int_{t_1}^{t_2}\left[\mu\ddot{q} + \frac{\gamma m M}{q^2(t)}\right]\delta q(t)\,dt = -\int_{t_1}^{t_2}\left[\mu\frac{d^2 q}{dt^2} + \frac{\gamma m M}{q^2(t)}\right]\delta q(t)\,dt \\
&= -\int_{t_1}^{t_2}\left[\mu\frac{d^2 r}{dt^2} + \frac{\gamma m M}{r^2(t)}\right]\delta r(t)\,dt,
\end{aligned} \tag{14.78}$$

where $\ddot{q} \equiv d^2 q(t)/dt^2$. Setting variation $\delta\mathcal{A}\left[q(t)\right]$ in (14.78) to zero and recalling that variation $\delta r(t)$ is *arbitrary* finally results in

$$\mu\frac{d^2 r}{dt^2} + \frac{\gamma m M}{r^2(t)} = 0. \tag{14.79}$$

This recovers condition (14.55) for $r(t)$ to correspond to an extremum of $I\left[r(t)\right]$, as defined by (14.50).

14.6.8 Summary

The Lagrangian, L, for the two-body, gravitational-attraction problem was obtained by constructing the kinetic (K) and potential (V) energies (both scalars) and taking their difference. The action, $\mathcal{A}$, then immediately followed. Applying Hamilton's principle of stationary action (by setting first variations of the action to zero, subject to the constraint of zero variation at the endpoints of the variation) then led to governing equation (14.79). This equation corresponds to Newton's second law for this problem.

This law was not imposed. It just dropped out (as expected) from constructing the Lagrangian and then applying Hamilton's principle.

14.7 A SYSTEM OF POINT PARTICLES

Following Salmon (1998), consider a system of n point particles, of mass m_i, and locations $\mathbf{r}_i = \mathbf{r}_i(t) = [x_i(t), y_i(t), z_i(t)]$ in three-dimensional Cartesian space.[16]

14.7.1 Coordinates

The vector of generalised coordinates, $\mathbf{q}(t)$, is now comprised of the n three-dimensional location (position) vectors, $\mathbf{q}_i(t) = \mathbf{r}_i(t)$, so that

$$\begin{aligned}\mathbf{q}(t) &= [\mathbf{q}_1(t), \mathbf{q}_2(t), \ldots, \mathbf{q}_n(t)] = [\mathbf{r}_1(t), \mathbf{r}_2(t), \ldots, \mathbf{r}_n(t)] \\ &= [x_1(t), y_1(t), z_1(t), x_2(t), y_2(t), z_2(t), \ldots, x_n(t), y_n(t), z_n(t)] = \mathbf{r}(t). \end{aligned} \tag{14.80}$$

A vector of generalised coordinates, $\mathbf{q}_i(t) \equiv [x_i(t), y_i(t), z_i(t)]$, is associated with each particle of index i. Since there are n such particles, there are $N = 3n$ independent, generalised coordinates. Thus configuration space is location space, with $3n$ independent, generalised coordinates.

14.7.2 Kinetic Energy (K)

From (14.80), the derivative of $\mathbf{q}(t)$ with respect to time, t, is

$$\dot{\mathbf{q}}(t) = \frac{d\mathbf{q}}{dt} = \left(\frac{d\mathbf{q}_1}{dt}, \frac{d\mathbf{q}_2}{dt}, \ldots, \frac{d\mathbf{q}_n}{dt}\right) = \left(\frac{d\mathbf{r}_1}{dt}, \frac{d\mathbf{r}_2}{dt}, \ldots, \frac{d\mathbf{r}_n}{dt}\right) = \dot{\mathbf{r}}(t). \tag{14.81}$$

Kinetic energy (K) is thus

$$K[\dot{\mathbf{q}}(t)] = K[\dot{\mathbf{r}}(t)] = K[\dot{\mathbf{r}}_1(t), \dot{\mathbf{r}}_2(t), \ldots, \dot{\mathbf{r}}_n(t)] = \frac{1}{2}\sum_{i=1}^{n} m_i \frac{d\mathbf{r}_i}{dt} \cdot \frac{d\mathbf{r}_i}{dt}, \tag{14.82}$$

where the last expression is just the sum of the kinetic energies of the individual point particles.

14.7.3 Potential Energy (V)

Potential energy (V) is assumed to be a prescribed (known) function of location only, and so

$$V[\mathbf{q}(t)] = V[\mathbf{q}_1(t), \mathbf{q}_2(t), \ldots, \mathbf{q}_n(t)] = V[\mathbf{r}_1(t), \mathbf{r}_2(t), \ldots, \mathbf{r}_n(t)] = V[\mathbf{r}(t)]. \tag{14.83}$$

Aside As an illustrative example, the n-body problem consists of n particles of mass $m_i, i = 1, 2, \ldots, n$, moving in three dimensions in an inertial frame of reference and under

[16] Because of the use of a subscript 'i' to identify a specific particle, the previous notation $\mathbf{r} = (x_1, x_2, x_3)$ is temporarily replaced in this section by $\mathbf{r} = (x, y, z)$. When we move (in the next section) from the discrete case of n particles to the continuous case of a fluid, we will revert to the original notation.

the influence of mutual gravitational attraction. Using Newton's law of gravitation, the potential energy is therefore

$$V\left[\mathbf{r}(t)\right] = V\left[\mathbf{r}_1(t), \mathbf{r}_2(t), \ldots, \mathbf{r}_n(t)\right] = -\frac{1}{2}\sum_{i=1}^{n}\sum_{j=1, j\neq i}^{n}\frac{\gamma m_i m_j}{\left|\mathbf{r}_i(t) - \mathbf{r}_j(t)\right|}, \tag{14.84}$$

where γ is Newton's universal gravitational constant.[17] Thus the more massive two particles are and the closer together they are, the stronger the inter-body interaction.

The number of particles can, in principle, vary from two – as in Section 14.6 – to as many as one wishes. Although the two-body problem can be solved exactly, analytic tractability very rapidly diminishes as the number of bodies increases.[18] Some exact solutions are nevertheless possible in restrictive circumstances. For example, for the three-body problem, it can be shown that a small object can maintain its position relative to two large bodies orbiting around their barycentre (i.e. their centre of mass). There are five such equilibrium positions (two stable to orbital perturbation, three unstable), known as Lagrange points. It is advantageous to place an observing platform at one of these points in the Earth-Sun-platform system; much less fuel is then needed to maintain the position of a platform there, even near a Lagrange point that is unstable to orbital perturbation.[19]

Classically, the n-body problem originated from examination of the motion of planets in the solar system. The Sun accounts for all but one or two per cent of the total mass of the solar system. To leading order, the number of bodies can therefore be reduced to two; the Sun plus any one of the planets. This is a two-body problem with an exact solution for the orbit of a planet around the Sun.[20] Although the Sun, because of its huge mass, dominates inter-body gravitational interaction within the solar system, the other planets (principally the giant gas planets Jupiter, Saturn, Neptune, and Uranus) do cause the Sun to measurably wobble a little (in an inertial reference frame) in response to their masses and motions. There are many other examples of the application of the n-body problem in celestial mechanics, including determination of artificial-satellite orbits around planets and interplanetary spacecraft navigation.

14.7.4 Lagrangian (L)

From definition (14.56) of a Lagrangian, and using (14.82) and (14.83) for K and V, respectively, the Lagrangian for a system of n point particles is

$$\begin{aligned} L\left[\mathbf{q}(t), \dot{\mathbf{q}}(t)\right] &= K\left[\dot{\mathbf{q}}(t)\right] - V\left[\mathbf{q}(t)\right] = \frac{1}{2}\sum_{i=1}^{n} m_i \frac{d\mathbf{r}_i}{dt}\cdot\frac{d\mathbf{r}_i}{dt} - V\left[\mathbf{r}(t)\right] \\ &= \frac{1}{2}\sum_{i=1}^{n} m_i \dot{\mathbf{r}}_i(t)\cdot\dot{\mathbf{r}}_i(t) - V\left[\mathbf{r}(t)\right] = L\left[\mathbf{r}(t), \dot{\mathbf{r}}(t)\right]. \end{aligned} \tag{14.85}$$

17 The factor of 1/2 in (14.84) removes the double counting of the interaction of particle j with particle i, which would otherwise occur. The condition $j \neq i$ is because a particle does not interact with itself, only with other particles; this also avoids the possibility of spurious division by zero.

18 Numerical solution is nevertheless possible.

19 Solar flares and the solar wind can adversely interfere with communications and weather satellites orbiting around Earth. They can also cause damaging surges to the networks used to distribute electrical power to homes and industry. Monitoring these flares using an observational platform located at a Lagrange point can help alleviate such problems. Furthermore, placement of a telescope there (in a vacuum) avoids the distortion that occurs when viewing the universe through Earth's atmosphere with telescopes located at Earth's surface.

20 Physically, however, the orbit is not exactly determined, due to neglect of the gravitational influence of bodies other than the Sun.

14.7.5 Action ($\mathcal{A}$) and Hamilton's Principle

Hamilton's principle of stationary action is to use (14.58) to take independent variations of the action, subject to endpoint constraints (14.59) on the variation, and to set these variations to zero. For the present problem this gives

$$\delta\mathcal{A}\left[\mathbf{q}(t)\right] = \delta\int_{t_1}^{t_2} L\left[\mathbf{r}(t),\dot{\mathbf{r}}(t)\right]dt = \int_{t_1}^{t_2}\delta L\left[\mathbf{r}(t),\dot{\mathbf{r}}(t)\right]dt = 0, \tag{14.86}$$

subject to the variation's endpoint constraints

$$\delta\mathbf{r}(t_1) = \delta\mathbf{r}(t_2) = 0, \tag{14.87}$$

where $L\left[\mathbf{r}(t),\dot{\mathbf{r}}(t)\right]$ is given by (14.85). Since there are $3n$ generalised coordinates for this problem – see (14.80) – there are $3n$ corresponding endpoint constraints (of the form (14.87)) at initial time $t = t_1$, plus a further $3n$ at final time $t = t_2$.

14.7.6 Variations of the Lagrangian

Variations, $\delta L\left[\mathbf{r}(t),\dot{\mathbf{r}}(t)\right]$, of the Lagrangian defined by (14.85) can be computed by using Taylor series or by using partial differentiation. It is instructive to do both for this problem and to verify that they do indeed lead to the same result.

14.7.6.1 Method 1, via Taylor Series

The variation, $\delta L\left[\mathbf{r}(t),\dot{\mathbf{r}}(t)\right]$, of the Lagrangian defined by (14.85) is

$$\begin{aligned}\delta L\left[\mathbf{r}(t),\dot{\mathbf{r}}(t)\right] &\equiv L\left\{\mathbf{r}(t)+\delta\mathbf{r}(t),\frac{d}{dt}\left[\mathbf{r}(t)+\delta\mathbf{r}(t)\right]\right\} - L\left[\mathbf{r}(t),\frac{d\mathbf{r}(t)}{dt}\right]\\ &= \frac{1}{2}\sum_{i=1}^{n} m_i\frac{d}{dt}\left[\mathbf{r}_i(t)+\delta\mathbf{r}_i(t)\right]\cdot\frac{d}{dt}\left[\mathbf{r}_i(t)+\delta\mathbf{r}_i(t)\right] - V\left[\mathbf{r}(t)+\delta\mathbf{r}(t)\right]\\ &\quad - \left\{\frac{1}{2}\sum_{i=1}^{n} m_i\frac{d\mathbf{r}_i(t)}{dt}\cdot\frac{d\mathbf{r}_i(t)}{dt} - V\left[\mathbf{r}(t)\right]\right\}.\end{aligned} \tag{14.88}$$

From (14.83), $V\left[\mathbf{r}(t)\right] = V\left[\mathbf{r}_1(t),\mathbf{r}_2(t),\ldots,\mathbf{r}_n(t)\right]$. Expanding $V\left[\mathbf{r}(t)+\delta\mathbf{r}(t)\right]$ in a Taylor series to first order then yields

$$\begin{aligned}V\left[\mathbf{r}(t)+\delta\mathbf{r}(t)\right] &= V\left[\mathbf{r}_1(t)+\delta\mathbf{r}_1(t),\mathbf{r}_2(t)+\delta\mathbf{r}_2(t),\ldots,\mathbf{r}_n(t)+\delta\mathbf{r}_n(t)\right]\\ &= V\left[\mathbf{r}_1(t),\mathbf{r}_2(t),\ldots,\mathbf{r}_n(t)\right] + \sum_{i=1}^{n}\frac{\partial V\left[\mathbf{r}(t)\right]}{\partial\mathbf{r}_i}\cdot\delta\mathbf{r}_i(t)\\ &= V\left[\mathbf{r}(t)\right] + \sum_{i=1}^{n}\frac{\partial V\left[\mathbf{r}(t)\right]}{\partial\mathbf{r}_i}\cdot\delta\mathbf{r}_i(t),\end{aligned} \tag{14.89}$$

where

$$\frac{\partial V\left[\mathbf{r}(t)\right]}{\partial\mathbf{r}_i} \equiv \left\{\frac{\partial V\left[\mathbf{r}(t)\right]}{\partial x_i},\frac{\partial V\left[\mathbf{r}(t)\right]}{\partial y_i},\frac{\partial V\left[\mathbf{r}(t)\right]}{\partial z_i}\right\} \tag{14.90}$$

denotes the three partial derivatives of V, with respect to the three Cartesian coordinates $\left(x_i, y_i, z_i\right)$; and

$$\delta\mathbf{r}_i(t) \equiv \left[\delta x_i(t),\delta y_i(t),\delta z_i(t)\right] \tag{14.91}$$

is the variation of $\mathbf{r}_i(t)$, with three independent components, $\delta x_i(t)$, $\delta y_i(t)$, and $\delta z_i(t)$. Substituting (14.89) into (14.88) and ignoring second-order terms in $\delta\mathbf{r}_i(t)$ finally gives

$$\begin{aligned}\delta L\left[\mathbf{r}(t),\dot{\mathbf{r}}(t)\right] &= \frac{1}{2}\sum_{i=1}^{n} m_i\left\{\cancel{\frac{d\mathbf{r}_i(t)}{dt}\cdot\frac{d\mathbf{r}_i(t)}{dt}} + 2\frac{d\mathbf{r}_i(t)}{dt}\cdot\frac{d\left[\delta\mathbf{r}_i(t)\right]}{dt}\right\} \\ &\quad - \frac{1}{2}\sum_{i=1}^{n} m_i\cancel{\frac{d\mathbf{r}_i(t)}{dt}\cdot\frac{d\mathbf{r}_i(t)}{dt}} - \sum_{i=1}^{n}\frac{\partial V\left[\mathbf{r}(t)\right]}{\partial\mathbf{r}_i}\cdot\delta\mathbf{r}_i(t) \\ &= \sum_{i=1}^{n} m_i\frac{d\mathbf{r}_i(t)}{dt}\cdot\frac{d\left[\delta\mathbf{r}_i(t)\right]}{dt} - \sum_{i=1}^{n}\frac{\partial V\left[\mathbf{r}(t)\right]}{\partial\mathbf{r}_i}\cdot\delta\mathbf{r}_i(t),\end{aligned} \tag{14.92}$$

for the n independent, *vector* variations, $\delta\mathbf{r}_i(t)$. (Associated with each vector variation $\delta\mathbf{r}_i(t)$, there are three independent *scalar* variations δx_i, δy_i, and δz_i; see the preceding discussion.)

14.7.6.2 Method 2, via Partial Differentiation

Let

$$\frac{\partial L\left[\mathbf{r}(t),\dot{\mathbf{r}}(t)\right]}{\partial\mathbf{r}_j} \equiv \left\{\frac{\partial L\left[\mathbf{r}(t),\dot{\mathbf{r}}(t)\right]}{\partial x_j}, \frac{\partial L\left[\mathbf{r}(t),\dot{\mathbf{r}}(t)\right]}{\partial y_j}, \frac{\partial L\left[\mathbf{r}(t),\dot{\mathbf{r}}(t)\right]}{\partial z_j}\right\}, \tag{14.93}$$

denote the three partial derivatives of L with respect to the three Cartesian coordinates x_j , y_j, and z_j, associated with particle j. This is simply taking partial derivatives, three at a time, for each particle j, of which there are n such particles. Similarly, let

$$\frac{\partial L\left[\mathbf{r}(t),\dot{\mathbf{r}}(t)\right]}{\partial\dot{\mathbf{r}}_j} \equiv \left\{\frac{\partial L\left[\mathbf{r}(t),\dot{\mathbf{r}}(t)\right]}{\partial\dot{x}_j}, \frac{\partial L\left[\mathbf{r}(t),\dot{\mathbf{r}}(t)\right]}{\partial\dot{y}_j}, \frac{\partial L\left[\mathbf{r}(t),\dot{\mathbf{r}}(t)\right]}{\partial\dot{z}_j}\right\} \tag{14.94}$$

denote the three partial derivatives of L with respect to $\dot{x}_j$, $\dot{y}_j$, and $\dot{z}_j$, associated with particle j.

Inserting Lagrangian (14.85) into the first component of (14.93) gives

$$\frac{\partial L\left[\mathbf{r}(t),\dot{\mathbf{r}}(t)\right]}{\partial x_j} = \frac{\partial}{\partial x_j}\left\{\frac{1}{2}\sum_{i=1}^{n} m_i\dot{\mathbf{r}}_i(t)\cdot\dot{\mathbf{r}}_i(t) - V\left[\mathbf{r}(t)\right]\right\} = -\frac{\partial V\left[\mathbf{r}(t)\right]}{\partial x_j}, \tag{14.95}$$

with similar expressions for the other two components. Thus

$$\frac{\partial L\left[\mathbf{r}(t),\dot{\mathbf{r}}(t)\right]}{\partial\mathbf{r}_j} = -\frac{\partial V\left[\mathbf{r}_1(t),\mathbf{r}_2(t),\ldots,\mathbf{r}_n(t)\right]}{\partial\mathbf{r}_j}, \tag{14.96}$$

where $\partial/\partial\mathbf{r}_j$ has the meaning of (14.93). Similarly inserting Lagrangian (14.85) into the first component of (14.94) gives

$$\begin{aligned}\frac{\partial L\left[\mathbf{r}(t),\dot{\mathbf{r}}(t)\right]}{\partial\dot{x}_j} &= \frac{\partial}{\partial\dot{x}_j}\left\{\frac{1}{2}\sum_{i=1}^{n} m_i\dot{\mathbf{r}}_i(t)\cdot\dot{\mathbf{r}}_i(t) - V\left[\mathbf{r}(t)\right]\right\} = \frac{\partial}{\partial\dot{x}_j}\left\{\frac{1}{2}m_j\dot{\mathbf{r}}_j(t)\cdot\dot{\mathbf{r}}_j(t)\right\} \\ &= \frac{1}{2}m_j\frac{\partial}{\partial\dot{x}_j}\left\{\left[\dot{x}_j(t)\right]^2 + \left[\dot{y}_j(t)\right]^2 + \left[\dot{z}_j(t)\right]^2\right\} = m_j\dot{x}_j,\end{aligned} \tag{14.97}$$

with similar expressions for the other two components. Thus

$$\frac{\partial L\left[\mathbf{r}(t),\dot{\mathbf{r}}(t)\right]}{\partial\dot{\mathbf{r}}_j} = m_j\left[\dot{x}_j(t),\dot{y}_j(t),\dot{z}_j(t)\right] = m_j\dot{\mathbf{r}}_j(t), \tag{14.98}$$

where $\partial/\partial\dot{\mathbf{r}}_j$ has the meaning of (14.94).

Rewriting (14.75) and then inserting (14.96) and (14.98) into the result finally yields

$$
\begin{aligned}
\delta L\left[\mathbf{r}(t), \dot{\mathbf{r}}(t)\right] &= \sum_{i=1}^{n} \frac{\partial L\left[\mathbf{r}(t), \dot{\mathbf{r}}(t)\right]}{\partial \mathbf{r}_i} \cdot \delta \mathbf{r}_i(t) + \sum_{i=1}^{n} \frac{\partial L\left[\mathbf{r}(t), \dot{\mathbf{r}}(t)\right]}{\partial \dot{\mathbf{r}}_i} \cdot \frac{d\left[\delta \mathbf{r}_i(t)\right]}{dt} \\
&= -\sum_{i=1}^{n} \frac{\partial V\left[\mathbf{r}(t)\right]}{\partial \mathbf{r}_i} \cdot \delta \mathbf{r}_i(t) + \sum_{i=1}^{n} m_i \dot{\mathbf{r}}_i(t) \cdot \frac{d\left[\delta \mathbf{r}_i(t)\right]}{dt} \\
&= \sum_{i=1}^{n} m_i \frac{d\mathbf{r}_i(t)}{dt} \cdot \frac{d\left[\delta \mathbf{r}_i(t)\right]}{dt} - \sum_{i=1}^{n} \frac{\partial V\left[\mathbf{r}(t)\right]}{\partial \mathbf{r}_i} \cdot \delta \mathbf{r}_i(t),
\end{aligned} \tag{14.99}
$$

for the n independent, vector variations, $\delta \mathbf{r}_i(t)$.

Equation (14.99), obtained via partial differentiation, agrees with (14.92), obtained via Taylor series. Phew.

14.7.7 Variations of the Action and Governing Equations

Variation $\delta \mathcal{A}\left[\mathbf{q}(t)\right]$ of action $\mathcal{A}\left[\mathbf{q}(t)\right]$ is given by (14.86). Substitution of (14.92) or, equivalently, (14.99), for variation $\delta L\left[\mathbf{q}(t), \dot{\mathbf{q}}(t)\right]$ of the Lagrangian, into (14.86) for variation $\delta \mathcal{A}\left[\mathbf{q}(t)\right]$ of the action, followed by integration by parts and use of endpoint constraints (14.87), then yields

$$
\begin{aligned}
\delta \mathcal{A}\left[\mathbf{r}(t)\right] &= \int_{t_1}^{t_2} \delta L\left[\mathbf{r}(t), \dot{\mathbf{r}}(t)\right] dt = \int_{t_1}^{t_2} \left\{ \sum_{i=1}^{n} m_i \frac{d\mathbf{r}_i(t)}{dt} \cdot \frac{d\left[\delta \mathbf{r}_i(t)\right]}{dt} - \sum_{i=1}^{n} \frac{\partial V\left[\mathbf{r}(t)\right]}{\partial \mathbf{r}_i} \cdot \delta \mathbf{r}_i(t) \right\} dt \\
&= -\int_{t_1}^{t_2} \left\{ \sum_{i=1}^{n} m_i \frac{d^2\mathbf{r}_i(t)}{dt^2} + \sum_{i=1}^{n} \frac{\partial V\left[\mathbf{r}(t)\right]}{\partial \mathbf{r}_i} \right\} \cdot \delta \mathbf{r}_i(t)\, dt + \cancel{\left[\sum_{i=1}^{n} m_i \frac{d\mathbf{r}_i(t)}{dt} \cdot \delta \mathbf{r}_i(t) \right]_{t_1}^{t_2}} \rightarrow 0 \\
&= -\int_{t_1}^{t_2} \sum_{i=1}^{n} \left\{ m_i \frac{d^2\mathbf{r}_i(t)}{dt^2} + \frac{\partial V\left[\mathbf{r}(t)\right]}{\partial \mathbf{r}_i} \right\} \cdot \delta \mathbf{r}_i(t)\, dt.
\end{aligned} \tag{14.100}
$$

Setting to zero variation $\delta \mathcal{A}\left[\mathbf{r}(t)\right]$ in (14.100), resulting from the n *arbitrary* independent vector variations, $\delta \mathbf{r}(t)$, finally gives the n governing equations

$$
m_i \frac{d^2\mathbf{r}_i(t)}{dt^2} + \frac{\partial V\left[\mathbf{r}(t)\right]}{\partial \mathbf{r}_i} = 0. \tag{14.101}
$$

These n vector equations (one per point particle) correspond to Newton's second law of motion for this problem. *This was not imposed.* It is instead a consequence of constructing a Lagrangian and then applying Hamilton's principle of stationary action.

14.8 GOVERNING EQUATIONS FOR GLOBAL FLUIDS: VECTOR FORM

14.8.1 Strategy

The traditional approach to fluid dynamics (as in Chapters 2–4) is to make the *continuum assumption* that a fluid is continuous at all scales, right down to infinitesimal scale.[21]

One can instead consider that a fluid is composed of molecules that mutually interact with one another like elastic billiard balls. This idealisation is fundamental to the development of thermodynamics. It is also conceptually useful for fluid dynamics. A fluid parcel can be considered to be composed of many molecules. Each molecule can be idealised as a point particle, whose evolution is governed by Newton's laws. One can then apply the theory for the mechanics of a system

[21] In reality, quantum mechanical effects become important at small enough scale. For the purposes of atmospheric and oceanic modelling, however, these effects can usually be neglected.

of point particles to obtain equations for the evolution of a parcel composed of many molecules; see Chapter 1 of Goldstein et al. (2001). This provides a means of transiting from a discrete representation to a continuous one; see Chapter 1 of Salmon (1998) and Chapter 13 of Goldstein et al. (2001).

Adopting this approach, we make the transition from a discrete system of n point particles to a continuous system as follows:

1. A suitable Lagrangian is constructed in location space in an inertial frame of reference (in Section 14.8.2).
2. The resulting Lagrangian is then successively transformed from an inertial frame of reference to a rotating one (in Section 14.8.3), and from location space to label space (in Section 14.8.4).
3. Hamilton's principle of stationary action is applied to obtain the governing equations in a rotating frame, first in vector form (in this section) and then in axial-orthogonal-curvilinear coordinates (in Section 14.9).

Herein, our main interest in applying Hamilton's principle of stationary action to a global atmosphere or ocean is to provide a systematic way of obtaining *dynamically consistent*, approximate models; see Chapters 16–18. To do so, it is sufficient to restrict attention to a fluid and its governing equations, *in the absence of any forcings*, and to only pay scant attention to boundary conditions. This avoids the complication of introducing constraints and associated Lagrange multipliers (Salmon, 1998). Thus the unapproximated equation sets of interest here are those given in Section 4.4 (in vector form) and Section 6.5 (in axial-orthogonal-curvilinear coordinates), with all forcings set to zero therein. Specifically:

- $\mathbf{F} = (F^{u_1}, F^{u_2}, F^{u_3}) = 0$ (no friction).
- $F^{\rho} = \dot{S}^i = 0$ (no change in composition).
- $\dot{Q} = 0$ (no heating) and $\dot{Q}_E = 0$ (no energy input).

Such a fluid is often termed a *perfect fluid*.

Applying Hamilton's principle of stationary action does not change the underlying physics in any way, only the way of looking at it. Anything and everything that can be obtained using Hamilton's principle can also be obtained by other, more traditional methods. What Hamilton's principle offers is a valuable tool for approximating equation sets in a systematic manner and *revealing equation sets and conservation properties that might otherwise remain hidden, or less well understood*. Thus it complements other methods rather than replacing them.

14.8.2 The Lagrangian in Location Space in an Inertial Frame

14.8.2.1 The Lagrangian (L) for a Conservative System of Point Particles

From (14.85), the Lagrangian for a conservative system of *n point particles* in location space in an inertial frame of reference is

$$L\left[\mathbf{r}(t), \dot{\mathbf{r}}(t)\right] = K\left[\dot{\mathbf{r}}(t)\right] - V\left[\mathbf{r}(t)\right], \tag{14.102}$$

where

$$K\left[\dot{\mathbf{r}}(t)\right] = \sum_{i=1}^{n} \frac{1}{2} m_i \dot{\mathbf{r}}_i(t) \cdot \dot{\mathbf{r}}_i(t) \equiv \sum_{i=1}^{n} \frac{1}{2} m_i \left(\frac{d\mathbf{r}_i}{dt} \cdot \frac{d\mathbf{r}_i}{dt} \right) \tag{14.103}$$

is kinetic energy, $V = V[\mathbf{r}(t)]$ is potential energy, $\mathbf{r} = \mathbf{r}(t)$ is position (location) vector, and n can be as large as we wish.

14.8.2.2 Transition from Point Particles to Fluid Parcels

The corresponding Lagrangian for a conservative system of *fluid parcels* (i.e. for a continuum) can be obtained from the discrete system of point particles by:

- Replacing the mass, m_i, of a point particle by an infinitesimal fluid parcel having the same mass.
- Distributing fluid parcels continuously throughout the domain using a mass-density distribution, $\rho = \rho\left[\mathbf{r}\left(t\right)\right]$.
- Summing the contributions to the Lagrangian of an infinity ($n \to \infty$) of infinitesimal fluid parcels that span the domain, via integration over the entire spatial domain.

Now m_i in (14.103) is the point mass of particle i at location $\mathbf{r} = \mathbf{r}_i\left(t\right)$. We can replace this point mass by a fluid parcel of the same mass, having infinitesimal volume $d\mathbf{r}_i \equiv \left.\left(dx_1 dx_2 dx_3\right)\right|_{\mathbf{r}=\mathbf{r}_i}$, and mass density $\rho_i \equiv \rho\left[\mathbf{r}_i\left(t\right)\right]$. Since, by definition, mass is equal to mass density multiplied by volume, the mass of the infinitesimal fluid parcel can be written as

$$m_i = \rho_i d\mathbf{r}_i \equiv \rho\left[\mathbf{r}_i\left(t\right)\right] d\mathbf{r}_i. \tag{14.104}$$

14.8.2.3 *Kinetic Energy (K) for Fluid Parcels*

Insertion of (14.104) into (14.103), noting from (14.13) that $\alpha \equiv 1/\rho$ is specific volume, and taking the limit of an infinite number (i.e. $n \to \infty$) of such infinitesimal fluid parcels, distributed continuously throughout the domain of the fluid and spanning it, yields the functional

$$K\left(\dot{\mathbf{r}}, \alpha\right) = \lim_{n\to\infty} \sum_{i=1}^{n} \left\{ \frac{1}{2} \rho\left[\mathbf{r}_i\left(t\right)\right] d\mathbf{r}_i \frac{d\mathbf{r}_i}{dt} \cdot \frac{d\mathbf{r}_i}{dt} \right\} = \iiint \frac{1}{2} \left[\frac{d\mathbf{r}\left(t\right)}{dt} \cdot \frac{d\mathbf{r}\left(t\right)}{dt} \right] \rho\left[\mathbf{r}\left(t\right)\right] d\mathbf{r}. \tag{14.105}$$

In (14.105), $\left(d\mathbf{r}/dt\right)^2/2 \equiv \dot{\mathbf{r}}^2/2$ is *specific kinetic energy* (i.e. kinetic energy per unit mass). The integral is over the entire three-dimensional domain of the global atmosphere or ocean.[22]

14.8.2.4 *Potential (V) for Fluid Parcels*

Transition from point particles to fluid parcels is less straightforward for the potential (V) than it is for the kinetic energy (K), since it brings in additional physics. This needs to be represented, and it needs additional variables to do so.

The potential, V, of a fluid parcel is then the sum of *internal* energy (namely microscopic energy contained *within* the parcel – which is absent from the system of point particles) and *external* potential energy (due to external forces acting *on* the fluid parcel); see Section 4.2. It is given by the functional

$$V\left(\mathbf{r}, \alpha, \eta, S^1, S^2, \ldots\right) = \iiint \left[\mathcal{E}\left(\alpha, \eta, S^1, S^2, \ldots\right) + \mathcal{V}\left(\mathbf{r}\right)\right] \rho\left(\mathbf{r}\right) d\mathbf{r}, \tag{14.106}$$

where:

- $\mathcal{E}\left(\alpha, \eta, S^1, S^2, \ldots\right)$ is *specific internal energy* (per unit mass of fluid, including all of its components), a prescribed function of its arguments that depends upon the properties of the fluid.[23]
- $\alpha \equiv 1/\rho$ is *specific volume*, and η is *specific entropy*.
- S^1, S^2, ... are *mass fractions* of the fluid's components (i.e. substances) of type i (these mass fractions collectively define the fluid's composition – see (4.1)).

[22] As a cross-check on the limiting procedure, instead of starting with the *discrete* $K\left[\dot{\mathbf{r}}\left(t\right)\right]$ of (14.103) and transforming to the *continuous* $K\left[\dot{\mathbf{r}}\left(t\right)\right]$ of (14.105), consider transforming in the opposite direction. To do so, define $\rho\left[\mathbf{r}\left(t\right)\right] = \sum_{i=1}^{n} m_i \delta\left(\mathbf{r}\text{-}\mathbf{r}_i\right)$, where $\delta\left(\mathbf{r}\text{-}\mathbf{r}_i\right)$ is the three-dimensional Dirac delta function, and $\mathbf{r} = \mathbf{r}_i, i = 1, 2, \ldots, n$, is a finite set of points in location space. Substitution of this definition of ρ into (14.105), with use of the sampling property $\iiint F\left(\mathbf{r}\right) \delta\left(\mathbf{r}\text{-}\mathbf{r}_i\right) d\mathbf{r} = F\left(\mathbf{r}_i\right)$, then recovers (14.103).

[23] Internal energy ($\mathcal{E}$) takes into account the work that *microscopic* energy, contained *within* a parcel, can do by expanding a parcel against the pressure imposed by its surrounding environment.

- $\mathcal{V} = \mathcal{V}(\mathbf{r}) = \mathcal{V}[\mathbf{r}(t)]$ is assumed to be a prescribed function of location, $\mathbf{r} = \mathbf{r}(t)$, *only*, with no explicit dependence on time, t. For the global atmosphere or ocean in the present context, $\mathcal{V}[\mathbf{r}(t)]$ is just the potential of Newtonian gravity (per unit mass of fluid).

Thus – see (14.106) – internal energy ($\mathcal{E}$) of a fluid parcel is treated in a similar manner to external potential energy ($\mathcal{V}$).

To ease the notation and to shorten later equations, (14.106) is rewritten as

$$V(\mathbf{r}, \alpha, \eta, S) = \iiint [\mathcal{E}(\alpha, \eta, S) + \mathcal{V}(\mathbf{r})] \rho(\mathbf{r})\, d\mathbf{r}. \tag{14.107}$$

In (14.107), S is to be understood as shorthand for

$$S = \left\{S^1, S^2, \ldots\right\}, \tag{14.108}$$

the set of mass fractions of substances (if present) of type i, however many there may be. Typically there are only two; dry air and water substance for the atmosphere, and water substance and salt for the oceans.

14.8.2.5 Lagrangian (L) for Fluid Parcels in Location Space in an Inertial Frame

Using (14.102), (14.105), and (14.107) for L, K, and V, respectively, the Lagrangian for the global atmosphere or ocean can now be written as:

The Lagrangian Functional in Location Space in an Inertial Frame

$$\begin{aligned} L(\mathbf{r}, \dot{\mathbf{r}}, \alpha, \eta, S) &= K(\dot{\mathbf{r}}, \alpha) - V(\mathbf{r}, \alpha, \eta, S) \\ &= \iiint \left[\frac{1}{2}\left(\frac{d\mathbf{r}}{dt} \cdot \frac{d\mathbf{r}}{dt}\right) - \mathcal{E}(\alpha, \eta, S) - \mathcal{V}(\mathbf{r})\right] \rho(\mathbf{r})\, d\mathbf{r}, \end{aligned} \tag{14.109}$$

where $\mathbf{r} = \mathbf{r}(t)$ and $\alpha(\mathbf{r}) \equiv 1/\rho(\mathbf{r})$.

14.8.3 The Lagrangian in Location Space in a Rotating Frame

In principle (and, with sufficient effort, also in practice), one can work entirely in an inertial frame of reference. This has the virtue of conceptual simplicity. However, Earth rotates in an inertial frame of reference. This has a profound impact on its shape and on the behaviour of its atmosphere and oceans.

As discussed in Chapters 7 and 8, Earth's shape (i.e. its 'Figure') is largely a consequence of its rotation rate. If Earth rotated faster, then it would be more oblate and, conversely, if it rotated slower, then it would be less oblate. Similarly, its geopotential surfaces (the geopotentials) would be more oblate or less oblate. The natural frame of reference for examining Earth's geopotential surfaces and the horizontal and vertical dynamical balances that hold on them and on normals to them, respectively – see Section 7.3 – are most evident, and simply expressed, in a rotating frame of reference. It is also much easier, for theoretical purposes, to develop simplified, approximate equation sets that well capture these important dynamical balances in a rotating frame of reference; see Chapters 16–18. These considerations motivate transformation of Lagrangian (14.109) to a rotating frame.

In (14.109), the Lagrangian is expressed in location space in an *inertial* frame of reference, where $\mathbf{r} = \mathbf{r}(t)$ denotes the location vector in this frame.[24] However, for the reasons just given, we

[24] For notational brevity, and to maintain the link with classical mechanics, $\mathbf{r}$ in the preceding analysis has been used to denote the location vector in an *inertial* frame of reference. However, in Chapter 2, $(\mathbf{r})_I$ denotes the location vector in an

wish to transform this Lagrangian into an equivalent one expressed in a frame rotating coaxially with Earth, with uniform angular velocity, Ω, relative to an inertial frame. To this end, we first rewrite (14.109) as

$$L\left(\widehat{\mathbf{r}}, \dot{\widehat{\mathbf{r}}}, \alpha, \eta, S\right) = \iiint \left[\frac{1}{2}\left(\frac{d\widehat{\mathbf{r}}}{dt}\right) \cdot \left(\frac{d\widehat{\mathbf{r}}}{dt}\right) - \mathcal{E}(\alpha, \eta, S) - \mathcal{V}\left(\widehat{\mathbf{r}}\right)\right] \rho\left(\widehat{\mathbf{r}}\right) d\widehat{\mathbf{r}}, \tag{14.110}$$

where $\widehat{\mathbf{r}} \equiv (\mathbf{r})_I$ denotes the location vector observed in an *inertial* frame of reference. This is simply a temporary change of notation ($\mathbf{r} \to \widehat{\mathbf{r}}$) to prepare the way for $\mathbf{r}$ to be used henceforth to instead denote the location vector $\mathbf{r} \equiv (\mathbf{r})_R$ in the *rotating* frame.

Equations (2.20) and (2.23) of Chapter 2 relate the velocity in an inertial frame of reference, $\widehat{\mathbf{u}} \equiv (\mathbf{u})_I$, to the velocity in a rotating frame, $\mathbf{u} \equiv (\mathbf{u})_R$, and to the rotation vector, $\mathbf{\Omega}$. Thus

$$(\mathbf{u})_I \equiv \widehat{\mathbf{u}} \equiv \frac{d\widehat{\mathbf{r}}}{dt} = \frac{d\mathbf{r}}{dt} + \mathbf{\Omega}\times\mathbf{r} = \mathbf{u} + \mathbf{\Omega}\times\mathbf{r}, \tag{14.111}$$

where, to be clear, $\mathbf{r}$ is now the location vector in the *rotating* frame, and $\widehat{\mathbf{r}}$ in the *inertial* frame.

Using (14.111) in (14.110), the Lagrangian in a *rotating* frame of reference, in *location* space, is given by

$$\begin{aligned} L(\mathbf{r}, \dot{\mathbf{r}}, \alpha, \eta, S) &= \iiint \left[\frac{1}{2}\left(\frac{d\mathbf{r}}{dt} + \mathbf{\Omega}\times\mathbf{r}\right) \cdot \left(\frac{d\mathbf{r}}{d\tau} + \mathbf{\Omega}\times\mathbf{r}\right) - \mathcal{E}(\alpha, \eta, S) - \mathcal{V}(\mathbf{r})\right] \rho(\mathbf{r})\, d\mathbf{r} \\ &= \iiint \left[\frac{1}{2}\frac{d\mathbf{r}}{dt} \cdot \frac{d\mathbf{r}}{dt} + (\mathbf{\Omega}\times\mathbf{r}) \cdot \frac{d\mathbf{r}}{dt} + \frac{|\mathbf{\Omega}\times\mathbf{r}|^2}{2} - \mathcal{E}(\alpha, \eta, S) - \mathcal{V}(\mathbf{r})\right] \rho(\mathbf{r})\, d\mathbf{r}. \end{aligned} \tag{14.112}$$

This may be rewritten as:

The Lagrangian Functional in Location Space in a Rotating Frame

$$L(\mathbf{r}, \dot{\mathbf{r}}, \alpha, \eta, S) = \iiint \left[\frac{1}{2}\left(\frac{d\mathbf{r}}{dt} \cdot \frac{d\mathbf{r}}{dt}\right) + (\mathbf{\Omega}\times\mathbf{r}) \cdot \frac{d\mathbf{r}}{dt} - \mathcal{E}(\alpha, \eta, S) - \Phi(\mathbf{r})\right] \rho(\mathbf{r})\, d\mathbf{r}, \tag{14.113}$$

where $\mathbf{r} = \mathbf{r}(t)$, and

$$\Phi(\mathbf{r}) \equiv \mathcal{V}(\mathbf{r}) - \frac{|\mathbf{\Omega}\times\mathbf{r}|^2}{2} \tag{14.114}$$

is the *potential of apparent gravity* (or *geopotential*, for short); see (2.29). As discussed in detail in Chapters 7 and 8, surfaces of constant Φ are of great importance in meteorology and oceanography; these are termed *geopotential surfaces*.

Comparing (14.113) with (14.109), it is seen that transformation to a rotating frame has resulted in two changes:

1. A new (rotational) term, $(\mathbf{\Omega}\times\mathbf{r}) \cdot \left(d\mathbf{r}/dt\right)$, now appears.
2. Newtonian potential of gravity, $\mathcal{V}(\mathbf{r})$, is replaced by potential of apparent gravity, $\Phi(\mathbf{r})$.

Note that (14.109) (for an inertial frame) can be recovered from (14.113) (for a rotating frame) by simply setting rotation rate $\mathbf{\Omega}$ identically to zero in (14.113) and (14.114).

inertial frame, whereas $\mathbf{r} \equiv (\mathbf{r})_R$ denotes the location vector in a *rotating* frame. To be notationally consistent with Chapter 2, $\mathbf{r}$ in the preceding analysis is to be interpreted as $\widehat{\mathbf{r}} \equiv (\mathbf{r})_I$.

14.8.4 The Lagrangian in Label Space in a Rotating Frame

14.8.4.1 *The Lagrangian (L)*

Because of the appearance of $\rho\,[\mathbf{r}\,(t)]$ in (14.113) – adjacent to $d\mathbf{r}$ – it often turns out to be easier to take variations if (14.112) is first transformed from location space to label space.[25]

Recall – from Section 14.2.3 – that *location* coordinates $\mathbf{r} = (x_1, x_2, x_3)$, plus time t, are related to *label* coordinates $\mathbf{a} = (a_1, a_2, a_3)$, plus time τ, via relations of the form[26]

$$\mathbf{r} = \mathbf{r}\,(\mathbf{a}, \tau)\,, \quad t = \tau, \tag{14.115}$$

$$\mathbf{a} = \mathbf{a}\,(\mathbf{r}, t)\,, \quad \tau = t. \tag{14.116}$$

Furthermore, labels $\mathbf{a} \equiv (a_1, a_2, a_3)$ are judiciously assigned – see (14.16) – so that

$$\rho d\mathbf{r} = \rho \frac{\partial\,(\mathbf{r})}{\partial\,(\mathbf{a})} d\mathbf{a} = d\mathbf{a}, \quad \alpha \equiv \frac{1}{\rho} = \frac{\partial\,(\mathbf{r})}{\partial\,(\mathbf{a})}, \tag{14.117}$$

where $d\mathbf{a} \equiv da_1 da_2 da_3$ is the infinitesimal volume element in *label* space. As shown in Section 14.3, mass conservation is implicitly embodied within (14.117) and leads to (14.22) and (14.23), that is, to

$$\frac{\partial \alpha}{\partial \tau} = \frac{D\alpha}{Dt} = \alpha \nabla \cdot \mathbf{u} \quad \Rightarrow \quad \frac{D\rho}{Dt} + \rho \nabla \cdot \mathbf{u} = 0. \tag{14.118}$$

Substituting (14.117) into (14.113) yields:

The Lagrangian Functional in Label Space in a Rotating Frame

$$L\,(\mathbf{r}, \dot{\mathbf{r}}, \alpha, \eta, S) = \iiint \left[\frac{1}{2} \left(\frac{\partial \mathbf{r}}{\partial \tau} \cdot \frac{\partial \mathbf{r}}{\partial \tau} \right) + (\boldsymbol{\Omega} \times \mathbf{r}) \cdot \frac{\partial \mathbf{r}}{\partial \tau} - \mathcal{E}\,(\alpha, \eta, S) - \Phi\,(\mathbf{r}) \right] d\mathbf{a}, \tag{14.119}$$

where, in the present context of label space, $d/dt = D/Dt \to \partial/\partial\tau$; see (14.9).

Equation (14.119) gives the Lagrangian for a (unforced) global atmosphere or ocean, expressed in vector form, in *label* space and in a *rotating frame of reference.*

Comparing (14.119) with (14.113), it is seen that transforming to label space results in a simpler form by absorbing density (ρ) into the volume element ($d\mathbf{a}$) for label coordinates. *This simplifies the taking of variations.*

14.8.4.2 *The Action ($\mathcal{A}$)*

Inserting Lagrangian (14.119) into (14.57) gives:

The Action Functional in Label Space in a Rotating Frame

$$\begin{aligned} \mathcal{A}\,(\mathbf{q}) = \mathcal{A}\,(\mathbf{r}, \alpha, \eta, S) &= \int_{\tau_1}^{\tau_2} L\,(\mathbf{r}, \dot{\mathbf{r}}, \alpha, \eta, S)\, d\tau \\ &= \int_{\tau_1}^{\tau_2} \left\{ \iiint \left[\frac{1}{2} \left(\frac{\partial \mathbf{r}}{\partial \tau} \cdot \frac{\partial \mathbf{r}}{\partial \tau} \right) + (\boldsymbol{\Omega} \times \mathbf{r}) \cdot \frac{\partial \mathbf{r}}{\partial \tau} - \mathcal{E}\,(\alpha, \eta, S) - \Phi\,(\mathbf{r}) \right] d\mathbf{a} \right\} d\tau. \end{aligned} \tag{14.120}$$

[25] Variations can nevertheless be taken in location space. One ends up with exactly the same results, as one must, but the intermediate calculations are often a lot messier!

[26] These are the abbreviated forms (14.3) and (14.4). The full ones are given by (14.1) and (14.2).

In (14.120), τ now appears instead of t, since the Lagrangian – $L(\mathbf{r}, \dot{\mathbf{r}}, \alpha, \eta, S)$, where $\mathbf{r} = \mathbf{r}(\mathbf{a}, \tau)$ – is now defined *in label space* rather than – see (14.57) – in location space.

14.8.4.3 *Configuration Space*

The introduction of additional physics – in the form of internal energy $\mathcal{E} = \mathcal{E}(\alpha, \eta, S)$ – in transitioning from a system of point particles to a fluid parcel enriches configuration space by introducing new variables. Whereas configuration space for a system of point particles is location space (with $\mathbf{q} = \mathbf{r}$), internal energy introduces the additional variables α, η, and S; see (14.120). The generalised coordinates are now the arguments of action $\mathcal{A}(\mathbf{r}, \alpha, \eta, S)$, as in (14.120); that is, $\mathbf{q} = (\mathbf{r}, \alpha, \eta, S)$. Note that $\dot{\mathbf{r}}$ is not a generalised coordinate since it is not independent of $\mathbf{r}$, and generalised coordinates are necessarily (by definition) mutually independent.

14.8.4.4 *Hamilton's Principle in Label Space*

From Section 14.5.3 – equations (14.58) and (14.59) –

Hamilton's Principle of Stationary Action in Label Space

Hamilton's principle of stationary action, but now expressed *in label space*, is to apply

$$\delta \mathcal{A}(\mathbf{r}, \alpha, \eta, S) \equiv \delta \int_{\tau_1}^{\tau_2} L(\mathbf{r}, \dot{\mathbf{r}}, \alpha, \eta, S)\, d\tau = \int_{\tau_1}^{\tau_2} \delta L(\mathbf{r}, \dot{\mathbf{r}}, \alpha, \eta, S)\, d\tau = 0, \tag{14.121}$$

for independent variations $\delta\mathbf{q}(\mathbf{a}, \tau)$ of generalised coordinates $\mathbf{q} = (\mathbf{r}, \alpha, \eta, S)$, such that

$$\delta\mathbf{q}(\mathbf{a}, \tau_1) = \delta\mathbf{q}(\mathbf{a}, \tau_2) = 0, \tag{14.122}$$

where Lagrangian $L(\mathbf{r}, \dot{\mathbf{r}}, \alpha, \eta, S)$ is given by (14.119), and action $\mathcal{A}(\mathbf{r}, \alpha, \eta, S)$, by (14.120).

14.8.4.5 *Variations of the Lagrangian in Label Space*

Taking variations of the Lagrangian, $L(\mathbf{r}, \dot{\mathbf{r}}, \alpha, \eta, S)$ – given by (14.119) – and using vector identities (A.3) and (A.8) of the Appendix to this book yields

$$\begin{aligned}
\delta L(\mathbf{r}, \dot{\mathbf{r}}, \alpha, \eta, S) &= \iiint \delta \left[\frac{1}{2}\left(\frac{\partial \mathbf{r}}{\partial \tau} \cdot \frac{\partial \mathbf{r}}{\partial \tau} \right) + (\boldsymbol{\Omega}\times\mathbf{r}) \cdot \frac{\partial \mathbf{r}}{\partial \tau} - \mathcal{E}(\alpha, \eta, S) - \Phi(\mathbf{r}) \right] d\mathbf{a} \\
&= \iiint \left[\frac{\partial \mathbf{r}}{\partial \tau} \cdot \frac{\partial (\delta\mathbf{r})}{\partial \tau} + (\boldsymbol{\Omega}\times\mathbf{r}) \cdot \frac{\partial (\delta\mathbf{r})}{\partial \tau} + (\Omega\times\delta\mathbf{r}) \cdot \frac{\partial \mathbf{r}}{\partial \tau} - \delta\Phi(\mathbf{r}) - \delta\mathcal{E}(\alpha, \eta, S) \right] d\mathbf{a} \\
&= \iiint \left[\left(\frac{\partial \mathbf{r}}{\partial \tau} + \boldsymbol{\Omega}\times\mathbf{r} \right) \cdot \frac{\partial (\delta\mathbf{r})}{\partial \tau} - \left(\boldsymbol{\Omega}\times\frac{\partial \mathbf{r}}{\partial \tau} \right) \cdot \delta\mathbf{r} - \frac{\partial \Phi}{\partial \mathbf{r}} \cdot \delta\mathbf{r} \right] d\mathbf{a} \\
&\quad - \iiint \left[\frac{\partial \mathcal{E}(\alpha, \eta, S)}{\partial \alpha}\delta\alpha + \frac{\partial \mathcal{E}(\alpha, \eta, S)}{\partial \eta}\delta\eta + \frac{\partial \mathcal{E}(\alpha, \eta, S)}{\partial S}\delta S \right] d\mathbf{a}.
\end{aligned} \tag{14.123}$$

In (14.123),

$$\frac{\partial \Phi}{\partial \mathbf{r}} \equiv \left(\frac{\partial \Phi}{\partial x_1}, \frac{\partial \Phi}{\partial x_2}, \frac{\partial \Phi}{\partial x_3} \right) = \nabla\Phi \tag{14.124}$$

results from taking variation $\delta\Phi(\mathbf{r})$ as in Section 14.7.6. Furthermore, the chain rule has been used to express variation $\delta\mathcal{E}(\alpha, \eta, S)$ in terms of variations $\delta\alpha$, $\delta\eta$, and δS^i, where (in shorthand)

$$\frac{\partial \mathcal{E}(\alpha, \eta, S)}{\partial S}\delta S \equiv \sum_i \frac{\partial \mathcal{E}(\alpha, \eta, S)}{\partial S^i}\delta S^i \equiv \sum_i \frac{\partial \mathcal{E}\left(\alpha, \eta, S^1, S^2, \ldots\right)}{\partial S^i}\delta S^i. \tag{14.125}$$

Now, from (4.17)–(4.19), the equations of state for $\mathcal{E}\left(\alpha, \eta, S^1, S^2, \ldots\right)$ are

$$p = -\frac{\partial \mathcal{E}(\alpha, \eta, S)}{\partial \alpha}, \quad T = \frac{\partial \mathcal{E}(\alpha, \eta, S)}{\partial \eta}, \quad \mu^i = \frac{\partial \mathcal{E}(\alpha, \eta, S)}{\partial S^i}. \tag{14.126}$$

Inserting (14.124) and (14.126) into (14.123) then yields

$$\begin{aligned}\delta L(\mathbf{r}, \dot{\mathbf{r}}, \alpha, \eta, S) = &\iiint \left[\left(\frac{\partial \mathbf{r}}{\partial \tau} + \mathbf{\Omega}\times\mathbf{r}\right) \cdot \frac{\partial (\delta\mathbf{r})}{\partial \tau} - \left(\mathbf{\Omega}\times\frac{\partial \mathbf{r}}{\partial \tau} + \nabla\Phi\right) \cdot \delta\mathbf{r}\right] d\mathbf{a} \\ &+ \iiint \left(p\delta\alpha - T\delta\eta - \mu\delta S\right) d\mathbf{a},\end{aligned} \tag{14.127}$$

where μdS is shorthand for

$$\mu\delta S \equiv \sum_i \mu^i \delta S^i \equiv \sum_i \frac{\partial \mathcal{E}\left(\alpha, \eta, S^1, S^2, \ldots\right)}{\partial S^i} \delta S^i. \tag{14.128}$$

$\delta\alpha$ Variation

For the first term in the last line of (14.127), using (14.117) and the chain rule (14.18) for Jacobians gives

$$\begin{aligned}\iiint (p\delta\alpha)\, d\mathbf{a} &= \iiint \left\{p\delta\left[\frac{\partial(\mathbf{r})}{\partial(\mathbf{a})}\right]\right\} d\mathbf{a} = \iiint \left\{p\delta\left[\frac{\partial(x_1, x_2, x_3)}{\partial(a_1, a_2, a_3)}\right]\right\} d\mathbf{a} \\ &= \iiint \left\{p\left[\frac{\partial(\delta x_1, x_2, x_3)}{\partial(a_1, a_2, a_3)} + \frac{\partial(x_1, \delta x_2, x_3)}{\partial(a_1, a_2, a_3)} + \frac{\partial(x_1, x_2, \delta x_3)}{\partial(a_1, a_2, a_3)}\right]\right\} d\mathbf{a} \\ &= \iiint \left\{p\left[\frac{\partial(\delta x_1, x_2, x_3)}{\partial(x_1, x_2, x_3)} + \frac{\partial(x_1, \delta x_2, x_3)}{\partial(x_1, x_2, x_3)} + \frac{\partial(x_1, x_2, \delta x_3)}{\partial(x_1, x_2, x_3)}\right]\right\} d\mathbf{r}.\end{aligned} \tag{14.129}$$

Now from (14.15),

$$\begin{aligned}\frac{\partial(\delta x_1, x_2, x_3)}{\partial(x_1, x_2, x_3)} &\equiv \det\begin{bmatrix} \frac{\partial(\delta x_1)}{\partial x_1} & \frac{\partial(\delta x_1)}{\partial x_2} & \frac{\partial(\delta x_1)}{\partial x_3} \\ \frac{\partial x_2}{\partial x_1} & \frac{\partial x_2}{\partial x_2} & \frac{\partial x_2}{\partial x_3} \\ \frac{\partial x_3}{\partial x_1} & \frac{\partial x_3}{\partial x_2} & \frac{\partial x_3}{\partial x_3} \end{bmatrix} \equiv \begin{vmatrix} \frac{\partial(\delta x_1)}{\partial x_1} & \frac{\partial(\delta x_1)}{\partial x_2} & \frac{\partial(\delta x_1)}{\partial x_3} \\ 0 & 1 & 0 \\ 0 & 0 & 1 \end{vmatrix} \\ &= \frac{\partial(\delta x_1)}{\partial x_1},\end{aligned} \tag{14.130}$$

and, similarly,

$$\frac{\partial(x_1, \delta x_2, x_3)}{\partial(x_1, x_2, x_3)} = \frac{\partial(\delta x_2)}{\partial x_2}, \quad \frac{\partial(x_1, x_2, \delta x_3)}{\partial(x_1, x_2, x_3)} = \frac{\partial(\delta x_3)}{\partial x_3}. \tag{14.131}$$

Substituting (14.130) and (14.131) into (14.129) and using divergence theorem (A.33) of the Appendix to this book and (14.117) yields

$$\begin{aligned}\iiint (p\delta\alpha)\, d\mathbf{a} &= \iiint \left\{p\left[\frac{\partial(\delta x_1)}{\partial x_1} + \frac{\partial(\delta x_2)}{\partial x_2} + \frac{\partial(\delta x_3)}{\partial x_3}\right]\right\} d\mathbf{r} = \iiint \left[p\nabla\cdot(\delta\mathbf{r})\right] d\mathbf{r} \\ &= -\iiint (\nabla p \cdot \delta\mathbf{r})\, d\mathbf{r} + \iiint \left[\nabla\cdot(p\delta\mathbf{r})\right] d\mathbf{r} = -\iiint (\nabla p\cdot\delta\mathbf{r})\, d\mathbf{r} \\ &\quad + \oiint_{\mathscr{S}} (p\delta\mathbf{r}\cdot d\mathbf{n})\, d\mathscr{S}\end{aligned}$$

$$= -\iiint \left(\frac{1}{\rho}\nabla p \cdot \delta\mathbf{r}\right) d\mathbf{a} + \oint\!\!\oint_{\mathscr{S}} \left(p\delta\mathbf{r}\cdot\mathbf{n}\right) d\mathscr{S}, \tag{14.132}$$

where $\mathbf{n}$ is the outward-pointing normal to the boundary $\mathcal{S}$ that encloses the domain.

$\delta\eta$ and δS^i Variations

In this section (Section 14.8), we have assumed that there are no forcings whatsoever. In particular, we assume that $\dot{Q} \equiv$ (0, i.e. fluid parcels are not heated nor cooled), and that $\dot{S}^i \equiv 0$ (i.e. there is no change in the composition of the substances within a fluid parcel or across its boundary); see Section 14.8.1. From (4.3), this means that, for a fluid parcel, $d\eta \equiv 0$ over any time period $d\tau$, and so – cf. (4.15) –

$$\frac{\partial\eta}{\partial\tau} = 0. \tag{14.133}$$

This implies that $\eta = \eta(\mathbf{a})$, with no dependence on τ, and that entropy, η, is materially conserved by a fluid parcel. It does not then matter which path is taken in location space since labels, $\mathbf{a}$, remain unchanged along any and all possible paths; see (14.5). Thus

$$\delta\eta = 0. \tag{14.134}$$

Similarly – see (4.5) – $S^i = S^i(\mathbf{a})$ is materially conserved by a fluid parcel in the absence of any change in composition, and – see (4.9) –

$$\frac{\partial S^i}{\partial\tau} = 0, \tag{14.135}$$

$$\delta S^i = 0 \quad \Rightarrow \quad \delta S = 0. \tag{14.136}$$

Thus, from (14.134) and (14.136), the second and third terms in the last line of (14.127) are zero.

Gathering these developments together, insertion of (14.132), (14.134), and (14.136) into (14.127) finally yields the variation of the Lagrangian, namely

$$\begin{aligned}\delta L(\mathbf{r},\dot{\mathbf{r}},\alpha,\eta,S) = &\iiint\left[\left(\frac{\partial\mathbf{r}}{\partial\tau} + \boldsymbol{\Omega}\times\mathbf{r}\right)\cdot\frac{\partial(\delta\mathbf{r})}{\partial\tau} - \left(\boldsymbol{\Omega}\times\frac{\partial\mathbf{r}}{\partial\tau} + \frac{1}{\rho}\nabla p + \nabla\Phi\right)\cdot\delta\mathbf{r}\right] d\mathbf{a}\\ &+ \oint\!\!\oint_{\mathcal{S}}\left(p\delta\mathbf{r}\cdot\mathbf{n}\right) d\mathcal{S}.\end{aligned} \tag{14.137}$$

14.8.4.6 *Variations of the Action and Governing Equations*

Applying Hamilton's principle of stationary action (14.121) and using (14.137) for the variations $\delta L(\mathbf{r},\dot{\mathbf{r}},\alpha,\eta,S)$ of the Lagrangian leads to

$$\begin{aligned}0 = \delta\mathcal{A}(\mathbf{r},\alpha,\eta,S) &= \int_{\tau_1}^{\tau_2}\delta L(\mathbf{r},\dot{\mathbf{r}},\alpha,\eta,S)\, d\tau\\ &= \int_{\tau_1}^{\tau_2}\left\{\iiint\left[\left(\frac{\partial\mathbf{r}}{\partial\tau} + \boldsymbol{\Omega}\times\mathbf{r}\right)\cdot\frac{\partial(\delta\mathbf{r})}{\partial\tau}\right] d\mathbf{a}\right\} d\tau\\ &- \int_{\tau_1}^{\tau_2}\left\{\iiint\left[\left(\boldsymbol{\Omega}\times\frac{\partial\mathbf{r}}{\partial\tau} + \frac{1}{\rho}\nabla p + \nabla\Phi\right)\cdot\delta\mathbf{r}\right] d\mathbf{a} - \oint\!\!\oint_{\mathscr{S}}\left(p\delta\mathbf{r}\cdot\mathbf{n}\right) d\mathscr{S}\right\} d\tau.\end{aligned} \tag{14.138}$$

Integrating the $\partial(\delta\mathbf{r})/\partial\tau$ terms in (14.138) by parts, applying constraints (14.122) on variations $\delta\mathbf{r}$, and noting that rotation vector Ω is constant then yields

$$0 = \delta\mathcal{A}(\mathbf{r},\alpha,\eta,S) = \left\{\iiint\left[\left(\frac{\partial\mathbf{r}}{\partial\tau} + \mathbf{\Omega}\times\mathbf{r}\right)\cdot\delta\mathbf{r}(\mathbf{a},\tau)\right]d\mathbf{a}\right\}\Bigg|_{\tau=\tau_1}^{\tau=\tau_2} \to 0$$
$$+\int_{\tau_1}^{\tau_2}\left\{\iiint\left[-\left(\frac{\partial^2\mathbf{r}}{\partial\tau^2} + \Omega\times\frac{\partial\mathbf{r}}{\partial\tau}\right)\cdot\delta\mathbf{r} - \left(\mathbf{\Omega}\times\frac{\partial\mathbf{r}}{\partial\tau} + \frac{1}{\rho}\nabla p + \nabla\Phi\right)\cdot\delta\mathbf{r}\right]d\mathbf{a}\right\}d\tau$$
$$+\int_{\tau_1}^{\tau_2}\left[\oiint_S\left(p\delta\mathbf{r}\cdot\mathbf{n}\right)dS\right]d\tau. \tag{14.139}$$

Rearranging this equation, we finally obtain

$$\int_{\tau_1}^{\tau_2}\left\{\iiint\left[\left(\frac{\partial^2\mathbf{r}}{\partial\tau^2} + 2\mathbf{\Omega}\times\frac{\partial\mathbf{r}}{\partial\tau} + \frac{1}{\rho}\nabla p + \nabla\Phi\right)\cdot\delta\mathbf{r}\right]d\mathbf{a}\right\}d\tau + \int_{\tau_1}^{\tau_2}\left[\oiint_S\left(p\delta\mathbf{r}\cdot\mathbf{n}\right)dS\right]d\tau = 0. \tag{14.140}$$

The last term of (14.140) depends upon conditions at the encompassing boundary of the domain. We briefly consider three typical kinds of boundary conditions:

1. *Rigid boundary*
   The boundary is rigid with outward-pointing normal, $\mathbf{n}$.
   Because it is rigid, $\delta\mathbf{r}\cdot\mathbf{n} = 0$ there, and so $\oiint\left(p\delta\mathbf{r}\cdot\mathbf{n}\right)dS = 0$ in (14.140).
   Also, because it is rigid, $\mathbf{u}\cdot\mathbf{n} \equiv (\partial\mathbf{r}/\partial t)\cdot\mathbf{n} = 0$ there.
   A rigid boundary is appropriate over land for the atmosphere, above bathymetry for an ocean, and at the lateral (horizontal) boundaries of a confined ocean.[27]
2. *Bi-periodic in the horizontal*
   Because the boundary is periodic in both horizontal directions, there are no lateral (horizontal) contributions from $\oiint\left(p\delta\mathbf{r}\cdot\mathbf{n}\right)dS = \iiint\left[\nabla\cdot\left(p\delta\mathbf{r}\right)\right]d\mathbf{r}$ in (14.140), only contributions from top and/or bottom boundaries.
   A bi-periodic lateral boundary is appropriate for a global atmosphere (e.g. for Earth) or for a global ocean (e.g. for some celestial bodies, such as Jupiter's moon, Europa, with a hypothesised global ocean, capped by thick, solid ice).
3. *Free surface*
   Because the boundary is a free surface, $\mathbf{r} = \mathbf{r}(t)$ and, in general, $\oiint\left(p\delta\mathbf{r}\cdot\mathbf{n}\right)dS \neq 0$ in (14.140), and must be determined by other means. (We examine such a situation, in a simple context, in Section 13.6 and Chapter 18.)
   A free surface is appropriate for the lower surface of a global atmosphere over an ocean, and for the upper surface of an ocean (although this surface can be considered to be approximately rigid under certain circumstances, e.g. if capped by solid ice).

From the preceding cursory discussion, we conclude that $\int_{\tau_1}^{\tau_2}\left[\oiint_S\left(p\delta\mathbf{r}\cdot\mathbf{n}\right)dS\right]d\tau$ in (14.140) is typically zero, except at a free surface.

Because the variation of $\delta\mathbf{r}$ in (14.140) is arbitrary, and because (14.140) holds for all possible variations of $\delta\mathbf{r}$,

$$\frac{\partial^2\mathbf{r}}{\partial\tau^2} + 2\mathbf{\Omega}\times\frac{\partial\mathbf{r}}{\partial\tau} + \frac{1}{\rho}\nabla p + \nabla\Phi = 0, \tag{14.141}$$

[27] Strictly speaking, this is not quite true. At small enough scale, oceanic tides become important, leading to semi-diurnal wetting and drying of land over beaches, or to changes in water level over coastal marshes and swamps. A lateral wall boundary condition is then no longer valid, and alternative boundary conditions have to be formulated. This complicates matters. It generally leads to the introduction of constraints and to the use of Lagrange multipliers to enforce them.

within the domain, and $p = 0$ at a boundary that is neither rigid nor periodic.[28]

Using $\mathbf{u} \equiv \partial \mathbf{r}/\partial \tau \equiv D\mathbf{r}/Dt$ – see (14.6)–(14.10) – (14.141) can be rewritten more conventionally as

$$\frac{D\mathbf{u}}{Dt} + 2\boldsymbol{\Omega}\times\mathbf{u} + \frac{1}{\rho}\nabla p + \nabla\Phi = 0. \tag{14.142}$$

Furthermore – from (14.118) –

$$\frac{D\rho}{Dt} + \rho\nabla\cdot\mathbf{u} = 0, \tag{14.143}$$

and (14.133) and (14.135) can be rewritten more conventionally as

$$\frac{D\eta}{Dt} = 0, \tag{14.144}$$

$$\frac{DS^i}{Dt} = 0. \tag{14.145}$$

It is seen that (14.142)–(14.145) recover (4.95)–(4.97), together with the second equation of (4.98), when all forcings are set to zero in these latter equations.

14.9 GOVERNING EQUATIONS FOR GLOBAL FLUIDS: CURVILINEAR FORM

One can make significant progress using vectors to both develop the governing equations for atmospheric and oceanic modelling (see Chapters 2–4), and to examine some of their fundamental properties (e.g. conservation of various physical quantities; see the first few sections of Chapter 15). Coordinate systems, however, are needed for both practical applications and theoretical investigations. In particular, axial-orthogonal-curvilinear coordinates facilitate the construction of dynamically consistent equation subsets of the full equations; see Chapters 16–19. This motivates the variational rederivation in this section of the governing equations given in Chapter 6 in axial-orthogonal-curvilinear coordinates. The rederivation is based on that described in Staniforth (2014a) for a global atmosphere. With a suitable prescription of specific internal energy, Staniforth (2014a)'s derivation also follows through for oceans.

14.9.1 The Lagrangian in Label Space in a Rotating Frame

The Lagrangian in label space, in a rotating frame of reference, and in vector form is given by (14.119), that is, by

$$L(\mathbf{r}, \dot{\mathbf{r}}, \alpha, \eta, S) = \iiint \left\{ \frac{1}{2}\left(\frac{\partial \mathbf{r}}{\partial \tau}\cdot\frac{\partial \mathbf{r}}{\partial \tau}\right) + (\boldsymbol{\Omega}\times\mathbf{r})\cdot\frac{\partial \mathbf{r}}{\partial \tau} - \mathcal{E}(\alpha, \eta, S) - \Phi(\mathbf{r}) \right\} d\mathbf{a}. \tag{14.146}$$

Using (5.5) of Chapter 5, together with (14.7), $\partial \mathbf{r}/\partial \tau$ in axial-orthogonal-curvilinear coordinates may be written as

$$\frac{\partial \mathbf{r}}{\partial \tau} \equiv \mathbf{u} \equiv u_1\mathbf{e}_1 + u_2\mathbf{e}_2 + u_3\mathbf{e}_3 \equiv h_1\frac{\partial \xi_1}{\partial \tau}\mathbf{e}_1 + h_2\frac{\partial \xi_2}{\partial \tau}\mathbf{e}_2 + h_3\frac{\partial \xi_3}{\partial \tau}\mathbf{e}_3, \tag{14.147}$$

where (by definition of an axial-orthogonal-curvilinear coordinate system – see (6.13))

$$h_i = h_i(\xi_2, \xi_3), \quad i = 1, 2, 3. \tag{14.148}$$

[28] The reason why $p = 0$ pops out of the preceding analysis for a boundary that is neither rigid nor periodic is because, for simplicity, boundary conditions were not taken into account when constructing the Lagrangian for this situation. For the atmosphere, $p = 0$ at the upper boundary is arguably quite reasonable. However, p is non-zero at the lower boundary of the atmosphere and at the upper boundary of an ocean. This needs to be considered when formulating boundary conditions there and incorporating them into the Lagrangian (as constraints). Furthermore, complicated important exchanges (of opposite sign) of energy and momentum between the atmosphere and an ocean also take place there, and these also have to be considered in practical applications. See Section 13.6.

In axial-orthogonal-curvilinear coordinates, $\boldsymbol{\Omega}\times\mathbf{r} = \Omega r_\perp \mathbf{e}_1$, where $r_\perp$ is perpendicular distance from the planet's (and the coordinate system's) rotation axis.[29] Now recall from Section 6.4.1 that, in axial-orthogonal-curvilinear coordinates, metric factor h_1 also measures perpendicular distance of a point from the coordinate system's rotation axis; see Fig. 6.1a. Thus $r_\perp = h_1$, and so

$$\boldsymbol{\Omega}\times\mathbf{r} = \Omega h_1 \mathbf{e}_1. \tag{14.149}$$

Inserting (14.147) and (14.149) into (14.146) yields:

The Lagrangian Functional in Label Space in a Rotating Frame

$$L\left(\boldsymbol{\xi},\dot{\boldsymbol{\xi}},\alpha,\eta,S\right) = \iiint \left[\frac{h_1^2}{2}\left(\frac{\partial \xi_1}{\partial \tau}\right)^2 + \frac{h_2^2}{2}\left(\frac{\partial \xi_2}{\partial \tau}\right)^2 + \frac{h_3^2}{2}\left(\frac{\partial \xi_3}{\partial \tau}\right)^2 + \Omega h_1^2 \frac{\partial \xi_1}{\partial \tau}\right] d\mathbf{a} - \iiint \left[\mathcal{E}\left(\alpha,\eta,S\right) + \Phi\left(\xi_2,\xi_3\right)\right] d\mathbf{a}. \tag{14.150}$$

Equation (14.150) is expressed in *axial-orthogonal-curvilinear coordinates* $\boldsymbol{\xi} = (\xi_1,\xi_2,\xi_3)$, where $\dot{\boldsymbol{\xi}} \equiv (\partial\xi_1/\partial\tau, \partial\xi_2/\partial\tau, \partial\xi_3/\partial\tau)$ and Φ is assumed to be zonally symmetric (i.e. independent of ξ_1).

As an illustrative example, setting $h_1 = r\cos\phi$, $h_2 = r$, and $h_3 = 1$ – see (5.15) – in (14.150) gives

$$L\left(\lambda,\phi,r,\alpha,\eta,S\right) = \iiint \left[\frac{r^2\cos^2\phi}{2}\left(\frac{\partial \lambda}{\partial \tau}\right)^2 + \frac{r^2}{2}\left(\frac{\partial \phi}{\partial \tau}\right)^2 + \frac{1}{2}\left(\frac{\partial r}{\partial \tau}\right)^2 + \Omega r^2\cos^2\phi \frac{\partial \lambda}{\partial \tau}\right] d\mathbf{a} - \iiint \left[\mathscr{E}\left(\alpha,\eta,S\right) + \Phi\left(r\right)\right] d\mathbf{a}, \tag{14.151}$$

in *spherical-polar coordinates* (λ,ϕ,r), where Φ is assumed to be spherically symmetric (by classical spherical-geopotential approximation (7.9)).

14.9.2 The Action ($\mathcal{A}$) and Hamilton's Principle

Using (14.150) in (14.57) gives:

The Action Functional in Label Space in a Rotating Frame

$$\begin{aligned}\mathcal{A}\left(\boldsymbol{\xi},\alpha,\eta,S\right) &= \int_{\tau_1}^{\tau_2} L\left(\boldsymbol{\xi},\dot{\boldsymbol{\xi}},\alpha,\eta,S\right) d\tau \\ &= \int_{\tau_1}^{\tau_2} \left\{\iiint \left[h_1^2\left(\frac{\partial \xi_1}{\partial \tau}\right)^2 + h_2^2\left(\frac{\partial \xi_2}{\partial \tau}\right)^2 + h_3^2\left(\frac{\partial \xi_3}{\partial \tau}\right)^2 + \Omega h_1^2 \frac{\partial \xi_1}{\partial \tau}\right] d\mathbf{a}\right\} d\tau \\ &\quad - \int_{\tau_1}^{\tau_2} \left\{\iiint \left[\mathcal{E}\left(\alpha,\eta,S\right) + \Phi\left(\xi_2,\xi_3\right)\right] d\mathbf{a}\right\} d\tau,\end{aligned} \tag{14.152}$$

where (14.152) is expressed in axial-orthogonal-curvilinear coordinates.

Applying equations (14.121) and (14.122) of Section 14.8.4.4 in axial-orthogonal-curvilinear coordinates leads to:

[29] For example, $r_\perp = r\cos\phi$ in spherical-polar coordinates, and $r_\perp = r$ in cylindrical-polar coordinates.

Hamilton's Principle of Stationary Action in Label Space and in Axial-Orthogonal-Curvilinear Coordinates

Hamilton's principle of stationary action, expressed *in label space* and *axial-orthogonal-curvilinear coordinates,* is to apply

$$\delta \mathcal{A}(\boldsymbol{\xi},\alpha,\eta,S) \equiv \delta \int_{\tau_1}^{\tau_2} L\left(\boldsymbol{\xi},\dot{\boldsymbol{\xi}},\alpha,\eta,S\right) d\tau = \int_{\tau_1}^{\tau_2} \delta L\left(\boldsymbol{\xi},\dot{\boldsymbol{\xi}},\alpha,\eta,S\right) d\tau = 0, \tag{14.153}$$

for independent variations $\delta\mathbf{q}(\mathbf{a},\tau)$ of generalised coordinates $\mathbf{q} = (\boldsymbol{\xi},\alpha,\eta,S)$, such that

$$\delta\mathbf{q}(\mathbf{a},\tau_1) = \delta\mathbf{q}(\mathbf{a},\tau_2) = 0, \tag{14.154}$$

where the Lagrangian, $L\left(\boldsymbol{\xi},\dot{\boldsymbol{\xi}},\alpha,\eta,S\right)$, is given by (14.150), and the action, $\mathcal{A}$, by (14.152).

14.9.3 Variations of the Action and Governing Equations

Taking variations of Lagrangian (14.150) and associated action (14.130) is a somewhat tedious exercise but a skill worth developing. Details may be found in the Appendix of this chapter. Assembling the variations of the components of the action from the results derived in the Appendix and maintaining the colour coding used therein (for later use in Chapter 16) leads to:

$$\begin{aligned}\delta_{\xi_1}\mathcal{A} &\equiv \delta_{\xi_1}\mathcal{A}^{(1)} + \delta_{\xi_1}\mathcal{A}^{(2)} + \delta_{\xi_1}\mathcal{A}^{(3)} \\ &= -\int_{\tau_1}^{\tau_2}\left[\iiint\left(\frac{\partial u_1}{\partial\tau} + \frac{u_1}{h_1}\frac{u_2}{h_2}\frac{\partial h_1}{\partial\xi_2} + 2\Omega\frac{u_2}{h_2}\frac{\partial h_1}{\partial\xi_2} + \frac{u_1}{h_1}\frac{u_3}{h_3}\frac{\partial h_1}{\partial\xi_3} + 2\Omega\frac{u_3}{h_3}\frac{\partial h_1}{\partial\xi_3}\right)h_1\delta\xi_1 d\mathbf{a}\right]d\tau \\ &\quad - \int_{\tau_1}^{\tau_2}\left[\iiint\left(\frac{1}{\rho h_1}\frac{\partial p}{\partial\xi_1}\right)h_1\delta\xi_1 d\mathbf{a}\right]d\tau,\end{aligned} \tag{14.155}$$

(from (14.204), (14.212), and (14.222));

$$\begin{aligned}\delta_{\xi_2}\mathcal{A} &\equiv \delta_{\xi_2}\mathcal{A}^{(1)} + \delta_{\xi_2}\mathcal{A}^{(2)} + \delta_{\xi_2}\mathcal{A}^{(3)} \\ &= -\int_{\tau_1}^{\tau_2}\left[\iiint\left(\frac{\partial u_2}{\partial\tau} - \frac{u_1^2}{h_1 h_2}\frac{\partial h_1}{\partial\xi_2} - 2\Omega\frac{u_1}{h_2}\frac{\partial h_1}{\partial\xi_2} - \frac{u_3^2}{h_2 h_3}\frac{\partial h_3}{\partial\xi_2} + \frac{u_2}{h_2}\frac{u_3}{h_3}\frac{\partial h_2}{\partial\xi_3}\right)h_2\delta\xi_2 d\mathbf{a}\right]d\tau \\ &\quad - \int_{\tau_1}^{\tau_2}\left[\iiint\left(\frac{1}{\rho h_2}\frac{\partial p}{\partial\xi_2} + \frac{1}{h_2}\frac{\partial\Phi}{\partial\xi_2}\right)h_2\delta\xi_2 d\mathbf{a}\right]d\tau,\end{aligned} \tag{14.156}$$

(from (14.207), (14.214), and (14.223)); and

$$\begin{aligned}\delta_{\xi_3}\mathcal{A} &\equiv \delta_{\xi_3}\mathcal{A}^{(1)} + \delta_{\xi_3}\mathcal{A}^{(2)} + \delta_{\xi_3}\mathcal{A}^{(3)} \\ &= -\int_{\tau_1}^{\tau_2}\left[\iiint\left(\frac{\partial u_3}{\partial\tau} + \frac{u_2}{h_2}\frac{u_3}{h_3}\frac{\partial h_3}{\partial\xi_2} - \frac{u_1^2}{h_1 h_3}\frac{\partial h_1}{\partial\xi_3} - 2\Omega\frac{u_1}{h_3}\frac{\partial h_1}{\partial\xi_3} - \frac{u_2^2}{h_2 h_3}\frac{\partial h_2}{\partial\xi_3}\right)h_3\delta\xi_3 d\mathbf{a}\right]d\tau \\ &\quad - \int_{\tau_1}^{\tau_2}\left[\iiint\left(\frac{1}{\rho h_3}\frac{\partial p}{\partial\xi_3} + \frac{1}{h_3}\frac{\partial\Phi}{\partial\xi_3}\right)h_3\delta\xi_3 d\mathbf{a}\right]d\tau,\end{aligned} \tag{14.157}$$

(from (14.210), (14.216), and (14.224)).

For *arbitrary* variations, with application of Hamilton's principle of stationary action (14.153), the sum of the round-bracketed terms for each variation ($\delta\xi_1$, $\delta\xi_2$, or $\delta\xi_3$) in (14.155)–(14.157), respectively, must be zero. The following governing equations for the three components of momentum then result:

$$\frac{\partial u_1}{\partial \tau} + \left(\frac{u_1}{h_1} + 2\Omega\right)\left(\frac{u_2}{h_2}\frac{\partial h_1}{\partial \xi_2} + \frac{u_3}{h_3}\frac{\partial h_1}{\partial \xi_3}\right) + \frac{1}{\rho h_1}\frac{\partial p}{\partial \xi_1} = 0, \tag{14.158}$$

$$\frac{\partial u_2}{\partial \tau} - \left(\frac{u_1}{h_1} + 2\Omega\right)\frac{u_1}{h_2}\frac{\partial h_1}{\partial \xi_2} - \frac{u_3^2}{h_2 h_3}\frac{\partial h_3}{\partial \xi_2} + \frac{u_2\, u_3}{h_2\, h_3}\frac{\partial h_2}{\partial \xi_3} + \frac{1}{\rho h_2}\frac{\partial p}{\partial \xi_2} + \frac{1}{h_2}\frac{\partial \Phi}{\partial \xi_2} = 0, \tag{14.159}$$

$$\frac{\partial u_3}{\partial \tau} + \frac{u_2\, u_3}{h_2\, h_3}\frac{\partial h_3}{\partial \xi_2} - \left(\frac{u_1}{h_1} + 2\Omega\right)\frac{u_1}{h_3}\frac{\partial h_1}{\partial \xi_3} - \frac{u_2^2}{h_2 h_3}\frac{\partial h_2}{\partial \xi_3} + \frac{1}{\rho h_3}\frac{\partial p}{\partial \xi_3} + \frac{1}{h_3}\frac{\partial \Phi}{\partial \xi_3} = 0, \tag{14.160}$$

where – from (14.195) –

$$\frac{\partial u_i}{\partial \tau} = \frac{\partial u_i}{\partial t} + \frac{u_1}{h_1}\frac{\partial u_i}{\partial \xi_1} + \frac{u_2}{h_2}\frac{\partial u_i}{\partial \xi_2} + \frac{u_3}{h_3}\frac{\partial u_i}{\partial \xi_3} = \frac{Du_i}{Dt}, \quad i = 1, 2, 3. \tag{14.161}$$

For a *geopotential coordinate system*, for which (by definition) $\Phi = \Phi(\xi_3)$, $\partial\Phi/\partial\xi_2 \equiv 0$. The term $(1/h_2)(\partial\Phi/\partial\xi_2)$ in (14.159) is then absent.

Taking into account a difference in notation ($\partial/\partial\tau \to D/Dt$, see (14.9) and (14.10)), and in the absence of forcing ($F^{u_1} = F^{u_2} = F^{u_3} = 0$), (14.158)–(14.160) agree (as they must) with (6.29)–(6.31) for the three components of the momentum equation. Equations (14.158)–(14.160) are also consistent with White and Wood (2012)'s (A.10)–(A.12) and (in the absence of the quasi-shallow terms therein) with Staniforth (2015a)'s (61)–(63).

From Section 14.8.4.5, momentum-component equations (14.158)–(14.160) are supplemented by mass-continuity equation (14.143), and by (14.133) and (14.135) for the material transport of entropy and substances, respectively. Thus, using identity (5.7) for divergence in axial-orthogonal-curvilinear coordinates,

$$\frac{\partial \rho}{\partial \tau} + \frac{\rho}{h_1 h_2 h_3}\left[\frac{\partial}{\partial \xi_1}(u_1 h_2 h_3) + \frac{\partial}{\partial \xi_2}(u_2 h_3 h_1) + \frac{\partial}{\partial \xi_3}(u_3 h_1 h_2)\right] = 0, \tag{14.162}$$

$$\frac{\partial \eta}{\partial \tau} = 0, \tag{14.163}$$

$$\frac{\partial S^i}{\partial \tau} = 0. \tag{14.164}$$

Again taking into account a difference in notation ($\partial/\partial\tau \to D/Dt$), and in the absence of forcing ($F^\rho = \dot{S}^i = \dot{Q} = 0$), (14.162)–(14.164) correspond to (6.32), to the second equation of (6.34), and to (6.33), respectively.

14.10 EULER–LAGRANGE EQUATIONS FOR GLOBAL FLUIDS

Momentum-component equations (14.158)–(14.160) were obtained using Hamilton's principle of stationary action. This involved taking variations of the action ($\mathcal{A}$) defined by (14.152), subject to conditions (14.154). An alternative, two-step procedure is now developed that has the virtue of being arguably simpler to apply in practice:

1. Derive (in Section 14.10.1) what are known as the Euler–Lagrange equations[30] associated with action (14.152). This still involves taking variations.[31] However, it only needs to be done once and for all. The result is expressed in terms of a Lagrangian density ($\mathscr{L}$), and it is generic.
2. Apply (in Section 14.10.2) the derived Euler–Lagrange equations. This then provides a complementary way, via partial differentiation of a Lagrangian density, of obtaining the three components of the momentum equation for a global atmosphere or ocean.

14.10.1 Derivation of the Euler–Lagrange Equations (Step 1)

Action (14.152) may be rewritten as

$$\mathcal{A}(\boldsymbol{\xi},\alpha,\eta,S)=\int_{\tau_1}^{\tau_2}\iiint \mathscr{L}\left(\boldsymbol{\xi},\dot{\boldsymbol{\xi}},\alpha,\eta,S\right)d\mathbf{a}d\tau, \tag{14.165}$$

where

$$\begin{aligned}\mathscr{L}&=\mathscr{L}\left[\boldsymbol{\xi}(\mathbf{a},\tau),\dot{\boldsymbol{\xi}}(\mathbf{a},\tau),\alpha,\eta,S\right]\\ &=\mathscr{L}\left[\xi_1(\mathbf{a},\tau),\xi_2(\mathbf{a},\tau),\xi_3(\mathbf{a},\tau),\dot{\xi}_1(\mathbf{a},\tau),\dot{\xi}_2(\mathbf{a},\tau),\dot{\xi}_3(\mathbf{a},\tau),\alpha,\eta,S\right]\\ &=\frac{h_1^2\dot{\xi}_1^2+h_2^2\dot{\xi}_2^2+h_3^2\dot{\xi}_3^2}{2}+\Omega h_1^2\dot{\xi}_1-\mathcal{E}(\alpha,\eta,S)-\Phi(\xi_2,\xi_3)\end{aligned} \tag{14.166}$$

is termed *Lagrangian density*, and

$$\dot{\boldsymbol{\xi}}(\mathbf{a},\tau)\equiv\frac{\partial\boldsymbol{\xi}(\mathbf{a},\tau)}{\partial\tau}. \tag{14.167}$$

(The dot notation in (14.165)–(14.167) is again the convenient abbreviation used in particle mechanics (Goldstein et al., 2001).)

Taking variations of action (14.165) and using the chain rule for differentiation then gives

$$\begin{aligned}\delta\mathcal{A}(\boldsymbol{\xi},\alpha,\eta,S)&=\int_{\tau_1}^{\tau_2}\iiint\left\{\delta\mathscr{L}\left(\boldsymbol{\xi},\dot{\boldsymbol{\xi}},\alpha,\eta,S\right)d\mathbf{a}\right\}d\tau\\ &=\int_{\tau_1}^{\tau_2}\left\{\iiint\left[\sum_{i=1}^{3}\left(\frac{\partial\mathscr{L}}{\partial\xi_i}\delta\xi_i+\frac{\partial\mathscr{L}}{\partial\dot{\xi}_i}\delta\dot{\xi}_i\right)+\frac{\partial\mathscr{L}}{\partial\alpha}\delta\alpha+\frac{\partial\mathscr{L}}{\partial\eta}\delta\eta+\frac{\partial\mathscr{L}}{\partial S}\delta S\right]d\mathbf{a}\right\}d\tau.\end{aligned} \tag{14.168}$$

For the second term on the last line of this equation, using definition (14.167), integrating by parts, and applying commutivity property (14.61) and endpoint conditions (14.154) leads to

$$\begin{aligned}\int_{\tau_1}^{\tau_2}\left[\iiint\left(\frac{\partial\mathscr{L}}{\partial\dot{\xi}_i}\delta\dot{\xi}_i\right)d\mathbf{a}\right]d\tau&=\int_{\tau_1}^{\tau_2}\left\{\iiint\left[\frac{\partial\mathscr{L}}{\partial\dot{\xi}_i}\delta\left(\frac{\partial\xi_i}{\partial\tau}\right)\right]d\mathbf{a}\right\}d\tau=\int_{\tau_1}^{\tau_2}\left\{\iiint\left[\frac{\partial\mathscr{L}}{\partial\dot{\xi}_i}\frac{\partial(\delta\xi_i)}{\partial\tau}\right]d\mathbf{a}\right\}d\tau\\ &=\cancelto{0}{\left\{\iiint\left[\frac{\partial\mathscr{L}}{\partial\dot{\xi}_i}(\delta\xi_i)\right]d\mathbf{a}\right\}_{\tau_1}^{\tau_2}}-\int_{\tau_1}^{\tau_2}\left\{\iiint\left[\frac{\partial}{\partial\tau}\left(\frac{\partial\mathscr{L}}{\partial\dot{\xi}_i}\right)(\delta\xi_i)\right]d\mathbf{a}\right\}d\tau\end{aligned}$$

[30] After Joseph-Louis Lagrange (1736–1813) and Leonhard Euler (1707–83).

[31] The Euler–Lagrange equations were, however, otherwise derived in the eighteenth century before the development of variational methods, the original motivation being to derive an equivalent, coordinate-free formulation of conservative Newtonian mechanics.

$$= -\int_{\tau_1}^{\tau_2} \left\{ \iiint \left[\frac{\partial}{\partial \tau} \left(\frac{\partial \mathscr{L}}{\partial \dot{\xi}_i} \right) (\delta \xi_i) \right] d\mathbf{a} \right\} d\tau. \tag{14.169}$$

Furthermore, from (14.126), (14.166), (14.134), and (14.136),

$$p = -\frac{\partial \mathcal{E}(\alpha, \eta, S)}{\partial \alpha} = \frac{\partial \mathscr{L}(\alpha, \eta, S)}{\partial \alpha}, \quad \delta \eta = \delta S = 0. \tag{14.170}$$

Inserting (14.169) and (14.170) into (14.168) then yields the variation of the action:

$$\delta \mathcal{A} = -\int_{\tau_1}^{\tau_2} \left\langle \iiint \left\{ \sum_{i=1}^{3} \left[\frac{\partial}{\partial \tau} \left(\frac{\partial \mathscr{L}}{\partial \dot{\xi}_i} \right) - \frac{\partial \mathscr{L}}{\partial \xi_i} \right] (\delta \xi_i) - p \delta \alpha \right\} d\mathbf{a} \right\rangle d\tau. \tag{14.171}$$

Now from (14.132), and with use of $\alpha \equiv 1/\rho$, (5.6), and (5.2), the last term in (14.171) can be rewritten as

$$\iiint p \delta \alpha d\mathbf{a} = -\iiint (\alpha \nabla p \cdot \delta \mathbf{r})\, d\mathbf{a} = -\iiint \left\{ \alpha \sum_{i=1}^{3} \left[\left(\frac{1}{h_i} \frac{\partial p}{\partial \xi_i} \right) (h_i \delta \xi_i) \right] \right\} d\mathbf{a}$$

$$= -\iiint \left\{ \sum_{i=1}^{3} \left[\alpha \frac{\partial p}{\partial \xi_i} (\delta \xi_i) \right] \right\} d\mathbf{a}. \tag{14.172}$$

For simplicity, the boundary contribution in (14.132) has been ignored (as before). (It does not anyway influence the governing equations *within the fluid*, which is the focus herein.) Inserting (14.172) into (14.171) and applying Hamilton's principle (14.153) of stationary action (by setting $\delta \mathcal{A} \equiv 0$) leads to

$$\delta \mathcal{A} = -\int_{\tau_1}^{\tau_2} \left\langle \iiint \left\{ \sum_{i=1}^{3} \left[\frac{\partial}{\partial \tau} \left(\frac{\partial \mathscr{L}}{\partial \dot{\xi}_i} \right) - \frac{\partial \mathscr{L}}{\partial \xi_i} + \alpha \frac{\partial p}{\partial \xi_i} \right] (\delta \xi_i) \right\} d\mathbf{a} \right\rangle d\tau = 0. \tag{14.173}$$

For arbitrary variations $\delta \xi_i, i = 1, 2, 3$, this then yields:

The Euler–Lagrange Equations

$$\frac{\partial}{\partial \tau} \left(\frac{\partial \mathscr{L}}{\partial \dot{\xi}_i} \right) - \frac{\partial \mathscr{L}}{\partial \xi_i} = -\alpha \frac{\partial p}{\partial \xi_i}, \quad i = 1, 2, 3, \tag{14.174}$$

for the three components of the momentum equation.

These equations correspond to (A.14) of White et al. (2005) (in the absence of forcing) and to (13) of Tort and Dubos (2014b).

Equations (14.174) are *generic* and were obtained under the assumptions that:

- The Lagrangian density has the form $\mathscr{L} = \mathscr{L}(\boldsymbol{\xi}, \dot{\boldsymbol{\xi}}, \alpha, \eta, S)$.
- The internal energy appearing therein has the form $\mathcal{E} = \mathcal{E}(\alpha, \eta, S)$.
- The geopotential (Φ) appearing therein is an axially symmetric function of position.
- Equations (14.163) and (14.164) hold for η and S.

This means that $\mathscr{L}$ does not have to have precisely the same definition as that given in the last line of (14.166). In particular, any of the terms on this line (within reason) can be approximated, and (14.174) will still hold for the approximated Lagrangian. This applies equally well to the contributions of kinetic, internal, and gravitational energies.

14.10.2 Application of the Euler–Lagrange Equations (Step 2)

We now move on to the second step, namely to obtain (via partial differentiation) the three components of the momentum equation from the Euler–Lagrange equations, (14.174). This is where the last line of (14.166) comes into play.

Recall, from (14.166), that

$$\mathscr{L}\left(\boldsymbol{\xi},\dot{\boldsymbol{\xi}},\alpha,\eta,S\right)=\frac{h_1^2\dot{\xi}_1^2+h_2^2\dot{\xi}_2^2}{2}+\frac{h_3^2\dot{\xi}_3^2}{2}+\Omega h_1^2\dot{\xi}_1-\mathcal{E}\left(\alpha,\eta,S\right)-\Phi\left(\xi_2,\xi_3\right), \tag{14.175}$$

where the colour coding of the Appendix to this chapter has been applied. The blue term is the non-hydrostatic contribution to the Lagrangian density. This term is absent when the *hydrostatic* or *quasi-hydrostatic* approximations are made. Colour coding allows this term to be tracked through to the three components of the momentum equation. Similarly, terms coded red are associated with a deep fluid. They are absent when *shallow* approximation is made. See Chapter 16 for further details.

Partially differentiating (14.175) with respect to ξ_1, ξ_2, ξ_3, $\dot{\xi}_1$, $\dot{\xi}_2$, and $\dot{\xi}_3$ and using definition (14.147) of $\mathbf{u}$ in orthogonal-curvilinear coordinates leads to

$$\frac{\partial\mathscr{L}}{\partial\xi_1}=0, \tag{14.176}$$

$$\begin{aligned}\frac{\partial\mathscr{L}}{\partial\xi_2}&=h_1\frac{\partial h_1}{\partial\xi_2}\dot{\xi}_1^2+h_2\frac{\partial h_2}{\partial\xi_2}\dot{\xi}_2^2+h_3\frac{\partial h_3}{\partial\xi_2}\dot{\xi}_3^2+2\Omega h_1\frac{\partial h_1}{\partial\xi_2}\dot{\xi}_1-\frac{\partial\Phi}{\partial\xi_2}\\&=\left(\frac{u_1}{h_1}+2\Omega\right)u_1\frac{\partial h_1}{\partial\xi_2}+\frac{u_2^2}{h_2}\frac{\partial h_2}{\partial\xi_2}+\frac{u_3^2}{h_3}\frac{\partial h_3}{\partial\xi_2}-\frac{\partial\Phi}{\partial\xi_2},\end{aligned} \tag{14.177}$$

$$\begin{aligned}\frac{\partial\mathscr{L}}{\partial\xi_3}&=h_1\frac{\partial h_1}{\partial\xi_3}\dot{\xi}_1^2+h_2\frac{\partial h_2}{\partial\xi_3}\dot{\xi}_2^2+h_3\frac{\partial h_3}{\partial\xi_3}\dot{\xi}_3^2+2\Omega h_1\frac{\partial h_1}{\partial\xi_3}\dot{\xi}_1-\frac{\partial\Phi}{\partial\xi_3}\\&=\left(\frac{u_1}{h_1}+2\Omega\right)u_1\frac{\partial h_1}{\partial\xi_3}+\frac{u_2^2}{h_2}\frac{\partial h_2}{\partial\xi_3}+\frac{u_3^2}{h_3}\frac{\partial h_3}{\partial\xi_3}-\frac{\partial\Phi}{\partial\xi_3},\end{aligned} \tag{14.178}$$

$$\frac{\partial\mathscr{L}}{\partial\dot{\xi}_1}=h_1^2\dot{\xi}_1+\Omega h_1^2=h_1\left(u_1+\Omega h_1\right),\quad\frac{\partial\mathscr{L}}{\partial\dot{\xi}_2}=h_2^2\dot{\xi}_2=h_2u_2,\quad\frac{\partial\mathscr{L}}{\partial\dot{\xi}_3}=h_3^2\dot{\xi}_3=h_3u_3. \tag{14.179}$$

Insertion of (14.176)–(14.179) into Euler–Lagrange equations (14.174) then yields

$$\frac{\partial}{\partial\tau}\left[h_1\left(u_1+\Omega h_1\right)\right]=-\alpha\frac{\partial p}{\partial\xi_1}, \tag{14.180}$$

$$\frac{\partial}{\partial\tau}\left(h_2u_2\right)-\left(\frac{u_1}{h_1}+2\Omega\right)u_1\frac{\partial h_1}{\partial\xi_2}-\frac{u_2^2}{h_2}\frac{\partial h_2}{\partial\xi_2}-\frac{u_3^2}{h_3}\frac{\partial h_3}{\partial\xi_2}+\frac{\partial\Phi}{\partial\xi_2}=-\alpha\frac{\partial p}{\partial\xi_2}, \tag{14.181}$$

$$\frac{\partial}{\partial\tau}\left(h_3u_3\right)-\left(\frac{u_1}{h_1}+2\Omega\right)u_1\frac{\partial h_1}{\partial\xi_3}-\frac{u_2^2}{h_2}\frac{\partial h_2}{\partial\xi_3}-\frac{u_3^2}{h_3}\frac{\partial h_3}{\partial\xi_3}+\frac{\partial\Phi}{\partial\xi_3}=-\alpha\frac{\partial p}{\partial\xi_3}. \tag{14.182}$$

Note that (14.180) expresses conservation of axial angular momentum.

Using useful relations (14.196)–(14.201), (14.180)–(14.182) may be rewritten as

$$\frac{\partial u_1}{\partial\tau}+\left(\frac{u_1}{h_1}+2\Omega\right)\left(\frac{u_2}{h_2}\frac{\partial h_1}{\partial\xi_2}+\frac{u_3}{h_3}\frac{\partial h_1}{\partial\xi_3}\right)=-\frac{\alpha}{h_1}\frac{\partial p}{\partial\xi_1}, \tag{14.183}$$

$$\frac{\partial u_2}{\partial\tau}-\left(\frac{u_1}{h_1}+2\Omega\right)\frac{u_1}{h_2}\frac{\partial h_1}{\partial\xi_2}-\frac{u_3^2}{h_2h_3}\frac{\partial h_3}{\partial\xi_2}+\frac{u_2u_3}{h_2h_3}\frac{\partial h_2}{\partial\xi_3}+\frac{1}{h_2}\frac{\partial\Phi}{\partial\xi_2}=-\frac{\alpha}{h_2}\frac{\partial p}{\partial\xi_2}, \tag{14.184}$$

$$\frac{\partial u_3}{\partial\tau}+\frac{u_2u_3}{h_2h_3}\frac{\partial h_3}{\partial\xi_2}-\left(\frac{u_1}{h_1}+2\Omega\right)\frac{u_1}{h_3}\frac{\partial h_1}{\partial\xi_3}-\frac{u_2^2}{h_2h_3}\frac{\partial h_2}{\partial\xi_3}+\frac{1}{h_3}\frac{\partial\Phi}{\partial\xi_3}=-\frac{\alpha}{h_3}\frac{\partial p}{\partial\xi_3}. \tag{14.185}$$

This recovers (14.158)–(14.160), but without (directly) taking variations. Of course, variations have indirectly been taken, since this was done in Step 1 to obtain the Euler–Lagrange equations. Now that we have the Euler–Lagrange equations, they are a very useful tool, particularly if one is not entirely comfortable taking variations!

14.11 CONCLUDING REMARKS

The present chapter has provided:

- An introduction to variational methods and Hamilton's principle of stationary action.
- A complementary derivation of the unified, governing, dynamical-core equations for an atmosphere or ocean overlying a rotating spheroidal planet.

The techniques developed in this chapter will prove to be very useful tools for the examination – in Part II of this book – of dynamically consistent approximation of the governing dynamical-core equations.

Now respect of fundamental conservation laws is key to dynamical consistency and to the fidelity of atmospheric and oceanic model forecasts with respect to reality. In the next (and final) chapter of Part I – and to complete the preparation for the analyses of Part II – we examine various different but equivalent ways of mathematically expressing these physical conservation laws. Since it is intrinsically linked to physical conservation principles and their expression, the variational methodology developed in this chapter is highly relevant to the developments of both Chapter 15 (on conservation) and Part II (on dynamical consistency).

APPENDIX: VARIATIONS IN AXIAL-ORTHOGONAL-CURVILINEAR COORDINATES

Preliminaries

Colour Coding

In this appendix, certain terms are colour coded for later use. Terms in blue are termed *non-hydrostatic*, and those in red *deep*.[32] The reasons for this will become apparent in Chapter 16. They are related to the physical significance and origin of these terms. For now, we can, if we wish, just ignore the colour coding and pretend that all terms are written in black.

Decomposition

Decompose Lagrangian (14.150) into three contributions so that

$$L\left(\boldsymbol{\xi},\dot{\boldsymbol{\xi}},\alpha,\eta,S\right) \equiv L\left(\xi_1,\xi_2,\xi_3,\dot{\xi}_1,\dot{\xi}_2,\dot{\xi}_3,\alpha,\eta,S\right) \equiv L^{(1)} + L^{(2)} + L^{(3)}, \tag{14.186}$$

where

$$L^{(1)} \equiv \iiint \left[\frac{h_1^2}{2}\left(\frac{\partial \xi_1}{\partial \tau}\right)^2 + \frac{h_2^2}{2}\left(\frac{\partial \xi_2}{\partial \tau}\right)^2 + \frac{h_3^2}{2}\left(\frac{\partial \xi_3}{\partial \tau}\right)^2\right] d\mathbf{a}, \tag{14.187}$$

$$L^{(2)} \equiv \iiint \Omega h_1^2 \frac{\partial \xi_1}{\partial \tau} d\mathbf{a}, \tag{14.188}$$

$$L^{(3)} \equiv \iiint \left[\mathcal{E}\left(\alpha,\eta,S\right) + \Phi\left(\xi_2,\xi_3\right)\right] d\mathbf{a}. \tag{14.189}$$

Similarly, decompose action (14.152) so that

$$\mathcal{A}\left(\boldsymbol{\xi},\alpha,\eta,S\right) \equiv \mathcal{A}\left(\xi_1,\xi_2,\xi_3\alpha,\eta,S\right) \equiv \mathcal{A}^{(1)} + \mathcal{A}^{(2)} + \mathcal{A}^{(3)}, \tag{14.190}$$

[32] Equivalently, blue terms are absent when the hydrostatic or quasi-hydrostatic approximations are made, and red terms when the shallow approximation is made.

where

$$\mathcal{A}^{(i)} \equiv \int_{\tau_1}^{\tau_2} L^{(i)} d\tau, \quad i = 1, 2, 3. \tag{14.191}$$

From (14.153), the variation of the decomposed actions is then

$$\delta \mathcal{A}^{(i)} = \int_{\tau_1}^{\tau_2} \delta L^{(i)} d\tau, \quad i = 1, 2, 3. \tag{14.192}$$

Useful Relations

For any scalar function f that can be written in location space as $f = f(\xi_1, \xi_2, \xi_3, t)$ and in label space as $f = f(a_1, a_2, a_3, \tau)$, applying the chain rule for partial differentiation gives

$$\frac{\partial f}{\partial \tau} = \frac{\partial t}{\partial \tau}\frac{\partial f}{\partial t} + \frac{\partial \xi_1}{\partial \tau}\frac{\partial f}{\partial \xi_1} + \frac{\partial \xi_2}{\partial \tau}\frac{\partial f}{\partial \xi_2} + \frac{\partial \xi_3}{\partial \tau}\frac{\partial f}{\partial \xi_3}. \tag{14.193}$$

Now $t = \tau$ and, from the definition (14.147) of $\mathbf{u}$ in an orthogonal-curvilinear coordinate system,

$$\frac{\partial \xi_i}{\partial \tau} = \frac{u_i}{h_i}, \quad i = 1, 2, 3. \tag{14.194}$$

Substitution of these into (14.193) yields

$$\frac{\partial f}{\partial \tau} = \frac{\partial f}{\partial t} + \frac{u_1}{h_1}\frac{\partial f}{\partial \xi_1} + \frac{u_2}{h_2}\frac{\partial f}{\partial \xi_2} + \frac{u_3}{h_3}\frac{\partial f}{\partial \xi_3} = \frac{Df}{Dt}. \tag{14.195}$$

Using axial-symmetry assumption (14.148) of an axial-orthogonal-curvilinear coordinate system and successively substituting $f = h_i(\xi_2, \xi_3)$ for $i = 1, 2, 3$ into (14.195) then leads to

$$\frac{\partial h_1}{\partial \tau} = \cancelto{0}{\frac{\partial h_1}{\partial t}} + \cancelto{0}{\frac{u_1}{h_1}\frac{\partial h_1}{\partial \xi_1}} + \frac{u_2}{h_2}\frac{\partial h_1}{\partial \xi_2} + \frac{u_3}{h_3}\frac{\partial h_1}{\partial \xi_3} = \frac{u_2}{h_2}\frac{\partial h_1}{\partial \xi_2} + \frac{u_3}{h_3}\frac{\partial h_1}{\partial \xi_3}, \tag{14.196}$$

$$\frac{\partial h_2}{\partial \tau} = \cancelto{0}{\frac{\partial h_2}{\partial t}} + \cancelto{0}{\frac{u_1}{h_1}\frac{\partial h_2}{\partial \xi_1}} + \frac{u_2}{h_2}\frac{\partial h_2}{\partial \xi_2} + \frac{u_3}{h_3}\frac{\partial h_2}{\partial \xi_3} = \frac{u_2}{h_2}\frac{\partial h_2}{\partial \xi_2} + \frac{u_3}{h_3}\frac{\partial h_2}{\partial \xi_3}, \tag{14.197}$$

$$\frac{\partial h_3}{\partial \tau} = \cancelto{0}{\frac{\partial h_3}{\partial t}} + \cancelto{0}{\frac{u_1}{h_1}\frac{\partial h_3}{\partial \xi_1}} + \frac{u_2}{h_2}\frac{\partial h_3}{\partial \xi_2} + \frac{u_3}{h_3}\frac{\partial h_2}{\partial \xi_3} = \frac{u_2}{h_2}\frac{\partial h_3}{\partial \xi_2} + \frac{u_3}{h_3}\frac{\partial h_3}{\partial \xi_3}. \tag{14.198}$$

These three equations then imply

$$\frac{\partial}{\partial \tau}\left(h_1 u_1\right) = h_1 \frac{\partial u_1}{\partial \tau} + u_1 \frac{\partial h_1}{\partial \tau} = h_1 \left[\frac{\partial u_1}{\partial \tau} + \frac{u_1}{h_1}\left(\frac{u_2}{h_2}\frac{\partial h_1}{\partial \xi_2} + \frac{u_3}{h_3}\frac{\partial h_1}{\partial \xi_3}\right)\right], \tag{14.199}$$

$$\frac{\partial}{\partial \tau}\left(h_2 u_2\right) = h_2 \frac{\partial u_2}{\partial \tau} + u_2 \frac{\partial h_2}{\partial \tau} = h_2 \left[\frac{\partial u_2}{\partial \tau} + \frac{u_2}{h_2}\left(\frac{u_2}{h_2}\frac{\partial h_2}{\partial \xi_2} + \frac{u_3}{h_3}\frac{\partial h_2}{\partial \xi_3}\right)\right], \tag{14.200}$$

$$\frac{\partial}{\partial \tau}\left(h_3 u_3\right) = h_3 \frac{\partial u_3}{\partial \tau} + u_3 \frac{\partial h_3}{\partial \tau} = h_3 \left[\frac{\partial u_3}{\partial \tau} + \frac{u_3}{h_3}\left(\frac{u_2}{h_2}\frac{\partial h_3}{\partial \xi_2} + \frac{u_3}{h_3}\frac{\partial h_3}{\partial \xi_3}\right)\right]. \tag{14.201}$$

For shallow approximation, $h_i(\xi_2, \xi_3) \to h_i^S(\xi_2), i = 1, 2, 3$, where superscript '$S$' denotes evaluation at the surface $\xi_3 = \xi_3^S =$ constant; see Section 5.3.1. Thus $\partial h_i / \partial \xi_3 \to \partial h_i^S / \partial \xi_3 \equiv 0, i = 1, 2, 3$, and red terms are absent when shallow approximation is made; that is, *they are only present for a deep atmosphere/ocean.*

Variations of $\mathcal{A}^{(1)}$

Variations of $\mathcal{A}^{(1)}$ with Respect to $\delta\xi_1$

Using definition (14.187) of $L^{(1)}$, axial-symmetry assumption (14.148), and definition (14.147) of $\mathbf{u}$, variation of $L^{(1)}$ with respect to $\delta\xi_1$ is given by

$$
\begin{aligned}
\delta_{\xi_1} L^{(1)} &= \delta_{\xi_1} \iiint \left[\frac{h_1^2}{2}\left(\frac{\partial \xi_1}{\partial \tau}\right)^2 + \frac{h_2^2}{2}\left(\frac{\partial \xi_2}{\partial \tau}\right)^2 + \frac{h_3^2}{2}\left(\frac{\partial \xi_3}{\partial \tau}\right)^2 \right] \mathbf{da} \\
&= \iiint \delta_{\xi_1} \left[\frac{h_1^2}{2}\left(\frac{\partial \xi_1}{\partial \tau}\right)^2 + \frac{h_2^2}{2}\left(\frac{\partial \xi_2}{\partial \tau}\right)^2 + \frac{h_3^2}{2}\left(\frac{\partial \xi_3}{\partial \tau}\right)^2 \right] \mathbf{da} \\
&= \iiint \left[h_1 \cancelto{0}{\frac{\partial h_1}{\partial \xi_1}} \left(\frac{\partial \xi_1}{\partial \tau}\right)^2 + h_2 \cancelto{0}{\frac{\partial h_2}{\partial \xi_1}} \left(\frac{\partial \xi_2}{\partial \tau}\right)^2 + h_3 \cancelto{0}{\frac{\partial h_3}{\partial \xi_1}} \left(\frac{\partial \xi_3}{\partial \tau}\right)^2 \right] (\delta\xi_1)\, \mathbf{da} \\
&\quad + \iiint h_1^2 \left(\frac{\partial \xi_1}{\partial \tau}\right) \frac{\partial (\delta\xi_1)}{\partial \tau} \mathbf{da} \\
&= \iiint \left[(h_1 u_1) \frac{\partial (\delta\xi_1)}{\partial \tau} \right] \mathbf{da}.
\end{aligned} \tag{14.202}
$$

Using (14.202) in (14.192), integrating by parts, and using constraints (14.154), the corresponding variation of $\mathcal{A}^{(1)}$ with respect to $\delta\xi_1$ is then

$$
\begin{aligned}
\delta_{\xi_1}\mathcal{A}^{(1)} &\equiv \delta_{\xi_1} \int_{\tau_1}^{\tau_2} L^{(1)} d\tau = \int_{\tau_1}^{\tau_2} \delta_{\xi_1} L^{(1)} d\tau = \int_{\tau_1}^{\tau_2} \left\{ \iiint \left[(h_1 u_1) \frac{\partial (\delta\xi_1)}{\partial \tau} \right] \mathbf{da} \right\} d\tau \\
&= -\int_{\tau_1}^{\tau_2} \left\{ \iiint \left[\frac{\partial}{\partial \tau} (h_1 u_1) \right] (\delta\xi_1)\, \mathbf{da} \right\} d\tau + \iiint \cancelto{0}{\left[(h_1 u_1 \delta\xi_1)_{\tau_1}^{\tau_2} \right]} \mathbf{da} \\
&= -\int_{\tau_1}^{\tau_2} \left\{ \iiint \left[\frac{\partial}{\partial \tau} (h_1 u_1) \right] (\delta\xi_1)\, \mathbf{da} \right\} d\tau.
\end{aligned} \tag{14.203}
$$

Insertion of useful relation (14.199) into (14.203) finally yields

$$
\begin{aligned}
\delta_{\xi_1}\mathcal{A}^{(1)} &= -\int_{\tau_1}^{\tau_2} \left\{ \iiint h_1 \left[\frac{\partial u_1}{\partial \tau} + \frac{u_1}{h_1} \left(\frac{u_2}{h_2}\frac{\partial h_1}{\partial \xi_2} + \frac{u_3}{h_3}\frac{\partial h_1}{\partial \xi_3} \right) \right] (\delta\xi_1)\, \mathbf{da} \right\} d\tau \\
&= -\int_{\tau_1}^{\tau_2} \left[\iiint \left(\frac{\partial u_1}{\partial \tau} + \frac{u_1}{h_1}\frac{u_2}{h_2}\frac{\partial h_1}{\partial \xi_2} + \frac{u_1}{h_1}\frac{u_3}{h_3}\frac{\partial h_1}{\partial \xi_3} \right) h_1 \delta\xi_1 \mathbf{da} \right] d\tau.
\end{aligned} \tag{14.204}
$$

This variation is the first contribution to $\delta_{\xi_1}\mathcal{A}$; see (14.155).

Variations of $\mathcal{A}^{(1)}$ with Respect to $\delta\xi_2$

Using definition (14.187) of $L^{(1)}$, axial-symmetry assumption (14.148) of the coordinate system, and definition (14.147) of $\mathbf{u}$, variation of $L^{(1)}$ with respect to $\delta\xi_2$ is given by

$$
\begin{aligned}
\delta_{\xi_2} L^{(1)} &= \delta_{\xi_2} \iiint \left[\frac{h_1^2}{2}\left(\frac{\partial \xi_1}{\partial \tau}\right)^2 + \frac{h_2^2}{2}\left(\frac{\partial \xi_2}{\partial \tau}\right)^2 + \frac{h_3^2}{2}\left(\frac{\partial \xi_3}{\partial \tau}\right)^2 \right] \mathbf{da} \\
&= \iiint \delta_{\xi_2} \left[\frac{h_1^2}{2}\left(\frac{\partial \xi_1}{\partial \tau}\right)^2 + \frac{h_2^2}{2}\left(\frac{\partial \xi_2}{\partial \tau}\right)^2 + \frac{h_3^2}{2}\left(\frac{\partial \xi_3}{\partial \tau}\right)^2 \right] \mathbf{da} \\
&= \iiint \left[h_1 \frac{\partial h_1}{\partial \xi_2} \left(\frac{\partial \xi_1}{\partial \tau}\right)^2 + h_2 \frac{\partial h_2}{\partial \xi_2} \left(\frac{\partial \xi_2}{\partial \tau}\right)^2 + h_3 \frac{\partial h_3}{\partial \xi_2} \left(\frac{\partial \xi_3}{\partial \tau}\right)^2 \right] (\delta\xi_2)\, \mathbf{da}
\end{aligned}
$$

$$+\iiint h_2^2\left(\frac{\partial\xi_2}{\partial\tau}\right)\frac{\partial(\delta\xi_2)}{\partial\tau}\mathbf{da}$$

$$=\iiint\left[\left(\frac{u_1^2}{h_1}\frac{\partial h_1}{\partial\xi_2}+\frac{u_2^2}{h_2}\frac{\partial h_2}{\partial\xi_2}+\frac{u_3^2}{h_3}\frac{\partial h_3}{\partial\xi_2}\right)(\delta\xi_2)+\left(h_2u_2\right)\frac{\partial(\delta\xi_2)}{\partial\tau}\right]\mathbf{da}. \tag{14.205}$$

Using (14.205) in (14.192), integrating by parts, and using constraints (14.154), the corresponding variation of $\mathcal{A}^{(1)}$ with respect to $\delta\xi_2$ is then

$$\delta_{\xi_2}\mathcal{A}^{(1)}\equiv\delta_{\xi_2}\int_{\tau_1}^{\tau_2}L^{(1)}d\tau=\int_{\tau_1}^{\tau_2}\delta_{\xi_2}L^{(1)}d\tau$$

$$=\int_{\tau_1}^{\tau_2}\left\{\iiint\left[\left(\frac{u_1^2}{h_1}\frac{\partial h_1}{\partial\xi_2}+\frac{u_2^2}{h_2}\frac{\partial h_2}{\partial\xi_2}+\frac{u_3^2}{h_3}\frac{\partial h_3}{\partial\xi_2}\right)(\delta\xi_2)+\left(h_2u_2\right)\frac{\partial(\delta\xi_2)}{\partial\tau}\right]\mathbf{da}\right\}d\tau$$

$$=\int_{\tau_1}^{\tau_2}\left\{\iiint\left[\frac{u_1^2}{h_1}\frac{\partial h_1}{\partial\xi_2}+\frac{u_2^2}{h_2}\frac{\partial h_2}{\partial\xi_2}+\frac{u_3^2}{h_3}\frac{\partial h_3}{\partial\xi_2}-\frac{\partial}{\partial\tau}\left(h_2u_2\right)\right](\delta\xi_2)\,\mathbf{da}\right\}d\tau$$

$$+\iiint\left[\cancelto{0}{\left(h_2u_2\delta\xi_2\right)_{\tau_1}^{\tau_2}}\right]\mathbf{da}$$

$$=\int_{\tau_1}^{\tau_2}\left\{\iiint\left[\frac{u_1^2}{h_1}\frac{\partial h_1}{\partial\xi_2}+\frac{u_2^2}{h_2}\frac{\partial h_2}{\partial\xi_2}+\frac{u_3^2}{h_3}\frac{\partial h_3}{\partial\xi_2}-\frac{\partial}{\partial\tau}\left(h_2u_2\right)\right](\delta\xi_2)\,\mathbf{da}\right\}d\tau. \tag{14.206}$$

Insertion of useful relation (14.200) into (14.206) finally yields

$$\delta_{\xi_2}\mathcal{A}^{(1)}=\int_{\tau_1}^{\tau_2}\left\langle\iiint\left\{\left[\frac{u_1^2}{h_1}\frac{\partial h_1}{\partial\xi_2}+\cancel{\frac{u_2^2}{h_2}\frac{\partial h_2}{\partial\xi_2}}+\frac{u_3^2}{h_3}\frac{\partial h_3}{\partial\xi_2}\right](\delta\xi_2)\right\}\mathbf{da}\right\rangle d\tau$$

$$-\int_{\tau_1}^{\tau_2}\left\{\iiint h_2\left[\frac{\partial u_2}{\partial\tau}+\frac{u_2}{h_2}\left(\cancel{\frac{u_2}{h_2}\frac{\partial h_2}{\partial\xi_2}}+\frac{u_3}{h_3}\frac{\partial h_2}{\partial\xi_3}\right)\right](\delta\xi_2)\,\mathbf{da}\right\}d\tau$$

$$=-\int_{\tau_1}^{\tau_2}\left[\iiint\left(\frac{\partial u_2}{\partial\tau}-\frac{u_1^2}{h_1h_2}\frac{\partial h_1}{\partial\xi_2}-\frac{u_3^2}{h_2h_3}\frac{\partial h_3}{\partial\xi_2}+\frac{u_2\,u_3}{h_2\,h_3}\frac{\partial h_2}{\partial\xi_3}\right)h_2\delta\xi_2\mathbf{da}\right]d\tau. \tag{14.207}$$

This variation is the first contribution to $\delta_{\xi_2}\mathcal{A}$; see (14.156).

Variations of $\mathcal{A}^{(1)}$ with Respect to $\delta\xi_3$

Using definition (14.187) of $L^{(1)}$, axial-symmetry assumption (14.148), and definition (14.147) of **u**, variation of $L^{(1)}$ with respect to $\delta\xi_3$ is given by

$$\delta_{\xi_3}L^{(1)}=\delta_{\xi_3}\iiint\left[\frac{h_1^2}{2}\left(\frac{\partial\xi_1}{\partial\tau}\right)^2+\frac{h_2^2}{2}\left(\frac{\partial\xi_2}{\partial\tau}\right)^2+\frac{h_3^2}{2}\left(\frac{\partial\xi_3}{\partial\tau}\right)^2\right]\mathbf{da}$$

$$=\iiint\delta_{\xi_3}\left[\frac{h_1^2}{2}\left(\frac{\partial\xi_1}{\partial\tau}\right)^2+\frac{h_2^2}{2}\left(\frac{\partial\xi_2}{\partial\tau}\right)^2+\frac{h_3^2}{2}\left(\frac{\partial\xi_3}{\partial\tau}\right)^2\right]\mathbf{da}$$

$$=\iiint\left[h_1\frac{\partial h_1}{\partial\xi_3}\left(\frac{\partial\xi_1}{\partial\tau}\right)^2+h_2\frac{\partial h_2}{\partial\xi_3}\left(\frac{\partial\xi_2}{\partial\tau}\right)^2+h_3\frac{\partial h_3}{\partial\xi_3}\left(\frac{\partial\xi_3}{\partial\tau}\right)^2\right](\delta\xi_3)\,\mathbf{da}$$

$$+\iiint h_3^2\left(\frac{\partial\xi_3}{\partial\tau}\right)\frac{\partial(\delta\xi_3)}{\partial\tau}\mathbf{da}$$

$$=\iiint\left[\left(\frac{u_1^2}{h_1}\frac{\partial h_1}{\partial\xi_3}+\frac{u_2^2}{h_2}\frac{\partial h_2}{\partial\xi_3}+\frac{u_3^2}{h_3}\frac{\partial h_3}{\partial\xi_3}\right)(\delta\xi_3)+\left(h_3u_3\right)\frac{\partial(\delta\xi_3)}{\partial\tau}\right]\mathbf{da}. \tag{14.208}$$

Using (14.208) in (14.192), integrating by parts, and using constraints (14.154), the corresponding variation of $\mathcal{A}^{(1)}$ with respect to $\delta\xi_3$ is then

$$\begin{aligned}
\delta_{\xi_3}\mathcal{A}^{(1)} &\equiv \delta_{\xi_3}\int_{\tau_1}^{\tau_2} L^{(1)}d\tau = \int_{\tau_1}^{\tau_2} \delta_{\xi_3}L^{(1)}d\tau \\
&= \int_{\tau_1}^{\tau_2}\left\{\iiint\left[\left(\frac{u_1^2}{h_1}\frac{\partial h_1}{\partial\xi_3}+\frac{u_2^2}{h_2}\frac{\partial h_2}{\partial\xi_3}+\frac{u_3^2}{h_3}\frac{\partial h_3}{\partial\xi_3}\right)(\delta\xi_3)+\left(h_3u_3\right)\frac{\partial\left(\delta\xi_3\right)}{\partial\tau}\right]\mathbf{da}\right\}d\tau \\
&= \int_{\tau_1}^{\tau_2}\left\{\iiint\left[\frac{u_1^2}{h_1}\frac{\partial h_1}{\partial\xi_3}+\frac{u_2^2}{h_2}\frac{\partial h_2}{\partial\xi_3}+\frac{u_3^2}{h_3}\frac{\partial h_3}{\partial\xi_3}-\frac{\partial}{\partial\tau}\left(h_3u_3\right)\right](\delta\xi_3)\,\mathbf{da}\right\}d\tau \\
&\quad+\iiint\left[\cancelto{0}{\left(h_3u_3\delta\xi_3\right)_{\tau_1}^{\tau_2}}\right]\mathbf{da} \\
&= \int_{\tau_1}^{\tau_2}\left\{\iiint\left[\frac{u_1^2}{h_1}\frac{\partial h_1}{\partial\xi_3}+\frac{u_2^2}{h_2}\frac{\partial h_2}{\partial\xi_3}+\frac{u_3^2}{h_3}\frac{\partial h_3}{\partial\xi_3}-\frac{\partial}{\partial\tau}\left(h_3u_3\right)\right](\delta\xi_3)\,\mathbf{da}\right\}d\tau.
\end{aligned}\tag{14.209}$$

Insertion of useful relation (14.201) into (14.209) finally yields

$$\begin{aligned}
\delta_{\xi_3}\mathcal{A}^{(1)} &= \int_{\tau_1}^{\tau_2}\left\langle\iiint\left\{\left[\frac{u_1^2}{h_1}\frac{\partial h_1}{\partial\xi_3}+\frac{u_2^2}{h_2}\frac{\partial h_2}{\partial\xi_3}+\cancel{\frac{u_3^2}{h_3}\frac{\partial h_3}{\partial\xi_3}}\right](\delta\xi_3)\right\}\mathbf{da}\right\rangle d\tau \\
&\quad-\int_{\tau_1}^{\tau_2}\left\{\iiint h_3\left[\frac{\partial u_3}{\partial\tau}+\frac{u_3}{h_3}\left(\frac{u_2}{h_2}\frac{\partial h_3}{\partial\xi_2}+\cancel{\frac{u_3}{h_3}\frac{\partial h_3}{\partial\xi_3}}\right)\right](\delta\xi_3)\,\mathbf{da}\right\}d\tau \\
&= -\int_{\tau_1}^{\tau_2}\left[\iiint\left(\frac{\partial u_3}{\partial\tau}+\frac{u_2}{h_2}\frac{u_3}{h_3}\frac{\partial h_3}{\partial\xi_2}-\frac{u_1^2}{h_1h_3}\frac{\partial h_1}{\partial\xi_3}-\frac{u_2^2}{h_2h_3}\frac{\partial h_2}{\partial\xi_3}\right)h_3\delta\xi_3\mathbf{da}\right]d\tau.
\end{aligned}\tag{14.210}$$

This variation is the first contribution to $\delta_{\xi_3}\mathcal{A}$; see (14.157).

Variations of $\mathcal{A}^{(2)}$

Variations of $\mathcal{A}^{(2)}$ with Respect to $\delta\xi_1$

Using definition (14.188) of $L^{(2)}$, and axial-symmetry assumption (14.148), variation of $L^{(2)}$ with respect to $\delta\xi_1$ is given by

$$\begin{aligned}
\delta_{\xi_1}L^{(2)} &= \delta_{\xi_1}\iiint\left(\Omega h_1^2\frac{\partial\xi_1}{\partial\tau}\right)\mathbf{da} = \iiint\Omega\delta_{\xi_1}\left(h_1^2\frac{\partial\xi_1}{\partial\tau}\right)\mathbf{da} \\
&= \iiint\left[2\Omega h_1\cancelto{0}{\frac{\partial h_1}{\partial\xi_1}}\frac{\partial\xi_1}{\partial\tau}(\delta\xi_1)+\Omega h_1^2\frac{\partial\left(\delta\xi_1\right)}{\partial\tau}\right]\mathbf{da} = \iiint\left[\Omega h_1^2\frac{\partial\left(\delta\xi_1\right)}{\partial\tau}\right]\mathbf{da}.
\end{aligned}\tag{14.211}$$

Using (14.211) in (14.192), integrating by parts, and using constraints (14.154), definition (14.147) of $\mathbf{u}$, and useful relation (14.196), the corresponding variation of $\mathcal{A}^{(2)}$ with respect to $\delta\xi_1$ is then

$$\begin{aligned}
\delta_{\xi_1}\mathcal{A}^{(2)} &= \int_{\tau_1}^{\tau_2}\delta_{\xi_1}L^{(2)}d\tau = \int_{\tau_1}^{\tau_2}\left\{\iiint\left[\Omega h_1^2\frac{\partial\left(\delta\xi_1\right)}{\partial\tau}\right]\mathbf{da}\right\}d\tau \\
&= \cancelto{0}{\left\{\iiint\Omega\left[h_1^2(\delta\xi_1)\right]\mathbf{da}\right\}_{\tau_1}^{\tau_2}}-\int_{\tau_1}^{\tau_2}\left\{\iiint\left[2\Omega h_1\frac{\partial h_1}{\partial\tau}(\delta\xi_1)\right]\mathbf{da}\right\}d\tau \\
&= -\int_{\tau_1}^{\tau_2}\left[\iiint\left(2\Omega\frac{u_2}{h_2}\frac{\partial h_1}{\partial\xi_2}+2\Omega\frac{u_3}{h_3}\frac{\partial h_1}{\partial\xi_3}\right)h_1\delta\xi_1\mathbf{da}\right]d\tau.
\end{aligned}\tag{14.212}$$

This variation is the second contribution to $\delta_{\xi_1}\mathcal{A}$; see (14.155).

Variations of $\mathcal{A}^{(2)}$ with Respect to $\delta\xi_2$

Using definition (14.188) of $L^{(2)}$ and definition (14.147) of **u**, variation of $L^{(2)}$ with respect to $\delta\xi_2$ is given by

$$\begin{aligned}\delta_{\xi_2} L^{(2)} &= \delta_{\xi_2} \iiint \left(\Omega h_1^2 \frac{\partial \xi_1}{\partial \tau}\right) \mathbf{da} = \iiint \delta_{\xi_2} \left(\Omega h_1^2 \frac{\partial \xi_1}{\partial \tau}\right) \mathbf{da} = \iiint \left[2\Omega h_1 \frac{\partial h_1}{\partial \xi_2} \frac{\partial \xi_1}{\partial \tau} (\delta\xi_2)\right] \mathbf{da} \\ &= \iiint \left[2\Omega \frac{u_1}{h_2} \frac{\partial h_1}{\partial \xi_2} (h_2 \delta\xi_2)\right] \mathbf{da}.\end{aligned} \tag{14.213}$$

Using (14.213) in (14.192), the corresponding variation of $\mathcal{A}^{(2)}$ with respect to $\delta\xi_2$ is then

$$\delta_{\xi_2} \mathcal{A}^{(2)} = \int_{\tau_1}^{\tau_2} \delta_{\xi_2} L^{(2)} d\tau = \int_{\tau_1}^{\tau_2} \left[\iiint \left(2\Omega \frac{u_1}{h_2} \frac{\partial h_1}{\partial \xi_2}\right) h_2 \delta\xi_2 \mathbf{da}\right] d\tau. \tag{14.214}$$

This variation is the second contribution to $\delta_{\xi_2}\mathcal{A}$; see (14.156).

Variations of $\mathcal{A}^{(2)}$ with Respect to $\delta\xi_3$

Using definition (14.188) of $L^{(2)}$ and definition (14.147) of **u**, variation of $L^{(2)}$ with respect to $\delta\xi_3$ is given by

$$\begin{aligned}\delta_{\xi_3} L^{(2)} &= \delta_{\xi_3} \iiint \left(\Omega h_1^2 \frac{\partial \xi_1}{\partial \tau}\right) \mathbf{da} = \iiint \Omega \delta_{\xi_3} \left(h_1^2 \frac{\partial \xi_1}{\partial \tau}\right) \mathbf{da} = \iiint \left[2\Omega h_1 \frac{\partial h_1}{\partial \xi_3} \frac{\partial \xi_1}{\partial \tau} (\delta\xi_3)\right] \mathbf{da} \\ &= \iiint \left[2\Omega \frac{u_1}{h_3} \frac{\partial h_1}{\partial \xi_3} (h_3 \delta\xi_3)\right] \mathbf{da}.\end{aligned} \tag{14.215}$$

Using (14.215) in (14.192), the corresponding variation of $\mathcal{A}^{(2)}$ with respect to $\delta\xi_3$ is then

$$\delta_{\xi_3} \mathcal{A}^{(2)} = \int_{\tau_1}^{\tau_2} \delta_{\xi_3} L^{(2)} d\tau = \int_{\tau_1}^{\tau_2} \left[\iiint \left(2\Omega \frac{u_1}{h_3} \frac{\partial h_1}{\partial \xi_3}\right) h_3 \delta\xi_3 \mathbf{da}\right] d\tau. \tag{14.216}$$

This variation is the second contribution to $\delta_{\xi_3}\mathcal{A}$; see (14.157).

Variations of $\mathcal{A}^{(3)}$

From (14.137), the variation of $\delta L^{(3)}$ with respect to $\delta\mathbf{r}$ is deduced to be

$$\delta L^{(3)} = -\iiint \left[\left(\frac{1}{\rho}\nabla p + \nabla\Phi\right) \cdot \delta\mathbf{r}\right] \mathbf{da} + \oint\!\!\oint_{\mathcal{S}} (p\delta\mathbf{r} \cdot \mathbf{n})\, d\mathcal{S}, \tag{14.217}$$

and, from (14.140), the corresponding variation of $\mathcal{A}^{(3)}$ is

$$\delta\mathcal{A}^{(3)} = -\int_{\tau_1}^{\tau_2} \left\{\iiint \left[\left(\frac{1}{\rho}\nabla p + \nabla\Phi\right) \cdot \delta\mathbf{r}\right] \mathbf{da} - \oint\!\!\oint_{\mathcal{S}} (p\delta\mathbf{r} \cdot \mathbf{n})\, d\mathcal{S}\right\} d\tau. \tag{14.218}$$

Now from (5.6) and (5.2),

$$\nabla f = \frac{1}{h_1}\frac{\partial f}{\partial \xi_1}\mathbf{e}_1 + \frac{1}{h_2}\frac{\partial f}{\partial \xi_2}\mathbf{e}_2 + \frac{1}{h_3}\frac{\partial f}{\partial \xi_3}\mathbf{e}_3, \tag{14.219}$$

$$\delta\mathbf{r} = h_1\delta\xi_1\mathbf{e}_1 + h_2\delta\xi_2\mathbf{e}_2 + h_3\delta\xi_3\mathbf{e}_3. \tag{14.220}$$

Inserting (14.219) and (14.220) into (14.218), and ignoring boundary conditions, leads to

$$\delta\mathcal{A}^{(3)} \equiv \delta_{\xi_1}\mathcal{A}^{(3)} + \delta_{\xi_2}\mathcal{A}^{(3)} + \delta_{\xi_3}\mathcal{A}^{(3)} = -\int_{\tau_1}^{\tau_2} \left[\iiint \left(\frac{1}{\rho}\nabla p + \nabla\Phi\right) \cdot (\delta\mathbf{r})\, \mathbf{da}\right] d\tau. \tag{14.221}$$

Decomposing (14.221) into its components using (14.219) and (14.220) and noting (from (14.150), (14.152), and (14.189)) that $\Phi = \Phi(\xi_2, \xi_3)$ then gives

$$\delta_{\xi_1}\mathcal{A}^{(3)} = -\int_{\tau_1}^{\tau_2}\left[\iiint\left(\frac{1}{\rho h_1}\frac{\partial p}{\partial \xi_1}\right)h_1\delta\xi_1 d\mathbf{a}\right]d\tau, \tag{14.222}$$

$$\delta_{\xi_2}\mathcal{A}^{(3)} = -\int_{\tau_1}^{\tau_2}\left[\iiint\left(\frac{1}{\rho h_2}\frac{\partial p}{\partial \xi_2} + \frac{1}{h_2}\frac{\partial \Phi}{\partial \xi_2}\right)h_2\delta\xi_2 d\mathbf{a}\right]d\tau, \tag{14.223}$$

$$\delta_{\xi_3}\mathcal{A}^{(3)} = -\int_{\tau_1}^{\tau_2}\left[\iiint\left(\frac{1}{\rho h_3}\frac{\partial p}{\partial \xi_3} + \frac{1}{h_3}\frac{\partial \Phi}{\partial \xi_3}\right)h_3\delta\xi_3 d\mathbf{a}\right]d\tau. \tag{14.224}$$

For a *geopotential coordinate system* (for which, by definition, $\Phi = \Phi(\xi_3)$), $\partial\Phi/\partial\xi_2 \equiv 0$ and the term $(1/h_2)(\partial\Phi/\partial\xi_2)$ in (14.223) is then absent.

Variations (14.222), (14.223), and (14.224) are the third contributions to $\delta_{\xi_1}\mathcal{A}$, $\delta_{\xi_2}\mathcal{A}$, and $\delta_{\xi_3}\mathcal{A}$, respectively; see (14.155), (14.156), and (14.157), respectively.

15

Conservation

ABSTRACT

Respect of conservation principles for various fundamental physical quantities is of crucial importance for weather and climate prediction. These principles may be expressed in different forms: Lagrangian or Eulerian; local or global; and using vectors or a coordinate system. The governing equations for motion of a planet's atmosphere and oceans are first summarised using vectors. From these equations, conservation principles for mass, axial angular momentum, total energy, and potential vorticity are derived and expressed in Lagrangian, Eulerian, local, and global forms. These principles, in their various forms, are then re-expressed in axial-orthogonal-curvilinear coordinates. This review of various ways of expressing conservation principles prepares the way for (a) construction of budget equations, to better understand the evolution of the atmosphere and oceans during model integrations; and (b) examination of the existence (or otherwise) of analogous conservation principles for mass, axial angular momentum, total energy, and potential vorticity for simplified sets of governing equations. Finally, a complementary way of examining conservation properties of a set of governing equations, using Noether's theorem and rotation, time-translation, and particle-relabelling symmetries of a Lagrangian, is summarised.

15.1 PREAMBLE

The governing equations for global atmospheric and oceanic modelling (see Chapters 2–4, 6, and 14) are based on fundamental physical conservation laws for mass, momentum, and total energy. An important – and rather surprising – consequence of these laws in the present context is the conservation of *potential vorticity* within a closed fluid parcel during Lagrangian transport by its enveloping flow. This conservation law – derived in Section 15.3.6 – is unique to fluid dynamics and has no analogue in particle dynamics (Salmon, 1998).

Since conservation principles are crucially important for weather and climate prediction, we now review them to:

1. Prepare the way for constructing budget equations to help improve understanding of the evolution of the atmosphere and oceans during model integrations.
2. Set the scene for Chapters 16–19 on the existence (or otherwise) of conservation principles for mass, axial angular momentum, total energy, and potential vorticity of various simplified sets of governing equations. An equation set that respects all four principles is – by the definition of Section 16.1.2 – dynamically consistent.

Strictly speaking, a quantity is only conserved if it is precisely constant during some time interval. In the presence of certain forcings, the quantity that would otherwise be conserved in their absence varies in time (i.e. its time tendency is *non-zero* instead of zero). The equation governing the evolution of the quantity is then not a *conservation* equation, but rather a *budget* equation that describes how the quantity varies in time due to prescribed forcing. In practice, however, many authors consider such an equation to nevertheless express a conservation principle, it being understood that conservation only precisely holds in the absence of forcing. Differences in authors' use of nomenclature is something to be aware of, but the meaning is usually clear from the context.

As will be seen, conservation principles may be expressed in many different ways; for example, Lagrangian versus Eulerian, local versus global, in vector form versus in a coordinate system (such as axial-orthogonal-curvilinear coordinates), and in the absence (as opposed to presence) of certain types of forcing.

The content of the remainder of this chapter, as a function of section, is:

15.2 Summary of the governing equations in vector form.
15.3 Derivation from these equations of conservation principles, expressed in *vector form*, for mass, axial angular momentum, total energy, and potential vorticity.
15.4 Expression of these principles in *axial-orthogonal-curvilinear coordinates*.
15.5 Use of Noether's theorem to relate rotation, time-translation, and particle-relabelling symmetries of a Lagrangian to conservation of axial angular momentum, total energy, and potential vorticity, respectively.
15.6 Concluding remarks.

15.2 GOVERNING EQUATIONS

A set of governing equations for general fluids, expressed in the notation of Chapter 4, and in vector form in a rotating frame of reference – see (4.95)–(4.102), (2.21), and (2.22) – may be written as follows:

Momentum

$$\frac{D\mathbf{u}}{Dt} = -2\boldsymbol{\Omega} \times \mathbf{u} - \alpha\nabla p - \nabla\Phi + \mathbf{F}. \tag{15.1}$$

Continuity of Total Mass Density

$$\frac{D\rho}{Dt} + \rho\nabla\cdot\mathbf{u} = F^{\rho}. \tag{15.2}$$

Substance Transport

$$\frac{DS^i}{Dt} = \dot{S}^i, \quad i = 1, 2, \ldots \tag{15.3}$$

Thermodynamic-Energy Equation

$$\frac{D\mathscr{E}}{Dt} + p\frac{D\alpha}{Dt} = \dot{Q}_E \qquad \text{or} \qquad \frac{D\eta}{Dt} = \frac{\dot{Q}}{T}. \tag{15.4}$$

Fundamental Equation of State

$$\mathscr{E} = \mathscr{E}\left(\alpha, \eta, S^1, S^2, \ldots\right) \qquad \text{or} \qquad \eta = \eta\left(\alpha, \mathscr{E}, S^1, S^2, \ldots\right). \tag{15.5}$$

Definition of Specific Volume

$$\alpha \equiv 1/\rho. \tag{15.6}$$

Definition of T, p, *and* μ^i

$$T = \frac{\partial \mathscr{E}\left(\alpha, \eta, S^1, S^2, \ldots\right)}{\partial \eta} \quad \text{or} \quad T = \left[\frac{\partial \eta\left(\alpha, \mathscr{E}, S^1, S^2, \ldots\right)}{\partial \mathscr{E}}\right]^{-1}, \tag{15.7}$$

$$p = -\frac{\partial \mathscr{E}\left(\alpha, \eta, S^1, S^2, \ldots\right)}{\partial \alpha} \quad \text{or} \quad p = T\frac{\partial \eta\left(\alpha, \mathscr{E}, S^1, S^2, \ldots\right)}{\partial \alpha}, \tag{15.8}$$

$$\mu^i = \frac{\partial \mathscr{E}\left(\alpha, \eta, S^1, S^2, \ldots\right)}{\partial S^i} \quad \text{or} \quad \mu^i = -T\frac{\partial \eta\left(\alpha, \mathscr{E}, S^1, S^2, \ldots\right)}{\partial S^i}. \tag{15.9}$$

Kinematic Equation

$$\mathbf{u} = \frac{D\mathbf{r}}{Dt}. \tag{15.10}$$

Material Derivative

$$\frac{D}{Dt} = \frac{\partial}{\partial t} + \mathbf{u} \cdot \nabla. \tag{15.11}$$

In these equations:

- S^i and $\dot{S}^i$ are as in (4.10) (for the atmosphere), and (4.11) (for oceans).
- $\dot{Q}$ is heating rate (per unit mass).
- $\dot{Q}_E \equiv \dot{Q} + \sum_i \mu^i \dot{S}^i$ is total rate of energy input (per unit mass).

15.3 CONSERVATION PRINCIPLES: VECTOR FORM

From (15.1)–(15.11), various useful equations and conservation laws may be deduced; see, for example, White et al. (2005) in the atmospheric context. Standard vector identities – see the Appendix on Vector Identities at the end of this book – facilitate the developments described herein.

15.3.1 Mass

Setting $F^\rho \equiv 0$ in mass-continuity equation (15.2) reduces it to

$$\frac{D\rho}{Dt} + \rho \nabla \cdot \mathbf{u} = 0. \tag{15.12}$$

This corresponds to a Lagrangian expression for the conservation of mass. To see this more readily, recall from (2.35) of Chapter 2 that, using Reynolds's transport theorem (A.53), (15.12) was derived from

$$\frac{D}{Dt} \iiint_V \rho dV = \iiint_V \left(\frac{D\rho}{Dt} + \rho \nabla \cdot \mathbf{u}\right) dV = 0. \tag{15.13}$$

This equation expresses conservation of the mass $\iiint_V \rho dV$ within a closed fluid parcel of volume V (with, by definition, no exchange of mass across the parcel's bounding surface), transported by its enveloping fluid environment. For an open parcel (with mass exchange across the parcel's bounding surface), a mass exchange (source when positive/sink when negative) term (F^ρ) is added to the right-hand side of (15.12) to obtain (4.96), that is, (15.2).

With use of vector identity (A.17), (15.2) can be rewritten in Eulerian form as

$$\frac{\partial \rho}{\partial t} + \nabla \cdot (\rho \mathbf{u}) = F^\rho. \tag{15.14}$$

This equation is written in terms of the fluxes of momentum ($\rho\mathbf{u}$), across the bounding surface of a volume *fixed in space*. As such, it is particularly useful for numerical models of the atmosphere and oceans discretised in an Eulerian manner. However, (15.2) is arguably the more natural form of the mass-continuity equation compared to (15.14), since its individual terms are all *frame invariant*. This is not so for (15.14) because $\partial\rho/\partial t$ is not frame invariant; see (2.21) of Vallis (2017).

15.3.2 Angular Momentum

Taking the vector product of position vector ($\mathbf{r}$) with momentum equation (15.1) gives

$$\mathbf{r}\times\frac{D\mathbf{u}}{Dt}=\mathbf{r}\times\left(\mathbf{F}-2\boldsymbol{\Omega}\times\mathbf{u}-\alpha\nabla p-\nabla\Phi\right). \tag{15.15}$$

Now the governing equations in Section 15.2 are expressed in a rotating frame of reference. The fixed origin for $\mathbf{r}$ is therefore assumed to lie on the frame's rotation axis. This origin is usually located at a planet's centre of mass; see Section 2.2.2.3. Using kinematic equation (15.10), together with vector identities (A.4) and (A.44), the left-hand side of (15.15) may be rewritten as

$$\mathbf{r}\times\frac{D\mathbf{u}}{Dt}=\frac{D\left(\mathbf{r}\times\mathbf{u}\right)}{Dt}-\frac{D\mathbf{r}}{Dt}\times\mathbf{u}=\frac{D\left(\mathbf{r}\times\mathbf{u}\right)}{Dt}-\overbrace{\mathbf{u}\times\mathbf{u}}^{\equiv 0}=\frac{D\left(\mathbf{r}\times\mathbf{u}\right)}{Dt}. \tag{15.16}$$

Substitution of this equation into (15.15) then yields:

The Budget Equation (Lagrangian Form) for Angular Momentum

$$\frac{D}{Dt}\left(\mathbf{r}\times\mathbf{u}\right)=\mathbf{r}\times\left(\mathbf{F}-2\boldsymbol{\Omega}\times\mathbf{u}-\alpha\nabla p-\nabla\Phi\right), \tag{15.17}$$

where $\mathbf{r}\times\mathbf{u}$ is *specific relative angular momentum* (i.e. relative angular momentum per unit mass) about the fixed origin for $\mathbf{r}$.

15.3.3 Axial Absolute Angular Momentum

Forming the scalar product of (15.17) with unit vector $\widehat{\boldsymbol{\Omega}}\equiv\boldsymbol{\Omega}/\left|\boldsymbol{\Omega}\right|$ (aligned with the planet's rotation vector $\boldsymbol{\Omega}$, assumed constant) and using kinematic condition (15.10) and material-derivative identity (A.46) results in

$$\begin{aligned}\frac{D}{Dt}\left[\widehat{\boldsymbol{\Omega}}\cdot\left(\mathbf{r}\times\mathbf{u}\right)\right]&=\widehat{\boldsymbol{\Omega}}\cdot\frac{D}{Dt}\left(\mathbf{r}\times\mathbf{u}\right)=\widehat{\boldsymbol{\Omega}}\cdot\left[\mathbf{r}\times\left(\mathbf{F}-\alpha\nabla p-\nabla\Phi\right)\right]-\widehat{\boldsymbol{\Omega}}\cdot\left[\mathbf{r}\times\left(2\boldsymbol{\Omega}\times\mathbf{u}\right)\right]\\&=\widehat{\boldsymbol{\Omega}}\cdot\left[\mathbf{r}\times\left(\mathbf{F}-\alpha\nabla p-\nabla\Phi\right)\right]-\widehat{\boldsymbol{\Omega}}\cdot\left[\mathbf{r}\times\left(2\boldsymbol{\Omega}\times\frac{D\mathbf{r}}{Dt}\right)\right].\end{aligned} \tag{15.18}$$

Now

$$\begin{aligned}\frac{D}{Dt}\left[\mathbf{r}\times\left(\boldsymbol{\Omega}\times\mathbf{r}\right)\right]&=\frac{D\mathbf{r}}{Dt}\times\left(\boldsymbol{\Omega}\times\mathbf{r}\right)+\left[\mathbf{r}\times\left(\cancelto{0}{\frac{D\boldsymbol{\Omega}}{Dt}}\times\mathbf{r}\right)\right]+\mathbf{r}\times\left(\boldsymbol{\Omega}\times\frac{D\mathbf{r}}{Dt}\right)\\&=-\boldsymbol{\Omega}\times\left(\mathbf{r}\times\frac{D\mathbf{r}}{Dt}\right)-\mathbf{r}\times\left(\frac{D\mathbf{r}}{Dt}\times\boldsymbol{\Omega}\right)+\mathbf{r}\times\left(\boldsymbol{\Omega}\times\frac{D\mathbf{r}}{Dt}\right)\\&=-\boldsymbol{\Omega}\times\left(\mathbf{r}\times\frac{D\mathbf{r}}{Dt}\right)+\mathbf{r}\times\left(\boldsymbol{\Omega}\times\frac{D\mathbf{r}}{Dt}\right)+\mathbf{r}\times\left(\boldsymbol{\Omega}\times\frac{D\mathbf{r}}{Dt}\right)\end{aligned}$$

$$= -\boldsymbol{\Omega} \times \left(\mathbf{r} \times \frac{D\mathbf{r}}{Dt} \right) + \mathbf{r} \times \left(2\boldsymbol{\Omega} \times \frac{D\mathbf{r}}{Dt} \right), \tag{15.19}$$

where use has been made of vector identities (A.50), (A.11), and (A.3). Taking the scalar product of (15.19) with unit vector $\widehat{\boldsymbol{\Omega}} \equiv \boldsymbol{\Omega}/|\boldsymbol{\Omega}|$ then gives

$$\begin{aligned} \widehat{\boldsymbol{\Omega}} \cdot \frac{D}{Dt} [\mathbf{r} \times (\boldsymbol{\Omega} \times \mathbf{r})] &= -\widehat{\boldsymbol{\Omega}} \cdot \left[\boldsymbol{\Omega} \times \left(\mathbf{r} \times \frac{D\mathbf{r}}{Dt} \right) \right]^{\to 0} + \widehat{\boldsymbol{\Omega}} \cdot \left[\mathbf{r} \times \left(2\boldsymbol{\Omega} \times \frac{D\mathbf{r}}{Dt} \right) \right] \\ &= \widehat{\boldsymbol{\Omega}} \cdot \left[\mathbf{r} \times \left(2\boldsymbol{\Omega} \times \frac{D\mathbf{r}}{Dt} \right) \right], \end{aligned} \tag{15.20}$$

where the crossed-out term is zero by virtue of vector identity (A.9) and the fact that $\widehat{\boldsymbol{\Omega}}$ and $\boldsymbol{\Omega}$ point in the same direction. Eliminating $\widehat{\boldsymbol{\Omega}} \cdot [\mathbf{r} \times (2\boldsymbol{\Omega} \times D\mathbf{r}/Dt)]$ between (15.18) and (15.20) and noting that $\widehat{\boldsymbol{\Omega}}$ is a constant vector delivers

$$\begin{aligned} \frac{D}{Dt} \left[\widehat{\boldsymbol{\Omega}} \cdot (\mathbf{r} \times \mathbf{u}) \right] &= \widehat{\boldsymbol{\Omega}} \cdot \left[\mathbf{r} \times \left(\mathbf{F} - \alpha \nabla p - \nabla \Phi \right) \right] - \widehat{\boldsymbol{\Omega}} \cdot \frac{D}{Dt} [\mathbf{r} \times (\boldsymbol{\Omega} \times \mathbf{r})] \\ &= \widehat{\boldsymbol{\Omega}} \cdot \left[\mathbf{r} \times \left(\mathbf{F} - \alpha \nabla p - \nabla \Phi \right) \right] - \frac{D}{Dt} \left\{ \widehat{\boldsymbol{\Omega}} \cdot [\mathbf{r} \times (\boldsymbol{\Omega} \times \mathbf{r})] \right\}. \end{aligned} \tag{15.21}$$

Rearrangement of this equation, after multiplication by ρ and use of definition (15.6), then yields

$$\rho \frac{D}{Dt} \left\{ \widehat{\boldsymbol{\Omega}} \cdot [\mathbf{r} \times (\mathbf{u} + \boldsymbol{\Omega} \times \mathbf{r})] \right\} = \widehat{\boldsymbol{\Omega}} \cdot \left[\mathbf{r} \times \left(\rho \mathbf{F} - \nabla p - \rho \nabla \Phi \right) \right]. \tag{15.22}$$

Using vector identity (A.8) for the triple scalar product in (15.22) finally yields:

The Budget Equation (Lagrangian Form) for Axial Angular Momentum

$$\rho \frac{DM}{Dt} = \widehat{\boldsymbol{\Omega}} \cdot \left[\mathbf{r} \times \left(\rho \mathbf{F} - \nabla p - \rho \nabla \Phi \right) \right] = \left(\rho \mathbf{F} - \nabla p - \rho \nabla \Phi \right) \cdot \left(\widehat{\boldsymbol{\Omega}} \times \mathbf{r} \right), \tag{15.23}$$

where

$$M \equiv \widehat{\boldsymbol{\Omega}} \cdot [\mathbf{r} \times (\mathbf{u} + \boldsymbol{\Omega} \times \mathbf{r})] \equiv (\mathbf{u} + \boldsymbol{\Omega} \times \mathbf{r}) \cdot \left(\widehat{\boldsymbol{\Omega}} \times \mathbf{r} \right), \tag{15.24}$$

is *specific axial absolute angular momentum* (i.e. axial absolute angular momentum per unit mass) about rotation axis $\widehat{\boldsymbol{\Omega}}$.

Axial-angular-momentum equation (15.23) could have been directly derived in an inertial frame of reference. It is reassuring that transforming to a rotating frame of reference and then working in this frame of reference recovers the original conservation principle for axial absolute angular momentum.

For an axially symmetric planet – see Chapters 7, 8, and 12 – $\nabla \Phi$ at a point lies in the meridional plane going through that point. Then, since the origin for $\mathbf{r}$ is assumed to lie on the planet's rotation axis, $\mathbf{r} \times \nabla \Phi$ is normal to this meridional plane. It is therefore normal to $\widehat{\boldsymbol{\Omega}}$, since this is aligned with the planet's rotation axis. Thus $\widehat{\boldsymbol{\Omega}} \cdot (\mathbf{r} \times \nabla \Phi) = 0$, and (15.23) reduces to

The Budget Equation (Lagrangian Form) for Axial Angular Momentum over an Axially Symmetric Planet

$$\rho\frac{DM}{Dt} = \widehat{\boldsymbol{\Omega}} \cdot \left[\mathbf{r} \times \left(\rho\mathbf{F} - \nabla p\right)\right] = \left(\rho\mathbf{F} - \nabla p\right) \cdot \left(\widehat{\boldsymbol{\Omega}} \times \mathbf{r}\right). \tag{15.25}$$

[For a planet that is not perfectly axially symmetric, the influence of orography (for the atmosphere) and of bathymetry (for an ocean) on conservation of axial absolute angular momentum is discussed in Section 13.6.]

In the absence of the physical forcing term $\mathbf{F}$, (15.25) further reduces to

The Budget Equation (Lagrangian Form) for Axial Angular Momentum over an Axially Symmetric Planet in the Absence of Physical Forcing

$$\rho\frac{DM}{Dt} = -\widehat{\boldsymbol{\Omega}} \cdot \left(\mathbf{r} \times \nabla p\right) = -\nabla p \cdot \left(\widehat{\boldsymbol{\Omega}} \times \mathbf{r}\right). \tag{15.26}$$

Note, though, that in realistic atmospheric and oceanic models, the forcing term $\mathbf{F}$, which includes important frictional effects, cannot be neglected.

The right-hand side of (15.26) is the *pressure-gradient torque*, per unit mass, about the planet's rotation axis. Since $\widehat{\boldsymbol{\Omega}}$ and $\mathbf{r}$ both lie in a meridional plane, $\widehat{\boldsymbol{\Omega}} \times \mathbf{r}$ is normal to the meridional plane and points in the zonal direction (usually denoted by $\mathbf{e}_\lambda$) of an axisymmetric coordinate system (obtained by rotating a 2D coordinate system about an axis). This means that it is only the *zonal* component of ∇p (usually denoted by $\nabla p \cdot \mathbf{e}_\lambda$) that contributes to the term $-\nabla p \cdot \left(\widehat{\boldsymbol{\Omega}} \times \mathbf{r}\right)$ appearing on the right-hand side of (15.26), and also in the two preceding equations. When integrated globally, as in Section 15.3.8, $-\nabla p \cdot \left(\widehat{\boldsymbol{\Omega}} \times \mathbf{r}\right)$ will then be zero due to periodicity of dependent variables in the zonal (λ) direction. Whilst the pressure-gradient torque term $-\nabla p \cdot \left(\widehat{\boldsymbol{\Omega}} \times \mathbf{r}\right)$ is non-zero *locally*, it does not contribute to the *global* axial-absolute-angular-momentum budget.

15.3.4 Absolute Vorticity

As shown in Section 2.2.2.4 of Chapter 2, momentum equation (15.1) may be equivalently written in vector-invariant form – see (2.34) – as

$$\frac{\partial \mathbf{u}}{\partial t} = -\mathbf{Z} \times \mathbf{u} - \nabla\left(K + \Phi\right) - \alpha\nabla p + \mathbf{F}, \tag{15.27}$$

where

$$\mathbf{Z} \equiv 2\boldsymbol{\Omega} + \nabla \times \mathbf{u} \equiv 2\boldsymbol{\Omega} + \boldsymbol{\zeta} \tag{15.28}$$

is *absolute vorticity* (i.e. the sum of *planetary vorticity*, $2\boldsymbol{\Omega}$, and *relative vorticity*, $\boldsymbol{\zeta} \equiv \nabla \times \mathbf{u}$), and

$$K \equiv \frac{\mathbf{u} \cdot \mathbf{u}}{2} \tag{15.29}$$

is *kinetic energy*. Both $\mathbf{Z}$ and K are specific quantities (i.e. amounts per unit mass).

Taking the curl of (15.27) – this is equivalent to taking the curl of (15.1) directly, so do this instead if you prefer – and using (15.28) together with vector identities (A.27), (A.22), and (A.26) then yields

$$
\begin{aligned}
\frac{\partial \zeta}{\partial t} &= -\nabla \times (\mathbf{Z} \times \mathbf{u}) - \nabla \times \cancelto{0}{[\nabla (K + \Phi)]} - \nabla \times (\alpha \nabla p) + \nabla \times \mathbf{F} \\
&= -\mathbf{Z}(\nabla \cdot \mathbf{u}) + \cancelto{0}{\mathbf{u}(\nabla \cdot \mathbf{Z})} - (\mathbf{u} \cdot \nabla)\,\mathbf{Z} + (\mathbf{Z} \cdot \nabla)\,\mathbf{u} - \alpha \cancelto{0}{\nabla \times (\nabla p)} - \nabla \alpha \times \nabla p + \nabla \times \mathbf{F}.
\end{aligned}
\tag{15.30}
$$

Rearranging (15.30) and noting that $D(2\boldsymbol{\Omega})/Dt = 0$ (since $\boldsymbol{\Omega}$ is a constant vector) finally leads to:

The Budget Equation (Lagrangian Form) for Absolute Vorticity

$$
\frac{D\mathbf{Z}}{Dt} = -\mathbf{Z}(\nabla \cdot \mathbf{u}) + (\mathbf{Z} \cdot \nabla)\,\mathbf{u} - \nabla \alpha \times \nabla p + \nabla \times \mathbf{F}, \tag{15.31}
$$

where $\mathbf{Z} \equiv 2\boldsymbol{\Omega} + \nabla \times \mathbf{u}$ is *specific absolute vorticity*, and $\alpha \equiv 1/\rho$ is specific volume.

The first three terms on the right-hand side of (15.31) are respectively known as:

- The *expansion/contraction* term.
- The *vortex tilting/stretching* term.
- The *solenoidal* (or *baroclinic generation*, or *non-homentropic*) term.

See Sections 8.4–8.6 of Hoskins and James (2014) and Section 4.3 of Vallis (2017) for physical interpretations of these terms.

15.3.5 Total Energy

Taking the scalar product of momentum equation (15.1) with $\rho\mathbf{u}$ and using vector identity (A.9) gives

$$
\rho\mathbf{u} \cdot \frac{D\mathbf{u}}{Dt} + \rho\mathbf{u} \cdot \nabla \Phi = -2\rho \cancelto{0}{\mathbf{u} \cdot (\boldsymbol{\Omega} \times \mathbf{u})} - \mathbf{u} \cdot \nabla p + \rho\mathbf{u} \cdot \mathbf{F} = -\mathbf{u} \cdot \nabla p + \rho\mathbf{u} \cdot \mathbf{F}. \tag{15.32}
$$

Using vector identity (A.17) in (15.32) and noting that $D\Phi/Dt \equiv \partial\Phi/\partial t + \mathbf{u} \cdot \nabla \Phi = \mathbf{u} \cdot \nabla \Phi$ (since $\Phi = \Phi(\mathbf{r})$ is independent of time) then leads to:

The Budget Equation (Lagrangian Form) for Kinetic + Gravitational Potential Energies $(K + \Phi)$

$$
\rho \frac{D}{Dt}(K + \Phi) = -\nabla \cdot (p\mathbf{u}) + p\nabla \cdot \mathbf{u} + \rho\mathbf{u} \cdot \mathbf{F}, \tag{15.33}
$$

where $K \equiv \mathbf{u} \cdot \mathbf{u}/2$ is specific kinetic energy, and Φ is specific gravitational potential energy.

Multiplication of internal-energy equation (15.4) by ρ, followed by use of mass-continuity equation (15.2), yields:

The Budget Equation (Lagrangian Form) for Internal Energy ($\mathscr{E}$)

$$\rho \frac{D\mathscr{E}}{Dt} = -\rho p \frac{D}{Dt}\left(\frac{1}{\rho}\right) + \rho \dot{Q}_E = \frac{p}{\rho}\frac{D\rho}{Dt} + \rho \dot{Q}_E = \frac{p}{\rho}\left(-\rho\nabla\cdot\mathbf{u} + F^{\rho}\right) + \rho\dot{Q}_E$$
$$= -p\nabla\cdot\mathbf{u} + \rho\dot{Q}_E + \frac{p}{\rho}F^{\rho}, \tag{15.34}$$

where $\mathscr{E}$ is specific internal energy.

Summing budget equations (15.33) and (15.34) and cancelling the $p\nabla\cdot\mathbf{u}$ terms of opposite sign finally delivers:

The Budget Equation (Lagrangian Form) for Total Energy ($E = K + \Phi + \mathscr{E}$)

$$\rho\frac{DE}{Dt} + \nabla\cdot(p\mathbf{u}) = \rho\left(\mathbf{u}\cdot\mathbf{F} + \dot{Q}_E\right) + \frac{p}{\rho}F^{\rho}, \tag{15.35}$$

where

$$E \equiv K + \Phi + \mathcal{E}, \tag{15.36}$$

is specific total energy, composed of the sum of specific kinetic (K), specific gravitational-potential (Φ), and specific internal ($\mathcal{E}$) energies, respectively.

15.3.6 Potential Vorticity

15.3.6.1 Derivation

Dividing absolute-vorticity equation (15.31) by ρ, eliminating $\nabla\cdot\mathbf{u}$ using mass-continuity equation (15.2), and using definition (15.6) of specific volume, leads to

$$\frac{1}{\rho}\frac{D\mathbf{Z}}{Dt} = -\frac{\mathbf{Z}}{\rho}\left(\frac{F^{\rho}}{\rho} - \frac{1}{\rho}\frac{D\rho}{Dt}\right) + \left(\frac{\mathbf{Z}}{\rho}\cdot\nabla\right)\mathbf{u} - \frac{1}{\rho}\nabla\left(\frac{1}{\rho}\right)\times\nabla p + \frac{\nabla\times\mathbf{F}}{\rho}$$
$$= -\frac{F^{\rho}\mathbf{Z}}{\rho^2} - \frac{D}{Dt}\left(\frac{1}{\rho}\right)\mathbf{Z} + \left(\frac{\mathbf{Z}}{\rho}\cdot\nabla\right)\mathbf{u} + \frac{1}{\rho^3}\nabla\rho\times\nabla p + \frac{\nabla\times\mathbf{F}}{\rho}. \tag{15.37}$$

Rearrangement of this equation and use of material-derivative identity (A.48) then gives

$$\frac{D}{Dt}\left(\frac{\mathbf{Z}}{\rho}\right) = \left(\frac{\mathbf{Z}}{\rho}\cdot\nabla\right)\mathbf{u} + \frac{1}{\rho^3}\nabla\rho\times\nabla p + \frac{\nabla\times\mathbf{F}}{\rho} - \frac{F^{\rho}}{\rho^2}\mathbf{Z}, \tag{15.38}$$

where the second term on the right-hand side is known as the *baroclinicity vector*.[1]

Consider now the material-conservation equation

$$\frac{D\Lambda}{Dt} = F^{\Lambda}, \tag{15.39}$$

[1] For a *barotropic fluid*, $p = p(\rho)$ from (2.48). The baroclinicity vector is then identically zero and absent from (15.38).

where Λ is a scalar field (to be discussed later), and F^{Λ} is a physical-forcing term for this equation. Taking the scalar product of $\nabla\Lambda$ and (15.38) yields

$$\left[\frac{D}{Dt}\left(\frac{\mathbf{Z}}{\rho}\right)\right]\cdot\nabla\Lambda = \left[\left(\frac{\mathbf{Z}}{\rho}\cdot\nabla\right)\mathbf{u}\right]\cdot\nabla\Lambda + \frac{1}{\rho^3}\nabla\Lambda\cdot\left(\nabla\rho\times\nabla p\right) + \frac{\nabla\Lambda}{\rho}\cdot\left(\nabla\times\mathbf{F} - \frac{F^{\rho}}{\rho}\mathbf{Z}\right). \quad (15.40)$$

Now – from identity (A.51) – for arbitrary vector $\mathbf{Z}$ and arbitrary scalars ρ and Λ,

$$\frac{\mathbf{Z}}{\rho}\cdot\frac{D}{Dt}\left(\nabla\Lambda\right) \equiv \left(\frac{\mathbf{Z}}{\rho}\cdot\nabla\right)\frac{D\Lambda}{Dt} - \left[\left(\frac{\mathbf{Z}}{\rho}\cdot\nabla\right)\mathbf{u}\right]\cdot\nabla\Lambda. \quad (15.41)$$

Summing these last two equations and using (15.39) and material-derivative identity (A.46) then leads to:

The Budget Equation (Lagrangian Form) for Potential Vorticity

$$\frac{D\Pi}{Dt} = \frac{1}{\rho^3}\nabla\Lambda\cdot\left(\nabla\rho\times\nabla p\right) + \frac{\mathbf{Z}}{\rho}\cdot\nabla F^{\Lambda} + \frac{\nabla\Lambda}{\rho}\cdot\left(\nabla\times\mathbf{F} - \frac{F^{\rho}}{\rho}\mathbf{Z}\right), \quad (15.42)$$

where

$$\Pi \equiv \frac{\mathbf{Z}\cdot\nabla\Lambda}{\rho} \equiv \frac{(2\boldsymbol{\Omega}+\boldsymbol{\zeta})\cdot\nabla\Lambda}{\rho} \equiv \frac{(2\boldsymbol{\Omega}+\nabla\times\mathbf{u})\cdot\nabla\Lambda}{\rho} \quad (15.43)$$

is termed *potential vorticity.*

This is a general result, true for both a moist atmosphere and a saline ocean.

In the absence of any forcings (i.e. $\mathbf{F} = F^{\rho} = F^{\Lambda} = 0$), (15.42) reduces to:

The Budget Equation (Lagrangian Form) for Potential Vorticity in the Absence of Forcings

$$\frac{D\Pi}{Dt} = \frac{1}{\rho^3}\nabla\Lambda\cdot\left(\nabla\rho\times\nabla p\right). \quad (15.44)$$

Strictly speaking, all forcings do not have to be absent from (15.42) for (15.44) to hold. For example, a non-zero, irrotational forcing $\mathbf{F}$ of the momentum equation does not survive (15.42), albeit any other example is likely to be contrived.

The issue now is:

- Under what conditions will the right-hand side (the solenoidal/baroclinicity term) of (15.44) vanish, to give Ertel's theorem[2] for conservation of potential vorticity, Π?

This all boils down to the precise choice of the as-yet-unspecified scalar field Λ. Recall – from (15.39) with F^{Λ} set to zero and from (15.44) – that there are two constraints on Λ that need to be simultaneously satisfied for potential vorticity to be conserved in the absence of physical forcing:

1. $D\Lambda/Dt = 0$, for some appropriately chosen Λ.
2. $\nabla\Lambda\cdot\left(\nabla\rho\times\nabla p\right) = 0$.

[2] After Hans Ertel, 1904–1971. This theorem is also known as the Rossby–Ertel theorem, since Carl-Gustaf Rossby (1898–1957) had earlier developed a less general version.

Simultaneous satisfaction of these two constraints is challenging. There are a restricted number of possibilities to satisfy either of them individually, and an even more restricted number to do so simultaneously. Let us now examine this further.

For the first condition, we continue to leave the precise definition of Λ open for now, and simply assume that Λ does indeed satisfy $D\Lambda/Dt = 0$ in the absence of physical forcing. We will return to this point after having developed Ertel's potential-vorticity theorem under this assumption.

For the second condition, this is trivially satisfied if pressure depends *solely* on density (and vice versa), that is, if $p = p(\rho)$. Then (independently of Λ)

$$\nabla\rho \times \nabla p(\rho) = \nabla\rho \times \left(\frac{dp}{d\rho}\nabla\rho\right) = \frac{dp}{d\rho}\nabla\rho \times \nabla\rho = 0, \tag{15.45}$$

since $\nabla\rho \times \nabla\rho \equiv 0$ from vector identity (A.4). This particular case, however, is of somewhat limited interest since $p = p(\rho)$ is the definition of a *barotropic* (or *homentropic*) fluid – see (2.48). The general, more interesting case of a *baroclinic* (or *non-homentropic*) fluid is then excluded. Boo hoo.

Not to worry though, since $\nabla\Lambda \cdot (\nabla\rho \times \nabla p) = 0$ is also satisfied when $\nabla\Lambda$, $\nabla\rho$, and ∇p are all *coplanar* (i.e. when all three lie in the same plane)[3]. This is a much weaker constraint than the barotropic (homentropic) one, $p = p(\rho)$. So we need to determine under what conditions (if any) this can happen.

Now if Λ is a function of (*only*) ρ and p, that is, $\Lambda = \Lambda(\rho, p)$, then (by the chain rule)

$$\nabla\Lambda(\rho, p) = \frac{\partial\Lambda}{\partial\rho}\nabla\rho + \frac{\partial\Lambda}{\partial p}\nabla p. \tag{15.46}$$

Thus, using vector identity (A.9),

$$\begin{aligned}\nabla\Lambda \cdot (\nabla\rho \times \nabla p) &= \left(\frac{\partial\Lambda}{\partial\rho}\nabla\rho + \frac{\partial\Lambda}{\partial p}\nabla p\right) \cdot (\nabla\rho \times \nabla p) \\ &= \frac{\partial\Lambda}{\partial\rho}\cancelto{0}{\nabla\rho \cdot (\nabla\rho \times \nabla p)} + \frac{\partial\Lambda}{\partial p}\cancelto{0}{\nabla p \cdot (\nabla\rho \times \nabla p)} \\ &= 0,\end{aligned} \tag{15.47}$$

and Condition 2 (i.e. $\nabla\Lambda \cdot (\nabla\rho \times \nabla p) = 0$) is satisfied, in addition to Condition 1 (i.e. $D\Lambda/Dt = 0$, which is assumed). Success – the right-hand side of (15.44) is guaranteed to be zero whenever Λ is chosen to be a function of (*only*) density and pressure, that is, when $\Lambda = \Lambda(\rho, p)$! Thus the previous restriction that the fluid be barotropic – that is, that $p = p(\rho)$ – has been lifted. Under these conditions, (15.44) reduces to

$$\frac{D\Pi}{Dt} \equiv \frac{D}{Dt}\left(\frac{\mathbf{Z} \cdot \nabla\Lambda}{\rho}\right) = 0, \tag{15.48}$$

and potential vorticity ($\Pi \equiv \mathbf{Z} \cdot \nabla\Lambda/\rho$) is materially conserved by a fluid parcel.

This analysis can be summarised as:

[3] Geometrically, $|\mathbf{A} \cdot \mathbf{B} \times \mathbf{C}|$ is the volume of a parallelepiped formed by vectors **A**, **B**, and **C**; see, for example, Spiegel and Lipschutz (2009). If these three vectors lie in the same plane, then the volume degenerates to zero.

Ertel's Theorem for Material Conservation of Potential Vorticity

If:

1. Λ is a materially conserved quantity that satisfies $D\Lambda/Dt = 0$.
2. Physical forcings are absent.
3. The fluid is;
   (a) *Either* barotropic (homentropic), that is, $p = p(\rho)$.
   (b) *Or* baroclinic (non-homentropic), with $\Lambda = \Lambda(\rho, p)$.

Then *potential vorticity*, defined by

$$\Pi \equiv \frac{\mathbf{Z} \cdot \nabla\Lambda}{\rho} \equiv \frac{(2\boldsymbol{\Omega} + \boldsymbol{\zeta}) \cdot \nabla\Lambda}{\rho} \equiv \frac{(2\boldsymbol{\Omega} + \nabla \times \mathbf{u}) \cdot \nabla\Lambda}{\rho}, \tag{15.49}$$

is materially conserved by a fluid parcel, that is,

$$\frac{D\Pi}{Dt} = 0. \tag{15.50}$$

This theorem is a remarkable and unexpected result. In classical particle mechanics, one frequently encounters conservation principles (e.g. for axial angular momentum and energy) for moving bodies. Since fluid mechanics is founded on classical particle mechanics, it is therefore not at all surprising, and is indeed expected, to find analogues of these conservation principles for the motion of fluid parcels.

Conservation of potential vorticity, however, is a very different phenomenon that has no analogue in classical particle mechanics. It provides a very strong local constraint on the evolution of fluid parcels by tightly coupling hydrodynamics with thermodynamics. This provides great insight into how a moving fluid evolves under the influence of physical forces. For detailed discussions of this phenomenon, see, for example, Salmon (1998), Hoskins and James (2014), and Vallis (2017).

15.3.6.2 *Two Related Questions*

Thus far, we have developed Ertel's theorem for conservation of potential vorticity (in the absence of physical forcings) by making:

Assumption 1 $D\Lambda/Dt = 0$ holds for some scalar field Λ.
Assumption 2 Λ has the functional form $\Lambda = \Lambda(\rho, p)$, where dependence on substances (S^i) is excluded.

Since $\Lambda = \Lambda(\rho, p)$ is assumed independent of S^i, the derivation of Ertel's theorem applies to a *dry atmosphere* and to pure-liquid-water oceans. The derivation does not apply to an atmosphere or ocean of variable composition. So far, so good, but this prompts two questions:

Question 1 In practice, what variable should one choose for Λ?
Question 2 Can Ertel's theorem be extended to also hold for a moist atmosphere, or for a saline ocean, of *variable* composition?

For the first question, a very natural choice for Λ is entropy, η; see governing equation (15.4) for η, with $\dot{Q}$ set to zero (i.e. no diabatic heating). Thus η is materially conserved and therefore satisfies Assumption 1. This choice is general. It applies equally well to both the atmosphere and the oceans; and governing equation (15.4) holds (in the absence of diabatic heating) for both a *moist* atmosphere and a *saline* ocean.

With this choice (i.e. $\Lambda = \eta$), we now consider the second question; does Ertel's theorem hold for a moist atmosphere, or for a saline ocean, of *variable* composition? The simplest generalisation of $\Lambda = \eta\left(\rho, p\right)$ is that η has the functional form $\eta = \eta\left(\rho, p, S\right)$, where S is specific humidity (for the atmosphere) or salinity (for the ocean). This is the minimum number (one) of substances (other than dry air for the atmosphere, or pure liquid water for the ocean) that allows the atmosphere to be moist, or the ocean to be saline. Then – using the chain rule and vector identity (A.9) –

$$\nabla\eta\left(\rho, p, S\right) = \frac{\partial\eta}{\partial\rho}\nabla\rho + \frac{\partial\eta}{\partial p}\nabla p + \frac{\partial\eta}{\partial S}\nabla S, \tag{15.51}$$

and so

$$\begin{aligned}
\nabla\eta\cdot\left(\nabla\rho\times\nabla p\right) &= \left(\frac{\partial\eta}{\partial\rho}\nabla\rho + \frac{\partial\eta}{\partial p}\nabla p + \frac{\partial\eta}{\partial S}\nabla S\right)\cdot\left(\nabla\rho\times\nabla p\right) \\
&= \frac{\partial\eta}{\partial\rho}\cancelto{0}{\nabla\rho\cdot\left(\nabla\rho\times\nabla p\right)} + \frac{\partial\eta}{\partial p}\cancelto{0}{\nabla p\cdot\left(\nabla\rho\times\nabla p\right)} + \frac{\partial\eta}{\partial S}\nabla S\cdot\left(\nabla\rho\times\nabla p\right) \\
&= \frac{\partial\eta}{\partial S}\nabla S\cdot\left(\nabla\rho\times\nabla p\right).
\end{aligned} \tag{15.52}$$

In general, the right-hand side of (15.52) is *non-zero* for a fluid of *variable* composition.[4] The right-hand side of (15.48) is then no longer zero, and potential vorticity (Π) is then no longer conserved (in the absence of physical forcings). Thus the answer to the second question is:

- Ertel's theorem does not generally hold exactly for a moist atmosphere, nor for a saline ocean, of *variable* composition.
- It does, however, do so for a moist atmosphere or saline ocean of *constant* composition.

Although this is a little disappointing, it does not really detract from the usefulness of potential vorticity in the atmospheric and oceanic sciences. Importantly, potential vorticity is materially conserved to a high degree of accuracy for many real-world situations; see, for example, Chapter 4 of Vallis (2017), and Chapters 10 and 17 of Hoskins and James (2014).

Having established that Ertel's theorem holds for both a dry atmosphere and an ocean of pure liquid water, we return to the first question. For both of these fluids, potential temperature and potential density[5] are materially conserved quantities.[6]

As an illustrative example, consider now the special case of a *dry* atmosphere, with the ideal-gas law ($p = \rho R^d T$) as the equation of state. Recall from Section 2.3.6 of Chapter 2 that then – from (2.71) with $\dot{Q}$ set identically zero –

$$\frac{D\theta}{Dt} = 0, \tag{15.53}$$

where:

- $\theta \equiv T\left(p/p_0\right)^{-R^d/c_p^d}$ is dry potential temperature.

[4] The special case $S = S\left(\rho, p\right)$, which does lead to the right-hand side being zero, is excluded since the composition is then not variable, but fixed.

[5] Potential temperature, θ, of a fluid parcel is defined to be the temperature that the parcel would have if moved adiabatically and reversibly (i.e. with $\dot{Q}$ held zero) and with fixed composition (i.e. with S held constant) to a standard reference pressure, p_0. Similarly, potential density, ρ_θ, of a fluid parcel is defined to be the density that the parcel would have if moved adiabatically and reversibly (i.e. with $\dot{Q}$ held zero) and with fixed composition (i.e. with S held constant) to a standard reference pressure, p_0. See Section 10.7 for more detail regarding these quantities.

[6] This is not the case for a moist atmosphere, nor for a saline ocean, of *variable* composition.

- p_0 is a constant reference pressure.
- R^d is the gas constant of dry air.
- c_p^d (a constant) is specific heat of dry air at constant pressure.

Furthermore, eliminating T from the ideal-gas law ($p = \rho R^d T$) and the definition $\theta \equiv T\left(p/p_0\right)^{-R^d/c_p^d}$ of dry potential temperature leads to

$$\theta = \frac{p_0}{\rho R^d}\left(\frac{p}{p_0}\right)^{1-\frac{R^d}{c_p^d}} = \theta\left(\rho, p\right), \tag{15.54}$$

for an ideal gas. Thus:

- For a dry atmosphere, instead of setting Λ to η (*entropy*), we can instead set it to θ (*dry potential temperature*).

Assumption 1 is satisfied – since $D\theta/Dt = 0$, from (15.53) – and so also is Assumption 2 – since $\theta = \theta\left(\rho, p\right)$, from (15.54).

Substitution of (15.54) into (15.53) allows the latter equation to be rewritten as

$$\frac{R^d}{p_0}\frac{D\theta}{Dt} = \frac{D}{Dt}\left[\frac{1}{\rho}\left(\frac{p}{p_0}\right)^{1-\frac{R^d}{c_p^d}}\right] = \frac{D}{Dt}\left(\frac{1}{\rho_\theta}\right) = -\frac{1}{\rho_\theta^2}\frac{D\rho_\theta}{Dt} = 0 \quad \Rightarrow \quad \frac{D\rho_\theta}{Dt} = 0, \tag{15.55}$$

where

$$\rho_\theta = \rho\left(\frac{p_0}{p}\right)^{1-\frac{R^d}{c_p^d}} = \rho_\theta\left(\rho, p\right), \tag{15.56}$$

is *potential density* (for an ideal gas, but not for an ocean). Thus:

- For a dry atmosphere, instead of setting Λ to η (*entropy*), we can instead set it to ρ_θ (*dry potential density*).

Assumption 1 is satisfied – since $D\rho_\theta/Dt = 0$, from (15.55) – and so also is Assumption 2 – since $\theta = \theta\left(\rho, p\right)$, from (15.56).

In practice, meteorologists usually set Λ to potential temperature, θ. From (15.54) and (15.56), it is seen that $\rho_\theta = \left(p_0/R^d\right)/\theta$ for an ideal gas, that is, (dry) potential density (ρ_θ) is inversely proportional to (dry) potential temperature (θ). Because of this simple relationship, and because of the convenience of using potential temperature, potential density is rarely used in meteorology.

For the ocean, the ideal-gas law is *not* an appropriate equation of state. This very much complicates matters. Nevertheless, both potential temperature and potential density are still materially conserved for an ocean of pure liquid water. However, they cannot be explicitly expressed by nice, simple formulae analogous to (15.54) and (15.56) for a dry atmosphere.

Oceanographers use both potential temperature and potential density, depending on application and on the accuracy of the equation of state employed. They also use potential enthalpy and conservative temperature (this is potential enthalpy scaled by a constant); see Section 10.7 for definition of these quantities.

$\mathbb{C}$	$F^{\mathbb{C}}$
$M \equiv (\mathbf{u} + \boldsymbol{\Omega} \times \mathbf{r}) \cdot (\widehat{\boldsymbol{\Omega}} \times \mathbf{r})$	$(\rho \mathbf{F} - \nabla p - \rho \nabla \Phi) \cdot (\widehat{\boldsymbol{\Omega}} \times \mathbf{r})$
$E \equiv K + \Phi + \mathcal{E}$	$-\nabla \cdot (p\mathbf{u}) + \rho (\mathbf{u} \cdot \mathbf{F} + \dot{Q}_E) + \frac{p}{\rho} F^{\rho}$
$\Pi \equiv \frac{\mathbf{Z} \cdot \nabla \Lambda}{\rho}$	$\frac{1}{\rho^2} \nabla \Lambda \cdot (\nabla \rho \times \nabla p) + \mathbf{Z} \cdot \nabla F^{\Lambda} + \nabla \Lambda \cdot \left(\nabla \times \mathbf{F} - \frac{F^{\rho}}{\rho} \mathbf{Z} \right)$

Table 15.1 Functional forms for $\mathbb{C}$ and $F^{\mathbb{C}}$ to put equations (15.23), (15.35), and (15.42) in the form $\rho D\mathbb{C}/Dt = F^{\mathbb{C}}$, for material conservation of axial absolute angular momentum (M), total energy (E), and potential vorticity (Π), respectively.

15.3.7 Eulerian Forms of Budget Equations for Conservation

15.3.7.1 *Conversion between Lagrangian and Eulerian Forms*

The preceding statements of conservation are expressed in Lagrangian form. They are applicable to fluid parcels (i.e. following the fluid flow). For modelling purposes, it is often convenient to have these statements available in Eulerian form. For example, in a numerical model based on an Eulerian formulation of the governing equations, it can be advantageous to design the discretisation so that it leads to an exact, discrete analogue of a conservation law. This can be achieved by computing fluxes of various quantities across the surfaces of a grid box (fixed in space) in such a way as to guarantee exact, local, discrete conservation of a quantity.

Using mass-continuity equation (15.2), it is straightforward to take a conservation statement expressed in Lagrangian form and to re-express it in Eulerian form. Consider a Lagrangian statement of conservation expressed in the form

$$\rho \frac{D\mathbb{C}}{Dt} = F^{\mathbb{C}}, \tag{15.57}$$

where $\mathbb{C}$ is a materially conserved quantity when $F^{\mathbb{C}}$ is zero. Equations (15.23), (15.35), and (15.42) for material conservation of axial absolute angular momentum, total energy, and potential vorticity, respectively, may all be written in this form. See Table 15.1 for the appropriate definitions of $\mathbb{C}$ and $F^{\mathbb{C}}$.

Recall – from (15.2) – that the mass-conservation equation in Lagrangian form is

$$\frac{D\rho}{Dt} + \rho \nabla \cdot \mathbf{u} = F^{\rho}. \tag{15.58}$$

Multiplying (15.58) by $\mathbb{C}$, summing the result with (15.57), and using material-derivative identity (A.43) then gives

$$\rho \frac{D\mathbb{C}}{Dt} + \frac{D\rho}{Dt}\mathbb{C} + \rho \mathbb{C} \nabla \cdot \mathbf{u} = F^{\mathbb{C}} + F^{\rho}\mathbb{C} \quad \Rightarrow \quad \frac{D(\rho \mathbb{C})}{Dt} + \rho \mathbb{C} \nabla \cdot \mathbf{u} = F^{\mathbb{C}} + F^{\rho}\mathbb{C}. \tag{15.59}$$

Using definition (15.11) for material derivative (D/Dt), (15.59) may be rewritten as

$$\frac{\partial (\rho \mathbb{C})}{\partial t} + \mathbf{u} \cdot \nabla (\rho \mathbb{C}) + \rho \mathbb{C} \nabla \cdot \mathbf{u} = F^{\mathbb{C}} + F^{\rho}\mathbb{C}. \tag{15.60}$$

With use of vector identity (A.17), this equation simplifies to the Eulerian statement of conservation

$$\frac{\partial(\rho\mathbb{C})}{\partial t}+\nabla\cdot(\rho\mathbb{C}\mathbf{u})=F^{\mathbb{C}}+F^{\rho}\mathbb{C}. \tag{15.61}$$

Thus, *combining Lagrangian (i.e. material) conservation law* (15.57) *with mass-continuity equation* (15.58) *leads to Eulerian conservation law* (15.61). It can be similarly shown that combining Eulerian conservation law (15.61) with mass-continuity equation (15.58) leads back to Lagrangian conservation law (15.57).

Summarising this analysis:

Lagrangian and Eulerian Forms of Conservation Statements

Using mass-continuity equation

$$\frac{D\rho}{Dt}+\rho\nabla\cdot\mathbf{u}=F^{\rho}, \tag{15.62}$$

conservation statements expressed in Lagrangian form

$$\rho\frac{D\mathbb{C}}{Dt}=F^{\mathbb{C}}, \tag{15.63}$$

may be equivalently expressed in Eulerian form as

$$\frac{\partial(\rho\mathbb{C})}{\partial t}+\nabla\cdot(\rho\mathbb{C}\mathbf{u})=F^{\mathbb{C}}+F^{\rho}\mathbb{C}, \tag{15.64}$$

and vice versa.

Eulerian forms equivalent to the corresponding Lagrangian forms (15.23)–(15.26), (15.35), and (15.42) (for the conservation principles of axial angular momentum, total energy, and potential vorticity, respectively) are explicitly given here.

15.3.7.2 *Eulerian Forms for Axial Absolute Angular Momentum*

The budget equation (15.23) for axial absolute angular momentum has the form of (15.57) when (see Table 15.1)

$$\mathbb{C}=M\equiv(\mathbf{u}+\boldsymbol{\Omega}\times\mathbf{r})\cdot\left(\widehat{\boldsymbol{\Omega}}\times\mathbf{r}\right),\quad F^{\mathbb{C}}=\left(\rho\mathbf{F}-\nabla p-\rho\nabla\Phi\right)\cdot\left(\widehat{\boldsymbol{\Omega}}\times\mathbf{r}\right). \tag{15.65}$$

Inserting (15.65) into (15.61) then gives:

The Budget Equation (Eulerian Form) for Axial Angular Momentum

$$\begin{aligned}\frac{\partial(\rho M)}{\partial t}&+\nabla\cdot\left\{\left[(\mathbf{u}+\boldsymbol{\Omega}\times\mathbf{r})\cdot\left(\widehat{\boldsymbol{\Omega}}\times\mathbf{r}\right)\right]\rho\mathbf{u}\right\}\\&=\left(\rho\mathbf{F}-\nabla p-\rho\nabla\Phi\right)\cdot\left(\widehat{\boldsymbol{\Omega}}\times\mathbf{r}\right)+F^{\rho}(\mathbf{u}+\boldsymbol{\Omega}\times\mathbf{r})\cdot\left(\widehat{\boldsymbol{\Omega}}\times\mathbf{r}\right).\end{aligned} \tag{15.66}$$

Equation (15.66) is the Eulerian form of the Lagrangian conservation principle (15.23) for axial absolute angular momentum.

Similarly, the Eulerian counterparts to (15.25) and (15.26) are:

The Budget Equation (Eulerian Form) for Axial Angular Momentum over an Axially Symmetric Planet

$$\frac{\partial(\rho M)}{\partial t} + \nabla \cdot \left\{ \left[(\mathbf{u} + \boldsymbol{\Omega} \times \mathbf{r}) \cdot \left(\widehat{\boldsymbol{\Omega}} \times \mathbf{r} \right) \right] \rho \mathbf{u} \right\}$$
$$= \left(\rho \mathbf{F} - \nabla p \right) \cdot \left(\widehat{\boldsymbol{\Omega}} \times \mathbf{r} \right) + F^{\rho} \left(\mathbf{u} + \boldsymbol{\Omega} \times \mathbf{r} \right) \cdot \left(\widehat{\boldsymbol{\Omega}} \times \mathbf{r} \right), \tag{15.67}$$

and

The Budget Equation (Eulerian Form) for Axial Angular Momentum over an Axially Symmetric Planet in the Absence of Physical Forcing

$$\frac{\partial(\rho M)}{\partial t} + \nabla \cdot \left\{ \left[(\mathbf{u} + \boldsymbol{\Omega} \times \mathbf{r}) \cdot \left(\widehat{\boldsymbol{\Omega}} \times \mathbf{r} \right) \right] \rho \mathbf{u} \right\} = -\nabla p \cdot \left(\widehat{\boldsymbol{\Omega}} \times \mathbf{r} \right). \tag{15.68}$$

15.3.7.3 *Eulerian Forms for Total Energy*

The budget equation (15.35) for total energy has the form of (15.57) when (see Table 15.1)

$$\mathbb{C} = E \equiv K + \Phi + \mathcal{E}, \quad F^{\mathbb{C}} = -\nabla \cdot \left(p\mathbf{u} \right) + \rho \left(\mathbf{u} \cdot \mathbf{F} + \dot{Q}_E \right) + \frac{p}{\rho} F^{\rho}. \tag{15.69}$$

Inserting (15.69) into (15.61) then gives:

The Budget Equation (Eulerian Form) for Total Energy

$$\frac{\partial}{\partial t}(\rho E) + \nabla \cdot \left[\left(E + \frac{p}{\rho} \right) \rho \mathbf{u} \right] = \rho \left(\mathbf{u} \cdot \mathbf{F} + \dot{Q}_E \right) + F^{\rho} \left(E + \frac{p}{\rho} \right), \tag{15.70}$$

where the contribution due to the $\nabla \cdot \left(p\mathbf{u} \right)$ term of $F^{\mathbb{C}}$ has been moved to the left-hand side of this equation and then combined with the energy-flux term, that is, with $\nabla \cdot (E\rho\mathbf{u})$. Equation (15.70) is the Eulerian form of the Lagrangian conservation principle (15.35) for total energy.

By definition, $h \equiv \mathcal{E} + p\alpha \equiv \mathcal{E} + p/\rho$, where h is *specific enthalpy*; see (3.93) and Table 9.2. Thus, using definition (15.36) of total energy, (15.70) may be written as:

An Alternative Budget Equation (Eulerian Form) for Total Energy

$$\frac{\partial}{\partial t}\left[\rho \left(K + \Phi + \mathcal{E} \right) \right] + \nabla \cdot \left[\left(K + \Phi + h \right) \rho \mathbf{u} \right] = \rho \left(\mathbf{u} \cdot \mathbf{F} + \dot{Q}_E \right) + F^{\rho} \left(E + \frac{p}{\rho} \right). \tag{15.71}$$

Examination of (15.71) shows that it is not the flux of total energy, $\nabla \cdot [(K + \Phi + \mathcal{E})\, \rho \mathbf{u}]$, that controls the rate of change of total (i.e. kinetic + potential + *internal*) energy, $\partial\, [\rho\, (K + \Phi + \mathcal{E})] / \partial t$, of an Eulerian fluid parcel (fixed in space for all time). It is instead the flux of kinetic energy + potential energy + *enthalpy*, $\nabla \cdot \left[(K + \Phi + h)\, \rho \mathbf{u} \right]$. This is due to the $\nabla \cdot (p\mathbf{u})$ term in (15.70), which term originates from the $pD\alpha/Dt$ compressibility term in internal-energy equation (15.4).

15.3.7.4 *Eulerian Form for Potential Vorticity*

The budget equation (15.42) for potential vorticity has the form of (15.57) when – see Table 15.1 –

$$\mathbb{C} = \Pi \equiv \frac{\mathbf{Z} \cdot \nabla \Lambda}{\rho}, \quad F^{\mathbb{C}} = \frac{1}{\rho^2} \nabla \Lambda \cdot (\nabla \rho \times \nabla p) + \mathbf{Z} \cdot \nabla F^{\Lambda} + \nabla \Lambda \cdot \left(\nabla \times \mathbf{F} - \frac{F^{\rho}}{\rho} \mathbf{Z} \right). \quad (15.72)$$

Inserting (15.72) into (15.61) then gives:

The Budget Equation (Eulerian Form) for Potential Vorticity

$$\frac{\partial\, (\rho \Pi)}{\partial t} + \nabla \cdot (\rho \Pi \mathbf{u}) = \frac{1}{\rho^2} \nabla \Lambda \cdot (\nabla \rho \times \nabla p) + \mathbf{Z} \cdot \nabla F^{\Lambda} + \nabla \Lambda \cdot \left(\nabla \times \mathbf{F} - \frac{F^{\rho}}{\rho} \mathbf{Z} \right) + F^{\rho} \Pi. \quad (15.73)$$

Equation (15.73) is the Eulerian form of the Lagrangian conservation principle (15.42) for potential vorticity.

For $\Lambda = \Lambda(\rho, p)$, that is, Λ is a function of *only* density and pressure (and not of composition), the first term on the right-hand side of (15.73) is absent, as discussed in Section 15.3.6. The remaining terms on the right-hand side are zero in the absence of physical forcing (i.e. when $\mathbf{F} = F^{\Lambda} = F^{\rho} = 0$), and (15.73) reduces to:

The Budget Equation (Eulerian Form) for the Potential Vorticity of a Dry Atmosphere or Pure Water Ocean in the Absence of Physical Forcing

$$\frac{\partial\, (\rho \Pi)}{\partial t} + \nabla \cdot (\rho \Pi \mathbf{u}) = 0. \quad (15.74)$$

15.3.8 Global Conservation

The Lagrangian and Eulerian conservation statements just developed are all *local* in nature. They are based on what happens locally to either a moving fluid parcel (the Lagrangian approach) or an elemental volume fixed in space (the Eulerian approach). What happens *globally* is also of interest. For example, one may wish to know under what conditions axial absolute angular momentum, total energy, and potential vorticity are conserved over the entire atmosphere or ocean, and how various physical forces influence this. The Eulerian conservation forms facilitate examination of these physical mechanisms, and assessment of their global, integrated impact.

Consider, therefore, an Eulerian conservation principle, expressed by (15.61) for some conserved quantity $\mathbb{C}$ (by which we mean conserved in the absence of forcings). Integrating this equation over the entire volume, V, of the atmosphere or of an ocean gives

$$\frac{d}{dt}\left[\iiint_V (\rho\mathbb{C})\, dV\right] + \iiint_V \nabla\cdot(\rho\mathbb{C}\mathbf{u})\, dV = \iiint_V \left(F^{\mathbb{C}} + F^{\rho}\mathbb{C}\right) dV. \tag{15.75}$$

By Gauss's divergence theorem (A.33), the second term on the left-hand side of this equation can be replaced by a surface integral over the (closed) surface, $\mathscr{S}$, that encloses V. Thus

$$\frac{d}{dt}\left[\iiint_V (\rho\mathbb{C})\, dV\right] + \oiint_{\mathscr{S}} \rho\mathbb{C}\mathbf{u}\cdot\mathbf{n}d\mathscr{S} = \iiint_V \left(F^{\mathbb{C}} + F^{\rho}\mathbb{C}\right) dV, \tag{15.76}$$

where $\mathbf{n}$ is the outward-pointing unit normal from $\mathscr{S}$. The surface integral on the left-hand side of this equation is often zero, due to periodicity in the horizontal (e.g. for a global atmosphere) and/or due to a rigid-wall boundary condition, $\mathbf{u}\cdot\mathbf{n}=0$, in the horizontal and/or in the vertical. Conservation law (15.76) then reduces to

$$\frac{d}{dt}\left[\iiint_V (\rho\mathbb{C})\, dV\right] = \iiint_V \left(F^{\mathbb{C}} + F^{\rho}\mathbb{C}\right) dV. \tag{15.77}$$

This situation, however, is not always the case; for example, for the free surface of an ocean below an atmosphere. Boundary conditions at the surface, $\mathscr{S}$, then determine any flux of $\mathbb{C}$ across it, and this must be taken into consideration in the formulation.

It may happen that the right-hand side of (15.61) contains the divergence of a quantity. This, for example, occurs for total energy; see the $\nabla\cdot\left(p\mathbf{u}\right)$ term in (15.69) and (15.70). The contribution of any divergence term to $\iiint_V F^{\mathbb{C}} dV$ in (15.75)–(15.77) may be re-expressed in terms of a surface integral, again using Gauss's divergence theorem (A.33). Depending on boundary conditions (such as periodicity and/or a rigid-wall condition $\mathbf{u}\cdot\mathbf{n}=0$), this contribution is also often zero.

In the absence of forcings $F^{\mathbb{C}}$ and F^{ρ}, (15.77) further reduces to

$$\frac{d}{dt}\left[\iiint_V (\rho\mathbb{C})\, dV\right] = 0. \tag{15.78}$$

15.4 CONSERVATION PRINCIPLES: CURVILINEAR FORM

15.4.1 Lagrangian and Eulerian Forms

In the preceding section, conservation principles were expressed in *vector* form. Using various expressions given in Section 5.2, conservation principles may instead be written in terms of *axial-orthogonal-curvilinear coordinates*. This is useful for both theoretical and practical purposes.

Using (5.5), (5.7), and (5.11), conservation equations (15.62)–(15.64) transform to

$$\frac{\partial\rho}{\partial t} + \frac{u_1}{h_1}\frac{\partial\rho}{\partial\xi_1} + \frac{u_2}{h_2}\frac{\partial\rho}{\partial\xi_2} + \frac{u_3}{h_3}\frac{\partial\rho}{\partial\xi_3} + \frac{\rho}{h_1h_2h_3}\left[\frac{\partial}{\partial\xi_1}\left(h_2h_3u_1\right) + \frac{\partial}{\partial\xi_2}\left(h_3h_1u_2\right) + \frac{\partial}{\partial\xi_3}\left(h_1h_2\right)u_3\right] = F^{\rho}, \tag{15.79}$$

$$\rho \frac{D\mathbb{C}}{Dt} = \rho \left(\frac{\partial \mathbb{C}}{\partial t} + \frac{u_1}{h_1} \frac{\partial \mathbb{C}}{\partial \xi_1} + \frac{u_2}{h_2} \frac{\partial \mathbb{C}}{\partial \xi_2} + \frac{u_3}{h_3} \frac{\partial \mathbb{C}}{\partial \xi_3} \right) = F^{\mathbb{C}}, \tag{15.80}$$

$$\frac{\partial (\rho \mathbb{C})}{\partial t} + \frac{1}{h_1 h_2 h_3} \left[\frac{\partial}{\partial \xi_1} (\rho \mathbb{C} u_1 h_2 h_3) + \frac{\partial}{\partial \xi_2} (\rho \mathbb{C} u_2 h_3 h_1) + \frac{\partial}{\partial \xi_3} (\rho \mathbb{C} u_3 h_1 h_2) \right] = F^{\mathbb{C}} + F^{\rho} \mathbb{C} \tag{15.81}$$

in axial-orthogonal-curvilinear coordinates, where

$$\mathbf{u} = u_1 \mathbf{e}_1 + u_2 \mathbf{e}_2 + u_3 \mathbf{e}_3 = (u_1, u_2, u_3) = \left(h_1 \frac{D\xi_1}{Dt}, h_2 \frac{D\xi_2}{Dt}, h_3 \frac{D\xi_3}{Dt} \right). \tag{15.82}$$

The Eulerian form of the mass-continuity equation in axial-orthogonal-curvilinear coordinates can be straightforwardly obtained by setting $\mathbb{C} \equiv 1$ and $F^{\mathbb{C}} \equiv 0$ in (15.81). Thus

$$\frac{\partial \rho}{\partial t} + \frac{1}{h_1 h_2 h_3} \left[\frac{\partial}{\partial \xi_1} (\rho u_1 h_2 h_3) + \frac{\partial}{\partial \xi_2} (\rho u_2 h_3 h_1) + \frac{\partial}{\partial \xi_3} (\rho u_3 h_1 h_2) \right] = F^{\rho}. \tag{15.83}$$

This corresponds to vector form (15.14) of the Eulerian continuity equation.

It remains to express $\mathbb{C}$ and $F^{\mathbb{C}}$ in terms of axial-orthogonal-curvilinear coordinates for $\mathbb{C} = M, E,$ and Π. This is, for the most part, straightforward. One simply expands the differential operations that appear in vector form in Table 15.1, in terms of axial-orthogonal-curvilinear coordinates, using various expressions given in Section 5.2 for this purpose.

It is not, however, immediately obvious how to represent $\widehat{\mathbf{\Omega}} \times \mathbf{r}$ in terms of axial-orthogonal-curvilinear coordinates. To do so, recall from Section 6.4.1 that, in these coordinates, metric factor h_1 measures perpendicular distance of a point from the rotation axis; see Fig. 6.1a, and also Fig. 20.2. Now $\widehat{\mathbf{\Omega}}$ is, by definition, a unit vector along this rotation axis. This means that $\left|\widehat{\mathbf{\Omega}} \times \mathbf{r}\right|$ also measures perpendicular distance of a point from the rotation axis, so that $\left|\widehat{\mathbf{\Omega}} \times \mathbf{r}\right| = h_1$. Since $\widehat{\mathbf{\Omega}}$ and $\mathbf{r}$ both lie in the meridional plane in which the point lies, $\widehat{\mathbf{\Omega}} \times \mathbf{r}$ is perpendicular to this plane and is in the direction of unit vector $\mathbf{e}_1$ in the zonal direction. Thus, putting this together,

$$\widehat{\mathbf{\Omega}} \times \mathbf{r} = h_1 \mathbf{e}_1, \tag{15.84}$$

and, similarly,

$$\mathbf{\Omega} \times \mathbf{r} = \Omega h_1 \mathbf{e}_1. \tag{15.85}$$

15.4.2 Axial Absolute Angular Momentum

From Table 15.1, specific axial absolute angular momentum in vector form is given by

$$M \equiv (\mathbf{u} + \mathbf{\Omega} \times \mathbf{r}) \cdot \left(\widehat{\mathbf{\Omega}} \times \mathbf{r}\right). \tag{15.86}$$

Introducing (15.82)–(15.85) into (15.86) then yields:

Axial Absolute Angular Momentum in Axial-Orthogonal-Curvilinear Coordinates

$$M \equiv (\mathbf{u} + \boldsymbol{\Omega} \times \mathbf{r}) \cdot \left(\widehat{\boldsymbol{\Omega}} \times \mathbf{r}\right) = \left(u_1 \mathbf{e}_1 + u_2 \mathbf{e}_2 + u_3 \mathbf{e}_3 + \Omega h_1 \mathbf{e}_1\right) \cdot \left(h_1 \mathbf{e}_1\right) = h_1 \left(u_1 + \Omega h_1\right), \tag{15.87}$$

since $\mathbf{e}_1$ is orthogonal to both $\mathbf{e}_2$ and $\mathbf{e}_3$ (and therefore $\mathbf{e}_1 \cdot \mathbf{e}_2 = \mathbf{e}_1 \cdot \mathbf{e}_3 = 0$).

15.4.3 Total Energy

From Table 15.1, specific total energy in vector form is given by

$$E \equiv K + \Phi + \mathcal{E} \equiv \frac{\mathbf{u} \cdot \mathbf{u}}{2} + \Phi + \mathcal{E}. \tag{15.88}$$

Introducing (15.82) into (15.88) then yields:

Total Energy in Axial-Orthogonal-Curvilinear Coordinates

$$E \equiv \frac{\mathbf{u} \cdot \mathbf{u}}{2} + \Phi + \mathcal{E} = \frac{(u_1 \mathbf{e}_1 + u_2 \mathbf{e}_2 + u_3 \mathbf{e}_3) \cdot (u_1 \mathbf{e}_1 + u_2 \mathbf{e}_2 + u_3 \mathbf{e}_3)}{2} + \Phi + \mathcal{E}$$
$$= \frac{u_1^2 + u_2^2 + u_3^2}{2} + \Phi + \mathcal{E} = \frac{1}{2} \left[h_1^2 \left(\frac{D\xi_1}{Dt} \right)^2 + h_2^2 \left(\frac{D\xi_2}{Dt} \right)^2 + h_3^2 \left(\frac{D\xi_3}{Dt} \right)^2 \right] + \Phi + \mathcal{E}. \tag{15.89}$$

15.4.4 Potential Vorticity

From Table 15.1, specific potential vorticity in vector form is given by

$$\Pi \equiv \frac{\mathbf{Z} \cdot \nabla \Lambda}{\rho}. \tag{15.90}$$

From (15.28),

$$\mathbf{Z} \equiv 2\boldsymbol{\Omega} + \nabla \times \mathbf{u} \equiv 2\boldsymbol{\Omega} + \boldsymbol{\zeta} \tag{15.91}$$

is *absolute vorticity*, 2Ω is *planetary vorticity*, and $\boldsymbol{\zeta} \equiv \nabla \times \mathbf{u}$ is *relative vorticity*, all of which are specific quantities. Now from (6.26) and (5.8), $2\boldsymbol{\Omega}$ and $\boldsymbol{\zeta} \equiv \nabla \times \mathbf{u}$ can be expressed in axial-orthogonal-curvilinear coordinates as

$$2\Omega = 2\Omega \left(\frac{1}{h_3} \frac{\partial h_1}{\partial \xi_3} \mathbf{e}_2 - \frac{1}{h_2} \frac{\partial h_1}{\partial \xi_2} \mathbf{e}_3 \right) = 2\Omega \left(0, \frac{1}{h_3} \frac{\partial h_1}{\partial \xi_3}, -\frac{1}{h_2} \frac{\partial h_1}{\partial \xi_2} \right), \tag{15.92}$$

$$\zeta = \zeta_1 \mathbf{e}_1 + \zeta_2 \mathbf{e}_2 + \zeta_3 \mathbf{e}_3 = \frac{1}{h_2 h_3} \left[\frac{\partial}{\partial \xi_2} (h_3 u_3) - \frac{\partial}{\partial \xi_3} (h_2 u_2) \right] \mathbf{e}_1$$
$$+ \frac{1}{h_3 h_1} \left[\frac{\partial}{\partial \xi_3} (h_1 u_1) - \frac{\partial}{\partial \xi_1} (h_3 u_3) \right] \mathbf{e}_2$$
$$+ \frac{1}{h_1 h_2} \left[\frac{\partial}{\partial \xi_1} (h_2 u_2) - \frac{\partial}{\partial \xi_2} (h_1 u_1) \right] \mathbf{e}_3. \tag{15.93}$$

Inserting (15.92) and (15.93) into (15.91) leads to:

Absolute Vorticity in Axial-Orthogonal-Curvilinear Coordinates

$$\mathbf{Z} = Z_1\mathbf{e}_1 + Z_2\mathbf{e}_2 + Z_3\mathbf{e}_3, \tag{15.94}$$

where

$$Z_1 = \frac{1}{h_2 h_3}\left[\frac{\partial}{\partial \xi_2}(h_3 u_3) - \frac{\partial}{\partial \xi_3}(h_2 u_2)\right], \tag{15.95}$$

$$Z_2 = \frac{2\Omega}{h_3}\frac{\partial h_1}{\partial \xi_3} + \frac{1}{h_3 h_1}\left[\frac{\partial}{\partial \xi_3}(h_1 u_1) - \frac{\partial}{\partial \xi_1}(h_3 u_3)\right], \tag{15.96}$$

$$Z_3 = -\frac{2\Omega}{h_2}\frac{\partial h_1}{\partial \xi_2} + \frac{1}{h_1 h_2}\left[\frac{\partial}{\partial \xi_1}(h_2 u_2) - \frac{\partial}{\partial \xi_2}(h_1 u_1)\right]. \tag{15.97}$$

With this preparation, introducing (5.6) and (15.94) into (15.90) yields:

Potential Vorticity in Axial-Orthogonal-Curvilinear Coordinates

$$\Pi \equiv \frac{\mathbf{Z}\cdot\nabla\Lambda}{\rho} = \frac{1}{\rho}\left(Z_1\mathbf{e}_1 + Z_2\mathbf{e}_2 + Z_3\mathbf{e}_3\right)\cdot\left(\frac{1}{h_1}\frac{\partial\Lambda}{\partial\xi_1}\mathbf{e}_1 + \frac{1}{h_2}\frac{\partial\Lambda}{\partial\xi_2}\mathbf{e}_2 + \frac{1}{h_3}\frac{\partial\Lambda}{\partial\xi_3}\mathbf{e}_3\right)$$
$$= \frac{1}{\rho}\left(\frac{Z_1}{h_1}\frac{\partial\Lambda}{\partial\xi_1} + \frac{Z_2}{h_2}\frac{\partial\Lambda}{\partial\xi_2} + \frac{Z_3}{h_3}\frac{\partial\Lambda}{\partial\xi_3}\right), \tag{15.98}$$

where Z_1, Z_2, and Z_3 are given by (15.95)–(15.97), respectively.

15.4.5 Expressions for $F^{\mathbb{C}}$

In a manner similar to that employed earlier for $\mathbb{C} = M, E$, and Π, expressions for $F^{\mathbb{C}}$ in axial-orthogonal-curvilinear coordinates may be obtained (using relations in Section 5.2) from the vector definitions given in Table 15.1. The resulting expressions are tabulated in Table 15.2.

To make things a little more tangible and familiar, expressions corresponding to those given in Table 15.2 for general axial-orthogonal-curvilinear coordinates (ξ_1, ξ_2, ξ_3) are given in Table 15.3 for the particular case of spherical-polar coordinates (λ, ϕ, r). They are obtained – see Section 5.2.2 – by making the transformations

$$(\xi_1, \xi_2, \xi_3) \rightarrow (\lambda, \phi, r), \quad (h_1, h_2, h_3) \rightarrow (h_\lambda, h_\phi, h_r) = (r\cos\phi, r, 1). \tag{15.99}$$

Note that $\Phi = \Phi(\xi_2, \xi_3)$ for a (axially symmetric) geopotential coordinate system; see, for example, the GREAT[7] coordinate system of Section 12.9. The term containing $\partial\Phi/\partial\xi_1$ in the top line of the second column of Table 15.2 is then absent (and similarly in Table 15.3).

15.4.6 Global Conservation

Global conservation principles, in vector form, are derived in Section 15.3.8. They are expressed in terms of a conserved quantity $\mathbb{C}$ (by which we mean conserved in the absence of forcings), and

[7] GREAT = Geophysically Realistic, Ellipsoidal, Analytically Tractable.

$\mathbb{C}$	$F^{\mathbb{C}}$
$M = h_1 \left(u_1 + \Omega h_1 \right)$	$h_1 \left(\rho F^{u_1} - \frac{1}{h_1} \frac{\partial p}{\partial \xi_1} - \frac{\rho}{h_1} \frac{\partial \Phi}{\partial \xi_1} \right)$
$E = \frac{u_1^2 + u_2^2 + u_3^2}{2} + \Phi + \mathcal{E}$	$-\frac{1}{h_1 h_2 h_3} \left[\frac{\partial \left(p u_1 h_2 h_3 \right)}{\partial \xi_1} + \frac{\partial \left(p u_2 h_3 h_1 \right)}{\partial \xi_2} + \frac{\partial \left(p u_3 h_1 h_2 \right)}{\partial \xi_3} \right] + \rho \left(u_1 F^{u_1} + u_2 F^{u_2} + u_3 F^{u_3} + \dot{Q}_E \right) + \frac{p}{\rho} F^{\rho}$
$\Pi = \frac{1}{\rho} \left(\frac{Z_1}{h_1} \frac{\partial \Lambda}{\partial \xi_1} + \frac{Z_2}{h_2} \frac{\partial \Lambda}{\partial \xi_2} + \frac{Z_3}{h_3} \frac{\partial \Lambda}{\partial \xi_3} \right)$	$\frac{1}{\rho^2 h_1 h_2 h_3} \left[\frac{\partial \left(\rho, p \right)}{\partial \left(\xi_2, \xi_3 \right)} \frac{\partial \Lambda}{\partial \xi_1} + \frac{\partial \left(\rho, p \right)}{\partial \left(\xi_3, \xi_1 \right)} \frac{\partial \Lambda}{\partial \xi_2} + \frac{\partial \left(\rho, p \right)}{\partial \left(\xi_1, \xi_2 \right)} \frac{\partial \Lambda}{\partial \xi_3} \right] + \frac{Z_1}{h_1} \left(\frac{\partial F^{\Lambda}}{\partial \xi_1} - \frac{F^{\rho}}{\rho} \frac{\partial \Lambda}{\partial \xi_1} \right) + \frac{Z_2}{h_2} \left(\frac{\partial F^{\Lambda}}{\partial \xi_2} - \frac{F^{\rho}}{\rho} \frac{\partial \Lambda}{\partial \xi_2} \right) + \frac{Z_3}{h_3} \left(\frac{\partial F^{\Lambda}}{\partial \xi_3} - \frac{F^{\rho}}{\rho} \frac{\partial \Lambda}{\partial \xi_3} \right) + \frac{1}{h_1 h_2 h_3} \left\{ \left[\frac{\partial \left(h_3 F^{u_3} \right)}{\partial \xi_2} - \frac{\partial \left(h_2 F^{u_2} \right)}{\partial \xi_3} \right] \frac{\partial \Lambda}{\partial \xi_1} + \left[\frac{\partial \left(h_1 F^{u_1} \right)}{\partial \xi_3} - \frac{\partial \left(h_3 F^{u_3} \right)}{\partial \xi_1} \right] \frac{\partial \Lambda}{\partial \xi_2} \right\} + \frac{1}{h_1 h_2 h_3} \left[\frac{\partial \left(h_2 F^{u_2} \right)}{\partial \xi_1} - \frac{\partial \left(h_1 F^{u_1} \right)}{\partial \xi_2} \right] \frac{\partial \Lambda}{\partial \xi_3}$

Table 15.2 Functional forms for $\mathbb{C}$ and $F^{\mathbb{C}}$ to express equations (15.80) and (15.81) in axial-orthogonal-curvilinear coordinates, for conservation of axial absolute angular momentum (M), total energy (E), and potential vorticity (Π), respectively, where $\frac{\partial (a,b)}{\partial (c,d)} \equiv \frac{\partial a}{\partial c}\frac{\partial b}{\partial d} - \frac{\partial a}{\partial d}\frac{\partial b}{\partial c}$. For an axially symmetric geopotential (of an axially symmetric planet), $\Phi = \Phi \left(\xi_2, \xi_3 \right)$, and the term containing $\partial \Phi / \partial \xi_1$ in the top line of the second column is then absent.

$\mathbb{C}$	$F^{\mathbb{C}}$
$M = r\cos\phi\left(u_\lambda + \Omega r\cos\phi\right)$	$\rho F^{u_\lambda} r\cos\phi - \frac{\partial p}{\partial\lambda} - \rho\frac{\partial\Phi}{\partial\lambda}$
$E = \frac{u_\lambda^2 + u_\phi^2 + u_r^2}{2} + \Phi + \mathcal{E}$	$-\frac{1}{r^2\cos\phi}\left[\frac{\partial\left(pu_\lambda r\right)}{\partial\lambda} + \frac{\partial\left(pu_\phi r\cos\phi\right)}{\partial\phi} + \frac{\partial\left(pu_r r^2\cos\phi\right)}{\partial r}\right] + \rho\left(u_\lambda F^{u_\lambda} + u_\phi F^{u_\phi} + u_r F^{u_r} + \dot{Q}_E\right) + \frac{p}{\rho}F^\rho$
$\Pi = \frac{1}{\rho}\left(\frac{Z_\lambda}{r\cos\phi}\frac{\partial\Lambda}{\partial\lambda} + \frac{Z_\phi}{r}\frac{\partial\Lambda}{\partial\phi} + Z_r\frac{\partial\Lambda}{\partial r}\right)$	$\frac{1}{\rho^2 r^2\cos\phi}\left[\frac{\partial\left(\rho,p\right)}{\partial\left(\phi,r\right)}\frac{\partial\Lambda}{\partial\lambda} + \frac{\partial\left(\rho,p\right)}{\partial\left(r,\lambda\right)}\frac{\partial\Lambda}{\partial\phi} + \frac{\partial\left(\rho,p\right)}{\partial\left(\lambda,\phi\right)}\frac{\partial\Lambda}{\partial r}\right] + \frac{Z_\lambda}{r\cos\phi}\left(\frac{\partial F^\Lambda}{\partial\lambda} - \frac{F^\rho}{\rho}\frac{\partial\Lambda}{\partial\lambda}\right) + \frac{Z_\phi}{r}\left(\frac{\partial F^\Lambda}{\partial\phi} - \frac{F^\rho}{\rho}\frac{\partial\Lambda}{\partial\phi}\right) + Z_r\left(\frac{\partial F^\Lambda}{\partial r} - \frac{F^\rho}{\rho}\frac{\partial\Lambda}{\partial r}\right) + \frac{1}{r^2\cos\phi}\left\{\left[\frac{\partial F^{u_r}}{\partial\phi} - \frac{\partial\left(rF^{u_\phi}\right)}{\partial r}\right]\frac{\partial\Lambda}{\partial\lambda} + \left[\frac{\partial\left(r\cos\phi F^{u_\lambda}\right)}{\partial r} - \frac{\partial F^{u_r}}{\partial\lambda}\right]\frac{\partial\Lambda}{\partial\phi}\right\} + \frac{1}{r^2\cos\phi}\left[\frac{\partial\left(rF^{u_\phi}\right)}{\partial\lambda} - \frac{\partial\left(r\cos\phi F^{u_\lambda}\right)}{\partial\phi}\right]\frac{\partial\Lambda}{\partial r}$

Table 15.3 Functional forms for $\mathbb{C}$ and $F^{\mathbb{C}}$ to express equations (15.80) and (15.81) in spherical-polar coordinates (λ, ϕ, r), for conservation of axial absolute angular momentum (M), total energy (E), and potential vorticity (Π), respectively, where $\frac{\partial\left(a,b\right)}{\partial\left(c,d\right)} \equiv \frac{\partial a}{\partial c}\frac{\partial b}{\partial d} - \frac{\partial a}{\partial d}\frac{\partial b}{\partial c}$. For the classical spherical-geopotential approximation, $\Phi = \Phi\left(r\right)$, and the term containing $\partial\Phi/\partial\lambda$ in the top line of the second column is then absent.

forcings $F^{\mathbb{C}}$ (the right-hand side of the material conservation equation (15.57)) and F^{ρ} (the right-hand side of the mass-conservation equation (15.58)). Vector forms for $\mathbb{C}$ and $F^{\mathbb{C}}$ are displayed in Table 15.1 for conservation of axial absolute angular momentum, total energy, and potential vorticity. The corresponding forms in axial-orthogonal-curvilinear coordinates are displayed in Table 15.2, and in Table 15.3 for the special case of spherical-polar coordinates.

From (5.3) and (6.13), the volume element in axial-orthogonal-curvilinear coordinates is

$$dV \equiv h_1 h_2 h_3 d\xi_1 d\xi_2 d\xi_3, \tag{15.100}$$

where (due to the assumed axial symmetry of the coordinate system)

$$h_i = h_i\left(\xi_2, \xi_3\right), \quad i = 1, 2, 3. \tag{15.101}$$

Furthermore, divergence in axial-orthogonal-curvilinear coordinates is given by (5.7), and so

$$\nabla \cdot \mathbf{A} = \frac{1}{h_1 h_2 h_3}\left[\frac{\partial}{\partial \xi_1}\left(A_1 h_2 h_3\right) + \frac{\partial}{\partial \xi_2}\left(A_2 h_3 h_1\right) + \frac{\partial}{\partial \xi_3}\left(A_3 h_1 h_2\right)\right], \tag{15.102}$$

where $h_i, i = 1, 2, 3$, satisfy (15.101).

Using these three equations, the global-conservation principles expressed in vector form in Section 15.3.8 may be re-expressed in axial-orthogonal-curvilinear coordinates. Consider, therefore, an Eulerian-conservation principle for some conserved quantity $\mathbb{C}$, expressed by (15.81) in axial-orthogonal-curvilinear coordinates. Integrating this equation over the entire volume, V, of the atmosphere or of an ocean and using (15.100)–(15.102) then gives (with cancellation of $h_1 h_2 h_3$ top and bottom in the divergence term)

$$\begin{aligned}
&\frac{d}{dt}\left[\iiint_V (\rho\mathbb{C})\, h_1 h_2 h_3 d\xi_1 d\xi_2 d\xi_3\right] \\
&\quad + \iiint_V \left[\frac{\partial}{\partial \xi_1}\left(\rho\mathbb{C} u_1 h_2 h_3\right) + \frac{\partial}{\partial \xi_2}\left(\rho\mathbb{C} u_2 h_3 h_1\right) + \frac{\partial}{\partial \xi_3}\left(\rho\mathbb{C} u_3 h_1 h_2\right)\right] d\xi_1 d\xi_2 d\xi_3 \\
&\quad = \iiint_V \left(F^{\mathbb{C}} + F^{\rho}\mathbb{C}\right) h_1 h_2 h_3 d\xi_1 d\xi_2 d\xi_3.
\end{aligned} \tag{15.103}$$

This equation is simply (15.75), but expressed in axial-orthogonal-curvilinear coordinates.

By Gauss' divergence theorem (A.33), the second term on the left-hand side of this equation can be replaced by a surface integral over the (closed) surface, $\mathscr{S}$, enclosing V. Thus

$$\frac{d}{dt}\left[\iiint_V (\rho\mathbb{C})\, h_1 h_2 h_3 d\xi_1 d\xi_2 d\xi_3\right] + \oiint_{\mathscr{S}} \rho\mathbb{C}\mathbf{u} \cdot \mathbf{n} d\mathscr{S} = \iiint_V \left(F^{\mathbb{C}} + F^{\rho}\mathbb{C}\right) h_1 h_2 h_3 d\xi_1 d\xi_2 d\xi_3, \tag{15.104}$$

where $\mathbf{n}$ is the outward-pointing unit normal from $\mathscr{S}$. This equation corresponds to (15.76), but expressed in axial-orthogonal-curvilinear coordinates. The surface integral on the left-hand side is often zero, due to periodicity in the horizontal (e.g. for a global atmosphere), and/or due to a rigid-wall boundary condition, $\mathbf{u} \cdot \mathbf{n} = 0$, in the horizontal and/or in the vertical. The conservation law (15.104) then reduces to

$$\frac{d}{dt}\left[\iiint_V (\rho\mathbb{C})\, h_1 h_2 h_3 d\xi_1 d\xi_2 d\xi_3\right] = \iiint_V \left(F^{\mathbb{C}} + F^{\rho}\mathbb{C}\right) h_1 h_2 h_3 d\xi_1 d\xi_2 d\xi_3. \tag{15.105}$$

This situation, however, is not always the case; for example, for the free surface of an ocean below an atmosphere. Boundary conditions at the surface, $\mathscr{S}$, then determine any flux of $\mathbb{C}$ across it, and this must be taken into consideration in the formulation.

It may happen that the right-hand side of (15.105) contains the divergence of a quantity. This, for example, occurs for total energy; see F^E (i.e. $F^{\mathbb{C}}$ with $\mathbb{C} = E$) in Table 15.2, with a term that corresponds to $\nabla \cdot (p\mathbf{u})$ expressed in axial-orthogonal-curvilinear coordinates. The contribution of any divergence term to $\iiint_V F^{\mathbb{C}} dV$ in (15.103)–(15.105) may be re-expressed in terms of a surface integral, again using Gauss' divergence theorem (A.33). Depending on boundary conditions (such as periodicity and/or a rigid-wall condition $\mathbf{u} \cdot \mathbf{n} = 0$), this contribution is also often zero.

In the absence of forcings $F^{\mathbb{C}}$ and F^{ρ}, (15.105) further reduces to the conservation law

$$\frac{d}{dt}\left[\iiint_V (\rho\mathbb{C})\, h_1 h_2 h_3 d\xi_1 d\xi_2 d\xi_3\right] = 0. \qquad (15.106)$$

15.5 NOETHER'S THEOREM, SYMMETRIES, AND CONSERVATION

15.5.1 Preliminaries

In the context of geophysical fluid dynamics, Noether's mathematical theorem[8] relates symmetry properties of a known Lagrangian to conservation laws for the physically important quantities of axial angular momentum, total energy, and potential vorticity:

Noether's Theorem

- For every continuous symmetry property of a Lagrangian, there is a corresponding quantity that is conserved in time, and vice versa.

This theorem provides a very powerful tool for examining the conservation properties of a conservative system, such as that described and discussed in Chapter 14 for a global atmosphere or ocean. Of particular interest herein is that:

Corollary

- If one approximates a Lagrangian whilst maintaining its symmetries, then quantities conserved in the original system remain conserved in the approximated one.

The mathematics for Noether's theorem is highly abstract, and salient results are stated herein rather than proved. Instead, we focus (in Part II of this book) on the *practical application* of Noether's theorem to obtain approximated governing equation sets for global atmospheric and oceanic flow, whereby important conservation principles of unapproximated equation sets are maintained under symmetry-preserving approximation. From here on, *we restrict attention to governing equation sets for single-substance fluids (i.e. for an ideal-gas atmosphere or a liquid-water ocean) in the absence of external physical forcing other than gravity.* This both simplifies things and recognises that Lagrangian formulations are most easily applied to conservative systems.

It is worth emphasising that applying Noether's theorem and Hamilton's principle of stationary action does not in any way change the underlying physics. Anything and everything, *once known*, that can be obtained this way can also be obtained by other, more traditional methods. What this approach offers is a valuable tool for:

- Approximating equation sets in a systematic, guided manner.
- Improving understanding of, and theoretical support for, known equation sets.
- Revealing new equation sets that might otherwise be overlooked.

[8] After Amalie Emmy Noether, 1882–1935.

Thus this approach *complements* other, more traditional methods.[9]

In the present context, there are three key symmetries of a Lagrangian that are of great interest. They are invariance of a Lagrangian with respect to:

1. Rotation about a planet's axis, together with assumed zonal symmetry of the gravitational potential, Φ (for *conservation of axial angular momentum*).
2. Time translation (for *conservation of total energy*).
3. Particle relabelling (for *conservation of potential vorticity*).

Recall from (14.119) that the Lagrangian, expressed in vector form *for a single-substance fluid in the absence of external physical forcing other than gravity*, is given by

$$L(\mathbf{r},\dot{\mathbf{r}},\alpha,\eta) = \iiint \left[\frac{1}{2}\left(\frac{\partial \mathbf{r}}{\partial \tau} \cdot \frac{\partial \mathbf{r}}{\partial \tau} \right) + (\boldsymbol{\Omega}\times\mathbf{r}) \cdot \frac{\partial \mathbf{r}}{\partial \tau} - \mathcal{E}(\alpha,\eta) - \Phi(\mathbf{r}) \right] d\mathbf{a}, \tag{15.107}$$

where $\mathbf{r} = \mathbf{r}(\mathbf{a},\tau)$ and $\mathbf{a} = \mathbf{a}(\mathbf{r},t)$ are position and label vectors, respectively, and t and τ are time in position and label space, respectively. From (14.150), the Lagrangian can be similarly and equivalently expressed in *axial-orthogonal-curvilinear coordinates,* $\boldsymbol{\xi} \equiv [\xi_1(\mathbf{a},\tau), \xi_2(\mathbf{a},\tau), \xi_3(\mathbf{a},\tau)]$, as

$$\begin{aligned} L(\boldsymbol{\xi},\dot{\boldsymbol{\xi}},\alpha,\eta) = & \iiint \left[\frac{h_1^2}{2}\left(\frac{\partial \xi_1}{\partial \tau}\right)^2 + \frac{h_2^2}{2}\left(\frac{\partial \xi_2}{\partial \tau}\right)^2 + \frac{h_3^2}{2}\left(\frac{\partial \xi_3}{\partial \tau}\right)^2 + \Omega h_1^2 \frac{\partial \xi_1}{\partial \tau} \right] d\mathbf{a} \\ & - \iiint \left[\mathcal{E}(\alpha,\eta) + \Phi(\xi_2,\xi_3) \right] d\mathbf{a}. \end{aligned} \tag{15.108}$$

In (15.107) and (15.108):

- $h_i = h_i(\xi_2,\xi_3)$, $i = 1,2,3$, are zonally symmetric (axially symmetric), metric (scale) factors.
- Internal energy ($\mathcal{E}$) is a *prescribed function* of specific volume (α) and entropy (η).
- Geopotential (Φ) is axially symmetric, that is, it is a *prescribed function* of coordinates ξ_2 and ξ_3 (but *not* of ξ_1). For an ellipsoidal planet, its functional form using GREAT coordinates (λ,χ,R) is given by (12.35) above the geoid and by (12.61) below it.

Labels $\mathbf{a} = \mathbf{a}(\mathbf{r},t)$ in (15.107) and (15.108) are assigned such that – see (14.16) –

$$\alpha = \alpha(\mathbf{a},\tau) = \alpha(a_1,a_2,a_3,\tau) \equiv \frac{1}{\rho} = \frac{\partial(\mathbf{r})}{\partial(\mathbf{a})} = \frac{\partial(x_1,x_2,x_3)}{\partial(a_1,a_2,a_3)} = \frac{\partial(x_1,x_2,x_3)}{\partial(\xi_1,\xi_2,\xi_3)}\frac{\partial(\xi_1,\xi_2,\xi_3)}{\partial(a_1,a_2,a_3)}. \tag{15.109}$$

As shown in Section 14.3, (15.109) then leads to mass-continuity equation (14.22), that is, to

$$\frac{\partial \alpha}{\partial \tau} = \alpha \nabla \cdot \mathbf{u} \quad \Rightarrow \quad \frac{D\rho}{Dt} + \rho \nabla \cdot \mathbf{u} = 0. \tag{15.110}$$

Furthermore – see (15.4) with $\dot{Q} \equiv 0$, and (14.133) –

$$\frac{\partial \eta}{\partial \tau} = 0 \quad \Rightarrow \quad \eta = \eta(\mathbf{a}) = \eta(a_1,a_2,a_3). \tag{15.111}$$

[9] A useful attribute of this complementarity is that results obtained with one approach can be cross-checked by the other. So if you are not confident that you have properly applied Noether's theorem to obtain a result, then you should anyway be able to confirm it using more traditional means!

Note that whilst α depends upon time τ and labels **a**, η is independent of τ with dependence only on **a**.

To help fix ideas, the prescribed functional form of $\mathcal{E}$ for an ideal gas (a good approximation for an atmosphere, but *not* for an ocean) may be obtained as a special (dry) case of (4.70) for a cloudy-air parcel. Specifically,

$$\mathcal{E}(\alpha,\eta) = C\alpha^{-\frac{R}{c_V}} \exp\left(\frac{\eta}{c_V}\right), \tag{15.112}$$

where:

- α is specific volume and η is specific entropy.
- C is an arbitrary (positive) constant (with no physical significance).
- R is the gas constant of the ideal gas.
- c_V is its (constant) specific heat at constant volume.

To further help fix ideas, setting $(h_1, h_2, h_3) = (r\cos\phi, r, 1)$ (see (5.15)) in (15.108) gives (see (14.151))

$$\begin{aligned} L(\lambda,\phi,r,\alpha,\eta) = &\iiint \left[\frac{r^2\cos^2\phi}{2}\left(\frac{\partial\lambda}{\partial\tau}\right)^2 + \frac{r^2}{2}\left(\frac{\partial\phi}{\partial\tau}\right)^2 + \frac{1}{2}\left(\frac{\partial r}{\partial\tau}\right)^2 + \Omega r^2\cos^2\phi\frac{\partial\lambda}{\partial\tau}\right]d\mathbf{a} \\ &- \iiint \left[\mathcal{E}(\alpha,\eta) + \Phi(r)\right] d\mathbf{a}, \end{aligned} \tag{15.113}$$

for *spherical-polar coordinates* (λ,ϕ,r). In (15.113), Φ is assumed to not only be axially symmetric, but also *spherically* symmetric (by the classical spherical-geopotential approximation (7.9)), so that

$$\Phi(r) = -\frac{\gamma M_P}{r}, \tag{15.114}$$

where γ is Newton's universal gravitational constant, and M_P is the mass of the planet (including its atmosphere and oceans).[10]

In what follows we go into some further detail regarding the three key symmetries of a Lagrangian and their link, through Noether's theorem, to the conservation of axial angular momentum, total energy, and potential vorticity.

15.5.2 Rotation Symmetry and Conservation of Axial Angular Momentum

15.5.2.1 Rotation Symmetry of a Lagrangian

We begin by defining:

Rotation Symmetry of a Lagrangian

A Lagrangian $L[\mathbf{r}(\mathbf{a},\tau)] \equiv L[x_1(\mathbf{a},\tau), x_2(\mathbf{a},\tau), x_3(\mathbf{a},\tau)]$ (in Cartesian coordinates) or $L[\boldsymbol{\xi}(\mathbf{a},\tau)] \equiv L[\xi_1(\mathbf{a},\tau), \xi_2(\mathbf{a},\tau), \xi_3(\mathbf{a},\tau)]$ (in axial-orthogonal-curvilinear coordinates) is said to have *rotation symmetry* (i.e. axial symmetry) if

[10] The mass of the planet, M_P, is not to be confused with specific axial angular momentum, M. They are completely different quantities.

$$L\left(\mathbf{r}'\right) = L(\mathbf{r}), \tag{15.115}$$

or

$$L\left(\xi_1', \xi_2, \xi_3\right) = L(\xi_1, \xi_2, \xi_3), \tag{15.116}$$

where

$$\mathbf{r}'(\mathbf{a}, \tau) \equiv \mathbf{r}(\mathbf{a}, \tau) + (\delta\xi_1)\,\widehat{\boldsymbol{\Omega}} \times \mathbf{r}(\mathbf{a}, \tau), \tag{15.117}$$

$$\xi_1'(\mathbf{a}, \tau) \equiv \xi_1(\mathbf{a}, \tau) + \delta\xi_1, \tag{15.118}$$

and $\delta\xi_1$ = constant is an infinitesimal rotation angle.

Equations (15.117) and (15.118) correspond to an infinitesimal rotation of a point, by constant angle $\delta\xi_1$, about a planet's rotation axis with unit vector $\widehat{\boldsymbol{\Omega}}$ directed along it.

15.5.2.2 *Rotation Symmetry of Lagrangians* (15.107) *and* (15.108)

Having defined rotation symmetry of a Lagrangian, we now show that Lagrangians (15.107) and (15.108) have this symmetry. Since Lagrangians (15.107) and (15.108) are equivalent, we only need to do this for one of them. Because of the rotation symmetry inherent in axial-orthogonal-curvilinear coordinates, it is natural to work with the second form of the Lagrangian, that is, (15.108), for this.

Partially differentiating infinitesimal rotation (15.118) with respect to τ and noting that $\delta\xi_1 =$ constant gives

$$\frac{\partial \xi_1'(\mathbf{a}, \tau)}{\partial \tau} = \frac{\partial\left[\xi_1(\mathbf{a}, \tau) + \delta\xi_1\right]}{\partial \tau} = \frac{\partial \xi_1(\mathbf{a}, \tau)}{\partial \tau}. \tag{15.119}$$

Now (see e.g. Spiegel and Lipschutz, 2009, p. 169)

$$\frac{\partial(x_1, x_2, x_3)}{\partial(\xi_1, \xi_2, \xi_3)} = h_1 h_2 h_3, \tag{15.120}$$

where (h_1, h_2, h_3) are the metric (or scale) factors of an orthogonal-curvilinear coordinate system with coordinates (ξ_1, ξ_2, ξ_3). Substituting (15.120) into (15.109), and using (15.101) for an assumed axial symmetry of the coordinate system, leads to

$$\alpha(\xi_1, \xi_2, \xi_3) = h_1 h_2 h_3 \frac{\partial(\xi_1, \xi_2, \xi_3)}{\partial(a_1, a_2, a_3)} = h_1(\xi_2, \xi_3)\, h_2(\xi_2, \xi_3)\, h_3(\xi_2, \xi_3) \begin{vmatrix} \dfrac{\partial \xi_1}{\partial a_1} & \dfrac{\partial \xi_1}{\partial a_2} & \dfrac{\partial \xi_1}{\partial a_3} \\ \dfrac{\partial \xi_2}{\partial a_1} & \dfrac{\partial \xi_2}{\partial a_2} & \dfrac{\partial \xi_2}{\partial a_3} \\ \dfrac{\partial \xi_3}{\partial a_1} & \dfrac{\partial \xi_3}{\partial a_2} & \dfrac{\partial \xi_3}{\partial a_3} \end{vmatrix}. \tag{15.121}$$

Consider now (15.121) subject to the infinitesimal rotation (15.118). Successively differentiating (15.118) with respect to the labels $\mathbf{a} = (a_1, a_2, a_3)$ gives

$$\frac{\partial \xi_1'(\mathbf{a}, \tau)}{\partial a_i} = \frac{\partial\left[\xi_1(\mathbf{a}, \tau) + \delta\xi_1\right]}{\partial a_i} = \frac{\partial \xi_1(\mathbf{a}, \tau)}{\partial a_i}, \quad i = 1, 2, 3. \tag{15.122}$$

Substituting $\xi_1^{'}$ for ξ_1 in (15.121), using (15.122) and reusing (15.121), then yields

$$\alpha\left(\xi_1^{'},\xi_2,\xi_3\right) = h_1\left(\xi_2,\xi_3\right) h_2\left(\xi_2,\xi_3\right) h_3\left(\xi_2,\xi_3\right) \begin{vmatrix} \frac{\partial \xi_1^{'}}{\partial a_1} & \frac{\partial \xi_1^{'}}{\partial a_2} & \frac{\partial \xi_1^{'}}{\partial a_3} \\ \frac{\partial \xi_2}{\partial a_1} & \frac{\partial \xi_2}{\partial a_2} & \frac{\partial \xi_2}{\partial a_3} \\ \frac{\partial \xi_3}{\partial a_1} & \frac{\partial \xi_3}{\partial a_2} & \frac{\partial \xi_3}{\partial a_3} \end{vmatrix}$$

$$= h_1\left(\xi_2,\xi_3\right) h_2\left(\xi_2,\xi_3\right) h_3\left(\xi_2,\xi_3\right) \begin{vmatrix} \frac{\partial \xi_1}{\partial a_1} & \frac{\partial \xi_1}{\partial a_2} & \frac{\partial \xi_1}{\partial a_3} \\ \frac{\partial \xi_2}{\partial a_1} & \frac{\partial \xi_2}{\partial a_2} & \frac{\partial \xi_2}{\partial a_3} \\ \frac{\partial \xi_3}{\partial a_1} & \frac{\partial \xi_3}{\partial a_2} & \frac{\partial \xi_3}{\partial a_3} \end{vmatrix} = \alpha\left(\xi_1,\xi_2,\xi_3\right). \tag{15.123}$$

Thus α is unchanged by infinitesimal rotation (15.118), that is, α has rotation symmetry. From (15.111), $\eta = \eta(\mathbf{a}) = \eta(a_1, a_2, a_3)$. Furthermore, it is a prescribed function of a fluid parcel's labels (whose values are determined by initial conditions), and unaffected by any change in the dependence of $\xi = (\xi_1, \xi_2, \xi_3)$ on $(\mathbf{a}, \tau)$. Thus it too has rotation symmetry.

Replacing ξ_1 by $\xi_1^{'}$ in (15.108), using (15.119) and (15.123), noting that η has rotational symmetry, and comparing the result to (15.108) (in unchanged form) then yields

$$\begin{aligned} L\left(\xi_1^{'},\xi_2,\xi_3\right) &= \iiint \left[\frac{h_1^2}{2}\left(\frac{\partial \xi_1^{'}}{\partial \tau}\right)^2 + \frac{h_2^2}{2}\left(\frac{\partial \xi_2}{\partial \tau}\right)^2 + \frac{h_3^2}{2}\left(\frac{\partial \xi_3}{\partial \tau}\right)^2 + \Omega h_1^2 \frac{\partial \xi_1^{'}}{\partial \tau}\right] d\mathbf{a} \\ &\quad - \iiint \left[\mathcal{E}(\alpha,\eta) + \Phi\left(\xi_2,\xi_3\right)\right] d\mathbf{a} \\ &= \iiint \left[\frac{h_1^2}{2}\left(\frac{\partial \xi_1}{\partial \tau}\right)^2 + \frac{h_2^2}{2}\left(\frac{\partial \xi_2}{\partial \tau}\right)^2 + \frac{h_3^2}{2}\left(\frac{\partial \xi_3}{\partial \tau}\right)^2 + \Omega h_1^2 \frac{\partial \xi_1}{\partial \tau}\right] d\mathbf{a} \\ &\quad - \iiint \left[\mathcal{E}(\alpha,\eta) + \Phi\left(\xi_2,\xi_3\right)\right] d\mathbf{a} \\ &= L\left(\xi_1,\xi_2,\xi_3\right). \end{aligned} \tag{15.124}$$

This equation satisfies condition (15.116) for rotation symmetry. Thus Lagrangian (15.108) (and, by equivalence, also Lagrangian (15.107)) has rotation symmetry. Note that it is implicit in the functional form $\Phi = \Phi(\xi_2, \xi_3)$ in (15.108) that Φ has rotation symmetry. If Φ were to be changed in such a way as to lose its rotation symmetry (i.e. if Φ were to depend on ξ_1, in addition to possible dependence on ξ_2 and ξ_3), then $L(\xi_1, \xi_2, \xi_3)$ would consequently also lose its rotation symmetry.

15.5.2.3 *Conservation of Global Axial Absolute Angular Momentum*

If Lagrangian L, defined by (15.107) or (15.108), has rotation symmetry according to criterion (15.115) or (15.116), then global axial absolute angular momentum

$$M = \widehat{\boldsymbol{\Omega}} \cdot \left[\mathbf{r} \times \left(\mathbf{u} + \boldsymbol{\Omega} \times \mathbf{r}\right)\right] = \left(\mathbf{u} + \boldsymbol{\Omega} \times \mathbf{r}\right) \cdot \left(\widehat{\boldsymbol{\Omega}} \times \mathbf{r}\right) = h_1\left(u_1 + \Omega h_1\right), \tag{15.125}$$

is conserved according to

$$\frac{d}{dt}\left[\iiint (\rho M)\, h_1 h_2 h_3 d\xi_1 d\xi_2 d\xi_3\right] = 0, \tag{15.126}$$

cf. (15.106) with $\mathbb{C} = M$. As mentioned earlier, Φ has to have rotation symmetry about the planet's rotation axis, $\boldsymbol{\Omega}$, for (15.126) to hold.

15.5.3 Time-Translation Symmetry and Conservation of Total Energy

15.5.3.1 *Time-Translation Symmetry of a Lagrangian*

We begin by defining:

Time-Translation Symmetry of a Lagrangian

A Lagrangian $L[\mathbf{r}(\mathbf{a},\tau)] \equiv L[x_1(\mathbf{a},\tau), x_2(\mathbf{a},\tau), x_3(\mathbf{a},\tau)]$ (in Cartesian coordinates) or $L[\boldsymbol{\xi}(\mathbf{a},\tau)] \equiv L[\xi_1(\mathbf{a},\tau), \xi_2(\mathbf{a},\tau), \xi_3(\mathbf{a},\tau)]$ (in axial-orthogonal-curvilinear coordinates) is said to have *time-translation symmetry* if

$$L\left[\mathbf{r}\left(\mathbf{a},\tau'\right)\right] = L[\mathbf{r}(\mathbf{a},\tau)], \tag{15.127}$$

or

$$L\left[\boldsymbol{\xi}\left(\mathbf{a},\tau'\right)\right] = L[\boldsymbol{\xi}(\mathbf{a},\tau)], \tag{15.128}$$

where

$$\tau' \equiv \tau + \delta\tau, \tag{15.129}$$

and $\delta\tau$ = constant is an infinitesimal time translation.

15.5.3.2 *Time-Translation Symmetry of Lagrangians* (15.107) *and* (15.108)

Having defined time-translation symmetry, we now show that Lagrangians (15.107) and (15.108) have this symmetry. Since Lagrangians (15.107) and (15.108) are equivalent to one another, we only need to do this for one of them. Because of the focus (in Part II of this book) on obtaining approximated governing equation sets in axial-orthogonal-curvilinear coordinates, we work here with the second form of the Lagrangian, that is, (15.108).

From (15.129)

$$\frac{\partial \xi_i\left(\mathbf{a},\tau'\right)}{\partial \tau'} = \frac{d\tau}{d\tau'}\frac{\partial \xi_i}{\partial \tau} = \frac{\partial \xi_i}{\partial \tau}, \quad i = 1,2,3. \tag{15.130}$$

Replacing τ by τ' in (15.108), using (15.130), and comparing the result to (15.108) (in unchanged form), then yields

$$\begin{aligned}
&L\left[\xi_1\left(\mathbf{a},\tau'\right), \xi_2\left(\mathbf{a},\tau'\right), \xi_3\left(\mathbf{a},\tau'\right)\right] \\
&= \iiint \left[\frac{h_1^2}{2}\left(\frac{\partial \xi_1}{\partial \tau'}\right)^2 + \frac{h_2^2}{2}\left(\frac{\partial \xi_2}{\partial \tau'}\right)^2 + \frac{h_3^2}{2}\left(\frac{\partial \xi_3}{\partial \tau'}\right)^2 + \Omega h_1^2 \frac{\partial \xi_1}{\partial \tau'}\right] d\mathbf{a} - \iiint \left[\mathcal{E}(\alpha,\eta) + \Phi(\xi_2,\xi_3)\right] d\mathbf{a} \\
&= \iiint \left[\frac{h_1^2}{2}\left(\frac{\partial \xi_1}{\partial \tau}\right)^2 + \frac{h_2^2}{2}\left(\frac{\partial \xi_2}{\partial \tau}\right)^2 + \frac{h_3^2}{2}\left(\frac{\partial \xi_3}{\partial \tau}\right)^2 + \Omega h_1^2 \frac{\partial \xi_1}{\partial \tau}\right] d\mathbf{a} - \iiint \left[\mathcal{E}(\alpha,\eta) + \Phi(\xi_2,\xi_3)\right] d\mathbf{a} \\
&= L\left[\xi_1(\mathbf{a},\tau), \xi_2(\mathbf{a},\tau), \xi_3(\mathbf{a},\tau)\right].
\end{aligned} \tag{15.131}$$

This equation satisfies condition (15.128) for time-translation symmetry. Thus Lagrangian (15.108) (and, equivalently, Lagrangian (15.107)) has time-translation symmetry.

15.5.3.3 *Conservation of Total Energy*

If Lagrangian L, defined by (15.107) or (15.108), has time-translation symmetry according to criterion (15.127) or (15.128), then

$$\frac{dE^{\text{total}}}{dt} = \frac{d}{dt}\left(E^{\text{kinetic}} + E^{\text{internal}} + E^{\text{gravitational}}\right) = 0, \tag{15.132}$$

where

$$E^{\text{total}} = \iiint \left[\frac{\mathbf{u}^2}{2} + \mathcal{E}(\alpha,\eta) + \Phi(\mathbf{r}) \right] d\mathbf{a} = \iiint \rho \left[\frac{\mathbf{u}^2}{2} + \mathcal{E}(\alpha,\eta) + \Phi(\mathbf{r}) \right] d\mathbf{r}, \quad (15.133)$$

$$E^{\text{kinetic}} = \iiint \left(\rho \frac{\mathbf{u}^2}{2} \right) d\mathbf{r} = \iiint \left(\rho \frac{\mathbf{u}^2}{2} \right) h_1 h_2 h_3 d\xi_1 d\xi_2 d\xi_3, \quad (15.134)$$

$$E^{\text{internal}} = \iiint \left[\rho \mathcal{E}(\alpha,\eta) \right] d\mathbf{r} = \iiint \left[\rho \mathcal{E}(\alpha,\eta) \right] h_1 h_2 h_3 d\xi_1 d\xi_2 d\xi_3, \quad (15.135)$$

$$E^{\text{gravitational}} = \iiint \left[\rho \Phi(\mathbf{r}) \right] d\mathbf{r} = \iiint \left[\rho \Phi(\xi_2,\xi_3) \right] h_1 h_2 h_3 d\xi_1 d\xi_2 d\xi_3, \quad (15.136)$$

$$\mathbf{u}^2 = \left(\frac{\partial \mathbf{r}}{\partial \tau} \right)^2 = \frac{h_1^2}{2} \left(\frac{\partial \xi_1}{\partial \tau} \right)^2 + \frac{h_2^2}{2} \left(\frac{\partial \xi_2}{\partial \tau} \right)^2 + \frac{h_3^2}{2} \left(\frac{\partial \xi_3}{\partial \tau} \right)^2. \quad (15.137)$$

To keep things simple, boundary conditions have been ignored here. (The focus herein is, for the most part, on the governing interior equations.) For most of the usual boundary conditions (rigid wall, horizontally periodic, free surface at zero pressure), boundary contributions are anyway identically zero; see Section 14.8.4.6.

15.5.4 Particle-Relabelling Symmetry and Conservation of Potential Vorticity

15.5.4.1 *Particle-Relabelling Symmetry*

We begin by defining:

Particle-Relabelling Symmetry of a Lagrangian

A Lagrangian $L[\mathbf{r}(\mathbf{a},\tau)]$ (in vector form) or $L[\xi_1(\mathbf{a},\tau), \xi_2(\mathbf{a},\tau), \xi_3(\mathbf{a},\tau)]$ (in axial-orthogonal-curvilinear coordinates) is said to have *particle-relabelling symmetry* if

$$L\left[\mathbf{r}\left(\mathbf{a}',\tau\right)\right] = L\left[\mathbf{r}(\mathbf{a},\tau)\right], \quad (15.138)$$

or

$$L\left[\xi_1\left(\mathbf{a}',\tau\right), \xi_2\left(\mathbf{a}',\tau\right), \xi_3\left(\mathbf{a}',\tau\right)\right] = L\left[\xi_1(\mathbf{a},\tau), \xi_2(\mathbf{a},\tau), \xi_3(\mathbf{a},\tau)\right], \quad (15.139)$$

where

$$\mathbf{a}' \equiv \left(a_1', a_2', a_3'\right) \equiv \mathbf{a} + \delta\mathbf{a}(a_1, a_2, a_3, \tau) \quad (15.140)$$

is a relabelling of vector label $\mathbf{a}$, and $\delta\mathbf{a}(a_1, a_2, a_3, \tau)$ is an infinitesimal vector function.

15.5.4.2 *Particle-Relabelling Symmetry of Lagrangians* (15.107) *and* (15.108)

Recall that particle labels $\mathbf{a} \equiv (a_1, a_2, a_3)$ have been assigned to satisfy (15.109) at time τ, but are otherwise arbitrary. Following Section 2 of Chapter 7 of Salmon (1998), assign label a_3 to be specific entropy, that is,

$$a_3 = \eta \quad \Rightarrow \quad \mathbf{a} = (a_1, a_2, \eta). \quad (15.141)$$

Now constrain the label variations $\delta\mathbf{a} \equiv (\delta a_1, \delta a_2, \delta a_3)$ to satisfy

$$\delta a_1 = -\frac{\partial}{\partial a_2}\left[\delta\psi(a_1, a_2, a_3, \tau)\right], \quad \delta a_2 = \frac{\partial}{\partial a_1}\left[\delta\psi(a_1, a_2, a_3, \tau)\right], \quad \delta a_3 = \delta\eta = 0, \quad (15.142)$$

where $\delta\psi(a_1, a_2, a_3, \tau)$ is an *arbitrary* infinitesimal function. That $\delta a_3 = 0$ in (15.142) follows from (15.111). Salmon (1998) shows that constraints (15.142) ensure (by his construction) that $\mathcal{E} + \Phi$ in Lagrangians (15.107) and (15.108) is invariant to fluid-particle relabelling (15.140)–(15.142). Lagrangians (15.107) and (15.108) are therefore invariant to this fluid-particle relabelling since the particle labels only appear therein through their presence in internal energy ($\mathscr{E}$); see (15.107)–(15.109).

15.5.4.3 *Conservation of Potential Vorticity*

Salmon (1998) also shows that this particle-relabelling symmetry leads to Ertel's theorem for conservation of potential vorticity. Thus

$$\frac{\partial}{\partial \tau}\left[\left(\frac{2\boldsymbol{\Omega} + \nabla \times \mathbf{u}}{\rho}\right) \cdot \nabla \eta\right] = \frac{\partial}{\partial \tau}\left[\frac{1}{\rho}\left(\frac{Z_1}{h_1}\frac{\partial \eta}{\partial \xi_1} + \frac{Z_2}{h_2}\frac{\partial \eta}{\partial \xi_2} + \frac{Z_3}{h_3}\frac{\partial \eta}{\partial \xi_3}\right)\right] = 0, \tag{15.143}$$

where (from (15.95)–(15.97))

$$Z_1 = \frac{1}{h_2 h_3}\left[\frac{\partial}{\partial \xi_2}\left(h_3 u_3\right) - \frac{\partial}{\partial \xi_3}\left(h_2 u_2\right)\right], \tag{15.144}$$

$$Z_2 = \frac{2\Omega}{h_3}\frac{\partial h_1}{\partial \xi_3} + \frac{1}{h_3 h_1}\left[\frac{\partial}{\partial \xi_3}\left(h_1 u_1\right) - \frac{\partial}{\partial \xi_1}\left(h_3 u_3\right)\right], \tag{15.145}$$

$$Z_3 = -\frac{2\Omega}{h_2}\frac{\partial h_1}{\partial \xi_2} + \frac{1}{h_1 h_2}\left[\frac{\partial}{\partial \xi_1}\left(h_2 u_2\right) - \frac{\partial}{\partial \xi_2}\left(h_1 u_1\right)\right]. \tag{15.146}$$

Note that Salmon (1998)'s derivation of Ertel's theorem is performed in an inertial frame of reference. A minor tweak to his analysis has therefore been made to obtain (15.143)–(15.146), which are valid in the planet's rotating frame. This gives rise to the planetary-vorticity contributions (associated with the explicit appearance of Ω) in these equations.

15.6 CONCLUDING REMARKS

The governing equations for atmospheric and oceanic dynamical cores are based on physical conservation principles. It is therefore only to be expected that, given a particular set of such governing equations, expressed in a particular form, one should be able to manipulate them to recover the conservation principles on which they are based. These physical principles may, however, be expressed in a variety of different, mathematical ways.

An unexpected result was the emergence in Section 15.3.6 of an *additional* conservation principle, namely Ertel's theorem for material conservation of potential vorticity. This fluid-dynamical principle is a consequence of other conservation principles with no analogue in classical particle mechanics. It provides a very strong local constraint on the evolution of fluid parcels by tightly coupling hydrodynamics with thermodynamics.

In this final chapter of Part I of this book, we have answered the question:

- What do various conservation principles look like, in both vector form and in axial-orthogonal-curvilinear coordinates, and in the presence of forcings, both internal and external?

A first practical benefit is that this facilitates the construction of budget equations to monitor and understand the evolution of the atmosphere and/or oceans during model integrations. A second, and important, further benefit is that this has prepared the way for Part II of this book, which examines dynamically consistent approximation of the governing dynamical-core equations.

Part II

DYNAMICALLY CONSISTENT EQUATION SETS

16

Deep and Shallow, Dynamically Consistent Equation Sets in 3D

ABSTRACT

An approximate equation set for motion of a global atmosphere or ocean is dynamically consistent if it implies conservation principles for mass, axial angular momentum, total energy, and potential vorticity, analogous to those of the unapproximated equations. Failure of an equation set to respect dynamical consistency makes results obtained using it unreliable. Initially negligible errors may spuriously increase to become non-negligible, and even dominant, during time integration. A quartet of approximate, dynamically consistent equation sets for global atmospheric and oceanic modelling is given in both spherical and spheroidal geometry. Each set is at least as accurate as the hydrostatic primitive equations. Historically, these equation sets were developed over a lengthy period of time. They may be written as a single, unified set of equations, with the option of jointly or severally applying two different kinds of approximation: (quasi-) hydrostatic and/or shallow. This provides insight. The classical derivation of the hydrostatic primitive equations is revisited and simplified. There is no need to separately invoke the 'traditional approximation' since it is shown to be a direct consequence of consistent application of shallow approximation. The four equation sets are alternatively derived – in a unified manner – via straightforward approximation of a Lagrangian. This is accomplished in two different but related ways: using Hamilton's principle of stationary action and taking variations, or using the Euler–Lagrange equations. This prepares the way for the derivation of other approximate, dynamically consistent equation sets in later chapters. A comparative discussion of the derivation methodologies is given in conclusion.

16.1 PREAMBLE

16.1.1 Introductory Remarks

Thus far we have considered the governing geophysical-fluid-dynamical equations for a global atmosphere or a global ocean overlying a rotating spheroidal planet. In particular we have maintained generality as much as is reasonably feasible. For global weather and climate prediction, it is desirable to make as few approximations as possible in order to maximise physical realism.[1] During the early decades of numerical weather prediction, the lack of powerful computers was a very serious problem indeed (Spekat, 2001). This very serious limitation necessarily resulted in the development and use of much simplified sets of governing equations to make quantitative atmospheric and (somewhat later) oceanic forecasting practicable.

Many such sets have been devised over the years, with early ones being necessarily rudimentary. The degree of approximation made depends on available computer capability and intended

[1] By way of contrast, the use of much simplified equation sets is often crucial for isolating and understanding fundamental physical phenomena such as wave propagation and instabilities.

application. Common to all such approximated equation sets is adoption of the classical spherical-geopotential approximation. See Sections 7.1.5 and 8.6 for a description of this approximation, its derivation as a zeroth-order *asymptotic limit*, and its consequent limitations.

Governing equation sets for global atmospheric and oceanic modelling are based on physical conservation laws for mass, momentum, and total energy; see Chapters 2–4, 6, 14, and 15. An important consequence of these laws in the present context is that potential vorticity is also conserved; see Section 15.3.6. The importance of preserving analogues of these fundamental conservation laws when formulating *approximate* governing equation sets for atmospheric and oceanic modelling is well established. Failure to do so risks making results obtained using such equation sets unreliable. Errors that are initially negligible can, if they are systematic, spuriously increase during time integration to become non-negligible, and even dominant, by final time (Phillips, 1966; Lorenz, 1967).

Scale analysis to determine what terms may be dropped or approximated in the unapproximated governing equations is a frequently used tool in geophysical fluid dynamics. Its use is, however, *insufficient* to guarantee the existence of conservation analogues after approximation; spurious sources and sinks may be introduced. Global equation sets given in early textbooks often did not preserve an analogue of the conservation principle for axial angular momentum when making the shallow-fluid approximation (Phillips, 1966, 1968; Veronis, 1968).

To maintain analogues of conservation laws when applying the shallow-fluid assumption, one has to omit some terms. Phillips (1966) showed (in spherical geometry) that conservation of axial angular momentum *with the shallow-fluid approximation* can, without prejudicing respect of other conservation principles, be recovered by omitting the so-called '$2\Omega\cos\phi$' terms (as well as four metric terms) in the components of the approximated momentum equation. Omitting the $2\Omega\cos\phi$ terms is termed the *traditional approximation*, a phrase introduced in Eckart (1960). It is, however, important to note that *recovering conservation of axial angular momentum does not physically justify neglect of the* $2\Omega\cos\phi$ *terms*. Respect of conservation principles is very important, but so is physical fidelity.

Over the years, various authors have questioned the validity of the traditional approximation for atmospheric and oceanic modelling, particularly for modelling equatorial flow; for example, Eckart (1960), Phillips (1966, 1968, 1973), Draghici (1987, 1989), Müller, 1989, White and Bromley (1995), Marshall et al. (1997), Kasahara (2003a,b), White et al. (2005), and Gerkema et al. (2008). In the deep oceans, the buoyancy frequency is typically an order of magnitude less than in the atmosphere (Gill, 1982, p. 52). The $2\Omega\cos\phi$ Coriolis terms are then correspondingly more important, since the ratio of planetary rotation frequency to buoyancy frequency is larger.[2]

16.1.2 Definition of Dynamical Consistency

In a global atmospheric context, White and Bromley (1995) coined the term *dynamically consistent* to describe approximate equation sets that possess analogues of the conservation principles for axial angular momentum, total energy, and potential vorticity that hold for the unapproximated equation set. Their designation has subsequently been adopted and adapted by White et al. (2005), White and Wood (2012), Dubos and Voitus (2014), Staniforth (2014a, 2015a), Tort and

[2] Since Gill (1982)'s excellent textbook is still frequently referenced – e.g. as herein – it is worth noting that there are a few errors regarding his description (at the bottom of his p. 93) of the application of the shallow-fluid approximation. His equations (4.12.14)–(4.12.16) for the components of the momentum equation for the deep case are correct. However, (only) setting $r = a$ in them, as described, violates conservation principles. Both of the $2\Omega\cos\phi$ terms, plus four metric terms, must also be omitted to properly apply the shallow-fluid approximation and thereby obtain conservation analogues. Also, his (4.12.11) is incorrect for the deep case; his implied shallow approximation of it is then also incorrect. Fortunately, none of these errors appear to have propagated elsewhere.

Dubos (2014a,b), and Tort et al. (2014). Herein, we adopt White and Bromley (1995)'s definition, but, additionally, we explicitly include conservation of mass rather than leaving this as being understood.[3] Thus:

Definition of Dynamical Consistency

An approximate set of governing equations for motion of a global atmosphere or ocean is considered to be *dynamically consistent* if it implies conservation principles (see Chapter 15) for:

- Mass.
- Axial angular momentum.
- Total energy.
- Potential vorticity.

analogous to those of the unapproximated equations.

Dynamical consistency focuses on analogues of fundamental conservation principles. Herein, these principles are considered to relate to the *macroscale* of fluid parcels; see Section 4.2.1.2 for discussion of microscopic versus macroscopic. As the name suggests, dynamical consistency intrinsically involves *consistent* representation of the time-dependent *dynamics* of fluid parcels, with emphasis on dynamics (as opposed to statics). Dynamical consistency is primarily influenced by the representations of the prognostic equations for momentum, continuity, and thermodynamic energy, and only secondarily by the representation of the fundamental equation of state. Herein, when examining dynamical consistency, it is implicitly assumed that thermodynamic representation of a fluid parcel is self-consistent and properly represents the *equilibrium* thermodynamic properties of the fluid. In other words, *it is assumed that internal energy ($\mathscr{E}$) of the fluid – as embodied in the fundamental equation of state – is well prescribed in a self-consistent manner.*

The complementary and equally important aspect of *thermodynamical consistency,* for consistent approximation of internal energy, is the focus of Chapter 9. In the present context, 'thermodynamical' is (in the author's view) something of a misnomer. When (as discussed in Section 4.2.1.2) a fluid parcel is viewed at the *microscale* (i.e. at fine scale within it), it certainly behaves *dynamically*; individually, the molecules of a fluid parcel are in random motion. However, when viewed at the *macroscale*, these individual random motions are essentially averaged out. The fluid parcel is then considered to be in a (quasi)*static* (as opposed to dynamic) state of thermodynamic *equilibrium*. *Dynamical consistency* focuses on the motion of fluid parcels (at the macroscale), whereas *thermodynamical consistency* focuses on the physical *equilibrium state* of fluid parcels (again at the macroscale); both are concerned with doing so in a consistent manner.

The concepts of dynamical and thermodynamical consistency (as defined herein) are linked through the need to respect *total*-energy conservation. The consistent representation of *internal* energy is embodied in the prescription of a thermodynamic potential. This is *assumed* when examining dynamical consistency. Dynamical consistency then ensures that, *with an assumed, self-consistent representation of internal energy*, an approximate model of the global atmosphere or of an ocean has conservation principles for mass, axial angular momentum, total energy, and potential vorticity, analogous to those of the unapproximated equations.

[3] The full, unapproximated continuity equation was retained in their work; mass conservation then implicitly and automatically follows from this.

16.1.3 A Guide to the Remainder of This Chapter

Having defined dynamical consistency, the remainder of this chapter is devoted to the description and derivation of a quartet (two deep, two shallow) of approximate equation sets, suitable for quantitative, global, atmospheric, and oceanic modelling. Attention here is focused on a model's dynamical core, that is, on its governing equations in the absence of forcings (other than gravity).

In Section 16.2, we begin by giving an historical overview of the development of these four equation sets. These equation sets are then expressed in a unified manner, firstly in spherical-polar coordinates (the coordinate system of choice in the literature, and in textbooks, for expressing global equation sets), and then (more generally) in axial-orthogonal-curvilinear coordinates. Colouration of terms facilitates comparison between different equation sets. For illustrative and reference purposes, the individual equation sets are given explicitly in two appendices (one for spherical geometry, the other for spheroidal geometry).

Having documented the mathematical forms for this quartet of equation sets, attention is turned in the remaining sections to how they may be derived. Three methodologies for doing so are identified in Section 16.3. Each methodology is then described and discussed in the following three sections:

- Classical Eulerian (in Section 16.4).
- Lagrangian using Hamilton's principle of stationary action (in Section 16.5).
- Lagrangian using Euler–Lagrange equations (in Section 16.6).

The classical, Eulerian derivation of the hydrostatic primitive equations is revisited in Section 16.7 and simplified. This then obviates the need to invoke the 'traditional approximation' since this is shown to be a direct consequence of consistent application of shallow approximation. Finally, concluding remarks are made in Section 16.8. This provides a comparative discussion of the three methodologies, plus a bridge to later chapters.

16.2 A UNIFIED QUARTET OF DYNAMICALLY CONSISTENT EQUATION SETS

16.2.1 Historical Overview

Assuming the classical *spherical*-geopotential approximation, a quartet of existing, approximate equation sets for the global atmosphere were consolidated within a unified framework in Staniforth (2001). All four sets are valid over a broad range of scales, from global down to the mesoscale, and are at least as accurate as the well-known, and frequently used, hydrostatic primitive equations. See Fig. 16.1 for a schematic of the interrelationships between individual members (two in dark blue, two in light blue) of this quartet.[4]

Dynamical consistency of these four equation sets was subsequently analysed in detail in White et al. (2005). It was also shown (in their Appendix A) that the component equations of momentum for each equation set have the form of Euler–Lagrange equations. Although Staniforth (2001)'s and White et al. (2005)'s analyses were presented in the context of *atmospheric* modelling, they are also applicable, almost unchanged, to the *oceanic* context. For illustrative and reference purposes, the governing equations for each of the equation sets analysed therein are summarised in Appendix A of the present chapter, together with their associated conservation principles.

White and Wood (2012) generalised the spherical quartet to *spheroidal* geometry by:

- Assuming a *spheroidal*-geopotential approximation instead of a *spherical* one.
- Using axial-orthogonal-curvilinear coordinates instead of spherical-polar ones.

[4] The blank space in the middle of this figure prepares the way for the insertion there of a pair of (quasi-shallow) equation sets, plus their interrelationships with the deep and shallow ones depicted here (in dark and light blue, respectively). See Chapter 17 and, in particular, Fig. 17.1. The quasi-shallow pair of equation sets are of intermediate complexity between the deep and shallow pairs.

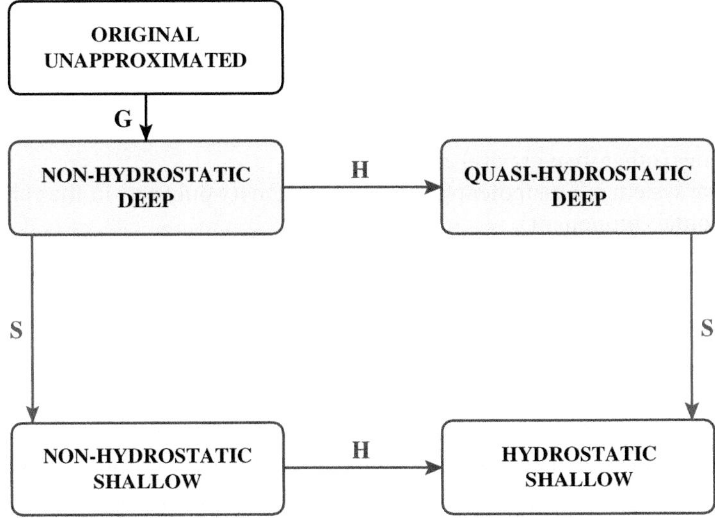

Figure 16.1 Interrelationships between four, dynamically consistent, approximate equation sets (deep ones in dark blue, shallow ones in light blue) derived from the original, unapproximated one (unshaded). **G** denotes application of a geopotential approximation; **H**, application of quasi-hydrostatic or hydrostatic approximation; and **S**, application of shallow-fluid approximation. See text for further details.

The governing equations and associated conservation principles for each of these four equation sets are similarly summarised for illustrative and reference purposes in Appendix B of the present chapter. The spheroidal quartet gives rise to additional metric terms when compared to the spherical one. An alternative, complementary derivation (using Hamilton's principle of stationary action) of the generalised quartet of equation sets is provided in Staniforth (2014a). In particular, components of the momentum equation for them are succinctly assembled in a unified manner.

The four, approximated equation sets (in both spherical and spheroidal geometry) depicted in Fig. 16.1 arise according to whether approximations are made of (quasi-)hydrostatic[5] and/or shallow nature. The associated equation sets are referred to as:

- Non-hydrostatic deep (or fully compressible).
- Quasi-hydrostatic deep (or quasi-hydrostatic).
- Non-hydrostatic shallow.
- Hydrostatic shallow (or hydrostatic primitive).

These sets can be organised in a hierarchy of successive approximation: the least-approximated ('original unapproximated') equation set at the top left of Fig. 16.1; and the most approximated ('hydrostatic shallow' = 'hydrostatic primitive') at the bottom right.

The original *unapproximated* set, despite the nomenclature, is not entirely unapproximated. When writing down the governing equations – see Chapters 2–4 – various assumptions (i.e. approximations) were necessarily made. These include (but are not limited to) the usual ones that:

- The atmosphere and oceans behave to leading order as fluids.
- Fluid composition can be well represented.
- Important physical forcings can also be well represented.

[5] For a deep fluid, this approximation is termed quasi-hydrostatic; whence the use here of the nomenclature *(quasi-)hydrostatic* to embrace both deep and shallow contexts.

Here the focus is on dynamically consistent approximation of the underlying geophysical-fluid-dynamical equations of a dynamical core. See Section 1.2 for definition and discussion of dynamical cores, and their important role in quantitative atmospheric and oceanic modelling. It is therefore assumed throughout this chapter that:

- Physical forcings (other than gravity) are absent.[6]
- The coordinate system is a geopotential one (so that gravity only acts in the vertical direction, with no horizontal component).
- The fluid is composed of a single substance (dry air for the atmosphere, pure liquid water for the oceans).

These assumptions not only shorten equations, they also enhance clarity in the present context by excluding extraneous detail.

At the other extreme – see bottom right of Fig. 16.1 – are the *hydrostatic-shallow* equations, which have brought so much joy to so many meteorologists and oceanographers over the past half century or so. For historical reasons, these equations are usually referred to as the *hydrostatic-primitive* equations: *hydrostatic* because they assume the hydrostatic approximation; and *primitive* (i.e. fundamental) because it was believed at the time that it would never be feasible to use a more general equation set for numerical modelling. How times change.

The least-approximated equation set – again see Fig. 16.1 – is the *non-hydrostatic-deep* one. It is obtained from the original, *unapproximated* set by applying a geopotential approximation (to represent a planet's gravitational attraction); this process is denoted by **G** on the schematic. Past practice (and also present at the time of writing) is to use the classical, *zeroth-order spherical* geopotential (together with spherical coordinates). More generally – as discussed in detail in Chapters 7, 8, and 12 – a *first-order spheroidal* approximation (together with a spheroidal coordinate system) is preferable.[7]

There are two ways to successively simplify the non-hydrostatic-deep equations to obtain the hydrostatic-shallow ones. These are to successively apply the (quasi-)hydrostatic (denoted by H on the schematic) and shallow (S) approximations, but in different order. See Fig. 16.1. One can first apply a hydrostatic-type (H) approximation to obtain the *quasi-hydrostatic-deep* equations,[8] and then further apply the shallow (S) approximation to obtain the hydrostatic-shallow ones. Alternatively, one can apply these two approximations in reverse order to instead obtain the *non-hydrostatic-shallow* equations as the intermediate equation set. The quasi-hydrostatic-deep equations are developed and discussed in White and Bromley (1995), and the non-hydrostatic-shallow ones in Tanguay et al. (1990).

To obtain the (quasi-)hydrostatic equation sets from their non-hydrostatic counterparts, various terms are omitted from the vertical momentum equation, resulting in a simpler, (quasi-)hydrostatic balance equation. The process of doing so is denoted by H on the schematic. To obtain the shallow equation sets from their deep counterpart – by applying the shallow-fluid assumption (S) – is more complicated due to the need to maintain dynamical consistency. Some terms have to be omitted. They are generally small, *but not always.* An underlying difficulty with applying the shallow-fluid approximation in a dynamically consistent manner is that this approximation distorts Euclidean geometry into non-Euclidean geometry, with the consequence

[6] Forcings are included in Chapter 15 for the conservation principles of the full equations. Guided by the full-equations analysis, forcings can be reinserted at any time.

[7] There is a subtlety to be aware of regarding geopotential approximation; see Section 8.6. For consistency reasons, when applying the *shallow-fluid approximation* (be it in spherical or spheroidal geometry), a *deep* geopotential representation must be appropriately approximated in a *shallow* manner. This then ensures that the Newtonian gravitational flux divergence outside a planet is identically zero. Doing so is part and parcel of correctly applying the shallow-fluid approximation.

[8] The *quasi-hydrostatic*-deep equations are not strictly hydrostatic due to the presence of some (typically small, but not always) terms that are absent in the shallow case; whence *quasi-hydrostatic*.

that intuitive Euclidean reasoning can break down (Thuburn and White, 2013). We will see in Section 16.7 that it is in fact possible to make the (dynamically consistent) transition from the deep-fluid pair of equation sets (non-hydrostatic-deep and quasi-hydrostatic-deep) to the corresponding pair of shallow-fluid ones (non-hydrostatic-shallow and hydrostatic-shallow) in a more straightforward but nevertheless still Eulerian manner than that usually employed in the literature.

The preceding discussion broadly describes how four approximated equation sets may be obtained by successive approximation of an 'unapproximated' equation set. Historically, however, for quantitative atmospheric and oceanic forecasting, and for practical reasons, the evolution of the quartet of equation sets has proceeded in the opposite sense; from *most* approximated (the hydrostatic-primitive equations) towards *least* approximated (the non-hydrostatic-deep equations), as computer power has dramatically increased over the years.

The reader may wonder why it has taken so long (fifty years, give or take) for this to happen, given that there is a relatively small difference (much less than a factor of two) in the number of terms that appear in any of these equation sets. The principal reasons are:

1. The serious stability constraints imposed by explicit time discretisation of the governing equations.
2. The fact that it is only relatively recently that it has been possible to integrate global models at high-enough resolution for some of the previously neglected terms to become non-negligible.

In the early days of atmospheric and oceanic modelling, time discretisations were *explicit*. Efficient *time-implicit* schemes had yet to be developed and, when they were, doubts (in retrospect, mostly unfounded) were expressed regarding their accuracy. For computational-stability reasons, maximum time-step length (Δt) of an *explicit* time scheme is proportional to spatial meshlength (Δs) and inversely proportional to signal speed (c_s). Without going into details, the most restrictive situation occurs for the vertical propagation of acoustic oscillations (i.e. sound waves). See Section 4.2.10 for representative values of c_s for the atmosphere and for the oceans. Basically, c_s is very large, and Δs is very small (particularly for the lowermost atmospheric layers, and uppermost oceanic ones), with the consequence that Δt is tiny for *computational-stability reasons* (only). But *not* for accuracy reasons. What to do?

Now vertically propagating acoustic oscillations carry very little energy. They therefore contribute negligibly to accuracy. Normal-mode analysis of the governing equations shows that such modes of energy propagation can be *entirely eliminated* by eliminating Du_3/Dt from the vertical component of the momentum equation.[9] Doing so then eliminates the severe time-step restriction (of an explicit time scheme) associated with vertically propagating acoustic oscillations! This, combined with the simplification afforded by the shallow-fluid assumption, reduces the vertical component of the momentum equation to an equation for *hydrostatic balance*, with just two terms.[10] These two assumptions then lead to the *hydrostatic primitive equations*. In summary, the preceding discussion explains why the hydrostatic primitive equations have been the workhorse equation set for so many years:

- The most troublesome modes of energy propagation are removed analytically.
- Hydrostatic balance is well respected physically in both the atmosphere and the oceans.

9 The *prognostic* vertical momentum equation is thus reduced to a *diagnostic* equation for (quasi-)hydrostatic balance. This means that the vertical component of velocity (u_3, *which is generally non-zero*) then has to be determined *diagnostically* from a consistently derived equation, known as *Richardson's equation* (after Lewis Fry Richardson, 1881–1953), instead of *prognostically*.

10 Hydrostatic balance was used extensively as a constraint in the early days of numerical weather prediction, not only in forecast models but also for processing observations (e.g. radiosonde profiles) and for initialisation (also known as balancing) of objectively analysed fields at initial time.

But, you might say, this sounds too good to be true. And indeed it is. Broadly speaking, and appropriately applied, everything works wonderfully well until horizontal scales are of the order of 10 km or so (for atmospheric models; Daley, 1988). However, previously neglected terms eventually become increasingly important as resolution is further increased. It is only relatively recently that computational power has been sufficient to run global models at such high resolutions.

To address this limitation then requires reintroduction of missing terms (e.g. to obtain the non-hydrostatic-deep equations) and, in particular, reintroduction of vertical acceleration. Even with today's computers, integrating these equations with an *explicit* time scheme would be computationally very expensive and inefficient. The very good news, however, is that computationally efficient time schemes have been developed over the past several decades. Using such schemes then permits the non-hydrostatic-deep equations to be integrated with a computational efficiency approaching that of integration of the workhorse hydrostatic primitive equations.

With this overview, we now assemble two unified quartets of equations sets; first in spherical geometry (in Section 16.2.2), then in spheroidal geometry (in Section 16.2.3). For both geometries, the corresponding conservation principles are also given.

16.2.2 A Unified Quartet in Spherical-Polar Coordinates

The four equation sets verbally described earlier are expressed in spherical-polar coordinates in Appendix A of the present chapter. They correspond to those given in White et al. (2005) for these coordinates. Examination reveals similarities and differences between these equation sets. By inspection they can be compactly unified in a manner similar to that given in Section 4 of Staniforth (2001), as follows.

16.2.2.1 *Governing Equations*

In the notation of Section 6.6 and with the introduction of some colour and the variable $\boxed{\mathrm{r}}$ of Staniforth (2001), the non-hydrostatic-deep equations (6.46)–(6.57), in spherical-polar coordinates (λ, ϕ, r) and in the absence of forcings, may be rewritten as:

A Unified Quartet of Equation Sets in Spherical-Polar Coordinates

$$\frac{D_{\boxed{\mathrm{r}}} u_\lambda}{Dt} - \left(\frac{u_\lambda}{\boxed{\mathrm{r}} \cos\phi} + 2\Omega \right) \left(u_\phi \sin\phi - \underbrace{u_r \cos\phi}_{\text{deep}} \right) + \frac{1}{\rho \boxed{\mathrm{r}} \cos\phi} \frac{\partial p}{\partial \lambda} = 0, \tag{16.1}$$

$$\frac{D_{\boxed{\mathrm{r}}} u_\phi}{Dt} + \left(\frac{u_\lambda}{\boxed{\mathrm{r}} \cos\phi} + 2\Omega \right) u_\lambda \sin\phi + \underbrace{\frac{u_\phi u_r}{r}}_{\text{deep}} + \frac{1}{\rho \boxed{\mathrm{r}}} \frac{\partial p}{\partial \phi} = 0, \tag{16.2}$$

$$\underbrace{\frac{D_{\boxed{\mathrm{r}}} u_r}{Dt}}_{\text{non-hydro}} - \underbrace{\left(\frac{u_\lambda^2}{r} + \frac{u_\phi^2}{r} + 2\Omega u_\lambda \cos\phi \right)}_{\text{deep}} + \frac{1}{\rho} \frac{\partial p}{\partial r} + \frac{d\Phi_{\boxed{\mathrm{r}}}}{dr} = 0, \tag{16.3}$$

$$\frac{D_{\boxed{\mathrm{r}}} \rho}{Dt} + \rho \nabla_{\boxed{\mathrm{r}}} \cdot \mathbf{u} = 0, \tag{16.4}$$

$$\frac{D_{\boxed{r}}\mathcal{E}}{Dt} + p\frac{D_{\boxed{r}}\alpha}{Dt} = 0, \tag{16.5}$$

where:

$$\alpha \equiv \frac{1}{\rho}, \tag{16.6}$$

$$\frac{D_{\boxed{r}}}{Dt} = \frac{\partial}{\partial t} + \frac{u_\lambda}{\boxed{r}\cos\phi}\frac{\partial}{\partial\lambda} + \frac{u_\phi}{\boxed{r}}\frac{\partial}{\partial\phi} + u_r\frac{\partial}{\partial r}, \tag{16.7}$$

$$\mathbf{u} = u_\lambda\mathbf{e}_\lambda + u_\phi\mathbf{e}_\phi + u_r\mathbf{e}_r \equiv \boxed{r}\cos\phi\frac{D_{\boxed{r}}\lambda}{Dt}\mathbf{e}_\lambda + \boxed{r}\frac{D_{\boxed{r}}\phi}{Dt}\mathbf{e}_\phi + \frac{D_{\boxed{r}}r}{Dt}\mathbf{e}_r, \tag{16.8}$$

$$\nabla_{\boxed{r}}\cdot\mathbf{A} = \frac{1}{\boxed{r}^2\cos\phi}\left[\frac{\partial}{\partial\lambda}\left(A_\lambda\boxed{r}\right) + \frac{\partial}{\partial\phi}\left(A_\phi\boxed{r}\cos\phi\right) + \frac{\partial}{\partial r}\left(A_r\boxed{r}^2\cos\phi\right)\right]. \tag{16.9}$$

	Non-Hydrostatic	(Quasi-)Hydrostatic
Deep *Additional terms retained*:	$\boxed{r} \to r$ • 'deep' • 'non-hydro'	$\boxed{r} \to r$ • 'deep'
Shallow *Additional terms retained*:	$\boxed{r} \to a$ • 'non-hydro'	$\boxed{r} \to a$

Table 16.1 Deep/shallow settings for $\boxed{r}$ in the unified equation set (16.1)–(16.89), plus additional terms retained in (16.1)–(16.3), as a function of the equation set. Black terms in (16.1)–(16.3) are present in all four sets.

Common to all four equation sets of the quartet, and needed to close them, is:

The Fundamental Equation of State

$$\mathcal{E} = \mathcal{E}(\alpha,\eta) \quad\Rightarrow\quad p \equiv -\frac{\partial\mathcal{E}(\alpha,\eta)}{\partial\alpha}, \quad T \equiv \frac{\partial\mathcal{E}(\alpha,\eta)}{\partial\eta}. \tag{16.10}$$

These three equations – see (6.35)–(6.37) – are *functionally* independent of whether the fluid is deep or shallow[11] and of whether the flow is (quasi-)hydrostatically balanced or not; they depend only upon the thermodynamic state of the fluid. By way of contrast, the forms of (16.1)–(16.5) and (16.7)–(16.9) *do* (explicitly) depend upon whether the fluid is deep or shallow.

The variable $\boxed{r}$ has been introduced to allow the quartet of equation sets given in Appendix A of the present chapter to be unified as the equation set (16.1)–(16.9). Setting $\boxed{r} = r$ leads to the

[11] Although $\mathcal{E} = \mathcal{E}(\alpha,\eta)$ does not *directly* depend on fluid depth (deep vs. shallow), there is *indirect* dependence. Evolution of $\alpha \equiv 1/\rho$ depends upon continuity equation (16.4), with dependence on fluid depth.

two *deep* equation sets (16.80)–(16.89), namely the *non-hydrostatic-deep* equations, by retaining $D_r u_r/Dt$; and to the *quasi-hydrostatic-deep* equations by omitting $D_r u_r/Dt$.

Recipes to obtain individual equation sets of the quartet from unified set (16.1)–(16.10) are given in Table 16.1. Application of these recipes then recovers the four equation sets given explicitly in Appendix A. For the deep terms in (16.2) and (16.3), it does not matter whether the division by r is as written, or alternatively as $\boxed{\mathrm{r}}$, since these terms only survive for the deep case (for which $\boxed{\mathrm{r}} = r$) *and are entirely absent* for the shallow one.

Setting $\boxed{\mathrm{r}} = a$ in (16.1)–(16.9) and neglecting 'deep' terms leads to the two shallow equation sets (16.96)–(16.105), namely the *non-hydrostatic-shallow* equations, by retaining $D_a u_r/Dt$; and to the *hydrostatic-shallow* (*hydrostatic-primitive*) equations, by omitting $D_a u_r/Dt$.

The form of the geopotential $\Phi_{\boxed{\mathrm{r}}}$ appearing in (16.3) depends upon whether the fluid is deep or shallow; see footnote 7. For the deep case, Φ is given by (5.26) of Section 5.2.2; and by (5.83) of Section 5.3.2 for the shallow case. Thus:

Geopotential Representation in Spherical-Polar Coordinates

Deep Fluid : $$\Phi_{\boxed{\mathrm{r}}} \rightarrow \Phi_r(r) = -g_a \frac{a^2}{r} \quad \Rightarrow \quad \mathbf{g}_r(r) = -\frac{d\Phi_r}{dr}\mathbf{e}_r = -g_a \frac{a^2}{r^2}\mathbf{e}_r, \tag{16.11}$$

Shallow Fluid : $$\Phi_{\boxed{\mathrm{r}}} \rightarrow \Phi_a(r) = g_a(r-a) \quad \Rightarrow \quad \mathbf{g}_a(r) = -\frac{d\Phi_a}{dr}\mathbf{e}_r = -g_a\mathbf{e}_r, \tag{16.12}$$

where $g_a \equiv g(r=a) =$ constant is the value of gravity at the planet's spherical surface $r = a$.

16.2.2.2 *Conservation Principles*

There are a number of ways of expressing conservation principles; see Chapter 15. The following principles, expressed in Lagrangian form, hold for the unified equation set in spherical-polar coordinates (Staniforth, 2001; White et al., 2005).

Unified Conservation Principles in Spherical-Polar Coordinates

Axial Angular Momentum

$$\frac{D_{\boxed{\mathrm{r}}}}{Dt}\left[\left(u_\lambda + \Omega \boxed{\mathrm{r}} \cos\phi\right)\boxed{\mathrm{r}}\cos\phi\right] = -\frac{1}{\rho}\frac{\partial p}{\partial \lambda}. \tag{16.13}$$

Total Energy

$$\frac{D_{\boxed{\mathrm{r}}}}{Dt}\left[\frac{1}{2}\left(u_\lambda^2 + u_\phi^2 + u_r^2\right) + \Phi_{\boxed{\mathrm{r}}} + \mathcal{E}\right] = -\frac{1}{\rho}\nabla_{\boxed{\mathrm{r}}} \cdot (p\mathbf{u}). \tag{16.14}$$

Potential Vorticity

$$\frac{D_{\boxed{\mathrm{r}}}}{Dt}\left(\frac{\mathbf{Z}_{\boxed{\mathrm{r}}} \cdot \nabla_{\boxed{\mathrm{r}}}\Lambda}{\rho}\right) = 0, \tag{16.15}$$

where:

$$\mathbf{Z}_{\boxed{r}} \equiv \nabla_{\boxed{r}} \times \mathbf{u} + 2\boldsymbol{\Omega} = \text{absolute vorticity}$$

$$= \frac{1}{\boxed{r}}\left[\frac{\partial u_r}{\partial \phi} - \frac{\partial}{\partial r}\left(\boxed{r}\, u_\phi\right)\right]\mathbf{e}_\lambda + \left[\frac{1}{\boxed{r}}\frac{\partial}{\partial r}\left(\boxed{r}\, u_\lambda\right) - \frac{1}{\boxed{r}\cos\phi}\frac{\partial u_r}{\partial \lambda} + 2\Omega\cos\phi\right]\mathbf{e}_\phi$$

$$+ \left\{\frac{1}{\boxed{r}\cos\phi}\left[\frac{\partial u_\phi}{\partial \lambda} - \frac{\partial}{\partial \phi}\left(u_\lambda \cos\phi\right)\right] + 2\Omega\sin\phi\right\}\mathbf{e}_r, \tag{16.16}$$

$$\frac{D_{\boxed{r}}\Lambda}{Dt} = 0, \tag{16.17}$$

$$\nabla_{\boxed{r}}\Lambda = \frac{1}{\boxed{r}\cos\phi}\frac{\partial \Lambda}{\partial \lambda}\mathbf{e}_\lambda + \frac{1}{\boxed{r}}\frac{\partial \Lambda}{\partial \phi}\mathbf{e}_\phi + \frac{\partial \Lambda}{\partial r}\mathbf{e}_r, \tag{16.18}$$

and $\Lambda = \Lambda(\alpha, \eta)$ is any materially conserved quantity that satisfies (16.17).

16.2.3 A Unified Quartet in Axial-Orthogonal-Curvilinear Coordinates

The four equation sets verbally described in Section 16.2.1 are expressed in axial-orthogonal-curvilinear coordinates in Appendix B of the present chapter. They correspond to those given in White and Wood (2012) for these coordinates. As for the spherical case, examination again reveals similarities and differences between these equation sets. By inspection, they can be compactly unified in a manner similar to that given in Section 4 of Staniforth (2015a) – but without the 'quasi-shallow' terms therein.

16.2.3.1 *Governing Equations*

In the notation of Section 6.5, and with the introduction of some colour, the non-hydrostatic-deep equations (6.29)–(6.40) may be written in the absence of sources and sinks as:

A Unified Quartet of Equation Sets in Axial-Orthogonal-Curvilinear Geopotential Coordinates

$$\frac{Du_1}{Dt} + \left(\frac{u_1}{h_1} + 2\Omega\right)\left(\frac{u_2}{h_2}\frac{\partial h_1}{\partial \xi_2} + \underbrace{\frac{u_3}{h_3}\frac{\partial h_1}{\partial \xi_3}}_{\text{deep}}\right) + \frac{1}{\rho h_1}\frac{\partial p}{\partial \xi_1} = 0, \tag{16.19}$$

$$\frac{Du_2}{Dt} - \left(\frac{u_1}{h_1} + 2\Omega\right)\frac{u_1}{h_2}\frac{\partial h_1}{\partial \xi_2} - \underbrace{\frac{u_3^2}{h_2 h_3}\frac{\partial h_3}{\partial \xi_2}}_{\text{non-hydro}} + \underbrace{\frac{u_2}{h_2}\frac{u_3}{h_3}\frac{\partial h_2}{\partial \xi_3}}_{\text{deep}} + \frac{1}{\rho h_2}\frac{\partial p}{\partial \xi_2} = 0, \tag{16.20}$$

$$\underbrace{\frac{Du_3}{Dt} + \frac{u_2}{h_2}\frac{u_3}{h_3}\frac{\partial h_3}{\partial \xi_2}}_{\text{non-hydro}} \underbrace{- \left[\left(\frac{u_1}{h_1} + 2\Omega\right)\frac{u_1}{h_3}\frac{\partial h_1}{\partial \xi_3} + \frac{u_2^2}{h_2 h_3}\frac{\partial h_2}{\partial \xi_3}\right]}_{\text{deep}} + \frac{1}{\rho h_3}\frac{\partial p}{\partial \xi_3} + \frac{1}{h_3}\frac{d\Phi}{d\xi_3} = 0, \tag{16.21}$$

$$\frac{D\rho}{Dt} + \rho\nabla\cdot\mathbf{u} = 0, \tag{16.22}$$

$$\frac{D\mathcal{E}}{Dt} + p\frac{D\alpha}{Dt} = 0, \tag{16.23}$$

where:

$$\alpha \equiv \frac{1}{\rho}, \tag{16.24}$$

$$\frac{D}{Dt} = \frac{\partial}{\partial t} + \frac{u_1}{h_1}\frac{\partial}{\partial\xi_1} + \frac{u_2}{h_2}\frac{\partial}{\partial\xi_2} + \frac{u_3}{h_3}\frac{\partial}{\partial\xi_3}, \tag{16.25}$$

$$\mathbf{u} \equiv u_1\mathbf{e}_1 + u_2\mathbf{e}_2 + u_3\mathbf{e}_3 = h_1\frac{D\xi_1}{Dt}\mathbf{e}_1 + h_2\frac{D\xi_2}{Dt}\mathbf{e}_2 + h_3\frac{D\xi_3}{Dt}\mathbf{e}_3, \tag{16.26}$$

$$\nabla\cdot\mathbf{A} = \frac{1}{h_1h_2h_3}\left[\frac{\partial}{\partial\xi_1}\left(A_1h_2h_3\right) + \frac{\partial}{\partial\xi_2}\left(A_2h_3h_1\right) + \frac{\partial}{\partial\xi_3}\left(A_3h_1h_2\right)\right]. \tag{16.27}$$

As noted on p. 983 of White and Wood (2012), the 'deep' terms (in red) in (16.19)–(16.21) form two pairwise sets of energy-conserving terms (one pair involving $\partial h_1/\partial\xi_3$, the other $\partial h_2/\partial\xi_3$); similarly, so do the pair of 'non-hydro' terms (in blue, involving $\partial h_3/\partial\xi_2$). These pairs of terms are three of the four energy-conserving pairs noted on p. 983 of White and Wood (2012).[12]

Common to all four equation sets of the quartet, and needed to close them, is again;

The Fundamental Equation of State

$$\mathcal{E} = \mathcal{E}(\alpha,\eta) \quad\Rightarrow\quad p \equiv -\frac{\partial\mathcal{E}(\alpha,\eta)}{\partial\alpha}, \quad T \equiv \frac{\partial\mathcal{E}(\alpha,\eta)}{\partial\eta}. \tag{16.28}$$

These three equations – see (6.35)–(6.37) – are functionally independent of whether the fluid is deep or shallow and of whether the flow is (quasi-)hydrostatically balanced or not; they depend only upon the thermodynamic state of the fluid. By way of contrast, the forms of (16.19)–(16.23) and (16.25)–(16.27) depend upon whether the fluid is deep or shallow. Retaining all terms in (16.19)–(16.28) recovers the *deep* equations (16.113)–(16.121).

The precise form of the geopotential Φ appearing in (16.21) depends upon:

- Whether the fluid is deep or shallow – see footnote 7.
- The composition and shape of the planet – see Chapters 7, 8, and 12.

In summary:

[12] White and Wood (2012)'s (A.3), (A.12), and (A.15) have a common typographic error in them. Their term $u_2^2\left(\partial h_1/\partial\xi_3\right)/\left(h_2h_3\right)$ is incorrect; it should be $u_2^2\left(\partial h_2/\partial\xi_3\right)/\left(h_2h_3\right)$ as in (16.21) herein. This error has also propagated to (3) and (13) of Staniforth and Wood (2013) but fortunately does not affect their subsequent analysis and conclusions.

Geopotential Representation in Axial-Orthogonal-Curvilinear Coordinates

Deep Fluid

For a deep fluid,

$$g \equiv |\mathbf{g}| = g(\xi_2, \xi_3) = \frac{1}{h_3(\xi_2)} \left| \frac{d\Phi(\xi_3)}{d\xi_3} \right|, \tag{16.29}$$

where $\Phi = \Phi(\xi_3)$ is a prescribed geopotential.

Shallow Fluid

For a shallow fluid,

$$g_S \equiv g\left(\xi_2, \xi_3 = \xi_3^S\right) = g_S(\xi_2). \tag{16.30}$$

Since

$$g_S(\xi_2) = \frac{1}{h_3^S(\xi_2)} \left| \frac{d\Phi_S(\xi_3)}{d\xi_3} \right|, \tag{16.31}$$

vertical integration of this equation gives the corresponding geopotential

$$\Phi_S(\xi_3) = \Phi_S\left(\xi_3^S\right) + g_S(\xi_2)\, h_3^S(\xi_2)\left(\xi_3 - \xi_3^S\right), \tag{16.32}$$

where $s_3 = h_3^S(\xi_2)\left(\xi_3 - \xi_3^S\right)$ is vertical distance from the planet's surface at constant ξ_2.

	Non-Hydrostatic	(Quasi-)Hydrostatic
Deep	Unapproximated h_i	Unapproximated h_i
Additional terms retained:	• 'deep' • 'non-hydro'	• 'deep'
Shallow	$h_i \to h_i^S$	$h_i \to h_i^S$
Additional terms retained:	• 'non-hydro'	

Table 16.2 Metric factor settings (deep h_i vs. shallow h_i^S) in the unified equation set (16.19)–(16.28), plus additional terms retained in (16.19)–(16.21), as a function of the equation set. Black terms in (16.19)–(16.21) are present in all four sets.

Recipes to obtain the individual equation sets of the quartet from unified set (16.19)–(16.28) are given in Table 16.2. Application of these recipes then recovers the two pairs of equation sets (four in all) given explicitly in Appendix B.

16.2.3.2 *Shallow Approximation*

A very nice property indeed of the expression of the governing equations in terms of axial-orthogonal-curvilinear coordinates is that application of the *shallow approximation* – in a dynamically consistent manner – is straightforward. All one has to do is to replace the *deep* metric factors $h_i = h_i(\xi_2, \xi_3), i = 1, 2, 3$, in (16.19)–(16.27) by their *shallow* analogues $h_i^S \equiv h_i\left(\xi_2, \xi_3 = \xi_3^S\right) = h_i^S(\xi_2), i = 1, 2, 3$, where $\xi_3 = \xi_3^S =$ constant corresponds to the planet's surface. The terms denoted 'deep' in (16.19)–(16.21) are then *automatically* absent since $\partial h_1^S(\xi_2)/\partial\xi_3 = \partial h_2^S(\xi_2)/\partial\xi_3 \equiv 0$.

This may be contrasted with the classical situation in spherical geometry, whereby metric factors are inserted *explicitly* in the governing equations (as opposed to only in the definition of the metric factors). 'Deep' terms in (16.1)–(16.3) have to then be *individually* removed; in a computer code this would normally be done using a logical switch. For a spherical model, it is beneficial to *first* rewrite (16.1)–(16.3) in the form of (16.19)–(16.21) *before* applying spherical metric factors, either deep ones or shallow ones according to the desired geometry. See Section 16.7.

16.2.3.3 Conservation Principles

There are a number of ways of expressing conservation principles; see Chapter 15. The following principles, expressed in Lagrangian form and in axial-orthogonal-curvilinear coordinates, hold for the preceding unified equation set. They hold in both deep and shallow geometry, subject to the recipes given in Table 16.2 for metric-factor settings, and for additional terms retained as a function of the equation set.

Unified Conservation Principles in Axial-Orthogonal-Curvilinear Coordinates

Axial Angular Momentum

$$\frac{D}{Dt}\left[h_1\left(u_1+\Omega h_1\right)\right]=-\frac{1}{\rho}\frac{\partial p}{\partial \xi_1}. \tag{16.33}$$

Total Energy

$$\frac{D}{Dt}\left[\frac{1}{2}\left(u_1^2+u_2^2+u_3^2\right)+\Phi+\mathcal{E}\right]=-\frac{1}{\rho}\nabla\cdot\left(p\mathbf{u}\right). \tag{16.34}$$

Potential Vorticity

$$\frac{D}{Dt}\left(\frac{\mathbf{Z}\cdot\nabla\Lambda}{\rho}\right)=\frac{D}{Dt}\left[\frac{1}{\rho}\left(\frac{Z_1}{h_1}\frac{\partial\Lambda}{\partial\xi_1}+\frac{Z_2}{h_2}\frac{\partial\Lambda}{\partial\xi_2}+\frac{Z_3}{h_3}\frac{\partial\Lambda}{\partial\xi_3}\right)\right]=0, \tag{16.35}$$

where:

$$\begin{aligned}\mathbf{Z}\equiv\nabla\times\mathbf{u}+2\boldsymbol{\Omega}&=\text{absolute vorticity}\\ &=\frac{1}{h_2h_3}\left[\frac{\partial}{\partial\xi_2}\left(h_3u_3\right)-\frac{\partial}{\partial\xi_3}\left(h_2u_2\right)\right]\mathbf{e}_1\\ &\quad+\left[\frac{1}{h_3h_1}\frac{\partial}{\partial\xi_3}\left(h_1u_1\right)-\frac{1}{h_1}\frac{\partial u_3}{\partial\xi_1}+\frac{2\Omega}{h_3}\frac{\partial h_1}{\partial\xi_3}\right]\mathbf{e}_2\\ &\quad+\left[\frac{1}{h_1}\frac{\partial u_2}{\partial\xi_1}-\frac{1}{h_1h_2}\frac{\partial}{\partial\xi_2}\left(h_1u_1\right)-\frac{2\Omega}{h_2}\frac{\partial h_1}{\partial\xi_2}\right]\mathbf{e}_3,\end{aligned} \tag{16.36}$$

$$\frac{D\Lambda}{Dt}=0, \tag{16.37}$$

$$\nabla\Lambda=\frac{1}{h_1}\frac{\partial\Lambda}{\partial\xi_1}\mathbf{e}_1+\frac{1}{h_2}\frac{\partial\Lambda}{\partial\xi_2}\mathbf{e}_2+\frac{1}{h_3}\frac{\partial\Lambda}{\partial\xi_3}\mathbf{e}_3, \tag{16.38}$$

and $\Lambda=\Lambda\left(\alpha,\eta\right)$ is any materially conserved quantity that satisfies (16.37).

Examination of (16.33) for conservation of axial angular momentum provides some insight regarding the absence of the '$2\Omega\cos\phi$' terms in shallow geometry (this corresponds to the 'traditional approximation'). In shallow geometry, $h_1(\xi_2,\xi_3) \to h_1^S(\xi_2)$. Thus

$$\frac{D}{Dt}\left(\Omega h_1^2\right) = 2\Omega h_1 \left(\frac{u_2}{h_2}\frac{\partial h_1}{\partial \xi_2} + \frac{u_3}{h_3}\frac{\partial h_1}{\partial \xi_3}\right) \to 2\Omega h_1^S \left(\frac{u_2}{h_2^S}\frac{dh_1^S}{d\xi_2}\right), \tag{16.39}$$

for the Coriolis contribution to axial-angular-momentum principle (16.33). For shallow geometry, note the absence in the rightmost expression of (16.39) of an analogue of the red (deep) term of the middle expression. This term corresponds to the absence of $2\Omega\left(u_3/h_3\right)\left(\partial h_1/\partial\xi_3\right)$ in (16.19) for shallow geometry (i.e. to the absence of a $2\Omega\cos\phi$ term). This is a consequence of respect of the conservation principle for axial angular momentum in shallow geometry. Furthermore, the $2\Omega\left(u_1/h_3\right)\left(\partial h_1/\partial\xi_3\right)$ term must also be absent from the vertical momentum component equation (16.35) to obtain a shallow analogue of total energy conservation. The absence of these $2\Omega\cos\phi$ terms from (16.33) and (16.35), respectively, corresponds to the traditional approximation.[13] Thus, in summary:

The Traditional Approximation

The traditional approximation is a (dynamically consistent) *consequence* of shallow approximation. Furthermore:

- That h_1^S is only a function of ξ_2 for shallow approximation, and not of ξ_3, automatically suppresses the $2\Omega\cos\phi$ terms of the deep pair of equation sets. As a side effect, it also simultaneously suppresses a pair of metric terms.
- Similarly for $h_2^S = h_2^S(\xi_2)$, which suppresses a pair of metric terms (only) but has no effect on the Coriolis terms.

16.3 DERIVATION METHODOLOGIES FOR APPROXIMATE EQUATION SETS

A quartet of dynamically consistent equation sets, drawn from the literature, is given for the case of spherical geometry in Appendix A of the present chapter and for the case of spheroidal geometry in Appendix B. In Section 16.2, each quartet, by inspection, was compactly unified into a single equation set from which the four individual sets may be obtained. The issue now is:

- How might one *derive* these individual equation sets from first principles?

Three methodologies for doing so are described in the sections that follow:

1. Classical Eulerian (in Section 16.4).
2. Lagrangian, using Hamilton's principle of stationary action (in Section 16.5).
3. Lagrangian, using Euler–Lagrange equations (in Section 16.6).

16.4 CLASSICAL EULERIAN DERIVATION

This methodology was the way in which the individual members of the spherical and spheroidal quartet of dynamically consistent equation sets were historically derived.

As discussed in Sections 16.1 and 16.2, the *spherical* quartet of equation sets gradually evolved over a period of time. Starting with the original equation set obtained from physical conservation principles, the classical, spherical-geopotential approximation was applied to obtain

[13] A similar argument implies the absence of the $\left(u_1/h_1\right)\left(u_3/h_3\right)\left(\partial h_1/\partial\xi_3\right)$ and $\left(u_1/h_1\right)\left(u_1/h_3\right)\left(\partial h_1/\partial\xi_3\right)$ metric terms in (16.33) and (16.35), respectively, for shallow geometry.

the non-hydrostatic-deep equations in spherical geometry; see Fig. 16.1. Application of the shallow-fluid, hydrostatic, and other approximations culminated in the dynamically consistent, hydrostatic-shallow (= hydrostatic-primitive) equations (Phillips, 1966; Lorenz, 1967). The intermediate (quasi-hydrostatic-deep and non-hydrostatic-shallow) equation sets independently emerged some years later. White et al. (2005) systematically verified the dynamical consistency of the individual members of the *spherical* quartet of equation sets.

Building on this work, the next evolutionary step was to generalise the quartet from spherical to *spheroidal* geometry (White and Wood, 2012) to determine:

- Which metric terms to omit to obtain dynamically consistent *spheroidal* analogues of the *spherical*, quasi-hydrostatic-deep, non-hydrostatic-shallow and hydrostatic-shallow (= hydrostatic-primitive) equation sets.
- How to represent the Coriolis force in terms of the metric factors of an axial-orthogonal-curvilinear coordinate system.

The methodology employed in Chapter 15 to obtain conservation principles for the non-hydrostatic-deep equations may be straightforwardly adapted to verify, a posteriori, the analogous principles for the other three equation sets. One simply follows the same steps, but using an approximated equation set instead of the non-hydrostatic-deep one. This approach, however, presupposes that the precise form of an approximated equation set is already known. To be clear, it *verifies* the dynamical consistency of an *existing* equation set as opposed to *deriving* an equation set that is dynamically consistent.

The question now is:

- Is there a more systematic methodology available to obtain dynamically consistent, approximate equation sets?

Indeed there is. It is the Lagrangian methodology, and it comes in two flavours using:

1. *Either* Hamilton's principle of stationary action (in Section 16.5).
2. *Or* the Euler–Lagrange equations (in Section 16.6).

16.5 LAGRANGIAN DERIVATION USING HAMILTON'S PRINCIPLE

This methodology was applied in Section 14.9 to derive the non-hydrostatic-deep equations for the global atmosphere or oceans, expressed in axial-orthogonal-curvilinear coordinates.

16.5.1 Hamilton's Principle of Stationary Action

We begin with a brief overview of the three steps of the methodology. Thus:

Hamilton's Principle of Stationary Action

1. From (14.150), and in the notation of Chapter 14, a Lagrangian is defined in label space in a rotating frame of reference. Thus

$$L\left(\xi,\dot{\xi},\alpha,\eta\right)=\iiint\mathscr{L}\left(\xi,\dot{\xi},\alpha,\eta\right)d\mathbf{a},\tag{16.40}$$

where $\xi=(\xi_1,\xi_2,\xi_3)$, $\dot{\xi}=\left(\dot{\xi}_1,\dot{\xi}_2,\dot{\xi}_3\right)$, $\mathbf{a}=(a_1,a_2,a_3)$, and

$$\mathscr{L}\left(\xi,\dot{\xi},\alpha,\eta\right)=\frac{h_1^2\dot{\xi}_1^2+h_2^2\dot{\xi}_2^2+h_3^2\dot{\xi}_3^2}{2}+\Omega h_1^2\dot{\xi}_1-\mathcal{E}\left(\alpha,\eta\right)-\Phi\left(\xi_3\right)\tag{16.41}$$

is *Lagrangian density*. Note that $h_i=h_i\left(\xi_2,\xi_3\right),i=1,2,3$, due to use of an axial-orthogonal-curvilinear coordinate system; and $\Phi=\Phi\left(\xi_3\right)$, since a geopotential coordinate system is assumed here. Furthermore – see (14.117) and (14.18) –

$$\alpha = \frac{\partial(\mathbf{r})}{\partial(\mathbf{a})} = \frac{\partial(\mathbf{r})}{\partial(\boldsymbol{\xi})}\frac{\partial(\boldsymbol{\xi})}{\partial(\mathbf{a})} = h_1 h_2 h_3 \frac{\partial(\boldsymbol{\xi})}{\partial(\mathbf{a})} \tag{16.42}$$

depends upon $h_i = h_i(\xi_2, \xi_3)$, $i = 1, 2, 3$. Equation (16.42) embodies mass conservation; see Section 14.3.

2. Using (14.152), the action is

$$\mathcal{A}(\boldsymbol{\xi}, \alpha, \eta) = \int_{\tau_1}^{\tau_2} L\left(\boldsymbol{\xi}, \dot{\boldsymbol{\xi}}, \alpha, \eta\right) d\tau, \tag{16.43}$$

where $L\left(\boldsymbol{\xi}, \dot{\boldsymbol{\xi}}, \alpha, \eta\right)$ is as in (16.40) and (16.41).

3. Using (14.153) and (14.154), Hamilton's principle of stationary action is applied. Taking variations of the action (and a deep breath) then gives

$$\delta\mathcal{A}(\boldsymbol{\xi}, \alpha, \eta) \equiv \delta \int_{\tau_1}^{\tau_2} L\left(\boldsymbol{\xi}, \dot{\boldsymbol{\xi}}, \alpha, \eta\right) d\tau = \int_{\tau_1}^{\tau_2} \delta L\left(\boldsymbol{\xi}, \dot{\boldsymbol{\xi}}, \alpha, \eta\right) d\tau = 0, \tag{16.44}$$

for independent variations $\delta\mathbf{q}(\mathbf{a}, \tau)$ of the generalised coordinates $\mathbf{q} = (\boldsymbol{\xi}, \alpha, \eta)$, such that

$$\delta\mathbf{q}(\mathbf{a}, \tau_1) = \delta\mathbf{q}(\mathbf{a}, \tau_2) = 0, \tag{16.45}$$

where the Lagrangian, $L\left(\boldsymbol{\xi}, \dot{\boldsymbol{\xi}}, \alpha, \eta\right)$, is given by (16.40) and (16.41), and the action, $\mathcal{A}$, by (16.43).

16.5.2 Approximation Strategy

The preceding methodology was applied in Section 14.9 – with variations taken in the Appendix of Chapter 14. This led to prognostic equations (14.158)–(14.163) for the non-hydrostatic-deep equation set. For a fluid composed of a single substance (dry air for the atmosphere, pure liquid water for the ocean), this equation set corresponds to (16.113)–(16.117) of Appendix B of the present chapter. The term $\partial\Phi/\partial\xi_2$ in (14.159) is absent from (16.114), due to use of a geopotential coordinate system here. Furthermore, thermodynamic-energy equation (14.163) is equivalent to (16.117); see (6.34) in the absence of forcing.

The question now is;

- How do we apply this methodology to obtain the other three equation sets of the quartet given in Appendix B of the present chapter?

From Sections 15.4 and 15.5, respectively, we know that:

- *The non-hydrostatic-deep equation set is dynamically consistent*, since all of the conservation principles for this were explicitly derived there.
- By Noether's theorem, *if one approximates a Lagrangian whilst maintaining its symmetries, then quantities conserved in the original system remain conserved in the approximated one.*

So all we have to do is to appropriately approximate the Lagrangian density ($\mathcal{L}$) – and thereby the Lagrangian (L). But how?

Two different kinds of approximation are made:

1. *(Quasi-)Hydrostatic*; by omitting the $h_3^2\dot{\xi}_3^2/2 \equiv u_3^2/2$ contribution to the kinetic-energy term in (16.41) for $\mathcal{L}$.
2. *Shallow Fluid*; by replacing *deep* metric factors $h_i = h_i(\xi_2, \xi_3)$, $i = 1, 2, 3$, everywhere they appear in (16.41) for $\mathcal{L}$ by their *shallow* counterparts $h_i^S = h_i^S(\xi_2) \equiv h_i\left(\xi_2, \xi_3^S\right)$, $i = 1, 2, 3$, where ξ_3^S is the value of ξ_3 at the planet's bounding surface.

These two approximations are then appropriately applied according to the simplifying approximations made for a particular equation set.

16.5.3 Derivation of the Spheroidal Quartet

The Lagrangian densities ($\mathscr{L}$) for the individual members of the spheroidal quartet of equation sets are then (from approximation of (16.41)):

1. $$\mathscr{L}_{\text{NHD}} = \frac{h_1^2\dot{\xi}_1^2 + h_2^2\dot{\xi}_2^2}{2} + \frac{h_3^2\dot{\xi}_3^2}{2} + \Omega h_1^2\dot{\xi}_1 - \mathcal{E}(\alpha,\eta) - \Phi(\xi_3), \tag{16.46}$$

2. $$\mathscr{L}_{\text{QHD}} = \frac{h_1^2\dot{\xi}_1^2 + h_2^2\dot{\xi}_2^2}{2} + \Omega h_1^2\dot{\xi}_1 - \mathcal{E}(\alpha,\eta) - \Phi(\xi_3), \tag{16.47}$$

3. $$\mathscr{L}_{\text{NHS}} = \frac{\left(h_1^2\right)^S\dot{\xi}_1^2 + \left(h_2^2\right)^S\dot{\xi}_2^2}{2} + \frac{\left(h_3^2\right)^S\dot{\xi}_3^2}{2} + \Omega\left(h_1^2\right)^S\dot{\xi}_1 - \mathcal{E}_S(\alpha,\eta) - \Phi_S(\xi_3), \tag{16.48}$$

4. $$\mathscr{L}_{\text{HS}} = \frac{\left(h_1^2\right)^S\dot{\xi}_1^2 + \left(h_2^2\right)^S\dot{\xi}_2^2}{2} + \Omega\left(h_1^2\right)^S\dot{\xi}_1 - \mathcal{E}_S(\alpha,\eta) - \Phi_S(\xi_3), \tag{16.49}$$

where:

1. NHD = Non-Hydrostatic Deep,
2. QHD = Quasi-Hydrostatic Deep,
3. NHS = Non-Hydrostatic Shallow,
4. HS = Hydrostatic Shallow (i.e. Hydrostatic Primitive).

Note that (deep) internal energy $\mathcal{E}(\alpha,\eta)$ depends upon (deep) $h_i = h_i(\xi_2,\xi_3), i = 1,2,3$, as noted earlier. When making the *shallow-fluid approximation*, the dependence is instead on (shallow) $h_i^S = h_i^S(\xi_2) \equiv h_i\left(\xi_2,\xi_3^S\right), i = 1,2,3$. To account for this, we denote the modified internal energy in (16.48) and (16.49) by $\mathcal{E}_S(\alpha,\eta)$.[14] Doing so is consistent with shallow-fluid approximation of (deep) mass-continuity equation (16.22), achieved by replacing its deep metric factors (h_1,h_2,h_3) by shallow ones $\left(h_1^S,h_2^S,h_3^S\right)$. Similar considerations also apply to (deep) geopotential Φ, which changes to (shallow) Φ_S under shallow approximation.

Terms in (16.46)–(16.49) have been arranged to allow us to see, at a glance, the subtle differences between the four Lagrangian densities, and how various terms are related to application of the two different kinds of approximation:

1. The (wide) blank spaces in (16.47) and (16.49) relate to *(quasi-)hydrostatic approximation.*
2. Terms in (16.48) and (16.49) bearing 'S' sub/superscripts relate to *shallow-fluid approximation.*

[14] The functional form of $\mathcal{E} = \mathcal{E}(\alpha,\eta)$ is *independent* of whether the shallow-fluid approximation is made or not. Strictly speaking, therefore, there is no need for subscript 'S' to appear on $\mathcal{E}$ in (16.48) and (16.49). We only do so to *presentationally* underline the fact that there is a *hidden* dependence/sensitivity of $\mathcal{E}$ under shallow-fluid approximation, due to dependence of α on this. This manifests itself when taking variations and, in particular, variations of α. Under shallow-fluid approximation, the metric factors appearing in (16.42) are the *shallow* ones (h_i^S), and not the unapproximated deep ones (h_i). In practice this subtlety is taken care of automatically provided that the deep metric factors h_i are replaced *everywhere* by shallow ones h_i^S; there will then, automatically, be no contributions involving $\partial h_i/\partial\xi_3$, since $\partial h_i^S(\xi_2)/\partial\xi_3 \equiv 0$.

For a particular Lagrangian density ($\mathscr{L}$), it remains to actually take variations of its associated action ($\mathscr{A}$); this is usually the hard part. Good news, however; most of the work for this has already been done – via colouration of terms – in Section 14.9 and in the Appendix of Chapter 14!

Colouration allows terms to be tracked during the taking of variations. Those coloured blue in the Appendix of Chapter 14 are associated with the presence of $h_3^2\dot{\xi}_3^2/2$ and $\left(h_3^2\right)^S\dot{\xi}_3^2/2$ in (16.46) and (16.48), respectively, for the non-hydrostatic-deep and non-hydrostatic-shallow equation sets. When taking variations, the presence of the blue terms in Lagrangian densities (16.46) and (16.48) then leads to terms materialising in the two associated equation sets. However, the $h_3^2\dot{\xi}_3^2/2$ and $\left(h_3^2\right)^S\dot{\xi}_3^2/2$ terms are absent in the Lagrangian densities (16.47) and (16.49) that lead to the quasi-hydrostatic-deep and hydrostatic-shallow equation sets, respectively. Setting blue terms identically zero during the course of taking variations then leads to the absence of blue terms in the two resulting equation sets.

Similarly, terms coloured red in the Appendix of Chapter 14 are associated with application of the shallow-fluid assumption. They only appear in the non-hydrostatic-deep and quasi-hydrostatic-deep equation sets obtained using Lagrangian densities (16.46) and (16.47), respectively, but are absent in the non-hydrostatic-shallow and hydrostatic-shallow equation sets obtained using (16.48) and (16.49), respectively. As formulated:

- Everything comes out straightforwardly when applying the shallow-fluid approximation. All $\partial h_i^S\left(\xi_2\right)/\partial\xi_3$ terms are then automatically zero, with no need for special measures.
- Notably, this includes the '$2\Omega\cos\phi$' Coriolis terms (because they are expressed in terms of metric factors h_i) which would otherwise need to be switched for dynamical consistency.
- Expressing Coriolis terms in terms of metric factors is not only mathematically elegant, but of practical benefit!

Equations (14.158)–(14.163) of Section 14.9, obtained variationally, are equivalent to unified prognostic equations (16.19)–(16.23) of the present chapter, and thus to those given explicitly in Appendix B, namely:

1. Non-Hydrostatic Deep, (16.113)–(16.117) with $\mathcal{N}=1$.
2. Quasi-Hydrostatic Deep, (16.113)–(16.117) with $\mathcal{N}=0$.
3. Non-Hydrostatic Shallow, (16.128)–(16.132) with $\mathcal{N}=1$.
4. Hydrostatic Shallow (= Hydrostatic Primitive), (16.128)–(16.132) with $\mathcal{N}=0$.

Contrast now the present variational approach with the 'classical Eulerian' one. Unification of the four equation sets with the variational approach drops out automatically. With the classical approach, all four equation sets have to be closely examined and compared before they can be judiciously unified.

16.5.4 Derivation of the Spherical Quartet

The spherical quartet is just a special case of the spheroidal one. The four equation sets can therefore be explicitly written down – as in Appendix A of the present chapter – simply by replacing the general metric factors of axial-orthogonal-curvilinear coordinates by the special ones of spherical coordinates; that is, $\left(h_1,h_2,h_3\right)\rightarrow\left(r\cos\phi,r,1\right)$ and $\left(h_1^S,h_2^S,h_3^S\right)\rightarrow\left(a\cos\phi,a,1\right)$.

Alternatively, these equation sets may be derived using the methodology presented earlier, but instead using the spherical analogues of (16.46)–(16.49), namely:

1. $$\mathscr{L}_{\text{NHD}} = \frac{r^2 \cos^2 \phi \dot{\lambda}^2 + r^2 \dot{\phi}^2}{2} + \frac{\dot{r}^2}{2} + \Omega \left(r^2 \cos^2 \phi \right) \dot{\phi} - \mathcal{E} \left(\alpha, \eta \right) \ - \Phi \left(r \right), \quad (16.50)$$

2. $$\mathscr{L}_{\text{QHD}} = \frac{r^2 \cos^2 \phi \dot{\lambda}^2 + r^2 \dot{\phi}^2}{2} + \Omega \left(r^2 \cos^2 \phi \right) \dot{\phi} - \mathcal{E} \left(\alpha, \eta \right) \ - \Phi \left(r \right), \quad (16.51)$$

3. $$\mathscr{L}_{\text{NHS}} = \frac{a^2 \cos^2 \phi \dot{\lambda}^2 + a^2 \dot{\phi}^2}{2} + \frac{\dot{r}^2}{2} + \Omega \left(a^2 \cos^2 \phi \right) \dot{\phi} - \mathcal{E}_S \left(\alpha, \eta \right) - \Phi_S \left(r \right), \quad (16.52)$$

4. $$\mathscr{L}_{\text{HS}} = \frac{a^2 \cos^2 \phi \dot{\lambda}^2 + a^2 \dot{\phi}^2}{2} + \Omega \left(a^2 \cos^2 \phi \right) \dot{\phi} - \mathcal{E}_S \left(\alpha, \eta \right) - \Phi_S \left(r \right). \quad (16.53)$$

Doing so leads to prognostic unified equations (16.1)–(16.5) and to the individual sets given explicitly in Appendix A, namely:

1. Non-Hydrostatic Deep, (16.80)–(16.84) with $\mathcal{N} = 1$.
2. Quasi-Hydrostatic Deep, (16.80)–(16.84) with $\mathcal{N} = 0$.
3. Non-Hydrostatic Shallow, (16.96)–(16.100) with $\mathcal{N} = 1$.
4. Hydrostatic Shallow (= Hydrostatic Primitive), (16.96)–(16.100) with $\mathcal{N} = 0$.

16.6 LAGRANGIAN DERIVATION USING EULER–LAGRANGE EQUATIONS

The Euler–Lagrange equations were derived variationally in Section 14.10 for the components of the momentum equation.[15] They were then applied to obtain the momentum components of the non-hydrostatic-deep equations for a global atmosphere or ocean. Here we outline how the same approach can be used to obtain the momentum components of the other three members of the quartet of equation sets examined in the present chapter. What follows is a generalisation to axial-orthogonal-curvilinear coordinates of the special case examined in Appendix A of White et al. (2005) for the quartet of equation sets in spherical-polar coordinates; see also Müller (1989) for the special case of the hydrostatic-primitive equation set (only) in spherical coordinates.

From either (6.41) or (14.174) – but with $\alpha \rightarrow 1/\rho$ and $\partial/\partial\tau \rightarrow D/Dt$ – the Euler–Lagrange equations for the components of the momentum equation are:

$$\frac{D}{Dt}\left(\frac{\partial \mathscr{L}}{\partial \dot{\xi}_i}\right) - \frac{\partial \mathscr{L}}{\partial \xi_i} = -\frac{1}{\rho}\frac{\partial p}{\partial \xi_i}, \quad i = 1, 2, 3, \quad (16.54)$$

where $\mathscr{L}$ is Lagrangian density. In the present context, and according to which equation set of the quartet is under consideration, $\mathscr{L}$ is defined by (16.46)–(16.49) in axial-orthogonal-curvilinear coordinates, and by (16.50)–(16.53) in spherical coordinates.

However, in a manner similar to that of the preceding section, we do not actually have to consider each equation set individually. It suffices to apply the Euler–Lagrange equations using Lagrangian density (16.46) of the non-hydrostatic-deep equation set. With a judicious use of colouration – as in Section 14.10 and the Appendix of Chapter 14 – the other three equation sets of the quartet are similarly subsumed and are simply special cases of the resulting unified equation set.

[15] As noted there, they can, however, be derived by non-variational methods.

Partially differentiating (16.46) with respect to ξ_1, ξ_2, ξ_3, $\dot{\xi}_1$, $\dot{\xi}_2$, and $\dot{\xi}_3$ and using definition (16.26) of $\mathbf{u}$ in axial-orthogonal-curvilinear coordinates leads to

$$\frac{\partial \mathscr{L}}{\partial \xi_1} = 0, \tag{16.55}$$

$$\frac{\partial \mathscr{L}}{\partial \xi_2} = \left(\frac{u_1}{h_1} + 2\Omega\right) u_1 \frac{\partial h_1}{\partial \xi_2} + \frac{u_2^2}{h_2}\frac{\partial h_2}{\partial \xi_2} + \frac{u_3^2}{h_3}\frac{\partial h_3}{\partial \xi_2}, \tag{16.56}$$

$$\frac{\partial \mathscr{L}}{\partial \xi_3} = \left(\frac{u_1}{h_1} + 2\Omega\right) u_1 \frac{\partial h_1}{\partial \xi_3} + \frac{u_2^2}{h_2}\frac{\partial h_2}{\partial \xi_3} + \frac{u_3^2}{h_3}\frac{\partial h_3}{\partial \xi_3} - \frac{d\Phi}{d\xi_3}, \tag{16.57}$$

$$\frac{\partial \mathscr{L}}{\partial \dot{\xi}_1} = h_1^2 \dot{\xi}_1 + \Omega h_1^2 = h_1 u_1 + \Omega h_1^2, \tag{16.58}$$

$$\frac{\partial \mathscr{L}}{\partial \dot{\xi}_2} = h_2^2 \dot{\xi}_2 = h_2 u_2, \tag{16.59}$$

$$\frac{\partial \mathscr{L}}{\partial \dot{\xi}_3} = h_3^2 \dot{\xi}_3 = h_3 u_3. \tag{16.60}$$

Substituting (16.55)–(16.60) into Euler–Lagrange equations (16.54) then gives

$$\frac{D}{Dt}\left[h_1\left(u_1 + \Omega h_1\right)\right] = -\alpha \frac{\partial p}{\partial \xi_1}, \tag{16.61}$$

$$\frac{D}{Dt}\left(h_2 u_2\right) - \left(\frac{u_1}{h_1} + 2\Omega\right) u_1 \frac{\partial h_1}{\partial \xi_2} - \frac{u_2^2}{h_2}\frac{\partial h_2}{\partial \xi_2} - \frac{u_3^2}{h_3}\frac{\partial h_3}{\partial \xi_2} = -\alpha \frac{\partial p}{\partial \xi_2}, \tag{16.62}$$

$$\frac{D}{Dt}\left(h_3 u_3\right) - \left(\frac{u_1}{h_1} + 2\Omega\right) u_1 \frac{\partial h_1}{\partial \xi_3} - \frac{u_2^2}{h_2}\frac{\partial h_2}{\partial \xi_3} - \frac{u_3^2}{h_3}\frac{\partial h_3}{\partial \xi_3} + \frac{d\Phi}{d\xi_3} = -\alpha \frac{\partial p}{\partial \xi_3}. \tag{16.63}$$

Useful relations (14.196)–(14.201) of the Appendix to Chapter 14 may be rewritten as

$$\frac{Dh_i}{Dt} = \frac{u_2}{h_2}\frac{\partial h_i}{\partial \xi_2} + \frac{u_3}{h_3}\frac{\partial h_i}{\partial \xi_3}, \qquad i = 1, 2, 3, \tag{16.64}$$

$$\frac{D}{Dt}\left(h_i u_i\right) = h_i \left[\frac{Du_i}{Dt} + \frac{u_i}{h_i}\left(\frac{u_2}{h_2}\frac{\partial h_i}{\partial \xi_2} + \frac{u_3}{h_3}\frac{\partial h_i}{\partial \xi_3}\right)\right], \qquad i = 1, 2, 3. \tag{16.65}$$

Using (16.64) and (16.65), (16.61)–(16.63) may then be rewritten as

$$\frac{Du_1}{Dt} + \left(\frac{u_1}{h_1} + 2\Omega\right)\left(\frac{u_2}{h_2}\frac{\partial h_1}{\partial \xi_2} + \frac{u_3}{h_3}\frac{\partial h_1}{\partial \xi_3}\right) = -\frac{1}{\rho h_1}\frac{\partial p}{\partial \xi_1}, \tag{16.66}$$

$$\frac{Du_2}{Dt} - \left(\frac{u_1}{h_1} + 2\Omega\right)\frac{u_1}{h_2}\frac{\partial h_1}{\partial \xi_2} - \frac{u_3^2}{h_2 h_3}\frac{\partial h_3}{\partial \xi_2} + \frac{u_2 u_3}{h_2 h_3}\frac{\partial h_2}{\partial \xi_3} = -\frac{1}{\rho h_2}\frac{\partial p}{\partial \xi_2}, \tag{16.67}$$

$$\frac{Du_3}{Dt} + \frac{u_2 u_3}{h_2 h_3}\frac{\partial h_3}{\partial \xi_2} - \left(\frac{u_1}{h_1} + 2\Omega\right)\frac{u_1}{h_3}\frac{\partial h_1}{\partial \xi_3} - \frac{u_2^2}{h_2 h_3}\frac{\partial h_2}{\partial \xi_3} + \frac{1}{h_3}\frac{d\Phi}{d\xi_3} = -\frac{1}{\rho h_3}\frac{\partial p}{\partial \xi_3}. \tag{16.68}$$

These three equations are identical to (16.19)–(16.21) for the momentum components of the unified equation set, expressed in axial-orthogonal-curvilinear coordinates, and thereby equivalent to (16.113)–(16.115) of the non-hydrostatic-deep equations.

16.7 EQUATION TRANSITION FROM DEEP FLUIDS TO SHALLOW FLUIDS

16.7.1 Shallow Approximation

The hydrostatic-primitive (i.e. hydrostatic-shallow) equations in spherical-polar coordinates (λ, ϕ, r) are usually derived using the classical Eulerian methodology. When doing so, *three* related approximations are generally made (as reported and articulated particularly clearly in Vallis, 2017, pp. 64–65):

1. The *shallow-fluid approximation*, whereby $r \to a$ everywhere it appears in undifferentiated form. Furthermore, two pairwise sets of energy-conserving metric terms are omitted.
2. The *traditional approximation*, whereby the '$2\Omega \cos \phi$' Coriolis terms are omitted.
3. The *hydrostatic approximation*, whereby the vertical component of the momentum equation is approximated by its two most important terms, so that

$$\frac{\partial p}{\partial r} = -\rho g_a. \tag{16.69}$$

Consistent with the classical *spherical-geopotential approximation*, g_a is taken to be the constant value of apparent gravity at Earth's assumed spherical surface $r = a$.

It is generally argued that the shallow-fluid and hydrostatic approximations (approximations 1 and 3 of the preceding list) simplify things. And indeed they do. *But*, it is *usually* further argued, one has to *additionally* invoke the traditional approximation (approximation 2) in order to achieve dynamical consistency, since shallow approximation alone is insufficient to do the job.[16] However, neither of the two alternative (Lagrangian) derivations invoked the traditional assumption! So:

- Why does this *appear* to be necessary for the classical (Eulerian) derivation in spherical-polar coordinates, but not for the two Lagrangian derivations?

In a nutshell, the traditional approximation turns out to be a *direct consequence* of (fully applying) shallow-fluid approximation;[17] it is *not* an additional, independent approximation. To shed further light on this,[18] we (classically) examine the more general problem of shallow approximation of the deep equations in *spheroidal* geometry. The spherical case then results as an immediate corollary.

The underlying reason why the traditional approximation *appears* to be an additional approximation (required for dynamical consistency) is due to *precipitately working in spherical-polar coordinates*. Things are actually a lot clearer if one maintains the generality of axial-orthogonal-curvilinear coordinates, before *eventual* specialisation (at the very end) to spherical-polar coordinates. The key to this development is the expression – introduced in White and Wood (2012)'s Appendix – of the Coriolis force in terms of metric (scale) factors; see (16.74) and (16.75) herein.

From (2.28) – with $\mathbf{F} \equiv 0$ – the vector form of the deep momentum equation is

$$\frac{D\mathbf{u}}{Dt} = -2\boldsymbol{\Omega} \times \mathbf{u} - \frac{1}{\rho}\nabla p - \nabla \Phi. \tag{16.70}$$

Shallow approximation of this equation then gives

$$\frac{D_S\mathbf{u}}{Dt} = -2\left(\boldsymbol{\Omega} \times \mathbf{u}\right)_S - \frac{1}{\rho}\nabla_S p - \nabla_S \Phi_S, \tag{16.71}$$

[16] Atypically, White et al. (2005) noted in their Introduction that shallow approximation 'involves *subsidiary* approximations not *overtly* associated with shallowness' (emphasis added here).

[17] This – see Chapter 17 – is not, however, the case for *quasi-shallow* approximation. Quasi-shallow approximation only *partially* applies shallow approximation. The traditional approximation then no longer follows as a consequence of shallow approximation. Indeed, this is the underlying reason why quasi-shallow approximation provides an excellent representation of the $2\Omega \cos \phi$ terms missing from the shallow-fluid equations.

[18] By 'further light', we mean beyond the analysis given in Section 16.2.3.3.

where subscript 'S' denotes evaluation at the planet's spheroidal surface. Using (5.69) and (5.63) and noting that $h_i^S = h_i^S(\xi_2), i = 1,2,3$ for axial-orthogonal-curvilinear coordinates leads to

$$\frac{D_S\mathbf{u}}{Dt} = \left(\frac{D_S u_1}{Dt} + \frac{u_1 u_2}{h_1^S h_2^S}\frac{dh_1^S}{d\xi_2}\right)\mathbf{e}_1 + \left(\frac{D_S u_2}{Dt} - \frac{u_3^2}{h_2^S h_3^S}\frac{dh_3^S}{d\xi_2} - \frac{u_1^2}{h_2^S h_1^S}\frac{dh_1^S}{d\xi_2}\right)\mathbf{e}_2 + \left(\frac{D_S u_3}{Dt} + \frac{u_2 u_3}{h_2^S h_3^S}\frac{dh_3^S}{d\xi_2}\right)\mathbf{e}_3, \tag{16.72}$$

$$\nabla_S f = \frac{1}{h_1^S}\frac{\partial f}{\partial \xi_1}\mathbf{e}_1 + \frac{1}{h_2^S}\frac{\partial f}{\partial \xi_2}\mathbf{e}_2 + \frac{1}{h_3^S}\frac{\partial f}{\partial \xi_3}\mathbf{e}_3. \tag{16.73}$$

Now for the crucially important part:

Expression of the Coriolis Force in Terms of Metric Factors

From (6.27), the Coriolis force may be expressed in terms of metric (scale) factors for axial-orthogonal-curvilinear coordinates as

$$-2\boldsymbol{\Omega}\times\mathbf{u} = -2\Omega\left[\left(\frac{u_2}{h_2}\frac{\partial h_1}{\partial \xi_2} + \frac{u_3}{h_3}\frac{\partial h_1}{\partial \xi_3}\right)\mathbf{e}_1 - \frac{u_1}{h_2}\frac{\partial h_1}{\partial \xi_2}\mathbf{e}_2 - \frac{u_1}{h_3}\frac{\partial h_1}{\partial \xi_3}\mathbf{e}_3\right], \tag{16.74}$$

$$\Downarrow$$

$$-2\left(\boldsymbol{\Omega}\times\mathbf{u}\right)_S = -2\Omega\left(\frac{u_2}{h_2^S}\mathbf{e}_1 - \frac{u_1}{h_2^S}\mathbf{e}_2\right)\frac{dh_1^S}{d\xi_2}, \quad \text{under shallow approximation.} \tag{16.75}$$

Using (16.72), (16.73), and (16.75) in (16.71); noting that $\Phi_S = \Phi_S(\xi_3)$ for a geopotential coordinate system, taking components of the resulting equation, and rearranging then yields

$$\frac{D_S u_1}{Dt} + \left(\frac{u_1}{h_1^S} + 2\Omega\right)\frac{u_2}{h_2^S}\frac{dh_1^S}{d\xi_2} + \frac{1}{\rho h_1^S}\frac{\partial p}{\partial \xi_1} = 0, \tag{16.76}$$

$$\frac{D_S u_2}{Dt} - \left(\frac{u_1}{h_1^S} + 2\Omega\right)\frac{u_1}{h_2^S}\frac{dh_1^S}{d\xi_2} - \frac{u_3^2}{h_2^S h_3^S}\frac{dh_3^S}{d\xi_2} + \frac{1}{\rho h_2^S}\frac{\partial p}{\partial \xi_2} = 0, \tag{16.77}$$

$$\frac{D_S u_3}{Dt} + \frac{u_2 u_3}{h_2^S h_3^S}\frac{dh_3^S}{d\xi_2} + \frac{1}{\rho h_3^S}\frac{\partial p}{\partial \xi_3} + \frac{1}{h_3^S}\frac{d\Phi_S}{d\xi_3} = 0. \tag{16.78}$$

Equations (16.76)–(16.78) recover the (purely) shallow equations (16.128)–(16.130) in spheroidal geometry. Importantly:

- The '$2\Omega\cos\phi$' Coriolis terms – when present, these are proportional to $\partial h_1/\partial\xi_3$ – *are entirely absent* from (16.75), and thereby also from (16.76) and (16.78), after application of the shallow mapping/approximation $h_i(\xi_2,\xi_3) \to h_i^S(\xi_2), i = 1,2,3$.
- *No recourse whatsoever to the 'traditional approximation' was required to apply shallow approximation.* The '$2\Omega\cos\phi$' Coriolis terms simply dropped out automatically as a direct consequence of shallow approximation.

The blue terms in (16.77) and (16.78), and also in (16.129) and (16.130), are the non-hydrostatic terms. In their absence (i.e. under hydrostatic approximation), the hydrostatic primitive equations in spheroidal geometry are recovered. The special case of the hydrostatic primitive equations (16.96)–(16.98) in *spherical* geometry (with $\mathcal{N} = 0$) then results as a corollary.

16.7.2 Conservation

The derivation of the shallow components (16.76)–(16.78) of the momentum equation in spheroidal geometry is more straightforward than the classical way of doing so used in most published works for the special case of spherical geometry. This is good news, but there is yet more good news. Proof of dynamical consistency is trivial!

How so, you may ask? The proof depends on two existing results:

1. The conservation properties for dynamical consistency of the deep equations to hold were derived in Chapter 15 from the deep equations expressed in vector form.
2. The usual vector identities involving gradient, divergence, and curl hold not only in 3D Euclidean space with (deep) orthogonal-curvilinear metric factors $h_i, i = 1,2,3$, but also in non-Euclidean space with shallow orthogonal-curvilinear metric factors $h_i^S, i = 1,2,3$ (White et al., 2005).

The second of these two results means that the vector derivation of conservation properties in deep spheroidal geometry of the first result also carries through for shallow geometry. As a special case, this property also holds when the governing equations are instead expressed in axial-orthogonal-curvilinear coordinates. Job done.

16.8 CONCLUDING REMARKS

A quartet of approximate, dynamically consistent equation sets, suitable for global atmospheric and oceanic modelling, has been given in this chapter in both spherical and spheroidal geometry. Three different methodologies to derive these equation sets have been described; one Eulerian, the other two Lagrangian. We now discuss their pros and cons. Factors to be considered are:

- Ease of application.
- Scientific rigour.
- Mathematical elegance.

Historically, all four sets were obtained using the popular 'classical-Eulerian' methodology, as discussed in Sections 16.2.1 and 16.4. For this methodology, the first step is to use scale analysis to identify terms that might be approximated or omitted in the unapproximated governing equations. *Only* doing this can then lead to (what is termed here) dynamically inconsistent formulations. The key issue for some of the early (and also a few later) derivations of shallow, hydrostatic governing-equation sets was not that the classical methodology is fundamentally flawed, but rather that *it was not fully applied.* It is fine to use scale analysis to identify terms that *might* be approximated or omitted. What is not fine is to then go ahead and do so *without verifying that the resulting approximated equation set remains dynamically consistent* (as defined in Section 16.1.2). This second step – verification of conservation principles – is *not an optional extra* of the classical methodology but rather an *integral component* of it.

If, for a proposed approximated equation set, it is found that one (or more) of the conservation principles is violated, then the cause needs to be identified and remedied. Whilst fixing the original problem, the revised formulation may, however, cause a different conservation law to be violated! The whole process then needs to be repeated until *all* conservation laws are verified for any revised formulation; or, failing that, defeat admitted.

A strength of the classical methodology is that it is relatively easy (albeit somewhat tedious) to apply it to determine the dynamical consistency of any proposed equation set. One just adapts the well-established steps used to verify the conservation properties of the unapproximated equation set; see Sections 15.3 and 15.4. If dynamical consistency is confirmed, then this is rigorously true.

A downside of the approach, however, is that one has to determine which combination(s) of (many) possible different approximations of terms – or possible omissions thereof – in multiple equations, will lead to preservation of dynamical consistency. The more possible combinations there are, the more work and insight that are required to examine the consequences for dynamical consistency/inconsistency. The labour required for this may, however, be substantially reduced – experience greatly helps here – by noting the following points:

1. Many terms occur in energy-conserving subsets (e.g. pairs).
2. Consistent approximation of metric factors distorts the geometry without adversely affecting conservation properties.
3. Axial-angular-momentum conservation and the zonal component of the momentum equation are closely related to one another.

This brings us to the two versions of the Lagrangian approach using:

1. *Either* Hamilton's principle of stationary action (as in Section 16.5).
2. *Or* the Euler–Lagrange equations (as in Section 16.6).

Both versions are based on judicious approximation of a Lagrangian. This gets to the heart of the issue. Instead of having to consider many different combinations of possible approximations to a variety of vector and scalar equations when applying the classical Eulerian methodology, one only has a *single, scalar* function to consider; namely the unapproximated Lagrangian. This generally has a very limited number of terms – see, for example, (16.41) – with much less scope for approximation. Furthermore, there are *symmetry constraints* on approximation of the Lagrangian that must be respected for dynamical consistency to be maintained; see Section 15.5. This further limits the scope for approximation and therefore the number of possibilities that need to be examined. See Section 17.3.1 for an illustration of how this simplifies things for the case of quasi-shallow approximation. Furthermore, one can examine each approximation *individually* (as opposed to in combination with other approximations); again a helpful simplification. Each approximation of the Lagrangian must individually respect the requisite symmetries. These considerations apply to both versions of the variational approach, since approximation of the Lagrangian is common to them.

The question now is which of the two Lagrangian methodologies should one use? This depends on one's experience and how deeply one wishes to delve into the subject. For reasons of simplicity, both versions described herein are somewhat simplified ones of more complete – yet more abstract – treatments. Although both versions lead to governing equation sets in a natural way, demonstration of dynamical consistency of an approximated equation set still needs to be done. This can be accomplished in two different ways:

1. *Explicitly*, by verifying conservation properties of the resulting equation set using traditional methods.
2. *Implicitly*, by verifying, a priori, the requisite symmetry properties of the approximated Lagrangian density.

The most elegant way is the latter. Specifically, the requisite symmetry properties are – see Section 15.5 – invariance of the (approximated) Lagrangian with respect to:

1. Rotation about a planet's axis, together with assumed zonal symmetry of the gravitational potential, Φ (for *conservation of axial angular momentum*).
2. Time translation (for *conservation of total energy*).
3. Particle relabelling (for *conservation of potential vorticity*).

In practice this is generally not too bad for conservation of axial angular momentum and total energy, but considerably more difficult for conservation of potential vorticity, due to label symmetry. Furthermore, in a more comprehensive treatment, the symmetries themselves lead to the actual conservation principles. This approach can, however, be highly challenging to apply in practice. So is there an easier practical, yet rigorous, alternative? Indeed there is!

It boils down to a mix-and-match approach. Taking variations to apply Hamilton's principle is hard work. It is far easier to use the Euler–Lagrange equations; to do so only involves partial differentiation and algebraic manipulation. But this latter approach (as simplified herein) only delivers the components of the momentum equation, albeit these are generally the crucial ones. Use of the Euler–Lagrange equations is recommended for *exploring* possible approximate forms of the Lagrangian. In practice, most 'reasonable' approximations do lead to dynamical consistency, provided one is careful to do things consistently in either deep or shallow geometry. In particular, this influences the (deep versus shallow) forms of the mass-continuity and thermodynamic-energy equations and the D/Dt, gradient, and divergence operators. Once one has a candidate approximated equation set, one can revert to the classical methodology fully applied to rigorously verify dynamical consistency. So a mix of two methodologies.

The Euler–Lagrange methodology allows one to relatively quickly identify, without necessarily requiring particularly deep insight, a small number of possible candidate approximation(s) together with the consequent approximated equation set. Each candidate approximation can then be examined individually, and its dynamical consistency verified a posteriori using the classical methodology. Any combination of individual, dynamically consistent approximations of the Lagrangian leads to a dynamically consistent equation set.

It was shown in Sections 16.5 and 16.6 how existing equation sets could have been *systematically* derived via suitable and quite straightforward approximation of a Lagrangian. However, as alluded to earlier, the two variants of the Lagrangian methodology offer much more than this. They can, for example, be used to discover *new*, dynamically consistent, equation sets, as we will see in Chapters 17 and 18.

The insight provided by the various developments reported in this chapter allowed us to revisit and significantly simplify the classical derivation of the (dynamically consistent) hydrostatic primitive equations. Contrary to conventional wisdom, it was shown (in Section 16.7) that there is no need to invoke the 'traditional approximation', since this approximation is actually a direct consequence of shallow approximation. It was also argued that the conservation properties of the deep pair of equation sets is automatically retained for the shallow pair of equation sets. This is because the usual vector identities involving gradient, divergence, and curl hold not only in 3D Euclidean space with (deep) orthogonal-curvilinear metric factors but also in non-Euclidean space with corresponding shallow, orthogonal-curvilinear metric factors. This amounts to a geometrical distortion that is invisible to the derivation of conservation properties.

APPENDIX A: FOUR EQUATION SETS IN SPHERICAL-POLAR COORDINATES

A quartet of equation sets (two deep, two shallow), expressed in spherical-polar coordinates, is assembled here for illustrative and reference purposes. The four sets are labelled:

1. Non-hydrostatic deep (or fully compressible).
2. Quasi-hydrostatic deep (or quasi-hydrostatic).
3. Non-hydrostatic shallow.
3. Hydrostatic shallow (or hydrostatic primitive).

See Sections 16.1 and 16.2.2 for description of their attributes.

To better focus on dynamical consistency, it is assumed that:

- Physical forcings (other than gravity) are absent, but see footnote 6.
- The coordinate system is a geopotential one, so $\Phi = \Phi(r)$.
- The fluid is composed of a single substance (dry air for the atmosphere, pure liquid water for the oceans).

Common to all four equation sets is the fundamental equation of state:

$$\mathcal{E} = \mathcal{E}(\alpha, \eta) \quad \Rightarrow \quad p \equiv -\frac{\partial \mathcal{E}(\alpha, \eta)}{\partial \alpha}, \quad T \equiv \frac{\partial \mathcal{E}(\alpha, \eta)}{\partial \eta}. \tag{16.79}$$

These three equations are functionally independent of whether the fluid is deep or shallow and of whether the flow is (quasi-)hydrostatically balanced or not; they only depend upon the thermodynamic state of the fluid. They do, however, have *indirect* dependence on depth since evolution of α depends on whether the fluid is deep or shallow.

Colour is used in the four equation sets to help identify links between them:

- Blue terms are associated with quasi-hydrostatic/hydrostatic approximation. Blue terms are present before application of this approximation (i.e. present in a non-hydrostatic equation set) and absent after application – that is, absent in a (quasi-)hydrostatic one.
  In what follows, $\mathcal{N}$ is the non-hydrostatic switch. When:
  - $\mathcal{N} = 1$, the term it multiplies is *retained*, and the equation set is then *non-hydrostatic.*
  - $\mathcal{N} = 0$, the term it multiplies is *omitted*, and the equation set is then *(quasi-)hydrostatic.*
- Red terms are associated with shallow approximation. Their presence in a deep-fluid equation set indicates that they are subsequently either eliminated or altered by shallow approximation.

Deep Equation Sets in Spherical-Polar Coordinates

Non-hydrostatic switch $\mathcal{N}$ determines the equation set:

$$\mathcal{N} = 1 \quad \Rightarrow \quad \textbf{non-hydrostatic-deep},$$
$$\mathcal{N} = 0 \quad \Rightarrow \quad \textbf{quasi-hydrostatic-deep}.$$

Equations (16.80)–(16.89) – with $\mathcal{N} = 0, 1$ – correspond to the two *deep* equation sets (in dark blue) on Fig. 16.1. Terms coloured red in (16.80)–(16.89) are affected – either by modification or omission – by subsequent application of shallow approximation. This facilitates detailed, term-by-term comparison of deep equations (16.80)–(16.95) with shallow equations (16.96)–(16.111).

Governing Equations

In the notation of Section 6.6, but with the introduction of some colour, non-hydrostatic-deep equations (6.46)–(6.57) in spherical-polar coordinates (λ, ϕ, r) may be written as:

$$\frac{D_r u_\lambda}{Dt} - \left(\frac{u_\lambda}{r\cos\phi} + 2\Omega\right)(u_\phi \sin\phi - u_r \cos\phi) + \frac{1}{\rho r \cos\phi}\frac{\partial p}{\partial \lambda} = 0, \tag{16.80}$$

$$\frac{D_r u_\phi}{Dt} + \left(\frac{u_\lambda}{r\cos\phi} + 2\Omega\right) u_\lambda \sin\phi + \frac{u_\phi u_r}{r} + \frac{1}{\rho r}\frac{\partial p}{\partial \phi} = 0, \tag{16.81}$$

$$\mathcal{N}\frac{D_r u_r}{Dt} - \left(\frac{u_\lambda^2}{r} + \frac{u_\phi^2}{r} + 2\Omega u_\lambda \cos\phi\right) + \frac{1}{\rho}\frac{\partial p}{\partial r} + \frac{d\Phi_r}{dr} = 0, \tag{16.82}$$

$$\frac{D_r \rho}{Dt} + \rho \nabla_r \cdot \mathbf{u} = 0, \tag{16.83}$$

$$\frac{D_r \mathcal{E}_r}{Dt} + p\frac{D_r \alpha}{Dt} = 0, \tag{16.84}$$

where:

$$\alpha \equiv \frac{1}{\rho}, \tag{16.85}$$

$$\frac{D_r}{Dt} \equiv \frac{\partial}{\partial t} + \frac{u_\lambda}{r\cos\phi}\frac{\partial}{\partial \lambda} + \frac{u_\phi}{r}\frac{\partial}{\partial \phi} + u_r\frac{\partial}{\partial r}, \tag{16.86}$$

$$\mathbf{u} = u_\lambda \mathbf{e}_\lambda + u_\phi \mathbf{e}_\phi + u_r \mathbf{e}_r \equiv r\cos\phi \frac{D_r\lambda}{Dt}\mathbf{e}_\lambda + r\frac{D_r\phi}{Dt}\mathbf{e}_\phi + \frac{D_r r}{Dt}\mathbf{e}_r, \tag{16.87}$$

$$\nabla_r \cdot \mathbf{A} = \frac{1}{r^2\cos\phi}\left[\frac{\partial}{\partial\lambda}(A_\lambda r) + \frac{\partial}{\partial\phi}\left(A_\phi r\cos\phi\right) + \frac{\partial}{\partial r}\left(A_r r^2\cos\phi\right)\right], \tag{16.88}$$

$$\Phi_r(r) = -g_a\frac{a^2}{r} \quad \Rightarrow \quad \mathbf{g}_r = -\frac{d\Phi_r}{dr}\mathbf{e}_r = -g_a\frac{a^2}{r^2}\mathbf{e}_r. \tag{16.89}$$

Equations (16.80)–(16.82) agree with White et al. (2005)'s (2.17)–(2.19), and (16.83) with their (2.6).

Conservation Principles

Associated with the preceding governing equations are the following conservation principles for axial angular momentum, total energy, and potential vorticity:

Axial Angular Momentum

$$\frac{D_r}{Dt}\left[(u_\lambda + \Omega r\cos\phi)\, r\cos\phi\right] = -\frac{1}{\rho}\frac{\partial p}{\partial\lambda}, \tag{16.90}$$

in agreement with White et al. (2005)'s (3.1).

Total Energy

$$\frac{D_r}{Dt}\left[\frac{1}{2}\left(u_\lambda^2 + u_\phi^2 + \mathcal{N}u_r^2\right) + \Phi_r + \mathcal{E}_r\right] = -\frac{1}{\rho}\nabla_r\cdot(p\mathbf{u}), \tag{16.91}$$

in agreement with White et al. (2005)'s (2.8) with $\mathcal{E} = c_v T$.

Potential Vorticity

$$\frac{D_r}{Dt}\left(\frac{\mathbf{Z}_r\cdot\nabla_r\Lambda}{\rho}\right) = 0, \tag{16.92}$$

where

$$\mathbf{Z}_r = \frac{1}{r}\left[\mathcal{N}\frac{\partial u_r}{\partial\phi} - \frac{\partial}{\partial r}\left(r u_\phi\right)\right]\mathbf{e}_\lambda + \left[\frac{1}{r}\frac{\partial}{\partial r}(r u_\lambda) - \frac{\mathcal{N}}{r\cos\phi}\frac{\partial u_r}{\partial\lambda} + 2\Omega\cos\phi\right]\mathbf{e}_\phi + \left\{\frac{1}{r\cos\phi}\left[\frac{\partial u_\phi}{\partial\lambda} - \frac{\partial}{\partial\phi}(u_\lambda\cos\phi)\right] + 2\Omega\sin\phi\right\}\mathbf{e}_r, \tag{16.93}$$

$$\frac{D_r\Lambda}{Dt} = 0, \tag{16.94}$$

$$\nabla_r\Lambda = \frac{1}{r\cos\phi}\frac{\partial\Lambda}{\partial\lambda}\mathbf{e}_\lambda + \frac{1}{r}\frac{\partial\Lambda}{\partial\phi}\mathbf{e}_\phi + \frac{\partial\Lambda}{\partial r}\mathbf{e}_r, \tag{16.95}$$

and $\Lambda = \Lambda(\alpha, \eta)$ is any materially conserved quantity that satisfies (16.94).

Equation (16.92) agrees with White et al. (2005)'s (2.9), and (16.93) with their (3.5)–(3.7).

Shallow Equation Sets in Spherical-Polar Coordinates

Non-hydrostatic switch $\mathcal{N}$ determines the equation set:

$$\mathcal{N}=1 \quad \Rightarrow \quad \textbf{non-hydrostatic-shallow},$$
$$\mathcal{N}=0 \quad \Rightarrow \quad \textbf{hydrostatic-shallow} \ (= \textbf{hydrostatic-primitive}).$$

Equations (16.96)–(16.105) – with $\mathcal{N} = 0, 1$ – correspond to the two *shallow* equation sets (in light blue) on Fig. 16.1.

Governing Equations

$$\frac{D_a u_\lambda}{Dt} - \left(\frac{u_\lambda}{a\cos\phi} + 2\Omega\right) u_\phi \sin\phi + \frac{1}{\rho a \cos\phi}\frac{\partial p}{\partial \lambda} = 0, \tag{16.96}$$

$$\frac{D_a u_\phi}{Dt} + \left(\frac{u_\lambda}{a\cos\phi} + 2\Omega\right) u_\lambda \sin\phi + \frac{1}{\rho a}\frac{\partial p}{\partial \phi} = 0, \tag{16.97}$$

$$\mathcal{N}\frac{D_a u_r}{Dt} + \frac{1}{\rho}\frac{\partial p}{\partial r} + \frac{d\Phi_a}{dr} = 0, \tag{16.98}$$

$$\frac{D_a \rho}{Dt} + \rho \nabla_a \cdot \mathbf{u} = 0, \tag{16.99}$$

$$\frac{D_a \mathcal{E}_a}{Dt} + p\frac{D_a \alpha}{Dt} = 0, \tag{16.100}$$

where

$$\alpha \equiv \frac{1}{\rho}, \tag{16.101}$$

$$\frac{D_a}{Dt} \equiv \frac{\partial}{\partial t} + \frac{u_\lambda}{a\cos\phi}\frac{\partial}{\partial \lambda} + \frac{u_\phi}{a}\frac{\partial}{\partial \phi} + u_r\frac{\partial}{\partial r}, \tag{16.102}$$

$$\mathbf{u} = u_\lambda \mathbf{e}_\lambda + u_\phi \mathbf{e}_\phi + u_r \mathbf{e}_r \equiv a\cos\phi\frac{D_a\lambda}{Dt}\mathbf{e}_\lambda + a\frac{D_a\phi}{Dt}\mathbf{e}_\phi + \frac{D_a r}{Dt}\mathbf{e}_r, \tag{16.103}$$

$$\nabla_a \cdot \mathbf{A} = \frac{1}{a\cos\phi}\left[\frac{\partial A_\lambda}{\partial \lambda} + \frac{\partial}{\partial \phi}\left(A_\phi \cos\phi\right)\right] + \frac{\partial A_r}{\partial r}, \tag{16.104}$$

$$\Phi_a(r) = g_a(r-a) \quad \Rightarrow \quad \mathbf{g}_a = -\frac{d\Phi_a}{dr}\mathbf{e}_r = -g_a\mathbf{e}_r. \tag{16.105}$$

Equations (16.96) and (16.97) correspond to White et al. (2005)'s (3.8) and (3.9), and (16.98) agrees with their (3.26).

Conservation Principles

Associated with the preceding governing equations are the following conservation principles for axial angular momentum, total energy, and potential vorticity:

Axial Angular Momentum

$$\frac{D_a}{Dt}\left[\left(u_\lambda + \Omega a\cos\phi\right) a\cos\phi\right] = -\frac{1}{\rho}\frac{\partial p}{\partial \lambda}, \tag{16.106}$$

in agreement with White et al. (2005)'s (3.18).

Total Energy

$$\frac{D_a}{Dt}\left[\frac{1}{2}\left(u_\lambda^2+u_\phi^2+\mathcal{N}u_r^2\right)+\Phi_a+\mathcal{E}_a\right]=-\frac{1}{\rho}\nabla_a\cdot(p\mathbf{u}), \tag{16.107}$$

in agreement with White et al. (2005)'s (3.29) with $\mathcal{E}=c_vT$.

Potential Vorticity

$$\frac{D_a}{Dt}\left(\frac{\mathbf{Z}_a\cdot\nabla_a\Lambda}{\rho}\right)=0, \tag{16.108}$$

where:

$$\mathbf{Z}_a=\left(\frac{\mathcal{N}}{a}\frac{\partial u_r}{\partial\phi}-\frac{\partial u_\phi}{\partial r}\right)\mathbf{e}_\lambda+\left(\frac{\partial u_\lambda}{\partial r}-\frac{\mathcal{N}}{a\cos\phi}\frac{\partial u_r}{\partial\lambda}\right)\mathbf{e}_\phi+\left\{\frac{1}{a\cos\phi}\left[\frac{\partial u_\phi}{\partial\lambda}-\frac{\partial}{\partial\phi}(u_\lambda\cos\phi)\right]+2\Omega\sin\phi\right\}\mathbf{e}_r, \tag{16.109}$$

$$\frac{D_a\Lambda}{Dt}=0, \tag{16.110}$$

$$\nabla_a\Lambda=\frac{1}{a\cos\phi}\frac{\partial\Lambda}{\partial\lambda}\mathbf{e}_\lambda+\frac{1}{a}\frac{\partial\Lambda}{\partial\phi}\mathbf{e}_\phi+\frac{\partial\Lambda}{\partial r}\mathbf{e}_r, \tag{16.111}$$

and $\Lambda=\Lambda(\alpha,\eta)$ is any materially conserved quantity that satisfies (16.110).

Equation (16.108) agrees with White et al. (2005)'s (3.30), and (16.109) with their (3.31) and (3.32).

APPENDIX B: FOUR EQUATION SETS IN AXIAL-ORTHOGONAL-CURVILINEAR COORDINATES

A quartet of equation sets (two deep, two shallow), expressed in axial-orthogonal-curvilinear coordinates, is assembled here for illustrative and reference purposes. The four sets are labelled:

1. Non-hydrostatic deep (or fully compressible).
2. Quasi-hydrostatic deep (or quasi-hydrostatic).
3. Non-hydrostatic shallow.
4. Hydrostatic shallow (or hydrostatic primitive).

See Sections 16.1 and 16.2.3 for description of their attributes.

To better focus on dynamical consistency, it is assumed that:

- Physical forcings (other than gravity) are absent, but see footnote 6.
- The coordinate system is a geopotential one, so $\Phi=\Phi(\xi_3)$.
- The fluid is composed of a single substance (dry air for the atmosphere, pure liquid water for the oceans).

Common to all four equation sets is the fundamental equation of state:

$$\mathcal{E}=\mathcal{E}(\alpha,\eta)\quad\Rightarrow\quad p\equiv-\frac{\partial\mathcal{E}(\alpha,\eta)}{\partial\alpha},\quad T\equiv\frac{\partial\mathcal{E}(\alpha,\eta)}{\partial\eta}. \tag{16.112}$$

These three equations are independent of whether the fluid is deep or shallow, and they depend only upon the thermodynamic state of the fluid. They do, however, have *indirect* dependence on depth since evolution of α depends on whether the fluid is deep or shallow.

Colour is used in the four equation sets to help identify links between them:

- Blue terms are associated with quasi-hydrostatic/hydrostatic approximation. Blue terms are present before application of this approximation (i.e. present in a non-hydrostatic equation set) and absent after application – that is, absent in a (quasi-)hydrostatic one.
  In what follows, $\mathcal{N}$ is the non-hydrostatic switch. When:
  - $\mathcal{N} = 1$, the term it multiplies is *retained*, and the equation set is then *non-hydrostatic*.
  - $\mathcal{N} = 0$, the term it multiplies is omitted, and the equation set is then *(quasi-)hydrostatic*.
- Red terms are associated with shallow approximation. They are present before application of this approximation (i.e. present for a deep fluid) and absent afterwards (i.e. absent for a shallow fluid).
  Various other terms are altered by shallow approximation due to replacement of deep metric factors, $h_i = h_i(\xi_2, \xi_3), i = 1, 2, 3$, by shallow metric factors, $h_i^S \equiv h_i(\xi_2, \xi_3^S) = h_i^S(\xi_2), i = 1, 2, 3$, where superscript '$S$' denotes evaluation at the planet's surface $\xi_3 = \xi_3^S =$ constant.

Deep Equation Sets in Axial-Orthogonal-Curvilinear Coordinates

Non-hydrostatic switch $\mathcal{N}$ determines the equation set:

$$\mathcal{N} = 1 \quad \Rightarrow \quad \textbf{non-hydrostatic-deep},$$
$$\mathcal{N} = 0 \quad \Rightarrow \quad \textbf{quasi-hydrostatic-deep}.$$

Equations (16.113)–(16.121) – with $\mathcal{N} = 0, 1$ – correspond to the two *deep* equation sets (in dark blue) on Fig. 16.1. Terms coloured red in (16.113)–(16.115) identify terms that automatically disappear under shallow approximation, since then $\partial h_1/\partial\xi_3 \rightarrow \partial h_1^S/\partial\xi_3 \equiv 0$ and $\partial h_2/\partial\xi_3 \rightarrow \partial h_2^S/\partial\xi_3 \equiv 0$. This facilitates comparison of deep equations (16.113)–(16.115) with shallow equations (16.128)–(16.130).

Governing Equations

In the notation of Section 6.5 and with the introduction of some colour, non-hydrostatic-deep equations (6.29)–(6.40) in axial-orthogonal-curvilinear coordinates (ξ_1, ξ_2, ξ_3) may be written as:

$$\frac{Du_1}{Dt} + \left(\frac{u_1}{h_1} + 2\Omega\right)\left(\frac{u_2}{h_2}\frac{\partial h_1}{\partial \xi_2} + \frac{u_3}{h_3}\frac{\partial h_1}{\partial \xi_3}\right) + \frac{1}{\rho h_1}\frac{\partial p}{\partial \xi_1} = 0, \tag{16.113}$$

$$\frac{Du_2}{Dt} - \left(\frac{u_1}{h_1} + 2\Omega\right)\frac{u_1}{h_2}\frac{\partial h_1}{\partial \xi_2} - \frac{\mathcal{N} u_3^2}{h_2 h_3}\frac{\partial h_3}{\partial \xi_2} + \frac{u_2}{h_2}\frac{u_3}{h_3}\frac{\partial h_2}{\partial \xi_3} + \frac{1}{\rho h_2}\frac{\partial p}{\partial \xi_2} = 0, \tag{16.114}$$

$$\mathcal{N}\left(\frac{Du_3}{Dt} + \frac{u_2}{h_2}\frac{u_3}{h_3}\frac{\partial h_3}{\partial \xi_2}\right) - \left(\frac{u_1}{h_1} + 2\Omega\right)\frac{u_1}{h_3}\frac{\partial h_1}{\partial \xi_3} - \frac{u_2^2}{h_2 h_3}\frac{\partial h_2}{\partial \xi_3} + \frac{1}{\rho h_3}\frac{\partial p}{\partial \xi_3} + \frac{1}{h_3}\frac{d\Phi}{d\xi_3} = 0, \tag{16.115}$$

$$\frac{D\rho}{Dt} + \rho \nabla \cdot \mathbf{u} = 0, \tag{16.116}$$

$$\frac{D\mathcal{E}}{Dt} + p\frac{D\alpha}{Dt} = 0, \tag{16.117}$$

where

$$\alpha \equiv \frac{1}{\rho}, \tag{16.118}$$

$$\frac{D}{Dt} = \frac{\partial}{\partial t} + \frac{u_1}{h_1}\frac{\partial}{\partial \xi_1} + \frac{u_2}{h_2}\frac{\partial}{\partial \xi_2} + \frac{u_3}{h_3}\frac{\partial}{\partial \xi_3}, \tag{16.119}$$

$$\mathbf{u} \equiv u_1\mathbf{e}_1 + u_2\mathbf{e}_2 + u_3\mathbf{e}_3 = h_1\frac{D\xi_1}{Dt}\mathbf{e}_1 + h_2\frac{D\xi_2}{Dt}\mathbf{e}_2 + h_3\frac{D\xi_3}{Dt}\mathbf{e}_3, \tag{16.120}$$

$$\nabla \cdot \mathbf{A} = \frac{1}{h_1 h_2 h_3}\left[\frac{\partial}{\partial \xi_1}(A_1 h_2 h_3) + \frac{\partial}{\partial \xi_2}(A_2 h_3 h_1) + \frac{\partial}{\partial \xi_3}(A_3 h_1 h_2)\right]. \tag{16.121}$$

Equations (16.113)–(16.115) agree with White and Wood (2012)'s equations (A.10)–(A.12) after correction of their (A.12) (see footnote 12 herein).

Conservation Principles

Associated with the preceding governing equations are the following conservation principles for axial angular momentum, total energy, and potential vorticity:

Axial Angular Momentum

$$\frac{D}{Dt}\left[h_1\left(u_1+\Omega h_1\right)\right]=-\frac{1}{\rho}\frac{\partial p}{\partial \xi_1}. \tag{16.122}$$

Total Energy

$$\frac{D}{Dt}\left[\frac{1}{2}\left(u_1^2+u_2^2+\mathcal{N}u_3^2\right)+\Phi+\mathcal{E}\right]=-\frac{1}{\rho}\nabla\cdot\left(p\mathbf{u}\right). \tag{16.123}$$

Potential Vorticity

$$\frac{D}{Dt}\left(\frac{\mathbf{Z}\cdot\nabla\Lambda}{\rho}\right)=\frac{D}{Dt}\left[\frac{1}{\rho}\left(\frac{Z_1}{h_1}\frac{\partial\Lambda}{\partial\xi_1}+\frac{Z_2}{h_2}\frac{\partial\Lambda}{\partial\xi_2}+\frac{Z_3}{h_3}\frac{\partial\Lambda}{\partial\xi_3}\right)\right]=0, \tag{16.124}$$

where:

$$\begin{aligned}
\mathbf{Z}\equiv\nabla\times\mathbf{u}+2\boldsymbol{\Omega}&=\text{absolute vorticity}\\
&=\frac{1}{h_2h_3}\left[\mathcal{N}\frac{\partial}{\partial\xi_2}\left(h_3u_3\right)-\frac{\partial}{\partial\xi_3}\left(h_2u_2\right)\right]\mathbf{e}_1\\
&\quad+\left\{\frac{1}{h_3h_1}\left[\frac{\partial}{\partial\xi_3}\left(h_1u_1\right)-\mathcal{N}\frac{\partial}{\partial\xi_1}\left(h_3u_3\right)\right]+\frac{2\Omega}{h_3}\frac{\partial h_1}{\partial\xi_3}\right\}\mathbf{e}_2\\
&\quad+\left\{\frac{1}{h_1h_2}\left[\frac{\partial}{\partial\xi_1}\left(h_2u_2\right)-\frac{\partial}{\partial\xi_2}\left(h_1u_1\right)\right]-\frac{2\Omega}{h_2}\frac{\partial h_1}{\partial\xi_2}\right\}\mathbf{e}_3,
\end{aligned} \tag{16.125}$$

$$\frac{D\Lambda}{Dt}=0, \tag{16.126}$$

$$\nabla\Lambda=\frac{1}{h_1}\frac{\partial\Lambda}{\partial\xi_1}\mathbf{e}_1+\frac{1}{h_2}\frac{\partial\Lambda}{\partial\xi_2}\mathbf{e}_2+\frac{1}{h_3}\frac{\partial\Lambda}{\partial\xi_3}\mathbf{e}_3, \tag{16.127}$$

and $\Lambda=\Lambda\left(\alpha,\eta\right)$ is any materially conserved quantity that satisfies (16.126).

Shallow Equation Sets in Axial-Orthogonal-Curvilinear Coordinates

Non-hydrostatic switch $\mathcal{N}$ determines the equation set;

$$\begin{aligned}
\mathcal{N}=1 &\quad\Rightarrow\quad \textbf{non-hydrostatic-shallow},\\
\mathcal{N}=0 &\quad\Rightarrow\quad \textbf{hydrostatic-shallow}\ (=\ \textbf{hydrostatic-primitive}).
\end{aligned}$$

Equations (16.128)–(16.136) – with $\mathcal{N}=0,1$ – correspond to the two *shallow* equation sets (in light blue) in Fig. 16.1.

Governing Equations

$$\frac{D_S u_1}{Dt} + \left(\frac{u_1}{h_1^S} + 2\Omega\right)\frac{u_2}{h_2^S}\frac{dh_1^S}{d\xi_2} + \frac{1}{\rho h_1^S}\frac{\partial p}{\partial \xi_1} = 0, \tag{16.128}$$

$$\frac{D_S u_2}{Dt} - \left(\frac{u_1}{h_1^S} + 2\Omega\right)\frac{u_1}{h_2^S}\frac{dh_1^S}{d\xi_2} - \frac{\mathcal{N} u_3^2}{h_2^S h_3^S}\frac{dh_3^S}{d\xi_2} + \frac{1}{\rho h_2^S}\frac{\partial p}{\partial \xi_2} = 0, \tag{16.129}$$

$$\mathcal{N}\left(\frac{D_S u_3}{Dt} + \frac{u_2}{h_2^S}\frac{u_3}{h_3^S}\frac{dh_3^S}{d\xi_2}\right) + \frac{1}{\rho h_3^S}\frac{\partial p}{\partial \xi_3} + \frac{1}{h_3^S}\frac{d\Phi_S}{d\xi_3} = 0, \tag{16.130}$$

$$\frac{D_S \rho}{Dt} + \rho \nabla_S \cdot \mathbf{u} = 0, \tag{16.131}$$

$$\frac{D_S \mathcal{E}_S}{Dt} + p\frac{D_S \alpha}{Dt} = 0, \tag{16.132}$$

where:

$$\alpha \equiv \frac{1}{\rho}, \tag{16.133}$$

$$\frac{D_S}{Dt} = \frac{\partial}{\partial t} + \frac{u_1}{h_1^S}\frac{\partial}{\partial \xi_1} + \frac{u_2}{h_2^S}\frac{\partial}{\partial \xi_2} + \frac{u_3}{h_3^S}\frac{\partial}{\partial \xi_3}, \tag{16.134}$$

$$\mathbf{u} \equiv u_1\mathbf{e}_1 + u_2\mathbf{e}_2 + u_3\mathbf{e}_3 = h_1^S\frac{D_S\xi_1}{Dt}\mathbf{e}_1 + h_2^S\frac{D_S\xi_2}{Dt}\mathbf{e}_2 + h_3^S\frac{D_S\xi_3}{Dt}\mathbf{e}_3, \tag{16.135}$$

$$\nabla_S \cdot \mathbf{A} = \frac{1}{h_1^S h_2^S h_3^S}\left[\frac{\partial}{\partial \xi_1}\left(A_1 h_2^S h_3^S\right) + \frac{\partial}{\partial \xi_2}\left(A_2 h_3^S h_1^S\right)\right] + \frac{1}{h_3^S}\frac{\partial A_3}{\partial \xi_3}. \tag{16.136}$$

Equations (16.128)–(16.130) agree with White and Wood (2012)'s equations (A.16)–(A.18).

Conservation Principles

Associated with the preceding governing equations are the following conservation principles for axial angular momentum, total energy, and potential vorticity:

Axial Angular Momentum

$$\frac{D_S}{Dt}\left[h_1^S\left(u_1 + \Omega h_1^S\right)\right] = -\frac{1}{\rho}\frac{\partial p}{\partial \xi_1}. \tag{16.137}$$

Total Energy

$$\frac{D_S}{Dt}\left[\frac{1}{2}\left(u_1^2 + u_2^2 + \mathcal{N} u_3^2\right) + \Phi_S + \mathcal{E}_S\right] = -\frac{1}{\rho}\nabla_S \cdot (p\mathbf{u}). \tag{16.138}$$

Potential Vorticity

$$\frac{D_S}{Dt}\left(\frac{\mathbf{Z}^S \cdot \nabla_S \Lambda}{\rho}\right) = \frac{D_S}{Dt}\left[\frac{1}{\rho}\left(\frac{Z_1^S}{h_1^S}\frac{\partial \Lambda}{\partial \xi_1} + \frac{Z_2^S}{h_2^S}\frac{\partial \Lambda}{\partial \xi_2} + \frac{Z_3^S}{h_3^S}\frac{\partial \Lambda}{\partial \xi_3}\right)\right] = 0, \tag{16.139}$$

where

$$\mathbf{Z}^S \equiv \nabla_S \times \mathbf{u} + 2\boldsymbol{\Omega} \cdot \mathbf{e}_3 = \text{absolute vorticity}$$

$$
\begin{aligned}
&= \left[\frac{\mathcal{N}}{h_2^S h_3^S} \frac{\partial}{\partial \xi_2} \left(h_3^S u_3 \right) - \frac{1}{h_3^S} \frac{\partial u_2}{\partial \xi_3} \right] \mathbf{e}_1 + \left(\frac{1}{h_3^S} \frac{\partial u_1}{\partial \xi_3} - \frac{\mathcal{N}}{h_1^S} \frac{\partial u_3}{\partial \xi_1} \right) \mathbf{e}_2 \\
&+ \left[\frac{1}{h_1^S} \frac{\partial u_2}{\partial \xi_1} - \frac{1}{h_1^S h_2^S} \frac{\partial}{\partial \xi_2} \left(h_1^S u_1 \right) - \frac{2\Omega}{h_2^S} \frac{d h_1^S}{d \xi_2} \right] \mathbf{e}_3, \qquad (16.140)
\end{aligned}
$$

$$
\frac{D_S \Lambda}{Dt} = 0, \qquad (16.141)
$$

$$
\nabla_S \Lambda = \frac{1}{h_1^S} \frac{\partial \Lambda}{\partial \xi_1} \mathbf{e}_1 + \frac{1}{h_2^S} \frac{\partial \Lambda}{\partial \xi_2} \mathbf{e}_2 + \frac{1}{h_3^S} \frac{\partial \Lambda}{\partial \xi_3} \mathbf{e}_3, \qquad (16.142)
$$

and $\Lambda = \Lambda\left(\alpha, \eta\right)$ is any materially conserved quantity that satisfies (16.141).

17

Quasi-Shallow, Dynamically Consistent Equation Sets in 3D

ABSTRACT

For many years it was conventional wisdom that to accurately represent the Coriolis force, whilst respecting dynamical consistency, there is no viable alternative but to adopt the deep-fluid equations. Recently, however, a pair (one non-hydrostatic, the other quasi-hydrostatic) of dynamically consistent quasi-shallow equation sets has been developed for global spheroidal (and, as a special case, spherical) atmospheres and oceans. This pair is of intermediate complexity between two standard pairs (one shallow, one deep) that are at least as accurate as the hydrostatic primitive equations. A feature of this new pair is that the fluid is assumed to be of essentially shallow depth, yet the Coriolis force is (almost) completely represented. This is in contradistinction to the traditional shallow pair, where the vertical component of the Coriolis force is entirely omitted in order to achieve dynamical consistency. To achieve this almost complete representation of the Coriolis force, planetary vorticity is approximated slightly less accurately than for the deep pair but much more accurately than for the shallow pair. The quasi-shallow pair is first derived using the classical Eulerian methodology. Next, it is rederived using (a) Hamilton's principle of stationary action, and (b) the Euler–Lagrange equations. The pair of quasi-shallow equations sets is then inserted into the framework that encompasses the standard quartet of equations sets (two deep, two shallow) described in Chapter 16. This results in a unified sextet of dynamically consistent equation sets. These are assembled and displayed in a comparative and informative manner, in both spheroidal and (as a special case) spherical geometry.

17.1 PREAMBLE

An approximate set of governing equations for motion of a global atmosphere or ocean is considered to be *dynamically consistent* if it implies conservation principles for mass, axial angular momentum, total energy, and potential vorticity, analogous to those of the unapproximated equations. See Section 16.1.2.

A quartet of approximate, dynamically consistent equation sets suitable for global atmospheric and oceanic modelling is considered in Chapter 16. See Fig. 16.1 for a schematic of the interrelationships between them. These equation sets are expressed in a unified manner in both spherical and spheroidal geometry. They arise according to whether approximations are made of (quasi-)hydrostatic and/or shallow nature, and the associated equation sets are referred to as:

- Non-hydrostatic deep (or fully compressible).
- Quasi-hydrostatic deep (or quasi-hydrostatic).
- Non-hydrostatic shallow.
- Hydrostatic shallow (or hydrostatic primitive).

To obtain the hydrostatic and quasi-hydrostatic equation sets from their non-hydrostatic counterparts, various terms are omitted from the vertical momentum equation, resulting in a simpler

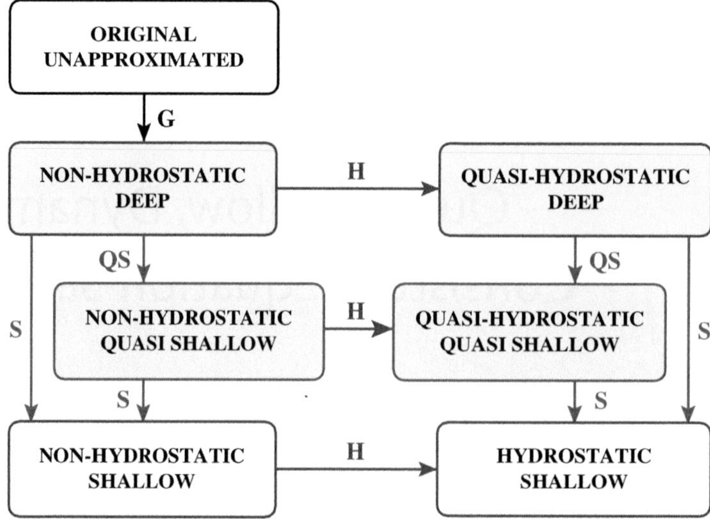

Figure 17.1 Interrelationships between six dynamically consistent, approximate equation sets (deep ones in dark blue, quasi-shallow ones in magenta, shallow ones in light blue) derived from the original, unapproximated one (unshaded). **G** denotes application of a geopotential approximation, **H** application of a quasi-hydrostatic or hydrostatic approximation, **QS** partial application of the shallow-fluid approximation, and **S** full application of the shallow-fluid approximation (or completion of it after partial application). See text for further details.

(hydrostatic or quasi-hydrostatic) balance equation. The process of doing so is denoted by H on the schematic.

For reasons of dynamical consistency – see Section 16.1.1 – the so-called '$2\Omega \cos \phi$' Coriolis terms are *entirely omitted* in the shallow pair, resulting in an incomplete representation of the Coriolis force. It was conventional wisdom that to accurately represent the Coriolis force, there is no viable alternative but to adopt the deep pair. However, Tort and Dubos (2014a) have recently derived a pair of dynamically consistent, quasi-shallow equation sets of intermediate complexity between these two standard pairs (one shallow, one deep) for global, *spherical* atmospheres and oceans. Thus the quartet became a *sextet* in *spherical* geometry (only). See Fig. 17.1 for a schematic of the interrelationships between individual members of this sextet. Those in dark blue (a deep pair) and in light blue (a shallow pair) are members of the original quartet – as depicted in Fig. 16.1. Those in magenta in Fig. 17.1 are the *additional*, intermediate pair of quasi-shallow equation sets to now be considered.

This advance then naturally led to the question:

- Can Tort and Dubos (2014a)'s quasi-shallow pair in *spherical* geometry be generalised to *spheroidal* geometry, or is it limited to spherical geometry?

Tort and Dubos (2014b) and Staniforth (2015a) independently answered this question by confirming that indeed it can. This then generalised the sextet from spherical to *spheroidal* geometry.

With this introduction, the goals of the present chapter[1] are to:

[1] This chapter contains public sector information licensed under the Open Government Licence v1.0. This information has been adapted and generalised for use herein.

1. Derive the pair of dynamically consistent, quasi-shallow equation sets – depicted schematically in magenta on Fig. 17.1 – in axial-orthogonal-curvilinear coordinates.
2. For comparative and informative purposes, display the equation sets in a unified manner with those of the quartet of equation sets considered in Section 16.

As in Chapter 16, the focus is again on dynamically consistent approximation of the underlying geophysical-fluid-dynamical equations of a dynamical core. It is therefore assumed throughout this chapter that:

- Physical forcings (other than gravity) are absent.
- The coordinate system is a geopotential one.
- The fluid is composed of a single substance (dry air for the atmosphere, pure liquid water for the oceans).

These assumptions not only shorten equations, they also enhance clarity in the present context by excluding extraneous detail. Notation is as in Chapter 16.

Three methodologies for deriving dynamically consistent equation sets were listed in Chapter 16.3:

1. Classical Eulerian.
2. Lagrangian, using Hamilton's principle of stationary action.
3. Lagrangian, using Euler–Lagrange equations.

These were then individually described in Sections 16.4–16.6, with a comparative discussion provided in Section 16.8.

The remainder of the present chapter is organised as follows. Since the standard way of deriving approximate equation sets is via the popular '*classical Eulerian*' approach, the quasi-shallow pair is first derived in this manner in Section 17.2. This involves a fair amount of intuitive reasoning, informed by experience, followed by a posteriori verification that dynamical consistency is indeed achieved. Two alternative, complementary, *Lagrangian* derivations are then developed in Section 17.3. They are based on approximation of a *single term* in the Lagrangian density; this greatly simplifies things. Applying either of the two Lagrangian methodologies then automatically leads – without further approximation – to the quasi-shallow pair of equation sets. The resulting unified sextet of equation sets is then displayed – in a comparative and informative manner – in spheroidal and spherical geometry in Sections 17.4 and 17.5, respectively. Conclusions are drawn in Section 17.6.

17.2 CLASSICAL EULERIAN DERIVATION

17.2.1 Axial Angular Momentum

To derive dynamically consistent equation sets, a key consideration is conservation of axial angular momentum (Phillips, 1966; White and Wood, 2012). From (16.122) (of Appendix B of Chapter 16), conservation of axial angular momentum of the deep equations in axial-orthogonal-curvilinear coordinates may be expressed as

$$\frac{D}{Dt}\left[h_1\left(u_1+\Omega h_1\right)\right]=-\frac{1}{\rho}\frac{\partial p}{\partial \xi_1}, \tag{17.1}$$

where

$$\frac{D}{Dt}\equiv\frac{\partial}{\partial t}+\frac{u_1}{h_1}\frac{\partial}{\partial \xi_1}+\frac{u_2}{h_2}\frac{\partial}{\partial \xi_2}+\frac{u_3}{h_3}\frac{\partial}{\partial \xi_3}, \tag{17.2}$$

is material derivative for a deep fluid. Guided by Tort and Dubos (2014a)'s functional form for axial angular momentum of their quasi-shallow pair of equation sets in *spherical* geometry, and by

Tort and Dubos (2014b)'s and Staniforth (2015a)'s forms in *spheroidal* geometry, we approximate the expression for specific axial angular momentum in (17.1) as

$$h_1\left(u_1+\Omega h_1\right)=h_1u_1+\Omega h_1^2\approx h_1^Su_1+\Omega\left[\left(h_1^S\right)^2+\left(\frac{\partial h_1^2}{\partial\xi_3}\right)^S\left(\xi_3-\xi_3^S\right)\right], \tag{17.3}$$

where $h_1^2\equiv\left(h_1\right)^2$ for brevity, and superscript 'S' denotes evaluation at the planet's axially symmetric, spheroidal, geopotential surface $\xi_3=\xi_3^S=$ constant.

The first two terms in the rightmost expression of (17.3) correspond to shallow representation of axial angular momentum; see (16.137). The expression in square brackets corresponds to a first-order expansion of h_1^2 in the vertical, about $\xi_3=\xi_3^S$; and $\left(\xi_3-\xi_3^S\right)$ plays the role here of $z\equiv r-a$ in Tort and Dubos (2014a) for the special case of spherical geometry. In (17.3) and onwards, terms in green denote terms associated with quasi-shallow approximation that are *additional* to those appearing in the pair of shallow-fluid equation sets. This facilitates tracking such terms during the ensuing analysis. Axial symmetry of the coordinate system is assumed so that scale factors $h_i=h_i\left(\xi_2,\xi_3\right), i=1,2,3$, and $h_i^S=h_i\left(\xi_2\right), i=1,2,3$ are independent of azimuthal (zonal) coordinate ξ_1.

Inserting (17.3) into (17.1) then gives the posited form for the

Axial Angular Momentum Evolution Equation

$$\rho\frac{D_S}{Dt}\left[h_1^S\left(u_1+\Omega h_1^S\right)+\Omega B\left(\xi_2\right)\left(\xi_3-\xi_3^S\right)\right]=-\frac{\partial p}{\partial\xi_1} \tag{17.4}$$

of the sought quasi-shallow pair of equation sets, where

$$\frac{D_S}{Dt}\equiv\frac{\partial}{\partial t}+\frac{u_1}{h_1^S}\frac{\partial}{\partial\xi_1}+\frac{u_2}{h_2^S}\frac{\partial}{\partial\xi_2}+\frac{u_3}{h_3^S}\frac{\partial}{\partial\xi_3} \tag{17.5}$$

is material derivative for a shallow fluid, and (for brevity and convenience)

$$B\left(\xi_2\right)\equiv\left(\frac{\partial h_1^2}{\partial\xi_3}\right)^S. \tag{17.6}$$

With use of shallow material derivative (17.5) and division by ρh_1^S, evolution equation (17.4) for axial angular momentum may be rewritten as

$$\frac{D_Su_1}{Dt}+\left(\frac{u_1}{h_1^S}+2\Omega\right)\frac{u_2}{h_2^S}\frac{dh_1^S}{d\xi_2}+\frac{\Omega}{h_1^S}\left[\frac{u_2}{h_2^S}\frac{dB\left(\xi_2\right)}{d\xi_2}\left(\xi_3-\xi_3^S\right)+B\left(\xi_2\right)\frac{u_3}{h_3^S}\right]=-\frac{1}{\rho h_1^S}\frac{\partial p}{\partial\xi_1}. \tag{17.7}$$

This equation is in the form of the first component of the momentum equation. Using definition (17.6), it is convenient to rewrite (17.7) in the form

$$\frac{D_Su_1}{Dt}+\left(\frac{u_1}{h_1^S}+2\Omega\right)\frac{u_2}{h_2^S}\frac{dh_1^S}{d\xi_2}+2\Omega\left[A\left(\xi_2\right)\left(\xi_3-\xi_3^S\right)u_2+\frac{u_3}{h_3^S}\left(\frac{\partial h_1}{\partial\xi_3}\right)^S\right]=-\frac{1}{\rho h_1^S}\frac{\partial p}{\partial\xi_1}, \tag{17.8}$$

where

$$A(\xi_2) \equiv \frac{1}{h_1^S h_2^S} \frac{d}{d\xi_2} \left[\frac{B(\xi_2)}{2} \right] = \frac{1}{h_1^S h_2^S} \frac{d}{d\xi_2} \left[\frac{1}{2} \left(\frac{\partial h_1^2}{\partial \xi_3} \right)^S \right]. \tag{17.9}$$

(The reader is advised that definitions (17.9) and (17.6) of $A(\xi_2)$ and $B(\xi_2)$, respectively, are used frequently in this chapter. Since $\left(\partial h_1^2/\partial \xi_3\right)^S \equiv 0$ and $A(\xi_2) \equiv B(\xi_2) \equiv 0$ in (purely) shallow geometry, green terms are then absent from the equations of the ensuing analysis.)

Note that $A(\xi_2) \to \left[h_2^S/\left(dh_1^S/d\xi_2\right)\right] A(\xi_2)$ and $B(\xi_2) \to \left[1/\left(h_1^S\right)^2\right] B(\xi_2)$ with respect to their definitions in Staniforth (2015a).[2] For the special, *spherical* case, $(h_1, h_2, h_3) = (r\cos\phi, r, 1)$ and then (from (17.6) and (17.9))

$$A(\phi) = -\frac{2}{a}\sin\phi, \quad B(\phi) = 2a\cos^2\phi, \tag{17.10}$$

in agreement with Tort and Dubos (2014a)'s corresponding explicit expressions.

17.2.2 Components of the Momentum Equation

By inspection (and guided by the analyses of White et al. (2005), White and Wood (2012), Tort and Dubos (2014a,b), and Staniforth (2015a)), the three components of the momentum equation are posited to be

$$\frac{D_S u_1}{Dt} + \left(\frac{u_1}{h_1^S} + 2\Omega\right) \frac{u_2}{h_2^S} \frac{dh_1^S}{d\xi_2} + 2\Omega \left[A(\xi_2)\left(\xi_3 - \xi_3^S\right) u_2 + \frac{u_3}{h_3^S} \left(\frac{\partial h_1}{\partial \xi_3} \right)^S \right] = -\frac{1}{\rho h_1^S} \frac{\partial p}{\partial \xi_1}, \tag{17.11}$$

$$\frac{D_S u_2}{Dt} - \left(\frac{u_1}{h_1^S} + 2\Omega\right) \frac{u_1}{h_2^S} \frac{dh_1^S}{d\xi_2} - \frac{\mathcal{N} u_3^2}{h_2^S h_3^S} \frac{dh_3^S}{d\xi_2} - 2\Omega A(\xi_2)\left(\xi_3 - \xi_3^S\right) u_1 = -\frac{1}{\rho h_2^S} \frac{\partial p}{\partial \xi_2}, \tag{17.12}$$

$$\mathcal{N}\left(\frac{D_S u_3}{Dt} + \frac{u_2}{h_2^S} \frac{u_3}{h_3^S} \frac{dh_3^S}{d\xi_2} \right) - 2\Omega \frac{u_1}{h_3^S} \left(\frac{\partial h_1}{\partial \xi_3} \right)^S = -\frac{1}{\rho h_3^S} \frac{\partial p}{\partial \xi_3} - \frac{1}{h_3^S} \frac{d\Phi_S}{d\xi_3}, \tag{17.13}$$

where $\mathcal{N}$ is the non-hydrostatic switch. When:

- $\mathcal{N} = 1$, the term it multiplies is retained.
- $\mathcal{N} = 0$, the term it multiplies is omitted.

Such terms are coloured in blue, again for tracking purposes.

A very nice feature of (17.11)–(17.13), developed in White and Wood (2012)'s Appendix and Section 6.4.1 herein, is the way in which $\cos\phi$ and $\sin\phi$ are alternatively expressed in terms of metric factors using (6.20) and (6.21) of Section 6.4.1.[3] Thus – for a shallow fluid –

[2] Redefinition of $A(\xi_2)$ avoids an unnecessary singularity for cylindrical geometry, and redefinition of $B(\xi_2)$ simplifies things a little.

[3] In Chapter 16, this allowed shallow approximation of the pair of deep equations to be achieved by the *purely geometric* mapping/approximation $h_i(\xi_2, \xi_3) \to h_i^S(\xi_2)$, $i = 1, 2, 3$.

$$\cos\phi = \frac{1}{h_3^S}\left(\frac{\partial h_1}{\partial \xi_3}\right)^S, \quad \sin\phi = -\frac{1}{h_2^S}\left(\frac{\partial h_1}{\partial \xi_2}\right)^S, \tag{17.14}$$

where ϕ is geographic latitude, that is, the angle between the local vertical (aligned with unit vector $\mathbf{e}_3$) and the equatorial plane; see, for example, Fig. 17.2 later in this chapter. This helps us to keep track of approximation of the Coriolis force.

A few words now about what was involved in the 'inspection'.

- Component equation (17.11) was developed in the immediately preceding subsection; this equation, *by construction*, embodies *conservation of axial angular momentum*.
- What *about total energy conservation*? Two additional (green) terms appear in (17.11). Now the budget equation for specific kinetic energy in axial-orthogonal-curvilinear coordinates is formed by successively multiplying the three component equations (17.11)–(17.13) by u_1, u_2, and u_3, respectively, and summing the results. See, for example, Section 17.2.5. To ensure that the term $2\Omega A(\xi_2)\left(\xi_3 - \xi_3^S\right)u_2$ in (17.11) does not create a spurious source of kinetic energy, a corresponding term $-2\Omega A(\xi_2)\left(\xi_3 - \xi_3^S\right)u_1$ must appear in (17.12) to counteract this. Similarly, to ensure that the term $2\Omega\left(u_3/h_3^S\right)\left(\partial h_1/\partial\xi_3\right)^S$ in (17.11) does not likewise create a spurious source of kinetic energy, a term $-2\Omega\left(u_1/h_3^S\right)\left(\partial h_1/\partial\xi_3\right)^S$ must appear in (17.13).
- What about *conservation of potential vorticity*? For now we just hope for the best, pending subsequent verification. Generally, the procedure to be followed is that if this works out (it does in the present instance), then great; otherwise we need to diagnose why not and examine how to fix things (assuming that this is indeed possible).

This analysis for the hypothesised form of the three component equations of momentum requires experience and insight regarding the fundamental properties of the governing equations in order to devise useful, dynamically consistent approximations of them. This is the downside of the 'classical Eulerian' approach, as discussed in Section 16.8. By way of contrast, the Lagrangian approach is more systematic and requires less insight, but employs less familiar methods.

17.2.3 The Mass-Continuity Equation

The mass-continuity equation in shallow, axial-orthogonal-curvilinear coordinates is given by the shallow form of (6.32) with $F^\rho \equiv 0$. Thus (and in agreement with the shallow mass-continuity equation (16.131) of Chapter 16)

$$\frac{D_S\rho}{Dt} + \rho\nabla_S\cdot\mathbf{u} = 0, \tag{17.15}$$

where (from (5.64) and (16.136))

$$\nabla_S\cdot\mathbf{A} \equiv \frac{1}{h_1^S h_2^S h_3^S}\left[\frac{\partial}{\partial\xi_1}\left(A_1 h_2^S h_3^S\right) + \frac{\partial}{\partial\xi_2}\left(A_2 h_3^S h_1^S\right)\right] + \frac{1}{h_3^S}\frac{\partial A_3}{\partial\xi_3}. \tag{17.16}$$

17.2.4 The Thermodynamic-Energy and State Equations

To complete the formulation of the governing equations for unforced shallow-fluid flow over a rotating spheroidal planet, the component equations (17.11)–(17.13) of momentum and mass-continuity equation (17.15) are supplemented by a thermodynamic-energy equation and a fundamental equation of state. Thus, from the shallow form of (6.34) (with $\dot{Q}_E \equiv 0$) and from (6.35)–(6.37) (with S^1 and S^2 omitted),

$$\frac{D_S \mathcal{E}_S}{Dt} + p \frac{D_S \alpha}{Dt} = 0, \tag{17.17}$$

$$\mathcal{E}_S = \mathcal{E}_S(\alpha, \eta) \quad \Rightarrow \quad p \equiv -\frac{\partial \mathcal{E}_S}{\partial \alpha}, \quad T \equiv \frac{\partial \mathcal{E}_S}{\partial \eta}, \tag{17.18}$$

respectively, where $\mathcal{E}_S$ is specific internal energy. The natural variables of $\mathcal{E}_S$ are α (specific volume) and η (specific entropy); see Section 9.2. The prescribed functional form for $\mathcal{E}_S$ in (17.18) defines the thermodynamic properties of the fluid. As in Section 16.5.3, specific internal energy $\mathcal{E}$ bears a subscript 'S' to signify that its first argument (α) is governed by the *shallow* mass-continuity equation (17.15), rather than by the analogous equation for a *deep* fluid. Strictly speaking this is not necessary, but it is convenient to do so in order to keep track of the influence, both direct and indirect, of shallow approximation. Thermodynamic-energy equation (17.17) has the usual form for a shallow fluid; see (16.132) of Chapter 16.

17.2.5 Total Energy Conservation

Multiplying (17.11)–(17.13) successively by ρu_1, ρu_2, and ρu_3 and summing the resulting equations leads to the combined kinetic-energy and gravitational-energy evolution equation

$$\rho \frac{D_S}{Dt}\left(\frac{u_1^2 + u_2^2 + \mathcal{N} u_3^2}{2}\right) + \rho \frac{u_3}{h_3^S}\frac{d\Phi_S}{d\xi_3} = -\left(\frac{u_1}{h_1^S}\frac{\partial p}{\partial \xi_1} + \frac{u_2}{h_2^S}\frac{\partial p}{\partial \xi_2} + \frac{u_3}{h_3^S}\frac{\partial p}{\partial \xi_3}\right), \tag{17.19}$$

that is,

$$\begin{aligned}
\rho \frac{D_S}{Dt}\left(K^{\mathcal{N}} + \Phi_S\right) &= -\frac{1}{h_1^S h_2^S h_3^S}\left[\frac{\partial}{\partial \xi_1}\left(p u_1 h_2^S h_3^S\right) + \frac{\partial}{\partial \xi_2}\left(p u_2 h_3^S h_1^S\right) + \frac{\partial}{\partial \xi_3}\left(p u_3 h_1^S h_2^S\right)\right] \\
&\quad + \frac{p}{h_1^S h_2^S h_3^S}\left[\frac{\partial}{\partial \xi_1}\left(u_1 h_2^S h_3^S\right) + \frac{\partial}{\partial \xi_2}\left(u_2 h_3^S h_1^S\right) + \frac{\partial}{\partial \xi_3}\left(p u_3 h_1^S h_2^S\right)\right] \\
&= -\nabla_S \cdot \left(p\mathbf{u}\right) + p\left(\nabla_S \cdot \mathbf{u}\right),
\end{aligned} \tag{17.20}$$

where

$$K^{\mathcal{N}} \equiv \frac{u_1^2 + u_2^2 + \mathcal{N} u_3^2}{2}. \tag{17.21}$$

This evolution equation for the sum of specific kinetic energy ($K^{\mathcal{N}}$) and specific gravitational potential energy (Φ_S) has the usual form for a shallow atmosphere/ocean, with the contributions of the $A(\xi_2)$ terms of (17.11) and (17.12) having (by design) cancelled one another. This cancellation ensures that the Coriolis force – as represented – does no work; see Section 2.2.2.3 regarding this fundamental property.

Multiplication of thermodynamic-energy equation (17.17) by $\rho \equiv 1/\alpha$, followed by use of mass-continuity equation (17.15), yields

$$\rho \frac{D_S \mathcal{E}_S}{Dt} = -\rho p \frac{D_S \alpha}{Dt} = -\rho p \frac{D_S}{Dt}\left(\frac{1}{\rho}\right) = \frac{p}{\rho}\frac{D_S \rho}{Dt} = -p \nabla_S \cdot \mathbf{u}. \tag{17.22}$$

This is an evolution equation for internal energy ($\mathcal{E}_S$).

Summing evolution equations (17.20) and (17.22) and cancelling the $p\nabla_S \cdot \mathbf{u}$ terms of opposite sign then delivers:

The Total Energy Evolution Equation

$$\rho \frac{D_S E^{\mathcal{N}}}{Dt} + \nabla_S \cdot \left(p \boldsymbol{u} \right) = 0. \tag{17.23}$$

Here

$$E^{\mathcal{N}} \equiv K^{\mathcal{N}} + \Phi_S + \mathcal{E}_S \tag{17.24}$$

is specific total energy, composed of the sum of specific kinetic ($K^{\mathcal{N}}$), specific gravitational-potential (Φ_S), and specific internal ($\mathcal{E}_S$) energies. Total-energy evolution equation (17.23) has the usual form for a shallow atmosphere/ocean – see (16.138) (for $\mathcal{N} = 0, 1$). Thus:

- The posited form of momentum equations (17.11)–(17.13) leads to total energy conservation.

17.2.6 Ertel's Theorem for the Quasi-Shallow Equations

Having, by construction, achieved conservation of mass, axial angular momentum and total energy, the only remaining conservation issue is whether or not (17.11)–(17.13) also lead to conservation of potential vorticity (i.e. to Ertel's theorem). An important property, exploited here, is that the usual vector identities involving gradient, divergence, and curl not only hold in 3D Euclidean space with (deep) orthogonal, curvilinear, metric factors $h_i, i = 1, 2, 3$, but also in non-Euclidean space with shallow, orthogonal, curvilinear, metric factors $h_i^S, i = 1, 2, 3$ (White et al., 2005). This simplifies the derivation by greatly reducing the amount of algebra.

Following Sections 15.3.4 and 15.3.6, but for the quasi-shallow equations instead of the deep ones, the posited momentum-component equations (17.11)–(17.13) can be rewritten in vector-invariant form as

$$\frac{\partial \mathbf{u}^{\mathcal{N}}}{\partial t} = -\mathbf{Z}^{QS} \times \mathbf{u} - \nabla_S \left(\Phi_S + \frac{\mathbf{u}^{\mathcal{N}} \cdot \mathbf{u}^{\mathcal{N}}}{2} \right) - \frac{1}{\rho} \nabla_S p, \tag{17.25}$$

where

$$\mathbf{u}^{\mathcal{N}} \equiv u_1 \mathbf{e}_1 + u_2 \mathbf{e}_2 + \mathcal{N} u_3 \mathbf{e}_3, \tag{17.26}$$

$$\begin{aligned} \mathbf{Z}^{QS} &\equiv \frac{2\Omega}{h_3^S} \left(\frac{\partial h_1}{\partial \xi_3} \right)^S \mathbf{e}_2 - 2\Omega \left[\frac{1}{h_2^S} \frac{d h_1^S}{d \xi_2} + A(\xi_2) \left(\xi_3 - \xi_3^S \right) \right] \mathbf{e}_3 + \left(\boldsymbol{\zeta}^S \right)^{\mathcal{N}} \\ &= \nabla_S \times \left\{ \Omega h_1^S \left[1 + \frac{B(\xi_2)}{\left(h_1^S \right)^2} \left(\xi_3 - \xi_3^S \right) \right] \mathbf{e}_1 + \mathbf{u}^{\mathcal{N}} \right\} \end{aligned} \tag{17.27}$$

is (quasi-shallow) absolute vorticity, and

$$\begin{aligned} \left(\boldsymbol{\zeta}^S \right)^{\mathcal{N}} &\equiv \left(\zeta_1^S \right)^{\mathcal{N}} \mathbf{e}_1 + \left(\zeta_2^S \right)^{\mathcal{N}} \mathbf{e}_2 + \left(\zeta_3^S \right)^{\mathcal{N}} \mathbf{e}_3 \equiv \nabla_S \times \mathbf{u}^{\mathcal{N}} \\ &\equiv \frac{1}{h_2^S h_3^S} \left[\mathcal{N} \frac{\partial}{\partial \xi_2} \left(h_3^S u_3 \right) - \frac{\partial}{\partial \xi_3} \left(h_2^S u_2 \right) \right] \mathbf{e}_1 \\ &+ \frac{1}{h_3^S h_1^S} \left[\frac{\partial}{\partial \xi_3} \left(h_1^S u_1 \right) - \mathcal{N} \frac{\partial}{\partial \xi_1} \left(h_3^S u_3 \right) \right] \mathbf{e}_2 \end{aligned}$$

$$+\frac{1}{h_1^S h_2^S}\left[\frac{\partial}{\partial \xi_1}\left(h_2^S u_2\right)-\frac{\partial}{\partial \xi_2}\left(h_1^S u_1\right)\right]\mathbf{e}_3 \tag{17.28}$$

is the usual (shallow) relative vorticity. Equation (17.27) is just the usual shallow expression for $\mathbf{Z}$, but with a modified planetary-vorticity term that now includes a vertical $\left(\xi_3-\xi_3^S\right)$ addition to the usual $\sin\phi$ $\left(\equiv -\left(\partial h_1/\partial \xi_2\right)^S/h_2^S\right.$; see (17.14)) contribution, plus a new $\cos\phi$ $\left(\equiv \left(\partial h_1/\partial \xi_3\right)^S/h_3^S\right.$; again see (17.14)) contribution.

Taking $\nabla_S\times$(17.25) and noting that

$$\frac{\partial}{\partial t}\left\{\frac{2\Omega}{h_3^S}\left(\frac{\partial h_1}{\partial \xi_3}\right)^S \mathbf{e}_2-2\Omega\left[\frac{1}{h_2^S}\frac{dh_1^S}{d\xi_2}+A\left(\xi_2\right)\left(\xi_3-\xi_3^S\right)\right]\mathbf{e}_3\right\}\equiv 0 \tag{17.29}$$

(because all terms within the braces are independent of time t) then gives

$$\frac{\partial \mathbf{Z}^{QS}}{\partial t}=-\nabla_S\times\left(\mathbf{Z}^{QS}\times\mathbf{u}\right)-\nabla_S\left(\frac{1}{\rho}\right)\times\nabla_S p. \tag{17.30}$$

Using vector identity (cf. (A.22) of the Appendix at the end of this book for the corresponding form in deep geometry)

$$\nabla_S\times(\mathbf{Z}\times\mathbf{u})\equiv\mathbf{Z}\left(\nabla_S\cdot\mathbf{u}\right)+\left(\mathbf{u}\cdot\nabla_S\right)\mathbf{Z}-\left(\mathbf{Z}\cdot\nabla_S\right)\mathbf{u}-\mathbf{u}\left(\nabla_S\cdot\mathbf{Z}\right), \tag{17.31}$$

and noting that $\nabla_S\cdot\mathbf{Z}^{QS}=0$, (17.30) can be rewritten as

$$\frac{1}{\rho}\frac{D_S\mathbf{Z}^{QS}}{Dt}=-\frac{\mathbf{Z}^{QS}}{\rho}\left(\nabla_S\cdot\mathbf{u}\right)+\frac{1}{\rho}\left(\mathbf{Z}^{QS}\cdot\nabla_S\right)\mathbf{u}-\frac{1}{\rho^3}\nabla_S\rho\times\nabla_S p. \tag{17.32}$$

The (shallow) mass-continuity equation (17.15) is now used to eliminate $\nabla_S\cdot\mathbf{u}$ from (17.32) to obtain the (quasi-shallow) absolute-vorticity equation

$$\frac{D_S}{Dt}\left(\frac{\mathbf{Z}^{QS}}{\rho}\right)=\left(\frac{\mathbf{Z}^{QS}}{\rho}\cdot\nabla_S\right)\mathbf{u}+\frac{1}{\rho^3}\nabla_S\rho\times\nabla_S p. \tag{17.33}$$

To complete the derivation of Ertel's theorem for the quasi-shallow equations, consider now the material-conservation equation

$$\frac{D_S\Lambda}{Dt}=0, \tag{17.34}$$

where Λ is any scalar field that satisfies (17.34). Some suitable possible candidates for Λ are discussed in Section 15.3.6. In the present context of a single-substance fluid (dry air for the atmosphere, and pure liquid water for the oceans), specific entropy (η), potential temperature (θ), potential enthalpy (h^0), conservative temperature (Θ), and potential density (ρ^θ) are all suitable candidates for Λ; see Section 10.7 for definitions of these quantities.

Now

$$\frac{\mathbf{Z}^{QS}}{\rho}\cdot\frac{D_S}{Dt}\left(\nabla_S\Lambda\right)\equiv\left(\frac{\mathbf{Z}^{QS}}{\rho}\cdot\nabla_S\right)\overbrace{\left(\frac{D_S\Lambda}{Dt}\right)}^{=0}-\left[\left(\frac{\mathbf{Z}^{QS}}{\rho}\cdot\nabla_S\right)\mathbf{u}\right]\cdot\left(\nabla_S\Lambda\right)$$

$$=-\left[\left(\frac{\mathbf{Z}^{QS}}{\rho}\cdot\nabla_S\right)\mathbf{u}\right]\cdot\left(\nabla_S\Lambda\right), \tag{17.35}$$

where (17.34) has been used to simplify the right-hand side of (17.35). Taking the scalar product of (17.33) with $\nabla_S\Lambda$ gives

$$\left(\nabla_S\Lambda\right)\cdot\frac{D_S}{Dt}\left(\frac{\mathbf{Z}^{QS}}{\rho}\right)=\left(\nabla_S\Lambda\right)\cdot\left[\left(\frac{\mathbf{Z}^{QS}}{\rho}\cdot\nabla_S\right)\mathbf{u}\right]+\left(\nabla_S\Lambda\right)\cdot\left(\frac{1}{\rho^3}\nabla_S\rho\times\nabla_S p\right). \tag{17.36}$$

Summing this equation with (17.35) then yields

$$\frac{D_S \Pi^{QS}}{Dt} = \nabla_S \Lambda \cdot \left(\frac{1}{\rho^3} \nabla_S \rho \times \nabla_S p \right), \tag{17.37}$$

where

$$\Pi^{QS} \equiv \frac{\mathbf{Z}^{QS} \cdot \nabla_S \Lambda}{\rho} \tag{17.38}$$

is quasi-shallow potential vorticity, and the quasi-shallow measure for absolute vorticity, $\mathbf{Z}^{QS}$, is defined by (17.27). Thus:

Conservation of Quasi-Shallow Potential Vorticity (Ertel's Theorem)

$$\frac{D_S \Pi^{QS}}{Dt} = 0, \tag{17.39}$$

where Π^{QS} is *quasi-shallow potential vorticity*, provided that:

1. *Either* $\nabla_S \rho \times \nabla_S p \equiv 0$ (i.e. the fluid is barotropic).
2. *Or* $\Lambda = \Lambda(\rho, p)$ (i.e. $\nabla_S \Lambda$, $\nabla_S \rho$ and $\nabla_S p$ are all coplanar).

Note that the second of the two conditions follows – with use of vector identities (A.3) and (A.9) of the Appendix at the end this book – from

$$\nabla_S \Lambda \cdot \left(\nabla_S \rho \times \nabla_S p \right) \equiv \left(\frac{\partial \Lambda}{\partial \rho} \nabla_S \rho + \frac{\partial \Lambda}{\partial p} \nabla_S p \right) \cdot \left(\nabla_S \rho \times \nabla_S p \right) \equiv 0. \tag{17.40}$$

17.2.7 Summary of the Eulerian Derivation

The Quasi-Shallow Pair of Equation Sets

The quasi-shallow pair of equation sets is comprised of:

- Component equations (17.11)–(17.13) of momentum.
- Mass-continuity equation (17.15).
- Thermodynamic-energy equation (17.17).
- Fundamental equation of state (17.18).

For reference purposes, this pair of equation sets is given in Appendix A of the present chapter, together with the associated conservation properties. Compared with the usual shallow-fluid equation sets (both non-hydrostatic and hydrostatic), quasi-shallow component equations (17.11)–(17.13) of momentum each include an *additional* Coriolis contribution. In particular, the presence of the Coriolis contribution in the vertical-component equation (17.13) means that classical hydrostatic balance is only achieved in the exceptional circumstance that the left-hand side of (17.13) fortuitously happens to be zero. The remaining equations (other than (17.11)–(17.13)) correspond to the usual ones obtained under shallow approximation. They do not contain any (green) quasi-shallow contributions, unlike (17.11)–(17.13), which do.

To obtain the quasi-shallow component equations of momentum, the functional form of the axial-angular-momentum equation was posited, and the first component of the momentum equation then deduced from it. The expression contained within the square brackets of (17.4) – where $B(\xi_2)$ is defined by (17.6) – is the quasi-shallow measure of axial angular momentum.

Next, the probable form of the two other momentum-component equations was inferred, and total energy conservation verified with these forms. The quasi-shallow measure of kinetic energy is $K^{\mathcal{N}} \equiv \left(u_1^2 + u_2^2 + \mathcal{N} u_3^2\right)/2$. This is identical to both the deep and shallow measures and subject to whether the (quasi-)hydrostatic assumption is made ($\mathcal{N} = 0$) or not made ($\mathcal{N} = 1$).

The final step for conservation was to verify that the three posited momentum-component equations (17.11)–(17.13) still lead to Ertel's theorem for conservation of potential vorticity. The measure of potential vorticity for the quasi-shallow equations is $\left(\mathbf{Z}^{QS} \cdot \nabla_S \Lambda\right)/\rho$, where $\mathbf{Z}^{QS}$ is defined by (17.27). This completed the demonstration that *the above-defined quasi-shallow equation set is dynamically consistent*; that is, it has conservation principles for mass, axial angular momentum, total energy, and potential vorticity analogous to those of the unapproximated equations.

17.2.8 Distance Metrics and Approximation of Planetary Vorticity

For the *deep*-fluid equations – expressed in (deep) axial-orthogonal-curvilinear coordinates – the distance metric is

$$\left(ds^2\right)^{\text{deep}} = \left(h_1\right)^2 d\xi_1^2 + \left(h_2\right)^2 d\xi_2^2 + \left(h_3\right)^2 d\xi_3^2. \tag{17.41}$$

This may be contrasted with the distance metric for *shallow*-fluid equations. This latter metric is straightforwardly obtained by setting deep metric factors h_1, h_2, and h_3 in (17.41) to their surface values h_1^S, h_2^S, and h_3^S, respectively. Thus

$$\left(ds^2\right)^{\text{shallow}} = \left(h_1^S\right)^2 d\xi_1^2 + \left(h_2^S\right)^2 d\xi_2^2 + \left(h_3^S\right)^2 d\xi_3^2. \tag{17.42}$$

The usual vector identities involving gradient, divergence, and curl continue to hold in shallow axial-orthogonal-curvilinear coordinates, but with use of shallow metric factors (h_1^S, h_2^S, and h_3^S) instead of deep ones (h_1, h_2, and h_3); (White et al., 2005).

Examination of quasi-shallow equation set (17.11)–(17.13), (17.15), and (17.17) shows that the appropriate distance metric is the shallow one, namely (17.42). All curvilinear operations – such as evaluating gradient, divergence, and curl – *with one exception*, can be seen to be computed using shallow metric (17.42), just as they are for the shallow equation set.

The single exception is due to the way in which the constant planetary-vorticity vector, $2\boldsymbol{\Omega}$ – aligned with the planet's rotation axis – is projected onto unit vectors $\mathbf{e}_2$ and $\mathbf{e}_3$. The usual way of doing this for the deep equations is to *exactly* express the planetary-vorticity vector as

$$2\boldsymbol{\Omega} = (2\Omega \cos\phi)\, \mathbf{e}_2 + (2\Omega \sin\phi)\, \mathbf{e}_3, \tag{17.43}$$

where ϕ is geographic latitude, that is, the angle between the local vertical (directed along $\mathbf{e}_3$) and the equatorial plane. See Fig. 17.2.

Using (17.14), (17.43) may be re-expressed for the deep and shallow situations as

$$(2\boldsymbol{\Omega})^{\text{deep}} = 2\Omega h_1 \nabla \times \mathbf{e}_1 = 2\Omega \left(\frac{1}{h_3}\frac{\partial h_1}{\partial \xi_3}\mathbf{e}_2 - \frac{1}{h_2}\frac{\partial h_1}{\partial \xi_2}\mathbf{e}_3 \right) = (2\Omega \cos\phi)\, \mathbf{e}_2 + (2\Omega \sin\phi)\, \mathbf{e}_3, \tag{17.44}$$

$$(2\boldsymbol{\Omega})^{\text{shallow}} = 2\Omega h_1^S \nabla_S \times \mathbf{e}_1 = 2\Omega \left(\frac{1}{h_3^S}\cancelto{0}{\frac{\partial h_1^S}{\partial \xi_3}}\mathbf{e}_2 - \frac{1}{h_2^S}\frac{\partial h_1^S}{\partial \xi_2}\mathbf{e}_3 \right) = -2\Omega \frac{1}{h_2^S}\frac{dh_1^S}{d\xi_2}\mathbf{e}_3 = (2\Omega \sin\phi)^S\, \mathbf{e}_3, \tag{17.45}$$

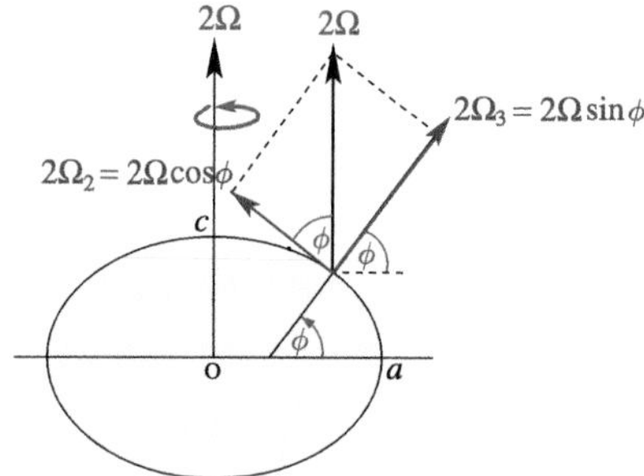

Figure 17.2 Exact projection of planetary-vorticity vector, $2\mathbf{\Omega}$, onto the ξ_2 (horizontal) and ξ_3 (vertical) axes for an axially symmetric, spheroidal planet; $2\mathbf{\Omega} = 2\Omega_2\mathbf{e}_2 + 2\Omega_3\mathbf{e}_3 = (2\Omega\cos\phi)\,\mathbf{e}_2 + (2\Omega_3\sin\phi)\,\mathbf{e}_3$, where ϕ is geographic latitude, and a and c are semi-major and semi-minor axes, respectively.

respectively. *Deep* projection (17.44) of the constant planetary-vorticity vector onto the unit vectors is *exact*, whereas *shallow* projection (17.45) is definitely *not*.

For *shallow* projection, the meridional component ($2\Omega\cos\phi$) is entirely lost. Quantitatively, the meridional component is particularly deficient in the tropics where $\cos\phi$ is of order unity. Contrastingly, the vertical component ($2\Omega\sin\phi$) is evaluated at the planet's surface as $(2\Omega\sin\phi)^S$ but, for a mildly oblate planet, the error in doing so is negligibly small.

For *quasi-shallow* projection, the constant planetary-vorticity vector is also approximated, but differently, and far more accurately, than for shallow projection. To see this, we extract the quasi-shallow projection of the planetary-vorticity vector from (17.27) by setting $\left(\zeta^S\right)^{\mathcal{N}} \equiv 0$ therein. Thus, using (17.14),

$$\begin{aligned}(2\mathbf{\Omega})^{\text{quasi-shallow}} &= \frac{2\Omega}{h_3^S}\left(\frac{\partial h_1}{\partial \xi_3}\right)^S \mathbf{e}_2 - 2\Omega\left[\frac{1}{h_2^S}\frac{dh_1^S}{d\xi_2} + A\left(\xi_2\right)\left(\xi_3 - \xi_3^S\right)\right]\mathbf{e}_3 \\ &= \left(2\Omega\cos\phi\right)^S \mathbf{e}_2 + 2\Omega\left[\left(\sin\phi\right)^S - A\left(\xi_2\right)\left(\xi_3 - \xi_3^S\right)\right]\mathbf{e}_3. \end{aligned} \tag{17.46}$$

Comparing quasi-shallow (17.46) with deep (17.44), reveals that *the quasi-shallow representation of planetary vorticity is almost exact*, provided that $\left|A\left(\xi_2\right)\left(\xi_3 - \xi_3^S\right)\right| \ll 1$. This condition corresponds to a shallow-fluid assumption. For example, insertion of (17.10) for the spherical case into this condition shows that it corresponds to $|r-a|/a \ll 1$, that is, to the usual shallow-fluid condition that the depth of the fluid is very small compared to the planet's mean radius.

The fact that shallow representation of planetary vorticity seriously misrepresents the actual behaviour has ramifications for the Coriolis force. This force (by convention placed on the right-hand side of the vector momentum equation) is defined to be $-2\mathbf{\Omega}\times\mathbf{u}$. Use of (17.44) leads to it being *completely* represented in the *deep* equations, and use of (17.46) to it being *almost* completely represented in the *quasi-shallow* equations. However, for the *shallow* equations, (17.45) is used instead. The inexactitude of (17.45) then leads to the well-known absence of the so-called '$2\Omega\cos\phi$' Coriolis terms in the shallow-fluid equations. The Coriolis force then acts *only* in the local horizontal plane.

Summarising the preceding discussion, the quasi-shallow equations are more accurate than the shallow ones, with a much better representation of both the constant planetary-vorticity vector

Equation set	$2\boldsymbol{\Omega}\cdot\mathbf{e}_2$	$2\boldsymbol{\Omega}\cdot\mathbf{e}_3$
Deep	$\frac{2\Omega}{h_3}\frac{\partial h_1}{\partial \xi_3} = 2\Omega\cos\phi$	$-\frac{2\Omega}{h_2}\frac{\partial h_1}{\partial \xi_2} = 2\Omega\sin\phi$
Quasi-Shallow	$\frac{2\Omega}{h_3^S}\left(\frac{\partial h_1}{\partial \xi_3}\right)^S = 2\Omega\left(\cos\phi\right)^S$	$-2\Omega\left[\frac{1}{h_2^S}\frac{dh_1^S}{d\xi_2} + A\left(\xi_2\right)\left(\xi_3 - \xi_3^S\right)\right] = 2\Omega\left[\left(\sin\phi\right)^S - A\left(\xi_2\right)\left(\xi_3 - \xi_3^S\right)\right]$
Shallow	0	$-\frac{2\Omega}{h_2^S}\frac{dh_1^S}{d\xi_2} = 2\Omega\left(\sin\phi\right)^S$

Table 17.1 Components of planetary-vorticity vector ($2\boldsymbol{\Omega}$) in the $\mathbf{e}_2$ and $\mathbf{e}_3$ directions (2nd and 3rd columns, respectively), as a function of equation set (1st column); $2\boldsymbol{\Omega}\cdot\mathbf{e}_1 \equiv 0$ for all equation sets.

and the Coriolis force. Also – see Sections 17.2.1–17.2.7 – this is achieved *without loss of dynamical consistency*.[4] The derivation of the quasi-shallow equations is, however, a little more complicated. It requires not only the adoption of a shallow metric *but also a carefully chosen approximation of planetary vorticity and, thereby, of the Coriolis force.*

For comparative convenience, the expressions for the individual components in the $\mathbf{e}_2$ and $\mathbf{e}_3$ directions of the planetary-vorticity vector ($2\boldsymbol{\Omega}$) are displayed in Table 17.1. The $\mathbf{e}_1$ component is identically zero for all three (namely deep, quasi-shallow and shallow) equation sets.

17.3 LAGRANGIAN DERIVATION

Three methodologies for deriving dynamically consistent equation sets were listed in Section 17.1. The first one ('classical Eulerian') has been applied earlier to derive a pair (one non-hydrostatic, the other hydrostatic) of dynamically consistent, *quasi-shallow*, equation sets.

In this section we describe use of the other two methodologies – both Lagrangian – to obtain the same pair of equation sets. Common to both methodologies is approximation of Lagrangian density. We therefore begin by considering how to do this in the present quasi-shallow context. We then describe how to use the approximated Lagrangian density to alternatively derive the two quasi-shallow equation sets: in Section 17.3.2, using Hamilton's principle of stationary action; and in Section 17.3.3, using Euler–Lagrange equations.

17.3.1 Approximation of Lagrangian Density

Lagrangian densities for the quartet of equation sets discussed in Chapter 16 are given in axial-orthogonal-curvilinear coordinates by (16.46)–(16.49). Recall from Section 17.2.2 that non-hydrostatic switch $\mathcal{N}$ is defined such that when $\mathcal{N} = 1$, the term it multiplies is retained, but when $\mathcal{N} = 0$, the term it multiplies is omitted. Using non-hydrostatic switch $\mathcal{N}$, Lagrangian densities (16.46) and (16.47) for the (non-hydrostatic-deep and quasi-hydrostatic-deep) pair of deep equation sets may be compactly combined as the *deep* Lagrangian density

$$\mathcal{L}_{\mathrm{D}}\left(\boldsymbol{\xi},\dot{\boldsymbol{\xi}},\alpha,\eta\right) = \frac{h_1^2\dot{\xi}_1^2 + h_2^2\dot{\xi}_2^2}{2} + \frac{\mathcal{N}h_3^2\dot{\xi}_3^2}{2} + \Omega h_1^2\dot{\xi}_1 - \mathcal{E}\left(\alpha,\eta\right) - \Phi\left(\xi_3\right), \tag{17.47}$$

[4] The *planetary* axial angular momentum is not defined with the same moment arm as the *relative* axial angular momentum. This means that the ability to *exactly* transform the quasi-shallow equations to other rotating frames, or to an inertial frame, has been sacrificed. This (in the author's view) is a small price to be paid for improved accuracy (compared to the shallow equation sets), without loss of dynamical consistency. The unapproximated deep representation $h_1\left(u_1 + h_1 u_1\right)$ of planetary axial angular momentum is, of course, even better, with the additional desirable attribute of being exactly transformable to other frames.

where

$$(u_1, u_2, u_3) = \left(h_1\dot{\xi}_1, h_2\dot{\xi}_2, h_3\dot{\xi}_3\right), \tag{17.48}$$

$\xi = (\xi_1, \xi_2, \xi_3)$, $\dot{\boldsymbol{\xi}} = \left(\dot{\xi}_1, \dot{\xi}_2, \dot{\xi}_3\right)$, and $h_i = h_i(\xi_2, \xi_3)$, $i = 1, 2, 3$. Quasi-hydrostatic approximation is achieved by omitting the $h_3^2\dot{\xi}_3^2/2$ (blue) term in (17.47).

Similarly, Lagrangian densities (16.48) and (16.49) for the (non-hydrostatic-shallow and hydrostatic-shallow/hydrostatic-primitive) pair of shallow equation sets may be compactly combined as the *shallow* Lagrangian density

$$\mathscr{L}_S\left(\boldsymbol{\xi}, \dot{\boldsymbol{\xi}}, \alpha, \eta\right) = \frac{\left(h_1^2\right)^S \dot{\xi}_1^2 + \left(h_2^2\right)^S \dot{\xi}_2^2}{2} + \frac{\mathcal{N}\left(h_3^2\right)^S \dot{\xi}_3^2}{2} + \Omega\left(h_1^2\right)^S \dot{\xi}_1 - \mathcal{E}_S(\alpha, \eta) - \Phi_S(\xi_3), \tag{17.49}$$

where 'S' super/subscripts denote application of shallow approximation at the planet's surface $\xi_3 = \xi_3^S =$ constant. In particular,

$$h_i^S \equiv h_i\left(\xi_2, \xi_3 = \xi_3^S\right) = h_i^S(\xi_2), \quad i = 1, 2, 3. \tag{17.50}$$

Hydrostatic approximation is achieved by omitting the $\left(h_3^2\right)^S \dot{\xi}_3^2/2$ (blue) term in (17.49).

As discussed in Section 16.1, and as is well known, a consequence of making the shallow-fluid approximation (whilst maintaining dynamical consistency) is that the '$2\Omega\cos\phi$' Coriolis terms are lost despite them being non-negligible in some contexts. An obvious solution is to not make this approximation. Practically speaking this does, however, complicate things somewhat, so we instead seek an approximation that is more accurate than the shallow-fluid one, but an approximation nevertheless.

Assume therefore that we instead wish to *minimally enhance* the pair of shallow equation sets (namely non-hydrostatic and hydrostatic) to well capture these missing terms, *whilst maintaining dynamical consistency*. One way of doing so is to apply the 'classical Eulerian' methodology – as we have already done in Section 17.2. This, as previously noted, is somewhat labour-intensive with no guarantee of success.[5] Let us imagine that we have been unsuccessful with this method ...or that we are impatient. Either scenario suggests we should consider the alternative, Lagrangian approach. The key to this approach is appropriate approximation of deep Lagrangian density ($\mathscr{L}_D$) or, put another way, minimal enhancement of shallow Lagrangian density ($\mathscr{L}_S$).

Examination of deep Lagrangian density (17.47) shows that, to achieve our goal, there is remarkably little scope for approximation. This is very good news indeed and greatly simplifies matters! We now reason as follows:

- The $2\Omega\cos\phi$ Coriolis terms in the components of the momentum equation can only originate from the $h_1^2\dot{\xi}_1$ term in (17.47), since it is the *only* term therein that depends upon parameter Ω. So a *single* term to approximate, instead of several/many in multiple combinations in multiple equations with the 'classical Eulerian' methodology. This greatly simplifies things.
- From (6.20)

$$\cos\phi = \frac{1}{h_3(\xi_2, \xi_3)} \frac{\partial h_1(\xi_2, \xi_3)}{\partial \xi_3}, \tag{17.51}$$

in (deep) axial-orthogonal-curvilinear coordinates. For $\cos\phi$ to be non-zero (to allow the $2\Omega\cos\phi$ Coriolis terms to materialise), h_1 *must depend on* ξ_3.[6]

[5] In hindsight, doing so often looks more straightforward than it actually was. Particularisation is usually straightforward; generalisation generally is not!

[6] This immediately explains why shallow-fluid approximation (17.49) fails to deliver the $2\Omega\cos\phi$ Coriolis terms; this approximation assumes $h_1^S = h_1^S(\xi_2) \Rightarrow \cos\phi = \left(\partial h_1^S/\partial\xi_3\right)/h_3^S = 0$. That h_1 must depend on ξ_3 is also clear from the dependence of planetary angular momentum on ξ_3.

- The *simplest* way[7] of approximating h_1^2 in the $h_1^2\dot{\xi}_1$ term of (17.47), whilst maintaining dependence on ξ_3, is by the first two terms of a Taylor series in ξ_3, that is,

$$h_1^2 \approx \left(h_1^2\right)^S + \left(\frac{\partial h_1^2}{\partial \xi_3}\right)^S \left(\xi_3 - \xi_3^S\right). \tag{17.52}$$

Taking just the first term in (17.52) is insufficient, since this is just shallow-fluid approximation and the $2\Omega \cos\phi$ Coriolis terms then fail to materialise.

- Furthermore, we know from the analysis of Chapter 16 that omitting or retaining the $h_3^2\dot{\xi}_3^2/2$ term in (17.47) amounts to applying (quasi-)hydrostaticity or not. There is no reason not to maintain that option here, and thereby obtain a *pair* of quasi-shallow equation sets.

Putting this reasoning together then strongly suggests that

$$\begin{aligned}\mathscr{L}_{QS}\left(\boldsymbol{\xi},\dot{\boldsymbol{\xi}},\alpha,\eta\right) = &\underbrace{\frac{\left(h_1^2\right)^S\dot{\xi}_1^2 + \left(h_2^2\right)^S\dot{\xi}_2^2}{2}}_{\text{horizontal kinetic energy}} + \underbrace{\frac{\mathcal{N}\left(h_3^2\right)^S\dot{\xi}_3^2}{2}}_{\text{non-hydrostatic}} - \underbrace{\Phi_S\left(\xi_3\right)}_{\text{gravity}} - \underbrace{\mathcal{E}_S\left(\alpha,\eta\right)}_{\text{internal energy}} \\ &+ \underbrace{\Omega\left(h_1^2\right)^S\dot{\xi}_1}_{\text{shallow Coriolis}} + \underbrace{\Omega\left(\frac{\partial h_1^2}{\partial \xi_3}\right)^S\left(\xi_3 - \xi_3^S\right)\dot{\xi}_1}_{\text{quasi-shallow Coriolis}}\end{aligned} \tag{17.53}$$

is the sought *quasi-shallow* approximation of deep Lagrangian density (17.47):

- The right-hand-side terms of (17.53), *with the exception of the last one*, are just those of shallow-fluid approximation (17.49).
- The last term (in green) is the *additional quasi-shallow contribution*. This is the term that allows the $2\Omega \cos\phi$ Coriolis terms to materialise.
- The blue, switched term is the usual non-hydrostatic term. In the absence of this term, the prognostic vertical-momentum equation reduces to a diagnostic quasi-hydrostatic equation.
- The *only way* that approximation (17.53) of the deep Lagrangian can fail to lead to dynamical consistency is if the (green) quasi-shallow contribution breaks a symmetry property; see the discussion of this possibility later in this section.

It remains to derive the pair of quasi-shallow equation sets using approximated Lagrangian density (17.53). We now examine two methodologies for doing so, using:

- Hamilton's principle of stationary action (Section 17.3.2).
- Euler–Lagrange equations (in Section 17.3.3).

17.3.2 Derivation using Hamilton's Principle of Stationary Action

Application of Hamilton's principle of stationary action to obtain a quartet of equation sets in axial-orthogonal-curvilinear coordinates, using approximated Lagrangian densities (16.46)–(16.49), is summarised in Section 16.5. It is based on the analysis given in Section 14.9 to similarly do so for the unapproximated non-hydrostatic-deep equation set.

[7] This is simpler than similarly expanding h_1 as $h_1 \approx h_1^S + \left(\partial h_1/\partial \xi_3\right)^S\left(\xi_3 - \xi_3^S\right)$, since h_1^2 then has an *additional* (quadratic) term in $\left(\xi_3 - \xi_3^S\right)^2$. Use of (17.52) avoids this complication.

The same procedure is used here to obtain the pair of quasi-shallow equation sets, the only difference being that approximated Lagrangian density (17.53) (with and without the blue $\left(h_3^2\right)^S \dot{\xi}_3^2/2$ term) is used instead of those of (16.46)–(16.49). We split taking variations of the associated action into two parts; the *purely shallow* part and the *quasi-shallow perturbation* away from it. This is because the labour for doing the first part has already been done in Section 14.9, and we can take advantage of it here.

The Lagrangian density of the *purely shallow* part is given by (17.49). From (14.155)–(14.157) of Section 14.9, the associated variations of the action ($\delta_{\xi_i} \mathcal{A}_S$) are (with red terms therein absent for the purely shallow case)[8]

$$\delta_{\xi_1} \mathcal{A}_S = -\int_{\tau_1}^{\tau_2} \left[\int \left(\frac{\partial u_1}{\partial \tau} + \frac{u_1}{h_1^S} \frac{u_2}{h_2^S} \frac{dh_1^S}{d\xi_2} + 2\Omega \frac{u_2}{h_2^S} \frac{dh_1^S}{d\xi_2} + \frac{1}{\rho h_1^S} \frac{\partial p}{\partial \xi_1} \right) h_1^S \delta \xi_1 \mathbf{da} \right] d\tau, \tag{17.54}$$

$$\delta_{\xi_2} \mathcal{A}_S = -\int_{\tau_1}^{\tau_2} \left[\int \left(\frac{\partial u_2}{\partial \tau} - \frac{u_1^2}{h_1^S h_2^S} \frac{dh_1^S}{d\xi_2} - \mathcal{N} \frac{u_3^2}{h_2^S h_3^S} \frac{dh_3^S}{d\xi_2} - 2\Omega \frac{u_1}{h_2^S} \frac{dh_1^S}{d\xi_2} + \frac{1}{\rho h_2^S} \frac{\partial p}{\partial \xi_2} \right) h_2^S \delta \xi_2 \mathbf{da} \right] d\tau, \tag{17.55}$$

$$\delta_{\xi_3} \mathcal{A}_S = -\int_{\tau_1}^{\tau_2} \left[\int \left(\mathcal{N} \frac{\partial u_3}{\partial \tau} + \mathcal{N} \frac{u_2}{h_2^S} \frac{u_3}{h_3^S} \frac{dh_3}{d\xi_2} + \frac{1}{\rho h_3^S} \frac{\partial p}{\partial \xi_3} + \frac{1}{h_3^S} \frac{d\Phi}{d\xi_3} \right) h_3^S \delta \xi_3 \mathbf{da} \right] d\tau. \tag{17.56}$$

The *additional quasi-shallow term* in Lagrangian density (17.53) is

$$\mathcal{L}'_{QS} \equiv \Omega \left[\left(\frac{\partial h_1^2}{\partial \xi_3} \right)^S \left(\xi_3 - \xi_3^S \right) \right] \frac{\partial \xi_1}{\partial \tau}. \tag{17.57}$$

This is just the *quasi-shallow perturbation* of the underlying *shallow* Lagrangian density, that is, of (17.53) without this term. The corresponding *quasi-shallow perturbation of the action* is then

$$\mathscr{A}'_{QS} \equiv \int_{\tau_1}^{\tau_2} \left(\int \mathcal{L}'_{QS} \mathbf{da} \right) d\tau. \tag{17.58}$$

From (17.121), (17.123), and (17.125) of Appendix B of the present chapter, the associated variations of the action ($\delta_{\xi_i} \mathscr{A}'_{QS}$) are

$$\delta_{\xi_1} \left(\mathscr{A}'_{QS} \right) = -\int_{\tau_1}^{\tau_2} \left\{ \int 2\Omega \left[A\left(\xi_2\right) \left(\xi_3 - \xi_3^S\right) u_2 + \frac{u_3}{h_3^S} \left(\frac{\partial h_1}{\partial \xi_3} \right)^S \right] h_1^S \delta \xi_1 \mathbf{da} \right\} d\tau, \tag{17.59}$$

$$\delta_{\xi_2} \left(\mathscr{A}'_{QS} \right) = -\int_{\tau_1}^{\tau_2} \left\{ \int \left[-2\Omega A\left(\xi_2\right) \left(\xi_3 - \xi_3^S\right) u_1 \right] h_2^S \delta \xi_2 \mathbf{da} \right\} d\tau, \tag{17.60}$$

$$\delta_{\xi_2} \left(\mathscr{A}'_{QS} \right) = -\int_{\tau_1}^{\tau_2} \left\{ \int \left[-2\Omega \frac{u_1}{h_3^S} \left(\frac{\partial h_1}{\partial \xi_3} \right)^S \right] h_3^S \delta \xi_3 \mathbf{da} \right\} d\tau, \tag{17.61}$$

where $A\left(\xi_2\right)$ is defined by (17.9).

Summing (17.54)–(17.56) with (17.59)–(17.61), respectively, yields the three components of the variation of the action $\mathscr{A}_{QS} = \mathscr{A}_S + \mathscr{A}'_{QS}$ associated with quasi-shallow Lagrangian density

[8] Notational note: for brevity in the present chapter, the volume integrals $\iiint (\quad) \mathbf{da}$ over labels $\mathbf{a}$ are more compactly written as $\int (\quad) \mathbf{da}$. This should not cause any confusion since no distinction is needed here between volume and area integrals.

(17.53). For arbitrary variations $\delta\xi_i, i = 1, 2, 3$, the summed integrands of each component of the variation of the action must be identically zero. Noting that $\partial/\partial\tau \rightarrow D_S/Dt$ in the present shallow geometry, the momentum components of the quasi-shallow equations are thus (with some minor rearrangement)

$$\frac{D_S u_1}{Dt} + \left(\frac{u_1}{h_1^S} + 2\Omega\right)\frac{u_2}{h_2^S}\frac{dh_1^S}{d\xi_2} + 2\Omega\left[A(\xi_2)\left(\xi_3 - \xi_3^S\right)u_2 + \frac{u_3}{h_3^S}\left(\frac{\partial h_1}{\partial \xi_3}\right)^S\right] = -\frac{1}{\rho h_1^S}\frac{\partial p}{\partial \xi_1}, \quad (17.62)$$

$$\frac{D_S u_2}{Dt} - \left(\frac{u_1}{h_1^S} + 2\Omega\right)\frac{u_1}{h_2^S}\frac{dh_1^S}{d\xi_2} - \mathcal{N}\frac{u_3^2}{h_2^S h_3^S}\frac{dh_3^S}{d\xi_2} - 2\Omega A(\xi_2)\left(\xi_3 - \xi_3^S\right)u_1 = -\frac{1}{\rho h_2^S}\frac{\partial p}{\partial \xi_2}, \quad (17.63)$$

$$\mathcal{N}\left(\frac{D_S u_3}{Dt} + \frac{u_3}{h_3^S}\frac{u_2}{h_2^S}\frac{dh_3^S}{d\xi_2}\right) - 2\Omega\frac{u_1}{h_3^S}\left(\frac{\partial h_1}{\partial \xi_3}\right)^S + \frac{1}{h_3^S}\frac{d\Phi}{d\xi_3} = -\frac{1}{\rho h_3^S}\frac{\partial p}{\partial \xi_3}. \quad (17.64)$$

As both desired and expected, (17.11)–(17.13) are recovered.

In the absence of the quasi-shallow (green) terms, we then recover the three components of momentum for the:

- Non-hydrostatic-shallow equations – see (16.128)–(16.130) of Chapter 16 with $\mathcal{N} = 1$.
- Hydrostatic-shallow (= hydrostatic-primitive) equations – see (16.128)–(16.130) of Chapter 16 with $\mathcal{N} = 0$.

The blue terms in (17.63) and (17.64) determine whether the quasi-shallow pair of equation sets is non-hydrostatic (when blue terms are present) or (almost-)[9]hydrostatic (in their absence).

Deriving the preceding equations does not, however, guarantee that the associated pair of equation sets are dynamically consistent. To demonstrate that they are, we need to (see Chapter 15 and Section 16.8):

1. *Either* classically construct analogues of the conservation principles for axial angular momentum, total energy, and potential vorticity from the derived quasi-shallow equation set.
2. *Or* apply Noether's theorem and verify that quasi-shallow perturbation (17.57) of the Lagrangian respects the symmetry properties necessary for conservation of axial angular momentum (rotation symmetry), total energy (time-translation symmetry), and potential vorticity (particle-relabelling symmetry).
3. *Or* judiciously combine elements of the first two alternatives.

Note that conservation of mass is assured by use of shallow mass-continuity equation (17.15).

For the thought experiment of the present context (Lagrangian development of a pair of quasi-shallow equation sets), we can straightforwardly apply the tried-and-true first option. In fact we essentially already have! Momentum-component equations (17.62)–(17.64) are equivalent to (17.11)–(17.13), with $A(\xi_2)$ defined by (17.9). We just have to systematically apply the steps used in Sections 17.2.1, 17.2.5, and 17.2.6 in essentially the same way. This then completes the proof that the pair of quasi-shallow equation sets are dynamically consistent.

But what about the other two alternatives (i.e. Items 2 and 3 of the list)? This depends upon how comfortable one is regarding verification of symmetries. For the application of Noether's theorem, one might be comfortable verifying the symmetries for the conservation analogues of axial angular momentum and total energy, but much less so verifying particle-relabelling symmetry (for conservation of potential vorticity). Particle-relabelling symmetry is quite abstract and challenging, with various subtleties; see, for example, Salmon (1998) and Badin and Crisciani (2018). In such a situation, one would adopt the third option, and verify the symmetries for the

[9] 'Almost' because the green term in (17.64) is (in general) small but nevertheless non-zero.

conservation analogues of axial angular momentum and total energy, but then classically verify conservation of potential vorticity as in Section 17.2.6.

In the end, *it does not matter which method one uses.* It only matters that *all* of the conservation principles for dynamical consistency are correctly verified by whichever method is most convenient.

17.3.3 Derivation Using the Euler–Lagrange Equations

From (16.54), the Euler–Lagrange equations for the components of the momentum equation for a (quasi-)shallow fluid are

$$\frac{D_S}{Dt}\left(\frac{\partial \mathscr{L}_{QS}}{\partial \dot{\xi}_i}\right) - \frac{\partial \mathscr{L}_{QS}}{\partial \xi_i} = -\frac{1}{\rho}\frac{\partial p}{\partial \xi_i}, \quad i = 1, 2, 3, \tag{17.65}$$

for Lagrangian density, $\mathscr{L}_{QS}$. Here D_S/Dt – the shallow material derivative – is defined by (17.5), and (from (17.48)) $\dot{\xi}_i \equiv D_S\xi_i/Dt = u_i/h_i^S, i = 1, 2, 3$. The Lagrangian density, $\mathscr{L}_{QS}$, for the pair of quasi-shallow equation sets is given by (17.53), that is, by

$$\begin{aligned}\mathscr{L}_{QS}\left(\boldsymbol{\xi}, \dot{\boldsymbol{\xi}}, \alpha, \eta\right) = & \underbrace{\frac{\left(h_1^2\right)^S \dot{\xi}_1^2 + \left(h_2^2\right)^S \dot{\xi}_2^2}{2}}_{\text{horizontal kinetic energy}} + \underbrace{\mathcal{N}\frac{\left(h_3^2\right)^S}{2}\dot{\xi}_3^2}_{\text{non-hydrostatic}} - \underbrace{\Phi_S\left(\xi_3\right)}_{\text{gravity}} - \underbrace{\mathcal{E}_S\left(\alpha, \eta\right)}_{\text{internal energy}} \\ & + \underbrace{\Omega\left(h_1^2\right)^S \dot{\xi}_1}_{\text{shallow Coriolis}} + \underbrace{\Omega\left(\frac{\partial h_1^2}{\partial \xi_3}\right)^S\left(\xi_3 - \xi_3^S\right)\dot{\xi}_1}_{\text{quasi-shallow Coriolis}},\end{aligned} \tag{17.66}$$

Recalling – from (17.50) – that $h_i^S = h_i^S(\xi_2)$, for $i = 1, 2, 3$, and also that $\left(\partial h_1^2/\partial\xi_3\right)^S$ is a function of ξ_2 only, appropriately partially differentiating quasi-shallow Lagrangian (17.66) leads to:

$$\frac{\partial \mathscr{L}_{QS}}{\partial \xi_1} = 0, \tag{17.67}$$

$$\begin{aligned}\frac{\partial \mathscr{L}_{QS}}{\partial \xi_2} &= \left(\frac{u_1}{h_1^S} + 2\Omega\right)u_1\frac{dh_1^S}{d\xi_2} + \frac{u_2^2}{h_2^S}\frac{dh_2^S}{d\xi_2} + \mathcal{N}\frac{u_3^2}{h_3^S}\frac{dh_3^S}{d\xi_2} + \Omega\frac{u_1}{h_1^S}\frac{d}{d\xi_2}\left[\left(\frac{\partial h_1^2}{\partial\xi_3}\right)^S\right]\left(\xi_3 - \xi_3^S\right) \\ &= \left(\frac{u_1}{h_1^S} + 2\Omega\right)u_1\frac{dh_1^S}{d\xi_2} + \frac{u_2^2}{h_2^S}\frac{dh_2^S}{d\xi_2} + \mathcal{N}\frac{u_3^2}{h_3^S}\frac{dh_3^S}{d\xi_2} + \Omega\frac{u_1}{h_1^S}\frac{dB\left(\xi_2\right)}{d\xi_2}\left(\xi_3 - \xi_3^S\right),\end{aligned} \tag{17.68}$$

$$\frac{\partial \mathscr{L}_{QS}}{\partial \xi_3} = \Omega\frac{u_1}{h_1^S}\left(\frac{\partial h_1^2}{\partial \xi_3}\right)^S - \frac{d\Phi}{d\xi_3} = \Omega\frac{u_1}{h_1^S}B\left(\xi_2\right) - \frac{d\Phi}{d\xi_3}, \tag{17.69}$$

$$\frac{\partial \mathscr{L}_{QS}}{\partial \dot{\xi}_1} = h_1^S\left(u_1 + \Omega h_1^S\right) + \Omega\left(\frac{\partial h_1^2}{\partial \xi_3}\right)^S\left(\xi_3 - \xi_3^S\right) = h_1^S\left(u_1 + \Omega h_1^S\right) + \Omega B\left(\xi_2\right)\left(\xi_3 - \xi_3^S\right), \tag{17.70}$$

$$\frac{\partial \mathscr{L}_{QS}}{\partial \dot{\xi}_2} = h_2^S u_2, \tag{17.71}$$

$$\frac{\partial \mathscr{L}_{QS}}{\partial \dot{\xi}_3} = \mathcal{N} h_3^S u_3, \tag{17.72}$$

where (from (17.50)) $u_i \equiv h_i^S\dot{\xi}_i, i = 1, 2, 3$, has been exploited, and $B\left(\xi_2\right)$ is defined by (17.6).

Substitution of (17.67)–(17.72) into Euler–Lagrange equations (17.65), for $i = 1, 2, 3$, gives

$$\frac{D_S}{Dt}\left[h_1^S\left(u_1 + \Omega h_1^S\right) + \Omega B(\xi_2)\left(\xi_3 - \xi_3^S\right)\right] = -\frac{1}{\rho}\frac{\partial p}{\partial \xi_1}, \tag{17.73}$$

$$\frac{D_S}{Dt}\left(h_2^S u_2\right) - \left(\frac{u_1}{h_1^S} + 2\Omega\right)u_1\frac{dh_1^S}{d\xi_2} - \frac{u_2^2}{h_2^S}\frac{dh_2^S}{d\xi_2} - \mathcal{N}\frac{u_3^2}{h_3^S}\frac{dh_3^S}{d\xi_2} - \Omega\frac{u_1}{h_1^S}\frac{dB(\xi_2)}{d\xi_2}\left(\xi_3 - \xi_3^S\right) = -\frac{1}{\rho}\frac{\partial p}{\partial \xi_2}, \tag{17.74}$$

$$\mathcal{N}\frac{D_S}{Dt}\left(h_3^S u_3\right) - \Omega\frac{u_1}{h_1^S}B(\xi_2) + \frac{d\Phi}{d\xi_3} = -\frac{1}{\rho}\frac{\partial p}{\partial \xi_3}. \tag{17.75}$$

Useful relations (14.196)–(14.201) of the Appendix to Chapter 14 may be rewritten in shallow geometry as

$$\frac{D_S h_i^S}{Dt} = \frac{u_2}{h_2^S}\frac{dh_i^S}{d\xi_2}, \quad i = 1, 2, 3, \tag{17.76}$$

$$\frac{D_S}{Dt}\left(h_i^S u_i\right) = h_i^S\left(\frac{D_S u_i}{Dt} + \frac{u_i}{h_i^S}\frac{u_2}{h_2^S}\frac{dh_i^S}{d\xi_2}\right), \quad i = 1, 2, 3. \tag{17.77}$$

Alternatively, they may be derived from first principles. Using (17.5), (17.6), (17.9), (17.76), and (17.77), (17.73)–(17.75) may then be rewritten as

$$\frac{D_S u_1}{Dt} + \left(\frac{u_1}{h_1^S} + 2\Omega\right)\frac{u_2}{h_2^S}\frac{dh_1^S}{d\xi_2} + 2\Omega\left[A(\xi_2)\left(\xi_3 - \xi_3^S\right)u_2 + \frac{u_3}{h_3^S}\left(\frac{\partial h_1}{\partial \xi_3}\right)^S\right] = -\frac{1}{\rho h_1^S}\frac{\partial p}{\partial \xi_1}, \tag{17.78}$$

$$\frac{D_S u_2}{Dt} - \left(\frac{u_1}{h_1^S} + 2\Omega\right)\frac{u_1}{h_2^S}\frac{dh_1^S}{d\xi_2} - \mathcal{N}\frac{u_3^2}{h_2^S h_3^S}\frac{dh_3^S}{d\xi_2} - 2\Omega A(\xi_2)\left(\xi_3 - \xi_3^S\right)u_1 = -\frac{1}{\rho h_2^S}\frac{\partial p}{\partial \xi_2}, \tag{17.79}$$

$$\mathcal{N}\left(\frac{D_S u_3}{Dt} + \frac{u_3}{h_3^S}\frac{u_2}{h_2^S}\frac{dh_3^S}{d\xi_2}\right) - 2\Omega\frac{u_1}{h_3^S}\left(\frac{\partial h_1}{\partial \xi_3}\right)^S + \frac{1}{h_3^S}\frac{d\Phi}{d\xi_3} = -\frac{1}{\rho h_3^S}\frac{\partial p}{\partial \xi_3}. \tag{17.80}$$

These three equations recover the three component equations (17.62)–(17.64) of momentum for the derivation of the quasi-shallow equations in Section 17.3.2 using Hamilton's principle of stationary action. The comments at the end of Section 17.3.2 regarding dynamical consistency therefore also apply here.

17.4 A UNIFIED SEXTET OF EQUATION SETS IN SPHEROIDAL GEOMETRY

A quartet of dynamically consistent equation sets – suitable for global atmospheric and oceanic modelling, and at least as accurate as the hydrostatic-primitive equations – was considered in Chapter 16. In particular, they were expressed in Section 16.2.3 in a unified manner and in axial-orthogonal-curvilinear coordinates. See Fig. 17.1 for a schematic of the interrelationships between the individual members of this quartet, and of the pair of quasi-shallow equation sets developed in this chapter. Taken together, these equation sets form a unified sextet.

It is instructive to assemble and display the prognostic equations of this sextet in a comparative manner. This allows us to see how individual terms successively simplify or vanish under quasi-shallow or shallow approximation. In what follows, the three forms – 'Deep', 'Quasi-Shallow' (abbreviated to 'QShallow'), and 'Shallow' – are displayed on successive lines. Notation is as in Chapter 16 and the present one. The deep and shallow equations given in this section are obtained from those given in Chapter 16, and the corresponding quasi-shallow ones from those given earlier in the present chapter.

17.4.1 Governing Equations

We start by assembling the component-momentum equations for u_1, u_2, and u_3, respectively. (In what follows, recall that $A(\xi_2)$ is defined by (17.9).) Thus:

Component Momentum Equations for u_1

Deep
$$\frac{Du_1}{Dt} + \left(\frac{u_1}{h_1} + 2\Omega\right)\frac{u_2}{h_2}\frac{\partial h_1}{\partial \xi_2} + \left(\frac{u_1}{h_1} + 2\Omega\right)\frac{u_3}{h_3}\frac{\partial h_1}{\partial \xi_3} = -\frac{1}{\rho h_1}\frac{\partial p}{\partial \xi_1},$$

QShallow
$$\frac{D_S u_1}{Dt} + \left(\frac{u_1}{h_1^S} + 2\Omega\right)\frac{u_2}{h_2^S}\frac{dh_1^S}{d\xi_2} + 2\Omega A(\xi_2)\left(\xi_3 - \xi_3^S\right)u_2 + 2\Omega\frac{u_3}{h_3^S}\left(\frac{\partial h_1}{\partial \xi_3}\right)^S = -\frac{1}{\rho h_1^S}\frac{\partial p}{\partial \xi_1},$$

Shallow
$$\frac{D_S u_1}{Dt} + \left(\frac{u_1}{h_1^S} + 2\Omega\right)\frac{u_2}{h_2^S}\frac{dh_1^S}{d\xi_2} = -\frac{1}{\rho h_1^S}\frac{\partial p}{\partial \xi_1}. \tag{17.81}$$

Component Momentum Equations for u_2

Deep
$$\frac{Du_2}{Dt} - \left(\frac{u_1}{h_1} + 2\Omega\right)\frac{u_1}{h_2}\frac{\partial h_1}{\partial \xi_2} - \frac{u_3^2}{h_2 h_3}\frac{\partial h_3}{\partial \xi_2} + \frac{u_2}{h_2}\frac{u_3}{h_3}\frac{\partial h_2}{\partial \xi_3} = -\frac{1}{\rho h_2}\frac{\partial p}{\partial \xi_2},$$

QShallow
$$\frac{D_S u_2}{Dt} - \left(\frac{u_1}{h_1^S} + 2\Omega\right)\frac{u_1}{h_2^S}\frac{dh_1^S}{d\xi_2} - \frac{u_3^2}{h_2^S h_3^S}\frac{dh_3^S}{d\xi_2} - 2\Omega A(\xi_2)\left(\xi_3 - \xi_3^S\right)u_1 = -\frac{1}{\rho h_2^S}\frac{\partial p}{\partial \xi_2},$$

Shallow
$$\frac{D_S u_2}{Dt} - \left(\frac{u_1}{h_1^S} + 2\Omega\right)\frac{u_1}{h_2^S}\frac{dh_1^S}{d\xi_2} - \frac{u_3^2}{h_2^S h_3^S}\frac{dh_3^S}{d\xi_2} = -\frac{1}{\rho h_2^S}\frac{\partial p}{\partial \xi_2}. \tag{17.82}$$

Component Momentum Equations for u_3

Deep
$$\frac{Du_3}{Dt} + \frac{u_2}{h_2}\frac{u_3}{h_3}\frac{\partial h_3}{\partial \xi_2} - \left(\frac{u_1}{h_1} + 2\Omega\right)\frac{u_1}{h_3}\frac{\partial h_1}{\partial \xi_3} - \frac{u_2^2}{h_2 h_3}\frac{\partial h_2}{\partial \xi_3} + \frac{1}{h_3}\frac{d\Phi}{d\xi_3} = -\frac{1}{\rho h_3}\frac{\partial p}{\partial \xi_3},$$

QShallow
$$\frac{D_S u_3}{Dt} + \frac{u_2}{h_2^S}\frac{u_3}{h_3^S}\frac{dh_3^S}{d\xi_2} - 2\Omega\frac{u_1}{h_3^S}\left(\frac{\partial h_1}{\partial \xi_3}\right)^S + \frac{1}{h_3^S}\frac{d\Phi_S}{d\xi_3} = -\frac{1}{\rho h_3^S}\frac{\partial p}{\partial \xi_3},$$

Shallow
$$\frac{D_S u_3}{Dt} + \frac{u_2}{h_2^S}\frac{u_3}{h_3^S}\frac{dh_3^S}{d\xi_2} + \frac{1}{h_3^S}\frac{d\Phi_S}{d\xi_3} = -\frac{1}{\rho h_3^S}\frac{\partial p}{\partial \xi_3}. \tag{17.83}$$

Next, component-momentum equations (17.81)–(17.83) are supplemented by two further prognostic equations; a mass-continuity equation and a thermodynamic-energy equation. Thus:

Mass-Continuity and Thermodynamic-Energy Equations

$$\left.\begin{array}{llll} & \text{Mass Continuity} & & \text{Thermodynamic Energy} \\ \text{Deep} & \dfrac{D\rho}{Dt}+\rho\nabla\cdot\mathbf{u}=0, & & \dfrac{D\mathcal{E}}{Dt}+p\dfrac{D\alpha}{Dt}=0, \\ \text{QShallow} & \dfrac{D_S\rho}{Dt}+\rho\nabla_S\cdot\mathbf{u}=0, & & \dfrac{D_S\mathcal{E}_S}{Dt}+p\dfrac{D_S\alpha}{Dt}=0, \\ \text{Shallow} & \dfrac{D_S\rho}{Dt}+\rho\nabla_S\cdot\mathbf{u}=0, & & \dfrac{D_S\mathcal{E}_S}{Dt}+p\dfrac{D_S\alpha}{Dt}=0. \end{array}\right\} \quad (17.84)$$

Recall that various quantities in prognostic equations (17.81)–(17.84) are defined as follows:

- $\alpha \equiv 1/\rho$.
- D/Dt and D_S/Dt by (17.2) and (17.5), respectively.
- $\mathbf{u}$ by (16.26) for the deep case, and similarly for the shallow case but with $h_i \rightarrow h_i^S$.
- $\nabla \cdot \mathbf{A}$ and $\nabla_S \cdot \mathbf{A}$ by (16.27) and (17.16), respectively.
- $\mathcal{E}$ and $\mathcal{E}_S$ by (16.28) and (17.18), respectively, and thereby p and T.

The precise forms for the geopotentials Φ and Φ_S appearing in (17.83) depend upon:

- Whether the fluid is deep or shallow.
- The composition and shape of the planet.

See Section 16.2.3.1 for further information regarding this.

From the preceding equations, we see that:

- Red (deep) terms in component-momentum equations (17.81)–(17.83) vanish under both quasi-shallow and shallow approximation.
- Blue terms appear in all three pairs of equation sets. They vanish under quasi-hydrostatic and hydrostatic approximation.
- Green terms are quasi-shallow terms that supplement the usual shallow ones. These terms are placed immediately below the deep terms from which they emanate. Green terms provide an excellent approximation of the $2\Omega\cos\phi$ terms that are absent from the (purely) shallow pair of equation sets.
- Shallow approximation can be applied directly to the pair of deep equation sets to obtain the corresponding pair of shallow ones – as shown schematically in Fig. 17.1. One simply applies the *purely geometric* mapping/approximation $h_i\left(\xi_2,\xi_3\right) \rightarrow h_i^S\left(\xi_2\right), i = 1,2,3$. This amounts to a non-Euclidean distortion of Euclidean geometry.
- Alternatively, one can apply quasi-shallow approximation (i.e. partial application of shallow approximation) to the pair of deep equation sets to obtain the corresponding quasi-shallow pair. Following this up with full application of the shallow approximation then yields the purely shallow pair of equation sets.
- Quasi-shallow approximation, however, is *not* purely geometric. It includes some additional approximated terms. These originate from approximation (17.52) of h_1^2 in axial angular

momentum equation (17.1). Shallow approximation, whereby $h_i \to h_i^S, i = 1,2,3$, treats the three metric factors in precisely the same manner everywhere they appear in the governing equations, and doing so is *purely geometric* (Thuburn and White, 2013). By way of contrast, approximation (17.52) of h_1^2 gives preferential treatment to only one metric factor (h_1), and only where it appears in terms related to the representation of the Coriolis force, but not elsewhere. See footnote 4 regarding an inconvenient consequence of this anomalous, additional approximation.

17.4.2 Conservation Principles

There are a number of ways of expressing conservation principles; see Chapter 15. The following principles, expressed in Lagrangian form and in axial-orthogonal-curvilinear coordinates, hold for the unified sextet of equation sets.

Conservation of Axial Angular Momentum

$$\left.\begin{array}{lll}
\text{Deep} & \dfrac{D}{Dt}\left[h_1\left(u_1+\Omega h_1\right)\right] & = -\dfrac{1}{\rho}\dfrac{\partial p}{\partial \xi_1}, \\
\text{Q Shallow} & \dfrac{D_S}{Dt}\left[h_1^S\left(u_1+\Omega h_1^S\right)+\Omega\left(\dfrac{\partial h_1^2}{\partial \xi_3}\right)^S\left(\xi_3-\xi_3^S\right)\right] & = -\dfrac{1}{\rho}\dfrac{\partial p}{\partial \xi_1}, \\
\text{Shallow} & \dfrac{D_S}{Dt}\left[h_1^S\left(u_1+\Omega h_1^S\right)\right] & = -\dfrac{1}{\rho}\dfrac{\partial p}{\partial \xi_1}.
\end{array}\right\} \tag{17.85}$$

Conservation of Total Energy

$$\left.\begin{array}{lll}
\text{Deep} & \dfrac{D}{Dt}\left[\dfrac{1}{2}\left(u_1^2+u_2^2+u_3^2\right)+\Phi+\mathcal{E}\right] & = -\dfrac{1}{\rho}\nabla\cdot\left(p\mathbf{u}\right), \\
\text{Q Shallow} & \dfrac{D_S}{Dt}\left[\dfrac{1}{2}\left(u_1^2+u_2^2+u_3^2\right)+\Phi_S+\mathcal{E}_S\right] & = -\dfrac{1}{\rho}\nabla_S\cdot\left(p\mathbf{u}\right), \\
\text{Shallow} & \dfrac{D_S}{Dt}\left[\dfrac{1}{2}\left(u_1^2+u_2^2+u_3^2\right)+\Phi_S+\mathcal{E}_S\right] & = -\dfrac{1}{\rho}\nabla_S\cdot\left(p\mathbf{u}\right).
\end{array}\right\} \tag{17.86}$$

Conservation of Potential Vorticity

$$\left.\begin{array}{lll}
\text{Deep} & \dfrac{D}{Dt}\left(\dfrac{\mathbf{Z}^D\cdot\nabla\Lambda}{\rho}\right) = \dfrac{D}{Dt}\left[\dfrac{1}{\rho}\left(\dfrac{Z_1^D}{h_1}\dfrac{\partial\Lambda}{\partial\xi_1}+\dfrac{Z_2^D}{h_2}\dfrac{\partial\Lambda}{\partial\xi_2}+\dfrac{Z_3^D}{h_3}\dfrac{\partial\Lambda}{\partial\xi_3}\right)\right] & = 0, \\
\text{Q Shallow} & \dfrac{D_S}{Dt}\left(\dfrac{\mathbf{Z}^{QS}\cdot\nabla_S\Lambda}{\rho}\right) = \dfrac{D_S}{Dt}\left[\dfrac{1}{\rho}\left(\dfrac{Z_1^{QS}}{h_1^S}\dfrac{\partial\Lambda}{\partial\xi_1}+\dfrac{Z_2^{QS}}{h_2^S}\dfrac{\partial\Lambda}{\partial\xi_2}+\dfrac{Z_3^{QS}}{h_3^S}\dfrac{\partial\Lambda}{\partial\xi_3}\right)\right] & = 0, \\
\text{Shallow} & \dfrac{D_S}{Dt}\left(\dfrac{\mathbf{Z}^S\cdot\nabla_S\Lambda}{\rho}\right) = \dfrac{D_S}{Dt}\left[\dfrac{1}{\rho}\left(\dfrac{Z_1^S}{h_1^S}\dfrac{\partial\Lambda}{\partial\xi_1}+\dfrac{Z_2^S}{h_2^S}\dfrac{\partial\Lambda}{\partial\xi_2}+\dfrac{Z_3^S}{h_3^S}\dfrac{\partial\Lambda}{\partial\xi_3}\right)\right] & = 0,
\end{array}\right\} \tag{17.87}$$

where

$$
\left.\begin{array}{llll}
\text{Deep} & \dfrac{D\Lambda}{Dt} = 0, & \nabla\Lambda & = \dfrac{1}{h_1}\dfrac{\partial\Lambda}{\partial\xi_1}\mathbf{e}_1 + \dfrac{1}{h_2}\dfrac{\partial\Lambda}{\partial\xi_2}\mathbf{e}_2 + \dfrac{1}{h_3}\dfrac{\partial\Lambda}{\partial\xi_3}\mathbf{e}_3, \\
\text{Q Shallow} & \dfrac{D_S\Lambda}{Dt} = 0, & \nabla_S\Lambda & = \dfrac{1}{h_1^S}\dfrac{\partial\Lambda}{\partial\xi_1}\mathbf{e}_1 + \dfrac{1}{h_2^S}\dfrac{\partial\Lambda}{\partial\xi_2}\mathbf{e}_2 + \dfrac{1}{h_3^S}\dfrac{\partial\Lambda}{\partial\xi_3}\mathbf{e}_3, \\
\text{Shallow} & \dfrac{D_S\Lambda}{Dt} = 0, & \nabla_S\Lambda & = \dfrac{1}{h_1^S}\dfrac{\partial\Lambda}{\partial\xi_1}\mathbf{e}_1 + \dfrac{1}{h_2^S}\dfrac{\partial\Lambda}{\partial\xi_2}\mathbf{e}_2 + \dfrac{1}{h_3^S}\dfrac{\partial\Lambda}{\partial\xi_3}\mathbf{e}_3,
\end{array}\right\} \tag{17.88}
$$

$\Lambda = \Lambda(\alpha, \eta)$ is any materially conserved quantity; and

$$
\left.\begin{array}{lll}
\text{Deep} & \mathbf{Z}^{\mathrm{D}} & = \left[\dfrac{\mathcal{N}}{h_2 h_3}\dfrac{\partial}{\partial\xi_2}(h_3 u_3) - \dfrac{1}{h_2 h_3}\dfrac{\partial}{\partial\xi_3}(h_2 u_2)\right]\mathbf{e}_1 \\
& & + \left[\dfrac{1}{h_3 h_1}\dfrac{\partial}{\partial\xi_3}(h_1 u_1) - \dfrac{\mathcal{N}}{h_1}\dfrac{\partial u_3}{\partial\xi_1} + \dfrac{2\Omega}{h_3}\dfrac{\partial h_1}{\partial\xi_3}\right]\mathbf{e}_2 \\
& & + \left[\dfrac{1}{h_1}\dfrac{\partial u_2}{\partial\xi_1} - \dfrac{1}{h_1 h_2}\dfrac{\partial}{\partial\xi_2}(h_1 u_1) - \dfrac{2\Omega}{h_2}\dfrac{\partial h_1}{\partial\xi_2}\right]\mathbf{e}_3, \\
\text{Q Shallow} & \mathbf{Z}^{\mathrm{QS}} & = \left[\dfrac{\mathcal{N}}{h_2^S h_3^S}\dfrac{\partial}{\partial\xi_2}(h_3^S u_3) - \dfrac{1}{h_3^S}\dfrac{\partial u_2}{\partial\xi_3}\right]\mathbf{e}_1 \\
& & + \left[\dfrac{1}{h_3^S}\dfrac{\partial u_1}{\partial\xi_3} - \dfrac{\mathcal{N}}{h_1^S}\dfrac{\partial u_3}{\partial\xi_1} + \dfrac{2\Omega}{h_3^S}\left(\dfrac{\partial h_1}{\partial\xi_3}\right)^S\right]\mathbf{e}_2 \\
& & + \left[\dfrac{1}{h_1^S}\dfrac{\partial u_2}{\partial\xi_1} - \dfrac{1}{h_1^S h_2^S}\dfrac{\partial}{\partial\xi_2}(h_1^S u_1) - \dfrac{2\Omega}{h_2^S}\dfrac{dh_1^S}{d\xi_2} - 2\Omega A(\xi_2)\left(\xi_3 - \xi_3^S\right)\right]\mathbf{e}_3, \\
\text{Shallow} & \mathbf{Z}^{\mathrm{S}} & = \left[\dfrac{\mathcal{N}}{h_2^S h_3^S}\dfrac{\partial}{\partial\xi_2}(h_3^S u_3) - \dfrac{1}{h_3^S}\dfrac{\partial u_2}{\partial\xi_3}\right]\mathbf{e}_1 \\
& & + \left[\dfrac{1}{h_3^S}\dfrac{\partial u_1}{\partial\xi_3} - \dfrac{\mathcal{N}}{h_1^S}\dfrac{\partial u_3}{\partial\xi_1}\right]\mathbf{e}_2 \\
& & + \left[\dfrac{1}{h_1^S}\dfrac{\partial u_2}{\partial\xi_1} - \dfrac{1}{h_1^S h_2^S}\dfrac{\partial}{\partial\xi_2}(h_1^S u_1) - \dfrac{2\Omega}{h_2^S}\dfrac{dh_1^S}{d\xi_2}\right]\mathbf{e}_3
\end{array}\right\} \tag{17.89}
$$

(Recall that $A(\xi_2)$ in (17.89) is defined by (17.9)).

17.5 A UNIFIED SEXTET OF EQUATION SETS IN SPHERICAL GEOMETRY

The prognostic equations of the unified sextet of equation sets in *spheroidal* geometry have been assembled in the preceding section in a comparative manner. It is also instructive to similarly do so in this section for the special, familiar case of *spherical* geometry. Similar comments to those given for spheroidal geometry also apply here.

17.5.1 Governing Equations

We begin by assembling the component-momentum equations for u_λ, u_ϕ, and u_r, respectively. Thus:

Component Momentum Equations for u_λ

$$\left.\begin{array}{lll}
\text{Deep} & \dfrac{D_r u_\lambda}{Dt} - \left(\dfrac{u_\lambda}{r\cos\phi} + 2\Omega\right) u_\phi \sin\phi + \left(\dfrac{u_\lambda}{r\cos\phi} + 2\Omega\right) u_r \cos\phi & = -\dfrac{1}{\rho r\cos\phi}\dfrac{\partial p}{\partial \lambda}, \\
\text{QShallow} & \dfrac{D_a u_\lambda}{Dt} - \left(\dfrac{u_\lambda}{a\cos\phi} + 2\Omega\right) u_\phi \sin\phi & \\
& \quad - 4\Omega\dfrac{(r-a)}{a} u_\phi \sin\phi + 2\Omega u_r \cos\phi & = -\dfrac{1}{\rho a\cos\phi}\dfrac{\partial p}{\partial \lambda}, \\
\text{Shallow} & \dfrac{D_a u_\lambda}{Dt} - \left(\dfrac{u_\lambda}{a\cos\phi} + 2\Omega\right) u_\phi \sin\phi & = -\dfrac{1}{\rho a\cos\phi}\dfrac{\partial p}{\partial \lambda}.
\end{array}\right\} \tag{17.90}$$

Component Momentum Equations for u_ϕ

$$\left.\begin{array}{lll}
\text{Deep} & \dfrac{D_r u_\phi}{Dt} + \left(\dfrac{u_\lambda}{r\cos\phi} + 2\Omega\right) u_\lambda \sin\phi + \dfrac{u_\phi u_r}{r} & = -\dfrac{1}{\rho r}\dfrac{\partial p}{\partial \phi}, \\
\text{QShallow} & \dfrac{D_a u_\phi}{Dt} + \left(\dfrac{u_\lambda}{a\cos\phi} + 2\Omega\right) u_\lambda \sin\phi & \\
& \quad + 4\Omega\dfrac{(r-a)}{a} u_\lambda \sin\phi & = -\dfrac{1}{\rho a}\dfrac{\partial p}{\partial \phi}, \\
\text{Shallow} & \dfrac{D_a u_\phi}{Dt} + \left(\dfrac{u_\lambda}{a\cos\phi} + 2\Omega\right) u_\lambda \sin\phi & = -\dfrac{1}{\rho a}\dfrac{\partial p}{\partial \phi}.
\end{array}\right\} \tag{17.91}$$

Component Momentum Equations for u_r

$$\left.\begin{array}{lll}
\text{Deep} & \dfrac{D_r u_r}{Dt} - \left(\dfrac{u_\lambda}{r\cos\phi} + 2\Omega\right) u_\lambda \cos\phi - \dfrac{u_\phi^2}{r} + \dfrac{d\Phi_r}{dr} & = -\dfrac{1}{\rho}\dfrac{\partial p}{\partial r}, \\
\text{QShallow} & \dfrac{D_a u_r}{Dt} - 2\Omega u_\lambda \cos\phi + \dfrac{d\Phi_a}{dr} & = -\dfrac{1}{\rho}\dfrac{\partial p}{\partial r}, \\
\text{Shallow} & \dfrac{D_a u_r}{Dt} + \dfrac{d\Phi_a}{dr} & = -\dfrac{1}{\rho}\dfrac{\partial p}{\partial r}.
\end{array}\right\} \tag{17.92}$$

Next, component-momentum equations (17.90)–(17.92) are supplemented by two further prognostic equations: a mass-continuity equation and a thermodynamic-energy equation. Thus:

Continuity and Thermodynamic-Energy Equations

$$
\left.
\begin{array}{llll}
 & \text{Mass Continuity} & & \text{Thermodynamic Energy} \\
\text{Deep} & \dfrac{D_r\rho}{Dt} + \rho\nabla_r\cdot\mathbf{u} = 0, & & \dfrac{D_r\mathcal{E}}{Dt} + p\dfrac{D_r\alpha}{Dt} = 0, \\
\text{QShallow} & \dfrac{D_a\rho}{Dt} + \rho\nabla_a\cdot\mathbf{u} = 0, & & \dfrac{D_a\mathcal{E}_S}{Dt} + p\dfrac{D_a\alpha}{Dt} = 0, \\
\text{Shallow} & \dfrac{D_a\rho}{Dt} + \rho\nabla_a\cdot\mathbf{u} = 0, & & \dfrac{D_a\mathcal{E}_S}{Dt} + p\dfrac{D_a\alpha}{Dt} = 0.
\end{array}
\right\} \quad (17.93)
$$

Recall that various quantities in (17.81)–(17.93) are defined as follows:

- $\alpha \equiv 1/\rho$.
- D_r/Dt and D_a/Dt by (16.86) and (16.102), respectively.
- $\mathbf{u}$ by (16.87) for the deep case, and by (16.103) for the shallow case.
- $\nabla_r \cdot \mathbf{A}$ and $\nabla_a \cdot \mathbf{A}$ by (16.88) and (16.104), respectively.
- $\mathcal{E}$ and $\mathcal{E}_S$ by (16.28) and (17.18), respectively, and thereby p and T.

The precise forms for the geopotentials Φ_r (deep) and Φ_a (shallow) appearing in (17.92) are given by (16.11) and (16.12), respectively.

17.5.2 Conservation Principles

Conservation of Axial Angular Momentum

$$
\left.
\begin{array}{lll}
\text{Deep} & \dfrac{D_r}{Dt}\left[r\cos\phi\left(u_1 + \Omega r\cos\phi\right)\right] & = -\dfrac{1}{\rho}\dfrac{\partial p}{\partial\lambda}, \\
\text{Q Shallow} & \dfrac{D_a}{Dt}\left[a\cos\phi\left(u_1 + \Omega a\cos\phi\right) + 2\Omega az\cos^2\phi\right] & = -\dfrac{1}{\rho}\dfrac{\partial p}{\partial\lambda}, \\
\text{Shallow} & \dfrac{D_a}{Dt}\left[a\cos\phi\left(u_1 + \Omega a\cos\phi\right)\right] & = -\dfrac{1}{\rho}\dfrac{\partial p}{\partial\lambda}.
\end{array}
\right\} \quad (17.94)
$$

Conservation of Total Energy

$$
\left.
\begin{array}{lll}
\text{Deep} & \dfrac{D_r}{Dt}\left[\dfrac{1}{2}\left(u_1^2+u_2^2+u_3^2\right) + \Phi_r + \mathcal{E}_r\right] & = -\dfrac{1}{\rho}\nabla_r\cdot(p\mathbf{u}), \\
\text{Q Shallow} & \dfrac{D_a}{Dt}\left[\dfrac{1}{2}\left(u_1^2+u_2^2+u_3^2\right) + \Phi_a + \mathcal{E}_a\right] & = -\dfrac{1}{\rho}\nabla_a\cdot(p\mathbf{u}), \\
\text{Shallow} & \dfrac{D_a}{Dt}\left[\dfrac{1}{2}\left(u_1^2+u_2^2+u_3^2\right) + \Phi_a + \mathcal{E}_a\right] & = -\dfrac{1}{\rho}\nabla_a\cdot(p\mathbf{u}).
\end{array}
\right\} \quad (17.95)
$$

Conservation of Potential Vorticity

$$
\left.\begin{array}{llll}
\text{Deep} & \dfrac{D_r}{Dt}\left(\dfrac{\mathbf{Z}^{\mathrm{D}}\cdot\nabla\Lambda}{\rho}\right) = \dfrac{D_r}{Dt}\left[\dfrac{1}{\rho}\left(\dfrac{Z_1^{\mathrm{D}}}{r\cos\phi}\dfrac{\partial\Lambda}{\partial\lambda}+\dfrac{Z_2^{\mathrm{D}}}{r}\dfrac{\partial\Lambda}{\partial\phi}+Z_3^{\mathrm{D}}\dfrac{\partial\Lambda}{\partial r}\right)\right] & = 0, \\
\text{Q Shallow} & \dfrac{D_a}{Dt}\left(\dfrac{\mathbf{Z}^{\mathrm{QS}}\cdot\nabla_{\mathrm{S}}\Lambda}{\rho}\right) = \dfrac{D_a}{Dt}\left[\dfrac{1}{\rho}\left(\dfrac{Z_1^{\mathrm{QS}}}{a\cos\phi}\dfrac{\partial\Lambda}{\partial\lambda}+\dfrac{Z_2^{\mathrm{QS}}}{a}\dfrac{\partial\Lambda}{\partial\phi}+Z_3^{\mathrm{QS}}\dfrac{\partial\Lambda}{\partial r}\right)\right] & = 0, \\
\text{Shallow} & \dfrac{D_a}{Dt}\left(\dfrac{\mathbf{Z}^{\mathrm{S}}\cdot\nabla_{\mathrm{S}}\Lambda}{\rho}\right) = \dfrac{D_a}{Dt}\left[\dfrac{1}{\rho}\left(\dfrac{Z_1^{\mathrm{S}}}{a\cos\phi}\dfrac{\partial\Lambda}{\partial\lambda}+\dfrac{Z_2^{\mathrm{S}}}{a}\dfrac{\partial\Lambda}{\partial\phi}+Z_3^{\mathrm{S}}\dfrac{\partial\Lambda}{\partial r}\right)\right] & = 0,
\end{array}\right\} \tag{17.96}
$$

$$
\left.\begin{array}{llll}
\text{Deep} & \dfrac{D_r\Lambda}{Dt} = 0, & \nabla_r\Lambda = \dfrac{1}{r\cos\phi}\dfrac{\partial\Lambda}{\partial\phi}\mathbf{e}_\lambda+\dfrac{1}{r}\dfrac{\partial\Lambda}{\partial\phi}\mathbf{e}_\phi+\dfrac{\partial\Lambda}{\partial r}\mathbf{e}_r, \\
\text{Q Shallow} & \dfrac{D_a\Lambda}{Dt} = 0, & \nabla_a\Lambda = \dfrac{1}{a\cos\phi}\dfrac{\partial\Lambda}{\partial\phi}\mathbf{e}_\lambda+\dfrac{1}{a}\dfrac{\partial\Lambda}{\partial\phi}\mathbf{e}_\phi+\dfrac{\partial\Lambda}{\partial r}\mathbf{e}_r, \\
\text{Shallow} & \dfrac{D_a\Lambda}{Dt} = 0, & \nabla_a\Lambda = \dfrac{1}{a\cos\phi}\dfrac{\partial\Lambda}{\partial\phi}\mathbf{e}_\lambda+\dfrac{1}{a}\dfrac{\partial\Lambda}{\partial\phi}\mathbf{e}_\phi+\dfrac{\partial\Lambda}{\partial r}\mathbf{e}_r,
\end{array}\right\} \tag{17.97}
$$

$\Lambda = \Lambda\left(\alpha, \eta\right)$ is any materially conserved quantity; and

$$
\left.\begin{array}{lll}
\text{Deep} & \mathbf{Z}^{\mathrm{D}} & = \dfrac{1}{r}\left[\mathcal{N}\dfrac{\partial u_r}{\partial\phi}-\dfrac{\partial}{\partial r}\left(ru_\phi\right)\right]\mathbf{e}_\lambda \\
& & \quad +\left[\dfrac{1}{r}\dfrac{\partial}{\partial r}\left(ru_\lambda\right)-\dfrac{\mathcal{N}}{r\cos\phi}\dfrac{\partial u_r}{\partial\lambda}+2\Omega\cos\phi\right]\mathbf{e}_\phi \\
& & \quad +\left[\dfrac{1}{r\cos\phi}\dfrac{\partial u_\phi}{\partial\lambda}-\dfrac{1}{r\cos\phi}\dfrac{\partial}{\partial\phi}\left(\cos\phi u_\lambda\right)+2\Omega\sin\phi\right]\mathbf{e}_r, \\
\text{Q Shallow} & \mathbf{Z}^{\mathrm{QS}} & = \dfrac{1}{a}\left[\mathcal{N}\dfrac{\partial u_r}{\partial\phi}-a\dfrac{\partial u_\phi}{\partial r}\right]\mathbf{e}_\lambda \\
& & \quad +\left[\dfrac{\partial u_\lambda}{\partial r}-\dfrac{\mathcal{N}}{a\cos\phi}\dfrac{\partial u_r}{\partial\lambda}+2\Omega\cos\phi\right]\mathbf{e}_\phi \\
& & \quad +\left[\dfrac{1}{a\cos\phi}\dfrac{\partial u_\phi}{\partial\lambda}-\dfrac{1}{a\cos\phi}\dfrac{\partial}{\partial\phi}\left(\cos\phi u_\lambda\right)+2\Omega\sin\phi+4\Omega\dfrac{z}{a}\sin\phi\right]\mathbf{e}_r, \\
\text{Shallow} & \mathbf{Z}^{\mathrm{S}} & = \dfrac{1}{a}\left[\mathcal{N}\dfrac{\partial u_r}{\partial\phi}-a\dfrac{\partial u_\phi}{\partial r}\right]\mathbf{e}_\lambda \\
& & \quad +\left[\dfrac{\partial u_\lambda}{\partial r}-\dfrac{\mathcal{N}}{a\cos\phi}\dfrac{\partial u_r}{\partial\lambda}\right]\mathbf{e}_\phi \\
& & \quad +\left[\dfrac{1}{a\cos\phi}\dfrac{\partial u_\phi}{\partial\lambda}-\dfrac{1}{a\cos\phi}\dfrac{\partial}{\partial\phi}\left(\cos\phi u_\lambda\right)+2\Omega\sin\phi\right]\mathbf{e}_r
\end{array}\right\} \tag{17.98}
$$

17.6 CONCLUDING REMARKS

A dynamically consistent pair of quasi-shallow equation sets – expressed in axial-orthogonal-curvilinear coordinates – has been derived using two contrasting but complementary approaches. The 'classical Eulerian' approach involves informed reasoning, coupled with a posteriori verification of dynamical consistency. The Lagrangian approach is more direct. Once the shallow assumption has been introduced into the Lagrangian density for all terms except the one dependent on planetary vorticity (2Ω), only this last term needs to be approximated to accurately (albeit not exactly) represent the '$2\Omega\cos\phi$' terms that are entirely missing from the pair of shallow-fluid equation sets.

The resulting pair of quasi-shallow equation sets is then combined with the quartet of (deep and shallow) equation sets described in Chapter 16. This leads to a unified sextet of equation sets in spheroidal geometry that are at least as accurate as the well-known hydrostatic primitive equations. This sextet is comprised of three pairs of equation sets: *deep*, *quasi-shallow*, and *shallow*. Each pair is itself comprised of a non-hydrostatic equation set and a (quasi-)hydrostatic one, depending on whether non-hydrostaticity is retained or not. The hydrostatic primitive equations are the only member of the sextet that are hydrostatic in the classical sense of an exact (two-term) balance between gravity and the vertical component of the pressure-gradient force.

An advantage of the Lagrangian approach, compared to the 'classical Eulerian' one, is that it facilitates identification of new and unsuspected equation sets via judicious approximation of Lagrangian density. This was indeed how the quasi-shallow pair of equation sets was originally discovered and developed (Tort and Dubos, 2014a,b; Staniforth, 2015a). Further examples of this approach are explored in Chapter 18 to develop some unconventional, global, shallow-water equation sets in 2D.

APPENDIX A: QUASI-SHALLOW EQUATION SETS IN AXIAL-ORTHOGONAL-CURVILINEAR COORDINATES

A pair of quasi-shallow equation sets (one non-hydrostatic, the other hydrostatic), expressed in axial-orthogonal-curvilinear coordinates, is assembled here for reference purposes. The two sets are labelled:

- Non-hydrostatic quasi-shallow.
- Hydrostatic quasi-shallow.

See Sections 17.1 and 17.2.8 for description of their attributes.

To better focus on dynamical consistency, it is assumed that:

- Physical forcings (other than gravity) are absent.[10]
- The coordinate system is a geopotential one, so $\Phi = \Phi(\xi_3)$.
- The fluid is composed of a single substance (dry air for the atmosphere, pure liquid water for the oceans).

Common to both equation sets is the fundamental equation of state:

$$\mathcal{E}_S = \mathcal{E}_S(\alpha, \eta) \quad \Rightarrow \quad p \equiv -\frac{\partial \mathcal{E}_S(\alpha, \eta)}{\partial \alpha}, \quad T \equiv \frac{\partial \mathcal{E}_S(\alpha, \eta)}{\partial \eta}. \tag{17.99}$$

These three equations are independent of whether the fluid is deep or shallow, and they depend only upon the thermodynamic state of the fluid. However, as noted in Section 6.35, it is convenient to append a subscript 'S' to $\mathcal{E}$ – as in (17.99) – to signify that its first argument (α) is governed by the *shallow* mass-continuity equation (17.15) rather than by the analogous equation

[10] Forcings are included in Chapter 15 for the full equations. Guided by the full-equations analysis, forcings can be reinserted at any time.

for a *deep* fluid. This facilitates keeping track of the influence, both direct and indirect, of shallow approximation.

Colour is used in the pair of *quasi-shallow* equation sets, summarised here, to help identify links between them and the analogous pair of *shallow* equation sets summarised in Appendix B of Chapter 16:

- Blue terms are associated with quasi-hydrostatic/hydrostatic approximation. They are present before application of this approximation (i.e. present for the non-hydrostatic-quasi-shallow equation set), and absent after application (i.e. absent in the hydrostatic-quasi-shallow equation set).
  In what follows, non-hydrostatic switch $\mathcal{N}$ determines the equation set;
  - $\mathcal{N}=1 \quad \Rightarrow$ **non-hydrostatic-quasi-shallow,**
  - $\mathcal{N}=0 \quad \Rightarrow$ **quasi-hydrostatic-quasi-shallow.**
- Green terms are the *supplementary terms* that generalise the pair of shallow equation sets to the quasi-shallow pair. In the absence of these terms, the shallow equations are retrieved.

Equations (17.100)–(17.109) – with $\mathcal{N}=0,1$ – correspond to the two *quasi-shallow* equation sets (in magenta) on Fig. 17.1.

Governing Equations

$$\frac{D_S u_1}{Dt} + \left(\frac{u_1}{h_1^S} + 2\Omega\right)\frac{u_2}{h_2^S}\frac{dh_1^S}{d\xi_2} + 2\Omega\left[A(\xi_2)\left(\xi_3 - \xi_3^S\right)u_2 + \frac{u_3}{h_3^S}\left(\frac{\partial h_1}{\partial \xi_3}\right)^S\right] = -\frac{1}{\rho h_1^S}\frac{\partial p}{\partial \xi_1}, \tag{17.100}$$

$$\frac{D_S u_2}{Dt} - \left(\frac{u_1}{h_1^S} + 2\Omega\right)\frac{u_1}{h_2^S}\frac{dh_1^S}{d\xi_2} - \frac{\mathcal{N} u_3^2}{h_2^S h_3^S}\frac{dh_3^S}{d\xi_2} - 2\Omega A(\xi_2)\left(\xi_3 - \xi_3^S\right)u_1 = -\frac{1}{\rho h_2^S}\frac{\partial p}{\partial \xi_2}, \tag{17.101}$$

$$\mathcal{N}\left(\frac{D_S u_3}{Dt} + \frac{u_2}{h_2^S}\frac{u_3}{h_3^S}\frac{dh_3^S}{d\xi_2}\right) - 2\Omega\frac{u_1}{h_3^S}\left(\frac{\partial h_1}{\partial \xi_3}\right)^S + \frac{1}{h_3^S}\frac{d\Phi_S}{d\xi_3} = -\frac{1}{\rho h_3^S}\frac{\partial p}{\partial \xi_3}, \tag{17.102}$$

$$\frac{D_S \rho}{Dt} + \rho \nabla_S \cdot \mathbf{u} = 0, \tag{17.103}$$

$$\frac{D_S \mathcal{E}_S}{Dt} + p\frac{D_S \alpha}{Dt} = 0, \tag{17.104}$$

where

$$A(\xi_2) = \frac{1}{h_1^S h_2^S}\frac{d}{d\xi_2}\left[\frac{1}{2}\left(\frac{\partial h_1^2}{\partial \xi_3}\right)^S\right], \tag{17.105}$$

$$\alpha \equiv \frac{1}{\rho}, \tag{17.106}$$

$$\frac{D_S}{Dt} = \frac{\partial}{\partial t} + \frac{u_1}{h_1^S}\frac{\partial}{\partial \xi_1} + \frac{u_2}{h_2^S}\frac{\partial}{\partial \xi_2} + \frac{u_3}{h_3^S}\frac{\partial}{\partial \xi_3}, \tag{17.107}$$

$$\mathbf{u} \equiv u_1\mathbf{e}_1 + u_2\mathbf{e}_2 + u_3\mathbf{e}_3 = h_1^S\frac{D_S\xi_1}{Dt}\mathbf{e}_1 + h_2^S\frac{D_S\xi_2}{Dt}\mathbf{e}_2 + h_3^S\frac{D_S\xi_3}{Dt}\mathbf{e}_3, \tag{17.108}$$

$$\nabla_S \cdot \mathbf{A} = \frac{1}{h_1^S h_2^S h_3^S}\left[\frac{\partial}{\partial \xi_1}\left(A_1 h_2^S h_3^S\right) + \frac{\partial}{\partial \xi_2}\left(A_2 h_3^S h_1^S\right)\right] + \frac{1}{h_3^S}\frac{\partial A_3}{\partial \xi_3}. \tag{17.109}$$

After taking account of the different definitions of $A(\xi_2)$, (17.100)–(17.102) agree with Staniforth (2015a)'s equations (11)–(13).

Conservation Principles

Associated with the governing equations are the following conservation principles for axial angular momentum, total energy, and potential vorticity:

Axial Angular Momentum

$$\frac{D_S}{Dt}\left\{h_1^S u_1 + \Omega\left[\left(h_1^S\right)^2 + \left(\frac{\partial h_1^2}{\partial \xi_3}\right)^S \left(\xi_3 - \xi_3^S\right)\right]\right\} = -\frac{1}{\rho}\frac{\partial p}{\partial \xi_1}. \tag{17.110}$$

Total Energy

$$\frac{D_S}{Dt}\left[\frac{1}{2}\left(u_1^2 + u_2^2 + \mathcal{N} u_3^2\right) + \Phi_S + \mathcal{E}_S\right] = -\frac{1}{\rho}\nabla_S \cdot \left(p\mathbf{u}\right). \tag{17.111}$$

Potential Vorticity

$$\frac{D_S}{Dt}\left(\frac{\mathbf{Z}^{QS}\cdot\nabla_S\Lambda}{\rho}\right) = \frac{D_S}{Dt}\left[\frac{1}{\rho}\left(\frac{Z_1^{QS}}{h_1^S}\frac{\partial\Lambda}{\partial\xi_1} + \frac{Z_2^{QS}}{h_2^S}\frac{\partial\Lambda}{\partial\xi_2} + \frac{Z_3^{QS}}{h_3^S}\frac{\partial\Lambda}{\partial\xi_3}\right)\right] = 0, \tag{17.112}$$

where

$$\begin{aligned}\mathbf{Z}^{QS} &\equiv \nabla_S \times \mathbf{u} + 2\boldsymbol{\Omega} = \text{absolute vorticity}\\ &= \left[\frac{\mathcal{N}}{h_2^S h_3^S}\frac{\partial}{\partial\xi_2}\left(h_3^S u_3\right) - \frac{1}{h_3^S}\frac{\partial u_2}{\partial \xi_3}\right]\mathbf{e}_1 + \left[\frac{1}{h_3^S}\frac{\partial u_1}{\partial\xi_3} - \frac{\mathcal{N}}{h_1^S}\frac{\partial u_3}{\partial \xi_1} + \frac{2\Omega}{h_3^S}\left(\frac{\partial h_1}{\partial\xi_3}\right)^S\right]\mathbf{e}_2\\ &\quad + \left[\frac{1}{h_1^S}\frac{\partial u_2}{\partial\xi_1} - \frac{1}{h_1^S h_2^S}\frac{\partial}{\partial\xi_2}\left(h_1^S u_1\right) - \frac{2\Omega}{h_2^S}\frac{dh_1^S}{d\xi_2} - 2\Omega A\left(\xi_2\right)\left(\xi_3 - \xi_3^S\right)\right]\mathbf{e}_3,\end{aligned} \tag{17.113}$$

$$\frac{D_S\Lambda}{Dt} = 0, \tag{17.114}$$

$$\nabla_S\Lambda = \frac{1}{h_1^S}\frac{\partial\Lambda}{\partial\xi_1}\mathbf{e}_1 + \frac{1}{h_2^S}\frac{\partial\Lambda}{\partial\xi_2}\mathbf{e}_2 + \frac{1}{h_3^S}\frac{\partial\Lambda}{\partial\xi_3}\mathbf{e}_3, \tag{17.115}$$

and $\Lambda = \Lambda(\alpha, \eta)$ is any materially conserved quantity that satisfies (17.114).

APPENDIX B: VARIATIONS FOR QUASI-SHALLOW CONTRIBUTIONS

The quasi-shallow contribution to Lagrangian density (17.53) is

$$\mathscr{L}'_{QS} \equiv \Omega\left(\frac{\partial h_1^2}{\partial\xi_3}\right)^S\left(\xi_3 - \xi_3^S\right)\frac{\partial\xi_1}{\partial\tau}. \tag{17.116}$$

This is the quasi-shallow perturbation of the shallow Lagrangian density (i.e. of (17.53) without this term). The corresponding perturbation of the action is then

$$\mathscr{A}'_{QS} \equiv \int_{\tau_1}^{\tau_2}\left(\int \mathscr{L}'_{QS}\,\mathbf{da}\right)d\tau, \tag{17.117}$$

with variations

$$\delta_{\xi_i}\left(\mathscr{A}'_{\text{QS}}\right) = \delta_{\xi_i}\int_{\tau_1}^{\tau_2}\left(\int \mathscr{L}'_{\text{QS}}\mathbf{da}\right)d\tau = \int_{\tau_1}^{\tau_2}\left[\int \delta_{\xi_i}\left(\mathscr{L}'_{\text{QS}}\right)\mathbf{da}\right]d\tau, \quad i = 1,2,3, \tag{17.118}$$

subject to the constraints

$$\delta\xi_i\left(\tau = \tau_1\right) = \delta\xi_i\left(\tau = \tau_2\right) = 0, \quad i = 1,2,3. \tag{17.119}$$

Variation of $\mathscr{A}'_{\text{QS}}$ with Respect to $\delta\xi_1$

Using definition (17.116) and axial-symmetry condition (17.50), the variation of $\mathscr{L}'_{\text{QS}}$ with respect to $\delta\xi_1$ is

$$\delta_{\xi_1}\left(\mathscr{L}'_{\text{QS}}\right) = \delta_{\xi_1}\left[\Omega\left(\frac{\partial h_1^2}{\partial\xi_3}\right)^S\left(\xi_3 - \xi_3^S\right)\frac{\partial\xi_1}{\partial\tau}\right] = \Omega\left(\frac{\partial h_1^2}{\partial\xi_3}\right)^S\left(\xi_3 - \xi_3^S\right)\frac{\partial\left(\delta\xi_1\right)}{\partial\tau}. \tag{17.120}$$

Inserting (17.120) into (17.118) with $i = 1$, integrating by parts, and using definitions (17.108) and (17.9) for u_1 and $A\left(\xi_2\right)$, respectively, axial-symmetry condition (17.50), and constraints (17.119) with $i = 1$, the corresponding variation of $\mathscr{A}'_{\text{QS}}$ with respect to $\delta\xi_1$ is then

$$\begin{aligned}
\delta_{\xi_1}\left(\mathscr{A}'_{\text{QS}}\right) &= \int_{\tau_1}^{\tau_2}\left\{\int\left[\Omega\left(\frac{\partial h_1^2}{\partial\xi_3}\right)^S\left(\xi_3 - \xi_3^S\right)\frac{\partial\left(\delta\xi_1\right)}{\partial\tau}\right]\mathbf{da}\right\}d\tau \\
&= \left\{\int\left[\Omega\left(\frac{\partial h_1^2}{\partial\xi_3}\right)^S\left(\xi_3 - \xi_3^S\right)\left(\delta\xi_1\right)\right]\mathbf{da}\right\}_{\tau_1}^{\tau_2} \rightarrow 0 \\
&\quad - \int_{\tau_1}^{\tau_2}\left\langle\int\left\{\Omega\frac{\partial}{\partial\tau}\left[\left(\frac{\partial h_1^2}{\partial\xi_3}\right)^S\left(\xi_3 - \xi_3^S\right)\right]\left(\delta\xi_1\right)\right\}\mathbf{da}\right\rangle d\tau \\
&= -\int_{\tau_1}^{\tau_2}\left\langle\int\Omega\left\{\frac{u_2}{h_2^S}\frac{\partial}{\partial\xi_2}\left[\left(\frac{\partial h_1^2}{\partial\xi_3}\right)^S\right]\left(\xi_3 - \xi_3^S\right) + \frac{u_3}{h_3^S}\left(\frac{\partial h_1^2}{\partial\xi_3}\right)^S\right\}\left(\delta\xi_1\right)\mathbf{da}\right\rangle d\tau \\
&= -\int_{\tau_1}^{\tau_2}\left\{\int 2\Omega\left[A\left(\xi_2\right)\left(\xi_3 - \xi_3^S\right)u_2 + \frac{u_3}{h_3^S}\left(\frac{\partial h_1}{\partial\xi_3}\right)^S\right]\left(h_1^S\delta\xi_1\right)\mathbf{da}\right\}d\tau.
\end{aligned} \tag{17.121}$$

Variation of $\mathscr{A}'_{\text{QS}}$ with Respect to $\delta\xi_2$

Using definition (17.116), definitions (17.48) and (17.9) for u_1 and $A\left(\xi_2\right)$, respectively, and axial-symmetry condition (17.50), the variation of $\mathscr{L}'_{\text{QS}}$ with respect to $\delta\xi_2$ is

$$\begin{aligned}
\delta_{\xi_2}\left(\mathscr{L}'_{\text{QS}}\right) &= \delta_{\xi_2}\left[\Omega\left(\frac{\partial h_1^2}{\partial\xi_3}\right)^S\left(\xi_3 - \xi_3^S\right)\frac{\partial\xi_1}{\partial\tau}\right] \\
&= \Omega\frac{d}{d\xi_2}\left[\left(\frac{\partial h_1^2}{\partial\xi_3}\right)^S\right]\left(\xi_3 - \xi_3^S\right)\frac{\partial\xi_1}{\partial\tau}\delta\xi_2 = \frac{2\Omega}{h_1^S h_2^S}\frac{d}{d\xi_2}\left[\left(\frac{\partial h_1^2}{\partial\xi_3}\right)^S\right]\left(\xi_3 - \xi_3^S\right)u_1\left(h_2^S\delta\xi_2\right) \\
&= \left[2\Omega A\left(\xi_2\right)\left(\xi_3 - \xi_3^S\right)u_1\right]\left(h_2^S\delta\xi_2\right).
\end{aligned} \tag{17.122}$$

Inserting (17.122) into (17.118) with $i = 2$, followed by rearrangement, the corresponding variation of $\mathscr{A}'_{\mathrm{QS}}$ with respect to $\delta\xi_2$ is

$$\delta_{\xi_2}\left(\mathscr{A}'_{\mathrm{QS}}\right) = -\int_{\tau_1}^{\tau_2}\left\{\iint\left[-2\Omega A\left(\xi_2\right)\left(\xi_3-\xi_3^S\right)u_1\right]\left(h_2^S\delta\xi_2\right)d\mathbf{a}\right\}d\tau. \tag{17.123}$$

Variation of $\mathscr{A}'_{\mathrm{QS}}$ with Respect to $\delta\xi_3$

Using definitions (17.48) and (17.9) for u_1 and $A\left(\xi_2\right)$, respectively, and axial-symmetry condition (17.50), the variation of $\mathscr{L}'_{\mathrm{QS}}$ with respect to $\delta\xi_3$ is

$$\begin{aligned}\delta_{\xi_3}\left(\mathscr{L}'_{\mathrm{QS}}\right) &= \delta_{\xi_3}\left[\Omega\left(\frac{\partial h_1^2}{\partial\xi_3}\right)^S\left(\xi_3-\xi_3^S\right)\frac{\partial\xi_1}{\partial\tau}\right] = \Omega\left(\frac{\partial h_1^2}{\partial\xi_3}\right)^S\delta_{\xi_3}\left(\xi_3-\xi_3^S\right)\frac{\partial\xi_1}{\partial\tau}\\ &= \Omega\frac{1}{h_1^S}\frac{u_1}{h_3^S}\left(\frac{\partial h_1^2}{\partial\xi_3}\right)^S\left(h_3^S\delta\xi_3\right) = 2\Omega\frac{u_1}{h_3^S}\left(\frac{\partial h_1}{\partial\xi_3}\right)^S\left(h_3^S\delta\xi_3\right).\end{aligned} \tag{17.124}$$

Inserting (17.124) into (17.118) with $i = 3$, followed by rearrangement, the corresponding variation of $\mathscr{A}'_{\mathrm{QS}}$ with respect to $\delta\xi_3$ is

$$\delta_{\xi_3}\mathscr{A}'_{\mathrm{QS}} = -\int_{\tau_1}^{\tau_2}\left\{\iint\left[-2\Omega\frac{u_1}{h_3^S}\left(\frac{\partial h_1}{\partial\xi_3}\right)^S\right]\left(h_3^S\delta\xi_3\right)d\mathbf{a}\right\}d\tau. \tag{17.125}$$

18

Shallow-Water Equation Sets in 2D

ABSTRACT

The 2D shallow-water equations, for a thin layer of incompressible fluid bounded below by a rigid boundary and above by a free surface, embody many terms and properties of the full 3D equations. These 2D equations have consequently been frequently used in spherical geometry, with constant gravity, both to advance theoretical understanding and to test prototypical numerical methods for quantitative modelling of Earth's atmosphere and oceans in 3D. Limitations to their wider use include:

1. Representation of Earth as a sphere.
2. Absence of meridional variation of gravity.
3. Incomplete representation of the Coriolis force.
4. Neglect of vertical acceleration.

These limitations are successively addressed in a step-by-step manner. The first step – this addresses the first two limitations – is to derive, from first principles using the traditional Eulerian methodology, a 'basic' equation set in spheroidal geometry with meridional variation of gravity. To close this equation set, a reference geopotential surface, three possible horizontal-coordinate systems, and five possible models of gravity are specified. The second and third steps instead use the complementary, Lagrangian methodology. For each of these, the Lagrangian density of the 'basic' equation set is enhanced by 'quasi-shallow' and 'non-hydrostatic' contributions, respectively. Doing so then leads to a unified quartet of equation sets, controlled by 'quasi-shallow' and 'non-hydrostatic' binary switches that selectively address the third and fourth limitations, respectively. The most complete set simultaneously addresses all four limitations. All members of the quartet are dynamically consistent, possessing conservation principles for mass, axial angular momentum, total energy, and potential vorticity.

18.1 PREAMBLE

The shallow-water equations have been used extensively in both meteorology and oceanography, dating back to the nineteenth century. The underlying reason for their enduring popularity and utility is that, *within a simplified, dynamically consistent,*[1] *2D framework*, they embody many of the fundamental terms and physical properties needed to realistically model Earth's atmosphere and oceans in 3D.

[1] Dynamical consistency is defined to be formal respect of conservation principles for mass, axial angular momentum, total energy, and potential vorticity. See Section 16.1.2.

Consequently, the shallow-water equations have been used to:

- Theoretically understand fundamental geophysical-fluid-dynamical processes.
- Test and validate, under well-controlled conditions, prototypical numerical methods for quantitative modelling of Earth's atmosphere and oceans.

18.1.1 The Shallow-Water Equations in Cartesian and Spherical Geometry

The governing equations for shallow-water flow are usually derived in the literature as a *constant-gravity specialisation* of the hydrostatic primitive equations for a thin layer of *incompressible* fluid. This fluid is assumed to be bounded below by a rigid orography/bathymetry, and above by a free surface. In a uniformly rotating coordinate system, and in either Cartesian or spherical geometry, the resulting shallow-water equations may then be compactly written in vector form (see e.g. Vallis, 2017, p. 109). Thus – in the present notation, and referring to Fig. 18.1 –

$$\frac{D_{\text{hor}}\mathbf{u}_{\text{hor}}}{Dt} + f\mathbf{e}_3 \times \mathbf{u}_{\text{hor}} + g\nabla_{\text{hor}}H = 0, \tag{18.1}$$

$$\frac{D_{\text{hor}}\widetilde{H}}{Dt} + \widetilde{H}\nabla_{\text{hor}} \cdot \mathbf{u}_{hor} = 0, \tag{18.2}$$

or

$$\frac{\partial \widetilde{H}}{\partial t} + \nabla_{\text{hor}} \cdot \left(\widetilde{H}\mathbf{u}_{hor}\right) = 0, \tag{18.3}$$

where (18.1) is the component equation for horizontal momentum, (18.2) and (18.3) are the Lagrangian and Eulerian forms of the mass-continuity equation, respectively, and

$$\frac{D_{\text{hor}}}{Dt} \equiv \frac{\partial}{\partial t} + \mathbf{u}_{\text{hor}} \cdot \nabla_{\text{hor}} \tag{18.4}$$

is the horizontal material derivative. Furthermore:

- $\mathbf{u}_{\text{hor}}$ is horizontal velocity.
- $\mathbf{e}_3$ is vertical unit vector.
- f is Coriolis parameter.
- Gravity, g, is constant (for dynamical consistency).
- $\widetilde{H} \equiv H - B > 0$ is fluid depth.
- H and B are the heights – above a horizontal reference surface – of the free surface and rigid bottom boundary, respectively.

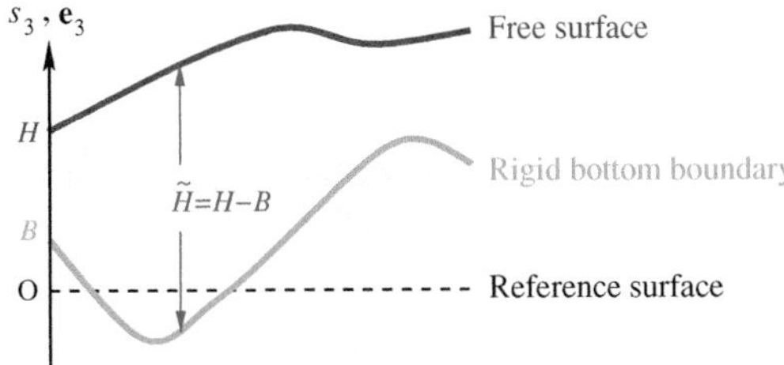

Figure 18.1 Vertical cross section of a shallow layer of fluid (shaded in light blue), confined between a rigid bottom boundary (solid green line) and an upper free surface (dark blue line), located at vertical distances B and H, respectively, from a horizontal reference surface (dotted line). The depth of the fluid is $\widetilde{H} \equiv H - B > 0$. See text for further details.

When viewed in the rotating frame of reference, the locations of the free surface (H) and the rigid bottom boundary (B) both vary spatially, but only that of the free surface varies temporally.

18.1.2 Some Recent Advances

Assumptions made, for reasons of simplicity, in the usual derivation of the shallow-water equations result, however, in various limitations on their applicability. These assumptions include:

1. Representation of Earth's form as (at best) a sphere, when it is better represented as a spheroid.
2. Representation of gravity as a constant everywhere, when it is known to vary meridionally.[2]
3. Incomplete representation of the Coriolis force.
4. Neglect of vertical acceleration.

For many applications of the shallow-water equations, particularly at large horizontal scales, the limitations arising from these assumptions are unimportant. It is nevertheless desirable to be able to relax these assumptions in the interest of widening the applicability of shallow-water equation sets as a paradigm for 3D, deep-fluid, geophysical modelling. But is this possible? Recent work collectively shows that indeed it is.

The first two limitations are intrinsically linked to one another and therefore need to be considered simultaneously. Starting with the more general *quasi-hydrostatic, deep* equations instead of the usual *hydrostatic-primitive, shallow* equations and assuming axial symmetry of the geopotential and independence of g on vertical coordinate, Staniforth and White (2015b) derived a dynamically consistent, shallow-water equation set in axial-orthogonal-curvilinear coordinates. Amongst other things, they showed that for a spheroidal planet,[3] and for reasons of dynamical consistency:

- g *must* vary meridionally and cannot be constant (except for the special case of a spherical planet, when g *must* be constant).
- $g\nabla_{\text{hor}}H$ in (18.1) herein *must* be rewritten as $\nabla_{\text{hor}}\left(gH\right)$ for the now-variable g.

Using the more general spheroidal equation set, Bénard (2015) showed that the potential impact of meridional variation of gravity for a mildly oblate, spheroidal planet may be greater than previously thought.

The third of the limitations was addressed by Tort et al. (2014) in *spherical* geometry and then further extended to *spheroidal* geometry by Staniforth (2015b). To respect dynamical consistency, a negligible approximation was, however, needed in both spherical and spheroidal geometry for restoration of the missing '$2\Omega\cos\phi$' contribution to the Coriolis force. To account for this approximation, the representation of the Coriolis force is more precisely termed 'almost complete' herein, rather than 'complete' as in Tort et al. (2014) and Staniforth (2015b).

The fourth of the limitations – neglect of vertical acceleration – was addressed in Cartesian geometry by Green and Naghdi (1976), Miles and Salmon (1985), Salmon (1998), and Dellar and Salmon (2005). It is then natural to wonder whether vertical acceleration can also be represented in the more-general context of a *spheroidal*, shallow-water equation set. This was confirmed in Staniforth (2015b), albeit to do so is quite complicated. A unified quartet of dynamically consistent, shallow-water equation sets, controlled by two binary switches, then results. The most complete of these sets, with all terms retained, simultaneously addresses all four of the limitations.

[2] It also varies vertically, but this variation is negligible in the context of a sufficiently shallow fluid.

[3] Nomenclature: The terms spheroid and ellipsoid in the literature are often used interchangeably, or with conflicting meanings between different authors. The convention adopted herein is that a *spheroid* is a solid of revolution that is approximately spherical, having an *almost*, or *precisely*, elliptic cross section in any meridional plane. When this cross section is *precisely* elliptic, the solid is then termed an *ellipsoid*.

The goals of the present chapter are to:

- Compare and contrast complementary, equivalent methodologies for the derivation, *in spheroidal geometry*,[4] of dynamically consistent, shallow-water equation sets of varying complexity.
- Describe a unified, switch-controlled quartet of such equation sets.
- For illustrative purposes, explicitly provide the governing equations and associated conservation principles for this quartet in spherical geometry[5].[6]

18.1.3 A Guide to the Remainder of this Chapter

As shown in Chapters 16 and 17, dynamically consistent equation sets in 3D may be derived in a variety of different but complementary ways. We adopt a similar approach in this chapter for the derivation of a quartet of 2D shallow-water equation sets of increasing complexity.[7]

To set the scene for later sections, and following Staniforth and White (2015b), we begin (in Section 18.2) by developing – from first principles using the classical Eulerian method – a 'basic' shallow-water equation set in axial-orthogonal-curvilinear coordinates. This set corresponds to the generalisation of (18.1)–(18.4) from Cartesian or spherical geometry to spheroidal geometry. To close this equation set and to apply it in practice requires the definition of:

- A reference geopotential surface.
- A specific horizontal coordinate system.
- A model of gravity – see Section 18.3.

Doing so then addresses the first two limitations listed in Section 18.1.2.

To instead derive an equation set by using a variant of the Lagrangian methodology requires the definition of a Lagrangian. To prepare the way for doing so and to provide some insight, the Lagrangian density for the 'basic' shallow-water equation set is obtained (in Section 18.4) in two different ways:

1. Mathematically – from the governing equations developed in Section 18.2.
2. From first physical principles.

The ensuing 'basic' Lagrangian density is then supplemented (in Section 18.5) by an additional term, in a similar manner to that used in Chapter 17 to obtain two, quasi-shallow equation sets in 3D. This yields a 2D 'quasi-shallow' Lagrangian density field and, with an associated, vertically averaged pressure field, prepares the way to address the third limitation – incomplete representation of the Coriolis force.

To accomplish this goal, application (in Section 18.6) of the Euler–Lagrange equations, using these two fields, then leads to the enhanced 'quasi-shallow' equation set in axial-orthogonal-curvilinear coordinates. This equation set subsumes the 'basic' one as a special case. Dynamical consistency of this equation set is demonstrated in an appendix, with the associated conservation principles summarised in Section 18.7. For illustrative and reference purposes, the 'quasi-shallow' equation set is explicitly given (in Section 18.8) in spherical geometry, together with the associated conservation principles. At this point, the third limitation of the list has been addressed.

It remains to address the fourth limitation. How to do so, using Hamilton's principle of stationary action, is described in Section 18.10. The key element for this is to supplement the Lagrangian density with an additional 'non-hydrostatic' term, obtained from vertical integration of vertical kinetic energy.

4 This naturally includes spherical geometry as a familiar, special case.

5 These results are a special case of the more general ones given in Staniforth (2015b) for *spheroidal* geometry.

6 Readers more interested in this quartet of equation sets than in its derivation should proceed directly to Section 18.10. They may, however, need to selectively consult the preceding sections regarding definitions and notation.

7 This chapter contains public sector information licensed under the Open Government Licence v1.0 and adapted for use herein.

The four combinations of switching the 'quasi-shallow' and 'non-hydrostatic' supplementary terms of Lagrangian density on or off then leads to the quartet of equation sets obtained in Staniforth (2015b) for spheroidal geometry. Since the form of the resulting equations, with all terms retained, is very complicated, they are not duplicated here. Instead, the equation set quartet is given more compactly (in Section 18.10) in spherical geometry, together with the associated conservation principles.

Finally, concluding remarks are given in Section 18.11.

18.2 EULERIAN DERIVATION OF THE BASIC SHALLOW-WATER EQUATIONS

18.2.1 The Governing Equations in 3D

18.2.1.1 *Axial-Orthogonal-Curvilinear Coordinates*

Consider an axial-orthogonal-curvilinear, geopotential coordinate system (ξ_1, ξ_2, ξ_3) in the zonal, meridional, and (upward) vertical directions, respectively, with associated unit vectors $(\mathbf{e}_1, \mathbf{e}_2, \mathbf{e}_3)$ and velocity components (u_1, u_2, u_3). Surfaces of constant ξ_3 represent surfaces of constant apparent geopotential, with apparent gravity acting only in the ξ_3 direction (i.e. the coordinate system is a geopotential one).

The metric (or scale) factors h_1, h_2, and h_3 are the quantities that appear in

$$ds^2 = h_1^2 d\xi_1^2 + h_2^2 d\xi_2^2 + h_3^2 d\xi_3^2, \tag{18.5}$$

where ds is infinitesimal distance. See Section 5.2.1. From (5.4), these factors are related to Cartesian coordinates (x_1, x_2, x_3) by the identities

$$h_i^2(\xi_2, \xi_3) \equiv \left(\frac{\partial x_1}{\partial \xi_i}\right)^2 + \left(\frac{\partial x_2}{\partial \xi_i}\right)^2 + \left(\frac{\partial x_3}{\partial \xi_i}\right)^2, \quad i = 1, 2, 3. \tag{18.6}$$

Because axial symmetry of the coordinate system is assumed (by definition of an axial-orthogonal-curvilinear coordinate system), h_1, h_2, and h_3 in (18.6) are independent of zonal (azimuthal) coordinate ξ_1. Standard expressions for gradient, curl, divergence, and material derivatives of scalars and vectors in axial-orthogonal-curvilinear coordinates may be found in Section 5.2.1 (subject to satisfaction of axial-symmetry constraint (18.6)).

From (5.5), velocity components in axial-orthogonal-curvilinear coordinates are given by

$$\mathbf{u} = u_1\mathbf{e}_1 + u_2\mathbf{e}_2 + u_3\mathbf{e}_3 = h_1\frac{D\xi_1}{Dt}\mathbf{e}_1 + h_2\frac{D\xi_2}{Dt}\mathbf{e}_2 + h_3\frac{D\xi_3}{Dt}\mathbf{e}_3 = h_1\dot{\xi}_1\mathbf{e}_1 + h_2\dot{\xi}_2\mathbf{e}_2 + h_3\dot{\xi}_3\mathbf{e}_3. \tag{18.7}$$

18.2.1.2 *Components of the Momentum Equation*

The three components of momentum balance for the *quasi-hydrostatic-deep equations* in axial-orthogonal-curvilinear geopotential coordinates are given by (16.113)–(16.115), with switch $\mathcal{N}$ set identically zero, that is, (blue) non-hydrostatic terms are omitted from these three equations. Thus

$$\frac{Du_1}{Dt} + \left(\frac{u_1}{h_1} + 2\Omega\right)\frac{u_2}{h_2}\frac{\partial h_1}{\partial \xi_2} + \underbrace{\left(\frac{u_1}{h_1} + 2\Omega\right)\frac{u_3}{h_3}\frac{\partial h_1}{\partial \xi_3}}_{\text{deep}} + \frac{1}{\rho h_1}\frac{\partial p}{\partial \xi_1} = 0, \tag{18.8}$$

$$\frac{Du_2}{Dt} - \left(\frac{u_1}{h_1} + 2\Omega\right)\frac{u_1}{h_2}\frac{\partial h_1}{\partial \xi_2} + \underbrace{\frac{u_2}{h_2}\frac{u_3}{h_3}\frac{\partial h_2}{\partial \xi_3}}_{\text{deep}} + \frac{1}{\rho h_2}\frac{\partial p}{\partial \xi_2} = 0, \tag{18.9}$$

$$\underbrace{-\left[\left(\frac{u_1}{h_1} + 2\Omega\right)\frac{u_1}{h_3}\frac{\partial h_1}{\partial \xi_3} + \frac{u_2^2}{h_2 h_3}\frac{\partial h_2}{\partial \xi_3}\right]}_{\text{deep}} + \frac{1}{\rho h_3}\frac{\partial p}{\partial \xi_3} + \frac{1}{h_3}\frac{d\Phi}{d\xi_3} = 0, \tag{18.10}$$

where

$$\frac{D}{Dt} \equiv \frac{\partial}{\partial t} + \frac{u_1}{h_1}\frac{\partial}{\partial \xi_1} + \frac{u_2}{h_2}\frac{\partial}{\partial \xi_2} + \frac{u_3}{h_3}\frac{\partial}{\partial \xi_3}, \tag{18.11}$$

is (deep) material derivative, expressed in axial-orthogonal-curvilinear coordinates. Notation is standard, and as in Chapter 16.

Equations (18.8) and (18.9) are the *horizontal* components of the momentum equation. *Vertical* component (18.10) is the *quasi-hydrostatic equation* itself. It is the *deep*-fluid analogue of the classical *shallow*-fluid hydrostatic approximation.[8] To obtain (18.10) in this form,

$$g \equiv \frac{1}{h_3}\frac{d\Phi}{d\xi_3} \tag{18.12}$$

has been used. This equation defines the acceleration of apparent gravity (g) in terms of geopotential (Φ). Because a geopotential coordinate system is assumed here, $\Phi = \Phi(\xi_3)$. Thus components of gravity are absent from horizontal momentum equations (18.8) and (18.9), since Φ is constant along any geopotential surface $\xi_3 =$ constant.

18.2.1.3 *Mass-Continuity Equation*

Only the three components of the momentum equation, plus the mass-continuity equation, are required to derive the shallow-water equations (see e.g. Vallis 2017, 2019). To complete the formulation of the relevant quasi-hydrostatic equations in axial-orthogonal-curvilinear geopotential coordinates, component equations (18.8)–(18.10) are thus supplemented by mass-continuity equation (16.116), that is, by

$$\frac{D\rho}{Dt} + \frac{\rho}{h_1h_2h_3}\left[\frac{\partial}{\partial \xi_1}\left(u_1h_2h_3\right) + \frac{\partial}{\partial \xi_2}\left(u_2h_3h_1\right) + \frac{\partial}{\partial \xi_3}\left(u_3h_1h_2\right)\right] = 0. \tag{18.13}$$

18.2.2 Assumptions and Preparatory Steps

To derive a shallow-water equation set, a number of assumptions need to be made. We assemble them here – together with a few preparatory steps common to both Eulerian and Lagrangian derivations – in a manner similar to that employed in Staniforth and White (2015b) and Staniforth (2015b). This avoids introducing them piecemeal during the derivations:

1. The fluid is assumed to be *homogeneous*, that is,

   $$\rho = \overline{\rho} = \text{constant}. \tag{18.14}$$

   Without loss of generality, $\overline{\rho}$ may be arbitrarily set to unity if desired.
2. The fluid is assumed to be either *shallow* or *quasi-shallow*.
   (a) The *shallow* assumption – see Chapter 16 – justifies neglect of the height variation of the metric factors, $h_i, i = 1, 2, 3$, in quasi-hydrostatic equations (18.8)–(18.10) and in (18.13). Because of the assumed axial symmetry of the coordinate system, they are then functions only of ξ_2. Thus

   $$h_1 = h_1^S(\xi_2), \quad h_2 = h_2^S(\xi_2), \quad h_3 = h_3^S(\xi_2), \tag{18.15}$$

   where superscript 'S' denotes evaluation at the planet's reference surface; see Fig. 18.2 and the following discussion. This assumption automatically eliminates the deep terms (in red) from (18.8)–(18.10), since $\partial h_i^S/\partial \xi_3, i = 1, 2, 3$, are then all identically zero. In particular, and noting (6.20), it automatically removes the '$2\Omega\cos\phi = 2\Omega\left(1/h_3\right)\left(\partial h_1/\partial \xi_3\right)$'

8 The presence of the deep (red) terms in (18.10) means that the balance is, in general, not precisely hydrostatic, only *quasi*-hydrostatic. In their absence, however, the balance *is* precisely hydrostatic; this occurs when making the shallow-fluid assumption.

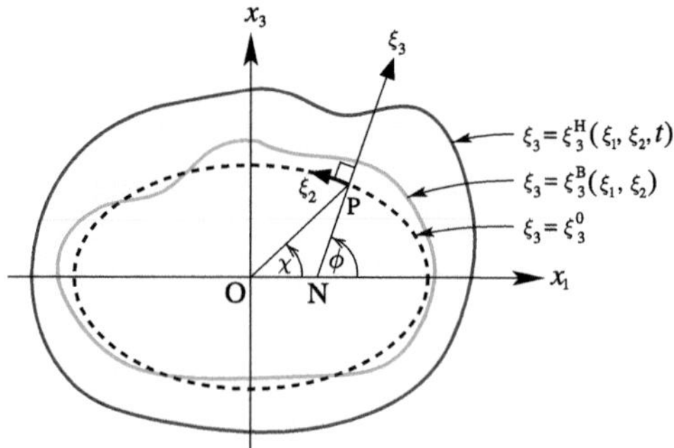

Figure 18.2 A cross section in the Ox_1x_3 meridional plane. The reference geopotential surface ($\xi_3 = \xi_3^0$) is depicted by the dashed curve: N is the point at which the normal to this surface at the general point P intersects Ox_1. χ is *geocentric* latitude, and ϕ is *geographic* latitude. ξ_2 and ξ_3 are the orthogonal coordinates in the meridional and outward-pointing (normal to the reference surface) directions, respectively: ξ_1, the azimuthal coordinate, points into the page for $x_1 > 0$; and out of it for $x_1 < 0$. The shallow layer of fluid is shaded in light blue. It is bounded below by the bottom orography/bathymetry (green contour), $\xi_3 = \xi_3^B(\xi_1, \xi_2)$, and above by its free surface (dark blue contour), $\xi_3 = \xi_3^H(\xi_1, \xi_2, t)$. See text for further details.

terms, with no need of a switch for this. Making the full shallow assumption (as opposed to the weaker quasi-shallow one, as in item 2(b)) – plus some further assumptions – ultimately leads to the classical, 'basic', shallow-water equations of this section.

(b) If the (weaker) *quasi-shallow* assumption is made instead – see Chapter 17 – height variation of the metric factors is again ignored, as in (18.15), *but with one exception*. For the computation of the Coriolis force, h_1^2 is exceptionally approximated using the two-term Taylor series expansion (17.52) in the vertical, namely

$$h_1^2 \approx \left(h_1^2\right)^S + \left(\frac{\partial h_1^2}{\partial \xi_3}\right)^S \left(\xi_3 - \xi_3^0\right). \tag{18.16}$$

Whenever h_1, h_2, and h_3 appear *undifferentiated* in the vertical, they nevertheless still obey shallow metric constraints (18.15). Making this assumption – plus some other assumptions – ultimately leads to a further ('*quasi-shallow*') shallow-water equation set; see Section 18.6. The quasi-shallow contribution in (18.16) is coloured green to facilitate tracking the influence of this term in the development of the quasi-shallow, shallow-water equations. In the absence of the quasi-shallow (green) term in (18.16), (purely) shallow assumption (18.15) is recovered as a special case.

3. An axially symmetric, reference, geopotential surface[9] – see Fig. 18.2 – is defined in such a way that it is representative of the location of the planet's surface. This reference surface corresponds to $\xi_3 = \xi_3^0 =$ constant (say), and so

$$\Phi_S\left(\xi_3 = \xi_3^0\right) \equiv \Phi_S^0 = \text{constant}. \tag{18.17}$$

4. As noted in White and Wood (2012) and elsewhere in the literature, a crucial step in deriving a shallow-fluid or a shallow-water model is to make gravity, g, *independent of* vertical

[9] This reference surface is graphically depicted in Fig. 18.2 as being elliptic. This is appropriate for terrestrial applications, with the virtue of being consistent with the World Geodetic System reference ellipsoid used for reporting observational data and for satellite navigation; see Section 12.2.4. However, for the formulation given herein, the shape of the reference surface (determined by the planet's internal mass distribution) is less restrictive: it is only assumed to be axially symmetric (i.e. independent of ξ_1).

coordinate ξ_3. Due to the use of an assumed, axially symmetric coordinate system, g is also independent of ξ_1. Furthermore, ξ_3 is a geopotential coordinate. Thus it is assumed that

$$g = g(\xi_2), \quad \Phi_S = \Phi_S(\xi_3). \tag{18.18}$$

However, from definition (18.12) of g in terms of Φ, (18.15) and (18.18) together imply that

$$\frac{d\Phi_S(\xi_3)}{d\xi_3} = h_3^S(\xi_2)\, g(\xi_2). \tag{18.19}$$

This separation of variables then means that the only way (18.19) can be satisfied is by setting

$$h_3^S(\xi_2)\, g(\xi_2) = \text{constant} = h_3^E g_S^E \ \text{(say)}, \tag{18.20}$$

where h_3^E and g_S^E are the values of h_3^S and g, respectively, at the planet's equator. Thus g can vary as a function of latitude, but then h_3^S must satisfy constraint (18.20) and vary as the reciprocal of g. Equation (18.20) is a simple consequence of the use of a geopotential coordinate system. Vertically integrating (18.19) from the reference geopotential surface $\xi_3 = \xi_3^0$ to ξ_3, subject to condition (18.17) on the reference geopotential surface, and using (18.20) then gives the explicit geopotential representation

$$\Phi_S(\xi_3) = \Phi_S^0 + h_3^E g_S^E \left(\xi_3 - \xi_3^0\right) = \Phi_S^0 + h_3^S(\xi_2)\, g(\xi_2)\left(\xi_3 - \xi_3^0\right). \tag{18.21}$$

5. Using shallow metric (18.15) and considering only changes along the ξ_3 direction (with ξ_1 and ξ_2 held fixed), (18.5) reduces to

$$ds_3 = h_3^S(\xi_2)\, d\xi_3. \tag{18.22}$$

Integration of (18.22) from the reference geopotential surface at $\xi_3 = \xi_3^0$, to ξ_3, with use of (18.20) and boundary condition $s_3\left(\xi_1, \xi_2, \xi_3 = \xi_3^0\right) = 0$, then gives

$$s_3(\xi_1, \xi_2, \xi_3) = h_3^S(\xi_2)\left(\xi_3 - \xi_3^0\right) = \frac{h_3^E g_S^E}{g(\xi_2)}\left(\xi_3 - \xi_3^0\right). \tag{18.23}$$

Thus $s_3 = s_3(\xi_1, \xi_2, \xi_3)$ measures distance from the reference geopotential surface (defined by $\xi_3 = \xi_3^0$) along a (generally curved) vertical coordinate line with ξ_1 and ξ_2 held fixed.

6. The layer of fluid is assumed to be bounded below by a rigid, impermeable, bottom orography/bathymetry located at a prescribed distance $B\left(\xi_1, \xi_2\right)$ along a vertical coordinate line from the reference geopotential surface ($\xi_3 = \xi_3^0$), that is, at $s_3^B \equiv B(\xi_1, \xi_2)$. From (18.20) and (18.23), the corresponding value of ξ_3 is given by

$$\xi_3^B\left(\xi_1, \xi_2\right) - \xi_3^0 \equiv \frac{s_3^B}{h_3^S(\xi_2)} = \frac{B(\xi_1, \xi_2)}{h_3^S(\xi_2)} = \frac{1}{h_3^E}\frac{g(\xi_2)}{g_S^E} B(\xi_1, \xi_2). \tag{18.24}$$

7. It is assumed that the deviation of the bottom orography/bathymetry from the reference geopotential surface is sufficiently small that the associated mass difference has negligible influence on Newtonian gravitational attraction. To guarantee negligible influence, it is therefore assumed that $\left|B\left(\xi_1, \xi_2\right)\right| \ll \overline{a}$, where $\overline{a}$ is the planet's mean radius.[10]

8. Similarly, the free surface of the shallow layer of fluid is assumed to be located at a distance $s_3^H \equiv H(\xi_1, \xi_2, t)$ along a vertical coordinate line from the reference geopotential surface (located at $s_3 = 0$, i.e. at $\xi_3 = \xi_3^0$). From (18.23), the corresponding value of ξ_3 is given by

$$\xi_3^H(\xi_1, \xi_2, t) - \xi_3^0 \equiv \frac{s_3^H}{h_3^S(\xi_2)} = \frac{H(\xi_1, \xi_2, t)}{h_3^S(\xi_2)} = \frac{1}{h_3^E}\frac{g(\xi_2)}{g_S^E} H(\xi_1, \xi_2, t), \tag{18.25}$$

[10] Clearly the thinner the layer of mass trapped between the reference geopotential surface and the bottom orography/bathymetry, the more negligible is its gravitational influence.

where $H(\xi_1, \xi_2, t)$ is referred to as the *free-surface height* and is one of the three dependent variables of the shallow-water equations. Whereas the bottom boundary is *fixed* in space, the *free*-surface boundary varies as a function of both space and time. As implied by the (*shallow-water*) terminology, it is assumed that the depth of the shallow layer of fluid is very small, that is,

$$s_3^H(\xi_1, \xi_2, t) - s_3^B(\xi_1, \xi_2) = H(\xi_1, \xi_2, t) - B(\xi_1, \xi_2) \ll \bar{a}. \tag{18.26}$$

The mass of this shallow layer of fluid is also assumed to make a negligible contribution to Newtonian gravitational attraction. Furthermore, to avoid singular behaviour, it is assumed that the planet, including its orography/bathymetry, is *entirely* covered by fluid, that is,

$$s_3^H(\xi_1, \xi_2, t) > s_3^B(\xi_1, \xi_2), \quad \text{for all time,} \tag{18.27}$$

with equality of these two distances strictly forbidden. Physically, this means that there is no 'wetting/drying' at the lower boundary (i.e. it is everywhere overlaid by fluid *for all time*).[11]

9. The other two dependent variables are the horizontal velocity components, u_1 and u_2. The fluid is constrained to move coherently in a wholly vertical, columnar manner[12] – see Section 7.6 of Salmon (1998) – and so

$$u_1 = u_1(\xi_1, \xi_2, t), \quad u_2 = u_2(\xi_1, \xi_2, t), \tag{18.28}$$

that is, u_1 and u_2 are assumed to be independent of ξ_3 for all time. For the situation examined in Staniforth and White (2015b) (this corresponds to the 'basic' equation set discussed in the present section), it was *deduced* that if (18.28) holds at initial time (i.e. at $t = 0$), then it holds for all time. Herein, (18.28) is *assumed* rather than deduced. Without this assumption, the quasi-shallow equation set of Section 18.6 would not materialise. This is because solutions of the 'basic' equation set are a *specialisation* of solutions of the 3D equations, whereas those of the 'quasi-shallow' one are not; they are instead *approximations*.

10. At the free surface of the fluid, it is assumed that the pressure is constant and positive, that is,

$$p\left[\xi_1, \xi_2, \xi_3^H(\xi_1, \xi_2, t), t\right] = p^H = \text{constant} \geq 0. \tag{18.29}$$

11. By definition (18.7), and with use of (18.15),

$$u_3(\xi_1, \xi_2, \xi_3, t) = h_3^S(\xi_2)\frac{D\xi_3}{Dt}. \tag{18.30}$$

Substituting for ξ_3 into (18.30) using (18.23), and also using definition (18.11) of material derivative, gives the following expression for the vertical component of velocity

$$u_3(\xi_1, \xi_2, \xi_3, t) = h_3^S\frac{D}{Dt}\left(\xi_3^0 + \frac{s_3}{h_3^S}\right) = \frac{Ds_3}{Dt} - \frac{s_3}{h_3^S}\frac{Dh_3^S(\xi_2)}{Dt} = \frac{Ds_3}{Dt} - \frac{s_3}{h_3^S}\frac{u_2}{h_2^S}\frac{dh_3^S}{d\xi_2}. \tag{18.31}$$

12. The planet's orography/bathymetry is assumed to be impermeable and stationary. Evaluating (18.31) at the fluid's bottom boundary, $s_3 = B(\xi_1, \xi_2)$, gives the kinematic condition there that

$$u_3^B(\xi_1, \xi_2) \equiv u_3\left[\xi_1, \xi_2, \xi_3^B(\xi_1, \xi_2)\right] = \frac{D_{\text{hor}}B(\xi_1, \xi_2)}{Dt} - \frac{B(\xi_1, \xi_2)}{h_2^S h_3^S}u_2\frac{dh_3^S}{d\xi_2}, \tag{18.32}$$

where

$$\frac{D_{\text{hor}}}{Dt} \equiv \frac{\partial}{\partial t} + \frac{u_1}{h_1^S}\frac{\partial}{\partial \xi_1} + \frac{u_2}{h_2^S}\frac{\partial}{\partial \xi_2}, \tag{18.33}$$

[11] This assumption implicitly limits the amplitude of any 'sloshing' of the fluid.

[12] The column is aligned with a vertical coordinate line (not necessarily straight) with ξ_1 and ξ_2 held fixed. Specific examples of such (analytically specified) curved, vertical, coordinate lines may be found in Staniforth and White (2015a) and Section 12.8 herein for ellipsoidal geometry with spatial variation of gravity.

is horizontal material derivative – in shallow geometry – along a geopotential surface $\xi_3 =$ constant. Note that $D/Dt \rightarrow D_{\text{hor}}/Dt$ in (18.32) since $B = B(\xi_1, \xi_2)$ is independent of ξ_3, and so $u_3 \partial B(\xi_1, \xi_2) / \partial \xi_3 \equiv 0$.

13. It is assumed that the kinematic condition also holds at the fluid's free surface $s_3 = H(\xi_1, \xi_2, t)$. Evaluating (18.31) there then gives

$$u_3^H(\xi_1, \xi_2, t) \equiv u_3\left[\xi_1, \xi_2, \xi_3^H(\xi_1, \xi_2, t), t\right] = \frac{D_{\text{hor}} H(\xi_1, \xi_2, t)}{Dt} - \frac{H(\xi_1, \xi_2, t)}{h_2^S h_3^S} u_2 \frac{dh_3^S}{d\xi_2}. \quad (18.34)$$

Similarly, $D/Dt \rightarrow D_{\text{hor}}/Dt$ in (18.34) since $H = H(\xi_1, \xi_2, t)$ is also independent of ξ_3.

14. Subtracting (18.32) from (18.34) yields

$$u_3^H(\xi_1, \xi_2, t) - u_3^B(\xi_1, \xi_2) = \frac{D_{\text{hor}} \widetilde{H}(\xi_1, \xi_2, t)}{Dt} - \frac{\widetilde{H}(\xi_1, \xi_2, t)}{h_2^S h_3^S} u_2 \frac{dh_3^S}{d\xi_2}, \quad (18.35)$$

where

$$\widetilde{H}(\xi_1, \xi_2, t) \equiv H(\xi_1, \xi_2, t) - B(\xi_1, \xi_2) > 0, \quad (18.36)$$

is *fluid depth*.

18.2.3 Derivation of the Basic Shallow-Water Equations

18.2.3.1 *Simplification of the Governing Equations for the Purely Shallow Homogeneous Case*

Consider now a *purely shallow, homogeneous* fluid. Introduction of (18.14), (18.15), and (18.21) into governing equations (18.8)–(18.10) and (18.13), with use of (18.28), then leads to

$$\frac{D_{\text{hor}} u_1}{Dt} + \left(\frac{u_1}{h_1^S} + 2\Omega\right) \frac{u_2}{h_2^S} \frac{dh_1^S}{d\xi_2} + \frac{1}{\overline{\rho} h_1^S} \frac{\partial p}{\partial \xi_1} = 0, \quad (18.37)$$

$$\frac{D_{\text{hor}} u_2}{Dt} - \left(\frac{u_1}{h_1^S} + 2\Omega\right) \frac{u_1}{h_2^S} \frac{dh_1^S}{d\xi_2} + \frac{1}{\overline{\rho} h_2^S} \frac{\partial p}{\partial \xi_2} = 0, \quad (18.38)$$

$$\frac{\partial p}{\partial \xi_3} = -\overline{\rho} \frac{d\Phi_S}{d\xi_3} = -\overline{\rho} h_3^E g_S^E = -\overline{\rho} h_3^S(\xi_2) g(\xi_2), \quad (18.39)$$

$$\frac{\partial}{\partial \xi_1}\left(u_1 h_2^S h_3^S\right) + \frac{\partial}{\partial \xi_2}\left(u_2 h_3^S h_1^S\right) + \frac{\partial}{\partial \xi_3}\left(u_3 h_1^S h_2^S\right) = 0. \quad (18.40)$$

Use of (18.28) in (18.37) and (18.38) simplified Du_1/Dt and Du_2/Dt to $D_{\text{hor}} u_1/Dt$ and $D_{\text{hor}} u_2/Dt$, respectively.

Equations (18.37)–(18.40) are essentially the hydrostatic primitive equations specialised to homogeneous, shallow, columnar, fluid flow; cf. (16.128)–(16.131), with $\mathcal{N} \equiv 0$. The significance and roles of these four equations are as follows:

- Equations (18.37) and (18.38) are *prognostic* equations for horizontal velocity components u_1 and u_2, respectively.
- Equation (18.39) originates from the *prognostic* equation for vertical component u_3 of velocity. Under the preceding assumptions, this equation reduces to a *diagnostic* equation for classical hydrostatic balance, whereby the vertical component of the pressure gradient exactly balances apparent gravity. Vertical integration of this first-order differential equation determines total pressure, $p = p(\xi_1, \xi_2, \xi_3, t)$.
- Equation (18.40), which originates from *prognostic* equation (18.13) for mass continuity, can be considered to be a *diagnostic* equation for u_3. Vertical integration of (18.40) then determines $u_3 = u_3(\xi_1, \xi_2, \xi_3, t)$.

Crucial to derivation of shallow-water equation sets is assumption (18.28) that u_1 and u_2 are *independent of vertical coordinate* ξ_3. Without this assumption, shallow-water equation sets fail to materialise.[13]

18.2.3.2 Determination of Total Pressure

Integrating hydrostatic equation (18.39) downwards from the free surface $\xi_3 = \xi_3^H$ to ξ_3 using (18.20) and (18.25), and applying constant-pressure condition (18.29) at the free surface, yields

$$\begin{aligned} p\left(\xi_1,\xi_2,\xi_3,t\right) &= p^H + \overline{\rho}\left[g\left(\xi_2\right)H\left(\xi_1,\xi_2,t\right) - h_3^E g_S^E\left(\xi_3-\xi_3^0\right)\right] \\ &= p^H + \overline{\rho}g\left(\xi_2\right)\left[H\left(\xi_1,\xi_2,t\right) - h_3^S\left(\xi_2\right)\left(\xi_3-\xi_3^0\right)\right]. \end{aligned} \tag{18.41}$$

Note that this equation for total pressure only holds for the purely shallow ('basic') case under present consideration. For the '*quasi*-shallow' case, (18.39) is missing a term, and so therefore is (18.41).

18.2.3.3 The Horizontal Momentum Components of the Basic Shallow-Water Equations

The functional form of (18.41) means that the two components of the horizontal pressure gradient in (18.37) and (18.38), namely $\left(\partial p/\partial\xi_1\right)/\left(\overline{\rho}h_1^S\right)$ and $\left(\partial p/\partial\xi_2\right)/\left(\overline{\rho}h_2^S\right)$, are both independent of vertical coordinate ξ_3, as in spherical coordinates (Vallis, 2017, 2019). Inserting (18.41) into (18.37) and (18.38) then yields the two, prognostic, horizontal components of the momentum equation in the *shallow-water* forms

$$\frac{D_{\text{hor}}u_1}{Dt} + \left(\frac{u_1}{h_1^S} + 2\Omega\right)\frac{u_2}{h_2^S}\frac{dh_1^S}{d\xi_2} + \frac{1}{h_1^S}\frac{\partial}{\partial\xi_1}\left[g\left(\xi_2\right)H\left(\xi_1,\xi_2,t\right)\right] = 0, \tag{18.42}$$

$$\frac{D_{\text{hor}}u_2}{Dt} - \left(\frac{u_1}{h_1^S} + 2\Omega\right)\frac{u_1}{h_2^S}\frac{dh_1^S}{d\xi_2} + \frac{1}{h_2^S}\frac{\partial}{\partial\xi_2}\left[g\left(\xi_2\right)H\left(\xi_1,\xi_2,t\right)\right] = 0. \tag{18.43}$$

Note that (18.42) and (18.43) also only hold for the purely shallow case, again due to the absence of a quasi-shallow term in (18.39) and in equations that follow from it.

18.2.3.4 The Shallow-Water Mass-Continuity Equation

To obtain the equation governing the evolution of fluid depth, $\widetilde{H}\left(\xi_1,\xi_2,t\right)$, mass-continuity equation (18.40) is first integrated from bottom boundary $\xi_3 = \xi_3^B$ to ξ_3, where ξ_3^B is defined by (18.24). Thus

$$u_3\left(\xi_1,\xi_2,\xi_3,t\right) - u_3^B\left(\xi_1,\xi_2\right) + \frac{\left[\xi_3 - \xi_3^B\left(\xi_1,\xi_2\right)\right]}{h_1^S h_2^S}\left[\frac{\partial}{\partial\xi_1}\left(u_1 h_2^S h_3^S\right) + \frac{\partial}{\partial\xi_2}\left(u_2 h_3^S h_1^S\right)\right] = 0. \tag{18.44}$$

The last square-bracketed term in (18.44) originates from horizontal divergence and is independent of ξ_3 by virtue of (18.15) and (18.28); this simplified the vertical integration and thereby the form of (18.44). It is seen from (18.44) that vertical velocity (u_3) *varies linearly* within a vertical column with respect to vertical coordinate ξ_3.

Evaluation of (18.44) at the free surface $\xi_3 = \xi_3^H$ and insertion of definitions (18.24), (18.25), and (18.36) into the result leads to

$$u_3^H\left(\xi_1,\xi_2,t\right) - u_3^B\left(\xi_1,\xi_2\right) + \frac{\widetilde{H}\left(\xi_1,\xi_2,t\right)}{h_1^S h_2^S h_3^S}\left[\frac{\partial}{\partial\xi_1}\left(u_1 h_2^S h_3^S\right) + \frac{\partial}{\partial\xi_2}\left(u_2 h_3^S h_1^S\right)\right] = 0. \tag{18.45}$$

[13] Solutions to (18.37)–(18.40) (but with Du_1/Dt and Du_2/Dt no longer simplified to $D_{\text{hor}}u_1/Dt$ and $D_{\text{hor}}u_2/Dt$, respectively) nevertheless exist. However, the fluid flow is then no longer columnar as it is for solutions of the shallow-water equations.

Using (18.35) to eliminate $u_3^H - u_3^B$ from this equation then yields the prognostic, shallow-water, mass-continuity equation for fluid depth ($\widetilde{H}$), namely

$$\frac{D_{\text{hor}}\widetilde{H}(\xi_1,\xi_2,t)}{Dt} + \frac{\widetilde{H}(\xi_1,\xi_2,t)}{h_1^S h_2^S h_3^S}\left[\frac{\partial}{\partial \xi_1}\left(u_1 h_2^S h_3^S\right) + \frac{\partial}{\partial \xi_2}\left(u_2 h_3^S h_1^S\right) - u_2 h_1^S \frac{dh_3^S}{d\xi_2}\right] = 0. \tag{18.46}$$

After some algebra, $h_3^S(\xi_2)$ drops out of this equation and it simplifies to

$$\frac{D_{\text{hor}}\widetilde{H}(\xi_1,\xi_2,t)}{Dt} + \frac{\widetilde{H}(\xi_1,\xi_2,t)}{h_1^S h_2^S}\left[\frac{\partial}{\partial \xi_1}\left(u_1 h_2^S\right) + \frac{\partial}{\partial \xi_2}\left(u_2 h_1^S\right)\right] = 0. \tag{18.47}$$

As noted in Staniforth and White (2015b), the disappearance of $h_3^S(\xi_2)$ from (18.47) is to be expected. This is because the requirement to satisfy (18.20) means that $g(\xi_2)$ would otherwise feature in the conservation of mass, which would be physically unreasonable. Of course, $g(\xi_2)$ remains in the shallow-water equations through its appearance in (18.42) and (18.43) for the components of momentum. Thus any physically reasonable ξ_2 variation is permitted provided that $h_3^S(\xi_2)$ varies as $1/g(\xi_2)$ in the 'parent' 3D, shallow model (from which the 'basic' shallow-water model is obtained by vertical integration).

Using definition (18.33) of D_{hor}/Dt, the shallow-water continuity equation (18.47) may be rewritten in Eulerian flux form as

$$\frac{\partial \widetilde{H}(\xi_1,\xi_2,t)}{\partial t} + \frac{1}{h_1^S h_2^S}\left[\frac{\partial}{\partial \xi_1}\left(u_1 h_2^S \widetilde{H}\right) + \frac{\partial}{\partial \xi_2}\left(u_2 h_1^S \widetilde{H}\right)\right] = 0. \tag{18.48}$$

18.2.3.5 Recapitulation

In axial-orthogonal-curvilinear geopotential coordinates, the 'basic' shallow-water equation set is comprised of component equations (18.42) and (18.43) for horizontal momentum, and mass-continuity equation (18.47) (in Lagrangian form) or (18.48) (in Eulerian form). Thus:

The 'Basic' Shallow-Water Equations in Axial-Orthogonal-Curvilinear Coordinates

Momentum

$$\frac{D_{\text{hor}}u_1}{Dt} + \left(\frac{u_1}{h_1^S} + 2\Omega\right)\frac{u_2}{h_2^S}\frac{dh_1^S}{d\xi_2} + \frac{1}{h_1^S}\frac{\partial}{\partial \xi_1}\left[g(\xi_2)\, H(\xi_1,\xi_2,t)\right] = 0, \tag{18.49}$$

$$\frac{D_{\text{hor}}u_2}{Dt} - \left(\frac{u_1}{h_1^S} + 2\Omega\right)\frac{u_1}{h_2^S}\frac{dh_1^S}{d\xi_2} + \frac{1}{h_2^S}\frac{\partial}{\partial \xi_2}\left[g(\xi_2)\, H(\xi_1,\xi_2,t)\right] = 0. \tag{18.50}$$

Mass-Continuity

$$\frac{D_{\text{hor}}\widetilde{H}(\xi_1,\xi_2,t)}{Dt} + \frac{\widetilde{H}}{h_1^S h_2^S}\left[\frac{\partial}{\partial \xi_1}\left(u_1 h_2^S\right) + \frac{\partial}{\partial \xi_2}\left(u_2 h_1^S\right)\right] = 0, \tag{18.51}$$

or

$$\frac{\partial \widetilde{H}(\xi_1,\xi_2,t)}{\partial t} + \frac{1}{h_1^S h_2^S}\left[\frac{\partial}{\partial \xi_1}\left(u_1 h_2^S \widetilde{H}\right) + \frac{\partial}{\partial \xi_2}\left(u_2 h_1^S \widetilde{H}\right)\right] = 0. \tag{18.52}$$

Material Derivative

$$\frac{D_{\text{hor}}}{Dt} \equiv \frac{\partial}{\partial t} + \frac{u_1}{h_1^S}\frac{\partial}{\partial \xi_1} + \frac{u_2}{h_2^S}\frac{\partial}{\partial \xi_2}. \tag{18.53}$$

This equation set in axial-orthogonal-curvilinear coordinates may be more compactly written in vector form as:

The 'Basic' Shallow-Water Equations in Vector Form

Momentum

$$\frac{D_{\text{hor}}\mathbf{u}_{\text{hor}}}{Dt} + f\mathbf{e}_3 \times \mathbf{u}_{\text{hor}} + \nabla_{\text{hor}}\left(gH\right) = 0. \tag{18.54}$$

Mass-Continuity

$$\frac{D_{\text{hor}}\widetilde{H}}{Dt} + \widetilde{H}\nabla_{\text{hor}} \cdot \mathbf{u}_{\text{hor}} = 0, \tag{18.55}$$

or

$$\frac{\partial \widetilde{H}}{\partial t} + \nabla_{\text{hor}} \cdot \left(\widetilde{H}\mathbf{u}_{\text{hor}}\right) = 0. \tag{18.56}$$

Material Derivative

$$\frac{D_{\text{hor}}}{Dt} \equiv \frac{\partial}{\partial t} + \mathbf{u}_{\text{hor}} \cdot \nabla_{\text{hor}}, \tag{18.57}$$

where

$$\mathbf{u}_{\text{hor}} \equiv u_1\mathbf{e}_1 + u_2\mathbf{e}_2, \tag{18.58}$$

$$\mathbf{u}_{\text{hor}} \cdot \nabla_{\text{hor}} \equiv \frac{u_1}{h_1^S}\frac{\partial}{\partial \xi_1} + \frac{u_2}{h_2^S}\frac{\partial}{\partial \xi_2}, \tag{18.59}$$

$$\nabla_{\text{hor}} \cdot \left(G\mathbf{u}_{\text{hor}}\right) \equiv \frac{1}{h_1^S h_2^S}\left[\frac{\partial}{\partial \xi_1}\left(h_2^S u_1 G\right) + \frac{\partial}{\partial \xi_2}\left(h_1^S u_2 G\right)\right], \tag{18.60}$$

and

$$f \equiv -\frac{2\Omega}{h_2^S}\frac{dh_1^S}{d\xi_2}, \tag{18.61}$$

is the Coriolis parameter.

Comparing (18.54)–(18.57) for shallow, incompressible, fluid flow in *spheroidal* geometry (with *variable* gravity, g) with their counterparts (18.1)–(18.4) in either Cartesian or spherical geometry (with *constant* g for dynamical consistency), we see that g in (18.54) can no longer be brought outside $\nabla_{\text{hor}}\left(gH\right)$ as it can for (18.1). If we were to do so, this would then violate dynamical consistency in the present *spheroidal* case.

18.2.3.6 Conservation Properties

The basic shallow-water equation set (18.49)–(18.53) or, equivalently, (18.54)–(18.57), has been derived in a consistent manner as a *specialised subset* of the quasi-hydrostatic equations. This latter equation set is dynamically consistent, that is, it formally respects conservation principles for mass, axial angular momentum, total energy, and potential vorticity; see (16.122)–(16.127) in the absence of the (blue) non-hydrostatic terms.[14] The only *approximation* that has been made is of the metric factors. Since, as noted in Appendix C of White et al. (2005), approximating metric factors leaves conservation principles of an equation set intact, the shallow-water *specialisation* in axial-orthogonal-curvilinear coordinates preserves the conservation principles of the quasi-hydrostatic equations. It is straightforward to verify, a posteriori, that this is indeed so. See Appendix B of the present chapter, with Q set identically zero everywhere therein.

18.3 HORIZONTAL COORDINATE SYSTEMS AND MODELS OF GRAVITY

To close the preceding formulation of the 'basic' shallow-water equations, we need to define:

1. A reference geopotential surface – characterised by parameter ξ_0.
2. A horizontal coordinate system – characterised by horizontal coordinates (ξ_1, ξ_2).
3. A model of gravity – characterised by functional form $g = g(\xi_2)$.

The purpose of this section is to illustrate how this may be accomplished in practice. In what follows, we do so with a focus on planet Earth, a planet of mild oblateness as discussed in Chapters 7, 8, and 12. This allows the use of fairly simple, (terrestrially) realistic representations of gravity. Much of what follows is, however, applicable to planets of much stronger ellipsoidal oblateness. In particular, the equation sets of Section 18.3.4 are valid for such planets provided that an appropriate model of gravity, valid for stronger oblateness, is used.[15]

18.3.1 A Reference Geopotential Surface

In the derivation of the 'basic' shallow-water equations, an axially symmetric *reference geopotential surface*, $\Phi_S\left(\xi_3 = \xi_3^0\right) \equiv \Phi_S^0 =$ constant, is assumed; see (18.17). It is also assumed that this surface is representative of that of the planet, and that it corresponds to $\xi_3 = \xi_3^0 =$ constant. In what follows, this reference surface is further assumed to be (precisely) *ellipsoidal* in form.[16] As discussed in Section 12.2, an important practical advantage of using an ellipsoidal geopotential reference surface is that this is consistent with use of the WGS 84 reference ellipsoid for reporting the locations of atmospheric and oceanic observational data for Earth.

In a geocentric, Cartesian reference frame, rotating synchronously about a planet's rotation axis, a point on the planet's assumed, reference, geopotential ellipsoid, having coordinates (x_1, x_2, x_3), satisfies – see (8.15) –

$$\frac{x_1^2 + x_2^2}{a^2} + \frac{x_3^2}{c^2} = 1, \tag{18.62}$$

where a and c are the ellipsoid's semi-major and semi-minor axes in any meridional plane (i.e. any plane that contains the planet's rotation axis).

[14] Conservation of axial angular momentum, however, requires the bottom orography/bathymetry to be axially symmetric, otherwise there is a (physical) bottom torque that has to be taken into account.

[15] The degree of oblateness of a planet is generally related to its rotation rate. All other things being equal, the faster a planet rotates, the more oblate it is likely to be. If it rotates too quickly, however, it may become dynamically unstable and break up due to the centrifugal force in the vicinity of its equator overpowering gravitational attraction there.

[16] Other forms are, however, possible, provided that they are axially symmetric and representative of a planet's shape. For example, the *spheroidal* surface defined in Section 8.2.1 would be suitable. There is no requirement that the reference geopotential surface be symmetric about the equatorial plane, and a 'pear-shaped' surface is permissible (Earth is in fact slightly 'pear shaped'). That said, simple, symmetric planetary shapes enhance analytical tractibility. Indeed, this helps explain the popularity of the classical spherical-geopotential approximation!

φ	$x_1(\lambda,\varphi)$	$x_2(\lambda,\varphi)$	$x_3(\varphi)$	$h_1^S(\varphi)$	$h_2^S(\varphi)$
χ	$\dfrac{ac\cos\lambda\cos\chi}{r_{ac}^{\chi}(\chi)}$	$\dfrac{ac\sin\lambda\cos\chi}{r_{ac}^{\chi}(\chi)}$	$\dfrac{ac\sin\chi}{r_{ac}^{\chi}(\chi)}$	$\dfrac{ac\cos\chi}{r_{ac}^{\chi}(\chi)}$	$\dfrac{ac}{r_{ac}^{\chi}(\chi)}$
θ	$a\cos\lambda\cos\theta$	$a\sin\lambda\cos\theta$	$c\sin\theta$	$a\cos\theta$	$a\left(1-e^2\cos^2\theta\right)^{\frac{1}{2}}$
ϕ	$\dfrac{a^2\cos\lambda\cos\phi}{r_{ac}^{\phi}(\phi)}$	$\dfrac{a^2\sin\lambda\cos\phi}{r_{ac}^{\phi}(\phi)}$	$\dfrac{c^2\sin\phi}{r_{ac}^{\phi}(\phi)}$	$\dfrac{a^2\cos\phi}{r_{ac}^{\phi}(\phi)}$	$\dfrac{a^2c^2}{\left[r_{ac}^{\phi}(\phi)\right]^3}$

Table 18.1 Expressions for Cartesian coordinates $x_1(\lambda,\varphi)$, $x_2(\lambda,\varphi)$, and $x_3(\varphi)$, and map-scale factors $h_1^S(\varphi)$ and $h_2^S(\varphi)$ for the representation of an ellipsoid using axial-orthogonal-curvilinear coordinates (λ,φ), where (generic) meridional coordinate φ is either: (i) χ (*geocentric* latitude); or (ii) θ (*parametric* latitude); or (iii) ϕ (*geographic* latitude). Furthermore, $r_{ac}^{\chi}(\chi)\equiv a\left(1-e^2\cos^2\chi\right)^{\frac{1}{2}}$ and $r_{ac}^{\phi}(\phi)\equiv a\left(1-e^2\sin^2\phi\right)^{\frac{1}{2}}$, where $e^2\equiv\left(a^2-c^2\right)/a^2$ and e is eccentricity.

18.3.2 Horizontal Coordinate Systems

Having defined the ellipsoidal, reference, geopotential surface (18.62), we now need to define an appropriate, horizontal, coordinate system to navigate over this surface in order to apply the 'basic' shallow-water equation set (18.49)–(18.53). In principle, any 2D axial-orthogonal-curvilinear coordinate system (ξ_1,ξ_2) may be used for this. Three illustrative examples of such a coordinate system are given in what follows, with pertinent quantities for them summarised in Table 18.1 for comparative and reference purposes. They are:

1. Geocentric-ellipsoidal coordinates (λ,χ).
2. Parametric-ellipsoidal coordinates (λ,θ).
3. Geographic-ellipsoidal coordinates (λ,ϕ).

Common to all three coordinate systems is zonal (azimuthal) coordinate $\xi_1\equiv\lambda$ (longitude). The distinguishing feature between these three axial-orthogonal-curvilinear coordinate systems is their meridional coordinate (latitude, which varies between $-\pi/2$ at the South Pole and $+\pi/2$ at the North Pole). In what follows, we consider $\xi_2\equiv\varphi$ to be a generic, meridional coordinate that can be set to one of the following:

1. χ (geocentric latitude).
2. θ (parametric latitude).
3. ϕ (geographic latitude).

See Fig. 18.3 for a (grossly exaggerated) graphical depiction of these three latitudes. All three are simply different but equivalent ways of specifying position in the meridional direction (with longitude λ held fixed). For Earth, all three latitudes are quantitatively very close to one another; do not be misled by Fig. 18.3 regarding this. There is less than 12 minutes of arc difference between any two of these latitudes, with maximum difference occurring at a latitude of approximately $\pm 45°$.

18.3.2.1 *Geocentric Ellipsoidal Coordinates* (λ,χ)

Definition (18.62) of the ellipsoidal, reference, geopotential surface in (rotating) Cartesian coordinates may be equivalently rewritten in terms of (rotating) *geocentric-ellipsoidal coordinates* (λ,χ) as (cf. (8.19)–(8.21))

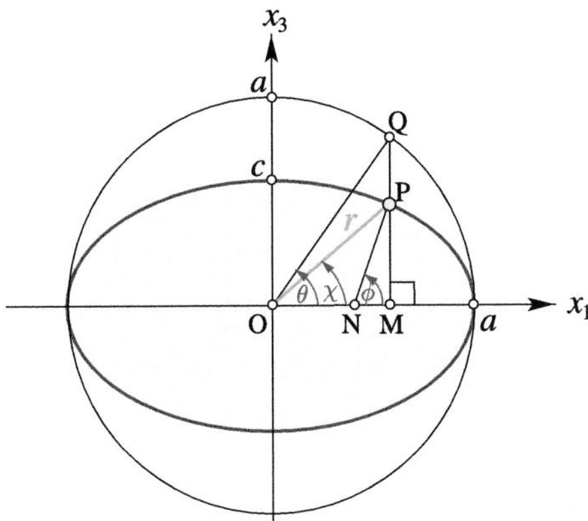

Figure 18.3 A cross section in the Ox_1x_3 meridional plane of an ellipsoid having semi-major and semi-minor axes a and c, respectively, that is inscribed in a circle of radius a. P is a general point on the ellipse with coordinates (x_1, x_3). N is the point at which the normal to the ellipse at the general point P (x_1, x_3) intersects the x_1 axis. θ, χ, and ϕ are *parametric, geocentric,* and *geographic* latitudes, respectively. For a spherical planet these latitudes are identical to one another, but for an ellipsoidal planet they are all distinct. The straight line QPM is parallel to Ox_3 and intersects the circle and Ox_1 at Q and M respectively. The equation of the ellipse is $\left(x_1^2/a^2\right) + \left(x_3^2/c^2\right) = 1$, and $r = \left(x_1^2 + x_3^2\right)^{1/2}$ is the distance of P from the origin O along the green line.

$$x_1 = \frac{ac \cos\lambda \cos\chi}{r_{ac}^{\chi}(\chi)}, \quad x_2 = \frac{ac \sin\lambda \cos\chi}{r_{ac}^{\chi}(\chi)}, \quad x_3 = \frac{ac \sin\chi}{r_{ac}^{\chi}(\chi)}, \tag{18.63}$$

where:

$$r_{ac}^{\chi}(\chi) \equiv \left(a^2 \sin^2\chi + c^2 \cos^2\chi\right)^{\frac{1}{2}} \equiv a\left(1 - e^2 \cos^2\chi\right)^{\frac{1}{2}}, \tag{18.64}$$

$$e^2 \equiv \left(a^2 - c^2\right)/a^2, \tag{18.65}$$

λ is longitude, χ is *geocentric latitude* (see Fig. 18.3), and e is eccentricity. Using (18.15), (18.63), and (18.64), the associated metric factors for geocentric-ellipsoidal coordinates (λ, χ) are

$$h_{\lambda}^{S}(\chi) = \frac{ac\cos\chi}{r_{ac}^{\chi}(\chi)}, \quad h_{\chi}^{S}(\chi) = \frac{ac}{r_{ac}^{\chi}(\chi)}. \tag{18.66}$$

18.3.2.2 *Parametric Ellipsoidal Coordinates* (λ, θ)

Equation (18.62) may be alternatively rewritten in terms of (rotating) *parametric-ellipsoidal coordinates* (λ, θ) as – see (8.22) –

$$x_1 = a\cos\lambda\cos\theta, \quad x_2 = a\sin\lambda\cos\theta, \quad x_3 = c\sin\theta, \tag{18.67}$$

where λ is longitude, and θ is *parametric latitude* (see Fig. 18.3). Using (18.15) and (18.67), the associated metric factors for parametric-ellipsoidal coordinates (λ, θ) are

$$h_{\lambda}^{S}(\theta) = a\cos\theta, \quad h_{\theta}^{S}(\theta) = \left(a^2\sin^2\theta + c^2\cos^2\theta\right)^{\frac{1}{2}} = a\left(1 - e^2\cos^2\theta\right)^{\frac{1}{2}}, \tag{18.68}$$

where e is eccentricity.

18.3.2.3 *Geographic Ellipsoidal Coordinates* (λ, ϕ)

Equation (18.62) may also be alternatively rewritten in terms of (rotating) *geographic-ellipsoidal coordinates* (λ, ϕ) as – see (8.24) and (8.25) –

$$x_1 = \frac{a^2 \cos\lambda \cos\phi}{r_{ac}^{\phi}(\phi)}, \quad x_2 = \frac{a^2 \sin\lambda \cos\phi}{r_{ac}^{\phi}(\phi)}, \quad x_3 = \frac{c^2 \sin\phi}{r_{ac}^{\phi}(\phi)}, \tag{18.69}$$

where

$$r_{ac}^{\phi}(\phi) \equiv \left(a^2 \cos^2\phi + c^2 \sin^2\phi\right)^{\frac{1}{2}} = a\left(1 - e^2 \sin^2\phi\right)^{\frac{1}{2}}, \tag{18.70}$$

λ is longitude, ϕ is *geographic latitude* (see Fig. 18.3), and e is eccentricity. Using (18.15) and (18.69), the associated metric factors for geographic-ellipsoidal coordinates (λ, ϕ) are

$$h_\lambda^S(\phi) = \frac{a^2 \cos\phi}{r_{ac}^{\phi}(\phi)}, \quad h_\phi^S(\phi) = \frac{a^2 c^2}{\left[r_{ac}^{\phi}(\phi)\right]^3}. \tag{18.71}$$

18.3.3 Models of Gravity

It remains to define some suitable models (i.e. representations) of gravity. Guided by the analyses of Chapters 7 and 8, a quintet of models of gravity, of varied simplicity and accuracy, are now assembled. For all five models, *the planet's reference surface coincides with a geopotential surface, as assumed in the derivation of the 'basic' shallow-water equations* (18.49)–(18.53). Respect of this important constraint is therefore considered to be *essential* in the present context.

18.3.3.1 *Ellipsoidal Planet with Variable Density and Realistic Meridional Variation of Gravity*

For mildly oblate, ellipsoidal planets, such as Earth, an accurate model of gravity, having realistic meridional variation, is the one developed in Section 8.5. Thus – from (8.101) and (8.102) –

$$g(\varphi) = g_S^E\left[1 + \left(\frac{5m}{2} - \widetilde{\varepsilon}\right)\sin^2\varphi\right], \tag{18.72}$$

at the planet's boundary, where

$$g_S^E = \frac{\gamma M}{a^2}\left(1 + \widetilde{\varepsilon} - \frac{3m}{2}\right) \tag{18.73}$$

is the constant value of g around the planet's equator. Furthermore:

- φ is generic latitude, that is, one of geocentric (χ), parametric (θ), or geographic (φ) latitudes.
- $m \equiv \Omega^2 a^3/(\gamma M)$.
- γ is universal gravitation constant.
- M is total planetary mass.
- $2\widetilde{\varepsilon} \equiv \left(a^2 - c^2\right)/c^2 \equiv a^2 e^2/c^2$, where e is *eccentricity*.

For Earth, $\widetilde{\varepsilon} \approx m \approx 1/300$ – see Table 7.1 – and both can be considered small. Although (18.72) was developed in terms of geocentric latitude (χ) as an $O(\widetilde{\varepsilon}, m)$ approximation for mildly oblate planets, it also holds when φ is parametric latitude (θ) or geographic latitude (ϕ). This, as previously mentioned, is because the difference between any two of these three latitudes is of $O\left(\widetilde{\varepsilon}^2\right)$ and therefore negligible.

From (18.72), Clairaut's fraction is – see (8.104) –

$$\frac{g_S^P - g_S^E}{g_S^E} = \frac{5m}{2} - \widetilde{\varepsilon}, \tag{18.74}$$

where

$$g_S^E \equiv g\left(\varphi = 0\right), \quad g_S^P \equiv g\left(\varphi = \frac{\pi}{2}\right), \tag{18.75}$$

and g_S^P is the value of gravity at the planet's poles.

This model of gravity can be considered in the present context to be the 'gold standard'. Four simpler, less accurate, models, developed in Chapters 7 and 8, and drawn from the literature, are assembled here. Formally, they can all be functionally expressed as special forms of the 'gold standard' one of (18.72).

18.3.3.2 *Ellipsoidal Planet with Constant Density*

It was found in Section 8.3 that for a mildly oblate, ellipsoidal planet of *constant density*, the resulting gravity at the planet's surface is given by (8.69) and (8.70). Thus

$$g\left(\varphi\right) = g_S^E\left(1 + \widetilde{\varepsilon}\sin^2\chi\right) = g_S^E\left(1 + \frac{5m}{4}\sin^2\chi\right), \tag{18.76}$$

where

$$g_S^E \equiv g\left(\varphi = 0\right) = \frac{\gamma M}{a^2}\left(1 - \frac{\widetilde{\varepsilon}}{5}\right) = \frac{\gamma M}{a^2}\left(1 - \frac{m}{4}\right). \tag{18.77}$$

This model of gravity corresponds to Newton's, for which $\widetilde{\varepsilon} = 5m/4$ in order for the planet's boundary to coincide with a geopotential surface; see (8.62). From (18.76), Clairaut's fraction is then – see (8.72) –

$$\frac{g_S^P - g_S^E}{g_S^E} = \widetilde{\varepsilon} = \frac{5m}{4}. \tag{18.78}$$

Although (18.76)–(18.78) were developed (in Section 8.3) independently of (18.72)–(18.74) (in Section 8.5), it is seen that formally setting $\widetilde{\varepsilon} = 5m/4$ in these latter three equations leads to (18.76)–(18.78). Thus, formally, the present case can be considered to be a special case of the more general one.

18.3.3.3 *Ellipsoidal Planet with Constant Gravity*

White and Inverarity (2012) investigated the possibility of using horizontal, *geodetic* coordinate surfaces to define geopotential surfaces (and thereby gravity); see Sections 5.2.4 and 12.2.5 herein for the definition of geodetic coordinates. Because, by definition, any two horizontal geodetic surfaces are uniformly spaced apart in the vertical, gravity is constant everywhere on any horizontal geodetic surface, including the one at a planet's boundary. Thus

$$g\left(\varphi\right) = g_S^E = \text{constant}. \tag{18.79}$$

Now the only horizontal, geodetic coordinate surface that is precisely *ellipsoidal* in shape is the one that represents the planet's boundary (White and Inverarity, 2012). In the present context, we only need to represent gravity at a planet's boundary, so the departure of geopotential surfaces elsewhere from ellipsoidal is of no concern here. This means that we can formally bring the present model of gravity (at a planet's boundary) under the mathematical umbrella of the model described in Section 18.3.3.1. This is achieved by setting $\widetilde{\varepsilon} = 5m/2$ in (18.72)–(18.74), leading to (18.79) and to

$$g_S^E = \frac{\gamma M}{a^2}\left(1 + m\right) = \text{constant}, \quad \frac{g_S^P - g_S^E}{g_S^E} = 0. \tag{18.80}$$

#	Shape	Characterisation	$\dfrac{g(\varphi)}{g_S^E}$	$\dfrac{g_S^E}{\gamma M/a^2}$	m	$\widetilde{\varepsilon}$	$\dfrac{g_S^P - g_S^E}{g_S^E}$
1	Ellipsoid	Realistic gravity/ Variable density	$1+\left(\dfrac{5m}{2}-\widetilde{\varepsilon}\right)\sin^2\varphi$	$1+\widetilde{\varepsilon}-\dfrac{3m}{2}$	m	$\widetilde{\varepsilon}$	$\dfrac{5m}{2}-\widetilde{\varepsilon}$
2	Ellipsoid	Constant density	$1+\dfrac{5m}{4}\sin^2\varphi$	$1-\dfrac{m}{4}$	m	$\dfrac{5m}{4}$	$\dfrac{5m}{4}$
3	Ellipsoid	Constant gravity	1	$1+m$	m	$\dfrac{5m}{2}$	0
4	Sphere	Variable density	$1+\dfrac{5m}{2}\sin^2\varphi$	$1-\dfrac{3m}{2}$	m	0	$\dfrac{5m}{2}$
5	Sphere	Classical	1	1	0	0	0

Table 18.2 Five models of gravity for mildly oblate, ellipsoidal, and spherical planets, with the planet's bounding surface coinciding with a geopotential surface. The last model, characterised as 'Classical', corresponds to the classical spherical-geopotential approximation. Clairaut's fraction $\equiv \left(g_S^P - g_S^E\right)/g_S^E$. See text for further details.

18.3.3.4 *Spherical Planet with Variable Density*

For a spherical planet with variable density, it was shown in Section 7.8 that – see (7.110) and (7.111) –

$$g(\varphi) = g_S^E\left(1+\frac{5m}{2}\sin^2\varphi\right), \tag{18.81}$$

where

$$g_S^E \equiv g(\varphi=0) = \frac{\gamma M}{a^2}\left(1-\frac{3m}{2}\right). \tag{18.82}$$

From (18.75) and (18.81), Clairaut's fraction is then – see (7.113) –

$$\frac{g_S^P - g_S^E}{g_S^E} = \frac{5m}{2}. \tag{18.83}$$

Setting $\widetilde{\varepsilon} \equiv 0$ in (18.72)–(18.74), that is, considering the special case of the ellipsoid being a sphere, recovers (18.81)–(18.83).

18.3.3.5 *Spherical Planet with the Classical Spherical Geopotential Approximation*

As discussed in Section 8.6, the classical spherical-geopotential approximation is not physically realisable. It is instead an *asymptotic limit* of the one given in Section 18.3.3.1. Setting $\widetilde{\varepsilon} \equiv m \equiv 0$ in (18.72)–(18.74) gives

$$g = g_S^E = \frac{\gamma M}{a^2} = \text{constant} \quad \Rightarrow \quad \frac{g_S^P - g_S^E}{g_S^E} = 0. \tag{18.84}$$

18.3.3.6 *Summary Table*

For convenience and comparative purposes, the quintet of models of gravity are summarised in Table 18.2.

18.3.3.7 *Two Excluded Spherical Models of Gravity*

As now discussed, two particular models of gravity for spherical planets have been excluded from the quintet of gravity models. Neither of them fit into the preceding mathematical and physical framework.

A Spherical Planet of Constant Density

The first such model is that developed in Section 7.6 for a spherical planet of constant density. It was found there – see (7.39) and (7.40) evaluated at $r = a$ – that the geopotential at the planet's spherical boundary (at $r = a$) varies as

$$\Phi_S(\chi) = -\frac{\gamma M}{a}\left(1 + \frac{m}{2}\cos^2\chi\right). \tag{18.85}$$

For $m \neq 0$ (i.e. for a rotating planet) this means that $\Phi_S(\chi)$ cannot be constant on the planet's boundary. In other words, *the planet's boundary cannot coincide with a geopotential surface*. However, in the present context of the development of the shallow-water equations *it is assumed that the planet's boundary is a geopotential surface*. For consistency reasons, this means that this model of gravity is not suitable for use in the shallow-water equations (except for the very special case of a non-rotating planet, which implies $m \equiv 0$).

A Spherical Planet with Variable Density and 'Realistic Meridional Variation of Gravity'

The second excluded model was introduced in Bénard (2015) and referenced in Staniforth and White (2015b) as being of interest. This model of gravity – for a *spherical* planet of variable density – is defined to have *exactly the same* quantitative, meridional variation of gravity[17] as the realistic one described in Section 18.3.3.1 for an *ellipsoidal* planet.[18] Although this 'realistic' representation of gravity for a *spherical* planet (as opposed to an ellipsoidal one) appears to be an attractive proposition, is it physically realisable? The following argument demonstrates that it is not, and corrects a misleading, incidental comment[19] to the contrary in Section 4.6 of Staniforth and White (2015b).

For a *spherical planet of variable density*, it was shown in Section 7.8 that it is possible to obtain a model of gravity *outside* the planet such that the planet's boundary coincides with a geopotential surface. For this model, gravity at the planet's boundary then varies according to (18.81) and (18.82); compare (7.110) and (7.111). It was also shown in Section 7.8 that – *for a spherical planet* – the constraint that *the planet's boundary coincides with a geopotential surface* means that there is *only one way* to represent gravity *outside* the planet. Consequently gravity at the planet's spherical boundary can only vary as (18.81) and therefore *cannot* vary as (18.72). Thus, by this argument, gravity cannot *physically* vary 'realistically' according to (18.72) at a planet's assumed spherical *geopotential* boundary (at $r = a$).[20]

17 Earth's physical values (of approximately 1/300) for $\widetilde{\varepsilon}$ and m are simply inserted into (18.72) when applying this representation.

18 The underlying motivation is to be able to compare two integrations of the shallow-water equations using a realistic (observed) variation of gravity, where the only difference between the two integrations is that one is for a spherical planet, the other for an ellipsoidal one. The impact of only changing the shape of the planet (whilst holding fixed its meridional variation of gravity) can then, in principle, be quantified.

19 The specific comment is 'Although this situation is physically realizable, . . . '.

20 The earlier derivation of the 'basic' shallow-water equations explicitly assumes that (for the purposes of representing gravity) a planet's boundary coincides with a geopotential surface. Consequently a model of gravity that violates this condition is *inconsistent* with the assumptions that lead to the 'basic' shallow-water equations (18.49)–(18.53).

18.3.4 The 'Basic' Shallow-Water Equations in Three Ellipsoidal Coordinate Systems

Having discussed how to close the formulation of the 'basic' shallow-water equations (18.49)–(18.53) by defining a reference geopotential surface (in Section 18.3.1), three horizontal coordinate systems (in Section 18.3.2), and five models of gravity (in Section 18.3.3), we now explicitly express this equation set in terms of:

1. Geocentric-ellipsoidal coordinates (λ, χ).
2. Parametric-ellipsoidal coordinates (λ, θ).
3. Geographic-ellipsoidal coordinates (λ, ϕ).

18.3.4.1 *Geocentric Ellipsoidal Coordinates* (λ, χ)

Introducing metric factors (18.66) into (18.49)–(18.53), the 'basic' shallow-water equations in geocentric-ellipsoidal coordinates (λ, χ) may be written as

$$\frac{D_{\text{hor}} u_\lambda}{Dt} - \left(\frac{r_{ac}^{\chi} u_\lambda}{ac \cos\chi} + 2\Omega \right) \frac{a^2}{\left(r_{ac}^{\chi}\right)^2} u_\chi \sin\chi + \frac{r_{ac}^{\chi}}{ac\cos\chi} \frac{\partial}{\partial\lambda} \left[g(\chi) H(\lambda, \chi, t) \right] = 0, \tag{18.86}$$

$$\frac{D_{\text{hor}} u_\chi}{Dt} + \left(\frac{r_{ac}^{\chi} u_\lambda}{ac \cos\chi} + 2\Omega \right) \frac{a^2}{\left(r_{ac}^{\chi}\right)^2} u_\lambda \sin\chi + \frac{r_{ac}^{\chi}}{ac} \frac{\partial}{\partial\chi} \left[g(\chi) H(\lambda, \chi, t) \right] = 0, \tag{18.87}$$

$$\frac{D_{\text{hor}} \widetilde{H}(\lambda, \chi, t)}{Dt} + \widetilde{H} \frac{\left(r_{ac}^{\chi}\right)^2}{ac\cos\chi} \left[\frac{\partial}{\partial\lambda} \left(\frac{u_\lambda}{r_{ac}^{\chi}} \right) + \frac{\partial}{\partial\chi} \left(\frac{u_\chi \cos\chi}{r_{ac}^{\chi}} \right) \right] = 0, \tag{18.88}$$

or

$$\frac{\partial \widetilde{H}(\lambda, \chi, t)}{\partial t} + \frac{\left(r_{ac}^{\chi}\right)^2}{ac\cos\chi} \left[\frac{\partial}{\partial\lambda} \left(\frac{u_\lambda}{r_{ac}^{\chi}} \widetilde{H} \right) + \frac{\partial}{\partial\chi} \left(\frac{u_\chi \cos\chi}{r_{ac}^{\chi}} \widetilde{H} \right) \right] = 0, \tag{18.89}$$

where

$$\frac{D_{\text{hor}}}{Dt} \equiv \frac{\partial}{\partial t} + \frac{r_{ac}^{\chi} u_\lambda}{ac\cos\chi} \frac{\partial}{\partial\lambda} + \frac{r_{ac}^{\chi} u_\chi}{ac} \frac{\partial}{\partial\chi}, \tag{18.90}$$

$$r_{ac}^{\chi}(\chi) \equiv \left(a^2 \sin^2\chi + c^2 \cos^2\chi\right)^{\frac{1}{2}} \equiv a\left(1 - e^2\cos^2\chi\right)^{\frac{1}{2}}. \tag{18.91}$$

18.3.4.2 *Parametric Ellipsoidal Coordinates* (λ, θ)

Introducing metric factors (18.68) into (18.49)–(18.53), the 'basic' shallow-water equations in parametric-ellipsoidal coordinates (λ, θ) may be written as

$$\frac{D_{\text{hor}} u_\lambda}{Dt} - \left(\frac{u_\lambda}{a\cos\theta} + 2\Omega \right) \frac{u_\theta \sin\theta}{\left(1 - e^2\cos^2\theta\right)^{\frac{1}{2}}} + \frac{1}{a\cos\theta} \frac{\partial}{\partial\lambda} \left[g(\theta) H(\lambda, \theta, t) \right] = 0, \tag{18.92}$$

$$\frac{D_{\text{hor}} u_\theta}{Dt} + \left(\frac{u_\lambda}{a\cos\theta} + 2\Omega \right) \frac{u_\lambda \sin\theta}{\left(1 - e^2\cos^2\theta\right)^{\frac{1}{2}}} + \frac{1}{a\left(1 - e^2\cos^2\theta\right)^{\frac{1}{2}}} \frac{\partial}{\partial\theta} \left[g(\theta) H(\lambda, \theta, t) \right] = 0, \tag{18.93}$$

$$\frac{D_{\text{hor}} \widetilde{H}(\lambda, \theta, t)}{Dt} + \frac{\widetilde{H}}{a\cos\theta} \left[\frac{\partial u_\lambda}{\partial\lambda} + \frac{1}{\left(1 - e^2\cos^2\theta\right)^{\frac{1}{2}}} \frac{\partial}{\partial\theta} \left(\cos\theta\, u_\theta\right) \right] = 0, \tag{18.94}$$

or

$$\frac{\partial \widetilde{H}(\lambda,\theta,t)}{\partial t}+\frac{1}{a\cos\theta}\left[\frac{\partial}{\partial\lambda}\left(u_\lambda\widetilde{H}\right)+\frac{1}{\left(1-e^2\cos^2\theta\right)^{\frac{1}{2}}}\frac{\partial}{\partial\theta}\left(\cos\theta u_\theta\widetilde{H}\right)\right]=0, \tag{18.95}$$

where

$$\frac{D_{\text{hor}}}{Dt}\equiv\frac{\partial}{\partial t}+\frac{u_\lambda}{a\cos\theta}\frac{\partial}{\partial\lambda}+\frac{u_\theta}{a\left(1-e^2\cos^2\theta\right)^{\frac{1}{2}}}\frac{\partial}{\partial\theta}. \tag{18.96}$$

18.3.4.3 *Geographic Ellipsoidal Coordinates* (λ,ϕ)

Introducing metric factors (18.71) into (18.49)–(18.53), the 'basic' shallow-water equations in geographic-ellipsoidal coordinates (λ,θ) may be written as

$$\frac{D_{\text{hor}}u_\lambda}{Dt}-\left(\frac{r_{ac}^{\phi}u_\lambda}{a^2\cos\phi}+2\Omega\right)u_\phi\sin\phi+\frac{r_{ac}^{\phi}}{a^2\cos\phi}\frac{\partial}{\partial\lambda}\left[g(\phi)H(\lambda,\phi,t)\right]=0, \tag{18.97}$$

$$\frac{D_{\text{hor}}u_\phi}{Dt}+\left(\frac{r_{ac}^{\phi}u_\lambda}{a^2\cos\phi}+2\Omega\right)u_\lambda\sin\phi+\frac{\left(r_{ac}^{\phi}\right)^3}{a^2c^2}\frac{\partial}{\partial\phi}\left[g(\phi)H(\lambda,\phi,t)\right]=0, \tag{18.98}$$

$$\frac{D_{\text{hor}}\widetilde{H}(\lambda,\phi,t)}{Dt}+\widetilde{H}\frac{r_{ac}^{\phi}}{a}\left[\frac{1}{a\cos\phi}\frac{\partial u_\lambda}{\partial\lambda}+\frac{\left(r_{ac}^{\phi}\right)^2}{ac^2}\frac{\partial u_\phi}{\partial\phi}-\frac{\tan\phi}{a}u_\phi\right]=0, \tag{18.99}$$

or

$$\frac{\partial\widetilde{H}(\lambda,\phi,t)}{\partial t}+\frac{r_{ac}^{\phi}}{a}\left[\frac{1}{a\cos\phi}\frac{\partial\left(u_\lambda\widetilde{H}\right)}{\partial\lambda}+\frac{\left(r_{ac}^{\phi}\right)^2}{ac^2}\frac{\partial\left(u_\phi\widetilde{H}\right)}{\partial\phi}-\frac{\tan\phi}{a}\left(u_\phi\widetilde{H}\right)\right]=0, \tag{18.100}$$

where

$$\frac{D_{\text{hor}}}{Dt}\equiv\frac{\partial}{\partial t}+\frac{r_{ac}u_\lambda}{a^2\cos\phi}\frac{\partial}{\partial\lambda}+\frac{\left(r_{ac}^{\phi}\right)^3}{a^2c^2}\frac{u_\phi}{}\frac{\partial}{\partial\phi}, \tag{18.101}$$

$$r_{ac}^{\phi}(\phi)\equiv\left(a^2\cos^2\phi+c^2\sin^2\phi\right)^{\frac{1}{2}}=a\left(1-e^2\sin^2\phi\right)^{\frac{1}{2}}. \tag{18.102}$$

18.3.5 Spherical-Polar Coordinates

The shallow-water equations in spherical-polar coordinates, with the classical spherical-geopotential approximation and constant g, may be obtained from any of the preceding three equations sets in ellipsoidal coordinates. Thus setting $\varepsilon = m = 0$ in (18.72) and (18.73); and setting $c = a \Rightarrow e = 0$ in either (18.86)–(18.91), or (18.92)–(18.96), or (18.97)–(18.102) yields

$$\frac{D_{\text{hor}}u_\lambda}{Dt}-\left(\frac{u_\lambda}{a\cos\varphi}+2\Omega\right)u_\varphi\sin\varphi+\frac{1}{a\cos\varphi}\frac{\partial}{\partial\lambda}\left[gH(\lambda,\varphi,t)\right]=0, \tag{18.103}$$

$$\frac{D_{\text{hor}}u_\varphi}{Dt}+\left(\frac{u_\lambda}{a\cos\varphi}+2\Omega\right)u_\lambda\sin\varphi+\frac{1}{a}\frac{\partial}{\partial\varphi}\left[gH(\lambda,\varphi,t)\right]=0, \tag{18.104}$$

$$\frac{D_{\text{hor}}\widetilde{H}(\lambda,\varphi,t)}{Dt}+\widetilde{H}\frac{1}{a\cos\varphi}\left[\frac{\partial u_\lambda}{\partial\lambda}+\frac{\partial}{\partial\varphi}\left(u_\varphi\cos\varphi\right)\right]=0, \tag{18.105}$$

or

$$\frac{\partial\widetilde{H}(\lambda,\varphi,t)}{\partial t}+\frac{1}{a\cos\varphi}\left[\frac{\partial}{\partial\lambda}\left(u_\lambda\widetilde{H}\right)+\frac{\partial}{\partial\varphi}\left(u_\varphi\cos\varphi\widetilde{H}\right)\right]=0, \tag{18.106}$$

where φ is generic latitude (i.e. indifferently one of χ, θ, and ϕ since they are identical in spherical geometry), and

$$\frac{D_{\text{hor}}}{Dt}\equiv\frac{\partial}{\partial t}+\frac{u_\lambda}{a\cos\varphi}\frac{\partial}{\partial\lambda}+\frac{u_\varphi}{a}\frac{\partial}{\partial\varphi}. \tag{18.107}$$

18.4 LAGRANGIAN DENSITY FOR THE BASIC SHALLOW-WATER EQUATIONS

To prepare the way for the development of the 'quasi-shallow' enhancement of the 'basic' shallow-water equations and to provide some insight, we now derive the Lagrangian density for the latter equations. We do so in two different but complementary ways from:

1. Mathematical manipulation of the basic shallow-water equations (in Section 18.4.1).
2. First physical principles (in Section 18.4.2).

The first way is entirely mathematical, with the physical principles already built into the 'basic' shallow-water equations. These equations are assumed to be known. This is true in the present context, since they have already been derived, but in general, it is, of course, a limitation of this approach. An advantage of this way of doing things is that no approximation is involved (beyond that already made to derive the 'basic' shallow-water equations in the first place). This then leads to the 'right answer' for the associated Lagrangian density. It thus provides a useful cross-check on what comes out of the second way of doing things (from first physical principles). A disadvantage of the first way, however, is that it does require some experience and insight to know exactly what manipulations of the 'basic' shallow-water equations will successfully lead to the end result.

The second way – from first physical principles – does not assume that the 'basic' shallow-water equations are known. Quite the contrary; the Lagrangian density is constructed from physical principles as a means of obtaining the 'basic' equation set through subsequent use of the Euler–Lagrange equations. This approach is based on a vertical averaging of kinetic and gravitational potential energies over the depth of the fluid (from the lower rigid boundary to the upper free-surface one). Proceeding in this way then provides a bridge to the derivation of the *quasi-shallow enhancement* of the 'basic' shallow-water equations via vertical averaging. The second way also provides some insight into why the manipulations of the first way of doing things are done in the manner shown.

18.4.1 Derivation from the Basic Shallow-Water Equations

Recall that component equations (18.42) and (18.43) for horizontal momentum of the 'basic' shallow-water equations in axial-orthogonal-curvilinear coordinates are

$$\frac{D_{\text{hor}}u_1}{Dt}+\left(\frac{u_1}{h_1^S}+2\Omega\right)\frac{u_2}{h_2^S}\frac{dh_1^S}{d\xi_2}+\frac{1}{h_1^S}\frac{\partial}{\partial\xi_1}\left[g(\xi_2)H(\xi_1,\xi_2,t)\right]=0, \tag{18.108}$$

$$\frac{D_{\text{hor}}u_2}{Dt}-\left(\frac{u_1}{h_1^S}+2\Omega\right)\frac{u_1}{h_2^S}\frac{dh_1^S}{d\xi_2}+\frac{1}{h_2^S}\frac{\partial}{\partial\xi_2}\left[g(\xi_2)H(\xi_1,\xi_2,t)\right]=0. \tag{18.109}$$

We now rewrite these two component equations in the form of the two Euler–Lagrange equations

$$\frac{D_{\text{hor}}}{Dt}\left(\frac{\partial \widehat{\mathscr{L}}}{\partial \dot{\xi}_i}\right) - \frac{\partial \widehat{\mathscr{L}}}{\partial \xi_i} = -\frac{1}{\overline{\rho}}\frac{\partial \widehat{p}}{\partial \xi_i}, \quad i = 1,2, \tag{18.110}$$

for some Lagrangian $\widehat{\mathscr{L}} = \widehat{\mathscr{L}}\left(\xi_1, \xi_2, \dot{\xi}_1, \dot{\xi}_2, t\right)$, and pressure $\widehat{p} = \widehat{p}\left(\xi_1, \xi_2, t\right)$, both of which are to be determined. To do so, the standard mathematical procedure used in Appendix A of White et al. (2005) – and also previously in Müller (1989) (following K. Hinkelmann) and Zdunkowski and Bott (2003) – is adapted to the present context. This approach assumes that component equations (18.108) and (18.109) are already known; for example, from the earlier Eulerian derivation. The object of the present exercise is thus to *deduce* the associated Lagrangian density ($\widehat{\mathscr{L}}$) and pressure ($\widehat{p}$) fields from them.

Now specific (relative), horizontal, kinetic energy is defined by

$$\mathscr{K} \equiv \frac{u_1^2 + u_2^2}{2} = \frac{\left(h_1^2\right)^S \dot{\xi}_1^2 + \left(h_2^2\right)^S \dot{\xi}_2^2}{2} = \frac{\left[h_1^2\left(\xi_2\right)\right]^S \dot{\xi}_1^2 + \left[h_2^2\left(\xi_2\right)\right]^S \dot{\xi}_2^2}{2}, \tag{18.111}$$

where – from (18.7) evaluated using metric factors (18.15) –

$$\left(u_1, u_2\right) = \left(h_1^S \frac{D\xi_1}{Dt}, h_2^S \frac{D\xi_2}{Dt}\right) = \left(h_1^S \dot{\xi}_1, h_2^S \dot{\xi}_2\right) = \left[h_1^S\left(\xi_2\right)\dot{\xi}_1, h_2^S\left(\xi_2\right)\dot{\xi}_2\right]. \tag{18.112}$$

Successively differentiating (18.111) with respect to $\dot{\xi}_1$ and $\dot{\xi}_2$, and using (18.112), then allows u_1 and u_2 to be expressed in terms of $\mathscr{K}$ as

$$\left(u_1, u_2\right) = \left(\frac{1}{h_1^S}\frac{\partial \mathscr{K}}{\partial \dot{\xi}_1}, \frac{1}{h_2^S}\frac{\partial \mathscr{K}}{\partial \dot{\xi}_2}\right). \tag{18.113}$$

Taking the (shallow, horizontal) material derivative of (18.113) with use of (18.33); using (shallow) metric factors (18.15); and reusing (18.113) yields

$$\begin{aligned}\frac{D_{\text{hor}}u_1}{Dt} &= \frac{D_{\text{hor}}}{Dt}\left(\frac{1}{h_1^S}\frac{\partial \mathscr{K}}{\partial \dot{\xi}_1}\right) = \frac{1}{h_1^S}\frac{D_{\text{hor}}}{Dt}\left(\frac{\partial \mathscr{K}}{\partial \dot{\xi}_1}\right) - \frac{1}{\left(h_1^S\right)^2}\frac{D_{\text{hor}}h_1^S}{Dt}\frac{\partial \mathscr{K}}{\partial \dot{\xi}_1} \\ &= \frac{1}{h_1^S}\frac{D_{\text{hor}}}{Dt}\left(\frac{\partial \mathscr{K}}{\partial \dot{\xi}_1}\right) - \frac{u_1}{h_1^S}\frac{u_2}{h_2^S}\frac{dh_1^S}{d\xi_2},\end{aligned} \tag{18.114}$$

$$\begin{aligned}\frac{D_{\text{hor}}u_2}{Dt} &= \frac{D_{\text{hor}}}{Dt}\left(\frac{1}{h_2^S}\frac{\partial \mathscr{K}}{\partial \dot{\xi}_2}\right) = \frac{1}{h_2^S}\frac{D_{\text{hor}}}{Dt}\left(\frac{\partial \mathscr{K}}{\partial \dot{\xi}_2}\right) - \frac{1}{\left(h_2^S\right)^2}\frac{D_{\text{hor}}h_2^S}{Dt}\frac{\partial \mathscr{K}}{\partial \dot{\xi}_2} \\ &= \frac{1}{h_2^S}\frac{D_{\text{hor}}}{Dt}\left(\frac{\partial \mathscr{K}}{\partial \dot{\xi}_2}\right) - \frac{u_2^2}{\left(h_2^S\right)^2}\frac{dh_2^S}{d\xi_2}.\end{aligned} \tag{18.115}$$

Substitution of (18.114) and (18.115) into component equations (18.108) and (18.109) of momentum, respectively, followed by multiplication by h_1^S and h_2^S, respectively, then leads to

$$\frac{D_{\text{hor}}}{Dt}\left(\frac{\partial \mathscr{K}}{\partial \dot{\xi}_1}\right) + 2\Omega h_1^S \frac{u_2}{h_2^S}\frac{dh_1^S}{d\xi_2} + \frac{\partial}{\partial \xi_1}\left[g\left(\xi_2\right)H\left(\xi_1, \xi_2, t\right)\right] = 0, \tag{18.116}$$

$$\frac{D_{\text{hor}}}{Dt}\left(\frac{\partial \mathscr{K}}{\partial \dot{\xi}_2}\right) - \left(\frac{u_1^2}{h_1^S}\frac{dh_1^S}{d\xi_2} + \frac{u_2^2}{h_2^S}\frac{dh_2^S}{d\xi_2}\right) - 2\Omega u_1 \frac{dh_1^S}{d\xi_2} + \frac{\partial}{\partial \xi_2}\left[g\left(\xi_2\right)H\left(\xi_1, \xi_2, t\right)\right] = 0. \tag{18.117}$$

Using (18.15) and (18.112), the Coriolis contributions in (18.116) and (18.117) may be rewritten as

$$2\Omega h_1^S \frac{u_2}{h_2^S}\frac{dh_1^S}{d\xi_2} = 2\Omega h_1^S \frac{D_{\text{hor}}h_1^S}{Dt} = \frac{D_{\text{hor}}}{Dt}\left[\Omega\left(h_1^2\right)^S\right] = \frac{D_{\text{hor}}}{Dt}\left\{\frac{\partial}{\partial \dot{\xi}_1}\left[\Omega\left(h_1^2\right)^S \dot{\xi}_1\right]\right\}, \tag{18.118}$$

$$2\Omega u_1 \frac{dh_1^S}{d\xi_2} = 2\Omega h_1^S \dot{\xi}_1 \frac{dh_1^S}{d\xi_2} = \frac{\partial}{\partial \xi_2}\left[\Omega \left(h_1^2\right)^S \dot{\xi}_1\right]. \tag{18.119}$$

Substitution of these two Coriolis contributions into component equations (18.116) and (18.117), respectively, then yields

$$\frac{D_{\text{hor}}}{Dt}\left\{\frac{\partial}{\partial \dot{\xi}_1}\left[\mathscr{K} + \Omega \left(h_1^2\right)^S \dot{\xi}_1\right]\right\} + \frac{\partial}{\partial \xi_1}\left[g\left(\xi_2\right) H\left(\xi_1, \xi_2, t\right)\right] = 0, \tag{18.120}$$

$$\frac{D_{\text{hor}}}{Dt}\left(\frac{\partial \mathscr{K}}{\partial \dot{\xi}_2}\right) - \left(\frac{u_1^2}{h_1^S}\frac{dh_1^S}{d\xi_2} + \frac{u_2^2}{h_2^S}\frac{dh_2^S}{d\xi_2}\right) - \frac{\partial}{\partial \xi_2}\left[\Omega \left(h_1^2\right)^S \dot{\xi}_1\right] + \frac{\partial}{\partial \xi_2}\left[g\left(\xi_2\right) H\left(\xi_1, \xi_2, t\right)\right] = 0. \tag{18.121}$$

Noting that

$$\frac{\partial}{\partial \xi_1}\left[\mathscr{K} + \Omega \left(h_1^2\right)^S \dot{\xi}_1\right] = \frac{\partial}{\partial \xi_1}\left\{\frac{1}{2}\left[\left(h_1^2\right)^S \dot{\xi}_1^2 + \left(h_2^2\right)^S \dot{\xi}_2^2\right] + \Omega \left(h_1^2\right)^S \dot{\xi}_1\right\} = 0, \tag{18.122}$$

$$\frac{\partial}{\partial \dot{\xi}_2}\left[\Omega \left(h_1^2\right)^S \dot{\xi}_1\right] = 0, \tag{18.123}$$

$$g\left(\xi_2\right) H\left(\xi_1, \xi_2, t\right) \equiv g\left(\xi_2\right)\left[B\left(\xi_1, \xi_2\right) + \frac{\widetilde{H}\left(\xi_1, \xi_2, t\right)}{2}\right] + \frac{g\left(\xi_2\right)\widetilde{H}\left(\xi_1, \xi_2, t\right)}{2}, \tag{18.124}$$

component equations (18.120) and (18.121) may be rewritten as

$$\begin{aligned}\frac{D_{\text{hor}}}{Dt}\left\{\frac{\partial}{\partial \dot{\xi}_1}\left[\mathscr{K} + \Omega \left(h_1^2\right)^S \dot{\xi}_1\right]\right\} &- \frac{\partial}{\partial \xi_1}\left\{\mathscr{K} + \Omega \left(h_1^2\right)^S \dot{\xi}_1 - g\left(\xi_2\right)\left[B\left(\xi_1, \xi_2\right) + \frac{\widetilde{H}\left(\xi_1, \xi_2, t\right)}{2}\right]\right\} \\ &= -\frac{\partial}{\partial \xi_1}\left[\frac{g\left(\xi_2\right)\widetilde{H}\left(\xi_1, \xi_2, t\right)}{2}\right],\end{aligned} \tag{18.125}$$

$$\begin{aligned}\frac{D_{\text{hor}}}{Dt}\left\{\frac{\partial}{\partial \dot{\xi}_2}\left[\mathscr{K} + \Omega \left(h_1^2\right)^S \dot{\xi}_1\right]\right\} &- \frac{\partial}{\partial \xi_2}\left\{\mathscr{K} + \Omega \left(h_1^2\right)^S \dot{\xi}_1 - g\left(\xi_2\right)\left[B\left(\xi_1, \xi_2\right) + \frac{\widetilde{H}\left(\xi_1, \xi_2, t\right)}{2}\right]\right\} \\ &= -\frac{\partial}{\partial \xi_2}\left[\frac{g\left(\xi_2\right)\widetilde{H}\left(\xi_1, \xi_2, t\right)}{2}\right],\end{aligned} \tag{18.126}$$

respectively. Since

$$\frac{\partial}{\partial \dot{\xi}_1}\left\{g\left(\xi_2\right)\left[B\left(\xi_1, \xi_2\right) + \frac{\widetilde{H}\left(\xi_1, \xi_2, t\right)}{2}\right]\right\} = \frac{\partial}{\partial \dot{\xi}_2}\left\{g\left(\xi_2\right)\left[B\left(\xi_1, \xi_2\right) + \frac{\widetilde{H}\left(\xi_1, \xi_2, t\right)}{2}\right]\right\} = 0, \tag{18.127}$$

(18.125) and (18.126) may be written in the form of the Euler–Lagrange equations (18.110) by defining

$$\mathscr{L}\left(\xi_1, \xi_2, \dot{\xi}_1, \dot{\xi}_2, t\right) \equiv \underbrace{\frac{\left(h_1^2\right)^S \dot{\xi}_1^2 + \left(h_2^2\right)^S \dot{\xi}_2^2}{2}}_{\text{horizontal KE}} + \underbrace{\Omega \left(h_1^2\right)^S \dot{\xi}_1}_{\text{Coriolis}} \underbrace{- g\left(\xi_2\right)\left[B\left(\xi_1, \xi_2\right) + \frac{\widetilde{H}\left(\xi_1, \xi_2, t\right)}{2}\right]}_{\text{gravity}}, \tag{18.128}$$

$$\widehat{p}\left(\xi_1, \xi_2, t\right) \equiv p^H + \overline{\rho}\frac{g\left(\xi_2\right)\widetilde{H}\left(\xi_1, \xi_2, t\right)}{2}, \tag{18.129}$$

where p^H = constant, and definition (18.111) of $\mathscr{K}$ has been used. This completes the determination of the Lagrangian density ($\widehat{\mathscr{L}}$) and pressure ($\widehat{p}$) that correspond to component equations (18.108) and (18.109) of momentum. The physical significance of the constant value p^H is explained in Section 18.4.2.5.

Aside We may now, if we wish, turn the problem around. Instead of posing the question:

- Given component momentum equations (18.108) and (18.109), what are the corresponding Lagrangian density and pressure fields (with answer (18.128) and (18.129), respectively)?

we can instead pose the question:

- Given Lagrangian density and pressure fields (18.128) and (18.129), respectively, what are the corresponding component momentum equations (with answer (18.108) and (18.109))?

When answering the latter question, we just need to:

1. Compute partial derivatives $\partial\widehat{\mathscr{L}}/\partial\xi_1$, $\partial\widehat{\mathscr{L}}/\partial\xi_2$, $\partial\widehat{\mathscr{L}}/\partial\dot{\xi}_1$, and $\partial\widehat{\mathscr{L}}/\partial\dot{\xi}_2$ from definition (18.128).
2. Insert these four partial derivatives, together with definition (18.129) of $\widehat{p}$, into Euler–Lagrange equations (18.110).

Manipulation of the two resulting Euler–Lagrange equations then leads to component momentum equations (18.108) and (18.109), respectively. Details for this may be found in Section 18.6, with quasi-shallow switch Q set identically zero therein; all quasi-shallow terms (in green) are then absent.

18.4.2 Derivation from First Physical Principles

The preceding derivation of Lagrangian density and pressure for the 'basic' shallow-water equations is entirely mathematical, with some of the manipulations being less than obvious; for example, decomposition (18.124) of $g(\xi_2)H(\xi_1,\xi_2,t)$, with one term on the right-hand side of this equation contributing to Lagrangian density ($\widehat{\mathscr{L}}$), the other to pressure ($\widehat{p}$). Furthermore, the derivation assumes that the component momentum equations are already known.

- But what if they are not?
- Can we instead obtain the Lagrangian density and pressure fields from first physical principles (from which the horizontal component momentum equations may subsequently be obtained using the Euler–Lagrange equations (18.110))?

Indeed we can!

Referring to Fig. 18.4, consider a vertical column of shallow fluid, of constant density $\overline{\rho}$ and of infinitesimal, horizontal cross section $dA \equiv h_1^S(\xi_2)\,h_2^S(\xi_2)\,d\xi_1 d\xi_2$, confined between a rigid lower boundary at $s_3 = B(\xi_1,\xi_2)$, and an upper free boundary at $s_3 = H(\xi_1,\xi_2,t)$. Note that dA is independent of vertical coordinate $s_3 \equiv h_3^S(\xi_2)\left(\xi_3 - \xi_3^0\right)$ – by virtue of (18.15) – and thus the column is of uniform cross section; see also a related discussion in Section 5.3.

18.4.2.1 *The Vertically Averaged Horizontal Kinetic Energy* $\widehat{\mathscr{K}}$

The horizontal (relative) kinetic energy (KE) of the arbitrary volume element (dV, depicted in red) within the fluid column (in medium blue) is $\overline{\rho}dV\left(u_1^2+u_2^2\right)/2 = \overline{\rho}ds_3 dA\left[\left(h_1^2\right)^S\dot{\xi}_1^2 + \left(h_2^2\right)^S\dot{\xi}_2^2\right]/2$. This is simply the elemental mass ($\overline{\rho}dV$), multiplied by specific horizontal KE [$\mathscr{K} \equiv \left(u_1^2+u_2^2\right)/2$]. The total (relative) horizontal KE of the whole column is then the (infinite) sum of all such (infinitesimal) horizontal KEs within the column, that is, it

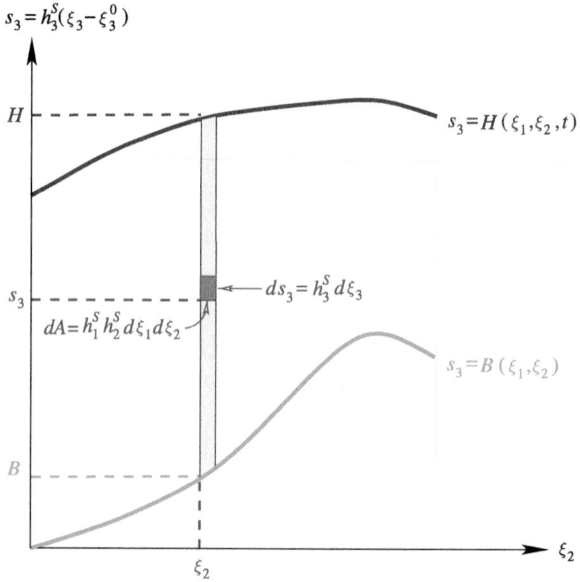

Figure 18.4 Vertical cross section in the $\xi_2 - s_3$ plane of a layer of shallow fluid (depicted in light blue), of constant density $\overline{\rho}$, confined between a lower rigid boundary (solid green line) located at $s_3 = B(\xi_1, \xi_2)$, and an upper free boundary (solid blue line) at $s_3 = H(\xi_1, \xi_2, t)$. Vertical coordinate $s_3 \equiv h_3^S(\xi_3 - \xi_3^0)$ measures vertical distance along a ξ_3 line, with ξ_1 and ξ_2 held fixed, from a reference geopotential surface located at $\xi_3 = \xi_3^0 \Rightarrow s_3 = 0$. A vertical column of shallow water, of infinitesimal horizontal cross section $dA \equiv h_1^S(\xi_2) h_2^S(\xi_2) d\xi_1 d\xi_2$, is shown in medium blue. Within this column, an infinitesimal volume element, $dV \equiv ds_3 dA \equiv h_3^S(\xi_2) d\xi_3 dA \equiv h_1^S(\xi_2) h_2^S(\xi_2) h_3^S(\xi_2) d\xi_1 d\xi_2 d\xi_3$, is depicted in red.

is the vertical integral of horizontal KE over the depth of the fluid column (from $s_3 = s_3^B \equiv B(\xi_1, \xi_2)$ to $s_3 = s_3^H \equiv H(\xi_1, \xi_2, t)$). Now $dA = h_1^S(\xi_2) h_2^S(\xi_2) d\xi_1 d\xi_2$ and (from (18.28)) $u_1 = u_1(\xi_1, \xi_2, t)$ and $u_2 = u_2(\xi_1, \xi_2, t)$, so that the column moves coherently and remains vertical. Exploiting the independence of these three quantities (dA, u_1, and u_2) on vertical coordinate $s_3 \equiv h_3^S(\xi_2)(\xi_3 - \xi_3^0)$ leads to

$$\begin{aligned}\text{Total horizontal KE of the column} &= \int_B^H \overline{\rho} dA \frac{\left(u_1^2 + u_2^2\right)}{2} ds_3 \\ &= \overline{\rho} dA \left[\frac{\left(h_1^2\right)^S \dot{\xi}_1^2 + \left(h_2^2\right)^S \dot{\xi}_2^2}{2}\right] \int_B^H ds_3 \\ &= \overline{\rho} dA \left[\frac{\left(h_1^2\right)^S \dot{\xi}_1^2 + \left(h_2^2\right)^S \dot{\xi}_2^2}{2}\right] \left[H(\xi_1, \xi_2, t) - B(\xi_1, \xi_2)\right],\end{aligned} \tag{18.130}$$

$$\begin{aligned}\text{Total mass of the column} &= \int_B^H \overline{\rho} dA ds_3 = \overline{\rho} dA \int_B^H ds_3 \\ &= \overline{\rho} dA \left[H(\xi_1, \xi_2, t) - B(\xi_1, \xi_2)\right].\end{aligned} \tag{18.131}$$

Thus, dividing (18.130) by (18.131), the horizontal KE of the column, *vertically averaged* over the depth of the fluid ($\widetilde{H} \equiv H - B$) within it, is

$$\widehat{\mathscr{K}} = \frac{\text{Total horizontal KE of the column}}{\text{Total mass of the column}} = \frac{\left(h_1^2\right)^S \dot{\xi}_1^2 + \left(h_2^2\right)^S \dot{\xi}_2^2}{2} = \mathscr{K}; \tag{18.132}$$

cf. (18.111). Note that $\widehat{\mathscr{K}}$ is a *specific* quantity, that is, it is an amount *per unit mass* (as indeed is $\mathscr{K}$).

Because specific horizontal KE is identical *everywhere within the column* – that is, entirely independent of vertical coordinate s_3 – this result could have been immediately written down; the average of a quantity that does not vary over the domain is just the value of the quantity at any point within that domain. So why didn't we just do so? It is because – see, for example, Sections 18.4.2.3 and 18.4.2.5 – we can follow the same procedure for quantities, such as gravitational-potential energy and pressure, that *do* vary within the column as a function of vertical coordinate s_3.

18.4.2.2 *The Vertically Averaged Coriolis Contribution* $\widehat{\Omega h_1^S u_1}$

In the preceding analysis, $\mathscr{K}$ is specific (relative) horizontal KE observed in a *rotating* frame of reference. It originates, however, from specific (absolute) horizontal KE observed in an *inertial* frame of reference. The Coriolis terms in the governing equations also originate from the same specific (absolute) horizontal KE observed in an *inertial* frame of reference, as does the centrifugal force (which contributes to apparent gravity – see Section 2.2.2.3). This results (in shallow geometry) in the appearance of a term

$$\widehat{\Omega h_1^S u_1} = \Omega h_1^S u_1 = \Omega \left(h_1^2\right)^S \dot{\xi}_1, \tag{18.133}$$

in Lagrangian density $\widehat{\mathscr{L}}$; see (18.128). Since $\Omega h_1^S u_1 = \Omega h_1^S(\xi_2)\, u_1(\xi_1, \xi_2, t)$ is independent of vertical coordinate $s_3 \equiv h_3^S(\xi_2)\left(\xi_3 - \xi_3^0\right)$, its vertical average $\widehat{\Omega h_1^S u_1}$ over the depth of the fluid $(H - B)$ is left unchanged as $\Omega h_1^S u_1$; just as vertical average $\widehat{\mathscr{K}}$ is left unchanged as $\mathscr{K}$.

18.4.2.3 *The Vertically Averaged Gravitational-Potential Energy* $\widehat{\Phi}$

The (gravitational-) potential energy (PE) of the arbitrary volume element (shown in red in Fig. 18.4) within the fluid column (in medium blue) is $\left(\overline{\rho} dV\right) g(\xi_2)\, s_3 = \left(\overline{\rho} dA ds_3\right) g(\xi_2)\, s_3$; this is simply the elemental mass $(\overline{\rho} dV)$ multiplied by gravity $(g(\xi_2))$ multiplied by the height (s_3) of the volume element above the reference surface (located at $\xi_3 = \xi_3^0 \Rightarrow s_3 = 0$). The PE of the whole column is then the (infinite) sum of all such (infinitesimal) PEs (i.e. it is the vertical integral of elemental PE over the depth of the fluid column). Thus

$$\begin{aligned}\text{Total PE of the column} &= \int_B^H \overline{\rho} dA g(\xi_2)\, s_3 ds_3 = \overline{\rho} dA g(\xi_2) \int_B^H s_3 ds_3 = \overline{\rho} dA g(\xi_2) \left(\frac{H^2 - B^2}{2}\right) \\ &= \overline{\rho} dA g(\xi_2) \left[H(\xi_1, \xi_2, t) - B(\xi_1, \xi_2)\right] \left[\frac{H(\xi_1, \xi_2, t) + B(\xi_1, \xi_2)}{2}\right],\end{aligned} \tag{18.134}$$

where independence of (shallow) $dA \equiv h_1^S(\xi_2)\, h_2^S(\xi_2)\, d\xi_1 d\xi_2$ on s_3 has again been exploited. This time, the integrand (in (18.134)) is no longer independent of s_3 (as it was in the previous two subsections) but *varies linearly* with respect to vertical coordinate s_3. Because the fluid is assumed to be shallow (with respect to the planet's mean radius), although the vertical variation of PE is non-zero, it is nevertheless very small. (Do not be misled by Fig. 18.4, which only depicts a minuscule window of the actual domain.)

Dividing (18.134) by (18.131) and using definition (18.36) of $\widetilde{H}$, the PE of the column, vertically averaged over the depth of the fluid ($\widetilde{H} \equiv H - B$) within it, is

$$\widehat{\Phi} = \frac{\text{Total PE of the column}}{\text{Total mass of the column}} = g(\xi_2) \left[\frac{H(\xi_1, \xi_2, t) + B(\xi_1, \xi_2)}{2}\right]$$

$$\equiv g(\xi_2)\left[B(\xi_1,\xi_2)+\frac{\widetilde{H}(\xi_1,\xi_2,t)}{2}\right]. \tag{18.135}$$

Note that $\widehat{\Phi}$ is also a *specific* quantity. A physical interpretation of (18.135) is that the vertically averaged PE is the average of the PEs at the lower and upper boundaries. This is a direct consequence of the *linear* vertical variation of Φ within the column and would not, for example, be true if the vertical variation were *quadratic.*

18.4.2.4 *The Vertically Averaged Lagrangian Density $\widehat{\mathscr{L}}$*

The vertically averaged Lagrangian density ($\widehat{\mathscr{L}}$) is obtained by summing (18.132) and (18.133) (for the horizontal kinetic-energy contributions $\widehat{\mathscr{K}}$ and $\widehat{\Omega h_1^S u_1}$ in an inertial frame of reference – with the centrifugal contribution accounted for in apparent gravity $g = g(\xi_2)$), and subtracting (18.135) (for the potential-gravitational energy contribution $\widehat{\Phi}$) from the result. This leads to:

The Vertically Averaged Lagrangian Density for the 'Basic' Shallow-Water Equations

$$\widehat{\mathscr{L}}=\widehat{\mathscr{K}}+\widehat{\Omega h_1^S u_1}-\widehat{\Phi}=\underbrace{\frac{\left(h_1^2\right)^S\dot{\xi}_1^2+\left(h_2^2\right)^S\dot{\xi}_2^2}{2}}_{\text{horizontal KE}}+\underbrace{\Omega\left(h_1^2\right)^S\dot{\xi}_1}_{\text{Coriolis}}\underbrace{-g(\xi_2)\left[B(\xi_1,\xi_2)+\frac{\widetilde{H}(\xi_1,\xi_2,t)}{2}\right]}_{\text{gravity}}, \tag{18.136}$$

and (18.128) is recovered. This verifies that derivation of the vertically averaged Lagrangian density from first physical principles is consistent with its mathematical derivation from the governing equations obtained in an Eulerian manner. So far, so very good, but what about the associated, vertically averaged pressure?

18.4.2.5 *The Vertically Averaged Pressure $\widehat{p}$*

The weight of an arbitrary volume element dV (shown in red in Fig. 18.4) of a fluid column with cross-sectional area dA is $\left(\overline{\rho}dV\right)g(\xi_2)=\left(\overline{\rho}dAds_3\right)g(\xi_2)$. This is simply the elemental mass $(\overline{\rho}dV)$ multiplied by gravity $(g(\xi_2))$. The total weight of fluid within the column *above* an areal element dA, located at a height s_3 above the ($s_3 = 0$) reference level, is then the (infinite) sum of the weights of all such infinitesimal volume elements located above s_3. In other words, it is the vertical integral of the fluid's weight confined within the column between $s_3' = s_3$ and $s_3' = H(\xi_1,\xi_2,t)$. Thus

$$\text{Total weight of the column of fluid above } s_3=\int_{s_3}^{H}\overline{\rho}dAg(\xi_2)\,ds_3'=\overline{\rho}dAg(\xi_2)\int_{s_3}^{H}ds_3'$$
$$=\overline{\rho}dAg(\xi_2)\left[H(\xi_1,\xi_2,t)-s_3\right]. \tag{18.137}$$

This column of fluid exerts a pressure on the (horizontal) areal element dA located at height s_3 above the reference level (at $s_3 = 0$). Thus (because pressure is force per unit area)

$$\text{Pressure at height } s \text{ due to the weight of the fluid column above it}$$
$$=\frac{\text{Weight of the fluid column above it}}{\text{Area of the cross section}}$$
$$=\frac{\overline{\rho}\cancel{dA}g(\xi_2)\left[H(\xi_1,\xi_2,t)-s_3\right]}{\cancel{dA}}=\overline{\rho}g(\xi_2)\left[H(\xi_1,\xi_2,t)-s_3\right]. \tag{18.138}$$

It is, however, assumed that a pressure p^H = constant ≥ 0 is externally applied at the fluid's free surface; see (18.29). To obtain the total pressure, acting at height s within the column, this pressure (p^H) must be added to that of (18.138) (which is due to the weight of the fluid alone). Thus

$$p\left(\xi_1, \xi_2, s_3, t\right) = p^H + \overline{\rho} g\left(\xi_2\right)\left[H\left(\xi_1, \xi_2, t\right) - s_3\right]. \tag{18.139}$$

Noting (from (18.23)) that $s_3\left(\xi_1, \xi_2, \xi_3\right) \equiv h_3^S\left(\xi_2\right)\left(\xi_3 - \xi_3^0\right)$, (18.139) recovers (18.41). The reason for this is that when deriving these two equations, the same physical assumptions were made; they were just couched in different language. Equation (18.41) was derived by integrating hydrostatic equation (18.39) downwards from the free surface using boundary condition (18.29). The derivation of (18.139) was expressed in terms of the weight of a column of fluid – this amounts to assuming hydrostatic balance within the fluid column – plus an applied pressure $p^H \geq 0$ at the free surface.[21] The well-known result for the pressure due to the weight of the fluid column (only) is often phrased as:

'*The hydrostatic relationship is equivalent to the statement that the pressure at any point in the fluid is equal to the weight of overlying fluid.*' (Hoskins and James, 2014, p. 90)

Having derived the pressure at a point anywhere within the fluid column of infinitesimal cross section $dA \equiv h_1^S\left(\xi_2\right) h_2^S\left(\xi_2\right) d\xi_1 d\xi_2$, we are now in a position to evaluate the vertical average (over the depth of the fluid column) of the pressure. Thus, integrating (18.139) over the depth of the fluid column and using definition (18.36) of $\widetilde{H}$ gives:

The Vertically Averaged Pressure for the 'Basic' Shallow-Water Equations

$$\begin{aligned}\widehat{p} &\equiv \frac{1}{(H-B)} \int_B^H p\, ds_3 = \frac{1}{(H-B)} \int_B^H \left[p^H + \overline{\rho} g\left(\xi_2\right)\left(H - s_3\right)\right] ds_3 \\ &= p^H + \frac{\overline{\rho} g\left(\xi_2\right)}{(H-B)} \int_B^H \left(H - s_3\right) ds_3 = p^H + \overline{\rho} g\left(\xi_2\right) \frac{(H-B)}{2} \\ &= p^H + \overline{\rho} \frac{g\left(\xi_2\right) \widetilde{H}\left(\xi_1, \xi_2, t\right)}{2}.\end{aligned} \tag{18.140}$$

This equation recovers (18.129) and verifies that derivation of the vertically averaged pressure ($\widehat{p}$) from first physical principles is consistent with its derivation from the governing equations obtained in an Eulerian manner.

18.5 QUASI-SHALLOW ENHANCEMENT OF LAGRANGIAN DENSITY

To respect dynamical consistency, the '$2\Omega \cos\phi$' Coriolis terms must be omitted in the 3D *shallow-fluid* equations – see Chapter 16 – and also therefore in the 2D 'basic' shallow-water equations derived from them. An incomplete representation of the Coriolis force then results in both contexts. The *quasi-shallow* approximation addresses this deficiency in 3D – see Chapter 17 – leading to an (almost) complete representation of the Coriolis force[22] in the context of a

[21] $p^H \equiv 0$ for a free surface underlying a vacuum.

[22] For brevity, Dellar and Salmon (2005), Tort and Dubos (2014a), Tort et al. (2014), and Staniforth (2015a) all characterise the inclusion of the $2\Omega \cos\phi$ terms using this approach as being '*with complete Coriolis force*'. A more precise terminology is '*with almost complete Coriolis force*'.

shallow fluid, *whilst still maintaining dynamical consistency*. The 3D quasi-shallow approximation is more accurate than the shallow one, particularly in the tropics. These considerations motivate the development of an analogous enhancement of the 2D 'basic' shallow-water equations to include a representation of the absent '$2\Omega\cos\phi$' Coriolis terms.

As discussed in Section 17.2.8, the shallow and quasi-shallow assumptions both employ a *shallow* metric, namely (18.15). Furthermore, the planetary-vorticity vector (and thereby the Coriolis terms) are approximated in both the shallow and quasi-shallow equations, and both of these 3D equation sets maintain dynamical consistency. The significant difference between them is the way in which the planetary-vorticity vector is approximated to achieve this important property. This also turns out to be the case for the 2D shallow-water equations.

Three methodologies for deriving dynamically consistent equation sets were listed in Section 16.3. The first one ('classical Eulerian') has been applied in Section 18.2 to derive the '*basic*' shallow-water equation set. In Sections 18.6 and 18.9, we describe use of the other two methodologies – both Lagrangian – to obtain a more general '*quasi-shallow*', shallow-water equation set that includes the 'basic' one as a special case. Increased generality is achieved by endowing the equations – in a dynamically consistent manner – with supplementary terms to *optionally* allow a more complete ('quasi-shallow') representation of the Coriolis force.

We begin by considering how to approximate Lagrangian density to obtain – in Section 18.6 – the more general quasi-shallow, equation set using the Euler–Lagrange equations. We then discuss – in Section 18.9 – an alternative Lagrangian methodology using Hamilton's principle of stationary action.

18.5.1 2D Quasi-Shallow Enhancement of Lagrangian Density

Two derivations of the 2D Lagrangian density for the basic shallow-water equations – one primarily mathematical, the other physical – have been given in Section 18.4. Both lead to the same Lagrangian density (i.e. to (18.136), alias (18.128)) and to the same associated pressure (i.e. to (18.140), alias (18.129)). For the purposes of obtaining the 'quasi-shallow', shallow-water equation set, we now enhance this semi-Lagrangian density with the addition of a term; we also obtain the associated enhanced, quasi-shallow pressure.

Recall from Section 17.3.1 that the *only term* in the 3D, purely shallow, Lagrangian density that needs to be modified for quasi-shallow enhancement is the one associated with planetary vorticity. This is the only term that depends upon the planet's rotation rate ($\mathbf{\Omega}$) and therefore the only one that can ultimately affect the Coriolis terms in the component equations of horizontal momentum. In the present context this strongly suggests that we need to enhance the Coriolis contribution of $\Omega\left(h_1^2\right)^S\dot{\xi}_1$ to the Lagrangian density (18.136) (alias (18.128)) of the 'basic' shallow-water equation set, by adding a term. To do so, we mimic/adapt what is done in Section 17.3.1 for the 3D case.

Now the Coriolis contribution $\Omega\left(h_1^2\right)^S\dot{\xi}_1$ to the *purely shallow* 3D Lagrangian density originates from the corresponding contribution $\Omega h_1^2\dot{\xi}_1$ to the *deep* 3D Lagrangian density. The only simplification made was the purely geometric one of evaluating this term in (purely) shallow geometry using shallow metric factors of the form (18.15). As shown in Section 17.3.1, all that needed to be done to obtain the 3D *quasi-shallow* Lagrangian density was to instead evaluate h_1^2 in its Coriolis contribution ($\Omega h_1^2\dot{\xi}_1$) as the first two terms of a Taylor series expansion in the vertical; that is, to use approximation (17.52). Noting (from (18.23)) that $s_3\left(\xi_1,\xi_2,\xi_3\right)\equiv h_3^S\left(\xi_2\right)\left(\xi_3-\xi_3^0\right)$, quasi-shallow, Taylor-series expansion/approximation (17.52) may be written in the present context as

$$\left(h_1^2\right)^Q=\left(h_1^2\right)^S+Q\left(\frac{\partial h_1^2}{\partial\xi_3}\right)^S\left(\xi_3-\xi_3^0\right)\equiv\left(h_1^2\right)^S+\frac{Q}{h_3^S}\left(\frac{\partial h_1^2}{\partial\xi_3}\right)^S s_3\equiv\left(h_1^2\right)^S+Q\sigma^S\left(\xi_2\right)s_3,\quad(18.141)$$

where (for brevity)

$$\sigma^S(\xi_2) \equiv \frac{1}{h_3^S}\left(\frac{\partial h_1^2}{\partial \xi_3}\right)^S, \tag{18.142}$$

and quasi-shallow dependence is coloured green to facilitate tracking its influence in the subsequent analysis.[23] Quasi-shallow switch Q has been introduced into (18.141) to be able to selectively switch quasi-shallow approximation on or off. When $Q = 0$, it is switched off; doing so then ultimately leads to the 'basic' shallow-water equations. When $Q = 1$, it is switched on; doing so then ultimately leads to the 'quasi-shallow' shallow-water equations with an (almost) complete representation of the Coriolis force. Using (18.141), *quasi-shallow approximation* of the (deep) Coriolis contribution $\Omega h_1^2 \dot{\xi}_1$ is thus

$$\left(\Omega h_1^2 \dot{\xi}_1\right)^Q = \Omega\left[\left(h_1^2\right)^S + Q\sigma^S(\xi_2)\, s_3\right]\dot{\xi}_1 = \Omega\left(h_1^2\right)^S \dot{\xi}_1 + Q\Omega\sigma^S(\xi_2)\, s_3\dot{\xi}_1. \tag{18.143}$$

Things are not, however, quite as simple in the present 2D context as in the 3D one due to the presence of vertical coordinate s_3 (which is a convenient proxy for ξ_3) in the additional, quasi-shallow term in (18.143).[24] This is where vertical averaging enters the quasi-shallow, shallow-water picture; an extra step is required to remove this dependence on s_3 from the green term in (18.143).

We saw in Section 18.4.2 that vertical averaging (over the depth $\widetilde{H}$ of a fluid column) of the kinetic-energy ($\mathscr{K}$) and Coriolis ($\Omega h_1^S u_1$) contributions to the *purely shallow*, vertically averaged, 2D, Lagrangian density ($\widehat{\mathscr{L}}$) left these contributions unchanged. This was for the simple reason that they are *independent* of vertical coordinate s_3. By way of contrast, this was not the case for the vertical averaging (again over the depth of a fluid column) of the gravitational-potential energy (Φ) – due to its (linear) dependence on s_3 – nor, for that matter, for the vertical averaging of the related pressure (p).

To derive the *quasi-shallow,* 2D, Lagrangian density, the only change needed to the purely shallow derivation of Section 18.4.2 is to vertically average the (additional) quasi-shallow term (in green) in (18.143) and to then add the result to Lagrangian density (18.136). Yes, it is indeed this simple – it takes longer to explain what to do, and why, than to actually do it!

Vertically averaging the additional, quasi-shallow contribution (in green in (18.143)) and using definition (18.36) of the depth $\widetilde{H}$ of the fluid column gives

$$\begin{aligned}\widehat{\Omega\sigma^S s_3\dot{\xi}_1} &= \frac{1}{H-B}\int_B^H \Omega\sigma^S(\xi_2)\, s_3\dot{\xi}_1\, ds_3 = \frac{\Omega\sigma^S(\xi_2)\,\dot{\xi}_1}{H-B}\int_B^H s_3\, ds_3 = \frac{\Omega\sigma^S(\xi_2)\,\dot{\xi}_1}{H-B}\left(\frac{H^2-B^2}{2}\right)\\ &= \Omega\sigma^S(\xi_2)\,\dot{\xi}_1\left[\frac{H(\xi_1,\xi_2,t)+B(\xi_1,\xi_2)}{2}\right] = \Omega\sigma^S(\xi_2)\,\dot{\xi}_1\left[B(\xi_1,\xi_2)+\frac{\widetilde{H}(\xi_1,\xi_2,t)}{2}\right].\end{aligned} \tag{18.144}$$

Inserting quasi-shallow contribution (18.144) into quasi-shallow Lagrangian density (18.136) then yields:

[23] Approximation (18.141) was in fact anticipated in Section 18.2.2 as (18.16).

[24] In the absence of the green term in (18.143), the remaining (leading-order) term, $\Omega\left(h_1^2\right)^S \dot{\xi}_1$, has no dependence on vertical coordinate (ξ_3, or its convenient proxy s_3).

The Vertically Averaged Lagrangian Density for the 'Quasi-Shallow' Shallow-Water Equations

$$\mathscr{L}_{SW}^{Q}\left(\xi_2,\xi_3,\dot{\xi}_1,\dot{\xi}_2\right) = \underbrace{\frac{\left(h_1^2\right)^S \dot{\xi}_1^2 + \left(h_2^2\right)^S \dot{\xi}_2^2}{2}}_{\text{horizontal KE}} \underbrace{-g\left(B+\frac{\widetilde{H}}{2}\right)}_{\text{gravity}} + \underbrace{\Omega\left(h_1^2\right)^S \dot{\xi}_1}_{\text{shallow Coriolis}} \underbrace{+Q\Omega\sigma^S\left(B+\frac{\widetilde{H}}{2}\right)\dot{\xi}_1}_{\text{quasi-shallow Coriolis}}, \tag{18.145}$$

where dependent variables have the functional forms

$$\left.\begin{aligned} \left(h_i^2\right)^S \equiv \left(h_i^S\right)^2 = \left[h_i^S\left(\xi_2\right)\right]^2, \ i=1,2; \quad g = g\left(\xi_2\right), \ \sigma^S = \sigma^S\left(\xi_2\right); \\ B = B\left(\xi_1,\xi_2\right); \quad \widetilde{H} = \widetilde{H}\left(\xi_1,\xi_2,t\right). \end{aligned}\right\} \tag{18.146}$$

The right-hand-side terms of (18.145), with the exception of the last one, are just those of Lagrangian density (18.136) for the 'basic' shallow-water equations.[25] The last term (in green) is the *additional quasi-shallow contribution*. This is the term that allows a very good approximation of the missing '$2\Omega\cos\phi$' Coriolis terms to materialise. The only way quasi-shallow approximation (18.145) of Lagrangian density can fail to lead to dynamical consistency (i.e. preservation of analogues of conservation principles for mass, axial angular momentum, total energy, and potential vorticity) is if the additional (green) quasi-shallow contribution causes this to happen. It is, however, shown (in Appendix B of the present chapter) that quasi-shallow approximation (18.145) of Lagrangian density does in fact lead to appropriate analogues of these conservation principles – and thereby to dynamical consistency of the 'quasi-shallow' shallow-water equation set.

18.5.2 An Alternative Derivation of 2D Quasi-Shallow Lagrangian Density

Derivation of 2D Lagrangian density (18.145) for the 'quasi-shallow' shallow-water equations was achieved by minimal enhancement of the corresponding 2D Lagrangian density (18.136) (alias (18.128)) for the 'basic' shallow-water equations. In Section 18.4.2, this involved vertical averaging of any terms with vertical variation.

Minimal enhancement of an existing 2D Lagrangian density is one way of obtaining the quasi-shallow, enhanced Lagrangian density. Another way, leading to the same result, is to instead vertically average an existing 3D, quasi-shallow Lagrangian density; see Appendix A of the present chapter for details.

18.5.3 The Vertically Averaged Pressure $\widehat{p}$

To classically derive the 'basic' shallow-water equations, hydrostatic equation (18.39) was vertically integrated downwards from the free surface at $\xi_3 = \xi_3^H$ to ξ_3, using upper-boundary condition (18.29), to determine p everywhere; see (18.41). A similar procedure for the quasi-shallow enhancement of the 'basic' shallow-water equations is followed here.

Inserting homogeneous condition (18.14) into quasi-hydrostatic equation (17.102) – with $\mathcal{N} \equiv 0$ therein; using explicit geopotential representation (18.21) and rearranging gives the diagnostic, first-order, quasi-hydrostatic, differential equation

$$\frac{\partial p}{\partial \xi_3} = -\overline{\rho}\frac{d\Phi_S}{d\xi_3} + 2Q\Omega\overline{\rho}u_1\left(\frac{\partial h_1}{\partial \xi_3}\right)^S = -\overline{\rho}h_3^E g_S^E + 2Q\Omega\overline{\rho}u_1\left(\frac{\partial h_1}{\partial \xi_3}\right)^S. \tag{18.147}$$

[25] This assertion follows as the special case $Q=0$ of the ensuing analysis.

This is the quasi-hydrostatic analogue of hydrostatic equation (18.39). Integrating (18.147) downwards from the free surface at $\xi_3 = \xi_3^H$ to ξ_3; applying upper-boundary condition (18.29); and using (18.23)–(18.25) then yields

$$p = p^H + \overline{\rho}\left[h_3^E g_S^E - 2Q\Omega u_1 \left(\frac{\partial h_1}{\partial \xi_3}\right)^S\right]\left(\xi_3^H - \xi_3\right)$$
$$= p^H + \overline{\rho}\left[h_3^S(\xi_2)\, g(\xi_2) - 2Q\Omega u_1 \left(\frac{\partial h_1}{\partial \xi_3}\right)^S\right]\left[\frac{H(\xi_1,\xi_2,t) - s_3}{h_3^S(\xi_2)}\right]. \quad (18.148)$$

In the absence of the (green) quasi-shallow term in this equation, and with use of (18.20), (18.23), and (18.25), (18.148) reduces to (18.41) as it must.

When (in Section 18.6) we use the Euler–Lagrange equations to obtain the horizontal components of momentum for the more general, 'quasi-shallow' version of the shallow-water equations, we will need the vertical average (mean value) $\widehat{p}$ of p, evaluated over the depth of the fluid between the lower, rigid boundary at $s_3 = s_3^B = B(\xi_1,\xi_2)$ and the upper, free-surface boundary at $s_3 = s_3^H = H(\xi_1,\xi_2,t)$. Anticipating this and using (18.25) leads to

$$\widehat{p} \equiv \frac{1}{(H-B)}\int_B^H p\,ds_3 = \frac{1}{(H-B)}\int_B^H \left\{p^H + \overline{\rho}\left[h_3^S(\xi_2)\, g(\xi_2) - 2Q\Omega u_1 \left(\frac{\partial h_1}{\partial \xi_3}\right)^S\right]\left[\frac{H - s_3}{h_3^S(\xi_2)}\right]\right\} ds_3$$
$$= p^H + \overline{\rho}\left[g(\xi_2) - \frac{2Q\Omega}{h_3^S(\xi_2)}\left(\frac{\partial h_1}{\partial \xi_3}\right)^S u_1\right]\frac{1}{(H-B)}\int_B^H (H - s_3)\, ds_3$$
$$= p^H + \overline{\rho}\left[g(\xi_2) - Q\Omega\sigma^S(\xi_2)\frac{u_1}{h_1^S}\right]\frac{(H-B)}{2} = p^H + \overline{\rho}\left[\frac{g(\xi_2)\widetilde{H}}{2} - \frac{Q\Omega u_1}{2h_1^S}\sigma^S(\xi_2)\widetilde{H}\right], \quad (18.149)$$

where p^H is the constant imposed pressure at the upper free boundary $s_3^H = H(\xi_1,\xi_2,t)$.

Partially differentiating (18.149) with respect to ξ_1 and ξ_2, respectively, then gives:

The Vertically Averaged Components of the Horizontal Pressure Gradient Force

$$\frac{1}{\overline{\rho}}\frac{\partial \widehat{p}}{\partial \xi_1} = \frac{\partial}{\partial \xi_1}\left[\frac{g(\xi_2)\widetilde{H}}{2}\right] - Q\Omega\frac{\partial}{\partial \xi_1}\left[\frac{u_1}{h_1^S}\sigma^S(\xi_2)\frac{\widetilde{H}}{2}\right], \quad (18.150)$$
$$\frac{1}{\overline{\rho}}\frac{\partial \widehat{p}}{\partial \xi_2} = \frac{\partial}{\partial \xi_2}\left[\frac{g(\xi_2)\widetilde{H}}{2}\right] - Q\Omega\frac{\partial}{\partial \xi_2}\left[\frac{u_1}{h_1^S}\sigma^S(\xi_2)\frac{\widetilde{H}}{2}\right]. \quad (18.151)$$

It remains to derive the 'quasi-shallow' shallow-water equations using approximated Lagrangian density (18.145), together with horizontal pressure-gradient terms (18.150) and (18.151), in the horizontal components of the Euler–Lagrange equations. This is now done in Section 18.6.

18.6 EULER–LAGRANGE DERIVATION OF THE QUASI-SHALLOW ENHANCED SET

18.6.1 The Euler–Lagrange Equations for Horizontal Momentum Components

From (18.110), the Euler–Lagrange equations for the horizontal components of the momentum equation for the 'quasi-shallow' enhanced shallow-water equations are

$$\frac{D_{\text{hor}}}{Dt}\left(\frac{\partial\widehat{\mathscr{L}}}{\partial\dot{\xi}_i}\right)-\frac{\partial\widehat{\mathscr{L}}}{\partial\xi_i}=-\frac{1}{\overline{\rho}}\frac{\partial\widehat{p}}{\partial\xi_i},\quad i=1,2, \tag{18.152}$$

where $\widehat{\mathscr{L}}=\mathscr{L}_{\text{SW}}^{Q}$ is the vertically averaged Lagrangian density (18.145); and $(1/\overline{\rho})\left(\partial\widehat{p}/\partial\xi_i\right), i=1,2$, are the vertically averaged pressure-gradient terms (18.150) and (18.151), respectively.

18.6.2 Partial Derivatives of Lagrangian Density $\mathscr{L}_{\text{SW}}^{Q}$

With use of functional forms (18.146), the relevant partial derivatives of Lagrangian density (18.145) for the horizontal Euler–Lagrange equations (18.152) are:

$$\begin{aligned}\frac{\partial\mathscr{L}_{\text{SW}}^{Q}}{\partial\xi_1}&=-\frac{\partial}{\partial\xi_1}\left[g\left(B+\frac{\widetilde{H}}{2}\right)\right]+Q\Omega\sigma^S\dot{\xi}_1\frac{\partial}{\partial\xi_1}\left(B+\frac{\widetilde{H}}{2}\right)\\&=-\frac{\partial}{\partial\xi_1}\left[g\left(B+\frac{\widetilde{H}}{2}\right)\right]+Q\Omega\sigma^S\frac{u_1}{h_1^S}\frac{\partial}{\partial\xi_1}\left(B+\frac{\widetilde{H}}{2}\right),\end{aligned} \tag{18.153}$$

$$\begin{aligned}\frac{\partial\mathscr{L}_{\text{SW}}^{Q}}{\partial\xi_2}&=h_1^S\frac{dh_1^S}{d\xi_2}\dot{\xi}_1^2+h_2^S\frac{dh_2^S}{d\xi_2}\dot{\xi}_2^2-\frac{\partial}{\partial\xi_2}\left[g\left(B+\frac{\widetilde{H}}{2}\right)\right]+\Omega\dot{\xi}_1\frac{\partial}{\partial\xi_2}\left[\left(h_1^2\right)^S+Q\sigma^S\left(B+\frac{\widetilde{H}}{2}\right)\right]\\&=\frac{u_1^2}{h_1^S}\frac{dh_1^S}{d\xi_2}+\frac{u_2^2}{h_2^S}\frac{dh_2^S}{d\xi_2}-\frac{\partial}{\partial\xi_2}\left[g\left(B+\frac{\widetilde{H}}{2}\right)\right]+\Omega\frac{u_1}{h_1^S}\left\{\frac{d\left(h_1^2\right)^S}{d\xi_2}+Q\frac{\partial}{\partial\xi_2}\left[\sigma^S\left(B+\frac{\widetilde{H}}{2}\right)\right]\right\},\end{aligned} \tag{18.154}$$

$$\frac{\partial\mathscr{L}_{\text{SW}}^{Q}}{\partial\dot{\xi}_1}=\left(h_1^2\right)^S\dot{\xi}_1+\Omega\left[\left(h_1^2\right)^S+Q\sigma^S\left(B+\frac{\widetilde{H}}{2}\right)\right]=h_1^Su_1+\Omega\left[\left(h_1^2\right)^S+Q\sigma^S\left(B+\frac{\widetilde{H}}{2}\right)\right], \tag{18.155}$$

$$\frac{\partial\mathscr{L}_{\text{SW}}^{Q}}{\partial\dot{\xi}_2}=\left(h_2^2\right)^S\dot{\xi}_2=h_2^Su_2. \tag{18.156}$$

18.6.3 Derivation of the Component Equation for u_1

Inserting (18.153), (18.155), and (18.150) into the first ($i=1$) Euler–Lagrange equation (18.152) gives

$$\begin{aligned}&\frac{D_{\text{hor}}}{Dt}\left\{h_1^Su_1+\Omega\left[\left(h_1^2\right)^S+Q\sigma^S\left(B+\frac{\widetilde{H}}{2}\right)\right]\right\}+\frac{\partial}{\partial\xi_1}\left[g\left(B+\frac{\widetilde{H}}{2}\right)\right]-Q\Omega\sigma^S\frac{u_1}{h_1^S}\frac{\partial}{\partial\xi_1}\left(B+\frac{\widetilde{H}}{2}\right)\\&=-\frac{\partial}{\partial\xi_1}\left(\frac{g\widetilde{H}}{2}\right)+Q\Omega\frac{\partial}{\partial\xi_1}\left(\frac{u_1}{h_1^S}\sigma^S\frac{\widetilde{H}}{2}\right).\end{aligned} \tag{18.157}$$

Rearranging (18.157) and using definition (18.36) of $\widetilde{H}$ yields

$$\begin{aligned}&\frac{D_{\text{hor}}}{Dt}\left[h_1^Su_1+\Omega\left(h_1^2\right)^S\right]+\frac{\partial}{\partial\xi_1}(gH)\\&-Q\Omega\left\{\frac{1}{2}\frac{\partial}{\partial\xi_1}\left(\frac{\sigma^S}{h_1^S}u_1\widetilde{H}\right)+\sigma^S\frac{u_1}{h_1^S}\frac{\partial}{\partial\xi_1}\left(B+\frac{\widetilde{H}}{2}\right)-\frac{D_S}{Dt}\left[\sigma^S\left(B+\frac{\widetilde{H}}{2}\right)\right]\right\}=0.\end{aligned} \tag{18.158}$$

Dividing this equation throughout by h_1^S; noting that $D_{\text{hor}}h_1^S/Dt=\left(u_2/h_2^S\right)\left(dh_1^S/d\xi_2\right)$ with use of (18.33) and $h_1^S=h_1^S(\xi_2)$; and further rearranging then leads to

$$\frac{D_{\text{hor}}u_1}{Dt}+\left(\frac{u_1}{h_1^S}+2\Omega\right)\frac{u_2}{h_2^S}\frac{dh_1^S}{d\xi_2}+\frac{1}{h_1^S}\frac{\partial}{\partial\xi_1}(gH)$$

$$-\frac{Q\Omega}{h_1^S}\left\{\frac{1}{2}\frac{\partial}{\partial\xi_1}\left(\frac{\sigma^S}{h_1^S}u_1\widetilde{H}\right)+\sigma^S\frac{u_1}{h_1^S}\frac{\partial}{\partial\xi_1}\left(B+\frac{\widetilde{H}}{2}\right)-\frac{D_S}{Dt}\left[\sigma^S\left(B+\frac{\widetilde{H}}{2}\right)\right]\right\}=0. \tag{18.159}$$

Now, again using definition (18.33) for D_{hor}/Dt:

$$\begin{aligned}\frac{D_{\text{hor}}}{Dt}\left[\sigma^S\left(B+\frac{\widetilde{H}}{2}\right)\right]&=\frac{\partial}{\partial t}\left[\sigma^S\left(B+\frac{\widetilde{H}}{2}\right)\right]+\frac{u_1}{h_1^S}\frac{\partial}{\partial\xi_1}\left[\sigma^S\left(B+\frac{\widetilde{H}}{2}\right)\right]\\&\quad+\frac{u_2}{h_2^S}\frac{\partial}{\partial\xi_2}\left[\sigma^S\left(B+\frac{\widetilde{H}}{2}\right)\right]\\&=\frac{\sigma^S}{2}\frac{\partial\widetilde{H}}{\partial t}+\sigma^S\frac{u_1}{h_1^S}\frac{\partial}{\partial\xi_1}\left(B+\frac{\widetilde{H}}{2}\right)+\frac{u_2}{h_2^S}\frac{\partial}{\partial\xi_2}\left[\sigma^S\left(B+\frac{\widetilde{H}}{2}\right)\right].\end{aligned} \tag{18.160}$$

Substituting (18.160) into component equation (18.159) then gives

$$\begin{aligned}&\frac{D_{\text{hor}}u_1}{Dt}+\left(\frac{u_1}{h_1^S}+2\Omega\right)\frac{u_2}{h_2^S}\frac{dh_1^S}{d\xi_2}+\frac{1}{h_1^S}\frac{\partial}{\partial\xi_1}(gH)\\&\quad-\frac{Q\Omega}{h_1^S}\left\{\frac{1}{2}\frac{\partial}{\partial\xi_1}\left(\frac{\sigma^S}{h_1^S}u_1\widetilde{H}\right)-\frac{u_2}{h_2^S}\frac{\partial}{\partial\xi_2}\left[\sigma^S\left(B+\frac{\widetilde{H}}{2}\right)\right]\right\}+\frac{Q\Omega}{h_1^S}\frac{\sigma^S}{2}\frac{\partial\widetilde{H}}{\partial t}=0.\end{aligned} \tag{18.161}$$

Now, from (18.48), the mass-continuity equation in Eulerian form may be written as

$$\frac{\partial\widetilde{H}}{\partial t}=-\frac{1}{h_1^Sh_2^S}\left[\frac{\partial}{\partial\xi_1}\left(h_2^Su_1\widetilde{H}\right)+\frac{\partial}{\partial\xi_2}\left(h_1^Su_2\widetilde{H}\right)\right]. \tag{18.162}$$

Substituting this equation into (18.161) to eliminate $\partial\widetilde{H}/\partial\widetilde{t}$ then yields

$$\begin{aligned}&\frac{D_{\text{hor}}u_1}{Dt}+\left(\frac{u_1}{h_1^S}+2\Omega\right)\frac{u_2}{h_2^S}\frac{dh_1^S}{d\xi_2}+\frac{1}{h_1^S}\frac{\partial}{\partial\xi_1}(gH)\\&\quad-\frac{Q\Omega}{h_1^S}\left\{\frac{1}{2}\frac{\partial}{\partial\xi_1}\left(\frac{\sigma^S}{h_1^S}u_1\widetilde{H}\right)-\frac{u_2}{h_2^S}\frac{\partial}{\partial\xi_2}\left[\sigma^S\left(B+\frac{\widetilde{H}}{2}\right)\right]\right\}\\&\quad-Q\frac{\Omega}{2}\frac{\sigma^S}{\left(h_1^S\right)^2h_2^S}\left[\frac{\partial}{\partial\xi_1}\left(h_2^Su_1\widetilde{H}\right)+\frac{\partial}{\partial\xi_2}\left(h_1^Su_2\widetilde{H}\right)\right]=0.\end{aligned} \tag{18.163}$$

This equation agrees with Staniforth (2015b)'s equation (64) (with $\mathcal{N}\equiv 0$ therein) for the ξ_1 component of the 'quasi-shallow' shallow-water equations.

18.6.4 Derivation of the Component Equation for u_2

Inserting (18.154), (18.156), and (18.151) into the second ($i=2$) Euler–Lagrange equation (18.152) gives

$$\begin{aligned}&\frac{D_{\text{hor}}}{Dt}\left(h_2^Su_2\right)-\frac{u_1^2}{h_1^S}\frac{dh_1^S}{d\xi_2}-\frac{u_2^2}{h_2^S}\frac{dh_2^S}{d\xi_2}+\frac{\partial}{\partial\xi_2}\left[g\left(B+\frac{\widetilde{H}}{2}\right)\right]\\&\quad-\Omega\frac{u_1}{h_1^S}\left\{\frac{d\left(h_1^2\right)^S}{d\xi_2}+Q\frac{\partial}{\partial\xi_2}\left[\sigma^S\left(B+\frac{\widetilde{H}}{2}\right)\right]\right\}\\&=-\frac{\partial}{\partial\xi_2}\left[\frac{g\left(\xi_2\right)\widetilde{H}}{2}\right]+Q\Omega\frac{\partial}{\partial\xi_2}\left(\frac{u_1}{h_1^S}\sigma^S\frac{\widetilde{H}}{2}\right).\end{aligned} \tag{18.164}$$

Rearranging (18.164) and using definition (18.36) of $\widetilde{H}$ yields

$$\frac{D_{\text{hor}}}{Dt}\left(h_2^S u_2\right) - \frac{u_1^2}{h_1^S}\frac{dh_1^S}{d\xi_2} - \frac{u_2^2}{h_2^S}\frac{dh_2^S}{d\xi_2} - \Omega\frac{u_1}{h_1^S}\frac{d\left(h_1^2\right)^S}{d\xi_2} + \frac{\partial}{\partial\xi_2}(gH) - Q\Omega\left\{\frac{1}{2}\frac{\partial}{\partial\xi_2}\left(\frac{\sigma^S}{h_1^S}u_1\widetilde{H}\right) + \frac{u_1}{h_1^S}\frac{\partial}{\partial\xi_2}\left[\sigma^S\left(B+\frac{\widetilde{H}}{2}\right)\right]\right\} = 0. \tag{18.165}$$

Dividing this equation throughout by h_2^S; noting that $D_{\text{hor}}h_2^S/Dt = \left(u_2/h_2^S\right)\left(dh_2^S/d\xi_2\right)$ with use of (18.33) and $h_2^S = h_2^S(\xi_2)$; and further rearranging then leads to

$$\frac{D_{\text{hor}}u_2}{Dt} - \left(\frac{u_1}{h_1^S} + 2\Omega\right)\frac{u_1}{h_2^S}\frac{dh_1^S}{d\xi_2} + \frac{1}{h_2^S}\frac{\partial}{\partial\xi_2}(gH) - \frac{Q\Omega}{h_2^S}\left\{\frac{1}{2}\frac{\partial}{\partial\xi_2}\left(\frac{\sigma^S}{h_1^S}u_1\widetilde{H}\right) + \frac{u_1}{h_1^S}\frac{\partial}{\partial\xi_2}\left[\sigma^S\left(B+\frac{\widetilde{H}}{2}\right)\right]\right\} = 0. \tag{18.166}$$

This agrees with Staniforth (2015b)'s equation (65) (with $\mathcal{N} \equiv 0$ therein) for the ξ_2 component of the 'quasi-shallow' shallow-water equations.

18.6.5 The 'Quasi-Shallow' Shallow-Water Equation Set

In axial-orthogonal-curvilinear geopotential coordinates, the 'quasi-shallow' shallow-water equation set is comprised of:

- Component equations (18.163) and (18.166) for horizontal momentum.
- Mass-continuity equation (18.47) (in Lagrangian form) or (18.48) (in Eulerian form).

This equation set is summarised in the highlighted box.

For the special case of Q set identically zero, the 'basic' shallow-water equation set (comprised of component equations (18.42) and (18.43) for horizontal momentum, and mass-continuity equation (18.47) or (18.48)) is recovered.

Setting $Q \equiv 1$ instead in (18.167), (18.168), and (18.170), and after taking into account minor differences in notation, equations (3.42)–(3.44) of Tort et al. (2014) are exactly recovered. This serves as a useful cross-check. Tort et al. (2014) show that their equations (3.42)–(3.44), that is, (18.167), (18.168), and (18.170) of the present chapter, can be used to recover Dellar and Salmon (2005)'s and Dellar (2011)'s shallow-water equation sets in Cartesian geometry on f and β-γ planes.

The 'Quasi-Shallow' Shallow-Water Equations in Axial-Orthogonal-Curvilinear Coordinates

Momentum

$$\underbrace{\frac{D_{\text{hor}}u_1}{Dt} + \left(\frac{u_1}{h_1^S} + 2\Omega\right)\frac{u_2}{h_2^S}\frac{dh_1^S}{d\xi_2} + \frac{1}{h_1^S}\frac{\partial}{\partial\xi_1}(gH)}_{\text{basic}} \underbrace{- \frac{Q\Omega}{h_1^S}\left\{\frac{1}{2}\frac{\partial}{\partial\xi_1}\left(\frac{\sigma^S}{h_1^S}u_1\widetilde{H}\right) - \frac{u_2}{h_2^S}\frac{\partial}{\partial\xi_2}\left[\sigma^S\left(B+\frac{\widetilde{H}}{2}\right)\right]\right\}}_{\text{quasi-shallow}}$$

$$\underbrace{-Q\frac{\Omega}{2}\frac{\sigma^S}{\left(h_1^S\right)^2 h_2^S}\left[\frac{\partial}{\partial \xi_1}\left(h_2^S u_1 \widetilde{H}\right)+\frac{\partial}{\partial \xi_2}\left(h_1^S u_2 \widetilde{H}\right)\right]}_{\text{quasi-shallow}} = 0, \qquad (18.167)$$

$$\underbrace{\frac{D_{\text{hor}} u_2}{Dt} - \left(\frac{u_1}{h_1^S}+2\Omega\right)\frac{u_1}{h_2^S}\frac{dh_1^S}{d\xi_2}+\frac{1}{h_2^S}\frac{\partial}{\partial \xi_2}(gH)}_{\text{basic}}$$

$$\underbrace{-\frac{Q\Omega}{h_2^S}\left\{\frac{1}{2}\frac{\partial}{\partial \xi_2}\left(\frac{\sigma^S}{h_1^S}u_1\widetilde{H}\right)+\frac{u_1}{h_1^S}\frac{\partial}{\partial \xi_2}\left[\sigma^S\left(B+\frac{\widetilde{H}}{2}\right)\right]\right\}}_{\text{quasi-shallow}} = 0. \qquad (18.168)$$

Mass-Continuity

$$\frac{D_{\text{hor}}\widetilde{H}(\xi_1,\xi_2,t)}{Dt}+\frac{\widetilde{H}(\xi_1,\xi_2,t)}{h_1^S h_2^S}\left[\frac{\partial}{\partial \xi_1}\left(u_1 h_2^S\right)+\frac{\partial}{\partial \xi_2}\left(u_2 h_1^S\right)\right] = 0, \qquad (18.169)$$

or

$$\frac{\partial \widetilde{H}(\xi_1,\xi_2,t)}{\partial t}+\frac{1}{h_1^S h_2^S}\left[\frac{\partial}{\partial \xi_1}\left(u_1 h_2^S \widetilde{H}\right)+\frac{\partial}{\partial \xi_2}\left(u_2 h_1^S \widetilde{H}\right)\right] = 0. \qquad (18.170)$$

Material Derivative

$$\frac{D_{\text{hor}}}{Dt} \equiv \frac{\partial}{\partial t}+\frac{u_1}{h_1^S}\frac{\partial}{\partial \xi_1}+\frac{u_2}{h_2^S}\frac{\partial}{\partial \xi_2}. \qquad (18.171)$$

18.7 'QUASI-SHALLOW' SHALLOW-WATER CONSERVATION PRINCIPLES

It is important to verify the dynamical consistency of the 'quasi-shallow' shallow-water equation set by deriving analogues of conservation principles for mass, axial angular momentum, total energy, and potential vorticity. There are a number of different ways of expressing conservation principles, from which other forms may be deduced; see Chapter 15. We now motivate the particular forms derived herein.

Historically, the 'basic' shallow-water equations have frequently been used as a simple prototype to develop and validate numerical methods for application in the broader context of comprehensive 3D atmospheric and oceanic models. The recent development of the enhanced 'quasi-shallow' shallow-water equations further extends the usefulness of the 'basic' set as a pathway towards the development of 3D atmospheric and oceanic dynamical cores. In particular, budget equations in Eulerian form for conservation of mass, axial angular momentum, and total energy are very useful tools for developing Eulerian numerical methods (e.g. using finite-difference, finite-element, and finite-volume techniques), and for understanding and assessing numerical integrations of governing equation sets. Contrastingly, material conservation of potential vorticity is most naturally expressed in Lagrangian form.

Bearing these comments in mind, the following conservation principles, expressed in axial-orthogonal-curvilinear coordinates, hold for 'quasi-shallow' shallow-water equations (18.167)–(18.171). See Appendix B of the present chapter for their detailed derivation. As a special case, setting $Q \equiv 0$ everywhere in what follows corresponds to use of the 'basic' shallow-water equations (without 'quasi-shallow' enhancement). Thus the conservation principles summarised in this section collectively imply dynamical consistency of both the 'quasi-shallow' shallow-water equations and, as a special case, the 'basic' shallow-water equations.

18.7.1 Mass

From (18.265) of Appendix B of the present chapter,

$$\frac{\partial \widetilde{H}}{\partial t}+\frac{1}{h_1^S h_2^S}\left[\frac{\partial}{\partial \xi_1}\left(h_2^S u_1 \widetilde{H}\right)+\frac{\partial}{\partial \xi_2}\left(h_1^S u_2 \widetilde{H}\right)\right]=0. \tag{18.172}$$

Global integration of (18.172), with application of periodicity, shows that mass is globally conserved, that is,

$$\frac{\partial}{\partial t}\left(\iint_{\text{globe}} \widetilde{H} h_1^S h_2^S d\xi_1 d\xi_2\right)=0. \tag{18.173}$$

18.7.2 Axial Angular Momentum

From (18.277) and (18.272) of Appendix B of the present chapter,

$$\frac{\partial M^{SW}}{\partial t}=-\nabla_{\text{hor}}\cdot\left\{\left[\underbrace{h_1^S\left(u_1+\Omega h_1^S\right)}_{\text{basic}}+\underbrace{Q\Omega\sigma^S\left(B+\frac{\widetilde{H}}{2}\right)}_{\text{quasi-shallow}}\right]\widetilde{H}\mathbf{u}_{\text{hor}}\right\}$$
$$-\frac{\partial}{\partial \xi_1}\left[\left(\underbrace{\frac{g}{2}}_{\text{basic}}-\underbrace{Q\frac{\Omega\sigma^S}{2}\frac{u_1}{h_1^S}}_{\text{quasi-shallow}}\right)\widetilde{H}^2\right]-\left(\underbrace{g}_{\text{basic}}-\underbrace{Q\Omega\sigma^S\frac{u_1}{h_1^S}}_{\text{quasi-shallow}}\right)\widetilde{H}\frac{\partial B}{\partial \xi_1}, \tag{18.174}$$

where $\nabla_{\text{hor}}\cdot(G\mathbf{u}_{\text{hor}})$ is defined by (18.60), and

$$M^{SW}\equiv\widetilde{H}\left[\underbrace{h_1^S\left(u_1+\Omega h_1^S\right)}_{\text{basic}}+\underbrace{Q\Omega\sigma^S\left(B+\frac{\widetilde{H}}{2}\right)}_{\text{quasi-shallow}}\right], \tag{18.175}$$

is (quasi-shallow) axial angular momentum. Equations (18.174) and (18.175) correspond to equations (78) and (79) of Staniforth, 2015b, respectively – with non-hydrostatic switch $\mathcal{N}\equiv 0$ and $\mathcal{A}\rightarrow M^{SW}$ therein. Note that definition (18.175) of axial angular momentum has an additional, quasi-shallow term (in green). The fluxes on the right-hand side of (18.174) are also influenced by quasi-shallow approximation.

Globally integrating (18.174), with application of biperiodicity and use of (18.15), gives

$$\frac{\partial}{\partial t}\left(\iint_{\text{globe}} M^{\text{SW}} h_1^S h_2^S d\xi_1 d\xi_2\right) = -\iint_{\text{globe}} \left(\underbrace{g}_{\text{basic}} - \underbrace{Q\Omega\sigma^S \frac{u_1}{h_1^S}}_{\text{quasi-shallow}}\right) \widetilde{H} \frac{\partial B}{\partial \xi_1} h_1^S h_2^S d\xi_1 d\xi_2. \quad (18.176)$$

It is seen from (18.176) that globally integrated axial angular momentum is conserved for a flat, rigid, lower boundary (i.e. when $B \equiv$ constant), and also for an axially symmetric one (i.e. when $B = B(\xi_2) \Rightarrow \partial B/\partial \xi_1 \equiv 0$). In the presence of azimuthal variation of the lower boundary (i.e. when $B = B(\xi_1) \Rightarrow \partial B/\partial \xi_1 \neq 0$), two bottom torque terms are present on the right-hand side of (18.176). The first of these is the usual one, with the second one (in green) owing its presence to quasi-shallow approximation.

18.7.3 Total Energy

From (18.285) and (18.286) of Appendix B of the present chapter,

$$\frac{\partial E^{\text{SW}}}{\partial t} = -\nabla_{\text{hor}} \cdot \left[\left(\underbrace{\frac{u_1^2 + u_2^2}{2} + gH}_{\text{basic}} - \underbrace{Q\frac{\Omega}{2}\frac{\sigma^S}{h_1^S} u_1 \widetilde{H}}_{\text{quasi-shallow}}\right) \widetilde{H} \mathbf{u}_{\text{hor}}\right], \quad (18.177)$$

where $\nabla_{\text{hor}} \cdot (G\mathbf{u}_{\text{hor}})$ is defined by (18.60), and

$$E^{\text{SW}} \equiv \underbrace{\left[\frac{u_1^2 + u_2^2}{2} + g\left(B + \frac{\widetilde{H}}{2}\right)\right] \widetilde{H}}_{\text{basic}}, \quad (18.178)$$

is the sum of kinetic and gravitational-potential energies. Equations (18.177) and (18.178) correspond to equations (82) and (83) of Staniforth, 2015b, respectively – with non-hydrostatic switch $\mathcal{N} \equiv 0$ and $\mathcal{E} \to E^{\text{SW}}$ therein. Note that total energy definition (18.178) is unaffected by quasi-shallow approximation, in contradistinction to the energy-flux terms on the right-hand side of (18.177).

As noted in Staniforth (2015b), (18.177) is in a similar form to, and essentially in agreement with,[26] Dellar and Salmon (2005)'s energy-evolution equation (20) for constant gravity in Cartesian β-γ-plane geometry. Equation (18.177) also agrees with Tort et al. (2014)'s equation (4.3) for constant gravity in the special case of *spherical* geometry.

Global integration of (18.177), with application of periodicity, shows that the sum of kinetic and gravitational-potential energies is globally conserved, that is,

[26] After correction of an inconsequential error in their equation; the $(u_1^2 + u_2^2)$ contribution to horizontal divergence is missing a division by 2.

$$\frac{\partial}{\partial t}\left(\iint_{\text{globe}} E^{SW} h_1^S h_2^S d\xi_1 d\xi_2\right) = 0. \tag{18.179}$$

18.7.4 Potential Vorticity

From (18.298), (18.299), and (18.289) of Appendix B of the present chapter,

$$\frac{D_{\text{hor}}\Pi^{SW}}{Dt} = 0, \tag{18.180}$$

where

$$\Pi^{SW} \equiv \frac{1}{\widetilde{H}}\left\{\underbrace{\zeta_3^S - \frac{2\Omega}{h_2^S}\frac{dh_1^S}{d\xi_2}}_{\text{basic}} \underbrace{- Q\frac{\Omega}{h_1^S h_2^S}\frac{\partial}{\partial \xi_2}\left[\sigma^S\left(B + \frac{\widetilde{H}}{2}\right)\right]}_{\text{quasi-shallow}}\right\} \tag{18.181}$$

is two-dimensional (quasi-shallow) potential vorticity and

$$\zeta_3^S \equiv \frac{1}{h_1^S h_2^S}\left[\frac{\partial}{\partial \xi_1}\left(h_2^S u_2\right) - \frac{\partial}{\partial \xi_2}\left(h_1^S u_1\right)\right] \tag{18.182}$$

is two-dimensional relative vorticity. Note that definition (18.181) of potential vorticity is influenced by quasi-shallow approximation through the green term, which originates from planetary vorticity. Potential vorticity is nevertheless materially conserved – in accordance with (18.180). Equations (18.180) and (18.181) correspond to equations (85) and (86) of Staniforth, 2015b, respectively – with non-hydrostatic switch $\mathcal{N} \equiv 0$ and $q \to \Pi^{SW}$ therein.

As noted in Staniforth (2015b), after taking into account differences in notation and geometry, (18.180) and (18.181) are in a similar form to, and essentially agree with, Dellar and Salmon (2005)'s potential vorticity equation (1) for constant gravity in Cartesian β-γ-plane geometry. Equations (18.180) and (18.181) are also in agreement with Tort et al. (2014)'s equation (B.24) for constant gravity in the special case of *spherical* geometry.

18.8 THE 'QUASI-SHALLOW' SHALLOW-WATER EQUATIONS IN SPHERICAL GEOMETRY

For illustrative and reference purposes, it is useful to express the enhanced 'quasi-shallow' shallow-water equations in spherical geometry, together with their conservation principles.

Now, shallow metric factors in spherical geometry $(\xi_1, \xi_2, \xi_3) = (\lambda, \phi, r)$ are given in standard notation by (5.72). Thus, in the present context,

$$(\xi_1, \xi_2, \xi_3) \to (\lambda, \phi, r - \bar{a}), \quad \left(h_1, h_2, h_3\right) \to \left(h_\lambda, h_\phi, h_r\right) = (r\cos\phi, r, 1), \tag{18.183}$$

$$\left(h_1^S, h_2^S, h_3^S\right) \to \left(h_\lambda^S, h_\phi^S, h_r^S\right) = (\bar{a}\cos\phi, \bar{a}, 1), \quad \sigma^S \equiv \frac{1}{h_3^S}\left(\frac{\partial h_1^2}{\partial \xi_3}\right)^S \to 2\bar{a}\cos^2\phi, \tag{18.184}$$

where $a \rightarrow \overline{a}$ is the planet's mean radius. Inserting (18.183) and (18.184) into (18.167)–(18.182) leads to the following governing equations, and associated conservation principles, for the 'quasi-shallow' shallow-water equations in spherical-polar coordinates:

The 'Quasi-Shallow' Shallow-Water Equations in Spherical-Polar Coordinates

Momentum Components

$$\underbrace{\frac{D_{hor}u_\lambda}{Dt} - \left(\frac{u_\lambda}{\overline{a}\cos\phi} + 2\Omega\right)u_\phi \sin\phi + \frac{1}{\overline{a}\cos\phi}\frac{\partial}{\partial\lambda}(gH)}_{\text{basic}}$$
$$\underbrace{-\frac{Q\Omega}{\overline{a}}\left\{\frac{\partial}{\partial\phi}\left(\cos\phi u_\phi \widetilde{H}\right) - \frac{2u_\phi}{\cos\phi}\frac{\partial}{\partial\phi}\left[\cos^2\phi\left(B + \frac{\widetilde{H}}{2}\right)\right] + 2\frac{\partial}{\partial\lambda}\left(u_\lambda \widetilde{H}\right)\right\}}_{\text{quasi-shallow}} = 0, \tag{18.185}$$

$$\underbrace{\frac{D_{hor}u_\phi}{Dt} + \left(\frac{u_\lambda}{\overline{a}\cos\phi} + 2\Omega\right)u_\lambda \sin\phi + \frac{1}{\overline{a}}\frac{\partial}{\partial\phi}(gH)}_{\text{basic}}$$
$$\underbrace{-\frac{Q\Omega}{\overline{a}}\left\{\frac{\partial}{\partial\phi}\left(\cos\phi u_\lambda \widetilde{H}\right) + \frac{2u_\lambda}{\cos\phi}\frac{\partial}{\partial\phi}\left[\cos^2\phi\left(B + \frac{\widetilde{H}}{2}\right)\right]\right\}}_{\text{quasi-shallow}} = 0. \tag{18.186}$$

Mass-Continuity

$$\frac{D_{hor}\widetilde{H}}{Dt} + \frac{\widetilde{H}}{\overline{a}\cos\phi}\left[\frac{\partial u_\lambda}{\partial\lambda} + \frac{\partial}{\partial\phi}\left(u_\phi\cos\phi\right)\right] = 0, \tag{18.187}$$

or

$$\frac{\partial\widetilde{H}}{\partial t} + \frac{1}{\overline{a}\cos\phi}\left[\frac{\partial}{\partial\lambda}\left(u_\lambda\widetilde{H}\right) + \frac{\partial}{\partial\phi}\left(\cos\phi u_\phi \widetilde{H}\right)\right] = 0. \tag{18.188}$$

Material Derivative

$$\frac{D_{hor}}{Dt} \equiv \frac{\partial}{\partial t} + \frac{u_\lambda}{\overline{a}\cos\phi}\frac{\partial}{\partial\lambda} + \frac{u_\phi}{\overline{a}}\frac{\partial}{\partial\phi}. \tag{18.189}$$

Associated Conservation Principles

Mass

$$\frac{\partial\widetilde{H}}{\partial t} = -\frac{1}{\overline{a}\cos\phi}\left[\frac{\partial}{\partial\lambda}\left(u_\lambda\widetilde{H}\right) + \frac{\partial}{\partial\phi}\left(\cos\phi u_\phi \widetilde{H}\right)\right], \tag{18.190}$$

$$\frac{\partial}{\partial t}\left(\int_0^{2\pi}\int_{-\frac{\pi}{2}}^{\frac{\pi}{2}} \widetilde{H}\overline{a}^2 \cos\phi d\lambda d\phi\right) = 0. \tag{18.191}$$

Axial Angular Momentum

$$\frac{\partial M^{SW}}{\partial t} = -\nabla_{hor} \cdot \left\{ \left[\underbrace{\overline{a}\cos\phi\left(u_\lambda + \Omega\overline{a}\cos\phi\right)}_{\text{basic}} + \underbrace{2Q\Omega\overline{a}\cos^2\phi\left(B + \frac{\widetilde{H}}{2}\right)}_{\text{quasi-shallow}} \right] \widetilde{H}\mathbf{u}_{hor} \right\}$$

$$- \frac{\partial}{\partial\lambda}\left[\left(\underbrace{\frac{g}{2}}_{\text{basic}} - \underbrace{Q\Omega\cos\phi u_\lambda}_{\text{quasi-shallow}}\right)\widetilde{H}^2\right] - \left(\underbrace{g}_{\text{basic}} - \underbrace{2Q\Omega\cos\phi u_\lambda}_{\text{quasi-shallow}}\right)\widetilde{H}\frac{\partial B}{\partial\lambda}, \tag{18.192}$$

$$\frac{\partial}{\partial t}\left(\int_0^{2\pi}\int_{-\frac{\pi}{2}}^{\frac{\pi}{2}} M^{SW}\overline{a}^2\cos\phi d\lambda d\phi\right)$$

$$= -\int_0^{2\pi}\int_{-\frac{\pi}{2}}^{\frac{\pi}{2}}\left(\underbrace{g}_{\text{basic}} - \underbrace{2Q\Omega\cos\phi u_\lambda}_{\text{quasi-shallow}}\right)\widetilde{H}\frac{\partial B}{\partial\lambda}\overline{a}^2\cos\phi d\lambda d\phi, \tag{18.193}$$

where

$$M^{SW} \equiv \widetilde{H}\left[\underbrace{\overline{a}\cos\phi\left(u_\lambda + \Omega\overline{a}\cos\phi\right)}_{\text{basic}} + \underbrace{2Q\Omega\overline{a}\cos^2\phi\left(B + \frac{\widetilde{H}}{2}\right)}_{\text{quasi-shallow}}\right], \tag{18.194}$$

$$\nabla_{hor}\cdot\left(G\mathbf{u}_{hor}\right) \equiv \frac{1}{\overline{a}\cos\phi}\frac{\partial}{\partial\lambda}\left(u_\lambda G\right) + \frac{\partial}{\partial\phi}\left(\cos\phi u_\phi G\right). \tag{18.195}$$

Total Energy

$$\frac{\partial E^{SW}}{\partial t} = -\nabla_{hor}\cdot\left[\left(\underbrace{\frac{u_\lambda^2 + u_\phi^2}{2} + gH}_{\text{basic}} - \underbrace{Q\Omega\cos\phi u_\lambda\widetilde{H}}_{\text{quasi-shallow}}\right)\widetilde{H}\mathbf{u}_{hor}\right], \tag{18.196}$$

$$\frac{\partial}{\partial t}\left(\int_0^{2\pi}\int_{-\frac{\pi}{2}}^{\frac{\pi}{2}} E^{SW}\overline{a}^2\cos\phi d\lambda d\phi\right) = 0, \tag{18.197}$$

where

$$E^{SW} \equiv \underbrace{\left[\frac{u_\lambda^2 + u_\phi^2}{2} + g\left(B + \frac{\widetilde{H}}{2}\right)\right]\widetilde{H}}_{\text{basic}}. \tag{18.198}$$

Potential Vorticity

$$\frac{D_{hor}\Pi^{SW}}{Dt} = 0, \tag{18.199}$$

where

$$\Pi^{SW} \equiv \frac{1}{\widetilde{H}} \left\{ \underbrace{\frac{1}{\bar{a}\cos\phi}\left[\frac{\partial u_\phi}{\partial \lambda} - \frac{\partial (u_\lambda \cos\phi)}{\partial \phi}\right] + 2\Omega \sin\phi}_{\text{basic}} - \underbrace{\frac{2Q\Omega}{\bar{a}\cos\phi}\frac{\partial}{\partial\phi}\left[\cos^2\phi\left(B + \frac{\widetilde{H}}{2}\right)\right]}_{\text{quasi-shallow}} \right\}. \tag{18.200}$$

18.9 DERIVATION OF A UNIFIED QUARTET OF EQUATION SETS

Of the variational approaches applied in this chapter, the one described in this section – applying Hamilton's principle of stationary action – is the most powerful. It leads naturally not only to the 'quasi-shallow' variant of the shallow-water equations but also to a recently developed 'non-hydrostatic' variant. However – due to the use of fluid-parcel labels – the present approach is also arguably the most complicated one! For this reason, we restrict ourselves here to giving an overview of the methodology, plus a few details regarding its application, before providing a particular example (in familiar spherical geometry) of a unified quartet of shallow-water equation sets in spheroidal geometry that may be obtained by using it. The reader interested in a more complete and detailed exposition may wish to consult Staniforth (2015b).

18.9.1 Methodology

In brief, the methodology consists of the following four steps:

1. Make essentially the same assumptions as those made in Section 18.2.2.
2. Derive the mass-continuity equation in a manner similar to that employed in Section 18.2.3.4, but with the added wrinkle that columnar flow is imposed slightly differently by using parcel labels. The end result – equation (18.47) – is the same.
3. Enhance Lagrangian density (18.145) with an optional, supplementary term. This term is derived via vertical integration, in label space, of *vertical* kinetic energy $(u_3^2/2)$.
4. Apply Hamilton's principle of stationary action (by using the enhanced Lagrangian density and taking variations in label space) to obtain a unified, switch-controlled quartet of shallow-water equation sets in axial-orthogonal-curvilinear coordinates. Two members of this quartet correspond to the 'basic' and 'quasi-shallow' sets derived earlier. Two further shallow-water equation sets – termed 'non-hydrostatic' – correspond to the optional inclusion of the supplementary term in the enhanced Lagrangian density.

18.9.2 Hamilton's Principle of Stationary Action

Recall from Section 14.9 that Hamilton's principle of stationary action was applied to derive the non-hydrostatic-deep equations for the global atmosphere or oceans, expressed in axial-orthogonal-curvilinear coordinates. To prepare the way for later developments, we first summarise the application of Hamilton's principle in the present context of incompressible fluid flow (for which $\alpha \equiv \overline{\alpha} =$ constant from (18.14)). Thus:

1. From (14.150), and in the notation of Chapter 14 and the present one, a 3D Lagrangian (L) is defined in a rotating frame of reference and in either location or label space. Thus

$$L\left(\xi, \dot{\xi}\right) = \iiint \mathscr{L}\left(\xi, \dot{\xi}\right) \overline{\rho} h_1 h_2 h_3 d\xi_1 d\xi_2 d\xi_3 = \iiint \mathscr{L}\left(\xi, \dot{\xi}\right) da_1 da_2 da_3, \tag{18.201}$$

where $\boldsymbol{\xi} = (\xi_1, \xi_2, \xi_3)$, $\dot{\boldsymbol{\xi}} = (\dot{\xi}_1, \dot{\xi}_2, \dot{\xi}_3)$, $\mathbf{a} = (a_1, a_2, a_3)$ are labels, $\overline{\rho} \equiv 1/\overline{\alpha} =$ constant, and

$$\mathscr{L}\left(\boldsymbol{\xi}, \dot{\boldsymbol{\xi}}\right) = \underbrace{\frac{h_1^2\dot{\xi}_1^2 + h_2^2\dot{\xi}_2^2}{2}}_{\text{horizontal KE}} + \underbrace{\frac{h_3^2}{2}\dot{\xi}_3^2}_{\text{vertical KE}} + \underbrace{\Omega h_1^2 \dot{\xi}_1}_{\text{Coriolis}} - \underbrace{\Phi\left(\xi_3\right)}_{\text{gravity}} \tag{18.202}$$

is *3D Lagrangian density*. The middle expression in (18.201) is given in location space, and the last one in label space. Note that $h_i = h_i\left(\xi_2, \xi_3\right), i = 1, 2, 3$, due to use of an axial-orthogonal-curvilinear coordinate system; and $\Phi = \Phi\left(\xi_3\right)$, since a geopotential coordinate system is assumed here. Furthermore – see (14.117), (14.18), and (18.14) –

$$\alpha = \overline{\alpha} = \frac{\partial\left(\mathbf{r}\right)}{\partial\left(\mathbf{a}\right)} = \frac{\partial\left(\mathbf{r}\right)}{\partial\left(\boldsymbol{\xi}\right)}\frac{\partial\left(\boldsymbol{\xi}\right)}{\partial\left(\mathbf{a}\right)} = h_1 h_2 h_3 \frac{\partial\left(\boldsymbol{\xi}\right)}{\partial\left(\mathbf{a}\right)} = \text{constant.} \tag{18.203}$$

This equation embodies mass conservation for incompressible fluid flow; see Section 14.3 with $\alpha \equiv \overline{\alpha} =$ constant.

2. Using (14.152), the action is

$$\mathcal{A}\left(\boldsymbol{\xi}\right) = \int_{\tau_1}^{\tau_2} L\left(\boldsymbol{\xi}, \dot{\boldsymbol{\xi}}\right) d\tau, \tag{18.204}$$

where $L\left(\boldsymbol{\xi}, \dot{\boldsymbol{\xi}}\right)$ is as in (18.201) and (18.202).

3. Using (14.153) and (14.154), Hamilton's principle of stationary action is applied. Taking variations of the action then gives

$$\delta\mathcal{A}\left(\xi\right) \equiv \delta \int_{\tau_1}^{\tau_2} L\left(\boldsymbol{\xi}, \dot{\boldsymbol{\xi}}\right) d\tau = \int_{\tau_1}^{\tau_2} \delta L\left(\boldsymbol{\xi}, \dot{\boldsymbol{\xi}}\right) d\tau = 0, \tag{18.205}$$

for independent variations $\delta\boldsymbol{\xi}\left(\mathbf{a}, \tau\right)$ of the generalised coordinates, $\boldsymbol{\xi}$, such that

$$\delta\boldsymbol{\xi}\left(\mathbf{a}, \tau_1\right) = \delta\boldsymbol{\xi}\left(\mathbf{a}, \tau_2\right) = 0. \tag{18.206}$$

Here τ is time; Lagrangian $L\left(\boldsymbol{\xi}, \dot{\boldsymbol{\xi}}\right)$ is given by (18.201) and (18.202); and action $\mathcal{A}\left(\xi\right)$ by (18.204).

18.9.3 Columnar Flow and Parcel Labels

The shallow layer of incompressible fluid is assumed to move coherently in a wholly vertical, columnar manner, as expressed by (18.28). To impose this columnar constraint within the variational framework of Hamilton's principle of stationary action, it is analogously assumed that horizontal coordinates ξ_1 and ξ_2 only depend on the first two of the three parcel labels $\mathbf{a} \equiv (a_1, a_2, a_3)$. Thus

$$\xi_1 = \xi_1\left(a_1, a_2, \tau\right), \quad \xi_2 = \xi_2\left(a_1, a_2, \tau\right), \tag{18.207}$$

where, recall, τ denotes time in label space; see Section 14.2.2.[27] But, concretely:

- How are labels assigned?

[27] Further recall that although $\tau \equiv t$, τ is intrinsically linked to *label space* (a_1, a_2, a_3, τ), whereas t is intrinsically linked to *location space* (x_1, x_2, x_3, t). Partial differentiation with respect to t in location space (x_1, x_2, x_3, t), holding (x_1, x_2, x_3) fixed, corresponds to taking the Eulerian derivative $\partial/\partial t$ at a fixed point in location space. Contrastingly, partial differentiation with respect to τ in label space (a_1, a_2, a_3, τ), holding (a_1, a_2, a_3) fixed, corresponds to taking the Lagrangian derivative $\partial/\partial\tau$ at a fixed point of a moving fluid parcel. $\partial/\partial\tau$ in label space is commonly written as D/Dt in location space, i.e. $\partial/\partial\tau = D/Dt$ are two different but equivalent ways of expressing the same Lagrangian derivative. Labels (a_1, a_2, a_3) of a fluid parcel, once assigned, are permanent and remain unchanged for all time.

This is, in general, quite arbitrary and a matter of convenience and context. Following Staniforth (2015b) – who followed Chapter 7.6 of Salmon (1998) – *vertical label* a_3 is assigned such that (for fixed a_1 and a_2, i.e. for a given column) a_3 varies monotonically from zero at the rigid lower boundary (where $s_3 = s_3^B = h_3(\xi_2)\left(\xi_3^B - \xi_3^0\right) = B(\xi_1, \xi_2)$) to an arbitrary constant ($\overline{H}$) at the free surface (where $s_3 = s_3^H = h_3(\xi_2)\left(\xi_3^B - \xi_3^0\right) = H(\xi_1, \xi_2, t)$). The present context is graphically depicted in Fig. 18.5:

- In location space – in panels (a) and (b); and in mixed location-label space – in panels (c) and (d).
- At two different times: at $t = \tau = t_1 = \tau_1$ – in panels (a) and (c); and at $t = \tau = t_2 = \tau_2$ – in panels (b) and (d).

Arbitrary constant $\overline{H}$ is simply a vertical scaling constant.[28]

The assignment of *horizontal labels* (a_1, a_2) is also quite arbitrary. Once assigned (at initial time $\tau = t = 0$), (a_1, a_2) are passively transported by the fluid flow, unchanged. To fix ideas, and following oft-used convention, the horizontal labels may be set to the values of the horizontal coordinates at initial time; that is, $(a_1, a_2) = (\xi_1, \xi_2)_{\tau=t=0}$.

A snapshot of a vertical cross section, at some time $t = t_1$, is depicted in location space $(\xi_2, s_3, t = t_1)$ in Fig. 18.5a.[29] This panel corresponds to Fig. 18.4. Assuming a left-to-right (incompressible) fluid flow, a snapshot at a later time ($t = t_2$) is displayed in Fig. 18.5b, again in location space. The free-surface height $s_3 = s_3^H = H(\xi_1, \xi_2, t)$ has evolved (think of a fluid sloshing) during the time interval $[t_1, t_2]$. A particular vertical column (in medium blue, and of infinitesimal cross section) has moved from its horizontal location at $\xi_2 = \xi_2(t = t_1)$ to that at $\xi_2 = \xi_2(t = t_2)$, as the column ascends the slope of the rigid lower boundary (in green). In doing so, the flow deforms the fluid column, including the infinitesimal volume element (in red) within it. The column contracts in the vertical, with a compensating expansion of its horizontal (infinitesimal) cross section to respect incompressibility of the fluid; cf. Fig. 18.5b with Fig. 18.5a.

The physical evolution depicted in Figs. 18.5a and b in location space is similarly depicted in Figs. 18.5c and d, but in *mixed* location-label space. Vertical coordinate s_3 in location space has been replaced by vertical coordinate a_3 in label space, whilst leaving unchanged horizontal coordinate ξ_2 in location space.[30] Vertical label coordinate a_3 remains unchanged in time at the rigid lower boundary and upper free surface, with values $a_3 = 0$ and $a_3 = \overline{H}$, respectively. When viewed in mixed location-label space $(\xi_2, a_3, \tau = t)$, although vertical label coordinate a_3 anywhere with the fluid column remains constant in time, the horizontal infinitesimal cross section of the column nevertheless expands or contracts as a function of horizontal location coordinate ξ_2. This is simply an artefact of viewing things in *mixed* location-label space, which is part of one thing and part of another.

At this point, the reader may wonder why one bothers at all with label coordinates, and why one does not always use arguably easier-to-understand location coordinates instead? The underlying reason is that certain mathematical operations that arise in applying Hamilton's principle proceed more easily in label coordinates. For example, $\partial/\partial\tau$ commutes with $\partial/\partial a_i, i = 1, 2, 3$, in label coordinates (a_1, a_2, a_3, τ), since a_i and τ are independent variables; this then means that $\partial^2 F/\partial\tau\partial a_i \equiv \partial^2 F/\partial a_i \partial\tau$. Now $\partial F/\partial\tau$ in label coordinates equates to DF/Dt in location coordinates (ξ_1, ξ_2, ξ_3, t). However, D/Dt does *not* analogously commute with $\partial/\partial\xi_i, i = 1, 2, 3$, in

[28] Setting $\overline{H}$ to a representative value of the free-surface height $s_3^H = H(\xi_1, \xi_2, t)$ above reference level $s_3^0 \equiv \xi_3 - \xi_3^0 = 0$ is a convenient choice that endows vertical label a_3 with the dimension of length, thereby aiding physical interpretation.

[29] It is convenient to use s_3 as a proxy for ξ_3, since s_3 corresponds to physical distance from reference surface $\xi_3 = \xi_3^0$. This facilitates physical visualisation and interpretation.

[30] One could, in both Figs. 18.5c and d, further replace horizontal coordinate ξ_2 in location space by horizontal coordinate a_2 in label space, to obtain a depiction wholly in label space. If one were to do so, then Fig. 18.5d would simply repeat Fig. 18.5c since fluid parcel labels, once assigned, remain unchanged for all time. In label space, the column does not move as τ changes.

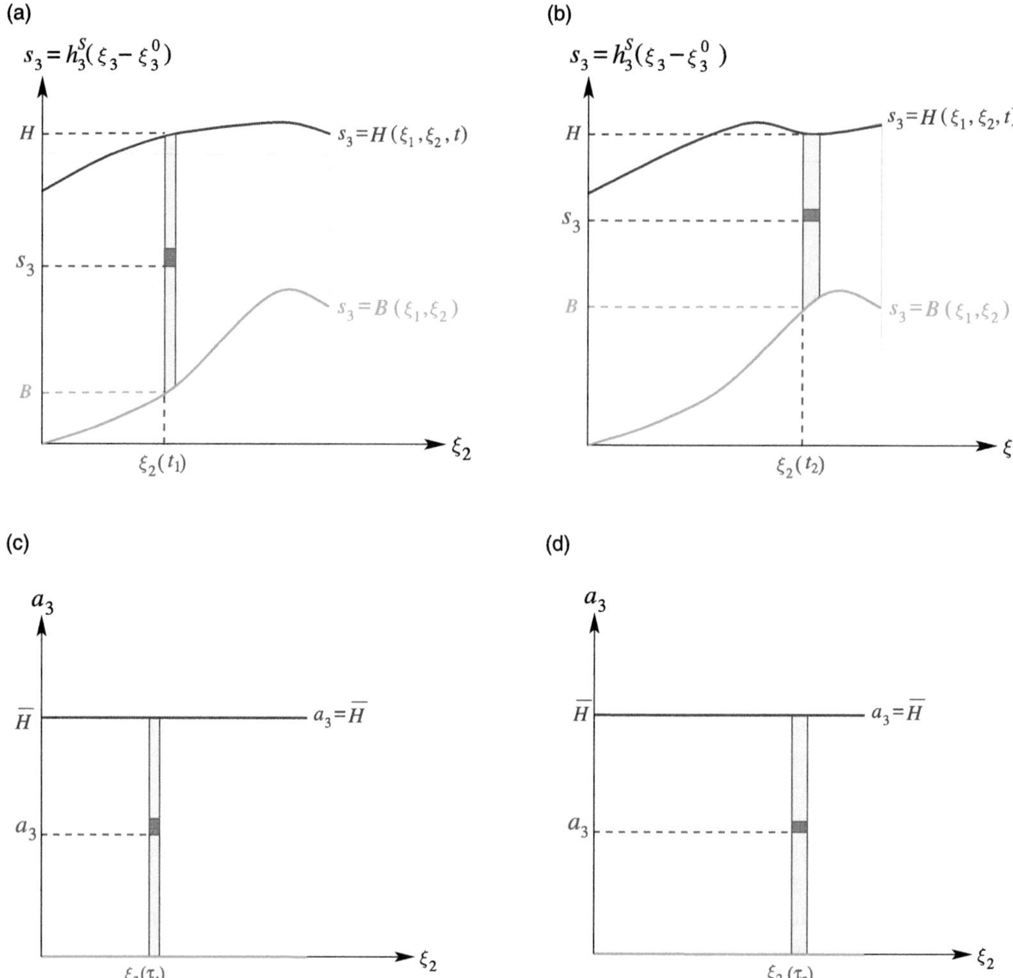

Figure 18.5 Snapshots of horizontal-vertical cross sections of a shallow layer of fluid (in light blue) showing the position at two different times of a given column of fluid (in darker blue) overlying a rigid boundary (in green). The vertical coordinate in panels (a) and (b) is s_3, the vertical coordinate of location space. For panels (c) and (d) it is a_3, the vertical coordinate of label space. The fluid column moves from left (at time $t = t_1$) to right (at later time $t = t_2$) in panels (a) and (b), respectively; and similarly in panels (c) and (d). The fluid's free-surface height (dark blue contour) *varies* as a function of time t for vertical (location) coordinate s_3 (cf. panel (b) with panel (a)), but it is *constant* in time $\tau = t$ for vertical (label) coordinate a_3 (cf. panel (d) with panel (c)). See text for further details.

location coordinates, and $\partial\,(DF/Dt)\,/\partial\xi_i \neq D\,(\partial F/\partial\xi_i)\,/Dt$. An important point here is that some operations happen to be easier to perform in one coordinate system than in another equivalent one and, in practice, one can exploit this. The end result must nevertheless be the same. Hamilton's principle is a *physical* principle, and the use of a different *mathematical* coordinate system is simply a different mathematical means to the same physical end.

18.9.4 Constant Density and the Mass-Continuity Equation

Thus far, vertical label a_3 has been defined to be a monotonic, increasing function from a lower-boundary value of $a_3 = 0$ to an upper-boundary one of $a_3 = \overline{H} > 0$. It remains to determine how

a_3 varies between these two prescribed, bounding values for a_3. How it does so is constrained by incompressibility of the fluid.

Now it was shown in Section 3.3 of Staniforth (2015b) that mass-conservation equation (18.203), together with the above vertical boundary conditions on a_3, lead to – cf. his equation (38) –

$$\xi_3 - \xi_3^0 = \frac{1}{h_3(\xi_2)}\left[B(\xi_1,\xi_2) + \frac{a_3}{\overline{H}}\widetilde{H}(\xi_1,\xi_2,t)\right]. \tag{18.208}$$

Solving this equation for a_3, with use of (18.23), then leads to

$$a_3 = \frac{\overline{H}}{\widetilde{H}(\xi_1,\xi_2,t)}\left[\left(\xi_3 - \xi_3^0\right)h_3(\xi_2) - B(\xi_1,\xi_2)\right] = \frac{\overline{H}}{\widetilde{H}(\xi_1,\xi_2,t)}\left[s_3 - B(\xi_1,\xi_2)\right]. \tag{18.209}$$

Thus a_3 is *linearly* related to s_3 (and also to ξ_3). This behaviour was *deduced* from various assumptions previously made, rather than being directly and arbitrarily *imposed* subsequently. To do otherwise would introduce an inconsistency.

It was also shown in Section 3.3 of Staniforth (2015b) that – his equation (39) –

$$u_3(\xi_1,\xi_2,t) = \left(\frac{\partial B}{\partial \tau} - \frac{B}{h_3}\frac{u_2}{h_2}\frac{dh_3}{d\xi_2}\right) + \frac{a_3}{\overline{H}}\left(\frac{\partial \widetilde{H}}{\partial \tau} - \frac{\widetilde{H}}{h_3}\frac{u_2}{h_2}\frac{dh_3}{d\xi_2}\right). \tag{18.210}$$

Noting that $\partial/\partial\tau = D/Dt$ and using (18.209), (18.210) may be rewritten as

$$u_3(\xi_1,\xi_2,t) = \frac{D_{\mathrm{hor}}B}{Dt^-} + \frac{a_3}{\overline{H}}\frac{D_{\mathrm{hor}}\widetilde{H}}{D\tau^-} = \frac{D_{\mathrm{hor}}B}{Dt^-} + \frac{\left[s_3 - B(\xi_1,\xi_2)\right]}{\widetilde{H}(\xi_1,\xi_2,t)}\frac{D_{\mathrm{hor}}\widetilde{H}}{D\tau^-}, \tag{18.211}$$

where[31]

$$\frac{D_{\mathrm{hor}}F}{Dt^-} \equiv \frac{D_{\mathrm{hor}}F}{Dt^-} - \frac{F}{h_3^S}\frac{u_2}{h_2^S}\frac{dh_3^S}{d\xi_2}. \tag{18.212}$$

It is seen from (18.210) and (18.211) that u_3 depends *linearly* on a_3 and s_3 (and also, through (18.23), on ξ_3) when viewed in label space and location space, respectively.

It was further shown in Section 3.3 of Staniforth (2015b) that mass-continuity equation (18.203) leads to (18.40) – see his equation (44). In a manner similar to that given in Section 18.2.3.4 herein, vertical integration of (18.40) was shown to lead to mass-continuity equation (18.47) herein – see his equation (49).

It remains to derive the two horizontal component equations for momentum using Hamilton's principle of stationary action.

18.9.5 The Shallow-Water Lagrangian and Lagrangian Density

To apply Hamilton's principle of stationary action – as summarised in Section 18.9.2 – one needs to suitably define the 3D Lagrangian density $\mathscr{L}\left(\boldsymbol{\xi},\dot{\boldsymbol{\xi}}\right)$ that appears in (18.201). Such a Lagrangian is essentially given by (17.53) of Section 17.3.1 for the Lagrangian derivation of a quartet of 3D, dynamically consistent, quasi-shallow equation sets for *compressible* flow. Since the flow in

[31] In general (i.e. for *spheroidal* geometry) the last term in (18.212) is non-zero. However, $h_3 \equiv 1$ for the special case of *spherical* geometry. The last term is then identically zero, and (18.212) reduces to $D_{\mathrm{hor}}F/Dt^- \equiv D_{\mathrm{hor}}F/Dt$. This property is exploited in Section 18.10.

the present context is assumed to be *incompressible*, we can drop the redundant internal-energy contribution ($\mathscr{E}_S$) from this Lagrangian. Thus:

$$\mathscr{L}_{QS}^{Q,\mathcal{N}}\left(\boldsymbol{\xi},\dot{\boldsymbol{\xi}}\right) = \underbrace{\frac{\left(h_1^2\right)^S \dot{\xi}_1^2 + \left(h_2^2\right)^S \dot{\xi}_2^2}{2}}_{\text{horizontal kinetic energy}} - \underbrace{\Phi_S\left(\xi_3\right)}_{\text{gravity}} + \underbrace{\Omega\left(h_1^2\right)^S \dot{\xi}_1}_{\text{shallow Coriolis}} + \underbrace{Q\Omega\left(\frac{\partial h_1^2}{\partial \xi_3}\right)^S\left(\xi_3 - \xi_3^S\right)\dot{\xi}_1}_{\text{quasi-shallow Coriolis}} + \underbrace{\frac{\mathcal{N}\left(h_3^2\right)^S \dot{\xi}_3^2}{2}}_{\text{non-hydrostatic}}, \tag{18.213}$$

where the two switches ($Q = 0, 1$, and $\mathcal{N} = 0, 1$) control the presence (for value unity) or absence (for value zero) of quasi-shallow and non-hydrostatic contributions, respectively.[32] The 'non-hydrostatic' contribution[33] corresponds to inclusion of vertical kinetic energy, $u_3 = \left(h_3^2\right)^S \dot{\xi}_3^2/2$.

18.9.5.1 *The Shallow-Water Lagrangian*

Having defined 3D Lagrangian density (18.213), and following Staniforth (2015b), insertion of (18.213) into (18.201) gives the shallow-water Lagrangian

$$\begin{aligned}
L_{SW}^{Q,\mathcal{N}} &= \iint \left\{ \int_0^{\overline{H}} \left[\frac{\left(h_1^2\right)^S \dot{\xi}_1^2 + \left(h_2^2\right)^S \dot{\xi}_2^2}{2} + \Omega\left(h_1^2\right)^S \dot{\xi}_1 \right] da_3 \right\} da_1 da_2 \\
&+ \iint \int_0^{\overline{H}} \left[-\Phi_S\left(\xi_3\right) + Q\Omega\left(\frac{\partial h_1^2}{\partial \xi_3}\right)^S \left(\xi_3 - \xi_3^S\right)\dot{\xi}_1 + \frac{\mathcal{N}\left(h_3^2\right)^S}{2}\dot{\xi}_3^2 \right] da_1 da_2 da_3 \\
&= \overline{H} \iint \left[\frac{\left(h_1^2\right)^S \dot{\xi}_1^2 + \left(h_2^2\right)^S \dot{\xi}_2^2}{2} + \Omega\left(h_1^2\right)^S \dot{\xi}_1 \right] da_1 da_2 \\
&+ \iint \left\{ \int_0^{\overline{H}} \left[-\Phi_S\left(\xi_3\right) + Q\Omega\left(\frac{\partial h_1^2}{\partial \xi_3}\right)^S \left(\xi_3 - \xi_3^S\right)\dot{\xi}_1 + \frac{\mathcal{N}\left(h_3^2\right)^S}{2}\dot{\xi}_3^2 \right] da_3 \right\} da_1 da_2.
\end{aligned} \tag{18.214}$$

The penultimate line in (18.214) is obtained from the first line by exploiting the facts that:

- The integrand is independent of the vertical.
- $\int_0^{\overline{H}} da_3 = \overline{H}$.

This simplification does not, however, hold for the terms in the second and fourth lines since their integrands depend on vertical label a_3. This complicates things a little. However, all that needs to be done is to explicitly introduce the (linear or quadratic, depending on the term) dependence on a_3 of the three individual terms and then integrate them in the vertical, as follows.

32 For $\mathcal{N} = 0$ and $\mathscr{E}_S = 0$, (18.213) corresponds to (18.248) of Appendix A to this chapter.

33 This contribution was termed 'non-hydrostatic' since it led to non-hydrostatic terms appearing in the 3D component equations (17.63) and (17.64) of momentum. These terms are absent in the hydrostatic primitive equations (i.e. when $\mathcal{N} = 0$).

The Gravity Contribution

Now – from (18.21) and (18.208) –

$$\Phi_S(\xi_3) = \Phi_S^0 + h_3^S(\xi_2)\,g(\xi_2)\left(\xi_3 - \xi_3^0\right) = \Phi_S^0 + g(\xi_2)\left[B(\xi_1,\xi_2) + \frac{a_3}{\overline{H}}\widetilde{H}(\xi_1,\xi_2,t)\right], \quad (18.215)$$

which is *linear* in a_3. Vertically integrating this equation then gives

$$\begin{aligned}\int_0^{\overline{H}} \Phi_S(\xi_3)\,da_3 &= \overline{H}\Phi_S^0 + g(\xi_2)\left[\overline{H}B(\xi_1,\xi_2) + \frac{1}{\overline{H}}\int_0^{\overline{H}} a_3 da_3 \widetilde{H}(\xi_1,\xi_2,t)\right] \\ &= \overline{H}\left\{\cancelto{0}{\Phi_S^0} + g(\xi_2)\left[B(\xi_1,\xi_2) + \frac{\widetilde{H}(\xi_1,\xi_2,t)}{2}\right]\right\} \\ &= \overline{H}g(\xi_2)\left[B(\xi_1,\xi_2) + \frac{\widetilde{H}(\xi_1,\xi_2,t)}{2}\right], \end{aligned} \quad (18.216)$$

where, for brevity and without loss of generality, Φ_S^0 has been set to zero since the geopotential is only ever defined to within an arbitrary additive constant.

The Quasi-Shallow Contribution

Using (18.208), it is seen that the second term in the integrand of the last line of (18.214) is also *linear* in a_3. Vertical integration of this (quasi-hydrostatic) term then gives

$$\begin{aligned}\int_0^{\overline{H}} Q\Omega\left(\frac{\partial h_1^2}{\partial \xi_3}\right)^S\left(\xi_3 - \xi_3^S\right)\dot{\xi}_1 da_3 &= \int_0^{\overline{H}} Q\Omega\frac{1}{h_3}\left(\frac{\partial h_1^2}{\partial \xi_3}\right)^S\left[B(\xi_1,\xi_2) + \frac{a_3}{\overline{H}}\widetilde{H}(\xi_1,\xi_2,t)\right]da_3 \\ &= \overline{H}Q\Omega\sigma^S\left[B(\xi_1,\xi_2) + \frac{\widetilde{H}(\xi_1,\xi_2,t)}{2}\right], \end{aligned} \quad (18.217)$$

where $\sigma^S = \sigma^S(\xi_2)$ is defined by (18.142).

The Non-Hydrostatic Contribution

Now using (18.210), it is seen that the last term in the integrand of the last line of (18.214) includes a *quadratic* dependence on a_3, instead of just the linear dependence of the preceding two contributions. Vertical integration of a quadratic contribution is nevertheless still straightforward to do. Thus

$$\begin{aligned}\int_0^{\overline{H}} \frac{\mathcal{N}\left(h_3^2\right)^S\dot{\xi}_3^2}{2}da_3 &= \int_0^{\overline{H}}\frac{\mathcal{N}u_3^2}{2}da_3 \\ &= \frac{\mathcal{N}}{2}\int_0^{\overline{H}}\left[\left(\frac{\partial B}{\partial \tau} - \frac{B}{h_3}\frac{u_2}{h_2}\frac{dh_3}{d\xi_2}\right) + \frac{a_3}{\overline{H}}\left(\frac{\partial \widetilde{H}}{\partial \tau} - \frac{\widetilde{H}}{h_3}\frac{u_2}{h_2}\frac{dh_3}{d\xi_2}\right)\right]^2 da_3 \\ &= \frac{\mathcal{N}\overline{H}}{2}\left[\left(\frac{\partial B}{\partial \tau^-}\right)^2 + \frac{\partial B}{\partial \tau^-}\frac{\partial \widetilde{H}}{\partial \tau^-} + \frac{1}{3}\left(\frac{\partial \widetilde{H}}{\partial \tau^-}\right)^2\right], \end{aligned} \quad (18.218)$$

where

$$\frac{\partial F}{\partial \tau^-} \equiv \frac{\partial F}{\partial \tau} - \frac{F}{h_3}\frac{u_2}{h_2}\frac{dh_3}{d\xi_2}, \quad (18.219)$$

for any differentiable scalar function F.

The Unified Shallow-Water Lagrangian

Inserting (18.216)–(18.218) into (18.214) and rearranging yields the unified, shallow-water Lagrangian (with switches Q and $\mathcal{N}$)

$$L_{\mathrm{SW}}^{Q,\mathcal{N}} = \overline{H} \iint \left[\frac{\left(h_1^2\right)^S \dot{\xi}_1^2 + \left(h_2^2\right)^S \dot{\xi}_2^2}{2} - g\left(B + \frac{\widetilde{H}}{2}\right) + \Omega \left(h_1^2\right)^S \dot{\xi}_1 \right] da_1 da_2$$
$$+ \overline{H} \iint \left\{ Q\Omega\sigma^S \left(B + \frac{\widetilde{H}}{2}\right) + \frac{\mathcal{N}}{2} \left[\left(\frac{\partial B}{\partial \tau^-}\right)^2 + \frac{\partial B}{\partial \tau^-}\frac{\partial \widetilde{H}}{\partial \tau^-} + \frac{1}{3}\left(\frac{\partial \widetilde{H}}{\partial \tau^-}\right)^2 \right] \right\} da_1 da_2. \tag{18.220}$$

This 2D shallow-water Lagrangian[34] corresponds to that obtained in Staniforth (2015b), that is,

$$L_{\mathrm{SW}}^{Q,\mathcal{N}} = \mathfrak{L} \equiv \mathfrak{L}^{(1)} + \mathfrak{L}^{(2)} + \mathfrak{L}^{(3)},$$

where $\mathfrak{L}$, $\mathfrak{L}^{(1)}$, $\mathfrak{L}^{(2)}$, and $\mathfrak{L}^{(3)}$ are defined by his equations (1), (52), (56), and (60), respectively – but without the (redundant) internal energy contribution of his (60).

18.9.5.2 *The Shallow-Water Lagrangian Density*

From (18.220), we can extract the corresponding 2D, shallow-water Lagrangian density

$$\mathscr{L}_{\mathrm{SW}}^{Q,\mathcal{N}}\left(\xi_1, \xi_2, \dot{\xi}_1, \dot{\xi}_2\right) = \underbrace{\frac{\left(h_1^2\right)^S \dot{\xi}_1^2 + \left(h_2^2\right)^S \dot{\xi}_2^2}{2}}_{\text{horizontal KE}} \underbrace{- g\left(B + \frac{\widetilde{H}}{2}\right)}_{\text{gravity}} + \underbrace{\Omega \left(h_1^2\right)^S \dot{\xi}_1}_{\text{shallow Coriolis}}$$
$$\underbrace{+ Q\Omega\sigma^S \left(B + \frac{\widetilde{H}}{2}\right)\dot{\xi}_1}_{\text{quasi-shallow Coriolis}} + \underbrace{\frac{\mathcal{N}}{2} \left[\left(\frac{\partial B}{\partial \tau^-}\right)^2 + \frac{\partial B}{\partial \tau^-}\frac{\partial \widetilde{H}}{\partial \tau^-} + \frac{1}{3}\left(\frac{\partial \widetilde{H}}{\partial \tau^-}\right)^2 \right]}_{\text{non-hydrostatic}}, \tag{18.221}$$

where, for simplicity, the multiplicative factor of $\overline{H}$ throughout has been ignored.[35] This is a slightly more general Lagrangian density than (18.145), since it includes an additional *non-hydrostatic* contribution (in blue) that originates from vertical kinetic energy $\left(h_3^2\right)^S \dot{\xi}_3^2/2$.

18.9.5.3 *An Alternative Derivation of Shallow-Water Lagrangian Density*

Noting that $\partial F/\partial \tau$ in label space is equivalent to DF/Dt in location space, shallow-water Lagrangian density (18.221) may be rewritten as

$$\mathscr{L}_{\mathrm{SW}}^{Q,\mathcal{N}}\left(\xi_1, \xi_2, \dot{\xi}_1, \dot{\xi}_2\right) = \underbrace{\frac{\left(h_1^2\right)^S \dot{\xi}_1^2 + \left(h_2^2\right)^S \dot{\xi}_2^2}{2}}_{\text{horizontal KE}} \underbrace{- g\left(B + \frac{\widetilde{H}}{2}\right)}_{\text{gravity}} + \underbrace{\Omega \left(h_1^2\right)^S \dot{\xi}_1}_{\text{shallow Coriolis}}$$

[34] All dependence on vertical label a_3 has been removed via vertical integration over a_3.

[35] As previously noted, the value of $\overline{H}$ is arbitrary, so there is no loss of generality by setting it to unity here. Alternatively, one could absorb $\overline{H}$ into the definition of the horizontal labels. Furthermore, any Lagrangian is only determined to within an arbitrary multiplicative constant (of either sign) since its value has no influence on the stationarity conditions of the Lagrangian.

$$
\underbrace{+ Q\Omega\sigma^{S}\left(B + \frac{\widetilde{H}}{2}\right)\dot{\xi}_1}_{\text{quasi-shallow Coriolis}}
$$

$$
\underbrace{+\frac{\mathcal{N}}{2}\left[\left(\frac{D_{\text{hor}}B}{Dt^{-}}\right)^2 + \frac{DB_{\text{hor}}}{Dt^{-}}\frac{D_{\text{hor}}\widetilde{H}}{Dt^{-}} + \frac{1}{3}\left(\frac{D_{\text{hor}}\widetilde{H}}{Dt^{-}}\right)^2\right]}_{\text{non-hydrostatic}}, \tag{18.222}
$$

where $D_{\text{hor}}F/Dt^{-}$ is defined by (18.212). Now for $\mathcal{N} = 0$, (18.222) reduces to (18.259). This latter Lagrangian density was obtained by setting $\mathcal{N} \equiv 0$ in 3D, quasi-hydrostatic, quasi-shallow Lagrangian density (17.53) and vertically averaging the result *in location space.* This suggests that one should be able to obtain the additional 'non-hydrostatic' term in (18.222) by instead setting $\mathcal{N} = 1$ in (17.53) and vertically averaging, *in location space*, the additional contribution $\left(h_3^2\right)^S \dot{\xi}_3^2/2$ for vertical kinetic energy. We now show that this is indeed the case.

Now – from (18.211) –

$$
\frac{\left(h_3^2\right)^S \dot{\xi}_3^2}{2} = \frac{u_3^2}{2} = \frac{1}{2}\left[\left(\frac{D_{\text{hor}}B}{Dt^{-}}\right)^2 + 2\frac{D_{\text{hor}}B}{Dt^{-}}\frac{D_{\text{hor}}\widetilde{H}}{Dt^{-}}\left(\frac{s_3 - B}{\widetilde{H}}\right) + \left(\frac{D_{\text{hor}}\widetilde{H}}{Dt^{-}}\right)^2\left(\frac{s_3 - B}{\widetilde{H}}\right)^2\right]. \tag{18.223}
$$

Vertically averaging (18.223) over the depth of the fluid $\widetilde{H} \equiv H - B$, as in Appendix A of the present chapter, gives

$$
\begin{aligned}
\frac{\widehat{\left(h_3^2\right)^S \dot{\xi}_3^2}}{2} &\equiv \frac{1}{\widetilde{H}}\int_B^H \frac{\left(h_3^2\right)^S \dot{\xi}_3^2}{2} ds_3 = \frac{1}{2\widetilde{H}}\left(\frac{D_{\text{hor}}B}{Dt^{-}}\right)^2 \int_B^H ds_3 + \frac{1}{\widetilde{H}}\frac{D_{\text{hor}}B}{Dt^{-}}\frac{D_{\text{hor}}\widetilde{H}}{Dt^{-}}\int_B^H \left(\frac{s_3 - B}{\widetilde{H}}\right) ds_3 \\
&\quad + \frac{1}{2\widetilde{H}}\left(\frac{D_{\text{hor}}\widetilde{H}}{Dt^{-}}\right)^2 \int_B^H \left(\frac{s_3 - B}{\widetilde{H}}\right)^2 ds_3 \\
&= \frac{1}{2}\left[\left(\frac{D_{\text{hor}}B}{Dt^{-}}\right)^2 + \frac{D_{\text{hor}}B}{Dt^{-}}\frac{D_{\text{hor}}\widetilde{H}}{Dt^{-}} + \frac{1}{3}\left(\frac{D_{\text{hor}}\widetilde{H}}{Dt^{-}}\right)^2\right].
\end{aligned} \tag{18.224}
$$

Adding this contribution to (18.259) then leads to (18.222), thereby verifying the conjecture.

18.9.6 Action, Variations and Governing Equations

Having constructed 2D Lagrangian (18.220) in label space – this corresponds to 2D Lagrangian density (18.221) – Step 1 of Section 18.9.2 for the derivation of the unified shallow-water equation set is now essentially complete. It remains to apply Steps 2 and 3. Insertion of (18.220) into (18.204) then defines the corresponding action $\mathcal{A}(\boldsymbol{\xi})$, thereby completing Step 2. The process of taking variations of this action – that is, applying Step 3 – is summarised in Appendix A of Staniforth (2015b). This is similar to what is described in the Appendix of Chapter 14 herein – for the derivation of the more general 3D equation set for atmospheric and oceanic modelling, expressed in axial-orthogonal-curvilinear coordinates.

Taking variations of the non-hydrostatic contributions (i.e. when $\mathcal{N} \equiv 1$) to the action – these originate from the non-hydrostatic terms in (18.220) – is a challenging and lengthy exercise. The resulting component equations for horizontal momentum, in axial-orthogonal-curvilinear coordinates, are neither concise nor pretty to behold – see equations (64) and (65) of Staniforth (2015b). We therefore restrict ourselves here to simply stating them here for the special case of (shallow) spherical-polar coordinates – see equations (73) and (74) of Staniforth (2015b). They are then a little more concise and a little less complicated. This gives the reader the opportunity to assess their possible interest in the augmented shallow-water equation set.

18.10 THE UNIFIED QUARTET IN SPHERICAL GEOMETRY

In what follows, the governing equations and associated conservation principles for the unified quartet of equation sets in spherical geometry are summarised in the format and notation of Section 18.8. This facilitates comparison of the complete, unified, shallow-water equation set (corresponding to $Q \equiv 1, \mathcal{N} \equiv 1$) with the equations of Section 18.8 for the 'quasi-shallow' subset (corresponding to $Q \equiv 1, \mathcal{N} \equiv 0$ of the unified set). In particular, it allows identification of the influence of the supplementary non-hydrostatic terms that only appear for $\mathcal{N} \equiv 1$.

The Unified Quartet of Shallow-Water Equations in Spherical-Polar Coordinates

Momentum Components

$$\underbrace{\frac{D_{\text{hor}} u_\lambda}{Dt} - \left(\frac{u_\lambda}{\overline{a} \cos\phi} + 2\Omega \right) u_\phi \sin\phi + \frac{1}{\overline{a}\cos\phi} \frac{\partial (gH)}{\partial \lambda}}_{\text{basic}}$$
$$\underbrace{- \frac{Q\Omega}{\overline{a}} \left\{ \frac{\partial}{\partial \phi} \left(\cos\phi u_\phi \widetilde{H} \right) - 2 \frac{u_\phi}{\cos\phi} \frac{\partial}{\partial \phi} \left[\cos^2\phi \left(B + \frac{\widetilde{H}}{2} \right) \right] + 2 \frac{\partial}{\partial \lambda} \left(u_\lambda \widetilde{H} \right) \right\}}_{\text{quasi-shallow}}$$
$$\underbrace{+ \frac{\mathcal{N}}{\overline{a}\cos\phi} \left\{ \frac{1}{\widetilde{H}} \frac{\partial}{\partial \lambda} \left[\widetilde{H}^2 \frac{D^2_{\text{hor}}}{Dt^2} \left(\frac{B}{2} + \frac{\widetilde{H}}{3} \right) \right] + \frac{\partial B}{\partial \lambda} \frac{D^2_{\text{hor}}}{Dt^2} \left(B + \frac{\widetilde{H}}{2} \right) \right\}}_{\text{non-hydrostatic}} = 0, \tag{18.225}$$

$$\underbrace{\frac{D_{\text{hor}} u_\phi}{Dt} + \left(\frac{u_\lambda}{\overline{a} \cos\phi} + 2\Omega \right) u_\lambda \sin\phi + \frac{1}{\overline{a}} \frac{\partial (gH)}{\partial \phi}}_{\text{basic}}$$
$$\underbrace{- \frac{Q\Omega}{\overline{a}} \left\{ \frac{\partial}{\partial \phi} \left(\cos\phi u_\lambda \widetilde{H} \right) + 2 \frac{u_\lambda}{\cos\phi} \frac{\partial}{\partial \phi} \left[\cos^2\phi \left(B + \frac{\widetilde{H}}{2} \right) \right] \right\}}_{\text{quasi-shallow}}$$
$$\underbrace{+ \frac{\mathcal{N}}{\overline{a}} \left\{ \frac{1}{\widetilde{H}} \frac{\partial}{\partial \phi} \left[\widetilde{H}^2 \frac{D^2_{\text{hor}}}{Dt^2} \left(\frac{B}{2} + \frac{\widetilde{H}}{3} \right) \right] + \frac{\partial B}{\partial \phi} \frac{D^2_{\text{hor}}}{Dt^2} \left(B + \frac{\widetilde{H}}{2} \right) \right\}}_{\text{non-hydrostatic}} = 0. \tag{18.226}$$

Mass-Continuity

$$\frac{D_{\text{hor}} \widetilde{H}}{Dt} + \frac{\widetilde{H}}{\overline{a}\cos\phi} \left[\frac{\partial u_\lambda}{\partial \lambda} + \frac{\partial}{\partial \phi} \left(u_\phi \cos\phi \right) \right] = 0, \tag{18.227}$$

or

$$\frac{\partial \widetilde{H}}{\partial t} + \frac{1}{\overline{a}\cos\phi} \left[\frac{\partial}{\partial \lambda} \left(u_\lambda \widetilde{H} \right) + \frac{\partial}{\partial \phi} \left(\cos\phi u_\phi \widetilde{H} \right) \right] = 0. \tag{18.228}$$

Material Derivative

$$\frac{D_{\text{hor}}}{Dt} \equiv \frac{\partial}{\partial t} + \frac{u_\lambda}{\overline{a}\cos\phi} \frac{\partial}{\partial \lambda} + \frac{u_\phi}{\overline{a}} \frac{\partial}{\partial \phi}. \tag{18.229}$$

Dynamical Consistency

Conservation principles for the general case (in axial-orthogonal-curvilinear coordinates) of the unified quartet of equation sets are derived in the online Supporting Information to Staniforth (2015b). The corresponding expressions in spherical-polar coordinates may be straightforwardly obtained by insertion of (18.183) and (18.184) into the appropriate equations of his Section 7. This results in (18.230)–(18.242). Collectively, these conservation principles demonstrate dynamical consistency of the unified quartet of shallow-water equation sets.

Associated Conservation Principles

Mass

$$\frac{\partial \widetilde{H}}{\partial t} = -\frac{1}{\overline{a}\cos\phi}\left[\frac{\partial}{\partial\lambda}\left(u_\lambda \widetilde{H}\right) + \frac{\partial}{\partial\phi}\left(\cos\phi u_\phi \widetilde{H}\right)\right], \tag{18.230}$$

$$\frac{\partial}{\partial t}\left(\int_0^{2\pi}\int_{-\frac{\pi}{2}}^{\frac{\pi}{2}} \widetilde{H}\overline{a}^2 \cos\phi d\lambda d\phi\right) = 0. \tag{18.231}$$

Axial Angular Momentum

$$\begin{aligned}
\frac{\partial M^{SW}}{\partial t} = &- \nabla_{hor} \cdot \left\{\left[\underbrace{\overline{a}\cos\phi\left(u_\lambda + \Omega\overline{a}\cos\phi\right)}_{\text{basic}} + \underbrace{2Q\Omega\overline{a}\cos^2\phi\left(B + \frac{\widetilde{H}}{2}\right)}_{\text{quasi-shallow}}\right]\widetilde{H}\mathbf{u}_{hor}\right\} \\
&- \frac{\partial}{\partial\lambda}\left\{\left[\underbrace{\frac{g}{2}}_{\text{basic}} - \underbrace{Q\Omega\cos\phi u_\lambda}_{\text{quasi-shallow}} + \underbrace{\mathcal{N}\frac{D_{hor}^2}{Dt^2}\left(\frac{B}{2} + \frac{\widetilde{H}}{3}\right)}_{\text{non-hydrostatic}}\right]\widetilde{H}^2\right\} \\
&- \left\{\left[\underbrace{g}_{\text{basic}} - \underbrace{2Q\Omega\cos\phi u_\lambda}_{\text{quasi-shallow}} + \underbrace{\mathcal{N}\frac{D_{hor}^2}{Dt^2}\left(B + \frac{\widetilde{H}}{2}\right)}_{\text{non-hydrostatic}}\right]\widetilde{H}\frac{\partial B}{\partial\lambda}\right\},
\end{aligned} \tag{18.232}$$

$$\begin{aligned}
&\frac{\partial}{\partial t}\left(\int_0^{2\pi}\int_{-\frac{\pi}{2}}^{\frac{\pi}{2}} M^{SW}\overline{a}^2\cos\phi d\lambda d\phi\right) \\
&= -\int_0^{2\pi}\int_{-\frac{\pi}{2}}^{\frac{\pi}{2}}\left[\underbrace{g}_{\text{basic}} - \underbrace{2Q\Omega\cos\phi u_\lambda}_{\text{quasi-shallow}} + \underbrace{\mathcal{N}\frac{D_{hor}^2}{Dt^2}\left(B + \frac{\widetilde{H}}{2}\right)}_{\text{non-hydrostatic}}\right]\widetilde{H}\frac{\partial B}{\partial\lambda}\overline{a}^2\cos\phi d\lambda d\phi,
\end{aligned} \tag{18.233}$$

where

$$M^{SW} \equiv \widetilde{H}\left[\underbrace{\overline{a}\cos\phi\left(u_\lambda + \Omega\overline{a}\cos\phi\right)}_{\text{basic}} + \underbrace{2Q\Omega\overline{a}\cos^2\phi\left(B + \frac{\widetilde{H}}{2}\right)}_{\text{quasi-shallow}}\right], \tag{18.234}$$

$$\nabla_{hor}\cdot\left(G\mathbf{u}_{hor}\right) \equiv \frac{1}{\overline{a}\cos\phi}\left[\frac{\partial}{\partial\lambda}\left(u_\lambda G\right) + \frac{\partial}{\partial\phi}\left(\cos\phi u_\phi G\right)\right]. \tag{18.235}$$

Total Energy

$$\frac{\partial E^{SW}}{\partial t} = -\nabla_{hor}\cdot\left\{\left[\underbrace{\left[\left(\frac{u_\lambda^2 + u_\phi^2}{2}\right) + gH\right.}_{\text{basic}} - \underbrace{Q\Omega\cos\phi u_\lambda\widetilde{H}}_{\text{quasi-shallow}}\right]\widetilde{H}\mathbf{u}_{hor}\right\}$$

$$-\nabla_{hor}\cdot\left\{\underbrace{\mathcal{N}\left[\frac{1}{6}\left(\frac{D_{hor}\widetilde{H}}{Dt}\right)^2 + \frac{1}{2}\frac{D_{hor}\widetilde{H}}{Dt}\frac{D_{hor}B}{Dt} + \frac{1}{2}\left(\frac{D_{hor}B}{Dt}\right)^2 + \widetilde{H}\frac{D_{hor}^2}{Dt^2}\left(\frac{B}{2} + \frac{\widetilde{H}}{3}\right)\right]\widetilde{H}\mathbf{u}_{hor}}_{\text{non-hydrostatic}}\right\}, \tag{18.236}$$

$$\frac{\partial}{\partial t}\left(\int_0^{2\pi}\int_{-\frac{\pi}{2}}^{\frac{\pi}{2}} E^{SW}\overline{a}^2\cos\phi d\lambda d\phi\right) = 0, \tag{18.237}$$

where

$$E^{SW} \equiv \underbrace{\left[\left(\frac{u_\lambda^2 + u_\phi^2}{2}\right) + g\left(B + \frac{\widetilde{H}}{2}\right)\right]\widetilde{H}}_{\text{basic}}$$

$$+ \underbrace{\frac{\mathcal{N}}{2}\left[\frac{1}{3}\left(\frac{D_{hor}\widetilde{H}}{Dt}\right)^2 + \frac{D_{hor}\widetilde{H}}{Dt}\frac{D_{hor}B}{Dt} + \left(\frac{D_{hor}B}{Dt}\right)^2\right]\widetilde{H}}_{\text{non-hydrostatic}}. \tag{18.238}$$

Potential Vorticity

$$\frac{D_{hor}\Pi^{SW}}{Dt} = 0, \tag{18.239}$$

where

$$\Pi^{SW} \equiv \frac{1}{\widetilde{H}}\left\{\underbrace{\frac{1}{\overline{a}\cos\phi}\left[\frac{\partial u_\phi}{\partial\lambda} - \frac{\partial\left(\cos\phi u_\lambda\right)}{\partial\phi}\right] + 2\Omega\sin\phi}_{\text{basic}} - \underbrace{\frac{2Q\Omega}{\overline{a}\cos\phi}\frac{\partial}{\partial\phi}\left[\cos^2\phi\left(B + \frac{\widetilde{H}}{2}\right)\right]}_{\text{quasi-shallow}}\right.$$

$$\left. + \underbrace{\mathcal{N}\zeta_*}_{\text{non-hydrostatic}}\right\} \tag{18.240}$$

is two-dimensional potential vorticity;

$$\zeta_* \equiv J\left[\frac{D_{\text{hor}}}{Dt}\left(\frac{B}{2}+\frac{\widetilde{H}}{3}\right),\widetilde{H}\right]+J\left[\frac{D_{\text{hor}}}{Dt}\left(B+\frac{\widetilde{H}}{2}\right),B\right] \tag{18.241}$$

is the spherical analogue of Miles and Salmon (1985)'s pseudovorticity ζ_* in Cartesian geometry; and

$$J\left[X_1,X_2\right] \equiv \frac{1}{\overline{a}^2\cos\phi}\frac{\partial\left(X_1,X_2\right)}{\partial\left(\lambda,\phi\right)} \tag{18.242}$$

is the 2D Jacobian in spherical-polar coordinates.

18.10.1 Influence of the 'Non-Hydrostatic' Terms

The four members of the unified quartet (18.225)–(18.229) of shallow-water equation sets *in spherical geometry* correspond to the following four combinations of the values of Q and $\mathcal{N}$:

1. $(Q,\mathcal{N}) \equiv (0,1)$: the 'basic' equation set, developed and discussed in Sections 18.2–18.4.
2. $(Q,\mathcal{N}) \equiv (1,0)$: the 'quasi-shallow' equation set, developed and discussed in Sections 18.5–18.8.
3. $(Q,\mathcal{N}) \equiv (0,1)$: *only* augmenting the 'basic' shallow-water equation set with additional 'non-hydrostatic' terms; 'quasi-shallow' contributions are entirely absent.
4. $(Q,\mathcal{N}) \equiv (1,1)$: the most complete shallow-water equation set.
   This set includes all 'quasi-shallow' and 'non-hydrostatic' contributions.

In Sections 18.5–18.8, we examined the influence of enhancing the 'basic' shallow-water equations with the addition of 'quasi-hydrostatic' contributions (in green). We now examine the influence of the supplementary 'non-hydrostatic' terms (in blue), included in component equations (18.185) and (18.186) of horizontal momentum for the unified quartet of equation sets in spherical geometry. This amounts to comparing equations (18.225)–(18.242) with (18.185)–(18.200).

18.10.1.1 Governing Equations

When $\mathcal{N} \equiv 1$ (the 'non-hydrostatic' case), two additional terms materialise in each of the component equations (18.225) and (18.226) in spherical geometry. More generally, in axial-orthogonal-curvilinear coordinates, two further non-hydrostatic terms appear in the second momentum equation – see equation (65) in Staniforth (2015b).

As noted in Staniforth (2015b):

- Setting $\mathcal{N} \equiv 1, Q \equiv 0$ and taking into account differences in notation and geometry, it can be shown that the non-hydrostatic terms in (18.225) and (18.226) agree with those given in Miles and Salmon (1985) for their case of constant gravity with $\Omega \equiv 0$. Miles and Salmon (1985) term these equations the *Green-Naghdi equations* since they are equivalent to those given in Green and Naghdi (1976) for Cartesian geometry, albeit derived differently by Miles and Salmon (1985).
- Using the variational framework, but applying it in Cartesian geometry (and accounting for differences in notation), generates some further equation sets. Setting g = constant, $Q \equiv 1$, and $\mathcal{N} \equiv 0$ or $\mathcal{N} \equiv 1$, leads to the two equation sets (one denoted shallow-water, the other Green-Naghdi) obtained in Dellar and Salmon (2005)'s Section V. Setting g = constant, $\Omega \equiv 0$ and $\mathcal{N} \equiv 1, Q \equiv 0$, and applying the variational framework in Cartesian geometry, leads to Miles and Salmon (1985)'s Green-Naghdi equation set. See also Salmon (1998), pp. 313–318.

- For $\mathcal{N} \equiv 1$, there is generally (in *spheroidal* geometry) a non-zero metric term (proportional to $dh_3^S/d\xi_2$) in the ξ_2 component of the momentum equation – see Staniforth (2015b)'s equation (65). This is, however, absent in *spherical* geometry – see (18.226) – since, from (18.183), $dh_3^S/d\xi_2 \equiv 0$. This is simply a manifestation of gravity being constant in spherical geometry to respect dynamical consistency (White et al., 2005): for the shallow-water equations, h_3^S is constrained by (18.20) to be the reciprocal of g and cannot therefore vary latitudinally.

18.10.1.2 Conservation Principles

Axial Angular Momentum

The presence (for $\mathcal{N} \equiv 1$) of non-hydrostatic terms in component equations (18.225) and (18.226) does not influence analogue definition (18.234) of axial angular momentum, which remains identical to (18.194). These non-hydrostatic terms do, however, modify the *fluxes* of angular momentum – cf. (18.232) with (18.192).

From (18.233), it is seen that global axial angular momentum remains conserved in the absence of orography/bathymetry (i.e. when $B \equiv 0$) and also for zonally symmetric orography/bathymetry (i.e. when $B = B(\phi)$). In the presence of azimuthally varying orography/bathymetry, bottom-torque terms appear on the right-hand side of (18.233), the first being the usual one, the second originating from the quasi-shallow term in (18.225) (with $Q \equiv 1$), and the last originating from the non-hydrostatic term in (18.225) (with $\mathcal{N} \equiv 1$).

Total Energy

The presence (for $\mathcal{N} \equiv 1$) of non-hydrostatic terms in component equations (18.225) and (18.226) modifies the definition of total energy. There is then an additional term – see (18.237). This term originates from the inclusion of *vertical* kinetic energy – see (18.218) – in both (18.220) and (18.221) for the unified Lagrangian and Lagrangian density, respectively. Unsurprisingly, the addition of vertical kinetic energy increases the total energy, as manifested in (18.237). Doing so also adds a contribution to the energy flux – see (18.236).

From (18.237), global total energy is conserved for all members of the quartet of equation sets, regardless of the presence or absence of quasi-shallow and/or non-hydrostatic terms in component momentum equations (18.225) and (18.226).

Potential Vorticity

The contributions to definition (18.240) of potential vorticity that depend on Ω originate from planetary vorticity, with a quasi-shallow contribution when $Q \equiv 1$. When $\mathcal{N} \equiv 1$, there is a non-hydrostatic contribution ζ_*, as defined by (18.241).

In the absence of non-hydrostatic terms (i.e. $\mathcal{N} \equiv 0$), and after taking account of differences in notation and geometry, (18.239)–(18.241) are in a similar form as, and essentially agree with potential-vorticity equation (1) of Dellar and Salmon (2005) for constant gravity in Cartesian β-γ-plane geometry. Similarly, in the absence of quasi-shallow terms (i.e. $Q \equiv 0$), but in the presence of non-hydrostatic terms (i.e. $\mathcal{N} \equiv 1$), and after again taking account of differences in notation and geometry, (18.239)–(18.241) are in a similar form as, and essentially agree with the corresponding expressions in Miles and Salmon (1985). In particular, (18.241) agrees with equation (C 3a) of Miles and Salmon (1985). Furthermore, these equations are also in agreement with equation (B.24) of Tort et al. (2014), for constant gravity in spherical geometry.

18.11 CONCLUDING REMARKS

18.11.1 A Brief Summary

Several complementary equivalent methodologies for the derivation of shallow-water equation sets in spheroidal geometry have been developed and contrasted in this chapter. This was accomplished in a step-by-step manner, culminating in a unified, switch-controlled quartet of equation sets of varying complexity. All members of the quartet are dynamically consistent, possessing conservation principles for mass, axial angular momentum, total energy, and potential vorticity. The simplest member of the quartet, the 'basic' equation set, was developed in spheroidal geometry using the classical Eulerian methodology and well-defined assumptions.

The remaining equation sets were developed using variants of the Lagrangian methodology. To do so, various approximations of Lagrangian density were made, followed by derivation of the corresponding equation sets from them. An advantage of the Lagrangian methodology (compared to the Eulerian one) is that it is easier to identify how one might approximate a Lagrangian density, with there only being a very limited number of possible 'sensible' ways of doing so, followed by a well-defined (albeit somewhat complicated and tedious) way of obtaining the corresponding equation sets.

The most complete member of the quartet of equation sets includes representations of the '$2\Omega \cos \phi$' Coriolis terms and vertical acceleration terms that are missing from traditional shallow-water equation sets. This equation set is then more broadly applicable than traditional ones, both in terms of phenomena represented and of geometry (spheroidal versus spherical).

18.11.2 A Simple Perspective on Model Design, Development, and Validation

Examination of (18.225)–(18.242) shows that the introduction of supplementary *non-hydrostatic* terms significantly complicates the 2D shallow-water equations, even with the simplification of spherical geometry. This consequently complicates and inhibits possible further investigation, either analytically or numerically. Let us assume that we nevertheless do wish to further explore, numerically, the influence of the non-hydrostatic terms.

- How might we proceed?

An obvious but high-risk way would be to design and build a 2D, global, numerical model using the full, switch-controlled, unified set of shallow-water equations (18.225)–(18.229) in spherical geometry. Doing so in one go would, however, be a challenging task, even with the assumed simplification of spherical (as opposed to spheroidal) geometry. It would require a horizontal (spatial) discretisation scheme, plus a time scheme. Traditional and well-understood techniques (such as spectral, finite-difference, and finite-element ones) would be natural candidates for spatial discretisation. However, the choice of a suitable time scheme is not obvious due to the appearance of second-order-in-time derivatives in (18.225) and (18.226). Building such a global, 2D, numerical model would be fraught with difficulty. Once formulated and coded, everything *might* work wonderfully well. However, it might very well not do so and, worse yet, after the expenditure of considerable effort. And if it does not perform as hoped for, is this due to a formulation error in the numerics, or to a coding error? So what should one do to reduce the risk factor, whilst increasing the probability of ultimate success?

The answer to this last question is, in a nutshell, to initially keep things as simple as possible, and to then gradually increase complexity in a step-by-step, controlled manner. If things go wrong, for whatever reason, it is then much easier to isolate the cause for this and to remedy it at source before progressing to the next step.[36] A significant advantage of having developed the unified quartet of shallow-water equation sets in a colour-coded manner is that it is straightforward, at a glance, to formulate *dynamically consistent* equation subsets.[37] The aspect of dynamical consistency is very important here. When simplifying an equation set, one definitely does not want to inadvertently end up with an equation set that, for example, includes a spurious source of energy leading to a numerical model being (inherently and spuriously) unstable. The colour coding also allows one to identify how, for example, certain terms in the governing equations affect the fluxes

[36] Or possibly admitting defeat in the face of insuperable problems!

[37] This is also true for the 3D equation sets developed in Chapters 16 and 17.

of axial angular momentum and total energy. These fluxes can then be quantified and monitored during integrations of a numerical model.

The above general discussion is all well and good theoretically, but how might one proceed practically in the present hypothetical context? In this context, we wish to focus attention on the non-hydrostatic terms (i.e. the blue terms) in (18.225) and (18.226). It therefore makes sense to (initially) eliminate quasi-hydrostatic terms by setting $Q \equiv 0$. Zap, the green terms in these two equations are gone! The presence of an underlying orography or bathymetry, $B = B(\lambda, \phi)$, significantly complicates the form of the (blue) non-hydrostatic terms, so we (initially) set B identically zero; zap, the B terms disappear! Doing so then not only eliminates the B terms, but also simplifies $\widetilde{H}$ to H (since $\widetilde{H} \equiv H - B \to H$). Whereas doing all this greatly simplifies (18.225) and (18.226), we are nevertheless still left with a global, 2D equation set. The next, natural simplification is to restrict attention to flows for which $\partial F/\partial\lambda \equiv 0$, where $F = \left(u_\lambda, u_\phi, H\right)$. This then reduces the (simplified) 2D global set to an even simpler 1D set. Introducing all of these simplifications into (18.225)–(18.229) then gives:

The 'Non-Hydrostatic' 1D Shallow-Water Equations in Spherical-Polar Coordinates

$$\underbrace{\frac{D_{\text{hor}} u_\lambda}{Dt} - \left(\frac{u_\lambda}{\overline{a}\cos\phi} + 2\Omega\right) u_\phi \sin\phi = 0,}_{\text{basic}} \tag{18.243}$$

$$\underbrace{\frac{D_{\text{hor}} u_\phi}{Dt} + \left(\frac{u_\lambda}{\overline{a}\cos\phi} + 2\Omega\right) u_\lambda \sin\phi + \frac{1}{\overline{a}}\frac{\partial (gH)}{\partial\phi}}_{\text{basic}} = \underbrace{-\frac{\mathcal{N}}{3\overline{a}H}\frac{\partial}{\partial\phi}\left(H^2\frac{D^2_{\text{hor}} H}{Dt^2}\right)}_{\text{non-hydrostatic}}, \tag{18.244}$$

$$\frac{D_{\text{hor}} H}{Dt} + \frac{H}{\overline{a}\cos\phi}\frac{\partial}{\partial\phi}\left(u_\phi\cos\phi\right) = 0, \tag{18.245}$$

or

$$\frac{\partial H}{\partial t} + \frac{1}{\overline{a}\cos\phi}\frac{\partial}{\partial\phi}\left(\cos\phi\, u_\phi H\right) = 0, \tag{18.246}$$

$$\frac{D_{\text{hor}}}{Dt} \equiv \frac{\partial}{\partial t} + \frac{u_\phi}{\overline{a}}\frac{\partial}{\partial\phi}. \tag{18.247}$$

In fact, for development purposes one can (and arguably should) further simplify things by initially setting the parameters Ω and $\mathcal{N}$ in (18.243) and (18.244) identically zero. Doing so then gives an even simpler 1D set of equations that can be solved using tried-and-true traditional techniques.[38] Switching $\mathcal{N}$ on (by setting its value to unity) then allows examination and validation of the approximation of the (single) non-hydrostatic term on the right-hand side of (18.244). Assuming that everything works satisfactorily for this, one can then introduce further terms in the 1D context and/or still keep things initially simple, but move to a corresponding 2D equation set.

Adopting the simplified, step-by-step strategy is both low-risk and robust. The effort required to initially formulate, code, and validate the numerical approximation of the preceding set of 1D equations is relatively small (compared to the 2D case), as is the incremental effort to increase complexity at a given stage in model development. Furthermore, if any problems are encountered,

[38] Alternatively, one could express the equations in yet simpler 1D Cartesian geometry.

then it should be relatively easy to identify why, because the incremental change is, by design, small.

Indeed, this approach is, in the author's view, the way one should develop 3D dynamical cores for quantitative atmospheric and oceanic modelling. The present context simply illustrates, by concrete example, how one can apply this approach in practice. The reader, however, is cautioned that when designing numerical models, one should not fall into the trap of developing methods that work wonderfully well in 1D but do not naturally generalise to 2D and 3D, particularly regarding accuracy and stability. The real world has three spatial dimensions – with many more ways for things to go wrong!

APPENDIX A: DERIVATION OF 2D QUASI-SHALLOW LAGRANGIAN DENSITY BY VERTICALLY AVERAGING THE 3D ONE

Lagrangian Density for the 3D Quasi-Hydrostatic Quasi-Shallow Equations

Recall that the Lagrangian density for the (quasi-hydrostatic and non-hydrostatic) pair of 3D quasi-shallow equation sets – developed in Chapter 17 and graphically depicted in magenta on Fig. 17.1 – is given in axial-orthogonal-curvilinear coordinates by (17.53). Here we restrict attention to the *quasi-hydrostatic* case – see the right-hand box in magenta on Fig. 17.1 – for which non-hydrostatic switch $\mathcal{N}$ is identically zero. Thus, in the notation of Chapter 17 and setting $\mathcal{N} \equiv 0$ in (17.53), the 3D, quasi-hydrostatic, quasi-shallow Lagrangian density is

$$\mathscr{L}^{Q}_{QS}\left(\boldsymbol{\xi},\dot{\boldsymbol{\xi}},\alpha,\eta\right) = \underbrace{\frac{\left(h_1^2\right)^S \dot{\xi}_1^2 + \left(h_2^2\right)^S \dot{\xi}_2^2}{2}}_{\text{horizontal KE}} - \underbrace{\Phi_S\left(\xi_3\right)}_{\text{gravity}} - \underbrace{\mathscr{E}_S\left(\alpha,\eta\right)}_{\text{internal energy}}$$
$$+ \underbrace{\Omega\left(h_1^2\right)^S \dot{\xi}_1}_{\text{shallow Coriolis}} + \underbrace{Q\Omega\left(\frac{\partial h_1^2}{\partial \xi_3}\right)^S \left(\xi_3 - \xi_3^0\right)\dot{\xi}_1}_{\text{quasi-shallow Coriolis}}, \tag{18.248}$$

where $\boldsymbol{\xi} \equiv (\xi_1,\xi_2,\xi_3)$, $\dot{\boldsymbol{\xi}} \equiv \left(\dot{\xi}_1,\dot{\xi}_2,\dot{\xi}_3\right)$, super/subscripts '$S$' denote evaluation at the planet's surface $\xi_3 = \xi_3^0 =$ constant, and h_1^S and h_2^S satisfy (18.15). Quasi-shallow switch Q has been introduced into (18.248) to be able to selectively switch quasi-shallow approximation on or off. When $Q \equiv 0$, it is switched *off*, leading to the (purely shallow) hydrostatic primitive equations, with an incomplete representation of the Coriolis force. But when $Q \equiv 1$, quasi-shallow approximation is switched *on*. The green term in (18.248) then materialises, leading to an (almost) complete representation of the Coriolis force.

Insertion of (18.248) into Euler–Lagrange equations (17.65) leads to the *quasi-hydrostatic quasi-shallow* component equations (17.100)–(17.102) – with $\mathcal{N} \equiv 0$ therein – for momentum; namely[39]

$$\frac{D_S u_1}{Dt} + \left(\frac{u_1}{h_1^S} + 2\Omega\right)\frac{u_2}{h_2^S}\frac{dh_1^S}{d\xi_2} + 2Q\Omega\left[A\left(\xi_2\right)\left(\xi_3 - \xi_3^S\right)u_2 + \frac{u_3}{h_3^S}\left(\frac{\partial h_1}{\partial \xi_3}\right)^S\right] = -\frac{1}{\rho h_1^S}\frac{\partial p}{\partial \xi_1}, \tag{18.249}$$

$$\frac{D_S u_2}{Dt} - \left(\frac{u_1}{h_1^S} + 2\Omega\right)\frac{u_1}{h_2^S}\frac{dh_1^S}{d\xi_2} - 2Q\Omega A\left(\xi_2\right)\left(\xi_3 - \xi_3^S\right)u_1 = -\frac{1}{\rho h_2^S}\frac{\partial p}{\partial \xi_2}, \tag{18.250}$$

[39] These three equations are part of the right-hand *quasi-shallow* equation set displayed in magenta on Fig. 17.1.

$$- 2Q\Omega \frac{u_1}{h_3^S} \left(\frac{\partial h_1}{\partial \xi_3} \right)^S + \frac{1}{h_3^S} \frac{d\Phi_S}{d\xi_3} = - \frac{1}{\rho h_3^S} \frac{\partial p}{\partial \xi_3}, \tag{18.251}$$

where

$$\frac{D_S}{Dt} \equiv \frac{\partial}{\partial t} + \frac{u_1}{h_1^S} \frac{\partial}{\partial \xi_1} + \frac{u_2}{h_2^S} \frac{\partial}{\partial \xi_2} + \frac{u_1}{h_3^S} \frac{\partial}{\partial \xi_3}, \tag{18.252}$$

is material derivative (in shallow geometry) and (from (17.105))

$$A(\xi_2) \equiv \frac{1}{h_1^S h_2^S} \frac{d}{d\xi_2} \left[\frac{1}{2} \left(\frac{\partial h_1^2}{\partial \xi_3} \right)^S \right]. \tag{18.253}$$

Before proceeding further, we note that the (in general) *prognostic* equation for vertical velocity (with a Du_3/Dt term) has (in the absence of this term) simplified to (18.251). This changes the role of this equation. It is no longer a *prognostic* equation for u_3 but rather a *diagnostic equation for p*.

Vertical Averaging of 3D Quasi-Shallow Lagrangian Density

The key to deriving the 2D Lagrangian density of the 'quasi-shallow', shallow-water equations from 3D, quasi-hydrostatic, quasi-shallow Lagrangian density (18.248) is to vertically average this latter equation. So that is what we now do.

The Non-Contribution of Internal Energy

There is no internal energy contribution to shallow-water Lagrangian density – see, for example, Salmon (1998), Chapter 7.6.[40]

The Vertically Averaged Horizontal Kinetic Energy $\widehat{\mathscr{K}}$

From (18.7), (18.15), and (18.28), the horizontal-kinetic-energy contribution $\mathscr{K} \equiv \left[\left(h_1^2\right)^S \dot{\xi}_1^2 + \left(h_2^2\right)^S \dot{\xi}_2^2 \right] /2$ to (18.248) is assumed to be independent of ξ_3. Thus $\widehat{\mathscr{K}} = \mathscr{K}$, where operator $\widehat{(\ )}$ denotes vertical average over the depth of the fluid $\widetilde{H}(\xi_1, \xi_2, t) \equiv H(\xi_1, \xi_2, t) - B(\xi_1, \xi_2)$, as defined by (18.36).

The Vertically Averaged Coriolis Contribution $\widehat{\Omega h_1^S u_1}$

For the 3D quasi-shallow approximation of the Coriolis terms, h_1^2 is approximated – see (17.52) and (18.16) – as

$$h_1^2 \approx \left(h_1^2\right)^S + Q \left(\frac{\partial h_1^2}{\partial \xi_3} \right)^S \left(\xi_3 - \xi_3^0\right) = \left(h_1^2\right)^S + \frac{Q}{h_3^S} \left(\frac{\partial h_1^2}{\partial \xi_3} \right)^S s_3 = \left(h_1^2\right)^S + Q\sigma^S(\xi_2)\, s_3, \tag{18.254}$$

where (18.23) and (for brevity)

$$\sigma^S(\xi_2) \equiv \frac{1}{h_3^S} \left(\frac{\partial h_1^2}{\partial \xi_3} \right)^S \tag{18.255}$$

have been used to obtain the two rightmost expressions in (18.254).

[40] Since $\rho = \overline{\rho}$ (from (18.14)), internal energy $\mathscr{E}_S = \mathscr{E}_S(\alpha \equiv 1/\rho, \eta)$ reduces to $\mathscr{E}_S = \mathscr{E}_S(\eta)$, and thermodynamic-energy equation (17.104) reduces to $D_S \mathscr{E}_S / Dt = 0$. Consequently internal energy, $\mathscr{E}_S$, does not contribute to conservation of total energy in the present, quasi-shallow, 2D, shallow-water context.

We now further approximate h_1^2 (solely for evaluation of the Coriolis terms) by vertically averaging (18.254) over the depth of the fluid ($\widetilde{H} \equiv H - B$). Thus

$$\widehat{h_1^2} \equiv \frac{1}{H-B}\int_B^H h_1^2 ds_3 \approx \frac{1}{H-B}\int_B^H \left[\left(h_1^2\right)^S + Q\sigma^S(\xi_2)\, s_3\right] ds_3$$
$$= \frac{1}{H-B}\left[\left(h_1^2\right)^S s_3 + Q\sigma^S(\xi_2)\frac{s_3^2}{2}\right]_B^H = \left(h_1^2\right)^S + Q\sigma^S(\xi_2)\left(\frac{H+B}{2}\right)$$
$$= \left(h_1^2\right)^S + Q\sigma^S(\xi_2)\left(B + \frac{\widetilde{H}}{2}\right). \tag{18.256}$$

Multiplying (18.256) through by $\Omega\dot{\xi}_1$ then takes care of vertically averaging the Coriolis contributions to the 3D, quasi-shallow, Lagrangian density defined by (18.248).

The Vertically Averaged Gravitational Potential Energy $\widehat{\Phi_S}$

It remains to vertically average the gravitational-potential-energy contribution to (18.248). Now from (18.21) and (18.23),

$$\Phi_S(\xi_3) = \Phi_S^0 + h_3^S(\xi_2)\, g(\xi_2)\left(\xi_3 - \xi_3^0\right) = \Phi_S^0 + \cancel{h_3^S(\xi_2)}\, g(\xi_2)\frac{s_3}{\cancel{h_3^S(\xi_2)}} = \Phi_S^0 + g(\xi_2)\, s_3. \tag{18.257}$$

We can similarly approximate $\Phi_S(\xi_3)$ by its vertical average $\widehat{\Phi_S}$ over the depth of the fluid $\widetilde{H}(\xi_1, \xi_2, t)$. Thus, vertically averaging (18.257) – and using definition (18.36) of fluid depth – gives

$$\Phi_S \approx \widehat{\Phi_S} = \frac{1}{H-B}\int_B^H \left[\Phi_S^0 + g(\xi_2)\, s_3\right] ds_3 = \frac{1}{H-B}\left[\Phi_S^0 s_3 + g(\xi_2)\frac{s_3^2}{2}\right]_B^H$$
$$= \Phi_S^0 + g(\xi_2)\left(\frac{H+B}{2}\right)$$
$$= \Phi_S^0 + g(\xi_2)\left(B + \frac{\widetilde{H}}{2}\right). \tag{18.258}$$

The Vertically Averaged Lagrangian Density $\mathscr{L}_{SW}^Q$

Assembling these developments strongly suggests that

$$\mathscr{L}_{SW}^Q\left(\xi_2, \xi_3, \dot{\xi}_1, \dot{\xi}_2\right) = \underbrace{\frac{\left(h_1^2\right)^S \dot{\xi}_1^2 + \left(h_2^2\right)^S \dot{\xi}_2^2}{2}}_{\text{horizontal KE}} \underbrace{- g\left(B + \frac{\widetilde{H}}{2}\right)}_{\text{gravity}} + \underbrace{\Omega\left(h_1^2\right)^S \dot{\xi}_1}_{\text{shallow Coriolis}} + \underbrace{Q\Omega\sigma^S\left(B + \frac{\widetilde{H}}{2}\right)\dot{\xi}_1}_{\text{quasi-shallow Coriolis}}, \tag{18.259}$$

is an appropriate, *'quasi-shallow', shallow-water* approximation of Lagrangian density (18.248). In (18.259), dependent variables have the functional forms

$$\left.\begin{array}{c}\left(h_i^2\right)^S \equiv \left(h_i^S\right)^2 = \left[h_i^S(\xi_2)\right]^2, \ i = 1, 2; \quad g = g(\xi_2), \ \sigma^S = \sigma^S(\xi_2); \\ B = B(\xi_1, \xi_2); \quad \widetilde{H} = \widetilde{H}(\xi_1, \xi_2, t).\end{array}\right\} \tag{18.260}$$

APPENDIX B: CONSERVATION PRINCIPLES FOR THE 'QUASI-SHALLOW' SHALLOW-WATER EQUATIONS

Conservation principles – stated without proof in Section 18.7 – for axial angular momentum, total energy, and potential vorticity of the 'quasi-shallow' shallow-water equations are derived here.[41]

Governing Equations

The two components of momentum balance are (18.167) and (18.168), namely

$$\frac{D_{hor}u_1}{Dt} + \left(\frac{u_1}{h_1^S} + 2\Omega\right)\frac{u_2}{h_2^S}\frac{dh_1^S}{d\xi_2} + \frac{1}{h_1^S}\frac{\partial (gH)}{\partial \xi_1} - Q\frac{\Omega}{h_1^S}\left\{\frac{1}{2}\frac{\partial}{\partial \xi_1}\left(\frac{\sigma^S}{h_1^S}u_1\widetilde{H}\right) - \frac{u_2}{h_2^S}\frac{\partial}{\partial \xi_2}\left[\sigma^S\left(B + \frac{\widetilde{H}}{2}\right)\right]\right\} - Q\frac{\Omega}{2}\frac{\sigma^S}{\left(h_1^S\right)^2 h_2^S}\left[\frac{\partial}{\partial \xi_1}\left(h_2^S u_1 \widetilde{H}\right) + \frac{\partial}{\partial \xi_2}\left(h_1^S u_2 \widetilde{H}\right)\right] = 0, \tag{18.261}$$

$$\frac{D_{hor}u_2}{Dt} - \left(\frac{u_1}{h_1^S} + 2\Omega\right)\frac{u_1}{h_2^S}\frac{dh_1^S}{d\xi_2} + \frac{1}{h_2^S}\frac{\partial (gH)}{\partial \xi_2} - Q\frac{\Omega}{h_2^S}\left\{\frac{1}{2}\frac{\partial}{\partial \xi_2}\left(\frac{\sigma^S}{h_1^S}u_1\widetilde{H}\right) + \frac{u_1}{h_1^S}\frac{\partial}{\partial \xi_2}\left[\sigma^S\left(B + \frac{\widetilde{H}}{2}\right)\right]\right\} = 0, \tag{18.262}$$

where (from (18.28)) $u_1 = u_1(\xi_1, \xi_2, t)$ and $u_2 = u_2(\xi_1, \xi_2, t)$; and (from (18.171))

$$\frac{D_{hor}}{Dt} \equiv \frac{\partial}{\partial t} + \frac{u_1}{h_1^S}\frac{\partial}{\partial \xi_1} + \frac{u_2}{h_2^S}\frac{\partial}{\partial \xi_2} \tag{18.263}$$

is horizontal material derivative.

The mass-continuity equation may be equivalently written – see (18.169) and (18.170) – in substantive form as

$$\frac{D_{hor}\widetilde{H}}{Dt} + \frac{\widetilde{H}}{h_1^S h_2^S}\left[\frac{\partial}{\partial \xi_1}\left(u_1 h_2^S\right) + \frac{\partial}{\partial \xi_2}\left(u_2 h_1^S\right)\right] = 0, \tag{18.264}$$

or in flux form as

$$\frac{\partial \widetilde{H}}{\partial t} + \frac{1}{h_1^S h_2^S}\left[\frac{\partial}{\partial \xi_1}\left(h_2^S u_1 \widetilde{H}\right) + \frac{\partial}{\partial \xi_2}\left(h_1^S u_2 \widetilde{H}\right)\right] = 0, \tag{18.265}$$

where, from (18.36), $\widetilde{H}(\xi_1, \xi_2, t) \equiv H(\xi_1, \xi_2, t) - B(\xi_1, \xi_2) \geq 0$ is fluid depth. Both forms are employed here.

Some Useful Relations

For generic scalars F and G

$$\nabla_{hor} \cdot (FG\mathbf{u}_{hor}) = \frac{1}{h_1^S h_2^S}\left[\frac{\partial}{\partial \xi_1}\left(h_2^S u_1 FG\right) + \frac{\partial}{\partial \xi_2}\left(h_1^S u_2 FG\right)\right]$$

[41] This appendix contains public sector information licensed under the Open Government Licence v1.0. It is an adapted, simplified, colourised version – with some minor changes of notation – of the online Supporting Information to Staniforth (2015b).

$$= G\left(\frac{u_1}{h_1^S}\frac{\partial}{\partial\xi_1} + \frac{u_2}{h_2^S}\frac{\partial}{\partial\xi_2}\right)F + F\frac{1}{h_1^S h_2^S}\left[\frac{\partial}{\partial\xi_1}\left(h_2^S u_1 G\right) + \frac{\partial}{\partial\xi_2}\left(h_1^S u_2 G\right)\right]$$
$$= G\mathbf{u}_{\text{hor}} \cdot \nabla_{\text{hor}} F + F\nabla_{\text{hor}} \cdot \left(G\mathbf{u}_{\text{hor}}\right), \qquad (18.266)$$

where (from (18.58))

$$\mathbf{u}_{\text{hor}} \equiv u_1\mathbf{e}_1 + u_2\mathbf{e}_2, \qquad (18.267)$$

∇_{hor} is the horizontal component (in shallow geometry) of ∇, and

$$\mathbf{u}_{\text{hor}} \cdot \nabla_{\text{hor}} \equiv \frac{u_1}{h_1^S}\frac{\partial}{\partial\xi_1} + \frac{u_2}{h_2^S}\frac{\partial}{\partial\xi_2}, \qquad (18.268)$$

$$\nabla_{\text{hor}} \cdot \left(G\mathbf{u}_{\text{hor}}\right) \equiv \frac{1}{h_1^S h_2^S}\left[\frac{\partial}{\partial\xi_1}\left(h_2^S u_1 G\right) + \frac{\partial}{\partial\xi_2}\left(h_1^S u_2 G\right)\right]. \qquad (18.269)$$

From (18.15), the metric factors are only functions of ξ_2, that is,

$$h_1^S = h_1^S(\xi_2), \quad h_2^S = h_2^S(\xi_2), \quad h_3^S = h_3^S(\xi_2). \qquad (18.270)$$

This property is frequently used in this appendix without explicit mention.

Conservation of Mass

Conservation of mass is embodied in (18.264) and (18.265), in Lagrangian and Eulerian forms, respectively.

Conservation of Axial Angular Momentum

Multiplying (18.261) by $\widetilde{H}h_1^S$ and using (18.263) and (18.265) yields

$$\widetilde{H}h_1^S\frac{\partial u_1}{\partial t} = -\widetilde{H}h_1^S\left(\frac{u_1}{h_1^S}\frac{\partial u_1}{\partial\xi_1} + \frac{u_2}{h_2^S}\frac{\partial u_1}{\partial\xi_2}\right) - \widetilde{H}\frac{u_1u_2}{h_2^S}\frac{dh_1^S}{d\xi_2} - \widetilde{H}\frac{\partial(gH)}{\partial\xi_1} - 2\Omega\widetilde{H}h_1^S\frac{u_2}{h_2^S}\frac{dh_1^S}{d\xi_2}$$
$$+ Q\Omega\frac{\widetilde{H}}{2}\frac{\partial}{\partial\xi_1}\left(\frac{\sigma^S}{h_1^S}u_1\widetilde{H}\right) - Q\Omega\widetilde{H}\frac{u_2}{h_2^S}\frac{\partial}{\partial\xi_2}\left[\sigma^S\left(B + \frac{\widetilde{H}}{2}\right)\right] - Q\Omega\frac{\widetilde{H}}{2}\sigma^S\frac{\partial\widetilde{H}}{\partial t}. \qquad (18.271)$$

We hypothesise that the measure for axial angular momentum is, as in Staniforth (2015b),

$$M^{\text{SW}} \equiv \widetilde{H}\left[h_1^S\left(u_1 + \Omega h_1^S\right) + Q\Omega\sigma^S\left(B + \frac{\widetilde{H}}{2}\right)\right], \qquad (18.272)$$

and we examine its (Eulerian) time tendency. Thus

$$\frac{\partial M^{\text{SW}}}{\partial t} \equiv \frac{\partial}{\partial t}\left\{\widetilde{H}\left[h_1^S\left(u_1 + \Omega h_1^S\right) + Q\Omega\sigma^S\left(B + \frac{\widetilde{H}}{2}\right)\right]\right\}$$
$$= \widetilde{H}\left(h_1^S\frac{\partial u_1}{\partial t} + Q\frac{\Omega}{2}\sigma^S\frac{\partial\widetilde{H}}{\partial t}\right) + \left[h_1^S\left(u_1 + \Omega h_1^S\right) + Q\Omega\sigma^S\left(B + \frac{\widetilde{H}}{2}\right)\right]\frac{\partial\widetilde{H}}{\partial t}$$
$$= \widetilde{H}h_1^S\frac{\partial u_1}{\partial t} + \left[h_1^S\left(u_1 + \Omega h_1^S\right) + Q\Omega\sigma^S\left(B + \widetilde{H}\right)\right]\frac{\partial\widetilde{H}}{\partial t}. \qquad (18.273)$$

Elimination of $\widetilde{H}h_1^S\partial u_1/\partial t$ between (18.271) and (18.273) then yields

$$\frac{\partial M^{\text{SW}}}{\partial t} = -\widetilde{H}h_1^S\left(\frac{u_1}{h_1^S}\frac{\partial u_1}{\partial\xi_1} + \frac{u_2}{h_2^S}\frac{\partial u_1}{\partial\xi_2}\right) - \widetilde{H}\frac{u_1u_2}{h_2^S}\frac{dh_1^S}{d\xi_2} - \widetilde{H}\frac{\partial(gH)}{\partial\xi_1} - 2\Omega\widetilde{H}h_1^S\frac{u_2}{h_2^S}\frac{dh_1^S}{d\xi_2}$$
$$+ Q\Omega\frac{\widetilde{H}}{2}\frac{\partial}{\partial\xi_1}\left(\frac{\sigma^S}{h_1^S}u_1\widetilde{H}\right) - Q\Omega\widetilde{H}\frac{u_2}{h_2^S}\frac{\partial}{\partial\xi_2}\left[\sigma^S\left(B + \frac{\widetilde{H}}{2}\right)\right] - Q\Omega\frac{\widetilde{H}}{2}\sigma^S\frac{\partial\widetilde{H}}{\partial t}$$

$$
\begin{aligned}
&+ \left[h_1^S\left(u_1+\Omega h_1^S\right)+Q\Omega\sigma^S\left(B+\widetilde{H}\right)\right]\frac{\partial\widetilde{H}}{\partial t}\\
&=-\widetilde{H}\left(\frac{u_1}{h_1^S}\frac{\partial}{\partial\xi_1}+\frac{u_2}{h_2^S}\frac{\partial}{\partial\xi_2}\right)\left(h_1^S u_1\right)-\frac{\partial}{\partial\xi_1}\left(g\frac{\widetilde{H}^2}{2}\right)-\widetilde{H}\frac{\partial\left(gB\right)}{\partial\xi_1}\\
&-\widetilde{H}\left(\frac{u_1}{h_1^S}\frac{\partial}{\partial\xi_1}+\frac{u_2}{h_2^S}\frac{\partial}{\partial\xi_2}\right)\left[\Omega\left(h_1^S\right)^2\right]+Q\widetilde{H}\frac{\partial}{\partial\xi_1}\left(\frac{\Omega}{2}\sigma^S\widetilde{H}\frac{u_1}{h_1^S}\right)\\
&+Q\widetilde{H}\frac{u_1}{h_1^S}\frac{\partial}{\partial\xi_1}\left[\Omega\sigma^S\left(B+\frac{\widetilde{H}}{2}\right)\right]-Q\widetilde{H}\left(\frac{u_1}{h_1^S}\frac{\partial}{\partial\xi_1}+\frac{u_2}{h_2^S}\frac{\partial}{\partial\xi_2}\right)\left[\Omega\sigma^S\left(B+\frac{\widetilde{H}}{2}\right)\right]\\
&+\left[h_1^S\left(u_1+\Omega h_1^S\right)+Q\Omega\sigma^S\left(B+\frac{\widetilde{H}}{2}\right)\right]\frac{\partial\widetilde{H}}{\partial t}\\
&=-\widetilde{H}\left(\frac{u_1}{h_1^S}\frac{\partial}{\partial\xi_1}+\frac{u_2}{h_2^S}\frac{\partial}{\partial\xi_2}\right)\left[h_1^S u_1+\Omega\left(h_1^S\right)^2+Q\Omega\sigma^S\left(B+\frac{\widetilde{H}}{2}\right)\right]\\
&+\left[h_1^S\left(u_1+\Omega h_1^S\right)+Q\Omega\sigma^S\left(B+\frac{\widetilde{H}}{2}\right)\right]\frac{\partial\widetilde{H}}{\partial t}-\frac{\partial}{\partial\xi_1}\left(g\frac{\widetilde{H}^2}{2}\right)-\widetilde{H}\frac{\partial\left(gB\right)}{\partial\xi_1}\\
&+Q\widetilde{H}\frac{\partial}{\partial\xi_1}\left(\frac{\Omega}{2}\sigma^S\widetilde{H}\frac{u_1}{h_1^S}\right)+Q\widetilde{H}\frac{u_1}{h_1^S}\frac{\partial}{\partial\xi_1}\left(\frac{\Omega}{2}\sigma^S\widetilde{H}\right)+Q\Omega\widetilde{H}\frac{u_1}{h_1^S}\frac{\partial}{\partial\xi_1}\left(\sigma^S B\right).
\end{aligned}
\tag{18.274}
$$

With use of (18.265), (18.274) leads to

$$
\begin{aligned}
\frac{\partial M^{SW}}{\partial t}=&-\widetilde{H}\left(\frac{u_1}{h_1^S}\frac{\partial}{\partial\xi_1}+\frac{u_2}{h_2^S}\frac{\partial}{\partial\xi_2}\right)\left[h_1^S\left(u_1+\Omega h_1^S\right)+Q\Omega\sigma^S\left(B+\frac{\widetilde{H}}{2}\right)\right]\\
&-\left[h_1^S\left(u_1+\Omega h_1^S\right)+Q\Omega\sigma^S\left(B+\frac{\widetilde{H}}{2}\right)\right]\frac{1}{h_1^S h_2^S}\left[\frac{\partial}{\partial\xi_1}\left(h_2^S\widetilde{H}u_1\right)+\frac{\partial}{\partial\xi_2}\left(h_1^S\widetilde{H}u_2\right)\right]\\
&+Q\frac{\partial}{\partial\xi_1}\left(\widetilde{H}\frac{\Omega}{2}\sigma^S\widetilde{H}\frac{u_1}{h_1^S}\right)+Q\Omega\widetilde{H}\frac{u_1}{h_1^S}\frac{\partial}{\partial\xi_1}\left(\sigma^S B\right)-\frac{\partial}{\partial\xi_1}\left(g\frac{\widetilde{H}^2}{2}\right)-\widetilde{H}\frac{\partial\left(gB\right)}{\partial\xi_1}.
\end{aligned}
\tag{18.275}
$$

Now setting $F=h_1^S\left(u_1+\Omega h_1^S\right)+Q\Omega\sigma^S\left[B+\left(\widetilde{H}/2\right)\right]$ and $G=\widetilde{H}$ in identity (18.266) gives

$$
\begin{aligned}
&\nabla_{\text{hor}}\cdot\left\{\left[h_1^S\left(u_1+\Omega h_1^S\right)+Q\Omega\sigma^S\left(B+\frac{\widetilde{H}}{2}\right)\right]\widetilde{H}\mathbf{u}_{\text{hor}}\right\}\\
&=\widetilde{H}\left(\frac{u_1}{h_1^S}\frac{\partial}{\partial\xi_1}+\frac{u_2}{h_2^S}\frac{\partial}{\partial\xi_2}\right)\left[h_1^S\left(u_1+\Omega h_1^S\right)+Q\Omega\sigma^S\left(B+\frac{\widetilde{H}}{2}\right)\right]\\
&+\left[h_1^S\left(u_1+\Omega h_1^S\right)+Q\Omega\sigma^S\left(B+\frac{\widetilde{H}}{2}\right)\right]\frac{1}{h_1^S h_2^S}\left[\frac{\partial}{\partial\xi_1}\left(h_2^S u_1\widetilde{H}\right)+\frac{\partial}{\partial\xi_2}\left(h_1^S u_2\widetilde{H}\right)\right].
\end{aligned}
\tag{18.276}
$$

Using (18.267) and (18.276), (18.275) may then be rewritten as

$$
\begin{aligned}
\frac{\partial M^{SW}}{\partial t}=&-\nabla_{\text{hor}}\cdot\left\{\left[h_1^S\left(u_1+\Omega h_1^S\right)+Q\Omega\sigma^S\left(B+\frac{\widetilde{H}}{2}\right)\right]\widetilde{H}\mathbf{u}_{\text{hor}}\right\}\\
&-\frac{\partial}{\partial\xi_1}\left[\left(\frac{g}{2}-Q\frac{\Omega\sigma^S}{2}\frac{u_1}{h_1^S}\right)\widetilde{H}^2\right]-\left(g-Q\Omega\sigma^S\frac{u_1}{h_1^S}\right)\widetilde{H}\frac{\partial B}{\partial\xi_1},
\end{aligned}
\tag{18.277}
$$

where M^{SW} is axial angular momentum.

Conservation of Total Energy

Multiplying the two components (18.261) and (18.262) of the momentum equation by u_1 and u_2, respectively, and summing the results yields

$$\frac{D_{\text{hor}}}{Dt}\left(\frac{u_1^2+u_2^2}{2}\right)+\frac{u_1}{h_1^S}\frac{\partial}{\partial\xi_1}\left(gH-Q\frac{\Omega}{2}\frac{\sigma^S}{h_1^S}u_1\widetilde{H}\right)+\frac{u_2}{h_2^S}\frac{\partial}{\partial\xi_2}\left(gH-Q\frac{\Omega}{2}\frac{\sigma^S}{h_1^S}u_1\widetilde{H}\right)$$
$$-Q\frac{\Omega}{2}\frac{\sigma^S}{h_1^S}\frac{u_1}{h_1^S h_2^S}\left[\frac{\partial}{\partial\xi_1}\left(h_2^S u_1\widetilde{H}\right)+\frac{\partial}{\partial\xi_2}\left(h_1^S u_2\widetilde{H}\right)\right]=0. \tag{18.278}$$

Equation (18.278) may be rewritten as

$$\left(\frac{\partial}{\partial t}+\frac{u_1}{h_1^S}\frac{\partial}{\partial\xi_1}+\frac{u_2}{h_2^S}\frac{\partial}{\partial\xi_2}\right)\left(\frac{u_1^2+u_2^2}{2}\right)+\left(\frac{u_1}{h_1^S}\frac{\partial}{\partial\xi_1}+\frac{u_2}{h_2^S}\frac{\partial}{\partial\xi_2}\right)(gH)$$
$$-Q\frac{\Omega}{2}\left\{\left(\frac{u_1}{h_1^S}\frac{\partial}{\partial\xi_1}+\frac{u_2}{h_2^S}\frac{\partial}{\partial\xi_2}\right)\left(\frac{\sigma^S}{h_1^S}u_1\widetilde{H}\right)+\frac{\sigma^S}{h_1^S}\frac{u_1}{h_1^S h_2^S}\left[\frac{\partial}{\partial\xi_1}\left(h_2^S u_1\widetilde{H}\right)+\frac{\partial}{\partial\xi_2}\left(h_1^S u_2\widetilde{H}\right)\right]\right\}=0. \tag{18.279}$$

Multiplying (18.279) by $\widetilde{H}$ and rearranging then leads to

$$\widetilde{H}\frac{\partial}{\partial t}\left(\frac{u_1^2+u_2^2}{2}\right)=-\widetilde{H}\left(\frac{u_1}{h_1^S}\frac{\partial}{\partial\xi_1}+\frac{u_2}{h_2^S}\frac{\partial}{\partial\xi_2}\right)\left(\frac{u_1^2+u_2^2}{2}+gH-Q\frac{\Omega}{2}\frac{\sigma^S}{h_1^S}u_1\widetilde{H}\right)$$
$$+Q\frac{\Omega}{2}\frac{\sigma^S}{h_1^S}\frac{u_1}{h_1^S h_2^S}\widetilde{H}\left[\frac{\partial}{\partial\xi_1}\left(h_2^S u_1\widetilde{H}\right)+\frac{\partial}{\partial\xi_2}\left(h_1^S u_2\widetilde{H}\right)\right]. \tag{18.280}$$

Now setting $F=\left(u_1^2+u_2^2\right)/2+gH-Q\Omega\sigma^S u_1\widetilde{H}/\left(2h_1^S\right)$ and $G=\widetilde{H}$ in identity (18.266) gives

$$-\widetilde{H}\left(\frac{u_1}{h_1^S}\frac{\partial}{\partial\xi_1}+\frac{u_2}{h_2^S}\frac{\partial}{\partial\xi_2}\right)\left(\frac{u_1^2+u_2^2}{2}+gH-Q\frac{\Omega}{2}\frac{\sigma^S}{h_1^S}u_1\widetilde{H}\right)$$
$$=-\nabla_{\text{hor}}\cdot\left[\left(\frac{u_1^2+u_2^2}{2}+gH-Q\frac{\Omega}{2}\frac{\sigma^S}{h_1^S}u_1\widetilde{H}\right)\widetilde{H}\mathbf{u}_{\text{hor}}\right]$$
$$+\left(\frac{u_1^2+u_2^2}{2}+gH-Q\frac{\Omega}{2}\frac{\sigma^S}{h_1^S}u_1\widetilde{H}\right)\nabla_{\text{hor}}\cdot\left(\widetilde{H}\mathbf{u}_{\text{hor}}\right). \tag{18.281}$$

Using (18.281) in (18.280) then yields

$$\widetilde{H}\frac{\partial}{\partial t}\left(\frac{u_1^2+u_2^2}{2}\right)=-\nabla_{\text{hor}}\cdot\left[\left(\frac{u_1^2+u_2^2}{2}+gH-Q\frac{\Omega}{2}\frac{\sigma^S}{h_1^S}u_1\widetilde{H}\right)\widetilde{H}\mathbf{u}_{\text{hor}}\right]$$
$$+\left(\frac{u_1^2+u_2^2}{2}+gH-Q\frac{\Omega}{2}\frac{\sigma^S}{h_1^S}u_1\widetilde{H}\right)\nabla_{\text{hor}}\cdot\left(\widetilde{H}\mathbf{u}_{\text{hor}}\right)$$
$$+Q\frac{\Omega}{2}\frac{\sigma^S}{h_1^S}u_1\widetilde{H}\nabla_{\text{hor}}\cdot\left(\widetilde{H}\mathbf{u}_{\text{hor}}\right)$$
$$=-\nabla_{\text{hor}}\cdot\left[\left(\frac{u_1^2+u_2^2}{2}+gH-Q\frac{\Omega}{2}\frac{\sigma^S}{h_1^S}u_1\widetilde{H}\right)\widetilde{H}\mathbf{u}_{\text{hor}}\right]$$
$$+\left(\frac{u_1^2+u_2^2}{2}+gH\right)\nabla_{\text{hor}}\cdot\left(\widetilde{H}\mathbf{u}_{\text{hor}}\right). \tag{18.282}$$

Now

$$gH\frac{\partial\widetilde{H}}{\partial t}=g\left(B+\widetilde{H}\right)\frac{\partial\widetilde{H}}{\partial t}=\frac{\partial}{\partial t}\left[g\widetilde{H}\left(B+\frac{\widetilde{H}}{2}\right)\right], \tag{18.283}$$

since $H \equiv B + \widetilde{H}$ (from (18.36)). Using (18.283) and (18.265), (18.282) can then be rewritten as

$$\frac{\partial}{\partial t}\left[\left(\frac{u_1^2+u_2^2}{2}+g\left(B+\frac{\widetilde{H}}{2}\right)\right)\widetilde{H}\right] = -\nabla_{\text{hor}}\cdot\left[\left(\frac{u_1^2+u_2^2}{2}+gH-Q\frac{\Omega}{2}\frac{\sigma^S}{h_1^S}u_1\widetilde{H}\right)\widetilde{H}\mathbf{u}\right]. \tag{18.284}$$

Rewriting (18.284) then yields the total-energy-conservation equation

$$\frac{\partial E^{\text{SW}}}{\partial t} = -\nabla_{\text{hor}}\cdot\left[\left(\frac{u_1^2+u_2^2}{2}+gH-Q\frac{\Omega}{2}\frac{\sigma^S}{h_1^S}u_1\widetilde{H}\right)\widetilde{H}\mathbf{u}\right], \tag{18.285}$$

where

$$E^{\text{SW}} \equiv \left[\frac{u_1^2+u_2^2}{2}+g\left(B+\frac{\widetilde{H}}{2}\right)\right]\widetilde{H} \tag{18.286}$$

is the sum of kinetic and gravitational-potential energies.

Conservation of Potential Vorticity

The two components (18.261) and (18.262) of the momentum equation can be rewritten as

$$\begin{aligned}\frac{\partial u_1}{\partial t} - \zeta_3^S u_2 + \frac{1}{h_1^S}\frac{\partial}{\partial \xi_1}\left(\frac{u_1^2+u_2^2}{2}\right) + \frac{1}{h_1^S}\frac{\partial (gH)}{\partial \xi_1} + 2\Omega\frac{u_2}{h_2^S}\frac{dh_1^S}{d\xi_2}\\ - Q\frac{\Omega}{2}\frac{1}{h_1^S}\frac{\partial}{\partial \xi_1}\left(\frac{\sigma^S}{h_1^S}u_1\widetilde{H}\right) + Q\frac{\Omega u_2}{h_1^S h_2^S}\frac{\partial}{\partial \xi_2}\left[\sigma^S\left(B+\frac{\widetilde{H}}{2}\right)\right]\\ - Q\frac{\Omega}{2}\frac{1}{h_1^S}\frac{\sigma^S}{h_1^S h_2^S}\left[\frac{\partial}{\partial \xi_1}\left(h_2^S u_1\widetilde{H}\right) + \frac{\partial}{\partial \xi_2}\left(h_1^S u_2\widetilde{H}\right)\right] = 0,\end{aligned} \tag{18.287}$$

$$\begin{aligned}\frac{\partial u_2}{\partial t} + \zeta_3^S u_1 + \frac{1}{h_2^S}\frac{\partial}{\partial \xi_2}\left(\frac{u_1^2+u_2^2}{2}\right) + \frac{1}{h_2^S}\frac{\partial (gH)}{\partial \xi_2} - 2\Omega\frac{u_1}{h_2^S}\frac{dh_1^S}{d\xi_2}\\ - Q\frac{\Omega}{2}\frac{1}{h_2^S}\frac{\partial}{\partial \xi_2}\left(\frac{\sigma^S}{h_1^S}u_1\widetilde{H}\right) - Q\frac{\Omega u_1}{h_1^S h_2^S}\frac{\partial}{\partial \xi_2}\left[\sigma^S\left(B+\frac{\widetilde{H}}{2}\right)\right] = 0,\end{aligned} \tag{18.288}$$

where

$$\zeta_3^S \equiv \frac{1}{h_1^S h_2^S}\left[\frac{\partial}{\partial \xi_1}\left(h_2^S u_2\right) - \frac{\partial}{\partial \xi_2}\left(h_1^S u_1\right)\right] \tag{18.289}$$

is relative vorticity.

Cross differentiating (18.287) and (18.288) gives

$$\begin{aligned}\frac{\partial}{\partial t}\left[\frac{\partial}{\partial \xi_2}\left(h_1^S u_1\right)\right] - \frac{\partial}{\partial \xi_2}\left(h_1^S\zeta_3^S u_2\right) + \frac{\partial^2}{\partial \xi_1\partial \xi_2}\left(\frac{u_1^2+u_2^2}{2}\right) + \frac{\partial^2 (gH)}{\partial \xi_1\partial \xi_2} + 2\Omega\frac{\partial}{\partial \xi_2}\left(h_1^S\frac{u_2}{h_2^S}\frac{dh_1^S}{d\xi_2}\right)\\ - Q\frac{\Omega}{2}\frac{\partial^2}{\partial \xi_1\partial \xi_2}\left(\frac{\sigma^S}{h_1^S}u_1\widetilde{H}\right) + Q\frac{\partial}{\partial \xi_2}\left\{\frac{\Omega u_2}{h_2^S}\frac{\partial}{\partial \xi_2}\left[\sigma^S\left(B+\frac{\widetilde{H}}{2}\right)\right]\right\}\\ - Q\frac{\Omega}{2}\frac{\partial}{\partial \xi_2}\left\{\frac{\sigma^S}{h_1^S h_2^S}\left[\frac{\partial}{\partial \xi_1}\left(h_2^S u_1\widetilde{H}\right) + \frac{\partial}{\partial \xi_2}\left(h_1^S u_2\widetilde{H}\right)\right]\right\} = 0,\end{aligned} \tag{18.290}$$

and

$$\frac{\partial}{\partial t}\left[\frac{\partial}{\partial \xi_1}\left(h_2^S u_2\right)\right] + \frac{\partial}{\partial \xi_1}\left(h_2^S\zeta_3^S u_1\right) + \frac{\partial^2}{\partial \xi_1\partial \xi_2}\left(\frac{u_1^2+u_2^2}{2}\right) + \frac{\partial^2 (gH)}{\partial \xi_1\partial \xi_2} - 2\Omega\frac{\partial}{\partial \xi_1}\left(h_2^S\frac{u_1}{h_2^S}\frac{dh_1^S}{d\xi_2}\right)$$

$$- Q\frac{\Omega}{2}\frac{\partial^2}{\partial\xi_1\partial\xi_2}\left(\frac{\sigma^S}{h_1^S}u_1\widetilde{H}\right) - Q\frac{\partial}{\partial\xi_1}\left\{\frac{\Omega u_1}{h_1^S}\frac{\partial}{\partial\xi_2}\left[\sigma^S\left(B+\frac{\widetilde{H}}{2}\right)\right]\right\} = 0. \tag{18.291}$$

Subtracting (18.290) from (18.291) then yields

$$\begin{aligned}
&\frac{\partial}{\partial t}\left[\frac{\partial}{\partial\xi_1}\left(h_2^S u_2\right) - \frac{\partial}{\partial\xi_2}\left(h_1^S u_1\right)\right] + \frac{\partial}{\partial\xi_1}\left(h_2^S\zeta_3^S u_1\right) + \frac{\partial}{\partial\xi_2}\left(h_1^S\zeta_3^S u_2\right) \\
&\quad - 2\Omega\left[\frac{\partial}{\partial\xi_1}\left(h_2^S\frac{u_1}{h_2^S}\frac{dh_1^S}{d\xi_2}\right) + \frac{\partial}{\partial\xi_2}\left(h_1^S\frac{u_2}{h_2^S}\frac{dh_1^S}{d\xi_2}\right)\right] \\
&\quad - Q\frac{\partial}{\partial\xi_1}\left\{\frac{\Omega u_1}{h_1^S}\frac{\partial}{\partial\xi_2}\left[\sigma^S\left(B+\frac{\widetilde{H}}{2}\right)\right]\right\} - Q\frac{\partial}{\partial\xi_2}\left\{\frac{\Omega u_2}{h_2^S}\frac{\partial}{\partial\xi_2}\left[\sigma^S\left(B+\frac{\widetilde{H}}{2}\right)\right]\right\} \\
&\quad + Q\frac{\Omega}{2}\frac{\partial}{\partial\xi_2}\left\{\frac{\sigma^S}{h_1^S h_2^S}\left[\frac{\partial}{\partial\xi_1}\left(h_2^S u_1\widetilde{H}\right) + \frac{\partial}{\partial\xi_2}\left(h_1^S u_2\widetilde{H}\right)\right]\right\} = 0.
\end{aligned} \tag{18.292}$$

Dividing (18.292) through by $h_1^S h_2^S$; using definition (18.289) of ζ_3^S, and mass-continuity equation (18.265); and noting that B, h_1^S, h_2^S, and σ^S are all independent of t, (18.292) can be rewritten as

$$\begin{aligned}
&\frac{\partial\zeta_3^S}{\partial t} + \frac{1}{h_1^S h_2^S}\left[\frac{\partial}{\partial\xi_1}\left(h_2^S\zeta_3^S u_1\right) + \frac{\partial}{\partial\xi_2}\left(h_1^S\zeta_3^S u_2\right)\right] \\
&\quad - 2\Omega\frac{1}{h_1^S h_2^S}\left[\frac{\partial}{\partial\xi_1}\left(h_2^S\frac{u_1}{h_2^S}\frac{dh_1^S}{d\xi_2}\right) + \frac{\partial}{\partial\xi_2}\left(h_1^S\frac{u_2}{h_2^S}\frac{dh_1^S}{d\xi_2}\right)\right] \\
&\quad - Q\frac{1}{h_1^S h_2^S}\frac{\partial}{\partial\xi_1}\left\{\frac{\Omega u_1}{h_1^S}\frac{h_1^S h_2^S}{h_1^S h_2^S}\frac{\partial}{\partial\xi_2}\left[\sigma^S\left(B+\frac{\widetilde{H}}{2}\right)\right]\right\} - Q\frac{\Omega}{h_1^S h_2^S}\frac{\partial}{\partial\xi_2}\left[\sigma^S\frac{\partial}{\partial t}\left(B+\frac{\widetilde{H}}{2}\right)\right] \\
&\quad - Q\frac{1}{h_1^S h_2^S}\frac{\partial}{\partial\xi_2}\left\{\frac{\Omega u_2}{h_2^S}\frac{h_1^S h_2^S}{h_1^S h_2^S}\frac{\partial}{\partial\xi_2}\left[\sigma^S\left(B+\frac{\widetilde{H}}{2}\right)\right]\right\} = 0.
\end{aligned} \tag{18.293}$$

Equation (18.293) can be further rewritten as

$$\begin{aligned}
&\frac{\partial\zeta_3^S}{\partial t} + \frac{u_1}{h_1^S}\frac{\partial\zeta_3^S}{\partial\xi_1} + \frac{u_2}{h_2^S}\frac{\partial\zeta_3^S}{\partial\xi_2} + \frac{\zeta_3^S}{h_1^S h_2^S}\left[\frac{\partial}{\partial\xi_1}\left(h_2^S u_1\right) + \frac{\partial}{\partial\xi_2}\left(h_1^S u_2\right)\right] \\
&\quad - \left[\frac{u_1}{h_1^S}\frac{\partial}{\partial\xi_1}\left(\frac{2\Omega}{h_2^S}\frac{dh_1^S}{d\xi_2}\right) + \frac{u_2}{h_2^S}\frac{\partial}{\partial\xi_2}\left(\frac{2\Omega}{h_2^S}\frac{dh_1^S}{d\xi_2}\right)\right] \\
&\quad - \left(\frac{2\Omega}{h_2^S}\frac{dh_1^S}{d\xi_2}\right)\frac{1}{h_1^S h_2^S}\left[\frac{\partial}{\partial\xi_1}\left(h_2^S u_1\right) + \frac{\partial}{\partial\xi_2}\left(h_1^S u_2\right)\right] - Q\frac{\partial}{\partial t}\left\{\frac{\Omega}{h_1^S h_2^S}\frac{\partial}{\partial\xi_2}\left[\sigma^S\left(B+\frac{\widetilde{H}}{2}\right)\right]\right\} \\
&\quad - Q\frac{u_1}{h_1^S}\frac{\partial}{\partial\xi_1}\left\{\frac{\Omega}{h_1^S h_2^S}\frac{\partial}{\partial\xi_2}\left[\sigma^S\left(B+\frac{\widetilde{H}}{2}\right)\right]\right\} - Q\frac{u_2}{h_2^S}\frac{\partial}{\partial\xi_2}\left\{\frac{\Omega}{h_1^S h_2^S}\frac{\partial}{\partial\xi_2}\left[\sigma^S\left(B+\frac{\widetilde{H}}{2}\right)\right]\right\} \\
&\quad - Q\frac{\Omega}{h_1^S h_2^S}\frac{\partial}{\partial\xi_2}\left[\sigma^S\left(B+\frac{\widetilde{H}}{2}\right)\right]\frac{1}{h_1^S h_2^S}\left[\frac{\partial}{\partial\xi_1}\left(h_2^S u_1\right) + \frac{\partial}{\partial\xi_2}\left(h_1^S u_2\right)\right] = 0,
\end{aligned} \tag{18.294}$$

that is, as

$$\begin{aligned}
&\frac{D_{\text{hor}}\zeta_3^S}{Dt} + \zeta_3^S\frac{1}{h_1^S h_2^S}\left[\frac{\partial}{\partial\xi_1}\left(h_2^S u_1\right) + \frac{\partial}{\partial\xi_2}\left(h_1^S u_2\right)\right] - \frac{D_{\text{hor}}}{Dt}\left(\frac{2\Omega}{h_2^S}\frac{dh_1^S}{d\xi_2}\right) \\
&\quad - \left(\frac{2\Omega}{h_2^S}\frac{dh_1^S}{d\xi_2}\right)\frac{1}{h_1^S h_2^S}\left[\frac{\partial}{\partial\xi_1}\left(h_2^S u_1\right) + \frac{\partial}{\partial\xi_2}\left(h_1^S u_2\right)\right] - Q\frac{D_{\text{hor}}}{Dt}\left\{\frac{\Omega}{h_1^S h_2^S}\frac{\partial}{\partial\xi_2}\left[\sigma^S\left(B+\frac{\widetilde{H}}{2}\right)\right]\right\} \\
&\quad - Q\left\{\frac{\Omega}{h_1^S h_2^S}\frac{\partial}{\partial\xi_2}\left[\sigma^S\left(B+\frac{\widetilde{H}}{2}\right)\right]\right\}\frac{1}{h_1^S h_2^S}\left[\frac{\partial}{\partial\xi_1}\left(h_2^S u_1\right) + \frac{\partial}{\partial\xi_2}\left(h_1^S u_2\right)\right] = 0,
\end{aligned} \tag{18.295}$$

and then as

$$\frac{D_{\text{hor}}}{Dt}\left\{\zeta_3^S - \frac{2\Omega}{h_2^S}\frac{dh_1^S}{d\xi_2} - Q\frac{\Omega}{h_1^S h_2^S}\frac{\partial}{\partial \xi_2}\left[\sigma^S\left(B+\frac{\widetilde{H}}{2}\right)\right]\right\}$$
$$+\left\{\zeta_3^S - \frac{2\Omega}{h_2^S}\frac{dh_1^S}{d\xi_2} - Q\frac{\Omega}{h_1^S h_2^S}\frac{\partial}{\partial \xi_2}\left[\sigma^S\left(B+\frac{\widetilde{H}}{2}\right)\right]\right\}\frac{1}{h_1^S h_2^S}\left[\frac{\partial}{\partial \xi_1}\left(h_2^S u_1\right)+\frac{\partial}{\partial \xi_2}\left(h_1^S u_2\right)\right]=0. \tag{18.296}$$

Using mass-continuity equation (18.264) in (18.296) gives

$$\frac{D_{\text{hor}}}{Dt}\left\{\zeta_3^S - \frac{2\Omega}{h_2^S}\frac{dh_1^S}{d\xi_2} - Q\frac{\Omega}{h_1^S h_2^S}\frac{\partial}{\partial \xi_2}\left[\sigma^S\left(B+\frac{\widetilde{H}}{2}\right)\right]\right\}$$
$$-\left\{\zeta_3^S - \frac{2\Omega}{h_2^S}\frac{dh_1^S}{d\xi_2} - Q\frac{\Omega}{h_1^S h_2^S}\frac{\partial}{\partial \xi_2}\left[\sigma^S\left(B+\frac{\widetilde{H}}{2}\right)\right]\right\}\frac{1}{\widetilde{H}}\frac{D_{\text{hor}}\widetilde{H}}{Dt}=0. \tag{18.297}$$

Now divide this equation by $\widetilde{H}$ and further rearrange to obtain

$$\frac{D_{\text{hor}}\Pi^{\text{SW}}}{Dt}=0, \tag{18.298}$$

where

$$\Pi^{\text{SW}} \equiv \frac{1}{\widetilde{H}}\left\{\zeta_3^S - \frac{2\Omega}{h_2^S}\frac{dh_1^S}{d\xi_2} - Q\frac{\Omega}{h_1^S h_2^S}\frac{\partial}{\partial \xi_2}\left[\sigma^S\left(B+\frac{\widetilde{H}}{2}\right)\right]\right\} \tag{18.299}$$

is 2D (quasi-shallow) potential vorticity.

A Barotropic Potential Vorticity (BPV) Equation for Flow over a Spheroidal Planet

ABSTRACT

A barotropic potential vorticity (BPV) equation is derived for a shallow global atmosphere or ocean confined between two rigid, spheroidal geopotential surfaces, the bottom one coinciding with a planet's surface. This equation is shown to be a specialisation of a more general (3D) potential vorticity (PV) equation given in Chapter 15. By solving a derived Poisson problem, the corresponding pressure field may, at any instant in time, be diagnostically obtained from the predicted PV and associated stream function. The fluid is assumed to be homogeneous and of shallow depth, and the flow to be purely horizontal. The governing equations that result from these simplifications (from which the BPV equation is derived) are shown to be dynamically consistent. In contradistinction to the classical (two-dimensionally) non-divergent, barotropic-vorticity equation, the flow is generally divergent in the horizontal. This then permits meridional variation of apparent gravity for a spheroidal planet. Exceptionally, apparent gravity is constant if the spherical-geopotential approximation is applied in spherical geometry. The classical (two-dimensionally) non-divergent barotropic-vorticity equation is then recovered as a special case. The form of the BPV equation provides useful physical insight into the influence of meridional variation of apparent gravity on fluid flow. Some exact, unsteady, non-linear solutions of the BPV equation in ellipsoidal geometry are given in Chapter 22.

19.1 PREAMBLE

The classical (non-divergent) barotropic-vorticity equation is well known. It is usually written (in a planet's rotating frame of reference) as

The Classical (Non-Divergent) Barotropic Vorticity Equation

$$\frac{D_{\text{hor}}}{Dt}(\zeta + f) = 0, \tag{19.1}$$

where:

$$\frac{D_{\text{hor}}}{Dt} = \frac{\partial}{\partial t} + \mathbf{u}_{\text{hor}} \cdot \nabla_{\text{hor}}, \tag{19.2}$$

$$\zeta = \nabla^2_{\text{hor}}\psi, \tag{19.3}$$

$$\mathbf{u}_{\text{hor}} = \mathbf{e}_3 \times \nabla_{\text{hor}}\psi, \tag{19.4}$$

$$\nabla_{\text{hor}} \cdot \mathbf{u}_{\text{hor}} = 0. \tag{19.5}$$

Furthermore:

- f is Coriolis parameter.
- $\zeta = \mathbf{e}_3 \cdot (\nabla_{\text{hor}} \times \mathbf{u}_{\text{hor}})$ is relative vorticity.
- $\mathbf{u}_{\text{hor}}$ is horizontal velocity.
- ψ is stream function.
- $\mathbf{e}_3$ is a unit vector in the local vertical direction.
- Subscript 'hor' denotes horizontal contributions only.

Written this way, and using appropriate coordinates, these equations can be expressed in different geometries, such as on f and β planes, and over a sphere.

Because of the way in which these equations are written, one might think that they can be straightforwardly applied in spheroidal[1] geometry by simply writing the spatial derivative operators in suitable spheroidal coordinates. Indeed they can, *but not generally in a physically realistic way*. There is an important subtlety that, at first sight, is not obvious. The resulting equations are then valid only for *constant* apparent gravity, g. Physically, however, g varies meridionally, and this is an important motivation for using spheroidal coordinates, as discussed in Chapters 7, 8, and 12.

We set ourselves three goals:

1. To identify the origin of this undesirable restriction on g (in this chapter).
2. To remove this restriction at source, and thereby obtain a more general (and also more accurate) form of the classical barotropic-vorticity equation that allows meridional variation of apparent gravity (also in this chapter).
3. To construct some exact, *unsteady*, non-linear solutions of this more general equation in ellipsoidal geometry (in Chapter 22).

To distinguish this more general equation from the classical barotropic-vorticity equation, it is termed the 'barotropic-*potential*-vorticity equation'. It has become standard practice in the meteorological and oceanographic literature to abbreviate potential vorticity to PV. This practice is followed here; thus *barotropic PV equation*, or *BPV equation* for short.

To set the scene, a set of governing equations for general fluids, expressed in axial-orthogonal-curvilinear coordinates, is recalled in Section 19.2. Various assumptions and preparatory steps for inviscid, horizontal, shallow, fluid flow between two rigid spheroidal geopotential surfaces are then given and applied in Section 19.3. Global conservation of mass, axial absolute angular momentum, and total energy for the resulting simplified equation set is established in Section 19.4. Using the developments of Section 19.3, the BPV equation for flow over a spheroidal planet, with meridional variation of apparent gravity, is derived in Section 19.5; an alternative derivation is also given in Section 19.6. Dynamical consistency of the simplified equation set is reviewed in Section 19.7. A diagnostic Poisson problem for the corresponding pressure field at any instant in time is obtained in Section 19.8, and the component equations of momentum are rederived variationally in Section 19.9. Concluding remarks are given in Section 19.10.

19.2 THE MOMENTUM AND MASS-CONTINUITY EQUATIONS IN CURVILINEAR FORM

Recall that a set of governing equations for general fluids is given in Section 6.5 for a planet rotating about its axis with angular frequency Ω. This set of equations provides the starting point for the derivation here of the BPV equation in axial-orthogonal-curvilinear coordinates.

[1] A *spheroid* is considered herein to be a solid of revolution that is approximately spherical, having an *almost* or *precisely* elliptic cross section in any meridional plane. When this cross section is *precisely* elliptic, the solid is then termed an *ellipsoid*.

19.2.1 Axial-Orthogonal-Curvilinear Coordinates

Let (ξ_1, ξ_2, ξ_3) be general, axisymmetric, orthogonal, curvilinear coordinates in the zonal, meridional, and (upward) vertical directions, respectively, with associated unit vectors $(\mathbf{e}_1, \mathbf{e}_2, \mathbf{e}_3)$ and velocity components (u_1, u_2, u_3). The metric (or scale) factors (h_1, h_2, h_3) are the quantities that appear in

$$ds^2 = h_1^2 d\xi_1^2 + h_2^2 d\xi_2^2 + h_3^2 d\xi_3^2, \tag{19.6}$$

where ds is infinitesimal distance. Standard expressions for gradient, curl, and divergence for these coordinates are given in Chapter 5. For *axial*-orthogonal-curvilinear coordinates – see (6.13) – h_1, h_2, and h_3 are, by definition, all independent of ξ_1, that is,

$$h_i = h_i(\xi_2, \xi_3), \quad i = 1, 2, 3. \tag{19.7}$$

19.2.2 Components of the Momentum Balance

Assume a *geopotential coordinate system* so that surfaces of constant ξ_3 represent surfaces of constant apparent geopotential (Φ), with apparent gravity acting only in the ξ_3 direction. Thus $\Phi = \Phi(\xi_3)$. Setting the right-hand-side forcing terms to zero in (6.29)–(6.31), the unforced components of the momentum balance in an axial-orthogonal-curvilinear, geopotential coordinate system are then

$$\frac{Du_1}{Dt} + \left(\frac{u_1}{h_1} + 2\Omega\right)\frac{u_2}{h_2}\frac{\partial h_1}{\partial \xi_2} + \underbrace{\left(\frac{u_1}{h_1} + 2\Omega\right)\frac{u_3}{h_3}\frac{\partial h_1}{\partial \xi_3}}_{\text{deep}} = -\frac{1}{\rho h_1}\frac{\partial p}{\partial \xi_1}, \tag{19.8}$$

$$\frac{Du_2}{Dt} - \left(\frac{u_1}{h_1} + 2\Omega\right)\frac{u_1}{h_2}\frac{\partial h_1}{\partial \xi_2} - \underbrace{\frac{u_3^2}{h_2 h_3}\frac{\partial h_3}{\partial \xi_2}}_{\text{non-hydro}} + \underbrace{\frac{u_2\, u_3}{h_2\, h_3}\frac{\partial h_2}{\partial \xi_3}}_{\text{deep}} = -\frac{1}{\rho h_2}\frac{\partial p}{\partial \xi_2}, \tag{19.9}$$

$$\underbrace{\left(\frac{Du_3}{Dt} + \frac{u_2\, u_3}{h_2\, h_3}\frac{\partial h_3}{\partial \xi_2}\right)}_{\text{non-hydro}} - \underbrace{\left[\left(\frac{u_1}{h_1} + 2\Omega\right)\frac{u_1}{h_3}\frac{\partial h_1}{\partial \xi_3} + \frac{u_2^2}{h_2 h_3}\frac{\partial h_2}{\partial \xi_3}\right]}_{\text{deep}} = -\frac{1}{\rho h_3}\frac{\partial p}{\partial \xi_3} - \frac{1}{h_3}\frac{d\Phi}{d\xi_3}, \tag{19.10}$$

where various terms have been colour coded for later convenience.

In (19.8)–(19.10), differentiation D/Dt following the fluid satisfies

$$\frac{D}{Dt} \equiv \frac{\partial}{\partial t} + \frac{u_1}{h_1}\frac{\partial}{\partial \xi_1} + \frac{u_2}{h_2}\frac{\partial}{\partial \xi_2} + \frac{u_3}{h_3}\frac{\partial}{\partial \xi_3}. \tag{19.11}$$

In (19.10),

$$g(\xi_2, \xi_3) \equiv \mathbf{e}_3 \cdot \nabla\Phi \equiv \frac{1}{h_3}\frac{d\Phi}{d\xi_3} \tag{19.12}$$

is apparent gravity (aligned with the normal to a geopotential surface ξ_3 = constant); and $\Phi = \Phi(\xi_3)$ is the potential of apparent gravity.

A very nice feature of (19.8)–(19.10), developed in White and Wood (2012) and in Section 6.4 herein, is – see (6.27) –

The Expression of the Coriolis Force in Terms of Metric Factors

$$-2\boldsymbol{\Omega} \times \mathbf{u} = -2\Omega\left[\left(\frac{u_2}{h_2}\frac{\partial h_1}{\partial \xi_2} + \frac{u_3}{h_3}\frac{\partial h_1}{\partial \xi_3}\right)\mathbf{e}_1 - \frac{u_1}{h_2}\frac{\partial h_1}{\partial \xi_2}\mathbf{e}_2 - \frac{u_1}{h_3}\frac{\partial h_1}{\partial \xi_3}\mathbf{e}_3\right]. \tag{19.13}$$

This allows the Coriolis terms to be grouped together in a neat way with the metric terms in (19.8)–(19.10). Importantly, it also greatly facilitates the optional application of the shallow-fluid assumption. Axial symmetry of the coordinate system – see (19.7) – is implicit in the derivation of (19.13).

19.2.3 The Mass-Continuity Equation

To derive the BPV equation, only the three components of the momentum balance and the mass-continuity equation are needed. There is no requirement for thermodynamic-energy and state equations. Thus, to close the set of governing equations, the mass-continuity equation is written as – see (6.32) with $F^{\rho} \equiv 0$ –

$$\frac{D\rho}{Dt} + \frac{\rho}{h_1 h_2 h_3}\left[\frac{\partial}{\partial \xi_1}\left(u_1 h_2 h_3\right) + \frac{\partial}{\partial \xi_2}\left(u_2 h_3 h_1\right) + \frac{\partial}{\partial \xi_3}\left(u_3 h_1 h_2\right)\right] = 0. \tag{19.14}$$

19.3 INVISCID, HORIZONTAL, SHALLOW FLOW IN SPHEROIDAL GEOMETRY

19.3.1 Assumptions and Preparatory Steps

To derive a set of shallow-water equations for unforced flow of a shallow layer of homogeneous, inviscid fluid over a spheroidal planet, with bottom orography and meridional variation of apparent gravity, a number of assumptions and preparatory steps were set out in Staniforth and White (2015b) and Chapter 18 herein. This approach is adapted here for the derivation of the BPV equation over a spheroidal planet.

A crucial difference between the two derivations, however, is the nature of the upper boundary. For the shallow-water equations the upper boundary is assumed to be a *free surface*, whereas for the BPV equation it is assumed to be a *rigid geopotential surface*. This latter assumption significantly restricts the associated fluid flow.

To derive the BPV equation, the following assumptions and preparatory steps are made:

1. The planet's spheroidal surface coincides with a zonally symmetric geopotential surface[2] – see Fig. 19.1. The vertical geopotential coordinate ξ_3 is chosen so that the planet's surface corresponds to $\xi_3 = \xi_3^B =$ constant.
2. The overlying layer of fluid is confined between bottom and top rigid geopotential surfaces located at $\xi_3 = \xi_3^B =$ constant and $\xi_3 = \xi_3^T =$ constant, respectively.
3. The fluid is homogeneous, that is,

$$\rho = \overline{\rho} = \text{constant}. \tag{19.15}$$

   This significantly simplifies mass-continuity equation (19.14); see (19.33).
4. The flow is purely horizontal for all time, that is,

$$u_3 \equiv 0, \quad \text{for all time,} \tag{19.16}$$

[2] This surface is graphically depicted in Fig. 19.1 as being elliptic. This is appropriate for terrestrial applications, with the virtue of being consistent with the World Geodetic System reference ellipsoid used for reporting observational data, and for satellite navigation. However, for the derivation given herein of the BPV equation, the shape of the surface (determined by the planet's internal mass distribution) is less restricted: it is only assumed to be zonally symmetric.

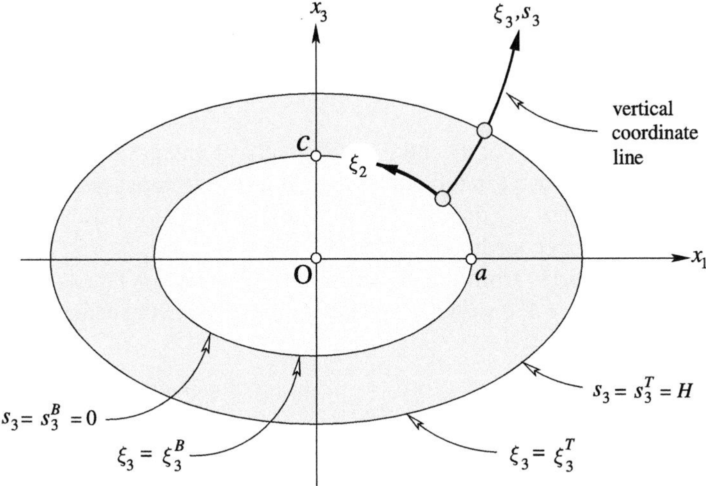

Figure 19.1 A cross section in the Ox_1x_3 meridional plane for horizontal barotropic flow over a spheroid. ξ_2 and ξ_3 are axial-orthogonal-curvilinear coordinates in the meridional and outward-pointing directions, respectively. ξ_1, the azimuthal coordinate, points into the page for $x_1 > 0$ and out of it for $x_1 < 0$. s_3 is physical distance from the spheroid's surface along a (curved) vertical coordinate line. The spheroid is light shaded. The overlying shallow layer of fluid (dark shaded) is confined between bottom and top rigid geopotential surfaces located at $\xi_3 = \xi_3^B$ and $\xi_3 = \xi_3^T$, respectively. When the spheroid is an ellipsoid, its semi-major and semi-minor axes are denoted by a and c, respectively. See text for further details.

everywhere. Thus infinitesimal fluid parcels can only move within geopotential surfaces (for which $\xi_3 =$ constant) and cannot cross them. This (quasi-hydrostatic) assumption removes the blue terms (denoted 'non-hydro') from (19.8)–(19.10), and constrains material derivative (19.11) to be purely horizontal, so that $D/Dt \to D_{\text{hor}}/Dt$.

5. The depth of the fluid is shallow. This assumption implies that scale factors $h_i, i = 1, 2, 3$ are independent of ξ_3, so that (19.7) (for the assumed zonal symmetry of the coordinate system) reduces to

$$h_i = h_i(\xi_2), \quad i = 1, 2, 3. \tag{19.17}$$

This then removes the red terms (denoted 'deep') from (19.8)–(19.10).

6. As noted in White and Wood (2012), Staniforth and White (2015b) and Chapter 18 herein, a crucial step in deriving shallow-fluid models is to make apparent gravity, g, *independent of height*. This is also the case for the present context. Due to the use of an assumed, zonally symmetric, coordinate system, g is also independent of ξ_1. Furthermore, ξ_3 is a geopotential coordinate. Thus it is assumed that

$$g = g(\xi_2), \quad \Phi = \Phi(\xi_3). \tag{19.18}$$

However, from definition (19.12) of g in terms of Φ, (19.17) and (19.18) imply that

$$\frac{d\Phi(\xi_3)}{d\xi_3} = h_3(\xi_2)\, g(\xi_2). \tag{19.19}$$

But the only way that (19.19) can be satisfied is by setting

$$h_3(\xi_2)\, g(\xi_2) = \text{constant} = h_3^E g^E \quad (\text{say}), \tag{19.20}$$

where h_3^E and g^E are the values of h_3 and g, respectively, at any point on the planet's equator. Thus g is permitted to vary as a function of latitude, but then h_3 must satisfy constraint (19.20) and vary as the reciprocal of g. Equation (19.20) is a simple but very important consequence of the use of a geopotential coordinate system.

7. Vertically integrating (19.19) from the planet's surface, $\xi_3 = \xi_3^B = \text{constant}$, to ξ_3, subject to the condition that $\Phi\left(\xi_3 = \xi_3^B\right) = \Phi^B = \text{constant}$ on the planet's surface, and using (19.20) gives

$$\Phi(\xi_3) = \Phi^B + h_3^E g^E \left(\xi_3 - \xi_3^B\right). \tag{19.21}$$

8. Using (19.17) and considering only changes along the ξ_3 direction, (19.6) reduces to

$$ds_3 = h_3(\xi_2)\, d\xi_3. \tag{19.22}$$

Integration of (19.22) from the planet's surface at $\xi_3 = \xi_3^B = \text{constant}$, to ξ_3, with use of (19.20) then gives

$$s_3(\xi_1, \xi_2, \xi_3) = h_3(\xi_2)\left(\xi_3 - \xi_3^B\right) = \frac{h_3^E g^E}{g(\xi_2)}\left(\xi_3 - \xi_3^B\right), \tag{19.23}$$

so that $s_3(\xi_1, \xi_2, \xi_3)$ measures distance from the planet's surface $\xi_3 = \xi_3^B$ along a (curved) vertical coordinate line with ξ_1 and ξ_2 held fixed.

9. The (rigid) geopotential surface $\xi_3 = \xi_3^T \equiv \text{constant}$ is located at a physical distance $s_3^T(\xi_2) \equiv H(\xi_2)$ along a (curved) vertical coordinate line from the planet's surface at $s_3 = 0$ (i.e. at $\xi_3 = \xi_3^B$). See Fig. 19.1. From (19.23),

$$s_3^T(\xi_2) \equiv H(\xi_2) = h_3(\xi_2)\left(\xi_3^T - \xi_3^B\right) = \frac{h_3^E g^E\left(\xi_3^T - \xi_3^B\right)}{g(\xi_2)}. \tag{19.24}$$

The depth of the shallow layer of fluid is very small, that is, $H(\xi_2) \equiv s_3^T(\xi_2) \ll \bar{a}$, where $\bar{a}$ is the planet's mean radius. Also the mass of this shallow layer of fluid is assumed to make a negligible contribution to Newtonian gravitational attraction.

10. The two horizontal momentum variables are

$$u_1 = u_1(\xi_1, \xi_2, \xi_3, t), \quad u_2 = u_2(\xi_1, \xi_2, \xi_3, t). \tag{19.25}$$

At initial time, they are assumed to be independent of vertical coordinate ξ_3, that is,

$$(u_1)_{t=0} = u_1(\xi_1, \xi_2, t = 0), \quad (u_2)_{t=0} = u_2(\xi_1, \xi_2, t = 0). \tag{19.26}$$

11. By definition (6.39) of u_3, and using horizontal-flow assumption (19.16),

$$u_3 \equiv h_3(\xi_2)\frac{D\xi_3}{Dt} = 0. \tag{19.27}$$

Substituting for ξ_3 into (19.27) using (19.23), and also using definition (19.11) and shallow metric factors (19.17), gives the following expression for the vertical component of velocity

$$h_3(\xi_2)\frac{D\xi_3}{Dt} = h_3\frac{D}{Dt}\left(\frac{s_3}{h_3}\right) = \frac{Ds_3}{Dt} - \frac{s_3}{h_3}\frac{Dh_3(\xi_2)}{Dt} = \frac{Ds_3}{Dt} - \frac{s_3}{h_3}\frac{u_2}{h_2}\frac{dh_3}{d\xi_2} = 0. \tag{19.28}$$

Evaluating (19.28) at the upper bounding surface $s_3 = s_3^T \equiv H(\xi_2)$ then gives

$$\frac{DH(\xi_2)}{Dt} - \frac{H(\xi_2)}{h_3(\xi_2)}\frac{u_2^T}{h_2(\xi_2)}\frac{dh_3(\xi_2)}{d\xi_2} = 0. \tag{19.29}$$

19.3.2 Simplification of the Governing Equations

Introduction of the assumptions and preparatory steps of Section 19.3.1 into governing equations (19.8)–(19.11) and (19.14) yields

$$\frac{D_{\text{hor}}u_1}{Dt} + \left(\frac{u_1}{h_1} + 2\Omega\right)\frac{u_2}{h_2}\frac{dh_1}{d\xi_2} = -\frac{1}{\overline{\rho}h_1}\frac{\partial p}{\partial \xi_1}, \tag{19.30}$$

$$\frac{D_{\text{hor}}u_2}{Dt} - \left(\frac{u_1}{h_1} + 2\Omega\right)\frac{u_1}{h_2}\frac{dh_1}{d\xi_2} = -\frac{1}{\overline{\rho}h_2}\frac{\partial p}{\partial \xi_2}, \tag{19.31}$$

$$\frac{\partial p}{\partial \xi_3} = -\overline{\rho}\frac{d\Phi}{d\xi_3} = -\overline{\rho}h_3^E g^E = \text{constant}, \tag{19.32}$$

$$\frac{\partial}{\partial \xi_1}\left(u_1 h_2 h_3\right) + \frac{\partial}{\partial \xi_2}\left(u_2 h_3 h_1\right) = 0, \tag{19.33}$$

where

$$\frac{D_{\text{hor}}}{Dt} \equiv \frac{\partial}{\partial t} + \frac{u_1}{h_1}\frac{\partial}{\partial \xi_1} + \frac{u_2}{h_2}\frac{\partial}{\partial \xi_2} \tag{19.34}$$

is horizontal material derivative along a geopotential surface, $\xi_3 =$ constant.[3]

Integrating hydrostatic equation (19.32) downwards from the upper, bounding geopotential surface $\xi_3 = \xi_3^T =$ constant then gives

$$p(\xi_1, \xi_2, \xi_3, t) = p^T(\xi_1, \xi_2, t) + \overline{\rho}h_3^E g^E\left(\xi_3^T - \xi_3\right). \tag{19.35}$$

[3] Multiplying (19.33) through by $\overline{\rho}/(h_1 h_2)$, and taking $\overline{\rho} =$ constant inside the partial derivatives, condition (19.33) may be physically interpreted as the mass flux between geopotential surfaces being horizontally non-divergent. This implies $h_3\mathbf{u}_{\text{hor}}$ is horizontally non-divergent, where $\mathbf{u}_{\text{hor}} \equiv u_1\mathbf{e}_1 + u_2\mathbf{e}_2$. It does not, in general, imply that $\mathbf{u}_{\text{hor}}$ is horizontally non-divergent, which only (exceptionally) occurs when $h_3 \equiv 1$.

This means that the horizontal pressure-gradient terms on the right-hand sides of (19.30) and (19.31) simplify, and (19.30) and (19.31) then reduce to

$$\frac{D_{\text{hor}}u_1}{Dt} + \left(\frac{u_1}{h_1} + 2\Omega\right)\frac{u_2}{h_2}\frac{dh_1}{d\xi_2} = -\frac{1}{\overline{\rho}h_1}\frac{\partial p^T(\xi_1,\xi_2,t)}{\partial\xi_1}, \tag{19.36}$$

$$\frac{D_{\text{hor}}u_2}{Dt} - \left(\frac{u_1}{h_1} + 2\Omega\right)\frac{u_1}{h_2}\frac{dh_1}{d\xi_2} = -\frac{1}{\overline{\rho}h_2}\frac{\partial p^T(\xi_1,\xi_2,t)}{\partial\xi_2}. \tag{19.37}$$

Examination of (19.36) and (19.37) shows that if (see (19.26)) u_1 and u_2 are initially independent of the vertical coordinate ξ_3, then this must be true for all time, that is,

$$u_1 = u_1(\xi_1,\xi_2,t), \quad u_2 = u_2(\xi_1,\xi_2,t), \quad \text{for all time.} \tag{19.38}$$

19.3.3 Recapitulation

Summarising the preceding developments for inviscid, horizontal, shallow, fluid flow, confined between two, rigid, spheroidal geopotential surfaces, and expressed in general, axial-orthogonal-curvilinear, geopotential coordinates:

- The two components of momentum balance are (19.36) and (19.37), respectively.
- The mass-continuity equation reduces to (19.33).
- The vertical component of momentum balance (19.32) has been used to simplify the two horizontal components and to obtain diagnostic equation (19.35) for total pressure; (19.32) is not needed for the further developments given here.
- Apparent gravity $g(\xi_2)$ and the vertical scale factor $h_3(\xi_2)$ are constrained by the reciprocal constraint (19.20).
- Flow at the upper, bounding geopotential surface is constrained by (19.29).

19.3.4 Vector Form

Equations (19.36), (19.37), and (19.33) may be more compactly written in vector form as follows.

19.3.4.1 *The Horizontal Momentum Balance*

From (5.11) and (5.12), with $u_3 \equiv 0$ and $h_i = h_i(\xi_2)$, and using horizontal material derivative (19.34),

$$\frac{D_{\text{hor}}\mathbf{u}_{\text{hor}}}{Dt} = \left(\frac{D_{\text{hor}}u_1}{Dt} + \frac{u_1u_2}{h_1h_2}\frac{dh_1}{d\xi_2}\right)\mathbf{e}_1 + \left(\frac{D_{\text{hor}}u_2}{Dt} - \frac{u_1^2}{h_1h_2}\frac{dh_1}{d\xi_2}\right)\mathbf{e}_2, \tag{19.39}$$

where

$$\mathbf{u}_{\text{hor}} \equiv u_1\mathbf{e}_1 + u_2\mathbf{e}_2, \tag{19.40}$$

is horizontal velocity. Now

$$f\mathbf{e}_3 \times \mathbf{u}_{\text{hor}} = -\frac{2\Omega}{h_2}\frac{dh_1}{d\xi_2}(-u_2\mathbf{e}_1 + u_1\mathbf{e}_2) = \frac{2\Omega}{h_2}\frac{dh_1}{d\xi_2}(u_2\mathbf{e}_1 - u_1\mathbf{e}_2), \tag{19.41}$$

where

$$f(\xi_2) \equiv -\frac{2\Omega}{h_2}\frac{dh_1}{d\xi_2}, \tag{19.42}$$

is the Coriolis parameter, expressed in terms of metric factors. Furthermore, from (5.6),

$$\frac{1}{\overline{\rho}}\nabla_{\text{hor}}p^T = \frac{1}{\overline{\rho}h_1}\frac{\partial p^T}{\partial \xi_1}\mathbf{e}_1 + \frac{1}{\overline{\rho}h_2}\frac{\partial p^T}{\partial \xi_2}\mathbf{e}_2, \tag{19.43}$$

where $\nabla_{\text{hor}}p^T$ is the horizontal gradient of p^T. Summing (19.39), (19.41), and (19.43) and rearranging, component equations (19.36) and (19.37) of horizontal momentum can be rewritten as

$$\frac{D_{\text{hor}}\mathbf{u}_{\text{hor}}}{Dt} + f\mathbf{e}_3 \times \mathbf{u}_{\text{hor}} = -\frac{1}{\overline{\rho}}\nabla_{\text{hor}}p^T. \tag{19.44}$$

19.3.4.2 *The Mass-Continuity Equation*

Using (19.17), mass-continuity equation (19.33) can be rewritten as

$$\begin{aligned} 0 &= \frac{H(\xi_2)}{h_1h_2h_3}\left[\frac{\partial}{\partial \xi_1}(u_1h_2h_3) + \frac{\partial}{\partial \xi_2}(u_2h_3h_1)\right] \\ &= H(\xi_2)\frac{u_2}{h_2h_3}\frac{dh_3}{d\xi_2} + \frac{H(\xi_2)}{h_1h_2}\left[\frac{\partial}{\partial \xi_1}(u_1h_2) + \frac{\partial}{\partial \xi_2}(u_2h_1)\right]. \end{aligned} \tag{19.45}$$

Thus – using (19.27), (19.29), and (19.38) in (19.45) –

$$\frac{D_{\text{hor}}H(\xi_2)}{Dt} + H(\xi_2)\nabla_{\text{hor}} \cdot \mathbf{u}_{\text{hor}} = 0, \tag{19.46}$$

where

$$\nabla_{\text{hor}} \cdot \mathbf{u}_{\text{hor}} \equiv \frac{1}{h_1h_2}\left[\frac{\partial}{\partial \xi_1}(u_1h_2) + \frac{\partial}{\partial \xi_2}(u_2h_1)\right] \tag{19.47}$$

is horizontal divergence of $\mathbf{u}_{\text{hor}}$.

In general, $H = H(\xi_2)$ and is *not* constant. It is therefore seen from (19.16), (19.33), (19.46), and (19.47) that although the flow is *three*-dimensionally non-divergent, it is not, in general, *two*-dimensionally non-divergent.

[Exceptionally, the flow is however *two*-dimensionally non-divergent *when using the classical spherical-geopotential approximation in spherical geometry*, since then $h_3 \equiv 1$ (from (5.15)) and (from (19.29)) H is then constant, which implies $D_{\text{hor}}H/Dt \equiv 0$ in (19.46).]

19.4 GLOBAL CONSERVATION

19.4.1 Mass

The fluid is assumed to be homogeneous; see item 19.3.1 of Section 19.3.1. Partially differentiating (19.15) with respect to time t gives

$$\frac{\partial \rho}{\partial t} = \frac{\partial \overline{\rho}}{\partial t} = 0. \tag{19.48}$$

Globally integrating (19.48) then yields:

Conservation of Globally Integrated Mass

$$\frac{d}{dt}\iiint_V (\overline{\rho})\, h_1 h_2 h_3 d\xi_1 d\xi_2 d\xi_3 = 0, \tag{19.49}$$

where the volume integral is over the entire 3D global domain.

Since the integrand in (19.49) is independent of ξ_3, and the values of ξ_3 at the bottom and top boundaries are both constant, the 3D conservation principle (19.49) can be re-expressed as the 2D one

$$\frac{d}{dt}\iint_A (\overline{\rho})\, h_1 h_2 h_3 d\xi_1 d\xi_2 = 0, \tag{19.50}$$

where the area integral is over the entire horizontal domain. Note the presence of h_3 in (19.50).

19.4.2 Axial Absolute Angular Momentum

From (19.17) and (19.34),

$$\frac{u_2}{h_2}\frac{dh_1}{d\xi_2} = \frac{D_{\text{hor}} h_1}{Dt}. \tag{19.51}$$

Multiplication of (19.36) by h_1, with use of (19.51) and (19.35), yields

$$\frac{D_{\text{hor}}\left(h_1 u_1\right)}{Dt} + 2\Omega h_1 \frac{D_{\text{hor}} h_1}{Dt} = -\frac{1}{\overline{\rho}}\frac{\partial p^T\left(\xi_1,\xi_2,t\right)}{\partial \xi_1} = -\frac{1}{\overline{\rho}}\frac{\partial p}{\partial \xi_1}. \tag{19.52}$$

Noting that $2\Omega h_1\left(D_{\text{hor}} h_1/Dt\right) = D_{\text{hor}}\left(\Omega h_1^2\right)/Dt$ and $D_{\text{hor}} M^{\text{BPV}}/Dt = DM^{\text{BPV}}/Dt$ (since $u_3 \equiv 0$ from (19.16)), (19.52) may be rewritten as

$$\overline{\rho}\frac{DM^{\text{BPV}}}{Dt} = \overline{\rho}\frac{D_{\text{hor}} M^{\text{BPV}}}{Dt} = -\frac{\partial p}{\partial \xi_1}, \tag{19.53}$$

where

$$M^{\text{BPV}} \equiv h_1\left(u_1 + \Omega h_1\right) \tag{19.54}$$

is axial absolute angular momentum per unit mass for the BPV (Barotropic Potential Vorticity) equation.

Since (from (19.16)) $u_3 \equiv 0 \Rightarrow D_{\text{hor}}/Dt = D/Dt$, and (from (19.15)) $\overline{\rho} = $ constant, the mass-continuity equation (19.33) can be rewritten as

$$\frac{D\overline{\rho}}{Dt} + \frac{\overline{\rho}}{h_1 h_2 h_3}\left[\frac{\partial}{\partial \xi_1}\left(u_1 h_2 h_3\right) + \frac{\partial}{\partial \xi_2}\left(u_2 h_3 h_1\right) + \frac{\partial}{\partial \xi_3}\left(u_3 h_1 h_2\right)\right] = 0. \tag{19.55}$$

Multiplying this equation by M^{BPV} and summing the resulting equation with (19.53) then gives

$$\frac{D\left(\overline{\rho} M^{\text{BPV}}\right)}{Dt} + \frac{\overline{\rho} M^{\text{BPV}}}{h_1 h_2 h_3}\left[\frac{\partial}{\partial \xi_1}\left(u_1 h_2 h_3\right) + \frac{\partial}{\partial \xi_2}\left(u_2 h_3 h_1\right) + \frac{\partial}{\partial \xi_3}\left(u_3 h_1 h_2\right)\right] = -\frac{\partial p}{\partial \xi_1}. \tag{19.56}$$

Using definition (19.11) of the material derivative in three dimensions, (19.56) may be rewritten as

$$\frac{\partial\left(\overline{\rho}M^{\mathrm{BPV}}\right)}{\partial t}+\frac{1}{h_1h_2h_3}\left[\frac{\partial}{\partial\xi_1}\left(\overline{\rho}M^{\mathrm{BPV}}u_1h_2h_3\right)+\frac{\partial}{\partial\xi_2}\left(\overline{\rho}M^{\mathrm{BPV}}u_2h_3h_1\right)+\frac{\partial}{\partial\xi_3}\left(\overline{\rho}M^{\mathrm{BPV}}u_3h_1h_2\right)\right]$$
$$=-\frac{\partial p}{\partial\xi_1}. \tag{19.57}$$

Equations (19.53) and (19.57) correspond to conservation principles (15.80) and (15.81), respectively, of Section 15.4 with $\rho=\overline{\rho}$ and $F^{\rho}\equiv 0$; and $\mathbb{C}$ and $F^{\mathbb{C}}$ given by the first row of Table 15.2, where $F^{u_1}=\partial\Phi/\partial\xi_1\equiv 0$ therein.

Globally integrating (19.57), applying periodicity in the horizontal, and noting that $u_3\equiv 0$ gives:

Conservation of Globally Integrated Axial Absolute Angular Momentum

$$\frac{d}{dt}\iiint_V\left(\overline{\rho}M^{\mathrm{BPV}}\right)h_1h_2h_3d\xi_1d\xi_2d\xi_3=0, \tag{19.58}$$

where the volume integral is over the entire 3D global domain.

Since the integrand in (19.58) is independent of ξ_3, and the values of ξ_3 at the bottom and top boundaries are both constant, the 3D conservation principle (19.58) can be re-expressed as the 2D one

$$\frac{d}{dt}\iint_A\left(\overline{\rho}M^{\mathrm{BPV}}\right)h_1h_2h_3d\xi_1d\xi_2=0, \tag{19.59}$$

where the area integral is over the entire horizontal domain.

This 2D principle may be obtained more directly by noting that the last term on the left-hand side of (19.57) is identically zero (since $u_3\equiv 0$) and (after multiplication by $h_1h_2h_3$) by only integrating (19.57) over the entire horizontal domain. Even though no vertical integration has been performed to obtain (19.59), vertical metric factor h_3 still makes its presence felt. This is because:

- h_3 varies meridionally (and therefore cannot be taken outside the horizontal integral).
- h_3 must appear for the horizontal integral to be over horizontal fluxes of the form $\partial\left(\overline{\rho}M^{\mathrm{BPV}}u_1h_2h_3\right)/\partial\xi_1+\partial\left(\overline{\rho}M^{\mathrm{BPV}}u_2h_3h_1\right)/\partial\xi_2$, and for these fluxes to then integrate to zero via application of horizontal periodicity.

Note that $h_3\equiv 1$ for the (special) spherical case, and h_3 is then essentially absent from both (19.58) and (19.59).

19.4.3 Total Energy

Multiplying (19.36) and (19.37) by u_1 and u_2, respectively, summing, and using (19.11), (19.27), (19.34), and (19.35) gives

$$\frac{DK}{Dt}=\frac{D_{\mathrm{hor}}K}{Dt}=-\frac{1}{\overline{\rho}}\left(\frac{u_1}{h_1}\frac{\partial}{\partial\xi_1}+\frac{u_2}{h_2}\frac{\partial}{\partial\xi_2}+\frac{u_3}{h_3}\frac{\partial}{\partial\xi_3}\right)p^T\left(\xi_1,\xi_2,t\right)$$
$$=-\frac{1}{\overline{\rho}}\left(\frac{u_1}{h_1}\frac{\partial}{\partial\xi_1}+\frac{u_2}{h_2}\frac{\partial}{\partial\xi_2}+\frac{u_3}{h_3}\frac{\partial}{\partial\xi_3}\right)p, \tag{19.60}$$

where

$$K \equiv \frac{u_1^2 + u_2^2 + u_3^2}{2} \tag{19.61}$$

is specific kinetic energy.

Since $\Phi = \Phi(\xi_3)$ (for a geopotential coordinate system) and $u_3 \equiv 0$ (from (19.27)),

$$\frac{D\Phi}{Dt} = \cancelto{0}{\frac{\partial \Phi}{\partial t}} + \frac{u_1}{h_1}\cancelto{0}{\frac{\partial \Phi}{\partial \xi_1}} + \frac{u_2}{h_2}\cancelto{0}{\frac{\partial \Phi}{\partial \xi_2}} + \frac{\cancelto{0}{u_3}}{h_3}\frac{\partial \Phi}{\partial \xi_3} = \frac{D_{\text{hor}}\Phi}{Dt} = 0. \tag{19.62}$$

To derive the BPV equation, there is no need to bring in the thermodynamic-energy equation as we will see. We can nevertheless, if we so desire, derive an evolution equation for internal energy $\mathscr{E}$. The fluid is assumed homogeneous – see (19.15) – and $\alpha \equiv 1/\rho$, so $\alpha = 1/\rho = 1/\overline{\rho} =$ constant $= \overline{\alpha}$ (say). Inserting $\alpha = \overline{\alpha} =$ constant into thermodynamic-energy equation (6.34), and setting $\dot{Q}_E \equiv 0$ (i.e. assuming no energy input) then gives

$$\frac{D\mathcal{E}}{Dt} = -p\frac{D\overline{\alpha}}{Dt} + \dot{Q}_E = 0, \tag{19.63}$$

that is, internal energy is materially conserved.

Summing (19.60), (19.62), and (19.63) yields (after multiplication by $\overline{\rho}$)

$$\overline{\rho}\frac{DE}{Dt} = \overline{\rho}\frac{D(K + \Phi + \mathscr{E})}{Dt} = -\left(\frac{u_1}{h_1}\frac{\partial}{\partial \xi_1} + \frac{u_2}{h_2}\frac{\partial}{\partial \xi_2} + \frac{u_3}{h_3}\frac{\partial}{\partial \xi_3}\right)p, \tag{19.64}$$

where $E \equiv K + \Phi + \mathscr{E}$ is total energy per unit mass.

Multiplying mass-continuity equation (19.33) by $p/\left(h_1 h_2 h_3\right)$, summing this with (19.64), and noting that $u_3 \equiv 0$ gives

$$\overline{\rho}\frac{DE}{Dt} = -\frac{1}{h_1 h_2 h_3}\left[\frac{\partial}{\partial \xi_1}(pu_1 h_2 h_3) + \frac{\partial}{\partial \xi_2}(pu_2 h_3 h_1) + \frac{\partial}{\partial \xi_2}(pu_3 h_1 h_2)\right]. \tag{19.65}$$

Multiplying (19.55) by E and summing it with (19.65) then delivers

$$\begin{aligned}\frac{D(\overline{\rho}E)}{Dt} &+ \frac{\overline{\rho}E}{h_1 h_2 h_3}\left[\frac{\partial}{\partial \xi_1}(u_1 h_2 h_3) + \frac{\partial}{\partial \xi_2}(u_2 h_3 h_1) + \frac{\partial}{\partial \xi_3}(u_3 h_1 h_2)\right] \\ &= -\frac{1}{h_1 h_2 h_3}\left[\frac{\partial}{\partial \xi_1}(pu_1 h_2 h_3) + \frac{\partial}{\partial \xi_2}(pu_2 h_3 h_1) + \frac{\partial}{\partial \xi_2}(pu_3 h_1 h_2)\right].\end{aligned} \tag{19.66}$$

Using definition (19.11) of the material derivative in three dimensions, (19.66) may be rewritten as

$$\begin{aligned}\frac{\partial}{\partial t}(\overline{\rho}E) &+ \frac{1}{h_1 h_2 h_3}\left[\frac{\partial}{\partial \xi_1}(\overline{\rho}Eu_1 h_2 h_3) + \frac{\partial}{\partial \xi_2}(\overline{\rho}Eu_2 h_3 h_1) + \frac{\partial}{\partial \xi_3}(\overline{\rho}Eu_3 h_1 h_2)\right] \\ &= -\frac{1}{h_1 h_2 h_3}\left[\frac{\partial}{\partial \xi_1}(pu_1 h_2 h_3) + \frac{\partial}{\partial \xi_2}(pu_2 h_3 h_1) + \frac{\partial}{\partial \xi_2}(pu_3 h_1 h_2)\right].\end{aligned} \tag{19.67}$$

Equations (19.65) and (19.67) correspond to conservation principles (15.80) and (15.81), respectively, of Section 15.4 with: $\rho = \overline{\rho}$ and $F^{\rho} \equiv 0$; and $\mathbb{C}$ and $F^{\mathbb{C}}$ given by the second row of Table 15.2, where $F^{u_1} = F^{u_2} = F^{u_3} = \dot{Q}_E \equiv 0$ therein.

Globally integrating (19.67), applying periodicity in the horizontal, and noting that $u_3 \equiv 0$ gives

Conservation of Globally Integrated Total Energy

$$\frac{d}{dt}\iiint_V (\overline{\rho}E)\, h_1 h_2 h_3 d\xi_1 d\xi_2 d\xi_3 = 0, \tag{19.68}$$

where the volume integral is over the entire 3D global domain.

In a manner similar to the re-expression of 3D conservation principle (19.58) for globally integrated axial absolute angular momentum as the 2D one (19.59), 3D conservation principle (19.68) for globally integrated total energy may be re-expressed as the 2D one

$$\frac{d}{dt}\iint_A (\overline{\rho}E)\, h_1 h_2 h_3 d\xi_1 d\xi_2 = 0. \tag{19.69}$$

Note the presence of vertical metric factor h_3 in the integrand of (19.69), for the same reasons that it similarly appears in the integrand of (19.59).

19.5 THE BPV EQUATION FOR A SPHEROIDAL PLANET

Everything is now in place to obtain a BPV equation for a *spheroidal* planet, analogous to (*but more general than*) the classical, non-divergent vorticity equation (19.1) for a *spherical* planet. This is accomplished by deriving an equation for the material conservation of PV from first principles.

19.5.1 Material Conservation of PV

Using (19.2) and the vector identity

$$(\mathbf{u}_{\text{hor}} \cdot \nabla)\, \mathbf{u}_{\text{hor}} \equiv \zeta \mathbf{e}_3 \times \mathbf{u}_{\text{hor}} + \nabla_{\text{hor}} \left(\frac{\mathbf{u}_{\text{hor}} \cdot \mathbf{u}_{\text{hor}}}{2} \right), \tag{19.70}$$

where

$$\zeta \equiv \mathbf{e}_3 \cdot (\nabla_{\text{hor}} \times \mathbf{u}_{\text{hor}}) = \frac{1}{h_1 h_2} \left[\frac{\partial}{\partial \xi_1}(h_2 u_2) - \frac{\partial}{\partial \xi_2}(h_1 u_1) \right], \tag{19.71}$$

horizontal momentum equation (19.44) may be rewritten in vector-invariant form as

$$\frac{\partial \mathbf{u}_{\text{hor}}}{\partial t} + (\zeta + f)\, \mathbf{e}_3 \times \mathbf{u}_{\text{hor}} + \nabla_{\text{hor}} \left(\frac{\mathbf{u}_{\text{hor}} \cdot \mathbf{u}_{\text{hor}}}{2} \right) = -\frac{1}{\overline{\rho}} \nabla_{\text{hor}} p^T. \tag{19.72}$$

Operating on this equation with $\mathbf{e}_3 \cdot \nabla_{\text{hor}} \times$ then gives

$$\frac{\partial (\zeta + f)}{\partial t} + \mathbf{e}_3 \cdot \left\{ \nabla_{\text{hor}} \times \left[(\zeta + f)\, \mathbf{e}_3 \times \mathbf{u}_{\text{hor}} \right] \right\} = 0, \tag{19.73}$$

where (from (19.42)) $\partial f(\xi_2)/\partial t \equiv 0$ has been exploited. Applying (5.65) with $\mathbf{A}_{\text{hor}} = (\zeta + f)\, \mathbf{e}_3 \times \mathbf{u}_{\text{hor}} = (\zeta + f)(-u_2 \mathbf{e}_1 + u_1 \mathbf{e}_2)$ gives

$$\begin{aligned}
\nabla_{\text{hor}} \times \mathbf{A}_{\text{hor}} &= \frac{1}{h_1 h_2 h_3} \begin{vmatrix} h_1 \mathbf{e}_1 & h_2 \mathbf{e}_2 & h_3 \mathbf{e}_3 \\ \frac{\partial}{\partial \xi_1} & \frac{\partial}{\partial \xi_2} & 0 \\ h_1 A_1 & h_2 A_2 & 0 \end{vmatrix} = \frac{1}{h_1 h_2} \left[\frac{\partial}{\partial \xi_1}(h_2 A_2) - \frac{\partial}{\partial \xi_2}(h_1 A_1) \right] \mathbf{e}_3 \\
&= \frac{1}{h_1 h_2} \left\{ \frac{\partial}{\partial \xi_1} \left[h_2 (\zeta + f)\, u_1 \right] + \frac{\partial}{\partial \xi_2} \left[h_1 (\zeta + f)\, u_2 \right] \right\} \mathbf{e}_3.
\end{aligned} \tag{19.74}$$

Taking the scalar product of this equation with $\mathbf{e}_3$ yields

$$\begin{aligned}\mathbf{e}_3 \cdot \nabla_{\text{hor}} \times \mathbf{A}_{\text{hor}} &= \left(\frac{u_1}{h_1}\frac{\partial}{\partial \xi_1} + \frac{u_2}{h_2}\frac{\partial}{\partial \xi_2}\right)(\zeta + f) + \frac{\zeta + f}{h_1 h_2}\left[\frac{\partial}{\partial \xi_1}(h_2 u_1) + \frac{\partial}{\partial \xi_2}(h_1 u_2)\right] \\ &= (\mathbf{u}_{\text{hor}} \cdot \nabla_{\text{hor}})(\zeta + f) + (\zeta + f)\, \nabla_{\text{hor}} \cdot \mathbf{u}_{\text{hor}}. \end{aligned} \tag{19.75}$$

Insertion of (19.75) into (19.73) then delivers

$$\frac{D_{\text{hor}}(\zeta + f)}{Dt} + (\zeta + f)\, \nabla_{\text{hor}} \cdot \mathbf{u}_{\text{hor}} = 0. \tag{19.76}$$

Elimination of the horizontal divergence $\nabla_{\text{hor}} \cdot \mathbf{u}_{\text{hor}}$ between this equation and (19.46) then gives:

The BPV (Barotropic Potential Vorticity) Equation for a Spheroidal Planet

$$\frac{D_{\text{hor}}\mathscr{P}}{Dt} = 0, \tag{19.77}$$

where

$$\mathscr{P} \equiv \frac{(\zeta + f)}{H(\xi_2)} \tag{19.78}$$

is PV (potential vorticity). Thus PV, $\mathscr{P}$, is materially conserved.

Using (19.24) and (19.78), BPV equation (19.77) may be rewritten as:

Two Alternative Forms of the BPV (Barotropic Potential Vorticity) Equation

$$\frac{D_{\text{hor}}}{Dt}\left[\frac{(\zeta + f)}{h_3(\xi_2)}\right] = 0, \tag{19.79}$$

$$\frac{D_{\text{hor}}}{Dt}\left[g(\xi_2)(\zeta + f)\right] = 0. \tag{19.80}$$

For the (special) *spherical* case; H, $h_3 \equiv 1$, and g are all non-zero *constants*. This allows their removal from (19.77), (19.79), and (19.80) to recover the classical form (19.1) of the non-divergent barotropic vorticity equation. This removal, *in general*, is not, however, permissible for a spheroid. The natural measure for PV is that given in (19.78), with those in (19.79) and (19.80) being equivalent surrogate forms. Whereas constant values of H, h_3, and g for the spherical case commute with the material derivative, D_{hor}/Dt, this is clearly not so when they have meridional variation, as they do for a general spheroid.

This explains the origin of the undesirable restriction on g, alluded to earlier, when naïvely applying the classical barotropic-vorticity equation for a spheroid. Whilst setting g constant is valid for the spherical case (and is indeed *required* by use of the classical spherical-geopotential assumption), it is generally *invalid* for a spheroid. Thus the first goal we set ourselves has now been met.

In essence, so also has the second goal; the restriction on g being constant is removed by retaining its meridional variation (and also that of its surrogates $1/H$ and $1/h_3$) *under* the material derivative D_{hor}/Dt, and by not (incorrectly in general) commuting it (or its surrogates $h_3(\xi_2)$ and $1/g(\xi_2)$) with D_{hor}/Dt (followed by its elimination by division). This principle is pursued and applied here to derive the BPV equation in terms of an appropriately defined stream function (of slightly different form to the classical one, to account for the flow no longer being two-dimensionally non-divergent; cf. (19.46) and (19.76) with (19.5) and (19.1)).

The three forms (19.77), (19.79), and (19.80) of the BPV equation may be contrasted:

Comparison of the Three Forms of the BPV Equation

1. Form (19.77) is basically the same as that of the shallow-water equations, except that whereas $H = H(\xi_1, \xi_2, t)$ for the shallow-water equations, $H = H(\xi_2)$ for the BPV equation examined here.
2. Form (19.79) naturally expresses PV conservation in terms of a metric factor, and it leads (in Chapter 22) to the development of an exact solution of the BPV equation for a homogeneous ellipsoid, with meridional variation of apparent gravity.
3. Form (19.80) provides useful physical insight into the influence of meridional variation of apparent gravity on fluid flow – see Section 19.10.

19.5.2 The BPV Equation in Stream-Function Form

The BPV equation for a spheroidal planet may be expressed in terms of a stream function. To accomplish this, note that (19.33) can be satisfied by:

The Definition of Stream Function Ψ

$$h_3 u_1 = -\frac{1}{h_2}\frac{\partial \Psi}{\partial \xi_2}, \quad h_3 u_2 = \frac{1}{h_1}\frac{\partial \Psi}{\partial \xi_1}. \tag{19.81}$$

For constant h_3, as in the (special) spherical case (for which $h_3 \equiv 1$), this corresponds to the classical definition (19.4) of the velocity components in terms of stream function ψ. The appearance of $h_3 = h_3(\xi_2)$ in (19.81) takes into consideration that h_3 *varies meridionally for the general spheroidal case.*

Using (19.81), the conservation equation for PV, in the form (19.79), can then be rewritten as

The BPV Equation in Stream-Function Form for a Spheroidal Planet

$$\begin{aligned}\frac{D_{hor}}{Dt}\left(\frac{\zeta+f}{h_3}\right) &= \frac{\partial}{\partial t}\left(\frac{\zeta+f}{h_3}\right) + \frac{u_1}{h_1}\frac{\partial}{\partial \xi_1}\left(\frac{\zeta+f}{h_3}\right) + \frac{u_2}{h_2}\frac{\partial}{\partial \xi_2}\left(\frac{\zeta+f}{h_3}\right)\\ &= \frac{\partial}{\partial t}\left(\frac{\zeta}{h_3}\right) + \frac{1}{h_1 h_2 h_3}\left[\frac{\partial(\Psi, \zeta/h_3)}{\partial(\xi_1,\xi_2)} + \frac{\partial(\Psi, f/h_3)}{\partial(\xi_1,\xi_2)}\right] = 0.\end{aligned} \tag{19.82}$$

Here: Coriolis parameter $f(\xi_2)$ is defined by (19.42);

$$\frac{\partial(F,G)}{\partial(\xi_1,\xi_2)} \equiv \frac{\partial F}{\partial \xi_1}\frac{\partial G}{\partial \xi_2} - \frac{\partial F}{\partial \xi_2}\frac{\partial G}{\partial \xi_1} \tag{19.83}$$

is the Jacobian of F and G with respect to ξ_1 and ξ_2; and, using (19.71) and (19.81),

$$\frac{\zeta}{h_3} = \frac{1}{h_1 h_2 h_3}\left[\frac{\partial}{\partial \xi_1}(h_2 u_2) - \frac{\partial}{\partial \xi_2}(h_1 u_1)\right]$$
$$= \frac{1}{h_1 h_2 h_3}\left[\frac{h_2}{h_3 h_1}\frac{\partial^2 \Psi}{\partial \xi_1^2} + \frac{\partial}{\partial \xi_2}\left(\frac{h_1}{h_2 h_3}\frac{\partial \Psi}{\partial \xi_2}\right)\right] \equiv \mathfrak{L}\Psi, \tag{19.84}$$

where $\mathfrak{L}$ is a linear, second-order differential operator. Note that $\mathfrak{L}$ for the (special) spherical case simplifies to the horizontal Laplacian operator. This is *not*, however, so for the (general) spheroidal case.

Using (19.24), BPV equation (19.82) may be written as:

Two Alternative Forms of the BPV Equation in Stream-Function Form

$$\frac{\partial}{\partial t}\left(\frac{\zeta}{H}\right) + \frac{1}{h_1 h_2 H}\left[\frac{\partial(\Psi, \zeta/H)}{\partial(\xi_1, \xi_2)} + \frac{\partial(\Psi, f/H)}{\partial(\xi_1, \xi_2)}\right] = 0, \tag{19.85}$$

$$\frac{\partial}{\partial t}(g\zeta) + \frac{g}{h_1 h_2}\left[\frac{\partial(\Psi, g\zeta)}{\partial(\xi_1, \xi_2)} + \frac{\partial(\Psi, fg)}{\partial(\xi_1, \xi_2)}\right] = 0, \tag{19.86}$$

where $H = H(\xi_2)$ and $g = g(\xi_2)$.

Since f, h_3, H, and g are all independent of time, the relative vorticity, ζ, in the time-tendency terms of (19.82), (19.85), and (19.86) may, if convenient for a particular purpose, be replaced by the absolute vorticity, $\zeta + f$.

19.6 AN ALTERNATIVE DERIVATION OF THE BPV EQUATION

In Section 19.3 we set out the assumptions and preparatory steps for the derivation in Section 19.5 of the BPV equation. It is instructive to rederive this equation in a complementary manner using the same assumptions and preparatory steps. This then highlights the fact that the 2D BPV equation is essentially a special case of the general 3D PV conservation law given in Section 15.4.4 for the unapproximated 3D equations of an atmospheric or oceanic fluid.

From Section 15.4.4, material conservation of PV (in the absence of forcings) for the unapproximated 3D equations is expressed as

$$\frac{D\Pi}{Dt} = 0, \tag{19.87}$$

where:

$$\Pi = \frac{1}{\rho}\left(\frac{Z_1}{h_1}\frac{\partial \Lambda}{\partial \xi_1} + \frac{Z_2}{h_2}\frac{\partial \Lambda}{\partial \xi_2} + \frac{Z_3}{h_3}\frac{\partial \Lambda}{\partial \xi_3}\right) \tag{19.88}$$

is specific potential vorticity in 3D,

$$Z_1 = \frac{1}{h_2 h_3}\left[\frac{\partial}{\partial \xi_2}(h_3 u_3) - \frac{\partial}{\partial \xi_3}(h_2 u_2)\right], \tag{19.89}$$

$$Z_2 = \frac{2\Omega}{h_3}\frac{\partial h_1}{\partial \xi_3} + \frac{1}{h_3 h_1}\left[\frac{\partial}{\partial \xi_3}(h_1 u_1) - \frac{\partial}{\partial \xi_1}(h_3 u_3)\right], \tag{19.90}$$

$$Z_3 = -\frac{2\Omega}{h_2}\frac{\partial h_1}{\partial \xi_2} + \frac{1}{h_1 h_2}\left[\frac{\partial}{\partial \xi_1}(h_2 u_2) - \frac{\partial}{\partial \xi_2}(h_1 u_1)\right] \tag{19.91}$$

are the three components of specific absolute vorticity $\mathbf{Z} = 2\mathbf{\Omega} + \nabla \times \mathbf{u}$; and Λ is an arbitrary scalar function that is materially conserved according to

$$\frac{D\Lambda}{Dt} = 0. \tag{19.92}$$

Applying assumptions (19.15)–(19.17) to (19.87)–(19.91) and also using (19.38) and (19.42) simplifies them to

$$\overline{\rho}\frac{D\Pi}{Dt} = \frac{D_{\text{hor}}}{Dt}\left[\frac{(f+\zeta)}{h_3}\frac{\partial \Lambda}{\partial \xi_3}\right] = 0, \tag{19.93}$$

$$Z_1 = Z_2 = 0, \quad Z_3 = f(\xi_2) + \frac{1}{h_1 h_2}\left[\frac{\partial}{\partial \xi_1}(h_2 u_2) - \frac{\partial}{\partial \xi_2}(h_1 u_1)\right] = f + \zeta. \tag{19.94}$$

It remains to suitably specify the scalar function Λ and, in particular, to do so in such a way that (19.92) is satisfied. We do so in three different ways, each of which leads to one of the forms of the BPV equation derived in Section 19.5:

1. Using (19.23),

$$\Lambda = \frac{s_3(\xi_1, \xi_2, \xi_3)}{H(\xi_2)} = \frac{h_3(\xi_2)}{H(\xi_2)}\left(\xi_3 - \xi_3^B\right) \quad \Rightarrow \quad \frac{1}{h_3}\frac{\partial \Lambda}{\partial \xi_3} = \frac{1}{H(\xi_2)}. \tag{19.95}$$

2. Using (19.21),

$$\Lambda = \frac{\Phi(\xi_3)}{h_3^E g^E} \quad \Rightarrow \quad \frac{1}{h_3}\frac{\partial \Lambda}{\partial \xi_3} = \frac{1}{h_3 h_3^E g^E}\frac{\partial \Phi(\xi_3)}{\partial \xi_3} = \frac{1}{h_3(\xi_2)}. \tag{19.96}$$

3. Using (19.19),

$$\Lambda = \Phi(\xi_3) \quad \Rightarrow \quad \frac{1}{h_3}\frac{\partial \Lambda}{\partial \xi_3} = \frac{1}{h_3}\frac{\partial \Phi(\xi_3)}{\partial \xi_3} = g(\xi_2). \tag{19.97}$$

Using the assumptions and preparatory steps of Section 19.3.1, it can be shown that each of these three forms satisfies (19.92) for material conservation of Λ, as they must. Insertion of (19.95)–(19.97) into (19.93) then yields

$$\frac{D_{\text{hor}}}{Dt}\left[\frac{(f+\zeta)}{H(\xi_2)}\right] = 0, \tag{19.98}$$

$$\frac{D_{\text{hor}}}{Dt}\left[\frac{(f+\zeta)}{h_3(\xi_2)}\right] = 0, \tag{19.99}$$

$$\frac{D_{\text{hor}}}{Dt}\left[(f+\zeta)\, g(\xi_2)\right] = 0. \tag{19.100}$$

Thus, as desired, (19.98)–(19.100) recover (19.77), (19.79) and (19.80), respectively, and the alternative derivation is now complete.

19.7 DYNAMICAL CONSISTENCY

By definition – see Section 16.1.2 – an approximate set of governing equations for motion of a global atmosphere or ocean is considered to be dynamically consistent if it implies conservation principles for *mass*, *axial angular momentum*, *total energy*, and *potential vorticity*, analogous to those of the unapproximated equations.

For the model considered in this chapter, conservation principles have been demonstrated in Section 19.4 for globally integrated mass, axial angular momentum, and total energy; and for material conservation of potential vorticity in Section 19.5. Thus the model considered in this chapter is dynamically consistent.

19.8 THE POISSON PROBLEM FOR PRESSURE

It is optional whether or not one computes the (diagnostic) pressure field (p) associated with the BPV equation, since this evolution equation can be integrated forward in time independently of p. It can, however, be beneficial to do so, not only for aesthetic satisfaction but also for practical reasons. For example, diagnosis of pressure (at any instant of time) from a velocity field is used in Chapter 23 to obtain exact, *unsteady* solutions of the full, unapproximated, 3D governing equations under certain conditions. This helps motivate the analysis of this section.

19.8.1 In Axial-Orthogonal-Curvilinear Coordinates

Partially differentiating mass-continuity equation (19.33) with respect to t and noting that h_1, h_2, and h_3 are all independent of time gives

$$\frac{\partial}{\partial \xi_1}\left(h_2 h_3 \frac{\partial u_1}{\partial t}\right) + \frac{\partial}{\partial \xi_2}\left(h_3 h_1 \frac{\partial u_2}{\partial t}\right) = 0. \tag{19.101}$$

Now the two components of horizontal momentum equation (19.72) are

$$\frac{\partial u_1}{\partial t} = (\zeta + f)\, u_2 - \frac{1}{h_1}\frac{\partial K}{\partial \xi_1} - \frac{1}{\overline{\rho} h_1}\frac{\partial p^T}{\partial \xi_1}, \tag{19.102}$$

$$\frac{\partial u_2}{\partial t} = -(\zeta + f)\, u_1 - \frac{1}{h_2}\frac{\partial K}{\partial \xi_2} - \frac{1}{\overline{\rho} h_2}\frac{\partial p^T}{\partial \xi_2}, \tag{19.103}$$

where K is given by (19.61) (with $u_3 \equiv 0$) and ζ by (19.71). Using (19.102) and (19.103) to eliminate $\partial u_1/\partial t$ and $\partial u_2/\partial t$ from (19.101) and dividing through by $h_1 h_2 h_3$ then yields:

The Diagnostic Poisson Equation for Top Pressure p^T Expressed in Axial-Orthogonal-Curvilinear Coordinates

$$\begin{aligned}
&\frac{1}{\overline{\rho} h_1 h_2 h_3}\left[\frac{\partial}{\partial \xi_1}\left(\frac{h_2 h_3}{h_1}\frac{\partial p^T}{\partial \xi_1}\right) + \frac{\partial}{\partial \xi_2}\left(\frac{h_3 h_1}{h_2}\frac{\partial p^T}{\partial \xi_2}\right)\right] \\
&= \frac{1}{h_1 h_2 h_3}\left\{\frac{\partial}{\partial \xi_1}\left[h_2 h_3 (\zeta + f)\, u_2\right] - \frac{\partial}{\partial \xi_2}\left[h_1 h_3 (\zeta + f)\, u_1\right]\right\} \\
&\quad - \frac{1}{h_1 h_2 h_3}\left[\frac{\partial}{\partial \xi_1}\left(\frac{h_2 h_3}{h_1}\frac{\partial K}{\partial \xi_1}\right) + \frac{\partial}{\partial \xi_2}\left(\frac{h_3 h_1}{h_2}\frac{\partial K}{\partial \xi_2}\right)\right].
\end{aligned} \tag{19.104}$$

Given horizontal wind components u_1 and u_2 (from which ζ can be computed) and appropriate boundary conditions (e.g. periodicity), (19.104) can be solved for the top pressure $p^T = p^T(\xi_1, \xi_2, t)$. The total pressure ($p$) elsewhere can then be obtained from (19.35).

19.8.2 In Vector Form

Since $p^T = p^T(\xi_1, \xi_2, t)$, and using (5.9),

$$\nabla^2 p^T = \frac{1}{h_1 h_2 h_3}\left[\frac{\partial}{\partial \xi_1}\left(\frac{h_2 h_3}{h_1}\frac{\partial p^T}{\partial \xi_1}\right) + \frac{\partial}{\partial \xi_2}\left(\frac{h_3 h_1}{h_2}\frac{\partial p^T}{\partial \xi_2}\right)\right]. \tag{19.105}$$

Furthermore, applying $u_3 \equiv 0$ and (5.9) with

$$\mathbf{A} = (A_1, A_2, A_3) = (\zeta + f)\,\mathbf{k} \times \mathbf{u} = \left[-(\zeta + f)\,u_2, (\zeta + f)\,u_1, 0\right] \tag{19.106}$$

gives

$$\nabla \cdot \left[(\zeta + f)\,\mathbf{k} \times \mathbf{u}\right] = -\frac{1}{h_1 h_2 h_3}\left\{\frac{\partial}{\partial \xi_1}\left[(\zeta + f)\,u_2 h_2 h_3\right] - \frac{\partial}{\partial \xi_2}\left[(\zeta + f)\,u_1 h_3 h_1\right]\right\}. \tag{19.107}$$

Inserting (19.105) and (19.107) into (19.104) then yields

The Diagnostic Poisson Equation for Top Pressure p^T Expressed in Vector Form

$$\nabla^2\left[\frac{p^T(\xi_1, \xi_2, t)}{\overline{\rho}}\right] = -\nabla \cdot \left[(\zeta + f)\,\mathbf{e}_3 \times \mathbf{u}_{\text{hor}}\right] - \nabla^2\left(\frac{\mathbf{u}_{\text{hor}} \cdot \mathbf{u}_{\text{hor}}}{2}\right). \tag{19.108}$$

This equation is the vector form of (19.104) for the top pressure $p^T = p^T(\xi_1, \xi_2, t)$. Given $\mathbf{u}_{\text{hor}}$ (from which ζ can be computed) and appropriate boundary conditions (e.g. periodicity), (19.108) can be solved for the top pressure $p^T = p^T(\xi_1, \xi_2, t)$. The total pressure ($p$) elsewhere can then be obtained from (19.35).

19.9 VARIATIONAL DERIVATION OF THE MOMENTUM EQUATIONS

In this section we show that momentum component equations (19.30)–(19.32) may be alternatively derived variationally – using the Euler–Lagrange equations – from the assumptions and preparatory steps of Section 19.3.1.

From (6.41) and (6.42), the Euler–Lagrange equations are

$$\frac{D}{Dt}\left(\frac{\partial \mathscr{L}}{\partial \dot{\xi}_i}\right) - \frac{\partial \mathscr{L}}{\partial \xi_i} = -\alpha\frac{\partial p}{\partial \xi_i}, \quad i = 1, 2, 3, \tag{19.109}$$

and Lagrangian density for the unapproximated equations is given by

$$\begin{aligned}\mathscr{L}&\left(\xi_1, \xi_2, \xi_3, \dot{\xi}_1, \dot{\xi}_2, \dot{\xi}_3, \alpha, \eta, S^1, S^2, \ldots\right)\\ &= \frac{h_1^2\dot{\xi}_1^2 + h_2^2\dot{\xi}_2^2 + h_3^2\dot{\xi}_3^2}{2} + \Omega h_1^2\dot{\xi}_1 - \mathscr{E}\left(\alpha, \eta, S^1, S^2, \ldots\right) - \Phi(\xi_2, \xi_3),\end{aligned} \tag{19.110}$$

where $\dot{\xi}_i \equiv D\xi_i/Dt$.

In the present context:

- $u_3 \equiv h_3\dot{\xi}_3 \equiv 0$ (from (19.27)) and so $D/Dt \to D_{\text{hor}}/Dt$.
- $\mathscr{E}$ and $S^1, S^2, \ldots$ are not needed.
- $\Phi = \Phi(\xi_3)$ (from (19.21)).

Thus Euler–Lagrange equations (19.109) simplify to

$$\frac{D_{\text{hor}}}{Dt}\left(\frac{\partial \mathscr{L}}{\partial \dot{\xi}_i}\right) - \frac{\partial \mathscr{L}}{\partial \xi_i} = -\alpha\frac{\partial p}{\partial \xi_i}, \quad i = 1, 2, 3, \tag{19.111}$$

and Lagrangian density (19.110) reduces to:

The Lagrangian Density for the BPV Equation in Axial-Orthogonal-Curvilinear Coordinates

$$\mathscr{L}_{\text{BPV}}\left(\xi_3, \dot{\xi}_1, \dot{\xi}_2\right) = \frac{h_1^2\dot{\xi}_1^2 + h_2^2\dot{\xi}_2^2}{2} + \Omega h_1^2\dot{\xi}_1 - \Phi\left(\xi_3\right), \tag{19.112}$$

where subscript 'BPV' signifies association with the 'Barotropic Potential Vorticity' equation.

Inserting Lagrangian density (19.112) into Euler–Lagrange equations (19.111), it is straightforward to verify that component-momentum equations (19.30)–(19.32) are recovered, as follows. Partially differentiating (19.112) with respect to $\xi_i, \dot{\xi}_i, i = 1, 2, 3$ and using definition (6.39) of $\mathbf{u}$ in axial-orthogonal-curvilinear coordinates leads to

$$\frac{\partial \mathscr{L}_{\text{BPV}}}{\partial \xi_1} = 0, \tag{19.113}$$

$$\frac{\partial \mathscr{L}_{\text{BPV}}}{\partial \xi_2} = h_1\frac{dh_1}{d\xi_2}\dot{\xi}_1^2 + h_2\frac{dh_2}{d\xi_2}\dot{\xi}_2^2 + 2\Omega h_1\frac{dh_1}{d\xi_2}\dot{\xi}_1 = \left(\frac{u_1}{h_1} + 2\Omega\right)u_1\frac{dh_1}{d\xi_2} + \frac{u_2^2}{h_2}\frac{dh_2}{d\xi_2}, \tag{19.114}$$

$$\frac{\partial \mathscr{L}_{\text{BPV}}}{\partial \xi_3} = -\frac{d\Phi}{d\xi_3}, \tag{19.115}$$

$$\frac{\partial \mathscr{L}_{\text{BPV}}}{\partial \dot{\xi}_1} = h_1^2\dot{\xi}_1 + \Omega h_1^2 = h_1\left(u_1 + \Omega h_1\right), \tag{19.116}$$

$$\frac{\partial \mathscr{L}_{\text{BPV}}}{\partial \dot{\xi}_2} = h_2^2\dot{\xi}_2 = h_2 u_2, \tag{19.117}$$

$$\frac{\partial \mathscr{L}_{\text{BPV}}}{\partial \dot{\xi}_3} = 0. \tag{19.118}$$

Substitution of (19.113)–(19.118) into Euler–Lagrange equations (19.111) then yields

$$\frac{D}{Dt}\left[h_1\left(u_1 + \Omega h_1\right)\right] = -\frac{1}{\overline{\rho}}\frac{\partial p}{\partial \xi_1}, \tag{19.119}$$

$$\frac{D}{Dt}\left(h_2 u_2\right) - \left(\frac{u_1}{h_1} + 2\Omega\right)u_1\frac{dh_1}{d\xi_2} - \frac{u_2^2}{h_2}\frac{dh_2}{d\xi_2} = -\frac{1}{\overline{\rho}}\frac{\partial p}{\partial \xi_2}, \tag{19.120}$$

$$\frac{d\Phi}{d\xi_3} = -\frac{1}{\overline{\rho}}\frac{\partial p}{\partial \xi_3}. \tag{19.121}$$

Manipulation of (19.119)–(19.121), with use of

$$\frac{Dh_i}{Dt} = \frac{u_2}{h_2}\frac{dh_i}{d\xi_2}, \quad i = 1, 2, \tag{19.122}$$

then recovers component-momentum equations (19.30)–(19.32).

Having alternatively and variationally rederived component-momentum equations (19.30)–(19.32) from Lagrangian density (19.112), everything else for the derivation of the BPV equation then goes through as before.

19.10 CONCLUDING REMARKS

A BPV equation, having several equivalent forms, has been derived for purely horizontal flow of a shallow, homogeneous, global atmosphere or ocean confined between two, rigid, spheroidal geopotential surfaces, the bottom one coinciding with the planet's. As the name suggests, this equation expresses material conservation of PV. Although $h_3\mathbf{u}_{\text{hor}}$ is horizontally non-divergent, $\mathbf{u}_{\text{hor}}$ itself generally is not. This then permits meridional variation of apparent gravity, which would otherwise be excluded. Because the derived BPV equation for a (general) spheroid includes meridional variation of apparent gravity, the present advance opens the way to examining this phenomenon in a simpler context than has hitherto been possible.

The form (19.80) of the BPV equation provides useful physical insight into the influence of meridional variation of apparent gravity on fluid flow. For example, consider a fluid parcel in the Northern Hemisphere, with positive relative vorticity, ζ, moving poleward. Physically, both f (Coriolis parameter) and g (apparent gravity) monotonically increase from the Equator to the North Pole. Consider first what happens for constant g. For poleward motion, the relative vorticity, ζ, must decrease in value to compensate for the increase of the Coriolis parameter, f, in order to maintain the constant value (when g is constant) of absolute vorticity, $\zeta + f$, along the path of the fluid parcel. Thus the fluid parcel must spin less rapidly relative to the planet as it moves poleward. Now consider what happens if g is no longer constant but varies meridionally. Since g increases as the fluid parcel moves poleward, the value of ζ must further decrease in order to preserve constancy of $g\left(\zeta + f\right)$ along the fluid parcel's path. Thus the impact of meridional variation of apparent gravity is to make the fluid parcel spin even less rapidly as it moves poleward. This reinforces the influence of the latitudinal variation of the Coriolis parameter on the flow.

To provide further physical insight, an *exact*, non-linear, unsteady solution, in closed form, of the BPV equation for a homogeneous ellipsoid is constructed in Chapter 22. This solution then provides the basis for exact, unsteady solutions (in Chapter 23) of more general equation sets in 3D.

Part III

EXACT STEADY AND UNSTEADY NON-LINEAR SOLUTIONS

20

Exact Steady Solutions of the Global Shallow-Water Equations

ABSTRACT

The 2D shallow-water equations embody many of the fundamental terms and physical properties needed to realistically model Earth's atmosphere and oceans in 3D. Exact solutions of these non-linear equations are consequently very useful. They provide theoretical insight into the physical properties of the equations and a means to evaluate and compare numerical methods for their approximation. A methodology is given for the construction of steady, vortical, axially symmetric, non-divergent solutions of the non-linear shallow-water equations in global, spheroidal geometry. This includes spherical and ellipsoidal geometries as special cases. An underlying, axially symmetric orography or bathymetry is optionally allowed. As illustrative examples, five families of such exact solutions, controlled by various parameters, are developed. It is shown how to rotate these solutions in spherical geometry to obtain solutions that are numerically more challenging. For numerical-experimentation purposes, it is important that an exact solution be physically stable to small-amplitude perturbation; this avoids drawing false conclusions. An existing stability analysis is generalised from spherical to spheroidal geometry. This leads to two criteria which, if satisfied by the solution (including specification of its parameters), guarantees physical stability of the solution when subjected to arbitrary, small-amplitude perturbation. This analysis is then applied to examine stability conditions for the illustrative example of twin, mid-latitude jets above a generalised cosine hill in spherical and ellipsoidal geometry.

20.1 PREAMBLE

Exact non-linear solutions of the governing equations for an atmospheric or oceanic dynamical core are useful for two reasons:

1. To provide theoretical insight into the underlying dynamical properties of atmospheric and oceanic fluid flows.
2. To help develop, test, and validate numerical discretisations of the dynamical-core equations, and simplified subsets of them.

Obtaining exact, non-linear solutions is, however, easier said than done due to:

- The complexity of the governing equations.
- Their inherent non-linearity.
- Geometrical complications.

These aspects very much inhibit analytic tractability and, consequently, the availability of exact solutions. In Part III, we assemble a number of known, exact, non-linear solutions. They are mostly drawn from the literature and, where applicable, generalised herein from *spherical* to *spheroidal/ellipsoidal* geometry. The described solutions come in two flavours.

1. Steady.
2. Unsteady.

We begin by developing exact, *steady*, non-linear solutions of the global, 2D, shallow-water equations (in Chapter 20) before similarly doing so for the global, 3D, dynamical-core equations (in Chapter 21). Next, exact, *unsteady*, non-linear, 2D solutions of the global, barotropic PV (potential vorticity) equation over an ellipsoid are developed (in Chapter 22). These solutions are then used to similarly obtain (in Chapter 23) exact, *unsteady*, non-linear solutions of the global, 3D, dynamical-core equations in ellipsoidal geometry.

Williamson et al. (1992) proposed a suite of seven test cases for the evaluation of discretisations of the global, shallow-water equations in *spherical* geometry. They also provided a variety of error measures to help model developers and users evaluate the various trade-offs associated with different discretisations. Their test cases have been extensively used for many years as a *de facto* standard for validating and comparing numerical discretisations of these equations. As noted in Staniforth and White (2007), whilst this suite has proven its worth, its scope is nevertheless somewhat limited. Of the seven cases:

- One is not a solution of the shallow-water equations, but of the simpler problem of passive advection.
- One has a closed-form analytic solution (for the simple case of solid-body rotation).
- One has a semi-analytic solution (for steady-state, zonal, geostrophic flow with compact support) that, although not in closed form, is computable to arbitrary precision.
- One specifies the solution and then constructs forcing terms to allow the specified solution to satisfy the forced, shallow-water equations.
- Three have no analytic solution.

These limitations simply reflect and illustrate the difficulty of developing exact, non-linear solutions of the shallow-water equations due to the challenges of analytic tractability. Simplification is therefore likely to be needed to make further progress. These considerations led Staniforth and White (2007) to describe a methodology for constructing steady, vortical, axially symmetric, non-divergent solutions of various, geophysical-fluid-dynamical equation sets in 2D and 3D Cartesian and spherical geometries. Of particular interest in the present chapter are the five solution families they obtained by applying this methodology to the shallow-water equations in *spherical* geometry. These five solution families may be used to enhance the diversity of the Williamson et al. (1992) suite of test cases in spherical geometry.

When testing a numerical model by initialising it with an exact, steady, non-linear solution, any significant departure from the initial state should be attributable to a deficiency in the model's discretisation scheme. In particular, it should *not* be due to an inherent instability of the exact, physical solution when subjected to arbitrary, small-amplitude perturbation. But:

- How does one determine whether or not an exact, steady, non-linear solution *is* physically stable?

In his seminal paper, Ripa (1983) derived sufficient conditions for the physical stability of general, exact, steady, axially symmetric, shallow-water flow in *spherical* geometry[1] in the *absence* of any underlying orography/bathymetry. Applying Ripa (1983)'s conditions, Staniforth and White (2008a) derived sufficient conditions – in terms of Rossby and Froude numbers – for physical stability of the five families of exact solutions developed in Staniforth and White (2007). However, consistent with Ripa (1983)'s analysis, these conditions are only valid for the special case of *spherical geometry in the absence of orography*. These conditions are, of course, very useful to have, but

[1] He also derived similar conditions in beta-plane geometry.

this still leaves open the following two questions. Can sufficient conditions for physical stability be obtained:

1. In the presence of an underlying, axially symmetric, rigid bottom boundary?
2. In spheroidal/ellipsoidal geometry?

To answer the first question, White and Staniforth (2009) generalised Ripa (1983)'s analysis to further include an underlying, axially symmetric, rigid bottom boundary, again in *spherical* geometry. To illustrate practical application of their analysis, they applied it to obtain sufficient conditions for physical stability of a pair of stationary jets overlying generalised cosine hills. They also noted that their analysis is an example of a Liapunov-Arnol'd stability treatment via energy arguments.

To answer the second question, White and Staniforth (2009)'s analysis is generalised herein (in Section 20.8) from *spherical* geometry to *spheroidal/ellipsoidal* geometry.

The plan of this chapter is as follows. To set the scene, the shallow-water equations in spheroidal geometry are recalled in Section 20.2. Staniforth and White (2007)'s derivation methodology for constructing steady, vortical, axially symmetric, non-divergent solutions of the shallow-water equations in *spherical* geometry is generalised in Section 20.3 to *spheroidal* geometry. To ease developments in later sections, physical interpretation of the metric (scale) factor in the zonal direction is reviewed in Section 20.4. Staniforth and White (2007)'s five families of exact steady solutions of the shallow-water equations in spherical geometry are generalised in Section 20.5 to spheroidal geometry. Axially symmetric, exact solutions are inherently one-dimensional, which makes it easier for a numerical model to well represent them. However, as discussed in Section 20.6, transforming them to a rotated coordinate system removes this spurious advantage to provide a more-challenging test problem, albeit only in spherical geometry.

In Section 20.7, we pause to collect our thoughts. In particular, we discuss the importance of an exact solution used for numerical-model-evaluation purposes being physically stable to arbitrary, small-amplitude perturbation. This then leads (in Section 20.8) to generalisation of White and Staniforth (2009)'s stability analysis in spherical geometry to spheroidal/ellipsoidal geometry. Having done so, this stability analysis is applied (in Section 20.9) to examine stability conditions for the illustrative example of twin, mid-latitude jets above a generalised cosine hill in ellipsoidal geometry. Finally, concluding remarks are made in Section 20.10.

20.2 THE SHALLOW-WATER EQUATIONS IN SPHEROIDAL GEOMETRY

From (18.49)–(18.51), (18.53), (18.15), (18.18), and (18.36) – and in the notation of Chapter 18 – the 'basic' shallow-water equations are

$$\frac{D_{\text{hor}}u_1}{Dt} + \left(\frac{u_1}{h_1^S} + 2\Omega\right)\frac{u_2}{h_2^S}\frac{dh_1^S}{d\xi_2} + \frac{1}{h_1^S}\frac{\partial}{\partial\xi_1}(gH) = 0, \tag{20.1}$$

$$\frac{D_{\text{hor}}u_2}{Dt} - \left(\frac{u_1}{h_1^S} + 2\Omega\right)\frac{u_1}{h_2^S}\frac{dh_1^S}{d\xi_2} + \frac{1}{h_2^S}\frac{\partial}{\partial\xi_2}(gH) = 0, \tag{20.2}$$

$$\frac{D_{\text{hor}}\widetilde{H}}{Dt} + \frac{\widetilde{H}}{h_1^S h_2^S}\left[\frac{\partial}{\partial\xi_1}\left(u_1 h_2^S\right) + \frac{\partial}{\partial\xi_2}\left(u_2 h_1^S\right)\right] = 0, \tag{20.3}$$

where

$$\frac{D_{\text{hor}}}{Dt} \equiv \frac{\partial}{\partial t} + \frac{u_1}{h_1^S}\frac{\partial}{\partial\xi_1} + \frac{u_2}{h_2^S}\frac{\partial}{\partial\xi_2}, \tag{20.4}$$

$$h_1^S = h_1^S(\xi_2), \quad h_2^S = h_2^S(\xi_2), \quad g = g(\xi_2), \tag{20.5}$$

$$\widetilde{H} = \widetilde{H}(\xi_1, \xi_2, t) \equiv H(\xi_1, \xi_2, t) - B(\xi_1, \xi_2). \tag{20.6}$$

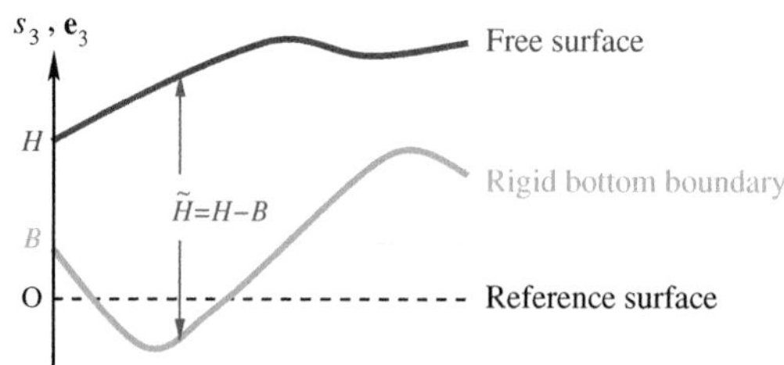

Figure 20.1 Vertical cross section of a shallow layer of fluid (in light blue), confined between a rigid bottom boundary (green line) and an upper free surface (dark blue line), located at vertical distances B and H, respectively, from a horizontal reference surface (dashed line). The depth of the fluid is $\widetilde{H} \equiv H - B$. See text for further details.

Furthermore:

- t is time.
- (ξ_1, ξ_2) are longitude and latitude, respectively.
- $(u_1, u_2) \equiv \left(h_1^S D\xi_1/Dt, h_2^S D\xi_2/Dt\right)$ are the zonal and meridional components of the flow, respectively.
- H and B are the elevations of the free surface and rigid bottom boundary, respectively, above a horizontal reference surface.
- $\widetilde{H} > 0$ is fluid depth.
- $g = g(\xi_2)$ is gravity.

See Fig. 20.1 for a graphical depiction of a shallow layer of fluid governed by the preceding equation set.

20.3 A DERIVATION METHODOLOGY

The methodology given in Staniforth and White (2007)'s Section 2 – to obtain exact steady solutions of the 'basic' shallow-water equations in spherical geometry – is generalised here from *spherical* to *spheroidal* geometry. Doing so then leads to the following six-step procedure for:

The Derivation of Exact Steady Axially Symmetric Solutions

1. Seek solutions that are both steady and axially symmetric. It is thus assumed that

$$u_1 = u_1(\xi_2), \quad u_2 = 0, \quad H = H(\xi_2), \quad B = B(\xi_2), \quad \widetilde{H} \equiv H - B = \widetilde{H}(\xi_2), \tag{20.7}$$

$$\Downarrow$$

$$\frac{D_{\text{hor}} u_1}{Dt} = \frac{D_{\text{hor}} u_2}{Dt} = \frac{D_{\text{hor}} \widetilde{H}}{Dt} = 0, \quad \frac{\partial}{\partial \xi_1}\left(u_1 h_2^S\right) + \frac{\partial}{\partial \xi_2}\left(u_2 h_1^S\right) = 0. \tag{20.8}$$

   Substitution of (20.7) and (20.8) into (20.1) (the first component-equation of momentum) and (20.3) (the mass-continuity equation) shows that they are both trivially satisfied.
2. Similarly, substitute (20.7) and (20.8) into the remaining governing equation (the second component equation of momentum), (20.2), to obtain

$$\frac{d}{d\xi_2}(gH) = \left(\frac{u_1}{h_1^S} + 2\Omega\right) u_1 \frac{dh_1^S}{d\xi_2}. \tag{20.9}$$

3. Decompose gH into the sum of its centripetal and Coriolis contributions – each being independent of the other – plus an arbitrary, constant contribution ($g_0 H_0$) that disappears under differentiation by ξ_2 in (20.9). Thus

$$g(\xi_2) H(\xi_2) = g_0 H_0 + g(\xi_2) H^{\text{cent}}(\xi_2) + g(\xi_2) H^{\text{Coriol}}(\xi_2), \tag{20.10}$$

where $H^{\text{Coriol}}(\xi_2)$ includes all contributions that depend on the planet's rotation parameter Ω. Dividing through by $g(\xi_2)$, (20.10) may be equivalently written as

$$H(\xi_2) = \frac{g_0 H_0}{g(\xi_2)} + H^{\text{cent}}(\xi_2) + H^{\text{Coriol}}(\xi_2). \tag{20.11}$$

4. Insert (20.10) – or, equivalently, (20.11) – into (20.9), and balance the two individual contributions $H^{\text{cent}}(\xi_2)$ and $H^{\text{Coriol}}(\xi_2)$ to obtain the first-order differential equations

$$\frac{d}{d\xi_2}\left(gH^{\text{cent}}\right) = \frac{u_1^2}{h_1^S}\frac{dh_1^S}{d\xi_2}, \tag{20.12}$$

$$\frac{d}{d\xi_2}\left(gH^{\text{Coriol}}\right) = 2\Omega u_1 \frac{dh_1^S}{d\xi_2}. \tag{20.13}$$

5. Specify a functional form for $u_1 = u_1(\xi_2)$ that is zero at the two poles (i.e. at the axially symmetric vortex centres).
6. Using this functional form, solve constraint equations (20.12) and (20.13) to obtain the individual contributions to $H(\xi_2)$, and thereby obtain the complete, exact, non-linear solution.

20.4 A PHYSICAL INTERPRETATION OF $h_1^S(\xi_2)$

In what follows, $h_1^S = h_1^S(\xi_2)$ appears frequently and plays a very important role. To better understand later developments, it is therefore of interest to provide a visual and physical interpretation of this quantity. Three prototypical planets are depicted in Fig. 20.2 using standard notation:

1. A *spherical* planet, in spherical-polar coordinates (λ, ϕ, r), with radius a and $h_1^S(\phi) = a\cos\phi$.
2. An *ellipsoidal* planet, in parametric-ellipsoidal coordinates (λ, θ, r), with equatorial radius a and $h_1^S(\theta) = a\cos\theta$.
3. A *contrived* planet, asymmetric about its (pseudo-) equatorial Ox_1x_2 plane,[2] and ellipsoidal northwards of it, with (pseudo-) equatorial radius a.

The spherical planet of panel (a) corresponds to the usual depiction of Earth as a perfect sphere. The ellipsoidal planet of panel (b) is a gross exaggeration (for pictorial clarity) of the more accurate representation of Earth as an ellipsoid. The analyses given here are not limited to planets that possess symmetry northwards and southwards of an equatorial plane; they need only respect symmetry about their rotation axis. By way of contrast, a contrived planet is depicted in panel (c). This planet is ellipsoidal northwards of the (pseudo-) equatorial plane (as in panel (b)), but distinctly different southwards of it.

[2] The nomenclature 'equatorial plane' implicitly assumes north-south symmetry about this plane. Because north-south symmetry is broken on panel (c), we instead use the terminology 'pseudo-equatorial plane'.

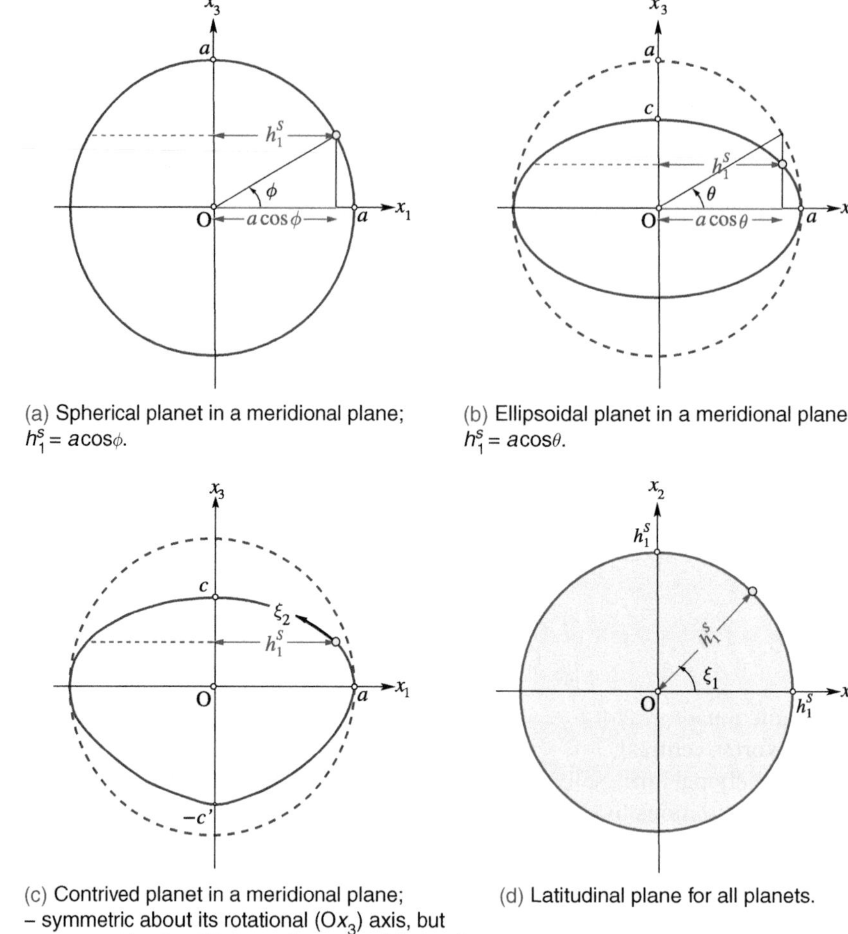

Figure 20.2 Visualisation of metric factor h_1^S, with depictions of meridional $(x_1 - x_3)$ cross sections for (a) a spherical planet; (b) an ellipsoidal planet; and (c) a contrived planet, asymmetric about its 'pseudo-equatorial plane'. For all three planets, panel (d) depicts a cross section in the latitudinal $(x_1 - x_2)$ plane that passes through a typical (yellow) point and the (red) dotted line on panels (a)–(c). (x_1, x_2, x_3) are the coordinates of the Cartesian system that rotates synchronously with the planet's rotation rate Ω. ϕ and θ are spherical and parametric-ellipsoidal latitudes, respectively, and ξ_1 is longitude. See text for further details.

Common to all three planets is a prototypical x_1-x_2 latitudinal plane that passes through a typical (yellow) point and the (red dashed) latitudinal radius shown on Figs. 20.2a–c. Physically, *all three radii are equal to perpendicular distance h_1^S from the planet's rotation axis, Ox_3*. The element of distance along the (red) latitudinal circle on Fig. 20.2d is $ds_1 = h_1^S d\xi_1$, where ξ_1 is longitude, and $h_1^S = h_1^S(\xi_2)$ is the metric (scale) factor in the ξ_1 direction with ξ_2 held fixed.

For an ellipsoidal planet, the use of parametric-ellipsoidal coordinates (λ, θ, r) is very natural since it gives rise to functional forms for $H^{\text{cent}}(\xi_2)$ and $H^{\text{Coriol}}(\xi_2)$ that are very similar indeed to those that occur in spherical geometry. This, as we will see, facilitates generalisation of previous work from spherical geometry to ellipsoidal geometry.

20.5 SOME ILLUSTRATIVE SOLUTIONS

To prepare the way for the first two illustrative examples (for equatorial and mid-latitude jets) let[3]

$$u_1(\xi_2) = u_0 \left(\frac{h_1^S}{a}\right)^m \left[1 - A^n \left(\frac{h_1^S}{a}\right)^n\right], \tag{20.14}$$

where:

- a is a representative scaling parameter, with dimension of length.
- u_0 is a representative value for $u_1(\xi_2)$.
- $m \geq 1$ and $n \geq 1$ are integers.

For all three ellipsoidal coordinates described in Section 18.3.2, a can be chosen to be equatorial radius, that is, $a = h_1^S(\xi_2 = 0)$, where ξ_2 is one of *geocentric*, *parametric*, or *geographic* latitude.[4] In *spherical* geometry, $h_1^S/a = \cos\phi$, where ϕ is latitude (all latitudes coincide for a sphere); see Fig. 20.2a. In *parametric-ellipsoidal* geometry, $h_1^S/a = \cos\theta$, where θ is parametric latitude (as opposed to geocentric and geographic latitudes, which have different functional forms); see Fig. 20.2b. It helps understanding to keep the *parametric-ellipsoidal*, functional form $h_1^S = a\cos\theta$ very much in mind in what follows. This form includes spherical geometry as a special case, so two geometries for the price of one!

Inserting (20.14) into (20.12) and (20.13) gives the two first-order, ordinary differential equations

$$\frac{d}{d\xi_2}\left(gH^{\text{cent}}\right) = \frac{u_1^2}{h_1^S}\frac{dh_1^S}{d\xi_2} = \frac{u_0^2}{a}\left(\frac{h_1^S}{a}\right)^{2m-1}\left[1 - 2A^n\left(\frac{h_1^S}{a}\right)^n + 2A^{2n}\left(\frac{h_1^S}{a}\right)^{2n}\right]\frac{dh_1^S}{d\xi_2}, \tag{20.15}$$

$$\frac{d}{d\xi_2}\left(gH^{\text{Coriol}}\right) = 2\Omega u_1\frac{dh_1^S}{d\xi_2} = 2\Omega u_0\left(\frac{h_1^S}{a}\right)^m\left[1 - A^n\left(\frac{h_1^S}{a}\right)^n\right]\frac{dh_1^S}{d\xi_2}. \tag{20.16}$$

Integrating these two equations then yields the centripetal and Coriolis contributions

$$H^{\text{cent}}(\xi_2) = \frac{u_0^2}{g(\xi_2)}\left[\frac{h_1^S(\xi_2)}{a}\right]^{2m}\left\{\frac{1}{2m} - \frac{2A^n}{(2m+n)}\left[\frac{h_1^S(\xi_2)}{a}\right]^n + \frac{A^{2n}}{(2m+2n)}\left[\frac{h_1^S(\xi_2)}{a}\right]^{2n}\right\}, \tag{20.17}$$

$$H^{\text{Coriol}}(\xi_2) = \frac{2\Omega a u_0}{g(\xi_2)}\left[\frac{h_1^S(\xi_2)}{a}\right]^{m+1}\left\{\frac{1}{m+1} - \frac{A^n}{m+n+1}\left[\frac{h_1^S(\xi_2)}{a}\right]^n\right\}. \tag{20.18}$$

These two contributions are seen to be independent of the precise functional form for $B(\xi_2)$, the height of the bottom boundary above the horizontal reference surface. Inserting these contributions into (20.11) leads to the free-surface height

$$H(\xi_2) = \frac{g_0 H_0}{g(\xi_2)} + \frac{u_0^2}{g(\xi_2)}\left[\frac{h_1^S(\xi_2)}{a}\right]^{2m}\left\{\frac{1}{2m} - \frac{2A^n}{(2m+n)}\left[\frac{h_1^S(\xi_2)}{a}\right]^n + \frac{A^{2n}}{(2m+2n)}\left[\frac{h_1^S(\xi_2)}{a}\right]^{2n}\right\}$$

[3] More generally, $u_1(\xi_2)$ may be defined to be h_1^S multiplied by any polynomial function of h_1^S, since (20.12) and (20.13) remain tractably integrable.

[4] For parametric-ellipsoidal coordinates, things are particularly simple since (from Table 18.1), $h_1^S = a\cos\theta$ (which has the same functional form as for spherical-polar coordinates) and then $h_1^S/a = \cos\theta$. See Figs. 18.3 and 20.2b.

$$+\frac{2\Omega a u_0}{g(\xi_2)}\left[\frac{h_1^S(\xi_2)}{a}\right]^{m+1}\left\{\frac{1}{m+1}-\frac{A^n}{m+n+1}\left[\frac{h_1^S(\xi_2)}{a}\right]^n\right\}. \tag{20.19}$$

Summarising:

- The zonal component (20.14) of the flow, $u_1 = u_1(\xi_2)$, is prescribed.
- (20.19) gives the corresponding steady-state response for the free-surface height, $H = H(\xi_2)$.

We now give two simple examples (for equatorial and mid-latitude jets) with specific choices of the parameters A, u_0, m, and n in (20.14) and (20.19). In what follows, we assume that $h_1^S(\xi_2) = h_1^S(-\xi_2)$ for $-\pi/2 \leq \xi_2 \leq \pi/2$ and that it varies monotonically from zero at a pole (where $\xi_2 = \pm\pi/2$) to its maximum value of a at the planet's equator (where $\xi_2 = 0$) – as in Figs. 20.2a and b for a sphere and for an ellipsoid, respectively. The assumptions of symmetry about the equator, and of monotonicity of meridional variation, are not essential. They are only made in the interests of simplifying the analysis, visualisation, and verbal description of the exact, non-linear solutions.

20.5.1 An Equatorial Jet

For our first example, we set $A = 0$, $u_0 = u_{max}$, and $m \geq 1$ in (20.14) and (20.19) to obtain:

The Equatorial Jet Solution

$$u_1(\xi_2) = u_{max}\left[\frac{h_1^S(\xi_2)}{a}\right]^m, \tag{20.20}$$

$$H(\xi_2) = \frac{1}{g(\xi_2)}\left\{g_0 H_0 + \frac{u_{max}^2}{2m}\left[\frac{h_1^S(\xi_2)}{a}\right]^{2m} + \frac{2\Omega a u_{max}}{m+1}\left[\frac{h_1^S(\xi_2)}{a}\right]^{m+1}\right\}. \tag{20.21}$$

For the special case ($\xi_2 = \phi$) of spherical geometry, and taking into account differences in notation, Staniforth and White (2007)'s equations (19) and (20) are recovered from (20.20) and (20.21) herein.

For the special case $m = 1$, the preceding solution corresponds to solid-body rotation. For $m \geq 2$, it corresponds to an equatorial jet – with maximum windspeed $|u_{max}|$ occurring at the planet's equator. Parameter m controls 'sharpness'. The greater its value, the narrower and more localised is the jet across the equator. Setting the parameters in (20.20) to $u_{max} = 10\,\mathrm{m\,s^{-1}}$ and $m = 10$ – and setting $\xi_2 = \theta$ for ellipsoidal geometry – leads to the black wind profile plotted in Fig. 20.3.

20.5.2 Twin Mid-Latitude Jets

For our second example, we set $A = 1$, $u_0 = 4u_{max}$, and $m = n = 1$ in (20.14) and (20.19) to obtain:

The Twin Mid-Latitude Jet Solution

$$u_1(\xi_2) = 4u_{max}\left[\frac{h_1^S(\xi_2)}{a}\right]\left\{1-\left[\frac{h_1^S(\xi_2)}{a}\right]\right\}, \tag{20.22}$$

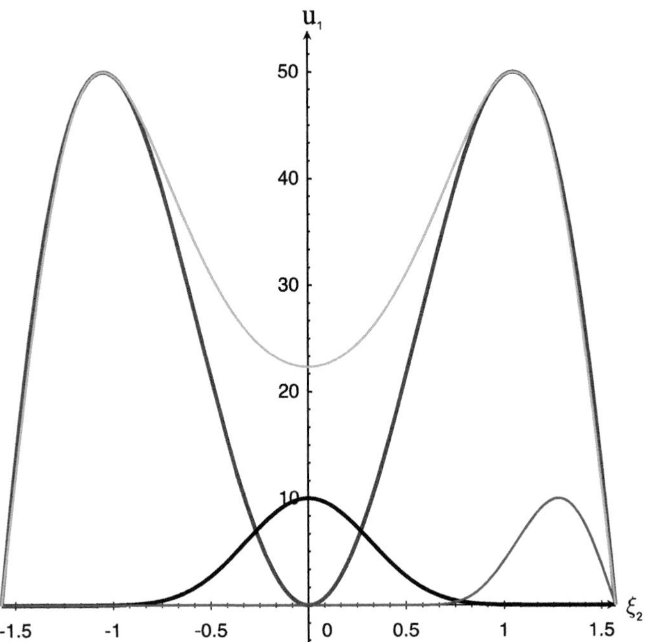

Figure 20.3 Wind profiles (ordinate), as a function of meridional coordinate ξ_2 (abscissa), for:
(a) An equatorial jet (in **black**), with $u_{\max} = 10\,\mathrm{m\,s^{-1}}, m = 10$.
(b) Twin, mid-latitude jets (in **blue**) at $\phi, \theta = \pm 60°$, with $u_{\max} = 50\,\mathrm{m\,s^{-1}}$.
(c) An isolated, polar vortex (in **red**), with $u_{\text{jet}} = 10\,\mathrm{m\,s^{-1}}, A = 2/\sqrt{3}$.
(d) Twin exponentially decaying, mid-latitude jets (in green), with $u_{\max} = 50\,\mathrm{m\,s^{-1}}, C = 2$.
ξ_2 corresponds to parametric latitude θ in ellipsoidal geometry, which includes ϕ in spherical geometry as a special case. See main text for further details.

$$H(\xi_2) = \frac{g_0 H_0}{g(\xi_2)} + \frac{8u_{\max}^2}{g(\xi_2)} \left[\frac{h_1^S(\xi_2)}{a}\right]^2 \left\{1 - \frac{4}{3}\left[\frac{h_1^S(\xi_2)}{a}\right] + \frac{1}{2}\left[\frac{h_1^S(\xi_2)}{a}\right]^2\right\} + \frac{4\Omega a u_{\max}}{g(\xi_2)} \left[\frac{h_1^S(\xi_2)}{a}\right]^2 \left\{1 - \frac{2}{3}\left[\frac{h_1^S(\xi_2)}{a}\right]\right\}. \tag{20.23}$$

For the special case ($\xi_2 = \phi$) of spherical geometry, and taking into account differences in notation, Staniforth and White (2007)'s equations (21) and (22) are recovered from (20.22) and (20.23) herein.

In ellipsoidal geometry, the preceding solution corresponds to a pair of mid-latitude jets, with maximum windspeed $|u_{\max}|$, located at $\theta = \pm 60°$. Setting the parameter in (20.22) to $u_{\max} = 50\,\mathrm{m\,s^{-1}}$ then leads to the blue wind profile plotted in Fig. 20.3.

20.5.3 An Isolated Polar Vortex

For our third example, we split the domain in two. The first subdomain covers the polar-cap region northwards of the limiting latitude obtained from solution of

$$h_1^S\left(\xi_2^{\lim}\right) = \frac{a}{A}, \tag{20.24}$$

with the second one covering the remainder of the planet's surface southward of it. To help fix ideas, $h_1^S(\theta) = a\cos\theta$ for an ellipsoidal planet, and then

$$\theta^{\text{lim}} = \cos^{-1}\left(\frac{1}{A}\right), \tag{20.25}$$

where $A > 1$ for θ^{lim} to be real. With this preparation:

The Zonal Component for an Isolated Polar Vortex

The zonal component of the isolated polar vortex is defined to be

$$u_1(\xi_2) = 0, \quad \text{for } -\frac{\pi}{2} \le \xi_2 \le \xi_2^{\text{lim}}, \tag{20.26}$$

$$= u_0 \left[\frac{h_1^S(\xi_2)}{a}\right]\left\{1 - A^2\left[\frac{h_1^S(\xi_2)}{a}\right]^2\right\}^n, \quad \text{for } \xi_2^{\text{lim}} \le \xi_2 \le \frac{\pi}{2}, \tag{20.27}$$

where it is assumed that $A > 1$ and $n \ge 1$.

Inserting (20.27) into (20.12) and (20.13) gives the two first-order, ordinary differential equations

$$\frac{d}{d\xi_2}\left(gH^{\text{cent}}\right) = \frac{u_1^2}{h_1^S}\frac{dh_1^S}{d\xi_2} = \frac{u_0^2}{a}\left(\frac{h_1^S}{a}\right)\left[1 - A^2\left(\frac{h_1^S}{a}\right)^2\right]^{2n}\frac{dh_1^S}{d\xi_2}, \tag{20.28}$$

$$\frac{d}{d\xi_2}\left(gH^{\text{Coriol}}\right) = 2\Omega u_1\frac{dh_1^S}{d\xi_2} = 2\Omega u_0\left(\frac{h_1^S}{a}\right)\left[1 - A^2\left(\frac{h_1^S}{a}\right)^2\right]^n\frac{dh_1^S}{d\xi_2}. \tag{20.29}$$

Integrating these two equations then yields the centripetal and Coriolis contributions

$$H^{\text{cent}}(\xi_2) = -\frac{u_0^2}{2(2n+1)A^2 g(\xi_2)}\left\{1 - A^2\left[\frac{h_1^S(\xi_2)}{a}\right]^2\right\}^{2n+1}, \tag{20.30}$$

$$H^{\text{Coriol}}(\xi_2) = -\frac{\Omega a u_0}{(n+1)A^2 g(\xi_2)}\left\{1 - A^2\left[\frac{h_1^S(\xi_2)}{a}\right]^2\right\}^{n+1}. \tag{20.31}$$

These two contributions are seen to be independent of the functional form for $B(\xi_2)$. Inserting them into (20.11) leads to:

The Free-Surface Height for an Isolated Polar Vortex

$$H(\xi_2) = \frac{g_0 H_0}{g(\xi_2)} - \frac{u_0^2}{2(2n+1)A^2 g(\xi_2)}\left\{1 - A^2\left[\frac{h_1^S(\xi_2)}{a}\right]^2\right\}^{2n+1}$$
$$- \frac{\Omega a u_0}{(n+1)A^2 g(\xi_2)}\left\{1 - A^2\left[\frac{h_1^S(\xi_2)}{a}\right]^2\right\}^{n+1}, \quad \text{for } \xi_2^{\text{lim}} \le \xi_2 \le \frac{\pi}{2}. \tag{20.32}$$

For the special case ($\xi_2 = \phi$) of spherical geometry, and taking into account differences in notation, Staniforth and White (2007)'s equations (23) and (24) are recovered from (20.27) and (20.32) herein.

Summarising:

- The zonal component (20.27) of the flow, $u_1 = u_1(\xi_2)$, is prescribed.
- (20.32) gives the corresponding steady-state response for the free-surface height, $H = H(\xi_2)$.

Now for physical interpretation of definition (20.27) for this example. Differentiating (20.27) with respect to ξ_2, and setting the result to zero determines the turning points of $u_1(\xi_2)$ within the subdomain $\xi_2^{\text{lim}} \leq \xi_2 \leq \pi/2$ for which u_1 is non-zero. Thus

$$\frac{du_1}{d\xi_2} = \frac{u_0}{a}\frac{dh_1^S}{d\xi_2}\left[1 - A^2\left(\frac{h_1^S}{a}\right)^2\right]^{n-1}\left[1 - (2n+1)A^2\left(\frac{h_1^S}{a}\right)^2\right] = 0. \tag{20.33}$$

Thus – noting that $h_1^S \geq 0$ and $dh_1^S/d\xi_2 \neq 0$ for $\xi_2^{\text{lim}} \leq \xi_2 < \pi/2$ – the turning points satisfy

$$h_1^S = \frac{a}{A}, \quad h_1^S = \frac{a}{\sqrt{2n+1}A}. \tag{20.34}$$

[To interpret (20.34), it helps here to consider an ellipsoidal planet, for which $h_1^S(\theta) = a\cos\theta$. Equation (20.34) then reduces to

$$\theta^{\text{jetmin}} = \cos^{-1}\left(\frac{1}{A}\right), \quad \theta^{\text{jetmax}} = \cos^{-1}\left(\frac{1}{\sqrt{2n+1}A}\right), \tag{20.35}$$

where θ^{jetmin} and θ^{jetmax} are the latitudes that correspond to the locations of the jet's minimum and maximum values. The first of these corresponds to the limiting latitude θ^{lim}; cf. (20.25). The second corresponds to the location of the maximum value of u_1.]

Substitution of (20.34) into (20.27) gives

$$u_1^{\text{jetmin}} = 0, \quad u_1^{\text{jetmax}} = \frac{u_0}{\sqrt{2n+1}A}\left(\frac{2n}{2n+1}\right)^n. \tag{20.36}$$

The second equation of (20.36) may be rewritten to express u_0 in terms of u^{jetmax} as

$$u_0 = u_1^{\text{jetmax}}\sqrt{2n+1}A\left(\frac{2n+1}{2n}\right)^n. \tag{20.37}$$

This then allows (20.27) to be alternatively written as

$$u_1(\xi_2) = u_1^{\text{jetmax}}\sqrt{2n+1}A\left(\frac{2n+1}{2n}\right)^n\left[\frac{h_1^S(\xi_2)}{a}\right]\left\{1 - A^2\left[\frac{h_1^S(\xi_2)}{a}\right]^2\right\}^n, \quad \text{for } \xi_2^{\text{lim}} \leq \xi_2 \leq \frac{\pi}{2}, \tag{20.38}$$

where parameter u_1^{jetmax} (the maximum value of u_1) in (20.38) is easier to physically interpret than parameter u_0 in (20.27). Setting the parameters in (20.38) to $u_1^{\text{jetmax}} = 10\,\text{m}\,\text{s}^{-1}$ and $A = 2/\sqrt{3}$ – and setting $\xi_2 = \theta$ for ellipsoidal geometry – leads to the red wind profile plotted in Fig. 20.3. This choice for A places the limiting latitude at $\theta^{\text{lim}} = 30°$.

The preceding exact, non-linear solution corresponds to an isolated polar vortex, confined northwards of the limiting latitude implicitly defined by (20.24) – and explicitly by (20.25) in ellipsoidal geometry. It has maximum windspeed u_1^{jetmax} defined by (20.36); in ellipsoidal geometry this occurs at latitude θ^{jetmax} defined by (20.35). The jet migrates poleward for increasing values of parameter A, and the jet becomes more sharply defined.[5] The wind is identically zero at the limiting latitude. For $n = 2$, $du_1/d\xi_2$ also vanishes there as do higher derivatives for $n \geq 3$. To maintain smoothness of the solution, the underlying orography should be chosen such that $B(\xi_2)$ and its first $n-2$ derivatives with respect to ξ_2 are also zero there.

20.5.4 Jovian Jets

Planet Jupiter has many more jets than Earth. These jets are both stronger and more stable than Earth's, with directional alternation between easterly and westerly (Heimpel et al., 2006). The strongest are located at low latitudes with broad structure, whereas those at higher latitudes are weaker and narrower. To construct a simplistic idealisation of multiple jets for a spheroidal planet such as Jupiter, let

$$\frac{du_1}{dh_1^S} = \frac{u_0}{a}\left(1 - \frac{h_1^S}{a}\right)\left(1 - B_1\frac{h_1^S}{a}\right)\cdots\left(1 - B_n\frac{h_1^S}{a}\right), \tag{20.39}$$

where $B_i > 1$ and $n \geq 1$.

Although (20.39) is written for a general axial-orthogonal-curvilinear coordinate system, with meridional coordinate ξ_2, things are easier to follow if we focus attention on a particular system for an ellipsoidal planet with semi-major and semi-minor axes a and c, respectively. Let that coordinate system be the parametric-ellipsoidal coordinate system (λ, θ) defined in Section 18.3.2.2. For this coordinate system (see (18.68)), $h_1^S = a\cos\theta$. Thus $h_1^S/a = \cos\theta$ in (20.39) and $-\pi/2 \leq \theta \leq +\pi/2$. Equation (20.39) can then be rewritten as

$$\frac{du_1}{d(\cos\theta)} = u_0(1-\cos\theta)(1-B_1\cos\theta)\cdots(1-B_n\cos\theta). \tag{20.40}$$

This has the same functional form as Staniforth and White (2007)'s analogous equation (25) for a *spherical* planet, but here, importantly, (20.40) is valid for an *ellipsoidal* planet. By our construction, the right-hand side of (20.40) is a polynomial of degree $n+1$ in $\cos\theta$, and its zeros correspond to the locations of the jets. The first zero is given by $\cos\theta = 1 \Rightarrow \theta_0 = 0$; this corresponds to an equatorial jet. The remaining $2n$ zeros then correspond to a further n jets per hemisphere, symmetrically located at

$$\theta_i = \pm\left|\cos^{-1}\left(\frac{1}{B_i}\right)\right|, i = 1, 2, \ldots n. \tag{20.41}$$

Since the right-hand side of (20.40) is a polynomial of degree $n+1$, and noting that $u_1(\theta = \pm\pi/2) = 0$ at the planet's poles, integration of (20.40) leads to u_1 being of the form

$$u_1(\theta) = u_0\cos\theta\,\mathbb{P}_{n+1}(\cos\theta), \tag{20.42}$$

where $\mathbb{P}_{n+1}(\cos\theta)$ is a (now-known) polynomial of degree $n+1$ in $\cos\theta$. Insertion of (20.42) into first-order differential equations (20.12) and (20.13), followed by integration, then yields

[5] For spherical geometry, Staniforth and White (2007) suggest that setting $A = \sqrt{2}/\sqrt{2n+1}$ or $A = 2/\sqrt{2n+1}$ places the jet maximum at $\phi_{\text{jet}} = 45°$ or at $\phi_{\text{jet}} = 60°$, respectively. This is incorrect, since these two values of A violate the assumption that $A > 1$. Only values of A greater than unity are admissible; this then places the jet maximum progressively further northward.

polynomials in $\cos\theta$ for $H^{\text{cent}}(\cos\theta)$ and $H^{\text{Coriol}}(\cos\theta)$, respectively. Insertion of these expressions into (20.11) finally leads to

$$H(\theta) = \frac{g_0 H_0}{g(\theta)} + \mathbb{Q}_{2n+4}(\cos\theta) + \mathbb{R}_{n+3}(\cos\theta), \tag{20.43}$$

where $\mathbb{Q}_{2n+4}(\cos\theta)$ and $\mathbb{R}_{n+3}(\cos\theta)$ are (now-known) polynomials in $\cos\theta$ of degree $2n+4$ and $n+3$, respectively.

For the special case of spherical geometry, and taking into account differences in notation, Staniforth and White (2007)'s equations (26) and (29) are recovered from (20.42) and (20.43) herein.

20.5.5 Exponentially Decaying Jet Solutions

In the preceding four examples, $u_1(\xi_2)$ has been conveniently expressed as a polynomial function. This greatly facilitates analytic tractability. However, $u_1(\xi_2)$ can also be specified otherwise, as now illustrated in our fifth example involving exponential decay:

The Zonal Component for Exponentially Decaying Jets

The zonal component for a pair of exponentially decaying jets is defined to be

$$u_1(\xi_2) = u_{\max} C\left(\frac{h_1^S}{a}\right)\exp\left\{\frac{1}{2}\left[1 - C^2\left(\frac{h_1^S}{a}\right)^2\right]\right\}, \tag{20.44}$$

where $u_{\max} > 0$ is maximum wind speed, and C is a sharpness parameter, both being prescribed constants.

Inserting (20.44) into (20.12) and (20.13) gives the two first-order, ordinary differential equations

$$\frac{d}{d\xi_2}\left(gH^{\text{cent}}\right) = \frac{u_1^2}{h_1^S}\frac{dh_1^S}{d\xi_2} = \frac{u_{\max}^2 C^2}{a}\left(\frac{h_1^S}{a}\right)\exp\left[1 - C^2\left(\frac{h_1^S}{a}\right)^2\right]\frac{dh_1^S}{d\xi_2}, \tag{20.45}$$

$$\frac{d}{d\xi_2}\left(gH^{\text{Coriol}}\right) = 2\Omega u_1\frac{dh_1^S}{d\xi_2} = 2\Omega u_{\max} C\left(\frac{h_1^S}{a}\right)\exp\left\{\frac{1}{2}\left[1 - C^2\left(\frac{h_1^S}{a}\right)^2\right]\right\}\frac{dh_1^S}{d\xi_2}. \tag{20.46}$$

Integrating these two equations then yields the centripetal and Coriolis contributions

$$H^{\text{cent}}(\xi_2) = -\frac{u_{\max}^2}{2g(\xi_2)}\exp\left\{1 - C^2\left[\frac{h_1^S(\xi_2)}{a}\right]^2\right\}, \tag{20.47}$$

$$H^{\text{Coriol}}(\xi_2) = -\frac{2\Omega a u_{\max}}{Cg(\xi_2)}\exp\left\langle\frac{1}{2}\left\{1 - C^2\left[\frac{h_1^S(\xi_2)}{a}\right]^2\right\}\right\rangle. \tag{20.48}$$

These two contributions are seen to be independent of the functional form for $B(\xi_2)$. Inserting them into (20.11) then leads to:

The Free-Surface Height for Exponentially Decaying Jets

$$H(\xi_2) = \frac{g_0 H_0}{g(\xi_2)} - \frac{u_{\max}^2}{2g(\xi_2)} \exp\left\{1 - C^2\left[\frac{h_1^S(\xi_2)}{a}\right]^2\right\} - \frac{2\Omega a u_{\max}}{Cg(\xi_2)} \exp\left\langle \frac{1}{2}\left\{1 - C^2\left[\frac{h_1^S(\xi_2)}{a}\right]^2\right\}\right\rangle. \tag{20.49}$$

For the special case ($\xi_2 = \phi$) of spherical geometry, and taking into account differences in notation, Staniforth and White (2007)'s equations (30) and (31) are recovered from (20.44) and (20.49) herein.

Summarising:

- The zonal component (20.44) of the flow, $u_1 = u_1(\xi_2)$, is prescribed.
- (20.49) gives the corresponding steady-state response for the free-surface height, $H = H(\xi_2)$.

For $C > 1$ – and setting $\xi_2 = \theta$ for ellipsoidal geometry – (20.44) attains its maximum value ($u_{\max}$) when

$$\theta = \cos^{-1}\left(\frac{1}{C}\right). \tag{20.50}$$

Thus (20.44) corresponds to a pair of symmetrically located mid-latitude jets (one per hemisphere). It is an alternative idealisation to (20.22) for large-scale zonal flow in the atmosphere. For example, $C = \sqrt{2}$ places the jet maxima at $\theta = \pm 45°$, whereas $C = 2$ places them at $\theta = \pm 60°$. As parameter C increases, so the two jets migrate poleward. Note that u remains positive at the planet's equator in contradistinction to twin-jet idealisation (20.22), for which it is zero. Setting the parameters in (20.44) to $u_{\max} = 50$ and $C = 2$ – and setting $\xi_2 = \theta$ for ellipsoidal geometry – leads to the green wind profile plotted in Fig. 20.3.

For $C < 1$, (20.44) instead corresponds to an equatorial jet, analogous to flow (20.20).

20.6 ROTATED SOLUTIONS IN SPHERICAL GEOMETRY

The five illustrative solution families just described may, with a suitable set of values for their parameters, be used to help validate and evaluate the accuracy of a shallow-water model in spheroidal geometry. A limitation, however, is that these solutions are all fundamentally *one dimensional*, since they all satisfy

$$u_1 = u_1(\xi_2), \quad u_2 = 0, \quad B = B(\xi_2), \quad H = H(\xi_2), \quad \widetilde{H} = \widetilde{H}(\xi_2). \tag{20.51}$$

This motivates the development of more challenging *two-dimensional* solutions for model-testing and validation purposes.

As discussed in Staniforth and White (2007), this may be accomplished in *spherical* geometry by simply transforming a known, exact, steady solution, with variation in only one dimension, to a *rotated coordinate system*. In the unrotated, spherical coordinate system (λ, ϕ), (20.51) becomes

$$u_1 = u_1(\phi), \quad u_2 = 0, \quad B = B(\phi), \quad H = H(\phi), \quad \widetilde{H} = \widetilde{H}(\phi). \tag{20.52}$$

Transforming now to the rotated, spherical coordinate system (λ', ϕ'), the 1D functional forms (20.52) then lead to the 2D forms

$$u_1' = u_1'(\lambda', \phi'), \quad u_2' = 0, \quad B' = B'(\lambda', \phi'), \quad H' = H'(\lambda', \phi'), \quad \widetilde{H}' = \widetilde{H}'(\lambda', \phi'). \tag{20.53}$$

Although the coordinate system has changed, the physical solution has not; only the way in which it is expressed has changed.

In Sections 20.5.1–20.5.5, five families of essentially 1D, exact, steady solutions in *spheroidal* geometry were derived. Each solution family generalises Staniforth and White (2007)'s analogous ones from *spherical* to *spheroidal* geometry. It would therefore be nice if we could also similarly generalise their procedure for obtaining rotated *spherical* solutions to *spheroidal* geometry. Unfortunately, this does not work out as one might hope.[6] Nevertheless, a spheroidal model's parameters can be trivially set to be spherical, since spheroidal geometry includes spherical geometry as a special case. One can then perform numerical validation and evaluation experiments using exact, *spherical*, rotated solutions.

All these illustrative solution families feature vortices centred on a pole. This means that there is no cross-polar flow. Such flows can be particularly challenging for numerical models that use a latitude-longitude grid; this is due to convergence of the meridians at the poles. However, the preceding solutions, as formulated, will not test how well a numerical model can accurately represent cross-polar flow. Now a useful test problem for assessing the performance of numerical models – particularly for those not based on latitude-longitude grids – is provided by the twin-jet solution developed in Section 20.5.2. Rotating this solution in such a way that the two jet maxima cross the unrotated geographical poles can then provide a test problem with cross-polar flow. This is particularly useful for testing models based on a latitude-longitude grid.

Details regarding rotated coordinate transformations and how they may be used to obtain exact, steady, spherical solutions from unrotated (λ, ϕ) coordinates to rotated (λ', ϕ') ones may be found in the Appendix of the present chapter. This appendix may also be of interest and use in other contexts.

20.7 INTERLUDE

Five families of exact, steady, non-linear solutions of the 'basic' shallow-water equations in spheroidal geometry were developed in Section 20.5 for zonally symmetric flow of a shallow layer of fluid overlying a rigid, zonally symmetric bottom boundary. Although these exact solutions are of interest in their own right, the underlying motivation for their development was to obtain test problems for:

- Validation of model codes.
- Evaluation of numerical methods for integrating the geophysical-fluid-dynamical equations governing atmospheric and oceanic flows.

When using one of these exact solutions for such purposes, an important issue is whether or not it is *physically* stable to small perturbations, such as those due to rounding errors arising from numerical computation, or to initial imbalances between discretely sampled, exact, dependent variables. Any significant departure from an exact initial state should be attributable to a deficiency in a model's numerics, and *not* due to an inherent physical instability to small-amplitude perturbation. This motivates the development (here) of sufficient conditions to *guarantee* that a particular solution, *with specific parameter values*, is indeed physically stable when slightly perturbed. Generally speaking, a family of exact solutions is only *guaranteed* to be physically stable within a certain range of parameter values, and it is valuable to determine a broad range of such values for this purpose.

The conditions derived in Section 20.8 are *sufficient* conditions; they guarantee that, if respected, the solution is physically stable to small-amplitude perturbation. They are not (in general) *necessary* conditions for stability; the solutions may or may not be physically stable to small-amplitude perturbation when these conditions are violated. What is important in the

[6] The underlying reason is related to geometrical symmetry. After *arbitrary* rotation about its centre, a rotated sphere occupies exactly the same physical space as it did before rotation. However, a spheroid (including an ellipsoid) fails to do so. This is a graphical manifestation of a spheroid failing to possess the isotropic symmetry of a sphere.

current context is that we can determine a useful, fairly broad range of parameter values that *guarantee* physical stability for model-validation and evaluation purposes.

The analysis that follows examines the physical stability of exact solutions to *small-amplitude* perturbation. It does so by using equations that are linearised about an exact solution. Non-linear terms are justifiably neglected provided that perturbations are of sufficiently small amplitude. If the exact solution is *physically* unstable to small-amplitude perturbation, then the amplitude of perturbations will grow in time. Although linear theory (as developed here) suggests that such perturbations will increase without bound, this does not mean that this would actually be the case for the full *non-linear* equations. In fact, the non-linear equations very much constrain what can happen due to their implied conservation laws.

For a *physically unstable*, exact solution, linear theory accurately predicts how the solution will, when subjected to small-amplitude perturbation, behave during a limited time period. After a certain period of time, the perturbations will grow sufficiently large that non-linear terms become important and possibly dominant. Linear theory then breaks down, and its predictions thereafter are rendered meaningless. This is simply a limitation of linear theory. It is great for predicting what will happen *for a limited time period*, when a flow is subjected to small-amplitude perturbation, but it is unreliable when perturbations grow to large amplitude.

In his seminal work, Ripa (1983) obtained sufficient conditions – in both β-plane and spherical geometry – for the physical stability of exact, steady, zonally symmetric, shallow-water flow over a purely horizontal (flat), rigid boundary. This is often referred to in the literature as Ripa's Theorem (Shepherd, 2003). These conditions were applied in Staniforth and White (2008a) to obtain sufficient conditions – *in spherical geometry with a flat, rigid boundary* – for the five families of illustrative exact solutions developed in Section 20.5 herein. The interested reader is referred to this latter paper for the detailed analysis and resulting parameter ranges that guarantee physical stability of the solutions to small-amplitude perturbation. Broadly speaking, it was observed that:

- Planetary rotation, Ω, plays essentially the same stabilising role in westerly flows governed by the (divergent) shallow-water equations as it does in westerly flows governed by the (non-divergent) barotropic-vorticity equation.
- Physical stability is enhanced by low Rossby number, $\text{Ro} \equiv u_{\max}/(\Omega a)$, and by low Froude number, $\text{Fr} \equiv u_{\max}/\sqrt{g_0 H_0}$, where $u_{\max}$ is maximum windspeed, a is spherical radius, g_0 is constant gravity (for a spherical planet), and H_0 is mean free-surface height of the shallow layer of fluid above the flat, rigid boundary.

Both Ripa (1983)'s analysis, and Staniforth and White (2008a)'s application of it, are limited by two assumptions;

1. A purely horizontal (flat) lower boundary.
2. Spherical geometry.

Practically speaking, forcing by a bottom boundary – be it by orography for the atmosphere or by bathymetry for an ocean – is important for atmospheric and oceanic modelling. It is therefore of interest to be able to test numerical models in the presence of an underlying, spatially varying, rigid boundary, and to do so using a *physically stable*, exact, non-linear solution. Sufficient conditions for stability of this case to small-amplitude perturbation is not, unfortunately, covered by Ripa (1983)'s analysis.

To address this (first) limitation, White and Staniforth (2009) generalised his analysis to include a zonally symmetric, rigid, horizontally varying bottom boundary, again in spherical geometry. For the twin mid-latitude-jets solution – developed in Section 20.5.2 but specialised to spherical geometry – they found that for guaranteed physical stability:

- The presence of such a bottom boundary reduces the range of parameter values.

- The higher the maximum height of the (zonally symmetric) bottom boundary is above the (horizontal) reference surface, and the smaller its horizontal scale, the more restrictive is the range of parameter values.

To address the second limitation, it is desirable to generalise White and Staniforth (2009)'s analysis in *spherical* geometry to *spheroidal* geometry.

- But is this possible?

Now the commutivity of certain operators in spherical geometry was exploited in their analysis, but the analogous operators in spheroidal geometry lose this property. At first blush it therefore seems unlikely that White and Staniforth (2009)'s analysis can be generalised to spheroidal geometry. However, although this lack of commutivity complicates matters, it fortunately turns out that this generalisation is indeed possible. With some work – and attention to detail – this is accomplished in what follows by closely following the logic of White and Staniforth (2009)'s analysis, whilst generalising it when needed.

20.8 THE STABILITY OF EXACT SOLUTIONS TO LINEAR PERTURBATION

20.8.1 The Non-Linear Equations

The zonal and meridional components (20.1) and (20.2) of the momentum equation for shallow-water flow on a rotating spheroid may be rewritten as

$$\frac{D_{\text{hor}}u_1}{Dt} - \left(f - \frac{u_1}{h_1^S h_2^S}\frac{dh_1^S}{d\xi_2}\right)u_2 + \frac{1}{h_1^S}\frac{\partial}{\partial \xi_1}(gH) = 0, \tag{20.54}$$

$$\frac{D_{\text{hor}}u_2}{Dt} + \left(f - \frac{u_1}{h_1^S h_2^S}\frac{dh_1^S}{d\xi_2}\right)u_1 + \frac{1}{h_2^S}\frac{\partial}{\partial \xi_2}(gH) = 0, \tag{20.55}$$

where

$$f(\xi_2) \equiv -\frac{2\Omega}{h_2^S}\frac{dh_1^S}{d\xi_2} \tag{20.56}$$

is the Coriolis parameter, and Ω is the rotation rate of the coordinate system about its polar axis. From (20.3), the associated mass-continuity equation is

$$\frac{D_{\text{hor}}\widetilde{H}}{Dt} + \frac{\widetilde{H}}{h_1^S h_2^S}\left[\frac{\partial}{\partial \xi_1}\left(u_1 h_2^S\right) + \frac{\partial}{\partial \xi_2}\left(u_2 h_1^S\right)\right] = 0, \tag{20.57}$$

where – from (20.6) – $\widetilde{H} \equiv H - B$ is fluid depth, and $H = H(\xi_1, \xi_2, t)$ and $B = B(\xi_1, \xi_2)$ are the elevations of the free-surface height and rigid bottom boundary above the horizontal reference surface, respectively. Recall Fig. 20.1 for a graphical depiction of B, H, and $\widetilde{H}$.

20.8.2 Potential Vorticity Conservation

As summarised in Section 18.7, with switch $Q \equiv 0$ everywhere therein, (20.54)–(20.57) imply the Lagrangian conservation law

$$\frac{D_{\text{hor}}\Pi}{Dt} = 0, \tag{20.58}$$

where

$$\Pi \equiv \frac{f + \zeta}{\widetilde{H}} \tag{20.59}$$

is potential vorticity (PV), and

$$\zeta \equiv \frac{1}{h_1^S h_2^S}\left[\frac{\partial}{\partial \xi_1}\left(h_2^S u_2\right) - \frac{\partial}{\partial \xi_2}\left(h_1^S u_1\right)\right] \tag{20.60}$$

is relative vorticity.[7] (For brevity in what follows, Π^{SW} in (18.180) and (18.181) has been abbreviated to Π in (20.58) and (20.59), and ζ_3^S in (18.181) and (18.182) to ζ in (20.59) and (20.60).) It will be found that the meridional gradient of PV plays an important role in the determination of sufficient conditions for the stability to perturbation of exact, axially symmetric, non-linear solutions.

20.8.3 Balanced Zonal Flow

Let u_1, u_2, B, H, and $\widetilde{H}$ have the functional forms

$$u_1 = \overline{u}(\xi_2), \quad u_2 = 0, \tag{20.61}$$

$$B = \overline{B}(\xi_2), \quad H = \overline{H}(\xi_2) \quad \Rightarrow \quad \widetilde{H} = \overline{\widetilde{H}}(\xi_2) = \overline{H}(\xi_2) - \overline{B}(\xi_2). \tag{20.62}$$

When $B = \overline{B}(\xi_2)$ – that is, when the bottom boundary is zonally symmetric – (20.61) and (20.62) define a zonally symmetric, basic-state solution of the non-linear governing equations (20.54)–(20.57), provided that $\overline{u}(\xi_2)$ and $\overline{H}(\xi_2)$ obey the balance condition – obtained from (20.55) –

$$\left(f - \frac{\overline{u}}{h_1^S h_2^S}\frac{dh_1^S}{d\xi_2}\right)\overline{u} + \frac{1}{h_2^S}\frac{\partial}{\partial \xi_2}\left(g\overline{H}\right) = 0. \tag{20.63}$$

This is easily verified by straightforward substitution of (20.61) and (20.62) into (20.54)–(20.57).

We now examine the linear stability of basic-state, steady flows that satisfy (20.54)–(20.57) and (20.61)–(20.63). These flows include the five illustrative exact solutions developed in Section 20.5. Key to establishing conditions for stability is to bound small-amplitude perturbations about an exact, non-linear, basic-state flow.

20.8.4 The Linearised Perturbation Equations

20.8.4.1 Expansion of (u_1, u_2, H) about $(u_1, u_2, H) = \left(\overline{u}, 0, \overline{H}\right)$

Holding the lower boundary $B = B(\xi_2)$ fixed, let u_1', u_2', and H' be *arbitrary, small-amplitude,* 2D perturbations about the basic-state solution (20.61) and (20.62); that is,

$$u_1 = \overline{u}(\xi_2) + u_1'(\xi_1, \xi_2, t), \quad u_2 = u_2'(\xi_1, \xi_2, t), \tag{20.64}$$

$$H = \overline{H}(\xi_2) + H'(\xi_1, \xi_2, t) \Rightarrow \widetilde{H} = \overline{H}(\xi_2) + H'(\xi_1, \xi_2, t) - B(\xi_2) = \overline{\widetilde{H}}(\xi_2) + H'(\xi_1, \xi_2, t). \tag{20.65}$$

[7] Detailed derivation of (20.58) may be found in Appendix B of Chapter 18.

20.8.4.2 *Prognostic Equations for u_1', u_2', and H'*

Substituting expansions (20.64) and (20.65) into (first) momentum-component equation (20.54) and neglecting terms with products of perturbations on the basis that they are negligible then gives

$$\frac{\overline{D}u_1'}{Dt} - \left[f - \frac{1}{h_1^S h_2^S} \frac{d\left(h_1^S \overline{u}\right)}{d\xi_2} \right] u_2' + \frac{1}{h_1^S} \frac{\partial}{\partial \xi_1} \left(gH'\right) = 0, \qquad (20.66)$$

where

$$\frac{\overline{D}}{Dt} \equiv \frac{\partial}{\partial t} + \frac{\overline{u}}{h_1^S} \frac{\partial}{\partial \xi_1}. \qquad (20.67)$$

From (20.59)–(20.62), the relative vorticity and PV of the balanced flow are given by

$$\overline{\zeta} \equiv \frac{1}{h_1^S h_2^S} \left[\cancelto{0}{\frac{\partial}{\partial \xi_1} \left(h_2^S \overline{u_2}\right)} - \frac{\partial}{\partial \xi_2} \left(h_1^S \overline{u}\right) \right] = -\frac{1}{h_1^S h_2^S} \frac{d\left(h_1^S \overline{u}\right)}{d\xi_2}, \qquad (20.68)$$

$$\overline{\Pi} \equiv \frac{f + \overline{\zeta}}{\overline{\overline{H}}} = \frac{1}{\overline{\overline{H}}} \left[f - \frac{1}{h_1^S h_2^S} \frac{d\left(h_1^S \overline{u}\right)}{d\xi_2} \right], \qquad (20.69)$$

respectively. With use of (20.69), (20.66) may be alternatively written as

$$\frac{\overline{D}u_1'}{Dt} - \overline{\Pi}\overline{\overline{H}}u_2' + \frac{1}{h_1^S} \frac{\partial \left(gH'\right)}{\partial \xi_1} = 0. \qquad (20.70)$$

Similarly linearising (second) momentum-component equation (20.55) – with use of balance condition (20.63) to eliminate terms involving $\overline{u}$ and/or $\overline{H}$ – yields

$$\frac{\overline{D}u_2'}{Dt} + \left(f - \frac{2\overline{u}}{h_1^S h_2^S} \frac{dh_1^S}{d\xi_2} \right) u_1' + \frac{1}{h_2^S} \frac{\partial \left(gH'\right)}{\partial \xi_2} = 0. \qquad (20.71)$$

With use of (20.64), (20.65), and (20.67), mass-continuity equation (20.57) linearises to

$$\frac{\overline{D}H'}{Dt} + \frac{1}{h_1^S h_2^S} \left[\frac{\partial}{\partial \xi_1} \left(\overline{\overline{H}}u_1' h_2^S \right) + \frac{\partial}{\partial \xi_2} \left(\overline{\overline{H}}u_2' h_1^S \right) \right] = 0. \qquad (20.72)$$

20.8.4.3 *Relative Vorticity and Potential Vorticity Perturbations (ζ' and Π')*

Now decompose relative vorticity (ζ) and potential vorticity (Π) as

$$\zeta = \overline{\zeta} + \zeta', \quad \Pi = \overline{\Pi} + \Pi', \qquad (20.73)$$

where ζ' and Π' are perturbations about the basic-state quantities $\overline{\zeta}$ and $\overline{\Pi}$, respectively. These four quantities are not arbitrary but are constrained by the specifications of the basic-state, exact solution and by the three arbitrary perturbations u_1', u_2', and H' about it. The two basic-state quantities, $\overline{\zeta}$ and $\overline{\Pi}$, are given by (20.68) and (20.69), respectively, in terms of (specified) $\overline{u}(\xi_2)$.

With use of (20.60), (20.68), and (20.73), perturbation relative vorticity (ζ') is related to perturbation velocity components (u_1' and u_2') by

$$\zeta' \equiv \frac{1}{h_1^S h_2^S}\left[\frac{\partial}{\partial \xi_1}\left(h_2^S u_2'\right) - \frac{\partial}{\partial \xi_2}\left(h_1^S u_1'\right)\right]. \tag{20.74}$$

Using (20.65), (20.69), (20.73), and (20.74) in definition (20.59) of PV gives, upon linearisation,

$$\Pi' = \frac{\zeta' - \overline{\Pi} H'}{\overline{\widetilde{H}}} = \frac{1}{h_1^S h_2^S \overline{\widetilde{H}}}\left[\frac{\partial}{\partial \xi_1}\left(h_2^S u_2'\right) - \frac{\partial}{\partial \xi_2}\left(h_1^S u_1'\right)\right] - \frac{\overline{\Pi}}{\overline{\widetilde{H}}} H'. \tag{20.75}$$

Equations (20.74) and (20.75) express and constrain the perturbations ζ' and Π' in terms of the three *arbitrary* perturbations u_1', u_2', and H' about $u_1 = \overline{u}$, $u_2 = 0$, and $H = \overline{H}$. Other perturbation quantities (such as Z', K', P', F', M', and E') are also not arbitrary, but dependent on u_1', u_2', and H' plus basic-state quantities.

20.8.4.4 *Prognostic Equations for PV and Potential Entrophy Perturbations (ζ' and Z')*

Examination of perturbation equations (20.74) and (20.75) reveals that the prognostic equation for the PV perturbation (Π') implied by the linearised zonal-momentum, meridional-momentum, and mass-continuity equations (i.e. (20.66) or (20.70), (20.71), and (20.72), respectively) may be obtained (without further approximation) by:

- Deriving an equation for $\overline{D}\zeta'/Dt$ from (20.66), (20.71), and (20.74), and hence an equation for $\overline{D}\left(\zeta'/\overline{\widetilde{H}}\right)/Dt$.
- Deriving an equation for $\overline{D}\left(\overline{\Pi}H'/\overline{\widetilde{H}}\right)/Dt$ from (20.72).
- Noting (20.75), constructing an equation for $\left(\overline{D}/Dt\right)\Pi'$.

This procedure is labour intensive, but the end result is exactly the same as that obtained by direct linearisation of (20.58), namely

$$\frac{\overline{D}\Pi'}{Dt} + \frac{1}{h_2^S}\frac{d\overline{\Pi}}{d\xi_2}u_2' = 0. \tag{20.76}$$

A key element in the derivation of stability criteria is to bound the sums of products (preferably squares) of perturbed quantities, one such quantity being

$$Z' \equiv \frac{\left(\Pi'\right)^2}{2}. \tag{20.77}$$

Multiplying (20.76) by Π', the prognostic equation for (non-negative) Z' is therefore

$$\frac{\overline{D}Z'}{Dt} + \frac{1}{h_2^S}\frac{d\overline{\Pi}}{d\xi_2}u_2'\Pi' = 0. \tag{20.78}$$

This equation plays an important role in enabling stability criteria to be derived; see Section 20.8.5.

20.8.4.5 *Prognostic Equation for Total Energy of Perturbations*

The desire to bound the sums of products of perturbed quantities motivates the development of a prognostic equation for total energy of the perturbations, where such quantities naturally occur. Multiplying momentum-component equations (20.66) and (20.71) by $\overline{\overline{H}}u_1'$ and $\overline{\overline{H}}u_2'$, respectively, summing the results, and rearranging gives

$$\frac{\overline{D}K'}{Dt} + \frac{1}{h_1^S h_2^S}\frac{d}{d\xi_2}\left(\frac{\overline{u}}{h_1^S}\right)\left[\left(h_1^S\right)^2\overline{\overline{H}}u_1'u_2'\right] + \overline{\overline{H}}\left[\frac{u_1'}{h_1^S}\frac{\partial\left(gH'\right)}{\partial\xi_1} + \frac{u_2'}{h_2^S}\frac{\partial\left(gH'\right)}{\partial\xi_2}\right] = 0, \tag{20.79}$$

where

$$K' \equiv \overline{\overline{H}}\left[\frac{\left(u_1'\right)^2 + \left(u_2'\right)^2}{2}\right]. \tag{20.80}$$

The factor $1/h_1^S$ that multiplies the $\left(h_1^S\right)^2\overline{\overline{H}}u_1'u_2'$ term in (20.79) has been retained without cancellation to ease later manipulation.

Defining

$$P' \equiv \frac{g\left(H'\right)^2}{2}, \tag{20.81}$$

and multiplying (20.72) by gH', leads to

$$\frac{\overline{D}P'}{Dt} + \frac{gH'}{h_1^S h_2^S}\left[\frac{\partial}{\partial\xi_1}\left(\overline{\overline{H}}u_1'h_2^S\right) + \frac{\partial}{\partial\xi_2}\left(\overline{\overline{H}}u_2'h_1^S\right)\right] = 0. \tag{20.82}$$

Summing (20.79) and (20.82), various terms combine into divergence form so that

$$\frac{\overline{D}}{Dt}(K'+P') + \frac{1}{h_1^S h_2^S}\frac{d}{d\xi_2}\left(\frac{\overline{u}}{h_1^S}\right)\left[\left(h_1^S\right)^2\overline{\overline{H}}u_1'u_2'\right] + \frac{1}{h_1^S h_2^S}\left[\frac{\partial}{\partial\xi_1}\left(\overline{\overline{H}}gH'u_1'h_2^S\right) + \frac{\partial}{\partial\xi_2}\left(\overline{\overline{H}}gH'u_2'h_1^S\right)\right] = 0. \tag{20.83}$$

Using advection operator (20.67), this equation may be written more compactly as

$$\frac{\partial}{\partial t}\left(K' + P'\right) + \frac{1}{h_1^S h_2^S}\frac{d}{d\xi_2}\left(\frac{\overline{u}}{h_1^S}\right)\left[\left(h_1^S\right)^2\overline{\overline{H}}u_1'u_2'\right] + \nabla\cdot\mathbf{A} = 0, \tag{20.84}$$

where **A** is a vector such that $\nabla \cdot \mathbf{A}$ vanishes when (after multiplication by areal element $h_1^S h_2^S d\xi_1 d\xi_2$) it is globally integrated.

20.8.5 Deriving Stability Criteria

20.8.5.1 *Balanced Solid Rotation Flow*

We are now in a position to answer the question:

- Is solid-rotation flow stable to small-amplitude perturbations?

For solid rotation, and noting that $h_1^S(\xi_2)/a$ measures perpendicular distance from the rotation axis – see Section 20.4,

$$\overline{u}(\xi_2) = u_0 \frac{h_1^S(\xi_2)}{a}, \tag{20.85}$$

where u_0 = constant. The $u_1^{'} u_2^{'}$ term in (20.84) then vanishes (since $u_1^{'} \equiv 0$ for solid rotation), and this equation simplifies (but only for this special case) to

$$\frac{\partial}{\partial t}\left(K' + P'\right) + \nabla \cdot \mathbf{A} = 0. \tag{20.86}$$

After multiplication by area element $h_1^S h_2^S d\xi_1 d\xi_2$, global integration of (20.86) over a spheroid gives

$$\frac{\partial}{\partial t} \iint \left(K' + P'\right) h_1^S h_2^S d\xi_1 d\xi_2 = 0, \tag{20.87}$$

since the divergence term integrates to zero. Substitution of definitions (20.80) and (20.81) into (20.87) shows that

$$\iint \left(K'+P'\right) h_1^S h_2^S d\xi_1 d\xi_2 \equiv \frac{1}{2} \iint \left\{ \overline{\overline{H}} \left[\left(u_1^{'}\right)^2 + \left(u_2^{'}\right)^2 \right] + g\left(H'\right)^2 \right\} h_1^S h_2^S d\xi_1 d\xi_2 = \text{constant in time.} \tag{20.88}$$

It follows that the flow must be stable to small-amplitude perturbations (as assumed in the preceding linearised analysis) since each individual term in (20.88) is non-negative. If the amplitude of one of the $u_1^{'}$, $u_2^{'}$, and $H^{'}$ perturbations increases anywhere within the domain at any instant in time, one or both of the other two perturbations must necessarily decrease somewhere to compensate for this. This guarantees boundedness of *small-amplitude* solutions to linear order.[8]

Thus the answer to the question posed earlier is – yes:

- Solid rotation is indeed stable to small-amplitude perturbations!

[8] It does not, however, guarantee boundedness of exact solutions when subjected to *large-amplitude* perturbations, since neglect of non-linear terms is then no longer justified.

20.8.5.2 *General Balanced Zonal Flow*

In an ideal world, the immediately preceding analysis for solid rotation would proceed just as smoothly for general, exact, balanced, zonal flow. Sadly things are not this simple due to the presence of the middle $(u_1^{'}u_2^{'})$ term in (20.84) because:

- This term does not globally integrate to zero.
- It can be of either sign.
- Conservation principle (20.88) is lost.

Further analysis is therefore called for if we are to derive stability criteria for the more general (and more interesting and useful) case. To do so is rather complicated, but nevertheless tractable. The necessary insight to proceed is provided by Ripa (1983)'s seminal paper (for *spherical* geometry, in the *absence* of spatial variation of a rigid bottom boundary, B), and by White and Staniforth (2009)'s extended analysis (to additionally include the *presence* of a meridionally varying bottom boundary, again in *spherical* geometry). The present analysis generalises these two previous ones to *spheroidal* geometry and subsumes them.

At this juncture, it is important to note that we are at liberty to add any combination of perturbation equations to (20.84) that we wish. In particular we are interested in combinations that ultimately lead to a quadratic function of perturbation quantities which can be exploited – in a manner somewhat analogous to that given previously for solid-rotation flow – to obtain stability criteria.

Examination of (20.84) shows that if the term

$$\frac{1}{h_1^S h_2^S}\left(\frac{\overline{u}}{h_1^S}\right)\frac{\partial}{\partial \xi_2}\left[\left(h_1^S\right)^2\overline{\widetilde{H}}u_1^{'}u_2^{'}\right] \tag{20.89}$$

were present on the left-hand side of (20.84), it could be combined with the existing term in $\overline{\widetilde{H}}u_1^{'}u_2^{'}$ to obtain

$$\begin{aligned}\frac{1}{h_1^S h_2^S}\frac{d}{d\xi_2}\left(\frac{\overline{u}}{h_1^S}\right)\left[\left(h_1^S\right)^2\overline{\widetilde{H}}u_1^{'}u_2^{'}\right]+\frac{1}{h_1^S h_2^S}\left(\frac{\overline{u}}{h_1^S}\right)\frac{\partial}{\partial \xi_2}\left[\left(h_1^S\right)^2\overline{\widetilde{H}}u_1^{'}u_2^{'}\right]\\ =\frac{1}{h_1^S h_2^S}\frac{\partial}{\partial \xi_2}\left[\left(\frac{\overline{u}}{h_1^S}\right)\left(h_1^S\right)^2\overline{\widetilde{H}}u_1^{'}u_2^{'}\right].\end{aligned} \tag{20.90}$$

This term would then vanish under ($h_1^S h_2^S$-weighted) global integration ... but at the expense of introducing another term, or terms, that would need to be handled! Undaunted by this observation, we nevertheless begin by constructing an equation containing a term of the form (20.90). We do so on the basis that:

- There is only a single problematic term in (20.84).
- Manipulating it into the sum of a flux term (that will globally integrate to zero), and a remaining (hopefully more benign) term, appears to be the most promising way forward in the absence of any obviously better option.

Multiplying (20.70) and (20.72) by $\overline{u}H'$ and $\overline{u}u_1^{'}$, respectively, and summing the two resulting equations (whilst noting (20.67)) gives

$$\frac{\overline{D}}{Dt}\left(\overline{u}H'u_1^{'}\right)+\frac{\overline{u}}{h_1^S h_2^S}\left[u_1^{'}\frac{\partial}{\partial \xi_1}\left(\overline{\widetilde{H}}u_1^{'}h_2^S\right)+u_1^{'}\frac{\partial}{\partial \xi_2}\left(\overline{\widetilde{H}}u_2^{'}h_1^S\right)+h_2^S H'\frac{\partial\left(gH'\right)}{\partial \xi_1}\right]-\overline{u}\overline{\widetilde{H}}\overline{\Pi}u_2^{'}H'=0. \tag{20.91}$$

Since $u_2'\Pi'$ can be expressed in terms of $\left(\overline{D}/Dt\right)\left(\Pi'\right)^2$ (via (20.77) and (20.78)), this motivates elimination of $u_2'H'$ in (20.91) in favour of $u_2'\Pi'$. Now – from (20.75) –

$$-\overline{\Pi}u_2'H' = \overline{\overline{H}}u_2'\Pi' - u_2'\zeta'. \tag{20.92}$$

With use of (20.74), (20.92) can be rewritten as

$$-\overline{u}\overline{\overline{H}}\overline{\Pi}u_2'H' = \overline{u}\left(\overline{\overline{H}}\right)^2 u_2'\Pi' + \frac{\overline{u}}{h_1^S h_2^S}\left[\overline{\overline{H}}u_2'\frac{\partial}{\partial\xi_2}\left(u_1'h_1^S\right) - \overline{\overline{H}}u_2'\frac{\partial}{\partial\xi_1}\left(u_2'h_2^S\right)\right]. \tag{20.93}$$

Substituting (20.93) into (20.91) gives (after quite some manipulation, and with use of (20.67))

$$\frac{\partial}{\partial t}\left(\overline{u}H'u_1'\right) + \frac{1}{h_1^S h_2^S}\left\{\left(\frac{\overline{u}}{h_1^S}\right)\frac{\partial}{\partial\xi_2}\left[\left(h_1^S\right)^2\overline{\overline{H}}u_1'u_2'\right] + \frac{\partial}{\partial\xi_1}\left(\overline{u}F'h_2^S\right)\right\} + \overline{u}\left(\overline{\overline{H}}\right)^2 u_2'\Pi' = 0, \tag{20.94}$$

where

$$F' \equiv \frac{g\left(H'\right)^2}{2} + \overline{\overline{H}}\left[\frac{\left(u_1'\right)^2 - \left(u_2'\right)^2}{2}\right] + \overline{u}H'u_1'. \tag{20.95}$$

Equation (20.94) contains a term of the desired form (20.89).[9]

Summing (20.84) with (20.94) then yields

$$\frac{\partial}{\partial t}\left(K' + P' + \overline{u}M'\right) + \overline{u}\left(\overline{\overline{H}}\right)^2 u_2'\Pi' + \frac{1}{h_1^S h_2^S}\left[\frac{\partial}{\partial\xi_2}\left(\overline{u}h_1^S\overline{\overline{H}}u_1'u_2'\right) + \frac{\partial}{\partial\xi_1}\left(\overline{u}F'h_2^S\right)\right] + \nabla\cdot\mathbf{A} = 0, \tag{20.96}$$

where

$$M' \equiv H'u_1'. \tag{20.97}$$

Bringing together (as $\nabla\cdot\mathbf{B}$) all terms that vanish when integrated over the global domain enables (20.96) to be written concisely as

$$\frac{\partial}{\partial t}\left(K' + P' + \overline{u}M'\right) + \overline{u}\left(\overline{\overline{H}}\right)^2 u_2'\Pi' + \nabla\cdot\mathbf{B} = 0. \tag{20.98}$$

The problematic (middle) term in (20.84) has effectively been decomposed into three contributions in (20.98):

- The time-tendency term $\partial\left(\overline{u}M'\right)/\partial t$.
- The (middle) term $\overline{u}\left(\overline{\overline{H}}\right)^2 u_2'\Pi'$.
- A term that has been absorbed into the divergence term ($\nabla\cdot\mathbf{B}$) that will globally integrate to zero.

It remains to manipulate key equation (20.98) to obtain stability criteria.

[9] There are three (inconsequential) typographic errors in White and Staniforth (2009); the last term in their equation (34) is missing a multiplicative factor of 2, and the $\overline{D}/Dt$ operator in their equations (38) and (40) should be replaced by $\partial/\partial t$.

To prepare for this, consider how the argument that led to (20.98) could be repeated by first multiplying (20.70) by $\overline{U}H'$ (instead of by $\overline{u}H'$) and (20.72) by $\overline{U}u_1'$ (instead of by $\overline{u}u_1'$), where $\overline{U} = \overline{U}(\xi_2)$ is a zonal flow *that is not equal to* $\overline{u}$. The analysis can be traced line-by-line up to (20.94), which now takes the form

$$\frac{\partial}{\partial t}\left(\overline{U}H'u_1'\right) + \frac{1}{h_1^S h_2^S}\left\{\frac{\overline{U}}{h_1^S}\frac{\partial}{\partial \xi_2}\left[\left(h_1^S\right)^2 \overline{\widetilde{H}} u_1' u_2'\right] + \frac{\partial}{\partial \xi_1}\left(\overline{U}F' h_2^S\right)\right\} + \overline{U}\left(\overline{\widetilde{H}}\right)^2 u_2'\Pi' = 0. \quad (20.99)$$

If $\overline{U}$ is a solid rotation, that is, if

$$\overline{U} = \alpha h_1^S, \quad \alpha \neq \frac{u_0}{a}, \quad (20.100)$$

where α is an *arbitrary* (real) *constant* (with unit of frequency), then (20.99) becomes

$$\frac{\partial}{\partial t}\left(\alpha h_1^S H' u_1'\right) + \frac{1}{h_1^S h_2^S}\left\{\frac{\partial}{\partial \xi_2}\left[\alpha\left(h_1^S\right)^2 \overline{\widetilde{H}} u_1' u_2'\right] + \frac{\partial}{\partial \xi_1}\left(\alpha F' h_1^S h_2^S\right)\right\} + \alpha h_1^S\left(\overline{\widetilde{H}}\right)^2 u_2'\Pi' = 0. \quad (20.101)$$

Bringing together (as $\nabla \cdot \mathbf{C}$) all terms that will vanish when integrated over the global domain and noting definition (20.97) enables (20.101) to be compactly written as

$$\frac{\partial}{\partial t}\left(\alpha h_1^S M'\right) + \alpha h_1^S\left(\overline{\widetilde{H}}\right)^2 u_2'\Pi' + \nabla \cdot \mathbf{C} = 0. \quad (20.102)$$

With this preparation we can again exploit our liberty to add or – as in the present case – subtract, a perturbation equation to/from key equation (20.98). Thus subtracting (20.102) from (20.98) gives

$$\frac{\partial}{\partial t}\left[K' + P' + \left(\overline{u} - \alpha h_1^S\right)M'\right] + \left(\overline{u} - \alpha h_1^S\right)\left(\overline{\widetilde{H}}\right)^2 u_2'\Pi' + \nabla \cdot \mathbf{D} = 0. \quad (20.103)$$

All terms that vanish when globally integrated have been gathered together in (20.103) as $\nabla \cdot \mathbf{D}$.

20.8.5.3 *Sufficient Conditions for Stability*

Area-weighted, global integration of (20.103) gives

$$\iint \left\{\frac{\partial}{\partial t}\left[K' + P' + \left(\overline{u} - \alpha h_1^S\right)M'\right] + \left(\overline{u} - \alpha h_1^S\right)\left(\overline{\widetilde{H}}\right)^2 u_2'\Pi'\right\} h_1^S h_2^S d\xi_1 d\xi_2 = 0. \quad (20.104)$$

Suppose now that PV gradient $\overline{\Pi}_{\xi_2} \equiv d\overline{\Pi}/d\xi_2$ does not vanish within the domain. (The criteria to be derived include this requirement, so there is no loss of generality in assuming it here.) Then, from (20.67) and (20.78),

$$u_2'\Pi' = -\frac{h_2^S}{\overline{\Pi}_{\xi_2}}\frac{\overline{D}Z'}{Dt} = -\frac{h_2^S}{\overline{\Pi}_{\xi_2}}\left(\frac{\partial Z'}{\partial t} + \frac{\overline{u}}{h_1^S}\frac{\partial Z'}{\partial \xi_1}\right) = -\frac{\partial}{\partial t}\left(\frac{h_2^S}{\overline{\Pi}_{\xi_2}}Z'\right) - \frac{\partial}{\partial \xi_1}\left(\frac{h_2^S \overline{u}}{h_1^S \overline{\Pi}_{\xi_2}}Z'\right). \quad (20.105)$$

From (20.104) and (20.105) it follows that

$$\iint \left[K' + P' + \left(\overline{u} - \alpha h_1^S\right) M' + \left(\alpha h_1^S - \overline{u}\right) \left(\overline{\widetilde{H}}\right)^2 \left(\frac{h_2^S}{\overline{\Pi}_{\xi_2}} Z' \right) \right] h_1^S h_2^S d\xi_1 d\xi_2 = \text{constant in time.} \tag{20.106}$$

Consider now the various contributors to the integral on the left-hand side of (20.106):

- From their definitions (20.80) and (20.81), K' and P' are non-negative throughout the domain. Locally, M' may take either sign (see definition (20.97)), but in the integrand of (20.106) it is multiplied by $\left(\overline{u} - \alpha h_1^S\right)$. For brevity, define

$$E' \equiv K' + P' + \left(\overline{u} - \alpha h_1^S\right) M'. \tag{20.107}$$

Use of definitions (20.80), (20.81), and (20.97) in (20.107) then leads to

$$E' = \frac{\overline{\widetilde{H}}}{2} \left(u_2'\right)^2 + \frac{\overline{\widetilde{H}}}{2} \left\{ \left[u_1' - \frac{1}{\overline{\widetilde{H}}} \left(\alpha h_1^S - \overline{u}\right) H' \right]^2 + \frac{1}{\left(\overline{\widetilde{H}}\right)^2} \left[g\overline{\widetilde{H}} - \left(\alpha h_1^S - \overline{u}\right)^2 \right] \left(H'\right)^2 \right\}. \tag{20.108}$$

This is certainly non-negative if

$$\left(\alpha h_1^S - \overline{u}\right)^2 \leq g\overline{\widetilde{H}}. \tag{20.109}$$

- Because Z' is non-negative throughout the domain (see definition (20.77)), the term in Z' in the integrand of (20.106) is non-negative provided its coefficient is positive. This is so if

$$\left(\alpha h_1^S - \overline{u}\right) \frac{1}{\overline{\Pi}_{\xi_2}} \geq 0, \quad \text{i.e. if} \quad \left(\alpha h_1^S - \overline{u}\right) \overline{\Pi}_{\xi_2} \geq 0, \tag{20.110}$$

since $\left(\overline{\Pi}_{\xi_2}\right)^2 \geq 0$, by assumption.

Sufficient Conditions for Stability (I)

Sufficient conditions for stability are therefore that *arbitrary parameter* α can be found such that, throughout the domain, both of the following two conditions simultaneously hold:

$$\left(\alpha h_1^S - \overline{u}\right) \overline{\Pi}_{\xi_2} \geq 0, \tag{20.111}$$

and

$$\left(\alpha h_1^S - \overline{u}\right)^2 \leq g\overline{\widetilde{H}}. \tag{20.112}$$

Real constant α, with the unit of frequency, in (20.111) and (20.112) is arbitrary, and one is therefore free to set its value as one wishes. Although arbitrary, some choices are, however, much better than others;[10] too large or too small a value can lead to violation of one or other of the preceding two conditions, or to more restrictive parameter ranges than would otherwise be the case. To facilitate practical application, three slightly weaker forms of sufficient conditions (20.111) and (20.112) are therefore derived here:

[10] The reciprocal of the reader's age has the right unit (of frequency), but this is unlikely to be a good choice.

1. Choosing $\alpha = \max(\overline{u}/h_1^S)$ means that criterion (20.111) is satisfied if $\overline{\Pi}_{\xi_2} \geq 0$ for all ξ_2. Condition (20.112) is also satisfied if

$$\max\left(\frac{\overline{u}}{h_1^S}\right) - \frac{\overline{u}}{h_1^S} \leq \frac{\sqrt{g\tilde{\overline{H}}}}{h_1^S}, \tag{20.113}$$

which is certainly the case if

$$\max\left(\frac{\overline{u}}{h_1^S}\right) \leq \min\left(\frac{\overline{u} + \sqrt{g\tilde{\overline{H}}}}{h_1^S}\right). \tag{20.114}$$

Thus stability is assured if $\overline{\Pi}_{\xi_2} \geq 0$ for all ξ_2 and (20.114) is obeyed.

2. By a similar argument, choosing $\alpha = \min(\overline{u}/h_1^S)$ means that criterion (20.111) is satisfied if $\overline{\Pi}_{\xi_2} \leq 0$ for all ξ_2. Criterion (20.112) is also satisfied if

$$\frac{\sqrt{g\tilde{\overline{H}}}}{h_1^S} \geq \frac{\overline{u}}{h_1^S} - \min\left(\frac{\overline{u}}{h_1^S}\right), \tag{20.115}$$

which is certainly the case if

$$\min\left(\frac{\overline{u}}{h_1^S}\right) \geq \max\left(\frac{\overline{u} - \sqrt{g\tilde{\overline{H}}}}{h_1^S}\right). \tag{20.116}$$

Stability is assured if $\overline{\Pi}_{\xi_2} \leq 0$ for all ξ_2 and (20.116) is obeyed.

3. Because $\min(A + B) \geq \min(A) + \min(B)$ for any real functions A and B, condition (20.114) is certainly obeyed if

$$\max\left(\frac{\overline{u}}{h_1^S}\right) - \min\left(\frac{\overline{u}}{h_1^S}\right) \leq \min\left(\frac{\sqrt{g\tilde{\overline{H}}}}{h_1^S}\right). \tag{20.117}$$

Because $\max(A - B) \leq \max(A) - \min(B)$ (as may be shown by setting $A = \max(A) - \lambda$ and $B = \min(B) + \mu$, where λ and μ are non-negative quantities), condition (20.116) is certainly obeyed if

$$\min\left(\frac{\overline{u}}{h_1^S}\right) \geq \max\left(\frac{\overline{u}}{h_1^S}\right) - \min\left(\frac{\sqrt{g\tilde{\overline{H}}}}{h_1^S}\right). \tag{20.118}$$

Condition (20.118) *is precisely equivalent to condition* (20.117). A third (weaker) version of the sufficient conditions for stability can therefore be stated, as follows.

Sufficient Conditions for Stability (II)

Stability is assured if, throughout the domain:

(a) The basic-state potential vorticity does not change sign, that is,

$$\overline{\Pi}_{\xi_2} \equiv \frac{d\overline{\Pi}}{d\xi_2} \text{ does not change sign,} \tag{20.119}$$

and

(b) Condition (20.117) is obeyed, that is,

$$\max\left(\frac{\overline{u}}{h_1^S}\right) - \min\left(\frac{\overline{u}}{h_1^S}\right) \leq \min\left(\frac{\sqrt{g\overline{\overline{H}}}}{h_1^S}\right). \tag{20.120}$$

In the (more limited) case of *spherical geometry* and a *flat (horizontal) bottom boundary,* this third set of sufficient conditions was applied in the stability analyses reported in Staniforth and White(2008a).

That $\overline{\overline{H}}$ appears in (20.120) instead of $\overline{H}$ (as in Staniforth and White (2008a)) means that *the sufficient conditions for stability are more restrictive in the presence of an underlying, meridionally varying bottom boundary* $B = B(\xi_2)$ *than in its absence.*

The third form is applied in Section 20.9 for the illustrative example – in ellipsoidal geometry – of the stability of stationary twin mid-latitude jets overlying zonally symmetric, generalised cosine hills.

20.9 ILLUSTRATIVE EXAMPLES OF THE APPLICATION OF THE STABILITY ANALYSIS

Having generalised White and Staniforth (2009)'s stability analysis, it is now feasible to apply it not only in spherical geometry but also in *spheroidal geometry.*

As a thought experiment and to set the scene, assume that we have formulated and carefully coded a shallow-water model in *ellipsoidal geometry* (a special case of spheroidal geometry) as a first step towards ultimately developing a comprehensive atmospheric or oceanic model in this geometry.[11] Having done so, we naturally wish to run this model using an exact, steady solution as a test problem. Any or (better still) all of the illustrative solutions developed in Section 20.5 are suitable candidates for this purpose. But, for any candidate solution, how might we specify its arbitrary parameter values in such a way as to guarantee that the solution is *physically stable*? To illustrate practical application of the stability analysis developed in Section 20.8, we now consider how to go about applying this analysis for the twin mid-latitude-jet solution developed in Section 20.5.2.

We begin by defining (in Section 20.9.1) this solution above a rigid, zonally symmetric, cosine hill in *ellipsoidal* coordinates. This includes both spherical coordinates and a flat bottom boundary as special cases. We then proceed – in a step-by-step manner, as in White and Staniforth (2009) – to examine the stability in *spherical* geometry of the twin-jet solution above the following:

- A flat bottom boundary (in Section 20.9.2).
- A zonally symmetric, cosine-squared hill (in Section 20.9.3).
- A zonally symmetric, generalised cosine hill (in Section 20.9.4).

These three steps prepare the way for discussion (in Section 20.9.5) of the stability of the twin-jet solution in *ellipsoidal geometry* over a *generalised cosine hill.*

20.9.1 Twin Mid-Latitude Jets above a Cosine Hill in Ellipsoidal Coordinates

The twin-jet solution is defined by (20.22) and (20.23) in terms of $h_1^S(\xi_2)/a$, where $h_1^S(\xi_2)$ is the metric factor in the ξ_1 direction of an axial-orthogonal-curvilinear coordinate system (ξ_1, ξ_2), and

[11] As discussed in Section 12.2.6, the choice of ellipsoidal geometry is a very natural one since the World Geodetic System, used for reporting data position using the Global Positioning System, is based on the WGS 84 reference *ellipsoid.*

a is a representative scaling parameter. By the preceding scenario, we wish to represent this exact solution in ellipsoidal geometry. But what ellipsoidal coordinate system, (ξ_1, ξ_2), should we use?

Three such ellipsoidal coordinate systems are described in Section 18.3.2. They are differentiated from one another by the specification of their meridional coordinate, namely:

- Geocentric latitude ($\xi_2 = \chi$).
- Parametric latitude ($\xi_2 = \theta$).
- Geographic latitude ($\xi_2 = \phi$).

Common to all three coordinate systems is longitude, $\xi_1 = \lambda$. We are free to choose whichever coordinate system is most convenient, since *physical* stability in ellipsoidal geometry is independent of the *mathematical* coordinate system used in any analysis of stability.[12] Examination of the fifth and sixth columns of Table 18.1 shows that the simplest forms for the metric factors $h_1^S(\xi_2)$ and $h_2^S(\xi_2)$ are obtained by use of *parametric-ellipsoidal coordinates* (λ, θ) for which

$$h_1^S(\theta) = a\cos\theta, \quad h_2^S(\theta) = a\left(1 - e^2\cos^2\theta\right)^{\frac{1}{2}}, \tag{20.121}$$

where $e^2 \equiv \left(a^2 - c^2\right)/a^2$, and e is eccentricity. The choice of parametric latitude (θ) as meridional coordinate has the virtues that:

- Metric factor h_1^S has exactly the same functional form as in familiar, spherical coordinates (with θ replaced by ϕ in the first equation of (20.121)).
- Spherical coordinates are included in (20.121) as the special case

$$\theta = \phi, \quad e \equiv 0 \quad \Rightarrow \quad h_1^S(\phi) = a\cos\phi, \quad h_2^S(\phi) = a. \tag{20.122}$$

These properties ease generalisation of White and Staniforth (2009)'s illustrative example in spherical geometry to the ellipsoidal case. The functional form of $h_2^S(\theta)$ – see the second equation of (20.121) – is nevertheless a complication when compared to the situation in spherical coordinates.

Inserting (20.121) into definitions (20.22) and (20.23) of the exact, twin-jet solution and noting that its meridional component of velocity is identically zero gives

$$\overline{u}(\theta) = 4u_{\max}\cos\theta\,(1 - \cos\theta), \tag{20.123}$$

$$\overline{v}(\theta) = 0, \tag{20.124}$$

$$\overline{\overline{H}}(\theta) = \frac{g_0 H_0}{g(\theta)} + \frac{8u_{\max}^2}{g(\theta)}\cos^2\theta\left(1 - \frac{4\cos\theta}{3} + \frac{\cos^2\theta}{2}\right) + \frac{4\Omega a u_{\max}}{g(\theta)}\cos^2\theta\left(1 - \frac{2\cos\theta}{3}\right) - B(\theta), \tag{20.125}$$

where $B = B(\theta)$ is an arbitrary, zonally symmetric, rigid bottom boundary that remains to be specified.

One of the problems used in Zerroukat et al. (2009) to test their shallow-water model in *spherical* geometry corresponds to (20.123)–(20.125), but with θ set to ϕ therein. For their problem:

[12] This means, for example, that the coordinate system used in the hypothetical numerical model does not have to be the same one as that used in the analysis of *physical* stability.

- Parameter values were set to

$$u_{\max} = 50\,\mathrm{m\,s^{-1}}, \quad g_0 H_0 = 10^5\,\mathrm{m^2\,s^{-2}}, \quad a = 6.37122 \times 10^6\,\mathrm{m}, \quad \Omega = 7.292 \times 10^{-5}\,\mathrm{s^{-1}}. \tag{20.126}$$

- The model of gravity (consistent with the use of spherical geometry) was

$$g(\theta) = g_0 = 9.80616\,\mathrm{m\,s^{-2}}. \tag{20.127}$$

- The rigid, bottom boundary was defined to be the *cosine-squared hill*

$$B(\theta) = B_{\max} \cos^2\left[\frac{\pi}{w}(\theta - \theta_c)\right] \quad \text{for } |\theta - \theta_c| \le \frac{w}{2}, \tag{20.128}$$

$$= 0, \quad \text{otherwise}, \tag{20.129}$$

where:
- $B_{\max} = 3.0 \times 10^3$ m is its maximum height.
- $w = \pi/3$ is its latitudinal width.
- $\theta = \theta_c = \pi/4$ is its latitudinal centring parameter.

In what follows, we adopt Zerroukat et al. (2009)'s parameter values for $u_{\max}$, $g_0 H_0$, a, Ω, and g_0 – as in (20.126) and (20.127) – for cases in both spherical and ellipsoidal geometry.

We consider three possible specifications of $B(\theta)$:

1. A flat (i.e. precisely horizontal) bottom boundary, for which $B(\theta) \equiv 0$.
2. A zonally symmetric, cosine-squared hill – as in Zerroukat et al. (2009) – defined by (20.128) and (20.129), where θ is parametric latitude.
3. A zonally symmetric, *generalised cosine hill* – as in White and Staniforth (2009) – defined by

$$B(\theta) = B_n(\theta) = (B_{\max})_n \cos^n\left[\frac{\pi}{w}(\theta - \theta_c)\right] \quad \text{for } |\theta - \theta_c| \le \frac{w}{2}, \quad n \ge 2, \tag{20.130}$$

$$= 0 \quad \text{otherwise}, \tag{20.131}$$

where θ is parametric latitude, and the latitudinal-width and hill-centring parameters are again set to $w = \pi/3$ and $\theta = \theta_c = \pi/4$, respectively.
(The special case $n = 2$ recovers the cosine-squared hill of (20.128).)

This completes the definition of the twin-jet solution above cosine hills in *ellipsoidal* coordinates. It includes spherical coordinates and/or a flat bottom boundary as special cases.

20.9.2 Stability of Twin Jets in Spherical Geometry above a Flat Boundary

For a precisely horizontal bottom boundary, that is, $B(\theta) \equiv 0$, Staniforth and White (2008a) showed analytically – by applying Ripa (1983)'s stability criteria – that the exact solution (20.123)–(20.127) in *spherical* geometry is guaranteed to be physically stable when subjected to small but otherwise arbitrary perturbations provided that

$$0 < \mathrm{Fr} \equiv \frac{u_{\max}}{\sqrt{g_0 H_0}} < \frac{1}{4}, \quad 0 \le \mathrm{Ro} \equiv \frac{u_{\max}}{\Omega a} \le \frac{1}{2}, \tag{20.132}$$

where Fr is Froude number, and Ro is Rossby number, both being non-dimensional.

Given the tractability of the analysis for the special case of spherical geometry with a flat bottom boundary, it is instructive to examine this case here in detail. For this special case, with $\theta \rightarrow \phi$, $g \rightarrow g_0$ and using stability conditions (20.119) and (20.120), plus metric factors (20.121) with $e \equiv 0$, stability is guaranteed if, throughout the domain $-\pi/2 \leq \phi \leq \pi/2$:

1. The meridional gradient of the exact, basic-state, potential vorticity does not change sign, that is,

$$\overline{\Pi}_\phi \equiv \frac{d\overline{\Pi}}{d\phi} \text{ does not change sign.} \tag{20.133}$$

and

2. The following inequality is satisfied,

$$\max\left(\frac{\overline{u}}{\cos\phi}\right) - \min\left(\frac{\overline{u}}{\cos\phi}\right) \leq \min\left(\frac{\sqrt{g_0\overline{\widetilde{H}}}}{\cos\phi}\right). \tag{20.134}$$

Both conditions must be satisfied for guaranteed stability of a solution to *small-amplitude* perturbations.

20.9.2.1 Preparation

For use in the preceding two conditions – namely (20.133) and (20.134) – the following relations are obtained from (20.56), (20.68), (20.69), (20.123), and (20.125) with $B \equiv 0$ and $g \equiv g_0$:

$$\overline{\Pi} \equiv \frac{f+\overline{\zeta}}{\overline{\widetilde{H}}}, \tag{20.135}$$

$$\overline{u} = 4u_{\text{max}}\cos\phi\left(1-\cos\phi\right), \tag{20.136}$$

$$\overline{\widetilde{H}} = H_0 + \frac{8u_{\text{max}}^2}{g_0}\cos^2\phi\left(1-\frac{4\cos\phi}{3}+\frac{\cos^2\phi}{2}\right) + \frac{4\Omega a u_{\text{max}}}{g_0}\cos^2\phi\left(1-\frac{2\cos\phi}{3}\right), \tag{20.137}$$

$$f = 2\Omega\sin\phi, \tag{20.138}$$

$$\overline{\zeta} = -\frac{1}{a\cos\phi}\frac{d\left(\overline{u}\cos\phi\right)}{d\phi}. \tag{20.139}$$

Now from (20.135),

$$\overline{\Pi}_\phi \equiv \frac{d\overline{\Pi}}{d\phi} \equiv \frac{d}{d\phi}\left(\frac{f+\overline{\zeta}}{\overline{\widetilde{H}}}\right) \equiv \frac{1}{\left(\overline{\widetilde{H}}\right)^2}\left[\overline{\widetilde{H}}\frac{d}{d\phi}\left(f+\overline{\zeta}\right) - \left(f+\overline{\zeta}\right)\frac{d\overline{\widetilde{H}}}{d\phi}\right] \equiv \frac{W\left(\phi\right)}{\left(\overline{\widetilde{H}}\right)^2}, \tag{20.140}$$

where

$$W\left(\phi\right) \equiv W\left[\overline{\widetilde{H}}\left(\phi\right), f\left(\phi\right)+\overline{\zeta}\left(\phi\right)\right] \equiv W\left(\overline{\widetilde{H}}, f+\overline{\zeta}\right) \equiv \overline{\widetilde{H}}\frac{d}{d\phi}\left(f+\overline{\zeta}\right) - \left(f+\overline{\zeta}\right)\frac{d\overline{\widetilde{H}}}{d\phi}, \tag{20.141}$$

is the Wronskian of $\overline{\widetilde{H}}$ and $f+\overline{\zeta}$ with respect to independent variable ϕ.

Since $1/\left(\overline{\widetilde{H}}\right)^2 > 0$ and is therefore single signed,[13] condition (20.133) is equivalent to $W\left(\phi\right)$ being single signed for $-\pi/2 \leq \phi \leq \pi/2$. Substitution of (20.137)–(20.139) into (20.141) then leads to

[13] In the derivation of the shallow-water equations it is assumed – see (18.36) – that $\overline{\widetilde{H}} > 0$. Physically, this means that no 'wetting/drying' can occur at the bottom boundary.

$$\frac{a}{2u_{\max}^3} g_0 W(\phi) = \left[\frac{1}{(\text{Fr})^2} + 8\mathbb{C}^2\left(1 - \frac{4\mathbb{C}}{3} + \frac{\mathbb{C}^2}{2}\right) + \frac{4}{\text{Ro}}\mathbb{C}^2\left(1 - \frac{2\mathbb{C}}{3}\right)\right]\left[\frac{\mathbb{C}}{\text{Ro}} + 2\left(3 + 2\mathbb{C} - 6\mathbb{C}^2\right)\right] + 8\mathbb{C}\left(1-\mathbb{C}\right)\left(1-\mathbb{C}^2\right)\left[\frac{1}{\text{Ro}} + 2\left(2 - 3\mathbb{C}\right)\right]\left[\frac{1}{\text{Ro}} + 2\left(1-\mathbb{C}\right)\right], \tag{20.142}$$

where $\mathbb{C} \equiv \cos\phi$ for conciseness, and – see (20.132) – $\text{Fr} \equiv u_{\max}/\sqrt{g_0 H_0}$ and $\text{Ro} \equiv u_{\max}/(\Omega a)$ are Froude and Rossby numbers, respectively.

20.9.2.2 *Application of Condition 1*

The first square-bracketed term on the top line of the right-hand side of (20.142) is everywhere non-negative for $0 \le \mathbb{C} \le 1$, as are $\mathbb{C}$, $(1-\mathbb{C})$, $(1-\mathbb{C}^2)$, and the last square-bracketed term on the second line. The second square-bracketed term on the top line, and the first square-bracketed term on the second line, are both non-negative for $0 \le \mathbb{C} \le 1$ provided that

$$0 \le \text{Ro} \equiv \frac{u_{\max}}{\Omega a} \le \frac{1}{2}. \tag{20.143}$$

Thus Condition 1 is satisfied provided that (20.143) is satisfied.

20.9.2.3 *Application of Condition 2*

From (20.136) and (20.137),

$$\max\left(\frac{\overline{u}}{\cos\phi}\right) = 4u_{\max}, \quad \min\left(\frac{\overline{u}}{\cos\phi}\right) = 0, \quad \min\left(\frac{\sqrt{g_0\widetilde{\overline{H}}}}{\cos\phi}\right) > \sqrt{g_0 H_0}. \tag{20.144}$$

Inserting these into (20.134) with use of the definition of Froude number (Fr) then gives

$$4u_{\max} < \sqrt{g_0 H_0} \quad \Rightarrow \quad 0 < \text{Fr} \equiv \frac{u_{\max}}{\sqrt{g_0 H_0}} < \frac{1}{4}. \tag{20.145}$$

20.9.2.4 *Sufficient Conditions for Physical Stability in Spherical Geometry over a Flat Boundary*

Putting together conditions (20.143) and (20.145) confirms the previously stated sufficient conditions (20.132) for physical stability of the examined twin-jet solution *in spherical geometry with a flat bottom boundary.*

Satisfaction of these conditions depends upon the specified values of the parameters. It is natural to wonder whether Zerroukat et al. (2009)'s specified values – see (20.126) – satisfy these conditions. To determine this, substitution of (20.126) into definitions (20.132) of Fr and Ro gives

$$\text{Fr} \equiv \frac{u_{\max}}{\sqrt{g_0 H_0}} = \frac{1}{2\sqrt{10}} \approx 0.158 < 0.25, \quad \text{Ro} \equiv \frac{u_{\max}}{\Omega a} = \frac{5}{7.292 \times 6.37122} \approx 0.108 < 0.5. \tag{20.146}$$

Comparing (20.146) with conditions (20.132) confirms that these conditions are indeed satisfied by Zerroukat et al. (2009)'s specified values, namely those in (20.126) herein.

20.9.2.5 *An Alternative Approach to Application of Condition 1*

It was analytically tractable to apply Condition 1 in the preceding context. However, it can be a lot more challenging to do so in more-complicated circumstances (e.g. in the presence of an

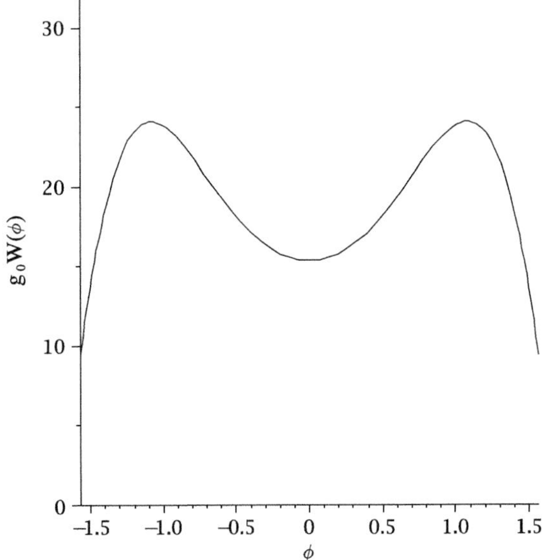

Figure 20.4 $g_0W(\phi)$ for stationary jets in the absence of orography with parameter values set as described in the text. Adapted from Fig. 1 of White and Staniforth (2009), and corrected here. (The labelling of the ordinate in their Figs. 1 and 2 was mistakenly missing scaling factor g_0. Mea culpa. This oversight is inconsequential since only the sign of $W(\phi)$ has significance.)

arbitrary, meridionally varying bottom boundary, $B = B(\phi)$). This motivates the following alternative numerical/graphical approach to application of Condition 1. This approach has broader applicability but provides less analytical insight. Which method to use in a given circumstance boils down to a matter of tractability, convenience, and objective.

The key quantity for application of Condition 1 is $W(\phi)$, defined by (20.141). It must be single signed for satisfaction of Condition 1. For a proposed set of parameter values – for example, to perform experiments with a numerical model – one can compute and plot $W(\phi)$ and then visually verify that it is indeed single signed over the entire domain $-\pi/2 \le \phi \le \pi/2$.[14] We now illustrate this with an example.

With parameter values set as in (20.126) and (20.127), and *with a flat bottom boundary* (i.e. $B(\phi) \equiv 0$), $g_0W(\phi)$ may be computed using (20.142). The result of doing so is plotted in Fig. 20.4.[15] Consistent with the preceding analysis, it is clear from this figure that $W(\phi)$ is positive definite everywhere within the domain ($-\pi/2 \le \phi \le \pi/2$), and therefore that Condition 1 is satisfied for the specified parameter values.

This alternative approach answers the question:

- Does a *specified* set of parameter values lead to satisfaction of Condition 1?

But what if the answer is that it does not? One can then, by trial and error, judiciously increase/decrease parameter values, rerun the computer program, and (hopefully) eventually obtain a set of parameter values that does lead to satisfaction of Condition 1. However, this is where the previous approach (*if analytically tractable*) is superior.

For a specified set of parameter values, it too answers the question. However, if the answer turns out to be that Condition 1 is not satisfied, *it identifies which parameter values to tweak, whether to*

[14] Use of a symbolic-algebra software package to take derivatives, if available, can reduce the labour for this.

[15] As a cross-check, evaluating (20.142) at the planet's poles (where $\mathbb{C} \equiv 0$) and using (20.126) and (20.146) gives $g_0W(\pi/2) = 12u_{\max}g_0H_0/a = 60/6.37122 \approx 9.42$.

increase or decrease them, and furthermore by roughly how much. Thus, for the present example, it is seen from (20.143) that sufficiently decreasing the value of u_{max} and/or increasing the value of Ωa will lead to satisfaction of Condition 1. This is valuable additional information.

20.9.2.6 *An Alternative Approach to Application of Condition 2*

The alternative approach to application of Condition 1, described earlier, can similarly be used for Application of Condition 2. This is not done here for the present example since it is straightforward to proceed analytically.

It is again – tractability permitting – similarly advantageous to proceed analytically. In the event of non-satisfaction of Condition 2, the analytic approach again identifies which parameter values to tweak, whether to increase or decrease them, and by how much. Thus, for the present example, it is seen from (20.145) that sufficiently decreasing the value of u_{max} and/or increasing the value of $g_0 H_0$ will lead to satisfaction of Condition 2.

20.9.3 Stability of Twin Jets in Spherical Geometry above a Cosine-Squared Hill

Having examined the physical stability of the twin-jet solution – in spherical geometry above a *flat bottom boundary* – we now complicate the situation by replacing this boundary by the *cosine-squared hill* defined by (20.128) and (20.129). In the presence of this hill, stability is still guaranteed by satisfaction of Conditions 1 and 2 (i.e. by satisfaction of (20.133) and (20.134), respectively). The only difference is that here the definition of $\widetilde{\widetilde{H}}$ contains a *non-zero* contribution from the bottom boundary, $B = B(\phi)$.

We now examine application of Conditions 1 and 2 for the new situation. Parameter values are all set to the values given in (20.126)–(20.129), except for the maximum hill height B_{max}, which is allowed to vary. The questions to be answered are:

- With this choice of parameter values, what range of maximum hill height guarantees that the exact solution is stable?
- Does this include the Zerroukat et al. (2009) choice of $B_{max} = 3.0 \times 10^3$ m?

20.9.3.1 *Application of Condition 1*

The presence of a *meridionally varying* bottom boundary adversely affects tractability of the application of Condition 1. To address this, we therefore adopt the alternative approach described earlier, for which $g_0 W(\phi)$ is *explicitly evaluated* as a function of ϕ.

With the preceding set of parameter values and with a flat bottom boundary (i.e. $B_{max} \equiv 0$), $g_0 W(\phi)$ has already been displayed in Fig. 20.4 as a function of ϕ over the complete domain. Consistent with the stability analysis of Section 20.9.2, it is seen that $g_0 W(\phi)$ – and therefore $W(\phi)$ – is positive everywhere within the domain $(-\pi/2 \leq \phi \leq \pi/2)$, and that Condition 1 is therefore satisfied. Contrastingly, we are interested here in what happens when B_{max} is increased from the zero value of a flat bottom boundary, thereby giving rise to the presence of a cosine-squared hill of maximum hill height B_{max}. We intuitively expect that stability will be maintained for small-enough hill heights, but how small/large is not evident.

For cosine-squared hill profile (20.128), and using the preceding set of parameter values, $g_0 W(\phi)$ is displayed in Fig. 20.5(a), over the subdomain $\pi/12 \leq \theta \leq 5\pi/12$ (i.e. where the bottom boundary is no longer flat), for four values of B_{max} ranging from zero to 4.95×10^3 m. [By varying B_{max} in the computer program used to plot $g_0 W(\phi)$, this latter (approximate) limiting value was determined experimentally. For larger values of B_{max}, Condition 1 is violated since $g_0 W(\phi)$ then has a change of sign over the domain.]

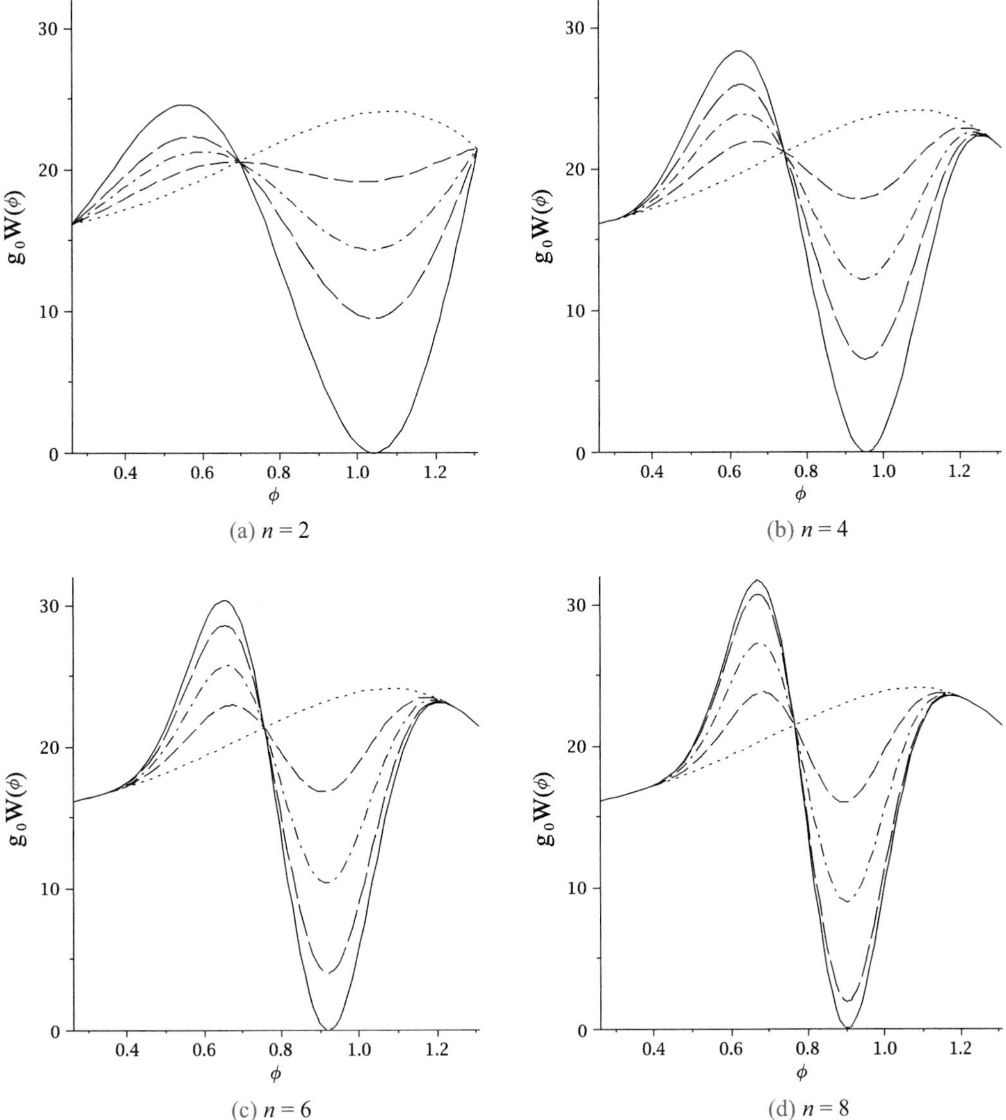

Figure 20.5 $g_0 W(\phi)$ over the subdomain $\pi/12 \leq \theta \leq 5\pi/12$ for stationary jets above an orography $B_n(\theta) = (B_{\text{max}})_n \cos^n [\pi (\theta - \theta_c)/w]$, where $w = \pi/3, \theta_c = \pi/4$, and panels (a)–(d) correspond to $n = 2, 4, 6$, and 8, respectively. For all panels, the dotted, dashed, dash-dotted, and long-dashed curves correspond to hill heights B_{max} of 0 (dotted); 1.0×10^3 m (dashed); 2.0×10^3 m (dash-dotted) and 3.0×10^3 m (long-dashed). The solid curves correspond to the maximum possible hill-height values for guaranteed stability, i.e.: (a) $(B_{\text{max}})_2 = 4.95 \times 10^3$ m, (b) $(B_{\text{max}})_4 = 4.15 \times 10^3$ m, (c) $(B_{\text{max}})_6 = 3.63 \times 10^3$ m, and (d) $(B_{\text{max}})_8 = 3.27 \times 10^3$ m. Adapted from Fig. 2 of White and Staniforth (2009), and corrected here to account for their erroneous scaling of the ordinate.

Thus, with the set of parameter values given in (20.126)–(20.129), Condition 1 is satisfied when the maximum hill height satisfies $0 \leq B_{\text{max}} \leq 4.95 \times 10^3$ m. This evidently includes Zerroukat et al. (2009)'s specific choice $B_{\text{max}} = 3.0 \times 10^3$ m.

20.9.3.2 *Application of Condition 2*

It remains to also apply Condition 2. From (20.123), with $\theta \to \phi$,

$$\max\left(\frac{\overline{u}}{\cos\phi}\right) = 4u_{\max}, \quad \min\left(\frac{\overline{u}}{\cos\phi}\right) = 0 \quad \Rightarrow \quad \max\left(\frac{\overline{u}}{\cos\phi}\right) - \min\left(\frac{\overline{u}}{\cos\phi}\right) = 4u_{\max}, \tag{20.147}$$

and, from (20.125)–(20.129) with $\theta \to \phi$,

$$\min\left(\frac{\sqrt{g_0\widetilde{\overline{H}}}}{\cos\phi}\right) > \sqrt{g_0H_0 - \max\left[g_0B(\phi)\right]} = \sqrt{g_0\left(H_0 - B_{\max}\right)}. \tag{20.148}$$

Using (20.147) and (20.148) together with the given values $u_{\max} = 50\,\mathrm{m\,s^{-1}}$ and $g_0H_0 = 10^5\,\mathrm{m^2\,s^{-2}}$ – see (20.126) – it is seen that (20.134) is satisfied provided that

$$g_0B_{\max} \leq g_0H_0 - 16u_{max}^2 = 6 \times 10^4\,\mathrm{m^2\,s^{-2}} \quad \Rightarrow \quad B_{\max} \leq 6.11 \times 10^3\,\mathrm{m}. \tag{20.149}$$

Thus Condition 2 is also satisfied for the preceding set of parameter values when the maximum hill height satisfies $0 \leq B_{\max} \leq 4.95 \times 10^3$ m (i.e. the range for $B_{\max}$ found for satisfaction of Condition 1).

20.9.3.3 *Sufficient Conditions for Physical Stability in Spherical Geometry over a Cosine-Squared Hill*

In the *presence* of a cosine-squared hill, both Conditions 1 and 2 are satisfied for the preceding set of parameter values when the maximum hill height satisfies $0 \leq B_{\max} \leq 4.95 \times 10^3$ m, and therefore the flow for stationary jets above a zonal cosine-squared hill is guaranteed to be physically stable when subjected to small but otherwise arbitrary perturbations. In particular, Zerroukat et al. (2009)'s choice of $B_{\max} = 3.0 \times 10^3$ m satisfies this condition. Both of the posed questions have thus been answered.

20.9.4 Stability of Twin Jets in Spherical Geometry above a Generalised Cosine Hill

The analysis given in Section 20.8 assumes that the zonally symmetric bottom boundary $B = B(\xi_2)$ is everywhere continuous, with continuous first derivative.[16] White and Staniforth (2009) noted that although the cosine-squared hill is sufficiently differentiable to satisfy these continuity conditions, additional differentiability of the bottom-boundary profile is desirable when defining a test problem to assess the accuracy of a finite-difference model of the shallow-water equations. This is because the derivation of the truncation errors of finite-difference formulae, using Taylor-series expansions, implicitly assumes further differentiability; for example when approximating ∇B by second-order or fourth-order finite differences in the momentum equation. This then motivated reconsideration of the Zerroukat et al. (2009) test problem, leading White and Staniforth (2009) to propose the alternative, more differentiable, generalised cosine hill profile defined by (20.130) and (20.131). Further motivation is that increasing the value of parameter n decreases the horizontal scale of the hill, thereby leading to a somewhat more challenging test problem.

Following this reasoning, we now further complicate the situation for the physical stability of the twin-jet solution by replacing the cosine-squared hill profile – defined by (20.128) and (20.129) and considered in the previous subsection – by the generalised cosine hill defined by (20.130) and (20.131). Setting $n = 2$ for this generalised cosine hill recovers the cosine-squared hill as a special case.

[16] If this were not so, then the mass-continuity equation could not be written in the form (20.57).

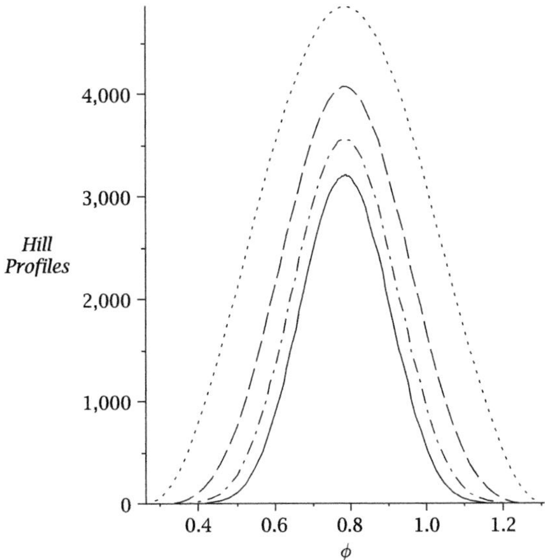

Figure 20.6 Generalised hill profiles of the form $B_n(\phi) = (B_{\max})_n \cos^n[\pi(\phi - \phi_c)/w]$ over the subdomain $\pi/12 \leq \theta \leq 5\pi/12$, where $(B_{\max})_n$ is the maximum value (in m) for given n that guarantees stability of the stationary jets problem with parameters set as in the text. Curves: $n = 2$ (dotted); $n = 4$ (long dashed); $n = 6$ (dash dotted); and $n = 8$ (solid).

The question posed now is:

- How large a value – for given order $n = 4, 6, 8$ – can one take for the maximum hill height $(B_{\max})_n$ for the generalised cosine-hill profile defined by (20.130) and (20.131), whilst maintaining stability of the flow to small but otherwise arbitrary perturbations?

Following the same methodology, with the same set of parameter values employed earlier for the cosine-squared definition (20.128) of the hill, but using definition (20.130) of the generalised hill – for $n = 4, 6, 8$ instead of (20.128) for the special case of $n = 2$ – then leads to the results shown in Fig. 20.5b–d for Condition 1; and, again, to (20.149) for Condition 2.

It is concluded from this figure and from (20.149) that both Conditions 1 and 2 are satisfied for this set of parameter values when the maximum hill height satisfies

$$\left.\begin{aligned} 0 &\leq (B_{\max})_2 \leq 4.95 \times 10^3 \text{ m}, \\ 0 &\leq (B_{\max})_4 \leq 4.15 \times 10^3 \text{ m}, \\ 0 &\leq (B_{\max})_6 \leq 3.63 \times 10^3 \text{ m}, \\ 0 &\leq (B_{\max})_8 \leq 3.27 \times 10^3 \text{ m}. \end{aligned}\right\} \tag{20.150}$$

Consequently the flow for stationary jets above a generalised cosine hill, when subjected to small but otherwise arbitrary perturbations, is guaranteed to be physically stable provided $B_{\max}$ is chosen to satisfy (20.150).

Setting $(B_{\max})_n$ for $n = 2, 4, 6, 8$ to the maximum values given in (20.150) and plotting the associated generalised cosine profiles of the form (20.131) over the subdomain $\pi/12 \leq \theta \leq 5\pi/12$ then gives the four profiles displayed in Fig. 20.6. It is seen that as n is increased (so that the horizontal scale of the hill is decreased) the maximum hill height for guaranteed stability diminishes accordingly. This answers the posed question.

20.9.5 Stability of Twin Jets in Ellipsoidal Geometry above a Generalised Cosine Hill

The preceding analysis of the stability of the twin-jet problem in *spherical* geometry has prepared the way for the discussion here of the corresponding problem in *ellipsoidal* geometry, as defined in Section 20.9.1.

To begin, we note that the more general formulation in ellipsoidal-parametric coordinates (λ, θ) is almost identical to that of its special case of spherical coordinates (λ, ϕ). It is almost but not quite as simple as replacing ϕ everywhere by the more general θ. There are, however, two exceptions:

1. Metric factor $h_2^S(\theta)$ has a more complicated functional form than $h_2^S(\phi)$.
2. Gravity $g(\theta)$ can vary meridionally, whereas $g(\phi)$ cannot (to respect dynamical consistency).

The questions now are:

- How do these two exceptions influence the conditions for stability?
- Practically speaking, how important are these exceptions?

20.9.5.1 *Conditions for Stability*

In parametric-ellipsoidal coordinates (λ, θ), and applying stability conditions (20.119) and (20.120) plus metric factors (20.121), stability in ellipsoidal geometry is assured if, throughout the domain $-\pi/2 \leq \theta \leq \pi/2$:

1. The basic-state, potential-vorticity gradient does not change sign, that is,

$$\overline{\Pi}_\theta \equiv \frac{d\overline{\Pi}}{d\theta} \text{ does not change sign,} \tag{20.151}$$

   where θ is parametric latitude; *and*
2. Condition (20.120) – with $\xi_1 = \theta$ and with use of (20.121) – is obeyed, that is,

$$\max\left(\frac{\overline{u}}{a\cos\theta}\right) - \min\left(\frac{\overline{u}}{a\cos\theta}\right) \leq \min\left(\frac{\sqrt{g(\theta)\,\widetilde{\overline{H}}(\theta)}}{a\cos\theta}\right). \tag{20.152}$$

20.9.5.2 *Preparation*

From (20.121), metric factor $h_2^S(\theta)$ satisfies

$$h_2^S(\theta) = a\left(1 - e^2\cos^2\theta\right)^{\frac{1}{2}}. \tag{20.153}$$

This has a more complicated functional form than its special, spherical form $h_2^S(\phi) = a$ for $e \equiv 0$. From (20.56) and (20.68), metric factor $h_2^S(\theta)$ influences both $f(\theta)$ and $\overline{\zeta}(\theta)$, and thereby – from (20.69) – $\overline{\Pi}(\theta)$. Thus – with use of (20.121), (20.123), (20.125), (20.128), and (20.129) –

$$f(\theta) + \overline{\zeta}(\theta) = \frac{1}{\left(1 - e^2\cos^2\theta\right)^{\frac{1}{2}}}\left[2\Omega + \frac{4u_{\max}}{a}(2 - 3\cos\theta)\right]\sin\theta, \tag{20.154}$$

$$\overline{\Pi}(\theta) = \frac{1}{\left(1 - e^2\cos^2\theta\right)^{\frac{1}{2}}}\left[2\Omega + \frac{4u_{\max}}{a}(2 - 3\cos\theta)\right]\frac{\sin\theta}{\widetilde{\overline{H}}(\theta)}, \tag{20.155}$$

where

$$\begin{aligned}\widetilde{\overline{H}}(\theta) = \frac{g_0}{g(\theta)}\Bigg[H_0 &+ \frac{8u_{\max}^2}{g_0}\cos^2\theta\left(1 - \frac{4\cos\theta}{3} + \frac{\cos^2\theta}{2}\right)\\ &+ \frac{4\Omega a u_{\max}}{g_0}\cos^2\theta\left(1 - \frac{2\cos\theta}{3}\right) - B_n(\theta)\Bigg].\end{aligned} \tag{20.156}$$

Equations (20.154)–(20.156) in parametric-ellipsoidal coordinates (λ, θ) may be compared to their counterparts for the special case of spherical-polar coordinates (λ, ϕ). Setting $\theta = \phi$ and $e = 0$ in (20.154)–(20.156) gives their simpler, spherical counterparts

$$f(\phi) + \overline{\zeta}(\phi) = \left[2\Omega + \frac{4u_{\max}}{a}(2 - 3\cos\phi)\right]\sin\phi, \tag{20.157}$$

$$\overline{\Pi}(\phi) = \left[2\Omega + \frac{4u_{\max}}{a}(2 - 3\cos\phi)\right]\frac{\sin\phi}{\overline{\widetilde{H}}(\phi)}, \tag{20.158}$$

$$\begin{aligned}\overline{\widetilde{H}}(\phi) = H_0 &+ \frac{8u_{\max}^2}{g_0}\cos^2\phi\left(1 - \frac{4\cos\phi}{3} + \frac{\cos^2\phi}{2}\right) \\ &+ \frac{4\Omega a u_{\max}}{g_0}\cos^2\phi\left(1 - \frac{2\cos\phi}{3}\right) - B_n(\phi),\end{aligned} \tag{20.159}$$

where $g(\phi) = g_0 =$ constant (for dynamical consistency in spherical geometry) has been exploited.

Comparing the functional forms of (20.154) and (20.155) with their simpler spherical counterparts (20.157) and (20.158), it is seen that they are the same except that the right-hand sides of (20.154) and (20.155) are multiplied by the factor

$$\left(1 - e^2\cos^2\theta\right)^{-\frac{1}{2}} = 1 + \frac{e^2}{2}\cos^2\theta + \cdots = 1 + O\left(e^2\right). \tag{20.160}$$

For mildly oblate planets, such as Earth, this factor is everywhere equal to unity to within less than one per cent.

Similarly comparing the functional form of (20.156) with its simpler spherical counterpart (20.159), the story is very similar. The functional form of (20.156) is the same except that the right-hand side of (20.156) is multiplied by the factor $g_0/g(\theta)$ and, for the models of gravity described in Section 18.3.3.1,

$$\frac{g_0}{g(\theta)} = 1 + O\left(e^2\right). \tag{20.161}$$

For mildly oblate planets, such as Earth, this factor is also everywhere equal to unity to within less than one per cent.

20.9.5.3 *Application of Condition 1*

The sign of the meridional gradient of potential vorticity $(d\overline{\Pi}/d\theta)$ is crucially important for the application of Condition 1 for stability. It is therefore important to examine how it is influenced by $h_2^S(\theta)$ and $g(\theta)$. Now – from (20.69) –

$$\overline{\Pi}_\theta \equiv \frac{d\overline{\Pi}}{d\theta} \equiv \frac{d}{d\theta}\left(\frac{f + \overline{\zeta}}{\overline{\widetilde{H}}}\right) \equiv \frac{1}{\left(\overline{\widetilde{H}}\right)^2}\left[\overline{\widetilde{H}}\frac{d}{d\theta}\left(f + \overline{\zeta}\right) - \left(f + \overline{\zeta}\right)\frac{d\overline{\widetilde{H}}}{d\theta}\right] \equiv \frac{W(\theta)}{\left(\overline{\widetilde{H}}\right)^2}, \tag{20.162}$$

where

$$W(\theta) \equiv W\left[\overline{\widetilde{H}}(\theta), f(\theta) + \overline{\zeta}(\theta)\right] \equiv W\left(\overline{\widetilde{H}}, f + \overline{\zeta}\right) \equiv \overline{\widetilde{H}}\frac{d}{d\theta}\left(f + \overline{\zeta}\right) - \left(f + \overline{\zeta}\right)\frac{d\overline{\widetilde{H}}}{d\theta} \tag{20.163}$$

is the Wronskian of $\overline{\widetilde{H}}$ and $f + \overline{\zeta}$ with respect to independent variable θ. The sign of $d\overline{\Pi}/d\theta$ is therefore determined by the sign of the right-hand side of (20.163).

Substitution of (20.157) and (20.158) into the right-hand side of (20.163) leads to an even more complicated expression in ellipsoidal geometry than it does for the previously examined case in spherical geometry. This again adversely affects analytic tractability for application of Condition 1 in ellipsoidal geometry and motivates use of the alternative approach to application of Condition

1. As in spherical geometry, this involves specification of parameter values, followed by explicit evaluation, using these values, of the right-hand side of (20.163) to obtain $W(\theta)$. By plotting Wronskian $W(\theta)$, it can be verified (or otherwise) that Condition 1 is satisfied for the set of specified parameter values.

If we were to do so using the same set of parameter values as the earlier one in spherical geometry, how would this affect things? The answer, in a nutshell, is by very little. This is because factors (20.160) and (20.161) are well-behaved functions that, for mildly oblate planets such as Earth, individually only vary by less than one per cent over the entire domain $-\pi/2 \leq \theta \leq \pi/2$. With some asymptotic analysis, it can be shown that the right-hand side of (20.163) can then only vary by $O\left(e^2\right)$ compared to values obtained in spherical geometry. This means that the functions plotted in Figs. 20.4–20.6 can only vary by a per cent or two at most, with the consequence that limiting values (such as for maximum hill height $B_{\max}$) can only change by a per cent or two at most. Results obtained in spherical geometry for Application of Condition 1 can thus be confidently used in ellipsoidal geometry provided one accounts for a possible per cent or two change in limiting values (e.g. for maximum hill height).

20.9.5.4 *Application of Condition 2*

It remains to also apply Condition 2. From (20.123),

$$\max\left(\frac{\overline{u}}{\cos\theta}\right) = 4u_{max}, \quad \min\left(\frac{\overline{u}}{\cos\theta}\right) = 0 \quad \Rightarrow \quad \max\left(\frac{\overline{u}}{\cos\theta}\right) - \min\left(\frac{\overline{u}}{\cos\theta}\right) = 4u_{\max}. \tag{20.164}$$

Now from (20.130), (20.131), and (20.156),

$$g(\theta)\,\overline{\overline{H}}(\theta) = g_0 H_0 + 8u_{\max}^2 \cos^2\theta\left(1 - \frac{4\cos\theta}{3} + \frac{\cos^2\theta}{2}\right) + 4\Omega a u_{\max}\cos^2\theta\left(1 - \frac{2\cos\theta}{3}\right) - g_0 B_n(\theta), \tag{20.165}$$

$$B_n(\theta) = (B_{\max})_n \cos^n\left[\frac{\pi}{w}(\theta - \theta_c)\right] \quad \text{for } |\theta - \theta_c| \leq \frac{w}{2},\ n \geq 2, \tag{20.166}$$

$$= 0 \quad \text{otherwise}. \tag{20.167}$$

Thus, using these three equations,

$$\min\left[\frac{\sqrt{g(\theta)\,\overline{\overline{H}}(\theta)}}{a\cos\theta}\right] > \sqrt{g_0 H_0 - \max\left[g_0 B_n(\theta)\right]} = \sqrt{g_0\left(H_0 - B_{\max}\right)}. \tag{20.168}$$

Insertion of (20.164) and (20.168) into Condition 2, that is, into (20.152), gives

$$4u_{\max} < \sqrt{g_0\left(H_0 - B_{\max}\right)}. \tag{20.169}$$

Inserting into (20.169) the given values $u_{\max} = 50\,\mathrm{m\,s^{-1}}$, $g_0 H_0 = 10^5\,\mathrm{m^2\,s^{-2}}$, and $g_0 = 9.80616\,\mathrm{m\,s^{-2}}$ – obtained from (20.126), (20.128), and (20.130) – it is seen that Condition 2, that is, (20.152), is satisfied provided that

$$B_{\max} < \frac{1}{g_0}\left(g_0 H_0 - 16u_{\max}^2\right) = \frac{6 \times 10^4}{9.80616}\,\mathrm{m} \approx 6.11 \times 10^3\,\mathrm{m}. \tag{20.170}$$

This is exactly the same condition obtained for the special case in spherical geometry – see (20.148) and (20.149). This – as noted in Section 20.9.4 – is true not only over a *cosine-squared hill* but also over a *generalised cosine hill*. Condition 2 is insensitive to the more general forms for $h_2^S(\theta)$ and $g(\theta)$ in ellipsoidal geometry, when compared to those in spherical geometry. It is also insensitive to the choice of a model of gravity.

20.9.5.5 *Sufficient Conditions for Stability of Twin Jets in Ellipsoidal Geometry above a Generalised Cosine Hill*

This is assured by the most stringent of Conditions 1 and 2. Since, for mildly oblate planets, ellipsoidal geometry is a small perturbation of spherical geometry, results from Sections 20.9.2–20.9.4 imply that Condition 1 will be the most stringent. So, in ellipsoidal geometry we expect the limitations of Condition 1 to apply within a per cent or two. For example, the maximum height of a cosine-squared hill would then need to satisfy $0 \leq B_{\max} \leq 4.95 \times 10^3\,\mathrm{m} \pm 0.1 \times 10^3\,\mathrm{m}$.

From the preceding analysis and discussion, we conclude that:

- The more complicated functional form of $h_2^S(\theta)$ compared to $h_2^S(\phi)$, and the possible meridional variation of gravity, collectively have relatively little influence on the stability conditions of the examined model problem over mildly oblate, ellipsoidal planets.

20.10 CONCLUDING REMARKS

A methodology has been developed to construct exact, steady, vortical, axially symmetric, non-divergent solutions of the 2D, non-linear, shallow-water equations in spheroidal geometry. Five families of such exact solutions have been constructed as illustrative examples. As part of the construction procedure, these solutions are constrained to respect a natural, non-linear, *horizontal* balance. An existing stability analysis has been generalised from spherical to spheroidal geometry, to obtain two criteria for the stability of such solutions when subjected to arbitrary, small-amplitude perturbations. Satisfaction of these two criteria then guarantees physical stability.

It is more challenging to develop a similar methodology for the construction of exact 3D solutions. There are then *two* non-linear, balance equations that need to be simultaneously satisfied – one for *horizontal* balance (as in the 2D case), the other for *vertical* balance. This complicates matters. It is, however, shown in Chapter 21 how to accomplish this balancing act. Three illustrative examples of exact, steady, vortical, axially symmetric solutions of the 3D, non-linear, governing equations are then constructed.

APPENDIX: ROTATED COORDINATE TRANSFORMATIONS

Closely following Staniforth and White (2007):

1. Various coordinate transformation relations between rotated (λ', ϕ') and unrotated (λ, ϕ) spherical-polar coordinate systems are developed.
2. These relations are then applied to transform exact, steady, spherical solutions to rotated spherical-polar coordinates.

Coordinate Transformation Relations

Let the north pole of the rotated (λ', ϕ') coordinate system be situated at the point $(\lambda, \phi) = (\lambda_P, \phi_P)$ of the unrotated, geographical coordinate system. The following relations – from Verkley (1984) – then hold between the rotated and unrotated coordinate systems:

$$\sin\phi' = \sin\phi_P \sin\phi + \cos\phi_P \cos\phi \cos(\lambda - \lambda_P), \tag{20.171}$$

$$\sin\phi = \sin\phi_P \sin\phi' - \cos\phi_P \cos\phi' \cos\lambda', \tag{20.172}$$

$$\cos\phi' \sin\lambda' = \cos\phi \sin(\lambda - \lambda_P). \tag{20.173}$$

Expressed in the unrotated (λ, ϕ) coordinate system, the wind components for the five exact, steady solutions developed in Section 20.5 satisfy

$$u_1(\phi) = a\cos\phi\frac{D\lambda}{Dt}, \quad u_2(\phi) = a\frac{D\phi}{Dt} = 0. \tag{20.174}$$

Similarly, the wind components expressed in the rotated (λ', ϕ') coordinate system satisfy

$$u_1^{'}(\lambda', \phi') = a\cos\phi'\frac{D\lambda'}{Dt}, \quad u_2^{'}(\lambda', \phi') = a\frac{D\phi'}{Dt} \neq 0. \tag{20.175}$$

With use of transformation relations (20.171)–(20.173), we now relate $u_1^{'}(\lambda', \phi')$ and $u_2^{'}(\lambda', \phi')$ to $u_1(\phi)$. Multiplying (20.171) by a, differentiating the result with respect to time, and using (20.174) and (20.175) gives

$$a\cos\phi'\frac{D\phi'}{Dt} = \cancel{a\left[\sin\phi_P\cos\phi - \cos\phi_P\sin\phi\cos(\lambda-\lambda_P)\right]\frac{D\phi}{Dt}} - a\cos\phi_P\cos\phi\sin(\lambda-\lambda_P)\frac{D\lambda}{Dt}$$

$$\Downarrow$$

$$u_2^{'}(\lambda', \phi')\cos\phi' = -\cos\phi_P\sin(\lambda-\lambda_P)\,u_1(\phi). \tag{20.176}$$

Multiplying (20.172) and (20.173) by a, differentiating the results with respect to time, and using (20.173)–(20.176) leads to

$$\cancel{a\cos\phi\frac{D\phi}{Dt}} = a\left[\sin\phi_P\cos\phi' + a\cos\phi_P\sin\phi'\cos\lambda'\right]\frac{D\phi'}{Dt} + a\cos\phi_P\cos\phi'\sin\lambda'\frac{D\lambda'}{Dt}$$

$$\Downarrow$$

$$0 = \left[\sin\phi_P\cos\phi' + \cos\phi_P\sin\phi'\cos\lambda'\right]u_2^{'}(\lambda', \phi') + u_1^{'}(\lambda', \phi')\cos\phi_P\sin\lambda'$$

$$\Downarrow$$

$$u_1^{'}(\lambda', \phi')\sin\lambda'\cos\phi' = \left[\sin\phi_P\cos\phi' + \cos\phi_P\sin\phi'\cos\lambda'\right]\sin(\lambda-\lambda_P)\,u_1(\phi), \tag{20.177}$$

and

$$a\cos\phi'\cos\lambda'\frac{D\lambda'}{Dt} - a\sin\phi'\sin\lambda'\frac{D\phi'}{Dt} = a\cos\phi\cos(\lambda-\lambda_P)\frac{D\lambda}{Dt} - \cancel{a\sin\phi\sin(\lambda-\lambda_P)\frac{D\phi}{Dt}}$$

$$\Downarrow$$

$$u_1^{'}(\lambda', \phi')\cos\lambda' - \sin\phi'\sin\lambda' u_2^{'}(\lambda', \phi') = \cos(\lambda-\lambda_P)\,u_1(\phi)$$

$$\Downarrow$$

$$u_1^{'}(\lambda', \phi')\cos\lambda'\cos\phi' = \left[\cos(\lambda-\lambda_P)\cos\phi' - \cos\phi_P\sin\phi'\sin\lambda'\sin(\lambda-\lambda_P)\right]u_1(\phi). \tag{20.178}$$

Multiplying (20.177) and (20.178) by $\sin\lambda'/\cos\phi'$ and $\cos\lambda'/\cos\phi'$, respectively, and summing finally yields

$$u_1^{'}(\lambda', \phi') = \left[\cos\lambda'\cos(\lambda-\lambda_P) + \sin\phi_P\sin\lambda'\sin(\lambda-\lambda_P)\right]u_1(\phi). \tag{20.179}$$

Transformation of Solutions to Rotated Coordinates

Everything is now in place to define the procedure to transform exact, steady, spherical solutions to rotated spherical-polar coordinates. Suppose that one of the exact, illustrative solutions in Section 20.5 is to be expressed in the rotated (λ', ϕ') coordinate system, whose north pole is located at the point (λ_P, ϕ_P) in the unrotated geographical coordinate system (λ, ϕ). The four-step procedure to express the corresponding solution at a specific point (λ', ϕ') of the rotated coordinate system is then:

1. Compute ϕ in the unrotated coordinate system using (20.172), and so

$$\phi = \sin^{-1}\left(\sin\phi_P \sin\phi' - \cos\phi_P \cos\phi' \cos\lambda'\right), \quad -\frac{\pi}{2} \leq \phi \leq \frac{\pi}{2}. \tag{20.180}$$

2. Using (20.180), compute $u(\phi)$ and $H(\phi)$ using the relevant expressions of Section 20.5. For the twin-jet solution, for example, this gives – from (20.22) and (20.23) –

$$u_1(\phi) = 4u_{\max}\cos\phi\,(1-\cos\phi), \tag{20.181}$$

$$H(\phi) = H_0 + \frac{8u_{\max}^2}{g_0}\cos^2\phi\left(1 - \frac{4\cos\phi}{3} + \frac{\cos^2\phi}{2}\right) + \frac{4\Omega a u_{\max}}{g_0}\cos^2\phi\left(1 - \frac{2\cos\phi}{3}\right). \tag{20.182}$$

3. Compute $\sin(\lambda - \lambda_P)$ and $\cos(\lambda - \lambda_P)$ in the unrotated coordinate system using (20.173) and (20.171), respectively, that is,

$$\sin(\lambda - \lambda_P) = \frac{\cos\phi' \sin\lambda'}{\cos\phi}, \tag{20.183}$$

$$\cos(\lambda - \lambda_P) = \frac{\sin\phi' - \sin\phi_P \sin\phi}{\cos\phi_P \cos\phi}. \tag{20.184}$$

4. Compute $u_1'(\lambda', \phi')$ and $u_2'(\lambda', \phi')$ from (20.179) and (20.176), respectively.

21

Exact 3D Steady Solutions of Global Equation Sets

ABSTRACT

Exact solutions of the governing equations for an atmospheric or oceanic dynamical core are very useful. A methodology for constructing exact, non-linear, steady, axially symmetric solutions of the 2D shallow-water equations was developed in Chapter 20. This methodology is extended to 3D in the present chapter under the assumptions that the flow is both steady and axially symmetric. Whereas the 2D solutions are constrained to respect a natural horizontal balance, the 3D ones have to simultaneously respect two balance conditions, one horizontal, the other vertical. These conditions non-linearly relate density, pressure, and the zonal component of velocity. A methodology for constructing 3D solutions that satisfy these two conditions is developed. This methodology is applicable not only in the atmospheric context but also in the oceanic one, thereby removing a restriction of previous work. With use of the two conditions, together with the ideal-gas law, the classical thermal-wind equation for zonal atmospheric flow is generalised. The construction methodology developed herein may be applied in different ways according to context and objectives. To illustrate this, three examples of increasing complexity are given. The third of these has found application as a basic steady state for a baroclinic-wave test problem that has been used in the literature for atmospheric model validation and assessment purposes. Depending upon interest and analytic tractability, further exact solutions may be obtained by applying and adapting the construction methodology developed herein.

21.1 PREAMBLE

As discussed in Chapter 20, exact, non-linear solutions of the governing equations for an atmospheric or oceanic dynamical core are very useful, both theoretically and for testing and validating numerical models. Developing such solutions is, however, highly challenging due to non-linearities and geometrical complications. A methodology for constructing steady, vertical, axially symmetric, non-divergent solutions of the non-linear, 2D, shallow-water equations in global, spheroidal geometry was given in Chapter 20. Five families of such solutions were then developed as illustrative examples of the application of the methodology:

1. An equatorial jet.
2. Twin mid-latitude jets.
3. An isolated polar vortex.
4. Multiple Jovian jets.
5. An exponentially decaying vortex.

In the present chapter we build on the 2D methodology of Chapter 20, and extend it to 3D.

The 2D steady solutions of Chapter 20 are constrained to respect a natural balance in the horizontal between the zonal component of velocity (u_1) and the free-surface height (H). The situation in 3D is more complicated. Not only do we need to respect a non-linear constraint on horizontal balance, we also need to simultaneously respect one on vertical balance. Furthermore, the exact, 3D, steady solutions of Staniforth and White (2007), White and Staniforth (2008), Staniforth and White (2011), and Staniforth and Wood (2013), developed over a period of several years and of increasing complexity, all assume the ideal-gas law as the equation of state. This then limits their methodology to the atmospheric context, since the ideal-gas law is not valid for ocean modelling. A key issue therefore, and a focus of the present chapter, is how to go about addressing this limitation. The work just cited is also, for the most part, restricted to spherical geometry. A further focus of the present chapter is thus to address this second limitation during the course of developing the theory.

The plan of this chapter is as follows. To set the scene, a unified quartet of governing equations for 3D atmospheric and oceanic dynamical cores – developed earlier in Chapter 16 – is recalled in Section 21.2. These equations are then simplified in Section 21.3 under the assumptions that the flow is both steady and axially symmetric. This gives rise to two key balance equations – one horizontal, the other vertical – that must be respected. These two first-order differential equations non-linearly relate the zonal component of velocity (u_1), density (ρ), and pressure (p). Three compatibility constraints are obtained from these two equations in Section 21.4 by elimination of one of these three variables in turn. In Section 21.5, a convenient change of dependent variable is then introduced to prepare the way for the subsequent analysis.

With these preparations, a methodology to construct exact steady solutions is developed in Section 21.6. Since no assumption is made regarding an equation of state, this methodology is valid not only in the atmospheric context but also in the oceanic one, thereby addressing the first limitation mentioned earlier. As a slight digression, a generalisation of the classical thermal-wind equation for zonal atmospheric flow is developed in Section 21.7 using the compatibility constraints of Section 21.4 and the ideal-gas law. Due to the latter assumption, this equation is valid only in the atmospheric context.

The theory developed in the preceding sections may be applied in a variety of different ways, depending upon the context and what one wishes to accomplish. Three illustrative examples of increasing complexity are given in Section 21.8. The first two illustrate how to generalise the five solution families of the 2D shallow-water equations – developed in Chapter 20 – to 3D. The third is a more complicated solution, with a temperature field of non-separable form. This solution is valid only in the atmospheric context due to use of the ideal-gas law. It has been used as a basic state in Ullrich et al. (2014)'s idealised baroclinic-wave test case in spherical geometry. Further exact, steady solutions may be developed using the theory developed herein, depending upon interest and analytic tractability. Finally, concluding remarks are made in Section 21.9.

21.2 A UNIFIED QUARTET OF GOVERNING EQUATIONS

To set the scene, recall the unified quartet of governing equations for atmospheric and oceanic dynamical cores, expressed in Chapter 16 in axial-orthogonal-curvilinear coordinates (ξ_1, ξ_2, ξ_3). Thus, from (16.19)–(16.28) and in the now-familiar notation of Chapter 16:

$$\frac{Du_1}{Dt} + \left(\frac{u_1}{h_1} + 2\Omega\right)\left(\frac{u_2}{h_2}\frac{\partial h_1}{\partial \xi_2} + \underbrace{\frac{u_3}{h_3}\frac{\partial h_1}{\partial \xi_3}}_{\text{deep}}\right) + \frac{1}{\rho h_1}\frac{\partial p}{\partial \xi_1} = 0, \tag{21.1}$$

$$\frac{Du_2}{Dt} - \left(\frac{u_1}{h_1} + 2\Omega\right)\frac{u_1}{h_2}\frac{\partial h_1}{\partial \xi_2} - \underbrace{\frac{u_3^2}{h_2 h_3}\frac{\partial h_3}{\partial \xi_2}}_{\text{non-hydro}} + \underbrace{\frac{u_2 u_3}{h_2 h_3}\frac{\partial h_2}{\partial \xi_3}}_{\text{deep}} + \frac{1}{\rho h_2}\frac{\partial p}{\partial \xi_2} = 0, \tag{21.2}$$

$$\underbrace{\frac{Du_3}{Dt} + \frac{u_2}{h_2}\frac{u_3}{h_3}\frac{\partial h_3}{\partial \xi_2}}_{\text{non-hydro}} - \underbrace{\left[\left(\frac{u_1}{h_1} + 2\Omega\right)\frac{u_1}{h_3}\frac{\partial h_1}{\partial \xi_3} + \frac{u_2^2}{h_2 h_3}\frac{\partial h_2}{\partial \xi_3}\right]}_{\text{deep}} + \frac{1}{\rho h_3}\frac{\partial p}{\partial \xi_3} + \frac{1}{h_3}\frac{d\Phi}{d\xi_3} = 0, \tag{21.3}$$

$$\frac{D\rho}{Dt} + \frac{\rho}{h_1 h_2 h_3}\left[\frac{\partial}{\partial \xi_1}(u_1 h_2 h_3) + \frac{\partial}{\partial \xi_2}(u_2 h_3 h_1) + \frac{\partial}{\partial \xi_3}(u_3 h_1 h_2)\right] = 0, \tag{21.4}$$

$$\frac{D\mathscr{E}}{Dt} + p\frac{D\alpha}{Dt} = 0, \tag{21.5}$$

where

$$\frac{D}{Dt} = \frac{\partial}{\partial t} + \frac{u_1}{h_1}\frac{\partial}{\partial \xi_1} + \frac{u_2}{h_2}\frac{\partial}{\partial \xi_2} + \frac{u_3}{h_3}\frac{\partial}{\partial \xi_3}, \tag{21.6}$$

$$\mathbf{u} \equiv u_1\mathbf{e}_1 + u_2\mathbf{e}_2 + u_3\mathbf{e}_3 = h_1\frac{D\xi_1}{Dt}\mathbf{e}_1 + h_2\frac{D\xi_2}{Dt}\mathbf{e}_2 + h_3\frac{D\xi_3}{Dt}\mathbf{e}_3, \tag{21.7}$$

$$\alpha \equiv \frac{1}{\rho}, \tag{21.8}$$

$$\mathscr{E} = \mathscr{E}(\alpha, \eta) \quad \Rightarrow \quad p \equiv -\frac{\partial \mathscr{E}(\alpha, \eta)}{\partial \alpha}, \quad T \equiv \frac{\partial \mathscr{E}(\alpha, \eta)}{\partial \eta}. \tag{21.9}$$

In these equations:

- (21.1)–(21.3) are the three component equations of momentum.
- (21.4) and (21.5) are the mass-continuity and thermodynamic-energy equations, respectively.
- (21.6)–(21.9) are the definitions of material derivative (D/Dt), velocity ($\mathbf{u}$), specific volume (α), and internal energy ($\mathscr{E}$), respectively.
- (21.9) also defines pressure (p) and density (ρ) diagnostically from internal energy ($\mathscr{E}$).

Terms labelled 'deep' (displayed in red) are absent when the shallow-fluid approximation is made, and those in blue are absent when the hydrostatic or quasi-hydrostatic approximation is made. Writing the equations in this manner allows us to simultaneously consider four equation sets rather than having to examine each one individually.

For a *deep* fluid, and by definition of an axial-orthogonal-curvilinear coordinate system, the deep metric factors satisfy

$$h_i = h_i(\xi_2, \xi_3), \quad i = 1, 2, 3. \tag{21.10}$$

For the special case of a *shallow* fluid, (21.10) then reduces to

$$h_i = h_i^S(\xi_2), \quad i = 1, 2, 3, \tag{21.11}$$

where superscript 'S' denotes evaluation at the planet's surface. Since $\partial h_i/\partial \xi_3 = \partial h_i^S/\partial \xi_3$ is then identically zero, it is seen that the 'deep' terms in (21.1)–(21.3) all automatically drop out in the shallow situation simply by using (21.11) instead of (21.10), with no need for a deep/shallow switch. As we will see, the non-hydrostatic terms all drop out in the present context of constructing exact, steady solutions, irrespective of whether the fluid is deep or shallow.

21.3 SIMPLIFICATION FOR STEADY, AXIALLY SYMMETRIC FLOW

For tractability reasons, steady, axially symmetric (i.e. zonally symmetric) solutions are sought such that

$$u_1 = u_1(\xi_2, \xi_3), \quad u_2 = u_3 = 0, \tag{21.12}$$

$$\rho = \rho(\xi_2, \xi_3), \quad \alpha = \alpha(\xi_2, \xi_3), \quad p = p(\xi_2, \xi_3), \quad T = T(\xi_2, \xi_3), \quad \Phi = \Phi(\xi_3). \tag{21.13}$$

From (21.6), (21.12) and (21.13), $D/Dt \equiv 0$ wherever the material derivative appears in prognostic equations (21.1)–(21.5). Exploiting this, together with functional forms (21.12) and (21.13), three of the prognostic equations are trivially satisfied, leaving only the two constraint equations

$$\frac{\partial h_1}{\partial \xi_2}\left(\frac{u_1}{h_1}+2\Omega\right)u_1 = \frac{1}{\rho}\frac{\partial p}{\partial \xi_2}, \tag{21.14}$$

$$\underbrace{\frac{\partial h_1}{\partial \xi_3}\left(\frac{u_1}{h_1}+2\Omega\right)u_1}_{\text{deep}} = \frac{1}{\rho}\frac{\partial p}{\partial \xi_3}+\frac{d\Phi}{d\xi_3}. \tag{21.15}$$

These two equations originate from the meridional and vertical component equations (21.2) and (21.3) of momentum, respectively. The 'deep' (red) terms in (21.15) are automatically absent when the shallow-fluid approximation is made due to (21.11), with the remaining two terms representing hydrostatic balance. The quartet of equation sets thus degenerates to a duo; deep or shallow according to whether $\partial h_1/\partial \xi_3$ is non-zero or zero, respectively.

Following White and Staniforth (2008) and Staniforth and White (2011), let

$$U \equiv \left(\frac{u_1}{h_1}+2\Omega\right)u_1, \tag{21.16}$$

where U has the dimensions of acceleration. With use of (21.16), (21.14) and (21.15) may be rewritten as

The Horizontal and Vertical Balance Equations

$$\frac{\partial h_1}{\partial \xi_2}\rho U = \frac{\partial p}{\partial \xi_2}, \tag{21.17}$$

$$\underbrace{\frac{\partial h_1}{\partial \xi_3}\rho U}_{\text{deep}} = \frac{\partial p}{\partial \xi_3}+\frac{d\Phi}{d\xi_3}\rho. \tag{21.18}$$

In contradistinction to the analyses of White and Staniforth (2008), Staniforth and White (2011), and Staniforth and Wood (2013), the ideal-gas law has *not* been assumed in the derivation of (21.17) and (21.18). These two equations therefore apply equally well for atmospheric flow *and* for oceanic flow. They are the key equations of the methodology employed herein for the construction of exact, steady solutions.

21.4 COMPATIBILITY CONSTRAINTS FOR BALANCE

In (21.17) and (21.18):

- $\Phi = \Phi(\xi_3)$ is a prescribed function that defines gravity.[1]
- Metric factor $h_1 = h_1(\xi_2, \xi_3)$ is a consequence of the specification of the assumed axial-orthogonal-curvilinear coordinate system.

Equations (21.17) and (21.18) can therefore be viewed as being a coupled pair of first-order, non-linear, differential equations that constrain the three dependent variables U, ρ, and p. *Specifying any one of these three dependent variables in an arbitrary way, the other two may be obtained from*

[1] For dynamical-consistency reasons, the precise functional form of $\Phi = \Phi(\xi_3)$ depends upon whether the fluid is deep or shallow; see Chapters 7 and 8. For the special case of spherical geometry, the deep and shallow functional forms are explicitly given by (8.167) and (8.169), respectively.

(21.17) *and* (21.18) *as a direct consequence.* This is the essence of the methodology employed here for constructing exact steady solutions.

An alternative way of viewing the pair of constraint equations (21.17) and (21.18) between the *three* dependent variables (U, ρ, and p) is that they also lead to three *compatibility constraints* between three pairs of dependent variables (namely ρ and U, p and ρ, and U and p). As we will see, this facilitates the development of various options for the construction of exact, steady solutions.

Explicitly, the three compatibility constraints are obtained by:

1. Cross differentiating (21.17) and (21.18) to eliminate p. This gives:

The First Compatibility Constraint

$$\frac{\partial}{\partial \xi_3}\left[\frac{\partial h_1}{\partial \xi_2}(\rho U)\right] - \underbrace{\frac{\partial}{\partial \xi_2}\left[\frac{\partial h_1}{\partial \xi_3}(\rho U)\right]}_{\text{deep}} = -\frac{d\Phi}{d\xi_3}\frac{\partial \rho}{\partial \xi_2}, \tag{21.19}$$

which relates ρ and U.

The 'deep' (red term) in (21.19) is automatically absent in the shallow case, since then $h_1 = h_1^S(\xi_2)$.

2. Eliminating ρU instead from (21.17) and (21.18). This gives:

The Second Compatibility Constraint

$$\frac{\partial (h_1, p)}{\partial (\xi_2, \xi_3)} = -\frac{\partial h_1}{\partial \xi_2}\frac{d\Phi}{d\xi_3}\rho, \tag{21.20}$$

which relates p and ρ.

In (21.20)

$$\frac{\partial (h_1, p)}{\partial (\xi_2, \xi_3)} \equiv \frac{\partial h_1}{\partial \xi_2}\frac{\partial p}{\partial \xi_3} - \underbrace{\frac{\partial h_1}{\partial \xi_3}\frac{\partial p}{\partial \xi_2}}_{\text{deep}} \tag{21.21}$$

is the Jacobian of h_1 and p with respect to ξ_2 and ξ_3. The 'deep' (red term) in (21.21) is automatically absent in the shallow case.

3. Eliminating ρ instead from (21.17) and (21.18). This gives:

The Third Compatibility Constraint

$$U = -\left[\frac{\partial (h_1, p)}{\partial (\xi_2, \xi_3)}\right]^{-1}\frac{d\Phi}{d\xi_3}\frac{\partial p}{\partial \xi_2}, \tag{21.22}$$

which relates U and p, where the Jacobian $\partial(h_1, p)/\partial(\xi_2, \xi_3)$ is defined by (21.21).

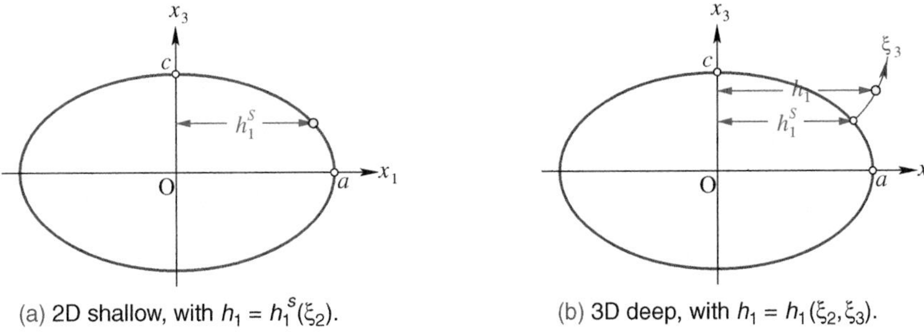

Figure 21.1 A spheroidal planet in an $x_1 - x_3$ meridional plane for (a) 2D shallow flow, with $h_1 = h_1^S(\xi_2)$; and (b) 3D deep flow, with $h_1 = h_1(\xi_2, \xi_3)$. A typical, arbitrary point is depicted on both panels as a yellow circle.

21.5 A CHANGE OF DEPENDENT VARIABLE

In Chapter 20, five families of exact steady solutions for the 2D shallow-water equations were constructed in terms of *shallow* metric factor $h_1^S(\xi_2)$. As discussed in Section 20.4, the physical significance of $h_1^S(\xi_2)$ is that it measures perpendicular distance of a planet's surface from its rotation axis. See Fig. 20.2 for a pictorial depiction of this. Things are, however, a bit more complicated for the construction of exact, steady solutions of the 3D dynamical-core equations. The corresponding metric factor in *deep* geometry depends not only on meridional coordinate ξ_2 but also on vertical coordinate ξ_3, that is, $h_1 = h_1(\xi_2, \xi_3)$ instead of $h_1 = h_1^S(\xi_2)$. See Fig. 21.1 for a graphical depiction of h_1. For 2D shallow flow (and also for 3D shallow flow) the fluid depth – see Fig. 21.1a – is sufficiently small that vertical variation of h_1 is negligible, that is, $h_1 = h_1^S(\xi_2)$. Contrastingly, for 3D deep flow – see Fig. 21.1b – this is no longer the case, that is, $h_1 = h_1(\xi_2, \xi_3)$.

As discussed in White and Staniforth (2008), there are two preferred directions in the problem. The first is that of gravity; this acts in a direction normal to geopotential surfaces (i.e. in the vertical with ξ_3 varying for fixed ξ_1 and ξ_2). The second is that of the planet's rotation axis, and it is planetary rotation that naturally brings in perpendicular distance from this axis, as measured by metric factor h_1. Now these two directions are not orthogonal to one another, except at the planet's equator. It turns out, however, that it can be convenient for the construction of exact, steady solutions of the 3D dynamical-core equations to use a mixed representation based on these two non-orthogonal directions. We now prepare the way for this.

Consider a generic scalar variable $F = F(\xi_2, \xi_3)$ where, for example, F could be pressure p. The two independent variables here are ξ_2 and ξ_3 of an axial-orthogonal-curvilinear coordinate system (ξ_1, ξ_2, ξ_3). Assuming differentiability of $F = F(\xi_2, \xi_3)$, we can compute its two partial derivatives $\partial F(\xi_2, \xi_3)/\partial\xi_2$ and $\partial F(\xi_2, \xi_3)/\partial\xi_3$, where:

- $\partial F(\xi_2, \xi_3)/\partial\xi_2$ is the derivative of F with respect to ξ_2 for fixed ξ_3; and, similarly,
- $\partial F(\xi_2, \xi_3)/\partial\xi_3$ is the derivative of F with respect to ξ_3 for fixed ξ_2.

Now consider the particular functional form $F = F[h_1(\xi_2, \xi_3), \xi_3] = F(h_1, \xi_3)$. This is a special case of the more general functional form $F = F(\xi_2, \xi_3)$ or (viewed another way), it is replacement of independent variable ξ_2 by surrogate independent variable $h_1(\xi_2, \xi_3)$. Assuming differentiability of $F = F(h_1, \xi_3)$, we can compute its two partial derivatives $\partial F(h_1, \xi_3)/\partial h_1$ and $\partial F(h_1, \xi_3)/\partial\xi_3$, where:

- $\partial F(h_1, \xi_3)/\partial h_1$ is the derivative of F with respect to h_1 for fixed ξ_3; and, similarly,
- $\partial F(h_1, \xi_3)/\partial\xi_3$ is the derivative of F with respect to ξ_3 for fixed h_1.

Using the chain rule for partial differentiation, we now relate the two pairs of partial derivatives to one another, that is, we relate $\partial F(\xi_2,\xi_3)/\partial\xi_2$ and $\partial F(\xi_2,\xi_3)/\partial\xi_3$ to $\partial F(h_1,\xi_3)/\partial h_1$ and $\partial F(h_1,\xi_3)/\partial\xi_3$. Thus

$$\frac{\partial F(\xi_2,\xi_3)}{\partial\xi_2} = \frac{\partial F[h_1(\xi_2,\xi_3),\xi_3]}{\partial h_1}\frac{\partial h_1(\xi_2,\xi_3)}{\partial\xi_2} = \frac{\partial F(h_1,\xi_3)}{\partial h_1}\frac{\partial h_1}{\partial\xi_2}, \tag{21.23}$$

$$\begin{aligned}\frac{\partial F(\xi_2,\xi_3)}{\partial\xi_3} &= \frac{\partial F[h_1(\xi_2,\xi_3),\xi_3]}{\partial h_1}\frac{\partial h_1(\xi_2,\xi_3)}{\partial\xi_3} + \frac{\partial F[h_1(\xi_2,\xi_3),\xi_3]}{\partial\xi_3}\\ &= \frac{\partial F(h_1,\xi_3)}{\partial h_1}\frac{\partial h_1}{\partial\xi_3} + \frac{\partial F(h_1,\xi_3)}{\partial\xi_3},\end{aligned} \tag{21.24}$$

where the last expressions in (21.23) and (21.24) are abbreviated forms of the middle ones.

With use of (21.23) and (21.24), the Jacobian defined by (21.21) may be rewritten in terms of $\partial p(h_1,\xi_3)/\partial h_1$ and $\partial p(h_1,\xi_3)/\partial\xi_3$, and simplified. Thus

$$\begin{aligned}\frac{\partial(h_1,p)}{\partial(\xi_2,\xi_3)} &\equiv \underbrace{\frac{\partial h_1(\xi_2,\xi_3)}{\partial\xi_3}\frac{\partial p(\xi_2,\xi_3)}{\partial\xi_2}}_{\text{deep}} - \frac{\partial h_1(\xi_2,\xi_3)}{\partial\xi_2}\frac{\partial p(\xi_2,\xi_3)}{\partial\xi_3}\\ &= \underbrace{\cancel{\frac{\partial h_1(\xi_2,\xi_3)}{\partial\xi_3}\frac{\partial p(h_1,\xi_3)}{\partial h_1}\frac{\partial h_1(\xi_2,\xi_3)}{\partial\xi_2}}}_{\text{deep}}\\ &\quad - \frac{\partial h_1(\xi_2,\xi_3)}{\partial\xi_2}\left[\cancel{\frac{\partial p(h_1,\xi_3)}{\partial h_1}\frac{\partial h_1(\xi_2,\xi_3)}{\partial\xi_3}} + \frac{\partial p(h_1,\xi_3)}{\partial\xi_3}\right]\\ &= -\frac{\partial h_1}{\partial\xi_2}\frac{\partial p(h_1,\xi_3)}{\partial\xi_3}.\end{aligned} \tag{21.25}$$

The two crossed-out terms in (21.25) essentially disappear regardless of whether the fluid is deep or shallow. For the deep case, they are non-zero but cancel; and for the shallow case, they are both identically zero (since then $\partial h_1/\partial\xi_3 = \partial h_1^S(\xi_2)/\partial\xi_3 \equiv 0$) and are therefore anyway absent.

Using (21.23) and (21.24), balance equations (21.17) and (21.18) may be rewritten as

$$\frac{\partial h_1}{\partial\xi_2}\rho U = \frac{\partial p(\xi_2,\xi_3)}{\partial\xi_2} = \frac{\partial p(h_1,\xi_3)}{\partial h_1}\frac{\partial h_1}{\partial\xi_2}, \tag{21.26}$$

$$\underbrace{\frac{\partial h_1}{\partial\xi_3}}_{\text{deep}}\rho U = \frac{\partial p(\xi_2,\xi_3)}{\partial\xi_3} + \frac{d\Phi}{d\xi_3}\rho = \frac{\partial p(h_1,\xi_3)}{\partial h_1}\frac{\partial h_1}{\partial\xi_3} + \frac{\partial p(h_1,\xi_3)}{\partial\xi_3} + \frac{d\Phi}{d\xi_3}\rho. \tag{21.27}$$

For later use, it is convenient to rewrite these two equations. Dividing (21.26) through by $\partial h_1/\partial\xi_2$ simplifies it to:

The Horizontal Balance Equation

$$\rho U = \frac{\partial p\left(h_1, \xi_3\right)}{\partial h_1}. \tag{21.28}$$

Insertion of (21.28) into (21.27) eliminates ρU in favour of p, and rearranging the resulting equation then yields:

The Vertical Balance Equation

$$\frac{d\Phi}{d\xi_3}\rho = \underbrace{\cancel{\frac{\partial h_1}{\partial \xi_3}\frac{\partial p\left(h_1, \xi_3\right)}{\partial h_1}}}_{\text{deep}} - \cancel{\frac{\partial p\left(h_1, \xi_3\right)}{\partial h_1}\frac{\partial h_1}{\partial \xi_3}} - \frac{\partial p\left(h_1, \xi_3\right)}{\partial \xi_3} = -\frac{\partial p\left(h_1, \xi_3\right)}{\partial \xi_3}. \tag{21.29}$$

The two crossed-out terms in (21.29) disappear regardless of whether the fluid is deep or shallow.

Equations (21.28) and (21.29) are a little simpler in form than are (21.17) and (21.18), from which they were obtained. This may be exploited in the construction of exact, steady solutions.

21.6 CONSTRUCTION OF EXACT STEADY SOLUTIONS

As previously mentioned, (21.17) and (21.18) can be viewed as being a pair of *non-linear*, first-order differential equations that constrain the dependent variables U, ρ, and p. However, they can be alternatively (and conveniently here) viewed as being a pair of *linear* equations for the three dependent variables (ρU), ρ, and p. Because p appears everywhere as a differentiated quantity in the governing equations, and in particular in (21.17) and (21.18), it is natural (albeit not at all essential) to make it the specified dependent variable from which all other dependent variables are obtained. This has the virtue of not having to integrate to obtain p, which would otherwise be the case, with all other variables obtained via differentiation and algebra *only*.

21.6.1 A Natural Construction Procedure

Taking advantage of these observations leads to:

A Natural Procedure to Construct Exact Steady Non-Linear Solutions in 3D

1. Specify p as any physically meaningful (according to context), axially symmetric, differentiable function of ξ_2 and ξ_3, so that

$$p = p\left(\xi_2, \xi_3\right). \tag{21.30}$$

2. Solve (21.17) for the composite variable (ρU), that is,

$$(\rho U) = \left(\frac{\partial h_1}{\partial \xi_2}\right)^{-1} \frac{\partial p}{\partial \xi_2}, \tag{21.31}$$

where $\partial h_1/\partial \xi_2 \neq 0$ is assumed.

3. Solve (21.18) for ρ to obtain

$$\rho = \left(\frac{d\Phi}{d\xi_3}\right)^{-1} \left[\underbrace{\frac{\partial h_1}{\partial \xi_3}(\rho U)}_{\text{deep}} - \frac{\partial p}{\partial \xi_3}\right] = \left(\frac{d\Phi}{d\xi_3}\right)^{-1} \left[\underbrace{\frac{\partial h_1}{\partial \xi_3}\left(\frac{\partial h_1}{\partial \xi_2}\right)^{-1} \frac{\partial p}{\partial \xi_2}}_{\text{deep}} - \frac{\partial p}{\partial \xi_3}\right], \tag{21.32}$$

where (21.31) has been used to obtain the rightmost expression.

4. Obtain U by eliminating ρ from (21.31) and (21.32). Thus

$$U = \frac{1}{\rho}\left(\frac{\partial h_1}{\partial \xi_2}\right)^{-1} \frac{\partial p}{\partial \xi_2} = \frac{d\Phi}{d\xi_3}\left[\underbrace{\frac{\partial h_1}{\partial \xi_3}\left(\frac{\partial h_1}{\partial \xi_2}\right)^{-1} \frac{\partial p}{\partial \xi_2}}_{\text{deep}} - \frac{\partial p}{\partial \xi_3}\right]^{-1} \left(\frac{\partial h_1}{\partial \xi_2}\right)^{-1} \frac{\partial p}{\partial \xi_2}, \tag{21.33}$$

where it is assumed that the square-bracketed term is non-zero.

5. Solve definition (21.16) of U to obtain u_1, so that

$$u_1 = -\Omega h_1 + \sqrt{\Omega^2 h_1^2 + h_1 U}, \tag{21.34}$$

where U is given by (21.33). As discussed in White and Staniforth (2008), the positive root is appropriate for terrestrial applications since the quadratic term in u_1 in (21.16) is typically much smaller than the linear term.

The restriction on application to the atmosphere *only* (due to the ideal gas assumption) of previous work (Staniforth and White, 2007; White and Staniforth, 2008; Staniforth and White, 2011; Staniforth and Wood, 2013) has been removed:

- Exact, non-linear, steady, axially symmetric solutions may therefore now also be constructed for oceanic flows using the procedure just presented.

21.6.2 Inclusion of Orography or Bathymetry

As noted in Staniforth and White (2011) for the special case of spherical geometry in an atmospheric context, the preceding solution procedure also holds in the presence of an underlying (rigid) orography/bathymetry of meridionally varying form $\xi_3^B = \xi_3^B(\xi_2)$. This can be shown by first transforming the governing equations using a generalised vertical coordinate $s = s(\xi_1, \xi_2, \xi_3, t)$ such that $(\xi_1, \xi_2, \xi_3, t) \rightarrow (\xi_1, \xi_2, s, t)$; see Section 13.2.2. From (13.22),

$$\dot{s} \equiv \frac{Ds}{Dt} = \left[\frac{u_3}{h_3} - \left(\frac{\partial \xi_3}{\partial t}\right)_s - \frac{u_1}{h_1}\left(\frac{\partial \xi_3}{\partial \xi_1}\right)_s - \frac{u_2}{h_2}\left(\frac{\partial \xi_3}{\partial \xi_2}\right)_s\right] \frac{\partial s}{\partial \xi_3}, \tag{21.35}$$

where $\dot{s}$ is *generalised vertical velocity*, and partial derivatives with respect to ξ_1, ξ_2, and ξ_3 are evaluated holding s constant. Since $u_2 = u_3 = 0$ everywhere (see (21.12)), (21.35) simplifies to

$$\dot{s} \equiv \frac{Ds}{Dt} = -\left[\left(\frac{\partial \xi_3}{\partial t}\right)_s + \frac{u_1}{h_1}\left(\frac{\partial \xi_3}{\partial \xi_1}\right)_s\right]\frac{\partial s}{\partial \xi_3}. \tag{21.36}$$

This equation holds *everywhere within the domain*, including at an underlying (rigid) orography/bathymetry of the form $\xi_3^B = \xi_3^B(\xi_1, \xi_2)$ where, to respect the condition of no normal flow, $\dot{s}$ must be identically zero. Since the underlying boundary is assumed to be rigid, $\partial \xi_3^B / \partial t \equiv 0$ there. Applying (21.36) at the rigid lower boundary for *non-zero* u_1, together with the condition of no normal flow there, then gives $\partial \xi_3^B / \partial \xi_1 = 0$.[2] Thus for the general case of non-zero u_1, the no-normal-flow condition is satisfied provided that $\xi_3^B = \xi_3^B(\xi_2)$, as declared earlier.

Physically, the preceding analysis corresponds to embedding a stationary, meridionally varying material surface – namely $\xi_3^B = \xi_3^B(\xi_2)$ – within the domain of an exact, steady solution above a *flat* lower boundary ($\xi_3 = a =$ constant) in such a way that:

- The governing equations are respected.
- No normal flow at the material surface is satisfied.

Doing so then leaves the steady flow undisturbed, and the flow below the embedded material surface (i.e. below the orography/bathymetry) can then be simply ignored.

Similar considerations also apply for an imposed upper boundary located at $\xi_3 = \xi_3^T(\xi_2)$.

21.6.3 A More Compact Construction Procedure

The preceding procedure for obtaining exact, steady solutions is by no means unique. Exploiting the change of dependent variable described in Section 21.5 leads to:

An Alternative More Compact Solution Procedure

1. Specify p as any physically meaningful (according to context), axially symmetric, differentiable function of $h_1 = h_1(\xi_2, \xi_3)$ and ξ_3, so that
$$p = p\left[h_1(\xi_2, \xi_3), \xi_3\right] = p(h_1, \xi_3). \tag{21.37}$$
2. Obtain the composite variable (ρU) from horizontal-balance equation (21.28), that is,
$$(\rho U) = \frac{\partial p(h_1, \xi_3)}{\partial h_1}. \tag{21.38}$$
3. Solve vertical-balance equation (21.29) for ρ to obtain
$$\rho = -\left(\frac{d\Phi}{d\xi_3}\right)^{-1}\frac{\partial p(h_1, \xi_3)}{\partial \xi_3}. \tag{21.39}$$
4. Obtain U by eliminating ρ from (21.38) and (21.39). Thus
$$U = \frac{1}{\rho}\frac{\partial p(h_1, \xi_3)}{\partial h_1} = -\frac{d\Phi}{d\xi_3}\left[\frac{\partial p(h_1, \xi_3)}{\partial \xi_3}\right]^{-1}\frac{\partial p(h_1, \xi_3)}{\partial h_1}, \tag{21.40}$$
where it is assumed that the square-bracketed term is non-zero.

[2] For the special case $u_1 \equiv 0$, $\partial \xi_3^B / \partial \xi_1$ can be non-zero, and variation of ξ_3^B with respect to ξ_1 is then allowed.

5. Solve definition (21.16) of U to obtain u_1, so that

$$u_1 = -\Omega h_1 + \sqrt{\Omega^2 h_1^2 + h_1 U}, \tag{21.41}$$

where U is given by (21.40).

21.7 A GENERALISED THERMAL-WIND EQUATION

Before giving some illustrative examples for the construction of exact, steady solutions, we embark on a slight digression. In the course of constructing some exact, non-separable, steady solutions in *spherical* geometry, White and Staniforth (2008) developed a generalised form of the classical thermal-wind equation for zonal atmospheric flow. The purpose of this section is to relate the present analysis to theirs, and to further generalise their thermal-wind equation from *spherical* geometry to *spheroidal* geometry. This is accomplished in a manner similar to that employed in Staniforth and Wood (2013).

From compatibility constraint (21.19):

$$\frac{\partial h_1}{\partial \xi_2}\frac{\partial}{\partial \xi_3}(\rho U) - \underbrace{\frac{\partial h_1}{\partial \xi_3}\frac{\partial}{\partial \xi_2}(\rho U)}_{\text{deep}} + \cancel{\frac{\partial^2 h_1}{\partial \xi_2 \partial \xi_3}(\rho U)} - \underbrace{\cancel{\frac{\partial^2 h_1}{\partial \xi_2 \partial \xi_3}(\rho U)}}_{\text{deep}} = -\frac{d\Phi}{d\xi_3}\frac{\partial \rho}{\partial \xi_2}, \tag{21.42}$$

where the two crossed-out terms mutually cancel in deep geometry and are altogether absent in shallow geometry. The thermal-wind equation, as the name implies, involves temperature, T. This variable, which is otherwise absent, enters the problem via an equation of state. In the atmospheric context examined in White and Staniforth (2008), the ideal-gas law,

$$p = \rho R T, \tag{21.43}$$

was employed, where R is the gas constant. Other equations of state may, however, be used instead. This would certainly be appropriate in an oceanic context, for which the ideal-gas law is certainly *not* ideal!

Elimination of ρ between (21.42) and (21.43), with use of compatibility condition (21.22), then leads to

$$\begin{aligned} p\left[\left(\frac{\partial h_1}{\partial \xi_2}\frac{\partial}{\partial \xi_3} - \underbrace{\frac{\partial h_1}{\partial \xi_3}\frac{\partial}{\partial \xi_2}}_{\text{deep}}\right)\left(\frac{U}{T}\right) + \frac{d\Phi}{d\xi_3}\frac{\partial}{\partial \xi_2}\left(\frac{1}{T}\right)\right] \\ = -\frac{1}{T}\left[U\left(\frac{\partial h_1}{\partial \xi_2}\frac{\partial p}{\partial \xi_3} - \underbrace{\frac{\partial h_1}{\partial \xi_3}\frac{\partial p}{\partial \xi_2}}_{\text{deep}}\right) + \frac{d\Phi}{d\xi_3}\frac{\partial p}{\partial \xi_2}\right] \\ = -\frac{1}{T}\left[U\frac{\partial\left(h_1, p\right)}{\partial\left(\xi_2, \xi_3\right)} + \frac{d\Phi}{d\xi_3}\frac{\partial p}{\partial \xi_2}\right] = 0. \end{aligned} \tag{21.44}$$

Assuming that p is non-zero, this reduces to

$$\Bigg(\frac{\partial h_1}{\partial \xi_2}\frac{\partial}{\partial \xi_3} - \underbrace{\frac{\partial h_1}{\partial \xi_3}\frac{\partial}{\partial \xi_2}}_{\text{deep}}\Bigg)\left[\frac{U(\xi_2,\xi_3)}{T(\xi_2,\xi_3)}\right] + \frac{d\Phi}{d\xi_3}\frac{\partial}{\partial \xi_2}\left[\frac{1}{T(\xi_2,\xi_3)}\right] = 0, \tag{21.45}$$

where the assumed functional dependence of U and T on ξ_2 and ξ_3 has been made explicit. Equation (21.45) corresponds to Staniforth and Wood (2013)'s (31), with $\mathfrak{R}$ set to h_1 therein. When U and T have the particular functional form $F = F\left[h_1(\xi_2,\xi_3),\xi_3\right] = F\left(h_1,\xi_3\right)$ discussed in Section 21.5, and with use of transformation relations (21.23) and (21.24), (21.45) may be more compactly written as

$$\frac{\partial}{\partial \xi_3}\left[\frac{U(h_1,\xi_3)}{T(h_1,\xi_3)}\right] + \frac{d\Phi}{d\xi_3}\frac{\partial}{\partial h_1}\left[\frac{1}{T(h_1,\xi_3)}\right] = 0. \tag{21.46}$$

This equation corresponds to Staniforth and Wood (2013)'s (32), with $\mathfrak{R}$ set to h_1 therein.

Now in deep spherical geometry, with spherical-polar coordinates $(\xi_1,\xi_2,\xi_3) = (\lambda,\phi,r)$,

$$\left(h_1,h_2,h_3\right) = (r\cos\phi, r, 1), \tag{21.47}$$

from (5.15). Insertion of (21.47) into (21.45) then yields, after some manipulation,

$$\left(\sin\phi\frac{\partial}{\partial r} + \frac{\cos\phi}{r}\frac{\partial}{\partial \phi}\right)\left(\frac{U}{T}\right) = -\frac{1}{r}\frac{d\Phi}{dr}\frac{\partial}{\partial \phi}\left(\frac{1}{T}\right). \tag{21.48}$$

This recovers White and Staniforth (2008)'s *deep*, generalised, thermal-wind equation (25) in spherical geometry, with deep geopotential $\Phi(r) = -ga^2/r$ (cf. (8.167)). We can thus conclude that (21.45) corresponds to the generalisation from spherical to spheroidal geometry of White and Staniforth (2008)'s deep, generalised, thermal-wind equation (25).

Similarly, in shallow spherical geometry, with spherical-polar coordinates $(\xi_1,\xi_2,\xi_3) = (\lambda,\phi,r)$,

$$\left(h_1,h_2,h_3\right) = (a\cos\phi, a, 1), \tag{21.49}$$

from (5.72). Insertion of (21.49) into (21.45) then yields, after some manipulation,

$$\frac{\partial}{\partial r}\left(\frac{U}{T}\right) = \frac{1}{a\sin\phi}\frac{d\Phi}{dr}\frac{\partial}{\partial \phi}\left(\frac{1}{T}\right). \tag{21.50}$$

This recovers White and Staniforth (2008)'s *shallow*, generalised, thermal-wind equation (28) in spherical geometry, with shallow geopotential $\Phi(r) = g_a(r-a)$ (cf. (8.169) herein).

The reader interested in further discussion of the generalised thermal-wind equation in spherical geometry is referred to White and Staniforth (2008), and in pseudo-Cartesian geometry to Staniforth and Wood (2013). Both of these analyses are limited to the atmospheric context due to use of the ideal-gas law. However, with an appropriate equation of state, the methodology described earlier may also be applied in an oceanic context. Indeed, and as noted in White and Staniforth (2008), a thermal-wind equation for zonal flow somewhat similar to that developed earlier was derived in Colin de Verdière and Schopp (1994) for incompressible flow in an oceanic context.

21.8 THREE ILLUSTRATIVE EXAMPLES

The preceding theory can be employed in a variety of different ways to obtain exact, steady solutions in 3D. Three illustrative examples are now described of increasing complexity. Simple solutions, such as solid-body rotation, are very helpful during the early stages of numerical model development as they help to identify and address fundamental coding and algorithmic flaws. More complicated solutions then facilitate further model development and validation.

We begin by illustrating how the five families of exact, steady solutions of the 2D shallow-water equations constructed in Chapter 20 may be adapted for use in 3D. We do so in two different ways. The first (in Section 21.8.1) makes no assumption regarding the functional form of an equation of state, whereas the second (in Section 21.8.2) assumes the ideal-gas law.

Our third example (in Section 21.8.3) is more complicated. It is drawn from Staniforth and Wood (2013) and related work, and again assumes the ideal-gas law.

21.8.1 Incompressible Flow

To recapitulate, the starting point for the development of the exact, 2D, steady solutions of Chapter 20 was the prescription of a functional form (including various parameters) for zonal velocity component $u_1 = u_1\left[h_1^S(\xi_2)\right] = u_1\left(h_1^S\right)$. The corresponding free-surface height $H = H(\xi_2)$ was then constructed to satisfy the balancing constraint – see (20.9) with use of definition (21.16) and the chain rule for differentiation – so that

$$U\left(h_1^S\right) \equiv \left[\frac{u_1\left(h_1^S\right)}{h_1^S} + 2\Omega\right] u_1\left(h_1^S\right) = \left(\frac{dh_1^S}{d\xi_2}\right)^{-1} \frac{d}{d\xi_2}(gH) = \frac{d}{dh_1^S}(gH). \tag{21.51}$$

In this equation, the *composite variable* $(gH) \equiv g(\xi_2)H(\xi_2)$ is expressed as a function of h_1^S *only* – instead of as a function of ξ_2 – so that $(gH) = (gH)\left(h_1^S\right)$. Satisfaction of constraint (21.51) could, however, have been accomplished the other way around, that is, by first prescribing composite variable (gH) as a function of h_1^S, and then deducing $U\left(h_1^S\right)$ and $u_1\left(h_1^S\right)$ from it via (21.51) and (21.16). This latter methodology is analogous to that described in Section 21.6. An advantage of doing things this way is that it only involves differentiation and algebra, without the complication (encountered in Chapter 20) of integration. For present purposes it does not matter which way around the construction was done. It only matters that it was done and that both u_1 and (gH) are known functions of h_1^S that together satisfy constraint (21.51).

For the 2D shallow-water equations, we only had to deal with the *single* (horizontal) constraint equation (21.51). Things are, however, generally more complicated in 3D since we instead have to satisfy *two* constraint equations; one horizontal, the other vertical. From (21.28) and (21.29), these may be written as

$$\rho U = \frac{\partial p(h_1, \xi_3)}{\partial h_1}, \tag{21.52}$$

$$\rho\frac{d\Phi}{d\xi_3} = -\frac{\partial p(h_1, \xi_3)}{\partial \xi_3}. \tag{21.53}$$

Furthermore, p in (21.52) and (21.53) is a function of *two* independent variables – namely h_1 and ξ_3 – whereas composite variable (gH) in (21.51) is a function of the *single* independent variable h_1^S. Having set ourselves the goal of adapting the exact, 2D, steady solutions of Chapter 20 to 3D, how should we proceed?

One way forward is to simplify things by restricting attention to flows for which density varies only in the vertical direction, and to decompose pressure as the sum of a hydrostatic term plus a perturbation that only depends on h_1. In other words, we seek solutions such that

$$\rho = \overline{\rho}(\xi_3), \quad p = \overline{p}(\xi_3) + p'(h_1), \tag{21.54}$$

where

$$\overline{\rho}(\xi_3)\frac{d\Phi}{d\xi_3} = -\frac{d\overline{p}(\xi_3)}{d\xi_3} \tag{21.55}$$

represents hydrostatic balance between $\overline{\rho}(\xi_3)$ and $\overline{p}(\xi_3)$. Proceeding this way enhances analytic tractability. Insertion of (21.54) and (21.55) into the two constraint equations (21.52) and (21.53) shows that (21.53) is identically satisfied, and that (21.52) transforms to

$$U(h_1, \xi_3) = \frac{1}{\overline{\rho}(\xi_3)}\frac{dp'(h_1)}{dh_1}. \tag{21.56}$$

This manoeuvre has nominally reduced *two* constraint equations (namely (21.52) and (21.53)) to the *single* constraint equation (21.56), albeit doing so has introduced some auxiliary equations (namely (21.54) and (21.55)). We are almost where we wish to be, but not quite. Comparing the functional form of (21.56) with that of (21.51), we see that U depends on *two* independent variables (h_1 and ξ_3) in (21.56) but only on a *single* one (h_1^S) in (21.51). To complete the analogy between (21.56) and (21.51), we therefore additionally make the *incompressible* assumption[3]

$$\rho = \overline{\rho}_0 = \text{constant}. \tag{21.57}$$

This assumption has two consequences:

1. It allows explicit integration of hydrostatic equation (21.55) to obtain

$$\overline{p}(\xi_3) = \overline{p}_0 - \overline{\rho}_0\Phi(\xi_3), \tag{21.58}$$

where $\overline{p}_0$ is the constant of integration; this constant provides a baseline value for pressure. Given a functional form $\Phi = \Phi(\xi_3)$ for the geopotential, plus values for the parameters $\overline{\rho}_0$ and $\overline{p}_0$, the unperturbed pressure $\overline{p}(\xi_3)$ is then a known function.
2. It eliminates the dependence of U on vertical variable ξ_3 – that is, $U(h_1, \xi_3) \to U(h_1)$. This (as we will see) then allows the five families of exact, 2D, steady solutions developed in Chapter 20 to be used to construct analogous families of exact, steady solutions in 3D.

Insertion of incompressibility assumption (21.57) into constraint equation (21.56) simplifies it to

$$U(h_1) = \frac{d}{dh_1}\left[\frac{p'(h_1)}{\overline{\rho}_0}\right]. \tag{21.59}$$

[3] This assumption could, of course, have been immediately introduced into (21.54) and (21.55), but it is instructive to delay this until its influence and benefit become clearer.

Comparison of (21.59) with (21.51) shows that they formally share the same functional form when

$$gH \to \frac{p'}{\overline{\rho}_0}, \quad h_1^S \to h_1. \tag{21.60}$$

This means that:

- The five 2D solution families constructed in Chapter 20 can be adapted to give a corresponding solution family in 3D.

For example, consider the equatorial and mid-latitude jet solutions constructed at the beginning of Section 20.5. These 2D solutions are defined by (20.14) and (20.19), that is, by

$$u_1\left(h_1^S\right) = u_0\left(\frac{h_1^S}{a}\right)^m\left[1 - A^n\left(\frac{h_1^S}{a}\right)^n\right], \tag{21.61}$$

$$\begin{aligned}(gH)\left(h_1^S\right) = &g_0H_0 + u_0^2\left(\frac{h_1^S}{a}\right)^{2m}\left[\frac{1}{2m} - \frac{2A^n}{(2m+n)}\left(\frac{h_1^S}{a}\right)^n + \frac{A^{2n}}{(2m+2n)}\left(\frac{h_1^S}{a}\right)^{2n}\right] \\ &+ 2\Omega a u_0\left(\frac{h_1^S}{a}\right)^{m+1}\left[\frac{1}{m+1} - \frac{A^n}{m+n+1}\left(\frac{h_1^S}{a}\right)^n\right],\end{aligned} \tag{21.62}$$

where:

- a is a representative scaling parameter, with dimension of length.
- u_0 is a representative value for $u_1\left(h_1^S\right)$.
- $m \geq 1$ and $n \geq 1$ are integers.

The equatorial and mid-latitude jet solutions are obtained by setting these parameters to appropriate values. In particular, the (very) special case of *solid-body rotation* is obtained by setting $m = 1, A = 0$ in (21.61) and (21.62).

From (21.12), (21.54), and (21.58)–(21.62), the analogous exact, steady solution of the deep, 3D equations for incompressible flow is thus:

$$u_1\left(h_1\right) = u_0\left(\frac{h_1}{a}\right)^m\left[1 - A^n\left(\frac{h_1}{a}\right)^n\right], \quad u_2 = u_3 = 0, \tag{21.63}$$

$$\overline{p}\left(\xi_3\right) = \overline{p}_0 - \overline{\rho}_0\Phi\left(\xi_3\right), \quad \rho = \overline{\rho}_0 = \text{constant}, \tag{21.64}$$

$$\begin{aligned}p'\left(h_1\right) = &\overline{\rho}_0u_0^2\left(\frac{h_1}{a}\right)^{2m}\left[\frac{1}{2m} - \frac{2A^n}{(2m+n)}\left(\frac{h_1}{a}\right)^n + \frac{A^{2n}}{(2m+2n)}\left(\frac{h_1}{a}\right)^{2n}\right] \\ &+ 2\Omega a\overline{\rho}_0u_0\left(\frac{h_1}{a}\right)^{m+1}\left[\frac{1}{m+1} - \frac{A^n}{m+n+1}\left(\frac{h_1}{a}\right)^n\right],\end{aligned} \tag{21.65}$$

$$p\left(h_1, \xi_3\right) = \overline{p}\left(\xi_3\right) + p'\left(h_1\right). \tag{21.66}$$

Since no assumption was made regarding an equation of state, this incompressible solution is valid in both atmospheric and oceanic contexts. Physically, this solution is more realistic for the oceanic case – since the ocean is incompressible to leading order, whereas the atmosphere is much less so – but it is still useful for the atmospheric case, particularly for code validation and troubleshooting purposes.

21.8.2 Atmospheric Isothermal Flow

We can be more adventurous in constructing exact, steady solutions by introducing an equation of state. This has the practical virtues of better tailoring the model problem to a situation of interest, and of more extensively exercising code and numerical algorithms. Here, following Staniforth and White (2007) but generalising from spherical to spheroidal geometry, we introduce an equation of state in the atmospheric context. Adopting the ideal-gas law

$$p = \rho R T, \tag{21.67}$$

– see (21.43) – key balance equations (21.28) and (21.29) may be rewritten as

$$U = RT \frac{\partial q\left(h_1, \xi_3\right)}{\partial h_1}, \tag{21.68}$$

$$\frac{d\Phi}{d\xi_3} = -RT \frac{\partial q\left(h_1, \xi_3\right)}{\partial \xi_3}, \tag{21.69}$$

where

$$q\left(h_1, \xi_3\right) \equiv \ln\left[p\left(h_1, \xi_3\right)\right]. \tag{21.70}$$

In (21.68) and (21.69), ρ has been eliminated in favour of T and, for convenience, p in favour of q.

By analogy with the incompressible analysis, we now seek solutions such that

$$T = \overline{T}\left(\xi_3\right), \quad q = \overline{q}\left(\xi_3\right) + q'\left(h_1\right), \tag{21.71}$$

where

$$\frac{d\Phi}{d\xi_3} = -R\overline{T}\left(\xi_3\right) \frac{d\overline{q}\left(\xi_3\right)}{d\xi_3} \tag{21.72}$$

represents hydrostatic balance between $\overline{T}(\xi_3)$ and $\overline{q}(\xi_3)$. Insertion of (21.71) and (21.72) into balance equations (21.68) and (21.69) shows that (21.69) is identically satisfied, and (21.68) transforms to

$$U\left(h_1, \xi_3\right) = RT\left(\xi_3\right) \frac{dq'\left(h_1\right)}{dh_1}. \tag{21.73}$$

Comparing the functional form of (21.73) with that of (21.51), we see that they are almost the same – but not quite, due to the dependence of T in (21.73) on ξ_3. All we have to do to make the two forms agree is to additionally make the *isothermal assumption*

$$T = \overline{T}_0 = \text{constant}, \tag{21.74}$$

and then – from (21.73) –

$$U\left(h_1\right) = \frac{d}{dh_1}\left[R\overline{T}_0 q^{'}\left(h_1\right)\right]. \tag{21.75}$$

Comparison of (21.75) with (21.51) shows that they formally have the same functional form when

$$gH \to R\overline{T}_0 q^{'}, \quad h_1^S \to h_1. \tag{21.76}$$

This means that:

- The five, 2D, solution families constructed in Chapter 20 can be adapted to give a corresponding solution family in 3D in the present context.

Setting $T = \overline{T}_0$ in hydrostatic-balance equation (21.72) and integrating leads to the explicit solution

$$\overline{q}\left(\xi_3\right) = \overline{q}_0 - \frac{\Phi\left(\xi_3\right)}{R\overline{T}_0}, \tag{21.77}$$

where $\overline{q}_0$ is the constant of integration; this provides a baseline value for the natural logarithm of pressure.

As specific examples, we again consider the equatorial and mid-latitude jet solutions constructed at the beginning of Section 20.5. These 2D solutions are defined by (21.61) and (21.62) of the present section. From (21.61), (21.62), (21.70), (21.71), (21.74), (21.76), and (21.77), the analogous, exact, steady solution of the deep 3D equations in the present context is thus:

$$u_1\left(h_1\right) = u_0\left(\frac{h_1}{a}\right)^m\left[1 - A^n\left(\frac{h_1}{a}\right)^n\right], \quad u_2 = u_3 = 0, \tag{21.78}$$

$$\overline{q}\left(\xi_3\right) = \overline{q}_0 - \frac{\Phi\left(\xi_3\right)}{R\overline{T}_0}, \quad T = \overline{T}_0 = \text{constant}, \tag{21.79}$$

$$\begin{aligned} \overline{q}^{'}\left(h_1\right) = {} & \frac{u_0^2}{R\overline{T}_0}\left(\frac{h_1}{a}\right)^{2m}\left[\frac{1}{2m} - \frac{2A^n}{(2m+n)}\left(\frac{h_1}{a}\right)^n + \frac{A^{2n}}{(2m+2n)}\left(\frac{h_1}{a}\right)^{2n}\right] \\ & + \frac{2\Omega a u_0}{R\overline{T}_0}\left(\frac{h_1}{a}\right)^{m+1}\left[\frac{1}{m+1} - \frac{A^n}{m+n+1}\left(\frac{h_1}{a}\right)^n\right]. \end{aligned} \tag{21.80}$$

$$q\left(h_1,\xi_3\right) = \overline{q}\left(\xi_3\right) + q^{'}\left(h_1\right), \quad p\left(h_1,\xi_3\right) = \exp\left[q\left(h_1,\xi_3\right)\right]. \tag{21.81}$$

For the special case of spherical geometry, this family of solutions corresponds to that given in Section 5.3 of Staniforth and White (2007).

21.8.3 Atmospheric Flow with Non-Separable Temperature

The construction procedure (21.37)–(21.41) to obtain exact, 3D, steady solutions has the great virtue that it is independent of any specification of an equation of state (i.e. the hydrodynamics is decoupled from the thermodynamics). The procedure is therefore valid in both the atmospheric context and the oceanic one. Depending upon what one wishes to accomplish, it is, however, often desirable to specialise to either the atmospheric context or to the oceanic one. This allows model problems (with associated exact, steady solutions) to be devised that are particularly relevant and useful in a specific context. To illustrate this, we describe one such exact, steady solution in the atmospheric context.[4] In this context, the ideal-gas law (21.67) is a very natural choice for the equation of state.

21.8.3.1 *Balance Equations*

Using ideal-gas law (21.67) to eliminate ρ in favour of T, key balance equations (21.28) and (21.29) may be rewritten as

$$\frac{U}{T} = R\frac{\partial q\left(h_1, \xi_3\right)}{\partial h_1}, \tag{21.82}$$

$$\frac{1}{T} = -R\left(\frac{d\Phi}{d\xi_3}\right)^{-1}\frac{\partial q\left(h_1, \xi_3\right)}{\partial \xi_3}, \tag{21.83}$$

where – from (21.70) – $q\left(h_1, \xi_3\right) \equiv \ln\left[p\left(h_1, \xi_3\right)\right]$.

21.8.3.2 *A Modified Construction Procedure*

Using ideal-gas law (21.67) and balance equations (21.82) and (21.83), construction procedure (21.37)–(21.41) may be modified to give the following sequence of steps:

1. Specify q as any physically meaningful (in the atmospheric context, since $p = \rho RT$ has been assumed), axially symmetric, differentiable function of $h_1 = h_1\left(\xi_2, \xi_3\right)$ and ξ_3, so that

$$q = q\left[h_1\left(\xi_2, \xi_3\right), \xi_3\right] = q\left(h_1, \xi_3\right). \tag{21.84}$$

2. Obtain composite variable (U/T) from (21.82), so that

$$\left(\frac{U}{T}\right) = R\frac{\partial q\left(h_1, \xi_3\right)}{\partial h_1}, \tag{21.85}$$

where $\partial q\left(h_1, \xi_3\right)/\partial h_1$ is obtained by partial differentiation of (21.84).

3. Solve (21.83) for T to obtain

$$T = -\frac{1}{R}\left(\frac{d\Phi}{d\xi_3}\right)\left[\frac{\partial q\left(h_1, \xi_3\right)}{\partial \xi_3}\right]^{-1}, \tag{21.86}$$

where $\partial q\left(h_1, \xi_3\right)/\partial \xi_3$ is obtained by partial differentiation of (21.84).

4. Obtain U by eliminating T from (21.85) and (21.86). Thus

[4] This section contains public sector information licensed under the Open Government Licence v1.0. This information has been adapted and generalised for use herein.

$$U = RT\frac{\partial q(h_1,\xi_3)}{\partial h_1} = -\frac{d\Phi}{d\xi_3}\left[\frac{\partial q(h_1,\xi_3)}{\partial \xi_3}\right]^{-1}\frac{\partial q(h_1,\xi_3)}{\partial h_1}, \tag{21.87}$$

where $\partial q(h_1,\xi_3)/\partial h_1$ and $\partial q(h_1,\xi_3)/\partial \xi_3$ are obtained by partial differentiation of (21.84).

5. Solve definition (21.16) of U to obtain u_1, so that

$$u_1 = -\Omega h_1 + \sqrt{\Omega^2 h_1^2 + h_1 U}, \tag{21.88}$$

where U is given by (21.87). As discussed in White and Staniforth (2008), the positive root is appropriate for terrestrial applications since the quadratic term in u_1 in (21.16) is typically much smaller than the linear term.

6. Obtain p and ρ from (21.67) and (21.70), so that

$$p = \exp q, \quad \rho = \frac{p}{RT}, \tag{21.89}$$

where T is given by (21.86).

21.8.3.3 *A Particular Functional Form for $q = q(h_1,\xi_3)$*

To apply the modified construction procedure, all that remains to be done is to specify a particular functional form for $q = q(h_1,\xi_3)$. As an illustrative example, and following Staniforth and Wood (2013) but generalising from spherical to spheroidal geometry, consider

$$\begin{aligned} q(h_1,\xi_3) = {} & q_0 - \frac{g_0}{R}A\left\{\exp\left[\frac{\Gamma}{T_0}(\xi_3 - a)\right] - 1\right\} \\ & + \frac{g_0}{R}(\xi_3 - a)\exp\left[-\left(\frac{\xi_3 - a}{bH_0}\right)^2\right]\left\{-B + C\left[\left(\frac{h_1}{a}\right)^k - \frac{k}{(k+2)}\left(\frac{h_1}{a}\right)^{k+2}\right]\right\}, \end{aligned}$$

$$k \geq 2. \tag{21.90}$$

In this equation (and anticipating the physical interpretation of Section 21.8.3.5):

- $q_0 \equiv \ln p_0$ is an arbitrary, constant value of the natural logarithm of pressure.
- g_0 is a constant related to gravity at the planet's surface.
- T_0 is a representative constant value of temperature.
- Γ is an assumed, constant lapse rate for temperature.
- $H_0 \equiv RT_0/g_0$ is a scale height of the atmosphere.
- b is a half-width parameter, and

$$A = \frac{1}{\Gamma}, \quad B = \frac{T_0 - T_0^P}{T_0 T_0^P}, \quad C = \left(\frac{k+2}{2}\right)\left(\frac{T_0^E - T_0^P}{T_0^E T_0^P}\right), \tag{21.91}$$

where T_0^P and T_0^E are surface values of temperature at the planet's poles and equator, respectively.

Specification (21.90) of $q(h_1,\xi_3)$ was inspired by the functional form used in Staniforth (2012) for exact solutions in z (geometric height) coordinates on a β-γ plane, which was itself inspired by that of Ullrich and Jablonowski (2012) for their channel-flow test problems in pressure-based coordinates on a β plane. The exact, 3D, steady solution that results from use of specification (21.90), expressed in spherical geometry, provides a basic state for Ullrich et al. (2014)'s idealised, baroclinic-wave test case. See what follows for further discussion of this basic state.

Further information regarding physical interpretation of the various parameters appearing in (21.90) and (21.91) is given in Section 21.8.3.5, following formal application (Section 21.8.3.4) of the modified construction procedure using specification (21.90) for $q\left(h_1,\xi_3\right)$.

21.8.3.4 *Application of the Modified Construction Procedure*

To prepare the way, partial differentiation of (21.90) with respect to h_1 and ξ_3 gives

$$\frac{\partial q\left(h_1,\xi_3\right)}{\partial h_1} = Ck\frac{g_0}{Ra}\left(\xi_3-a\right)\exp\left[-\left(\frac{\xi_3-a}{bH_0}\right)^2\right]\left[\left(\frac{h_1}{a}\right)^{k-1}-\left(\frac{h_1}{a}\right)^{k+1}\right], \tag{21.92}$$

$$\begin{aligned}\frac{\partial q\left(h_1,\xi_3\right)}{\partial \xi_3} = &-\frac{g_0}{R}A\frac{\Gamma}{T_0}\exp\left[\frac{\Gamma}{T_0}\left(\xi_3-a\right)\right]\\ &+\frac{g_0}{R}\left[1-2\left(\frac{\xi_3-a}{bH_0}\right)^2\right]\exp\left[-\left(\frac{\xi_3-a}{bH_0}\right)^2\right]\\ &\times\left\{-B+C\left[\left(\frac{h_1}{a}\right)^{k}-\frac{k}{(k+2)}\left(\frac{h_1}{a}\right)^{k+2}\right]\right\}.\end{aligned} \tag{21.93}$$

With this preparation, application of the modified construction procedure to the present illustrative example yields the following sequence of steps to obtain a specific, exact, 3D, steady solution:

1. Let q be specified by functional form (21.90).
2. Obtain composite variable (U/T) from (21.85) and (21.92), so that

$$\left(\frac{U}{T}\right)=R\frac{\partial q\left(h_1,\xi_3\right)}{\partial h_1}=Ckg_0\frac{\left(\xi_3-a\right)}{a}\exp\left[-\left(\frac{\xi_3-a}{bH_0}\right)^2\right]\left[\left(\frac{h_1}{a}\right)^{k-1}-\left(\frac{h_1}{a}\right)^{k+1}\right]. \tag{21.94}$$

3. Obtain T from (21.86), so that

$$T=-\frac{1}{R}\left(\frac{d\Phi}{d\xi_3}\right)\left[\frac{\partial q\left(h_1,\xi_3\right)}{\partial \xi_3}\right]^{-1}, \tag{21.95}$$

   where $\partial q\left(h_1,\xi_3\right)/\partial\xi_3$ is given by (21.93).
4. Obtain U from (21.94) and (21.95), so that

$$U=RT\frac{\partial q\left(h_1,\xi_3\right)}{\partial h_1}=-\frac{d\Phi}{d\xi_3}\left[\frac{\partial q\left(h_1,\xi_3\right)}{\partial \xi_3}\right]^{-1}\frac{\partial q\left(h_1,\xi_3\right)}{\partial h_1}, \tag{21.96}$$

   where $\partial q\left(h_1,\xi_3\right)/\partial h_1$ and $\partial q\left(h_1,\xi_3\right)/\partial\xi_3$ are given by (21.92) and (21.93), respectively.
5. Obtain u_1 from (21.88), so that

$$u_1=-\Omega h_1+\sqrt{\Omega^2h_1^2+h_1U}, \tag{21.97}$$

   where U is given by (21.96).
6. Obtain p and ρ from (21.89), so that

$$p=\exp q,\quad \rho=\frac{p}{RT}, \tag{21.98}$$

   where T is given by (21.95).

This six-step procedure illustrates how an exact, 3D, steady solution may be constructed from the explicit specification (21.90) of $q\left(h_1, \xi_3\right)$. Furthermore, this solution is obtained via partial differentiation and algebraic manipulation *only*, without recourse to integration. However, this implicitly assumes that a suitable specification of $q\left(h_1, \xi_3\right)$ is available for application in a particular context of interest. For particularly simple contexts, such as solid-body rotation, it is relatively straightforward to suitably specify $q\left(h_1, \xi_3\right)$. However, this is not the case more generally. For example, specification (21.90) of $q\left(h_1, \xi_3\right)$ was not obtained this way. It was instead obtained from a suitable specification of *temperature* $T = T\left(h_1, \xi_3\right)$ rather than $q = q\left(h_1, \xi_3\right)$. This then introduced the need for integration to obtain the complete solution, which, in turn, placed restrictions on the functional form of $T = T\left(h_1, \xi_3\right)$ for analytic tractability. See White and Staniforth (2008), Staniforth and White (2011), and Staniforth and Wood (2013) for details regarding how $T = T\left(h_1, \xi_3\right)$ was developed in the context of *spherical* geometry and how this ultimately led to specification (21.90) of $q\left(h_1, \xi_3\right)$.

It is important to understand that, depending on context, there are many different ways of constructing exact, steady solutions. However, *they all share the need to respect the key balance constraints imposed by the governing equations and simplifying assumptions.*

21.8.3.5 *Physical Interpretation of Parameters*

We are now in a position to physically interpret the parameters that appear in (21.90) and (21.91). Using (21.93), and by analogy with Ullrich et al. (2014)'s corresponding equation (13) for T in spherical geometry, (21.95) may be explicitly rewritten in *non-separable* form as

$$T\left(h_1, \xi_3\right) = \frac{1}{g_0}\frac{d\Phi}{d\xi_3}\left\{\widetilde{\tau}_1\left(\xi_3\right) - \widetilde{\tau}_2\left(\xi_3\right)\left[\left(\frac{h_1}{a}\right)^k - \frac{k}{(k+2)}\left(\frac{h_1}{a}\right)^{k+2}\right]\right\}^{-1}, \tag{21.99}$$

where[a]

$$\widetilde{\tau}_1\left(\xi_3\right) = A\frac{\Gamma}{T_0}\exp\left[\frac{\Gamma}{T_0}\left(\xi_3 - a\right)\right] + B\left[1 - 2\left(\frac{\xi_3 - a}{bH_0}\right)^2\right]\exp\left[-\left(\frac{\xi_3 - a}{bH_0}\right)^2\right], \tag{21.100}$$

$$\widetilde{\tau}_2\left(\xi_3\right) = C\left[1 - 2\left(\frac{\xi_3 - a}{bH_0}\right)^2\right]\exp\left[-\left(\frac{\xi_3 - a}{bH_0}\right)^2\right]. \tag{21.101}$$

[a] $\widetilde{\tau}_1\left(\xi_3\right)$ and $\widetilde{\tau}_2\left(\xi_3\right)$ in Ullrich et al. (2014) originate from similar but scaled expressions $\tau_1\left(\xi_3\right)$ and $\tau_2\left(\xi_3\right)$ that appear in Staniforth and White (2011)'s similar equation (8) for $T\left(h_1, \xi_3\right)$. To avoid possible confusion, we maintain the tilde notation here. The reason why Ullrich et al. (2014) (equivalently) rewrote Staniforth and White (2011)'s similar equation (8) as their (13) is that it simplified their parallel treatment of the deep- and shallow-atmosphere cases.

Equation (21.99) is the generalisation of Ullrich et al. (2014)'s (13) in *spherical* geometry to *spheroidal* geometry herein.

Evaluation of (21.99) at a planet's surface at $\xi_3 = a$, with use of (21.100) and (21.101), gives

$$\begin{aligned} T\left(h_1^S, \xi_3 = a\right) &= \frac{1}{g_0}\left(\frac{d\Phi}{d\xi_3}\right)_{\xi_3 = a}\left\{\widetilde{\tau}_1\left(\xi_3 = a\right) - \widetilde{\tau}_2\left(\xi_3 = a\right)\left[\left(\frac{h_1^S}{a}\right)^k - \frac{k}{(k+2)}\left(\frac{h_1^S}{a}\right)^{k+2}\right]\right\}^{-1} \\ &= \frac{1}{g_0}\left(\frac{d\Phi}{d\xi_3}\right)_{\xi_3 = a}\left\{A\frac{\Gamma}{T_0} + B - C\left[\left(\frac{h_1^S}{a}\right)^k - \frac{k}{(k+2)}\left(\frac{h_1^S}{a}\right)^{k+2}\right]\right\}^{-1}, \end{aligned} \tag{21.102}$$

where $h_1^S \equiv (h_1)_{\xi_3=a}$ is h_1 evaluated at the planet's surface. Setting the arbitrary value of g_0 to[5]

$$g_0 = \left(\frac{d\Phi}{d\xi_3}\right)_{\xi_3=a}, \tag{21.103}$$

(21.102) simplifies to

$$T\left(h_1^S, \xi_3 = a\right) = \left\{A\frac{\Gamma}{T_0} + B - C\left[\left(\frac{h_1^S}{a}\right)^k - \frac{k}{(k+2)}\left(\frac{h_1^S}{a}\right)^{k+2}\right]\right\}^{-1}. \tag{21.104}$$

For the purposes of physical interpretation, consider now the special case when $B = C = 0$. For this special case, (21.99)–(21.101) and (21.104) then reduce to

$$\widetilde{\tau}_1\left(\xi_3; B = C = 0\right) = A\frac{\Gamma}{T_0}\exp\left[\frac{\Gamma}{T_0}\left(\xi_3 - a\right)\right], \quad \widetilde{\tau}_2\left(\xi_3; B = C = 0\right) = 0, \tag{21.105}$$

$$T\left(\xi_3; B = C = 0\right) = T_0\frac{1}{g_0}\frac{d\Phi}{d\xi_3}\frac{1}{A\Gamma}\exp\left[-\frac{\Gamma}{T_0}\left(\xi_3 - a\right)\right], \tag{21.106}$$

$$T\left(\xi_3 = a; B = C = 0\right) = \frac{1}{A\Gamma}T_0, \tag{21.107}$$

respectively. Note that setting $B = C = 0$ has removed horizontal variation of T from (21.101) and (21.104). It is natural to set

$$A = \frac{1}{\Gamma}, \tag{21.108}$$

– in agreement with (21.91) – so that (21.106) and (21.107) further reduce to

$$T\left(\xi_3; B = C = 0\right) = T_0\frac{1}{g_0}\frac{d\Phi}{d\xi_3}\exp\left[-\frac{\Gamma}{T_0}\left(\xi_3 - a\right)\right], \tag{21.109}$$

$$T\left(\xi_3 = a; B = C = 0\right) = T_0. \tag{21.110}$$

From (21.110), T_0 *can now be interpreted as being a representative value of surface temperature.*

- But how may we interpret Γ physically?

Now close to the planet's surface, the vertical variation of $d\Phi/d\xi_3$ may be considered negligible. Thus, to a very good approximation and using (21.103), $d\Phi/d\xi_3 \approx \left(d\Phi/d\xi_3\right)_{\xi_3=a} = g_0$. Inserting this approximation into (21.109) then gives

$$T\left(\xi_3; B = C = 0\right) \approx T_0\exp\left[-\frac{\Gamma}{T_0}\left(\xi_3 - a\right)\right] \quad \Rightarrow \quad \left[\frac{dT\left(\xi_3; B = C = 0\right)}{d\xi_3}\right]_{\xi_3=a} \approx -\Gamma. \tag{21.111}$$

Thus Γ *may be interpreted as being a representative lapse rate of temperature at the planet's surface.*

Having examined the special case when $B = C = 0$ to physically interpret parameters T_0 and Γ, and to representatively relate A to Γ – see (21.108) – we now return to the general case for which B and C are both non-zero. The objective now is to relate parameters B and C to physical quantities.

[5] For a mildly oblate ellipsoidal planet, $g\left(\xi_2, \xi_3\right) = h_3\left(\xi_2, \xi_3\right)\left(d\Phi/d\xi_3\right)$, $h_3\left(\xi_2, \xi_3 = a\right) \approx 1$, and $g\left(\xi_2, \xi_3 = a\right) = h_3\left(\xi_2, \xi_3 = a\right)\left(d\Phi/d\xi_3\right)_{\xi_3=a} \approx \left(d\Phi/d\xi_3\right)_{\xi_3=a}$. For the special case of a spherical planet, $h_3 \equiv 1$ and $g\left(\xi_3 = a\right) = \left(d\Phi/d\xi_3\right)_{\xi_3=a}$ precisely.

Parameter	Value	Units	Description
a	6.371229×10^6	m	Mean radius of Earth
b	2	–	Half-width parameter
g_0	9.80616	$\mathrm{m\,s^{-2}}$	Gravitational acceleration at Earth's surface
k	3	–	Power used for temperature field
p_0	10^5	Pa	Surface pressure
R	287.0	$\mathrm{J\,kg^{-1}\,K^{-1}}$	Ideal gas constant
T_0^E	310	K	Surface equatorial temperature
T_0^P	240	K	Surface polar temperature
Γ	0.005	$\mathrm{K\,m^{-1}}$	Lapse rate
Ω	7.29212×10^{-5}	$\mathrm{s^{-1}}$	Earth's angular velocity

Table 21.1 Parameter values for Ullrich et al. (2014)'s exact, steady, basic state.

For an ellipsoidal planet, $h_1^S = 0$ at the planet's two poles, and $h_1^S = a$ at its equator; see Section 20.4. Evaluating (21.104) at its two poles and at its equator, with use of (21.108), thus gives

$$T_0^P \equiv T\left(h_1^S = 0, \xi_3 = a\right) = \left(\frac{1}{T_0} + B\right)^{-1}, \tag{21.112}$$

$$T_0^E \equiv T\left(h_1^S = a, \xi_3 = a\right) = \left(\frac{1}{T_0} + B - \frac{2C}{k+2}\right)^{-1}, \tag{21.113}$$

respectively. Solving these two equations for B and C then yields – in agreement with (21.91) –

$$B = \frac{T_0 - T_0^P}{T_0 T_0^P}, \quad C = \left(\frac{k+2}{2}\right)\left(\frac{T_0^E - T_0^P}{T_0^E T_0^P}\right), \tag{21.114}$$

where T_0 is an arbitrary, representative surface value for T. Following Staniforth and Wood (2013) and Ullrich et al. (2014), this arbitrary value is set to

$$T_0 = \frac{1}{2}\left(T_0^E + T_0^P\right), \tag{21.115}$$

for simplicity. Using this value in (21.114), B and C are then expressed in terms of T_0^E and T_0^P as

$$B = \frac{\left(T_0^E - T_0^P\right)}{\left(T_0^E + T_0^P\right) T_0^P}, \quad C = \left(\frac{k+2}{2}\right)\left(\frac{T_0^E - T_0^P}{T_0^E T_0^P}\right). \tag{21.116}$$

21.8.3.6 *Visualisation in Spherical Geometry*

The particular, exact, steady solution constructed in Sections 21.8.3.3 and 21.8.3.4 was expressed in spherical geometry in Staniforth and Wood (2013).[6] This solution was subsequently used as a basic state for Ullrich et al. (2014)'s idealised, baroclinic-wave test case with parameter values set to those in Table 21.1 herein.

Using

$$(\xi_1, \xi_2, \xi_3) = (\lambda, \phi, r) \quad \Rightarrow \quad (h_1, h_2, h_3) = (r\cos\phi, r, 1), \tag{21.117}$$

in general specification (21.90) of q then yields the corresponding specification

[6] This section contains public sector information licensed under the Open Government Licence v1.0.

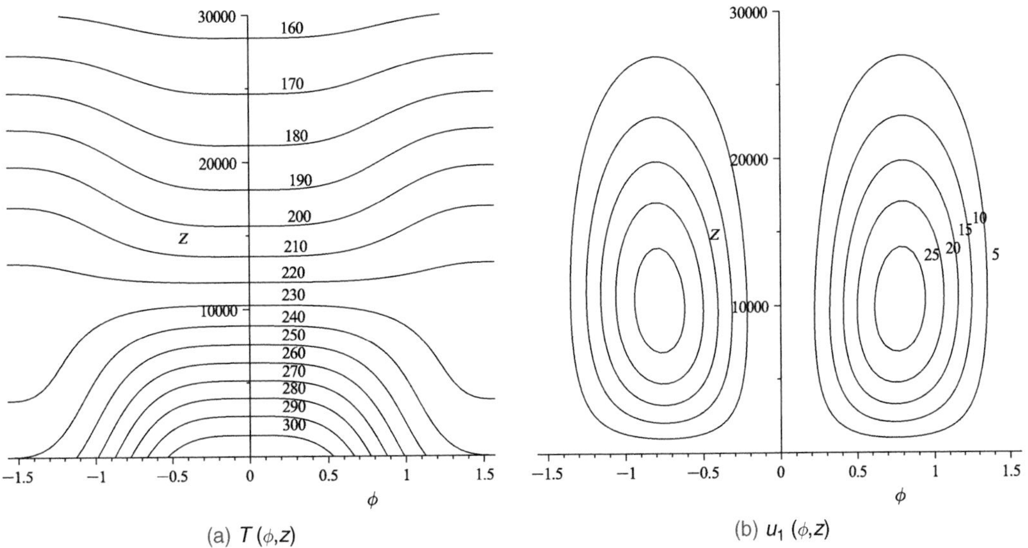

Figure 21.2 Meridional cross sections for the deep-atmosphere, non-separable, steady solution described in the text: (a) temperature $T(\phi, z)$, contour interval 10 K; and zonal wind $u_1(\phi, z)$, contour interval 5 m s^{-1}. Height $z \equiv r - a$ is in metres. From Fig. 1 of Staniforth and Wood (2013).

$$\begin{aligned} q(r\cos\phi, r) = q_0 &- \frac{g_0}{R} A \left\{ \exp\left[\frac{\Gamma}{T_0}(r-a) \right] - 1 \right\} \\ &+ \frac{g_0}{R}(r-a)\exp\left[-\left(\frac{r-a}{bH_0} \right)^2 \right] \\ &\times \left\{ -B + C\left[\left(\frac{r\cos\phi}{a} \right)^k - \frac{k}{(k+2)}\left(\frac{r\cos\phi}{a} \right)^{k+2} \right] \right\}, \end{aligned} \tag{21.118}$$

in spherical geometry. The corresponding temperature (T), zonal wind (u_1), pressure (p), and density (ρ) fields may be similarly obtained by evaluating (21.92)–(21.98) in spherical geometry using (21.117) and – from (8.167) – the (dynamically consistent) geopotential in deep spherical geometry

$$\Phi(r) = -g_0 \frac{a^2}{r}. \tag{21.119}$$

The temperature (T) and zonal wind (u_1) fields for a deep atmosphere, corresponding to the parameter values of Table 21.1, are displayed in Fig. 21.2, and some vertical profiles in Fig. 21.3. The depicted fields approximate the primary dynamic characteristics of Earth's atmosphere, with a single westerly jet of magnitude ≈ 28 m s^{-1} in each hemisphere. See Staniforth and Wood (2013) and Ullrich et al. (2014) for additional fields and further information.

21.9 CONCLUDING REMARKS

This concludes our review of the construction of exact, steady, vortical, axially symmetric solutions of the 3D, non-linear, governing equations of global, atmospheric and oceanic, dynamical cores in spheroidal geometry.

For the analogous but simpler 2D solutions constructed in Chapter 20, two criteria were obtained, respect of which guarantees stability of the solutions when subjected to arbitrary small-amplitude perturbations in 2D. It is an open question as to whether it is analytically tractable to

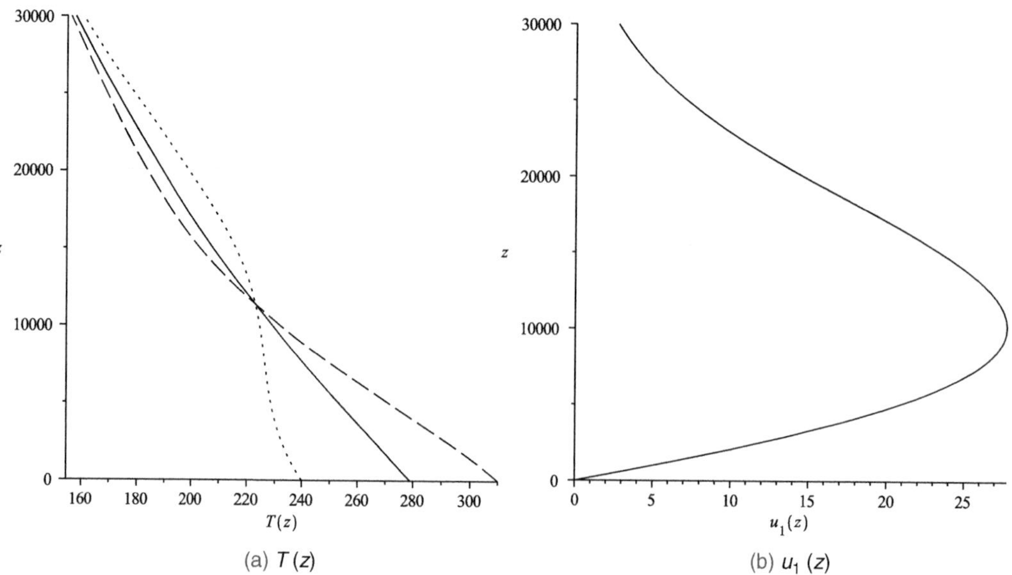

Figure 21.3 Profiles for the deep-atmosphere, non-separable, steady solution described in the text: (a) polar (dotted), mid-latitude (solid), and equatorial (dashed) temperature $T(z)$, in K; and (b) mid-latitude zonal wind $u_1(z)$, in m s^{-1}. Height $z \equiv r - a$ is in metres. From Fig. 2 of Staniforth and Wood (2013).

generalise the 2D stability analysis of Chapter 20 to 3D. Whilst this is certainly desirable, it is not, however, essential. It just means that the usefulness of the exact, steady, 3D solutions are limited to shorter integration periods.

The 2D and 3D exact, *steady* solutions constructed in Chapter 20 and the present chapter, respectively, are particularly useful for assessing the ability of dynamical cores to accurately represent large-scale horizontal and vertical balances. This is an important attribute for physically realistic atmospheric and oceanic models in spheroidal geometry. It is, however, similarly important that such models also accurately represent *unsteady* fluid flow. This challenging subject is taken up in Chapter 20 for some unsteady 2D flows, and in Chapter 21 for some unsteady 3D flows. In both contexts, the solutions are generalisations to spheroidal geometry of Rossby–Haurwitz solutions in spherical geometry.

Exact Unsteady Solutions of the Barotropic Potential Vorticity Equation over an Ellipsoid

ABSTRACT

A family of exact, unsteady, non-linear solutions of the barotropic potential vorticity (BPV) equation is obtained for confined flow between two rigid, ellipsoidal, geopotential surfaces. These solutions are a generalisation of Rossby–Haurwitz waves from the spherical context to the ellipsoidal one. They are based on the eigenfunctions of a linear, second-order, partial differential operator. Their meridional structure is that of associated Legendre functions. It is shown that such solutions are only possible for a particular model of apparent gravity. This model corresponds to a constant-density ellipsoidal planet. Approximately two thirds of the observed meridional variation of apparent gravity is represented. This can be compared to no variation whatsoever for the spherical-geopotential approximation employed in today's global atmospheric and oceanic models. Meridional variation of apparent gravity changes the propagation frequency via the introduction of dependency on both eccentricity and zonal wave number. These changes correspond to a systematic (small) increase in the beta effect. The evolution of potential vorticity (PV) can be computed independently of the pressure field. A balancing pressure is nevertheless needed in the underlying equations used to obtain the BPV equation. It can be diagnosed at any instant in time from solution of an elliptic equation. The balancing pressure is first obtained for a particular exact solution over an ellipsoid. It corresponds to the ellipsoidal generalisation of a particular solution frequently used for numerical experimentation in a spherical context. This result is then generalised to obtain a family of exact, unsteady, non-linear solutions of the BPV equation.

22.1 PREAMBLE

The goal of this chapter is to derive some *exact, unsteady, non-linear solutions* of the barotropic potential vorticity equation – abbreviated to 'BPV equation' from here onwards – for confined flow between two rigid, *ellipsoidal* geopotential surfaces. Several equivalent forms of this equation were derived in Chapter 19 in axial-orthogonal-curvilinear coordinates (ξ_1, ξ_2, ξ_3). Associated with these coordinates are unit vectors $(\mathbf{e}_1, \mathbf{e}_2, \mathbf{e}_3)$, velocity components (u_1, u_2, u_3), and metric (or scale) factors (h_1, h_2, h_3). The (shallow-fluid) metric factors are assumed to only vary in the meridional direction, that is,

$$h_i = h_i(\xi_2), \quad i = 1, 2, 3. \tag{22.1}$$

See Chapter 5 and Section 19.2.1 for further details of shallow axial-orthogonal-curvilinear coordinates. Superscript 'S' on h_i – to denote 'shallow' – is dropped in the present chapter for brevity.

For present purposes, the most convenient form of the BPV equation is the stream-function one developed in Section 19.5.2. A stream function, Ψ, is defined – see (19.81) – such that

$$h_3 u_1 = -\frac{1}{h_2}\frac{\partial \Psi}{\partial \xi_2}, \quad h_3 u_2 = \frac{1}{h_1}\frac{\partial \Psi}{\partial \xi_1}. \tag{22.2}$$

Horizontal velocity components u_1 and u_2 then satisfy the 3D non-divergence condition – see (19.33) –

$$\frac{\partial}{\partial \xi_1}\left(u_1 h_2 h_3\right) + \frac{\partial}{\partial \xi_2}\left(u_2 h_3 h_1\right) = 0, \tag{22.3}$$

where $u_3 \equiv 0$, that is, the flow is purely horizontal, so that only u_1 and u_2 appear in (22.3).[1] Note the presence of $h_3 = h_3(\xi_2)$ in (22.2), (22.3), and (22.4); this is absent from traditional formulations in spherical-polar coordinates using the classical spherical-geopotential approximation, since then $h_3 \equiv 1$.

Using (22.2), the BPV equation has the form – see (19.82) –

$$\begin{aligned}\frac{D_{\text{hor}}}{Dt}\left(\frac{\zeta+f}{h_3}\right) &= \frac{\partial}{\partial t}\left(\frac{\zeta+f}{h_3}\right) + \frac{u_1}{h_1}\frac{\partial}{\partial \xi_1}\left(\frac{\zeta+f}{h_3}\right) + \frac{u_2}{h_2}\frac{\partial}{\partial \xi_2}\left(\frac{\zeta+f}{h_3}\right)\\ &= \frac{\partial}{\partial t}\left(\frac{\zeta}{h_3}\right) + \frac{1}{h_1 h_2 h_3}\left[\frac{\partial\left(\Psi, \zeta/h_3\right)}{\partial\left(\xi_1, \xi_2\right)} + \frac{\partial\left(\Psi, f/h_3\right)}{\partial\left(\xi_1, \xi_2\right)}\right] = 0. \end{aligned} \tag{22.4}$$

In this equation: ζ is relative vorticity;

$$f(\xi_2) \equiv -\frac{2\Omega}{h_2}\frac{dh_1}{d\xi_2} \tag{22.5}$$

is the Coriolis parameter;

$$\frac{\partial(F,G)}{\partial(\xi_1,\xi_2)} \equiv \frac{\partial F}{\partial \xi_1}\frac{\partial G}{\partial \xi_2} - \frac{\partial F}{\partial \xi_2}\frac{\partial G}{\partial \xi_1} \tag{22.6}$$

is the definition of the Jacobian of F and G with respect to ξ_1 and ξ_2; and

$$\begin{aligned}\frac{\zeta}{h_3} &= \frac{1}{h_1 h_2 h_3}\left[\frac{\partial}{\partial \xi_1}\left(h_2 u_2\right) - \frac{\partial}{\partial \xi_2}\left(h_1 u_1\right)\right]\\ &= \frac{1}{h_1 h_2 h_3}\left[\frac{h_2}{h_3 h_1}\frac{\partial^2 \Psi}{\partial \xi_1^2} + \frac{\partial}{\partial \xi_2}\left(\frac{h_1}{h_2 h_3}\frac{\partial \Psi}{\partial \xi_2}\right)\right] \equiv \mathfrak{L}\Psi, \end{aligned} \tag{22.7}$$

[1] By multiplying (22.3) through by $\overline{\rho}/\left(h_1 h_2\right)$ and taking $\overline{\rho}$ = constant inside the partial derivatives, condition (22.3) may be physically interpreted as the mass flux between geopotential surfaces being horizontally non-divergent. This implies $h_3\mathbf{u}_{\text{hor}}$ is horizontally non-divergent, where $\mathbf{u}_{\text{hor}} \equiv u_1\mathbf{e}_1 + u_2\mathbf{e}_2$.

where $\mathfrak{L}$ is a linear, second-order, partial-differential operator. Equations (22.4)–(22.7) correspond to (19.82), (19.42), (19.83), and (19.84), respectively, of Chapter 19.

In deriving BPV equation (22.4), it was assumed as part of the shallow formulation (item 6 of Section 19.3.1) that apparent gravity is independent of vertical coordinate ξ_3. It was shown there that apparent gravity (g) and vertical metric factor (h_3) *cannot be independently prescribed.* Instead they must satisfy the constraint – see (19.20) –

$$h_3(\xi_2)\, g(\xi_2) = \text{constant} = h_3^E g^E, \tag{22.8}$$

where h_3^E and g^E are the values of h_3 and g, respectively, at the planet's equator. Thus g is permitted to vary as a function of latitude, but then h_3 must satisfy the constraint (22.8) and vary as the reciprocal of g; and vice versa.

The challenge now is to develop some exact, unsteady, non-linear solutions of the BPV equation (22.4) in ellipsoidal geopotential coordinates. This is met in Section 22.2 by developing a family of exact, unsteady, Rossby–Haurwitz-like solutions for confined flow between two rigid, ellipsoidal geopotential surfaces, with an alternative derivation given in Section 22.3. The balancing-pressure field for a particular such solution of practical interest is obtained in Section 22.4 as a prelude to developing (in Section 22.5) the corresponding field for the general family of solutions. Concluding remarks are made in Section 22.6.

22.2 DERIVATION OF EXACT UNSTEADY SOLUTIONS

22.2.1 The Eigenproblem for the Linear Differential Operator $\mathfrak{L}$

The eigenproblem associated with linear differential operator $\mathfrak{L}$ – defined by (22.7) – is at the very heart of the development of exact, unsteady solutions of BPV equation (22.4) over an ellipsoid. This 2D eigenproblem may be written as

$$\mathfrak{L}\Psi \equiv \frac{1}{h_1 h_2 h_3}\left[\frac{h_2}{h_3 h_1}\frac{\partial^2 \Psi}{\partial \xi_1^2} + \frac{\partial}{\partial \xi_2}\left(\frac{h_1}{h_2 h_3}\frac{\partial \Psi}{\partial \xi_2}\right)\right] = \Lambda \Psi, \tag{22.9}$$

where Λ is an eigenvalue, and the (shallow) metric factors (h_1, h_2, h_3) are known functions of ξ_2 (only) – as in (22.1).

To solve this eigenproblem, we seek separable solutions of (22.9) of the form

$$\Psi = L(\xi_1)\, M(\xi_2). \tag{22.10}$$

Inserting (22.10) into (22.9) and rearranging gives

$$\frac{h_1 h_3}{h_2 M(\xi_2)}\frac{d}{d\xi_2}\left[\frac{h_1}{h_2 h_3}\frac{dM(\xi_2)}{d\xi_2}\right] - \Lambda h_1^2 h_3^2 = -\frac{1}{L(\xi_1)}\frac{d^2 L(\xi_1)}{d\xi_1^2} = m^2 \ (\text{say}). \tag{22.11}$$

Separating variables then yields

$$\frac{d^2 L(\xi_1)}{d\xi_1^2} + m^2 L(\xi_1) = 0 \quad \Rightarrow \quad L^m(\xi_1) = A^m \cos(m\xi_1) + B^m \sin(m\xi_1), \tag{22.12}$$

$$\frac{d}{d\xi_2}\left[\frac{h_1}{h_2 h_3}\frac{dM(\xi_2)}{d\xi_2}\right] = \frac{\left(\Lambda h_1^2 h_3^2 + m^2\right) h_2}{h_1 h_3} M(\xi_2), \tag{22.13}$$

where A^m and B^m are constant coefficients and m is integer.

The second equation of (22.12) explicitly defines the zonal structure of an eigenfunction of (22.9), whilst (22.13) governs its meridional structure. Whereas determination of the zonal structure function $L(\xi_1)$ was straightforward, determination of the meridional structure function $M(\xi_2)$ is a lot more challenging. This is because of the appearance of the three metric factors $h_1(\xi_2)$, $h_2(\xi_2)$, and $h_3(\xi_2)$ in (22.13).

22.2.2 Geopotential Representation and Coordinate Systems

Metric factors h_1 and h_2 together describe the shape of the planet. Since, via constraint (22.8), h_3 is proportional to the reciprocal of apparent gravity, g, metric factor h_3 reflects the meridional variation of g. Because of apparent gravity, the three metric factors are *not* independent of one another, and there is something of a chicken-and-egg issue. The shape of the planet is determined by apparent gravity, but apparent gravity depends upon the shape of the planet (and its internal mass distribution), which depends on apparent gravity, and so on *ad infinitum*. It is therefore a coupled problem with consistency constraints; see Chapters 7 and 8. This very much complicates matters.

To proceed further, one needs a representation of apparent gravity and, in particular, of the geopotential, Φ.

In a self-consistent formulation – see Chapter 12 – the representation of Φ leads to:

- The equation for the associated geopotential surfaces.
- The precise shape of the planet.
- The coordinate lines orthogonal to geopotential surfaces.
- The acceleration due to apparent gravity, g.
- The ingredients for definition of an orthogonal-curvilinear-coordinate system, with metric factors h_1, h_2, and h_3.

The geopotential representation and GREAT (Geophysically Realistic, Ellipsoidal, Analytically Tractable) coordinate system, developed in Staniforth and White (2015a) and further developed in Chapter 12, are highly recommended[2] for the dynamical core of a physically realistic, 3D, atmospheric or oceanic prediction model. However, as formulated, they are not directly applicable to the present situation. In particular, they are designed for *deep*-fluid applications, whereas BPV equation (22.4) has been derived using *shallow*-fluid assumption (22.1). To address this issue, and to also ease tractability, a similar, but simpler, approach is taken here.

An ellipsoidal-geopotential representation is therefore constructed that is intermediate in complexity between the GREAT one and the simple, ubiquitous, classical, spherical-geopotential approximation.[3] The important rationale for this strategy is that it facilitates, and goes hand-in-hand with, the development of an *exact, unsteady, non-linear* solution of BPV equation (22.4) in ellipsoidal coordinates. This solution is a generalisation, to ellipsoidal geopotential coordinates, of the Rossby–Haurwitz[4] one in spherical-polar coordinates.

22.2.3 The Functional Form for an Exact Solution and Compatibility Conditions

Inspired by the Rossby–Haurwitz solution in spherical-polar coordinates with the classical spherical-geopotential approximation – see Rochas (1984, 1986) and references therein – exact solutions of BPV equation (22.4) in spheroidal geopotential coordinates are sought of the form

[2] At least by its authors!

[3] The resulting ellipsoidal, geopotential representation nevertheless turns out to be a shallow-fluid analogue of a special simplified case of the GREAT one.

[4] After Carl-Gustaf Rossby (1898–1957) and Bernhard Haurwitz (1905–86).

$$\Psi(\xi_1, \xi_2, t) = -\omega F(\xi_2) + K_n^m \Psi_n^m\left(\xi_1 - \sigma_n^m t, \xi_2\right). \tag{22.14}$$

Here,

$$\Psi_n^m(\xi_1, \xi_2) \equiv L^m\left(\xi_1 - \sigma_n^m t\right) M_n^m(\xi_2) \tag{22.15}$$

is the eigenfunction of (22.9) associated with eigenvalue Λ_n^m, and:

- $L^m(\xi_1)$ satisfies zonal-structure equation (22.12).
- $M_n^m(\xi_2)$ satisfies meridional-structure equation (22.13).
- m is zonal wave number.
- n is total wave number.
- K_n^m is a constant coefficient.
- ω and σ_n^m are *constant* (real) frequencies.

By analogy with the solution in spherical-polar coordinates,[5] the first term in (22.14) should correspond to a super (solid-body) rotation. Using decomposition (22.2) and the fact that solid-body rotation about a planet's axis corresponds to $u_1 = \omega h_1$ (since h_1 measures perpendicular distance from the rotation axis – see Sections 6.4.1 and 20.4), the following constraint on $F(\xi_2)$ should be respected:

$$\frac{dF}{d\xi_2} = h_1 h_2 h_3. \tag{22.16}$$

As developed in Section 19.3.1 and summarised in Section 22.1 of the present chapter, metric factor h_3 appearing in (22.16) is intimately related to the assumed model of apparent gravity, through satisfaction of (22.8). Usually h_3 (and also h_1 and h_2) results from specifying the geopotential Φ. Here, however, Φ is *not specified*, and h_3 cannot therefore be obtained from it. Instead, *h_3 is deduced by imposing the constraint that* (22.14) *be an exact solution of BPV equation* (22.4) *in ellipsoidal geopotential coordinates.*[6] The representation of Φ consistent with h_3 then follows.

From (22.7), $\zeta/h_3 = \mathfrak{L}\Psi$. Substituting this into BPV equation (22.4) then gives

$$\frac{\partial}{\partial t}(\mathfrak{L}\Psi) + \frac{1}{h_1 h_2 h_3}\left[\frac{\partial\left(\Psi, f/h_3\right)}{\partial(\xi_1, \xi_2)} + \frac{\partial(\Psi, \mathfrak{L}\Psi)}{\partial(\xi_1, \xi_2)}\right] = 0. \tag{22.17}$$

Inserting assumed form (22.14) for an exact solution into definition (22.9) of $\mathfrak{L}\Psi$ and using constraint (22.16) and structure equations (22.12) and (22.13) yields

5 There is some variation in the literature regarding definition of the precise form of Ψ for a Rossby–Haurwitz wave. In spherical geometry, with the classical spherical-geopotential approximation, form (22.14) consists of the superposition of a surface spherical-harmonic perturbation on a super (solid-body) background rotation. This captures the essence of the underlying physics. Although more general forms for Ψ than (22.14) exist in the *spherical* context, the present author is somewhat pessimistic regarding their possible generalisation to the ellipsoidal one. Attention here is therefore focused on form (22.14), for which exact, *ellipsoidal* solutions are obtained here.

6 One could specify h_3 but, in general, this leads to analytical intractability. This motivates the present *deductive* approach, guided by analogy and analytical tractability.

$$\begin{aligned}\mathfrak{L}\Psi &= \mathfrak{L}\left[-\omega F(\xi_2) + K_n^m \Psi_n^m\left(\xi_1 - \sigma_n^m t, \xi_2\right)\right] = -\omega \mathfrak{L} F(\xi_2) + K_n^m \mathfrak{L}\Psi_n^m\left(\xi_1 - \sigma_n^m t, \xi_2\right) \\ &= -\frac{\omega}{h_1 h_2 h_3}\left[\underbrace{\frac{h_2}{h_3 h_1}\frac{\partial^2 F(\xi_2)}{\partial \xi_1^2}}_{0} + \frac{d}{d\xi_2}\left(\frac{h_1}{h_2 h_3}\frac{dF}{d\xi_2}\right)\right] + K_n^m \Lambda_n^m \Psi_n^m\left(\xi_1 - \sigma_n^m t, \xi_2\right) \\ &= -\frac{2\omega}{h_2 h_3}\frac{dh_1}{d\xi_2} + K_n^m \Lambda_n^m \Psi_n^m\left(\xi_1 - \sigma_n^m t, \xi_2\right). \end{aligned} \tag{22.18}$$

Using (22.14), (22.18), and definitions (22.5) and (22.6), various terms appearing in (22.17) may be rewritten as follows:

$$\frac{\partial}{\partial t}(\mathfrak{L}\Psi) = K_n^m \Lambda_n^m \frac{\partial \Psi_n^m\left(\xi_1 - \sigma_n^m t, \xi_2\right)}{\partial t}, \tag{22.19}$$

$$\frac{\partial\left(\Psi, f/h_3\right)}{\partial\left(\xi_1, \xi_2\right)} = \frac{\partial \Psi}{\partial \xi_1}\frac{d}{d\xi_2}\left(\frac{f}{h_3}\right) = -2\Omega K_n^m \frac{d}{d\xi_2}\left(\frac{1}{h_2 h_3}\frac{dh_1}{d\xi_2}\right)\frac{\partial \Psi_n^m\left(\xi_1 - \sigma_n^m t, \xi_2\right)}{\partial \xi_1}, \tag{22.20}$$

$$\begin{aligned}\frac{\partial(\Psi, \mathfrak{L}\Psi)}{\partial\left(\xi_1, \xi_2\right)} &= \underbrace{\omega^2 \frac{\partial\left[F(\xi_2), 2\left(dh_1/d\xi_2\right)/\left(h_2 h_3\right)\right]}{\partial\left[\xi_1, \xi_2\right]}}_{0} - \omega K_n^m \Lambda_n^m \frac{\partial\left[F(\xi_2), \Psi_n^m\left(\xi_1 - \sigma_n^m t, \xi_2\right)\right]}{\partial\left[\xi_1, \xi_2\right]} \\ &\quad - \omega K_n^m \frac{\partial\left[\Psi_n^m\left(\xi_1 - \sigma_n^m t, \xi_2\right), 2\left(dh_1/d\xi_2\right)/\left(h_2 h_3\right)\right]}{\partial\left[\xi_1, \xi_2\right]} \\ &\quad + \underbrace{\left(K_\nu^m\right)^2 \Lambda_n^m \frac{\partial\left[\Psi_n^m\left(\xi_1 - \sigma_n^m t, \xi_2\right), \Psi_n^m\left(\xi_1 - \sigma_n^m t, \xi_2\right)\right]}{\partial\left[\xi_1, \xi_2\right]}}_{0}. \end{aligned} \tag{22.21}$$

Applying definition (22.6) of the Jacobian, the first and last terms on the right-hand side of (22.21) are identically zero. With use of constraint (22.16), (22.21) consequently further simplifies to

$$\frac{\partial(\Psi, \mathfrak{L}\Psi)}{\partial\left(\xi_1, \xi_2\right)} = \omega K_n^m\left[h_1 h_2 h_3 \Lambda_n^m - 2\frac{d}{d\xi_2}\left(\frac{1}{h_2 h_3}\frac{dh_1}{d\xi_2}\right)\right]\frac{\partial \Psi_n^m\left(\xi_1 - \sigma_n^m t, \xi_2\right)}{\partial \xi_1}. \tag{22.22}$$

Inserting (22.19), (22.20), and (22.22) into (22.17) then gives

$$\begin{aligned}&K_n^m \Lambda_n^m \frac{\partial \Psi_n^m\left(\xi_1 - \sigma_n^m t, \xi_2\right)}{\partial t} - \frac{2\Omega K_n^m}{h_1 h_2 h_3}\frac{d}{d\xi_2}\left(\frac{1}{h_2 h_3}\frac{dh_1}{d\xi_2}\right)\frac{\partial \Psi_n^m\left(\xi_1 - \sigma_n^m t, \xi_2\right)}{\partial \xi_1} \\ &\qquad + \omega K_n^m\left[\Lambda_n^m - \frac{2}{h_1 h_2 h_3}\frac{d}{d\xi_2}\left(\frac{1}{h_2 h_3}\frac{dh_1}{d\xi_2}\right)\right]\frac{\partial \Psi_n^m\left(\xi_1 - \sigma_n^m t, \xi_2\right)}{\partial \xi_1} = 0. \end{aligned} \tag{22.23}$$

Noting from (22.12) and (22.15) that

$$\begin{aligned}\frac{\partial \Psi_n^m\left(\xi_1 - \sigma_n^m t, \xi_2\right)}{\partial t} &= \frac{\partial L^m\left(\xi_1 - \sigma_n^m t\right)}{\partial t} M_n^m(\xi_2) = -\sigma_n^m \frac{\partial L^m\left(\xi_1 - \sigma_n^m t\right)}{\partial \xi_1} M_n^m(\xi_2) \\ &= -\sigma_n^m \frac{\partial \Psi_n^m\left(\xi_1 - \sigma_n^m t, \xi_2\right)}{\partial \xi_1}, \end{aligned} \tag{22.24}$$

(22.23) simplifies to

$$\left(\sigma_n^m - \omega\right)\Lambda_n^m = -\frac{2(\Omega + \omega)}{h_1 h_2 h_3}\frac{d}{d\xi_2}\left(\frac{1}{h_2 h_3}\frac{dh_1}{d\xi_2}\right), \tag{22.25}$$

provided that

$$K_n^m \frac{\partial \Psi_n^m\left(\xi_1 - \sigma_n^m t, \xi_2\right)}{\partial t} = -\sigma_n^m K_n^m \frac{\partial \Psi_n^m\left(\xi_1 - \sigma_n^m t, \xi_2\right)}{\partial \xi_1} \neq 0. \tag{22.26}$$

Admissible values for eigenvalue Λ_n^m appearing in (22.25) are determined by the precise shape of the planet, which remains to be specified. For example, values are determined in Section 22.2.7 for an ellipsoidal planet. At this point it is assumed that Λ_n^m is of known value.

Equation (22.25) is a *dispersion relation* for the frequency σ_n^m. It has been derived under the assumption that σ_n^m is *constant* (to enable exact solutions of the BPV equation to be found). Noting that Ω, ω, and Λ_n^m are mutually independent constants, examination of (22.25) shows that the following condition must hold for consistency with this assumption:

$$\frac{1}{h_1 h_2 h_3} \frac{d}{d\xi_2} \left(\frac{1}{h_2 h_3} \frac{dh_1}{d\xi_2} \right) = \text{constant} = -C_1 \ (\text{say}). \tag{22.27}$$

This equation can be considered to be a *compatibility condition* for finding solutions in spheroidal geopotential coordinates analogous to Rossby–Haurwitz solutions in spherical-polar coordinates.

22.2.4 Application of the Compatibility Condition for the General Case

Provided compatibility condition (22.27) is satisfied, insertion into (22.25) then gives the dispersion relation in the form

$$\sigma_n^m = \omega + \frac{2\left(\Omega + \omega\right) C_1}{\Lambda_n^m}, \tag{22.28}$$

where C_1 (to be determined) depends on the shape of the planet (which itself is intimately linked to the representation of apparent gravity).

Eliminating $h_1 h_2 h_3$ between (22.27) and (22.16) gives

$$\frac{d}{d\xi_2} \left(\frac{1}{h_2 h_3} \frac{dh_1}{d\xi_2} + C_1 F \right) = 0. \tag{22.29}$$

Integrating this then yields

$$F\left(\xi_2\right) = -\frac{1}{C_1 h_2 h_3} \frac{dh_1}{d\xi_2}, \tag{22.30}$$

where the integration constant has been arbitrarily set to zero since Ψ in (22.14), and therefore in $F\left(\xi_2\right)$, is only determined to within an arbitrary additive constant.

Multiplying through by $2h_1 \, dh_1/d\xi_2$, compatibility condition (22.27) may be rewritten as

$$\frac{d}{d\xi_2} \left[\left(\frac{1}{h_2 h_3} \frac{dh_1}{d\xi_2} \right)^2 + C_1 h_1^2 \right] = 0 \tag{22.31}$$

and integrated to give

$$\left(\frac{1}{h_2 h_3} \frac{dh_1}{d\xi_2} \right)^2 = C_2 - C_1 h_1^2, \tag{22.32}$$

where C_2 is constant. Solving (22.32) for h_3 yields

$$h_3 = \frac{1}{\left(C_2 - C_1 h_1^2\right)^{\frac{1}{2}} h_2} \frac{dh_1}{d\xi_2}. \tag{22.33}$$

Now – from (22.8) – metric factor $h_3(\xi_2)$ is related to apparent gravity $g(\xi_2)$ by

$$h_3(\xi_2) = \frac{h_3^E g^E}{g(\xi_2)}. \tag{22.34}$$

Eliminating $h_3(\xi_2)$ between (22.33) and (22.34) then gives

$$g(\xi_2) = \frac{h_3^E g^E h_2}{dh_1/d\xi_2} \left(C_2 - C_1 h_1^2\right)^{\frac{1}{2}}. \tag{22.35}$$

An important condition that remains to be applied is that *both* $h_3(\xi_2)$ *and* $g(\xi_2)$ *must be non-singular everywhere over the spheroid.* If attention is restricted to planets of small ellipticity, such as Earth, the metric factors are then small perturbations of those for a sphere, for which $(h_1, h_2) = (a\cos\phi, 1)$, where ϕ is latitude. From this one can infer that:

- h_2 in (22.33) and (22.35) is well-behaved.
- There is a potential singularity at the planet's equator associated with $dh_1/d\xi_2$.
- $dh_1/d\xi_2$ must be proportional to $\left(C_2 - C_1 h_1^2\right)^{\frac{1}{2}}$ in the vicinity of the equator.

This is about as far as one can proceed for the general case. The analysis to this point is valid for *any* zonally symmetric spheroid of arbitrary meridional cross-section. Application of compatibility condition (22.27) has led to form (22.33) for the vertical metric factor h_3, and thence to functional form (22.35) for apparent gravity, g. To proceed further in a tractable manner, a specific choice for the shape of the spheroid needs to be made. The choice of an ellipsoid is of particular interest since the World Geodetic System (WGS 84), used for reporting data position using the Global Positioning System (GPS), is based on a reference ellipsoid; see Section 12.2.4. An ellipsoid is therefore now examined in some detail, using parametric-ellipsoidal coordinates to represent it.

22.2.5 Parametric Ellipsoidal Coordinates

For parametric-ellipsoidal coordinates, $\xi_1 \equiv \lambda$ is longitude, and $\xi_2 \equiv \theta$ is parametric latitude; see Fig. 22.1 for a depiction of a cross section in a meridional plane.

From Section 18.3.2.2, the equations that define the ellipsoid's surface in rotating Cartesian coordinates (x_1, x_2, z_2) may be written as – see (18.67) –

$$x_1 = a\cos\lambda\cos\theta, \quad x_2 = a\sin\lambda\cos\theta, \quad x_3 = c\sin\theta, \tag{22.36}$$

where a and c are the semi-major and semi-minor axes, respectively, of the ellipsoid's cross section in a meridional plane. It is easily verified that (22.36) lead to the following two equivalent equations (in Cartesian and polar forms, respectively) for the ellipsoid's surface:

$$\frac{x_1^2 + x_2^2}{a^2} + \frac{x_3^2}{c^2} = 1, \tag{22.37}$$

$$r(\theta) \equiv \left(x_1^2 + x_2^2 + x_3^2\right)^{\frac{1}{2}} = \left(a^2\cos^2\theta + c^2\sin^2\theta\right)^{\frac{1}{2}} = a\left(1 - e^2\sin^2\theta\right)^{\frac{1}{2}}, \tag{22.38}$$

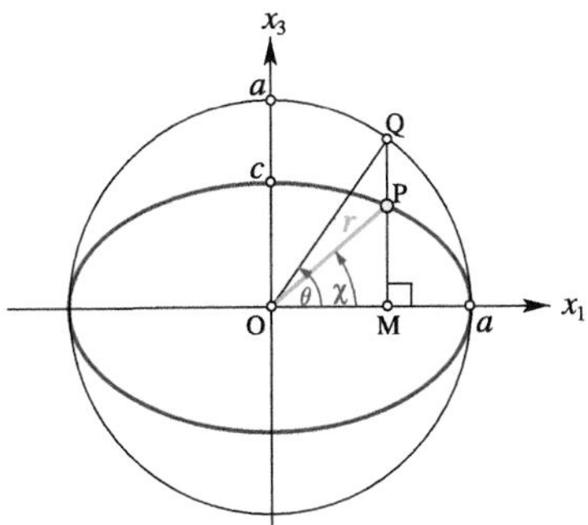

Figure 22.1 An ellipse (shaded) in the Ox_1x_3 meridional plane, having semi-major and semi-minor axes a and c, respectively, and inscribed in a circle of radius a. P is a general point on the ellipse with coordinates (x_1, x_3). θ and χ are *parametric* and *geocentric* latitudes, respectively. The line QPM is parallel to Ox_3 and intersects the circle and Ox_1 at Q and M respectively. The equation of the ellipse is $\left(x_1^2/a^2\right) + \left(x_3^2/c^2\right) = 1$, and $r = \left(x_1^2 + x_3^2\right)^{1/2}$ is the distance of P from origin O along the green line.

where

$$e \equiv \frac{\left(a^2 - c^2\right)^{\frac{1}{2}}}{a} \tag{22.39}$$

is *eccentricity*. The associated metric factors h_1 and h_2 are then – see Section 18.3.2.2, or by insertion of (22.36) into (5.4) –

$$h_1 = a\cos\theta, \quad h_2 = \left(a^2\sin^2\theta + c^2\cos^2\theta\right)^{\frac{1}{2}} = c\left(1 + 2\widetilde{\varepsilon}\sin^2\theta\right)^{\frac{1}{2}}, \tag{22.40}$$

where

$$2\widetilde{\varepsilon} \equiv \frac{a^2 - c^2}{c^2} \equiv e^2\frac{a^2}{c^2}. \tag{22.41}$$

22.2.6 The Implied Model of Apparent Gravity for an Ellipsoid

Using horizontal metric factors (22.40),

$$\frac{1}{h_2}\frac{dh_1}{d\xi_2} = \frac{1}{h_2}\frac{dh_1}{d\theta} = -\frac{a\sin\theta}{c\left(1 + 2\widetilde{\varepsilon}\sin^2\theta\right)^{\frac{1}{2}}}. \tag{22.42}$$

Substitution of (22.42) into (22.33), with use of (22.40), then gives

$$h_3^2 = \frac{a^2\sin^2\theta}{c^2\left(C_2 - C_1a^2\cos^2\theta\right)\left(1 + 2\widetilde{\varepsilon}\sin^2\theta\right)}. \tag{22.43}$$

Since both h_3 *and its reciprocal (which is proportional to g) must be non-singular,* $C_2 - C_1a^2\cos^2\theta$ *must be proportional to* $\sin^2\theta$ *to avoid any singularity at the planet's equator, where* $\theta = 0$. Thus

$$C_1a^2 = C_2, \tag{22.44}$$

and (22.43) simplifies to

$$h_3^2 = \frac{a^2}{C_2c^2\left(1 + 2\widetilde{\varepsilon}\sin^2\theta\right)} = \frac{\left(h_3^E\right)^2}{\left(1 + 2\widetilde{\varepsilon}\sin^2\theta\right)} \quad \Rightarrow \quad h_3(\theta) = \frac{h_3^E}{\left(1 + 2\widetilde{\varepsilon}\sin^2\theta\right)^{\frac{1}{2}}}, \tag{22.45}$$

where

$$h_3^E = \frac{a}{\sqrt{C_2}c}, \tag{22.46}$$

is the constant value of h_3 at the planet's equator. Substitution of (22.45) into (22.34), with use of (22.41) and (19.21), finally gives:

The Implied Model of Gravity

$$g(\theta) = \frac{1}{h_3(\theta)} \frac{d\Phi}{d\xi_3} = g^E \left(1 + 2\widetilde{\varepsilon} \sin^2\theta\right)^{\frac{1}{2}} = g^E \left(1 + e^2 \frac{a^2}{c^2} \sin^2\theta\right)^{\frac{1}{2}}, \tag{22.47}$$

where

$$\Phi(\xi_3) = \Phi^B + h_3^E g^E \left(\xi_3 - \xi_3^B\right). \tag{22.48}$$

This model of apparent gravity, by construction, turns out to be the *only* one that leads to analogue Rossby–Haurwitz solutions of the (non-linear) BPV equation in parametric-ellipsoidal coordinates. (Other models of apparent gravity may be considered, but they will not lead to analogue Rossby–Haurwitz solutions of the *non-linear* BPV equation that propagate with constant phase speed and without change of shape.[7]) This model of apparent gravity corresponds to an ellipsoidal planet of uniform density, as assumed by Newton, but restricted here to a shallow fluid; see Section 22.3.1 and Section 12.8 herein, and Staniforth and White (2015a)'s Section 5, with associated footnote. As such, this model represents approximately two thirds of the observed meridional variation of apparent gravity. This is much better than the zero variation of existing global models of Earth's atmosphere and oceans, all of which (to the author's knowledge) employ the classical spherical-geopotential approximation. For present purposes, and also for constructing test problems for dynamical cores of global models, this limitation is therefore perfectly acceptable.

The question now is:

- What value should one take for h_3^E?

The answer is that this is arbitrary; taking a different value simply amounts to measuring distance in the vertical using different units – see (19.6). A convenient and natural choice is $h_3^E = 1$. This corresponds to setting $R = a$ (to obtain values at the bottom surface) and $x_3 = 0$ (to obtain the equatorial value at the bottom surface) in (12.104) of Section 12.9.3. It also corresponds to the value obtained in the limiting case of spherical-polar coordinates (when $c \to a$). Setting $h_3^E = 1$ in (22.46) gives

$$C_2 = \frac{a^2}{c^2} \quad \Rightarrow \quad C_1 = \frac{C_2}{a^2} = \frac{1}{c^2}, \tag{22.49}$$

where use has been made of (22.44). Using this expression for C_1 in (22.30) and using (22.42) and (22.45) with $h_3^E = 1$ then gives

$$F(\theta) = ac\sin\theta. \tag{22.50}$$

22.2.7 Solution of the Meridional Structure Equation for an Ellipsoid

Using separation of variables, 2D eigenproblem (22.9) was reduced in Section 22.2.1 to the 1D problem (22.13), that is, to

$$\frac{d}{d\xi_2}\left[\frac{h_1}{h_2 h_3} \frac{dM_n^m(\xi_2)}{d\xi_2}\right] = \left(\Lambda_n^m h_1 h_2 h_3 + m^2 \frac{h_2}{h_3 h_1}\right) M_n^m(\xi_2), \tag{22.51}$$

7 Other models of gravity could, however, conceivably lead to analogue Rossby–Haurwitz solutions of a *linearised* form of the BPV equation.

where Λ_n^m is the 2D eigenvalue. The crucial question now is:

- Can the associated eigenfunctions be tractably obtained for parametric-ellipsoidal coordinates? (If this cannot be done, then the last term in assumed form (22.14) for the exact solutions remains undetermined.)

Fortunately they can, as now shown.

From the preceding analysis, $\xi_1 = \lambda$, $\xi_2 = \theta$, and – from metric factors (22.40) and (22.45) with $h_3^E = 1$ – we obtain:

The Metric Factors for an Exact Solution

$$h_1 = a\cos\theta, \quad h_2 = c\left(1 + 2\widetilde{\varepsilon}\sin^2\theta\right)^{\frac{1}{2}}, \quad h_3 = \left(1 + 2\widetilde{\varepsilon}\sin^2\theta\right)^{-\frac{1}{2}}, \tag{22.52}$$

where $2\widetilde{\varepsilon} \equiv \left(a^2 - c^2\right)/c^2 \equiv e^2a^2/c^2$ from (22.41).

Thus

$$\frac{h_1}{h_2h_3} = \frac{a}{c}\cos\theta, \quad h_1h_2h_3 = ac\cos\theta, \quad \frac{h_2}{h_3h_1} = \frac{c\left(1 + 2\widetilde{\varepsilon}\sin^2\theta\right)}{a\cos\theta}. \tag{22.53}$$

Using (22.53), meridional-structure equation (22.51) may be rewritten as

$$\frac{d}{d\theta}\left[\cos\theta\frac{dM_n^m(\theta)}{d\theta}\right] - \left[\Lambda_n^m c^2\cos\theta + m^2\frac{c^2\left(1 + 2\widetilde{\varepsilon}\sin^2\theta\right)}{a^2\cos\theta}\right]M_n^m(\theta) = 0. \tag{22.54}$$

By making the transformation of independent variable

$$\mu \equiv \sin\theta, \tag{22.55}$$

and defining n such that

$$\Lambda_n^m \equiv \frac{-n(n+1) + e^2m^2}{c^2}, \tag{22.56}$$

(22.54) may then be rewritten as

$$\left(1 - \mu^2\right)\frac{d^2M_n^m(\mu)}{d\mu^2} - 2\mu\frac{dM_n^m(\mu)}{d\mu} + \left[n(n+1) - \frac{m^2}{\left(1 - \mu^2\right)}\right]M_n^m(\mu) = 0. \tag{22.57}$$

This equation is the associated Legendre equation, whose solutions are the associated Legendre functions $P_n^m(\mu)$ and $Q_n^m(\mu)$ of the first and second kinds, respectively, of order m and degree n. For the present problem, only the solutions of the first kind, $P_n^m(\mu)$, are admissible since those of the second kind, $Q_n^m(\mu)$, are singular at the poles. As a cross-check, taking the limit $c \to a\ (\Rightarrow e \to 0)$ in (22.56) recovers the usual value $\Lambda_n^m \equiv -n(n+1)/a^2$ for the eigenproblem $\mathfrak{L}\Psi = \nabla_{\text{hor}}^2\Psi = \Lambda\Psi$ in spherical-polar coordinates.

22.2.8 A Family of Exact Unsteady Solutions over an Ellipsoid

An exact solution in parametric-ellipsoidal coordinates of BPV equation (22.4) may be obtained from (22.14) and (22.15) by using:

- (22.12) for $L^m(\xi_1)$, with $A^m = ac$ and $B^m = 0$.
- (22.28) and (22.56) for σ_n^m.

- (22.50) for $F(\theta)$.
- The analysis of Section 22.2.7 to identify $M_n^m(\mu)$ with the associated Legendre function $P_n^m(\mu)$.
- (22.2), (22.7), and (22.52) – for u_1, u_2, and ζ/h_3 corresponding to stream function (22.60).

Thus:

A Family of Exact Rossby–Haurwitz-Like Solutions of the BPV Equation in Parametric-Ellipsoidal Coordinates

The BPV equation (22.4) in parametric-ellipsoidal coordinates (22.36), with associated scale factors (22.52), is

$$\frac{D_{\text{hor}}}{Dt}\left(\frac{\zeta+f}{h_3}\right)=\frac{\partial}{\partial t}\left(\frac{\zeta+f}{h_3}\right)+\frac{u_1}{h_1}\frac{\partial}{\partial\lambda}\left(\frac{\zeta+f}{h_3}\right)+\frac{u_2}{h_2}\frac{\partial}{\partial\theta}\left(\frac{\zeta+f}{h_3}\right)$$
$$=\frac{\partial}{\partial t}\left(\frac{\zeta}{h_3}\right)+\frac{1}{h_1h_2h_3}\left[\frac{\partial\left(\Psi,\zeta/h_3\right)}{\partial\left(\lambda,\theta\right)}+\frac{\partial\left(\Psi,f/h_3\right)}{\partial\left(\lambda,\theta\right)}\right]=0, \quad (22.58)$$

where (from (22.5) with $\xi_2=\theta$ and (22.52))

$$f(\theta)\equiv-\frac{2\Omega}{h_2}\frac{dh_1}{d\theta}=-\frac{2\Omega a\sin\theta}{c\left(1+2\tilde{\varepsilon}\sin^2\theta\right)^{\frac{1}{2}}}$$

is the Coriolis parameter.
For a homogeneous, ellipsoidal planet, apparent gravity is (from (22.47))

$$g(\mu)=g^E\left(1+2\tilde{\varepsilon}\mu^2\right)^{\frac{1}{2}}=g^E\left(1+e^2\frac{a^2}{c^2}\mu^2\right)^{\frac{1}{2}}, \quad (22.59)$$

where $\mu\equiv\sin\theta$ (from (22.55)), and $2\tilde{\varepsilon}\equiv\left(a^2-c^2\right)/c^2\equiv e^2a^2/c^2$ (from (22.41)).
An exact, unsteady family of solutions of (22.58) is then given by

$$\Psi(\lambda,\mu,t)=-\omega ac\mu+acK_n^m\cos\left[m\left(\lambda-\sigma_n^mt\right)\right]P_n^m(\mu)$$
$$=-\omega acP_1^0(\mu)+acK_n^m\cos\left[m\left(\lambda-\sigma_n^mt\right)\right]P_n^m(\mu), \quad (22.60)$$

$$u_1(\lambda,\mu,t)=a\omega\left(1-\mu^2\right)^{\frac{1}{2}}-aK_n^m\cos\left[m\left(\lambda-\sigma_n^mt\right)\right]\left(1-\mu^2\right)^{\frac{1}{2}}\frac{dP_n^m(\mu)}{d\mu}, \quad (22.61)$$

$$u_2(\lambda,\mu,t)=-cmK_n^m\sin\left[m\left(\lambda-\sigma_n^mt\right)\right]\frac{\left(1+2\tilde{\varepsilon}\mu^2\right)^{\frac{1}{2}}P_n^m(\mu)}{\left(1-\mu^2\right)^{\frac{1}{2}}}, \quad (22.62)$$

$$\frac{\zeta(\lambda,\mu,t)}{h_3(\mu)}=\frac{2\omega a}{c}P_1^0(\mu)+ac\Lambda_n^mK_n^m\cos\left[m\left(\lambda-\sigma_n^mt\right)\right]P_n^m(\mu), \quad (22.63)$$

where

$$\frac{dP_n^m(\mu)}{d\mu}\equiv\frac{(n+1)(n+m)P_{n-1}^m(\mu)-n(n-m+1)P_{n+1}^m(\mu)}{(2n+1)\left(1-\mu^2\right)} \quad (22.64)$$

relates $dP_n^m(\mu)/d\mu$ to $P_{n-1}^m(\mu)$ and $P_{n+1}^m(\mu)$ via a classical recurrence relation for associated Legendre functions; and (from (22.28), (22.49), and (22.56))

$$\sigma_n^m=\omega+\frac{2(\Omega+\omega)}{c^2\Lambda_n^m}=\omega-\frac{2(\Omega+\omega)}{\left[n(n+1)-e^2m^2\right]}, \quad (22.65)$$

$$\Lambda_n^m = \frac{-n(n+1) + e^2 m^2}{c^2}. \tag{22.66}$$

Parameters ω and K_{m+1}^m have units of frequency (i.e. of time^{-1}).

The third goal identified in Section 19.1 – to construct some exact, physically meaningful, unsteady non-linear solutions of the BPV equation in ellipsoidal geometry – has now also been met.

As a first cross-check, setting $e \equiv 0$ ($\Rightarrow c \equiv a$) in (22.65) reduces it to

$$\sigma_n = \omega - \frac{2(\Omega + \omega)}{n(n+1)}, \tag{22.67}$$

with no dependence then on zonal wave number m. This agrees with the usual result for Rossby–Haurwitz waves in spherical-polar coordinates; see the second equation of Rochas (1986). Comparison of (22.65) in the ellipsoidal context with its counterpart (22.67) in the spherical one shows that meridional variation of apparent gravity changes the propagation frequency (σ) via the introduction of dependency on both eccentricity (e) and zonal wave number (m). This serves to speed up the east-to-west propagation (relative to the background flow) of Rossby–Haurwitz waves in the ellipsoidal context when compared to the spherical one.

Similarly, setting $e \equiv 0$ ($\Rightarrow c \equiv a$) in (22.59) confirms that apparent gravity loses its meridional variation in the (special) spherical context; gravity is then constant everywhere, both in the horizontal and in the vertical.

22.2.9 A Particular Exact Unsteady Solution over an Ellipsoid

The exact solution given in (22.60) is somewhat abstract due to the appearance of the associated Legendre function $P_n^m(\sin\theta)$. Following Phillips (1959), it has become customary to set $n \equiv m+1$ in (22.60) and (22.65) when using Rossby–Haurwitz waves (in the spherical context) for numerical experimentation (Hoskins, 1973; Williamson et al., 1992). This has the virtue of making things more tangible.

Aside How does setting $n \equiv m+1$ simplify matters? To answer this question we need to examine some properties of associated Legendre functions $P_n^m(\mu)$, where $\mu = \sin\theta$ here. Since (to within a constant normalisation factor)

$$P_n^m(\mu) \sim \left(1-\mu^2\right)^{\frac{m}{2}} \frac{d^m P_n(\mu)}{d\mu^m}, \tag{22.68}$$

where $P_n(\mu)$ is a Legendre polynomial of degree n, so (setting $n \equiv m+1$)

$$P_{m+1}^m(\mu) \sim \left(1-\mu^2\right)^{\frac{m}{2}} \frac{d^m P_{m+1}(\mu)}{d\mu^m}. \tag{22.69}$$

Now $P_{m+1}(\mu)$ is a *polynomial* of degree $m+1$, with the property that $d^m P_{m+1}(\mu)/d\mu^m \sim \mu$ to within a multiplicative constant. Thus (with $\mu = \sin\theta$), $P_{m+1}^m(\sin\theta)$ has the simple form

$$P_{m+1}^m(\sin\theta) \sim \cos^m\theta \sin\theta. \tag{22.70}$$

Setting $n = m+1$ in (22.60), using (22.70) to express $P_{m+1}^m(\sin\theta)$ explicitly, and renormalising K_{m+1}^m, (22.60) and (22.65) then take the particular forms

$$\Psi(\lambda,\theta,t) = -\omega ac\sin\theta + acK_{m+1}^{m}\cos\left[m\left(\lambda - \sigma_{m+1}^{m}t\right)\right]\cos^{m}\theta\sin\theta, \tag{22.71}$$

$$\sigma_{m+1}^{m} = \omega - \frac{2(\Omega+\omega)}{\left[(m+1)(m+2) - e^2m^2\right]}. \tag{22.72}$$

Equation (22.71) is the generalisation to parametric-ellipsoidal coordinates of Phillips (1959)'s equation (36); a^2 in his (36) changes to ac in (22.71), with the spherical case recovered when $a \equiv c$.

Using (22.2), (22.7), and (22.52), the expressions for u_1, u_2 and ζ/h_3 corresponding to stream function (22.71) may be determined. They are displayed in the summary box. It is straightforward to verify, by direct substitution, that BPV equation (22.73) in parametric-ellipsoidal coordinates, with metric factors defined by (22.52), and with model (22.74) of apparent gravity is exactly satisfied by (22.75)–(22.80). This provides a further cross-check on the analysis.

A Particular Exact Rossby–Haurwitz-Like Solution of the BPV Equation in Parametric-Ellipsoidal Coordinates

The BPV equation (22.4) in parametric-ellipsoidal coordinates (22.36), with associated scale factors (22.52), is

$$\begin{aligned}\frac{D_{\text{hor}}}{Dt}\left(\frac{\zeta+f}{h_3}\right) &= \frac{\partial}{\partial t}\left(\frac{\zeta+f}{h_3}\right) + \frac{u_1}{h_1}\frac{\partial}{\partial\lambda}\left(\frac{\zeta+f}{h_3}\right) + \frac{u_2}{h_2}\frac{\partial}{\partial\theta}\left(\frac{\zeta+f}{h_3}\right)\\ &= \frac{\partial}{\partial t}\left(\frac{\zeta}{h_3}\right) + \frac{1}{h_1h_2h_3}\left[\frac{\partial(\Psi,\zeta/h_3)}{\partial(\lambda,\theta)} + \frac{\partial(\Psi,f/h_3)}{\partial(\lambda,\theta)}\right] = 0, \end{aligned}\tag{22.73}$$

where (from (22.5) with $\xi_2 = \theta$ and (22.52))

$$f(\theta) \equiv -\frac{2\Omega}{h_2}\frac{dh_1}{d\theta} = -\frac{2\Omega a\sin\theta}{c\left(1+2\widetilde{\varepsilon}\sin^2\theta\right)^{\frac{1}{2}}}$$

is the Coriolis parameter.

For a homogeneous, ellipsoidal planet, apparent gravity is (from (22.47))

$$g(\theta) = g^E\left(1+2\widetilde{\varepsilon}\sin^2\theta\right)^{\frac{1}{2}} = g^E\left[1+e^2\frac{a^2}{c^2}\sin^2\theta\right]^{\frac{1}{2}}, \tag{22.74}$$

where (from (22.41)) $2\widetilde{\varepsilon} \equiv \left(a^2-c^2\right)/c^2 \equiv e^2a^2/c^2$.

A particular, exact, unsteady solution of (22.73) is then given by

$$\Psi(\lambda,\theta,t) = -\omega ac\sin\theta + acK_{m+1}^{m}\cos\left[m\left(\lambda-\sigma_{m+1}^{m}t\right)\right]\cos^{m}\theta\sin\theta, \tag{22.75}$$

$$u_1(\lambda,\theta,t) = a\omega\cos\theta + aK_{m+1}^{m}\cos\left[m\left(\lambda-\sigma_{m+1}^{m}t\right)\right]\left[m\cos^{m-1}\theta - (m+1)\cos^{m+1}\theta\right], \tag{22.76}$$

$$u_2(\lambda,\theta,t) = -cmK_{m+1}^{m}\sin\left[m\left(\lambda-\sigma_{m+1}^{m}t\right)\right]\left(1+2\widetilde{\varepsilon}\sin^2\theta\right)^{\frac{1}{2}}\cos^{m-1}\theta\sin\theta, \tag{22.77}$$

$$\frac{\zeta(\lambda,\theta,t)}{h_3(\theta)} = \frac{2\omega a}{c}\sin\theta + ac\Lambda_{m+1}^{m}K_{m+1}^{m}\cos\left[m\left(\lambda-\sigma_{m+1}^{m}t\right)\right]\cos^{m}\theta\sin\theta, \tag{22.78}$$

where

$$\sigma_{m+1}^{m} = \omega - \frac{2(\Omega+\omega)}{\left[(m+1)(m+2)-e^2m^2\right]}, \tag{22.79}$$

$$\Lambda_{m+1}^{m}=\frac{-(m+1)(m+2)+e^2m^2}{c^2}=-\left[\frac{m^2}{a^2}+\frac{(3m+2)}{c^2}\right]. \tag{22.80}$$

This solution is a generalisation to parametric-ellipsoidal coordinates of a particular Rossby–Haurwitz solution in spherical-polar ones. See the preceding discussion for further details.

22.3 A COMPLEMENTARY DERIVATION OF EXACT UNSTEADY SOLUTIONS

22.3.1 Geopotential Representation

The GREAT geopotential representation (12.35) (exterior to a planet) of Chapter 12 is

$$\begin{aligned}\Phi(\chi,r)=&-\frac{\gamma M_P}{r}+\frac{\gamma M_P}{R}\left\{1+\left[\left(\frac{8\widetilde{\varepsilon}-5\widetilde{m}}{2}\right)+\left(\frac{5\widetilde{m}-4\widetilde{\varepsilon}}{2}\right)\frac{R^2}{a^2}\right]\sin^2\chi\right\}^{\frac{1}{2}}\\&-\frac{\gamma M_P}{R}\left[1+\left(\frac{8\widetilde{\varepsilon}-7\widetilde{m}}{12}\right)+\left(\frac{11\widetilde{m}-4\widetilde{\varepsilon}}{12}\right)\frac{R^2}{a^2}\right],\end{aligned} \tag{22.81}$$

where:

- γ is the universal gravitational constant.
- M_P is the mass of the planet.
- R is the equatorial radius of the geopotential surface $\Phi=\Phi_R$.
- $\widetilde{\varepsilon}$ is defined by (22.41).
- $\widetilde{m}$ is a ratio of centrifugal and mass gravitational forces.
- χ is geocentric latitude.
- r is distance from the ellipsoid's centre of mass.

Note that m in (12.35) has been replaced in (22.81) by $\widetilde{m}$; this is to distinguish it from use of m here to denote zonal wave number. Similarly M in (12.35) has been replaced in (22.81) by M_P to distinguish it from use of M here to denote meridional structure function.

A special case of (22.81) is when $\widetilde{m}=4\widetilde{\varepsilon}/5$, which removes the R^2/a^2 term from its first line. This value corresponds to the apparent gravitational field of a homogeneous ellipsoidal planet of small eccentricity (as first examined by Newton), and the geopotential surfaces reduce to an infinite set of concentric similar ellipsoids (White et al., 2008). Setting $\widetilde{m}=4\widetilde{\varepsilon}/5$ in (22.81) then simplifies it to

$$\Phi(\chi,r)=-\frac{\gamma M_P}{r}+\frac{\gamma M_P}{R}\left(1+2\widetilde{\varepsilon}\sin^2\chi\right)^{\frac{1}{2}}-\frac{\gamma M_P}{R}\left[1+\frac{\widetilde{\varepsilon}}{5}\left(1+2\frac{R^2}{a^2}\right)\right]. \tag{22.82}$$

Equation (22.82) is expressed in terms of geocentric latitude, χ. However, elsewhere in this chapter, parametric-ellipsoidal coordinates are used to navigate in the horizontal, and the meridional coordinate is instead parametric latitude, θ. Using the coordinate definitions given in Section 18.3.2, it can be shown that these two latitudes are related to one another by

$$\sin\chi=\frac{c\sin\theta}{a\left(1-e^2\sin^2\theta\right)^{\frac{1}{2}}}, \tag{22.83}$$

where e^2 is defined by (22.39). Thus, using (22.41) and (22.83) in (22.82), the last of these equations may be rewritten in terms of parametric latitude as

$$\Phi(\theta,r)=-\frac{\gamma M_P}{r}+\frac{\gamma M_P}{R}\left(1-e^2\sin^2\theta\right)^{-\frac{1}{2}}-\frac{\gamma M_P}{R}\left[1+\frac{\widetilde{\varepsilon}}{5}\left(1+2\frac{R^2}{a^2}\right)\right]. \tag{22.84}$$

This geopotential representation can be considered to be the starting point of the alternative derivation of exact solutions. It corresponds to the apparent gravitational field of a homogeneous ellipsoidal planet of small eccentricity (e).

22.3.2 Location of the Planet's Surface

Setting $r = R = a$ and $\theta = 0$ in (22.84) gives the value of the geopotential Φ_a that defines the planet's surface. Thus

$$\Phi_a = -\frac{\gamma M_P}{a}\left(1 + \frac{3\widetilde{\varepsilon}}{5}\right). \tag{22.85}$$

The equation for the planet's surface is obtained by setting $\Phi(\theta, r, R = a) = \Phi_a$. Thus

$$-\frac{\gamma M_P}{r} + \frac{\gamma M_P}{a}\left(1 - e^2 \sin^2\theta\right)^{-\frac{1}{2}} - \cancel{\frac{\gamma M_P}{a}\left(1 + \frac{3\widetilde{\varepsilon}}{5}\right)} = -\cancel{\frac{\gamma M_P}{a}\left(1 + \frac{3\widetilde{\varepsilon}}{5}\right)}, \tag{22.86}$$

which simplifies to

$$r(\theta) = a\left(1 - e^2 \sin^2\theta\right)^{\frac{1}{2}}. \tag{22.87}$$

This agrees with (22.38) for an ellipsoid. It also agrees with (12.36), with $m = 4\widetilde{\varepsilon}/5$ (but with $\widetilde{m} = 4\widetilde{\varepsilon}/5$ in the present notation) inserted into it. This verifies that geopotential representation (22.82) does indeed lead to an ellipsoidal planet. Thus the representation of the geopotential – and therefore of apparent gravity – and the shape of the planet (an ellipsoid) are all mutually consistent.

22.3.3 Metric Factors

Expressions for the metric factors (h_1, h_2, h_3) of the GREAT coordinate system are given by (12.109)–(12.111). These can be straightforwardly adapted to the present situation by evaluating them at the planet's surface (consistent with shallow assumption (22.1), made here), and by setting $R = a$ and $m = 4\widetilde{\varepsilon}/5$ everywhere in (12.109)–(12.111). For this value of m (denoted by $\widetilde{m}$ here), μ and ν reduce to $\mu = 2\widetilde{\varepsilon}$ and $\nu = 0$, respectively.[8] Using these values, and also the expressions given in Tables 12.1 and 12.2 for parametric-ellipsoidal coordinates, then leads (in the present notation) to

$$\left.\begin{aligned} h_1(\theta) = a\cos\theta, \quad h_2(\theta) = c\left(1 + 2\widetilde{\varepsilon}\sin^2\theta\right)^{\frac{1}{2}}, \\ h_3(\theta) = \left[1 + 2\widetilde{\varepsilon}\left(1 + 2\widetilde{\varepsilon}\right)\frac{x_3^2}{a^2}\right]^{-\frac{1}{2}} = \left(1 + 2\widetilde{\varepsilon}\sin^2\theta\right)^{-\frac{1}{2}}, \end{aligned}\right\} \tag{22.88}$$

where (22.39), (22.41), and (22.45) have been used. These expressions for the metric factors agree with (22.52).

22.3.4 Apparent Gravity

Apparent gravity g is obtained from (22.8), with h_3^E arbitrarily set to unity (as in Section 22.2.6). Thus

$$g(\theta) = \frac{g^E}{h_3(\theta)} = g^E\left(1 + 2\widetilde{\varepsilon}\sin^2\theta\right)^{\frac{1}{2}}, \tag{22.89}$$

which agrees with (22.47).

[8] μ here should not be confused with the μ appearing in Section 22.2, where it denotes $\sin\theta$.

22.3.5 The BPV Equation

Using the metric factors defined by (22.88), and definition (22.5) of the Coriolis parameter f, BPV equation (22.4) becomes

$$\frac{\partial}{\partial t}(\mathfrak{L}\Psi) + \frac{1}{ac\cos\theta}\frac{\partial(\Psi, \mathfrak{L}\Psi)}{\partial(\lambda, \theta)} + \frac{2\Omega}{c^2\cos\theta}\frac{\partial(\Psi, \sin\theta)}{\partial(\lambda, \theta)} = 0, \tag{22.90}$$

where – from (22.7) –

$$\mathfrak{L}\Psi = \left(1 + 2\tilde{\varepsilon}\sin^2\theta\right)^{\frac{1}{2}}\zeta = \frac{1}{\cos\theta}\left[\frac{\left(1 + 2\tilde{\varepsilon}\sin^2\theta\right)}{a^2\cos\theta}\frac{\partial^2\Psi}{\partial\lambda^2} + \frac{1}{c^2}\frac{\partial}{\partial\theta}\left(\cos\theta\frac{\partial\Psi}{\partial\theta}\right)\right]. \tag{22.91}$$

22.3.6 Meridional Structure Functions

Since the metric factors in this alternative derivation are now known, the meridional structure functions are explicitly determined as in Section 22.2.7. The eigenfunctions are therefore

$$M_n^m(\theta) = P_n^m(\sin\theta), \tag{22.92}$$

where $P_n^m(\sin\theta)$ are associated Legendre functions of the first kind, of order m and degree n, and the corresponding eigenvalues are

$$\Lambda_n^m \equiv \frac{-n(n+1) + e^2m^2}{c^2}. \tag{22.93}$$

22.3.7 A Family of Exact Solutions

Solutions are sought of the form (22.14), where (now that the meridional structure functions are known) $\Psi_n^m(\lambda - \sigma t, \theta)$ is a known function, but $F(\theta)$ has to be determined. This solution form then leads to (22.16), so that

$$\frac{dF}{d\theta} = h_1h_2h_3 = ac\cos\theta, \tag{22.94}$$

where (22.88) for the metric factors has been used. Integrating (22.94) then gives

$$F(\theta) = ac\sin\theta, \tag{22.95}$$

where the integration constant has been arbitrarily set to zero since Ψ, and therefore F, is only determined to within an arbitrary additive constant. Equation (22.95) agrees with (22.50).

Substitution of the assumed solution form (22.14), where $F(\theta)$ and $\Psi_n^m(\lambda - \sigma t, \theta)$ are now known functions, into BPV equation (22.90) then leads (after a fair amount of algebra) to

$$\sigma_n^m = \omega + \frac{2(\Omega + \omega)}{c^2\Lambda_n^m} = \omega - \frac{2(\Omega + \omega)}{\left[n(n+1) - e^2m^2\right]}. \tag{22.96}$$

This dispersion relation agrees with (22.65). The alternative derivation of a family of exact solutions is now complete.

22.4 DIAGNOSIS OF PRESSURE FOR A PARTICULAR SOLUTION

A particular, exact, Rossby–Haurwitz-like solution of the BPV equation was constructed in Section 22.2.9 for flow over a homogeneous ellipsoid. As noted in Section 19.8, it is optional whether or not one computes the corresponding (diagnostic) pressure field (p), since the evolution equation for PV can be integrated forward in time independently of p. It can however be beneficial to do so, not only for aesthetic satisfaction but also for practical reasons. For example, diagnosis of pressure (at any instant of time) from a velocity field is used in Chapter 23 to obtain exact, *unsteady* solutions of the full, unapproximated governing equations under certain conditions. This helps motivate the analysis of this section.

Written in parametric-ellipsoidal coordinates (22.36), diagnostic Poisson equation (19.104) for top pressure $p^T(\lambda,\theta,t)$ is

$$\left[\frac{\partial}{\partial\lambda}\left(\frac{h_2h_3}{h_1}\frac{\partial}{\partial\lambda}\right)+\frac{\partial}{\partial\theta}\left(\frac{h_3h_1}{h_2}\frac{\partial}{\partial\theta}\right)\right]\left[\frac{p^T(\lambda,\theta,t)}{\overline{\rho}}\right] \\ =\frac{\partial}{\partial\lambda}\left[h_2h_3(\zeta+f)u_2\right]-\frac{\partial}{\partial\theta}\left[h_1h_3(\zeta+f)u_1\right] \\ -\left[\frac{\partial}{\partial\lambda}\left(\frac{h_2h_3}{h_1}\frac{\partial K}{\partial\lambda}\right)+\frac{\partial}{\partial\theta}\left(\frac{h_3h_1}{h_2}\frac{\partial K}{\partial\theta}\right)\right], \tag{22.97}$$

where $K=\left(u_1^2+u_2^2\right)/2$ since (from (19.16)) $u_3\equiv 0$ is assumed in the derivation of the BPV equation.[9] For a homogeneous, ellipsoidal planet, apparent gravity is given by (22.74), and the metric (scale) factors in (22.97) by (22.40), that is, by

$$h_1=a\cos\theta,\quad h_2=\left(a^2\sin^2\theta+c^2\cos^2\theta\right)^{\frac{1}{2}}=c\left(1+2\widetilde{\varepsilon}\sin^2\theta\right)^{\frac{1}{2}}. \tag{22.98}$$

Given solution (22.75) for $\Psi=\Psi(\lambda,\theta,t)$, the right-hand side of (22.97) can be computed using (22.76)–(22.78) at any instant in time, and hence the top pressure $p^T=p^T(\lambda,\theta,t)$ can be obtained by solving (22.97). The pressure elsewhere in the domain, $p=p(\lambda,\theta,\xi_3,t)$, can then be determined from (19.35), that is, from

$$p(\lambda,\theta,\xi_3,t)=p^T(\lambda,\theta,t)+\overline{\rho}h_3^Eg^E\left(\xi_3^T-\xi_3\right). \tag{22.99}$$

This is one way of obtaining the pressure field. The downside is the need to solve the 2D, second-order, partial-differential equation (22.97), which is a non-trivial task.

Alternatively, one can obtain $p^T=p^T(\lambda,\theta,t)$ from the horizontal components of momentum balance used to derive (22.97). Horizontal-component equations (19.36) and (19.37) can be rewritten in parametric-ellipsoidal coordinates as

$$\frac{1}{h_1}\frac{\partial}{\partial\lambda}\left[\frac{p^T(\lambda,\theta,t)}{\overline{\rho}}\right]=-\frac{D_{\text{hor}}u_1}{Dt}-\left(\frac{u_1}{h_1}+2\Omega\right)\frac{u_2}{h_2}\frac{dh_1}{d\theta}, \tag{22.100}$$

$$\frac{1}{h_2}\frac{\partial}{\partial\theta}\left[\frac{p^T(\lambda,\theta,t)}{\overline{\rho}}\right]=-\frac{D_{\text{hor}}u_2}{Dt}+\left(\frac{u_1}{h_1}+2\Omega\right)\frac{u_1}{h_2}\frac{dh_1}{d\theta}, \tag{22.101}$$

where (from (19.34))

$$\frac{D_{\text{hor}}}{Dt}=\frac{\partial}{\partial t}+\frac{u_1}{h_1}\frac{\partial}{\partial\lambda}+\frac{u_2}{h_2}\frac{\partial}{\partial\theta}, \tag{22.102}$$

is horizontal material derivative – expressed in parametric-ellipsoidal coordinates – along a geopotential surface $\xi_3=$ constant. Using (22.76) and (22.77) for u_1 and u_2, respectively, the

[9] K here is specific kinetic energy and should not be confused with constant coefficient K_n^m elsewhere in this chapter.

right-hand sides of (22.100) and (22.101) can be evaluated. Equations (22.100) and (22.101) then comprise two coupled *first-order*, partial-differential equations which can be simultaneously integrated to obtain $p^T(\lambda,\theta,t)/\overline{\rho}$. This is the approach adopted here to diagnostically deliver $p^T = p^T(\lambda,\theta,t)/\overline{\rho}$, and thence $p = p(\lambda,\theta,\xi_3,t)$, from solutions of the BPV equation.[10]

Without going into too much detail, the solution procedure is as follows:

1. Insertion of (22.76) and (22.77) for u_1 and u_2 into the right-hand sides of (22.100) and (22.101), respectively, defines the latter explicitly.
2. Examination of the resulting right-hand sides, and inspired by Phillips (1959)'s functional form for the spherical case, solutions are sought of the form

$$\frac{p^T(\lambda,\theta,t)}{\overline{\rho}} = \frac{p^T_{\text{pole}}}{\overline{\rho}} + a^2 A(\theta) + a^2 \cos\left[m\left(\lambda - \sigma^m_{m+1}t\right)\right] B(\theta) + a^2 \cos\left[2m\left(\lambda - \sigma^m_{m+1}t\right)\right] C(\theta). \quad (22.103)$$

Partial differentiation of (22.103) with respect to λ and θ, respectively, leads to

$$\frac{1}{h_1}\frac{\partial}{\partial\lambda}\left(\frac{p^T}{\overline{\rho}}\right) = -\frac{am}{\cos\theta}B(\theta)\sin\left[m\left(\lambda - \sigma^m_{m+1}t\right)\right] - \frac{2am}{\cos\theta}C(\theta)\sin\left[2m\left(\lambda - \sigma^m_{m+1}t\right)\right], \quad (22.104)$$

$$\frac{1}{h_2}\frac{\partial}{\partial\theta}\left(\frac{p^T}{\overline{\rho}}\right) = -\frac{a^2}{c\left(1+2\widetilde{\varepsilon}\sin^2\theta\right)^{\frac{1}{2}}}\frac{dA}{d\theta} - a^2\frac{\cos\left[m\left(\lambda - \sigma^m_{m+1}t\right)\right]}{c\left(1+2\widetilde{\varepsilon}\sin^2\theta\right)^{\frac{1}{2}}}\frac{dB}{d\theta} - a^2\frac{\cos\left[2m\left(\lambda - \sigma^m_{m+1}t\right)\right]}{c\left(1+2\widetilde{\varepsilon}\sin^2\theta\right)^{\frac{1}{2}}}\frac{dC}{d\theta}. \quad (22.105)$$

3. Comparison of the coefficients of $\sin\left[m\left(\lambda - \sigma^m_{m+1}t\right)\right]$ and $\sin\left[2m\left(\lambda - \sigma^m_{m+1}t\right)\right]$, respectively, between (22.104) and the computed right-hand side of (22.100) determines $B(\theta)$ and $C(\theta)$.
4. Differentiation of the now-known expressions for $B(\theta)$ and $C(\theta)$, with insertion of the results into (22.105), means that the only unknown remaining on the right-hand side of (22.105) is $dA/d\theta$. Comparison of coefficients between the recomputed right-hand side of (22.105) and the computed right-hand side of (22.101) then leads to a first-order differential equation for $dA/d\theta$.
5. Integration of this ordinary differential equation determines $A(\theta)$.
6. Insertion of the computed expressions for $A(\theta)$, $B(\theta)$, and $C(\theta)$ into (22.103) yields pressure $p^T = p^T(\lambda,\theta,t)$ at the top (geopotential) boundary $\xi_3 = \xi_3^T$. Total pressure $p = p(\lambda,\theta,\xi_3,t)$ elsewhere in the domain then follows from (22.99).

To simplify various expressions when applying this procedure, use is made of identity (22.41), namely

$$2\widetilde{\varepsilon} \equiv \frac{a^2 - c^2}{c^2} \equiv e^2\frac{a^2}{c^2}. \quad (22.106)$$

[10] For this alternative solution procedure to be equivalent to solving diagnostic pressure equation (22.97), u_1 and u_2 must additionally satisfy the 3D non-divergence condition (22.3) (with $(\xi_1,\xi_2) \to (\lambda,\theta)$). This condition is automatically satisfied since, by construction, u_1 and u_2 are expressed in (22.2) (again with $(\xi_1,\xi_2) \to (\lambda,\theta)$) in terms of a stream function, Ψ. If u_1 and u_2 were not constrained in this way, then the equivalence would be lost, and the alternative procedure would be invalid.

Note that solutions $p^T(\lambda, \theta, t)$ of (second-order) elliptic-boundary-value problem (22.97) and coupled (first-order) problems (22.100) and (22.101) are only determined to within an arbitrary additive constant. As noted in Phillips (1959), the prescribed value of p^T_{pole} determines the horizontal average of p^T.

Summarising, results from application of this procedure are:

The Balancing Pressure for a Particular Exact Rossby–Haurwitz-Like Solution of the BPV Equation in Parametric Ellipsoidal Coordinates

The balancing pressure for the exact Rossby–Haurwitz-like solution, given in Section 22.2.9, of the BPV equation in parametric-ellipsoidal coordinates is

$$p(\lambda, \theta, \xi_3, t) = p^T(\lambda, \theta, t) + \overline{\rho} h^E_3 g^E \left(\xi^T_3 - \xi_3\right), \tag{22.107}$$

where:

$$\begin{aligned}\frac{p^T(\lambda, \theta, t)}{\overline{\rho}} = \frac{p^T_{\text{pole}}}{\overline{\rho}} &+ a^2 A(\theta) + a^2 \cos\left[m\left(\lambda - \sigma^m_{m+1} t\right)\right] B(\theta) \\ &+ a^2 \cos\left[2m\left(\lambda - \sigma^m_{m+1} t\right)\right] C(\theta)\end{aligned} \tag{22.108}$$

defines the pressure at the top (geopotential) boundary $\xi_3 = \xi^T_3$; and

$$\begin{aligned}A(\theta) = &\left(\Omega + \frac{\omega}{2}\right)\omega \cos^2\theta - \frac{\left(K^m_{m+1}\right)^2 m^2}{2} \cos^{2m-2}\theta \\ &+ \frac{\left(K^m_{m+1}\right)^2}{4}\left[\left(2m^2 - m - 2\right) + 2e^2 m^2\right] \cos^{2m}\theta \\ &+ \frac{\left(K^m_{m+1}\right)^2}{4}\left(m + 1 - 2e^2 m^2\right) \cos^{2m+2}\theta,\end{aligned} \tag{22.109}$$

$$\begin{aligned}B(\theta) = \; &2(\Omega + \omega) K^m_{m+1}\left[\frac{m^2 + 2m + 2 - e^2 m^2}{(m+1)(m+2) - e^2 m^2}\right] \cos^m\theta \\ &- 2(\Omega + \omega) K^m_{m+1}\left[\frac{(m+1)^2 - e^2 m^2}{(m+1)(m+2) - e^2 m^2}\right] \cos^{m+2}\theta,\end{aligned} \tag{22.110}$$

$$C(\theta) = \frac{\left(K^m_{m+1}\right)^2}{4}\left[-(m+2)\cos^{2m}\theta + (m+1)\cos^{2m+2}\theta\right]. \tag{22.111}$$

Taking into account differences in notation and setting $c \equiv a \Rightarrow e \equiv 0$, expressions (22.109)–(22.111) for $A(\theta)$, $B(\theta)$, and $C(\theta)$ agree with the corresponding ones given in Phillips (1959) for the (special) spherical case (with constant apparent gravity).

The arbitrary constant in (22.108) is p^T_{pole}, since $A(\theta)$, $B(\theta)$, and $C(\theta)$ are all identically zero at a pole, where $\theta = \pm\pi/2$.

22.5 DIAGNOSIS OF PRESSURE FOR A FAMILY OF SOLUTIONS

A family of exact, Rossby–Haurwitz-like solutions of the BPV equation was constructed in Section 22.2.8 for flow confined between two rigid, ellipsoidal geopotential surfaces. The balancing pressure for a particular solution of this family was then obtained in Section 22.4. Adapting the solution procedure used for this, it is also possible to obtain the corresponding balancing pressure for the more general family of solutions. Thus:

The Balancing Pressure for a Family of Exact Rossby–Haurwitz-Like Solutions over an Ellipsoid

The balancing pressure for a family of exact Rossby–Haurwitz-like solutions, given in Section 22.2.8, of the BPV equation in parametric-ellipsoidal coordinates is

$$p\left(\lambda,\mu,\xi_3,t\right)=p^T\left(\lambda,\mu,t\right)+\overline{\rho}h_3^E g^E\left(\xi_3^T-\xi_3\right), \tag{22.112}$$

where

$$\frac{p^T\left(\lambda,\mu,t\right)}{\overline{\rho}}=\frac{p^T_{\text{pole}}}{\overline{\rho}}+a^2A\left(\mu\right)+a^2\cos\left[m\left(\lambda-\sigma^m_{m+1}t\right)\right]B\left(\mu\right) + a^2\cos\left[2m\left(\lambda-\sigma^m_{m+1}t\right)\right]C\left(\mu\right) \tag{22.113}$$

defines the pressure at the top (geopotential) boundary $\xi_3=\xi_3^T$; and

$$\mu\equiv\sin\theta \quad\Rightarrow\quad \left(1-\mu^2\right)^{\frac{1}{2}}\equiv\cos\theta,\quad -\frac{\pi}{2}\le\theta\le\frac{\pi}{2}, \tag{22.114}$$

$$A\left(\mu\right)=\left(\Omega+\frac{\omega}{2}\right)\omega\left(1-\mu^2\right)+C\left(\mu\right)-\frac{\left(K^m_n\right)^2}{2}\left[\frac{m^2}{\left(1-\mu^2\right)}-e^2m^2\right]\left[P^m_n\left(\mu\right)\right]^2, \tag{22.115}$$

$$B\left(\mu\right)=2\left(\Omega+\omega\right)K^m_n\left\{\frac{\left(1-\mu^2\right)}{\left[n\left(n+1\right)-e^2m^2\right]}\frac{dP^m_n\left(\mu\right)}{d\mu}+\mu P^m_n\left(\mu\right)\right\}, \tag{22.116}$$

$$C\left(\mu\right)=-\frac{\left(K^m_n\right)^2}{4}\left\{\left(1-\mu^2\right)\left[\frac{dP^m_n\left(\mu\right)}{d\mu}\right]^2+\left[n\left(n+1\right)-\frac{m^2}{\left(1-\mu^2\right)}\right]\left[P^m_n\left(\mu\right)\right]^2\right\}, \tag{22.117}$$

$$\frac{dP^m_n\left(\mu\right)}{d\mu}\equiv\frac{\left[\left(n+1\right)\left(n+m\right)P^m_{n-1}\left(\mu\right)-n\left(n-m+1\right)P^m_{n+1}\left(\mu\right)\right]}{\left(2n+1\right)\left(1-\mu^2\right)}. \tag{22.118}$$

To obtain the preceding results, only two recurrence relations for solutions of the associated Legendre equation

$$\left(1-\mu^2\right)\frac{d^2P^m_n\left(\mu\right)}{d\mu^2}-2\mu\frac{dP^m_n\left(\mu\right)}{d\mu}+\left[n\left(n+1\right)-\frac{m^2}{\left(1-\mu^2\right)}\right]P^m_n\left(\mu\right)=0 \tag{22.119}$$

were needed; specifically (for fixed m and variable n)

$$\left(n-m+1\right)P^m_{n+1}\left(\mu\right)=\left(2n+1\right)\mu P^m_n\left(\mu\right)-\left(n+m\right)P^m_{n-1}\left(\mu\right), \tag{22.120}$$

$$\left(1-\mu^2\right)\frac{dP^m_n\left(\mu\right)}{d\mu}=\frac{1}{\left(2n+1\right)}\left[\left(n+1\right)\left(n+m\right)P^m_{n-1}\left(\mu\right)-n\left(n-m+1\right)P^m_{n+1}\left(\mu\right)\right]. \tag{22.121}$$

Suitably shifting indices of n by one in these two equations and combining the resulting relations leads to the two further useful relations (again for fixed m)

$$\left(1-\mu^2\right)\frac{dP^m_{n+1}\left(\mu\right)}{d\mu}=-\left(n+1\right)\mu P^m_{n+1}\left(\mu\right)+\left(n+m+1\right)P^m_n\left(\mu\right), \tag{22.122}$$

$$\left(1-\mu^2\right)\frac{dP_{n-1}^m(\mu)}{d\mu}=n\mu P_{n-1}^m(\mu)-(n-m)\,P_n^m(\mu)\,. \tag{22.123}$$

Expressions (22.115)–(22.117) for the coefficients of the balancing pressure of the general case are pleasingly compact. The spherical result may be obtained by setting eccentricity e identically zero. The balancing pressure, given in Section 22.4, for a particular member of the general family is recovered by:

- Setting $n \equiv m+1$ in (22.113)–(22.118).
- Using identity

$$P_{m+1}^m(\mu)=\mu\,(2m+1)\,P_m^m(\mu)\,. \tag{22.124}$$

- Using identity (22.120) with $n=m+1$.
- Using (22.70) with $\mu \equiv \sin\theta$.

This is reassuring.

22.6 CONCLUDING REMARKS

To provide further physical insight, a family of *exact* unsteady solutions, in closed form, of the *non-linear* BPV equation has been constructed in parametric-ellipsoidal coordinates. Due to the strong mutual coupling between the planet's shape and its apparent gravity, it was not obvious, a priori, that this is possible. In the spherical context, such exact solutions certainly do exist; they are the well-known Rossby–Haurwitz waves. To address the tractability challenge in spheroidal coordinates, the problem was turned around. Instead of posing the question:

- For a chosen geopotential representation (i.e. representation of apparent gravity), can exact closed-form solutions of the BPV equation be obtained?

it was effectively reposed as:

- Can any exact closed-form solutions be obtained in spheroidal coordinates that are analogous to Rossby–Haurwitz waves in spherical-polar coordinates?

This builds in the strong coupling between the shape of the planet and the representation of its apparent gravity.

Happily, although not trivial to apply, and it could have conceivably turned out otherwise, this tactic led to tractable solution of the BPV equation in parametric-ellipsoidal coordinates in Section 22.2. The solution procedure was carried through almost to its conclusion for an *arbitrary*, zonally symmetric spheroid. It is only at the last step that a *particular* form (i.e. an ellipsoid) for the spheroid had to be assumed.

For an ellipsoidal planet, it was found that there is one *and only one* model of apparent gravity for which exact solutions exist that are analogous to Rossby–Haurwitz waves in spherical-polar coordinates.[11] This model corresponds to an ellipsoidal planet of uniform density, as originally assumed by Newton. Approximately two thirds of the observed meridional variation of apparent gravity is represented by this model. This is much better than the zero variation of existing, 3D, global models of Earth's atmosphere and oceans, all of which (to the author's knowledge) employ the classical spherical-geopotential approximation.

The family of exact solutions is based on the eigenfunctions of a linear, second-order, partial-differential operator (even though the solution itself satisfies the *non-linear* BPV equation). Somewhat surprisingly, these eigenfunctions turn out to be associated Legendre functions.[12]

[11] If any other model is chosen, then the solution procedure breaks down.

[12] The naïve expectation was that they would be some kind of elliptic function. However, although associated Legendre functions are commonly associated with spherical geometry, they do also occur naturally in other geometries (Lebedev, 1972).

The propagation frequency of an exact solution is given by (22.65). Examination of this equation reveals that meridional variation of apparent gravity (via the planet's associated non-zero ellipticity) changes the frequency of propagation. This introduces a dependency on ellipticity, e, and zonal wave number, m, both of which are absent from the corresponding equation in spherical-polar coordinates. The change corresponds to an increase of the beta effect (given that $c < a$). This is qualitatively what one would expect given that the geopotentials are farther apart at the equator than they are at the poles. Poleward motion leads to vortex compression and reduction of relative vorticity, and this reinforces the beta effect (as planetary vorticity increases from south pole to north pole, through zero at the planet's equator).

With the benefit of hindsight, a more direct derivation of exact solutions has also been outlined. It does, however, presuppose that one knows, a priori, what model of apparent gravity will lead to exact solutions! This is something of a guessing game. If one guesses correctly (as implicitly done in Section 22.3), then everything will nicely fall into place. But if one guesses incorrectly then, after quite some wasted effort, an impasse will be reached. Furthermore, even if one does guess correctly, one still does not know whether exact solutions exist for other models of gravity (e.g. for an inhomogenous distribution of the planet's mass). By way of contrast, the present approach is *deductive*; one determines under what conditions (if any) it is possible to find an exact solution that is analogous to Rossby–Haurwitz waves in spherical-polar coordinates. It is therefore a surer approach.

It is customary when using Rossby–Haurwitz waves for numerical experimentation in spherical-polar coordinates to set $n \equiv m + 1$, where m is zonal wave number and n is total wave number. An analogous solution of the BPV equation in parametric-ellipsoidal coordinates has been developed here, together with the corresponding balancing pressure. Having shown that it is possible to obtain the balancing pressure for the *particular* case of a family of exact solutions of the BPV equation, the corresponding balancing pressure for the *general* family was then given.

These solutions are used in Chapter 23 to construct *unsteady* solutions of Rossby–Haurwitz type that exactly satisfy the 3D, *non-linear*, dynamical-core equations of global atmospheric and oceanic models over an ellipsoidal planet of homogeneous composition. This then leads to the definition of some *unsteady* test problems to quantitatively measure the accuracy of 3D atmospheric and oceanic dynamical cores for the geophysically important problem of (unsteady) Rossby–Haurwitz-like solutions.

Finally, it is possible in *spherical-polar coordinates* – see Rochas (1986) – to enrich the solution form (22.60) for Ψ by:

- Adding a term in $P_n\left(\sin\theta\right)$.
- And/or taking a sum over m of $K_n^m L^m\left(\lambda - \sigma t\right) P_n^m\left(\sin\theta\right)$ (instead of taking just a single term), whilst keeping n fixed.

It is natural to wonder if such an enrichment is also possible in parametric-ellipsoidal coordinates. This possibility is an open question.[13]

[13] The reader is advised, however, that cursory examination suggests (but certainly does not prove) that neither enrichment is possible. If this speculation is indeed true, it might be due to the classical spherical-geopotential approximation creating spurious 'axial supersymmetries' that do not exist in the more general, ellipsoidal context. A sphere is symmetrical about *any* axis passing through its centre, but this is not so for an ellipsoid. Any solution that depends on an 'axial supersymmetry' for its existence will then no longer hold.

23

Exact Unsteady Solutions in 3D over an Ellipsoidal Planet

ABSTRACT

Solutions of Rossby–Haurwitz type are developed that exactly satisfy the 3D, non-linear, dynamical-core equations of global atmospheric and oceanic models over an ellipsoidal planet of homogeneous composition. Apparent gravity varies meridionally, but not zonally nor vertically. The fluid is homogeneous, of shallow depth, and confined between two rigid geopotential surfaces. That the top boundary is rigid is crucial to obtaining exact solutions. The flow is purely horizontal for all time and barotropic/homentropic. An ongoing challenge to numerical modellers is to develop global dynamical cores that run efficiently on evolving, massively parallel computer architectures. These cores generally use an underlying quasi-uniform global grid. They are susceptible to 'grid imprinting', whereby the structure of an underlying grid spuriously manifests itself during model forecasts, thereby adversely affecting accuracy. Test problems to identify and quantify these errors are therefore valuable. Three exploratory avenues are outlined for the design of unsteady test problems. The first is applicable in both ellipsoidal and spherical geometry, whereas the other two are limited to spherical geometry. The first defines an individual propagating wave. It is particularly simple; this facilitates diagnosis of the source of problems observed during model integrations. The second defines a sum of multiple waves, each of which propagates with its own constant amplitude and constant angular phase speed. The third defines a sum of multiple rotated waves. These waves individually propagate with variable amplitude and angular phase speed and, contrary to the first two avenues, exhibit cross-polar flow.

23.1 PREAMBLE

Test problems are very valuable for developing, validating and assessing dynamical cores of atmospheric models (Williamson et al., 1992; Lauritzen et al., 2010; Ullrich et al., 2014). Analytical solutions of the governing, 3D, dynamical-core equations are attractive candidates for these test problems. Most such solutions are, however, of a very basic character and leave important aspects of model formulation untried. In particular there is a dearth of *exact*, *unsteady*, *non-linear* solutions.

In this regard, Staniforth and White (2008b) extended Läuter et al. (2005)'s exact, unsteady solution for the shallow-water equations to 3D. Whilst useful, their 3D solution nevertheless has two limitations (shared with the original, 2D, Läuter et al. (2005) one):

1. It necessarily employs a forcing, without which time dependence is lost.
2. The frequency is inherently inertial.

A more realistic and more challenging test problem would be one that:

- *Exactly* represents geophysically important wave propagation.
- Can be configured for a broad range of wavelengths and frequencies.
- Can be applied to flow over ellipsoidal planets as well as over spherical ones.
- Can be adapted to test both global atmospheric and oceanic dynamical cores.

It is well known that Rossby–Haurwitz waves are exact solutions of the classical, 2D, non-divergent, barotropic-vorticity equation in spherical geometry (Rochas, 1984, 1986). However, it is also well known that these waves are *not* exact solutions of the more general, 2D, shallow-water equations (Phillips, 1959; Hoskins, 1973; Williamson et al., 1992), but only approximate ones. At first sight it therefore appears unlikely that analogous Rossby–Haurwitz waves can be found that are exact solutions of 3D, dynamical-core, equation sets in spherical-polar coordinates, let alone in ellipsoidal ones. The purpose of this chapter is to demonstrate that, happily, it is in fact possible with:

- An appropriate choice of fluid composition.
- An appropriate choice of bottom and top boundary conditions.
- The application of some further configuration conditions.

This then provides the foundation for a set of unsteady test problems.

This chapter is organised as follows. To set the scene, a quartet of 3D dynamical-core equation sets in axial-orthogonal-curvilinear coordinates is presented in Section 23.2; they correspond to those examined in Chapter 16. Some preparatory steps are described in Section 23.3. Exact Rossby–Haurwitz-like solutions of the *barotropic potential-vorticity* (BPV) equation are then recalled in Section 23.4, together with their corresponding balancing pressure. With this preparation, a family of exact unsteady solutions of 3D dynamical-core equation sets is developed in Section 23.5, with a particularly simple member of this family given in Section 23.6. The crucial role that the top boundary condition plays in the feasibility of obtaining exact Rossby–Haurwitz-like solutions is discussed in Section 23.7. Three possible avenues for designing unsteady test problems for 3D dynamical cores are outlined in Section 23.8. Finally, concluding remarks are given in Section 23.9.

23.2 A QUARTET OF EQUATION SETS FOR UNFORCED 3D FLUID FLOW OVER A ROTATING ELLIPSOIDAL PLANET

23.2.1 Unforced 3D Flow over a Rotating Ellipsoidal Planet

Consider unforced fluid flow, confined between two geopotential surfaces, over a rotating ellipsoidal planet. See Fig. 23.1 for a cross section of an ellipsoidal planet with overlying fluid. The governing equations for this flow will first be expressed in general axial-orthogonal-curvilinear coordinates and later specialised to parametric-ellipsoidal coordinates for solution tractability.

23.2.2 Axial-Orthogonal-Curvilinear Coordinates

Let (ξ_1, ξ_2, ξ_3) be general axial-orthogonal-curvilinear coordinates in the zonal, meridional, and (upward) vertical directions, respectively, with associated unit vectors $(\mathbf{e}_1, \mathbf{e}_2, \mathbf{e}_3)$ and velocity components (u_1, u_2, u_3). The metric (or scale) factors (h_1, h_2, h_3) are the quantities that appear in

$$ds^2 = h_1^2 d\xi_1^2 + h_2^2 d\xi_2^2 + h_3^2 d\xi_3^2, \tag{23.1}$$

where ds is infinitesimal distance. Standard expressions for gradient, curl, and divergence for these coordinates are given in Chapter 5. For axial-orthogonal-curvilinear coordinates – see (6.13) – h_1, h_2, and h_3 are, by definition, all independent of ξ_1, that is,

$$h_i = h_i(\xi_2, \xi_3), \quad i = 1, 2, 3. \tag{23.2}$$

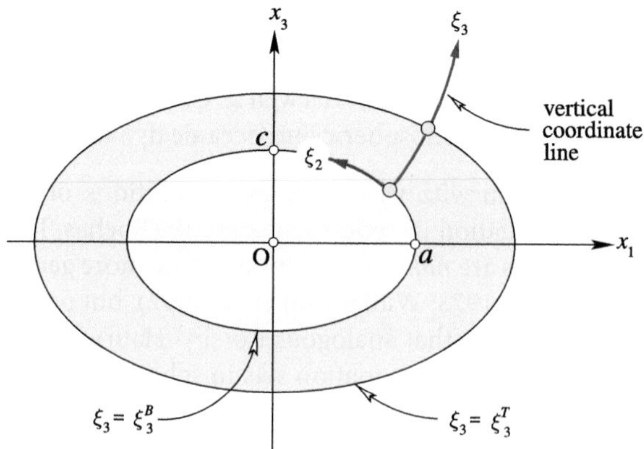

Figure 23.1 A cross section (not to scale) in the Ox_1x_3 meridional plane for flow over an ellipsoid. ξ_2 and ξ_3 are axial-orthogonal-curvilinear coordinates in the meridional and outward-pointing directions, respectively. ξ_1, the azimuthal coordinate, points into the page for $x_1 > 0$, and out of it for $x_1 < 0$. The ellipsoid is light shaded and has semi-major and semi-minor axes a and c, respectively. The overlying shallow layer of fluid (dark shaded) is confined between the bottom and top geopotential surfaces located at $\xi_3 = \xi_3^B$ and $\xi_3 = \xi_3^T$, respectively. See text for further details.

23.2.3 Components of the Momentum Balance

Assume a geopotential coordinate system so that, by definition, $\Phi = \Phi(\xi_3)$. For unforced flow, expressed in an axial-orthogonal-curvilinear, *geopotential* coordinate system, (16.19)–(16.21) give:

The Components of the Momentum Balance

$$\frac{Du_1}{Dt} + \left(\frac{u_1}{h_1} + 2\Omega\right)\frac{u_2}{h_2}\frac{\partial h_1}{\partial \xi_2} + \underbrace{\left(\frac{u_1}{h_1} + 2\Omega\right)\frac{u_3}{h_3}\frac{\partial h_1}{\partial \xi_3}}_{\text{deep}} = -\frac{1}{\rho h_1}\frac{\partial p}{\partial \xi_1}, \tag{23.3}$$

$$\frac{Du_2}{Dt} - \left(\frac{u_1}{h_1} + 2\Omega\right)\frac{u_1}{h_2}\frac{\partial h_1}{\partial \xi_2} - \underbrace{\frac{u_3^2}{h_2 h_3}\frac{\partial h_3}{\partial \xi_2}}_{\text{non-hydro}} + \underbrace{\frac{u_2}{h_2}\frac{u_3}{h_3}\frac{\partial h_2}{\partial \xi_3}}_{\text{deep}} = -\frac{1}{\rho h_2}\frac{\partial p}{\partial \xi_2}, \tag{23.4}$$

$$\underbrace{\left(\frac{Du_3}{Dt} + \frac{u_2}{h_2}\frac{u_3}{h_3}\frac{\partial h_3}{\partial \xi_2}\right)}_{\text{non-hydro}} - \underbrace{\left[\left(\frac{u_1}{h_1} + 2\Omega\right)\frac{u_1}{h_3}\frac{\partial h_1}{\partial \xi_3} + \frac{u_2^2}{h_2 h_3}\frac{\partial h_2}{\partial \xi_3}\right]}_{\text{deep}} + \frac{1}{h_3}\frac{d\Phi}{d\xi_3} = -\frac{1}{\rho h_3}\frac{\partial p}{\partial \xi_3}, \tag{23.5}$$

where various terms have been colour coded for later convenience. In (23.3)–(23.5), ρ is density, p is pressure, and differentiation D/Dt following the fluid satisfies

$$\frac{D}{Dt} = \frac{\partial}{\partial t} + \frac{u_1}{h_1}\frac{\partial}{\partial \xi_1} + \frac{u_2}{h_2}\frac{\partial}{\partial \xi_2} + \frac{u_3}{h_3}\frac{\partial}{\partial \xi_3}. \tag{23.6}$$

Due to the use of a geopotential coordinate system, no horizontal components of apparent gravity appear in (23.3) and (23.4), since the geopotential (Φ), by definition, is constant everywhere on a geopotential surface.

Metric Factors for a Shallow Fluid

For a *shallow fluid*, metric factors (h_1, h_2, h_3) are not only independent of ξ_1 (due to axial symmetry of the coordinate system) but also independent of vertical coordinate ξ_3, and (23.2) reduces to

$$h_i = h_i(\xi_2), \quad i = 1, 2, 3. \tag{23.7}$$

The red ('deep') terms in (23.3)–(23.5) are then all identically zero (and therefore absent).

A very nice feature of (23.3)–(23.5) – developed in Section 6.4 – is the way in which the Coriolis terms are alternatively expressed in terms of metric factors. This has allowed the Coriolis terms to be grouped together in a neat way with the metric terms. Importantly, it also greatly facilitates optional application of the shallow-fluid assumption in a deep-fluid model. Thus, if a deep-fluid model is formulated in terms of metric factors (e.g. as in Wood et al., 2014), optionally replacing the deep metric factors by shallow ones at all model levels automatically changes the deep model into a corresponding shallow one, *with no changes or switches needed elsewhere.* This facilitates controlled experimentation to measure the impact of deep terms, without the need for extensive reformulation and recoding of a model.

Application of the Hydrostatic or Quasi-Hydrostatic Approximation

To apply the hydrostatic or quasi-hydrostatic approximation, the blue ('non-hydro') terms are set identically zero in (23.5) (and are therefore absent).

Note that for a deep fluid, the *quasi*-hydrostatic approximation does not correspond to *exact* hydrostatic balance as usually defined (due to the presence of the 'deep' terms in (23.5)). Furthermore, application of the hydrostatic or quasi-hydrostatic approximation means that the vertical velocity (u_3) is no longer a *prognostic* variable – it is then absent from (23.5) – but must be determined *diagnostically*.

23.2.4 The Mass-Continuity Equation

For unforced flow, expressed in an axial-orthogonal-curvilinear, geopotential coordinate system, (16.22) – with $\nabla \cdot \mathbf{u}$ defined by (16.27) – gives:

The Mass-Continuity Equation

$$\frac{D\rho}{Dt} + \frac{\rho}{h_1 h_2 h_3}\left[\frac{\partial}{\partial \xi_1}(u_1 h_2 h_3) + \frac{\partial}{\partial \xi_2}(u_2 h_3 h_1) + \frac{\partial}{\partial \xi_3}(u_3 h_1 h_2)\right] = 0. \tag{23.8}$$

For a *deep* fluid, the metric factors in (23.8) satisfy (23.2). For a *shallow* fluid, they instead satisfy (23.7).

23.2.5 The Thermodynamic-Energy and State Equations

To complete the formulation of the governing equations for unforced fluid flow over a rotating spheroidal planet (including an ellipsoidal one as a special case), (23.3)–(23.5) and (23.8) are supplemented by:

The Thermodynamic-Energy and State Equations

$$\frac{D\mathscr{E}}{Dt} + p\frac{D\alpha}{Dt} = 0, \tag{23.9}$$

$$\mathscr{E} = \mathscr{E}\left(\alpha, \eta\right), \tag{23.10}$$

where $\mathscr{E}$ is specific internal energy. These two equations correspond to (16.23) and (16.28), respectively. The natural variables of $\mathscr{E}$ are α (specific volume) and η (specific entropy); see Section 9.2.

It is important to note that the prescribed functional form for $\mathscr{E}$ in (23.10) defines the thermodynamic properties of the fluid. For the purposes of constructing exact solutions for testing dynamical cores, one can therefore (within reason) choose *any* convenient functional form that leads to exact solutions. This property is exploited herein. See Section 23.8.5.1 for two prototypical, functional forms for $\mathscr{E}$; one atmospheric, the other oceanic.

23.2.6 A Quartet of Equation Sets

Equations (23.3)–(23.10) collectively govern unforced fluid flow over a rotating spheroidal planet. As described in Chapter 16, combining the deep/shallow option with the non-hydrostatic/(quasi-)hydrostatic option leads to:

A Quartet of Equation Sets

1. The *non-hydrostatic deep* (or fully compressible) equations; by retaining all terms in (23.3)–(23.5).
2. The *quasi-hydrostatic deep* (or quasi-hydrostatic) equations; by omitting the 'non-hydro' (blue) terms.
3. The *non-hydrostatic shallow* equations; by omitting the 'deep' (red) terms.
4. The *hydrostatic shallow* (or hydrostatic primitive) equations; by omitting all 'non-hydro' (blue) and 'deep' (red) terms.

Most quantitative global atmospheric and oceanic prediction models are built on and around one or other of these four equation sets.

23.3 PREPARATORY STEPS

A BPV (barotropic potential vorticity) equation was obtained in Chapter 19 for a shallow global atmosphere or ocean confined between two rigid, spheroidal, geopotential surfaces, with the bottom one coinciding with a planet's surface. To accomplish this, a number of simplifying assumptions were made regarding the composition of the fluid and the characteristics of its flow. A family of exact, unsteady, non-linear solutions of this equation was then developed in Chapter 22 for an ellipsoidal planet.

In Section 23.5, we exploit these solutions to obtain, under somewhat similar conditions, a family of exact solutions for testing 3D atmospheric and oceanic dynamical cores. To obtain the exact solutions of Chapter 22, it was sufficient to simultaneously satisfy the momentum balance equations (23.3)–(23.5) and the mass-continuity equation (23.8). Here *we additionally need to simultaneously satisfy thermodynamic-energy and state equations* (23.9) *and* (23.10), *respectively*. See Section 23.8.5.1 for how this may be done in the present context of developing exact solutions for testing 3D dynamical cores.

23.3.1 Fluid Composition and Flow Characteristics

Guided by the developments of Chapters 19 and 22, we make the following

Assumptions for Fluid Composition and Flow

1. The planet's surface is restricted to be *ellipsoidal*, and it coincides with a geopotential surface – see Fig. 23.1. Vertical geopotential coordinate ξ_3 is chosen so that the planet's surface corresponds to $\xi_3 = \xi_3^B = \text{constant}$.
2. An overlying layer of fluid is confined between bottom and top, rigid geopotential surfaces located at $\xi_3 = \xi_3^B = \text{constant}$ and $\xi_3 = \xi_3^T = \text{constant}$, respectively.
3. The fluid is homogeneous, that is,

$$\rho = \overline{\rho} = \text{constant} \quad \Rightarrow \quad \alpha \equiv \frac{1}{\rho} = \frac{1}{\overline{\rho}} = \overline{\alpha} = \text{constant}. \tag{23.11}$$

4. The flow is purely horizontal for all time, that is,

$$u_3 \equiv 0, \quad \text{for all time,} \tag{23.12}$$

and the horizontal velocity components u_1 and u_2 are independent of vertical coordinate ξ_3 for all time, that is,

$$u_1 = u_1(\xi_1, \xi_2, t), \quad u_2 = u_2(\xi_1, \xi_2, t), \quad \text{for all time.} \tag{23.13}$$

5. The depth of the fluid layer is shallow and the (shallow) metric factors satisfy (23.7). This assumption is crucial for the flow to remain horizontal if initially so. It prevents the 'deep' terms in (23.5) generating vertical motion. For the flow to remain horizontal, it is crucial that hydrostatic balance be maintained: cf. (23.5) and (23.22).

Assumption (23.12) should be considered to be a working hypothesis, to be verified retrospectively. Solutions are sought that satisfy $u_3 \equiv 0$. When putative solutions with this property have been obtained, it may then be verified by direct substitution that they do indeed satisfy all relevant equations.

23.3.2 Parametric-Ellipsoidal Coordinates and Model of Gravity

To obtain a family of exact Rossby–Haurwitz-like solutions, it was seen in Chapter 22 that the parametric-ellipsoidal coordinate system (λ, θ, ξ_3) and the model of gravity are intrinsically linked by:

The Constraint between the Vertical Metric Factor (h_3) and Gravity (g)

$$h_3(\theta)\, g(\theta) = g^E = \text{constant}, \tag{23.14}$$

where λ is longitude, θ is parametric latitude,[1] and g^E is the value of g at any point on the planet's equator.[2] It was also found – in Section 22.2.6 – that exact Rossby–Haurwitz-like solutions are only possible for a particular model of apparent gravity, namely (22.47). This gives:

The Representation of Gravity (g)

$$g(\theta) = g^E \left(1 + 2\widetilde{\varepsilon} \sin^2 \theta\right)^{\frac{1}{2}} = g^E \left(1 + e^2 \frac{a^2}{c^2} \sin^2 \theta\right)^{\frac{1}{2}}, \tag{23.15}$$

where – from (22.106) –

$$2\widetilde{\varepsilon} \equiv \frac{a^2 - c^2}{c^2} = e^2 \frac{a^2}{c^2}, \tag{23.16}$$

and e is eccentricity.[3] The corresponding geopotential for this model of gravity is – from (22.48) –

The Representation of Geopotential (Φ)

$$\Phi(\xi_3) = \Phi^B + g^E \left(\xi_3 - \xi_3^B\right), \tag{23.17}$$

where $\Phi\left(\xi_3 = \xi_3^B\right) = \Phi^B =$ constant on the planet's surface $\xi_3 = \xi_3^B =$ constant, and h_3^E has been set to unity in (22.48).

Representation (23.15) corresponds to the gravitational attraction exerted by an ellipsoidal planet of uniform density on a shallow layer of fluid overlying it. From (22.52), this corresponds to:

The Shallow Metric Factors for Parametric Ellipsoidal Coordinates

$$h_1 = a\cos\theta, \quad h_2 = c\left(1 + 2\widetilde{\varepsilon} \sin^2 \theta\right)^{\frac{1}{2}}, \quad h_3 = \left(1 + 2\widetilde{\varepsilon} \sin^2 \theta\right)^{-\frac{1}{2}}. \tag{23.18}$$

Recalling (23.15), it is seen that metric factors (23.18) do satisfy constraint (23.14).

23.3.3 The Decomposition of Total Pressure p

It was found in Section 22.4 that integrating hydrostatic equation (19.32) – compare (23.22) – downwards[4] from the top, bounding geopotential surface at $\xi_3 = \xi_3^T =$ constant leads to:

[1] For a spherical planet, parametric latitude (θ) coincides with geographic latitude (ϕ), and $h_3 \equiv 1 \Rightarrow g =$ constant.

[2] Scaling constant h_3^E in (22.8) has been arbitrarily set to unity as justified in Section 22.2.6. This amounts to using equatorial radius (a) as the unit of length.

[3] For a spherical planet, $\widetilde{\varepsilon} = e = 0$ and (23.15) reduces to $g = g^E =$ constant.

[4] Hydrostatic equation (23.22) could instead be integrated *upwards* from $\xi_3 = \xi_3^B$. This would give $p(\lambda, \theta, \xi_3, t) = p^B(\lambda, \theta, t) - \overline{\rho} g^E \left(\xi_3 - \xi_3^B\right)$. Doing so would, however, unnecessarily impose a limit on the depth of the fluid. (Otherwise p^T could become negative, which is unphysical.) Formulating the problem in terms of p^T (≥ 0), as in (23.19), avoids this limitation. For increasing ξ_3^T, and using (23.22), p^B then simply increases correspondingly according to $p^B(\lambda, \theta, t) = p^T(\lambda, \theta, t) + \overline{\rho} g^E \left(\xi_3^T - \xi_3^B\right) > p^T(\lambda, \theta, t) \geq 0$.

The Decomposition of Total Pressure (p)

$$p(\lambda,\theta,\xi_3,t) = p^T(\lambda,\theta,t) + \overline{\rho}g^E\left(\xi_3^T - \xi_3\right), \tag{23.19}$$

where:

- $p^T(\lambda,\theta,t) \equiv p\left(\lambda,\theta,\xi_3 = \xi_3^T,t\right)$ is pressure at the model's top (rigid) surface, $\xi_3 = \xi_3^T =$ constant.
- $\Phi^T \equiv \Phi\left(\xi_3 = \xi_3^T\right) =$ constant.
- h_3^E has again been set to unity.

Total pressure p therefore decomposes into the sum of a *horizontally varying and temporally varying* contribution, $p^T(\lambda,\theta,t)$, and a *vertically varying (only)* contribution, $\overline{\rho}g^E\left(\xi_3^T - \xi_3\right)$.

23.3.4 Application of the Preceding Developments

Introduction of the developments of Sections 23.3.1–23.3.3 into governing equations (23.3)–(23.5) and (23.8)–(23.10) simplifies them. Doing so then yields

$$\frac{D_{\text{hor}}u_1}{Dt} + \left(\frac{u_1}{h_1} + 2\Omega\right)\frac{u_2}{h_2}\frac{dh_1}{d\theta} = -\frac{1}{\overline{\rho}h_1}\frac{\partial p^T}{\partial\lambda}, \tag{23.20}$$

$$\frac{D_{\text{hor}}u_2}{Dt} - \left(\frac{u_1}{h_1} + 2\Omega\right)\frac{u_1}{h_2}\frac{dh_1}{d\theta} = -\frac{1}{\overline{\rho}h_2}\frac{\partial p^T}{\partial\theta}, \tag{23.21}$$

$$\frac{\partial p}{\partial\xi_3} = -\overline{\rho}\frac{d\Phi}{d\xi_3} = -\overline{\rho}g^E = \text{constant}, \tag{23.22}$$

$$\frac{\partial}{\partial\lambda}\left(u_1h_2h_3\right) + \frac{\partial}{\partial\theta}\left(u_2h_3h_1\right) = 0, \tag{23.23}$$

$$\frac{D_{\text{hor}}\mathscr{E}}{Dt} = 0, \tag{23.24}$$

where

$$\mathscr{E} = \mathscr{E}(\eta), \tag{23.25}$$

$$\frac{D_{\text{hor}}}{Dt} \equiv \frac{\partial}{\partial t} + \frac{u_1}{h_1}\frac{\partial}{\partial\lambda} + \frac{u_2}{h_2}\frac{\partial}{\partial\theta}. \tag{23.26}$$

Thus – see (23.12) and (23.23) – the flow is horizontal and three-dimensionally non-divergent.[5] Because of homogeneous assumption (23.11), $\mathscr{E}(\alpha,\eta)$ in (23.10) reduces to $\mathscr{E}(\eta)$ in (23.25). Thus the flow is assumed to be barotropic/homentropic.

Comparison of (23.20)–(23.22) with (23.3)–(23.5) shows that no matter which member of equation set quartet (23.3)–(23.5) one starts with, (23.20)–(23.22) still result. This is because, under the assumptions of Section 23.3.1, the terms denoted 'deep' and 'non-hydro' are all identically zero and therefore cannot contribute. The exact solutions developed in this chapter are therefore applicable to *all* members of the quartet of equation sets.

Simplification here of the governing equations is only done for the purpose of developing exact solutions:

[5] Furthermore, vector $h_3\mathbf{u}$ is horizontally non-divergent. The flow is therefore also two-dimensionally non-divergent if $h_3 =$ constant. This is so for a spherical planet, but not (in general) for an ellipsoidal one.

- Once developed, the exact solutions of the simplified equations are also exact solutions of the quartet of equation sets.

Caveats

- This is true provided that the Coriolis terms of a deep model (i.e. Models 1 and 2 of Section 23.2.6) *are expressed in terms of metric factors*, as described and discussed in Section 23.2.3. Replacement of deep metric factors by shallow ones at all model levels then automatically changes the deep model into a corresponding shallow one; that is, Models 1 and 2 of Section 23.2.6 automatically change to Models 3 and 4, respectively.
- However, *if the $2\Omega\cos\phi$ Coriolis terms of a deep model are coded explicitly*, replacement of deep metric factors by shallow ones everywhere in the model then no longer automatically deactivates these terms. Their continued presence in the model then makes it dynamically inconsistent. To avoid this inconsistency, these terms would need to be manually deactivated in the code.

23.3.5 The BPV (Barotropic Potential Vorticity) Equation

Recall from Chapter 19 that several forms of the BPV equation were derived from governing equations that are equivalent to those of Section 23.3.4. For present purposes, the most convenient form is the stream function one of Section 19.5.2. For this form, setting $(\xi_1, \xi_2) = (\lambda, \theta)$ in (19.81) gives:

The Definition of (u_1, u_2) in Terms of Stream Function Ψ

$$h_3 u_1 = -\frac{1}{h_2}\frac{\partial \Psi}{\partial \theta}, \quad h_3 u_2 = \frac{1}{h_1}\frac{\partial \Psi}{\partial \lambda}. \tag{23.27}$$

Velocity components u_1 and u_2 then satisfy the 3D non-divergence condition (23.23), that is, (19.33) with $(\xi_1, \xi_2) \rightarrow (\lambda, \theta)$.[6]

Setting $(\xi_1, \xi_2) = (\lambda, \theta)$ in (19.82) – or similarly doing so in (22.4) – yields:

The BPV Equation for an Ellipsoidal Planet

$$\begin{aligned}\frac{D_{\text{hor}}}{Dt}\left(\frac{\zeta + f}{h_3}\right) &= \frac{\partial}{\partial t}\left(\frac{\zeta + f}{h_3}\right) + \frac{u_1}{h_1}\frac{\partial}{\partial \lambda}\left(\frac{\zeta + f}{h_3}\right) + \frac{u_2}{h_2}\frac{\partial}{\partial \theta}\left(\frac{\zeta + f}{h_3}\right) \\ &= \frac{\partial}{\partial t}\left(\frac{\zeta}{h_3}\right) + \frac{1}{h_1 h_2 h_3}\left[\frac{\partial\left(\Psi, \zeta/h_3\right)}{\partial\left(\lambda, \theta\right)} + \frac{\partial\left(\Psi, f/h_3\right)}{\partial\left(\lambda, \theta\right)}\right] = 0,\end{aligned} \tag{23.28}$$

where ζ is relative vorticity. In this equation:

$$f\left(\xi_2\right) \equiv -\frac{2\Omega}{h_2}\frac{dh_1}{d\theta} \tag{23.29}$$

[6] Equation (23.23) only reduces to 2D non-divergence if $h_3 = h_3(\theta)$ reduces to $h_3 \equiv$ constant. This is so in a spherical context but not in a spheroidal (including ellipsoidal) one.

is the Coriolis parameter;

$$\frac{\partial (F,G)}{\partial (\lambda,\theta)} \equiv \frac{\partial F}{\partial \lambda}\frac{\partial G}{\partial \theta} - \frac{\partial F}{\partial \theta}\frac{\partial G}{\partial \lambda} \tag{23.30}$$

is the Jacobian of F and G with respect to λ and θ, and

$$\frac{\zeta}{h_3} = \frac{1}{h_1h_2h_3}\left[\frac{\partial}{\partial \lambda}(h_2u_2) - \frac{\partial}{\partial \theta}(h_1u_1)\right] = \frac{1}{h_1h_2h_3}\left[\frac{h_2}{h_3h_1}\frac{\partial^2 \Psi}{\partial \lambda^2} + \frac{\partial}{\partial \theta}\left(\frac{h_1}{h_2h_3}\frac{\partial \Psi}{\partial \theta}\right)\right] \equiv \mathfrak{L}\Psi, \tag{23.31}$$

where $\mathfrak{L}$ is a linear, second-order, partial-differential operator. Equations (23.29)–(23.31) correspond to (19.42), (19.83), and (19.84), respectively, with $(\xi_1,\xi_2) \to (\lambda,\theta)$.

23.3.6 The Diagnosis of Top Pressure p^T

Recall from Section 19.8 – and also from (22.97) of Section 22.4 – that top pressure, $p^T(\xi_1,\xi_2,t)$, expressed in axial-orthogonal-curvilinear coordinates, may be obtained from solution of (19.104). Setting $(\xi_1,\xi_2) = (\lambda,\theta)$ in this latter equation then gives:

The Diagnostic Differential Equation for Top Pressure (p^T) in Parametric Ellipsoidal Coordinates

$$\left[\frac{\partial}{\partial \lambda}\left(\frac{h_2h_3}{h_1}\frac{\partial}{\partial \lambda}\right) + \frac{\partial}{\partial \theta}\left(\frac{h_3h_1}{h_2}\frac{\partial}{\partial \theta}\right)\right]\left[\frac{p^T(\lambda,\theta,t)}{\overline{\rho}}\right]$$
$$= \frac{\partial}{\partial \lambda}\left[h_2h_3(\zeta+f)u_2\right] - \frac{\partial}{\partial \theta}\left[h_1h_3(\zeta+f)u_1\right] - \left[\frac{\partial}{\partial \lambda}\left(\frac{h_2h_3}{h_1}\frac{\partial K}{\partial \lambda}\right) + \frac{\partial}{\partial \theta}\left(\frac{h_3h_1}{h_2}\frac{\partial K}{\partial \theta}\right)\right], \tag{23.32}$$

where $K = \left(u_1^2+u_2^2\right)/2$, since $u_3 \equiv 0$ from (23.12). Given stream-function Ψ at any instant in time – and thereby u_1 and u_2 from (23.27) – the right-hand side of (23.32) can be computed. Top pressure $p^T(\lambda,\theta,t)$ can be obtained by solving this equation. Using the resulting solution, pressure elsewhere in the domain, $p(\lambda,\theta,\xi_3,t)$ can then be retrieved using (23.19).

This is one way of diagnosing top pressure $p^T = p^T(\lambda,\theta,t)$. As noted in Section 22.4, the downside of this method is the need to solve a 2D, *second-order*, partial-differential equation, namely (23.32), which is a non-trivial task. As outlined there, an alternative procedure is to obtain $p^T = p^T(\lambda,\theta,t)$ from the horizontal components of momentum used to derive (23.32), namely (22.100) and (22.101). These two equations correspond to (23.20) and (23.21) here, respectively. Given u_1 and u_2 at any instant in time, the right-hand sides of (23.20) and (23.21) may be computed. Top pressure $p^T = p^T(\lambda,\theta,t)$ may then be obtained by simultaneously solving *two* coupled, *first-order* differential equations.

Since the two methods are equivalent,[7] one can use whichever one is most convenient. The second one was employed in Section 22.4 to obtain $p^T = p^T(\lambda, \theta, t)$ for a family of exact, Rossby–Haurwitz-like solutions of the BPV equation in parametric-ellipsoidal coordinates, and for a particular member of this family.

23.4 EXACT BAROTROPIC SOLUTIONS OVER AN ELLIPSOID

23.4.1 Exact Solutions of the BPV Equation

23.4.1.1 The Model of Gravity

Recall that exact unsteady solutions of the BPV equation over an ellipsoid were obtained in Section 22.2 in parametric-ellipsoidal coordinates. They are a generalisation of Rossby–Haurwitz waves from a *spherical* context to an *ellipsoidal* one. To achieve this generalisation, it was found that a particular model of apparent gravity must be adopted. This model – see Section 23.3.2 – corresponds to the gravitational attraction exerted by an ellipsoidal planet of uniform density. Thus (23.17) holds for $\Phi = \Phi(\xi_3)$ and – from (23.15) –

$$g(\mu) = g^E \left(1 + 2\widetilde{\varepsilon}\mu^2\right)^{\frac{1}{2}} = g^E \left(1 + e^2 \frac{a^2}{c^2} \mu^2\right)^{\frac{1}{2}}, \tag{23.33}$$

where $\mu \equiv \sin\theta$ (cf. (22.55)), and $2\widetilde{\varepsilon} \equiv \left(a^2 - c^2\right)/c^2 \equiv e^2 a^2/c^2$ (from (23.16)).

23.4.1.2 A Family of Exact Solutions

BPV equation (23.28) then admits – see (22.60)–(22.66) –

A Family of Exact Rossby–Haurwitz-Like Solutions of the BPV Equation over an Ellipsoid

$$\begin{aligned}\Psi(\lambda, \mu, t) &= -\omega a c \mu + a c K_n^m \cos\left[m\left(\lambda - \sigma_n^m t\right)\right] P_n^m(\mu) \\ &= -\omega a c P_1^0(\mu) + a c K_n^m \cos\left[m\left(\lambda - \sigma_n^m t\right)\right] P_n^m(\mu),\end{aligned} \tag{23.34}$$

$$u_1(\lambda, \mu, t) = a\omega\left(1 - \mu^2\right)^{\frac{1}{2}} - aK_n^m \cos\left[m\left(\lambda - \sigma_n^m t\right)\right]\left(1 - \mu^2\right)^{\frac{1}{2}} \frac{dP_n^m(\mu)}{d\mu}, \tag{23.35}$$

$$u_2(\lambda, \mu, t) = -cmK_n^m \sin\left[m\left(\lambda - \sigma_n^m t\right)\right] \frac{\left(1 + 2\widetilde{\varepsilon}\mu^2\right)^{\frac{1}{2}} P_n^m(\mu)}{\left(1 - \mu^2\right)^{\frac{1}{2}}}, \tag{23.36}$$

$$\frac{\zeta(\lambda, \mu, t)}{h_3(\mu)} = \frac{2\omega a}{c} P_1^0(\mu) + ac\Lambda_n^m K_n^m \cos\left[m\left(\lambda - \sigma_n^m t\right)\right] P_n^m(\mu), \tag{23.37}$$

[7] As noted in Section 22.4, for this alternative solution procedure to be equivalent to solving diagnostic pressure equation (23.32), u_1 and u_2 must satisfy the 3D non-divergence condition (23.23). This condition is automatically satisfied here since, by construction, u_1 and u_2 are expressed in (23.27) in terms of a stream function, Ψ. If u_1 and u_2 were not constrained this way, then the equivalence would be lost, and the alternative procedure would be invalid.

where

- ω, a, c, and K_n^m are prescribed constants.
- $P_n^m(\mu)$ is an associated Legendre function of order m and degree n.
- $$\frac{dP_n^m(\mu)}{d\mu} \equiv \frac{(n+1)(n+m)P_{n-1}^m(\mu) - n(n-m+1)P_{n+1}^m(\mu)}{(2n+1)\left(1-\mu^2\right)}, \tag{23.38}$$

  relates $dP_n^m(\mu)/d\mu$ to $P_{n-1}^m(\mu)$ and $P_{n+1}^m(\mu)$ via a classical recurrence relation for associated Legendre functions.

The (constant) frequency σ_n^m in (23.34)–(23.37) is related to m, n, ω, and Ω by:

The Dispersion Relation

$$\sigma_n^m = \omega + \frac{2(\Omega+\omega)}{c^2\Lambda_n^m} = \omega - \frac{2(\Omega+\omega)}{\left[n(n+1) - e^2m^2\right]}, \tag{23.39}$$

where

$$\Lambda_n^m = \frac{-n(n+1) + e^2m^2}{c^2} \tag{23.40}$$

is an eigenvalue of eigenproblem (22.9) for the structure functions of the BPV equation, expressed in parametric-ellipsoidal coordinates.

The first term on the right-hand side of (23.34) corresponds to a background superrotation. The second is a superimposed perturbation that propagates east to west relative to the background flow, without change of shape.

23.4.1.3 *A Particular Exact Solution*

Following Phillips (1959), it has become customary to set $n \equiv m+1$ in (23.34)–(23.40) when using Rossby–Haurwitz waves for numerical experimentation in the *spherical* context (Hoskins, 1973; Williamson et al., 1992). Similarly doing so in the present *ellipsoidal* context leads to:

A Particular Exact Rossby–Haurwitz-Like Solution of the BPV Equation over an Ellipsoid

$$\Psi(\lambda,\theta,t) = -\omega ac\sin\theta + acK_{m+1}^m\cos\left[m\left(\lambda - \sigma_{m+1}^m t\right)\right]\cos^m\theta\sin\theta, \tag{23.41}$$

$$u_1(\lambda,\theta,t) = a\omega\cos\theta + aK_{m+1}^m\cos\left[m\left(\lambda - \sigma_{m+1}^m t\right)\right]\left[m\cos^{m-1}\theta - (m+1)\cos^{m+1}\theta\right], \tag{23.42}$$

$$u_2(\lambda,\theta,t) = -cmK_{m+1}^m\sin\left[m\left(\lambda - \sigma_{m+1}^m t\right)\right]\left(1+2\tilde{\varepsilon}\sin^2\theta\right)^{\frac{1}{2}}\cos^{m-1}\theta\sin\theta, \tag{23.43}$$

$$\frac{\zeta(\lambda,\theta,t)}{h_3(\theta)} = \frac{2\omega a}{c}\sin\theta + ac\Lambda_{m+1}^m K_{m+1}^m\cos\left[m\left(\lambda - \sigma_{m+1}^m t\right)\right]\cos^m\theta\sin\theta, \tag{23.44}$$

where

$$\sigma_{m+1}^{m} = \omega - \frac{2(\Omega+\omega)}{\left[(m+1)(m+2) - e^2 m^2\right]}, \tag{23.45}$$

$$\Lambda_{m+1}^{m} = \frac{-(m+1)(m+2) + e^2 m^2}{c^2} = -\left[\frac{m^2}{a^2} + \frac{(3m+2)}{c^2}\right]. \tag{23.46}$$

For testing shallow-water models in a *spherical* context (for which $c = a = \overline{a} \Rightarrow \widetilde{\varepsilon} = e = 0$, where $\overline{a}$ is mean radius), Williamson et al. (1992) recommended using (23.41)–(23.46) with parameter values set to

$$m = 4, \quad \omega = K_{m+1}^{m} = 7.848 \times 10^{-6}\,\mathrm{s}^{-1}, \quad g = \text{constant} = 9.80616\,\mathrm{m\,s}^{-2}, \tag{23.47}$$

$$\overline{a} = 6.37122 \times 10^{6}\,\mathrm{m}, \quad \Omega = 7.292 \times 10^{-5}\,\mathrm{s}^{-1}. \tag{23.48}$$

In (23.48), $\overline{a}$ (in the *spherical* context) represents Earth's *mean* radius. In the *ellipsoidal* context, a represents *equatorial* radius, and c represents *polar* radius. Appropriate values for a and c in the ellipsoidal context are – from Table 7.1 – the WGS 84 (World Geodetic System 84) values

$$a = 6378.1370000 \times 10^{6}\,\mathrm{m}, \quad c = 6356.7523142 \times 10^{6}\,\mathrm{m}. \tag{23.49}$$

At the time of the Williamson et al. (1992) work, it was commonly believed that barotropic Rossby–Haurwitz waves, with zonal wave numbers less than or equal to $m = 5$, are *dynamically* (i.e. physically, as opposed to computationally) stable when subjected to small initial perturbations. Since then, however, this has been found to be untrue (Thuburn and Li, 2000). In particular the $m = 4$ wave (as recommended in Williamson et al. (1992)) is in fact *unstable* to a triad interaction that was not included in the Hoskins (1973) analysis (albeit it was discussed by Baines (1976) in the context of the classical, nondivergent, barotropic-vorticity equation). As indicated by the Thuburn and Li (2000) study – and also later studies – a better choice for the zonal wave number would, in retrospect, be $m = 3$. This is particularly important for long integrations in time, since initially negligible errors (e.g. due to round-off error) then have sufficient time to grow and no longer be negligible.

23.4.2 The Balancing Pressure for Solutions of the BPV Equation

The balancing pressure for the preceding solutions of the BPV equation are obtained from decomposition (23.19) of p into a vertically varying part $\overline{\rho} g^E \left(\xi_3^T - \xi_3\right)$ and a horizontally varying and temporally varying top-pressure part $p^T(\lambda,\theta,t) \equiv p\left(\lambda,\theta,\xi_3 = \xi_3^T,t\right)$. For the family of exact Rossby–Haurwitz-like solutions of the BPV equation, decomposition (23.19) is equivalently written as

$$p(\lambda,\mu,\xi_3,t) = p^T(\lambda,\mu,t) + \overline{\rho} g^E \left(\xi_3^T - \xi_3\right), \tag{23.50}$$

where

$$\mu \equiv \sin\theta \quad \Rightarrow \quad \left(1-\mu^2\right)^{\frac{1}{2}} \equiv \cos\theta, \quad -\frac{\pi}{2} \le \theta \le \frac{\pi}{2}. \tag{23.51}$$

It remains to give the balancing top pressure that corresponds to the preceding family of exact solutions of the BPV equation, and the balancing top pressure that corresponds to the particular exact solution.

23.4.2.1 *The Balancing Top Pressure for the Family of Solutions*

Equations (22.113) and (22.115)–(22.117) give:

The Balancing Top Pressure for the Family of Exact Rossby–Haurwitz-Like Solutions of the BPV Equation over an Ellipsoid

$$\frac{p^T(\lambda,\mu,t)}{\overline{\rho}} = \frac{p^T_{\text{pole}}}{\overline{\rho}} + a^2 A(\mu) + a^2 \cos\left[m\left(\lambda - \sigma^m_{m+1}t\right)\right] B(\mu) + a^2 \cos\left[2m\left(\lambda - \sigma^m_{m+1}t\right)\right] C(\mu), \tag{23.52}$$

where

$$A(\mu) = \left(\Omega + \frac{\omega}{2}\right)\omega\left(1-\mu^2\right) + C(\mu) - \frac{\left(K^m_n\right)^2}{2}\left[\frac{m^2}{\left(1-\mu^2\right)} - e^2m^2\right]\left[P^m_n(\mu)\right]^2, \tag{23.53}$$

$$B(\mu) = 2\left(\Omega + \omega\right) K^m_n \left\{\frac{\left(1-\mu^2\right)}{\left[n(n+1) - e^2m^2\right]} \frac{dP^m_n(\mu)}{d\mu} + \mu P^m_n(\mu)\right\}, \tag{23.54}$$

$$C(\mu) = -\frac{\left(K^m_n\right)^2}{4}\left\{\left(1-\mu^2\right)\left[\frac{dP^m_n(\mu)}{d\mu}\right]^2 + \left[n(n+1) - \frac{m^2}{\left(1-\mu^2\right)}\right]\left[P^m_n(\mu)\right]^2\right\}, \tag{23.55}$$

and p^T_{pole} is the (constant and arbitrary – see Section 22.4) value of p^T evaluated over a pole (so that $\cos\left(\theta_{pole}\right) \equiv \cos\left(\pm\pi/2\right) \equiv 0$).[a]

[a] Evaluating (23.19) at a pole on the bottom geopotential surface $\xi_3 = \xi_3^B$ gives $p^B_{\text{pole}} = p^T_{\text{pole}} + \overline{\rho} g^E\left(\xi_3^T - \xi_3^B\right)$. Thus one could equivalently prescribe p^B_{pole} instead of p^T_{pole}.

In (23.54), $dP^m_n(\mu)/d\mu$ can be expressed in terms of $P^m_{n-1}(\mu)$ and $P^m_{n+1}(\mu)$ using (23.38).

23.4.2.2 *The Balancing Top Pressure for a Particular Solution*

Similarly, (22.108)–(22.111) give:

The Balancing Top Pressure for the Particular Exact Rossby–Haurwitz-Like Solution of the BPV Equation over an Ellipsoid

$$\frac{p^T(\lambda,\theta,t)}{\overline{\rho}} = \frac{p^T_{\text{pole}}}{\overline{\rho}} + a^2 A(\theta) + a^2 \cos\left[m\left(\lambda - \sigma^m_{m+1}t\right)\right] B(\theta) + a^2 \cos\left[2m\left(\lambda - \sigma^m_{m+1}t\right)\right] C(\theta), \tag{23.56}$$

where the value of p^T_{pole} is again arbitrary, and the coefficients are given by

$$A(\theta) = \left(\Omega + \frac{\omega}{2}\right)\omega\cos^2\theta - \frac{\left(K^m_{m+1}\right)^2 m^2}{2}\cos^{2m-2}\theta + \frac{\left(K^m_{m+1}\right)^2}{4}\left[\left(2m^2 - m - 2\right) + 2e^2m^2\right]\cos^{2m}\theta$$

$$+\frac{\left(K_{m+1}^{m}\right)^{2}}{4}\left(m+1-2e^{2}m^{2}\right)\cos^{2m+2}\theta, \tag{23.57}$$

$$B(\theta) = 2(\Omega+\omega)K_{m+1}^{m}\left[\frac{m^{2}+2m+2-e^{2}m^{2}}{(m+1)(m+2)-e^{2}m^{2}}\right]\cos^{m}\theta - 2(\Omega+\omega)K_{m+1}^{m}\left[\frac{(m+1)^{2}-e^{2}m^{2}}{(m+1)(m+2)-e^{2}m^{2}}\right]\cos^{m+2}\theta, \tag{23.58}$$

$$C(\theta) = \frac{\left(K_{m+1}^{m}\right)^{2}}{4}\left[-(m+2)\cos^{2m}\theta+(m+1)\cos^{2m+2}\theta\right]. \tag{23.59}$$

23.5 A FAMILY OF EXACT, UNSTEADY 3D SOLUTIONS

The preceding development of a family of exact, unsteady solutions of 3D, dynamical-core equation sets and the conditions under which these solutions hold may be summarised as follows:

A Family of Exact, Unsteady Solutions of the 3D Dynamical-Core Equations over an Ellipsoid

1. The fluid is homogeneous, as in (23.11), with prescribed value:
   (a) $\overline{\alpha} \equiv 1/\overline{\rho} = 0.8016\ \text{m}^3\ \text{kg}^{-1}$ (for the *atmosphere*; from Section 23.8.5.4);
   (b) $\overline{\alpha} \equiv 1/\overline{\rho} = 9.738 \times 10^{-4}\ \text{m}^3\ \text{kg}^{-1}$ (for the *oceans*; from Section 23.8.5.3).
2. This fluid is contained between two rigid, concentric, ellipsoidal geopotential surfaces, the bottom one located at the planet's reference surface $\xi_3 = \xi_3^B =$ constant, and the top one at $\xi_3 = \xi_3^T =$ constant, where ξ_3 is vertical coordinate.
3. This fluid is of shallow depth, with metric factors $h_i = h_i(\mu), i = 1, 2, 3$, defined by (23.18) in parametric-ellipsoidal coordinates with $\mu \equiv \sin\theta$.
4. Apparent gravity $g = g(\mu)$ is defined by (23.33) (alias (23.15)), with $g^E = 9.80616\ \text{m s}^{-2}$, and the corresponding geopotential $\Phi = \Phi(\xi_3)$ by (23.17).[a]
5. A time-dependent, Rossby–Haurwitz-like stream function $\psi = \psi(\lambda, \mu, t)$ is defined by (23.34), with:
   (a) prescribed values for parameters ω, K_n^m, m, n, a, c, and Ω;
   (b) e and σ_n^m evaluated using (23.16) and (23.39), respectively.
6. Horizontal components of momentum $u_1(\lambda, \mu, t)$ and $u_2(\lambda, \mu, t)$ – obtained from ψ, and independent of vertical coordinate (ξ_3) – are given by (23.35) and (23.36).
7. Vertical component of momentum (u_3) is identically zero, as in (23.12).
8. Top pressure $p^T(\lambda, \mu, t)$ is obtained from (23.52)–(23.55) and (23.38); 3D pressure $p(\lambda, \mu, \xi_3, t)$ then follows from (23.19).[b]
9. Temperature $T(\lambda, \mu, \xi_3, t)$ is obtained from:
   (a) (23.65) for the *atmosphere* (with $R^d = 287\text{J kg}^{-1}\ \text{K}^{-1}, \overline{\alpha} = 0.8016\ \text{m}^3\ \text{kg}^{-1}$);
   (b) (23.72) for the *oceans* (with $\beta_T = 1.67 \times 10^{-4}\ \text{K}^{-1}, \beta_p = 4.39 \times 10^{-10}\ \text{Pa}^{-1}$).

a Setting $g^E = 9.80616\ \text{m s}^{-2}$ has the virtue of being consistent with the Williamson et al. (1992) value for the *spherical* case when $c = a = \overline{a} \Rightarrow e = 0 \Rightarrow g = g^E =$ constant, everywhere.

b The value of p^T_{pole} in (23.52) is arbitrary and may be set to any convenient value; e.g. to 101 325 Pa for an ocean.

The 'tunable' (i.e. prescribable) parameters of this solution family are ω, K, m, n, Ω, a, c, ξ_3^B, ξ_3^T, $\overline{\rho}$, R^d, β_T, β_p, p^T_{pole}, and g^E. For the purposes of testing dynamical cores, one is at liberty, within reason, to set these parameters as one wishes. For example, one could set them to correspond with a planet having a smaller radius than Earth's but with a faster rotation rate, as in Wedi and Smolarkiewicz (2009). Similarly, one could make a planet more oblate by increasing the value of e.

Setting $c = a = \overline{a} \Rightarrow e = 0$ in the *ellipsoidal* family of solutions yields a *spherical* family.

By direct substitution into the governing equations, it can be verified that each and every one of them is exactly satisfied by the solution given earlier (i.e. the derived solution is indeed exact).

23.6 A PARTICULAR EXACT, UNSTEADY, 3D SOLUTION

A particular exact, unsteady solution, developed in the preceding discussion, together with the conditions under which it holds, may be similarly summarised by appropriately referring to different equations in several places. For the convenience of readers only interested in the particular solution, but wishing to see a 'stand-alone' summary without reference to the more general one, the particular exact, unsteady solution, developed earlier, together with the conditions under which it holds may be summarised as follows:

A Particular Exact, Unsteady Solution of the 3D Dynamical-Core Equations over an Ellipsoid

1. The fluid is homogeneous, as in (23.11), with prescribed value:
   (a) $\overline{\alpha} \equiv 1/\overline{\rho} = 0.8016\ \text{m}^3\ \text{kg}^{-1}$ (for the *atmosphere*; from Section 23.8.5.4);
   (b) $\overline{\alpha} \equiv 1/\overline{\rho} = 9.738 \times 10^{-4}\ \text{m}^3\ \text{kg}^{-1}$ (for the *oceans*; from Section 23.8.5.3).
2. This fluid is contained between two rigid, concentric, ellipsoidal geopotential surfaces, the bottom one located at the planet's reference surface $\xi_3 = \xi_3^B =$ constant, and the top one at $\xi_3 = \xi_3^T =$ constant, where ξ_3 is vertical coordinate.
3. This fluid is of shallow depth, with metric factors $h_i = h_i(\theta), i = 1, 2, 3$, defined by (23.18) in parametric-ellipsoidal coordinates.
4. Apparent gravity $g = g(\theta)$ is defined by (23.33) (alias (23.15)), with $g^E = 9.80616\ \text{m s}^{-2}$, and the corresponding geopotential $\Phi = \Phi(\xi_3)$ by (23.17).[a]
5. A time-dependent, Rossby–Haurwitz-like stream function $\psi = \psi(\lambda, \theta, t)$ is defined by (23.41), with:
   (a) prescribed values for the parameters ω, K^m_{m+1}, m, a, c, and Ω; and
   (b) e and σ^m_{m+1} evaluated using (23.16) and (23.45), respectively.
6. Horizontal components of momentum $u_1(\lambda, \theta, t)$ and $u_2(\lambda, \theta, t)$ – obtained from ψ, and independent of vertical coordinate ξ_3 – are given by (23.42) and (23.43).
7. Vertical component of momentum (u_3) is identically zero, as in (23.12).
8. Top pressure $p^T(\lambda, \theta, t)$ is obtained from (23.56)–(23.59); 3D pressure $p(\lambda, \theta, \xi_3, t)$ then follows from (23.19).[b]
9. Temperature $T(\lambda, \theta, \xi_3, t)$ is obtained from:
   (a) (23.65) for the *atmosphere* (with $R^d = 287\text{J kg}^{-1}\ \text{K}^{-1}, \overline{\alpha} = 0.8016\ \text{m}^3\ \text{kg}^{-1}$);
   (b) (23.72) for the *oceans* (with $\beta_T = 1.67 \times 10^{-4}\ \text{K}^{-1}, \beta_p = 4.39 \times 10^{-10}\ \text{Pa}^{-1}$).

a Setting $g^E = 9.80616\ \text{m s}^{-2}$ has the virtue of being consistent with the Williamson et al. (1992) value for the *spherical* case when $c = a = \overline{a} \Rightarrow e = 0 \Rightarrow g = g^E =$ constant, everywhere.
b The value of p^T_{pole} in (23.52) is arbitrary and may be set to any convenient value; e.g. to 101 325 Pa for an ocean.

The 'tunable' (i.e. prescribable) parameters of this solution family are ω, K, m, Ω, a, c, ξ_3^B, ξ_3^T, $\overline{\rho}$, R^d, β_T, β_p, p^T_{pole}, and g^E. For the purposes of testing dynamical cores, one is at liberty, within reason, to set these parameters as one wishes. For example, one could set them to correspond with a planet having a smaller radius than Earth's but with a faster rotation rate, as in Wedi and Smolarkiewicz (2009). Similarly, one could make a planet more oblate by increasing the value of e.

Setting $c = a = \overline{a} \Rightarrow e = 0$ in the *ellipsoidal* family of solutions yields a *spherical* family.

By direct substitution into the governing equations, it can be verified that each and every one of them is exactly satisfied by the solution derived earlier (i.e. the derived solution is indeed exact).

23.7 THE TOP BOUNDARY CONDITION

Recall that Rossby–Haurwitz waves are not exact solutions of the general shallow-water equations, but only *approximate* ones (Phillips, 1959; Hoskins, 1973; Williamson et al., 1992). Yet, as described earlier, it has been shown that it is possible to construct *exact* Rossby–Haurwitz solutions in 3D. It is therefore natural to ask why this is so. The resolution of this seeming paradox is as follows.

The difference in behaviour is not due to a property of the governing equations *per se*, but rather to a particular choice of *top boundary condition*. See Sections 13.6 and 13.9 for a related analysis, including discussion of the importance of boundary conditions in the vertical and their impact on conservation of axial angular momentum and total energy.

Common to both the general shallow-water equations and the 3D dynamical-core equations used in the preceding analysis, the bottom surface is assumed to be rigid. For the shallow-water equations, the top surface is assumed to be a *free* surface that varies in both time and space. For the present 3D solutions, however, the upper surface is assumed to be *rigid*, as in, for example, Davies et al. (2005), Satoh et al. (2008), Skamarock et al. (2012), Ullrich and Jablonowski (2012), Gassmann (2013), and Wood et al. (2014). Why is this difference between top boundary conditions crucial? It is because the existence of a free, moving surface (as in the general shallow-water equations) necessarily gives rise to a *non-zero* vertical component of velocity, u_3. By way of contrast, the property that u_3 is identically zero everywhere and for all time is an *essential* element of the present development of *exact* (as opposed to almost exact) solutions in 3D.

23.8 TEST CASES FOR VALIDATING 3D DYNAMICAL CORES

23.8.1 Preamble

The exact, unsteady, non-linear solutions (summarised in Section 23.5) of the 3D dynamical-core equations for a shallow layer of fluid overlying an ellipsoidal planet are of theoretical interest. They are an exact, 3D generalisation to ellipsoidal geometry of 2D Rossby–Haurwitz waves in spherical geometry. Furthermore, this generalisation to ellipsoidal geometry modifies the (classical) angular phase speed (σ) of Rossby–Haurwitz waves in spherical geometry by introducing a dependence on both eccentricity (e) and zonal wave number (m); see (23.39). Not only are the exact solutions of *theoretical* interest, they are also of *practical* interest. The purpose of this section is to explore how these solutions could be used as test cases for the validation of 3D dynamical cores. We start by giving some further background regarding this possibility.

To exploit massively parallel computer architectures, a great deal of effort has been, and is continuing to be invested in developing accurate and efficient discretisations of 3D dynamical-core equation sets on quasi-uniform global grids. See Staniforth and Thuburn (2012) for a review of global horizontal grids[8] and associated discretisation issues. The geometrical properties of quasi-uniform grids can, however, adversely influence the accuracy of numerical simulations. In

[8] This review shows why grids are termed 'quasi-uniform' rather than 'uniform'. Only five truly uniform global grids are possible, and they all have extremely low resolution (with between 4 and 20 points for an entire sphere). They correspond to the gnomonic projection onto a sphere of the Platonic solids, namely the tetrahedron, octahedron, cube, icosahedron, and dodecahedron.

particular, the structure of the underlying grid can become visible in the solution in the form of noise or systematic errors, a phenomenon known as 'grid imprinting' (Lauritzen et al., 2010; Staniforth and Thuburn, 2012). Such behaviour can also *spuriously* trigger baroclinic instability. The most unstable baroclinic waves have zonal wave numbers in the range of 4 to 10, and this correlates well with the various symmetries of popular quasi-uniform grids, based on cubed spheres and icosahedrons.

It is therefore highly desirable to have model problems, *with exact solutions*, to be able to quantify any grid imprinting of a dynamical core. Plotting difference fields between exact and numerical solutions then reveals the signature of any such grid imprinting. The family of exact, *unsteady* solutions developed herein thus has the potential to provide a valuable, complementary tool for testing and validating 3D, non-linear, dynamical cores. Three exploratory avenues are suggested in Sections 23.8.2–23.8.4.

The 3D solutions developed earlier and summarised in Sections 23.5 and 23.6 are inherently for *horizontal* flow, and they are expressed in terms of spherical harmonics, $\exp{(im\lambda)}\, P_n^m\, (\sin\theta)$.[9] This means that so-called spectral discretisation methods, based on spherical-harmonic expansion in the horizontal, are ideally suited for representation of these 3D solutions. For test problems (such as those described here), with solutions expressed in terms of spherical harmonics, and with all other things being equal (e.g. horizontal resolution and time scheme), it is inevitable that spectrally discretised dynamical cores will perform better than those that use alternative, horizontal-discretisation methods (such as finite differences and finite elements).

Spectral discretisation methods are not, however, ideally suited to massively parallel computer architectures, due primarily to the global (non-local) nature of spherical harmonics. Broadly speaking, non-locality results in high-volume, relatively slow, inter-processor communication; this inhibits computational efficiency. By way of contrast, locally defined, finite-difference, and finite-element methods benefit from relatively fast, 'nearest neighbour' (i.e. local), inter-processor communication; this enhances computational efficiency.

The prime purpose of the test problems outlined here is to provide a tool to quantitatively measure the accuracy of *non-spectral* dynamical cores for the geophysically important problem of (unsteady) Rossby–Haurwitz-like solutions of the 3D dynamical-core equations. It is emphasised that susceptibility of these methods to grid imprinting (i.e. to *systematic* errors) is a particularly important aspect.

From the preceding discussion, one might conclude that these test problems are of little help in testing and validating a spectral dynamical core. This, however, is not so. They are useful not only for detecting coding errors, but also for detecting deficiencies in vertical discretisation. For example, if the initial, horizontal flow is not maintained during a model integration, one needs to understand why and to investigate what can be done to eliminate or mitigate the cause.

In what follows, Section 23.8.2 is applicable for unsteady flow in the *ellipsoidal* context and also (as a special case) to unsteady flow in the spherical one. However, Sections 23.8.3 and 23.8.4 are restricted to the *spherical* context. This is because the solutions are only known to be valid in this context; pending confirmation, preliminary analysis indicates that they do not generalise to the ellipsoidal context. The test problems outlined in Sections 23.8.3 and 23.8.4 can nevertheless be used as part of the validation process of an *ellipsoidal* dynamical core simply by setting eccentricity to zero. An ellipsoidal dynamical core then simplifies, as a special case, to a spherical one. These (spherical) test problems may then reveal underlying deficiencies in the more general, ellipsoidal formulation.

[9] As noted in Lebedev (1972), p. 161, 'spherical harmonic' is something of a misnomer since spherical harmonics also arise naturally for applications in more general, non-spherical domains; e.g. for the ellipsoidal domains considered in this chapter.

23.8.2 Individual Waves in Spherical and Ellipsoidal Models

An obvious candidate for a test problem is to use the particular, exact, unsteady solution (summarised in Section 23.6) of the dynamical-core equations over an ellipsoid. This solution is for an *individual* Rossby–Haurwitz wave, defined by stream function (23.41). In the *spherical* context, parameter values could be set to those given in (23.47) and (23.48), that is, to those recommended by Williamson et al. (1992) for testing shallow-water models (albeit in their case using an approximate solution rather than an exact one). Alternatively, (23.49) could be used instead of (23.48) in the *ellipsoidal* context. This test problem (in either context) would enable modellers to assess the extent to which this exact solution is preserved during the integration of a dynamical core and, in particular, the extent to which grid imprinting is or is not observed during the integration, and purely horizontal flow is or is not preserved. For this configuration of exact solution, the zonal wave number has been arbitrarily set to $m = 4$, and the meridional wave number to $n = m + 1 = 5$.[10] Noting the discussion at the end of Section 23.4.1.3, a better choice would probably be $m = 3, n = m + 1 = 4$.

Since popular quasi-uniform grids, based on cubed spheres and icosahedrons, have symmetries associated with wave numbers in the range $m, n = 3, 4, 5, 6$, setting $(m, n) = (3, 4)$ or $(m, n) = (4, 5)$ could conceivably favour dynamical cores having a particular grid with a particular grid orientation. This would then restrict the generality of any conclusions drawn from comparative integrations using different dynamical cores, with different grids and different grid orientations. It therefore seems desirable to also define test problems with exactly the same parameters as in (23.47) and (23.48) or (23.49), but to instead (say) set $(m, n) = (3, 4), (4, 5), (5, 6), (6, 7)$, to create a range of pertinent test problems.[11] Performing comparative experiments with all four suggested test problems would then lead to more general and robust conclusions.

An important virtue of test problems based on an individual wave, as just described, is their simplicity. This facilitates pinpointing the source of any issues observed when integrating a dynamical core for these test problems, and may even suggest remedies for them.

At this point, the reader is reminded of the stability issue raised in Section 23.4.1.3. Many choices of zonal wave number m for Rossby–Haurwitz waves are dynamically (i.e. physically) unstable to small initial perturbations (e.g. due to unavoidable round-off error), no matter how small these perturbations may be. This does not mean that integrations with such choices of m are not useful; quite the contrary. But what it does mean is that the period in time for which they are useful is much shorter because initially negligible errors grow much faster. For the quartet of suggested test problems, the first one ($(m, n) = (3, 4)$) can be expected to give accurate results for a much longer period in time than for the other three. It is very important to understand these kinds of issues when assessing integrations of test problems in order to avoid drawing incorrect conclusions; see Thuburn and Li (2000) for a related discussion. Similar considerations also apply to the possible test problems outlined here.

23.8.3 Multiple Waves in Spherical Models

A vice, however, of test problems based on an *individual* Rossby–Haurwitz-like wave is that one needs to run an ensemble of them (having four members for the earlier proposal), with different horizontal scales, to increase the robustness of conclusions drawn from experimentation. This then raises the question of whether one can set up a test problem to cover a *range of scales* in the initial conditions and to also do so in such a way that dynamical cores with particular grids

[10] This choice was first used in Phillips (1959) and has subsequently been adopted by the numerical modelling community for testing shallow-water models (Williamson et al., 1992).

[11] Setting $n = m + 1$ avoids the need for software to compute associated Legendre functions $P_n^m(\sin\theta)$, since $P_{m+1}^m(\sin\theta) \sim \cos^m\theta\sin\theta$, as noted in Section 22.2.9.

and grid orientations would not be unduly advantaged or disadvantaged. A possible step in this direction is now proposed.

To this point herein, for simplicity and to maintain continuity with the existing literature on test problems, Rossby–Haurwitz waves of the form (23.34) (including particular form (23.41)) for stream function ψ have been considered. There are, however, more general forms available than this one. Specifically, Ertel showed that the more general specification

$$\psi = -\omega a^2 \sin\theta + \sum_{m=-n}^{n} K_n^m a^2 \exp\left[im\left(\lambda - \sigma_n t\right)\right] P_n^m\left(\sin\theta\right), \tag{23.60}$$

also leads to a solution of the classical, 2D, non-divergent barotropic-vorticity equation in *spherical* geometry (Rochas, 1984).[12] Here K_n^m is complex, and K_n^{-m} is the complex conjugate of K_n^m; and σ_n is defined by (23.39) herein, but with $e \equiv 0$ (i.e. for the spherical case). For this form, the solutions again correspond to a superrotation plus a superimposed perturbation with total wavenumber n. However, instead of an *individual* wave associated with a specific zonal wave number, m, there is now a sum of *multiple* waves, each having a different zonal wave number, m, but sharing a common total wave number, n.

It is suggested that multiple-wave form (23.60) could be exploited to define a test problem that would not unduly advantage or disadvantage dynamical cores based on particular grids with particular orientations. For example, one might consider using

$$\psi = -\omega a^2 \sin\theta + \sum_{m=0}^{10} K_{10}^m a^2 \cos\left[m\left(\lambda - \sigma_{10} t\right)\right] P_{10}^m\left(\sin\theta\right), \tag{23.61}$$

where the K_{10}^m are real and (say) equal.[13]

The procedure summarised in Section 23.5 to obtain the exact solution of the dynamical-core equations then goes through virtually unchanged. In summary:

1. The momentum components u_1 and u_2 are computed from the new stream function defined by (23.60).
2. These components are used to compute the right-hand side of the pressure-balance equation (23.32).
3. This equation is then solved for the top pressure, p^T.[14]
4. Pressure (p) elsewhere then follows from p^T, and temperature (T) from p.

23.8.4 Multiple Rotated Waves in Spherical Models

Although form (23.60) for stream function, ψ, is more general than that of (23.34), Rochas (1986) developed a yet more general form. The development of this latter form is somewhat complicated and involves coordinate transformations; the reader is referred to Rochas (1986) for details. Natural questions are then:

12 Caveat: This result was derived in a spherical context. It probably does not generalise to ellipsoids.

13 Note, however, that there remains some residual arbitrariness in the choice of the relative weightings, K_{10}^m, since there are different ways of normalising associated Legendre functions $P_n^m(\sin\theta)$. This would need to be thought through.

14 Alternatively, p^T could be obtained using the procedure outlined in Section 23.3.6.

1. What property distinguishes Rochas (1986)'s form from that of (23.61)?
2. What benefits might accrue from using his form to develop a test problem?

Rochas (1986) showed that a key property of forms (23.34) and (23.60) is that non-linear wave interactions automatically vanish from the vorticity equation, so that solutions of the analogous linear problem are also solutions of the non-linear one. This then means that waves propagate with constant amplitude and constant angular phase speed. The novelty in Rochas (1986) was to introduce a rotation of coordinates into the solution procedure in such a way that the non-linear wave interactions no longer vanish, whilst simultaneously retaining solution tractability. The resulting solution is then a modification of form (23.60) for ψ. This answers the first question, albeit only descriptively and without going into detail.

To answer the second question, it suffices for present purposes to simply quote from Rochas (1986)'s conclusion:

The main advantages of the new solutions are

(i) the amplitude of the new solutions is not constant, i.e. they do not represent patterns which rotate bodily around the Earth's axis, as was the case in all of the former solutions;

(ii) the mean zonal velocity on a latitude circle is not constant with time; these solutions represent phenomena like the oscillation of the zonal index; and

(iii) they include solutions in which cross-polar flows are allowed, a particularity which may be interesting to test numerical methods.

The trade-off here is between increased generality (albeit limited to the spherical context) and simplicity. Increased generality comes at the price of coordinate transformations.

23.8.5 Some Practical Considerations

23.8.5.1 *Prescription of Internal Energy and Equation of State*

Specific internal energy ($\mathscr{E}$) completely defines a fluid's thermodynamic-equilibrium state, and its precise functional form only depends on the physical properties of the fluid; see Sections 4.2.3 and 9.2. The natural variables of $\mathscr{E}$ are specific volume (α) and entropy (η); consequently $\mathscr{E} = \mathscr{E}(\alpha, \eta)$, in general. In the present, restricted context, however, $\alpha = \overline{\alpha} =$ constant; see assumption (23.11). This means that $\mathscr{E} = \mathscr{E}(\eta)$ – see (23.25) – so that $\mathscr{E}$ now only depends on a single variable (η) rather than two (α and η); the flow is then said to be *barotropic* or *homentropic*. The assumption that $\alpha = \overline{\alpha} =$ constant not only eliminates dependence of $\mathscr{E}$ on α, it also simplifies thermodynamic-energy equation (23.9) to (23.24) by eliminating the term $pD\alpha/Dt = pD\overline{\alpha}/Dt = 0$ from it.

Realistic atmospheric and oceanic forecast models generally do not prescribe $\mathscr{E}$ in terms of α and η for observational-data reasons; in particular, η is not a measured quantity. Instead, pressure (p) and temperature (T) are frequently used as thermodynamic state variables in both the thermodynamic-energy and state equations. In the present context of formulating test problems for atmospheric and oceanic models, it therefore makes sense to mimic the thermodynamic formulation of these models by using p and T, so that is what we do. How best to mimic what is done in a particular model very much depends on its prescribed forms for the state and thermodynamic-energy equations. Three mimicry strategies are outlined here.

Assuming $\alpha = \overline{\alpha} =$ constant not only facilitates the development of exact solutions, it also *analytically* 'slaves' the evolution of temperature (T) to the *independent* evolution of velocity ($\mathbf{u}$) and pressure (p). T is then passively advected according to the consequently simplified thermodynamic-energy equation $DT/Dt = \partial T/\partial t + \mathbf{u} \cdot \nabla T = 0$. Analytically speaking, this means that *any* prescription of internal energy and equation of state *that respects* $\alpha = \overline{\alpha} =$ constant is admissible, since the evolution of **u** and p is independent of this prescription. In a numerical model, however, and depending upon its numerics, this 'slaving' property may very well not be exactly respected. Diagnosing why can then lead to improved numerics and a better numerical model.

There are basically three strategies one can follow to adapt a dynamical core[15] to run the test problems outlined in Sections 23.8.2–23.8.4. These are:

1. Make no changes to the formulation and code of the existing dynamical core.
2. Minimally change the formulation and code for the existing state and thermodynamic-energy equations, to impose $\alpha = \overline{\alpha} =$ constant.
3. Replace the formulation and code for the existing state and thermodynamic-energy equations, to impose $\alpha = \overline{\alpha} =$ constant.

All three strategies use an exact solution – for example, evaluated at a model's gridpoints – as initial conditions. For some models it may be possible to use all three strategies, each one of which can provide valuable information. A further, desirable, possibility is to use an exact solution to critically examine and possibly improve the formulation of a dynamical core *before it is coded.*

Strategy 1 is the easiest to implement in practice. One simply deactivates any forcings (if present) and provides an exact solution as initial conditions (which includes $\alpha = \overline{\alpha} =$ constant, ideally imposed exactly, otherwise approximately). Doing so, however, takes no account of whether dynamic balances are exactly respected by the provided initial conditions; they may not be (e.g. due to the discretisation of non-linear terms). If dynamic balances are not respected at initial time, then they are unlikely to be respected thereafter. So, for example, the numerical solution may develop small, non-zero values for u_3, whereas $u_3 \equiv 0$ for the exact solution (i.e. the flow should be purely horizontal). It is sometimes possible to tweak the initial conditions to numerically respect a dynamic balance. How to do so, and whether this is indeed possible is, however, model dependent and beyond the scope of the present work. The important point here is to be aware that examining this issue, and consequently tweaking the numerics, may help isolate the source of a problem observed during a test integration.

Strategy 2 preserves and exercises most of the numerics of the dynamical core during a test integration. Minimally tweaking the formulation and code (as opposed to leaving it as is for Strategy 1) may then better isolate the source of any problems observed during a test integration. For example, it may prevent non-zero values of u_3 developing, which might otherwise obscure the source of an observed problem.

Strategy 3 exercises less of the numerics of the dynamical core during a test integration. However, if applied with sufficient care, it can be used to eliminate the thermodynamic formulation of the dynamical core as being the source of any problems observed during test integrations. This then focuses attention on the numerical discretisation of the three components of the momentum equation and of the mass-continuity equation, which collectively control the evolution of **u** and p.

For all three strategies, it is beneficial to start off by integrating the dynamical core for a single timestep, and then for a second one, and to carefully examine the fields during the time steps to see whether they behave as expected. This is particularly useful for identifying coding errors during the initial development of a dynamical core. Longer integrations can then be performed to examine other performance issues, such as numerical stability.

These fairly general considerations, whilst useful, are far from being exhaustive. They have been introduced to provide some illustrative examples of how exact solutions can be used to diagnose weaknesses in the formulation of a dynamical core and, possibly, to identify errors in its code. If, for example, the formulation should lead to u_3 being zero to machine precision, and this is not

[15] Implicit in the definition of a dynamical core is the absence of parametrised forcing terms; i.e. $F(X) \equiv 0$ in equation (1.1).

the case in a test integration, one obviously needs to investigate why and remedy the problem at source.

To further help fix ideas we give two illustrative, prototypical examples here regarding adaptation of the thermodynamics of a dynamical core for testing purposes; one for the atmosphere, the other for the ocean. For both, we first express the thermodynamic-energy and state equations in terms of the *three* thermodynamic-state variables p, T, and α – this mimics what is done in an unmodified dynamical core – before then applying assumption (23.11), that is, setting $\alpha = \overline{\alpha} =$ constant which mimics the exact solution. The number of *independent* thermodynamic state variables then reduces to *two* (p and T) and, analytically *but not necessarily numerically*, the flow is then barotropic/homentropic.

23.8.5.2 *The Atmospheric Prototype*

To leading order, and as discussed in Section 2.3, the atmosphere behaves as a (dry) ideal gas. Now for an ideal gas, specific internal energy and the equation of state may be obtained from Table 9.4. Doing so then leads to:

The Internal Energy and Equation of State for the Atmospheric Prototype

$$\mathscr{E}(T) = c_v^d T, \tag{23.62}$$

$$p\alpha(p, T) = R^d T. \tag{23.63}$$

Here, both c_v^d and R^d are assumed constant[16] and:

- $c_v^d = 717 \text{ J kg}^{-1} \text{ K}^{-1}$ is specific heat at constant volume of the gas.
- $R^d = 287 \text{J kg}^{-1} \text{ K}^{-1}$ is its gas constant.

Substitution of (23.62) into thermodynamic-energy equation (23.9) – before application of constant-density assumption (23.11) – then gives

$$c_v^d \frac{DT}{Dt} + p\frac{D\alpha}{Dt} = 0. \tag{23.64}$$

Many other equivalent forms are also possible; see Section 2.3.6 for examples.

Application of constant-density assumption (23.11) simplifies (23.63) (equation of state) and (23.64) (thermodynamic-energy equation) to

$$R^d T = \overline{\alpha} p, \tag{23.65}$$

$$\frac{DT}{Dt} = 0. \tag{23.66}$$

Specification of a representative value for $\overline{\alpha}$ for the atmospheric prototype is deferred until Section 23.8.5.4.

[16] The values quoted here are taken from Table 9.3.

If, for example, the equation of state of a dynamical core has the form of (23.63), and its thermodynamic-energy equation the form of (23.64), then it is relatively straightforward to adapt it to run a test problem that assumes $\alpha = \overline{\alpha} =$ constant. Not only do (23.63) and (23.64) analytically reduce to (23.65) and (23.66), respectively, they will also do so numerically for any reasonable discretisation of $D\alpha/Dt$. If, however, the forms are very different, then more work may be required, for example, to apply Strategy 3; or perhaps only Strategy 1 may be viable without extensive reformulation and coding.

Substitution of (23.65) into (23.66) gives

$$\frac{Dp}{Dt} = 0. \tag{23.67}$$

Since the flow is purely horizontal – by (23.12) – and using pressure-decomposition (23.19), (23.67) then reduces to

$$\frac{D_{\text{hor}}p^T}{Dt} = 0. \tag{23.68}$$

Equations (23.67) and (23.68) are an *analytic* consequence of applying $\alpha = \overline{\alpha} =$ constant. Depending on the numerics of a dynamical core, they may or may not be respected numerically. The extent to which they do so is one measure of the accuracy of a dynamical core.

23.8.5.3 *The Oceanic Prototype*

To leading order, the oceans behave as pure liquid water. From Table 9.5 – with $L_0^f \equiv 0$ – this then gives:

The Internal Energy for the Oceanic Prototype

$$\mathscr{E}(T) = c^l T, \tag{23.69}$$

where c^l is (constant) specific heat capacity at constant pressure.

In the absence of salinity, Vallis (2017)'s linearised state equation (1.56) for the oceans reduces to

$$\alpha\left(p, T\right) = \overline{\alpha} + \beta_T\left(T - T_0\right) - \beta_p\left(p - p_0\right). \tag{23.70}$$

This equation is valid for *small* variations around constant reference values $(\alpha, p, T) = (\overline{\alpha}, p_0, T_0)$; it corresponds to bilinear expansion of α about $(p, T) = (p_0, T_0)$. Representative values for the parameters in (23.69) and (23.70) are (from his Table 1.1 and Table 11.1 herein):

- $c^l = 3986\ \text{J kg}^{-1}\ \text{K}^{-1}$ (specific heat capacity at constant pressure).
- $\overline{\alpha} = 9.738 \times 10^{-4}\ \text{m}^3\ \text{kg}^{-1}$ (reference specific volume).
- $T_0 = 283$ K (reference temperature).

- $\beta_T = 1.67 \times 10^{-4}\ \mathrm{K}^{-1}$ (thermal expansion coefficient).
- $\beta_p = 4.39 \times 10^{-10}\ \mathrm{Pa}^{-1}$ (compressibility coefficient).

For a representative value of pressure, the IOC et al. (2010) reference value is reasonable, and then

- $p_0 = p_r = 101\ 325$ Pa.

Substitution of (23.69) into thermodynamic-energy equation (23.9) (before application of constant-density assumption (23.11)) then gives

$$c^l \frac{DT}{Dt} + p\frac{D\alpha}{Dt} = 0. \tag{23.71}$$

Application of constant-density assumption (23.11) simplifies (23.70) (equation of state) and (23.71) (thermodynamic-energy equation) to

$$\beta_T (T - T_0) = \beta_p (p - p_0), \tag{23.72}$$

$$\frac{DT}{Dt} = 0. \tag{23.73}$$

Substitution of (23.72) into (23.73) gives

$$\frac{Dp}{Dt} = 0. \tag{23.74}$$

Since the flow is purely horizontal (by (23.12)), and using pressure-decomposition (23.19), (23.74) then reduces to

$$\frac{D_{\mathrm{hor}} p^T}{Dt} = 0. \tag{23.75}$$

Considerations similar to those described for the atmospheric prototype to adapt a dynamical core for testing purposes also apply to the oceanic prototype.

23.8.5.4 *Comparison of the Atmospheric and Oceanic Prototypes*

We can now compare and contrast the specified forms of internal energy and equation of state for the atmospheric and oceanic prototypes. But first, we answer the deferred question:

- What would a representative value of $\overline{\alpha}$ be for the atmospheric prototype?

For simplicity (and also physical realism), we apply continuity at the atmospheric/oceanic interface. Taking the *oceanic* values $T_0 = 283$ and $p_0 = 101325$ Pa and substituting them into the *atmospheric* equation of state (23.65) then gives

- $\overline{\alpha} = R^d T_0 / p_0 = 287 \times 283/101325\ \mathrm{m}^3\ \mathrm{kg}^{-1} = 0.8016\ \mathrm{m}^3\ \mathrm{kg}^{-1}$.

Comparing now this atmospheric value for $\overline{\alpha}$ with the corresponding oceanic one of $\overline{\alpha} = 9.738 \times 10^{-4}\ \mathrm{m}^3\ \mathrm{kg}^{-1}$, we see that it is three orders of magnitude larger than the oceanic one. This difference between the atmospheric and oceanic contexts simply reflects the fact that liquid water is three orders of magnitude denser than atmospheric gas. Nevertheless, setting $\alpha = \overline{\alpha} =$ constant results in both contexts in a linear relation between p and T (only). Thus, given a value of p or T, the value of the other one immediately follows, with no other thermodynamic state variable involved, that is, $T = T(p)$ rather than $T = T(p, \alpha)$. This loss of a degree of freedom (due to setting $\overline{\alpha} =$ constant) simply reflects the fact that the flow is barotropic/homentropic.

Despite physically very different state equations, in both atmospheric and oceanic contexts:

- $\mathscr{E}(T) \propto T$ (cf. (23.62) with (23.69)).
- T and p are linearly related to one another (cf. (23.65) with (23.72)).
- $DT/Dt = 0$ (see (23.66) and (23.73)).
- $Dp/Dt = 0$ (see (23.67) and (23.74)).
- $Dp^T/Dt = 0$ (see (23.68) and (23.75)).

Although thermodynamic state variables p and T are continuous between the atmosphere and ocean at their interface, their densities there (expressed here as specific volumes, α) are very different (by three orders of magnitude).

The exact solutions, developed earlier, assume a shallow layer of fluid – see assumption 5 of Section 23.3.1. With parameter values set to those given in the preceding subsections, the atmospheric prototype characterises a shallow atmosphere above mean sea level, and the oceanic one a shallow ocean below. By adapting an atmospheric dynamical core to use the oceanic prototype instead of the atmospheric one, one could see how it might behave as an oceanic dynamical core, and vice versa!

23.9 CONCLUDING REMARKS

A quartet of 3D, dynamical-core equation sets for geophysical-fluid-dynamical flow over a spheroidal planet has been given. The following assumptions are then made:

- The planet is ellipsoidal and of homogeneous composition.
- The fluid is homogeneous, of shallow depth, and confined between two rigid, ellipsoidal geopotential surfaces.
- The flow is purely horizontal for all time.

Under these conditions, a family of exact, unsteady, Rossby–Haurwitz-like solutions has been developed. It has been verified, by substitution, that these solutions do indeed satisfy the governing equations. The family of exact solutions is based on solution of the BPV equation. With an appropriate setting of parameter values, this solution family may be used to test global atmospheric and oceanic dynamical cores.

Although Rossby–Haurwitz waves are exact solutions of the classical, 2D, non-divergent, barotropic-vorticity equation, they are *not* exact solutions of the more general 2D, shallow-water equations. It therefore appears paradoxical that Rossby–Haurwitz waves can provide the basis for exact solutions of even more general, 3D, dynamical-core equations. The resolution of this seeming paradox is that for the 2D shallow-water equations, the upper surface is assumed to be a *free* surface, whereas for the solutions developed herein, it is instead assumed to be *rigid*. This difference in top boundary condition is crucial. If, for the 3D dynamical-core equations, the upper surface were free, then the vertical velocity would no longer be zero everywhere, and the present solution procedure would break down.

An important current challenge of atmospheric and oceanic model development is how to develop dynamical cores that can run efficiently on evolving, massively parallel, computer architectures. These dynamical cores generally use an underlying quasi-uniform global grid of one kind or another. They are susceptible to 'grid imprinting', whereby the structure of an underlying grid manifests itself in model integrations, adversely affecting accuracy.

The motivation for the present development of exact, unsteady solutions of 3D dynamical-core equation sets is to provide the foundation for designing challenging and useful, *unsteady* test problems for the development, validation, and improvement of the dynamical cores of comprehensive atmospheric and oceanic forecast models. This would then facilitate examination of issues such as grid imprinting, not only for a particular dynamical core, but also in a comparative manner for several dynamical cores. Three exploratory avenues are suggested for the design of such test problems. The first is applicable in ellipsoidal geometry, whereas the other two are limited to spherical geometry.

The first avenue is to define an *individual propagating wave* (i.e. to use a single spherical harmonic to represent the perturbation of the stream function about a super solid-body rotation of the fluid). This has the important virtue of simplicity and also facilitates pinpointing the source of any issues observed during model integrations. A limitation, however, is that if one wishes to perform comparative experiments using dynamical cores having different grids, then an ensemble of such test problems needs to be defined to avoid inadvertently favouring one dynamical core, with its associated grid, over another due to the definition of the test problem.

To address this limitation, the second avenue is to define a sum of *multiple waves*, each one having a different horizontal spatial form. This approach exploits the fact that it is possible to define such a wave sum in such a way as to still lead to an exact solution (in the spherical context). Specifically, the perturbation of the stream function is represented as the sum of spherical harmonics having a common total wave number, n, but having different zonal wave numbers, m.

A third avenue is to define a sum of *multiple, rotated waves*. This builds on and exploits a more general and more complicated solution of the classical, 2D, non-divergent, barotropic-vorticity equation (again in the spherical context). It results in a more challenging flow problem with waves that propagate with variable amplitude and angular phase speed rather than with the usual constant amplitude and angular phase speed.

For tractability reasons, the exact solutions developed herein have no vertical motion. As such, test problems based on them cannot strongly test those parts of model formulation that describe vertical motions.[17] The exact solutions given herein would nevertheless be valuable for establishing whether these parts of model formulation are spuriously excited when they should not be. In this regard, the exact solutions decouple the hydrodynamics from the thermodynamics; the momentum and mass-continuity equations can be integrated in time independently of the state and thermodynamic-energy equations. The converse is not true, however; evolution (i.e. transport) of thermodynamic energy depends upon the velocity field provided by integration of the momentum and mass-continuity equations. Integrations of a numerical model using initial conditions of the exact solutions may not necessarily respect this decoupling property. The reasons for this would then need to be investigated.

Finally, it is anticipated that the methodology used in the present work to develop exact, unsteady solutions for dynamical-core equation sets in ellipsoidal and spherical geometry could also be used to obtain analogous exact, unsteady solutions on β planes with similar assumptions. Although this idea has not been explored, there appears to be no obvious reason why it would not work.

[17] All test problems have their limitations, including the ones outlined here. To properly test a dynamical core, one needs an ensemble of test problems that collectively exercise the discretisations of all terms in the governing equations.

Appendix: Vector Identities

The standard vector identities given in this appendix may be found in textbooks on vector analysis; see, for example, Spiegel and Lipschutz (2009). The identities for material derivatives of integrals over volumes may also be found in textbooks; see, for example, Section 1.1 of Vallis (2017).

In what follows:

- f and g are generic scalars; and
- **A**, **B**, **C**, and **D** are generic vectors.

Scalar and Vector

$$\mathbf{A}+\mathbf{B} \equiv \mathbf{B}+\mathbf{A}, \tag{A.1}$$

$$\mathbf{A}\cdot\mathbf{B} \equiv \mathbf{B}\cdot\mathbf{A}, \tag{A.2}$$

$$\mathbf{A}\times\mathbf{B} \equiv -\mathbf{B}\times\mathbf{A}, \tag{A.3}$$

$$\mathbf{A}\times\mathbf{A} \equiv 0, \tag{A.4}$$

$$\mathbf{A}+(\mathbf{B}+\mathbf{C}) \equiv (\mathbf{A}+\mathbf{B})+\mathbf{C} \equiv \mathbf{A}+\mathbf{B}+\mathbf{C}, \tag{A.5}$$

$$\mathbf{A}\cdot(\mathbf{B}+\mathbf{C}) \equiv \mathbf{A}\cdot\mathbf{B}+\mathbf{A}\cdot\mathbf{C}, \tag{A.6}$$

$$\mathbf{A}\times(\mathbf{B}+\mathbf{C}) \equiv \mathbf{A}\times\mathbf{B}+\mathbf{A}\times\mathbf{C}, \tag{A.7}$$

$$\mathbf{A}\cdot(\mathbf{B}\times\mathbf{C}) \equiv \mathbf{B}\cdot(\mathbf{C}\times\mathbf{A}) \equiv \mathbf{C}\cdot(\mathbf{A}\times\mathbf{B}), \tag{A.8}$$

$$\mathbf{A}\cdot(\mathbf{A}\times\mathbf{C}) \equiv 0, \tag{A.9}$$

$$\mathbf{A}\times(\mathbf{B}\times\mathbf{C}) = (\mathbf{A}\cdot\mathbf{C})\,\mathbf{B}-(\mathbf{A}\cdot\mathbf{B})\,\mathbf{C}, \tag{A.10}$$

$$\mathbf{A}\times(\mathbf{B}\times\mathbf{C})+\mathbf{B}\times(\mathbf{C}\times\mathbf{A})+\mathbf{C}\times(\mathbf{A}\times\mathbf{B}) \equiv 0, \tag{A.11}$$

$$(\mathbf{A}\times\mathbf{B})\cdot(\mathbf{C}\times\mathbf{D}) \equiv (\mathbf{A}\cdot\mathbf{C})(\mathbf{B}\cdot\mathbf{D})-(\mathbf{A}\cdot\mathbf{D})(\mathbf{B}\cdot\mathbf{C}), \tag{A.12}$$

$$(\mathbf{A}\times\mathbf{B})\times(\mathbf{C}\times\mathbf{D}) \equiv [(\mathbf{A}\times\mathbf{B})\cdot\mathbf{D}]\,\mathbf{C}-[(\mathbf{A}\times\mathbf{B})\cdot\mathbf{C}]\,\mathbf{D}. \tag{A.13}$$

Gradient, Divergence, Curl, and Laplacian

$$\nabla\left(f+g\right) \equiv \nabla f+\nabla g, \tag{A.14}$$

$$\nabla\left(fg\right) \equiv \nabla\left(gf\right) \equiv f\nabla g+g\nabla f, \tag{A.15}$$

$$\nabla\cdot(\mathbf{A}+\mathbf{B}) \equiv \nabla\cdot\mathbf{A}+\nabla\cdot\mathbf{B}, \tag{A.16}$$

$$\nabla\cdot\left(f\mathbf{A}\right) \equiv f\nabla\cdot\mathbf{A}+\mathbf{A}\cdot\nabla f, \tag{A.17}$$

$$\nabla\times\left(f\mathbf{A}\right) \equiv f\left(\nabla\times\mathbf{A}\right)+\left(\nabla f\right)\times\mathbf{A}, \tag{A.18}$$

$$\nabla(\mathbf{A}\cdot\mathbf{B}) \equiv \mathbf{A}\times(\nabla\times\mathbf{B}) + \mathbf{B}\times\big(\nabla\times\mathbf{A})\big) + (\mathbf{A}\cdot\nabla)\,\mathbf{B} + (\mathbf{B}\cdot\nabla)\,\mathbf{A}, \tag{A.19}$$

$$\nabla\cdot(\mathbf{A}\times\mathbf{B}) \equiv \mathbf{B}\cdot(\nabla\times\mathbf{A}) - \mathbf{A}\cdot(\nabla\times\mathbf{B}), \tag{A.20}$$

$$\nabla\times(\mathbf{A}+\mathbf{B}) \equiv \nabla\times\mathbf{A} + \nabla\times\mathbf{B}, \tag{A.21}$$

$$\nabla\times(\mathbf{A}\times\mathbf{B}) \equiv \mathbf{A}(\nabla\cdot\mathbf{B}) - \mathbf{B}(\nabla\cdot\mathbf{A}) + (\mathbf{B}\cdot\nabla)\,\mathbf{A} - (\mathbf{A}\cdot\nabla)\,\mathbf{B}, \tag{A.22}$$

$$(\mathbf{A}\cdot\nabla)\,\mathbf{A} \equiv \frac{1}{2}\nabla(\mathbf{A}\cdot\mathbf{A}) + (\nabla\times\mathbf{A})\times\mathbf{A}, \tag{A.23}$$

$$\nabla^2 f \equiv \nabla\cdot(\nabla f), \tag{A.24}$$

$$\nabla^2\mathbf{A} \equiv \nabla(\nabla\cdot\mathbf{A}) - \nabla\times(\nabla\times\mathbf{A}), \tag{A.25}$$

$$\nabla\cdot(\nabla\times\mathbf{A}) \equiv 0, \tag{A.26}$$

$$\nabla\times(\nabla f) \equiv 0. \tag{A.27}$$

Helmholtz Decomposition

Let

$$\mathbf{u} = \mathbf{u}_\chi + \mathbf{u}_\Psi, \tag{A.28}$$

be a three-dimensional vector field, where $\mathbf{u}_\chi$ and $\mathbf{u}_\Psi$ are irrotational (curl-free) and solenoidal (divergence-free) vector fields, respectively, so that

$$\nabla\times\mathbf{u}_\chi = 0, \quad \nabla\cdot\mathbf{u}_\Psi = 0. \tag{A.29}$$

Then:

$$\mathbf{u}_\chi = \nabla\chi, \quad \mathbf{u}_\Psi = -\nabla\times\boldsymbol{\Psi}, \tag{A.30}$$

where

$$\nabla^2\chi = \nabla\cdot\mathbf{u}_\chi = \nabla\cdot\mathbf{u}, \quad \nabla^2\boldsymbol{\Psi} = \nabla\times\mathbf{u}_\Psi = \nabla\times\mathbf{u}, \tag{A.31}$$

and χ and $\boldsymbol{\Psi}$ are scalar and vector potentials, respectively. Note, however, that the decomposition may not be unique, depending upon the domain and boundary conditions.

Volume and Surface Integrals

Let V be a volume entirely enclosed by a closed surface S, with surface element $d\mathbf{S} \equiv \mathbf{n}dS$, where $\mathbf{n}$ is the outward pointing normal from S. Then:

$$\iiint_V (\nabla f)\; dV \equiv \oiint_S f\; d\mathbf{S}, \tag{A.32}$$

$$\iiint_V (\nabla\cdot\mathbf{A})\; dV \equiv \oiint_S \mathbf{A}\cdot d\mathbf{S}, \tag{A.33}$$

$$\iiint_V (\nabla\times\mathbf{A})\; dV \equiv -\oiint_S \mathbf{A}\times d\mathbf{S}, \tag{A.34}$$

$$\iiint_V \left[f\nabla^2 g + (\nabla f)\cdot(\nabla g)\right]\; dV \equiv \oiint_S (f\nabla g)\cdot d\mathbf{S}, \tag{A.35}$$

$$\iiint_V \left(f\nabla^2 g - g\nabla^2 f\right)\; dV \equiv \oiint_S (f\nabla g - g\nabla f)\cdot d\mathbf{S}, \tag{A.36}$$

$$\iiint_V \{\mathbf{A}\cdot[\nabla\times(\nabla\times\mathbf{B})] - \mathbf{B}\cdot[\nabla\times(\nabla\times\mathbf{A})]\}\; dV \equiv \oiint_S [\mathbf{B}\times(\nabla\times\mathbf{A}) - \mathbf{A}\times(\nabla\times\mathbf{B})]\cdot d\mathbf{S}. \tag{A.37}$$

Identities (A.33), (A.35), and (A.36) are known, respectively, as Gauss's divergence theorem[1], and Green's first and second identities[2].

Integral Definitions of Gradient, Divergence, and Curl

Identities (A.32)-(A.34) lead to the following integral definitions of gradient, divergence, and curl, respectively:

$$\nabla f \equiv \lim_{\Delta V \to 0} \frac{1}{\Delta V} \oiint_S f d\mathbf{S}, \tag{A.38}$$

$$\nabla \cdot \mathbf{A} \equiv \lim_{\Delta V \to 0} \frac{1}{\Delta V} \oiint_S \mathbf{A} \cdot d\mathbf{S}, \tag{A.39}$$

$$\nabla \times \mathbf{A} \equiv - \lim_{\Delta V \to 0} \frac{1}{\Delta V} \oiint_S \mathbf{A} \times d\mathbf{S}. \tag{A.40}$$

Material Derivatives of Scalars and Vectors

Let

$$\frac{D}{Dt} \equiv \frac{\partial}{\partial t} + \mathbf{u} \cdot \nabla, \tag{A.41}$$

be the material derivative operator that operates on either a scalar or a vector. Then:

$$\frac{D}{Dt}(f+g) = \frac{Df}{Dt} + \frac{Dg}{Dt}, \tag{A.42}$$

$$\frac{D}{Dt}(fg) = \frac{Df}{Dt} g + f \frac{Dg}{Dt}, \tag{A.43}$$

$$\frac{D}{Dt}(\mathbf{A} \times \mathbf{B}) = \frac{D\mathbf{A}}{Dt} \times \mathbf{B} + \mathbf{A} \times \frac{D\mathbf{B}}{Dt}, \tag{A.44}$$

$$\frac{D}{Dt}(\mathbf{A} + \mathbf{B}) = \frac{D\mathbf{A}}{Dt} + \frac{D\mathbf{B}}{Dt}, \tag{A.45}$$

$$\frac{D}{Dt}(\mathbf{A} \cdot \mathbf{B}) = \frac{D\mathbf{A}}{Dt} \cdot \mathbf{B} + \mathbf{A} \cdot \frac{D\mathbf{B}}{Dt}, \tag{A.46}$$

$$\frac{D}{Dt}(\mathbf{A} \times \mathbf{B}) = \frac{D\mathbf{A}}{Dt} \times \mathbf{B} + \mathbf{A} \times \frac{D\mathbf{B}}{Dt}, \tag{A.47}$$

$$\frac{D}{Dt}(f\mathbf{A}) = \frac{Df}{Dt} \mathbf{A} + f \frac{D\mathbf{A}}{Dt}, \tag{A.48}$$

$$\frac{D}{Dt}[\mathbf{A} \cdot (\mathbf{B} \times \mathbf{C})] = \frac{D\mathbf{A}}{Dt} \cdot (\mathbf{B} \times \mathbf{C}) + \mathbf{A} \cdot \left(\frac{D\mathbf{B}}{Dt} \times \mathbf{C}\right) + \mathbf{A} \cdot \left(\mathbf{B} \times \frac{D\mathbf{C}}{Dt}\right), \tag{A.49}$$

$$\frac{D}{Dt}[\mathbf{A} \times (\mathbf{B} \times \mathbf{C})] = \frac{D\mathbf{A}}{Dt} \times (\mathbf{B} \times \mathbf{C}) + \mathbf{A} \times \left(\frac{D\mathbf{B}}{Dt} \times \mathbf{C}\right) + \mathbf{A} \times \left(\mathbf{B} \times \frac{D\mathbf{C}}{Dt}\right), \tag{A.50}$$

$$\mathbf{A} \cdot \frac{D}{Dt}(\nabla f) = (\mathbf{A} \cdot \nabla) \frac{Df}{Dt} - [(\mathbf{A} \cdot \nabla)\mathbf{u}] \cdot \nabla f. \tag{A.51}$$

Material Derivatives of Integrals over Volumes

Let ρ be the density (i.e. mass per unit volume) of a fluid parcel of finite volume V, entirely contained within a closed material surface S, with surface element $d\mathbf{S} \equiv \mathbf{n} dS$, where $\mathbf{n}$

[1] After Johann Carl Friedrich Gauss (1777–1855).

[2] After George Green (1793–1841).

is the outward pointing normal from S. Let $\mathbf{u}$ be the velocity at a point within or on the surface of the fluid parcel. Then:

$$\frac{D}{Dt}\iiint_V dV \equiv \iiint_V \nabla\cdot\mathbf{u}dV \equiv \oiint_S \mathbf{u}\cdot d\mathbf{S}, \tag{A.52}$$

$$\frac{D}{Dt}\iiint_V fdV \equiv \iiint_V \left(\frac{Df}{Dt} + f\nabla\cdot\mathbf{u}\right) dV, \tag{A.53}$$

$$\frac{D}{Dt}\iiint_V \rho fdV \equiv \iiint_V \rho\frac{Df}{Dt}dV, \tag{A.54}$$

$$\frac{D}{Dt}\iiint_V \mathbf{A}dV \equiv \iiint_V \left(\frac{D\mathbf{A}}{Dt} + \mathbf{A}\nabla\cdot\mathbf{u}\right) dV, \tag{A.55}$$

$$\frac{D}{Dt}\iiint_V \rho\mathbf{A}dV = \iiint_V \rho\frac{D\mathbf{A}}{Dt}dV. \tag{A.56}$$

Identities (A.53) (for a scalar) and (A.56) (for a vector) are collectively known as the Reynolds' transport theorem.[3]

Surface and Contour Integrals

Let S now instead be an *open* surface, with surface element $d\mathbf{S} \equiv \mathbf{n}dS$, and let S be bounded by the *closed* contour C, with line element $d\mathbf{l}$. Then:

$$\iint_S (\nabla\times\mathbf{A})\cdot d\mathbf{S} \equiv \oint_C \mathbf{A}\cdot d\mathbf{l}, \tag{A.57}$$

$$\iint_S (\nabla f\times\nabla g)\cdot d\mathbf{S} \equiv \oint_C f\,dg \equiv -\oint_C g\,df, \tag{A.58}$$

$$\iint_S (d\mathbf{S}\times\nabla f) \equiv \oint_C f\,d\mathbf{l}, \tag{A.59}$$

$$\iint_S (d\mathbf{S}\times\nabla)\times\mathbf{A} \equiv \oint_C d\mathbf{l}\times\mathbf{A}. \tag{A.60}$$

Identity (A.57) is known as Stokes's theorem.[4]

Position Vector

Let $\mathbf{r} \equiv x\mathbf{i} + y\mathbf{j} + z\mathbf{k}$ be the position vector from the origin to the point (x, y, z) with magnitude r. Then:

$$\nabla r \equiv \frac{\mathbf{r}}{r}, \tag{A.61}$$

$$\nabla\cdot\mathbf{r} \equiv 3, \tag{A.62}$$

$$\nabla\times\mathbf{r} \equiv 0, \tag{A.63}$$

$$\nabla\left(\frac{1}{r}\right) \equiv -\frac{\mathbf{r}}{r^3} \equiv -\frac{1}{r^2}\left(\frac{\mathbf{r}}{r}\right). \tag{A.64}$$

[3] After Osborne Reynolds (1842–1912).
[4] After George Gabriel Stokes (1819–1903).

References

Akmaev, R. A., 2011: Whole atmosphere modeling: connecting terrestrial and space weather. *Rev. Geophys.*, **49**, RG4004, doi: https://doi.org/10.1029/2011RG000364.

Akmaev, R. A., and H.-M. H. Juang, 2008: Using enthalpy as a prognostic variable in atmospheric modelling with variable composition. *Q. J. R. Meteorol. Soc.*, **134**, 2193–7, doi: https://doi.org/10.1002/qj.345.

Ambaum, M. H. P., 2010: *Thermal Physics of the Atmosphere*. Wiley, 239 pp.

Arakawa, A., and C. S. Konor, 1996: Vertical differencing of the primitive equations based on the Charney-Phillips grid in hybrid σ-p vertical coordinates. *Mon. Wea. Rev.*, **124**, 511–28.

Arakawa, A., and V. R. Lamb, 1977: Computational design of the basic dynamical processes of the UCLA general circulation model. *Methods in Comp. Phys.*, **17**, 174–265.

Badin, G., and F. Crisciani, 2018: *Variational Formulation of Fluid and Geophysical Fluid Dynamics*. Advances in Geophysical and Environmental Mechanics and Mathematics, Springer, 218 pp., doi: https://doi.org/10.1007/978-3-319-59695-2.

Baines, P. G., 1976: The stability of planetary waves on a sphere. *J. Fluid. Mech.*, **73**, 193–213.

Barnes, R. T. H., R. Hide, A. A. White, and C. A. Wilson, 1983: Atmospheric angular momentum fluctuations, length of day changes and polar motion. *Proc. R. Soc. Lond. A*, **387**, 31–73.

Bénard, P., 2014: An oblate-spheroid geopotential approximation for global meteorology. *Q. J. R. Meteorol. Soc.*, **140**, 170–184, doi: https://doi.org/10.1002/qj.2141.

Bénard, P., 2015: An assessment of global forecast errors due to spherical geopotential approximation in the shallow-water case. *Q. J. R. Meteorol. Soc.*, **141**, 195–206, doi: https://doi.org/10.1002/qj.2349.

Bryan, G. H., and J. M. Fritsch, 2002: A benchmark simulation for moist nonhydrostatic numerical models. *Mon. Wea. Rev.*, **130**, 2917–28, doi: https://doi.org/10.1175/1520-0493(2002)130<2917:ABSFMN>2.0.CO;2.

Callen, H. B., 1985: *Thermodynamics and an Introduction to Thermostatistics*. 2nd ed., Wiley, 493 pp.

Chandrasekhar, S., 1967: Ellipsoidal figures of equilibrium – an historical account. *Commun. Pure Appl. Math.*, **20**, 251–65.

Chandrasekhar, S., 1969: *Ellipsoidal Figures of Equilibrium*. Yale University Press, 253 pp.

Clairaut, A. C., 1743: *Théorie de la Figure de la Terre tirée des principes de l'hydrostatique*. Kessinger Publishing (facsimile reproduction), P.O. Box 1404, Whitefish, MT 59937 USA. Also accessed online at https://archive.org/details/thoriedelafigur00claigoog.

Colin de Verdière, A., and R. Schopp, 1994: Flows in a rotating spherical shell: the equatorial case. *J. Fluid Mech.*, **276**, 233–60.

Curry, J. A., and P. J. Webster, 1999: *Thermodynamics of Atmospheres and Oceans*. Academic Press, 467 pp.

Daley, R., 1988: The normal modes of the spherical non-hydrostatic equations with applications to the filtering of acoustic modes. *Tellus*, **40A**, 96–106.

Davies, T., M. Cullen, A. Malcolm et al., 2005: A new dynamical core for the Met Office's global and regional modelling of the atmosphere. *Q. J. R. Meteorol. Soc.*, **131**, 1759–82, doi: https://doi.org/10.1256/qj.04.101.

Dellar, P. J., 2011: Variations on a beta-plane: derivation of non-traditional beta-plane equations from Hamilton's principle on a sphere. *J. Fluid Mech.*, **674**, 174–95, doi: https://doi.org/10.1017/S0022112010006464.

Dellar, P. J., and R. Salmon, 2005: Shallow water equations with a complete Coriolis force and topography. *Phys. Fluids*, **17**, 106 601, doi: https://doi.org/10.1063/1.2116747.

de Szoeke, R. A., 2004: An effect of the thermobaric nonlinearity of the equation of state: a mechanism for sustaining solitary Rossby waves. *J. Phys. Oceanogr.*, **34**, 2042–56, doi: https://doi.org/10.1175/1520-0485(2004)034<2042:AEOTTN>2.0.CO;2.

Draghici, I., 1987: Non-hydrostatic Coriolis effects in an isentropic coordinate frame. *Meteorol. Hydrol.*, **19**, 13–27.

Draghici, I., 1989: The hypothesis of a marginally shallow atmosphere. *Meteorol. Hydrol.*, **19**, 13–27.

Dubos, T., and F. Voitus, 2014: A semi-hydrostatic theory of gravity-dominated compressible flow. *J. Atmos. Sci.*, **71**, 4621–38, doi: https://doi.org/10.1175/JAS-D-14-0080.1.

Dutton, J. A., 1986: *The Ceaseless Wind: An Introduction to the Theory of Atmospheric Motion*. 1st ed., McGraw-Hill, 579 pp.

Dziewonski, A. M., and D. L. Anderson, 1981: Preliminary reference Earth model. *Phys. Earth Planet. Inter.*, **25**, 297–356.

Eckart, C., 1960: *The Hydrodynamics of Oceans and Atmospheres*. Pergamon Press, 290 pp.

Feistel, R., 2003: A new extended Gibbs thermodynamic potential of seawater. *Progress in Oceanography*, **58**, 43–114, doi: https://doi.org/10.1016/S0079-6611(03)00088-0.

Feistel, R., 2008: A Gibbs function for seawater thermodynamics for −6 to 80 °C and salinity up to 120 g kg^{-1}. *Deep-Sea Res. I*, **55**, 1639–71, doi: https://doi.org/10.1016/j.dsr.2008.07.004.

Feistel, R., D. G. Wright, K. Miyagawa et al., 2008: Mutually consistent thermodynamic potentials for fluid water, ice and seawater: a new standard for oceanography. *Ocean Sci.*, **4**, 275–91, doi: https://doi.org/10.5194/os-4-275-2008.

Feistel, R., D. G. Wright, H. J. Kretzscmar et al., 2010a: Thermodynamic properties of sea air. *Ocean Sci.*, **6**, 91–141, doi: https://doi.org/10.5194/os-6-91-2010.

Feistel, R., D. G. Wright, H. J. Kretzscmar et al., 2010b: Numerical implementation and oceanographic application of the thermodynamic potentials of liquid water, water vapour, ice, seawater and humid air. Part 1: Background and equations. *Ocean Sci.*, **6**, 633–77, doi: https://doi.org/10.5194/os-6-633-2010.

Gassmann, A., 2013: A global hexagonal C-grid non-hydrostatic dynamical core (ICON-IAP) designed for energy consistency. *Q. J. R. Meteorol. Soc.*, **139**, 152–75, doi: https://doi.org/10.1002/qj.1960.

Gates, W. L., 2004: Derivation of the equations of atmospheric motion in oblate spheroidal coordinates. *J. Atmos. Sci.*, **61**, 2478–87, doi: https://doi.org/10.1175/1520-0469(2004)061<2478:DOTEA>2.0.CO;2.

Gerkema, T., J. T. F. Zimmerman, L. R. M. Maas, and H. van Haren, 2008: Geophysical and astrophysical fluid dynamics beyond the traditional approximation. *Rev. Geophys.*, **46**, RG2004–2033, doi: https://doi.org/10.1029/2006RG000220.

Gill, A. E., 1982: *Atmosphere-Ocean Dynamics*. Academic Press, 662 pp.

Goldstein, H., C. P. Poole, and J. L. Safko, 2001: *Classical Mechanics*. 3rd ed., Addison-Wesley, 638 pp.

Green, A. E., and P. M. Naghdi, 1976: A derivation of equations for wave propagation in water of variable depth. *J. Fluid Mech.*, **78**, 237–46.

Haltiner, G. J., and F. L. Martin, 1957: *Dynamical and Physical Meteorology*. McGraw-Hill, 470 pp.

Hawking, S., 2011: *A Brief History of Time: From the Big Bang to Black Holes*. Bantam Books, 272 pp.

Heimpel, M., J. Aurnou, and J. Wicht, 2006: Simulation of equatorial and high-latitude jets on Jupiter in a deep convection model. *Nature*, **438**, 193–196, doi: https://doi.org/10.1038/nature04208.

Hoskins, B. J., 1973: Stability of the Rossby–Haurwitz wave. *Q. J. R. Meteorol. Soc.*, **99**, 723–45.

Hoskins, B. J., and I. N. James, 2014: *Fluid Dynamics of the Midlatitude Atmosphere*. Wiley, 408 pp.

Ingersoll, A. P., 2013: *Planetary Climates*. Princeton University Press, 278 pp.

IOC, SCOR, and IAPSO, 2010: The international thermodynamic equation of seawater – 2010: calculations and use of thermodynamic properties. Intergovernmental Oceanographic Commission, Manuals and Guides No. 56, UNESCO (English), 196 pp. available online at http:teos-10.org/pubs/TEOS-10_Manual.pdf.

Jeffreys, H., 1976: *The Earth: Its Origin, History and Physical Constitution*. 6th ed., Cambridge University Press, 586 pp.

Kasahara, A., 1974: Various vertical coordinate systems used for numerical weather prediction. *Mon. Wea. Rev.*, **102**, 509–22.

Kasahara, A., 2003a: On the nonhydrostatic atmospheric models with inclusion of the horizontal component of the Earth's angular velocity. *J. Meteorol. Soc. Japan*, **81**, 935–50, doi: https://doi.org/10.2151/jmsj.81.935.

Kasahara, A., 2003b: The roles of the horizontal component of the Earth's angular velocity in nonhydrostatic linear models. *J. Atmos. Sci.*, **60**, 1085–95, doi: https://doi.org/10.1175/1520-0469(2003)60<1085:TROTHC>2.0.CO;2.

Klemp, J., 2011: A terrain-following coordinate with smoothed coordinate surfaces. *Mon. Wea. Rev.*, **139**, 2163–9, doi: https://doi.org/10.1175/MWR-D-10-05046.1.

Lanczos, C., 1970: *The Variational Principles of Mechanics*. 4th ed., Dover, 418 pp.

Laprise, R., 1992: The Euler equations of motion with hydrostatic pressure as an independent variable. *Mon. Wea. Rev.*, **120**, 197–207.

Laprise, R., and C. Girard, 1990: A spectral general circulation model using a piecewise-constant finite-element representation on a hybrid vertical coordinate system. *J. Climate*, **3**, 32–52.

Lauritzen, P. H., C. Jablonowski, M. A. Taylor, and R. D. Nair, 2010: Rotated versions of the Jablonowski steady-state and baroclinic wave test cases: a dynamical core intercomparison. *J. Adv. Model. Earth Syst.*, **2**, Art. #15, 34 pp, doi: https://doi.org/10.3894/JAMES.2010.2.15.

Läuter, M., D. Handorf, and K. Dethloff, 2005: Unsteady analytical solutions of the spherical shallow water equations. *J. Comput. Phys.*, **210**, 535–53, doi: https://doi.org/10.1016/j.jcp.2005.04.022.

Lebedev, N. N., 1972: *Special Functions and Their Applications*. Dover, 308 pp., translated by R. A. Silverman.

Leuenberger, D., M. Koller, O. Fuhrer, and C. Schär, 2010: A generalization of the SLEVE vertical coordinate. *Mon. Wea. Rev.*, **138**, 3683–9, doi: https://doi.org/10.1175/2010MWR3307.1.

Lorenz, E. N., 1967: *The Nature and Theory of the General Circulation of the Atmosphere*, vol. 218. World Meteorological Organization, 161 pp.

Lynch, P., 2014: *The Emergence of Numerical Weather Prediction: Richardson's Dream*. Cambridge University Press, 292 pp.

Marshall, J., C. Hill, L. Perelman, and A. Adcroft, 1997: Hydrostatic, quasi-hydrostatic, and non-hydrostatic ocean modeling. *J. Geophys. Res.*, **102**, 5733–52.

Miles, J., and R. Salmon, 1985: Weakly dispersive nonlinear gravity waves. *J. Fluid Mech.*, **157**, 519–31.

Müller, R., 1989: A note on the relation between the 'traditional approximation' and the metric of the primitive equations. *Tellus*, **41A**, 175–8.

Neuenschwander, D. E., 2017: *Emmy Noether's Wonderful Theorem*. 2nd ed., Johns Hopkins University Press, 321 pp.

Pavlis, N. K., S. A. Holmes, S. C. Kenyon, and J. K. Factor, 2012: The development and evaluation of the Earth Gravitational Model 2008 (EGM2008). *J. Geophys Res: Solid Earth*, **117**, B04 406, doi: https://doi.org/10.1029/2011JB008916.

Pavlis, N. K., S. A. Holmes, S. C. Kenyon, and J. K. Factor, 2013: Correction to: The development and evaluation of the Earth Gravitational Model 2008 (EGM2008). *J. Geophys Res: Solid Earth*, **118**, 2633, doi: https://doi.org/10.1002/jgrb.50167.

Pawlowicz, R., T. McDougall, R. Feistel, and R. Tailleux, 2012: An historical perspective on the development of the Thermodynamic Equation of Seawater – 2010. *Ocean Sci.*, **8**, 161–74, doi: https://doi.org/10.5194/os-8-161-2012.

Phillips, N. A., 1957: A coordinate system having some special advantages for numerical forecasting. *J. Meteor.*, **14**, 184–5.

Phillips, N. A., 1959: Numerical integration of the primitive equations on the hemisphere. *Mon. Wea. Rev.*, **87**, 333–5.

Phillips, N. A., 1966: The equations of motion for a shallow rotating atmosphere and the 'traditional' approximation. *J. Atmos. Sci.*, **23**, 626–8.

Phillips, N. A., 1968: Reply to 'Comments on Phillips' proposed simplification of the equations of motion for a shallow rotating atmosphere' by G. Veronis. *J. Atmos. Sci.*, **25**, 1155–7.

Phillips, N. A., 1973: Principles of large-scale numerical weather prediction. *Dynamic Meteorology*, P. Morel, ed., Reidel, 1–96.

Press, W., S. Teukolsky, W. Vetterling, and B. Flannery, 1992: *Numerical Recipes in FORTRAN: The Art of Scientific Computing*. 2nd ed., Cambridge University Press, 963 pp.

Ramsey, A. S., 1940: *An Introduction to the Theory of Newtonian Attraction*. Cambridge University Press, 184 pp.

Rasch, P. J., 1986: Toward atmospheres without tops: absorbing upper boundary conditions for numerical models. *Q. J. R. Meteorol. Soc.*, **112**, 1195–1218.

Ripa, P., 1983: General stability conditions for zonal flows in a one-layer model on the beta-plane or sphere. *J. Fluid Mech.*, **126**, 463–89.

Rochas, M., 1984: Comments on 'A generalized class of time-dependent solutions of the vorticity equation for nondivergent barotropic flow'. *Mon. Wea. Rev.*, **112**, 390.

Rochas, M., 1986: A new class of exact time-dependent solutions of the vorticity equation. *Mon. Wea. Rev.*, **114**, 961–6.

Roulstone, I., and J. Norbury, 2013: *Invisible in the Storm: The Role of Mathematics in Understanding the Weather*. Princeton University Press, 325 pp.

Salmon, R., 1998: *Lectures on Geophysical Fluid Dynamics*. Oxford University Press, 378 pp.

Satoh, M., T. Matsuno, H. Tomita et al., 2008: Nonhydrostatic icosahedral atmospheric model (NICAM) for global cloud resolving simulations. *J. Comput. Phys.*, **227**, 3486–514, doi: https://doi.org/10.1016/j.jcp.2007.02.006.

Schär, C., D. Leuenberger, O. Fuhrer, D. Lüthi, and C. Girard, 2002: A new terrain-following vertical coordinate formulation for atmospheric prediction models. *Mon. Wea. Rev.*, **130**, 2459–80, doi: https://doi.org/10.1175/1520-0493(2002)130<2459:ANTFVC>2.0.CO;2.

Shepherd, T. G., 2003: Ripa's theorem and its relatives. *Nonlinear Processes in Geophysical Fluid Dynamics*, O. U. Velasco-Fuentes, J. Sheinbaum, and J. Ochoa, eds., Kluwer Academic, 1–14, doi: https://doi.org/10.1007/978-94-010-0074-1.

Simmons, A. J., and D. M. Burridge, 1981: An energy and angular-momentum conserving vertical finite-difference scheme and hybrid vertical coordinates. *Mon. Wea. Rev.*, **109**, 758–66.

Skamarock, W., J. Klemp, M. Duda et al., 2012: A multiscale nonhydrostatic atmospheric model using centroidal Voronoi tesselations and C-grid staggering. *Mon. Wea. Rev.*, **140**, 3090–105, doi: https://doi.org/10.1175/MWR-D-11-00215.1.

Spekat, A., ed., 2001: *Proceedings of the 50th Anniversary of Numerical Weather Prediction Commemorative Symposium, Potsdam, Germany, 9–10 March 2000*, Deutsche Meteorologische Gesellschaft e.V., Berlin.

Spiegel, M. R., and S. L. Lipschutz, 2009: *Vector Analysis*. 2nd ed., McGraw-Hill, 253 pp.

Staniforth, A., 2001: Developing efficient unified nonhydrostatic models. *Proceedings of the 50th Anniversary of Numerical Weather Prediction Commemorative Symposium, Potsdam, Germany, 9–10 March 2000*, A. Spekat, ed., Deutsche Meteorologische Gesellschaft e.V., 185–200.

Staniforth, A., 2012: Exact stationary axisymmetric solutions of the Euler equations on beta-gamma planes. *Atmos. Sci. Letters*, **13**, 79–87, doi: https://doi.org/10.1002/asl.375.

Staniforth, A., 2014a: Deriving consistent approximate models of the global atmosphere using Hamilton's principle. *Q. J. R. Meteorol. Soc.*, **140**, 2383–7, doi: https://doi.org/10.1002/qj.2273.

Staniforth, A., 2014b: Spheroidal and spherical geopotential approximation. *Q. J. R. Meteorol. Soc.*, **140**, 2685–92, doi:https://doi.org/10.1002/qj.2324.

Staniforth, A., 2015a: Consistent quasi-shallow models of the global atmosphere in non-spherical geopotential coordinates with complete Coriolis force. *Q. J. R. Meteorol. Soc.*, **141**, 979–86, doi: https://doi.org/10.1002/qj.2399.

Staniforth, A., 2015b: Dynamically consistent shallow-water equation sets in non-spherical geometry with latitudinal variation of gravity. *Q. J. R. Meteorol. Soc.*, **141**, 2429–43, doi: https://doi.org/10.1002/qj.2533.

Staniforth, A., and J. Côté, 1991: Semi-Lagrangian integration schemes for atmospheric models – a review. *Mon. Wea. Rev.*, **119**, 2206–23.

Staniforth, A., and J. Thuburn, 2012: Horizontal grids for global weather prediction and climate models: a review. *Q. J. R. Meteorol. Soc.*, **138**, 1–26, doi: https://doi.org/10.1002/qj.958.

Staniforth, A., and A. White, 2015a: Geophysically Realistic, Ellipsoidal, Analytically Tractable (GREAT) coordinates for atmospheric and oceanic modelling. *Q. J. R. Meteorol. Soc.*, **141**, 1646–57, doi: https://doi.org/10.1002/qj.2467.

Staniforth, A., and A. White, 2019: Forms of the thermodynamic energy equation for moist air. *Q. J. R. Meteorol. Soc.*, **145**, 386–93, doi: https://doi.org/10.1002/qj.3421.

Staniforth, A., A. White, N. Wood et al., 2006: The Joy of U.M. 6.3 – Model Formulation. Unified Model Documentation Paper 15, UK Meteorological Office.

Staniforth, A., and A. A. White, 2007: Some exact solutions of geophysical fluid dynamics equations for testing models in spherical and plane geometry. *Q. J. R. Meteorol. Soc.*, **133**, 1605–14, doi: https://doi.org/10.1002/qj.122.

Staniforth, A., and A. A. White, 2008a: Stability of some exact solutions of the shallow-water equations for testing numerical models in spherical geometry. *Q. J. R. Meteorol. Soc.*, **134**, 771–8, doi: https://doi.org/10.1002/qj.240.

Staniforth, A., and A. A. White, 2008b: Unsteady exact solutions of the flow equations for three-dimensional spherical atmospheres. *Q. J. R. Meteorol. Soc.*, **134**, 1615–26, doi: https://doi.org/10.1002/qj.300.

Staniforth, A., and A. A. White, 2011: Further non-separable exact solutions of the deep- and shallow-atmosphere equations. *Atmos. Sci. Letters*, **12**, 356–61, doi: https://doi.org/10.1002/asl.349.

Staniforth, A., and A. A. White, 2015b: The shallow water equations in non-spherical geometry with latitudinal variation of gravity. *Q. J. R. Meteorol. Soc.*, **141**, 655–62, doi: https://doi.org/10.1002/qj.2394.

Staniforth, A., and N. Wood, 2003: The deep-atmosphere equations in a generalized vertical coordinate. *Mon. Wea. Rev.*, **131**, 1931–8, doi: https://doi.org/10.1175//2564.1.

Staniforth, A., and N. Wood, 2013: Exact axisymmetric solutions of the deep- and shallow-atmosphere Euler equations in curvilinear and plane geometries. *Q. J. R. Meteorol. Soc.*, **139**, 1113–20, doi: https://doi.org/10.1002/qj.2018.

Staniforth, A., N. Wood, and C. Girard, 2003: Energy and energy-like invariants for deep non-hydrostatic atmospheres. *Q. J. R. Meteorol. Soc.*, **129**, 3495–9, doi: https://doi.org/10.1256/qj.03.18.

Tanguay, M., A. Robert, and R. Laprise, 1990: A semi-implicit semi-Lagrangian fully compressible regional forecast model. *Mon. Wea. Rev.*, **118**, 1970–80.

Tassoul, J.-L., 1978: *Theory of Rotating Stars*. Princeton University Press, 506 pp.

Thuburn, J., 2017: Use of the Gibbs thermodynamic potential to express the equation of state in atmospheric models. *Q. J. R. Meteorol. Soc.*, **236**, 1185–96, doi: https://doi.org/10.1002/qj.3020.

Thuburn, J., and Y. Li, 2000: Numerical simulations of Rossby–Haurwitz waves. *Tellus*, **52A**, 181–9, doi: https://doi.org/10.3402/tellusa.v2i2.12258.

Thuburn, J., and A. A. White, 2013: A geometrical view of the shallow-atmosphere approximation, with application to the semi-Lagrangian departure point calculation. *Q. J. R. Meteorol. Soc.*, **139**, 261–8, doi: https://doi.org/10.1002/qj.1962.

Todhunter, I., 1873: *A History of the Mathematical Theories of Attraction and the Figure of the Earth from the Time of Newton to that of Laplace. Volume I.* Macmillan, 476 pp.

Torge, W., 2001: *Geodesy*. 3rd ed., Walter de Gruyter, 416 pp.

Tort, M., and T. Dubos, 2014a: Dynamically consistent shallow-atmosphere equations with a complete Coriolis force. *Q. J. R. Meteorol. Soc.*, **140**, 2388–92, doi: https://doi.org/10.1002/qj.2274.

Tort, M., and T. Dubos, 2014b: Usual approximations to the equations of atmospheric motion: a variational perspective. *J. Atmos. Sci.*, **71**, 2452–66, doi: https://doi.org/10.1175/JAS-D-13-0339.1.

Tort, M., T. Dubos, F. Bouchut, and V. Zeitlin, 2014: Consistent shallow-water equations on the rotating sphere with complete Coriolis force and topography. *J. Fluid Mech.*, **748**, 789–821, doi: https://doi.org/10.1017/jfm.2014.172.

Ullrich, P. A., and C. Jablonowski, 2012: Operator-split Runge-Kutta-Rosenbrock methods for non-hydrostatic atmospheric models. *Mon. Wea. Rev.*, **140**, 1257–84, doi: https://doi.org/10.1175/MWR-D-10-05073.1.

Ullrich, P. A., T. Melvin, C. Jablonowski, and A. Staniforth, 2014: A proposed baroclinic wave test case for deep and shallow atmosphere dynamical cores. *Q. J. R. Meteorol. Soc.*, **140**, 1590–1602, doi: https://doi.org/10.1002/qj.2241.

Vallis, G. K., 2017: *Atmospheric and Oceanic Fluid Dynamics: Fundamentals and Large-Scale Circulation*. 2nd ed., Cambridge University Press, 946 pp., doi: https://doi.org/10.1017/9781107588417.

Vallis, G. K., 2019: *Essentials of Atmospheric and Oceanic Dynamics*. Cambridge University Press, 356 pp., doi: https://doi.org/10.1017/9781107588431.

van der Toorn, R., and J. T. F. Zimmerman, 2008: On the spherical approximation of the geopotential in geophysical fluid dynamics and the use of a spherical coordinate system. *Geophys. Astrophys. Fluid*, **102**, 349–71, doi: https://doi.org/10.1080/03091920801900674.

Velasco, S., and C. Fernández-Pineda, 2007: Thermodynamics of a pure substance at the triple point. *Am. J. Phys.*, **75**, 1086–91, doi: https://doi.org/10.1119/1.2779880.

Verkley, W. T. M., 1984: The construction of barotropic motions on a sphere. *J. Atmos. Sci.*, **41**, 2492–504.

Veronis, G., 1968: Comments on Phillips' proposed simplification of the equations of motion for a shallow rotating atmosphere. *J. Atmos. Sci.*, **25**, 1154–5.

Wedi, N. P., and P. K. Smolarkiewicz, 2009: A framework for testing global non-hydrostatic models. *Q. J. R. Meteorol. Soc.*, **135**, 469–84, doi: https://doi.org/10.1002/qj.377.

WGS 84, 2004: World Geodetic System 1984, Its Definition and Relationships with Local Geodetic Systems. NIMA Technical Report TR8350.2, 3rd edition, amendment 2, 151 pp.

White, A., 2016: Comments on: Is the Coriolis effect an 'optical illusion'? by Anders Persson. *Q. J. R. Meteorol. Soc.*, **142**, 2585–9, doi: https://doi.org/10.1002/qj.2853.

White, A. A., and R. A. Bromley, 1995: Dynamically consistent, quasi-hydrostatic equations for global models with a complete representation of the Coriolis force. *Q. J. R. Meteorol. Soc.*, **121**, 399–418.

White, A. A., B. J. Hoskins, I. Roulstone, and A. Staniforth, 2005: Consistent approximate models of the global atmosphere: shallow, deep, hydrostatic, quasi-hydrostatic and non-hydrostatic. *Q. J. R. Meteorol. Soc.*, **131**, 2081–107, doi: https://doi.org/10.1256/qj.04.49.

White, A. A., and G. W. Inverarity, 2012: A quasi-spheroidal system for modelling global atmospheres: geodetic coordinates. *Q. J. R. Meteorol. Soc.*, **138**, 27–33, doi: https://doi.org/10.1002/qj.885.

White, A. A., and A. Staniforth, 2008: A generalized thermal wind equation and some non-separable exact solutions of the flow equations for three-dimensional spherical atmospheres. *Q. J. R. Meteorol. Soc.*, **134**, 1931–9, doi: https://doi.org/10.1002/qj.323.

White, A. A., and A. Staniforth, 2009: Stability criteria for shallow water flow above zonally symmetric orography on the sphere. *Q. J. R. Meteorol. Soc.*, **135**, 1897–905, doi: https://doi.org/10.1002/qj.504.

White, A. A., A. Staniforth, and N. Wood, 2008: Spheroidal coordinate systems for modelling global atmospheres. *Q. J. R. Meteorol. Soc.*, **134**, 261–70, doi: https://doi.org/10.1002/qj.208.

White, A. A., and N. Wood, 2012: Consistent approximate models of the global atmosphere in non-spherical geopotential coordinates. *Q. J. R. Meteorol. Soc.*, **138**, 980–8, doi: https://doi.org/10.1002/qj.972.

Williamson, D. L., J. B. Drake, J. J. Hack, R. Jakob, and P. N. Swarztrauber, 1992: A standard test set for numerical approximations to the shallow-water equations in spherical geometry. *J. Comp. Phys.*, **102**, 211–24.

Wood, N., and A. Staniforth, 2003: The deep-atmosphere Euler equations with a mass-based vertical coordinate. *Q. J. R. Meteorol. Soc.*, **129**, 1289–1300, doi: https://doi.org/10.1256/qj.02.153.

Wood, N., A. Staniforth, A. White et al., 2014: An inherently mass-conserving semi-implicit semi-Lagrangian discretisation of the deep-atmosphere global nonhydrostatic equations. *Q. J. R. Meteorol. Soc.*, **140**, 1505–20, doi: https://doi.org/10.1002/qj.2235.

Wright, D. G., R. Feistel, J. H. Reissmann et al., 2010: Numerical implementation and oceanographic application of the thermodynamic potentials of liquid water, water vapour, ice, seawater and humid air. Part 2: The library routines. *Ocean Sci.*, **6**, 695–718, doi: https://doi.org/10.5194/os-6-695-2010.

Zdunkowski, W., and A. Bott, 2003: *Dynamics of the Atmosphere: A Course in Theoretical Meteorology*. Cambridge University Press, 738 pp., doi: https://doi.org/10.1017/CBO9780511805462.

Zerroukat, M., N. Wood, A. Staniforth, A. White, and J. Thuburn, 2009: An inherently mass-conserving semi-implicit semi-Lagrangian discretisation of the shallow water equations on the sphere. *Q. J. R. Meteorol. Soc.*, **135**, 1104–16, doi: https://doi.org/10.1002/qj.458.

Index